T0140561

Advances in Intelligent Systems and Computing

Volume 1231

The series "Advances in Intelligent Systems and Computing" contains publications on theory, applications, and design methods of Intelligent Systems and Intelligent Computing. Virtually all disciplines such as engineering, natural sciences, computer and information science, ICT, economics, business, e-commerce, environment, healthcare, life science are covered. The list of topics spans all the areas of modern intelligent systems and computing such as: computational intelligence, soft computing including neural networks, fuzzy systems, evolutionary computing and the fusion of these paradigms, social intelligence, ambient intelligence, computational neuroscience, artificial life, virtual worlds and society, cognitive science and systems, Perception and Vision, DNA and immune based systems, self-organizing and adaptive systems, e-Learning and teaching, human-centered and human-centric computing, recommender systems, intelligent control, robotics and mechatronics including human-machine teaming, knowledge-based paradigms, learning paradigms, machine ethics, intelligent data analysis, knowledge management, intelligent agents, intelligent decision making and support, intelligent network security, trust management, interactive entertainment, Web intelligence and multimedia.

The publications within "Advances in Intelligent Systems and Computing" are primarily proceedings of important conferences, symposia and congresses. They cover significant recent developments in the field, both of a foundational and applicable character. An important characteristic feature of the series is the short publication time and world-wide distribution. This permits a rapid and broad dissemination of research results.

**** Indexing: The books of this series are submitted to ISI Proceedings, EI-Compendex, DBLP, SCOPUS, Google Scholar and Springerlink ****

More information about this series at http://www.springer.com/series/11156

Michael E. Auer · Dominik May
Editors

Cross Reality and Data Science in Engineering

Proceedings of the 17th International Conference on Remote Engineering and Virtual Instrumentation

Editors
Michael E. Auer
Carinthia University of Applied Sciences
Villach, Austria

Dominik May
College of Engineering
University of Georgia
Athens, GA, USA

ISSN 2194-5357 ISSN 2194-5365 (electronic)
Advances in Intelligent Systems and Computing
ISBN 978-3-030-52574-3 ISBN 978-3-030-52575-0 (eBook)
https://doi.org/10.1007/978-3-030-52575-0

This Springer imprint is published by the registered company Springer Nature Switzerland AG
The registered company address is: Gewerbestrasse 11, 6330 Cham, Switzerland

Preface

The REV conference is the annual conference of the International Association of Online Engineering (IAOE) and the Global Online Laboratory Consortium (GOLC).

REV2020 on "Cross Reality and Data Science in Engineering" was the 17th in a series of annual events concerning the area of Remote Engineering and Virtual Instrumentation.

In a globally connected world, the interest in online collaboration, teleworking, remote services and other digital working environments is rapidly increasing. In response to that, the general objective of this conference is to contribute and discuss fundamentals, applications and experiences in the fields of online and remote engineering, virtual instrumentation and other related new technologies like cross reality, data science & big data, Internet of Things & Industrial Internet of Things, Industry 4.0, cybersecurity, and M2M & smart objects. Another objective of the conference is to discuss guidelines and new concepts for engineering education in higher and vocational education institutions, including emerging technologies in learning, MOOCs & MOOLs, and open resources.

REV2020 has been organized in cooperation with the Engineering Education Transformations Institute and the Georgia Informatics Institutes for Research and Education at the College of Engineering at the University of Georgia, Athens, GA, USA, from February 26 to 28, 2020.

REV2020 offered again an exciting technical program as well as networking opportunities. Outstanding scientists and industry leaders accepted the invitation for keynote speeches:

- **Hans J. Hoyer**
  - Secretary General of the International Federation for Engineering Education Societies (IFEES) and Resident Scholar in Global Engineering at George Mason University, Fairfax VA, USA

- **Isa Jahnke**
  - Director of the Information Experience Lab and Associate Professor of Information Science and Learning Technologies at the University of Missouri, Columbia, MO, USA
- **Max Hoffmann**
  - Chair of Information Management in Mechanical Engineering at RWTH Aachen University, Aachen, Germany
- **Francis Limousy, Paul Miller, and Basil Ittiavira**
  - UL, Atlanta, GA, USA
- **Melanie Spare and John DeTellem**
  - Siemens, Atlanta, GA, USA
- **Dan Schaffer**
  - Phoenix Contact USA, Middletown, PA, USA

It was in 2004 when we started this conference series in Villach, Austria, together with some visionary colleagues and friends from around the world. When we started our REV endeavor, the Internet was just 10 years old! Since then the situation regarding Online Engineering and Virtual Instrumentation has radically changed. Both are today typical working areas of most of the engineers and are inseparable connected with

- Cross Reality
- Data science & big data
- Internet of Things & Industrial Internet of Things,
- Industry 4.0 & cyberphysical systems,
- Collaborative networks and grids
- Cybersecurity, and
- M2M & smart objects

to name only a few.

With our conference in 2004, we already tried to focus on the upcoming use of the Internet for engineering tasks and the opportunities as well as challenges around it. And as we can see very successful.

The REV2020 conference takes up the following topics in its variety and discusses the state-of-the-art and future trends under the global theme "Cross Reality and Data Science in Engineering":

- Applications & Experiences
- Artificial intelligence
- Augmented reality
- Big data–data science
- Cyberphysical system

- Cybersecurity
- Collaborative work in virtual environments
- Cross-reality applications
- Evaluation of online labs
- Human–machine interaction & usability
- Internet of Things
- Industry 4.0
- M2M concepts
- Virtual and mixed reality
- Networking, edge & cloud technology
- Online and biomedical engineering
- Process visualization
- Remote control & measurements
- Remote & crowd sensing
- Smart objects
- Smart world (city, buildings, home, etc.)
- Standards & standardization proposals
- Teleworking environments
- Virtual instrumentation
- Virtual & remote laboratories

As submission types have been accepted:

- Full Paper, Short Paper
- Work in progress, poster
- Special sessions
- Workshops, tutorials

All contributions were subject to a double-blind review. The review process was very competitive. We had to review near to 200 submissions. A team of over 90 reviewers and program committee members did this terrific job. Our special thanks go to all of them.

Due to the time and conference schedule restrictions, we could finally accept only the best 70 submissions for presentation or demonstration. The conference had again over 90 participants from 25 countries from all continents.

REV2021 will be held in Hong Kong.

Michael E. Auer
REV General Chair

Dominik May
REV2020 Chair

Organization

Committees

General Chair

Michael E. Auer	Founding President and CEO of the IAOE, CTI Frankfurt/Main New York, Vienna, Germany

REV2020 Chair

Dominik May	University of Georgia, Athens, GA, USA & IAOE Vice-President

Program Co-chairs

Joachim Walther	University of Georgia, Athens, GA, USA & Engineering Education Transformations Institute
Kyle Johnsen	University of Georgia, Athens, GA, USA & Georgia Informatics Institutes
Kalyan Ram B.	IAOE President & Electrono Solutions Pvt. Ltd., India

International Advisory Board

Abul Azad	President Global Online Laboratory Consortium, USA
Denis Gillet	EPFL Lausanne, Switzerland
Bert Hesselink	Stanford University, USA
Zorica Nedic	University of South Australia

Teresa Restivo	University of Porto, Portugal
Hamadou Saliah-Hassane	Université TÉLUQ, Montreal, Canada
Cornel Samoila	University of Brasov, Romania
Franz Schauer	Tomas Bata University, Czech Republic
Tarek Sobh	University of Bridgeport, USA
Claudius Terkowsky	TU Dortmund University, Germany
Vasant Honavar	Penn State University, USA
Valerie Varney	RWTH Aachen, Germany
Krishna Vedula	IUCEE, India

Technical Program Co-chairs

Sebastian Schreiter	IAOE, France
Virginia R. Bacon Talati	University of Georgia, Athens, GA, USA

Publication Chair and Web Master

Sebastian Schreiter	IAOE, France

IEEE Liaison

Manuel Castro	MIT Madrid, Spain

Workshop and Tutorial Chairs

Andreas Pester	Carinthia University of Applied Sciences, Austria
Tobias R. Ortelt	TU Dortmund University, Germany

Special Session Chairs

Valerie Varney	RWTH Aachen, Germany
Teresa Restivo	University of Porto, Portugal

International Program Committee

Akram Abu-Aisheh	Hartford University, USA
Yacob Astatke	Morgan State University, USA
Gustavo Alves	ISEP Porto, Portugal
Nael Bakarad	Grand Valley State University, USA
David Boehringer	University of Stuttgart, Germany
Michael Callaghan	University of Ulster, Northern Ireland

Contents

Virtual and Remote Laboratories

Augmented Reality

Virtual and Remote Laboratories

Digital Laboratories for Educating the IoT-Generation Heatmap for Digital Lab Competences

Valentin Kammerlohr(✉), Anke Pfeiffer, and Dieter Uckelmann

Hochschule Für Technik Stuttgart, Stuttgart, Germany
Valentin.Kammerlohr@hft-stuttgart.de

Abstract. The Sharing Economy is the basis for challenging existing business models and the creation of new cooperation's. The BMBF-funded project Open Digital Lab 4 You (DigiLab4U) promotes the sharing of laboratory (lab) infrastructures for Internet of Things (IoT)/Industry 4.0 (I4.0) related education. Through the digitalisation in education and research, new forms of cross-location networking of lab infrastructures arise. The paper illustrates innovation options for IoT/I4.0 lab infrastructure, exemplary based on a RFID measuring chamber using (I) a Virtual Reality (VR) Mockup, (II) a Service Robot Arm and (III) VR with live data as combination of (I) + (II), to realise cross-location networking. Related IoT-/I4.0-competences within lab education are raised and applied to the curriculum for educating undergraduate engineering. A matrix containing a competence evaluation system for digital labs is provided. Considerations for the evaluation of labs are presented to allow the classification of labs as well as to provide a quick orientation for users with regard to the learning objectives to be achieved.

Researchers and participants of digital labs are invited to explore the use of clustering lab experiences according to the introduced method. Clustering real, remote and virtual labs in this way can help to find out which learning outcomes can be achieved in specific labs.

Keywords: Digital lab · Industry 4.0 · IoT-/I4.0-Education · Internet of Things · Engineering education · IoT-/I4.0-Competences · Virtual reality · RFID · Sharing economy

1 Introduction

The further development of the information and consumer society from having to sharing (technical term "Sharing Economy") offers opportunities for new cooperation models. New concepts for sharing goods and services between private individuals and companies (B2B & B2C) are emerging worldwide [4]. Well-known representatives such as ShareNow and Airbnb are seizing new business areas with innovative business models. In the university environment, sharing of (lab-)infrastructures is yet uncommon. Therefore, we see numerous university labs for IoT and I4.0 with cost-intensive duplicate infrastructures. I4.0 refers to the German government program (2011) for maintain and strengthen the industrial location Germany, while IoT encompass I4.0

M. E. Auer and D. May (Eds.): REV 2020, AISC 1231, pp. 3–20, 2021.
https://doi.org/10.1007/978-3-030-52575-0_1

beside further areas like smart cities or smart heath [5]. In the further document the term IoT is used to cover I4.0 and areas like smart cities.

Educating students on IoT, requires practical and industry-oriented (lab-) experience [2]. In general, experiments in lab environments with different specializations play an important role in education of engineers and scientists [23, 33]. Traditional labs offer students the opportunity to experiment with real systems [7, 28], but these cause high costs for lab infrastructure, space and maintenance staff with marginal utilization and rapid loss of innovation [35]. Sharing of lab-infrastructures can help to overcome these issues.

In this context, the cross-institutional Open Digital Lab 4you[1] (brief: DigiLab4U) project intends to offer a digitalized lab environment. DigiLab4U pursues the objective to provide students, researchers and professors within the IoT-environments with practical, digitalized and networked labs – independent of their location. Figure 1 illustrates the distributed lab environment with local and distant access possibility for users. The participating institutions across international borders work toward common goals concerning teaching, learning and researching in the field of IoT. The cooperation of the universities and research institutions allows to pool the resources so that faculties, learners and researchers have access to a larger variety of digital courses based on different IoT labs.

Essential steps in the project are the technical integration of the partner institutions into the DigiLab4U learning environment. The large variance of the different labs requires a variety of digitalization technologies and didactical concepts that needs to be evaluated.

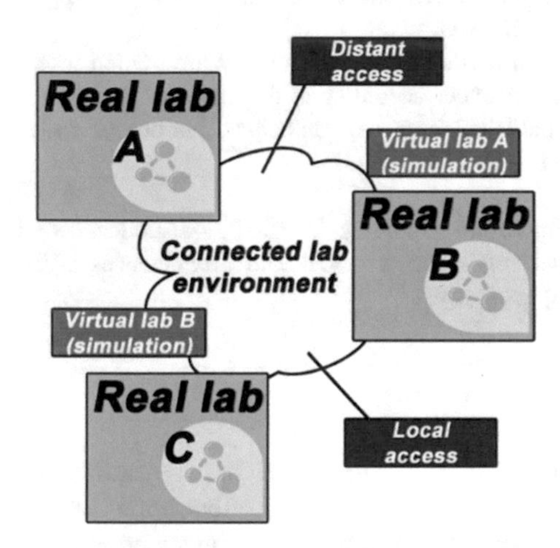

Fig. 1. DigiLab4U – digitalized lab environment in the field of IoT, based on [31]

[1] https://digilab4u.com/

Structure of the Document

The paper compares three different digitalized lab-scenarios. Therefore, the relevance of labs in IoT-education are described in a first step in Sect. 2. In a next step a typical lab scenario – a Radio-frequency Identification (RFID) measurement chamber and the lab competences for a digital environment are explained. Building on this, the digitalization stages, starting with the Virtual Reality mockup, the service robot arm for automation of handling activities and the integration of live data are explained further in Sect. 3. This paragraph concludes by the evaluation of the stages by using a matrix and a grid-diagram which are based on the lab competences for digital environments. Finally, the paper is completed by a discussion and an outlook in Sect. 4. The goal is not only to reflect different options for digitalizing a lab in engineering education, but also to include the judgement of the users (teachers and learners) for the ongoing development of the DigiLab4U platform.

Research Methodology

As research methodology Design-based Research [12] has been applied. The iterative approach supports a continuous development process. As a first step, the problem is specified and the research questions are developed.

- **Which different innovation stages support the development from real to digital laboratories? (RQ1)**
- **Which competences can be acquired in a specific digital laboratory? (RQ2)**

The next step includes the evaluation of literature and experiences and this builds the theoretical framework for the design considerations. This leads to a further step where a first design of a learn- and competence-matrix for labbased-learning was developed and will be cyclical re-designed in later stages of the process. In a next step, the design will be tested and formatively evaluated prospectively in the summer semester 2020 in order to finally develop a learn- and competence-matrix for lab-based-learning in networked environments. At this point, if necessary, it is planned to go back to the design (re-design) and repeat the cycle until the matrix has a degree of maturity that allows a more summative evaluation.

2 Why Labs Are Needed for Educating the IoT-Generation?

Engineering education requires application-based tasks and problematics to foster conceptual understanding as well as practical knowledge and experiences [3]. In general, the use of labs in engineering education can provide huge benefits such as opportunities to engage in processes of investigation and inquiry by relating theory to practice and vice versa [13, 19]. Lab-based learning offers students the opportunity to enhance their theoretical knowledge with practical skills. In the course of digitalization, engineering sciences are confronted with a number of challenges that require in-depth knowledge of IoT and I4.0 [29]. Lab-based learning offers here many possibilities to embed the IoT-topic in the engineering education for example via practical lab-exercises, where students can experience the integration of mobile technology with the internet [27, 32].

2.1 Labs in Education

Currently most universities use real labs for the mediation of practical knowledge, such as Maker Spaces for engineering courses. Those hands-on labs are implemented *"[...] in almost every university in the USA"* [18]. Other universities have a focus on remote access. For example the TU Dortmund University runs a material characterization tests with a robot arm [17]. The Institute of Machine Tools and Production Technology at TU Braunschweig operates a teaching and learning factory for Energy and Resource Efficiency, Digitalization and Urban Factories [18]. The Digital Capability Center (DCC) Aachen is a learning factory for the I4.0 value chain [18] and the Industry 4.0 Learning Factory/4.0 Pilot Factory (I40PF) at TU Wien runs a platform for research, teaching and training with regard to future production [11].

The term **digital labs** describe virtual and remote labs in which personal presence in the lab is not necessary, but access via the Internet is possible. Figure 2 illustrates the possible types of access, laboratories and organizations. A virtual or simulated lab is an online lab that offers software simulations or applications [23, 34]. On the other hand, a remote lab is an online lab that allows real experiments with real hardware by performing real measurements [23, 34]. The difficulty here is that every real lab needs an individual solution for digitalization, such as robotics or augmented reality. For the DigiLab4u project, the dimension of involved organisations is visualized. Hereby, the provision of the lab environment is subdivided for the usage in one/multiple organisations.

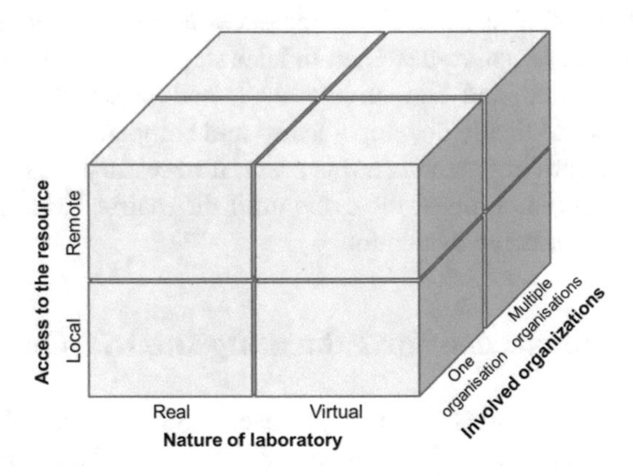

Fig. 2. Classification of laboratories, based on [1, 23, 34]

In general, the different kinds of labs have been investigated so far within their added value for the engineering education, but the results seem to be inconsistent with regard to the comparison of real, remote and virtual labs [6, 23]. The provision of a lab alone does not clarify which competences can be acquired [23]. [6] recommends further empirical studies related to learning outcomes to answer the question how efficient a lab can foster learning objectives or rather competences.

2.2 IoT-Competences Within Laboratory Education

In addition to the question why labs are important in the engineering sciences, arises also the question which competences and goals should be addressed. One overall goal of labs in engineering education according to [13] "*[…] is to prepare students to practice engineering and, in particular, to deal with the forces and materials of nature*" [13]. The Accreditation Board for Engineering and Technology (ABET) extracted thirteen learning objectives to clarify the meaning of lab education and to specify the learning outcome for their use in the practice of engineering education (see Table 1) [13].

Against the background of digitalization, we are currently experiencing a discussion, which competencies are required with regard to innovations in the IoT sectors [2, 18]. What exactly distinguishes IoT-competences is not quite clear and a view on research studies of IoT-competences shows a rather uneven picture. A representative survey on I4.0 differentiates the corporate competencies between technology/data-oriented (e.g. "ability to exchange with machines"); process/customer-oriented (e.g. """participation in innovation processes") and infrastructure/organization-oriented (e.g. "social/communication competence") [2]. While [11] for example differentiates between personal competencies, social/interpersonal competencies, action-related competencies and domain-related competencies [11]. A qualitative analyse of chosen competence profiles for I4.0 of [30] confirms this assessment. The qualitative analyse of five well-founded studies in this context, showed no agreement as to which competencies will be required in the future for the I4.0. This overview does not claim to be complete and sees itself at this point in time as work in progress. It serves as an orientation for the targeted competences and learning objectives in the DigiLab4U project. This pursues two concerns in the project context. First, to build a competence framework for the didactical considerations of the lab environment that could give the users a quick overview of the desired competencies in the respective lab. Second, it should be helpful to differentiate competences and objectives regarding the three different lab options (see below) that differ strongly in the degree of digitalization.

The table presented below maps general competencies with relevant lab-based goals and possible IoT-competencies (see Table 1). In doing so, the focus lies on the competencies and goals that can be promoted in labs. The ABET objectives serve as perspective and decision-criterion for the selection of corresponding requirements. For this first draft, two studies were selected which explicitly address essential competences for engineering studies:

- Competency development study Industry 4.0 of the German Academy of Science and Engineering, 2016 [2].
- VDI-Study (Verein Deutscher Ingenieure): Engineer training for digital transformation of the Association of German Engineers, 2019 [15].

The German Qualifications Framework for Lifelong Learning (DQR) is used as the higher-level reference frame in order to allocate the teaching objectives and the engineering-scientific qualifications [8].

In the current studies a series of requirements can be identified that can be well addressed in lab-based teaching and learning settings. In general, it should be noted that the more recent competence requirements in the course of digitalization (incl. IoT, I4.0)

Table 1. IoT-competences within laboratory education, based on [2, 8, 15]

German qualification framework (DQR)		ABET objectives for laboratory- based learning	IoT-requirements in current studies
Professional competence	**Knowledge** means the totality of facts, fundamentals, theories and practice of the lab-based work area as a result of learning and understanding	(1) Instrumentation (4) Data (9) Safety (13) Sensory awareness	IT-Security (acatech)
	Skills are the ability to apply knowledge and use know-how to perform tasks and solve problems. The skills include cognitive skills (logical, intuitive and creative thinking) and practical skills (dexterity and use of methods, materials, tools and instruments)	(2) Models (3) Experiment (5) Design (7) Creativity (8) Psychomotor	Process know-how (acatech) Data analysis, Data literacy (acatech, VDI) Ability to perform analysis (VDI) Problem solving and optimization competences (acatech, VDI)
Personal competence	**Social Competence** refers to the ability and willingness to work with others in a goal-oriented manner, to grasp their interests and social situations, to deal with and communicate with them rationally and responsibly, and to help shape the world of work and life	(10) Communication (11) Teamwork	Interdisciplinary thinking and acting (acatech, VDI) Communication and cooperation skills (VDI)
	Autonomy means the ability and willingness to act independently and responsibly, to reflect on one's own actions and those of others, and to further develop one's own ability to act	(6) Learn from failure (12) Ethics in the laboratory	Autonomous decisions (acatech) Self-learning competence (VDI) Ethical beliaviour (VDI)

are formulated in very general terms. Therefore, it is important to decide didactically in the respective lab scenario how corresponding lab competences and objectives can be promoted.

In the following sections, this will be illustrated exemplarily for the lab scenario (RFID measuring chamber) with regard to different degrees of digitalization.

3 Target Scenario: RFID Competencies for the Undergraduate Engineering Curriculum in a Digital Environment

RFID is one of the key-technologies for IoT and I4.0 [14] therefore understanding the RFID-basics is mandatory for students of IoT related programs. Within the bachelor degree program of Information Logistics at the HFT-Stuttgart, RFID competencies are required in the corresponding lab curriculum. In the corresponding lab, researchers and students can use extensive equipment for testing and researching RFID. In addition to various read/write systems for industrial and logistic applications, a measuring chamber is available for precise signal strength measurements. Measurements are carried out both in the area of teaching by students, in research by scientific staff, and in application-oriented research with industrial companies in the region. Up to now, samples have been placed in the measuring chamber by hand under the supervision of a technical employee of HFT-Stuttgart.

The RFID measuring chamber of the manufacturer Voyantic Ltd measures[2] $160 \times 106 \times 120$ cm and is used for the performance measurement of RFID transponders/tags (see Fig. 3). In addition to the chamber, an automatic rotation system for three-dimensional measurement and the Tagformance software for measurement and evaluation are installed. Within the software the test procedure threshold, readrange or orientation are available. The RFID tags to be measured are mounted on carrier plates made of different materials for test purposes.

Fig. 3. RFID measuring chamber as real laboratory, at HFT-Stuttgart. The RFID-tag is positioned on a substrate (e.g. glass, cardboard, metal) in the middle of a platform, which can be rotated.

[2] https://voyantic.com/products/tagformance-pro/accessories/rfid-measurement-cabinet

The students need to open the RFID measuring chamber manually. A substrate with a RFID tag is chosen from the stock and inserted into the foam device of the RFID measuring chamber. Then, the measuring chamber is closed manually. The user starts the software and selects the appropriate test procedure. At the end of the test, another test can be started in the software or another RFID tag can be measured. For the test of another RFID tag the process starts again. In order to provide a fully remote operation, existing manual procedures need to be automated or virtualized.

3.1 Lab Objectives: RFID Measuring Chamber

The labs in DigiLab4U are to be evaluated to determine the extent to which they support the achievement of learning objectives and thus the acquisition of competencies for the field IoT. Depending on the type of lab (real, remote or virtual), certain tendencies may emerge in the research field, but ultimately the underlying didactic concept seems to have an impact factor, that needs to be determined for the assessment of which learning objectives and competences can be achieved in which lab. In accordance with the objectives of engineering labs of ABET the following learning objectives for the RFID-lab scenario can be allocated:

1. *Instrumentation and digitization:* Utilize appropriate substrates and tags in the RFID measuring chamber and apply the required software to make measurements of the corresponding range. Measure the effect of rotation angles on read-range and received and transmitted signal strength. Being able to work at and configure the interface between lab and digital performance.
2. *Models:* Identify the strength and limitations of RFID-technology for real world settings. Evaluating and assessing which substrate with which tag can be adequately used in a scenario.
3. *Experiment:* Devise an experimental approach, specify appropriate equipment and procedures, implement these procedures, and interpret the resulting data to characterize an engineering material, component or system (is not relevant in the RFID Chamber, but in a project work, which follows the lab work in the chamber).
4. *Data Analysis:* Analyse the impact of RFID-tag antennas on read-range, of chip-designs and generations on read-range and the impact of different substrates including documentary and reading skills in dealing with measurement data in the lab-based environment.
5. *Design:* Design and assemble a RFID measurement, where an authentic test report has to be filled and the measurement has to be aligned to authentic conditions.
6. *Learn from failure:* Identify measurement errors and not optimal measurement results, reconsider those results in order to optimize measurement processes.
7. *Creativity and independent critical thoughts:* Depending on the lab task, different appropriate levels of independent thought, creativity and the capability in solving authentic exercises are required.
8. *Psychomotor:* Identifying and assessing a “best RFID-tag” for a random field of application (e.g. identifying a bottle of water with a read-range of 1 m in the European frequency range).

9. *Safety:* Identify health, safety, and environmental issues in real, remote and virtual labs related to technological and digital processes and activities, and deal with them responsibly.
10. *Communication:* Communicate, argue and present the results and professional relevance of the RFID lab, including peer-review and coaching sessions.
11. *Teamwork:* Teamwork in an (inter-)national and cross-institutional context (through the DigiLab4U-network).
12. *Ethics in the laboratory:* Scientific integrity will be considered and reflected, as well as the collection of Learning Analytics for lab-based learning purposes.
13. *Sensory Awareness:* Implementing coaching sessions with teachers to give the students the opportunity to review, reflect and to present and argue results.

3.2 Innovation Options of the RFID Measuring Chamber

The utilization-rate of the chamber in education currently is quite low, as students use the chamber once a year for a time of about two weeks. The objective for the RFID-chamber is to provide access to the infrastructure beyond the boundaries of individual institutes as part of the DigiLab4U-project. The RFID measuring chamber shall be digitalized by enhancing the experiment through other technologies used in IoT, such as robotics, Augmented/Virtual Reality (AR/VR) and remote access.

Figure 4 describes the stages of innovation that will be achieved using the Reality-Virtuality (RV) continuum model. The required technical enhancements and constructions at the chamber are currently being implemented by different teams within the DigiLab4u project. The following section provides the concepts and the process sequence of the three innovation stages from Virtual Reality Mockup (Option 1), Service robot arm for automation (Option 2) to Digital Lab – Virtual Reality with live data (Option 3).

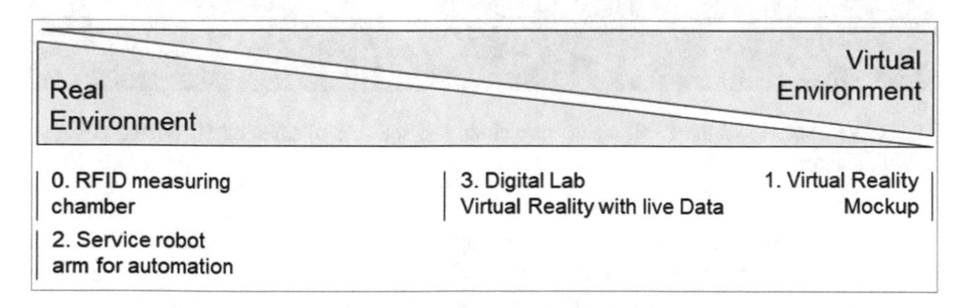

Fig. 4. Reality-Virtuality (RV) continuum, innovation stages (based on [24]) of the RFID measuring chamber as real lab, at HFT-Stuttgart.

Option 1: Virtual Reality Mockup

The concept of VR mock-ups was taken from the industrial design for visualization of scale models, it can refer to purely visual dummies or semi-functional prototypes [10]. A VR mockup, ultra-concurrent lab or fixed remote lab [22] has a set of sample scenarios, based on real lab results. The results are visualized through VR for students to have a similar experience as with real lab experiments but without any direct remote access to the lab.

For the RFID measuring chamber a 3D object (see Fig. 5), surface model with realistic dimensions and visuals is reconstructed [10]. This includes exact size [10], identical surface (e.g. colouring) [10], realistic mechanics and animation (e.g. door opener) [10], program and process flow and identical operating sequence (incl. timing of sequences) of the RFID measuring chamber. In addition, accessories such as RFID-tags are to be realistically visualized (e.g. size, shape, colour). Students should learn how to manipulate the chamber and receive test results of "sample solutions" based on real experiment data.

Fig. 5. VR mockup of the RFID measuring chamber, selected pictures of 3D object, designed by Nils Höhner

The VR-RFID measuring chamber application can be started within the HTC VIVE. The user starts in the VR environment in front of the chamber. The environment is visualized as realistically as possible, with the RFID-tags to the left of the chamber and the computer and RFID reader to the right. The functions of the simulation are operated via the two VIVE controllers. A digital assistant explains (voice via HTC VIVE speaker) the basics of RFID technology, the measuring chamber and the practical relevance. First, the user opens the door of the chamber with the controller, then he can view the interior and get an explanation of the functions, such as the automatic rotation system. On the left of the chamber is a selection of different RFID-tags (based on the predefined mockup), of those one can be inserted into the chamber. After the chamber has been closed, the test procedure of the Tagformance software can be selected and started on the computer. Based on the pre-programmed mockups, the results are displayed in the window on the computer. The procedures can be repeated in the appropriate combination. Second, the test results can be transferred via interfaces for further use.

Option 2: Service Robot Arm for Automation
The concept of using an already existing conventional lab remotely by integration of automation technology is not entirely new. The Institute for Forming Technology and Lightweight Construction (IUL) of the University of Dortmund is currently undertaking a similar project. They are using a KUKA robot, a real-time control system, a safety system and a camera system within their Tube Bending Lab for the automation [21].

A comparable concept shall be applied in the RFID measuring chamber. A multi-axis robot arm, an integrated safety system, an automated locking mechanism for the chamber door, a tag magazine and a camera system will be applied (see Fig. 6). With the help of the equipment, the manual handling activities can be automated and are the prerequisites for the digitalization of the lab.

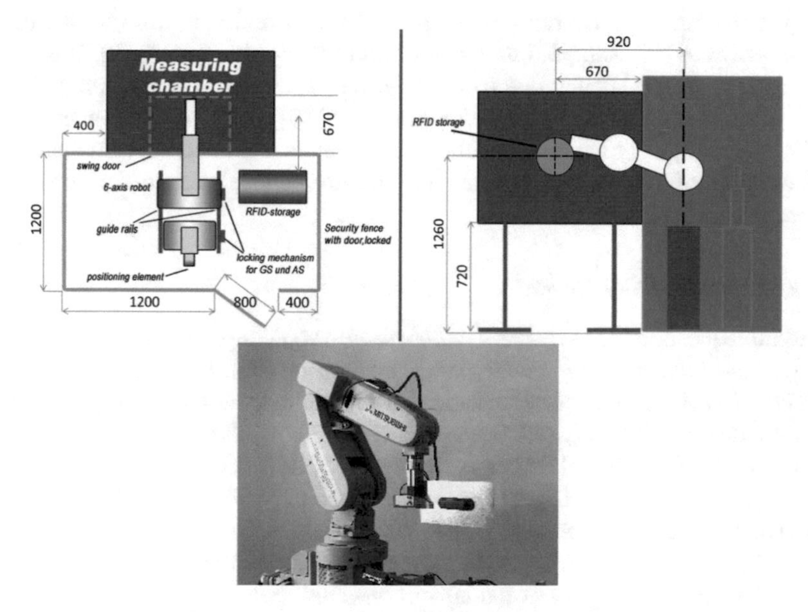

Fig. 6. Technical setup of the service robot arm for automation (top view, front view, robot arm)

The RFID measuring chamber lab with the service robot arm for automation application can be accessed by the user via an internet browser. The user starts on an introduction side for the RFID technology, the measuring chamber and the related equipment. The main activity of the user is the selection of the test parameters like tag type, substrate, orientation and test procedure within a "light" Tagformance Software. After the selection the chamber door opens automatically by the locking mechanism, the robot arm picks the selected tag from the tag magazine and places it in the chamber according to the selected parameter. The physical movements in the lab can be seen by the user via a camera system. The door is closing after the robot has left the chamber. The test procedure is started and the live test results are displayed to the user in the light software. In order to guarantee the necessary safety standards and to improve the security awareness on site, a safety system is installed in front of the measuring chamber and integrated in the program flow. As a result, the entire manual process is automated on site. The user can repeat in the appropriate combination and the test results also can be transferred via interfaces for further use.

Option 3: Digital Lab – Virtual Reality with Live Data
The Digital Lab – Virtual Reality with live data harmonize the virtual and real world and takes the advantages of both like interaction with real equipment calibration; realistic data; no time and place restrictions and medium costs [25]. The VR mockup and the service robot arm are to be combined for this purpose in order to have a live and real experience in a VR environment (see, e.g. Open Remote Lab [22]).

The flow for the user will be similar to the VR mockup but the live data of the service robot arm is transferred to the VR headset via a web interface to enable a remote access. The interactions in the headset are processed, interpreted and recommended for action using Learning Analytics (LA) on a central server ("*Data collection and pre-processing*", "*Analytics and action*" and "*Post-processing*") [9]. It is expected that there is a strong latency due to the physical movement ("*Transport-*", "*Simulations-*", "*Generierungs-*" and "*Darstellungslatenz*" [10]) in the VR system compared to the VR simulation, which is what the evaluation is intended to demonstrate [10]. A video stream of the webcam gives an insight into the local activities of the robot and the measuring chamber.

3.3 Findings and Conclusion

The research question "*Which different innovation stages support the development from real to digital laboratories?*" (**RQ1**) is to be taken up first. Based on the RFID measuring chamber three exemplary scenarios for the digitalization of labs were presented. Within the Reality-Virtuality (RV) continuum (see Fig. 4) the degree of virtual and real components in the Virtual Reality Mockup (Option 1), Service robot arm for automation (Option 2) and Digital Lab – Virtual Reality with live data (Option 3) are visualized. The literature (see Sect. 2.1) also shows, that these scenarios/procedures are different for each lab typ. For this reason, the following objective matrix (see Fig. 8) and grid-diagram (see Fig. 9) of the lab options shall help the reader a) to understand the competence objectives based on the degree of digitalization and b) to use it as a handling tool for the digitalization of their specific IoT labs.

In general all three innovation levels are improving data analysis (Objective 4 [13]) and design (Objective 5 [13]).

Findings of the Virtual Reality Mockup (see Fig. 8.-1).

- Experiment (Objective 3 [13]): Preconfigured scenarios do not provide an opportunity to develop an experimental approach with appropriate devices and methods.

- Learn from Failure (Objective 6 [13]): Due to the isolated simulation, it is not possible to identify unsuccessful outcomes and then re-engineer effective solutions.
- Creativity (Objective 7 [13]): Due to the isolated simulation, independent thought, creativity, and capability in real-world problem solving is completely hemmed.
- Teamwork (Objective 11 [13]): The use of VR glasses restricts the teamwork, including structure individual and joint accountability; assign roles, responsibilities, and tasks; monitor progress; meet deadlines; and integrate individual contributions into a final deliverable.
- Sensory Awareness (Objective 13 [13]): The use of VR glasses prevent the sense of the skin/touch and the sense of smell and cannot be used to formulating conclusions about real-world problems.

Findings of the Service Robot Arm for Automation (see Fig. 8.-2).

- Instrumentation (Objective 1 [13]): The user is applying no appropriate sensors, instrumentation, and/or software tools to make measurements of physical quantities.
- Creativity (Objective 7 [13]): Due to the preconfigured components and processes, independent thought, creativity, and capability in real-world problem solving is mostly hemmed.

Findings of the Digital Lab – Virtual Reality with Live Data (see Fig. 8.-3).

- Learn from Failure (Objective 6 [13]): Due to the isolated simulation, it is not possible to identify unsuccessful outcomes and then re-engineer effective solutions.
- Creativity (Objective 7 [13]): Due to the preconfigured components and processes, independent thought, creativity, and capability in real-world problem solving is mostly hemmed.
- Teamwork (Objective 11 [13]): The use of VR glasses restrict the teamwork, including structure individual and joint accountability; assign roles, responsibilities, and tasks; monitor progress; meet deadlines; and integrate individual contributions into a final deliverable.
- Sensory Awareness (Objective 13 [13]): The use of VR glasses prevent the sense of the skin/touch and the sense of smell and cannot be used to formulating conclusions about real-world problems.

In order to concretize the results more precisely, an objective matrix (see Fig. 8) and a grid-diagram (see Fig. 9) for the better visualization has been created. Each digital lab scenario has been evaluated by the authors so far on a theoretical basis based on the collected competence objectives. The matrix aims to be a useful tool for verifying the laboratories objectives and can be evaluated by several qualitative and quantitative research methods. Thereby each competence can be evaluated with 0 to 3 points (see Fig. 7). *Not fulfilled* (0 points): goal is not funded; *partially fulfilled* (1 points): goal is partially achieved; *fulfilled* (2 points): goal is fully achieved; *outperformed* (3 points): goal is fully achieved and goes beyond conventional/traditional approaches (research potential).

The second research question *"Which competences can be acquired in a specific digital laboratory?"* (**RQ2**) cannot be answered in general terms. It can only be answered with regard to the didactical concept and the pursued learning objectives on which the lab environment is based. The evaluation of a lab scenario, which achieved the corresponding learning outcomes can also support self-reflection learning processes by teachers and learners [16].

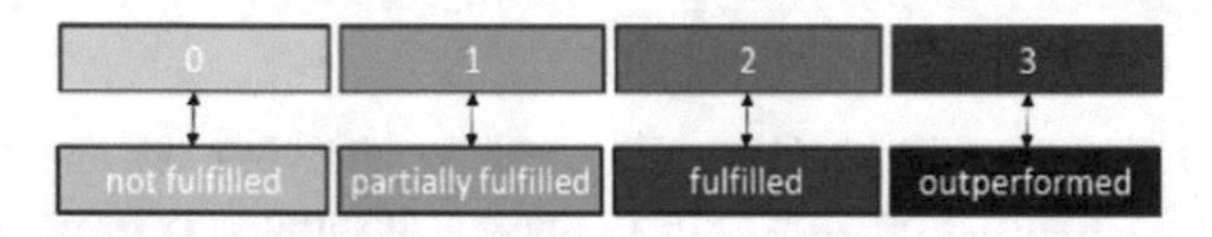

Fig. 7. Objective meter to measure and assess digital labs

Objective	1. Virtual Reality Mockup	2. Service robot arm for automation	3. Digital Lab: Virtual Reality with live Data
1. Instrumentation	1	0	1
2. Models	1	1	1
3. Experiment	1	2	2
4. Data Analysis	3	3	3
5. Design	3	3	3
6. Learn from Failure	1	2	2
7. Creativity	0	1	1
8. Psychomotor	2	2	2
9. Safety	2	2	2
10. Communication	2	2	2
11. Teamwork	1	2	1
12. Ethics in the Laboratory	2	2	2
13. Sensory Awareness	1	2	1
Total	20	24	23

Fig. 8. Evaluation Matrix of the RFID measuring chamber innovation stages, based on the lab competences for a digital environment

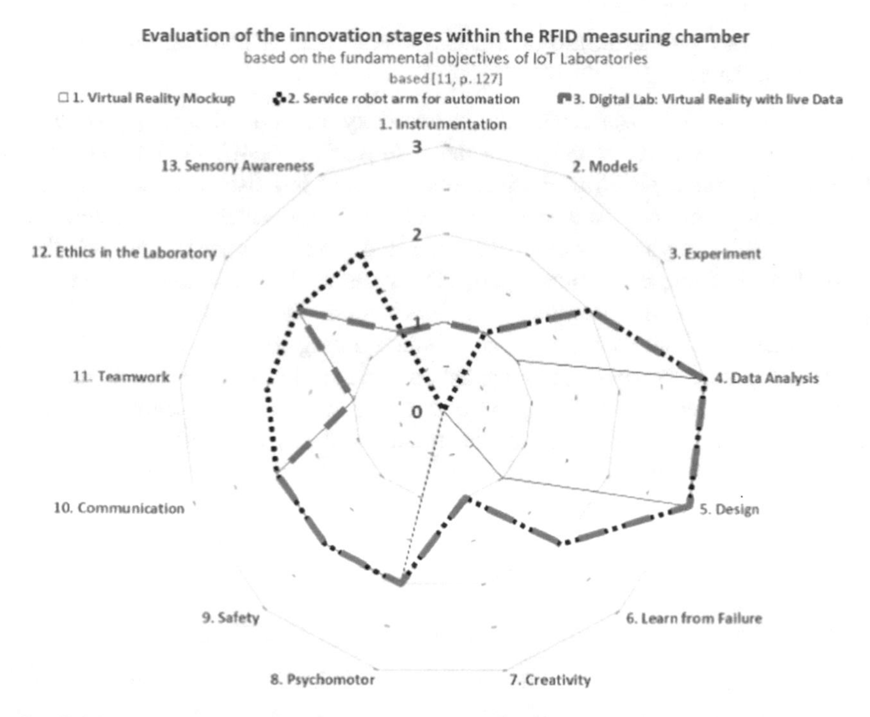

Fig. 9. Grid-diagram of the RFID measuring chamber innovation stages, based on the lab competences for a digital environment

4 Discussion and Outlook

The innovation levels shown can not only be applied to this specific lab but can also be used as a handling tool for other labs to further expand the Sharing Economy in the university environment. The digital labs allow a distributed, cost distributed, time-independent (synchronous/asynchronous) use of the lab infrastructure in parallel of multiple uses worldwide [25]. These advantages have also been shown during evaluation of the case study. However, the use of a digital environment also implements some disadvantages, as a virtual lab will not be able to fully replace a real lab. But the combination of digital labs, VR and live data is nearly closing the gap and meets the required skills demanded by the market [2].

The developed matrix with the corresponding graphical evaluation shows the advantages and disadvantages of different digitalization stages related to the didactical competences. Based on the achieved lab competences the didactic setting can be adapted to create a balanced learning scenario. On the other hand, it also can be considered which competencies my students/researchers should acquire and accordingly the lab should be designed and digitalized. From the DigiLab4U point of view the matrix offers two advantages:

- Users can use the matrix to quickly find out which competencies can be achieved in which lab scenario.

- For laboratory providers, the matrix is a useful tool for defining the objectives of their own laboratory.

However, the digitalization of labs respectively offers another major advantage – the integration of additional media. In the case of DigiLab4U, the labs will be integrated into a learning platform in which not only labs but also entire learning content from different media can be formed and collected. The latter with the intention of gaining learning analytics to support learning and teaching processes as well the (re-) design of the learning environment [20]. The main challenge in the lab-context is to capture and analyse the learning processes within the lab-learning environment, because multiple signal sources that accumulate during the students work in the lab, have to be captured, processed, aggregated and analysed to produce traces that outline and describe the learning activities and user interactions of the learners within the learning processes. Therefore the characteristics and properties of a combination of several modes and sources – and also those sources that do not necessarily provide digital traces like oral presentations or gestures - have to be recorded and described to understand and support particular learning processes [26]. The data generated in the learning platform can also be clustered according to the anticipated lab-achievements of the users.

The authors intend to further develop the findings as follows:

- *Methodical outlook:* Use the concept in multiple lab innovation levels and in different labs. Collection of statistically relevant data from user groups. Compare with several user groups if the competences for IoT in digital and real labs are different.
- *Technical outlook:* Implementation of the concept within DigiLab4U for teaching and research purpose. Development of an architecture that allows the providers of the labs to easily integrate labs into the DigiLab4U environment. Scale with the system worldwide.
- *Didactical outlook:* Development of a transparent IoT certification process for providers and users. We are planning to integrate Open Badges that cover corresponding competences for professional and social competences.

Acknowledgement. This work has been funded by the German Federal Ministry of Education and Research (BMBF) through the project DigiLab4U (No. 16DHB2112). The authors wish to acknowledge the BMBF for their support. We also wish to acknowledge our gratitude to all DigiLab4U project partners for their contribution.

References

1. Abele, E., Metternich, J., Tisch, M., Chryssolouris, G., Sihn, W., ElMaraghy, H., et al.: Learning factories for research, education, and training. Procedia CIRP 32 (2015). https://doi.org/10.1016/j.procir.2015.02.187
2. Acatech (ed.) Kompetenzentwicklungsstudie Industrie 40: Erste Ergebnisse und Schlussfolgerungen. München (2016)

3. Balamuralithara, B., Woods, P.C.: Virtual laboratories in engineering education: the simulation lab and remote lab. Comput. Appl. Eng. Educ. **17**(1), 108–118 (2009). https://doi.org/10.1002/cae.20186
4. Beutin, N.: Share Economy 2017: The New Business Model (2018)
5. BMWi. Digitale Strategie 2025: Bundesministerium für Wirtschaft und Energie. Hirschen Group GmbH, Gutenberg Beuys Feindruckerei GmbH, Berlin (2016)
6. Brinson, J.R.: Learning outcome achievement in non-traditional (virtual and remote) versus traditional (hands-on) laboratories: a review of the empirical research. Comput. Educ. **87**, 218–237 (2015). https://doi.org/10.1016/j.compedu.2015.07.003
7. Bruchmüller, H.-G., Haug, A.: Labordidaktik für Hochschulen: Eine Hinführung zum praxisorientierten Projekt-Labor (Schriftenreihe Report/ Lenkungsausschuss der Studienkommission für Hochschuldidaktik an den Fachhochschulen in Baden-Württemberg, vol. 40. Leuchtturm-Verl, Alsbach/Bergstraße (2001)
8. Bund-Länder-Koordinierungsstelle für den Deutschen Qualifikationsrahmen für lebenslanges Lernen. Handbuch zum Deutschen Qualifikationsrahmen: Struktur - Zuordnungen - Verfahren - Zuständigkeiten (2013). https://www.dqr.de/media/content/DQR_Handbuch_01_08_2013.pdf. Accessed 17 Oct 2019
9. Chatti, M.A., Dyckhoff, A.L., Schroeder, U., Thüs, H.: A reference model for learning analytics. Int. J. Technol. Enhan. Learn. **4**(5/6), 318 (2012). https://doi.org/10.1504/IJTEL.2012.051815
10. Dörner, R., Broll, W., Grimm, P., Jung, B.: Virtual und Augmented Reality (VR/AR). Springer, Heidelberg (2013)
11. Erol, S., Jäger, A., Hold, P., Ott, K., Sihn, W.: Tangible industry 4.0: a scenario-based approach to learning for the future of production. Procedia CIRP **54**, 13–18 (2016). https://doi.org/10.1016/j.procir.2016.03.162
12. Euler, D., Sloane, P.F.E. (eds.) Design-based research (Zeitschrift für Berufs- und Wirtschaftspädagogik Beiheft, vol. 27). Franz Steiner Verlag, Stuttgart (2014)
13. Feisel, L.D., Rosa, A.J.: The role of the laboratory in undergraduate engineering education. J. Eng. Educ. **94**(1), 121–130 (2005)
14. Fleisch, E., Mattern, F. (eds.): Das Internet der Dinge: Ubiquitous Computing und RFID in der Praxis: Visionen, Technologien, Anwendungen, Handlungsanleitungen. Springer, Heidelberg
15. Gottburgsen, A., Wannemacher, K., Wernz, J., Willige, J.: Ingenieurausbildung für die digitale Transformation: Zukunft durch Veränderung. VDI-Studie (2019)
16. Govaerts, S., Verbert, K., Duval, E., Pardo, A.: The student activity meter for awareness and self-reflection. In: Konstan, J.A., Chi, E.H., Höök, K. (eds.) The 2012 ACM Annual Conference Extended Abstracts, Austin, Texas, USA, 05–10 May 2012, p. 869. ACM Press, New York (2012). https://doi.org/10.1145/2212776.2212860
17. Grodotzki, J., Ortelt, T.R., Tekkaya, A.E.: Remote and virtual labs for engineering education 4.0. Procedia Manuf. **26**, 1349–1360 (2018). https://doi.org/10.1016/j.promfg.2018.07.126
18. Haertel, T., Terkowsky, C., Dany, S. (eds.): Hochschullehre & Industrie 4.0: Herausforderungen - Lösungen - Perspektiven (2019)
19. Hofstein, A., Lunetta, V.N.: The laboratory in science education: foundations for the twenty-first century. Sci. Educ. **88**(1), 28–54 (2004). https://doi.org/10.1002/sce.10106
20. Ifenthaler, D., Schumacher, C.: Learning Analytics im Hochschulkontext. WiSt Wirtschaftswissenschaftliches Studium **45**(4), 176–181 (2016). https://doi.org/10.15358/0340-1650-2016-4-176

21. Kruse, D., Kuska, R., Ferich, S., May, D., Ortelt, T.R., Tekkaya, E.A.: More than "did you read the script?": different approaches for preparing students for meaningful experimentation processes in remote and virtual laboratories. In: Auer, M.E., Zutin, D.G. (eds.) Online Engineering & Internet of Things: Proceedings of the 14th International Conference of Remote Engineering and Virtual Instrumentation, pp. 160–169, vol. 22. Springer, New York (2018)
22. Langmann, R.: Industrial internet of things und remote labs in der Lehre für Automatisierungsingenierinnen und -ingeniere. In: Haertel, T., Terkowsky, C., Dany, S. (eds.) Hochschullehre & Industrie 4.0: Herausforderungen - Lösungen – Perspektiven, pp. 105–125 (2019)
23. Ma, J., Nickerson, J.V.: Hands-on, simulated, and remote laboratories. ACM Comput. Surv. **38**(3), 7-es (2006). https://doi.org/10.1145/1132960.1132961
24. Milgram, P., Takemura, H., Utsumi, A., Kishino, F.: Augmented reality: a class of displays on the reality-virtuality continuum. In: Das, H. (ed.) Photonics for Industrial Applications, Boston, MA, 31 October 1994, SPIE Proceedings, pp. 282–292 (1995). https://doi.org/10.1117/12.197321
25. Nedic, Z., Machotka, J., Nafalski, A.: Remote laboratories versus virtual and real laboratories. In: 33rd Annual Frontiers in Education, 5–8 November 2003, pp. T3E_1–T3E_6+T3 E-6. Boulder, CO. (2003)
26. Ochoa, X.: Multimodal learning analytics. In: Lang, C., Siemens, G., Wise, A., Gasevic, D. (eds.) Handbook of Learning Analytics, pp. 129–141. Society for Learning Analytics Research (SoLAR) (2017)
27. Parkhomenko, A., Tulenkov, A., Sokolyanskii, A., Zalyubovskiy, Y., Parkhomenko, A., Stepanenko, A.: The application of the remote lab for studying the issues of smart house systems power efficiency, safety and cybersecurity. In: Auer, M.E., Langmann, R. (eds.) Smart Industry & Smart Education, Lecture Notes in Networks and Systems, vol. 47, pp. 395–402. Springer, Cham (2019)
28. Pleul, C.: Das Labor als Lehr-Lern-Umgebung in der Umformtechnik. Dissertation. Technische Universität Dortmund, Shaker Verlag GmbH (2016)
29. Raikar, M.M., Desai, P.M.V., Narayankar, P.: Upsurge of IoT (internet of things) in engineering education: a case study, pp. 191–197. IEEE (2018). https://doi.org/10.1109/icacci.2018.8554546
30. Terkowsky, C., May, D., Frye, S.: Labordidaktik: Kompetenzen für die Arbeitswelt 4.0. In: Haertel, T., Terkowsky, C., Dany, S. (eds.) Hochschullehre & Industrie 4.0: Herausforderungen - Lösungen – Perspektiven, pp. 89–103 (2019)
31. Uckelmann, D., Scholz-Reiter, B., Rügge, I., Hong, B., Rizzi, A. (eds.): The Impact of Virtual, Remote and Real Logistics Labs: Communications in Computer and Information Science. Springer, Heidelberg (2012)
32. Vujović, V., Maksimović, M.: The impact of the "Internet of Things" on engineering education. In: Yuen, K.S., Li, C.K. (eds.) Making Learning Mobile and Ubiquitous, pp. 135–144. Open University of Hong Kong, Hong Kong (2015)
33. Zubía, J.G., Alves, G. (eds.): Using Remote Labs in Education: Two Little Ducks in Remote Experimentation. University of Deusto, Bilbao (2011)
34. Zutin, D.G., Auer, M.E., Maier, C., Niederstatter, M.: Lab2go – a repository to locate educational online laboratories. In: 2010 IEEE Education Engineering 2010 – The Future of Global Learning Engineering Education (EDUCON 2010), Madrid, 14–16 Apr 2010, pp. 1741–1746. IEEE (2010). https://doi.org/10.1109/educon.2010.5492412
35. Zvacek, S.M.: Preface: University of Kansas, (USA). In: Zubía, J.G., Alves, G. (eds.) Using Remote Labs in Education: Two Little Ducks in Remote Experimentation, pp. 11–16. University of Deusto, Bilbao (2011)

Safety and Security in Federated Remote Labs – A Requirement Analysis

Dieter Uckelmann[1(✉)], Davide Mezzogori[2], Giovanni Esposito[2], Mattia Neroni[2], Davide Reverberi[2], and Maria Ustenko[2]

[1] Faculty of Surveying, Informatics and Mathematics, Hochschule für Technik, Stuttgart, Germany
dieter.uckelmann@hft-stuttgart.de

[2] Department of Engineering and Architecture, University of Parma, Parma, Italy
{davide.mezzogori,giovanni.esposito, mattia.neroni,maria.ustenko}@unipr.it, davide.reverberi1@studenti.unipr.it

Abstract. Recently, the interest of the educational community in laboratory-(lab-) based education has grown steadily. As remote and virtual labs have started to be a reliable alternative to traditional hands-on labs, security and safety issues are becoming increasingly important, especially in the case of remote laboratories, as their interconnected nature raises new and challenging issues. When multiple institutions are involved in a federated labs infrastructure, the complexity increases. Since a structured approach to assess safety and security for federated remote labs is missing, this paper aims at clarifying the general requirements to be considered and proposes a general concept for assessing safety and security in federated labs. Firstly, we analyze the current state on safety and security in remote labs by means of a literature review. Secondly, we investigate existing requirements and define operational requirements for a safety and security guideline in federated remote labs. Thirdly, we provide an overview about standardization approaches and existing guidelines and suggest a guideline, which matches our requirements analysis.

Keywords: Safety · Security · Remote labs · Federated labs

1 Introduction and Context

The interest of research community in lab-based education has been growing [1]. Remote and virtual labs within networks have started to be a reliable alternative to the traditional hands-on laboratories, because of sharing of costs, efficient control of user access to the experiment environment, and availability of a more diverse range of experimentation [2]. Safety and security in labs are important issues to avoid both malfunctioning behavior and intentionally caused harms. This is especially true for remote labs, since they are highly-connected systems [3]. Remote labs in public institutions such as universities and schools are extremely vulnerable to security issues [4]. Federated remote labs including multiple independent institutions and network structures are yet increasing the security risks involved.

M. E. Auer and D. May (Eds.): REV 2020, AISC 1231, pp. 21–36, 2021.
https://doi.org/10.1007/978-3-030-52575-0_2

In literature, there has been little focus on safety except for e.g. chemistry and biology labs. Although Scopus lists more than 2,400 works on remote and networked labs, few of them discuss safety and security issues in detail. However, a structured approach to access safety and security for federated remote labs is missing. Furthermore, most researchers focus either just on safety or security, and they lack work approaching both the issues concurrently [3] except for Henke et al., who have developed a corresponding system for their Grid of Online Lab Devices Ilmenau (GOLDi) to *"protect the physical systems in the lab against wrong control algorithms or malicious trying to sabotage or to destroy the system"* [5]. However, their work focusses a single institution with its remote labs. Further research is needed, especially considering multiple federated labs consisting of different institutions, lab infrastructures and networks.

While safety and security are required, there are other conflicting objectives. Henke et al. for example see flexibility for different student solution designs in labs as a key objective [5]. From a lab-management perspective, the additional safety and security requirements for remote labs should not cause huge efforts or costs as otherwise, they will not be interested to implement their labs in a federated structure.

Federated remote infrastructures lead to new safety and security issues. We see similar issues in connected Industry 4.0 environments. *"Industry 4.0 is more vulnerable to cyber-espionage because of the smart and connected business processes"* [6]. Remote access is among the Top10 threats for industrial control systems [7]. The Industrial Internet consortium sees the *"increased security risks due to increased attack surface"* as the number one key safety challenge for the Industrial Internet of Things (IIRA), followed by the convergence of Information Technology (IT) and Operational Technology (OT) [8]. Concerning its impact, the main problem is that security issues in the "virtual world" can cause harm in the "real world".

In federated lab-structures, a common guideline for participating labs needs to be established. Based on our analysis, we propose to use the VDI/VDE 2182 on "IT-security for industrial automation" as basic guideline, which can be extended to fulfill the specific needs for federated remote labs. This work is based on a funded project for sharing lab resources called Open Digital Lab for You (DigiLab4U[1]), currently including German and Italian labs.

2 Research Methodology

We utilize different methodologies. Firstly, we perform a literature research (1) on scientific papers related to safety for remote labs and (2) on legal requirements. For the research on existing papers, we use a query on Scopus (scientific database) and identify relevant papers, based on title, keywords and abstract. Secondly, we investigate existing requirements and define operational requirements for a safety and security guideline in federated remote labs. Legal safety, security and privacy requirements as well as organization needs are considered. Thirdly, we provide an overview about

[1] https://digilab4u.com/.

standardization approaches and existing guidelines and identify a guideline, which matches the main criteria of our requirements analysis.

3 Literature Review

With the aim of performing a review of the existing material on safety and security concerns in remote-labs literature, we made use of Scopus, *the largest abstract and citation database of peer-reviewed literature.* We queried the database by using the string *(TITLE-ABS-KEY ("remote lab*")) AND (safety OR security).* The string complies with the Scopus rules: (i) combinatory rules of *AND* and *OR* operators, and (ii) use of asterisk *"*"* in order to cut the suffix of the word "*laboratory*" and then consider all its possible declination. Furthermore, the tool is not case sensitive. The query string, as set, gathered 244 documents. We worked on titles and abstracts to only select the papers, which are strictly focused on safety and security issues. For instance, thirty-three documents, propose a holistic description of the developed and implemented solution, however they do not really discuss safety and security issues in-depth.

We identified twenty-three papers of interest including the work of Henke et al. [5] already discussed in paragraph 1, hence we do not consider this paper for the literature review. Because of language issues we could only analyze English documents available on the web (fifteen papers). Four of them relate to "*safety*", eleven relate to "*security*". In the followings, we briefly describe what these papers deal with, distinguishing them by the focus (i.e. safety or security).

3.1 Safety

The problem of safety in remote laboratories is well discussed by Maiti, Kist & Maxwell [6]. Their work relates to laboratories where multiple users develop experiments and share them as part of collaborative systems (i.e. Peer-to-Peer Remote Access Lab). The items on which the reliability of the whole system depends are (i) components and design of experimental rigs, and (ii) network and users (developers) characteristics. Casini, Prattichizzo & Vicino [7] identified the reduction of time to fix software / hardware failure as a key need and solved the issue by means of a bootable (live) device (CD) on the server side. Kozík T., Šimon M. [8] suggested to implement suitable authentication mechanism as first step to achieve the access control. Furthermore, since the remote laboratory is connected to the Internet, it is necessary to protect it by firewall and by an Intrusion Detection System (IDS), which role is to identify an abuse unauthorized or improper use of a computer system, thus also addressing security issues. Marangé, Gellot & Riera [9] as well as Maiti, Kist & Maxwell [6] proposed an approach using two validation filters to guarantee the safety of the operators and the equipment. It is based on the definition of logical constraints which should in no case to be violated. One filter called "system validation filter" validates outputs before sending them to the plant. The second filter called "functional validation filter" validates the use of the functions with regard to the autonomy mode selected. This filter reduces the use of safety constrains which could be violated in the system validation filter.

3.2 Security

Gerža, Schauer & Jašek [10] focused on the security of remote labs against malign attack. The authors analyze the general and specific software and hardware risks and provide the necessary behavior and practices to implement for preventing them. According to these authors, we have identified three main concerns over security in retrieved papers: (i) users' authentication and authorization (ii) access to and (iii) communication with the server. Ocaya [11] proposed a simplified authentication of permitted clients built-into the server, through alphanumeric username and password (issued at registration and periodically changeable). Chellaiah et al. [12] suggested to secure the users' authentication by means of narrative constructs using a sequence of cartoon images to generate an image-based password system. Krbeček & Schauer [13] also proposed to assure the security of the system by means of users' registration and reservation through username-and-password access-system, however they focused on securing the storage of data into the Learning Management System (LMS) [14]. The solution provided is twofold. The outside communication is based on TCP/IP protocol, which assures the reliability of data transmission. The inner communication between the experiment and Remote Laboratory Management System (RLMS) "*ensures the transmission of the measured data and preserves them for later use*", using Java language to provide communication and diagnostic services. The use of Java application and tools for security issues concerning the communication between equipment and server have had a widespread development in recent years, since the most security applications in virtual and remote labs (VRLs) are developed with high level programming tools using Java [15]. Unfortunately, smart devices usually do not run Java. For this reason Sáenz et al. [15] proposed a structure that allows a remote connection with hardware devices by means of a client-server configuration that serve to the client a Javascript application; leaving to the server the task of running the Java part of the virtual and remote lab. A similar solution was deployed in Herrera et al. [16]. They made use of a software solution to provide the security of machineries within their remote bench for testing electrical machine based on Easy Java Simulation (EJS) to connect the real hardware to the user interface for controlling (i) the load voltage and frequency when the machine is connected to the load in an islanded way, and (ii) the active and reactive power injected to the net when the machine is directly connected. In order to pledge the security of the network of both the single institution and the federated labs (i.e. the universities) a main approach seems to be the use of virtual machines. Border [17] presented the Remote Laboratory Emulation Systems (RLES) solution for accessing and scheduling the labs based on "*read-only libraries of virtual servers that can easily be copied, stored and deployed*'". Li & Mohammed [18] installed the virtual machines on students' personal computer with the guest operating systems and their applications run concurrently on a single physical machine. Richter et al. [19] assured the security of the network by using virtual machines on the users' side, while at the server side they split the virtual machine into two virtual network cards, the former managing the access to the system from the server itself, and the latter managing the host system making it possible to reach the virtual machine from the outside: "*this small virtual network is otherwise unconnected to the rest of the university system and any potentially malicious programs could not be passed to any other machine on the university campus*".

Finally, a solution summarizing several concepts analyzed by means of a remote-lab architecture, seems to be the one proposed by Pálka & Schauer [20]. The authors divided the infrastructure into multiple security zones that provide different levels of protection on the basis of whether a user should be granted access to specific resources. Furthermore, to increase flexibility and the ability to recover from a successful attack authors proposed a balanced control and an increased focus on user awareness as well as data protection anchored in the information assets. Table 1 provides an overview of all the reviewed papers.

Table 1. Overview of existing papers related to remote labs and safety or security

Ref.	Authors and year (list in alphabetical order)	Issue	Content (proposed solutions)
[17]	Border C., 2007	Security	usage of read-only libraries and virtual servers by means of virtual machines for accessing the lab of their RLES
[7]	Casini M., Prattichizzo D., Vicino A., 2007	Safety	technique based on the usage of a bootable (live) device on the server side of a remote laboratory is designed
[12]	Chellaiah P., Nair B., Achuthan K., Diwakar S., 2017	Security	authentication system based on image-based password setting is proposed
[10]	Gerža M., Schauer F., Jašek R., 2014	Security (Review)	focus is on security of remote labs against malign attack; the general and specific software and hardware risks are analyzed; the behavior and practices to implement for preventing them are provided
[16]	Herrera M.R.S., Márquez J.M. A., Borrero A.M., Sánchez M.A. M., 2013	Security	usage of EJS to allow the communication between equipment and user interface for monitoring activities
[8]	Kozík T., Šimon M., 2012	Safety	research on (1) a suitable authentication mechanism, (2) firewall (for the internet protection), (3) IDS (for detecting improper use of computer systems)
[13]	Krbeček M., Schauer F., Jašek R., 2013	Security	research on (1) creation of automatic “log-file” where all the users’ data are stored, (2) security of the lab is provided by means of registration-and-reservation system, (3) use of UPS power supply ensuring the restart and accessibility after a power blackout

(*continued*)

Table 1. (*continued*)

Ref.	Authors and year (list in alphabetical order)	Issue	Content (proposed solutions)
[14]	Krbeček M., Schauer F., 2015	Security	research on (1) use of TCP/IP protocol to assure reliability of communication, (2) Java language to provide application for communication interface and diagnostic service
[18]	Li P., Mohammed T., 2008	Security	security of the lab is provided by means of virtual machine for performing the experiment
[6]	Maiti A., Kist A.A., Maxwell A. D., 2015	Safety (Review)	conditions to satisfy the locating and running laboratories to get the desired results, especially, on the characteristics and components in remote access laboratories where multiple users develop experiments and share them as part of a collaborative system
[9]	Marangé P., Gellot F., Riera B., 2007	Safety	two validation filters based on the definition of logical constraints to assure the safety of lab, (1) the "system validation filter" validates outputs before sending them to the plant, (2) the "functional validation filter" validates the use of the functions with regard to the autonomy mode selected
[11]	Ocaya R.O., 2011	Security	authentication through username and passwords allows to secure unauthorized access for low cost labs
[20]	Pálka L., Schauer F., 2016	Security	research on security of the data storage of the Data Center, with remote laboratories working under the Laboratory Management System (LMS); aims at present a security solution against data risks when they are shared through different users (of the same network or among networks too); solution deployed relies on a new architecture with a dynamic, multitiered trust model that e.g. for an individual user, the level of access provided may vary dynamically over time, depending on a variety of factors-such as

(*continued*)

Table 1. (*continued*)

Ref.	Authors and year (list in alphabetical order)	Issue	Content (proposed solutions)
			whether the user is accessing the network from a highly secure managed device or an untrusted unmanaged device
[19]	Richter Th., Watson R., Kassavetis S., Kraft M., Grube P., Boehringer D., De Vries P., Hatzikraniotis E., Logothetidis S., 2012	Security	usage of two virtual machines, the former managing the access to the server, and the latter managing the access to the virtual machine from the outside
[15]	Sáenz J., Esquembre F., García F.J., Torre L.D.L., Dormido S., 2016	Security	proposed system relies on Java and Javascript applications allowing the communication between device side and client-server side

The literature research shows, that there still is a need for further domain-specific research on security and safety. While 5 out of 71 papers (two are not available on the web) have focus on "safety", 15 out of 172 (four papers not available on the web) papers are focused on "security" (Fig. 1). This suggests that the "security" word is very inflated but poorly stressed. On the other hand, the only paper that considers both "safety" and "security" issues together is the work of Henke et al. [5] (out of 244!).

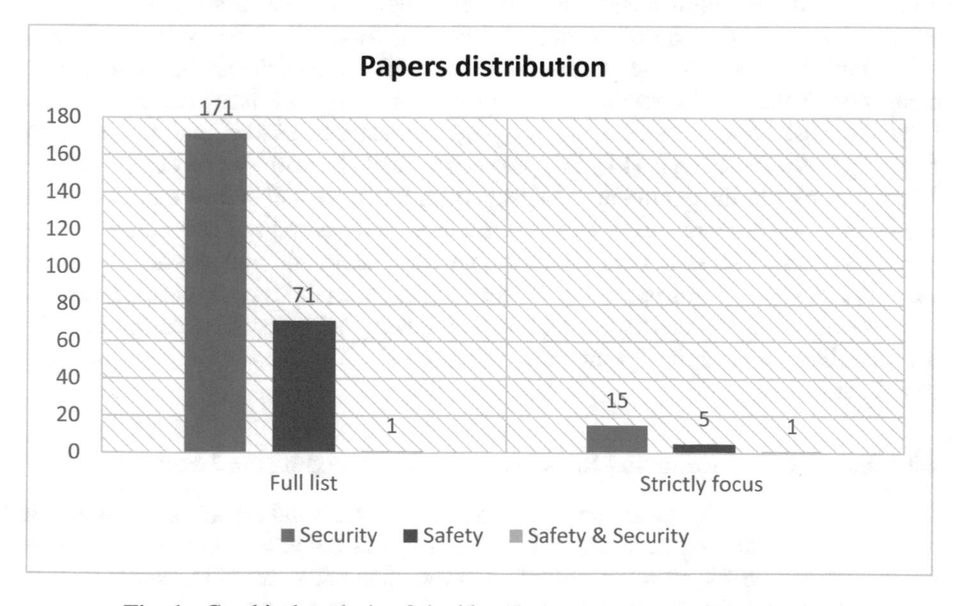

Fig. 1. Graphical analysis of the identified papers on security and safety

4 Requirement Analysis

In this paragraph, we identify existing requirements and future needs for federated remote labs.

4.1 Legal Safety Requirements

Each county or region has different requirements and procedures on safety. Legal requirements may need to be respected in relation to "product safety" (e.g. Produktsicherheitsgesetz in Germany [12]) and "occupational safety and health" (e.g. Directive 89/391 in Europe, Arbeitsschutzgesetz [13] in Germany, D.Lgs. 81/08 "Testo unico sulla salute e sicurezza sul lavoro" in Italy). Further specific laws, for example on chemicals or electromagnetic fields may apply in certain lab scenarios. In remote labs, the product users (e.g. students of another university) are usually not at risk, as they access the infrastructure through an Internet-connection. However, if physical components are used by the students (e.g. processor-boards connected to the federated lab infrastructure) the requirements concerning product safety may apply. The European Directive 89/391 defines minimum requirements, which have been implemented in national laws (Table 2). The European Directive 89/391 is based on a list of general principles. As currently, only German and Italian labs are participating in DigiLab4U, we have focused on these regulations.

Directive 89/391 defines not only the principles but also the obligations and actions for employers and workers in every situation.

4.2 Privacy/GDPR Requirements

The federated labs infrastructure in DigiLab4U will collect data about students and experiments, with the aim of creating a Learning Analytics (LA) system. As such, being initially designed and developed in European universities, the system must comply with the GDPR regulation 2016/679. As in remote labs, the risk of data breaches must be considered both: for proper counter-measures, and for promptly notifying users. As stated in [14], data protection should be provided by design and as a default. Moreover, the definition and design of the network and software infrastructure should take into account such issues as centralized or decentralized data collection and storage systems. The issues are relevant both from economic and practical points of view. For example, if processed data is sent back to each lab, it would certainly require a greater technical and economical efforts than having only a centralized solution. On the other hand, a centralized solution will impose significant costs for managing federated network.

4.3 Existing Organizational Requirements and Procedures in Local Labs

While working in a lab, several safety issues can arise. Safe and reliable systems should prevent harm to lab assistants, students, machines, as well as protect user data. These goals are common for each institution; however, different universities can have their own norms on security and safety. As an example, students at the University of Parma

Table 2. Directive 89/391 principles implemented into German and Italian law

Directive principles	German law	Italian law
Avoiding risks	*work shall be shaped so as to avoid, as far as possible, any risk to life and physical and mental health and to keep the remaining risk as low as possible*	*risks must be eliminated and, where not possible, the impact must be reduced at minimum according with the acquired knowledge based on the technical progress*
Evaluating the risks	*consideration shall be given to special risks to groups of workers requiring particular protection*	*preparation of DVR (Document of Risk Evaluation) – is the evaluation of all the risks for the health and the safety of workers, including the risks for group of workers subjected to particular risks, related-work stress, pregnant workers and the risks concerning the difference of gender, age, and workers with origin from different countries*
Adapting the work to the individual	*regulations with direct or indirect gender-specific effects shall be permissible only where this is imperative on biological grounds*	
Combating the risks at source	*risks shall be combated at their source*	*elimination of the risks at source*
Adapting the technical progress	*measures shall be planned with the objective of properly linking technology, labor organization, other conditions of work, social relations and the environmental influence on the work place*	*schedule the prevention, with an integration of technical and productive conditions and in relation with the work environment and organizational factors*
Replacing the dangerous by the non- or the less dangerous		*substitution of everything dangerous with what is not, or that are less dangerous*
Developing a coherent overall prevention policy	*when implementing the measures, consideration shall be given to the state of the art, occupational medicine and hygiene, as well as other established findings of ergonomics*	*scheduling the necessary measures to have a continuous improvement in safety levels, also with adoption of behavior code and good practices*
Prioritizing collective measures (over individual protective measures)	*individual protective measures shall be subordinate to other measures*	*collective prevention measures have to be preferred to individual measures*
Giving appropriate instructions to the workers	*workers shall be given appropriate training instructions*	*information and formation for workers, manager, person in charge instructions for the workers and utilization of warning and safety signal*

are required to take a Moodle-course, while students at HFT Stuttgart must attend a face-to-face class session at the beginning of their studies.

Current safety practices at HFT Stuttgart (faculty C): in order to achieve a 100% training rate, new students at HFT Stuttgart do not get their account data for the university network, before they have attended the basic safety instruction course. In a lab-scenario, where students from University of Parma access remote labs at HFT Stuttgart, the need to have a "common denominator" to let the students conduct experiments in an easy and safe manner occurs.

Current safety practices for the RFID lab at the University of Parma (based on an interview with the corresponding lab manager): the access to the lab is forbidden to students without supervision by official full-time staff (i.e. professors, teaching assistants, etc.). The main source of risk is a conveyor, and other handling equipment. No harmful substances are present in the lab. If an experiment requires the usage of any special equipment for which the students have never been trained before, such training will be provided by a professor in charge. To operate lab in a safe manner not more than 7 students are allowed to stay in a lab at the same time. Students are considered equally to employees, as stated by Italian laws D.M. 363/98 and D.Lgs.81/08. It is stated that "*students of university courses, PhD students, postgraduates, trainees, scholarship holders and similar subjects are equal to employees if they attend educational, research or service laboratories where machinery, equipment and work equipment in general are used, and chemical, physical and biological agents are present...*". According to an Article 37 of D.Lgs 81/08, the employer is required to provide training for all the employees. It is mandatory for students of Parma University to pass a Moodle-based course, which is comprised of three modules. The first two modules are mandatory for everyone, regardless of which specific course is attended, the third one is only required for those students who have any lab activities in their courses. The first module consists of a generic training, related to situations of risk, possible damage and injuries, and the consequent measures and procedures for prevention and protection. The second module is related to low risk activities (e.g. manual handling of loads, work-related stress and organizational wellness, etc.), and the last module is concerned with medium risk (e.g. dangerous substances) tasks. Students must access with the university access credentials and attend the lessons in order. Lessons are administered through audio and videos, which have to be fully watched in the given order. After that the platform unlocks a multiple-choice test, which the student must successfully pass, in order to unlock the next module. Once all tests are positively passed, the system produces a certificate attesting the training.

Current safety at BIBA Lab: BIBA is obliged to follow the safety and security guidelines of the University of Bremen. This implies a yearly safety and security training for all staff members including assistants. The University has a person responsible for this and in addition, BIBA has its own safety and security manager, who carries out the training of each new staff member at the beginning.

In addition, there are specific guidelines for the BIBAgamingLab, that are based on a risk analysis carried out in 2015 based on different ISO guidelines and regularly updated every February. The reason for this, it is a lab with limited access for employees and thus there are restricted working hours. Even trained personnel are only allowed to be alone in the lab, if they have announced this to the researchers outside the lab. Outside standard working hours for BIBA, any work requires that 3 persons are available and that it is approved by the head of the BIBAgamingLab. Student or other visitors are not allowed to be unsupervised in the lab.

BIBA personnel who do not have access on regularly basis to the lab are allowed to carry out work if it is requested beforehand and are aware of the gamingLab guidelines.

Most of the work carried out in the lab is related to computer games. Thus, there are guidelines in-line with the GDPR with comprised ethical consideration (for more informations see[2]). These guidelines cover all the aspects related to ethical issues, data management and privacy, which the Lab mostly deals with, including the specific requirements. The guidelines are to be well known to all employees involved in relevant activities. Furthermore, since many of the games we offer, can be accesses online of externals, these guidelines also need to be followed by those and consent hereto is required before (if we store data) as well as a document that it corresponds to national legislation.

4.4 Organizational Needs for Lab-Networks

Federated remote lab environments deal with different regional, national and organizational requirements. To find a common guideline simplicity and flexibility are necessary. Derived from the need for simplicity, we also identify a need to look at safety and security issues jointly. The three needs are described in more detail in the following.

4.5 The Need for Simplicity

In a federated remote lab-networks, multiple institutions open their lab-environments to other institutions. However, these activities need to pay off through money, reputation, or access to other labs in a sharing society. Any extra burden, including additional activities for safety and security implied by making labs remotely accessible, will negatively impact the economic balance. Additionally, the limited number of human resources for small university labs (e.g. one professor and one technician) are to be considered. Therefore, the safety and security assessment are expected to be easily applied and maintained. This rules out some of the standards and guidelines in paragraph 5, which are more focused on large enterprises. VDI/VDE 2182 provides a simple approach based on tables. The document is available in German and English. A software-support is favored in DigiLab4U – as its core-focus is digitalization. The Federal Office for Information Security (BSI) has provided a lightweight software tool utilizing the same asset-based approach (LARS [9]). However, currently it is only available in German and it is still more complex than the VDI/VDE 2182 tables.

[2] https://zenodo.org/record/1256626#.Xcsu71dKg2w; https://beaconing.eu/wp-content/uploads/deliverables/D1.8.pdf; https://beaconing.eu/wp-content/uploads/deliverables/D1.9.pdf.

4.6 The Need to Look Jointly at Safety and Security

In highly digitalized and networked scenarios such as the Internet of Things (IoT) or federated lab infrastructures, a separate investigation of safety and security issues would be time- and money-consuming. The need on integrating safety and security has been mentioned in literature, as *"safety and security can negatively influence each other, analyzing their interplay in an efficient manner means reducing the effort that needs to be invested in achieving a safe and secure system"* [4]. Security for safety in the context of Cyber-Physical Systems (CPS) is a current topic in research [9, 10]. Safety and security may even benefit from each other. For example, in IT security, logging mechanisms have been used for decades [11].

4.7 The Need for a Flexible and Iterative Approach

University labs are constantly changing, except for fundamental lab-classes, where the same experiments are performed year after year. In relation to current topics such as the IoT and Industry 4.0 however, technology and concepts are under constant development. Researcher as well as a student in research-oriented lectures requires a certain flexibility [5]. Current research approaches "*...lack evaluation of their support for efficient system update handling.*" [4]. Iterative safety and security measures need to address the whole lifecycle of the remote labs including development, testing, maintenance, and operation. The Deming Cycle – PDCA (Plan, Do, Check, Act) is one of the most used iterative approaches. It has been adopted in a wide number of fields and in standards (e.g. ISO 9001:2015, VDI/VDE 2182). The PDCA is an iterative approach for continuous improvement for product, process and services and in problem solving scenarios.

5 Safety and Security Related Standards and Guidelines

The following table provides a non-comprehensive overview about standards related to safety and security, as standards are provided from numerous different national and international institutions, industries and interest groups (Table 3).

Table 3. An excerpt of current standards and initiatives concerning safety and security

Standard/Initiative	Title	Remarks
Guidelines of the German Federal Office for Information Security (BSI, Germany)	IT-Grundschutz (IT Baseline Protection Manual)	Compatible with ISO/IEC 27001; contains security recommendation on a wide variety of topics (see: https://www.bsi.bund.de/EN/Topics/ITGrundschutz/itgrundschutz_node.html

(*continued*)

Table 3. (*continued*)

Standard/Initiative	Title	Remarks
HSG 65 (UK)	Managing for Health and Safety	British guideline for leaders and line managers, helps to implement the PDCA approach in Safety and Security, it's a guide to implement a Safety and Security system in different situation
IEC 62443	Industrial communication networks	Series of standards on security for Industrial Automation and Control Systems (IACS), based on ISO 27000, complex set of standards
ISO 45001	Occupational health and safety management systems – Requirements with guidance for use	Enables organizations to provide safe and healthy workplaces, identification of potential risks, continuous process improvement and risk reduction
ISO/IEC 27000 Series	Security techniques – Information Security Management Systems (ISMS)	Composed of 46 standards to cover the issues concerning IT, continuous feedback and improvement
ISO/IEC 15048	Security techniques – Evaluation Criteria for IT security	A resource used for the evaluation of the security of IT product and system, the security function of IT product and system are evaluated by a common set of requirements, same common set of requirements is used during the security evaluation of the IT products and system
Information Security Forum (UK)	The ISF Standard for Good Practice for Information Security 2018	Organized into six categories (agile system development, alignment of information risk with operational risk, collaboration platforms, Industrial Control Systems (ICS), information privacy and threat Intelligence), primarily published with the scope of eliminate redundancy

(*continued*)

Table 3. (*continued*)

Standard/Initiative	Title	Remarks
North American Electric Reliability Corporation (US)	NERC 1300: Cyber Security Standard	Cyber security and compliance practices, *"Ensure to prevent cyber threats and protect critical cyber assets that can affect the reliability of bulk electric systems"*
NIST (US)	NIST SP 800-53	Database developed for the US federal agencies, fix the Information Security Framework, provides an overview on security controls based on impact (low, high, medium)
VDI/VDE 2182 (Germany)	IT-security for industrial automation	Clarifies the IT Security topic for production plant, uses examples for manufacturing and process industry; series of recommendations, safety and security topic for multiple actors (manufacturers, operators, integrators), asset-based, iterative approach based on Deming Cycle

6 Conclusion and Outlook

We have assessed literature as well as legal and operational requirements for safety and security in federated remote labs. The legal requirements have been investigated from an engineering point of view – this does not replace a legal consultation. Therefore, in DigiLab4U, we will use an external auditing to ensure our proposed approach. Simplicity, flexibility and the joint security and safety assessment have been identified as the key needs for a federated lab-network safety and security guideline. Relevant national and international guidelines have been investigated. A general guideline (VDI/VDE 2182), providing a simple and flexible approach has been identified. However, we need to adjust the general guideline to the specific requirements for federated labs in general and more specifically for DigiLab4U. This guideline will be made available for further labs, which are interested in participating in the network. Last, not least, we need to implement the guidelines in the individual labs and institutions – and match the current local procedures.

References

1. Heradio, R., de la Torre, L., Galan, D., Cabrerizo, F.J., Herrera-Viedma, E., Dormido, S.: Virtual and remote labs in education: a bibliometric analysis. Comput. Educ. **98**, 14–38 (2016)
2. Tawfik, M., Sancristobal, E., Ros, S., Hernandez, R., Robles, A., et al.: Middleware solutions for service-oriented remote laboratories: a review. In: Proceedings of IEEE Global Engineering Education Conference (EDUCON), pp. 74–82. IEEE (2014)
3. Lisova, E., Šljivo, I., Čaušević, A.: Safety and security co-analyses: a systematic literature review. IEEE Syst. J. **13**(3), 2189–2200 (2018)
4. Orduña, P., Rodriguez-Gil, L., Garcia-Zubia, J., Angulo, I., Hernandez, U., Azcuenaga, E.: Increasing the value of remote laboratory federations through an open sharing platform: LabsLand. In: Auer, M.E., Zutin, D.G. (eds.) Online Engineering & Internet of Things: Lecture Notes in Networks and Systems, vol. 22, pp. 859–873. Springer, Cham (2018)
5. Henke, K., Vietzke, T., Wuttke, H.-D., Ostendorff, S.: Safety in interactive hybrid online labs. Int. J. Online Eng. (iJOE) **11**(3), 56–61 (2015)
6. Maiti, A., Kist, A., Maxwell, A.D.: Design and operational reliability of a Peer-to-Peer distributed remote access laboratory. In: Proceedings of 2015 12th International Conference on Remote Engineering and Virtual Instrumentation (2015)
7. Casini, M., Prattichizzo, D., Vicino, A.: Operating remote laboratories through a bootable device. IEEE Trans. Industr. Electron. **54**(6), 3134–3140 (2007)
8. Kozík, T., Šimon, M.: Preparing and managing the remote experiment in education. In: 2012 15th International Conference on Interactive Collaborative Learning (ICL) (2012)
9. Marangé, P., Gellot, F., Riera, B.: Control validation of DES systems: application to remote laboratories. In: 2nd International Conference on Digital Information Management (2007)
10. Gerža, M., Schauer, F., Jašek, R.: Security of ISES measureserver® module for remote experiments against malign attacks. Int. J. Online Eng. **10**(3), 4–10 (2014)
11. Ocaya, R.O.: A framework for collaborative remote experimentation for a physical laboratory using a low cost embedded web server. J. Network Comput. Appl. **34**(4), 1408–1415 (2011)
12. Chellaiah, P., Nair, B., Achuthan, K., Diwakar, S.: Using theme-based narrative construct of images as passwords: implementation and assessment of remembered sequences. Int. J. Online Eng. (iJOE) **13**(11), 77–93 (2017)
13. Krbeček, M., Schauer, F., Jasek, R.: Security aspects of remote e-laboratories. Int. J. Online Eng. **9**(3), 34–39 (2013)
14. Krbeček, M., Schauer, F.: Communication and diagnostic interfaces in remote laboratory management systems. Int. J. Online Eng. **11**(5), 43–49 (2015)
15. Sáenz, J., Esquembre, F., Garcia, F.J., de la Torre, L., Dormido, S.: A new model for a remote connection with hardware devices using Javascript. IFAC-PapersOnLine **49**(6), 133–137 (2016)
16. Herrera, M.S., Märquez, J.A., Borrero, A.M., Sänchez, M.M.: Testing bench for remote practical training in electric machines. IFAC Proc. Volumes **46**(17), 357–362 (2013)
17. Border, C.: The development and deployment of a multi-user, remote access virtualization system for networking, security, and system administration classes. ACM SIGCSE Bull. **39** (1), 576–580 (2007)
18. Li, P., Mohammed, T.: Integration of virtualization technology into network security laboratory. In: 38th Annual Frontiers in Education Conference, S2A (2008)

19. Richter, T., Watson, R., Kassavetis, S., Kraft, M., Grube, P., Boehringer, D., Logothetidis, S.: The WebLabs of the University of Cambridge: a study of securing remote instrumentation. In: 9th International Conference on Remote England Virtual Instrumentation (REV) (2012)
20. Pálka, L., Schauer, F.: Safety of communication and neural networks for security enhancement in data warehouse for remote laboratories and Laboratory Management System. In: 6th International Conference on Computing, Communication and Networking Technologies (ICCCNT) (2015)

Analog Electronic Experiments in Ultra-Concurrent Laboratory

Narasimhamurthy K. C.[1(✉)], Pablo Orduna[2,3], Luis Rodríguez-Gil[2], Bharath G. C.[1], C. N. Susheen Srivatsa[1], and Karthik Mulamuttal[1]

[1] Siddaganga Institute of Technology, Tumkur, India
kcnmurthy@sit.ac.in,
bharathgc8@gmail.com, susheensrivatsa@gmail.com,
karthikmulamuttal@gmail.com

[2] LabsLand Inc., Bilbao, Spain
{Pablo, luis}@labsland.com

[3] University of Deusto, Bilbao, Spain

Abstract. Conventionally in the classrooms, concepts related to Electronic courses like Electronic circuits will be conveyed by writing circuit diagram, deriving the design equations & showing the waveforms. Sometimes simulations are also used for the purpose in the classroom. However, most of the times students will not be convinced by this pedagogical approach as they have not seen the real time implementation of the circuit and real time typical values of the experiment. Therefore students can't connect to the concepts and they fail to analyze the circuit and hence the learning become incomplete. In this paper an innovative approach towards experiential learning is introduced for Analog electronic circuits.

Keywords: Ultra-concurrent Laboratory · High pass filter · CE amplifier · CC amplifier · Voltage gain

1 Introduction

In this digital-era it's difficult to keep student's attention in the classroom for a long time by the conventional chalk & talk. So it's necessary to have disturbed classroom by creating the facility to make students getting engaged in the learning process. One such way is to create laboratory environment in the classroom itself. However, there are infrastructural issues involved and moreover it is impractical to think that all the students conduct experiment during the class hour and get the result for better understanding of the concepts that are being discussed simultaneously in the class. In such case, student's confidence will not boost rather it may go low to overcome this, we propose ultra-concurrent laboratory - "A remote laboratory allows students to interact online with real time circuits and equipments. All data and multimedia are real and pre-recorded". Thus the laboratory provides a real time laboratory experience to many students simultaneously with a high reliability [1–3]. This remote laboratory can be conducted by students simultaneously and observe the response of the circuit for the selected combination of circuit parameters. So ultra-concurrent laboratory, for which the results are guaranteed will help students to get involved in continuous learning process [4–8].

M. E. Auer and D. May (Eds.): REV 2020, AISC 1231, pp. 37–45, 2021.
https://doi.org/10.1007/978-3-030-52575-0_3

2 Development of Ultra-Concurrent Laboratory

Experiments considered for the discussion of ultra concurrent laboratory in this paper are High pass filter and Common emitter (CE) amplifier. Each laboratory includes the schematic of the circuit diagram of the experiment with variable parameters in terms of component and input values. The ultra concurrent laboratory consists of video records of Proto-type circuit implementation, oscilloscope view and the Bode plot views. In Ultra-concurrent laboratory, the change in any parameter in the schematic will be immediately reflected in all the videos. So using ultra concurrent laboratory, faculty can convey the concepts with more confidence and proofs. Students can also verify it with many possible combinations of parameters during the class hour or beyond. Figure 1 shows the schematics of the experiments that are implemented in ultra concurrent laboratory.

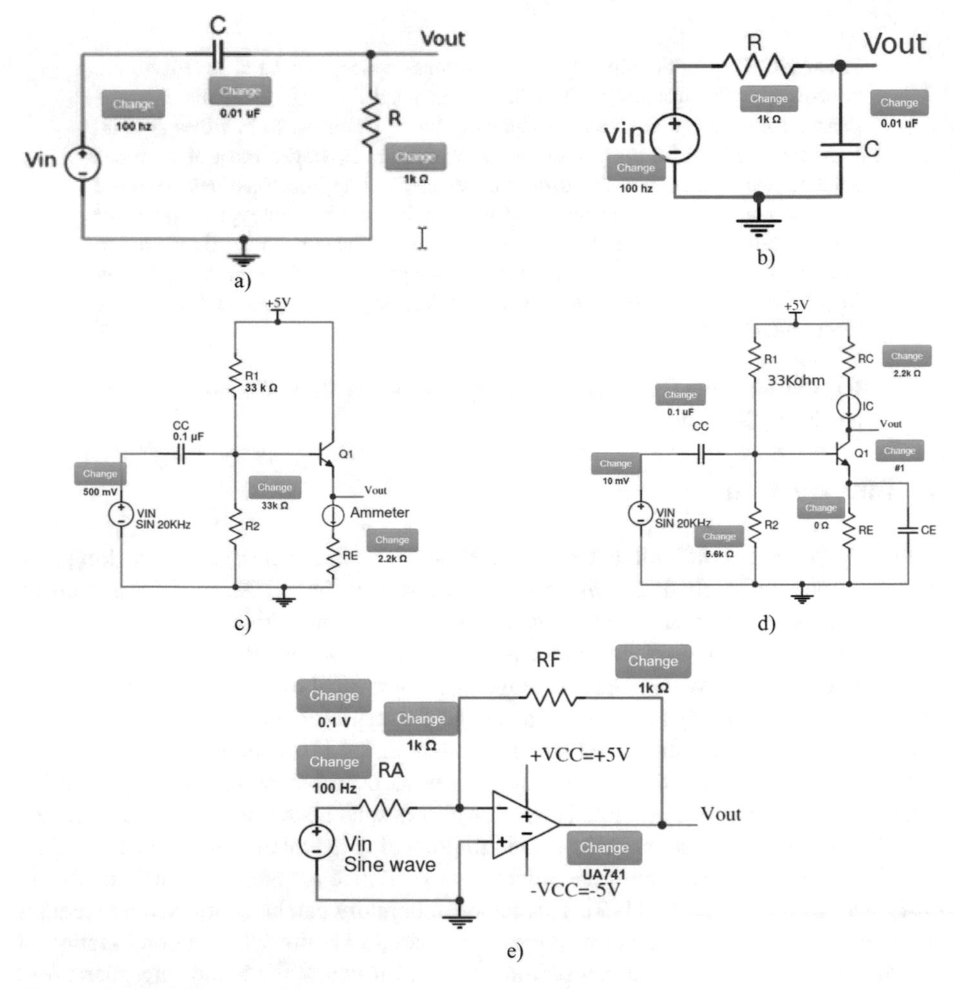

Fig. 1. Schematic of circuit diagrams of the experiments implemented in ultra-concurrent laboratory a) high pass filter b) low pass filter c) common collector amplifier d) common emitter amplifier e) Opamp based inverting amplifier.

In High pass filter shown in Fig. 1a and Low pass filter in Fig. 1b there is option of selecting input signal of five different frequencies 100, 500, 1000, 2000, 4000 Hz at constant amplitude of 1 V. There is option of selecting Resistor R from 1 kΩ, 10 kΩ and parallel combination of 1 kΩ with 10 kΩ, similarly the capacitor can be chosen from 0.1 μF, 0.01 μF and parallel combination of 0.1 μF with 0.01 μF. This will enable user to conduct HPF and LPF experiments with 45 time responses.

Similarly CC amplifier in Fig. 1c has four different values for R_2 (33 kΩ, 38 kΩ, 33 kΩ parallel with 38 kΩ and High impedance), four values of R_E (3.3 kΩ, 2.2 kΩ, 3.3 kΩ parallel with 2.2 kΩ and High impedance) this will result in 16 different combinations of CC amplifier.. Ultra concurrent laboratory provides option to select one of the three input signal amplitude 500 mV, 750 mV, 1000 mV at constant frequency of 20 kHz. This results in 48 distinct time responses. Users get enough data to analyze the working of CC amplifier.

CE amplifier in Fig. 1d has two different values for C_c (0.1 and 0.01 μF), R_2 (4.7 and 5.6 kΩ), R_C (2.2 and 1 kΩ) and $R_{E'}$(0 and 50 Ω) this will result in 16 different combinations of CE amplifier. As there are three BJTs available for selection, during the conduction it results in 48 different possible distinct circuit combinations CE amplifier. Ultra concurrent laboratory provides option to select one of the three input signal amplitude 0, 10 mV, 20 mV at constant frequency of 20 kHz. This results in 144 distinct time responses. This will give user huge data base to analyze the concepts of CE amplifiers. There are 48 different Bode plots possible for all CE amplifier hardware circuits.

Another recently added experiment, Opamp based inverting amplifier shown in Fig. 1e has two different circuit configurations. First one is **Standard configuration** and the second **Special configuration**. In standard configuration three different values for R_A (1 kΩ, 10 kΩ, 1 kΩ parallel with 10 kΩ) and similar combinations for R_F will result in 9 different combinations of inverting amplifier. In this experiment there is also an option of applying input signal of different amplitude and frequencies.

In **Special configuration** the major difference compared to Standard configuration is the feedback resistor is 100 kΩ and R_A (1 kΩ, 1 kΩ parallel with 10 kΩ). This adds few more combination for inverting amplifier circuits. In case of Special configuration as the voltage gain of the amplifier is more, the input signal amplitude can be 5 mV, 10 mV and 20 mV. To differentiate between Opamp with different open loop gains two Opamp μA 741 and OP37 are used. Out of which OP37 has better specifications compared to μA 741. Total number of time responses possible in the Opamp inverting amplifier is 216. Such huge amount of data is made possible probably for the first time for a single experiment.

The Ultra-concurrent laboratory is co-developed with LabsLand[1], the global network of remote laboratories, which offers real laboratories from over a dozen countries from all the continents. This provides a set of advantages to the development of the ultra-concurrent laboratory:

1. Relying on LabsLand experience on the development of laboratories: LabsLand has developed many ultra-concurrent laboratories and already optimize the videos in

[1] https://labsland.com.

different qualities for different network bandwidths and host the lab in optimized cloud services in Amazon Web Services.

2. LabsLand platform: SIT has its own space at LabsLand[2], where SIT students use the laboratories and SIT obtains analytics on the usage. LabsLand spaces also count with integration of major Learning Management Systems such as Canvas, Moodle, Sakai, Google Classroom and others, so students do not need to register in Labs-Land to use the laboratory.
3. Sustainability: once the laboratory is developed, anyone can use these and other laboratories by registering in LabsLand for a small subscription fee; which is shared with SIT using different mechanisms. This allows SIT to use laboratories of other universities and to re-invest in future SIT labs.

3 Results and Discussions

Ultra-concurrent Laboratory will enable the user to have the similar experience as though experiment is conducted in the conventional laboratory, because the Circuit view shows the real time prototype of the HPF.

Oscilloscope view shows the input and output waveform of the filter circuit and Bode plots provide the magnitude and phase responses as shown in Fig. 2. One typical reading of HPF is shown in Table 1.

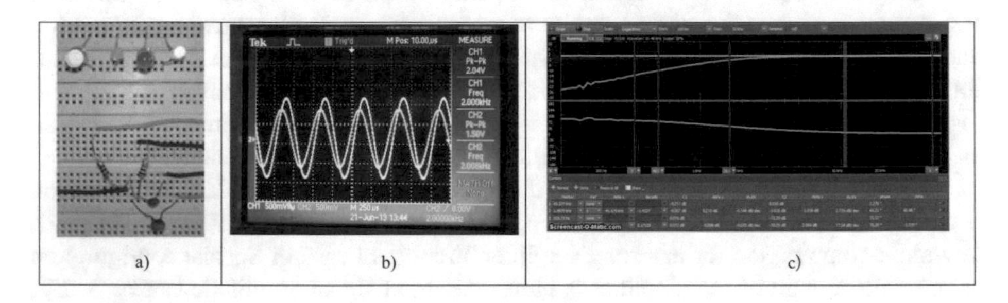

Fig. 2. Real time views of high pass filter a) circuit b) oscilloscope c) bode plot

Table 1. HPF time response analysis

Sl No	R	C	f, Hz	Vin,V	Vout, V	Voltage gain	Voltage gain (dB)
1	1 KΩ	0.1 uF	100	2.04	0.16	0.078	−22.15

Figure 3 shows the real time responses of the CE amplifier. These are the screenshots taken during the conduction of the experiment. The glowing LEDs on the breadboard

[2] https://labs.land/sit.

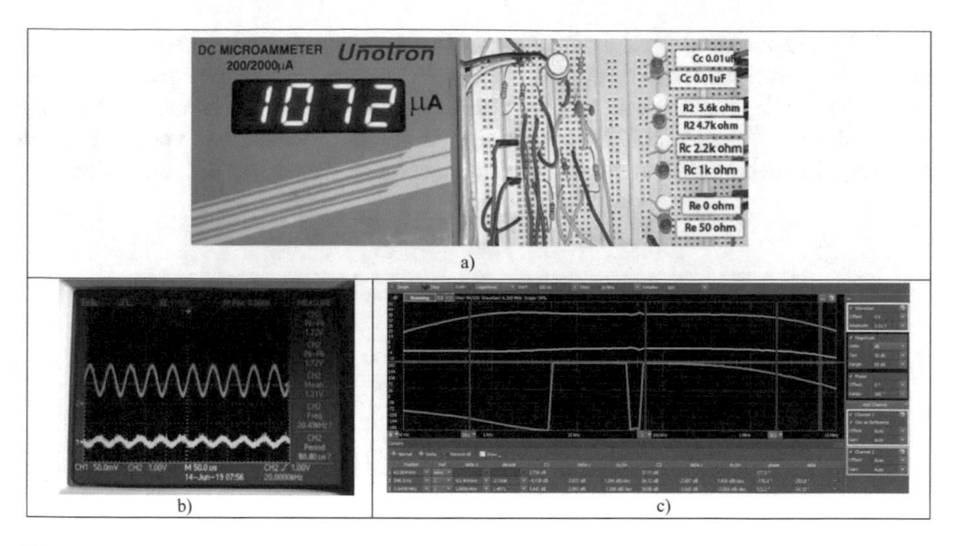

Fig. 3. Real time views of CE amplifier a) circuit and am meter b) oscilloscope c) bode plots

Indicate which component is being used during the conduction of experiment at that time. In CE amplifier, in time response for sixteen different possible circuits with three input amplitude, for three BJTs, totally 144 distinct oscilloscope views are available.

Using Bode plots frequency response of 16 different CE amplifiers can be analyzed. Concepts like constant mid band gain, 3 dB bandwidth, gain bandwidth product, movement of dominant poles, Phase and gain margin of CE amplifiers can be analyzed quickly.

The results of CC amplifier are shown in Fig. 4. The Am meter in Fig. 4a shows the DC emitter current of the CC amplifier for the selected biasing conditions and is independent of the input voltage amplitude.

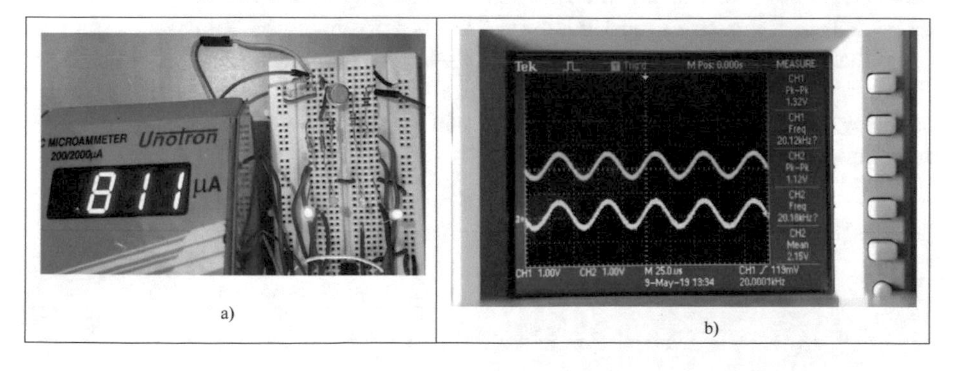

Fig. 4. Real time views of CC amplifier a) circuit and am meter b) oscilloscope

Table 2 shows few typical readings obtained during the conduction of the CC amplifier experiment. It can be observed that as R_2 is increased for the given R_E value, emitter current I_E increases. In another possible variations, as R_E is increased for the given value of $R_{2,}$ I_E reduces. However the voltage gain remains less than unity in all the cases. Likewise 48 readings can be obtained for analyzing the CC amplifier.

Table 2. Typical readings of CC amplifier.

R_2 kΩ	R_E kΩ	I_E (μA)	Vout (mV)	V_E DC (V)	Voltage gain (V/V)
33	2.2	811	1240	2.3	0.95
33‖38	3.3	434	1320	1.41	0.87
High impedance	3.3‖2.2	1117	1640	4.33	0.93
38	High impedance	80	1320	1.83	0.78

In the case of Opamp Inverting amplifier an important observation that can be made is bandwidth of amplifier for same gain but circuit being realized using two different Opamps μA741 and OP37. These observations in Table 3 give an insight into selection of Opamps based on the applications.

Table 3. Typical readings of inverting amplifier

Opamp	R_A kΩ	R_F k Ω	Voltage gain dB	Bandwidth kHz	Remarks
μA741	1	100	40.86	2.59	Suitable for only low frequency applications
OP37	1	100	39.98	192.08	Suitable for high frequency applications

The advantage of using Ultra-concurrent lab is not only in performing the experiments from anywhere at anytime with any one and any number of times, Teacher can keep track of the uses of the laboratory. This is very much essential to monitor the individual usage of the laboratory especially when assignments are given to students. Figure 5, 6, 7 and 8 gives the details of the lab usage that will be very useful in awarding grades. Figure 5 gives the day wise usage of the laboratory, indicating maximum 280 uses on a particular day.

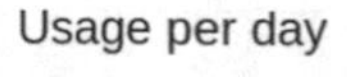

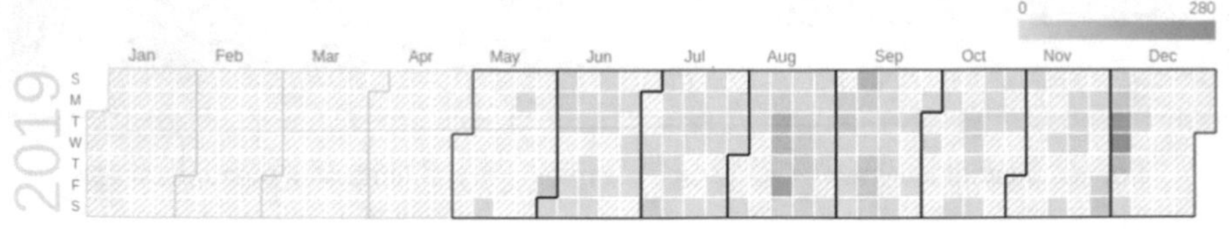

Fig. 5. Usage of the Laboratory

Laboratories used

- Electronics Deusto: 47 uses
  - Experiment with resistors: 10 uses
  - Experiment Ohm and Kirchhoff Laws: 6 uses
  - Analogical electronic experiments: 31 uses
- Arduino robot (code): 26 uses
  - Montmeló: 2 uses
  - 8-circuit: 2 uses
  - Any circuit: 2 uses
  - Arduino IDE: 13 uses
  - Robot example: 7 uses
- Kinematics: 4 uses
- Pendulum: 3 uses
- Arduino Board (code): 2 uses
  - Arduino Board IDE: 1 use
  - circuit: 1 use
- Common Collector Amplifier: 468 uses
- High-Pass Filter: 360 uses
- Low-Pass Filter: 455 uses
- Intel DE2-115: 10 uses
  - DE2-115 IDE VHDL: 5 uses
  - Intel DE2-115: 5 uses
- FPGA Laboratory: 1 use
- Common Emitter Amplifier: 637 uses
- Inverting Operational Amplifier: 116 uses
- Non-Inverting Operational Amplifier: 86 uses

Fig. 6. Statistics of laboratories used

Siddaganga Institute of Technology / Harshini Ns (harshinins2000)

Summary

Basic information

Property	Students	Instructors
Number of laboratory sessions	55	0

Groups

- Example group: 29 uses
- 4 TCE 2019-20 : 26 uses

Laboratories used

- Common Collector Amplifier: 7 uses
- High-Pass Filter: 10 uses
- Low-Pass Filter: 10 uses
- Common Emitter Amplifier: 17 uses
- Inverting Operational Amplifier: 5 uses
- Non-Inverting Operational Amplifier: 6 uses

Fig. 7. Detailed information about students laboratory usage

Uses per time of day

	Monday	Tuesday	Wednesday	Thursday	Friday	Saturday	Sunday
00:00	5	4	10	2			2
01:00							1
02:00							
03:00		1					
04:00							
05:00			7	1		2	
06:00	1		16	9	3		
07:00	2	4	17	23	9	4	
08:00	8	13	31	18		5	6
09:00	14	14	5	10	6	16	14
10:00	16	27	34	15	16	14	16
11:00	6	19	23	31	46	10	21
12:00	12	22	33	15	45	18	11
13:00	6	29	44	18	17	5	11
14:00	5	77	37	11	2	11	9
15:00	10	75	64	10	8	13	9
16:00	29	94	50	11	41	8	15
17:00	5	23	7	13	30	10	11
18:00	13	17	16	30	39	13	6
19:00	18	28	26	26	33	15	14
20:00	12	9	27	4	22	22	8
21:00	20	17	18	5	22	10	5
22:00	19	12	29	2	1	3	14
23:00	14	25	18	3		6	12

Fig. 8. Hour wise split out of laboratory uses.

Figure 6 shows the details of the number of times a particular lab being used by the students. Using this statistics one can analyze which lab is popular and which lab need to improved for better usage.

The system also helps to know the interest among students in using this innovative approach toward experiential learning. Figure 7 shows the details of a student who has used different laboratories 55 times.

Such time of information will help faculty members to know the students interest and modify their lesson plans accordingly. As the system allows the access 24/7, Fig. 8 gives the details of the laboratory being used over the entire day. This will help us to know the active period of students.

4 Conclusions

In an innovative way to convey the concepts of Analog electronic circuits, Ultra-concurrent laboratory is developed keeping in view the possible parameters that can be varied which will have an impact on the circuit response. Further the real time videos of circuit, oscilloscope and Bode plots along with the values of the key parameters help the user to understand the concepts by experiential learning. Useful report on the lab usage will help the faculty to keep track of the learning process of the students.

Acknowledgment. Authors of this paper express their gratitude for all those who helped in developing the Remote Laboratory system. We also thank officials of Siddaganga Institute of Technology (SIT), Tumakuru, the Management, Director, CEO and Principal for their support in establishing Ultra-concurrent Laboratory system.

References

1. http://vlab.co.in/
2. Hesselink, L., Rizal, D., Bjornson, E., Paik, S., Batra, R., Catrysse, P., Savage, D., Wong, A.: Stanford cyberlab: internet assisted laboratories. Int. J. Distance Educ. Technol. **1**(1), 21–39 (2003)
3. http://www.uml.edu/IT/Services/vLabs/
4. de la Torre, L., Guinaldo, M., Heradio, R., Dormido, S.: The ball and beam system: a case study of virtual and remote lab enhancement with moodle. IEEE Trans. Industr. Inf. **11**(4), 934–945 (2015)
5. de la Torre, L., Sanchez, J.P., Dormido, S.: What remote labs can do for you. Phys. Today **69**, 48–53 (2016)
6. Sanchez-Herrera, M.R.S., Mejias, A., Marquez, M., Andujar, J.M.: A fully integrated open solution for the remote operation of pilot plants. IEEE Trans. Ind. Inform. **15**(7), 3943–3951 (2018)
7. Heradio, R., de la Torre, L., Dormido, S.: Virtual and remote labs in control education: a survey. Annu. Rev. Control **42**, 1–10 (2016)
8. IEEE Std 1876-2019: IEEE Standard for Networked Smart Learning Objects for Online Laboratories, 2019, section 4: First layer of standardization: Lab as a Service (LaaS) and Section 5: Recommended practices for using online labs as learning objects (2019)

STEM Education on Equal Terms Through the Flipped Laboratory Approach

Lena Claesson[1(✉)], Mirka Kans[2], Lars Håkansson[2], and Kristian Nilsson[1]

[1] Blekinge Institute of Technology, Karlskrona, Sweden
{lena.claesson,Kristian.nilsson}@bth.se
[2] Linnaeus University, Växjö, Sweden
{mirka.kans,lars.hakansson}@lnu.se

Abstract. The educational phenomena studied in this paper is remote-controlled physical laboratory environments and their applicability in upper secondary school physics education. In order to gain a better understanding of the situation and needs regarding laboratory activities in the upper secondary school, eight physics teachers were interviewed at six different schools. This revealed that the resources for laboratory activities vary between schools and may be inconsistent with the Swedish National Agency for Education curriculum. Furthermore, 165 upper secondary school students answered a questionnaire survey regarding subject preferences, program choices, views on technology and self-ability, and approach to technology and technology-related situations. The acquired knowledge provides a basis concerning the needs and conditions of teaching and learning within the subject of physics. This new knowledge motivates the development of the Flipped laboratory concept that is introduced in this paper, based on remote-controlled physical laboratories, for upper secondary school.

Keywords: Remote laboratory · STEM education · Physics education · Upper secondary school · Flipped laboratory

1 Introduction

There is a substantial shortage of technicians and engineers in Sweden [1]. Attracting more women to engineering is one way to approach this problem, and research shows that equal recruitment and staffing contribute to positive business results [2]. Furthermore, the level of competence in the technology-related sector needs to be increased, but the number of engineers graduated annually is not sufficient to meet the needs of industry. The number of students in Sweden who graduated with a bachelor or master's degree in engineering was 6,440 in the academic year 2017/2018 [3]. Of these, 32% were female. The total number of students enrolled in university engineering programs was approximately 39,500 in 2018. The finishing rate is thus low, and only about half of the enrolled students receive a degree [4]. The proportion of students in upper secondary STEM programs has remained stable over a five-year period: of the total upper secondary school students, about 12% are enrolled in Natural

M. E. Auer and D. May (Eds.): REV 2020, AISC 1231, pp. 46–62, 2021.
https://doi.org/10.1007/978-3-030-52575-0_4

science (N) programs and 8% in Technical (T) programs, a total of about 74,000 students [5]. In N and T programs, approximately 83% and 68% respectively enters post-secondary education. Still, there is a substantial potential to increase the proportion of students who start STEM related programs and thereafter move on to post-secondary education. At the same time, there is a downward trend in young people's general interest in technology-related education, and in particular, it is observed that students' interest in technology is decreasing with age [6]. It is thus important to ensure that students choose STEM programs also in the future, and to encourage their interest in STEM subjects at an early stage. Potential areas for improvement can be found in the core business itself, i.e. the STEM education, and in particular laboratory and other kinds of practical work.

Equal laboratory resources, regardless of school or geographical location, is one of the necessities for increasing the number of young people and adults who have relevant knowledge, skills and abilities in the STEM field. The laboratory teaching must also be attractive and understandable for all students regardless of background and gender. One way to achieve this is to link technology with emerging societal challenges, i.e. putting technology understanding in a societal perspective. Flexible learning in the form of distance-based teaching, flipped classrooms and student-centered learning is seen by many as the future of learning [7–9], and are thus enablers for reaching these goals.

This paper reports on findings from a nine-month long feasibility project, in which the current state of Swedish STEM education on upper secondary school level and future possibilities to improve the education by the means of technical and digital solutions was investigated. The overall aim was to enable STEM education on equal terms for upper secondary schools in Sweden, e.g. by ensuring laboratory resources for all schools, teachers and students. The overall goal was to reach good and relevant education for all, and increased technology-related competence. A conceptual solution, Flipped laboratory, is proposed that fulfil the aim through modern pedagogy and technology. Remote-controlled physical laboratory environments based on the VISIR (Virtual Instruments Systems In Reality) concept comprise an important part of the solution, as they will arouse interest via computer and/or smartphone for experimenting with measuring instruments, and experimental objects, and provides students the opportunity to prepare for scheduled laboratory exercises. Learning becomes less space dependent; the students can do the laboratory work at home, before and after a lesson, which strengthens the students' self-esteem in technology related learning activities. This, in turn, can contribute to increase students' interest for higher education in STEM. A critical factor is that teachers feel comfortable with and can use the support effectively, which is why we also focus on upper secondary school teachers' skills development in the intended solution.

The educational phenomena studied in this paper is remote-controlled physical laboratory (RL) environments and their applicability in upper secondary school education. Educational remote labs are real laboratories that can be controlled over the Internet via a computer, mobile device, or tablet, without installing any new software, or requiring physical laboratory facilities. The RL can be used as support and complement to regular teaching in upper secondary schools. International initiatives that address similar areas to our project are LabsLand, VISIR+, PILAR and REMLABNET:

- LabsLand [10] offers different top-quality laboratories from different providers across the globe, for many educational levels and subjects.
- The VISIR+ [11] project aims to define and develop a set of educational modules comprising hands-on, virtual and VISIR remote lab, combined with calculus, following an enquiry-based teaching and learning methodology, in electrical and electronic circuits' theory and practice.
- The PILAR [12] project suggests a solution that aims to interconnect all VISIR systems with each other, in order to create a grid of laboratories shared and accessed by all participants, expanding and empowering the existing systems to a new level of service and capacity.
- REMLABNET [13] is a Remote Laboratory Management System for the integrating and management of remote experiments for starting university level and secondary schools.

In this paper, the focus is on understanding the needs and conditions of STEM education in Swedish upper secondary school, and especially teaching and learning conditions within the subject of physics. The understanding is necessary for further design of the Flipped laboratory concept. The main goal is to propose a solution design that addresses the needs and conditions of laboratory-intensive STEM education within Swedish upper secondary school.

2 Study Description

The feasibility project used a theory-informed, iterative and interdisciplinary approach. Statistics from Statistics Sweden and the Swedish National Agency for Education, analysis of similar concepts, interviews with teachers and a questionnaire survey have been used as a basis for the concept development. In order to gain a better understanding of the situation and needs regarding laboratory activities in the upper secondary school, eight physics teachers were interviewed at six different schools. The gender distribution amongst physics teachers was rather equal, see Table 1. All schools had both male and female teachers and the ratio of female teachers ranged between 29% and 67%. The average of female physics teachers in Sweden is 28%.

Table 1. Gender distribution for the interview objects

	School 1	School 2	School 3	School 4	School 5	School 6
Female physics teachers	2	1	1	4	4	6
Male physics teachers	5	1	2	2	5	13
% female students	52	63	10	60	51	33
Program	N	N	T	N	N	N/T

A questionnaire survey was designed for upper secondary school students regarding subject preferences, program choices, views on technology and self-ability and approach to technology and technology-related situations. The questionnaire was distributed in classes on second year in four different programs in the schools referred to as 1, 3 and 5 in the interview study. A total of 165 upper secondary school students answered the questionnaire (originally, 168 responses were recorded, but three of these were blank except for the first question regarding gender, and therefore excluded from the study). The quantitative results were statistically processed using variance analysis while qualitative results were processed using content analysis. Table 2 describes the study participants and the response rate for the questionnaire study.

Table 2. Participants in the questionnaire study

	Students, total	Female students (in %)	Students in year 2	Study participants			Response rate (in %)
				Male	Female	Other	
Technical program, School 3	217	10	79	53	7	3	80
Natural science program, School 1	358	52	127	10	18	0	22
Natural science program, School 5	551	51	142	10	8	0	13
Social science program, School 5	402	63	182	13	35	1	27
Humanistic program, School 5	41	88	8	1	5	0	75
School/program not stated				1	0	0	
Total			538	88	73	4	31

In addition, a market survey was made, in which relevant stakeholders and similar projects were investigated using structured database searches as well as internet searches. Findings from the reviews, the interview study and the questionnaire study were summarized in the form of a SWOT analysis for the Flipped laboratory concept. The Flipped laboratory concept has thereafter been developed and designed to meet identified needs. The development was made by an interdisciplinary project team and included functional as well as non-functional design dimensions of the solution. First, the area of focus was determined, i.e. which course and learning objectives to address. Thereafter, the main pedagogic approach was designed in the form of thematic scenarios. Finally, the design of remote laboratories to be included in the scenarios was determined.

3 Feasibility Analysis for the Flipped Laboratory Approach

3.1 Results from the Interview and Questionnaire Studies

Upper secondary school education is regulated in the syllabuses for the upper secondary school, Gy11, which describe the prerequisites for teaching in general and in syllabuses that define content as well as learning objectives for each course. In Gy11, laboratory parts in the natural science courses have received a greater emphasis. The natural science courses perceived by teachers and students as hard because of the dense content and required mathematical competences. An example is the course Physics1, which is compulsory for natural science and technology programs. The interviewed teachers describe that the students do not have sufficient foundations for mathematics as well as laboratory and experimental activities, but also that they lack study technique to set up their studies effectively. Some students also lack motivation in general to study the subject. In other words, there is a need to create better conditions for students to acquire knowledge and skills in the subject of physics.

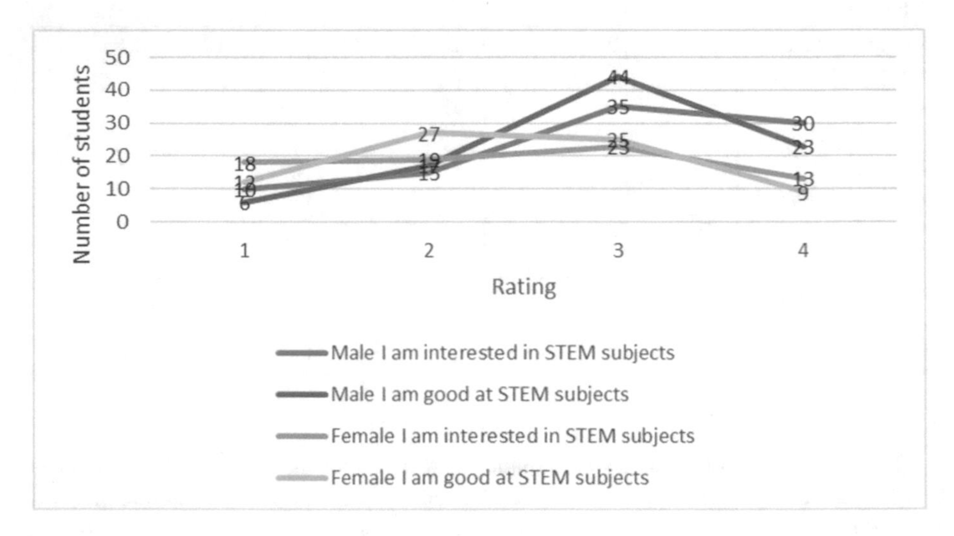

Fig. 1. Interest and perceived self-efficacy

The interviewed teachers did not recognize any direct differences in activity or cognitive competences with respect to gender in the student group. The students on T and N programs who answered our survey expressed that they are both interested in, and good at, STEM subjects, see Fig. 1. The ANOVA variance analysis showed no significant differences between genders in terms of interest or perceived ability, although the mean for male students was higher as compared to the female students for both questions. However, differences were found between programs: students in the natural science program were more interested in, and good at, STEM subjects than other students, while social science students were less interested and valued their knowledge lower for these subjects.

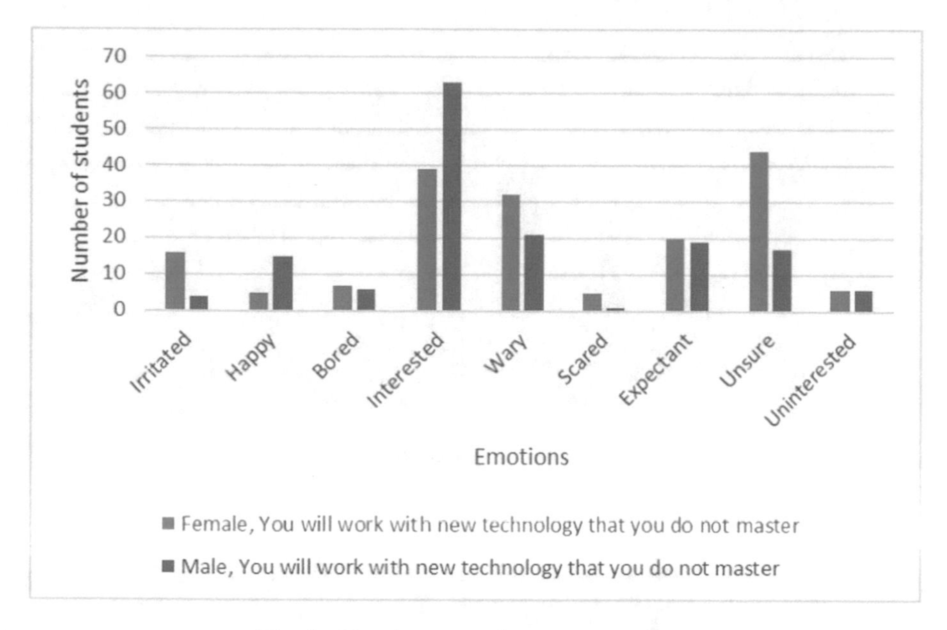

Fig. 2. Emotions regarding new technology

Favorite subjects, on the other hand, are in the STEM sciences for male students and in the humanities and social sciences for female students. Male students appreciate practical, theoretical, logical, structured, and scientific subjects that have distinct answers and that create understanding. Future usability is also important. Female students appreciate creative, challenging and analytical subjects and subjects that require problem-solving ability. In other words, female students do not associate characteristics such as analytical and problem-solving with STEM subjects, suggesting that a different approach is needed to support especially female students' interest and learning for STEM. In addition, female students experience negative emotions (typical feelings are cautious, insecure and afraid) when they face STEM related situations to a greater extent than male ones (typical feelings are happy and interested), see Figs. 2, 3 and 4. This is one reason for the thematic approach to the solution.

The interview study revealed that resources for laboratory work vary between schools. Some schools have good access to laboratory classrooms and equipment, while others only have very basic equipment. Especially expensive or dangerous equipment may be missing. Although the curriculum defines the total number of hours for the course, the number of hours students receive for laboratory teaching varies. Causes of variation include access to laboratory classrooms and equipment, but also the opportunity to split the class into smaller groups. Besides the lack of resources, a number of different problems with laboratory activities were described by the teachers. Technical problems were mentioned, such as computers that are malfunctioning or complicated equipment, as well as problems with installation. Moreover, it takes time to develop and design suitable laboratory work, time that some teachers do not feel they have as the number of hours for preparation and teaching is limited: "I must cover all the material in the book". Some subject areas are perceived as very theoretical, and it

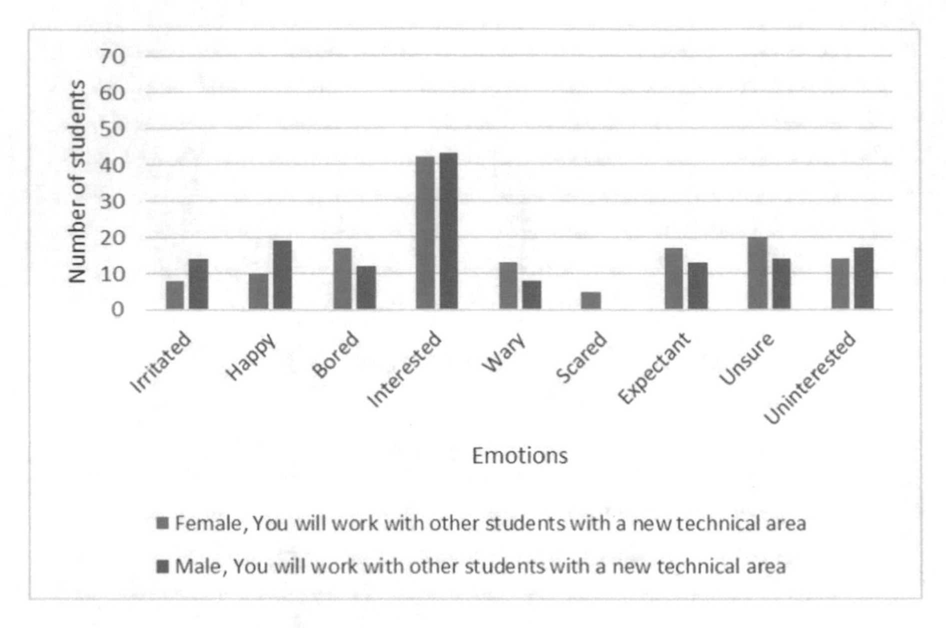

Fig. 3. Emotions regarding group work

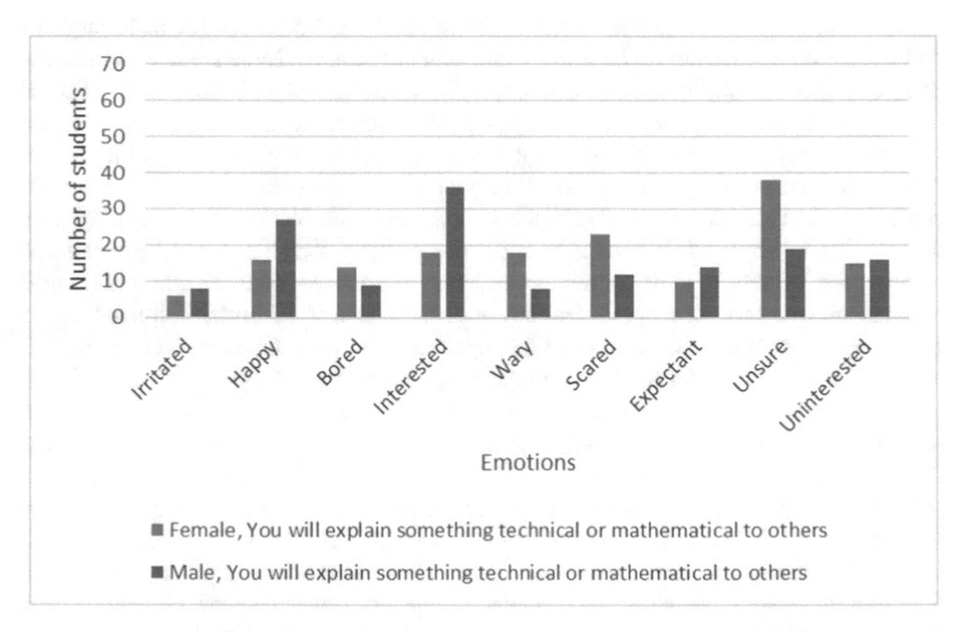

Fig. 4. Emotions regarding explanation or presentation

could be difficult to develop practical work to cover the content. In smaller schools, the lack of colleagues to discuss and cooperate with was a problem. There are also problems associated with assessment of laboratory work.

In other words, in order to achieve the learning objectives, support is especially needed for laboratory education; students need to have adequate time for laboratory work as well as help to acquire knowledge and skills, and teachers need to be able to offer adequate laboratory work even with limited resources, as well as help in assessing the outcomes. In the interviews, the Flipped laboratory concept was described, and the respondents were asked to give examples of suitable areas for the concept. Following areas were mentioned:

- Electricity - Electrical Power.
- Radiation theory - Radioactive radiation.
- Acoustics - Interference and standing waves.
- Mechanics - Distance, velocity and acceleration to describe motions.
- Atomic physics - spectroscope, lattice, spectral lamp.

3.2 Results from the Market Analysis

There are a number of publishers on the Swedish market with the focus on education material for upper secondary schools. Main part offer digital learning materials, often in the form of an electronic version of the paper book with additional services such as speech synthesis or digital diagnoses. Three publishers currently have digital learning materials in the area of Science and Technology aimed at the T and N programs, of which two actors are partly focused on the relevant area (N/T) and have a technical solution that is suitable for the development of the Flipped laboratory concept. One publisher offers digital education materials, integrated with an online-based encyclopedia. Another publisher offers a digital physics book, but also provides a digital comprehensive solution for teaching containing theoretical material, pictures, films, diagnoses and laboratory exercises. The laboratories are in the form of tutorials for physical labs, with the possibility of submitting a digital laboration report. None of these actors currently provides digital laboratory exercises in the form of simulating or remotely controlling the laboratories through the online solution. Pasco, Vernier, Texas instruments and National Instruments are examples of companies that supply physical laboratory equipment, for example in the form of components and data logging systems. Several of the companies provide complete environments and supporting materials for teachers. A trend seen in several solutions is wireless communication. Suppliers with complete environments are in first hand interesting for the implementation of the Flipped laboratory concept.

Online-based laboratory environments can be either simulated or real-time based, of which remotely controlled environments are of the latter type. A number of European and international universities have developed online laboratories, including Blekinge Institute of Technology (BTH), University of Deusto, Universitat Politecnica de Catalunya and The Open University (UK). Main part of the remote laboratory environments are research or development projects, such as PILAR, the Virtual Instruments Systems In Reality (VISIR) and VISIR+ [11, 12, 14]. The electronic laboratory Openlabs [15] developed by VISIR at BTH has been used and evaluated at upper secondary school in Sweden. Evaluations show that the laboratory is easy to use and the design appeals to the students. Internationally, there are 18 VISIR labs used by both colleges and upper secondary schools. An ongoing EU project, PILAR [12], proposes a solution that aims to connect all VISIR systems with each other. Another project, VISIR+ [11], is covering the broad field of Electrical Engineering. The aim is to define, develop and evaluate a set of training modules containing practical, virtual and remote controlled experiments, the latter supported by a VISIR lab. The VISIR electronics lab has already been developed into a platform for open laboratories, and has experience built up via the e-learning laboratory. Collaboration partners for this project can therefore be found in existing consortia without any competitive situation.

There are several other interesting initiatives internationally such as Olabs [16] in India. OLabs provides experiments in physics, chemistry and biology for students in classes 9 to 12, with content adapted to NCERT/CBSE and the national syllabus. A number of lessons in biology, chemistry and physics are offered, with simulated laboratory exercises as part of the learning material. Another is the Remote Laboratory Management System, REMLABNET [13], which integrates and maintains remote controlled laboratories for upper secondary and university level in collaboration with Go-Lab [17]. The Remlab was developed when the demand and use of equipment from the Internet School Experimental System, ISES [18], was very large. The EU project Open Discovery of STEM Laboratories [19] has developed MOOCS aimed at students aged 12–18, one of which uses the VISIR environment for laboratory work. Most laboratory environments have been developed for post-secondary education. There are few commercial actors, but LabsLand [10] is one initiative that offers remote real-time laboratories. LabsLand offers laboratory access to educational organizations and private individuals through a platform service. A real laboratory might be a small arduino-powered robot in Spain, a study of a particle's movement in Brazil, or a lab that can measure radioactivity in Australia. You can connect and share a lab from your own university or just buy the service and get access to all labs. LabsLand does not own and manage the labs, but the organization or company that shares their labs through the platform has full operational and maintenance responsibilities. LabsLand is an actor that is of interest to this project, in that their platform enables international dissemination.

3.3 SWOT Analysis for the Flipped Laboratory Approach

The flipped laboratory is an extension of the pedagogical method flipped classroom, where information is transferred to the students before they come to class. In the flipped laboratory the students can access theoretical content that allows them to understand the practical work in the remote laboratory. The content can be reading text or view a recorded part of a teachers lecture. The flipped laboratory is a student-centered environment.

The feasibility analysis revealed a number of obstacles and improvement areas in STEM education in general, and for physics laboratory education in specific. The market analysis showed that several approaches to strengthen the STEM and laboratory education has been developed, and that digital technology seems to be a good way to overcome some of the obstacles. These results were used as the basis for evaluating the Flipped laboratory concept in the form of a SWOT (Strengths, Weaknesses, Opportunities and threats) analysis. Figure 5 summarizes the findings.

	Positive	Negative
Internal	**Strengths with the concept:** • Provides students with time and support to understand complex laboratory tasks • Gives teachers the opportunity to prepare for and perform complex laboratory tasks • Gives teachers in training the opportunity to practice their own laboratory skills • Does not require any physical equipment or laboratory equipment • The technology is managed and maintained by third party • Focuses on context, not on technology • Relevant team of researchers/developers has been identified	**Weaknesses with the concept:** • Expensive primary investment; a sustainable business model for its realization is necessary • Accessibility and security must be guaranteed; otherwise, the service may not be available when users want access • Can be seen as a substitute for, instead of complementary to, regular teaching
External	**Opportunities with the concept:** • Focuses on Swedish conditions and the upper secondary school curriculum; easy to spread nationally • Similar service is not available commercially on the Swedish market; great market potential • Educational material publishers with a flipped approach exist in the market; appropriate collaborators • A platform and business model for distribution (especially internationally) already exists; appropriate collaborator • Can be developed in many directions subject/content/market wise • Has great potential to be launched internationally; can be scaled up/down and existing networks/international stakeholders are a good basis for dissemination	**Threats with the concept:** • Stakeholders who develops and launches their own solution based on, or similar to, our concept • Economic conditions in the municipality/school do not allow procurement of the concept • Public opinions, politicians and/or decision makers do not put STEM education and skills provision on the agenda

Fig. 5. SWOT analysis

Strengths with the Concept

Online-based teaching is asynchronous, i.e. independent of modes of time and space and space, and therefore allows students as well as teachers to prepare for and/or complete the task at the appropriate time and place [20, 21]. This is especially evident for the laboratory work: a student who needs more time to complete the laboratory task, or who want to do it more than once, can do so. Measurement errors, uncertainty about what happened at the laboratory or other uncertainty factors can be handled by running the laboratory once again. Similarly, teachers can carry out the laboratories whenever and wherever they want and can thus prepare themselves before the actual class. This strength can be used in teacher training as well, to allow the students to practice their laboratory skills.

Flipped laboratory means that laboratory setup, measuring equipment and objects are physically available at one or more locations (inside or outside Sweden). The individual school does not have to invest in expensive, bulky, complex or dangerous equipment. In addition, it does not have to operate, manage and maintain the equipment. Thus, several schools can use the same set of laboratories, which provides a resource efficiency that can be seen in the price as well as the opportunity to reach resources that might otherwise not be reachable when investment space is not available. The teaching material itself is not technology-focused, but the focus is on con-text; to put different technical and/or scientific phenomena in a context. In this way, interest can be aroused in student groups who do not have a STEM interest, but rather see problem solving and analysis as attractive characteristics for a school subject. This is also emphasized in the upper secondary school curriculum as well as in the physics course curricula. A relevant team of researchers/developers has been identified for the development of the concept, and the work on designing the concept has already begun.

Weaknesses with the Concept

Without a sustainable business model, there is the risk that the concept cannot be developed into a full-scale solution. Although the idea is to be able to reuse as much as possible of existing solutions, the concept represents a costly investment when developing the laboratory environments, and where robots, control systems and programming are examples of large expenses. A major focus must not only be on functional requirements, but accessibility and security must be guaranteed, which requires a good basic design, otherwise there is a risk that the service is not available when users want access. The basic idea is that Flipped laboratory should be a complement to regular classes. The concept will be a major improvement in schools where resource problems exist (time, equipment, teachers, rooms etc.), but there is a risk that it will replace normal laboratory teaching in other schools as well. The concept therefore focuses primarily on areas that currently include expensive, complex, bulky and/or dangerous laboratory work.

Opportunities with the Concept

The Flipped laboratory concept has been developed to suit the Swedish upper secondary school and the Swedish market. It is therefore easy to spread nationally, especially through a teaching material supplier, or a similar stakeholder. There are already a couple of interesting stakeholders in the market that currently do not provide a similar solution but have the commercial and technological maturity and capacity to do so. The market potential is great, especially if the concept is marketed together with other digital solutions that are developed for the upper secondary school. The concept is also suitable to be disseminated internationally either as a comprehensive solution (adjustments to each nation's curriculum may be needed) or by disseminating the laboratories separately. For the latter, there is a stakeholder that provides a business model and technology platform for this. The scalability allows the concept to be adapted to specific stakeholders, markets or contexts. The project team is a part of existing national and international networks, which facilitates dissemination.

Threats with the Concept
An imminent threat is that another stakeholder develops and launches their own solution that is based on, or is similar to, our concept. The project must therefore balance confidentiality with research transparency, and we see that this is best done by linking possible stakeholders to the project already in the development process. Another threat is that the financial situation of the municipality/school, which constitutes the primary market, does not allow procurement of the concept. In addition, shifts in political and overall public debate must be considered; will STEM skills requirements and development be on the agenda in the future?

4 Conceptual Design for the Flipped Laboratory Approach

In Fig. 6 a tentative web interface illustrates a descriptive of the concept with the principal conceptual elements included. The central functionality in the Flipped laboratory is the remote-controlled physical laboratory environments as exemplified in the middle of the figure in form of two themes, "Radioactivity" and "Electronics". In the fully developed version, the themes will be represented by illustrations of the context, and not the laboratories and experimental setups used to study the phenomena in the context.

In addition to the remote-controlled laboratories, several functionalities will enable a flipped learning approach and problem centered investigation, where the technology is put in a societal perspective and context. Theoretical introductions and a Wiki support the inquiry of knowledge. Self-tests, quizzes, and online lab report tools support the assessment and feedback. The possibility to collaborate between students and teachers, both peer-to-peer (student to student) and teacher-student communication, creates additional opportunities to activate students and enable student-centered learning.

In addition, a number of design criteria have to be considered when developing the solution, such as:

- Usability: The system should be easy to learn, it should be easy to find help, it should provide a standardized look, etc.
- Reliability: High accessibility when the system is used, backup if something happens, easy to get support if needed, etc.
- Performance: Supporting a high number of concurrent users, fast response time, etc.
- Security: Login procedure, handling of sensitive data, etc.
- Interoperability: Could be integrated with other systems, platform independent, etc.

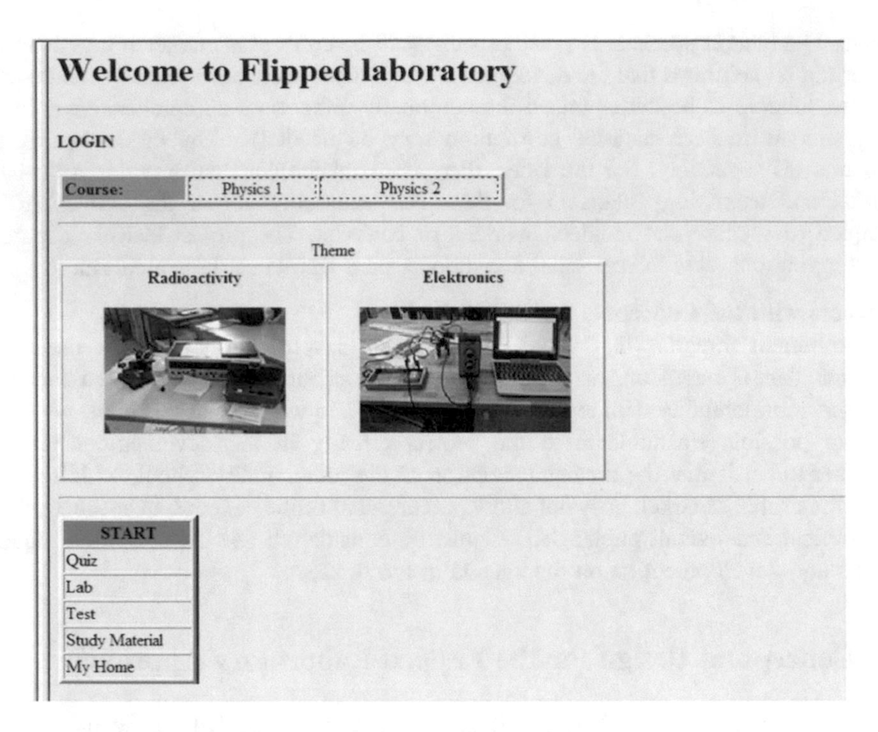

Fig. 6. Principal descriptive of Flipped laboratory concept

4.1 Overall Process Design

In the following, an example of the overall design and process flow is given for one theme; Radiation theory with focus on medical applications. The theme consists of two parts. Part one "A visit to the dentist", is described below, while part two, "Cancer treatment", is not described.

A Visit to the Dentist

- The purpose of the scenario: *Introduce the students to the field of radiation via familiar and relevant application areas. Problems that are being addressed are risks and safety associated with the use of radioactive materials as well as medical treatment options.*
- Theory included: *Radioactive decay, ionizing radiation, particle radiation, half-life and activity, interaction between different types of radiation and biological systems, absorbed and equivalent dose, radiation safety.*
- Laboratory work included: *Radioactive radiation range and half-life, absorbents.*

Overall Process

1. The students are introduced to the subject area and specific problem statements, such as “Is it dangerous to go to the dentist or to work as a dentist?” via a short text and a short video introduction to the theme.
2. The scenario “A visit to the dentist” is presented in an easy-to-understand way that raises questions. Students can reflect on where and how radioactivity comes into the context, and its possible impact. The opportunity to search more information is available through a Wiki function.
3. Laboratory “Dentist visit”. Laboratory instructions are given to the students, and the laboratory set-up is clearly explained. Students log into the specific laboratory environment. In the laboratory, a number of different radiation sources and absorbents are investigated. The source, absorbent, distance to source, measurement time and number of measurements to be done are selected. In a theme area when a part is selected; predefined experiments are available for that particular theme and part. The radiation level is measured with Geiger meters.
4. The students write a laboration report. The laboration report is written in online forms with predefined fields. All students are required to provide well-motivated answers to the questions “Is it dangerous to go to the dentist?” and “Is it dangerous to work as a dentist?” based on their results from the experiments.

4.2 Implementation Plan

The results from the feasibility study and the syllabuses for Physics 1 and 2 will form the basis for the further development. During the feasibility study, areas that for practical reasons are difficult to carry out laboratory work in a school environment were identified. One such area is radioactive decay. The ambition is to be able to provide an environment where students can, through a browser, do experiments with radioactive isotopes far from the harmful environment that they bring. Another area is electricity, and especially power current that requires special equipment for laboratory work. Radioactive decay and electricity forms part of the curriculum for the course Physics1 under the headlines Radiation in medicine and technology and energy and energy resources respectively. Two themes will be developed based on these headlines. In the further implementation, following steps will be taken (see Fig. 7):

1. Detailed concept development
2. Laboratory development
3. Testing and evaluation
4. Business model development

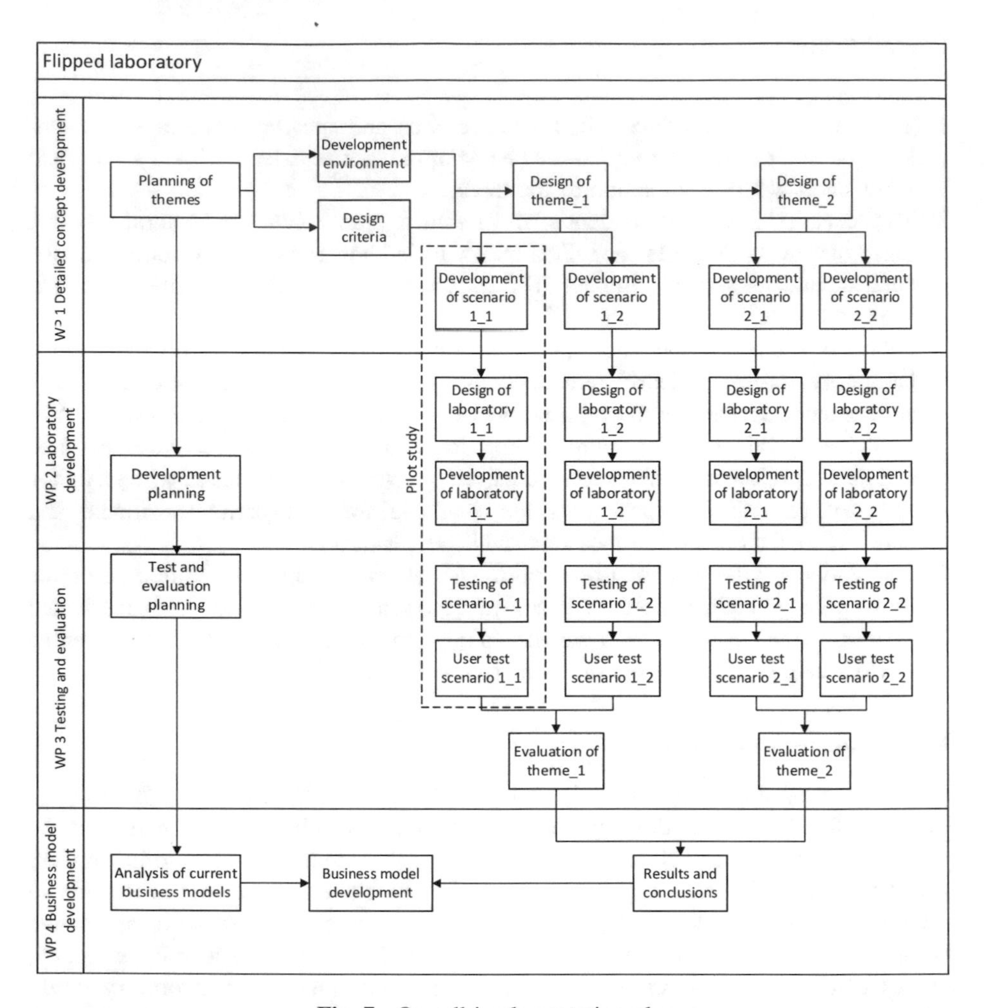

Fig. 7. Overall implementation plan

The identified areas for thematic and remote controlled laboratory work will further be clarified in the concept development phase. The overall design of the technical solutions will be developed together with relevant business partners. Thematic online-based teaching environments based on scenarios will be designed in dialogue with researchers and representatives from upper secondary schools. Once the themes are defined and the technical solution feasible, a first scenario with the associated remote laboratory will be built and tested. This will be made available to our main school partner in the project for validation (test pilot version), while the scenario is expanded with another laboratory (one scenario contains one or two laboratories). After the first test pilot, a limited number of upper secondary classes/students will have access to Flipped laboratory for evaluation. After pilot testing, iterations and validations, up to four thematic scenarios with associated laboratories will be built with the knowledge we have gathered during the project. Evaluation of the concept compared to traditional teaching will also take place. The overall goal is to increase students' interest and understanding of STEM. Evaluation towards this goal is included in the test and evaluation phase. This is enabled through the Flipped laboratory concept, and consequently a practical goal is to be able to offer upper secondary schools' nationwide access to the concept. For this, an appropriate business model must be designed. Ownership may be fully or partly at an educational institution that runs laboratories, but ownership may also lie within third parties such as education materials publishers or actors like LabsLand.

5 Conclusions

In conclusion, in order to achieve STEM education on equal terms and increased technology-related competence we need to reform the education with modern technology and pedagogy. Remote technology solutions can make the learning less time and space dependent and provide equal access to laboratories, particularly if they are complex, risky or costly. Flipped classroom pedagogy gives the opportunity for efficient lectures, and lectures that focus on understanding rather than on knowledge replication. The Flipped laboratory concept combines both. By the concept we want to make laboratory teaching attractive and understandable for all students by integrating laboratory elements into specific thematic areas, and thus putting technology understanding in a societal perspective. The innovation level lies both in a technology solution that is currently unavailable and in a new way of teaching technology-intensive elements. A number of similar initiatives exist, both in the field of digital learning and in remote-controlled laboratories, but at present there is no solution that provides thematic teaching with integrated laboratories that can be carried out online by connecting to a physical laboratory environment, and which is adapted to the Swedish upper secondary school curricula and conditions.

Since no physical equipment is needed at the individual school, both planning and learning become time and place independent, which supports students as well as teachers. With the Flipped laboratory concept, students can access the learning material and carry out laboratories remotely (as reinforcement to regular teaching), which strengthens individual learning. Teachers get support to instruct experimental/laboratory-related work methods in an integrated way, where the technology is clearly put in a societal perspective, and where physical resources are not required to perform the laboratory work.

When the concept is validated, it can be expanded in terms of content (covering both the course Physics1 and Physics2), subject (courses in Technology or Natural science), scope (added functionality) or level (towards primary school and/or post-graduate level). The concept can moreover be disseminated as a total concept or as selected parts in an international market (to suit national curricula, for example only the remotely controlled laboratories without associated learning material).

References

1. Statistics Sweden: Arbetskraftsbarometern 2018 (2018). [E document]. https://www.scb.se/hitta-statistik/statistik-efter-amne/utbildning-och-forskning/analyser-och-prognoser-om-utbildning-och-arbetsmarknad/arbetskraftsbarometern/
2. Nordström, E.: Jakten på kompetens driver jämställdheten framåt. Svenskt näringsliv (2019). https://www.svensktnaringsliv.se/fragor/kvotering/jakten-pa-kompetens-driver-jamstalldheten-framat_732472.html
3. Statistics Sweden: Higher Education. Students and graduates at first and second cycle studies 2017/18. UF20 - Universitet och högskolor. Grundutbildning (2019). [E document]. https://www.scb.se/publikation/38155
4. Statistics Sweden: Higher Education. Throughput at first and second cycle studies up to and including 2015/16. UF20 - Universitet och högskolor. Grundutbildning (2017). [E document]. https://www.scb.se/publikation/31515
5. Swedish National Agency for Education: Sök statistik om förskola, skola och vuxenutbildning – Skolverket (2019). https://www.skolverket.se/skolutveckling/statistik/sok-statistik-om-forskola-skola-och-vuxenutbildning. Accessed 05 Nov 2019
6. Svenningsson, J., Hultén, M., Hallström, J.: Understanding attitude measurement: exploring the meaning and use of the PATT short questionnaire. Int. J. Technol. Des. Educ. **28**, 67–83 (2018)
7. Gavin, K.G.: Design of the curriculum for a second-cycle course in civil engineering in the context of the Bologna framework. Eur. J. Eng. Educ. **35**(2), 175–185 (2010)
8. Mirriahi, N., Alonzo, D., McIntyre, S., Kligyte, G., Fox, B.: Blended learning innovations: leadership and change in one Australian institution. Int. J. Educ. Dev. Inf. Commun. Technol. **11**(1), 4–16 (2015)
9. De Hei, M.S.A., Strijbos, J.-W., Sjoer, E., Admiraal, W.: Collaborative learning in higher education: lecturers' practices and beliefs. Res. Pap. Educ. **30**(2), 232–247 (2015)
10. Labsland. https://www.labsland.com
11. Visir+. http://www2.isep.ipp.pt/visir/
12. Pilar. http://www.ieec.uned.es/pilar-project/news_events.html?lng=en
13. Remlabnet. http://www.remlabnet.eu/
14. Tawfik, M., Sancristobal, E., Martin, S., Gil, R., Diaz, G., Colmenar, A., Peire, J., Castro, F. M., Nilsson, K., Zackrisson, J., Håkansson, L., Gustavsson, I.: Virtual instrument systems in reality (VISIR) for remote wiring and measurement of electronic circuits on breadboard. IEEE Trans. Learn. Technol. **6**(01), 60–72 (2013)
15. Openlabs. https://openlabs.bth.se/electronics/index.php/en?page=StartPage#
16. Olabs. https://www.olabs.edu.in/
17. Go-Lab. https://www.golabz.eu/
18. Internet School Experimental System. http://www.ises.info/index.php/en/ises
19. Open Discovery of STEM Laboratories. http://opendiscoverylabs.eu/
20. Hrastinski, S.: Asynchronous and synchronous e-learning. Educause Q. **31**(4), 51–55 (2008)
21. Kearney, M., Schuck, S., Burden, K., Aubusson, P.: Viewing mobile learning from a pedagogical perspective. Res. Learn. Technol. **20** (2012)

Utilizing User Activity and System Response for Learning Analytics in a Remote Lab

Andrea Schwandt(✉), Marco Winzker, and Markus Rohde

Bonn-Rhein-Sieg University, Sankt Augustin, Germany
{andrea.schwandt,marco.winzker,markus.rohde}@h-brs.de

Abstract. Providers of learning material have a genuine interest in feedback about their learning resources. For a remote lab, providers like to know how students use the lab environment, which experiments they perform and what results they get from the experiments. This contribution describes an approach to perform learning analytics by recording the user interactions and the behavior of the hardware inside the remote lab.

Keywords: Remote lab · Learning analytics · Digital design

1 Introduction

According [1], learning analytics is defined as "the measurement, collection, analysis and reporting of data about learners and their contexts for purposes of understanding and optimizing learning and the environments in which it occurs". This information is not only interesting and helpful for providers of learning material and platforms like the remote lab, which is analyzed in this contribution. Ideally, the learning analysis is used to provide individual feedback to learners of their current learning progress [2, 3]. When students have predefined tasks to solve, this goal can be reached straightforward by checking if the outcome of a student matches the expected result. However, regarding open defined tasks in research-based learning, this represents a major challenge. Still, learning analytics provide valuable feedback to teachers to adjust their learning material and provide a maximum on learning outcome [4]. Our work performs learning analytics for a remote lab and, in this context, especially monitoring the user actions, which provide valuable information about how students exploit the system.

Another point is data protection and privacy. The general learning analytics without the direct, individual feedback to students enables an anonymization of the collected data and thus meets the GDPR (**G**eneral **D**ata **P**rotection **R**egulation) in Europe.

The following paper starts with a short outline of the motivation for learning analytics in the FPGA Vision Remote Lab in Sect. 2, followed by the definition of the methodology in Sect. 3, its appliance on the FPGA Vision Remote Lab and results in Sect. 4. Section 5 continues with a recommendation on further analysis and finally a conclusion is presented in Sect. 6.

M. E. Auer and D. May (Eds.): REV 2020, AISC 1231, pp. 63–74, 2021.
https://doi.org/10.1007/978-3-030-52575-0_5

2 Motivation

In January 2018, the authors released a new teaching concept [5] consisting of online lectures and a remote lab to support these online lectures and their hands-on labs in the field of image processing with FPGAs (**F**ield **P**rogrammable **G**ate **A**rrays) [6]. Students can perform image processing on an input image and observe the effects of the algorithm in the output image and the core current of the FPGA. The core current of the FPGA corresponds to the power dissipation for performing the image processing task.

Students upload an FPGA binary and can interact with the experiment with three switches, the choice of predefined images and the option to upload user specific jpg-images. Two full experiments are available which differ in the provided FPGA, the Cyclone IV and the Cyclone V from Intel [7]. Additionally, the two laboratory setups can be accessed in a demo-mode for demonstration purposes.

Figure 1 illustrates the scenario for students working with the remote lab. For performing an experiment, students write VHDL (**V**ery **H**igh Speed Integrated Circuit Hardware **D**escription **L**anguage) code and compile it on their computers. Then they login to the remote lab, reserve the experiment according to the FPGA the code is compiled for, upload their binary file and observe the system response. After that, students can perform further activities like operating a switch connected to an input pin of the FPGA or change the input image for signal processing. Additionally, students can upload a new FPGA binary and perform another experiment as long as there is still reservation time available.

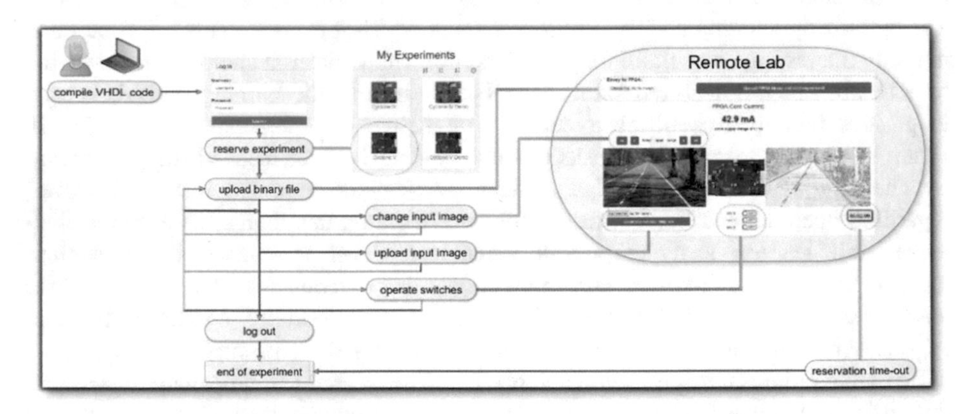

Fig. 1. Typical design flow of a remote lab experiment

The learning analytics information is intended to reach several goals. One purpose is an interest in understanding the different ways students exploit the remote lab, e.g. if they primarily use it to work on the proposed lab tasks or if they take advantage of the remote lab to perform additional experiments. Beyond the possibility to provide feedback to students of their learning progress, this information shall be used to adjust the design of the remote lab and learning material.

Modules in engineering studies often require that students solve a predefined number of tasks as a prerequisite for examination. Therefore, another goal is to evaluate students' work with regard to this. However, as described earlier, considering open defined tasks to motivate research-based learning, this goal equates a notable challenge.

3 Methodology

The WebLab-Deusto [8] RLMS (**R**emote **L**ab **M**anagement **S**ystem) is used to implement the FPGA Vision Remote Lab. It generates a log file as well as entries in an SQL-database storing the activity performed by all users. The gathered data contains the country the users are in, when users start and finish an experiment, which experiment is used, which actions they perform and the measurement of the FPGA core currents [9].

To interpret the learning analytics data it is categorized as follows:

(1) General data like country, date, start and end of experiment
(2) User actions like "switch operation" and "file upload"
(3) Measurement results from the system

3.1 Analyzation of General Data

General data can be retrieved directly from the logged data of the remote lab. Figure 2 displays this general data retrieved from the dashboard of the RLMS. It contains the

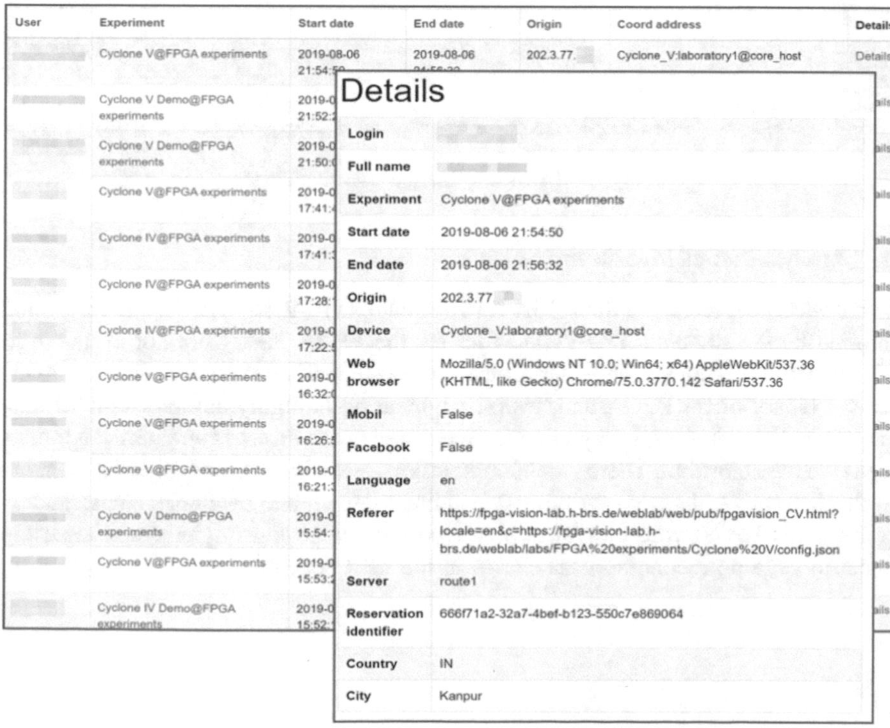

Fig. 2. User logs with general user data

login name, start and end date of experiment and further information. This data will be anonymized while processing learning analysis.

3.2 Analyzation of User Actions

Data about the user actions can be retrieved also directly from the logged data of the remote lab. Figure 3 shows an excerpt of the user actions logged in the RLMS. The user is operating the switches of the experiment and the system answers by providing the output image of the signal processing and the measurement of the FPGA core current. This data is still personalized and will be anonymized in the course of analyzation.

2019-10-22 10:59:52.132334	2019-10-22 10:59:53.147373	sw1_on	command received switch11	N/A
2019-10-22 10:59:54.398399	2019-10-22 10:59:54.607702	update	Image update processed	N/A
2019-10-22 10:59:54.681654	2019-10-22 10:59:55.949633	measure	92.3 mA	N/A
2019-10-22 10:59:56.756366	2019-10-22 10:59:57.774313	sw2_on	command received switch21	N/A
2019-10-22 10:59:59.041478	2019-10-22 10:59:59.247463	update	Image update processed	N/A
2019-10-22 10:59:59.291216	2019-10-22 11:00:00.749351	measure	92.5 mA	N/A
2019-10-22 11:00:03.127726	2019-10-22 11:00:04.143721	sw1_off	command received switch10	N/A
2019-10-22 11:00:05.386484	2019-10-22 11:00:05.581489	update	Image update processed	N/A
2019-10-22 11:00:05.658382	2019-10-22 11:00:06.748912	measure	91.8 mA	N/A

Fig. 3. Excerpt of logged user actions

3.3 Analyzation of Measurement Data

The measurement results of the FPGA core current can be used to deduce the algorithm compiled by the students. The VHDL files for the FPGA binary are not available for analysis because compilation is performed on the student's computer. Students upload only the binary of the FPGA and reverse engineering is highly complex and virtually impossible. Nevertheless, interpreting the measured core current is an indirect approach to retrieve the algorithm. There are different tasks proposed to the students for exercises and for these tasks, data on core current is already known and available. Consequently, the measured FPGA core current allows a correlation to the implemented algorithm subject to certain preconditions as shown in the next chapter.

4 Learning Analytics of the FPGA Vision Remote Lab

The data for learning analytics of the FPGA Vision Remote Lab has been collected since the official start in Jan. 2018. This data is analyzed respecting the categories defined in Sect. 3.

4.1 Analyzation of General Data

At the first stage, the total worldwide accesses to the FPGA Vision Remote Lab from Jan. 2018 to Aug. 2019 are analyzed as shown in Fig. 4. Most of these experiments are located in the country of our university, in Germany. The remote lab is used by partner universities in Argentina and Ukraine and consequently these countries have high access numbers of the lab, too. Additionally, experiment accesses from individual learners located in over 30 other countries can be noted.

Fig. 4. Worldwide accesses (last updated Sept. 2, 2019)

These accesses include experiments of our administration team for testing purposes as well as the reservations of the demo experiment. However, the learning analytics should only cover uses of the FPGA Vision Remote Lab from students performing experiments including the upload of a compiled FPGA code. Thus, the analyzed data has been filtered to leave out accesses to the remote lab from the administration team as well as reservations of the demo experiments.

Figure 5 shows the filtered data and depicts the use of the two available experiments split up between the different student groups. We distinguish between four groups: Germany, the two countries of our partner universities, Argentina and Ukraine, and one group for all other countries. For our analysis we assume, that accesses from Germany are mainly by our own students, accesses from Argentina and Ukraine are by students from partner universities and accesses from other countries are by individual learners who use the material without affiliation to a participating universities.

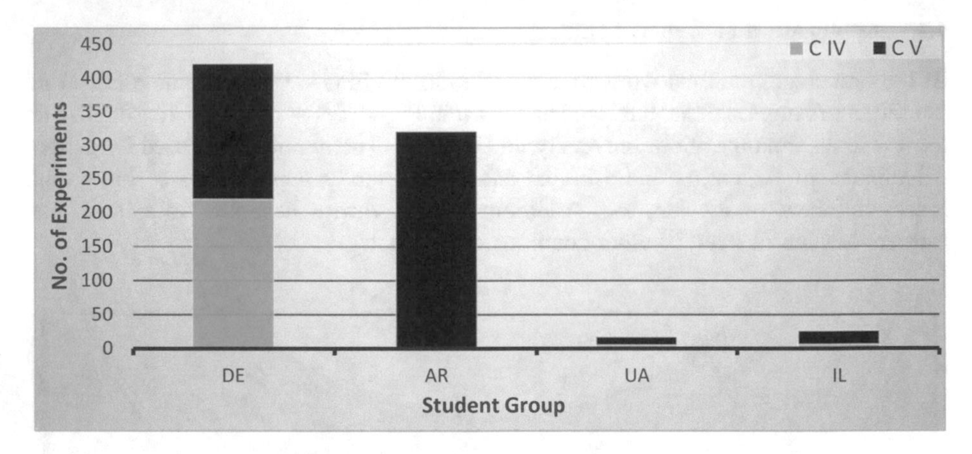

Fig. 5. Use of cyclone IV (C IV) and cyclone V (C V) FPGAs by different student groups *DE: Germany, AR: Argentina, UA: Ukraine, IL: Individual Learners*

Analyzing the data, the first noticeable aspect is that experiments with the older Cyclone IV FPGA are performed mainly by students from our university in Germany and only five times by students from Ukraine as well as six times by individual learners. However, the reason why students from the authors' university use the Cyclone IV experiment can be explained: The laboratory setup for the onsite hands-on labs uses the Cyclone IV FPGA, too. Therefore, many students from the Bonn-Rhein-Sieg University exploit the remote lab to finish experiments they run out of time during the onsite hands-on labs or to prepare their onsite hands-on lab. Students from other institutions prefer the newer Cyclone V FPGA.

The next, deeper analysis examines the monthly use of the experiments by each student group as shown in Fig. 6.

This analysis correlates with the fact that the German course, in which the remote lab is used, takes place every summer semester (April to July). Small numbers of experiments from Jan. 2018 to March 2018 can be correlated to colleagues who evaluate the remote lab for teaching purposes. They come from Germany and in March 2018 from Argentina when one of the authors visited our partner universities there and promoted the FPGA Vision Remote Lab. In December 2018 and February 2019 our partner universities in Argentina used the remote lab in a course, and our partner university in Ukraine did so in December 2018.

The remote lab was released in Jan. 2018 with basic functionality supporting the Cyclone V FPGA. In April 2018, the remote lab provided nearly complete functionality by adding power measurement, switches and the possibility to upload own images.

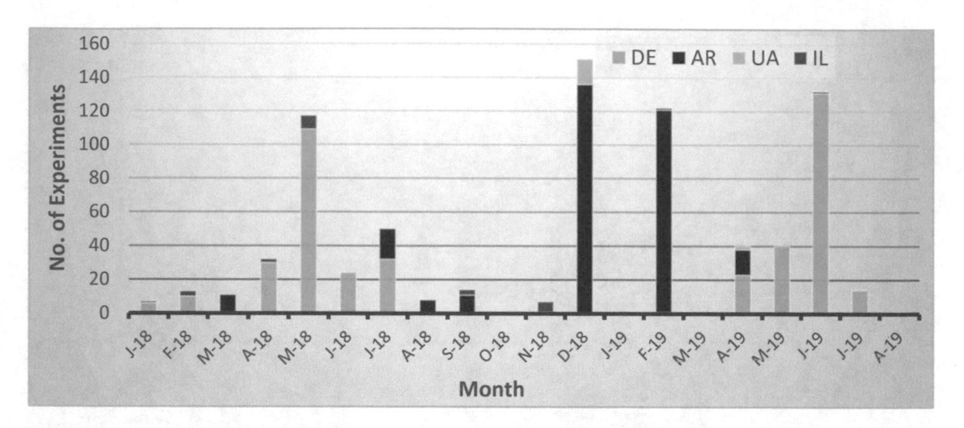

Fig. 6. Experiment count for each student group by month *DE: Germany, AR: Argentina, UA: Ukraine, IL: Individual Learners*

Finally, the experiment with the Cyclone IV FPGA was released in May 2018. Respecting this and the start of the summer term in April, further analysis will be applied to data from April 2018 to August 2019.

4.2 Analyzation of User Actions

The analyzations described in Sect. 4.1 examine the basic data like the origin of the student groups, reserved experiments and time of reservation. The following chapter intends to provide a deeper insight in how students exploit the FPGA Vision Remote Lab. Consequently, the analysis of the basic data is combined with the examination of the conducted user actions.

The FPGA Remote Lab provides different possibilities for user actions. First, students have to upload their own binary file to be able to execute the experiment.

As a measure of activity the total amount of binary file uploads has been retrieved from the logged data. We calculate the ratio between the numbers of uploaded binary files and the amount of reserved experiments. It has been observed that not every experiment reservation is followed by a binary upload. Thus, a second value is calculated where only active experiments are taken into account, i.e. experiments where the students uploads at least one binary file.

Figure 7 depicts the ratio between amount of uploaded binary files and the total amount of reservations (all) as well as merely for the reservations of an experiment resulting in at least one binary file upload (active).

The first ratio demonstrates students who reserved an experiment without uploading their own binary file namely did not really perform an experiment. This may have various reasons. On the one hand, e.g. one student attending the course in the summer term 2018 reported he had connection problems at his home unrelated to the remote lab. Accordingly, poor internet connections might be the reason for the dropouts. On the other hand, some students might not watch the provided YouTube video explaining how to operate the remote lab and reserve an experiment without knowing about the proper handling.

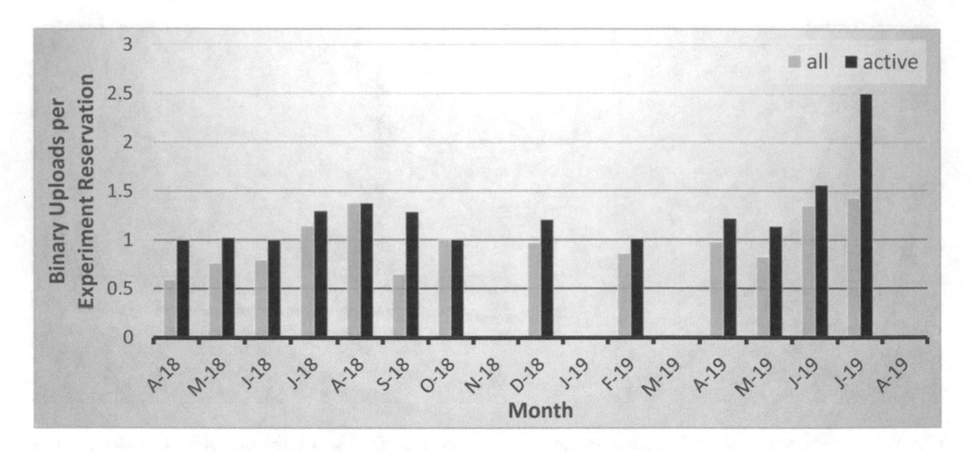

Fig. 7. Uploads of binary files per reserved experiment by month

The second ratio shows how intensive students exploit the available time of up to five minutes. Here, it can be noted that several students upload more than one binary file during an experiment session.

The next user activity option is the choice of the input image for image processing. Ten different images are provided, one of those being a wildcard that can be replaced by an individual uploaded image. The different images enable several purposes. One purpose is to offer various switching activities to evaluate the corresponding power consumption behavior. Another purpose is to generate different street scenes to evaluate lane detection algorithms. The exploitation of the different images is a measure how students work with the remote lab and their interests. Accordingly, the authors are interested to learn how often which images are used and if user-specific images are uploaded.

Figure 8 shows the percentage usage of the different images. Nearly 50% of the students stick to the default image (0) every experiment starts with. Only about 10% switch to the next image (1) and an even smaller number of 5% switch to image (2). The remaining values of the images 3 to 8 are even lower. Image 9 is the wildcard image and it is selected by 24% of the students. However, not all students do really upload an image. Only 16% of the usage is performed with an individual image.

The experiments of the remote lab also provide three switches for user actions. This allows parametric programming to have some possibilities of variations with only one binary upload. As a first analysis, the usage numbers of switches is evaluated. All three switches are used in total more than a hundred times. This quantity does not reveal how the switches are used inside the FPGA design. Users might operate the switches while exploring the user interface without using them inside the FPGA. However, analyzing measurement data gives the opportunity to learn about functional usage of the switches. If the measurement results change significantly after a user operates a switch, it validates the switch is implemented in the design. This analysis will be presented in the next section.

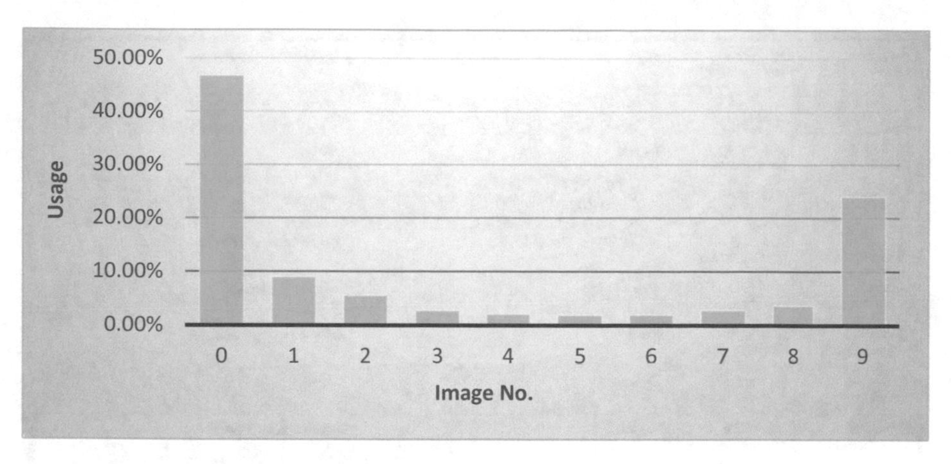

Fig. 8. Percentage usage of provided and uploaded images *0: Motorway in Germany 1, 1: Motorway in Germany 2, 2: Road in Germany 1, 3: Road in Germany 2, 4: Road on Canary Islands, Spain, 5: Road in Andes mountains, Argentina, 6: Road in Germany 3, 7: Bridge in Denmark, 8: Test card, 9: Wildcard and upload image*

4.3 Analyzation of Measurement Data

Analyzing user actions combining with deeper analysis provides a good insight of students' progress. The measured core current e.g. allows, under certain preconditions, a sufficiently precise correlation to the image processing algorithm and its complexity, especially in combination with other user actions.

Figure 9 shows an excerpt of the logged data of one experiment as displayed in the administration panel of the WebLab-Deusto RLMS.

The experiment uses all possible combination of switch states. According to switching actions, the excerpt of the data starts at a binary equivalent of an integer 3 and the measured core current of 50.2 mA. During the experiment, the binary switch value increases by one, while the measured core current increases by about an average of 8.8 mA with each step. This implies that the corresponding VHDL program consists of identical function blocks, which can be activated and chained one by one.

One of the suggested algorithms to start with is a register chain of 700 registers in total, where portions of 100 registers should be activated one by one by using the switches. Given that algorithm and comparing the known results of this known experiment with the results given in Fig. 9, it could be assumed that this is the experiment performed.

For the hands-on labs, of the course, several image processing algorithms are suggested to our students. Measuring series performed while preparing the hands-on lab [10] proof the reproducibility of the results and show the parameters influencing the measured core current, like temperature, routing inside the FPGA, and so on. Even different FPGAs of the same device show different power consumptions with the identical binary. Additionally, students' programs are supposed to be slightly different. Thus, it is not possible to define the expected values within their tolerances. However, the different tasks differ significantly in power consumption and in the behavior or

2019-10-28 14:48:29.317222	2019-10-28 14:48:30.330044	sw0_on	command received switch01
2019-10-28 14:48:31.559230	2019-10-28 14:48:31.839194	update	Image update processed
2019-10-28 14:48:31.868300	2019-10-28 14:48:33.023964	measure	50.2 mA
2019-10-28 14:48:36.252292	2019-10-28 14:48:37.264476	sw2_on	command received switch21
2019-10-28 14:48:37.265724	2019-10-28 14:48:38.284965	sw1_off	command received switch10
2019-10-28 14:48:38.287025	2019-10-28 14:48:39.300853	sw0_off	command received switch00
2019-10-28 14:48:40.949714	2019-10-28 14:48:41.218476	update	Image update processed
2019-10-28 14:48:41.253219	2019-10-28 14:48:42.622521	measure	58.7 mA
2019-10-28 14:48:44.044296	2019-10-28 14:48:45.051671	sw0_on	command received switch01
2019-10-28 14:48:46.287725	2019-10-28 14:48:46.552662	update	Image update processed
2019-10-28 14:48:46.579458	2019-10-28 14:48:48.022536	measure	67.5 mA
2019-10-28 14:48:49.126584	2019-10-28 14:48:50.140424	sw1_on	command received switch11
2019-10-28 14:48:50.207517	2019-10-28 14:48:51.218403	sw0_off	command received switch00
2019-10-28 14:48:52.855380	2019-10-28 14:48:53.128929	update	Image update processed
2019-10-28 14:48:53.151660	2019-10-28 14:48:54.621760	measure	76.4 mA
2019-10-28 14:48:56.893450	2019-10-28 14:48:57.906102	sw0_on	command received switch01
2019-10-28 14:48:59.149705	2019-10-28 14:48:59.420134	update	Image update processed
2019-10-28 14:48:59.446861	2019-10-28 14:49:00.620276	measure	85.6 mA
2019-10-28 14:49:07.868241	2019-10-28 14:49:17.917214	@@@finish@@@	None

Fig. 9. Excerpt of logged measurement data and user actions

sequence of their power consumption, so it is possible to keep them apart. Accordingly, most of students' experiments can be compared to them and consequently identified. However, this implies that students implement one of the proposed algorithms. Nevertheless, assuming the course is successful and students are motivated to program their own algorithms it is possible that students implement other applications with core measurements similar to a proposed algorithm. In addition, the authors might encounter an algorithm with an unknown core current and might not be able to determine it.

5 Recommended Further Analysis

Further work will investigate dependencies between different activities. Such analysis can include:

- Core current measurement before and after activating a switch.

- Core current measurement after upload of an individual image.
- Usage of different FPGAs be the same user, i.e. do students perform the same experiment on both FPGAs provided?

6 Conclusions

The general data allows a conclusion which users (student from our university, a partner university or individual learner) used the remote lab and when. It can be seen that students' log in times correspond with respective course times.

User actions give insight how intense the remote lab is used by individual users. Additional, they show whether the provided possibilities for user actions are accepted or not.

Most measurement results can be correlated to the implemented image processing algorithm. Consequently, the information can also be used for assessment of students, e.g. to determine if prerequisites for exams are fulfilled.

The results show that measurement results of a remote lab has offers potential for learning analytics which can be exploited to monitor students' activity and analyze which exercises they perform with the lab.

References

1. Peña-Ayala, A.: Learning Analytics: Fundaments, Applications, and Trends, vol. 94. Springer International Publishing, Cham (2017)
2. Lang, C., Siemens, G., Wise, A., et al.: Handbook of Learning Analytics. Society for Learning Analytics Research (SoLAR). Canada: Society for Learning Analytics Research (2017)
3. Wuttke, H.D., Hamann, M., Henke, K.: Learning analytics in online remote labs. In: Conference E@I (ed.) 3rd Experiment International Conference: Online Experimentation : June 2nd-4th, 2015, Ponta Delgada, São Miguel Island, Azores, Portugal : exp.at'15, pp. 255–260. IEEE, Piscataway, NJ (2015)
4. de Brandão, Damasceno, Andr´e Luiz, dos Santos Ribeiro, D., Barbosa, S.D.J.: What the literature and instructors say about the analysis of student interaction logs on virtual learning environments. In: Frontiers in Education 2019: Bridging Education to the Future. IEEE (2019)
5. Winzker, M., Schwandt, A.: Open education teaching unit for low-power design and FPGA image processing. In: Frontiers in Education 2019: Bridging Education to the Future. IEEE (2019)
6. Schwandt, A., Winzker, M.: Make it open - improving usability and availability of an FPGA remote lab. In: Ashmawy, A.K., Schreiter, S. (eds.) Proceedings of 2019 IEEE Global Engineering Education Conference (EDUCON) 9–11 April, 2019, Dubai, UAE, pp. 232–236. IEEE, Piscataway, New Jersey (2019)
7. Intel® Cyclone® Series FPGAS and SoCs-Intel® FPGA: (2018). https://www.intel.de/content/www/de/de/products/programmable/cyclone-series.html. Accessed 01 Nov 2019
8. WebLab–Deusto: (2019) http://weblab.deusto.es/website/index.html. Accessed 08 Dec 2018

9. Orduna, P., Almeida, A., Lopez-de-Ipina, D., et al.: Learning analytics on federated remote laboratories: tips and techniques. In: 2014 IEEE Global Engineering Education Conference (EDUCON), pp. 299–305. IEEE (2014)
10. Schwandt, A., Winzker, M., Abu Shanab, S.: Design of lab exercises for teaching energy-efficient digital design. In: IEEE Global Engineering Education Conference (EDUCON), 2015: 18–20 March 2015, Tallinn University of Technology (TUT), Tallinn, Estonia, pp. 112–117. IEEE, Piscataway, NJ (2015)

Programmable Remote Laboratory for Mobile Robots

Rafael Franco-Vera(✉), Xuemin Chen, and Wei Wayne Li

Texas Southern University, Texas, USA
rjfranco0812@gmail.com

Abstract. As mobile robots become more popular, more remote laboratories (RL) using mobile robots are being developed for online education. However, in most current RL for mobile robots, the control which users have over the robot is limited by the pre-programmed controlled algorithms. This limitation creates a gap in the learning experience of the students. This paper proposes an implementation of a fully programmable remote laboratory for mobile robots into a web application. In the RL user-interface, students are able to implement and execute their C++ source code and see the results in real-time through a live feed webcam. The AJAX technology is used to transfer data between the web application and web server, making it possible to execute the code. This application is designed to provide students with interactive tools and a contextual learning scene, making this research of significant relevance to online engineering education by providing students with direct programming experience on remote laboratory website.

Keywords: Programmable · Remote laboratory · Mobile robot · Flexible framework

1 Introduction

Laboratory work has long been identified as an important element of undergraduate degree courses in many disciplines, especially engineering and applied sciences [1]. It is impossible to imagine science without laboratories. To most, the concept of science is defined by those special buildings (laboratories) in which experts investigate natural phenomena and processes [2]. However, with the rapid development and availability of advanced telecommunications infrastructure and internetbased/wireless technological applications, it has made it possible to evolve the method of using a traditional laboratory for educational purposes into the development of remote laboratories. A remote laboratory (RL) is used to identify laboratories with physical equipment that performs the tests locally, but the user can have access remotely through an interface. A traditional laboratory has practical sessions on a fixed schedule regularly and with a limited time for the completion of the activity. Students sometimes feel that they did not reach all objectives and that work could not be done regularly because the laboratory is occupied with other classes in contrast to a remote laboratory that can be accessed equally as well from home and school, and can reduce logistical issues related to scheduling lab time for

M. E. Auer and D. May (Eds.): REV 2020, AISC 1231, pp. 75–81, 2021.
https://doi.org/10.1007/978-3-030-52575-0_6

students. Because of its effectiveness, flexibility, and cost-saving, remote laboratories have become an important component of online learning [3].

As the demand for RL increases, new innovative platforms/frameworks for RL are developed, [3–5, 7]; therefore, creating opportunities to implement new forms of robotics-related experiments into a RL, such as mobile robots. Currently, there is a variety of web-enabled robot platforms that allow the user to remotely control their sensors and other related components. Typically, in common implementation of these systems, the user interface (UI) only allows control of certain elements of the robot. Two great examples, of such system are: 1) the remote lab for mobile robot application [6], which permits the user to control the robot's movement with a series of buttons to move left, right, forward, and backwards or by pre-programmed controlled algorithms implemented in the robot, 2) the remote lab for multi-robot experiments with virtual obstacles [8], which, focuses on defining virtual obstacles and uses a Matlab simulator for visualization.

The systems described above are an excellent tool to captivate students' interest by providing the opportunity to perform various experiments and allowing them to familiarize themselves with the concept of robotics using mobile robots. However, they do not provide students with the fundamental capability of implementing and executing their program, therefore, falling short in giving students the required programming practice needed for their education and limiting their learning experience. To address this vital issue, the programmable remote laboratory using AJAX technology is presented in this paper. This programmable remote laboratory will enable students to fully program the mobile robots through a user interface without the need for software plug-ins or third-party software. This will not only captivate the student's interest; it will also enhance their learning by giving them a first-hand programming experience when running experiments in a remote laboratory.

This new programmable remote laboratory for mobile robots will be implemented into the Virtual Remote Laboratory (VR-Lab) developed at Texas Southern University (TSU) [7]. To demonstrate the effectiveness of this new programmable remote laboratory, the Amigobot mobile robot manufactured by Adept will be used. The Amigobot uses the Advanced Robotics Interface for Applications (ARIA) object-oriented, application programming interface (API) [11], which, includes pre-build libraries written in C++ language. However, this new programmable remote laboratory can be used for different engineering RL experiments and programming languages, e.g. Python.

The rest of the paper is organized as follows. Description of the components that constitute the programmable remote laboratory in Sect. 2. Section 3 discusses the implementation of the experiment. Concluding remarks are drawn in Sect. 4.

2 Methodology

In this paper, the objective is to build a new programmable remote laboratory (PRL) platform that will give engineering students the required programming experience needed in their education. To make this possible, we designed the PRL for mobile robots using AJAX technology and implemented it into the VR-Lab framework developed at TSU [7]. The PRL for mobile robot platform is a client-server application and uses Web 2.0 technology to reduce or eliminate the need for browser plug-ins or third-party software. The structure of the laboratory is shown in Fig. 1.

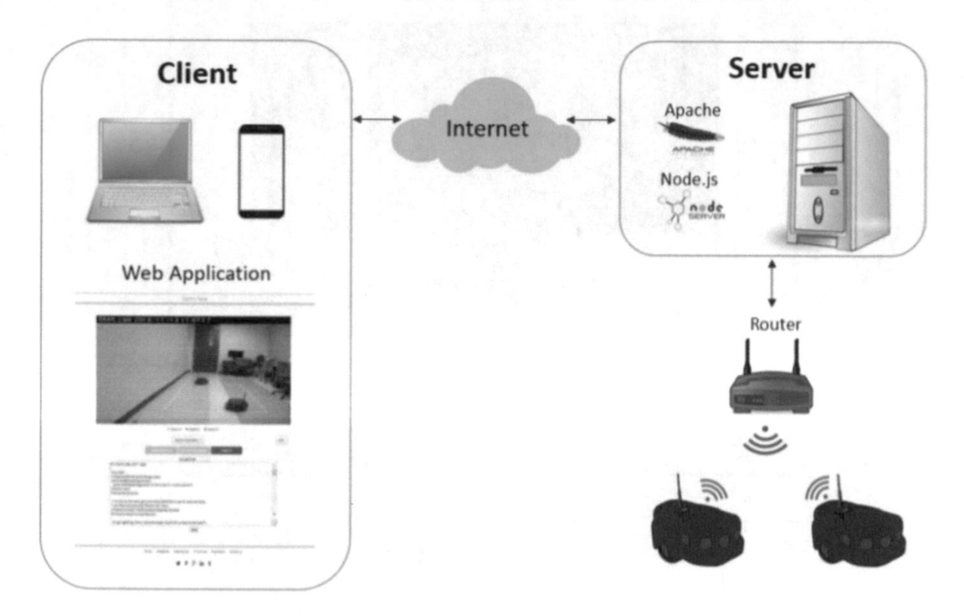

Fig. 1. Remote laboratory architecture

2.1 Client Web Application

The client web application design and implementation are based on Hypertext Preprocessing (PHP), HyperText Markup Language (HTML), Cascading Style Sheets (CSS), and JavaScript, along with server-based Mashup technology for UI development. The UI consists of a live stream of the mobile robots, a light switch to turn lab's lights on, a simple operation option, and a text-area with a pre-loaded C++ program example. A screenshot of the UI is shown in Fig. 2 and a closer look of the C++ example code is shown in Fig. 4.

If the student has little to no programming experience, he/she can get familiarized with the mobile robot programming style by selecting the "Simple Operation" control option, which simply allows them to control the mobile robot through a series of inputs, such as, 1st distance, angle for turning, 2nd distance, and speed. Although, watching the mobile robot move can be quite impressive for a beginner, the actual purpose of this option correlates with the pre-loaded C++ program example in the text-area. Because the programming logic implemented in the inputs of the "Simple Operation" option is the same as the example program displayed in the text-area. The example is set up to be easily understood by using the simple option and reading the code along with the movement of the robot. On the other hand, if the student is an advanced programmer, he/she can clear the pre-loaded example and start writing their C++ program. Furthermore, with the use of AJAX technology when the student clicks the run button their code will be sent to the server to be compiled and executed in the background.

Fig. 2. User interface

2.2 Server Application

To improve the web service technology performance issue for the server application implementation, the combined solution of both Apache HTTP web engine and Node.js web engine is implemented for real-time communication. Apache HTTP web engine is the most used web server software around the world and supports a variety of UI features development [9]. Node.js is an open-source, cross-platform runtime environment for developing server-side web applications. It also enables web developers to create an entire web application in JavaScript which are both server-side and browser-side. Thus, the server application built on the top of a MySQL database, an Apache webserver engine, and a Node.js web server engine.

Using AJAX

The approach taken for this programmable remote lab was to use AJAX (Asynchronous JavaScript and XML) technologies to communicate with the server without the need for extra plugins. It is not a single technology, nor is it a programming language. It is a set of web development techniques that allow web applications to send and retrieve data from a server asynchronously (behind the scenes) without interfering with the display and behavior of the existing page [12]. This system involves:

- **HTML5** for the main language and CSS for the presentation of the remote lab Web page.
- The **Document Object Model (DOM)** to dynamically access and update the content, structure, and style of a document.
- **XML/JSON** for the interchange of data.

- The **XMLHttpRequest** object for the asynchronous communication by exchanging data with the server behind the scenes.
- Lastly, **JavaScript** programming language bring it all together.

The general procedure of how AJAX works is quite simple. Ajax applications create a JavaScript-based engine that runs on the browser so instead of loading a traditional HTML page, the browser loads the engine, which then displays or performs the requested action [10]. Figure 3 gives a better understanding of this process along with the executing process that will be explained next.

In the browser, the user creates a JavaScript call by clicking the "Run" button on the UI which activates the XMLHttpRequest. Then behind the scene, the browser creates an HTTP POST method request to the server. The request is to open a PHP script as a POST method and sends the user's source code to the server, encoded by the "encodedURIComponent()" function. This function is used to encode user-entered fields from forms POST'd to the server. Once the request is made the server receives the data (user's code) allowing the PHP script to retrieve it using the "$_POST" variable, which collects the values sent by the HTTP POST method. The PHP script continues by adding the appropriate libraries needed to complete the user's program then followed by the program execution functions "system" and "exce" used to compile and execute the program. In the "system" function we use the g++ compiler to compile the program and the "exec" function simply executes the code [13].

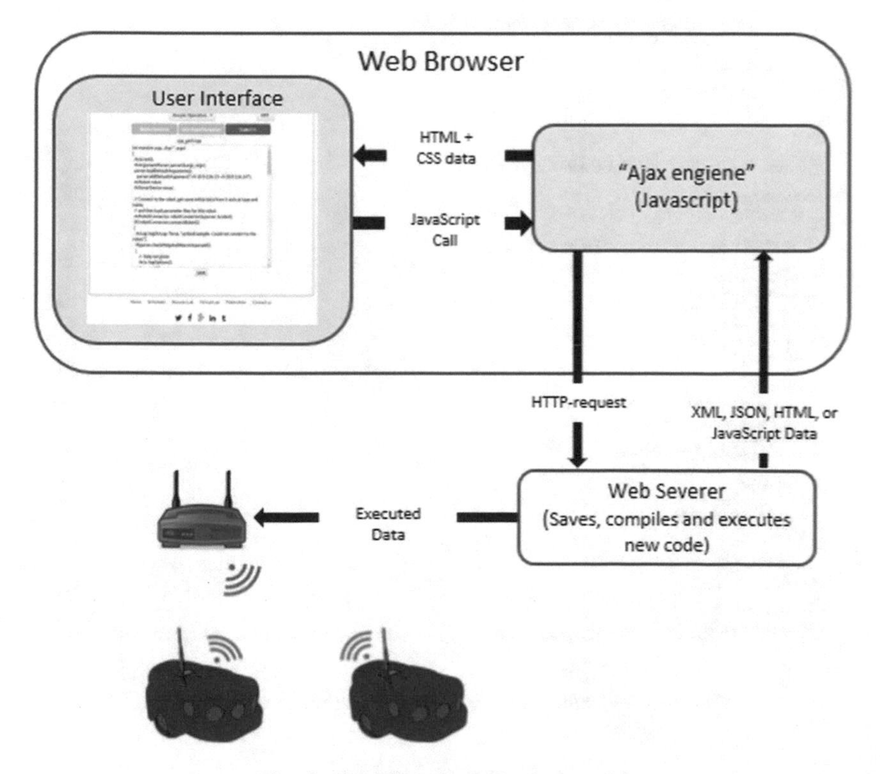

Fig. 3. AJAX web application model

3 Experiment Implementation Discussion

To demonstrate the effectiveness of this new programmable remote laboratory, the Amigobot mobile robot manufacture by Adept is used. The Amigobot uses the Advanced Robotics Interface for Applications (ARIA) object-oriented, application programming interface (API) [11], which, includes pre-build libraries written in C++ language. The ARIA software is provided by the Pioneer manufacturers, MobileRobots Inc. (formerly ActivMedia) and can be used to control any of their models, e.g. AmigoBots, PeopleBots, and Pioneers.

All the required libraries needed to communicate with the Amigobot have been downloaded into the server, such as the ARIA and ArNetworking libraries. Therefore, the students can now log into the VR-Lab select their experiment and start practicing their coding on the Amigobot mobile robot and see the results through the live video stream as shown in Fig. 2.

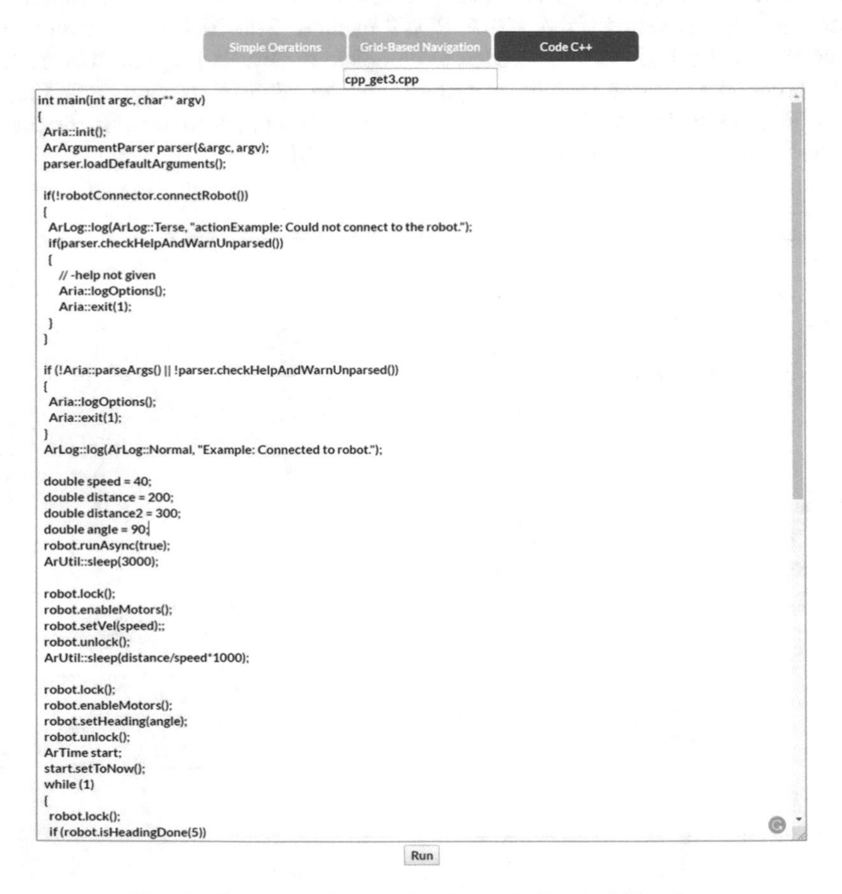

Fig. 4. C++ example code for the Amigobot mobile robot.

4 Conclusion

This paper proposed a new programmable remote laboratory application that was successfully implemented, providing students a meaningful platform in order effectively program mobile robots through the user interface without the need for software plug-ins or third-party software. The programmable remote laboratory application was implemented into the VR-Lab with the use of AJAX technologies, utilizing the Amigobot mobile robot in order to test its effectiveness of enhancing the programing skills of engineering students by letting them connect with the mobile robots and allowing them to practice their programming skills by learning through example source codes or by creating and implement their own programs.

Acknowledgement. This work was supported in part by RENEW which is a technology partner to POWDER. RENEW/POWDER is part of the National Science Foundation's PAWR program and is funded in part by NSF award CNS-1827940 and the PAWR Industry Consortium.

References

1. Feisel, L.D., Peterson, G.D., Arnas, O., Carter, L., Rosa, A., Worek, W.: Learning objectives for engineering education laboratories. In: Proceedings of the 32nd Annual Conference Frontiers in Education, FIE 2002, vol. 2, pp. F1D-1–F1D-24 (2002)
2. Schidgen, H.: History beginnings of the laboratory in early modern world. Professor of the Theory of Media Worlds (2011)
3. Wang, N., Chen, X., Song, G., Alan, Q., Parsaei, H.R.: Design of a new mobile optimized remote laboratory application architecture for M-learning. IEEE Trans. Ind. Electron. 64(3), 2382–2391 (2017)
4. Olmi, C., Cao, B., Chen, X., Song, G.: A unified framework for remote laboratory experiments. In: Proceedings of ASEE Annual Conference & Exposition, Vancouver, BC, Canada (2011)
5. Wang, N., Song, G.B., Chen, X.: Framework for rapid integration offline experiments into remote laboratory. Int. J. Online Eng. (iJOE) **13**(12), 192–205 (2017)
6. Neamtu, D.V., Fabregas, E., Wyns, B., De Keyser, R., Dormido, S., Ionescu, C.M.: A remote laboratory for mobile robot applications. Int. Fed. Autom. Control (IFAC) **44**, 7280–7285 (2011)
7. Wang, N., Chen, X., Lan, Q., Song, G., Parsaei, H., Ho, S.C.: A novel wiki-based remote laboratory platform for engineering education. IEEE Trans. Learn. Technol, 99 (2016). https://doi.org/10.1109/TLT.2016.2593461
8. Casini, M., Garulli, A., Giannitrapani, A., Vicino, A.: A remote lab for multi-robot experiments with virtual obstacles. Int. Fed. Autom. Control (IFAC) **45**(11), 354–359 (2012)
9. Fielding, R.T., Kaiser, G.: The apache HTTP server project. IEEE Internet Comput. **1**(4), 88–90 (1997)
10. Paulson, L.D.: Building rich web applications with Ajax. Computer **38**(10), 14–17 (2005)
11. Whitbrook, A.: Programming Mobile Robots with Aria and Player: A Guide to C Object-Oriented Control, pp. 1–15. Springer-Verlag, London (2010)
12. Crane, D., Sonneveld, J., Bibeault, B., Goddard, T., Gray, C., Venkataraman, R., Walker, J.: Ajax in Practice, pp. 4–15. Manning Publications Co., Shelter Island (2007)
13. Hagen, W.V.: The Definitive Guide to GCC. 2nd ed. (2006)

Using VISIR Remote Lab in the Classroom: Case of Study of the University of Deusto 2009–2019

Javier Garcia-Zubia[1(✉)], Jordi Cuadros[2], Unai Hernandez-Jayo[1], Susana Romero[1], Vanessa Serrano[2], Ignacio Angulo[1], Gustavo Alves[3], Andre Fidalgo[3], Pablo Orduña[4], and Luis Rodríguez-Gil[4]

[1] Faculty of Engineering, University of Deusto, Bilbao, Spain
{zubia, unai.hernandez, ignacio.angulo}@deusto.es
[2] IQS, Barcelona, Spain
{jordi.cuadros, vanessa.serrano}@iqs.url.edu
[3] Instituto Superior de Engenharia do Porto, (IPP), Porto, Portugal
{gca, anf}@isep.ipp.com
[4] LabsLand, S.L., Bilbao, Spain
{pablo, luis}@labsland.com

Abstract. During the last ten years, the University of Deusto is using VISIR remote lab for analog electronics in the Faculty of Engineering. Present paper shows how we are using the VISIR with students, and its results. It is a catalog of experiments to invite other universities to join the VISIR consortium. Ongoing and future research in VISIR are also presented.

Keywords: VISIR · Remote laboratory

1 Introduction

VISIR is probably the most awarded, shared, referenced and longest-running remote laboratory in the world. Professor Ingvar Gustavsson, after some prototypes development (he started his work in the VISIR in 1999), released the version 3 of the VISIR remote lab in 2009. The main improvement and advance respect similar initiatives was the designed relay switching matrix that allowed students to create over it real implementations of circuits using real components, by previously the design of these circuits in a virtual breadboard.

The completed and detailed description of VISIR was first included in the work published in IEEE Transactions on Learning Technologies [1]. This paper, according with IEEExplore has been cited 103 times and has received 1391 full text views. According with Google Scholar has been cited 208 times, so it is possible to observe the impact that this publication had on the area of knowledge of remote labs. In fact, if a search is performed in (only) IEEExplore using the word “VISIR”, 74 conference papers, 10 papers in journals and 2 early access articles are shown, a scientific production generated by multiple authors from all around the world. It is probably not

M. E. Auer and D. May (Eds.): REV 2020, AISC 1231, pp. 82–102, 2021.
https://doi.org/10.1007/978-3-030-52575-0_7

possible to find a technological development in the field of distance, remote or virtual education that has had such an impact like VISIR.

This extensive scientific production is possible thanks to the kindness of Professor Ingvar Gustavsson, who always facilitated that any institution could deploy an instance of the VISIR laboratory. Then, it is possible to find VISIR instances in Sweden (1), Spain (2), Portugal (1), Austria (2), Germany (1), India (1), Morocco (1), Argentina (2), Brazil (2), Costa Rica (1) and USA (1). VISIR has been also the core and engine on multiple European projects as VISIR+ [2] and PILAR [3], which have allowed the spread of VISIR around the world, increasing the number of potential users of this lab. Even more, the International Association of Online Engineering (IAOE) count with a Special Group of Interest denoted as VISIR Federation which goal is to provide a uniform system where students can register and use the federated VISIR based laboratories and learning materials from different institutions belonging to the federation (http://online-engineering.org/VISIR-Federation_about.php).

In 2015, the Global Online Laboratory Consortium awarded the VISIR as the best remote laboratory in the world. In addition, in 2018, the European Society for Engineering Education (SEFI) posthumously awarded the Maffioli Award Professor Ingvar Gustavsson for his work about the importance of performing lab experiments in different ways in engineering education.

VISIR is a remote lab that also nowadays provides a scenario for developing new ideas, concepts and technologies. In this way, new approaches as the VISIR-LXI developed by University of Deusto [4], or new tools as the VISIR Dashboard for learning analytics have been created [5]. Many other research topics have been carried out in the framework of at least 6 PhD works.

Finally, VISIR remote lab is part of the available portfolio of remote labs traded by LabsLand, the spin-off of the WebLab-Deusto research group of the University of Deusto. This has been possible because VISIR is being a stable, effective and reliable remote lab for more than 10 years. This makes possible to offer this remote lab professionally and as a final product to institutions and users interested in having a VISIR instance or having access to some of the instances currently available on the LabsLand portal [6].

This paper focuses on describing the instance of the VISIR remote lab provided by the University of Deusto, detailing all the available experiments that can be performed over it and that are available on the LabsLand website (https://labsland.com/).

2 VISIR at the University of Deusto

The available version of the VISIR at the University of Deusto is based on the last available distribution of the HTML5 web client, providing access to 14 component boards in the switching matrix. Two of them are boards where two components can be connected to multiple nodes of a circuit, increasing the number of combinations in the circuits.

VISIR is used or have been used by ten teachers during the last ten years in six different subjects of five different engineering degrees. During this time, the students have opened more than 80.000 VISIR sessions in which they have performed more than 2.000.000 actions and experiments, what is for sure the highest number of actions for any remote lab in the world.

Attending only to 2018–2019, more than 200 students accessed VISIR to open 10.881 sessions with 282.051 required actions without remarkable technical problems, except that we had to replace a relay due to a bad configuration of an LC circuit that burned one of them.

3 VISIR: Learning Process

VISIR remote lab is a complex hardware-software tool, but its use in the classroom or at home is very easy. VISIR's interface reproduces a work place in a classical hands-on laboratory: a breadboard with different devices (resistor, capacitors, coils, diodes, transistors, etc.) and instruments (power source, function generator, multimeter and oscilloscope). This interface is the main and powerful aspect from the student point of view (see Fig. 1). The student pick and place the devices in the breadboard, then they connected them using coloured wires and the mouse, power (DC or function generation) and multimeter or oscilloscope are connected, and finally Perform Experiment is clicked to see the measurement results.

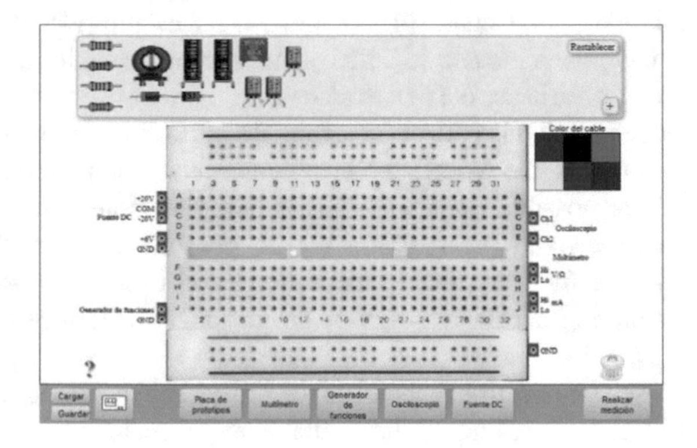

Fig. 1. VISIR interface

A remote lab can be used at the classroom as a classical hands-on lab. With VISIR a student can perform experiments for demonstrating a physical law like Ohm's Law, but also it can be used for discovering the law under an inquiry approach. In the first strategy, teacher has a central position because he explains the law and organizes the experiment, but using an inquiry approach. But a remote lab as VISIR offers the teacher a new opportunity: he can use the VISIR in the classroom with the students, so the teacher and the students can interact among them during a session using the VISIR as a central element. The VISIR also fosters the collaborative work.

VISIR has probed to have a positive effect in the students' learning process [7, 8]. This paper presents the set of experiments designed and used at the University of Deusto in different subjects, but it will not discuss again the effectiveness of VISIR.

The teachers at the University of Deusto organize the experiments with VISIR in four different scenarios:

- DC circuits.
- AC circuits.
- Characterization circuits.
- Active circuits.

3.1 DC Circuits

In this scenario the VISIR offers the students a basic "box" with four resistors: 2×1 kΩ resistors and 2×10 kΩ. With four resistors the student can create all the possible connections and circuits. Then, the student is not restricted to use only some connections recommended by the teacher. Our approach is to leave the student work without restrictions or impositions.

Circuits with Resistors
At this scenario, the student can create any circuit with these four resistors and measure it with the multimeter (see Fig. 2). During this session, the teacher starts showing the student how to use the multimeter and how to use the breadboard.

Again we have two strategies: teacher explains the mathematical model of the parallel-series connection or the student discovers (guided by the teacher) this mathematical model.

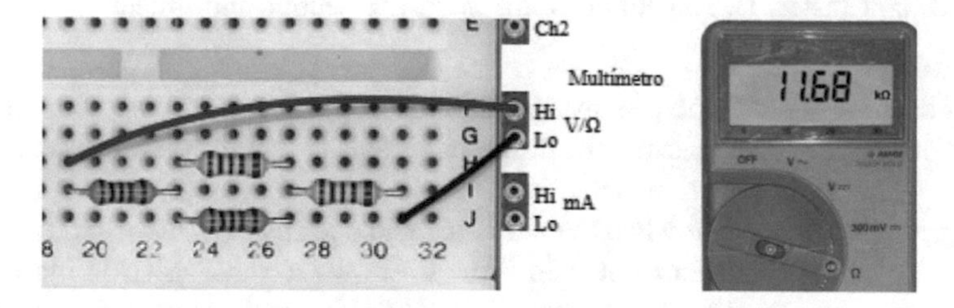

Fig. 2. Measuring resistors with VISIR

After the session the students has to complete a test.

An additional experiment is to measure the resistance of the multimeter. Ideally, it is infinite, but this is not the reality. In this case the expected value was 11,91 k, but the obtained real value is 11,68 k, so we can calculate the resistance of the multimeter.

Ohm's and Kirchhoff's Laws
The main objective of this session is to test or discover the Ohm's and Kirchhoff's laws. Starting with a basic circuit with a single resistor, then we will add new resistors in parallel and in series to measure the voltages and currents (see Fig. 3).

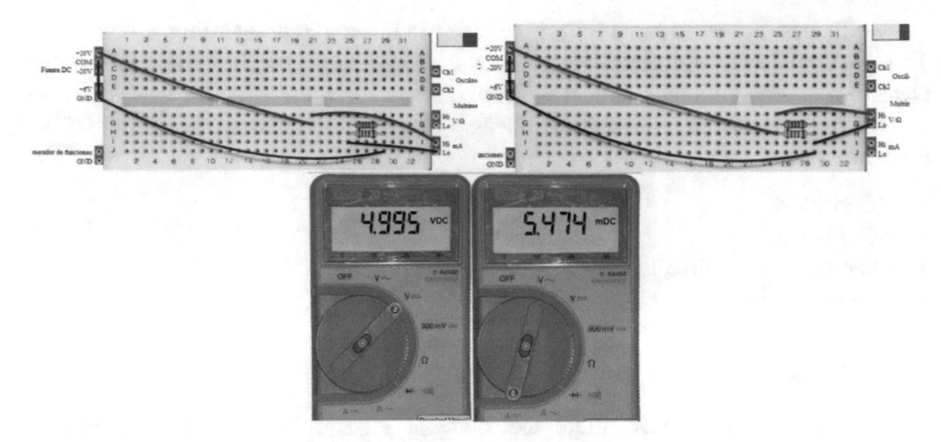

Fig. 3. Measuring voltage and current in a parallel circuit

A secondary and instrumental objective is to learn how to measure voltage (in parallel) and current (in series).

The circuit is powered with 5 V and the circuit is a parallel connection of two resistors 1 kΩ and 10 kΩ. Theoretically the voltage must 5 V and the current must be 5.5 mA, but these are not the values obtained in VISIR because they are real values.

Again we can measure the error introduced by the multimeter (and the breadboard, etc.) in the measurement, and we can compare if this new value is similar to the previous obtained.

Some VISIR platforms (i.e., ISEP en Portugal and UFSC in Brazil) have two multimeters, in this case they can measure the voltage and the current at the same time. It is easier and faster. U. Deusto does not have the second multimeter.

DC Circuits

In this scenario the student can mathematically solved any circuit with these four resistors, and then he can measure the different signals to test if the two results match or not.

In Fig. 4 we can see that when the student makes a mistake (the cable is not in the correct hole), the VISIR does not help him. It shows a value what is incorrect, but nothing is recommended to the student. This situation replicates the real work in the lab, so there is not difference between working with VISIR and working with a real lab, and because of that the student is correctly trained with VISIR.

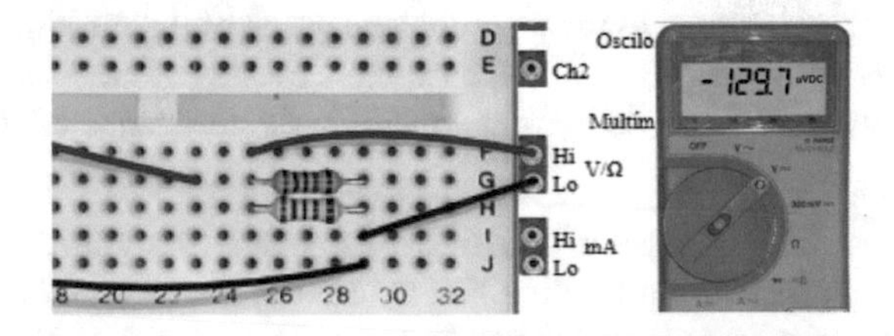

Fig. 4. Error during resistance measurement in VISIR

Sometimes the VISIR shows a message that says that the circuit cannot be created or it is unsafe. That means that something is wrong in the circuit. In Fig. 5, the multimeter is trying to measure the current in parallel. This avoids the break of the devices, multimeter or VISIR.

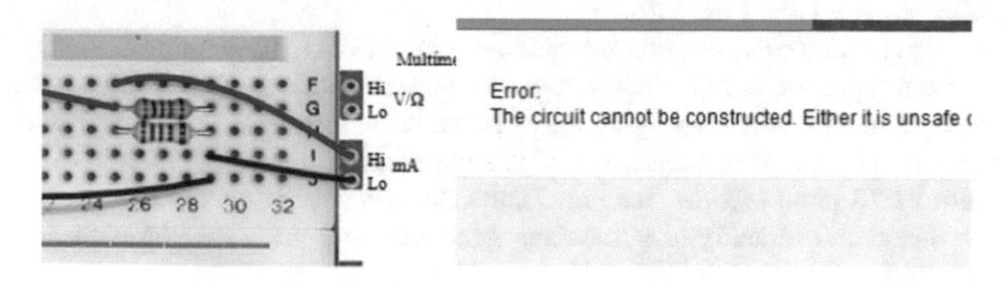

Fig. 5. Error during current measurement in VISIR

3.2 AC Circuits

AC circuits are less complex because in our case we need only one resistor and one capacitor or coil.

With AC circuits, the students learn how to use the function generator and the oscilloscope. In addition, the students learn how the frequency signals are characterized (rms value, average value, etc.) and how to do it with the oscilloscope and the multimeter.

The main objective is to create RC and RL circuits to analyze their behaviors.

Measuring Frequency Signals
Our first experiment (see Fig. 6) consists on measuring different values (minimum, maximum, peak to peak, average, and rms) of different signals (sinusoidal, square and triangle).

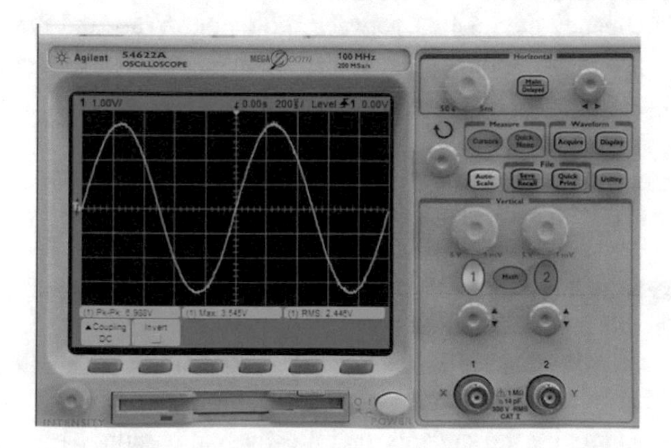

Fig. 6. VISIR oscilloscope

The main objective is to compare the different values and to explain why the rms value is the most significant value for representing AC signals. A secondary objective is to correlate the maximum and the rms values of different signals, that is, to obtain for a sinusoidal signal that $V_{rms} = \frac{V_{max}}{\sqrt{2}}$.

RC Circuit as a Low Pass Filter

An RC circuit depends on different variables: type of signal, amplitude, frequency, resistor and capacitor. A RC circuit is a good opportunity to explain to the student how to organize an experiment changing only one variable each time. The organization of data in tables is one of the objectives of this experiment.

From VISIR point of view (see Fig. 7), the analysis of a RC circuit is easy. Simply create the circuit, modify the function generator and measure the values with oscilloscope.

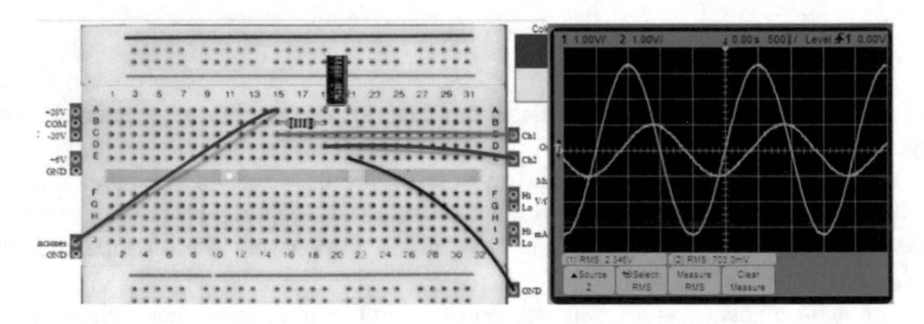

Fig. 7. RC circuit analysis

The data analysis allows the student to demonstrate that the RC circuit is a low pass filter.

Cut-off Frequency and Other Experiments

At the cut-off frequency the relation between the output Vrms and the input Vrms is $\sqrt{2}$, that is, $V_{rms,o} = \frac{V_{rms,i}}{\sqrt{2}}$, or the output signal is around 70% of the input signal. The value of the cut-off frequency is mathematically obtained as $f_c = \frac{1}{2 \cdot \pi R \cdot C}$. For R = 1 kΩ and C = 1 uF, $f_c \cong 160\,\text{Hz}$.

At this moment, the teacher can suggest an experiment to test if this true: create the RC circuit, excite it with a 160 Hz sinusoidal signal and measure the Vrms values (see Fig. 8).

After the experiment, $\frac{V_{rms,i}}{V_{rms,o}} = \frac{2{,}665}{1.870} = 1{,}42$, what more or less is equal to the expected value.

The student should repeat the experiment for 1 kΩ, 10 kΩ, 0.1 μF, 1 μF and 10 μF to see that the formula remains correct.

After we use to provide the students the following expression, $\frac{V_{rms,f1}}{V_{rms,f2}} = \frac{f2}{f1}$. They have to test if it is correct or not (it was "discovered" by two students two years ago).

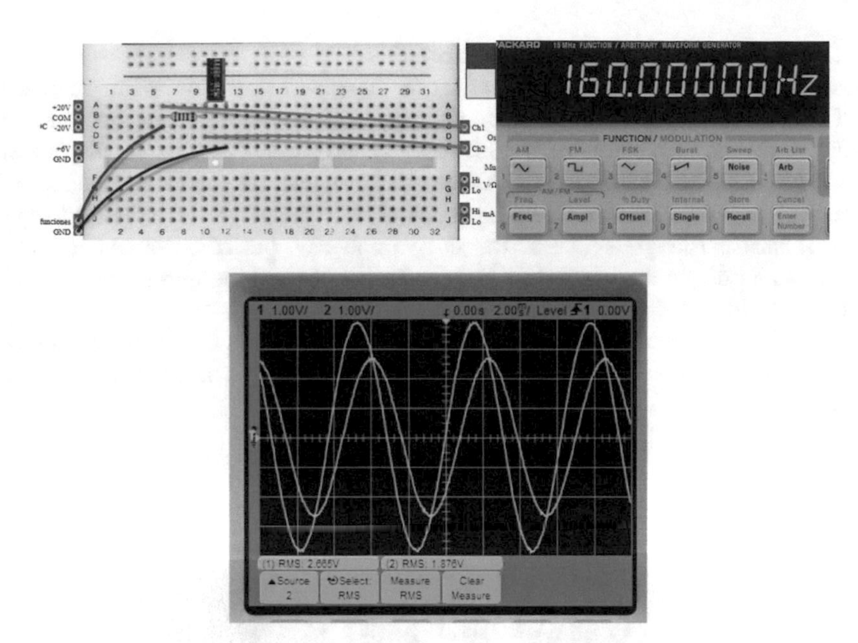

Fig. 8. Experimenting with a RC circuit

It is established that if an input signal has a frequency equal to 5 · fc, then it will be 80% filtered, that is, the output will be an 20% of the input. This can easily be tested by the student. For example, for the previous RC circuit and a 800 Hz sinusoidal input (see Fig. 9).

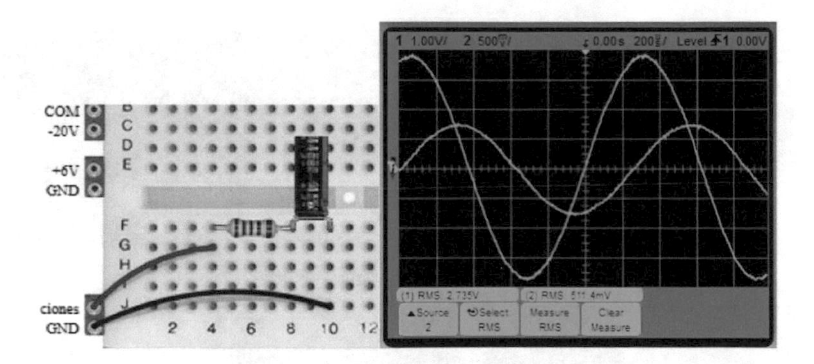

Fig. 9. Test of the cut-off frequency

The obtained result is $\%f = 100 - \dfrac{0.511\,\text{V}}{2.735\,\text{V}} = 100 - 18{,}2 = 81{,}8\%$. It seems that the design rule is correct, but this must be tested for different values of R and C.

After this, the student should be able to obtain the rule for 90%.

RC Circuit with Non-sinusoidal Input Signals

The student has seen that in a RC circuit if the input is a sinusoidal with A amplitude and f frequency, the output will be another sinusoidal with the same frequency f, a smaller amplitude (depending of the frequency f) and a time delay (some displacement to the right in the oscilloscope). At this moment, we can ask the students about what happens if we change the sinusoidal input signal by a triangle. Maybe the students will expect a smaller triangle signal of the same frequency, but it is not. After this, the teacher can introduce the concept of Fourier Transform and the misconceptions when making experiments.

Figure 10 shows the output of an RC circuit (R = 1 kΩ and C = 1 μF) excited with a 6 Vpp and 500 Hz triangle signal. Clearly, the output is not a triangle, it is a "sinusoidal".

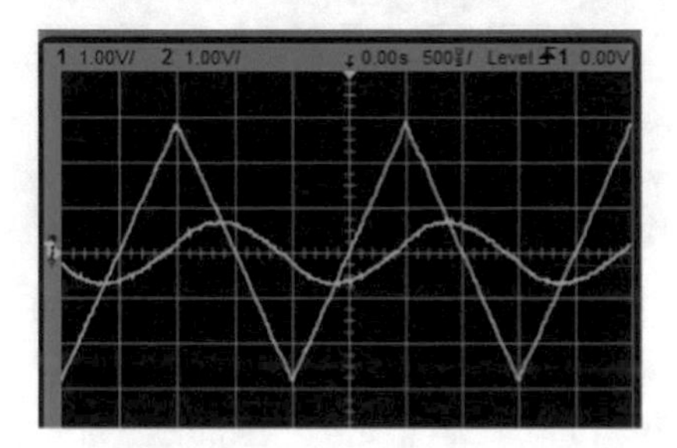

Fig. 10. RC circuit with a triangle input signal

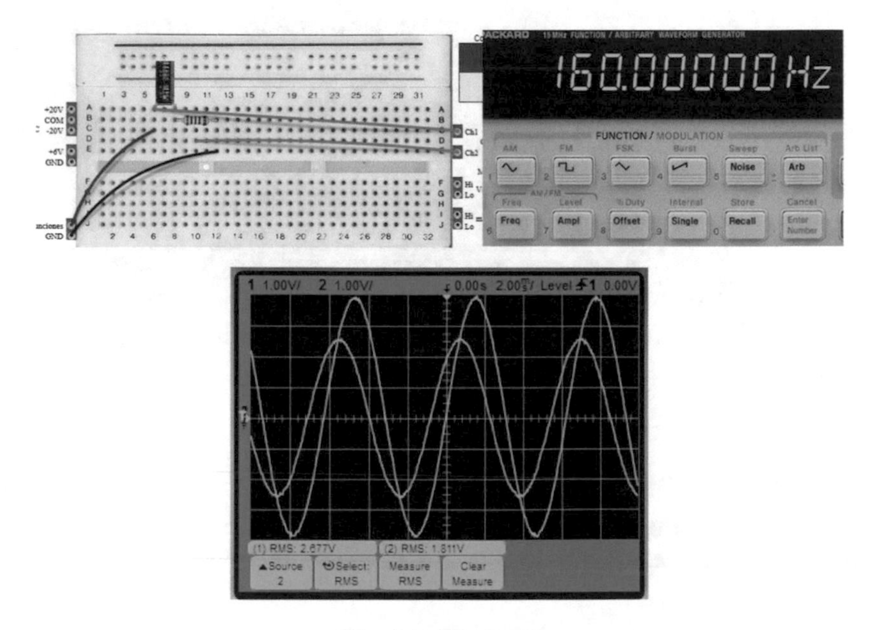

Fig. 11. CR circuit

CR Circuit as a High-Pass Filter

At this moment, the students can face alone the analysis of the CR circuit to see that only changing the order of the C and R, the behavior of the circuit is different (see Fig. 11).

Time Constant in a RC Circuit

When we excite a RC circuit with a sinusoidal input signal what is displayed in the oscilloscope, it represents the steady state response. To see the transient state of the RC circuit is common to use a square signal. Figure 12 shows the output of a RC circuit (R = 1 kΩ and C = 1 μF) excited with a 100 Hz and 6 Vpp square signal.

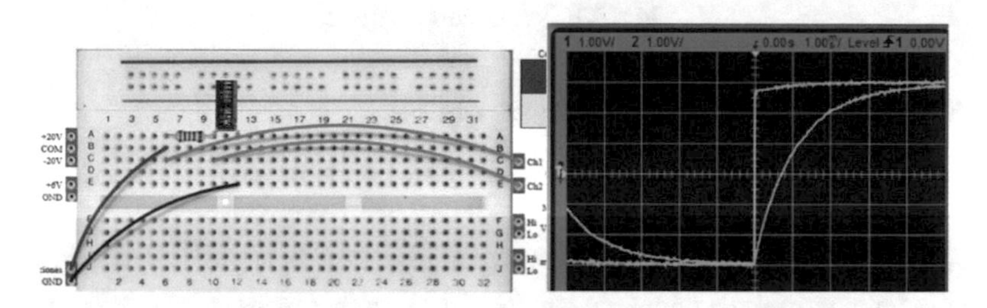

Fig. 12. Experiment with time constant of a RC circuit

The time constant,τ is defined as the time in which the signal reaches the 63% of its total excursion, Vpp. In the experiment: at what time the output reaches 3.78 V (6 V × 0.63)? Looking at the Fig. 12 this is at 1 ms.

The following expression obtains the time constant of a RC circuit, $\tau = R \cdot C = 1000 \cdot 0.000001 = 1\,\text{ms}$. This obtained value is equal to the experimental value.

Also there are other important indexes for a RC circuit:

- at 0.7 · τ seconds, the output is 50% of the input,
- at 5 · τ seconds, the capacitor is charged,
- and at 2,2 · τ seconds, the output grows from 10% to 90% of its final value.

All these values can be tested by the student with different experiments.

Analysis of RL Circuit

In this case, the student analyzes a RL circuit to analyze its behavior. With R = 1 kΩ, L = 100 mH and a 160 Hz sinusoidal (see Fig. 13).

If the students have obtained the mathematical expression of the V_L, they can see that the obtained signal in VISIR is not equal to the expression. This can be tested for different R and L values.

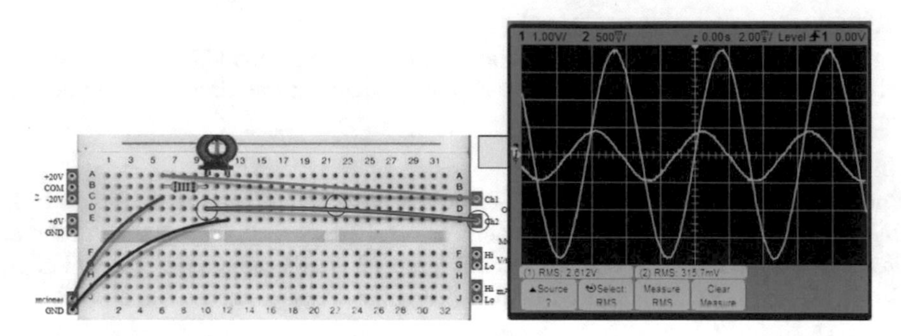

Fig. 13. RL circuit analysis

3.3 Experiments for Devices Characterization

The different devices used in VISIR circuits can be characterized using circuits and instruments

Resistor Characterization

In this case, the student can power a simple circuit with a resistor with different values to measure the voltage and the current in the resistor. After this, the student can draw the characteristic curve. In the figure, we can see a 1 kΩ resistor powered with 1 V (see Fig. 14).

Fig. 14. Resistor characterization experiment

The student will register all the data using a table to draw the curve.

Capacitor Characterization

Using a RC circuit the student can measure the $V_{C,rms}$ and $I_{C,rms}$ and then he can obtain X_C as the V/I, $X_C = \frac{V_{C,rms}}{I_{C,rms}}$. The student can discover or demonstrate the mathematical expression $X_C = \frac{1}{2 \cdot \pi \cdot f}$.

In Fig. 15 we can see a RC circuit with R = 1 kΩ and C = 1 μF excite with a 160 Hz sinusoidal of 8 Vpp.

With these values, $X_C = \dfrac{1.910\,\text{V}}{1.962\,\text{mA}} = 973,5\,\Omega$, and using the expression $X_C = \frac{1}{2 \cdot \pi \cdot f \cdot 0,000001} = 994,7\,\Omega$. At this moment, the teacher can introduce the concept of absolute and relative errors.

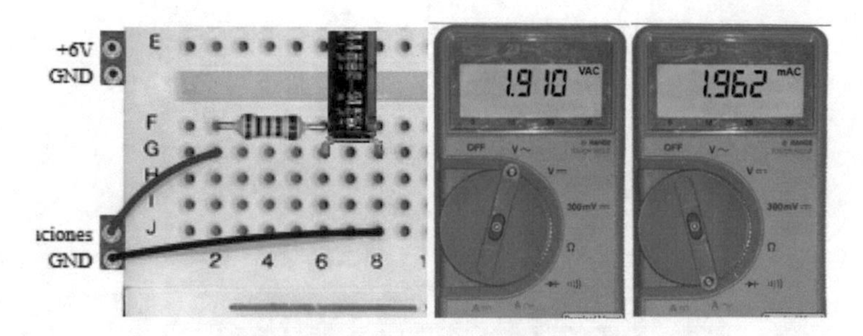

Fig. 15. Capacitor characterization experiment

Coil Characterization
Using the same previous approach, the student can repeat the experiment with a RL circuit. In this case the expression is $X_L = 2 \cdot \pi \cdot f \cdot L$.

With R = 100 Ω and L = 10 mL and a 160 Hz sinusoidal signal we obtain the following results in Fig. 16.

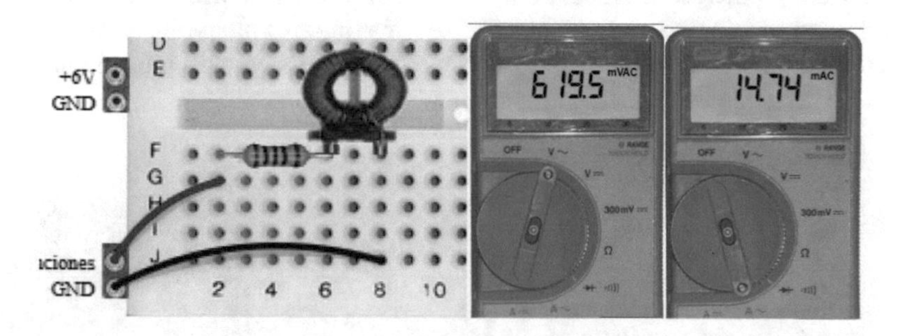

Fig. 16. Coil characterization experiment

Attending to the previous results, $X_L = \dfrac{0.620\,\mathrm{V}}{14.74\,\mathrm{mA}} = 42\,\Omega$. Using the expression, $X_L = 2 \cdot \pi \cdot 160 \cdot 0.01 = 10.05\,\Omega$. The two results are not similar, maybe because they are small.

If the input signal was a 500 Hz sinusoidal signal (see Fig. 17), then using the expression, $X_L = 2 \cdot \pi \cdot 500 \cdot 0.01 = 31.4\,\Omega$, and looking to the figure below $X_L = \frac{0.953}{14.20} = 67.1\,\Omega$.

Again, the differences are high, and they cannot be assigned to measurement errors. The problem is due to the coils have an additional resistor, which cannot be removed, it is intrinsic to the own coil. Using VISIR, the value of this r_l can be measured.

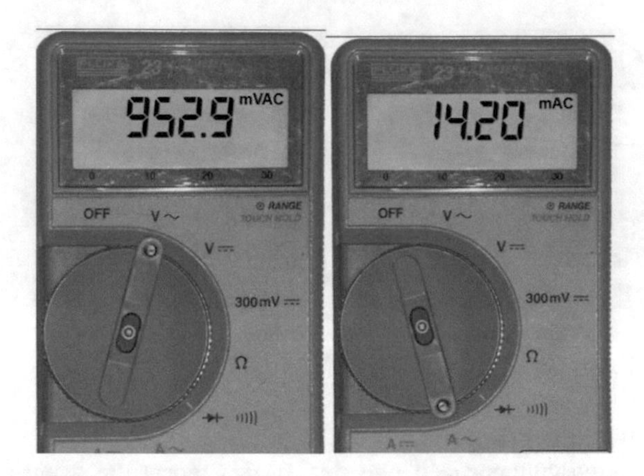

Fig. 17. New measurements for the coil characterization

Diode Characterization

A diode is characterized using the DC power, the diode and the multimeter. Figure 18 shows the experiment for an input of 0.9 V. The problem with this circuit is that VISIR limits the current to safe the equipment. The maximum voltage input is 1 V, if the student power the diode with more than 1 V, then a message will inform him that the current has exceeded the limit.

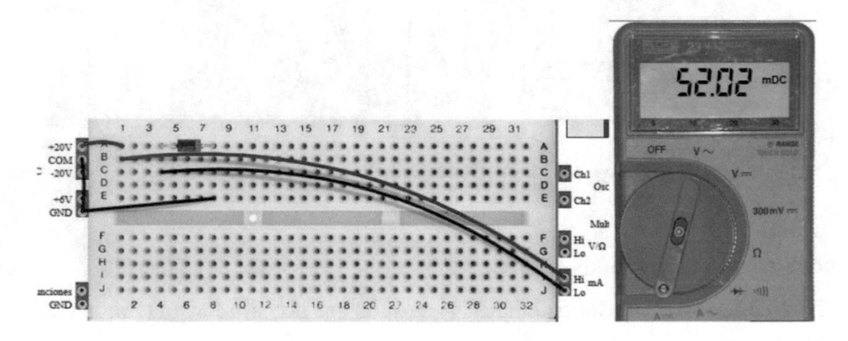

Fig. 18. Diode characterization experiment

To overcome the previous problem we have included a 6 Ω (25 W) in the set of devices. Using this limit resistor we can reach higher values of input voltage (see Fig. 19).

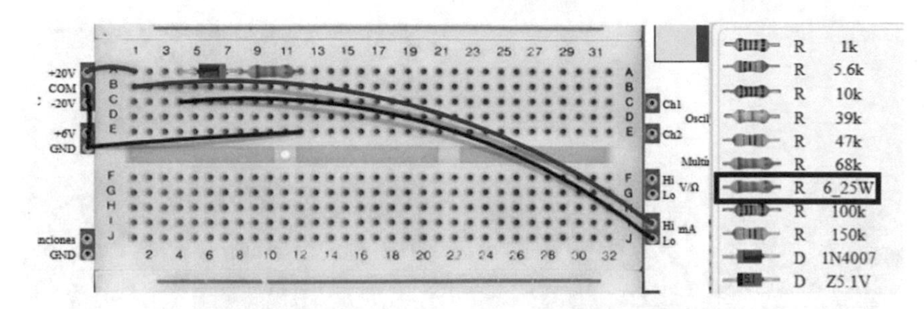

Fig. 19. Diode characterization experiment with a limit resistor

Using the two previous circuits the student can obtain the V-I characteristic curve. For doing this, the student needs to make a lot of boring measurements, but also the students can share a doc in the cloud, that is, they can work collaboratively.

Also the student can estimate the value of the rd resistor in the two curves, and both values must be similar.

Again, if VISIR had two multimeters (not in U. Deusto) the obtention of the V-I and Vi-Vo curves would be easier and better.

If we want to obtain the diode transfer curve, we have two options. The first option is to measure the voltage drop in the resistor (see Fig. 20), and then draw the Vi-Vo transfer curve. Also we can make the same with the voltage drop in the diode.

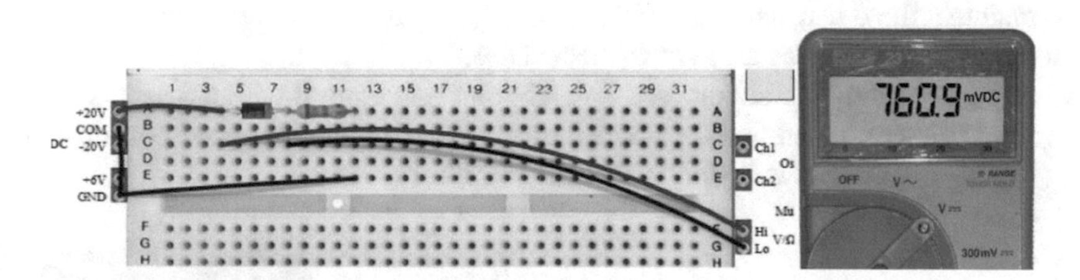

Fig. 20. Measuring the voltage drop in the diode

The second option is to power the circuit with a sinusoidal signal (see Fig. 21), i.e., 3 Vpp and 100 Hz and use the oscilloscope. In this case, we should use a 1 kΩ output resistor.

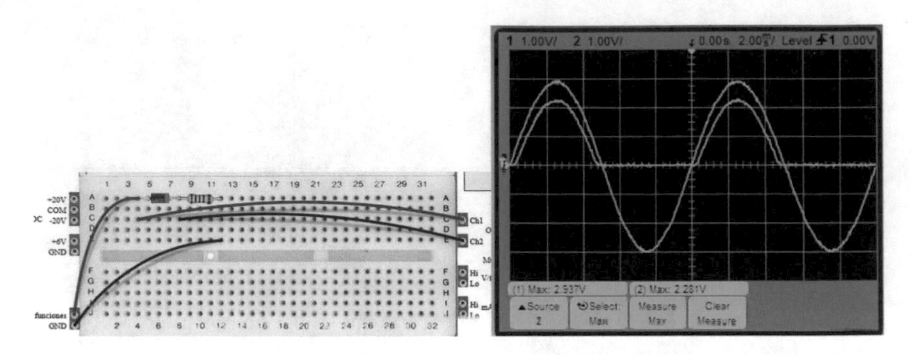

Fig. 21. Diode transfer curve with an oscilloscope

In a hands-on lab, the oscilloscope has an X-Y button to see the transfer curve. This button was in the Flash version of VISIR, but it was not in the html5 version of it. Now this button has been recently implemented by UNR in Argentina, but still we have not deployed it in U. Deusto.

Anyway, the student should test if the behavior of the diode is the same for different frequencies and amplitudes of the sinusoidal input signal.

Characterization of the Function Generator and the Oscilloscope Using the Maximum Power Transfer Theorem

It was recommended in the previous circuit to change the 6 Ω resistor by one of 1 kΩ. If we maintain the 6 Ω resistor we will see that the behavior is not the expected because the function generator has an associated resistor.

Creating the following circuit (see Fig. 22) and powering it with a 6 Vpp sinusoidal input signal, it is expected that the voltage drop in the 100 Ω output resistor must be the input signal, 6 Vpp, but it is not, it is 4.06 V. Why?

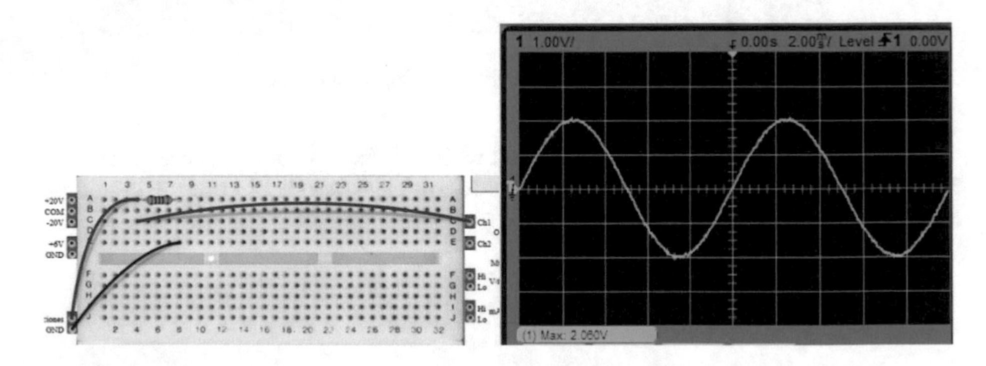

Fig. 22. Function generator characterization with an oscilloscope

Using the Ohm's Law it is clear that there is another resistor, rs, and it is in the source input, in the function generator. This resistor can be calculated and it is 50 Ω.

$$Vpp = Vrs + VR = I \cdot rs + I \cdot R = I(rs + R), 6\,\mathrm{V} = Vrs + 2\,\mathrm{V},$$

$$so\ I = {}^{4.06\,\mathrm{V}}/_{100\,\Omega} = 40\,\mathrm{mA},\ also\ Vrs = 6 - 4 = 2\,\mathrm{V},\ so\ rs = {}^{2\,\mathrm{V}}/_{40\,\mathrm{mA}} = 50\,\Omega$$

The function generator presents a rs 50 Ω resistor, and attending to the maximum power transfer theorem, so the function generator expects a 50 Ω output resistor. But the oscilloscope presents a high input resistance, so when the function generator is connected to the oscilloscope, it shows in the screen a signal that is the double of the selected in the function generator (see Fig. 23).

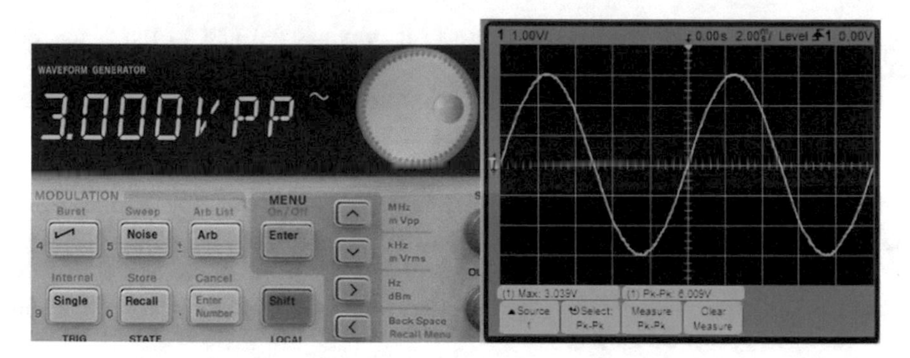

Fig. 23. Voltage drop in the function generator

Consequently, if we need a 3 Vpp input signal, we must select 1.5 Vpp in the function generator.

3.4 AC-DC Converter Circuit

An AC-DC circuit is the combination of a rectifier + filter. This circuit is simple and it combines active and passive devices for implementing a common circuit in engineering.

The objective of an AC-DC converter is to convert an alternate sinusoidal signal into a constant signal, in a "flat line". This means to convert a signal with 0 V average value (Vcc = 0 V) into another with a higher value (Vcc > 0 V).

Half-Wave Rectifier Circuit

First the student should analyzes, understands and measures a half-wave rectifier circuit (see Fig. 24).

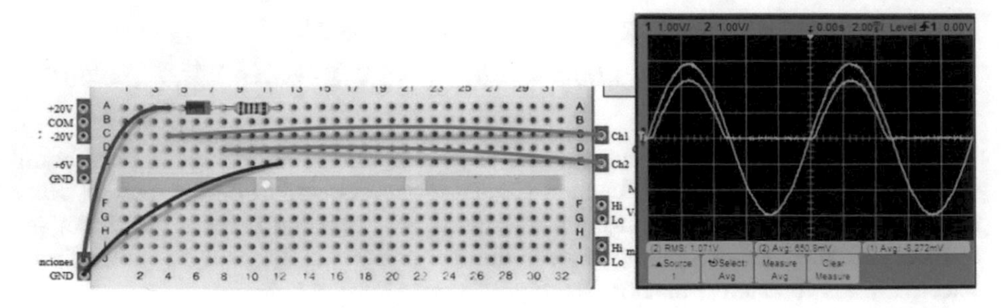

Fig. 24. Half-wave rectifier circuit

This circuit is called "half-wave" because it removes the half of the input, or it remains a half of the input.

It is clear that the output is not flat, but it is also clear that voltage average value is not 0 V, it is 0.651 V.

Two indexes describe the performance of the AC-DC converter: the ripple factor and the form factor. $Ripple factor = \frac{V'_{rms}}{V_{cc}}$ *and* $Form factor = \frac{V_{rms}}{V_{cc}}$. The ideal values of the ripple factor and the form factor are 0 and 1 (0% and 100%), respectively

The V'_{rms} is the rsm voltage value of the alternate part of the output signal (AC signal), that is, $v'(t) = v(t) - Vcc$. The student has to learn how to measure this value in the oscilloscope (see Fig. 25).

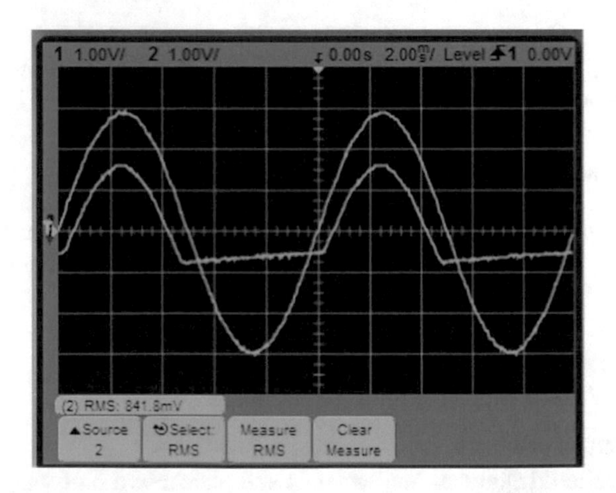

Fig. 25. Measuring the rms voltage value of the AC signal

For this rectifier, the indexes values are $Ripple factor = \frac{0.842}{0.651} = 129\%$ *and* $Form factor = \frac{1.07}{0.651} = 164\%$.

An additional activity for the student is to discover if the values of the ripple and form factors of a half-wave rectifier depend on the frequency and on the Vpp of the sinusoidal input signals. The same can be proposed for other signals like triangles, square, etc.

Half-Wave Rectifier + Filter: AC-DC Converter

The next objective is to improve the quality of the two indexes, mainly the ripple factor. Simply adding a capacitor in parallel with the load resistor, the behavior is better; the output is close to a "flat signal".

Figure 26 shows a circuit with a 1 μF capacitor and 1 kΩ load resistor powered with a 1 kHz and 6 Vpp sinusoidal signal.

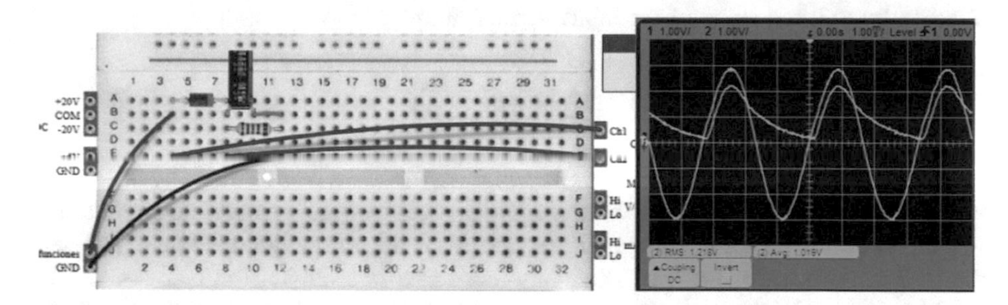

Fig. 26. AC-DC converter based on a half-wave rectifier

The output signal presents $Ripple\,factor = \frac{0.652}{1.019} = 64\%$ *and* $Form\,factor = \frac{1.218}{1.019} = 120\%$. These values are not good enough, but are better than the obtained values without the rectifier.

What can be done to increase these indexes? The answer is increase the capacitor value. Figure 27 shows the effect of a 10 μF capacitor.

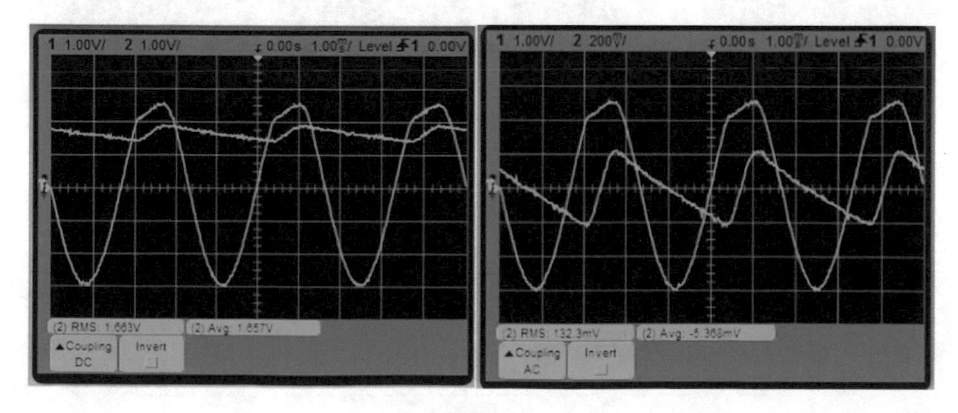

Fig. 27. AC-DC converter with a 10 μF capacitor

Using the previous values $Ripple factor = \frac{0.132}{1.667} = 8\%$ *and* $Form factor = \frac{1.663}{1.667} = 100\%$. Now the output signal in the left is “flat”, but how we can improve it? How we can improve the AC-DC rectifier? Maybe increasing the value of the capacitor?

At this moment we can include the concept of efficiency, how much of the input signal is presented in the output signal in terms of power. In this case we have to measure the voltage and currents using the oscilloscope and the multimeter, and even estimating some values because they cannot be measured using the VISIR mutimeter.

An additional task is to ask the student if the frequency of the sinusoidal input signal affects the behavior of the AC-DC converter. That is, are the ripple and form factors affected by the frequency of the input signal? Why?

4 Use of VISIR and VISIR Dashboard

In the previous section it has been shown what experiments can be made by the teacher and the students. This set of experiments is designed to provide active learning opportunities within the learning process. However, it is not only important to see what the students can make, but also what they are doing in reality.

The VISIR Dashboard, designed by the IQS in Barcelona (Spain), allows the teacher to see the detailed activity of the courses and the students.

The tool manages thousands of data and extract from them different graphical information (see for instance Fig. 28) to help the teacher with the evaluation of the students [5].

Figure 28 indicates how many students have completed a number of circuits. For instance, four students have implemented more than 200 circuits, and one student created 650 circuits.

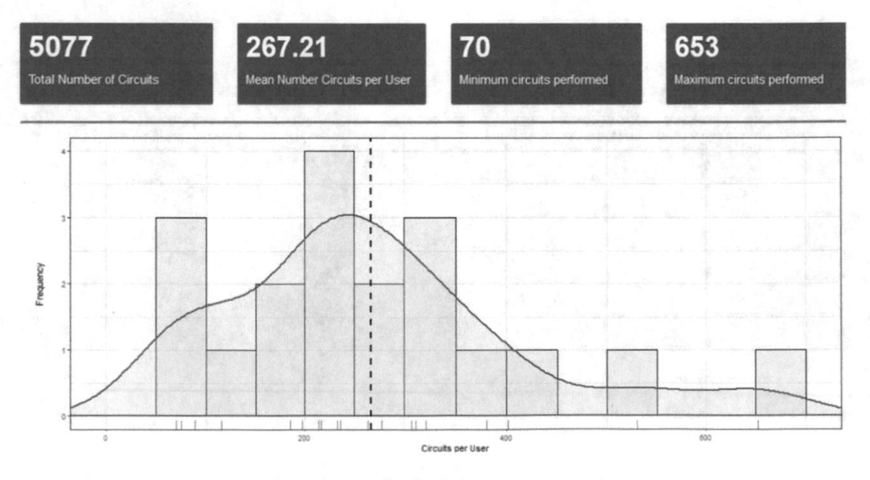

Fig. 28. Distribution of number of circuits created by students

VISIR Dashboard also offers the teacher to see the errors made by the students when measuring resistance, voltage and current.

Figure 29 shows the behavior of each student. Each bullet means one student, and if a student is in the top-right corner, he has made a lot of experiments in a lot of time. The best student should be in the top-left corner (no students there), because it means that the student has been able to do a lot of circuits in not so much time. For instance, one student (the teacher can see the name) tested almost 120 circuits and he/she spent around four hours for this task. The teacher should analyses these values to see how was the teaching/learning process.

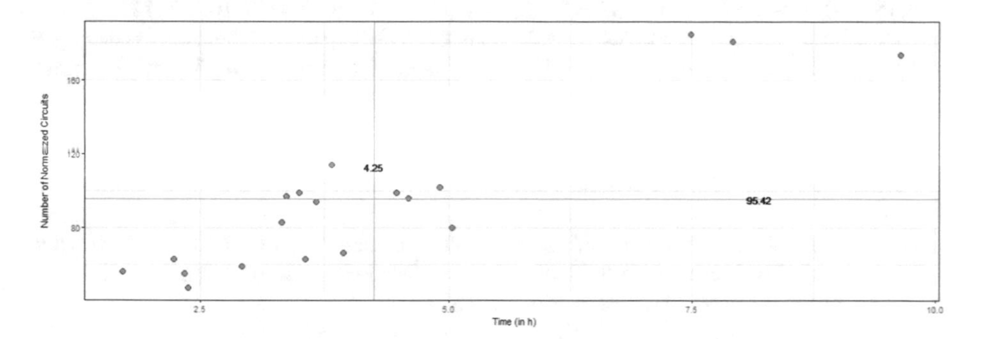

Fig. 29. Number of circuits versus time

5 Conclusion and Future Work

University of Deusto deployed VISIR ten years ago, and during this time, we have been using it intensively in different subjects. VISIR promotes the collaborative work among the students during the lessons, including an inquiry approach for teaching/learning process. The whole experience, including teachers, students, technicians and administrators, is very positive because it allows each participant to work like in a classical lab.

The paper has shown the experiments catalog implemented, used and tested by the University of Deusto. This catalog is ready to be used by others universities, schools and partners.

The future and ongoing research and activities are related with different topics:

- Increase the number of countries and institutions that are members of VISIR consortium.
- Federation of VISIR consortium. Federation means to group all the available VISIRs to be offered to the user as a unique one.
- Online tutoring. Using the VISIR Dashboard we can track the activity of the students in real time, so we can implement an online tutoring system to recommend the students with circuits and tasks. Also VISIR could be an adaptive learning platform.
- New features should be added to the html5 interface that are not still available.

Acknowledgement. The authors acknowledge the support provided by the ACM2019-20 Project "Ampliación de las herramientas LA en VISIR".

References

1. Gustavsson, I., et al.: On objectives of instructional laboratories, individual assessment, and use of collaborative remote laboratories. IEEE Trans. Learn. Technol. **2**(4), 263–274 (2009). https://doi.org/10.1109/tlt.2009.42
2. Alves, G.R., et al.: Spreading remote lab usage a system — a community — a federation. In: 2016 2nd International Conference of the Portuguese Society for Engineering Education (CISPEE), Vila Real, pp. 1–7 (2016). https://doi.org/10.1109/cispee.2016.7777722
3. Garcia-Loro, F., et al.: PILAR: a federation of VISIR remote laboratory systems for educational open activities. In: 2018 IEEE International Conference on Teaching, Assessment, and Learning for Engineering (TALE), Wollongong, NSW, pp. 134–141 (2018). https://doi.org/10.1109/tale.2018.8615277
4. Hernandez-Jayo, U., Garcia-Zubia, J.: Remote measurement and instrumentation laboratory for training in real analog electronic experiments. Measurement **82**, 123–134 (2016). ISSN 0263-2241
5. García-Zubía, J., et al.: Dashboard for the VISIR remote lab. In: 2019 5th Experiment International Conference (exp.at 2019), Funchal (Madeira Island), Portugal, pp. 42–46 (2019). https://doi.org/10.1109/expat.2019.8876527
6. Orduña, P., Rodriguez-Gil, L., Garcia-Zubia, J., Angulo, I., Hernandez, U., Azcuenaga, E.: LabsLand: a sharing economy platform to promote educational remote laboratories maintainability, sustainability and adoption. In: 2016 IEEE Frontiers in Education Conference (FIE), Erie, PA, USA, pp. 1–6 (2016). https://doi.org/10.1109/fie.2016.7757579
7. Garcia-Zubia, J., Cuadros, J., Romero, S., Hernandez-Jayo, U., Orduña, P., Güenaga, M.L., Gonzalez-Sabate, L., Gustavsson, I.: Empirical analysis of the use of the VISIR remote lab in teaching analog electronics. Trans. Educ. **7**(2), 149–156 (2017). https://doi.org/10.1109/te.2016.2608790
8. Marques, M., Viegas, M., Costa-Lobo, M., Fidalgo, A., Alves, G., Rocha, J.: How remote labs impact on course outcomes: various practices using VISIR. Trans. Educ. **57**(3), 151–159 (2014). https://doi.org/10.1109/TE.2013.2284156

How to Design Digitalized Laboratories?

Lessons Learned from Implementing Virtual and Remote Labs

Natascha Strenger and Sulamith Frerich(✉)

Ruhr-University, Bochum, Germany
strenger@fvt.rub.de, frerich@vvp.rub.de

Abstract. This contribution is showing ways to overcome issues on the way to digitalized laboratories in engineering education. The results presented in this paper were gained throughout a long-term study: 10 different laboratories were surveyed over a time period of 8 years. A non-standardized survey method was chosen for this evaluation, including a semi-structured guideline with open questions developed during several phases of pre-testing. The didactics and technical concepts of the laboratories are addressed, as well as challenges encountered during implementation and operation.

Key findings of the whole study were identified by looking at didactical set-ups, technical aspects, and project managing topics. Although some aspects of hands-on experiments on-site were easily conveyed into virtual or remote laboratories, others needed to overcome severe impairments. However, personal commitment and financial support were identified as important success factors.

By addressing original learning objectives as well as technical challenges that arose during set-up and digitalization of the laboratories, the results of this contribution clearly emphasize the connection between didactical purposes and technical realization. In some cases, both virtual and remote laboratories needed additional assistance in rephrasing learning objectives and adapting them throughout the process.

Keywords: Digitalization · Engineering education · Virtual learning · Remote laboratories

1 Introductory Background

1.1 Types of Virtual and Remote Laboratories

Currently, the implementation and set-up of virtual and remote laboratories enhances engineering education. There are several already existing types of experiments and systems containing remote or virtual laboratories. This contribution uses the term remote laboratory once there is a remotely operated experimental setup involved. With regard to providers of remote laboratories, there are either providers of infrastructural or management systems or providers of hardware components. In general, there is a wide range of different experimental setups. Those can be categorized either according to disciplines

M. E. Auer and D. May (Eds.): REV 2020, AISC 1231, pp. 103–111, 2021.
https://doi.org/10.1007/978-3-030-52575-0_8

or according to the complexity of the experimental set-up. However, complexity is usually closely linked to the objectives of the respective experimental setting.

Most of the laboratories found provide visualizations [1–4]. The observation of physical phenomena is allowed, which is why they are aiming primarily at pupils and undergraduate students. Often, the number of variable parameters is limited. In addition, the corresponding effects are shown in a small time interval. As a result, the time needed is rather short, usually within the range of a few minutes.

Remote laboratories are particularly established in the field of electronics [5, 6], and their design is rather compact. Possible applications may be explanations of electrical phenomena or verifications of control units in order to transfer them to microcontrollers of mechatronic systems. Instead of real plants on a miniaturized level, model plants are often used [6, 7]. Larger process plants and production systems are currently rarely available for remote access [8–11].

1.2 Diversity of Laboratories Investigated

The project ELLI (Excellent Teaching and Learning in Engineering Sciences), based on a cooperation between RWTH Aachen University, TU Dortmund and Ruhr-University Bochum (RUB), aims at improving engineering education. One of the main aspects of ELLI in its key area “Virtual Learning” is considering new ways of using laboratories in lectures and seminars as means of enhancing students’ practical skills. The implementation of such laboratories into educational concepts at all three universities started in 2011, when ELLI was launched.

In total, 9 pilot plants were implemented and provided with control facilities to make them remotely accessible, in the sense of online operability. Some of the pre-existing facilities were re-modelled for remote access, but in most cases, new testing facilities were designed. In addition, one completely virtual laboratory was developed to provide a “digital twin” to a real test stand. Thus, the actual experiment is represented as a simulation model of measurement data.

The didactic spectrum of all 10 digitalized laboratories shows a great variety of options. In most cases, different types of experiments can be conducted. Thus, in addition to demonstrational and illustrational purposes, experiments aim at facilitating research-led learning possibilities.

The ideas for the laboratories and test settings are anchored in the subject matters of the operating chairs or disciplines. In many cases, the implementation was exploring ways of success in “remotizing” a specific experimental set-up.

The only standardized requirement was the use of the control software LabVIEW, in order to set an identical framework. Thus, the open approach has enabled a variety of developments, and various technical challenges had to be overcome.

In the course of the cooperation throughout the ELLI project, a special relationship of trust developed between all participants. Thus, a constructive way of dealing with challenges and issues came into being. This proved to be a success factor for both the mutual consultation on the implementation of laboratories, and the conduct of this study.

2 Research Design of the Laboratory Evaluation

2.1 Continuous Evaluation of the Laboratory Implementation

The process of planning and implementation of the 10 different laboratories at RUB was documented and evaluated with various tools during the last years. The purpose was to keep track of the findings and possible challenges experienced by the scientific and teaching staff responsible for the laboratories.

During the first years from 2012 to 2016, the laboratory operators wrote reports about their progress of developing the respective virtual or remote lab and its build-up. At the same time, regular meetings were organized by the project ELLI with all laboratory operators that took place twice per year. These meetings were held to keep track of the individual progress and illustrate subsequent steps for each accompanied lab. In addition, it was made sure that individual experiences of the laboratory operators could be shared and discussed with the project staff and other group members. Difficulties they encountered during the implementation phase could thus be debated in a collegial environment. Usually, issues and problems could be addressed rather rapidly.

2.2 Qualitative Study with Expert Interviews

After the implementation phase of the remote laboratories had been completed, a qualitative study was designed in order to focus attention on the concluding observations of the laboratory operators. Furthermore, an investigation into their own experiences during the process of laboratory build-up and operation was intended to lead to recommendations for others working on similar projects in the form of a guideline.

The research design for this study included a thorough evaluation via qualitative expert interviews with representatives of all chairs who operated a virtual or remote lab at the three engineering faculties of RUB. In this non-standardized, qualitative research method of expert interviews, the participants are regarded as experts in their field of work or research due to their own experience in or with the respective field of interest [12, p. 35]. The study took place from June 2018 to March 2019, including two months of preparation and pretests and a further three months for analyzing and categorizing the research results.

During the interviews (which each had a duration of 45 to 90 min), the faculty employees entrusted with the implementation and operation of the laboratories were questioned about their personal experiences alongside a semi-structured interview guideline with open questions. After several rounds of preliminary testing, the final interview guideline contained five categories which covered all stages of the laboratories life cycle - from the planning and development phases to their application in practical laboratories as well as in lectures and seminars at the departments of civil, mechanical and electrical engineering. A list of these categories is shown in Table 1.

Table 1. Categories of the final interview guideline

1	General questions and history of the remote laboratory
2	Planning phase of the laboratory
3	Development and implementation
4	Conducting experiments
5	Conclusions and outlook

In the final section of the interview, participants were invited to formulate summarizing conclusions regarding (the future of) remote laboratories in general and about the laboratory they had been working on in particular. The interviewees were generally asked to answer questions narratively, according to their own opinion. Follow-up questions were asked in some cases in order to complete the picture of a certain aspect. Recording of all ten interviews was done via a voice recorder and additional notes taken by the interviewers. Afterwards, the interviews were transcribed and analyzed with qualitative content analysis [13, p. 49f], which is a common qualitative method in social sciences.

3 Research Results/Lessons Learned

The results of the qualitative interview study were considered with regard to their beneficial effect to others who would like to implement similar laboratories. They comprise aspects of project management, didactic aspects, and technical challenges. In the following, the lessons derived from study results will be listed. Based on these results, Sect. 4 will then present recommendations the authors would like to give to colleagues in engineering education who are working on similar projects.

3.1 Project, Process and Knowledge Management

With regard to project management and securing process knowledge about the implementation of the remote laboratories, three major challenges were identified during the interviews.

First, the development of the test stands required more time than originally planned in most cases. Many interviewees reported that, at the time of the original call for applications (in 2011), remote-controlled systems for application in higher education contexts could not simply be purchased as standard products. Progress in almost all projects was often slowed down due to the fact that many hardware and software components had to be developed by the research assistants. This was also the case for the connecting systems between hardware and software as well as in many cases for the entire back-up and safety systems. Furthermore, the reduction of complexity of some remote test stands for the purpose of remote learning was a new task for all of the research staff involved. This meant that newly designed and sometimes rather complex learning settings had to be downsized and recalibrated multiple times, making many interim results redundant.

Second, the implementation often demanded interdisciplinary skills, covering the fields of electrical engineering and programming or information technology. As many projects were based at chairs whose research originally specializes in other fields of engineering (mechanical engineering, civil engineering), these skills were not always readily available. The necessity to either integrate external expertise or to train internal staff accordingly caused further delays in the implementation. Many interviewees described the lack of IT skills as a regularly occurring problem. Not surprisingly, when asked about things that could be done to improve the working process, the first thing to be mentioned was additional and readily available expertise and capacity in this field. In several cases, the research staff themselves had to gain the necessary expertise and thus enhance their competences, which led to increased flexibility and problem-solving skills as well as to an overall spirit of innovation within the project, which could certainly be considered to be a beneficial (albeit time consuming) side-effect.

The final challenge occurred due to the fact that frequent staff changes during the years of the project sometimes impeded continuous knowledge management. Due to the staffing structure at university chairs in Germany, the contract duration for research assistants is typically limited to 1–3 years, during which they have to complete the research for their individual dissertation project as well. For many remote laboratories, this led to the situation that staff working on the project changed for multiple times between the project start in 2012 and the evaluation in 2018. These changes sometimes made an accompanying documentation difficult, which in turn posed challenges to keeping the laboratories operational without any interruptions. At some chairs, however, permanent university staff was in charge of the remote laboratories and these were the projects which experienced the fewest problems with knowledge management.

3.2 Challenges in the Didactic Set-Up of Laboratory Experiments

In accordance with the original call for applications, the target group of the remote laboratories comprises engineering students at Bachelors' as well as Masters' level. With regard to the didactic concept of the lab experiments, the interviews revealed that input from experts in didactics was rarely consulted. Instead, the experiments were contrived by research assistants and teaching staff at the engineering chairs. Several interviewees pointed out that they could rely on their own experience in teaching engineering subjects while designing the tasks. According to them, it is crucial to know the set-up of a specific experiment in order to comprehend the teaching approaches suitable to impart the knowledge behind the experiment – a knowledge which external didactic experts from other disciplines than engineering might lack.

The laboratories generally need to be highly automatized. In some cases, this prerequisite imposed strict limitations on the allowable actions students might perform and reduced the complexity of the experimental set-up. As a result, it became possible to execute an originally complex task without understanding the underlying principles. This necessitated a shift of the focus of the learning objectives towards the results and their successive analyse.

Concerning the targeted learning outcome that students are supposed to gain, a number of interviewees expressed that, in their opinion, the on-site and hands-on laboratory experience cannot be replaced. They base this estimation on the experience that the perception (sounds, smells, etc.) of being at the real test stand cannot be represented in equal intensity in a digitally recreated environment. Another difference according to them lies in the perception of time it takes to conduct individual steps of an experiment, which is also hard to convey while sitting on a computer at home instead of standing in the experimental hall.

3.3 Technical Aspects

Technical challenges of implementing virtual and remote laboratories occurred during acquisition of components, construction of plants, setting up the control and virtual interfaces, and implementation of safety precautions. While assessing the set-up of the virtual interfaces and designing the control units, many interviewees mentioned the complexity issue again. Since the experimental set-up needed to be simplified in many cases, some lab operators questioned the benefit of the remote operability. In their opinion, an easier set-up plant would have served similarly well, without causing so much effort in its adaption to being a remote lab.

Since the variety of tasks to be fulfilled by the designed user interfaces was quite large, their implementation also varied in a range of very simple to very complex, respectively. Figure 1 gives an example of a LabVIEW interface setup for one of the remote laboratories, as given in [14].

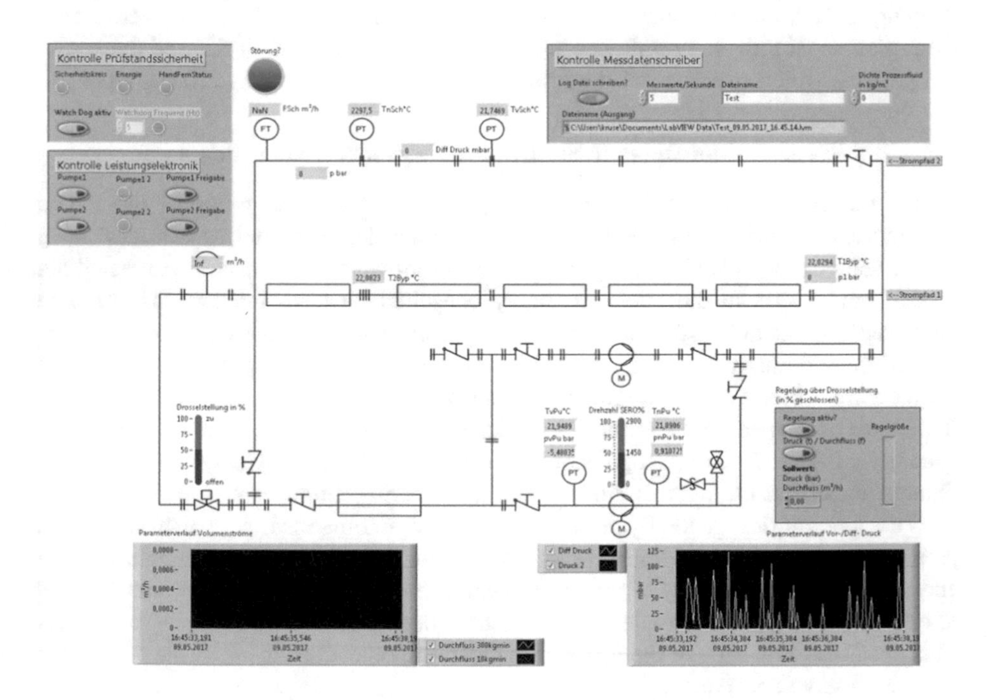

Fig. 1. Example of LabVIEW interface setup [14]

While some setups implemented only a few buttons, diagrams or video streaming options, others established several options of managing experimental data. In general, it could be observed that the learning objectives are directly found within the design of the user interface. It was found that some virtual or remote labs were planned to bring students in touch with new topics in an inspiring way, without high thresholds. Other virtual or remote labs were aiming at preparing students for a real hands-on laboratory on-site, resulting in rather complex user interface designs.

The implementation of safety precautions was a major technical aspect, to ensure the safety of the equipment as well as the safety of people working in the laboratory. Especially in remotely operated conditions, all plants need to be able to shut down safely at any time. Many setups implemented automated shut-downs using their respective user interfaces. However, many interviewees reported that additional limitations of the experimental design became necessary, to avoid human or technical failure and damages to the equipment. It was mentioned as well that some remote laboratories need to go back to their initial operating state at regular time intervals. Thus, their availability was tied to prior requests and reservations.

It was stated by many interviewees that implementing feedback onto actions undertaken by users facilitates the realistic impression created by remote laboratories. Videos, sounds, and visualized measurements are illustrating the real experiment remotely operated, and they pass on detailed information about its effects. However, most of these elements generate a substantial amount of data to be transferred to the user. Therefore, it remains a major task to maintain a robust software for administrating user schedules and data storage.

4 Recommendations Based on Lessons Learned from ELLI

As a summary, the experiences gained throughout 8 years of implementing virtual and remote laboratories lead to some recommendations for others interested in similar tasks.

Concerning the aspects of project management, it has to be stated that the creation of innovative virtual and remote laboratories was supported to a high amount by a regular exchange among colleagues who gained insight into the creation process, and developed adapted solutions. Laboratories with permanent staff in charge were the ones running most smoothly. Therefore, it is of great importance that standardized documentation for all aspects regarding virtual labs or laboratory plants with remote operability is pursued if staff is changing more frequently.

Addressing didactical challenges of virtual and remote laboratories, it is important to focus on learning outcomes and a specific didactic approach is certainly required to make good use of the laboratories. Most digitalized labs are not fully able to convey all aspects of hands-on and on-site experiments. Therefore, the main goal cannot be to "copy" the entire reality of a laboratory experiment, but to make use of the specific conditions of the remote and virtual settings in order to achieve adequately designed teaching-learning objectives. Thus, the laboratories can be used to enhance real set-ups by showing logical causalities between operating parameters, visualizing underlying scientific phenomena, and demonstrating different plant operating states. In order to define the learning objectives in accordance with the laboratories' potential, it takes an intensive and continuous dialogue between people with specialized (technical) expertise and didactic expertise.

Finally, technical challenges occur mainly in creating user interfaces and implementing safety precautions. The design of user interfaces is important, since it constitutes the individual learning outcomes by illustrating the plant and enabling its control. Therefore, its complexity is linked to the complexity of the experiment. In addition, the user interface usually links operating modes to safety requirements. The realistic impression of the experiment is enhanced by implementing videos, sounds or data visualizations. Thus, data transfer requires a stable online network.

With regard to the didactic expertise, as well as concerning IT-skills that could often not be provided by the employees at the engineering chairs themselves, flexible funds that are available even after the installation of the laboratories would be very helpful. While formative evaluation and exchange between the lab operators was already induced at several stages, it might be useful for other similar projects to additionally provide some sort of "stand-by" staff resources for assisting with technical problems, especially concerning IT aspects, and for providing didactic expertise when required.

Acknowledgement. This work was supported by the German Federal Ministry of Education and Research under Grant 01PL16082B in the Teaching Quality Pact.

References

1. GO-LAB (2019). https://www.golabz.eu/. Accessed 23 Aug 2019
2. LabsLand (2019). https://labsland.com/en/labs. Accessed 23 Aug 2019
3. TU Berlin: Remote Farm (2019). https://remote.physik.tu-berlin.de/experimente/. Accessed 23 Aug 2019
4. iLab (2019). https://icampus.mit.edu/projects/ilabs/. Accessed 23 Aug 2019
5. Winzker, M.: Vision Remote Lab (2019). https://fpga-vision-lab.h-brs.de/weblab/login. Accessed 23 Aug 2019
6. Lustig, F., Dvořák, J., Kuriščák, P.: Professional and hobby hands-on-remote experiments. In: AIP Conference Proceedings, vol. 2152, p. 030020 (2019). https://doi.org/10.1063/1.5124764
7. Technical University Ilmenau (2019). https://www.goldi-labs.net/index.php?Site=2. Accessed 23 Aug 2019
8. University of Cambridge (2019). https://como.ceb.cam.ac.uk/research/weblabs/. Accessed 23 Aug 2019
9. Institute of Information Engineering, Automation, and Mathematics (2019). http://www.ies.stuba.sk/index.php?menu=3&page_id=22#kolona. Accessed 23 Aug 2019
10. ELL-ELLI Lab Library (2019). http://www.elli-lab-library.de/index.php/de/. Accessed 23 Aug 2019
11. Otto-von-Guericke University Magdeburg. Industrial eLab (2019). http://www.elab.ovgu.de/Das+Projekt/Arbeitsplan.html. Accessed 23 Aug 2019

12. Kaiser, R.: Qualitative Experteninterviews. EP. Springer, Wiesbaden (2014). https://doi.org/10.1007/978-3-658-02479-6
13. Mayring, P.: Qualitative Inhaltsanalyse. Grundlagen und Techniken. 11., aktualisierte und überarb. Aufl. Weinheim: Beltz (Studium Paedagogik) (2010)
14. Kruse, D.: [online resource] Virtualisierung eines verfahrenstechnischen Prozesses als remote Labor für die Aus- und Weiterbildung in Industrie 4.0., Diss., Ruhr-University Bochum (2017). http://nbn-resolving.de/urn/resolver.pl?urn=urn:nbn:de:hbz:294-55103

FPGA Remote Laboratory: Experience in UPNA and UNIFESP

Cándido Aramburu Mayoz[1(✉)], Ana Lúcia da Silva Beraldo[2], Aitor Villar-Martinez[3,4], Luis Rodriguez-Gil[3], Wilson Francisco Moreira de Souza Seron[2], Tiago de Oliveira[2], and Pablo Orduña[3,4]

[1] Universidad Pública de Navarra, Pamplona, Spain
candido@unavarra.es
[2] Universidade Federal de São Paulo, São José dos Campos, Brazil
{ana.beraldo,wilson.seron,tiago.oliveira}@unifesp.br
[3] LabsLand, Bilbao, Spain
{aitor,luis,pablo}@labsland.com
[4] University of Deusto, Bilbao, Spain
{aitor.v,pablo.orduna}@deusto.es

Abstract. FPGAs are a powerful technology for teaching hardware design, through which students can learn how to design hardware using common hardware description languages such as VHDL or Verilog. However, the use of FPGAs in the classroom has several different problems, such as availability or associated costs of acquiring and maintaining the FPGAs. Also, relying on FPGA boards in the classroom is typically problematic because students cannot use the boards for doing projects after classes or during weekends. To solve this problem, remote laboratories can help allow more flexibility to students. A remote laboratory is a hardware and software system that allows students to access a real FPGA located somewhere else on the Internet. In this article, a cross-national remote laboratory is presented. This remote laboratory has multiple copies (17 devices at the time of this writing) deployed both in UPNA (Spain) and UNIFESP (Brazil), and using LabsLand (Spain) for technology and management of the laboratory. Students of both institutions access these laboratories transparently. This provides automatic fault tolerance and increases the potential number of concurrent students using the laboratories by sharing the boards among both institutions.

Keywords: FPGA · Altera · VHDL · Verilog · Hardware design · Remote laboratory · LabsLand

1 Introduction

Universities are increasingly interested in the curricular integration of practical skills that have so far been primarily present in environments such as Fab labs [1, 2] or Maker initiatives [3] and in the development of flexible online learning platforms such as Massive Open Online Courses (MOOCs) [4].

M. E. Auer and D. May (Eds.): REV 2020, AISC 1231, pp. 112–127, 2021.
https://doi.org/10.1007/978-3-030-52575-0_9

An effective tool to accomplish such an objective are remote laboratories [5]. This tool allows students to access real physical educational laboratories from anywhere and at any time, using only a web browser. Examples of such a laboratory are electronic instruments, programmable boards and embedded computers.

In a non-face-to-face manner, by connecting the student to these laboratories through a local network or through the internet network, accessing the practical station through a web interface installed on the user's computer, with the advantage of being able to resort to those remote laboratories that because of their cost or complexity cannot be developed locally in the study center, workshop or user premises. The platform is of interest to professionals in the technological environment who use the resources for a specific project as well as teaching centers as a practical tool in the curriculum of their curricula. Interest in competitive FPGA accelerators with classic processing units has led cloud service companies to offer FPGA remote resources such as Amazon Web Service F1 (AWS) instances developed on their Amazon Elastic Compute Cloud (EC2) platforms.

The paradigm shift of teaching methods totally dependent on local resources in the school towards a more flexible and personal educational system leads us to hybrid learning with both local resources (face-to-face interactivity with the teacher, seminars, etc.) as LMS online remote tools (access management, video readings, self-assessment systems, discussion forums, videoconferences, etc. … and also towards the need for remote laboratories for the development of practical skills).

This need in learning together with geographical issues [6] has led UPNA and UNIFESP universities to develop their own and shared remote laboratories. The remote laboratory shared between UPNA and UNIFESP universities is accessible through a web browser through an interface developed by LabsLand. UPNA has the Sakai learning management platform and UNIFESP has Moodle LMS platform that allows the connection to the web interface that integrates the development tools necessary for the remote programming of the FPGA device.

Section 2 describes the most important aspects in the evolution of remote laboratories and the teaching environment for which the remote laboratory for universities in both countries has been developed. Section 3 indicates both the common hardware and software characteristics of both laboratories and then details each one of them. Section 4 shows the mode of operation of the students. Sections 5 and 6 publish the results of the survey of students on the use of the remote laboratory. In Sect. 7, characteristics observed during the performance of the laboratory practices are discussed and finally in Sect. 8 the conclusions of the experience obtained are presented.

2 Background

2.1 Remote Laboratories

Throughout the last decades, significant research efforts have been dedicated to the field of virtual and remote laboratories [7]. Remote laboratories provide students with the means to experiment remotely with real equipment, but through the Internet, in a way that is similar to hands-on labs [8]. Research suggests that they are effective teaching tools, and that they can be just as effective as their hands-on equivalents [9–11].

There are many fields to which educational remote laboratories are applicable, including physics, biology, and others. However, the focus of this work are programmable embedded system and specifically FPGA devices [12–15].

The research literature describes various embedded systems and FPGA devices remote laboratories (e.g., [16–28]). Though the specific state of the art advancements that the works describe, the particular architectures that they use and the particular characteristics of the remote laboratories vary significantly, we observe that certain components tend to be present in many remote laboratories. Most of the laboratories provide access to a particular hardware (e.g., an educational FPGA board). They often have certain remotization hardware (e.g., a raspberry device or a FitPC) to control that board and to provide input or output peripheral simulation.

There is often some kind of "laboratory server" that provides a software layer for the laboratory, and which sometimes is located in the same "device server" that remotizes that hardware. Often, the laboratories feature a web camera, which lets students observe how the remote experiment is behaving, providing an interactive live-stream that serves them as their eyes would in a hands-on lab [29].

Most laboratories assume that users have a specific toolset (e.g., Intel Quartus Prime, Xilinx Vivado) installed on their local PCs, and accept locally-compiled binary files as an input. Others accept source files as well, or even provide an IDE environment, and synthesize the logic remotely. That way users do not need to install any specific software on their local devices.

Throughout the last decades, many of the base technological challenges of remote laboratories have been covered by researchers. However, many of the remote labs that are described in the literature were prototypes in nature, not necessarily intended for scalable [30], reliable, production-level usage, and not necessarily accessible online anymore. One of the key challenges in the remote labs field today is to attain that kind of remote laboratories, and several initiatives aim for that goal [31, 32]. It is also one of the key focuses of the particular scheme described in this work.

2.2 Sharing Remote Laboratories

One interesting feature of remote laboratories is that they can be shared across different institutions. While this can be done manually (e.g., students of one university register in the remote laboratory of other university), Remote Laboratory Management Systems (RLMS) have implemented an automated way to exchange laboratories in what in the literature has commonly be referred as remote laboratory federations [33]. The iLab initiative was the first one allowing this feature, and other architectures as WebLab-Deusto have also implemented it with extended features, such as federated load balance [34] which is that if two universities have copies of the same laboratory, students can go first to their university and only if it is not available they are automatically redirected to the other copy in the other university, and if both copies are busy or unavailable, then a shared queue is created.

At a higher level of abstraction, repositories were created listing or providing access to different remote and virtual laboratories. Examples include Go-Lab, LiLa or Lab2go. In this case, the approach is providing a curated list of resources and teachers can find them and access them in a more or less integrated way.

At the same time, modern initiatives were taken to provide uniform access to a curated list of remote laboratories. LabsLand[1] [35], for example, provides a list of laboratories from over 20 universities in over 11 countries. It constantly tracks which laboratories are available and which ones have technical issues, removing them from the shared queues in an automatic fashion. LabsLand enforces federated load balance by providing real-time remote laboratories only when the same laboratory is available in multiple institutions. Additionally, LabsLand provides universities and high schools without remote laboratories the opportunity to have a LabsLand space (where they can see what their students did, integrate the laboratories in a number of Learning Management Systems such as Moodle, Canvas, Sakai, Google Classroom and others), and access laboratories from the growing LabsLand network of laboratory provider. Under this model, LabsLand allows universities to share their laboratories with other institutions while at the same time consuming curated laboratories from third parties in a uniform way.

2.3 FPGA Teaching in UPNA and UNIFESP

At UPNA, in the first course of the Telecommunications Engineering degree, the four-month subject of "Digital Systems I" is taught, whose agenda includes the design and analysis of basic digital systems such as combinational and sequential circuits. Practical designs are made with the Intel Quartus II version 17.1 tool. The practices contemplate for the design of the digital circuits both their graphic description by means of schemes and their description by means of the VHDL language, designing by means of VHDL a finite state machine sequencer described with a functional model.

Due to the unavailability of the remote laboratory until the beginning of the course, practices were started in a laboratory with 20 local instances where the Quartus II design tool is installed in each of them. The final design practice using the VHDL language was carried out in the remote laboratory with 4 instances formed by the altera DE1-SoC development kit that integrates an ARM Cortex-A9 processor and an FPGA into the Altera Cyclone V 5CSEMA5F31C6 SoC. In the subject "Digital Systems I", 80 students organized in 4 groups of 20 students were enrolled in guided practices in the presence of the teacher in 6 sessions of 2 h each group.

At UNIFESP, at the Institute of Science and Technology, there is the Digital Circuit Course, basic digital systems, combinational and sequential circuits are taught. Students develop their projects using the Quartus II system by means of schematic designs and Verilog language, synthesizing them on the FPGA DE2-115 board. The 51 students enrolled in the course were able to access the remote lab throughout the semester from August to November of 2019.

[1] https://labsland.com.

3 Technical Solution

3.1 LabsLand and WebLabLib

The described remote laboratory is designed from the beginning to be integrated into the LabsLand network of remote laboratories. It is also designed to leverage much of the technology and common functionalities that LabsLand provides (e.g., the web-based administration platform, web-based IDE, distributed compilation environment, laboratory scalability and management, LMS integration, real-time quality assurance).

To meet those goals, the remote laboratory has been designed and implemented using WebLabLib [36]. This is an open source framework, maintained by LabsLand, for the development of remote laboratories. Laboratories that rely on WebLabLib are almost automatically integrable into WebLab-Deusto or into the LabsLand network, thus gaining access to features such as LMS integration.

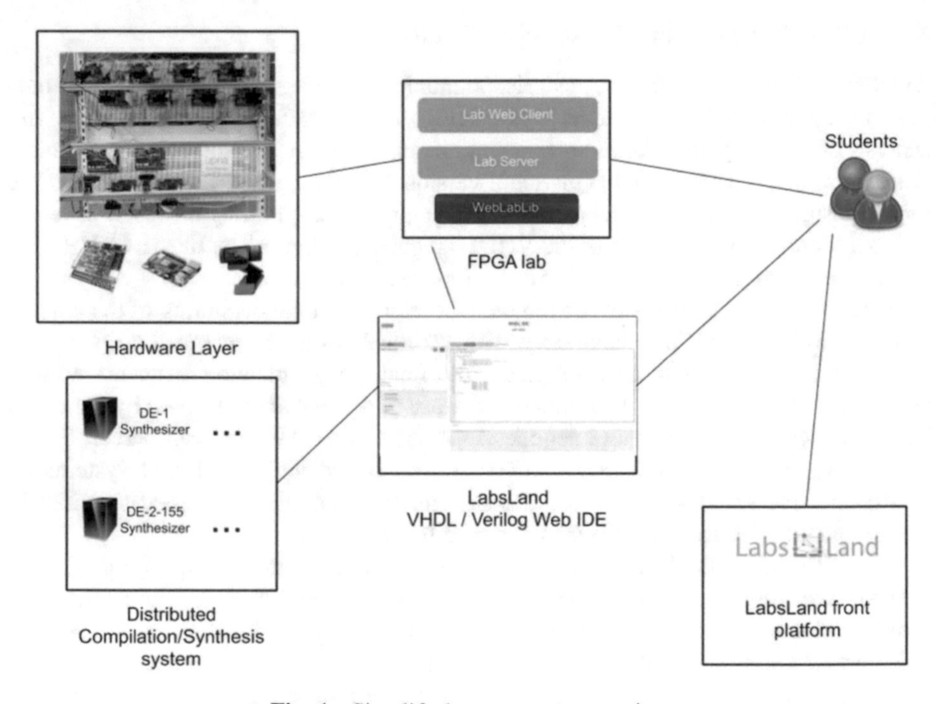

Fig. 1. Simplified components overview.

Figure 1 provides a simplified overview of the components that form the described system. After accessing LabsLand through its portal or through an integrated LMS, students will typically access the LabsLand Web IDE.

The LabsLand Web IDE is the platform to which students access to create and edit their VHDL or Verilog logics. Once they are ready to test it, the IDE internally queues these logics for remote synthesis. This is done through the Distributed Compilation/Synthesis system. The core of this system relies on LabsLand and cloud technologies, but it is formed by numerous "compilers". Those are deployed in different institutions and places (such as in UPNA, in UNIFESP and LabsLand premises) and support both VHDL and Verilog. In this case they turn the input logic into a binary file.

This binary file is programmed into a physical, remote FPGA. Students now enter the FPGA lab itself. It provides an interface to view and interact with that FPGA, which can now run their logic. The FPGA lab server and UI relies on WebLabLib and related technologies.

The hardware layer, which will be described in more detail in the following subsections, is formed by the educational FPGA boards themselves and the remotization hardware (mainly a raspberry for each board, and a web camera). There are currently several of these instances deployed at UPNA and several at UNIFESP.

3.2 Deployment at UPNA

The remote laboratory developed at UPNA is based on 14 DE1-SoC development kits and 2 DE2-115 development kits purchased from the Terasic provider. The compilation of the designs in the VHDL language and the programming of the FPGA devices of each instance is done with the Intel Quartus II version 17.1 development tool installed on the main server, which gives access to two private networks, an ethernet network for instances based on the DE1-SoC kit and a usb network for DE2-115 kits. For the interconnection of the DE1-SoC kits to the ethernet network, the switch TP-Link 34 ports was used and for the DE2-115 kits the usb hub Orico 12 ports.

The DE1-SoC kit has a programmable unit a Cyclone V 5CSEMA5F31C6 SoC which integrates two ARM Cortex A9 and FPGA cores. The Kit has a wide variety of peripherals that should be highlighted: switches, buttons, SD driver, ethernet communications ports, usb, jtag, etc. The DE1-SoC kit allows different types of boot by configuring an array of microswitches and in our case the kit is configured to boot from the SD card where an image of the linux kernel is located that is loaded into the DDR3 memory of the HPS processor ARM.

The selected configuration allows the bitstream to be transferred through the private network to the ARM embedded processor, the operating system being responsible for loading the bitstream in the FPGA, instead of selecting the configuration that allows the transfer of the bitstream through the USB cable connected to the JTAG interface responsible for loading the program into the FPGA.

Each instance consists of the Terasic programmable kit, a secondary server consisting of a Raspberry 3B+ embedded system and a Logitech camera with USB output. The secondary Rpi server performs two basic functions: emulate the physical activation of the Terasic kit inputs such as switches and buttons and on the other hand control the video camera. Each Rpi server is in communication with the main server through the same ethernet network to which the Terasic kits are connected, giving a total of 28 + 1 units connected to the ethernet network, through which it receives from the main server the orders that emulate the input peripherals such as switches and buttons and send the video bitstream to the main server. The connection of the Rpi server with the Terasic kit is made through the GPIO bus and the connection with the video camera through the USB port. Figure 2 shows the UPNA remote laboratory.

Fig. 2. UPNA remote lab

3.3 Deployment at UNIFESP

The described remote laboratory is designed from the beginning to be integrated into the LabsLand network of remote laboratories.

The remote laboratory developed at UNIFESP is based on 2 DE2-115 development kits purchased from the Terasic provider. The compilation of the designs in schematic files or in Verilog is done with the Intel Quartus II version 17.1 development tool installed on the main server, using two Raspberry Pi 3B+ and two webcams with USB output.

Raspberry Pi emulates the physical activation of inputs such as switches and push buttons from FPGA DE2-115. Figure 3 shows the UNIFESP remote laboratory.

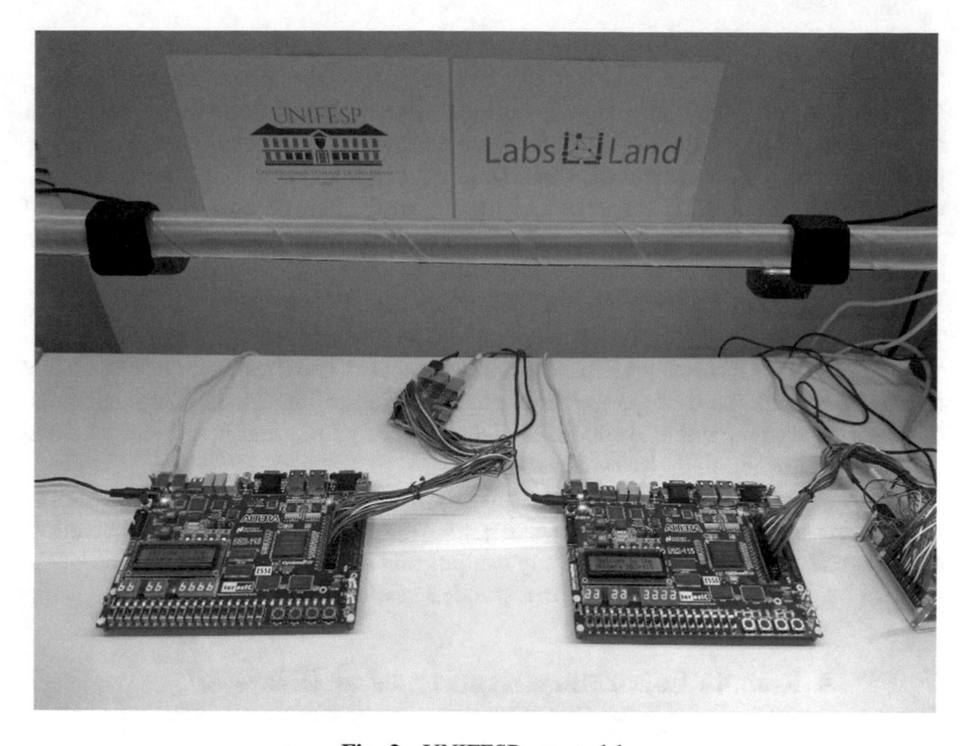

Fig. 3. UNIFESP remote lab

4 Procedure from a Student Perspective

The UPNA relies on the Sakai learning management system. It allows the control of student access to the remote laboratory by activating a work session exclusively for enrolled students and a menu with links to external applications such as the web interface for accessing the remote laboratory.

The IDE web is organized in different windows. The work window allows the editing of the VHDL module through a custom editor that simplifies the student's task through help functions, templates, autocompletion, etc. Once the compilation phase of the source module has been successfully completed, it is allowed to enter the programming phase that one of the available instances needs to reserve.

The functions for the development of the reconfigurable FPGA system through the compilation of the VHDL source module and programming of the FPGA unit are performed through the commands and scripts of the Intel Quartus II Prime Lite version software installed on the main server.

Once access to one of the instances has been obtained, the FPGA programming window opens displaying a real-time image of the entire kit, allowing you to see its outputs as LEDs, displays, etc. and the icons of the virtual inputs such as switches and buttons. The student has a limited time of 2 min to program the FPGA, activate the virtual inputs and observe the actual output in the image captured by the camera.

5 FPGA Remote Laboratory Experience at UPNA

UPNA used the remote laboratory in the spring semester of 2019 in the subject of Digital Systems I for the teaching of the master classes of the syllabus corresponding to programmable logic devices and description of combinational and sequential logic circuits using the VHDL language and for the realization of a practical session of design of a digital lock by means of the programming language VHDL. Because the remote laboratory was finished assembling at the beginning of May, it was not possible to use it throughout the entire course, as it will be the case in the 2019–2020 course.

The real-time availability of a remote laboratory without the need for any infrastructure other than a computer with access to the network and a projector made possible its use in the classroom of the master classes allowing to immediately implement the concepts of programming and circuits digital explained getting easier to understand and getting their use for study by students.

It was possible to analyze the use made by the students of the remote laboratory through the monitoring records, observing the use of it at times outside the school calendar and from outside the university campus and even the partial realization of the practice before accessing the classroom practice session with the teacher.

6 FPGA Remote Laboratory Experience at UNIFESP

6.1 Contextualization

UNIFESP has a conventional laboratory for Computer Systems courses, which includes the practices of Digital Circuits, Architecture and Organization of Computers, Compilers, Operating Systems, and Computer Networks. Besides, this lab offers twenty-five computers with 25 FPGA DE2-115 from Intel kits for use with Quartus Prime II Software. However, the demand for students to use this equipment is higher than currently offered. Due to the small number of kits and the need for more time to carry out the proposed projects in these subjects, not all students can be attended during the class period. Thus, the implementation of a remote FPGA laboratory results in higher student turnover and better use of laboratory equipment, which may be accessed at all times, including weekends and holidays.

UNIFESP has used the remote laboratory as part of the Digital Circuit Design classes from August to November 2019. The students were using Moodle to access the lab.

6.2 Survey and Results

Remote lab use began in August 2019, and students were able to access the lab by the end of the semester. The remote lab was available to 3 groups of students from the Digital Circuit Design subject totaling 51 students. However, students could also use the conventional laboratory to perform their activities, so not all students chose to make use of the remote laboratory.

As we can see in Fig. 4, the remote FPGA laboratory was widely used at UNIFESP.

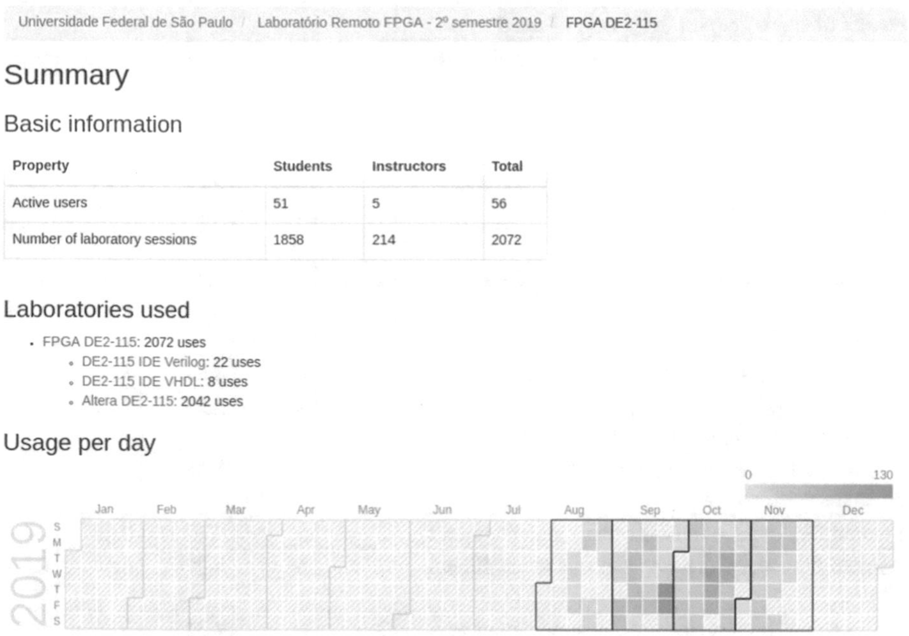

Universidade Federal de São Paulo / Laboratório Remoto FPGA - 2º semestre 2019 / FPGA DE2-115

Summary

Basic information

Property	Students	Instructors	Total
Active users	51	5	56
Number of laboratory sessions	1858	214	2072

Laboratories used

- FPGA DE2-115: 2072 uses
  - DE2-115 IDE Verilog: 22 uses
  - DE2-115 IDE VHDL: 8 uses
  - Altera DE2-115: 2042 uses

Usage per day

0 130

Jan Feb Mar Apr May Jun Jul Aug Sep Oct Nov Dec

2019 S M T W T F S

Fig. 4. Remote lab uses at UNIFESP from August to November/2019

The students were asked to complete a survey adapted from [5], as described in Table 1. In the survey, the student should indicate their degree of agreement or disagreement on a 5-point Likert scale [37] and the possible answers were: 1 - Strongly Disagree, 2 - Partially Disagree, 3 - Neither Agree nor Disagree, 4 - Partially Agree, 5 - Strongly Agree. The questions were grouped into dimensions related to learning, lab acceptance, immersion and usability, teacher guidance, and technical restrictions, and there were three open-ended questions so that students could explain what they liked and disliked and also they could give suggestions.

The total number of students enrolled in the Digital Circuit Projects discipline was 51, 42 accessed the remote lab at least once, and 35 of them answered the survey presented in Table 1.

Table 1. Online survey, items, and results. Source: table adapted from [5].

Item	Min	Max	Mean	Std. Dev.	Var.
Learning					
Using the remote lab helped me learn better how to design digital circuits	1	5	**4.0**	1.32	1.76
I used the lab several times and noticed strange results	1	5	**3.6**	1.26	1.60
I believe I can implement several real digital circuit designs	2	5	**4.34**	0.8	0.64
I could use the scientific concepts to explain the results of digital circuit designs performed	2	5	**4.28**	0.56	0.75
Using the remote lab has improved my ability to apply theoretical concepts to practice	1	5	**4.25**	0.98	0.96
Using the remote lab strengthened my practical skills	1	5	**4.20**	1.13	1.28
Using the remote lab strengthened my theoretical knowledge	1	5	**3.68**	1.13	1.28
Lab acceptance					
I was able to use the remote lab 24 h/7days	2	5	**4.20**	1.07	1.16
I would rather do activities in traditional labs than use a remote lab	1	5	**3.77**	1.11	1.24
I always shared the results with my colleagues	1	5	**3.65**	1.23	1.52
I was less afraid of damaging remote equipment than when working in the traditional lab	1	5	**3.97**	1.22	1.49
I performed activities or projects different from those assigned to me	1	5	**1.80**	1.20	1.45
I would like to have remote labs for other subjects	3	5	**4.62**	0.68	0.47
Remote labs serve as a complement to traditional labs	3	5	**4.77**	0.54	0.29
While using the remote lab, I was motivated to continue carrying out the activities	1	5	**4.31**	0.96	0.92
I used the remote lab more often than I needed to on basis of the assignment out of curiosity	1	5	**2.65**	1.51	2.29
Immersion and usability					
I think I can handle the remote lab well	1	5	**4.28**	0.89	0.79
I found that the remote lab and its devices are easy to use	2	5	**4.28**	0.95	0.91
I can see similarities using the remote lab and the traditional lab	1	5	**4.25**	0.98	0.96
The equipment used in the remote laboratory is identical to their real equivalence	1	5	**3.88**	1.18	1.39
Although I was far from the remote lab equipment, I felt myself in control of it	1	5	**3.82**	1.15	1.32

(*continued*)

Table 1. (*continued*)

Item	Min	Max	Mean	Std. Dev.	Var.
Teacher guidance					
The instructions for carrying out projects were always clear	2	5	**4.71**	0.66	0.44
I consulted the remote lab manual to learn more about its use	1	5	**4.71**	0.75	0.56
I didn't need the assistance of the monitor or teacher to do my activities	1	5	**3.74**	1.33	1.78
The objectives of the activities were clear to me all the time	2	5	**4.65**	0.63	0.40
Technical restrictions					
I have had many difficulties accessing the remote lab	1	5	**1.80**	1.15	1.34
The response time of the system was adequate	1	5	**4.0**	1.18	1.41
I found it difficult to find time to carry out the activities proposed by the teacher	1	5	**2.48**	1.26	1.61
The remote lab worked without any problems	1	5	**3.17**	1.33	1.79

Open-Ended Questions:

1. What did you find most interesting when using the remote lab? The most relevant items cited by the students were: being able to access the same equipment used in class, perform activities outside of class, be able to access the equipment anytime and anywhere, practicality and accessibility.
2. What drawbacks have you found when using the remote lab? Some drawbacks cited by the students were: lab unavailability due to technical issues, response delay between commands and display results, having to change pins in code to execute in the remote lab, short lab access time and to wait in line to access lab.
3. If you could change anything about the remote lab and its usage in course contexts, what would that be? Students would like to use the same code in both the conventional lab and the remote lab, so their most cited item was that they didn't need to change their pinouts to use the remote lab. Include a key-activated mapping. Enlarge board output image. The evaluation performed by the teacher could be by the remote laboratory. Map keys to buttons and change the switches layout.

7 Discussion

Most students from UNIFESP have accessed the remote lab at extra-class times. In this period of use, some students reported inconsistencies about the project developed in the conventional laboratory and in the remote laboratory.

Analyzing some of these projects, reported as incoherent, we found that in reality, there were no inconsistencies, the project was executed according to the Verilog and schematic designs elaborated. For example, in one of these cases, the student failed to assign output pins. Then, in the conventional lab, the pin was not mapped to any FPGA output, while in the remote lab, the FPGA has assigned the pins to one segment of the 7-segments display. When executing the project, a signal appeared on the screen that the student thought was "random", but in reality, it was the student's project execution, so there were no errors between remote and conventional lab.

Other reported cases involve timing. In one project, the students had to develop a state machine that produced a numerical sequence on a 7-segments display. Due to the speed of the internet used by the student, the viewing time was different from the conventional laboratory. Although in all cases analyzed, the project executed was the same, the student had the impression that the remote laboratory, due to the differences mentioned, had instabilities that did not occur in the conventional laboratory. In the survey, some students argued that their project was correct and stated that the remote lab had instabilities. Thus, it is clear that part of the students are suspicious that the remote laboratory works the same way as the conventional laboratory.

Although some students inherently have this distrust about the instability of remote lab operation, there are already some students who are using the remote lab during class. Some students prefer to use the remote lab exclusively, even while in the classroom when the student can use the physical FPGA kit.

During the course work presentations in the evaluation process, some of the students preferred to present their projects using only the remote laboratory.

We believe, therefore, that with their implementation and regular use in the classroom and evaluation processes, remote laboratories will gain more and more space and earn the trust and credibility of all students.

Specifically, regarding the cooperation with UPNA and UNIFESP, four advantages can be mentioned:

- Related to infrastructure

During the semester, the internet or power outage made the use of UNIFESP's FPGAs unfeasible, and the students could continue using the remote laboratory through the remote laboratory allocated at UPNA. Similarly, when UPNA fails, students can use the UNIFESP lab transparently.

- Related to demand

The cooperation allows for a higher number of kits, thus reducing the waiting time in line, which can also be enhanced by the timezone between the two countries (4 h).

- Related to FPGA diversification

UNIFESP only has the FPGA DE2-115 board, while UPNA also has the FPGA DE1-SoC board, which uses ARM microcontrollers. With cooperation, students will be able to use any of the FPGAs.

- Related to the culture of use

Finally, the cooperation between both universities and the advantages previously listed will allow the dissemination of a culture of use of remote laboratories among students, encouraging the development of work inside and outside the classroom. It even enables students to use the extra-class time to work on their projects toward more meaningful and student-centered learning.

8 Conclusions

In the Online Learning of Engineering studies in subjects such as Digital System Prototyping with FPGA and Embedded Computer Architectures, remote laboratories are necessary to promote motivation in individual study and develop skills such as creativity and initiative in students. The remote laboratories for reconfigurable FPGA systems, jointly developed by UPNA, UNIFESP, and the LabsLand company, has demonstrated a distributed architecture with shared resources. It has achieved a very high availability for the FPGA cards, and its friendly web-based IDE interface contributes to a remarkable performance in student learning.

Acknowledgments. Thanks the financial contribution made by UPNA and UNIFESP Universities for the acquisition of the necessary material for the development of the laboratories and the donation of the Terasic DE1-SoC kits made by Intel. Also thank the technical staff who have worked on the installation of the necessary resources for this project as well as the valuable contributions made by the teaching staff.

References

1. Mikhak, B., Lyon, C., Gorton, T., Gershenfeld, N., McEnnis, C., Taylor, J.: Fab Lab: an alternate model of ICT for development. In: 2nd International Conference on Open Collaborative Design for Sustainable Innovation, pp. 1–7, December 2002
2. Troxler, P., Wolf, P.: Bending the rules. The Fab Lab innovation ecology. In: 11th International CINet Conference, Zurich, Switzerland, pp. 5–7, September 2010
3. Blikstein, P., Krannich, D.: The makers' movement and FabLabs in education: experiences, technologies, and research. In: Proceedings of the 12th International Conference on Interaction Design and Children, pp. 613–616. ACM, June 2013
4. Malchow, M., Bauer, M., Meinel, C.: Embedded smart home—remote lab MOOC with optional real hardware experience for over 4000 students. In: 2018 IEEE Global Engineering Education Conference (EDUCON) (2018)
5. May, D.: Introducing Remote Laboratory Equipment to Circuits-Concepts, Possibilities, and First Experiences. American Society for Engineering Education (2019)
6. Fotopoulos, V., Orphanoudakis, T., Fanariotis, A., Skodras, A.: Remote FPGA laboratory course development based on an open multimodal laboratory facility. In: Proceeding PCI 2015 Proceedings of the 19th Panhellenic Conference on Informatics, pp. 447–452 (2015)
7. Bose, R.: Virtual labs project: a paradigm shift in internet-based remote experimentation. IEEE Access **1**, 718–725 (2013). https://doi.org/10.1109/access.2013.2286202
8. Corter, J.E., Nickerson, J.V., Esche, S.K., Chassapis, C.: Remote versus hands-on labs: a comparative study. In: 34th Annual Frontiers in Education, FIE 2004, p. F1G-17. IEEE, October 2004

9. Brinson, J.R.: Learning outcome achievement in non-traditional (virtual and remote) versus traditional (hands-on) laboratories: a review of the empirical research. Comput. Edu. **87**, 218–237 (2015). https://doi.org/10.1016/j.compedu.2015.07.003
10. Feisel, L.D., Peterson, G.D., Arnas, O., Carter, L., Rosa, A., Worek, W.: Learning objectives for engineering education laboratories. In: Proceedings 32nd Annual Frontiers in Education (FIE), vol. 2, p. F1D (2002). https://doi.org/10.1109/fie.2002.1158127
11. de Jong, T., Linn, M.C., Zacharia, Z.C.: Physical and virtual laboratories in science and engineering education. Science **340**(6130), 305–308 (2013). https://doi.org/10.1126/science.1230579
12. Zhang, K., Chang, Y., Chen, M., Bao, Y., Xu, Z.: Computer organization and design course with FPGA cloud. In: 50th ACM Technical Symposium on Computer Science Education, 27 February–2 March 2019 (2019)
13. Vandorpe, J., Vliegen, J., Smeets, R., Mentens, N., Drutarovsky, M., Varchola, M., Lemke-Rust, K., Plöger, P., Samarin, P., Koch, D., Hafting, Y., Tørresen, J.: Remote FPGA design through eDiViDe—European digital virtual design lab. In: 2013 23rd International Conference on Field programmable Logic and Applications (2013)
14. Torrensen, J., Wold, A., Hafting, Y., Nygaard, T., Skogstrom, R., Norenda, J.: Flexible teaching of reconfigurable logic design including a remote cloud lab. In: Proceedings of the Sixth International Conference on E-Learning and E-Technologies in Education (ICEEE 2017), Lodz, Poland (2017)
15. Petrvalský, M., Petura, O., Drutarovský, M.: Remote FPGA laboratory for testing VHDL implementations of digital FIR filters. Acta Electrotechnica et Informatica **15**(2), 3–8 (2015)
16. Gilibert, M., Picazo, J., Auer, M.E., Pester, A., Cusidó, J.A., Ortega, J.A.: 80C537 microcontroller remote lab for E-learning teaching. iJOE Int. J. Online Eng. **2**(4), 1–3 (2006)
17. de Moraes, A.G., de Sales, A.K.M.: Aplicacao de laboratorios remotos em microcontroladores PIC. In: Proceedings 19th Congresso Brasileiro de Automática (CBA), pp. 3634–3641 (2012)
18. Zenzerović, P., Sučić, V.: Remote laboratory for microcontroller systems design. In: Proceedings 34th International Convention MIPRO, pp. 1685–1688, May 2011
19. Fotopoulos, V., Anastasios, I.S., Anastasios, F.: Preparing a remote conducted course for microcontrollers based on Arduino. In: Proceedings 7th International Conference Open Distance Learning (ICODL), pp. 1–6 (2013)
20. Ferreira, J., Nedić, Z., Machotka, J., Nafalski, A., Göl, Ö.: International collaborative learning using remote workbenches. In: Proceedings Annual Conference Engineering Technology Education, pp. 47–51 (2010)
21. Tawfik, M., Sancristobal, E., Martin, S., Diaz, G., Peire, J., Castro, M.: Expanding the boundaries of the classroom: implementation of remote laboratories for industrial electronics disciplines. IEEE Ind. Electron. Mag. **7**(1), 41–49 (2013). https://doi.org/10.1109/mie.2012.2206872
22. Tawfik, M., Sancristobal, E., Martin, S., Diaz, G., Castro, M.: State-of-the-art remote laboratories for industrial electronics applications. In: Proceedings Technology Application Electronic Technology Conference (TAEE), June 2012, pp. 359–364 (2012). https://doi.org/10.1109/taee.2012.6235465
23. Reichenbach, M., Schmidt, M., Pfundt, B., Fey, D.: A new virtual hardware laboratory for remote FPGA experiments on real hardware. In: Proceedings 2011 International Conference e-Learning, e-Business, Enterprise Information Systems, and e-Government (EEE), pp. 17–23 (2011)
24. Morgan, F., Cawley, S., Newell, D.: Remote FPGA lab for enhancing learning of digital systems. ACM Trans. Reconfigurable Technol. Syst. **5**(3) (2012). Art. No. 18. https://doi.org/10.1145/2362374.2362382

25. Morgan, F., et al.: ViciLogic: online learning and prototyping platform for digital logic and computer architecture. In: Proceedings eChallenges (e), pp. 1–9, October 2014
26. Hercog, D., Gergic, B., Uran, S., Jezernik, K.: A DSP-based remote control laboratory. IEEE Trans. Ind. Electron. **54**(6), 3057–3068 (2007). https://doi.org/10.1109/tie.2007.907009
27. Rodriguez-Gil, L., García-Zubia, J., Orduña, P., López-de-Ipiña, D.: Towards new multiplatform hybrid online laboratory models. IEEE Trans. Learn. Technol. **10**(3), 318–330 (2016)
28. Villar-Martínez, A., Rodríguez-Gil, L., Angulo, I., Orduña, P., García-Zubía, J., López-De-Ipiña, D.: Improving the scalability and replicability of embedded systems remote laboratories through a cost-effective architecture. IEEE Access **7**, 164164–164185 (2019)
29. Rodríguez-Gil, L., García-Zubia, J., Orduña, P., López-de-Ipiña, D.: An open and scalable web-based interactive live-streaming architecture: the WILSP platform. IEEE Access **5**, 9842–9856 (2017)
30. Angulo, I., Rodriguez-Gil, L., García-Zubia, J.: Scaling up the lab: an adaptable and scalable architecture for embedded systems remote labs. IEEE (2018)
31. De Jong, T., Sotiriou, S., Gillet, D.: Innovations in STEM education: the Go-Lab federation of online labs. Smart Learn. Environ. **1**(1), 3 (2014)
32. Orduña, P., Rodriguez-Gil, L., Garcia-Zubia, J., Angulo, I., Hernandez, U., Azcuenaga, E.: LabsLand: a sharing economy platform to promote educational remote laboratories maintainability, sustainability and adoption. In: 2016 IEEE Frontiers in Education Conference (FIE), pp. 1–6. IEEE, October 2016
33. Harward, V.J., Del Alamo, J.A., Lerman, S.R., Bailey, P.H., Carpenter, J., DeLong, K., Long, P.D.: The iLab shared architecture: a web services infrastructure to build communities of internet accessible laboratories. Proc. IEEE **96**(6), 931–950 (2008)
34. Orduña, P.: Transitive and scalable federation model for remote laboratories. Universidad De Deusto, Bilbao, Spain, Thesis doctoral, vol. 86, pp. 78–91 (2013)
35. Orduña, P., Rodriguez-Gil, L., Garcia-Zubia, J., Angulo, I., Hernandez, U., Azcuenaga, E.: Increasing the value of remote laboratory federations through an open sharing platform: LabsLand. In: Online Engineering & Internet of Things, pp. 859–873. Springer, Cham (2018)
36. Orduña, P., Rodriguez-Gil, L., Angulo, I., Hernandez, U., Villar, A., Garcia-Zubia, J.: WebLabLib: new approach for creating remote laboratories. In: International Conference on Remote Engineering and Virtual Instrumentation, pp. 477–488. Springer, Cham, February 2019
37. Sullivan, G.M., Artino Jr., A.R.: Analyzing and interpreting data from Likert-type scales. J. Grad. Med. Educ. **5**(4), 541–542 (2013)

Real Laboratories Available Online: Establishment of ReVEL as a Conceptual Framework for Implementing Remote Experimentation in South African Higher Education Institutions and Rural-Based Schools – A Case Study at the University of Fort Hare

Pumezo Mzoxolo Kwinana[1(✉)], Phumzile Nomnga[2], Mncedi Rani[1], and Mantile L. Lekala[3]

[1] Physics Department, University of Fort Hare, Alice, South Africa
pkwinana@ufh.ac.za, mncedirani@gmail.com

[2] Computer Science Department, University of Fort Hare, Alice, South Africa
pnomnga@ufh.ac.za

[3] Physics Department, University of South Africa, Pretoria, South Africa
lekalml@unisa.ac.za

Abstract. The University of Fort Hare (UFH) in South Africa counts with its own laboratories as part of ReVEL (Remote and Virtual Education Laboratory), and intends to use them both for the University students and for rural-based schools. One example of these laboratories by UFH is the LabsLand Arduino Robot. LabsLand is the global network of real online laboratories, and counts with technologies to make laboratories available online. One Labsland Arduino Robot was deployed at UFH by the end of October 2019. The full article includes details of the deployed system. The aim of this paper is to present a conceptual framework for the establishment of ReVEL for implementation in South Africa, a developing country. It is also aimed at giving a qualitative account of relevant stakeholders' insights on the implementation of ReVEL model in the South African context. A survey and a series of in-depth, semi-structured individual and focus group interviews on the perception of implementation of remote labs were conducted for data collection guided by a thematic analysis approach to the UFH community (instructors and students) and to rural-based science schools. UFH instructors and students and learners from schools participated in the survey. Responses were analyzed accordingly. Survey responses were received from both instructors and students. 87% of instructors applauded this technology and strongly agreed on its advantages of enabling sharing of larger variety of specialized equipment while reducing the financial burden and improving the educational experience. Others were cautious about the lack of hands-on interaction with the real equipment in the remote laboratory and preferred the feeling of the actual equipment. 96.7% of Students praised its convenience and flexibility for obtaining results for analysis. 50% preferred to work in both face-to-face and remote lab. The initiative of

M. E. Auer and D. May (Eds.): REV 2020, AISC 1231, pp. 128–142, 2021.
https://doi.org/10.1007/978-3-030-52575-0_10

ReVEL as a platform for delivering the real laboratories through use of internet to a developing country such as South Africa has received a resounding acceptance and appreciation for its potential deliverables. The results suggested that, stakeholders from different communities such academia, industry, and government praised ReVEL for prospects of providing teaching and learning cutting-edge interactive technologies such robotics, high performance computing ranger and other teaching, learning and research benefits in various of study and to rural based schools.

Keywords: Remote laboratories · Arduino · Robotics

1 Introduction

There is a general agreement by numerous researchers that laboratory work plays a vital role in education since it contributes greatly to the development of students' understanding of scientific concepts and processes of science in practical subjects [1, 2, 3]. In a globally connected world where researchers need to update their technical skills and share knowledge continuously, remote and virtual laboratories have emerged as excellent complementary models to share specialized skills and resources over wide geographical areas, and to reduce overall costs while improving educational experience. A remote laboratory is a software and hardware tool that enables students to access real equipment located somewhere else on the internet. Figure 1 shows a graphical representation of remote and virtual laboratories:

For example, students learning how to program a small Arduino-based robot can write the code online. Then, in the web browser, they can connect to a remote laboratory that shows through a webcam how a real robot behaves in a real environment running their code. In the literature there is a wide range of remote laboratories in many fields (e.g., robotics, electronics, physics, chemistry). Software frameworks have been developed to make the development of remote laboratories more affordable (e.g., Remote Laboratory Management Systems such as WebLabDeusto (http://weblab.deusto.es) [4], Relle (http://relle.ufsc.br) [5, 6], iLab Shared Architecture (http://ilab.mit.edu) [7] or RemLabNet (http://www.remlabnet.eu) [8]) and tools (e.g., gateway4labs (http://gateway4labs.readthedocs.org) [9]) to provide. Integrations with

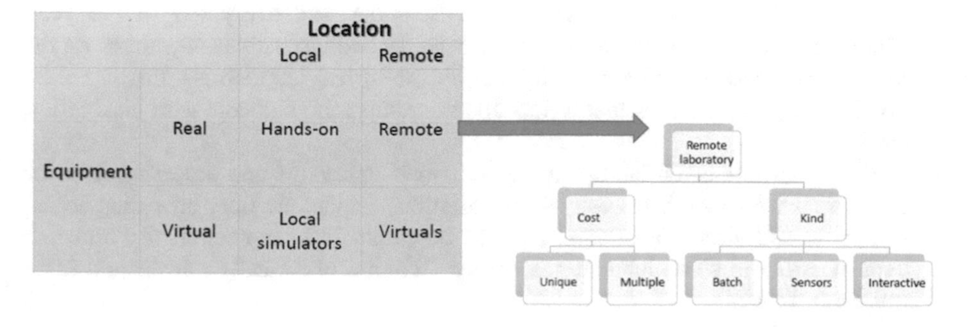

Fig. 1. Graphical description of remote laboratory operations in respect to its location and type

educational tools (such as Moodle, Sakai or other LMS, both through ad hoc solutions and through standards such as IMS LTI) or repositories linking remote and virtual laboratories (such as Go-Lab (http://www.golabz.eu) [10, 11]).

In particular, the WebLab-Deusto research group has created different open source technologies (WebLab-Deusto (http://weblabdeusto.readthedocs.org), weblablib (https://docs.labsland.com/weblablib/)) for developing remote laboratories and many types of laboratories, as well as different remote laboratories for over a decade. However, just like most of the research on the literature, most laboratories have been mostly used by the same university that created the laboratory, or by others but with no reliability guarantee. For this reason, the group has recently started a spin-off startup called LabsLand (https://labsland.com), which provides access to laboratories from different providers (higher institutions) to different consumers (universities like UFH and schools). This initiative has led to the creation new solutions to enable the provision of availability guarantees; which is something not addressed in the literature for professional support.

South African higher institutions, mostly historically disadvantaged institutions (HDIs) have limited research equipment that will offer the required hands-on experimental expertise to students due to the past political imbalances. These challenges among higher education institutions in our country has resulted to the migration of students from the HDIs to Historically Advantaged Institutions (HAIs). It is thus imperative that institutions should share useful teaching and research equipment in a cost-effective way. It is for these reasons that ReVEL (Remote and Virtual Education Laboratory) facility was initiated at the University of Fort Hare (UFH) in South Africa. Its vision is to enable effective sharing of research and teaching equipment through the internet to promote online instrumentation among higher institutions and provide science experimentation resources to needy high schools.

In this paper, a conceptual framework for remote and virtual experiment laboratory is presented. Also, the paper will present perceptions of UFH lecturers, technicians, teachers and learners from surrounding schools. ReVEL facility is the first of its kind in South Africa. The main objective of the paper is to present a number of learning approaches that ReVEL will support at UFH, both within the institution and to the surrounding rural-based schools in the Eastern Cape of South Africa. Educational values for the introduction of ReVEL will include, among others, the following advantages:

- *Flexibility:* ReVEL will be accessible 24 h a day and every day of the year. Moreover, students will connect from their respective institutions anywhere via the internet, whereas in the case of physical access to real laboratories, this is not be possible (e.g., institutions that lack adequate equipment, students with disabilities, etc.);
- *Maximal use:* Research laboratory equipment is expensive and accessing it to the institutions who own them costly (transportation, renting, etc.). By providing access to remote and virtual laboratories, it will be available to more students, and it is even possible to share laboratories among different historically advantaged institutions in future.

- *Real experimentation experience and skills development:* Even though remote and virtual laboratories are accessed through the internet, they provide valuable instrumentation experience; nothing can be compared to the interaction with an experimental setup, although it is performed in a remote way.
- *Interactive learning:* The learning process requires students to be active by performing experiments in available laboratories worldwide interactively. In that way, they develop techniques on sharing specialized skills and resources over a wide geographical area and thus reducing overall costs and improving the educational experience. Perception

The technological advances that characterizes the Fourth Industrial Revolution (4IR) such as artificial intelligence, big data and robotics amongst others, poise as key enablers of economic development and participation. The UFH has an indelible reputation of playing a leading role, spanning over 100 years, as the producer and disseminator of new knowledge, not just in the academia but across all spheres of life. As a modernized institution, promotion of teaching and learning strategies that are responsive to local and wider context [12] have been prioritized to define the new academic trajectory. One of the teaching and learning strategies is what is Venant et al., 2016 termed "Practical Activities" which refers to "any learning and teaching activity that engages learners in manipulating and analyzing real and physical objects", the involving of learners and teachers in a laboratory, a spatial and temporal space hosting devices used for experiments [13]. The establishment of ReVEL at UFH in the small town of Alice, Eastern Cape Province of South Africa, served as a potent instrument for community development and empowerment. Effectively, this means the beneficiaries of ReVEL will now be empowered to learn, live and interact with their peers from around the globe.

2 ReVEL as a Conceptual Framework for Implementing Remote Experimentation in South African Higher Education Institutions (HEIs) and Rural-Based Schools

The establishment of ReVEL at UFH serves as a conceptual framework for implementing remote experimentation in the South African Higher Education Institutions (HEIs) and rural-based schools. More often than not, 'theoretically' and 'conceptual' frameworks are two terms that are used interchangeably in literature although there is a distinction between the two. According to [6], the distinction originates from research the approach employed in a particular study in terms of whether or not it is 'seductive' or 'inductive.' Effectively, this distinction clarifies that "a deductive approach to literature review typically makes use of theories and theoretical frameworks whereas the inductive approach tends to lead to the development of a conceptual framework – which may take the form of a (conceptual) model."

Acknowledging the impact the Remote and Virtual Laboratories can make in society beyond academic boundaries but to socio-economic levels, it was imperative for ReVEL establishment at UFH to take an inductive approach. The aim of taking this approach was to ensure that relevant theories are incorporated and integrated to provide a methodology that is suitable to all concepts underpinning this initiative. These concepts range from above mentioned educational values, to historical context of our social setting and the intend use of ReVEL by University Students and rural-based schools. Figure 2 depicts all the concepts that are incorporated to implement ReVEL:

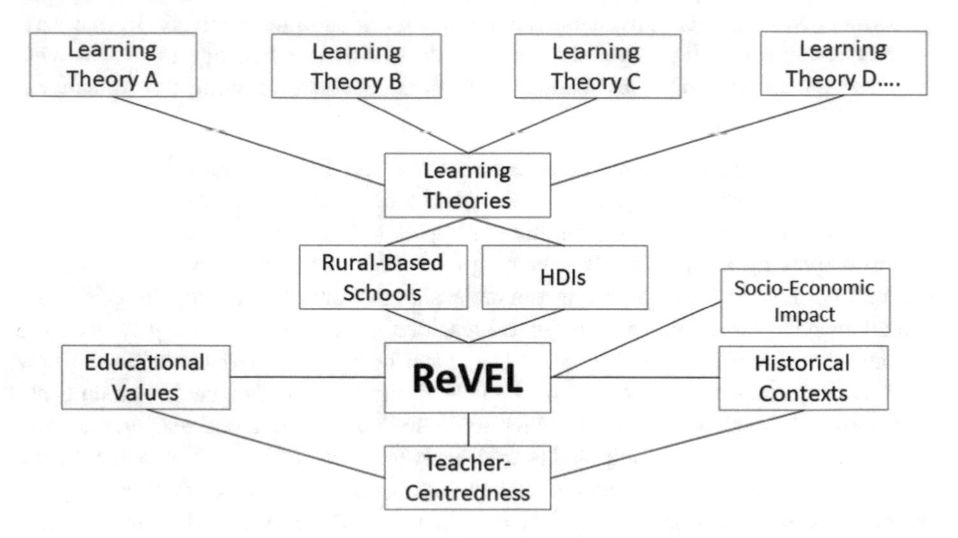

Fig. 2. ReVEL as a conceptual framework for implementing remote experimentation in South African Higher Education Institutions (HEIs) and rural-based schools

3 Aims and Objectives of Establishing ReVEL

The main aims and objectives for the establishment of ReVEL are as follows:

- To introduce technology that enables South African higher institutions to widen their scope of research by accessing and interacting with research equipment worldwide.
- To share techniques, specialized skills and resources over a wide geographical area and thus reducing overall costs while improving the educational experience in instrumentation.
- To introduce modern skills like coding and robotics to students to enable them to perform their experiments anywhere and at any time.
- To establish essential assessment models authenticating remote and virtual laboratories work that will yield comparative results to that of conventional (face-to-face) laboratories.
- To simplify explanations of complex physical phenomena like nuclear fission reactors and other scientific concepts which can be understood more easily if using computerized information techniques instead of traditional educational tools. Virtual laboratory techniques allow the simulations of physical phenomena, which can be subsequently visualized through attractive and allusive animations.
- To enable students to gain competence in acquiring the necessary skills to operate physical or real experimental setup equipment that is remotely located.

- To mitigate movement of postgraduate students from Historically Disadvantaged Institutions (HDIs) to Historically Advantaged Institutions (HAIs) due to lack of research equipment.
- To empower rural-based science schools with modern instrumentation skills so as to enable them grasp scientific concepts at an early educational stage.

4 Remote Laboratory Architecture

As depicted in Fig. 3, the remote laboratories deployed at the University of Fort Hare have a WebLab-Deusto server and the particular laboratories of two Arduino robots and four Arduino boards. Additionally, LabsLand services are located in Amazon Web Services, providing a space for UFH[1] where students can be managed, and they can either enter directly or through Moodle, Google Classroom or other technologies that LabsLand [14] supports.

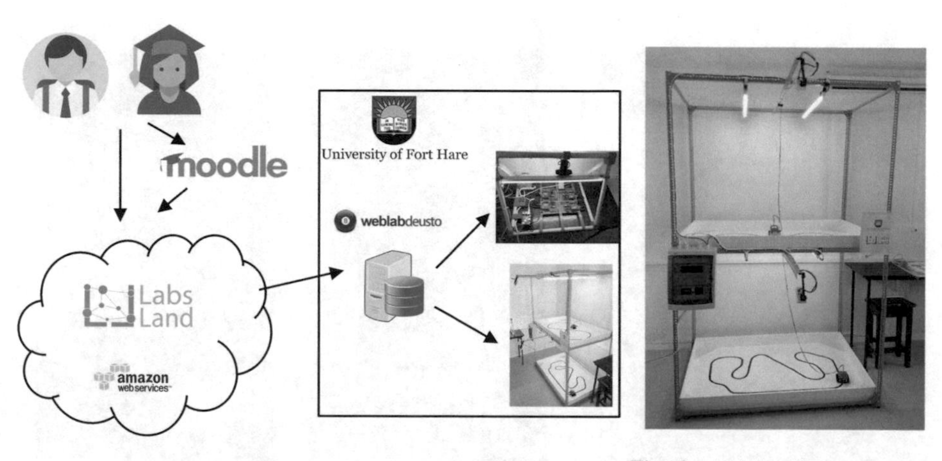

Fig. 3. Architecture of the Remote Laboratories at UFH

Fig. 4. Arduino Robot Remote Laboratory

Both types of laboratories (Arduino robot and Arduino board) are duplicated in multiple locations, including Colombia [15] and Spain (see Fig. 7). This allows students to access the Arduino robot laboratory at UFH and, if it is busy (other student using them) or unavailable (due to any network or power error), automatically students will be using the other deployments in other countries. Additionally, students from all over the world using the LabsLand network can also access the laboratories at UFH, creating a global educational platform.

In summary, the architecture and network of remote laboratories worldwide allows users to access the labs at anytime and anywhere in an internet-enabled environment.

---

[1] https://labs.land/ufh.

Fig. 5. Arduino Board Remote Laboratory

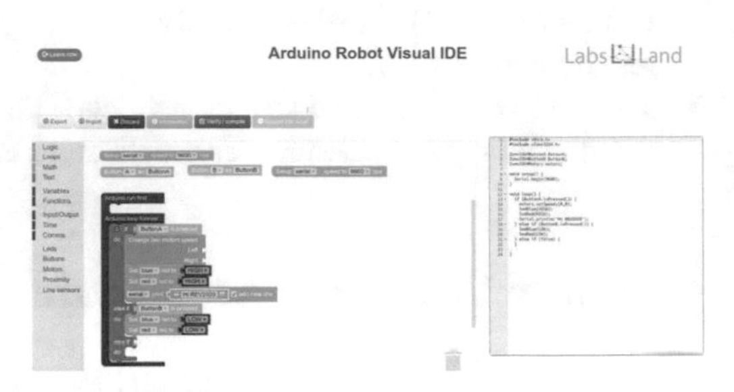

Fig. 6. Programming environment for Arduino for schools

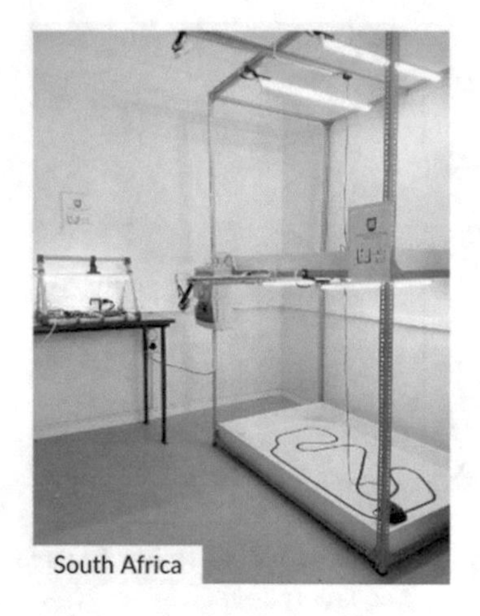

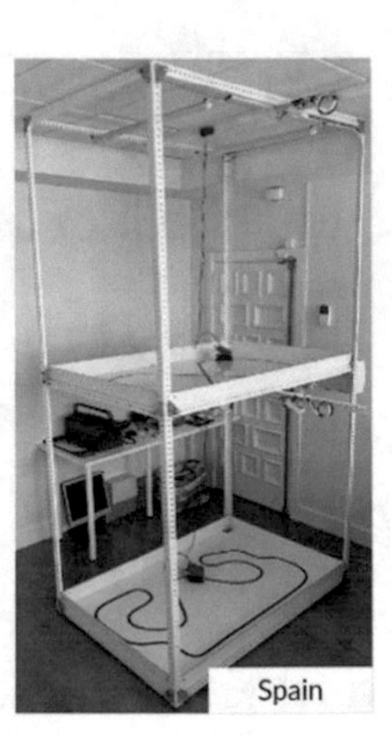

Fig. 7. Different deployments of the robot laboratory in UFH (South Africa), UNAD (Colombia) and LabsLand (Spain)

5 Official Deployment of ReVEL Technology

5.1 Official Launch of ReVEL at the University of Fort Hare

UFH community, industry partners and school learners attended the deployment of ReVEL on 31 October 2019. Dr Pablo Orduna led the deployment team to introduce this type of technology, which is the first of its kind in South Africa. Figure 3 illustrates the demonstrations by Dr Pablo on how the remote laboratories will work. Learners with their educators also had time to interact with the Labsland specialists. Learners had time to interact with the equipment. The community generally welcomed the

Official Launch of ReVEL - 31 October 2019

Fig. 8. Deployment of ReVEL by Labsland specialist on 31 October 2019

technology. They this technology has a huge potential to revolutionize instrumentation in both higher institutions and schools. It is no doubt that research questions:

- What are effective pedagogical characteristics for the use of remote virtual laboratories as a technological teaching tool to yield comparative outcomes to that of traditional (*face-to-face*) laboratory systems?
- Can virtual laboratories assist students to gain competence in acquiring the necessary skills to operate the physical equipment? In other words, does the virtual lab help the student in understanding the real experimental setup?

and many more of this nature will be addressed. It is also no doubt that the students will gain skills from performing experiments from the facility of this nature.

5.2 LabsLand Network Effect

LabsLand[2] is a global network of real laboratories available through the internet. UFH counts with its own space in LabsLand[3]. LabsLand [7] counts with over 20 remote

[2] https://labsland.com.

[3] https://labs.land/unad.

laboratories in a dozen countries in all the continents, and in particular, the Arduino Robotics laboratory deployed at UFH is also available in UNAD in Colombia, as well as in the University of Deusto in Spain and in LabsLand facilities also in Spain. The Arduino board laboratory is also deployed in University of Deusto and LabsLand premises.

The fact of having the same remote laboratory in multiple locations is a great advantage for all the nodes in the network for two reasons:

1. Students of UFH and rural schools by default access the remote laboratories in UFH. However, if for any reason (maintenance, internet failure, etc.) the remote laboratory is broken, or it is busy with other students using it, UFH students will automatically go to the robotics laboratory in other locations. This improves considerably the user experience since students do not need to wait long to access the laboratory.
2. Students of other parts of the world can access the UFH remote laboratory. In general, and especially taking into account the time difference between Colombia and Spain and South Africa, the negative impact on the UFH students is minimal. However, the positive impact is impactful for UFH: firstly, it improves the UFH visibility in schools and universities worldwide, showing how UFH is an innovative leader in distance education, and secondly, whenever students from other institutions access UFH laboratories, UFH obtains, through a compensation mechanism, access to other LabsLand laboratories (not only Arduino Robotics), so this is translated that at the end of the year, UFH and UFH partner schools obtains access to other laboratories for free for its students.

For these reasons, these remote laboratories are not only remote laboratories deployed in a University for its own students, but, thanks to the LabsLand network, it becomes globally available through LabsLand, and the students of UFH also access other laboratories from other global institutions.

6 Perceptions of the Remote Laboratory Deployment by UFH Community and Schools

A survey was conducted after the deployment to see how the UFH community and schools view the remote technology introduced. Structured questionnaires were distributed to all the UFH instructors who attended ReVEL launch and interacted with Labsland engineers. The other group was the one of school learners that attended the ReVEL launch and performed some exercises thereafter using the ReVEL remote laboratory. The questionnaire was based on wanting to know about their perceptions on the technology's *flexibility* (how accessible is it in comparison with the face-to-face equipment, etc.); *Maximum use* and *real experimentation experience and skills development* (whether the remote and virtual technology can provide a similar experience as that of face-to-face one, etc.). Results were analyzed accordingly to assess their perception about the technology. Table 1 tabulates results of the UFH community responses:

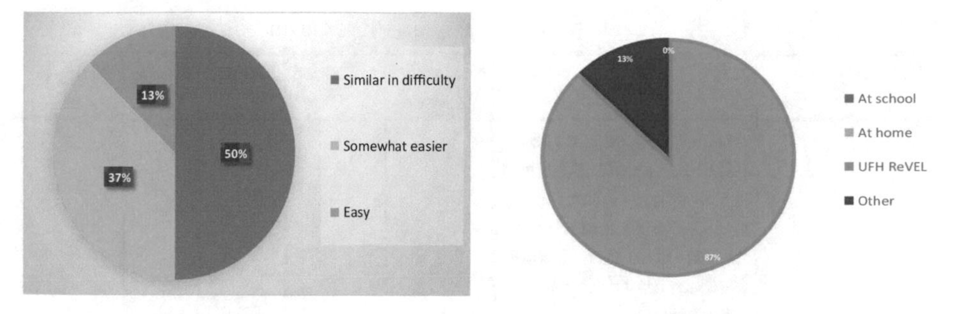

Fig. 9. Responses from the UFH community on the use of remote laboratory

Table 1. Responses from the UFH community on the use of remote laboratory

Category	Item	Strongly disagree	Disagree	Neutral	Agree	Strongly agree
Maximal use	The material covered and the degree of difficulty is appropriate for the level of the course	–	1	1	5	1
	I would recommend the remote lab experiment to others as a supplementary tool for practical sessions	–	–	1	4	3
Real experimentation experience and skills development	I felt like the data being generated was real	–	–	1	7	–
	I felt like I was completing a laboratory using real equipment	–	–	–	8	–
	I felt like I was in an environment created by a computer	–	1	3	4	–
	I felt like I was physically in the environment logging data	–	–	–	8	–
	Performing the remote access lab has helped me to develop practical and computer skills	–	–	2	5	1
Interactive learning	I found the remote laboratory intellectually stimulating	–	–	–	6	2

(*continued*)

Table 1. (*continued*)

Category	Item	Strongly disagree	Disagree	Neutral	Agree	Strongly agree
	Performing the remote lab experiment has helped me understand the theory and concepts underlying the experiments as I could perform experiments repeatedly	–	–	2	5	1
	I learned more in performing the experiments via remote access rather than in personal experimentation	–	–	4	4	–

Table 2. Responses from the school learners on the use of remote laboratory

	Item	Strongly disagree	Disagree	Neutral	Agree	Strongly agree
Real laboratories & Skills acquisition	a) I felt like I was completing a laboratory using a real equipment	0	0	0	15	5
	b) I felt like the data being generated was real	0	1	0	15	4
	c) I felt like I was in an environment created by a computer	1	2	2	5	10
	d) I felt like I was physically in the environment logging data	0	4	3	6	7
	e) Performing the remote access lab experiments has helped me to develop practical and computer skills	0	4	1	11	4

(*continued*)

Table 2. (*continued*)

	Item	Strongly disagree	Disagree	Neutral	Agree	Strongly agree
Interactive learning	a) I found the remote laboratory intellectually \stimulating	0	1	11	5	3
	b) Performing the remote lab experiment has helped me understand the theory and concepts underlying the experiment	0	0	10	8	2
	c) I learned more in performing the experiments via remote access rather than in-person experimentation	0	4	0	10	6
	d) Compared to physical performance of experiments, the remote operation of the instrument was	0	2	12	6	0
	e) Performing the remote access lab experiment has helped me understand the theory and concepts underlying the experiments as I could perform experiments repeatedly	0	0	9	9	0

(*continued*)

Table 2. (*continued*)

	Item	Strongly disagree	Disagree	Neutral	Agree	Strongly agree
Maximal use	a) Where did you access the remote lab website?	1	0	4	4	11
	b) I would recommend the remote experiment to others as a supplementary tool for practices	0	0	4	13	3

Figure 9 (a) and (b) summarizes results tabulated in Table 1. From the graphs, it can be seen that instructors view this instrumentation as almost similar to that of face-face one. They also see it as a facility that is accessible and will make learning interactive. The most important thing is that they valued was the fact that students can now do research worldwide while at UFH, thereby minimizing transportation costs etc.

The responses in Fig. 10 from the school learners are almost similar to those of UFH community. The learners were so receptive of this technology. It seemed to have appealed to them as a younger generation. Due to lack of internet availability, they had an impression that they could only access this when they are inside ReVEL. These are some matters that will need to clarified in future.

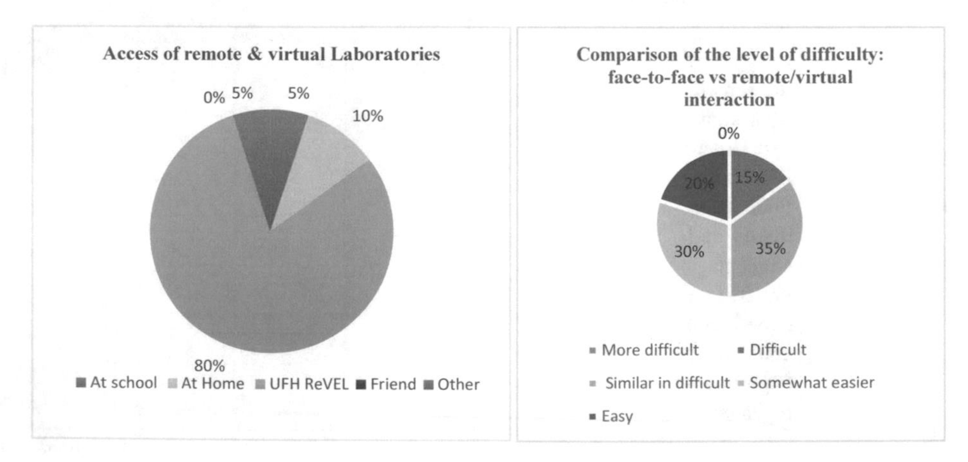

Fig. 10. Responses from the school learner on the use of remote laboratory

7 Conclusion and Recommendations

It can be concluded that the new instrumentation technology has broadened internet-based forms of learning to promote ideas of inter-institutional joint laboratory assets. This is crucial to the developing countries like South Africa as it will close the gap between HDIs and HAIs. The student's remote and virtual experience, despite the limited perception confined to computer screen and the surrounding sound at lab test site [16], can simulate the in-lab one in the remote laboratory environment. In this paper, we have described a novel platform for remote-controlled labs equipment in a collaborative web space that involved more than twenty institutions in the network with the assistance of Labsland. The technology will promote research in our institution and will assist to link other sister institutions to improve their research outputs. It will also assist many South African rural-based schools that lack adequate equipment to perform their syllabus-based experiments.

Acknowledgements. The authors wish to express their sincere gratitude to Labsland Laboratory Providers/Engineers, Dr Pablo Orduna and Mr Aitor Villar Martinez for the successful deployment of the UFH remote laboratory technology. Their unflinching support even during post deployment is highly appreciated. We would like to acknowledge industry and Government partners, viz. Albert Wessels and Armscor for their generous sponsorship to enable UFH to purchase launch the remote and virtual technology in South Africa. Our sincere acknowledgement to the FOSST Discovery Centre colleagues at the University of Fort Hare for valuable contributions. Lastly, we would like to recognize the Institutional Advancement at University of Fort Hare for publishing the official launch of ReVEL, a facility that is the first of its kind in South Africa.

References

1. Ma, J., Nickerson, J.V.: Hands-on, simulated, and remote laboratories: a comparative literature review. ACM Comput. Surv. **38**(3), 7-es (2006). Article 7
2. Hanson, B., Culmer, P., et al.: Real laboratories operated at a distance. IEEE Trans. Learn. Technol. **2**(4), 331 (2009)
3. Corter, J.E., Nickerson, J.V., et al.: Constructing reality: a study of remote, hands-on, and simulated laboratories. ACM Trans. Comput. Hum. Interact. **14**(2), 7-es (2007). Article 7
4. Orduña, P., Bailey, P.H., DeLong, K., L'opez-De-Ipiña, D., Garc'ia-Zubia, J.: Towards federated interoperable bridges for sharing educational remote laboratories. Comput. Hum. Behav. **30**, 389–395 (2014). http://www.sciencedirect.com/science/article/pii/S0747563213001416
5. Simao J.P.S., et al.: Relle: Sistema de gerenciamento de experimentos remotos (2016)
6. Carlos, L.M., Bento da Silva, J., Sommer Bilessimo, S., et al.: Estratégias de Integração de Tecnologia no Ensino: Uma Solução Baseada em Experimentação Remota Móvel (2017)
7. Harward, V.J., et al.: The ilab shared architecture: a web services infrastructure to build communities of internet accessible laboratories. Proc. IEEE **96**(6), 931–950 (2008)
8. Schauer, F., Krbecek, M., Beno, P., Gerza, M., Palka, L., Spilakov, P., Tkac, L.: REMLABNET III – federated remote laboratory management system for university and secondary schools. In: 2016 13th International Conference on Remote Engineering and Virtual Instrumentation (REV), pp. 238–241. IEEE (2016)

9. Orduna, P., Zutin, D.G., Govaerts, S., Zorrozua, I.L., Bailey, P.H., Sancristobal, E., Salzmann, C., Rodriguez-Gil, L., DeLong, K., Gillet, D., et al.: An extensible architecture for the integration of remote and virtual laboratories in public learning tools. IEEE Revista Iberoamericana de Tecnologias del Aprendizaje **10**(4), 223–233 (2015)
10. de Jong, T., Linn, M.C., Zacharia, Z.C.: Physical and virtual laboratories in science and engineering education. Science **340**(6130), 305–308 (2013)
11. Gillet, D., de Jong, T., Sotirou, S., Salzmann, C.: Personalised learning spaces and federated online labs for stem education at school. In: 2013 IEEE Global Engineering Education Conference (EDUCON), pp. 769–773. IEEE (2013)
12. Bouabid, M.E.A.: De la Conception à l'Exploitation des Travaux Pratiques en ligne (2012)
13. Imenda, S.: Is there a conceptual difference between theoretical and conceptual frameworks? J. Soc. Sci. **38**(2), 185–195 (2014)
14. Orduña, P., Rodriguez-Gil, L., Garcia-Zubia, J., Angulo, I., Hernandez, U., Azcuenaga, E.: Labsland: a sharing economy platform to promote educational remote laboratories maintainability, sustainability and adoption. In: 2016 IEEE Frontiers in Education Conference (FIE), pp. 1–6. IEEE, October 2016
15. Orduña, P., Rodriguez-Gil, L., Angulo, I., Martinez, G., Villar, A., Hernandez, Garcia-Zubia, J.: Addressing technical and organizational pitfalls of using remote laboratories in a commercial environment. In: 2018 IEEE Frontiers in Education Conference (FIE), pp. 1–7. IEEE, October 2018
16. Harward, V.J., del Alamo, J.A., Lerman, S.R., Bailey, P.H., Carpenter, J., DeLong, K., Felknor, C., Hardison, J., Harrison, B., Jabbour, I., Long, P.D., Mao, T., Naamani, L., Northridge, J., Schulz, M., Talavera, D., Varadharajan, C., Wang, S., Yehia, K., Zbib, R., Zych, D.: The iLab shared architecture: a web service infrastructure to build communities of internet accessible laboratories. Proc. IEEE **96**(6), 931–950 (2008). https://doi.org/10.1109/JPROC.2008.921607

Remote Labs in Germany—An Overview About Similarities and Variations

Tobias R. Ortelt(✉), Tobias Haertel, and Silke Frye

TU Dortmund University — Engineering Education, Dortmund, Germany
tobias.ortelt@tu-dortmund.de

Abstract. Remote Labs can be divided into different kind of categories like subject area, learning scenario, state of development and availableness. In Germany, several remote labs were developed over the last years or are currently under development. This paper list the different remote labs as well as the projects behind these labs and characterizes the similarities and variations of these remote labs. Therefore, a survey on remote labs in Germany was carried out to give an overview about the ecosystem of remote labs in Germany.

Keywords: Remote labs · Mechanical Engineering · Electrical Engineering

1 Introduction

Remote labs are a common solution to combine theory with practical relevance. Since more than ten years and remote labs are developed all over the world. New developments like smartphones or mobile computers speed up this development.

2 HFD Community Working Group "Remote Labs in Germany"

In 2018, a so called "Community Working Group Remote Labs in Germany" was founded by different institutions like universities, universities for applied science and research associations to promote remote labs in Germany. The buildup of this network was sponsored by the Hochschulforum Digitalisierung (German Forum for Higher Education in the Digital Age, abbreviated HFD). Currently, 14 institutions are involved in this network and are listed in Table 1. Figure 1 shows the different institutions all over the German landscape.

M. E. Auer and D. May (Eds.): REV 2020, AISC 1231, pp. 143–153, 2021.
https://doi.org/10.1007/978-3-030-52575-0_11

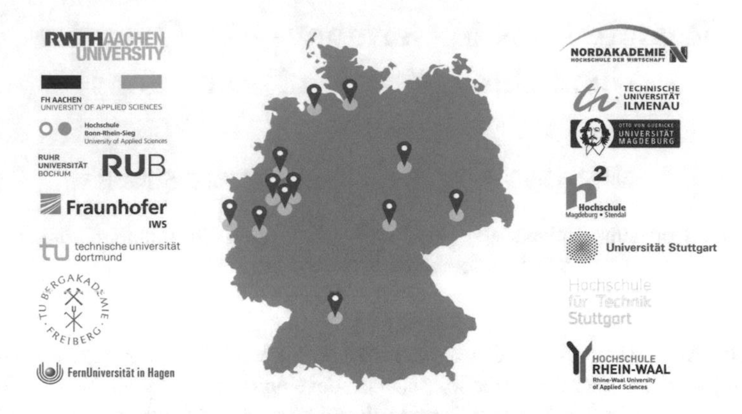

Fig. 1. Involved institutions in Germany

Table 1. Involved institutions/ chairs of the Community Working Group "Remote Labs in Germany"

Institutions	Chairs
RWTH Aachen University	Cybernetics Lab
FH Aachen University of Applied Science	Airplane electrics and electronics
Bonn-Rhine-Sieg University of Applied Sciences	Electrical Engineering, Mechanical Engineering and Technical Journalism (EMT)
Ruhr-Universität Bochum	Virtualization of Process Technology – Experimental Experiences in Engineering Education
Fraunhofer Institute	Wind Energy Systems
TU Dortmund University	1) Center of Higher Education 2) Engineering Education 3) Institute of Forming Technology and Lightweight Components
TU Bergakademie Freiberg - University of Resources	Software Technology and Robotics
Nordakademie Graduate School	Engineering Department
Ilemnau University of Technology	Integrated Communication Systems Group
Otto von Guericke University Magdeburg	
Magdeburg-Stendal University of Applied Science	
University of Stuttgart	
Hochschule für Technik Stuttgart University of Applied Sciences	Information logistics
Rhine-Waal University of Applied Science	Computer Engineering

3 Remote Labs in Germany

3.1 DigiLab4U

The project “Open Digital Lab 4 you”, short version “DigiLab4U” is a joint project of HFT Stuttgart, BIBA - Bremer Institut für Produktion und Logistik, IWM Koblenz-Landau, RTWH Aachen University and University of Palermo. The project is funded by the German Ministry for Education and Research and is leaded by HFT Stuttgart. The project started in October 2018 and is focusing on the combination of real experiment environment and remote labs in the fields of production and logistics. Due to the short project duration no remote lab was developed yet.

3.2 ELLI Lab Library

ELL Platform. The ELLI Lab Library (ELL) platform was developed in the ongoing project “ELLI – Excellent Teaching and Learning in Engineering Science” at the RWTH-Aachen University, Ruhr-Universität Bochum and TU Dortmund University to make their different remote labs available [1]. Figure 2 shows the two different approaches of Bochum and Dortmund. On the one hand, the team of Bochum developed up to nine remote labs with more than nine experiments. These experiments are in different subject areas – from thermodynamics to robot programming. On the other hand, the team of TU Dortmund University developed four different experiments in one remote lab in the field of material characterization in forming technology. Both approaches use iLab as a remote management system and NI LabVIEW to control the different remote labs or experiments.

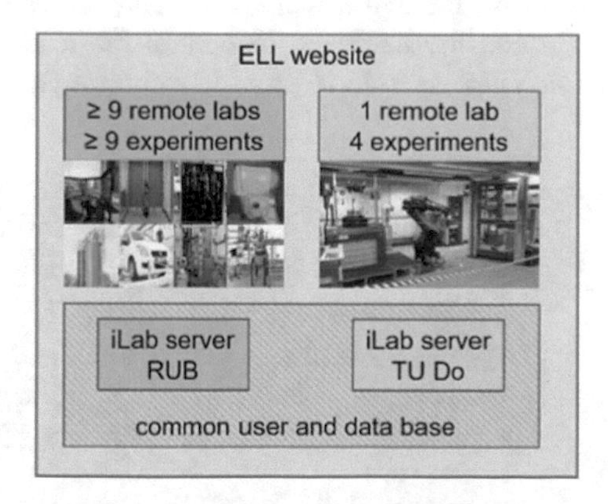

Fig. 2. Concept of the ELL platform [1]

Ruhr-Universität Bochum. At Ruhr-Universität Bochum more than nine remote labs in different engineering areas were developed. These remote labs were developed by different chairs or instates in interaction with the project team in Bochum and cover different engineering disciplines like thermodynamics, process engineering or automation engineering.

For example, a remote lab for flow measuring was developed to determine different properties of fluids. Different sensors like temperature sensors, pressure sensors were integrated. The combination of a pump and a throttle is used to generate different pressure levels during the experiment [2] (Fig. 3).

Fig. 3. Remote lab for flow measuring at Ruhr-Universität Bochum [2]

TU Dortmund University. At the moment there are remote labs at four different institutions. On the one hand, there are remote labs in the field of mechanical engineering at the Faculty of Mechanical Engineering. On the other hand, a remote lab for electronic circuits is installed at the Center of Higher Education. At the Faculty of Mechanical Engineering, more precisely at the Institute of Forming Technology and Lightweight Components a remote lab for material characterization in forming technology were developed [3]. With this remote lab different material properties, like Youngs modulus or deep drawing limit, can be determined. A robot is used to interact with two testing machines for different experiments to handle the specimens [4]. Figure 4 shows this robot (4), the two machines for material testing (1, 2), an optical measurement system (3) and the hardware to control the testing cell (5, 6). In all different experiments LabVIEW is used to provide a control panel.

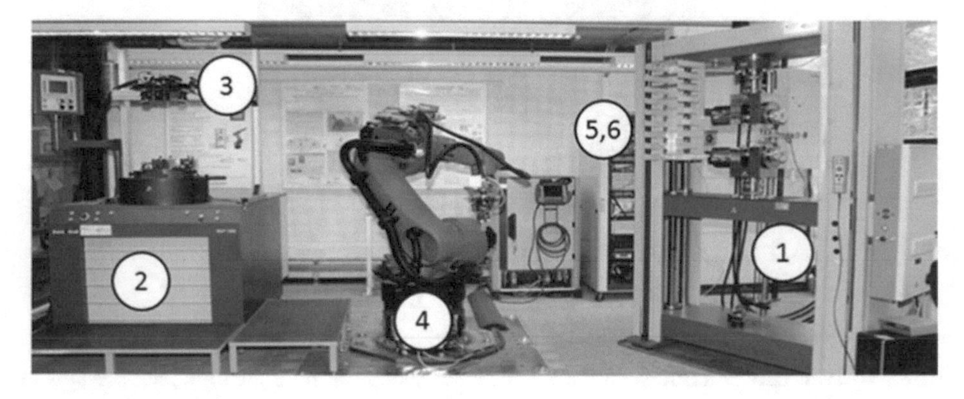

Fig. 4. Teleoperated testing cell at TU Dortmund University [3]

At the current stage of development four different experiments are implemented:

1. Uniaxial tensile test
   An uniaxial tensile test, according to DIN EN ISO 6892-1, was developed as a remote lab
2. Tensile test at high temperatures
   To determine material properties at elevated temperatures, up to 1000 °C, an induction heater was integrated.
3. Compression Test
   A compression test was automated to determine material properties for bulk metal forming (Fig. 5).

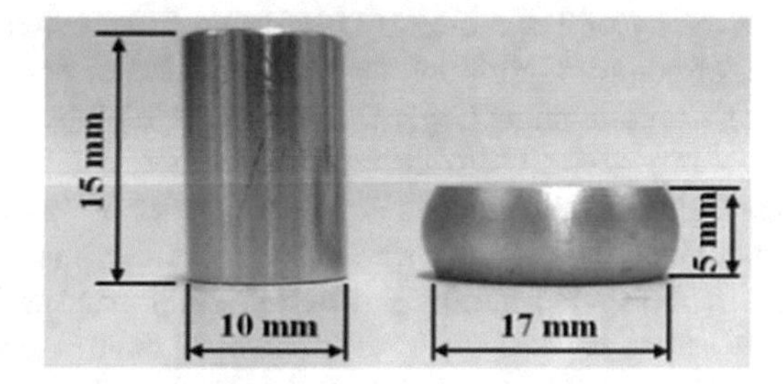

Fig. 5. Specimen before the compression test (left) and after (right) [5]

4. CuppingTest
   A remote experiment for a deep drawing cupping test was developed to determine sheet metal material properties. Users can control the punch speed and the clamping force during the process to construct a process window (Fig. 6).

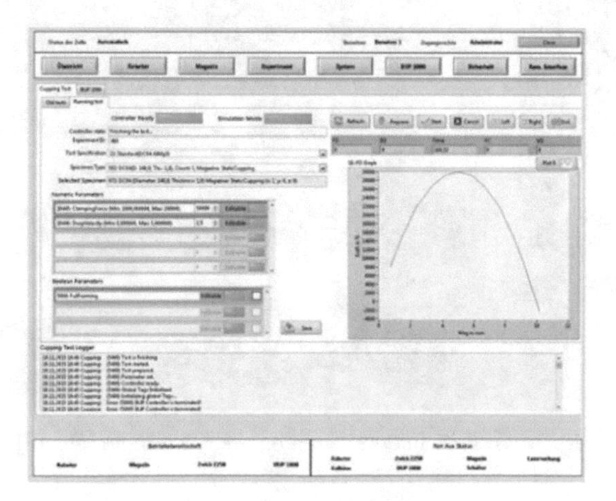

Fig. 6. User-Interface for the cupping test [6]

Next to the development of remote labs in the field of forming technology two other remote labs are currently under development. One remote lab focus on the machining process. The other remote lab is dealing with material testing.

At the Center of Higher Education a remote lab for electrical circuits was installed in 2019. A so called VISIR system (Virtual Instruments Systems In Reality) enables a 24 h access to experiments in the field of electrical and electronical engineering. The different experiments can be modified for different experience levels – from primary school pupils to undergrad students.

3.3 GOLDi - Grid of Online Laboratory Devices Ilmenau

A "Grid of Online Laboratory Devices Ilmenau – GOLDi" was developed over the past years at Ilmenau University of Technology. The focus of the developed remote laboratories is the design of control algorithms for different kind of controllers, like Finite State Machine (FSM), Programmable Logic Controller (PLC), Microcontroller or Field Programmable Gate Array (FPGA), to control electromechanical hardware models. These electromechanical hardware models are online available as a physical system (remote lab) or as a simulation system (virtual lab). Therefore, the GOLDi labs are so called hybrid online labs [7]. Furthermore, different combinations between the controllers and the hardware can be used to test the developed control algorithms. Figure 7 shows the internal architecture for remote labs using physical hardware. Two different protections units are interacting with the Bus Control Unit to secure the controller on the one hand and the hardware models on the other hand.

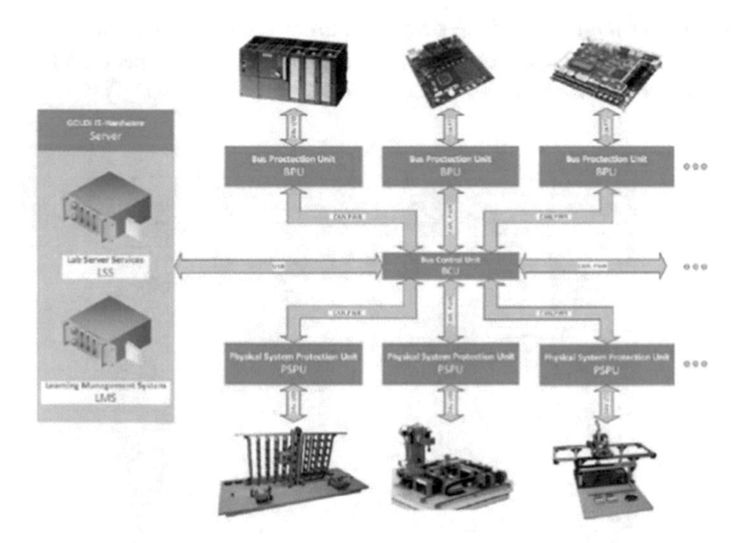

Fig. 7. GOLDi internal architecture for remote labs [8]

Ongoing developments aim on a cloud service for remote laboratories where users can interact with different independent labs [9]. Figure 8 shows the different involved entities. On one side the users, which can be located all over the world, and on the other side the current GOLDi partners from Australia, Ukraine, Georgia and Germany.

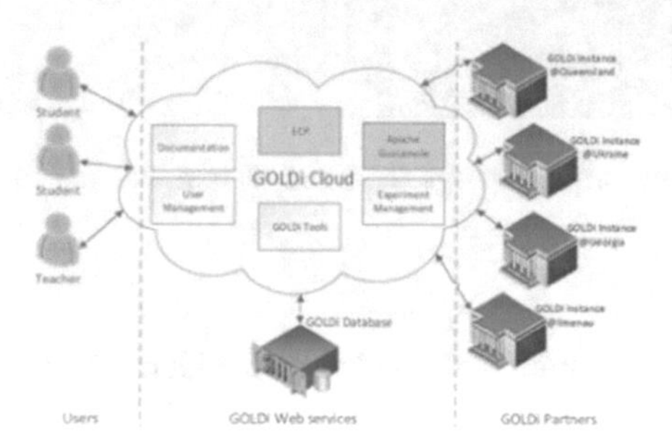

Fig. 8. GOLDi cloud architecture

3.4 FPGA Vision Remote Lab

At Bonn-Rhine-Sieg University of Applied Sciences a remote lab for low-power design with a FPGA system was developed. Figure 9 shows the setup of the developed remote lab. Students implement their design in a free of charge software using the own computer in a first step. Afterwards they upload the compiled bitfile to the server, select an image to check their design and the code is running on the FPGA. After approximate 60 s the recorded output is available for the students. The remote lab is integrated to an online course with video lectures and additional informations.

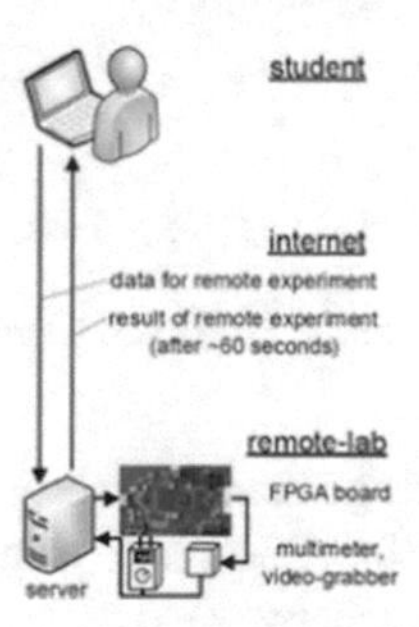

Fig. 9. Setup of the FPGA Remote-Lab [10]

Figure 10 shows the selected image of a road scene and the calculated image after line detection.

Fig. 10. Input: original image (left) and Output: calculated image (right)

3.5 Industrial eLab

An interdisciplinary research project, funded by the German Ministry of Education and Research, started in 2017 at Ottovon-Guericke University of Magdeburg and Magdeburg-Stendal University of Applied Sciences. The project Industrial eLab developed a remote lab for computer science. A platform and a software were developed to provide access to the remote lab. Users design software for an embedded system to control movement of small robots. The movement of the robots can be observed by video streams (Fig. 11).

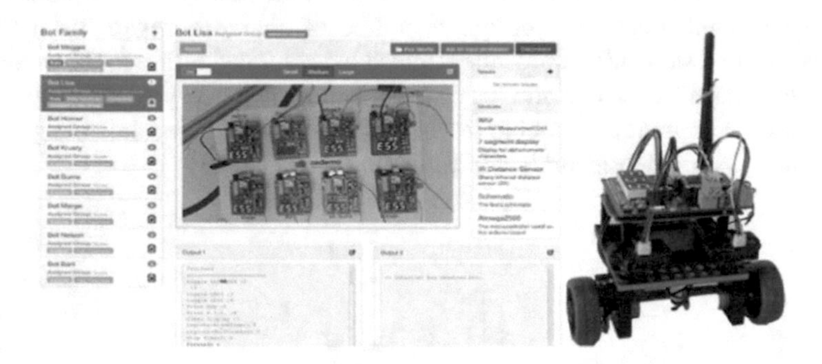

Fig. 11. Interface of the remote lab (left), Controllable robot (right) [11]

Two main benefits of the developed remote lab were documented. On the one hand, students enjoyed the freedom of time and place to access the remote lab. 74% of the access of the remote lab occurred between 6 PM in the evening and 8 AM in the morning, where regular hands-on laboratories are not available [11]. On the other hand, the teachers enjoyed that no regular maintenance jobs are needed, because the system can return to a defined setup without any interaction of human beings [11].

4 Similarities and Variations of German Remote Labs

4.1 Categories for Remote Labs

Currently, there are countless remote labs all over the world available, which allow time- and location-independent experimentation. The eco-system of remote labs in Germany is raising for more than 10 years now. These laboratories can be classified according to different characteristics. Therefore, categories have been created with which remote labs can be subdivided so that similarities and variations can be recognized.

The following categories were chosen to describe the presented remote labs:

- Subject area
  The category "Subject area" names the field of the target group of the remote lab or the subject area in which the remote lab was developed. Of course, all remote labs can also be used interdisciplinary.
- Start of development
  This is the beginning of the development of the shown remote labs.
- State of development
  The state of development states the current status about the development or further development of the remote laboratories.
- Funding
  The funding of remote labs is an important factor for the development. All described remote labs of the German eco system were developed by project funding.
- Availability
  The availability is an important or the most important factor for remote labs. In the best case, the laboratory can be reached 24 h a day, 365 days a year. At the other extreme, the laboratory can only be made available on request or on demand.
- Parallel utilization
  Remote labs in general allow a higher degree of utilization of equipment. Therefore, parallel utilization by more than one user are desirable. The investigated remote labs offers different kind of parallel utilization. On the one hand, there is just a one-to-one access possible. On the user hand, several users can have access to the remote lab at the same time. Another approach to improve availability is the parallel deployment of the same laboratories.
- Visual observability
  There are several ways in which remote laboratories can be monitored. Firstly, live video streams are often transmitted via webcam. For example, users can watch a video to see how the experiment is proceeding. Another extreme is the output of pure measured values without video or photo.
- Learning Scenario
  Just like the high number of different remote laboratories, the possible scenarios for integrating remote laboratories into teaching are just as high. Different kind of scenarios are possible with the listed remote labs. On the lowest level the remote lab is used in the lecture hall. In this case the educator controls the remote lab in

interaction with the students. On the highest level the remote lab is accessible 24/7 and the users can create their own experiments with no or just a few regulations.

- Remote Lab Management System (RLMS)
  Due to the large variety of remote laboratories and the associated requirements, different systems are used to integrate them. Therefore, some remote labs are available with standard software tools. For some remote labs special software tools were developed to aim the needed specification of its remote lab.

4.2 Outcome

Table 2 shows the listed remote labs, the different categories and the outcomes. In some cases the remote laboratories differ in other cases the results are the same. The outcomes were summarized by literature review, interviews with the developers or testing the remote labs.

Table 2. Facts about the studied remote labs

Categories	Remote Labs					
	Ruhr-Universität Bochum	TU Dortmund University		GOLDi	FPGA Vision Remote-Lab	Industrial eLab
		Faculty of Mechanical Engineering	Center of Higher Education			
Subject area	Mechanical Engineering, Electrical Engineering	Mechanical Engineering	Electrical and Electronical Engineering	Electrical Engineering	Electrical Engineering	Computer Science
Start of development	2011	2011	2019 (Installation)	2005	2016	2017
State of development	Ongoing	Ongoing	Ongoing	Ongoing	Ongoing	Ongoing
Funding	Project funding	Project funding	Project funding	Project funding	Project funding	Project funding
Availability	On demand	On demand	24 h a day	24 h a day	24 h a day	24 h a day
Parallel utilization	No	No	Yes	Yes	No	Yes (different robots)
Visual observability	Video stream, plots and control panel	Video stream, plots and control panel	Measurements (no video stream)	Video stream, plots and control panel	Input image and output image	Video stream, plots and control panel
Learning scenario	Integrated to lab courses	Integrated to lectures, tutorials, lab courses and online courses	Integrated to lectures, tutorials, lab courses and online courses	Integrated to lectures and tutorials	Integrated to lectures, tutorials, and online courses [10]	Integrated to lectures
Remote Lab Management System (RLMS)	iLab	iLab (New software under development)	WeblabDeusto/LabsLand	GOLDi	WeblabDeusto	Own developed tool

5 Summary

Different remote labs of the Community Working Group "Remote Labs in Germany" were investigated to address similarities and variations. The results reflect the wide range of remote laboratories and their respective characteristics. An important common feature of all investigated remote labs is the fact that they were all developed by project funding.

References

1. Kruse, D., et al.: Remote labs in ELLI: lab experience for every student with two different approaches. In: 2016 IEEE Global Engineering Education Conference (EDUCON) (2016)
2. Kruse, D.: Virtualisierung eines verfahrenstechnischen Prozesses als remote Labor für die Aus- und Weiterbildung in Industrie 4.0. Ruhr-Universität Bochum (2017)
3. Ortelt, T.R., et al.: Development of a tele-operative testing cell as a remote lab for material characterization. In: Proceedings of 2014 International Conference on Interactive Collaborative Learning, ICL 2014 (2014)
4. Sadiki, A., et al.: The challenge of specimen handling in remote laboratories for Engineering Education. In: Proceedings of 2015 12th International Conference on Remote Engineering and Virtual Instrumentation, REV 2015 (2015)
5. Selvaggio, A., et al.: Development of a remote compression test lab for engineering education. In: 16th International Conference on Remote Engineering and Virtual Instrumentation (REV 2019), Bengaluru, India (2019)
6. Selvaggio, A., et al.: Development of a cupping test in remote laboratories for engineering education. In: Frerich, S., et al. (eds.) Engineering Education 4.0: Excellent Teaching and Learning in Engineering Sciences, pp. 465–476. Springer, Cham (2016)
7. Henke, K., et al.: Fields of applications for hybrid online labs. In: 2013 10th International Conference on Remote Engineering and Virtual Instrumentation (REV) (2013)
8. Henke, K., et al.: GOLDi — grid of online lab devices Ilmenau: demonstration of online experimentation. In: 2015 3rd Experiment International Conference (exp.at 2015) (2015)
9. Henke, K., et al.: GOLDi-Lab as a Service – Next Step of Evolution. Springer, Cham (2019)
10. Winzker, M., et al.: Teaching across the ocean with video lectures and remote-lab. In: 2018 IEEE World Engineering Education Conference (EDUNINE) (2018)
11. Zug, S., et al.: Poster - Industrial eLAB, in BMBF-Fachtagung "Hochschule im digitalen Zeitalter", Berlin, Germany (2017)

An Implementation of Microservices Based Architecture for Remote Laboratories

Mohammed Moussa(✉), Abdelhalim Benachenhou, Smail Belghit, Abderrahmane Adda Benattia, and Abderrahmane Boumehdi

Laboratoire Electromagnétisme et Optique Guidée, Université Abdelhamid Ibn Badis Mostaganem, Mostaganem, Algérie
{mohamed.moussa, abdelhalim.benachenhou, abderrahmane.addabenattia, abderrahmane.boumehdi}@univ-mosta.dz, liamssi.dz@gmail.com

Abstract. In the fields of science, technology, engineering and mathematics, remote laboratories offer a new opportunity to increase the number of hours devoted to scientific experiments. Universities facing the growing number of students need to pool their equipment. This article describes an implementation, based on microservices, of a remote lab management and federation solution. This solution uses the ThingsBoard platform interfaced by a REST API. The feasibility has been demonstrated by the deployment of eight instances of practical work used by 120 students.

Keywords: Remote laboratories · Federation · Microservices · ThingsBoard · REST

1 Introduction

According to Feisel and Rosa, laboratories are essential elements of any engineering training program [1]. Until now, the dominant approach has been to use dedicated hands-on labs or simulation-based tools [2]. Universities with a growing number of undergraduate students face the cost of acquiring laboratory equipment. These facilities are not always profitable because they are only accessible during business hours. That is why Web-based remote laboratories (RL), allowing students access from anywhere at any time, are attracting more interest. To maximize efficiency, solutions are developed to allow the sharing of expensive equipment between institutions.

The development of RLs presents many technical challenges, as experiments differ from one RL to another, each having very different characteristics (in terms of components, complexity, types of measuring instruments, etc.).

Several studies have proposed the design of RLs. Tawfik and al. [3] propose a service-oriented architecture to solve the problem of interoperability and standardization of RLs. In [4], authors proposes an open source framework based on (Flask for Python) for the simplification of the development of the RL.

The aim of this study is to reduce the complexity of the management of electronics RL and to allow their federation through a microservice architecture. This paper is

M. E. Auer and D. May (Eds.): REV 2020, AISC 1231, pp. 154–161, 2021.
https://doi.org/10.1007/978-3-030-52575-0_12

structured as follows: Sect. 2 presents the methodological framework for achieving the objectives of this work. Section 3 presents the proposed architecture, while Sect. 4 describes the validation process by a case study. Finally, Sect. 5 presents the conclusions and future work.

2 Process for RL Design

The RL design process consists of a series of steps that teachers take to build hands-on labs fulfilling one or more educational objectives and accessible from anywhere. This process is different from the traditional design of a practical training session. To describe this process, we take the example of a passive filters practical work (PW) that can be generalized to any other PW.

The first step is to design a circuit allowing all the possible combinations during a PW session. Switches allow selection of a particular configuration. Figure 1 shows an exemplary circuit for selecting the R-R, R-C, C-R, L-R and R-L sub-circuits.

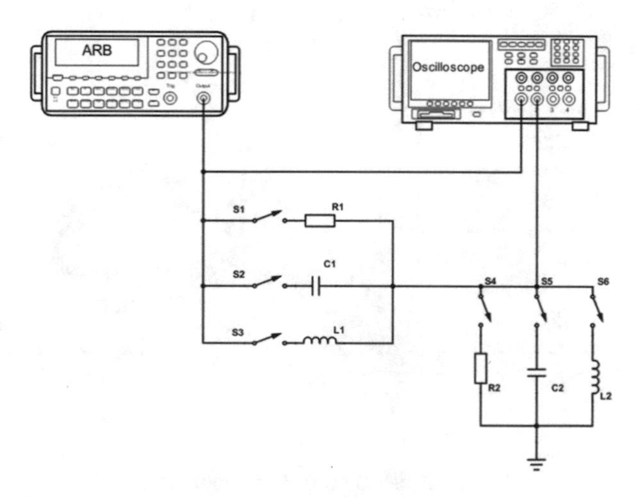

Fig. 1. Example of a multi-configuration passive filter circuit.

The second step is to develop the printed circuits by separating the switches from the other components. The first printed circuit including the switches is called switching board (SB). The second including the other elements is called the PW board. Putting side by side the two boards allows reconstituting the global circuit. The SB described in [5] is reusable. When designing the PW it is necessary to take into account the constraints of the SB: number and nature of the switches (SPST or SPDT), number of components such as digital potentiometers requiring an SPI (Serial Peripheral Interface) link etc.

3 System Architecture

This architecture called Most@lab extends a previous work described in [6]. The heart of this architecture, as given in Fig. 2, is based on several web-services. The architecture is modular to allow reuse of services in a variety of scenarios and practical works. This architecture is based on open sources technologies, and standard such as RESTful and JSON. Because web-services promise better isolation, this does not force developers to use the same programming language for the different *Application Programming Interface* (API).

The Gateway API is the intermediate component between the client interface and internal services (They are related to the GPIOs and the SPI bus of the single board computer such as a Raspberry Pi): it is first responsible for sending to the registry the request for client (1) and return the list of available RLs (2). Then, the Gateway API is responsible for passing on to the client the full description of the components involved in the hands-on experience.

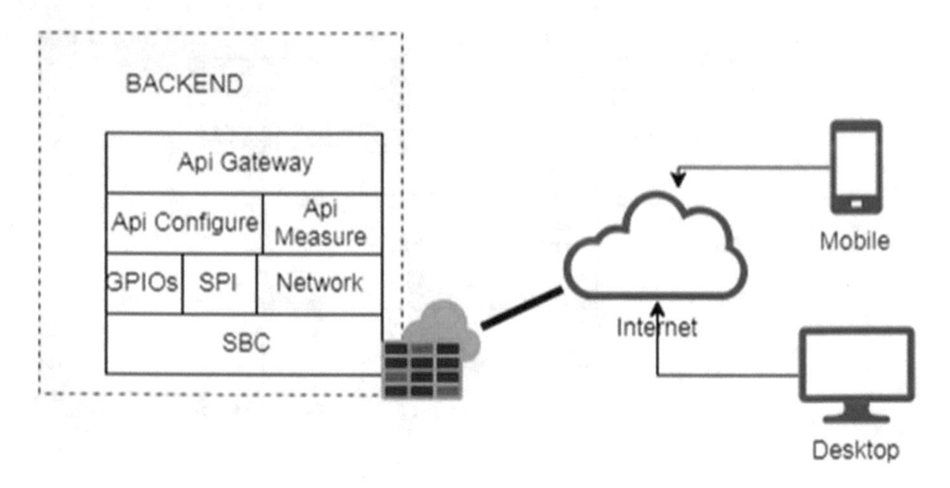

Fig. 2. Most@Lab system architecture

3.1 API Configure

Upon receiving a REST GET request from the client, this microservice interacts with the General Purpose Input Output (GPIO) of the Single Board Computer (SBC) that controls the SB. It makes it possible to select a particular configuration of the circuit by acting on the relays and by setting the values of the digital potentiometers if the circuit contains them.

3.2 API Measure

The API microservice Measure controls signal shaping equipments like Digital Power Supply and Arbitary Waveform Generator, and brings back measurements from instruments such as Digital Multimeter and oscilloscope.

3.3 Multi-lab Architecture

Figure 3 shows a remote lab federation model based on connected object concepts ThingsBoard [7].

This federation model has the following advantages:

1) The integration of a ThingsBoard platform for RL management with minimal administrator intervention;
2) The implementation of a Resty intermediate service to collect only the information necessary for the creation of user interfaces (LabsUI).
3) The development of future services.

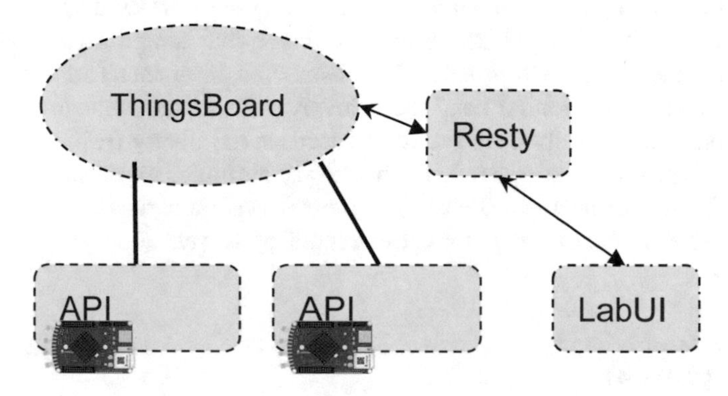

Fig. 3. Most@Lab multi-lab architecture.

3.4 ThingsBoard for Laboratory Management

Most@lab architecture integrates with the ThingsBoard platform. The ThingsBoard platform is defined with the following rules:

The first part of the regulation concerns the definition attributes of laboratories under ThingsBoard:

R1. Under ThingsBoard, each instance of a lab is represented as one device;
R2. Any device representing an instance of a lab must be of the “LAB” type;
R3. Each instance of a lab is identified by its ThingsBoard ID;
R4. Each instance of a lab has a laboratory type that is defined by an attribute client-side named “labTypeId” which must be published by the lab itself;
R5. Instances of the same type of laboratory have the same “labTypeId”;

R6. Each lab must publish its remote (web) address as a client-side attribute named "URL";
R7. Each instance of a lab connects to ThingsBoard with a JavaScript Web Token (JWT);
R8. Each client must publish a JWT named "Session Creation Token";
R9. Each lab must maintain its activity status (active or not) under ThingsBoard;
R10. To achieve the goal R9, a lab instance periodically updates its status (every minute) by updating the status attribute.

The second part of the rules concerns attributes of federation:

- A lab provider (university for example) is represented as a Customer on ThingsBoard;
- An administrative agent is represented as a user ThingsBoard.

3.5 API Resty

Resty is a REST API developed on NodeJS. Its main role is to mediate between ThingsBoard and the various micro-services of the system. Resty does not really offer new capabilities, it is simply an abstraction layer that simplifies features already available in ThingsBoard and in labs. This abstraction layer therefore depends on the ThingsBoard administration API and the gateway APIs of the laboratories.

Resty sends periodically a GET request based on the library (restify-clients) to the Thingsboard server and extracts only the necessary attributes to prepare them for other service i.e. Labs user interface (LabsUi) or the reservation service.

Below, an example of a data format extracted by Resty.

```
lab = {
id: entity.id.id,
name: entity.name,
type: entity.type,
labTypeId:findAttributes(obj, "labTypeId"),
labType: findAttributes(obj, "labType"),
tenantId: entity.tenantId.id,
createdTime: entity.createdTime,
active: findAttributes(obj, "active"),
state:findAttributes(obj, "state"),
url: findAttributes(obj, "url"),
img: findAttributes(obj, "img"),
token:findAttributes(obj, "sessionCreationToken")
};
```

3.6 Scheduling Microservice

The scheduling microservice (SM) is an extension of the solution proposed in [6]. The SM provides exclusive access for an instance of a RL. It works according to the following steps:

1) The client sends a request to the microservice Resty to book a given lab.
2) The Rest microservice, which represents the entry point of Most@lab architecture, checks if an instance of this lab is available.
3) Resty provides the lab scheduler with the list of connected instances.
4) The scheduler selects a free instance, generates the GUI using the metadata provided by Resty and declares this instance as busy.
5) In the case where all the instances of a requested lab are busy, the scheduler returns the estimated waiting time for the release of an instance.

4 Use Case: Introduction to the Use of the Oscilloscope

The Most@lab solution was tested with the PW "introduction to the Redpitaya STEMLab oscilloscope" [8]. Redpitaya STEMLab is a two channels waveform generator and a two channels oscilloscope.

The objective of this training is to be able to shape a signal (to select a waveform, to configure the amplitude, the frequency the phase, ….), to characterize by the oscilloscope a signal (measurement amplitude, rms value, phase shift,). The student can perform an experiment by accessing the oscilloscope as shown by Fig. 4.

Step 1: A PW provider registers on the Most@lab platform which assigns him an id_Lab_Provider identifier.
Step 2: The lab provider declares on Most@lab a new PW. The platform assigns him an Id_PW identifier.
Step 3: The lab provider declares on Most @ lab an instance of the PW. Among the information declared is the time required to perform the PW. The platform assigns him an Id_PW_inst identifier.
Step 3 is repeated for the number of instances implemented. In this study eight instances of the PW "Introduction to the Redpitaya STEMLab Oscilloscope" were installed to be used by 120 students.
Step 4: On each RedPitaya board, a "connect" function has been implemented according to the MQTT protocol. This function publishes the device attributes on the Thingsboard platform. The device attribute is the concatenation of the attributes mentioned in steps 1 to 3.
Step 5: The graphical user interface of the PW requests the scheduler access to the PW having the id_PW. The scheduler assigns a free instance of the PW for the nominal duration.

In case an instance is no longer available, the Thingsboard is automatically updated by MQTT protocol. The scheduler removes this instance from the available PW list.

In cases where all instances of a PW are in use. The new request receives from the scheduler the estimated waiting time.

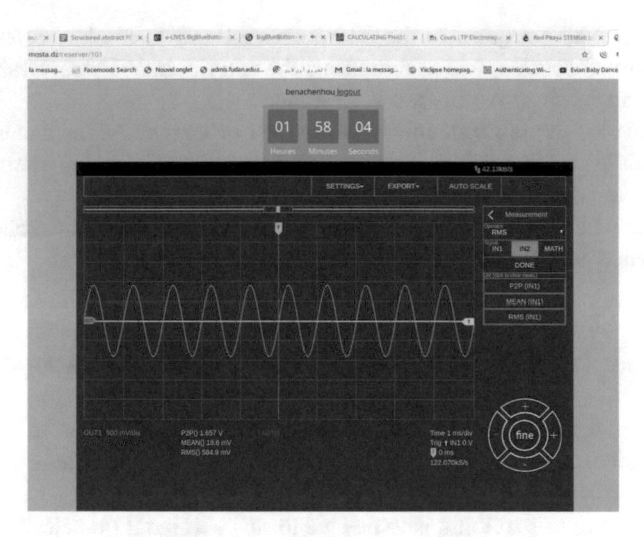

Fig. 4. Practical work user interface.

5 Conclusion and Future Work

This article presents a new method for managing and federating remote labs using the concept of microservices. This architecture uses the ThingsBoard platform that interfaces with a REST API.

The feasibility was tested by the implementation of PW comprising 8 instances and used by 120 students.

Our future work will focus on the implementation of several microservices for the monitoring and exploitation of the logs generated by the Resty and Thingsboard platform.

Acknowledgment. This work is partially supported by the ReLaTraPE project funded by the Agence Universitaire de la Francophonie (AUF) and by the e-LIVES project funded by the European Commission, under agreement number 2017-2891/001-001. The European Commission support for the production of this publication does not constitute an endorsement of the contents which reflects the views only of the authors, and the Commission cannot be held responsible for any use which may be made of the information contained therein.

References

1. Feisel, L.D., Rosa, A.J.: The role of the laboratory in undergraduate engineering education. J. Eng. Educ. **94**(1), 121–130 (2005)
2. Ma, J., Nickerson, J.V.: Hands-on, simulated, and remote laboratories: a comparative literature review. ACM Comput. Surv. (CSUR) **38**(3), 7 (2006)
3. Tawfik, M., Salzmann, C., Gillet, D., Lowe, D., Saliah-Hassane, H., Sancristobal, E., Castro, M.: Laboratory as a Service (LaaS): a novel paradigm for developing and implementing modular remote laboratories. Int. J. Online Eng. **10**(4), 13–21 (2014)

4. Orduña, P., Rodriguez-Gil, L., Angulo, I., Hernandez, U., Villar, A., Garcia-Zubia, J.: WebLabLib: new approach for creating remote laboratories. In: Auer, M.E., Ram B., K. (eds.) REV2019 2019. LNNS, vol. 80, pp. 477–488. Springer, Cham (2020). https://doi.org/10.1007/978-3-030-23162-0_43
5. Benattia, A.A., Moussa, M., Benachenhou, A., Mebrouka, A.: Design of a low cost switching board enabling a reconfigurable remote experiment. iJOE **15**(12), 33 (2019)
6. Moussa, M., Benachenhou, A., Adda Benatia, A.: Work-in-Progress: a smart scheduling system for shared interactive remote laboratories. In: Auer, M.E., Guralnick, D., Simonics, I. (eds.) ICL 2017. AISC, vol. 716, pp. 601–606. Springer, Cham (2018). https://doi.org/10.1007/978-3-319-73204-6_65
7. https://thingsboard.io/docs/
8. https://www.redpitaya.com/

Experience with the VISIR Remote Laboratory at the Universidad Estatal a Distancia (UNED)

Carlos Arguedas-Matarrita[1(✉)], Marco Conejo-Villalobos[1], Fernando Ureña Elizondo[1], Oscar Barahona-Aguilar[1], Pablo Orduña[2,3], Luis Rodriguez-Gil[3], Unai Hernandez-Jayo[2], and Javier García-Zubia[2]

[1] Universidad Estatal a Distancia, San José, Costa Rica
carguedas@uned.ac.cr
[2] Universidad de Deusto, Bilbao, Spain
[3] LabsLand, Bilbao, Spain
pablo@labsland.com

Abstract. The Universidad Estatal a Distancia (UNED) in Costa Rica is the public distance-education university in Costa Rica. It has been regularly using different types of remote laboratories since 2017, and is in the process of deploying new remote laboratories, as well as providing high quality simulations for improving the experimentation process in distance education; in the particular fields of Physics and Engineering. LabsLand is the global network of educational real laboratories available online. In 2018, UNED and LabsLand deployed the VISIR electronics remote laboratory in UNED. Since then, at the time of this writing, 167 students have had 979 laboratory sessions, including both in class and in multiple workshops with school teachers and international events with other researchers and professors.

Keywords: Remote laboratories · Electronics · Online education · VISIR · LabsLand

1 Introduction

Education is undergoing major changes influenced by the advancement of Information and Communication Technologies (ICT), which have generated educational resources in all areas of knowledge. In the specific field of physics education, one of these advances is the Remote Laboratories (RL). They allow real experiments to be carried out through the Internet, they are tools that support face-to-face teaching [1], and in turn constitute an ideal complement for distance education [2], because displacement to a physical space is not required and can be accessed at the time the student considers appropriate. Distance Education represents an effort to break spatio-temporal barriers or pedagogical limitations more typical of face-to-face education and requires resources that provide that flexibility.

An essential aspect for education in a current context is educational flexibility, understood as the possibilities of re-organization of education, depending on the

M. E. Auer and D. May (Eds.): REV 2020, AISC 1231, pp. 162–170, 2021.
https://doi.org/10.1007/978-3-030-52575-0_13

various interests or needs. Along these lines, the characteristics of the RLs are presented as educational resources to strengthen experimental work in this educational modality.

1.1 Context

The Universidad Estatal a Distancia of Costa Rica (UNED), is the pioneer institution of higher education in this modality, was created in 1977, it was conceived "as a project to innovate in higher education" [3, p. 182] This higher education center comes to support the work carried out by the other public universities in Costa Rica, with UNED having a particularity as it is conceived as "an institution of higher education specialized in teaching through the media social" [4, p. 33], a concept that has characterized it from its beginnings to the present day.

This higher education center comes to support the work carried out by the other public universities in Costa Rica, with UNED having a particularity as it is conceived as a higher education institution specialized in teaching through social media, a concept that has characterized it from its beginnings to the present day.

Currently, UNED has 37 university centers, distributed throughout the country (see Fig. 1), which makes it the university with the greatest coverage in Costa Rica, and using technology as a means to break with the space-time barrier.

Fig. 1. University centers of the UNED.

1.2 VISIR the Remote Laboratory

The VISIR remote laboratory is an Electronics remote laboratory well established in the literature and developed by the Blekinge Institute of Technology (BTH) in Sweden, which allows students to create circuits and interact with instruments (multimeter, function generator, oscilloscope and power supply) in a real-time way and using a very efficient mechanism: students have access to the real hardware for a fraction of a second, allowing the system to be used with dozens of concurrent students. It is the remote laboratory that has been deployed in more universities in the world.

In the case of the VISIR remote laboratory at UNED [6], the deployment is described in Fig. 2:

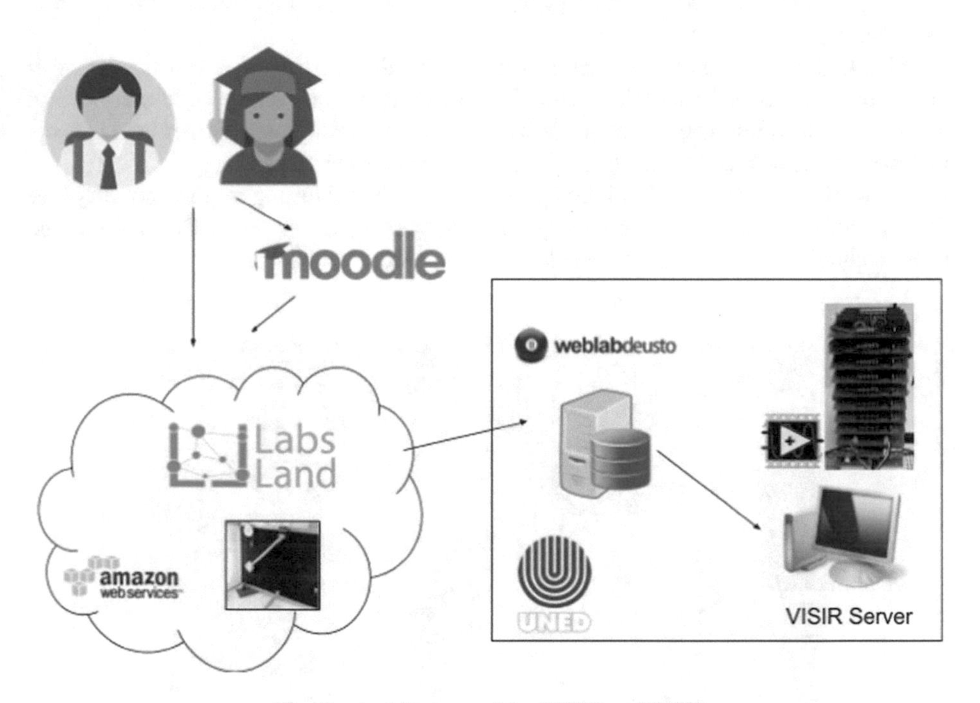

Fig. 2. Architecture of the VISIR at UNED.

1. The VISIR server is a Windows machine that contains the VISIR software (Measurement Server -C++ native application- and Equipment Server -LabVIEW application-, as described in [5]) and it is directly connected to the hardware: both a National Instruments PXI equipment with the instruments (Oscilloscope, Function Generator, Multimeter, DC Power Supply), as well as a the switching matrix. This machine is in the same room as the equipment, as it is directly connected to it.
2. The WebLab-Deusto server is in a Linux server, hosted by the IT Services of UNED. This software is in charge of the low-level management of the remote laboratory: the priority queue of the laboratory, automatic checks in the local

network to confirm that the laboratory is working and notification management to the administrators when the equipment is not available due to some planned or unplanned failure.

3. Outside UNED, in Amázon Web Services (AWS), the LabsLand portal [7] manages the authentication, authorization, group management of UNED for using the VISIR laboratory (Fig. 3), as well as other laboratories that students have access, such as the Inclined Plane and Pendulum. UNED uses Moodle as a Learning Management System, and the LabsLand portal, through the UNED portal[1], integrates Moodle so students do not need to register in LabsLand but they can go directly to the laboratories. However, optionally students (especially users who are not enrolled in UNED as students: e.g., participants in workshops) can also use LabsLand directly without going through Moodle. Also in Amazon Web Services, LabsLand has remote laboratories such as the Pendulum laboratory, used in UNED.

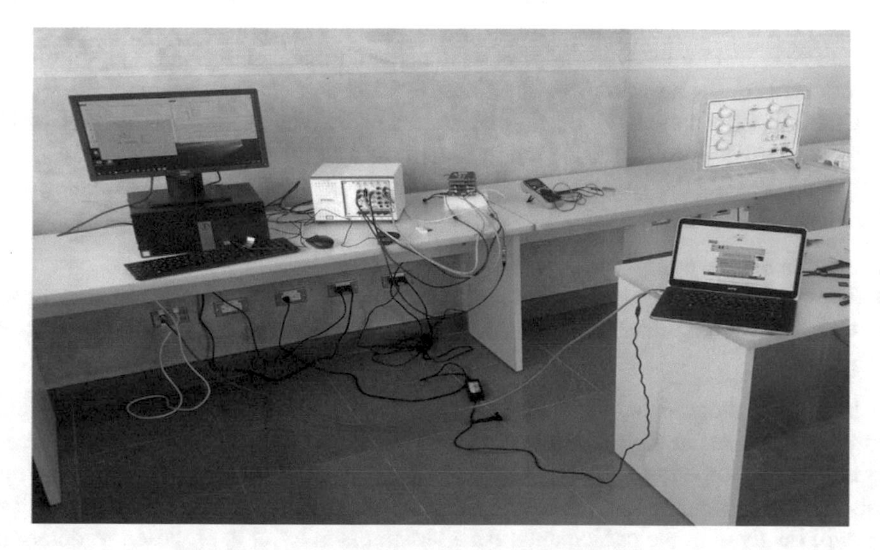

Fig. 3. VISIR system deployed in UNED (Costa Rica).

2 Methodology

This work refers to two workshops offered in Costa Rica and Argentina, which were coordinated by members of the Remote Experimentation Laboratory team of the UNED.

The workshops sought to meet two objectives: a) Familiarize teachers with the use of the VISIR RL, b) Use the VISIR by performing laboratory practices to experimentally verify Ohm's Law, c) Use other remote laboratories available at LabsLand.

[1] https://labs.land/unedcr/.

The design of each workshop included, in chronological order: diagnostic questions that the participants had to answer, they focused on indicating the differences between a series circuit and a parallel circuit, writing Ohm's law indicating each of the variables involved and finally the resolution of an exercise that involves a resistive electrical circuit. Then the facilitator gave a brief review of concepts and laws related to series, parallel and combined resistive circuits (Ohm's Law, equivalent resistance); a brief introduction to the use of the Protoboard, explaining the way in which the serial connections are made and then the equivalent resistance was measured with a multimeter in several arrangements of proposed resistors.

Having made that first part of the traditional laboratory, the use of RL VISIR was introduced: it was clearly explained that it is not a simulation, but a real experiment that is operated remotely, through the Internet. The operation of each of the components of interest of the interface (component area, assembly area and instrument area) was explained. Special emphasis was placed on the correct use of the multimeter explaining each of its functions; with the detailed detail of the most important functions for the planned activities for the workshop, as well as the correct connection of the multimeter in the VISIR for the measurement of difference of electric potential and electric current.

Once developed the basic disciplinary concepts related to the contents that will be addressed, and after having made the introduction for the use of the different resources offered by VISIR, the participants were admitted in the RL environment and were given the following instructions to work in the RL:

1) Access the page https://labs.land/unedcr/register.
2) If they were not previously registered on the LabsLand platform, register for it.
3) Once registered, enter the given code and access the laboratory assigned to the group.

Following these steps, the following activities were carried out: resistance measurement exercises in various arrangements, calculation of the equivalent resistance of the circuit and experimental determination of the equivalent resistance value using a multimeter, determination of the experimental error and recording of results in the form that will be given to the participants. As a closing activity, it is proposed to determine the potential difference and the current through each of the resistors arranged in different circuits by calculation and measure the values of these quantities with a multimeter.

2.1 Data Collection

To assess the participants' prior knowledge about RL and their perception of RL VISIR, an opinion survey was developed as a data collection instrument that was applied in a self-administered way with the Google Forms tool.

3 Results

Results were found referring to the use of remote laboratories in the UNED and data of the instruments applied to the assistants of the two workshops on the assessment of VISIR and other RLs.

3.1 Laboratories Used in UNED

The inclined plane laboratory has been used by 93 students 294 times in total in 2018 and 2019 (Fig. 4).

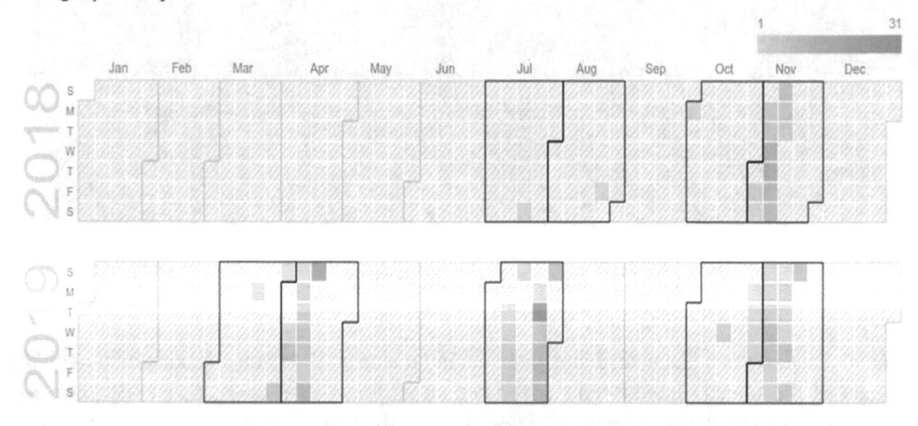

Fig. 4. Uses inclined plane.

The pendulum laboratory has been used 233 times by 38 students, especially in July and August (Fig. 5).

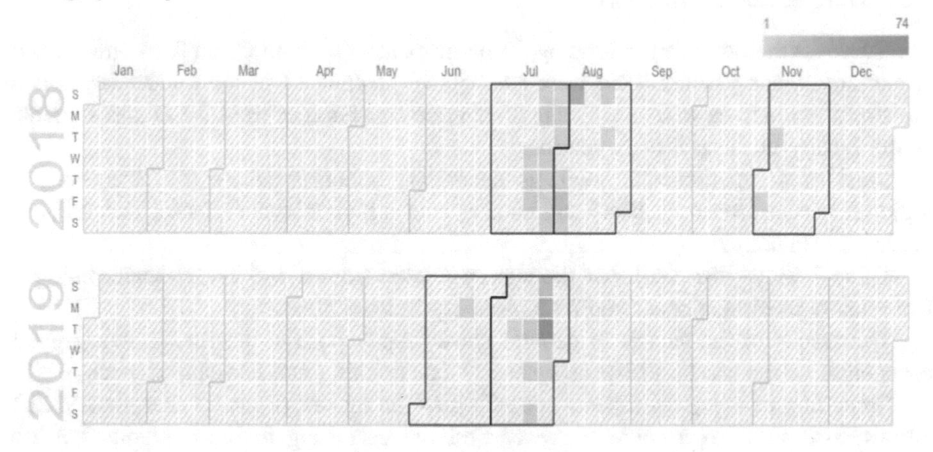

Fig. 5. Uses pendulum.

The most used laboratory is VISIR, used by 167 students at UNED (Fig. 6).

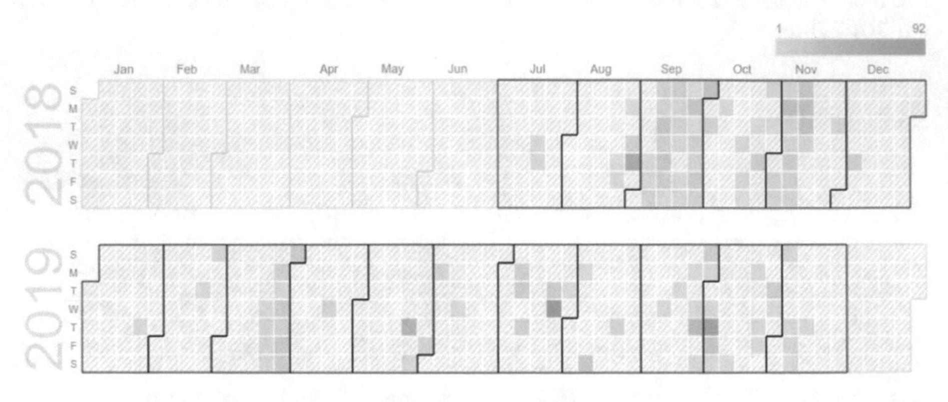

Fig. 6. Uses of VISIR.

In both the year 2018 and the year 2019 the use of remote laboratories by the students of the UNED have been carried out mostly in the evening and early morning hours, an aspect that was expected since due to the nature of the distance education many of the students work during the day and must use part of the night to carry out their learning activities, in this sense this type of laboratory will strengthen the experimental component in this modality [8], benefiting the student since it allows him to choose the moment of entering and carrying out the practice assigned to him, an essential aspect in distance education.

3.2 Data of the Instruments

The survey was completed in total by 25 participants, (16 from Costa Rica and 9 from Argentina), among whom there were teachers in practice 15 of the middle level as a university student, as well as students of science education of the UNED and students of the faculty in Physics of Argentina.

Most of the participants work as secondary school teachers (57,70%), one a regional science advisor (3,80%) and another one at higher level (7,70%), also 26,90% university students.

Table 1 shows the results of the survey, stated from a 1–5 Likert scale, being 1 Completely in disagreement and 5 Completely in agreement.

The results of the survey in general show highly positive results, showing great acceptance of the use of VISIR as resources for teaching and learning physics.

Regarding pedagogical aspects according to the results, it can be seen that VISIR allows laboratory experiments to be carried out according to the contents that are developed in the subject of electrical and electronic circuits and the results that are congruent with the conceptual body of physics.

Table 1. Likert scale averages

Question: Usability	Min.	Max.	Average
1. Does the simple result use the RL?	3	5	4,71
2. Was the interface friendly?	3	5	4,50
3. Would you be interested in using it in your institution?	3	5	4,71
4. Would you be interested in recommending it in your school to other teachers?	3	5	4,71
Question: Pedagogical aspects			
5. It allows real experimentation without having institutions that do not have laboratory enclosures.	3	5	4,71
6. 24 h availability	3	5	4,71
7. RL promote autonomous and collaborative work.	3	5	4,54
8. Promote technological skills.	3	5	4,63
9. Lets test the experiments several times if the results seem strange.	3	5	4,63
10. Was the experiment used appropriate to address the proposed experimental content?	3	5	4,67

4 Conclusions

According to the results mentioned before, the general assessment on the use of VISIR in this workshop was highly positive, both as regards the participants' point of view about the potentiality of the resource as about their interests and/or motivations to use it in their lessons in the future.

This article will provide a clear overview of the use of the VISIR remote laboratory in the Universidad Estatal a Distancia of Costa Rica; used both by UNED students and in workshops by other researchers and professors, in which the use of RLs was widely accepted by users.

References

1. Navarro, E., Tizón, J.M.: Docencia presencial y laboratorio remoto: una unión idónea para lãs prácticas de motores alternativos. Model. Sci. Educ. Learn. **9**(1), 129–138 (2016)
2. García, F., Macho, A., San Cristóbal, E., Rodríguez, M., Díaz, G., Castro, M.: Remote laboratories for electronics and new steps in learning process integration. In: REV 2016 13th International Conference on Remote Engineering and Virtual Instrumentation, pp. 106–111. Madrid, España (2016)
3. Dengo, E.: Educación Costarricense. EUNED, San José (2004)
4. Molina, S.: La joven Benemérita. Universidad Estatal a Distancia. Editorial Universidad Estatal a Distancia, San José (2008)
5. Gustavsson, I., et al.: The VISIR project–an open source software initiative for distributed online laboratories. In: REV 2007 (2007)

6. Arguedas-Matarrita, C., Concari, S.B., Rodriguez-Gil, L., Orduña, P., Elizondo, F.U., Hernandez-Jayo, U., Carlos, L.M., Bento da Silva, J., Marchisio, S.T., Conejo-Villalobos, M., García-Zubía, Alves, J.B.M.: Remote experimentation in the teaching of physics in Costa Rica: first steps. In: 2019 5th Experiment International Conference (exp. at 2019), pp. 208–212. IEEE (2019)
7. Orduña, P., Rodríguez-Gil, L., García-Zubia, J., Angulo, I., Hernández, U., Azcuenaga, E.: Labsland: a sharing economy platform to promote educational remote laboratories maintainability, sustainability and adoption. In: 2016 IEEE Frontiers in Education Conference (FIE), pp. 1–6. IEEE (2016)
8. Arguedas, C., Ureña, F., Conejo, M.: Laboratorios remotos: Herramientas para fomentar el aprendizaje experimental de la Física en educación a distancia. Lat. Am. J. Phys. Educ. **10**(3), 3309–3311 (2016)

Mobile Arduino Robot Programming Using a Remote Laboratory in UNAD: Pedagogic and Technical Aspects

Experience Using a Remote Mobile Robotics Laboratory at UNAD

Paola Andrea Buitrago[1](✉), Raúl Camacho[1], Harold Esneider Pérez[1], Oscar Jaramillo[1], Aitor Villar-Martinez[2,3], Luis Rodríguez-Gil[3], and Pablo Orduna[2,3]

[1] Universidad Nacional Abierta y A Distancia - UNAD, Dosquebradas, Colombia
{paola.buitrago, raul.camacho, harold.perez, oscar.jaramillo}@unad.edu.co
[2] Universidad de Deusto, Bilbao, Spain
{aitor.v, pablo.orduna}@deusto.es
[3] LabsLand, Bilbao, Spain
luis@labsland.com

Abstract. The National Open and Distance University - UNAD, is an autonomous university entity of the national order, with special regime. It has 8 zones and 64 education centers. By using the first remote mobile robotics laboratory based on the Arduino platform and incorporating the LabsLand Arduino Robot, an alternative that in addition to reducing costs allows coverage for students who are located in distant areas and who have difficulty accessing laboratory practices. It is planned to replace the use of simulators or loose microcontrollers, without any robotic structure, by incorporating this new technological tool. The educational approach it is due to the need to incorporate "e-learning" activities and laboratory practices in the course of microprocessors and microcontrollers during the month of October and the first week of November and measure the technological and pedagogical impact on open and distance education. For this purpose, 110 students participated and from those, 78 students conducted the survey subsequently the results evaluated in the activities are compared with the rest of the students who carry out traditional activities and a matrix of indicators is obtained, that allows measuring the educational impact on the use of technological tools compared to traditional laboratory activities and practices in the course of microprocessors and microcontrollers.

Keywords: Remote laboratories · Arduino · Microcontrollers · Distance education · LabsLand · UNAD · CEAD · CCAV · CERES y UDR

M. E. Auer and D. May (Eds.): REV 2020, AISC 1231, pp. 171–183, 2021.
https://doi.org/10.1007/978-3-030-52575-0_14

1 Introduction

The Universidad Nacional Abierta y a Distancia – UNAD, is equipped with a school of Basic Sciences, Technological Studies in Engineering ECBTI, where it is found subscribed to the Cadena de Formations de Electronica, Telecomunicaciones Y Redes – ETR, conformed by one specialization program [1], two Engineering programs [2] and two Technological programs [2]. For the present study realized the students of Electrical Engineering, program offered through distance learning and Telecommunications Engineering, program offered through our online method, were taken into account; these programs are founded in the educational model of UNAD supported through E-learning, that focuses its actions in the process of learning and highly recognizes it in three forms (self-directed learning, meaningful and collaborative) which mobilizes in the students, as an active protagonist, the appropriation and comprehension of the reality.

It is important to keep in mind that the designed course contents of the program of UNAD, are conceived as a permanent space of curricular innovation, educational and didactic. In them are included formative sessions and educational practices and learning, which taken together, permit the students to advance in the construction of their academic process in each one of the disciplines of the determined formative program.

For UNAD the Practical Component is the opportunity to construct the learning process from the perspective experience, interaction and dialog in the context of that area of specific knowledge. This construction is nurtured through the use of the technology of the information and the communication as well as the utilization of the physical and remote space [1].

The UNAD uses the following scenarios to develop practical components: with technological support, physical and remote; where the student comprehends the strategies for the academic accompaniment of the formative process that include examples, demonstrations or construction of the knowledge in real or simulated situations [1].

It is precisely on this scenario that remote laboratories focus the investigation realized between the UNAD and the LabsLand global network; where the first remote laboratory has implemented the Programming of Mobile Robot Arduino and that during the development of the article the impact that it has had according to the trial conducted with the participation of the students from the mentioned programs, will be evidenced; analyzing situations like these: student mobilization; taking into account that the programs are offered on a national level in Colombia through Distance learning or on our online program. Offering real devices through the Internet improves the laboratory availability, development of the established skills of the courses that articulate the practical component in remote form and the improvements of the curriculum.

2 Background

The programs of telecommunication engineering and electronics engineering of the Universidad Nacional Abierta y a Distancia UNAD, are offered in Colombia since the year 2006. For the second academic period of 2019, the Telecommunications Engineering program enrolled 1,654 students and the electronics engineering program enrolled 2,650 students (registered source and academic control of the UNAD). Both programs in their curriculum articulate the three scenarios of the development of the practical component; defining the remote scenario as that which is made possible to work at a distance in the physical laboratories of the UNAD, through the use of the Internet, cameras and specific equipment for data management [1].

The remote labs are technological tools with physical equipment, that can be used to test locally giving the user access, in a remote form, through an interface implemented through a software, that permits the students to do their practices as if they were in a traditional laboratory, usually the access is made through the Internet [5].

A remote lab is a solution of hardware and software that permit the students to interact with real equipment located in other places in the Internet. Likewise, in this manner, students interact with a real lab as if they were in a practice session in a real laboratory. Once the equipment is remote, it is also possible to share it between institutions so that students from one school or Universities can access a lab of another University [17].

Presently this type of practical component (Remote Laboratories) is implemented in the courses of Cisco Systems, through the Remote Lab Smart Lab; where in the second semester of 2019 close to 500 students have enrolled to perform the respective laboratory practices; where they have demonstrated the advantages that are precisely described in our educational model, Online Program and Distance Learning, like: promote and strengthen the self-directed learning, provide real equipment with simulation tools, reduce the damage by incorrect use of equipment, generate reports of the use of the laboratory, optimization of human resource and traditional lab material, availability of lab 24/7 and flexibility of the curriculum.

The Universidad Nacional Abierta y a Distancia UNAD, provides to establish in their Academic Pedagogical Solidary Project, 6 substantial responsibilities (Integral Formation – Investigation – Regional Development – Technological Innovation - Internationalization – Inclusion, Participation and Cooperation), that each academic program should include in their annual operative plan and respond before the National Ministry of Education. Taking in account the experience of the LabsLand global network in the elaboration and implementation of remote laboratories; the UNAD signs with said company in a joint agreement in the year 2018, to strengthen the investigation, the internationalization and innovation. The remote lab of Programming of Robot Mobile Arduino, is the result of work with said company and its' infrastructure is located in the CEAD Medellin of the UNAD.

2.1 Remote Laboratories

A remote laboratory is a software and hardware tool that allows students to remotely access real equipment located in the university. Users access this equipment as if they were in a traditional hands-on-lab session, but through the Internet. To show a clear example, Fig. 1 shows a mobile low cost robot laboratory described in [10]. Students learn to program a Microchip PIC microcontroller, and they write the code at home, compile it with the proper tools, and then submit the binary file to a real robot through the Internet. Then, students can see how the robot performs with their program through the Internet (e.g., if it follows the black line according to the submitted program, etc.) in a real environment.

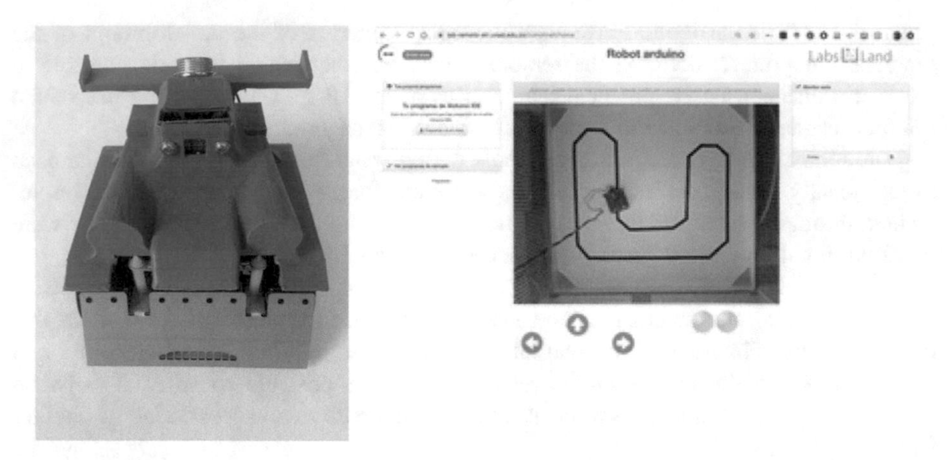

Fig. 1. At the left, the mobile robot itself in the UNAD. At the right, the user interface once the program has been submitted.

3 Deployment of the LabsLand Robotics Lab in UNAD

3.1 Robotics Laboratory

The remote mobile robotics laboratories based on Arduino platform at UNAD allowed students to have remote access not only in the existing laboratory, if not also through the alliance and collaboration with the LabsLand global network. The students of the UNAD were able to work collaboratively with other remote mobile robotics laboratories of various universities, thus allowing availability when the UNAD laboratory was occupied by some student.

The operation of the current platform of the mobile robotics remote laboratory based on the Arduino platform, it is managed based on the sharing of access to remote laboratories by different universities [7].

In this way, UNAD offers its laboratory in the LabsLand network through the exchange of uses of people in Europe or other continents, you will have free access to other mobile robotics laboratories in other universities around the world. Likewise, if the UNAD laboratory, in addition to being used by UNAD, is used by other universities, UNAD students will be able to access other laboratories in other countries free of charge for UNAD.

The Robot Mobile Arduino zumo is controlled by a Raspberry Pi 3 B+; that is in charge of downloading the program in the Arduino and additionally permits the access to the robot's switches (A, B, C and reset), and monitors other functions of the robot, this Raspberry Pi communicates via wi-fi with a wireless router and this also communicates with the server via Ethernet cable of the UNAD, the server is connected to the UNAD network and has access to Internet through a public IP. On the other hand, the WebLab-Deusto server communicates with the Labsland network and is in charge of operating the remote laboratories network of the Mobile Robotic of the UNAD and other universities, besides LabsLand's own robots. The robot's infrastructure required to be protected with a polyethylene net of high density with the purpose of protecting the internal structure of the drawer and the robot Pololu Zumo, as is observed in the structure of Fig. 2 [16] every time that said structure is found in a lab where the students have access, for the security of the equipment it is required that the same not be physically manipulated inside the facilities.

The robot used in the remote laboratory of mobile robotics is based in the platform Arduino of the UNAD, it's shown on Fig. 3. The robot Zumo 32U4 used in the UNAD, is a robot controlled by a microcontroller ATmega 32U4 compatible with Arduino. In the main board of Pololu Zumo 32U4 there is a microcontroller AVR ATmega 32u4 integrated of Atmel, along with two Bridge H drivers that feed the robots motors. The robot also presents a variety of sensors, including quadrature encoder and sensors of inertia (accelerometer and a gyroscope) in the principal plate, along with reflectance and proximity sensors in the matrix of the frontal sensors. The integrated switch buttons offer a convenient interface for the entrance of the user, a buzzer and LED indicators that permit the robot to provide feedback.

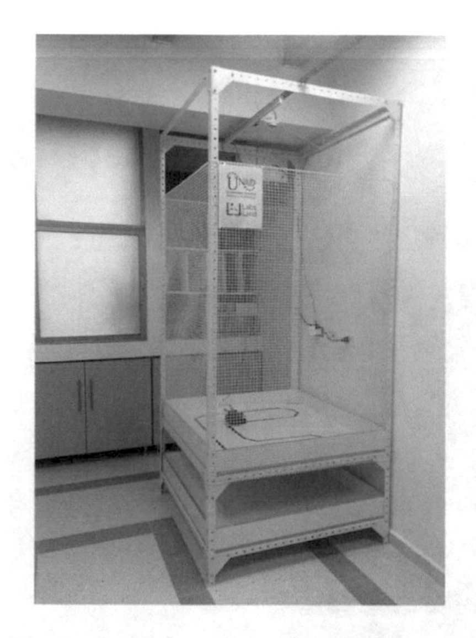

Fig. 2. Implementation of the laboratory of mobile robotics of LabsLand global network.

Fig. 3. Robot Pololu zumo 32U4 implemented in the UNAD

3.2 Integration of the Robotics Lab in UNAD

The students of the UNAD can enter the mobile robotic laboratory based in the Arduino platform through the Moodle, hence not requiring the registration in another application, but within the same UNAD Campus, they can access said resource with their campus username and password that was given to them upon university enrollment. The remote laboratory of mobile robotics based on the Arduino platform of the UNAD is equipped with a graphic interface of the user that permits the student to enter his work environment. Consequently, the users graphic setting is presented on Moodle in the remote laboratory based in Arduino. To access the remote laboratory, it is necessary to keep in mind the courses accepted norms and conditions, which are an informed consent in which the participant decides to participate in the trial of said investigation. Subsequently the student enters the integrated developing environment of Arduino (IDE), that permits the student to program in "C/C++" code, when the code is compiled by the student, this can be sent to the robot in the UNAD to be programmed by the same as is shown in Fig. 4 and Fig. 5 the architecture implemented.

The Robotics Remote Lab based in the Arduino platform has been presented pilot project for the students in electronic and telecommunication engineering with around 350 students of which 110 students participated in the experiment in pairs, which used the Robotics Remote Lab with the actual existing physical hardware of the CEAD of Medellin. The incorporation of this educational tool permitted the student the opportunity to analyze theoretical concepts of microcontrollers and microprocessors with the purpose of preparing without previous help of the virtual campus tutor. Furthermore, those who participated had access to didactic material of the management and programming of the robot. During the lab practices in the university initially the students performed the experiment online to get familiar with the tools and concepts learned in the lab completing a practical exercise by himself.

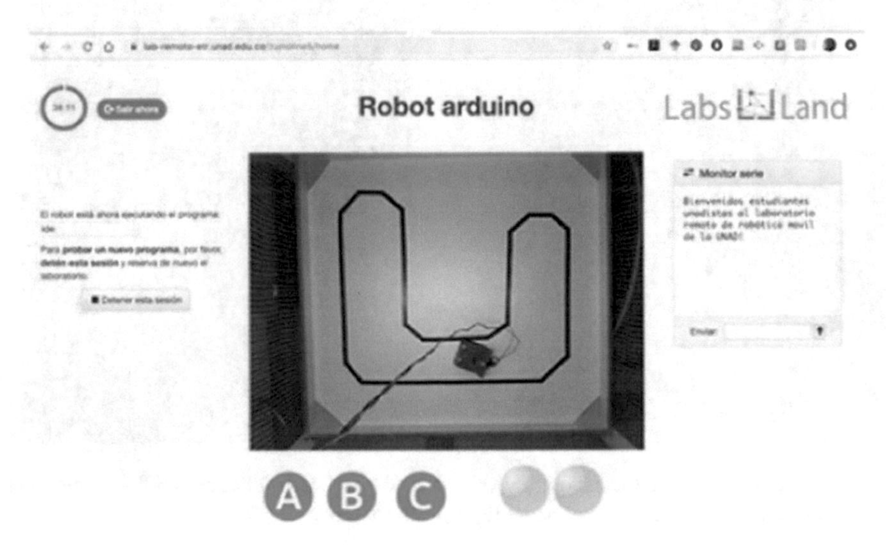

Fig. 4. Remote Laboratory of mobile robotic based in the Arduino platform, overhead view of the camera, programmed and executing.

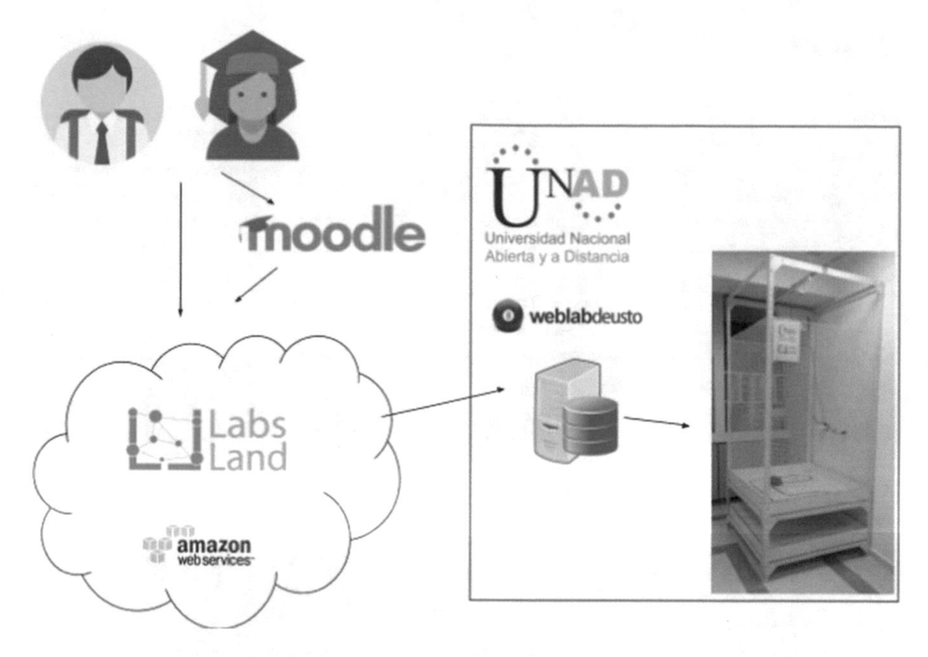

Fig. 5. Architecture of the Robotics Remote Lab at UNAD

Although not all 110 students that participated developed the activity to its entirety, this document will discuss the experience obtained by the students. Furthermore, it being a tool in trial it allows to evidence the capacity and ability of the student to interact with the Robotic Remote Lab, in continuation are the analyses obtained:

All in all, 248 experiments by the 110 students (in 55 pairs) have been done in course context using the remote mobile robotics laboratory based on the Arduino platform at UNAD. The LabsLand environment gives the opportunity to examine the lab uses with regard to time of the day and day of the week, on which the experiments have been conducted. Figure 6 shows the respective data. The data clearly shows, that the students have been using the tool mostly in direct context with the lab class on Saturday mornings and afternoons. Hence, most of the experiments have been conducted during class time. However, the data also shows, that experiments also have been carried out late in the afternoon on week or even on Sundays. This supports the aim to give the students more flexibility in performing own experiments and give them access to such equipment at uncommon times of the day or week. Even though the numbers are not overwhelming high, there can be a demand detected.

Uses per time of day

	Lunes	Martes	Miércoles	Jueves	Viernes	Sábado	Domingo
00:00					1		
01:00				1			
02:00				3			
03:00				1			
04:00							
05:00							
06:00				1			
07:00							
08:00						17	
09:00		1				23	
10:00		2	2	1	3	3	
11:00		2		1	4	3	
12:00				4		2	
13:00						5	
14:00				2	4	3	2
15:00				7	9	2	
16:00		2		3	2	4	1
17:00				3	2	6	
18:00		2		1	4	10	1
19:00	2	6	2		1	15	
20:00	8	10	2			12	2
21:00			4		2		
22:00			3			6	
23:00			2			21	

Fig. 6. Measurements of real-time use of the system for the remote mobile robotics laboratory at UNAD

3.3 LabsLand Network Effect

LabsLand is a global network of real laboratories available through the Internet. UNAD counts with its own space in LabsLand. LabsLand counts with multiple remote laboratories in many locations in all the continents, and in particular, the Arduino Robotics laboratory deployed at UNAD is also available in University of Fort Hare in South Africa, as well as in the University of Deusto in Spain and in LabsLand facilities also in Spain.

The fact of having the same remote laboratory in multiple locations is a great advantage for all the nodes in the network for two reasons:

1. Students of UNAD by default access the remote laboratory in UNAD. However, if for any reason (maintenance, Internet failure, etc.) the remote laboratory is not working, or it is busy with other students using it, UNAD students will automatically go to the robotics laboratory in other locations. This improves considerably the user experience since students do not need to wait long to access the laboratory.
2. Students of other parts of the world can access the UNAD remote laboratory. In general, and especially taking into account the time difference between Colombia and Spain and South Africa, the negative impact on the UNAD students is minimal.

However, the positive impact is impactful for UNAD: firstly, it improves the UNAD visibility in schools and universities worldwide, showing how UNAD is an innovative leader in distance education, and secondly, whenever students from other institutions access UNAD laboratories, UNAD obtains, through a compensation mechanism, access to other LabsLand laboratories (not only Arduino Robotics), so this is translated that at the end of the year, UNAD obtains access to other laboratories for free for its students.

For these reasons, this Arduino Robot remote laboratory is not only one remote laboratory deployed in a University for its own students, but, thanks to the LabsLand network, it becomes globally available through LabsLand, and the students of UNAD also access other laboratories from other global institutions. In the Fig. 7 shows the different deployments made in South Africa, Colombia and Spain [18].

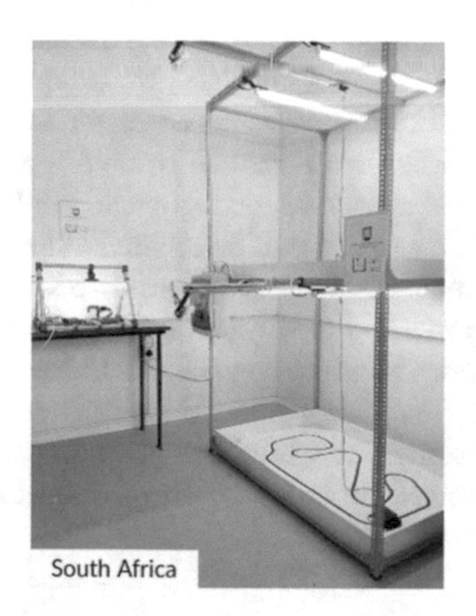

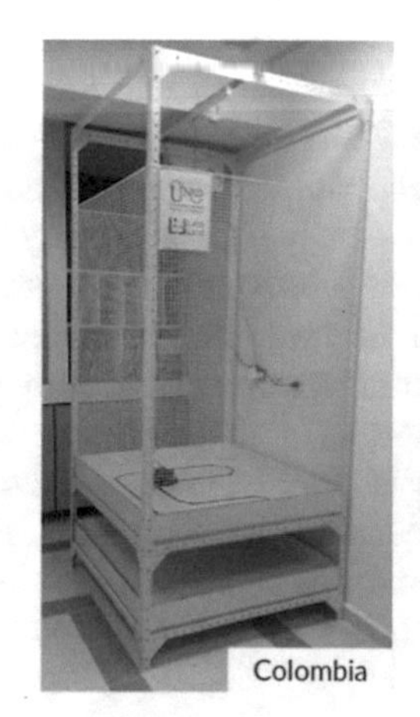

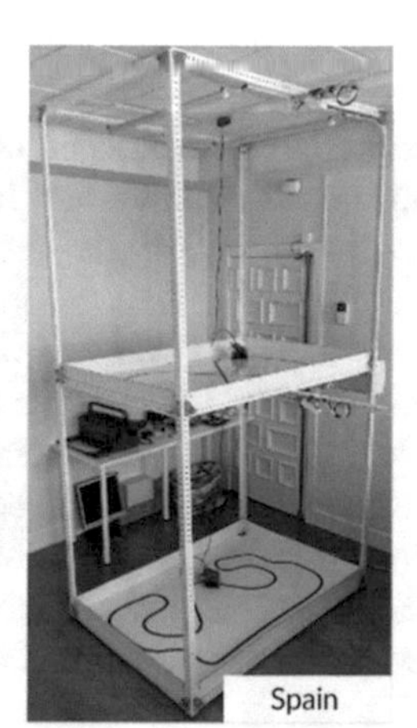

Fig. 7. Different deployments of the robot laboratory in UFH (South Africa), UNAD (Colombia) and LabsLand (Spain)

4 Experience and Perception of a First Pilot of the Robot in UNAD

4.1 Integration of the Robotics Lab in UNAD

Initially the pilot test was conducted with 350 students of the course of microprocessors and microcontrollers of which 110 students participated in the test. Data collection it was done through an online survey, the students were asked to conduct the survey after conducting the didactic tests within the pilot course, and 78 surveys with informed consent were obtained.

Some used the remote laboratory individually at home and others in pairs in the laboratory. The survey included closed questions which allowed evaluating specific aspects of the remote laboratory interface. However, it is worth mentioning that the work presented is in its testing stage. In order to develop the pilot test evaluation tool, the impact on the following indicator variables was taken into account: training quality (focused on the student through the appropriation of knowledge), technological tool (focused on the technology and infrastructure of the remote laboratory), learning strategy (focused on the development of student competencies) and finally curricular integration (Focused on evaluation of meaningful learning). At present there are no standardized referents that allow establishing own evaluation mechanisms of a remote laboratory as a teaching tool, therefore, it is crucial to have statistical results that allow us to establish whether the use of these tools allows improving teaching processes in distance education.

Referents in the literature include studies where it is observed whether remote and virtual laboratories are pedagogically effective. One of the first studies was by Ma and Nickerson [19], in which he analyzes virtual, remote and face-to-face laboratories, concluding that its use can be equivalent at the pedagogical level. Another study in the same line and with similar conclusions was the one presented by [20]. Along the same lines [21] made a meta-study of 50 remote laboratories and simulators, in which it measured the incidence of which learning characteristics are equal to or better than those used in traditional laboratories. This does not mean that a remote or traditional laboratory, due to the fact that it is a laboratory, has a specific pedagogical result; everything will depend on the laboratory and in what context it will be applied. Along these lines, [22] uses as an example the basic electronics laboratory to see its educational impact in the context of electronics, recently published in IEEE Transactions on Education. The use of remote vs. traditional laboratories in educational impact, due to the fact that it is a remote laboratory, it has a determining pedagogical result; everything will depend on the specific laboratory and in a context to which it applies. In this sense, according to [22] he uses as an example the basic electronics laboratory, that was used to measure the educational impact in the context of electronics.

4.2 Survey Performed Among Students

The survey was conducted online to 78 students after the end of the pilot test. The R statistics software was used to analyze the results of the survey. For the survey, students were asked to rate on a scale of 1 (strongly disagree) to 5 (strongly agree). Of the four categories in which the questions were classified as evidenced in Table 1. The survey system was conducted randomly. From 110 students we received a complete set of 78 responses, which form the basis for the following analysis of results. For the analysis of the 15 questions a test of significant differences of Tukey was performed. The questions were classified the four categories “training quality”, “technological tool”, “learning strategy” and “curriculum integration”.

Table 1. Indicators evaluated in the survey by question and type of scale used (N = 78).

Questions	1	2	3	4	5
Training quality					
Do you think that the remote mobile robotics laboratory based on the Arduino platform encourages and stimulates my learning?	0 (0%)	0 (0%)	4 (5%)	19 (24%)	**55 (71%)**
Does the use of learning tools based on remote mobile robotics laboratories in Arduino generate academic and professional added value?	0 (0%)	0 (0%)	4 (5%)	15 (19%)	**59 (76%)**
Do you consider that the use of technological tools based on remote mobile robotics laboratories is a methodology that improves the distance learning process?	0 (0%)	1 (1%)	8 (10%)	12 (15%)	**57 (73%)**
Technological tool					
How do you like the graphical user interface of the Arduino-based mobile robotics remote laboratory platform (icons, buttons, screens, navigation structure)?	0 (0%)	9 (12%)	17 (22%)	**32 (41%)**	20 (26%)
How did you find the operation of the tool?	0 (0%)	1 (1%)	12 (15%)	**40 (51%)**	25 (32%)
What is your opinion about the material used as a study guide for the use of the remote mobile robotics laboratory based on the Arduino platform?	0 (0%)	0 (0%)	7 (9%)	**50 (64%)**	21 (27%)
Learning strategy					
In general terms, do you consider that the practical exercise proposed improves your learning?	0 (0%)	1 (1%)	14 (18%)	**35 (45%)**	28 (36%)
Do you consider that the practical activity through the use of a remote mobile robotics laboratory based on the Arduino platform represents an important load of work?	7 (9%)	5 (6%)	**58 (74%)**	6 (8%)	2 (3%)
Curriculum integration					
Do you consider that the use of development tools based on remote laboratories favors the process of updating courses of the Electronics, Telecommunications and networks chain?	0 (0%)	0 (0%)	5 (6%)	31 (40%)	**42 (54%)**
Do you think that the creation of pedagogical materials based on technological tools based on the use of remote mobile robotics laboratories on the Arduino platform of the UNAD, favors the quality standards of the academic programs of the ETR chain at UNAD?	0 (0%)	0 (0%)	5 (6%)	26 (33%)	**47 (60%)**
Do you think that the learning environment based on the incorporation of remote mobile robotics laboratories in Arduino in the UNAD favors learning?	0 (0%)	1 (1%)	4 (5%)	26 (33%)	**47 (60%)**

5 Conclusions and Future Work

With the implementation of this remote mobile robotics laboratory as a pedagogical strategy in the UNAD through the use of a structured environment, an initial pilot test was carried out; So, UNAD students gained practical experience and applied the knowledge acquired in other subjects and areas.

It was identified if the tool could serve as a learning alternative by measuring the perception of students through a survey, and in this way improve the processes of availability of the educational resources of the UNAD, this was observed in the amount of Experiments done by the students of the remote laboratory.

The integration into Moodle of a remote mobile robotics laboratory based on the Arduino platform as a pilot test in laboratory practices in Arduino-based programming courses such as the microprocessors and microcontrollers course will allow the future incorporation of the learning tool into the learning processes, evaluation and accreditation of UNAD courses. In general terms, the initial experience of the pilot test was successful because it allows new educational resources to be available to students, and they constitute a pedagogical tool that allows the student to develop autonomous, meaningful and collaborative learning within the UNAD.

As future work, these remote mobile robotics laboratories at UNAD will provide the opportunity for students and teachers at UNAD to carry out degree work, courses and/or projects from the different approaches that a mobile robotics environment can provide; from areas of systems, electronics, mechatronics, biology, physics, mathematics, etc. For example, in mathematics and physics: the design of odometric systems (which are still used, despite the appearance of GPS systems), in physics: real examples of mobile behavior with uniform rectilinear motion and uniformly accelerated movement; in biology: experimentation in genetic and neuronal algorithms (pattern recognition: "Image identification"), collaborative behavior, etc.

References

1. Abadía García, C., Montero Vargas, R.: Lineamientos académicos para el desarrollo del componente práctico. Bogotá, Colombia (2015)
2. Lineamientos generales del currículo en la UNAD. Aspectos del trabajo colaborativo y acompañamiento docente, Bogotá D.C, Colombia, July 2014
3. Lineamientos generales del currículo en la UNAD. Serie lineamientos microcurriculares en la UNAD V2, Bogotá D.C, Colombia, December 2016
4. Orduña, P., Rodriguez Gil, L., Angulo, I., Martinez, G., Villar, A., Hernández, U., et al.: Abordar las dificultades técnicas y organizativas del uso de laboratorios remotos en un entorno comercial. IEEE (2019)
5. Soria, M., Fernández, R., Gómez, M., Paz, H., Pozzo, M., Dobboletta, E., et al.: Perspectivas de los Laboratorios Remotos en la Educación Media y Superior de Santiago del Estero. In: 1ER. CONGRESO LATINOAMERICANO DE INGENIERÍA, ENTRE RÍOS, ARGENTINA, pp. 13–15 (2017)
6. García-Zubía, J., Angulo, I., Martínez-Pieper, G., Orduña, P., Rodríguez-Gil, L., Hernandez-Jayo, U.: Learning to program in K12 using a remote controlled robot: RoboBlock. In: REV Conference 2017, New York (2017)

7. Orduna, P., Rodriguez-Gil, L., Garcia-Zubia, J., Angulo, I., Hernandez, U., Azcuenaga, E.: Increasing the value of remote laboratory federations through an open sharing platform: LabsLand. In: REV Conference 2017, New York (2017)
8. Wang, N., Ho, M., Lan, Q., Chen, X., Song, G., Parsaei, H.: Developing a remote laboratory at TAMUQ based on a novel unified framework. Age **26**, 1 (2015)
9. Islamgozhayev, T.U., Mazhitov, S.S., Zholmyrzayev, A.K., Toishybek, E.T.: IICT-bot: educational robotic platform using omni-directional wheels with open source code and architecture. In: 2015 International Siberian Conference on Control and Communications (SIBCON), pp. 1–3. IEEE, May 2015
10. Kalúz, M., Čirka, Ľ., Valo, R., Fikar, M.: ArPi lab: a low-cost remote laboratory for control education. IFAC Proc. Vol. **47**(3), 9057–9062 (2014)
11. Orduna, P., Caminero, A., Lequerica, I., Zutin, D.G., Bailey, P., Sancristobal, E., Rodriguez-Gil, L., Robles-Gomez, A., Latorre, M., DeLong, K., Tobarra, L.: Generic integration of remote laboratories in public learning tools: organizational and technical challenges. In: 2014 IEEE Frontiers in Education Conference (FIE), pp. 1–7. IEEE, October 2014
12. Saliah-Hassane, H., Reuzeau, A.: Mobile open online laboratories: a, October 2014
13. Garcia-Zubia, J., Cuadros, J., Romero, S., Hernandez-Jayo, U., Orduña, P., Guenaga, M., Gustavsson, I.: Empirical analysis of the use of the VISIR remote lab in teaching analog electronics. IEEE Trans. Educ. **60**(2), 149–156 (2017)
14. Harward, V.J., Del Alamo, J.A., Lerman, S.R., Bailey, P.H., Carpenter, J., DeLong, K., Felknor, C., Hardison, J., Harrison, B., Jabbour, I., Long, P.D.: The iLab shared architecture: a web services infrastructure to build communities of internet accessible laboratories. Proc. IEEE **96**(6), 931–950 (2008)
15. Maalouf, E., Saad, M., Saliah, H.: A higher level path tracking controller for a four-wheel differentially steered mobile robot. Robot. Auton. Syst. **54**(1), 23–33 (2006)
16. Buitrago, P., et al.: Use of remote laboratories in engineering as an alternative to pedagogical mediation and social inclusion in distance education. In: 2018 Congreso Internacional de Innovación y Tendencias en Ingeniería (CONIITI), Bogota, pp. 1–6 (2018)
17. Orduña, P., et al.: Addressing technical and organizational pitfalls of using remote laboratories in a commercial environment. In: 2019 IEEE Frontiers in Education Conference (FIE), pp. 1–7 (2019)
18. Orduña, P., Rodriguez-Gil, L., Garcia-Zubia, J., Angulo, I., Hernandez, U., Azcuenaga, E.: LabsLand: a sharing economy platform to promote educational remote laboratories maintainability, sustainability and adoption. In: 2016 IEEE Frontiers in Education Conference (FIE), pp. 1–6. IEEE (2016)
19. Ma, J., Nickerson, J.V.: Hands-on, simulated, and remote laboratories: a comparative literature review. ACM Comput. Surv. (CSUR) **38**(3), 7 (2006)
20. De Jong, T., Linn, M.C., Zacharia, Z.C.: Physical and virtual laboratories in science and engineering education. Science **340**(6130), 305–308 (2013)
21. Brinson, J.R.: Learning outcome achievement in non-traditional (virtual and remote) versus traditional (hands-on) laboratories: a review of the empirical research. Comput. Educ. **87**, 218–237 (2015)
22. Garcia-Zubia, J., Cuadros, J., Romero, S., Hernandez-Jayo, U., Orduña, P., Guenaga, M., Gonzalez-Sabate, L., Gustavsson, I.: Empirical analysis of the use of the VISIR remote lab in teaching analog electronics. IEEE Trans. Educ. **60**(2), 149–156 (2016)

Multi-phase Flowloop Remote Laboratory

Martin Sierra Apel(✉), Felix J. C. Odebrett, Carlos Paz, and Nelson Perozo

Clausthal Technical University of Technology, Institute of Petroleum Engineering, Clausthal-Zellerfeld, Germany
{martin.sierra.apel,felix.odebrett, carlos.a.p.carvajal,nelson.perozo}@tu-clausthal.de

Abstract. Multi-phase systems are found in various industrial scenarios, not being an exception the oil and gas industry. The prediction and determination of the flow pattern in the production tubing is of great importance since this is directly involved with the optimization of the production itself. The study and comprehension of this fundamental branch is of great importance to become a petroleum engineer at the Clausthal University of Technology. Mixtures of hydrocarbons, water, sediments and gases found in reservoirs manifest different behaviors during their transportation to the ground's surface. Predicting such behaviors enable a better understanding for adequate choice of pumping machinery and pipeline setups for a successful transport to their final processing facility.

In order to achieve a better predictiveness of how multi-phase systems behave in real life, the Clausthal University of Technology developed and built an industrial-scale Flowloop test unit for further investigation. This enables the engineering students a better comprehension and visualization of how multi-phase transportations react under variable production rates and mixture proportions.

Since the number of students is elevated, the utilization of the physical test unit is limited, not making it available to all of them all the time. The university decided therefore to develop an ultra-concurrent remote laboratory of the Flowloop in collaboration with LabsLand in order to revert this situation.

The goal of this remote laboratory is to provide access to all interested parties to the Multi-Phase Flowloop via a web interface, enabling them to experiment probable scenarios and using it as a complementary tool to their theoretical material provided in class. Besides, such an interface will let lecturers and petroleum experts show remotely different scenarios that make the class's content more visualizable and understandable without recurring to the physical unit itself.

This is a digital world; millennials are used to technology and learn easily with their modern technological devices. Given the lack of availability to experiment with the physical unit itself, the professor of the Drilling and Production Department of the Institute of Petroleum Engineering at TUC encouraged his staff to search for alternatives where students would be comfortable to work with. The advantage of such multi-phase systems is the easy replicability of their experiments, making a documentation of each test very easy. However, for an appropriate and accurate documentation of each experiment, the Flowloop

M. E. Auer and D. May (Eds.): REV 2020, AISC 1231, pp. 184–192, 2021.
https://doi.org/10.1007/978-3-030-52575-0_15

needs adequate equipment such as pressure transducers, positive displacement units, variable flow pumps and high-speed cameras that record the flow pattern.

With this remote laboratory, the Institute awaits an increase of interest from the students and hopes the content in class will be more understandable. This will be reflected in future exam results.

Besides, it's hoped to encourage other institutes at the TUC and third parties to consider creating such interfaces via remote laboratory to provide full time availability to their students and give access to others in need of such information for further research.

It is time to adapt lectures and find alternatives that make learning more interesting and foment the increase of knowledge without enormous investment.

Keywords: Multi-phase · Flowloop · Petroleum · Engineering · Clausthal · Technology · Remote laboratory · LabsLand

1 New Paths in Teaching Petroleum Engineering

The digital transformation is not only a much-discussed topic in today's media but has a great impact on our society as well as on industries and the way how future work will be carried out. Automation, cloud computing, big data, machine-learning and remote control are already rapidly changing the working world and open up a vast amount of opportunities. The oil and gas industry, although said to be traditionally only slowly adapting to new technologies, is acknowledging this major shift and puts forward many projects in digitalizing the oil field by implementing these new technologies.

The geographical locations of drilling rigs and production facilities are naturally governed by the geological existence of an oil or natural gas reservoir. These locations are often in remote places like the open sea or the back country far-off supporting infrastructure. Thus, high investments and efforts must be undertaken to build the existing facilities and operate the drilling and production activities while specialized personnel must be transported and accommodated. Automation and remote control are therefore especially in the discipline of deep drilling of increasing importance as they could provide the means to decrease the number of personnel needed on-site. Repetitive and dangerous tasks like pipe handling on the rig-floor could be automated while others like borehole measurements could be carried out centralized.

These changes in the petroleum industry directly influence the requirements on the skill set of future petroleum engineers. Universities therefore need to react to new developments and adapt their curriculum. "The next generation of petroleum engineers will have to address demands for sustainability, lower carbon intensity, and needs for radical productivity improvements, which only artificial intelligence (AI) and digital can drive. This suggests that we will need to revisit university education for petroleum engineers and all aspects of career development and training." [1] The Drilling and Production department of the Institute of Petroleum Engineering of the Clausthal University of Technology oversees the specialization "Drilling and Production" within the master's course "Petroleum Engineering". With respect to the current developments in the oil and gas industry the teaching staff of the department is currently working on new ways to implement the digitalization into the student's education.

On the one hand it is planned to use existing as well as create different multimedia tools to support the lectures. Web-based applications just like remote laboratories or drilling simulations provide a low cost and flexible way to give students the possibility to comprehend new topics. By using this many-sided approach of teaching it is intended to create a better and deeper understanding of the drilling environment for the students. On the other hand, the new educational concept aims to sensitize students to the increasing importance of digital technologies in the oil and gas industry by making them more comfortable in using and understanding mechatronic systems as well as computer science approaches.

In that scope a digital drilling laboratory was established at the department. As a teaching and researching lab, it features two scale-model drilling machines set up to simulate realistic rig equipment and able to perform drilling activities. The digital drilling lab is provided with computer resources suitable for the use of software-based drilling simulators to emulate rig floor operations. The lab is designed to offer students a complete learning experience of the entire drilling process through a hands-on approach. The department hopes to reinforce and complement the learning process of drilling engineering students by using different multimedia learning tools and allow students to use internet-based learning platforms for developing programmatic contents. The department aims to upgrade the laboratory drilling machines that students will not only have on-site but also remote access to autonomous and full operative drilling facilities to perform the experimental work associated with the lectures.

To drive the digitalization in the drilling department and to get acquainted with the development and installation of web-based learning applications, the multi-phase flowloop remote laboratory was created, which serves as a kick-off point to support the learning progress of the master's students.

2 Multi-phase Flow in Oil and Gas Applications

The movement of fluids dictates all petroleum engineering considerations: Being it the flow of hydrocarbons in the geological reservoir deep in the underground, the flow of drilling mud during drilling operations or the production of oil and gas from the reservoir to the surface. In all these processes multi-phase flow occurs.

During drilling operations, a special liquid based on water or oil with several additives is constantly pumped into the borehole. This drilling mud is circulated through the drill pipe, leaves the drill-bit through nozzles and then rises back to the surface in the annulus between the drill pipe and the borehole wall. The main purpose of the drilling mud is to transport the crushed rocks, so-called cuttings, from the borehole bottom to the surface in order to drill further. Hence, when the cuttings are suspended in the drilling mud a two-phase flow of liquid and solid particles occur in the annulus. In the undesired event of a kick, gas from an underground formation is flowing into the wellbore and thus introduces even a third phase to the flow [2].

After a well is successfully drilled, pipes are placed in the wellbore as an effective way to connect the reservoir with the surface and by this way to produce the hydrocarbons. During production, multi-phase flow will always form in the pipe as oil-water, gas-water or even oil-gas-water flow. Depending on the flow rates and velocities of the different phases a variety of flow patterns can form in the pipe ranging from gas bubbles or gas slugs in the liquid phase to mist in the gas phase. If sand control measures at the borehole bottom are not sufficient, sand will also be produced in the pipe which will add another phase to the flow [3].

Understanding the different flow patterns, when and how they occur is of great importance for drilling and production procedures to safely and efficiently conduct operations. To simulate and investigate multi-phase flow a Multi-Phase Flowloop Laboratory can be used.

3 The Flowloop Multi-phase Laboratory

The Clausthal University of Technology developed and built therefore an industrial-scale Multi-Phase Flowloop test unit in order to increase the understanding and predictiveness of multi-phase flow patterns in possible real-life scenarios. The unit itself, see Fig. 1, is not a complex device.

- The test section consists mainly of a transparent 0.04 m internal diameter straight Plexiglas pipeline (1) with a length of 6 m.
- This pipe is supported by an hydraulic screw-jack (2) and a carbon steel beam, the system can be inclined up to an angle of 90° from its horizontal ground level.
- A 6 KW stainless steel centrifugal pump (3) connected to a main tank (4) provides water to the system. Equipped with an electromagnetic induction flow meter and by the rotational speed of the pump, the rate of the water flow can be adjusted to the desired values.
- In order to develop various phase-flow patterns, air is injected to the water stream via an air-water mixer (5).
- Supplementary devices, such as an industrial air compressor with filtering and cooling systems, provides condensation-free air to the unit.
- A hopper (6) fitted to the top of the water tank serves to feed another phase, representing eventual solid sediment.
- An electric solid feeder with different vibration frequency-setups, allows the loading of the sediment to be adjusted and set to a desired feeding rate.
- The behavior of the flow is sometimes unrecognizable by the human eye capacities. Therefore, a high-speed camera (7) was mounted at the Plexiglas pipeline to record the flow. Saving its recordings serve for further slow-motion analysis.
- To control the complete unit and save all recorded information, a customized LabView software was programmed. This interface accesses each element of the unit allowing to configure and save all desirable scenarios.

With this setup, it's possible to reproduce different multi-phase flow patterns that were described in the previous chapter in the pipe of the flowloop. When water and air are flowing simultaneously, the behavior of a liquid-gas flow can be studied, mimicking the flow of an oil-gas phase during production. If the flowloop is run with water and particles the behavior of unwanted sand production during oil recovery can be reproduced. This setup could also depict the transportation of cuttings dissolved in the drilling mud. As the flowloop does not provide the opportunity to install a second

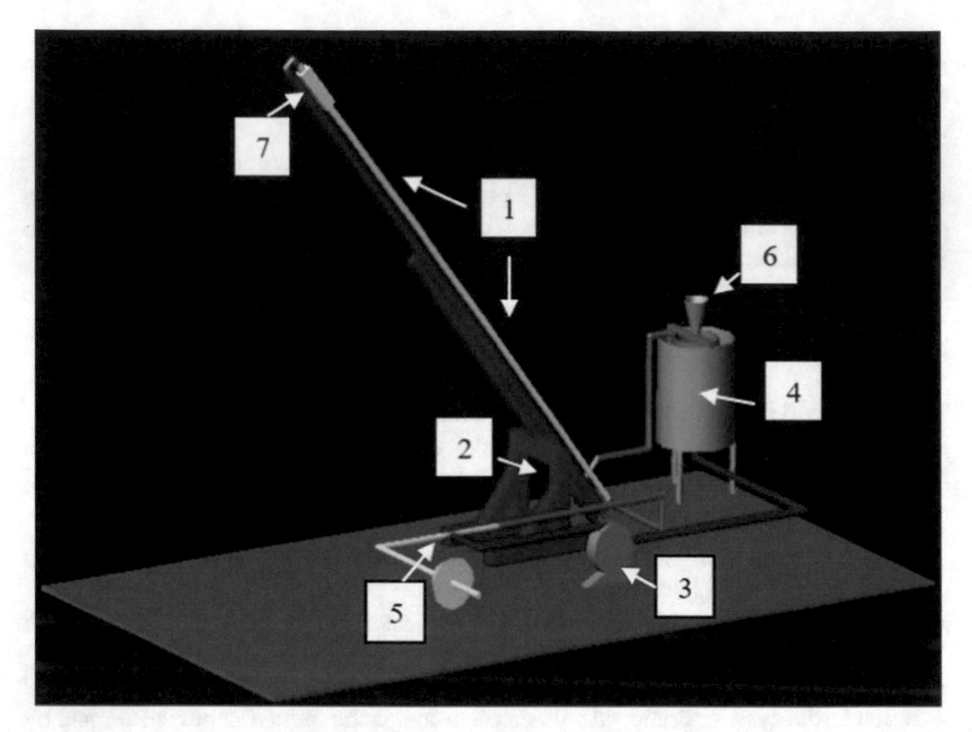

Fig. 1. Overview of the multi-phase flowloop test unit, [4]

smaller pipe to create an annulus, the setup has only limited capabilities to adequately simulate the transport of cuttings during drilling. However, different flow behaviors of the particles can be recognized that represent at least to certain extent the flow of cuttings in real drilling applications. Well boreholes for oil and gas production are not only drilled vertically but can also be deviated up to being completely horizontal. The possibility to change the angle of the flowloop continuously between 0° and 90° grants the opportunity to reproduce all different angles that could occur in a real-life borehole.

4 The Ultra-concurrent Remote Laboratory

A first prototype of the Multi-Phase Flowloop remote laboratory was created in collaboration with LabsLand. This global platform specializes in creating web interfaces for remote laboratories providing the necessary feedback to start a first concept.

Most remote laboratories are usually “real time laboratories”. Via web interfaces, users can control physical units with help of telemetry devices in real time. The Flowloop has the huge advantage of an easy replicability of its experiments. Same input parameters setups of each experiment lead to same results, showing a repeatable behavior of the multi-phase flow. It was therefore decided to base the laboratory on an ultra-concurrent laboratory.

For this first concept of the laboratory, a three-phase flow with gases was excluded. Still, for an increased number of different setups, three different flow rates of 70, 90 and 110 l/min, and intervals of 5° from its horizontal configuration of 0° up to 45° were chosen. A total of 30 experiments were carried out at the real lab and pre-recorded.

The interface created by LabsLand allows the user to configure these two variables in a very logic and easy way, see Fig. 2.

Depending on the values chosen, the pre-recorded documentation is shown. All the data is still completely real, but this way, several users can use the digital lab at the same time and get to see all pre-recorded scenarios, see Fig. 3.

The global network of remote laboratories LabsLand[1], offers real laboratories from over a dozen countries in all continents. This provides a set of advantages to the development of the ultra-concurrent laboratory:

- Relying on LabsLand experience [5] on the development of laboratories: LabsLand has developed many ultra-concurrent laboratories and already optimized videos in different qualities for different network bandwidths and host the lab in optimized cloud services in Amazon Web Services.
- LabsLand platform: TU Clausthal has its own space at LabsLand[2], where TU Clausthal students use the laboratories and TU Clausthal obtains analytics on the usage. LabsLand spaces also count with integration of major Learning Management Systems such as Canvas, Moodle, Sakai, Google Classroom and others, so students do not need to register in LabsLand to use the laboratory.
- Sustainability: once the laboratory is developed, anyone can use these and other laboratories by registering in LabsLand for a subscription fee; which is shared with TU Clausthal using different mechanisms. This allows TU Clausthal to use laboratories of other universities and to re-invest in future TU Clausthal labs.

5 Expectations and Probable Upgrades

With this remote laboratory it is intended to increase the student's interest in the matter and improve their academical performance. The lecturer will be using the interface during class as well, keeping students at safe locations. With the remote laboratory, students can conduct little experiments at any time from their own computers. It is hoped that this leads to an increase of student's interest in the matter and a better understanding of the underlaying flow principles. This will hopefully be reflected in future exam results.

This first prototype of the remote laboratory does not include all possible scenarios that can be simulated by the actual flowloop. It is therefore planned to further upgrade the online application on LabsLand. At first, new videos will be produced that document the flow of water and sediment at angles above 45° in 5° steps until 90° for all three different flow rates that are currently implemented. Secondly, two new flow rates

[1] https://labsland.com.

[2] https://labs.land/tu-clausthal/.

Fig. 2. Configuration of input parameters at the LabsLand-interface

Flowloop

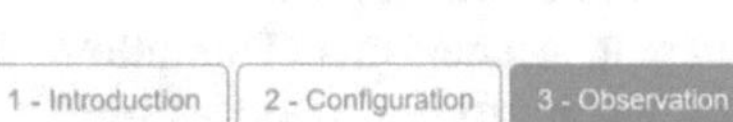

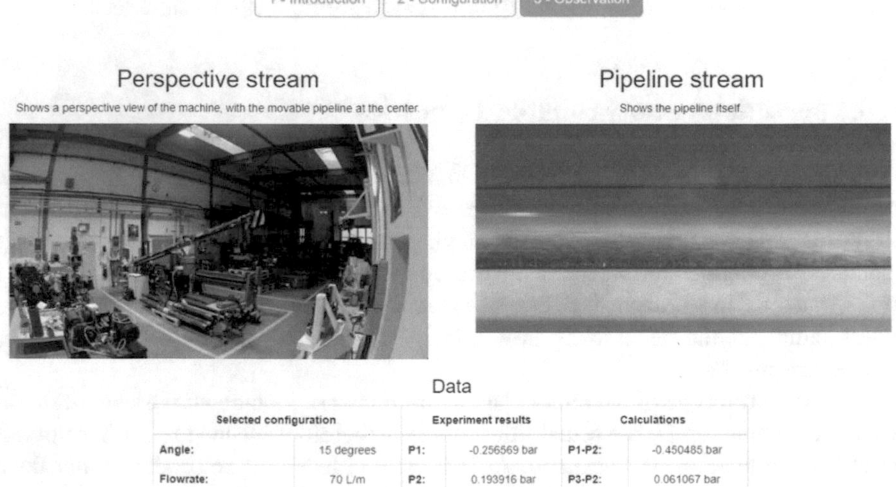

Selected configuration		Experiment results		Calculations	
Angle:	15 degrees	P1:	-0.256569 bar	P1-P2:	-0.450485 bar
Flowrate:	70 L/m	P2:	0.193916 bar	P3-P2:	0.061067 bar
Sensor flowrate:	69.4319 L/m	P3:	0.254983 bar		

Fig. 3. Pre-recorded information and images shown per user's request

of 80 l/min and 100 l/min will be introduced. New videos for all angles between 0° and 90° (again in 5° steps) will hence be recorded for these flow rates. After uploading the new videos and upgrading the interface of the remote laboratory, a wider range of angles and flow rates will be available for the user. In a next step, the remote laboratory will be extended with a new experiment in which water and air are flowing simultaneously in the flowloop. Analogous to the existing experiment setup, videos will be produced from the actual flowloop for all already described testing conditions (at angles between 0° and 90° and five different flow rates: 70, 80, 90, 100 and 110 l/min) which are than uploaded into the website.

Simultaneously, the drilling and production department will work on further remote laboratory setups using different experimental installations available at the institute. The goal in the near future is implementing the drilling machines from the above-mentioned digital drilling laboratory firstly in an ultra-concurrent web-based laboratory and finally in a remote accessible real-life drilling application.

Besides, it's hoped to encourage other Institutes from the Clausthal University of Technology and third parties to consider creating such interfaces via remote laboratory to increase the availability to their students and give access to others in need of such information for further research.

6 Conclusions

Educational institutions must realize the need to adapt to a new generation of students, not leaving behind the application of new existing technologies. In the 21st century, students grew up with electronic devices as a given part of their lives and are used to easy access to information and knowledge via the internet by just a click. The Institute of Petroleum Engineering realizes it must develop a wider range of teaching material by using multimedia approaches. Furthermore, the education of future petroleum engineers needs to drastically adapt to the new challenges that are awaiting the industry.

By developing and implementing the Multi-Phase Flowloop remote laboratory in cooperation with LabsLand, the drilling and production department of the Institute of Petroleum Engineering goes the first step in offering students the possibility to simulate their own experiments online and deepen their understanding of the underlaying physical principles of multi-phase flows. During the development of the remote laboratory the teaching staff of the department had the chance to enhance their knowledge and get valuable practice in setting up digital learning environments. With the digital drilling lab and the Multi-Phase Flowloop remote laboratory the first steps are taken in modernizing and digitalizing the education of future petroleum engineers at the Clausthal University of Technology.

References

1. Mathieson, D., Meehan, N., Potts, J.: The end of petroleum engineering as we know it. In: SPE-194746-MS, Bahrain (2019)

2. Bourgoyne, A., Millheim, K., Chenevert, M., Young, F.: Applied Drilling Engineering. Society of Petroleum Engineers, Richardson (1984)
3. Economides, M., et al.: Petroleum Production Systems. Pearson Education, Upper Saddle River (2019)
4. Bello, O.: Modelling particle transport in gas-oil-sand multiphase flows and its applications to production operations, Clausthal-Zellerfeld (2008)
5. Orduña, P., Rodriguez-Gil, L., Garcia-Zubia, J., Angulo, I., Hernandez, U., Azcuenaga, E.: Increasing the value of remote laboratory federations through an open sharing platform: LabsLand. In: Auer, M.E., Zutin, D.G. (eds.) Online Engineering & Internet of Things. LNNS, vol. 22, pp. 859–873. Springer, Cham (2018). https://doi.org/10.1007/978-3-319-64352-6_80

Manufacturing and Developing Remote Labs in Physics for Practical Experiments in the University

Zineb Laouina[1(✉)], Lynda Ouchaouka[1], Ali Elkebch[2], Mohamed Moussetad[1], Mohamed Radid[3], Yassine Khazri[1], and Ahmed Asabri[1]

[1] Laboratory of Engineering and Materials, Ben M'sik Faculty of Sciences, Hassan II University Casablanca (UH2C), Boulevard CdtDriss El Harti, B.P. 7955 Casablanca, Morocco
laouina.zinebl@gmail.com

[2] Laboratory of Structural Engineering, Intelligent Systems and Electric Energy, National School of Arts and Crafts (ENSAM Casablanca), (UH2C), Casablanca, Morocco

[3] Laboratory of Physical Chemistry of Materials LCPM, Ben M'sik Faculty of Sciences, (UH2C), Casablanca, Morocco

Abstract. Laboratory experimentation plays a crucial role in scientific education and engineering in higher education. The traditional hands-on labs are the natural scenarios where practical skills can be improved, but, thanks to Information and Communication Technologies (ICT), virtual and remote labs can provide a framework where Science, Technology, Engineering and Mathematics (STEM) disciplines can also be developed. This paper deals particularly with remote access to a real laboratory equipment using contemporary computer and network technology for creating the environment that will enable a remote user to perform the required practical experiments in physics. The main goal of this paper is to build an interactive system to control practical experiments in physics by the students enrolled in the first year of undergraduate program of Physics and Chemistry at the University.

The experiment adopted for this work is the simple pendulum manipulation.

In this work, we have focused on the design and creation of a prototype to make the equipment of the practical experiments of laboratories accessible to students via the Internet network. In order to eliminate time and space constraints through affording flexibility, protect students while conducting dangerous experiments and reduce the cost and the maintenance of laboratory equipment, which is expensive.

Keywords: Remote lab · Simple pendulum · 3D design · Practical experiments

M. E. Auer and D. May (Eds.): REV 2020, AISC 1231, pp. 193–204, 2021.
https://doi.org/10.1007/978-3-030-52575-0_16

1 Introduction

Recent years have witnessed major transformations in information and communication technology systems called ICT (Information and Communication Technologies), and since their emergence, there has been a revolution in the field of education, particularly in distance learning contexts. ICT in Education cover all digital tools and products that can be used in education, and it includes a set of hardware and software designed and used for distance learning and training purposes.

The strategy of setting up remote experimental laboratories is no longer to be demonstrated nowadays, both for the trainer and for the learner. Various works exist in the field of distance learning, but few are those concerning online training platforms integrating practical work.

However, these training environments support several pedagogical objectives since they allow adaptation to the learning pace of each learner, free learners from geographical constraints, allow economies of scale to be achieved. Also, they help improving the quality of experiment, effectiveness of time spent at lab by rehearsal and also safety and security by minimizing risks.

In addition, real-time remote control, manipulation and synchronization of communication between the server and the client currently represent a challenge and an added value for remote experimentation platforms. To this end, an appropriate and thoughtful choice must be made for the development software and communications protocol adopted between two remote computers.

The purpose of this study is to build an interactive remote laboratory for learners, which allows them to use, control and interact with the measuring instruments used in the real laboratories based on control interface. The chosen manipulation is the simple pendulum, which is part of the practical work of students enrolled in the first year of the bachelor's degree course in physical sciences. The aims of this work is to provide learners with access to measuring instruments of remote laboratory by delivering different activities related to simple pendulum and measurement experiments.

2 State of the Art

Remote Laboratories have become part of current teaching and learning practices, particularly in engineering, robotics and electronics. They are emerging as a valuable alternative to conventional hands-on labs. Remote Laboratories are flexible, distributed environments which enable a learner to perform real experiments via the web, by allowing them to collect real results, not only simulated, to use them [1]. The strategy that is often used in building a remote laboratory, which has been published, is using a computer-based server.

Web technology based on embedded system has grown rapidly in the post PC-Era [2]. Implementing a web enabled controller to the system replaces the computer as a web server, thus lowering the cost of the system [3]. Also, various technologies in web programming have been applied to provide a comfortable remote lab environment, such as Socket, Applet, Ajax, Corba, LabVIEW, etc. [4].

Simulated and remote experiments problematic has been the topic of scientific activities of our team since 2010, with the objective of providing the Ben M'sik Faculty of Science with a range of simulated and/or online manipulation in order to enhance the faculty's distance learning, by "developing a remote practice for laboratory experiments on measuring instruments" [5] and by "Creating online leading University laboratories" [6] in addition to a range of participation in European capacity building projects to enhance the quality of higher education Tempus Project Escience, and Erasmus Project Express.

3 Experiments

This paper is organized as follows: a first section contains the mechanical design and the development of our prototype; a second section contains the hardware and software design involving an embedded web server and lab module, and a last section contains the uploading of the manipulation.

3.1 Architecture of Remote Laboratory

The architecture of remote laboratories is based on the existence of a classical control engineering laboratories where usually students and researchers carry out experiments in the presence form [9]. There exist many experiments using this platform [6, 8, 10]. Through the design of client server application that is Web based, these experiments can be reached by any Internet user [7]. This opens laboratories to remote users that can experiment from the familiar environment of their browsers. The client server application communicates through TCP/IP protocol (Fig. 1).

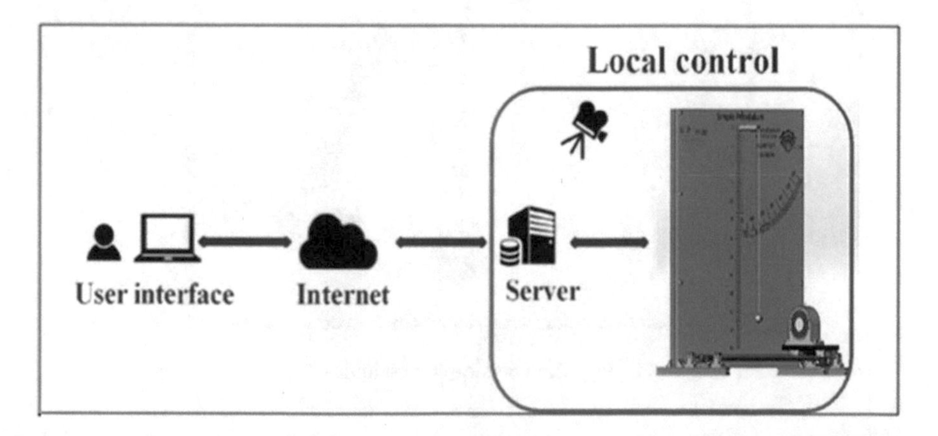

Fig. 1. Remote laboratory architecture

3.2 Description of the Experiments

The simple pendulum is characterized by three physical quantities, in particular the mass, the length and the angle.

Simple pendulum by definition consists of a mass m hanging from a string of length l and fixed at a point of suspension (Fig. 2).

The period of oscillation for small displacements is given by the following formula:

$$T = 2\pi\sqrt{\frac{l}{g}} \tag{1}$$

T: Time required for one cycle of periodic motion (sec)
g: Acceleration due to gravity.
l: Length of the pendulum is a distance from a point of suspension to the center of a bob.
m: Mass of the pendulum bob.
θ: The angular displacement.

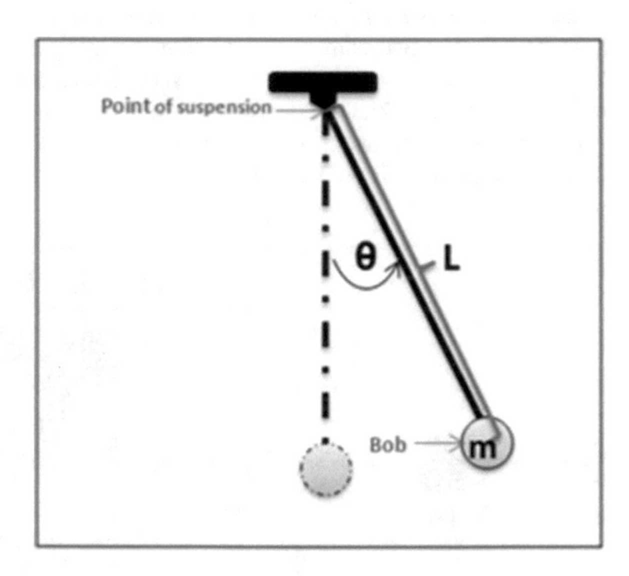

Fig. 2. The simple pendulum

The purpose of the experiment is to determine, the period T of the simple pendulum for amplitudes ranging from 5° to 30°, also to determine the gravity acceleration g and the influence of angle θ on the value of **g**. Finally, it is necessary to compare between the obtained and the theoretical values of **g**.

3.3 Mechanical Work

In this work, we firstly designed the prototype of the practical experiment, and then we developed our prototype.

Mechanical Design

All the pieces of the experiment were designed in the SolidWorks Software as shown in Fig. 3 and Fig. 4.

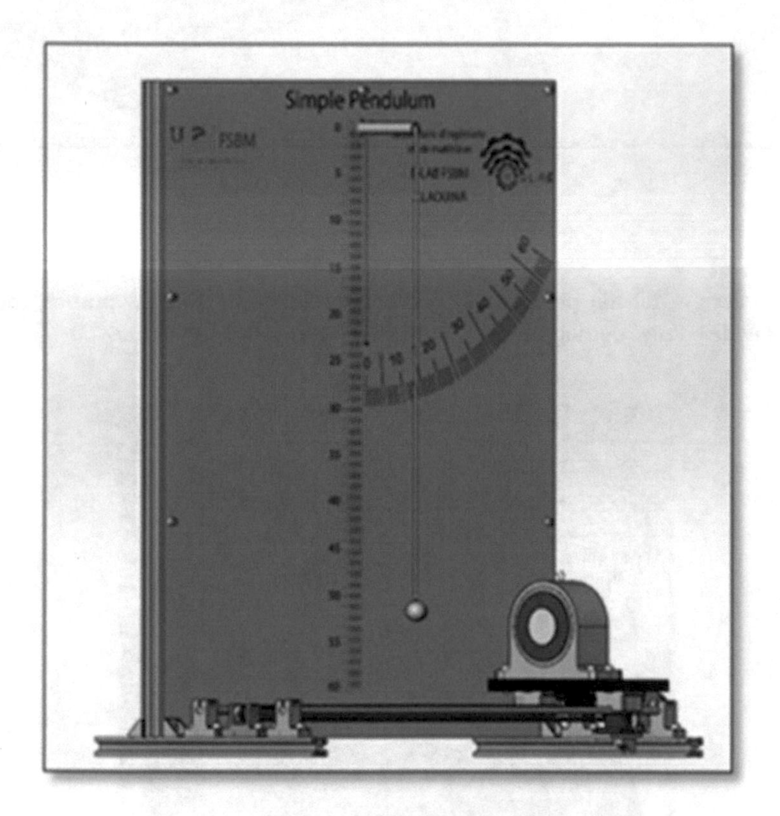

Fig. 3. 3D design of simple pendulum experiment

Figure 4 below represents the belt-pulley mechanism of our prototype.

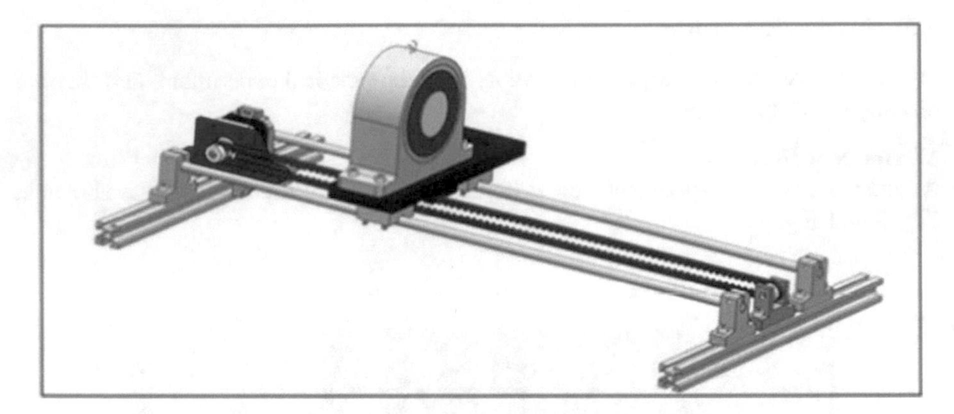

Fig. 4. 3D design of the belt-pulley mechanism

Development

After the design, all the previous elements were made by the 3D printer, the laser cutting and the vinyl cutting in order to produce the following prototype, Fig. 5.

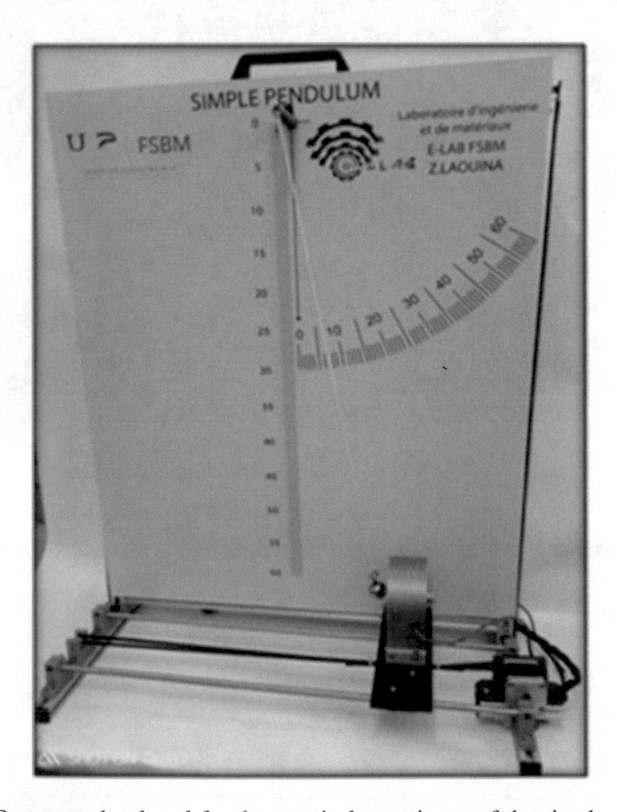

Fig. 5. Prototype developed for the practical experiment of the simple pendulum

3.4 Electronic Work

The electronic work focused on the design and development of the control card.

Software

Figures 6 and 7 show the 3D design and electrical schematic of the electronic card used in our work:

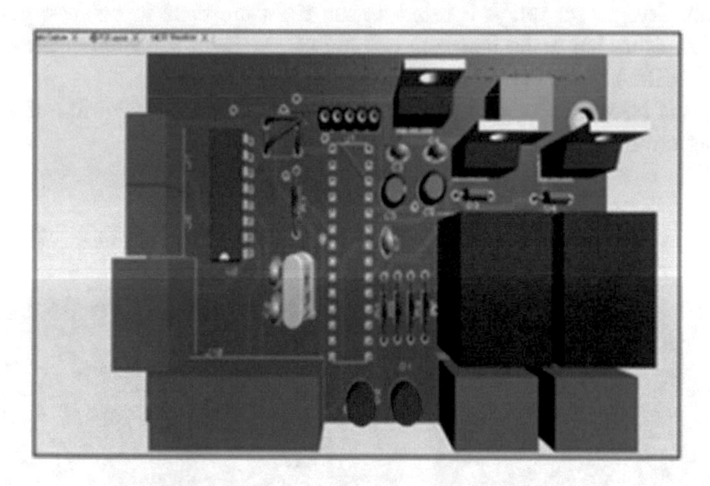

Fig. 6. 3D design of the control card

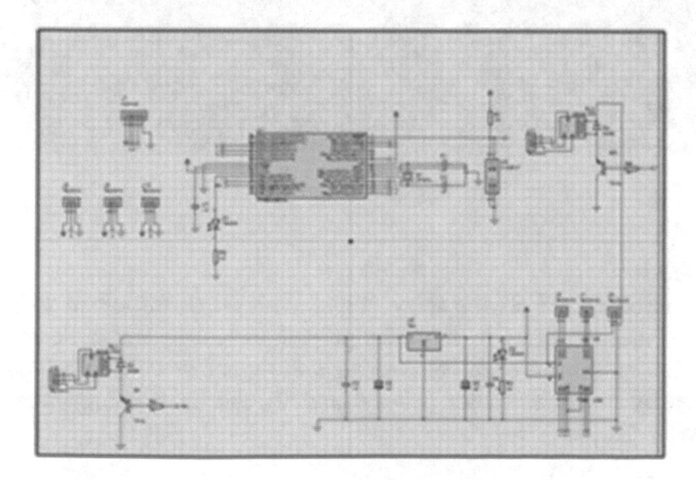

Fig. 7. 3D electrical schematic

Hardware

Description

The printed circuit board called "control board" is populated with passive and active components that allow the expected functionalities to be achieved. In the center of the board, the ATMEGA328 microcontroller of Microchip manufacturer's AVR family is chosen to manage the operating logic by being clocked at 16 MHz.

For the power part of the circuit where the loads are higher than the microcontroller's capacity, power relays located on the right are used to achieve galvanic isolation. On the other hand, the integrated circuit on the right is a stepper motor driver. It is from the NEMA17 family 17HS4401.

Finally, the power demand in the board (12 V) requires the addition of a regulation and filtering circuit to protect the components (Fig. 8).

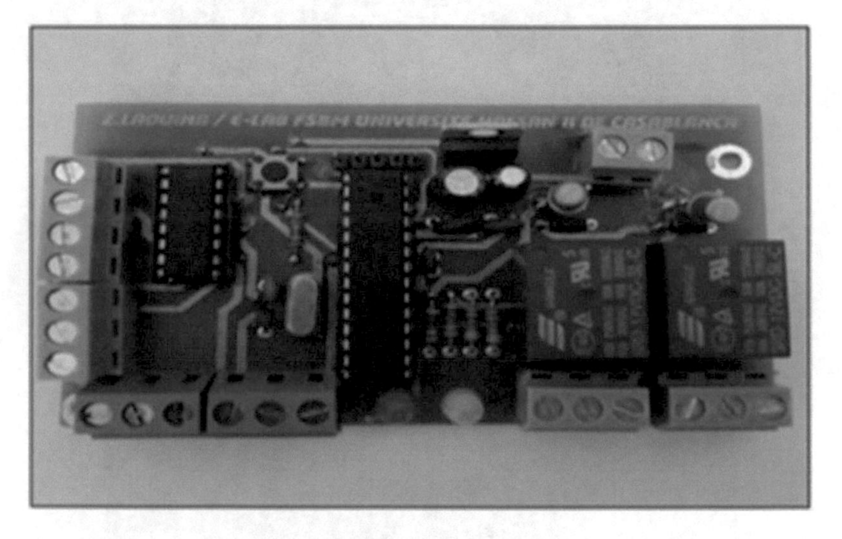

Fig. 8. Control card

Operating Mode

The microcontroller controls the relay output with an on/off signal to activate and deactivate the electromagnet to drive the pendulum ball by magnetic effect. On the other hand, the movement of the electromagnet is guaranteed by a belt-pulley mechanism driven by a stepper motor (Figs. 4 and 9), the latter being interfaced to the microcontroller via the driver.

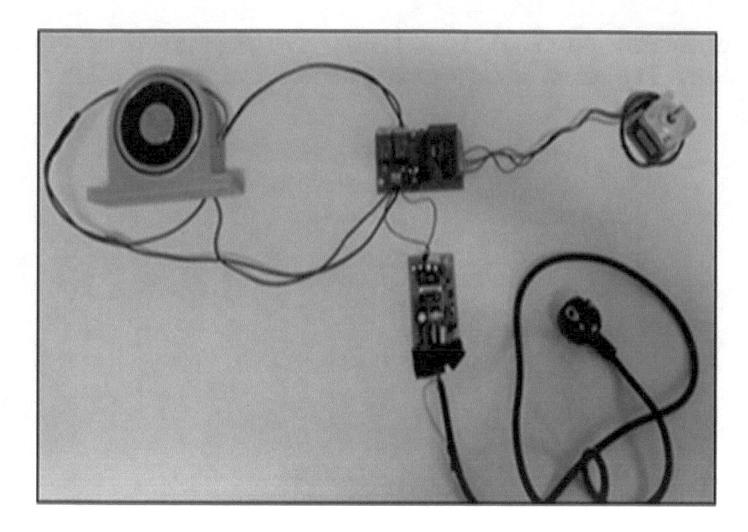

Fig. 9. Control card with component

3.5 Uploading of the Handling

The User Interface

The manual board control software is based on LabVIEW software. It is ergonomically designed to guide the user through the experience, including the operating mode; an IP camera is used to provide a real-time viewing experience. The server and the camera were connected to the Internet through a router switch. The user has the option of inserting one of the set point angle configurations and then sending it to the control card (Fig. 10).

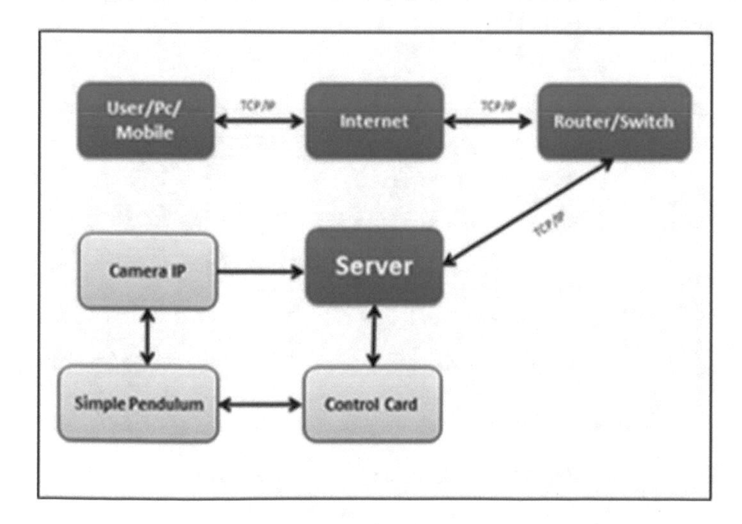

Fig. 10. Synoptic of remote labs

The user interface on the PC is an HTML web page provided by a LabVIEW software server. The access to the page is done in server [13] (Fig. 11).

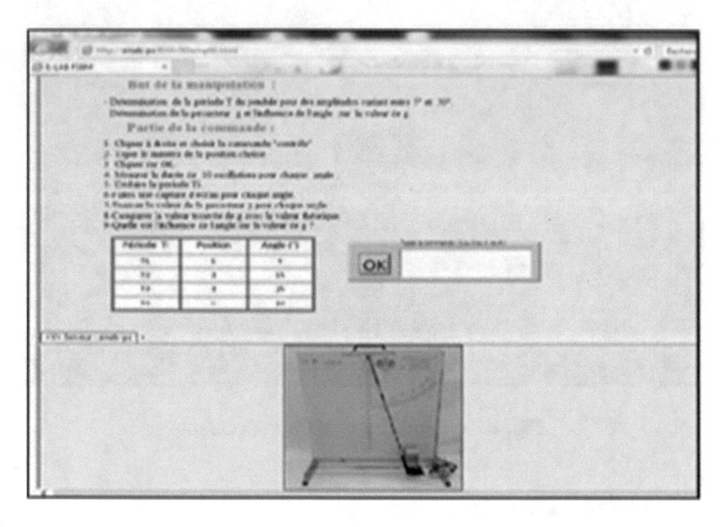

Fig. 11. User interface

This is the user interface that will be exposed to clients who will remotely connect to our web server. The Web server is a service provided by LabView software housed in a computer (or physical server). The web page (user interface) is presented in an ergonomic manner to guide students through the practical work, in this case in French language which is the learning language in the current context.

This example shows the operating mode of the pendulum experiment:

The user has the possibility to control the pivot angle of the pendulum by entering the appropriate field (Fig. 12):

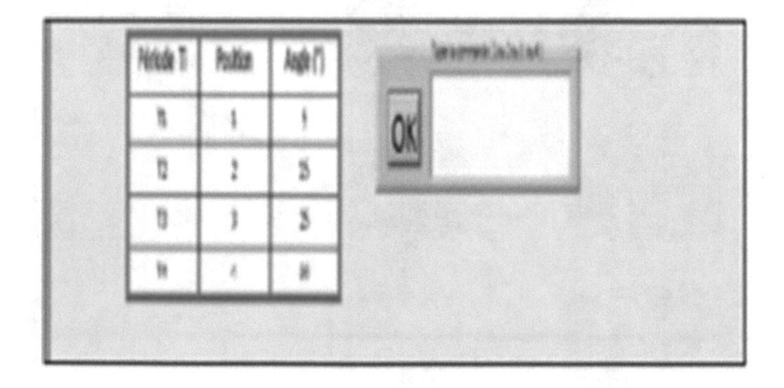

Fig. 12. Positions angle of the pendulum

The positions are provided in the table; then the clock is released, and the user is led to time 10 oscillations by viewing on the camera interface in order to deduce the period T (Fig. 13).

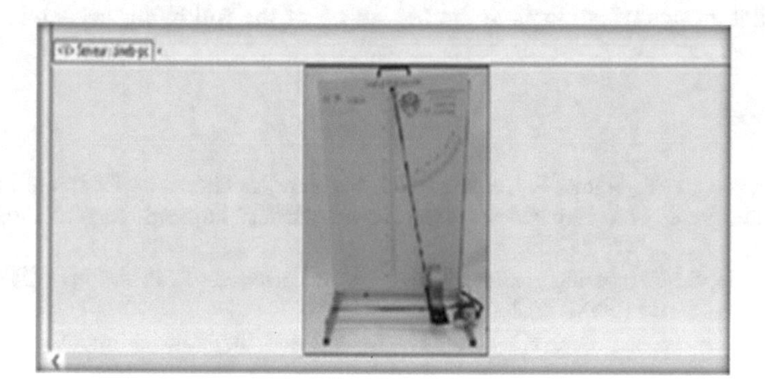

Fig. 13. Camera view user interface

Afterwards, the student must calculate the value of the gravitational constant g corresponding to each position, and finally compare it with the known theoretical value.

In order to determine the learning outcomes, the student is invited to make an analysis and deductions about the effect of the angle on the value of g following the results obtained by the experiment.

4 Conclusion

In this paper, we have worked on the design and creation of a prototype to make the equipment of the practical experiments of laboratories accessible to students via a server. Our work is in line with the work of our team, which has carried out several research projects on practical work at distance [11, 12].

In this work we have presented a global architecture (software and hardware) for the remote implementation of practical laboratory works in the field of physics.

The progress of our work will allow students not only to do simulations and virtual instrumentation, but they can also do real-world experiments linked with the physics of the phenomenon under study through observation or control. This is done remotely with TCP/IP network connection.

5 Perspectives

The prospects of our work are to put our prototype online and carry out tests to compare the measurements obtained with a conventional manipulation; then we will proceed with the implementation of a management platform for students and users.

Concerning the manipulation itself, we plan to introduce variations using a manipulator arm to vary the masses used in the simple pendulum, as well as a sensor to count the pendulum's oscillations.

Finally, we have planned a procedure for evaluating learning through this system on two pilot groups of students at the beginning of the following academic year.

References

1. Limpraptono, F.Y., Ratna, A.A.P., Sudibyo, H.: New Architecture of Remote Laboratories Multiuser based on Embedded Web Server (2012). https://doi.org/10.1109/rev.2012.6293113
2. Murugesan, S.: Harnessing green IT: principles and practices. IT Prof. **10**(1), 24–33 (2008). https://doi.org/10.1109/MITP.2008.10
3. Yang, L., Jiang, L., Yue, K., Pang, H.: Design and implementation of the lab remote monitoring system based on embedded web technology. In: 2010 International Forum on Information Technology and Applications (IFITA), vol. 2, pp. 172–175 (2010)
4. Garcia-Zubia, J., Lopez-de-Ipiña, D., Orduña, P.: Evolving towards better architectures for remote laboratories: a practical case. Int. J. Online Eng. (iJOE) **1**(2), 1 (2005)
5. Khazri, Y., Rouane, M., Fahli, A., Moussetad, M., Khaldouni, A., Naddami, A.: Developing a Remote Practice for Laboratory Experiments on Measuring Instruments (2015)
6. Khazri, Y., Al Sabri, A., Sabir, B., Toumi, H., Moussetad, M., Fahli, A.: Development and Management of a Remote Laboratory in Physics for Engineering Education (E-LAB FSBM), pp. 1–6 (2017). https://doi.org/10.1145/3090354.3090460
7. Rojko, A., Hercog, D., Jezernik, K.: E-training in mechatronics using innovative remote laboratory. Math. Comput. Simul. **82**, 508–516 (2011)
8. Barros, C., Leão, C.P., Soares, F., Minas, G., Machado, J.: RePhyS: a multidisciplinary experience in remote physiological systems laboratory iJOE. In: EDUCON2013, vol. 9, 5, June 2013
9. Fabregas, E., Farias, G., Dormido-Canto, S., Dormido, S., Esquembre, F.: Developing a remote laboratory for engineering education. Comput. Educ. **57**, 1686–1697 (2011)
10. Markan, C.M., Gupta, S., Mittal, S., Kumar, G.: Remote laboratories – a cloud based model for teleoperation of real laboratories. iJOE **9**(2), 36–43 (2013)
11. Sabir, B., Khazri, Y., Moussetad, M., Touri, B.: Hardware and software co-design of Arabic alphabets recognition platform for blind and visually impaired persons. Open Electr. Electron. Eng. J. **11**, 193–200 (2017). https://doi.org/10.2174/1874129001711010193
12. Al Sabri, A., Moussetad, M., Youness, A., Khazri, Y., Ahmed, F., Saeed, Al.: Create and develop remote labs for practical experiments, pp. 1–6 (2017). https://doi.org/10.1109/icca-ticet.2017.8095306
13. National Instruments, "Distance-Learning Remote Laboratories using LabVIEW"

Multimodal Data Representation Based on Multi-image Concept for Immersive Environments and Online Labs Development

Andreas Pester[1] and Yevgeniya Sulema[2](✉)

[1] Carinthia University of Applied Sciences, Villach, Austria
a.pester@cti.ac.at
[2] Igor Sikorsky Kyiv Polytechnic Institute, Kiev, Ukraine
sulema@pzks.fpm.kpi.ua

Abstract. In this paper we present an approach of multimodal data complex representation and processing based on the mathematical abstractions, such as aggregates and multi-images, as well as operations and relations between these mathematical objects, which are defined in the Algebraic System of Aggregates. This approach can be employed for online laboratories and immersive environments development for complex representation and processing of multimodal data which describe an object (process, phenomenon) of study. Such complex representation is resulted in obtaining a Digital Twin of the object of study that enables effective solving such tasks as modelling, prediction, classification, etc. In the paper we also propose the general architecture of an online laboratory and an immersive environment based on the presented mathematical background. A possible approach to software development for multimodal data processing is discussed in the paper as well.

Keywords: Immersive environment · Online lab · Multi-image concept · Algebraic system of aggregates

1 Introduction

Learning in engineering can require representation of a large amount of multimodal data describing a real-world object (subject, process, phenomenon, event) to be studied. The way of data presenting to a learner depends on an educational concept and its technical realization. In this paper we consider two cases: immersive environments and online labs. An immersive environment is a simulation which gives the sensation of physical presence of a learner in a certain realistic environment where he or she can interact with the object of study. An online lab does not suppose any immersion and it propose less spectacular but more practical way of learning the object of study by providing a learner with technical tools which allow to obtain (generate, measure, control, record) various characteristics of the object. In both cases we face with a necessity to represent and process multimodal data in a computer system used for hardware and software realization of a certain education approach. Since any software is based on specific algorithmic and mathematical methods, we focus our paper on this low-level stage of the development.

M. E. Auer and D. May (Eds.): REV 2020, AISC 1231, pp. 205–222, 2021.
https://doi.org/10.1007/978-3-030-52575-0_17

Our research objective is to provide developers of immersive environments and online labs with the mathematical apparatus which enables complex representation and easy processing of multimodal data describing a real-world object (subject, process, phenomenon, event) of studying. The motivation of this research is based on that there is a lack of both mathematical background and common approaches to multimodal data representation and processing. The majority of existing standards and methods for data representation, processing, and transmission are focused on specific media and they mostly do not consider combination of data modalities. Besides, there is a gap between physical nature of data (*modality*) and algorithms and methods of data processing. In our opinion, this gap should be filled in by a certain mathematical background for multimodal data representation and processing.

In this paper we propose a general structure of a mixed online laboratory. We suppose that such laboratory uses remote equipment for measuring different technical characteristics of a real object of study by using wide range of technical tools (sensors). It means that in the online lab we operate with multimodal data sequences which describe the object and its behaviour in course of time. To process such complex data defined in time domain, we need to present them in an appropriate way. In this paper we propose an overall approach for complex data representation in online labs based on the multi-image concept. The processing of a multi-image requires specific operations to be applied on it according to the purpose of data processing.

The same approach is applied to an immersive environment. In spite of that immersive environments (in contrast with online labs), suppose the use of artificially generated data, such multimodal data are also defined in time domain and therefore they also can be represented as a multi-image (i.e. mathematical and digital representation) of the object (subject, process, phenomenon, event) to be studied by learners in such immersive environment.

2 Related Work

In [1] a course, which offers a few virtual and remote laboratories based on automatic control, is proposed. Elio Sancristobal et al. [2] show the importance of the use of virtual and remote labs in education as well as the necessity of integrating them in learning scenarios, and show how modern universities have developed e-learning tools. The authors in [3] present a cost-effective approach to remote experimentation and introduce ArPi Lab, which is the remote laboratory for education in area of process control. The authors of [4] describe a completely functional Operational Amplifier iLab using an Android-based mobile platform that will allow students all over the world to conduct experiments remotely with a mobile device. Zoricu Nedic et al. [5] present remote laboratory NetLab with the specially designed graphical user interface. The augmented remote laboratory which is a new concept of virtual and remote laboratories and allows students to experience sensations and explore learning experiences that may exceed those offered by traditional laboratory classes to a certain extent, is proposed in [6]. A state-of-the-art remote laboratory project called 'Virtual Instrument Systems in Reality', which allows wiring and measuring of electronic circuits remotely on a virtual workbench that replicates physical circuit breadboards, is described in [7]. More examples of both virtual labs and remote labs are presented in [8–10] and in [11, 12] respectively.

The authors of [13] conduct a literature overview of modern remote laboratories and identify possible paths of development for the next generation of remote laboratories. Internet functions, which include video file generation and real-time control as MWS features useful for undergraduate courses, are described in [14]. In [15] Francisco Esquembre presents the study on creation and use of virtual and remote laboratories for improving Science and Engineering teaching and learning. In [16] the authors propose a review of the literature selection to compare the value of physical and virtual investigations and offer recommendations on combining them in order to strengthen science learning.

Veljko Potkonjak et al. [17] outline problems in existing concepts and technologies in the field of fully-software-based virtual laboratories. [18] presents Embedded Systems' Hardware-Software CoDesign as well as an overview of the approach based on using ready platforms. The authors of [19] introduce the design and implementation of Networked Control System Laboratory 3D, which is a web-based 3D control laboratory for remote real-time experimentation.

In [20] Christophe Salzmann et al. present the Smart Device specification aimed at interfacing with remote labs as well as the extensible and platform-agnostic specification of the Smart Device services and internal functionalities. The authors of [21] review the various online delivery methods and discuss possible future directions for virtual and remote engineering laboratory development. Ruben Heradio et al. [22] analyze the literature on virtual and remote labs identifying the most influential publications and the most researched topics in order to study the evolution of interest in these topics along the way.

Thus, we can conclude that there is a need in such educational tools as immersive environments and online labs; at the same time, new mathematical and software approaches should be developed to make representation and processing of multimodal data in these tools more productive.

3 Mathematical Background

3.1 Aggregates and Multi-images

In this paper we employ both the Algebraic System of Aggregates (ASA) [23–25] and the concept of multi-image, which is based on ASA, for complex representation of information about an object (subject, process, phenomenon, event) to be studied by learners in engineering education process.

ASA is an algebraic system [26, 27] which operates with special mathematical objects called *aggregates*. An aggregate is a data structure which consists of ordered data sets – tuples. Tuple elements can belong to any data type, including numbers (integer, single, double, etc.; floating point, fixed point), symbolic values (characters, strings, etc.), logical values, linguistic variables, etc. These values can be both sharp and fuzzy; they also can be undefined for a certain moment of time.

Formally, an *aggregate* A can be defined as a tuple of arbitrary tuples, elements of which belong to predefined sets [23–25]:

$$A = \left[\left[M_j | \langle a_i^j \rangle_{i=1}^{n_j}\right]\right]_{j=1}^{N} = [\![\{A\}|\langle A\rangle]\!] \quad (1)$$

where $\{A\}$ is a tuple of sets M_j, $\langle A\rangle$ is a tuple of elements tuples $\langle a_i^j \rangle_{i=1}^{n_j}$ corresponding to the tuple of sets ($a_i^j \in M_j$).

From practical point of view, any tuple in an aggregate is a sequence of data values obtained (generated, measured, recorded) in course of time. For example, we can measure such data values by using a number of sensors which allow us to observe the object of study by registering its characteristics. Some of sensors can register the same type of data and other sensors register different modalities of the object's nature and behaviour. In some cases, not only time but also a coordinate stamp is important.

To obtain an overall description of the object of study, we use the multi-image concept. A *multi-image* is an aggregate the first tuple of which is time values tuple. These values can be natural numbers or values of any other type which can be used for evident and monosemantic representation of time; they define moments when data values of other modalities are obtained (generated, measured, recorded). Thus, the multi-image of an object is its "digital" representation as a complex data structure which describes the object based on a number of its data modalities with respect to time. In fact, a multi-image can be considered as a data model for a Digital Twin concept [28].

Formally, a *multi-image* [23–25] is a non-empty aggregate such as:

$$I = \left[\left[T, M_1, \ldots M_N | \langle t_1, \ldots t_\tau \rangle, \left\langle a_1^1, \ldots, a_{n_1}^1 \right\rangle, \ldots, \left\langle a_1^N, \ldots, a_{n_N}^N \right\rangle \right]\right] \quad (2)$$

where T is a set of time values; $\tau \geq n_i, i \in [1, \ldots, N]$.

The main advantage of the multi-image concept is that we can operate with multiple values defined in a certain time moment as with a complex value of a function of several variables. Thus, we always have an overall view on the object's nature and behaviour in every moment of time. It allows us to compare multi-images, predict further state and behaviour of the object, model this behaviour, etc.

Moreover, having the object's multi-image for a certain time duration, we can reproduce the object virtually by using hardware for its 2D or 3D visualization, monophonic, stereophonic or spatial sound reproduction as well as reconstruction of other features of the object of study by using certain actuators.

3.2 Operations and Relations in ASA

There are three main types of operations on aggregates in ASA: arithmetical operations, logical operations, and ordering operations.

The logical operations on aggregates are *union*, *intersection*, *difference*, *symmetric difference*, *exclusive intersection* [23]. The result of logical operations depends on aggregates compatibility: aggregates can be compatible, quasi-compatible, incompatible, or hiddenly compatible. Aggregates A_1 and A_2 are *compatible* ($A_1 \div A_2$) if they have equal lengths as well as both the sets type and sequence order in these aggregates are the same. Aggregates A_1 and A_2 are *quasi-compatible* ($A_1 \doteq A_2$) if both the sets type and sequence order in these aggregates coincide partly. Aggregates A_1 and A_2 are *incompatible* ($A_1 \stackrel{\circ}{=} A_2$) if neither sets type nor sets sequence order in these aggregates coincide. Aggregates A_1 and A_2 are *hiddenly compatible* ($A_1 (\div) A_2$) if both aggregates have the same set of sets but the order of these sets differs. Hiddenly compatible aggregates can be transformed into compatible ones by applying ordering operations [24].

Let us consider the following aggregates:

$$A_1 = [[M_1, M_2, \ldots, M_N | \langle a_1^1, a_2^1, \ldots, a_l^1 \rangle, \langle b_1^1, b_2^1, \ldots, b_m^1 \rangle, \ldots, \langle w_1^1, w_2^1, \ldots, w_n^1 \rangle]] \quad (3)$$

$$A_2 = \left[\left[M_1, M_2, \ldots, M_N | \langle a_1^2, a_2^2, \ldots, a_r^2 \rangle, \left\langle b_1^2, b_2^2, \ldots, b_q^2 \right\rangle, \ldots, \left\langle w_1^2, w_2^2, \ldots, w_p^2 \right\rangle\right]\right] \quad (4)$$

$$A_3 = [[M_1^1, M_2^1, \ldots, M_N^1 | \langle a_1, a_2, \ldots, a_l \rangle, \langle b_1, b_2, \ldots, b_m \rangle, \ldots \langle w_1, w_2, \ldots, w_n \rangle]] \quad (5)$$

$$A_4 = [[M_1^2, M_2^2, \ldots, M_K^2 | \langle c_1, c_2, \ldots, c_r \rangle, \langle d_1, d_2, \ldots, d_q \rangle, \ldots \langle z_1, z_2, \ldots, z_p \rangle]] \quad (6)$$

$$\begin{aligned} A_5 = [[M_1, M_2^1, \ldots, M_x, \ldots, M_N^1 | \langle a_1^1, a_2^1, \ldots, a_l^1 \rangle, \langle b_1, b_2, \ldots, b_m \rangle, \ldots, \\ \langle f_1^1, f_2^1, \ldots, f_t^1 \rangle, \ldots, \langle w_1, w_2, \ldots, w_n \rangle]] \end{aligned} \quad (7)$$

$$\begin{aligned} A_6 = [[M_1, M_2^2, \ldots, M_x, \ldots, M_K^2 | \langle a_1^2, a_2^2, \ldots, a_r^2 \rangle, \langle d_1, d_2, \ldots, d_q \rangle, \ldots, \\ \langle f_1^2, f_2^2, \ldots, f_v^2 \rangle, \ldots, \langle z_1, z_2, \ldots, z_p \rangle]] \end{aligned} \quad (8)$$

Then, a *union* of compatible aggregates ($A_1 \div A_2$) is the aggregate which includes components of both aggregates, with elements of i-tuple of A_2 added into the end of i-tuple of A_1:

$$\begin{aligned} A_1 \cup A_2 = [[M_1, M_2, \ldots, M_N | \langle a_1^1, a_2^1, \ldots, a_l^1, a_1^2, a_2^2, \ldots, a_r^2 \rangle, \\ \langle b_1^1, b_2^1 \ldots, b_m^1, b_1^2, b_2^2, \ldots, b_q^2 \rangle, \ldots \langle w_1^1, w_2^1, \ldots, w_n^1, w_1^2, w_2^2, \ldots, w_p^2 \rangle]]. \end{aligned} \quad (9)$$

If aggregates are incompatible ($A_3 \stackrel{\circ}{=} A_4$), their *union* is an aggregate which includes components of both aggregates, with tuples of A_4 added into the end of the tuple of tuples of A_3 and the corresponding sets of A_4 added into the end of the sets sequence of A_3:

$$\begin{aligned} A_3 \cup A_4 = [[M_1^1, M_2^1, \ldots, M_N^1, M_1^2, M_2^2, \ldots, M_K^2 | \langle a_1, a_2, \ldots, a_l \rangle, \langle b_1, b_2, \ldots, b_m \rangle, \\ \ldots, \langle w_1, w_2, \ldots, w_n \rangle, \langle c_1, c_2, \ldots, c_r \rangle, \langle d_1, d_2, \ldots, d_q \rangle, \ldots, \langle z_1, z_2, \ldots, z_p \rangle]]. \end{aligned} \quad (10)$$

If aggregates are quasi-compatible ($A_5 \doteq A_6$), their *union* is an aggregate which includes components of both aggregates, with elements of i-tuple of A_6 added into the end of i-tuple of A_5 if elements of these tuples belong to the same set; otherwise, the tuples of A_6 are added into the end of the tuple of tuples of A_5 and the corresponding sets of A_6 are added into the end of the sets sequence of A_5:

$$\begin{aligned} A_5 \cup A_6 = [\![& M_1, M_2^1, \ldots, M_x, \ldots, M_N^1, M_2^2, \ldots, M_K^2 | \langle a_1^1, a_2^1, \ldots, a_l^1, a_1^2, a_2^2, \ldots, a_r^2 \rangle \\ & \langle b_1, b_2, \ldots, b_m \rangle, \ldots, , \langle f_1^1, f_2^1, \ldots, f_t^1, f_1^2, f_2^2, \ldots, f_v^2 \rangle, \ldots, \\ & \langle w_1, w_2, \ldots, w_n \rangle, \langle d_1, d_2, \ldots, d_q \rangle, \ldots, \langle z_1, z_2, \ldots, z_p \rangle]\!]. \end{aligned} \tag{11}$$

An *intersection* of compatible aggregates ($A_1 \doteqdot A_2$) is an aggregate which includes only common components of both aggregates in every tuple:

$$\begin{aligned} A_1 \cap A_2 = [\![& M_1, M_2, \ldots, M_N | \langle a_{l_1}^1, \ldots, a_{l_\alpha}^1, a_{r_1}^2, \ldots, a_{r_\beta}^2 \rangle, \langle b_{m_1}^1, \ldots, b_{m_\gamma}^1, b_{q_1}^2, \ldots, b_{q_\delta}^2 \rangle, \\ & \ldots, \langle w_{n_1}^1, \ldots, w_{n_\lambda}^1, w_{p_1}^2, \ldots, w_{p_\mu}^2 \rangle]\!], \end{aligned} \tag{12}$$

where $a_{l_i}^1 \in \overline{a^1}, a_{l_i}^1 \in \overline{a^2}, i \in \langle 1, \ldots, \alpha \rangle; a_{r_j}^2 \in \overline{a^1}, a_{r_j}^2 \in \overline{a^2}, j \in \langle 1, \ldots, \beta \rangle$; $b_{m_k}^1 \in \overline{b^1}, b_{m_k}^1 \in \overline{b^2}, k \in \langle 1, \ldots, \gamma \rangle; b_{q_s}^2 \in \overline{b^1}, b_{q_s}^2 \in \overline{b^2}, s \in \langle 1, \ldots, \delta \rangle$;
$w_{n_u}^1 \in \overline{w^1}, w_{n_u}^1 \in \overline{w^2}, u \in \langle 1, \ldots, \lambda \rangle; w_{p_y}^2 \in \overline{w^1}, w_{p_y}^2 \in \overline{w^2}, y \in \langle 1, \ldots, \mu \rangle$.

If aggregates are incompatible ($A_3 \stackrel{\circ}{=} A_4$), their *intersection* is the null-aggregate, and if aggregates are quasi-compatible ($A_5 \doteq A_6$), their *intersection* is an aggregate which includes only common components of both aggregates only in the tuples of common sets:

$$A_5 \cap A_6 = \left[\!\!\left[M_1, \ldots, M_x | \left\langle a_{l_1}^1, \ldots, a_{l_\alpha}^1, a_{r_1}^2, \ldots, a_{r_\beta}^2 \right\rangle, \ldots, \left\langle f_{t_1}^1, \ldots, f_{t_\rho}^1, f_{v_1}^2, \ldots, f_{v_\omega}^2 \right\rangle \right]\!\!\right] \tag{13}$$

where $a_{l_i}^1 \in \overline{a^1}, a_{l_i}^1 \in \overline{a^2}, i \in \langle 1, \ldots, \alpha \rangle; a_{r_j}^2 \in \overline{a^1}, a_{r_j}^2 \in \overline{a^2}, j \in \langle 1, \ldots, \beta \rangle$;

$$f_{t_e}^1 \in \overline{f^1}, f_{t_e}^1 \in \overline{f^2}, e \in \langle 1, \ldots, \rho \rangle; f_{v_h}^2 \in \overline{f^1}, f_{v_h}^2 \in \overline{f^2}, h \in \langle 1, \ldots, \omega \rangle$$

A *difference* of compatible aggregates, $A_1 \doteqdot A_2$, is an aggregate which includes only components present in A_1 and absent in A_2 in every tuple:

$$A_1 \backslash A_2 = \left[\!\!\left[M_1, M_2, \ldots, M_N | \left\langle a_{i_1}^1, \ldots, a_{i_\alpha}^1 \right\rangle, \left\langle b_{m_1}^1, \ldots, b_{m_\gamma}^1 \right\rangle, \ldots, \left\langle w_{n_1}^1, \ldots, w_{n_\lambda}^1 \right\rangle \right]\!\!\right] \tag{14}$$

where $a_{l_i}^1 \in \overline{a^1}, a_{l_i}^1 \notin \overline{a^2}, i \in \langle 1, \ldots, \alpha \rangle; b_{m_k}^1 \in \overline{b^1}, b_{m_k}^1 \notin \overline{b^2}, k \in \langle 1, \ldots, \gamma \rangle; w_{n_u}^1 \in \overline{w^1}$, $w_{n_u}^1 \notin \overline{w^2}, u \in \langle 1, \ldots, \lambda \rangle$.

If aggregates are incompatible ($A_3 \stackrel{\circ}{=} A_4$), their *difference* is an aggregate equal to the first of them (A_3), and if aggregates are quasi-compatible ($A_5 \doteq A_6$), their *difference* is an aggregate which includes elements of A_5, which are absent in A_6, in tuples of common sets and all tuples of sets defined only in A_5:

$$A_5 \backslash A_6 = [\![M_1, M_2^1, \ldots, M_x, \ldots, M_N^1 | \langle a_{i_1}^1, \ldots, a_{i_\alpha}^1 \rangle, \langle b_1, b_2, \ldots, b_m \rangle, \ldots, \\ \langle f_{t_1}^1, \ldots, f_{t_\rho}^1 \rangle, \ldots, \langle w_1, w_2, \ldots, w_n \rangle]\!] \tag{15}$$

where $a_{l_i}^1 \in \overline{a^1}, a_{l_i}^1 \notin \overline{a^2}, i \in \langle 1, \ldots, \alpha \rangle; f_{t_e}^1 \in \overline{f^1}, f_{t_e}^1 \notin \overline{f^2}, e \in \langle 1, \ldots, \rho \rangle$.

A *symmetric difference* of compatible aggregates, $A_1 \div A_2$, is an aggregate which includes both components present in A_1 and absent in A_2 and components present in A_2 and absent in A_1 in every tuple:

$$A_1 \Delta A_2 = [\![M_1, M_2, \ldots, M_N | \langle a_{l_1}^1, \ldots, a_{l_\alpha}^1, a_{r_1}^2, \ldots, a_{r_\beta}^2 \rangle, \langle b_{m_1}^1, \ldots, b_{m_\gamma}^1, b_{q_1}^2, \ldots, b_{q_\delta}^2 \rangle, \\ \ldots, \langle w_{n_1}^1, \ldots, w_{n_\lambda}^1, w_{p_1}^2, \ldots, w_{p_\mu}^2 \rangle]\!], \tag{16}$$

where $a_{l_i}^1 \in \overline{a^1}, a_{l_i}^1 \notin \overline{a^2}, i \in \langle 1, \ldots, \alpha \rangle; a_{r_j}^2 \notin \overline{a^1}, a_{r_j}^2 \in \overline{a^2}, j \in \langle 1, \ldots, \beta \rangle;$

$$b_{m_k}^1 \in \overline{b^1}, b_{m_k}^1 \notin \overline{b^2}, k \in \langle 1, \ldots, \gamma \rangle; b_{q_s}^2 \notin \overline{b^1}, b_{q_s}^2 \in \overline{b^2}, s \in \langle 1, \ldots, \delta \rangle;$$

$$w_{n_u}^1 \in \overline{w^1}, w_{n_u}^1 \notin \overline{w^2}, u \in \langle 1, \ldots, \lambda \rangle; w_{p_y}^2 \notin \overline{w^1}, w_{p_y}^2 \in \overline{w^2}, y \in \langle 1, \ldots, \mu \rangle$$

A *symmetric difference* of incompatible aggregates ($A_3 \mathrel{\mathring{=}} A_4$) is equal to the *union* of this aggregates. A *symmetric difference* of quasi-compatible aggregates, $A_5 \doteq A_6$, is an aggregate which includes elements of A_5, which are absent in A_6, and elements of A_6, which are absent in A_5, in tuples of common sets, all tuples of sets defined only in A_5, and all tuples of sets defined only in A_6:

$$A_5 \Delta A_6 = [\![M_1, M_2^1, \ldots, M_x, \ldots, M_N^1, M_2^2, \ldots, M_K^2 | \langle a_{l_1}^1, \ldots, a_{l_\alpha}^1, a_{r_1}^2, \ldots, a_{r_\beta}^2 \rangle, \\ \langle b_1, b_2, \ldots, b_m \rangle, \ldots, \langle f_{t_1}^1, \ldots, f_{t_\rho}^1, f_{v_1}^2, \ldots, f_{v_\omega}^2 \rangle, \ldots, \\ \langle w_1, w_2, \ldots, w_n \rangle, \langle d_1, d_2, \ldots, d_q \rangle, \ldots, \langle z_1, z_2, \ldots, z_p \rangle]\!] \tag{17}$$

where $a_{l_i}^1 \in \overline{a^1}, a_{l_i}^1 \notin \overline{a^2}, i \in \langle 1, \ldots, \alpha \rangle; a_{r_j}^2 \notin \overline{a^1}, a_{r_j}^2 \in \overline{a^2}, j \in \langle 1, \ldots, \beta \rangle;$

$$f_{t_e}^1 \in \overline{f^1}, f_{t_e}^1 \notin \overline{f^2}, e \in \langle 1, \ldots, \rho \rangle; f_{v_h}^2 \notin \overline{f^1}, f_{v_h}^2 \in \overline{f^2}, h \in \langle 1, \ldots, \omega \rangle$$

Finally, an *exclusive intersection* of compatible aggregates ($A_1 \div A_2$) is an aggregate which includes only components of the first aggregate (A_1) common for both aggregates in every tuple:

$$A_1 \neg A_2 = \left[\!\left[M_1, M_2, \ldots, M_N | \left\langle a_{l_1}^1, \ldots, a_{l_\alpha}^1 \right\rangle, \left\langle b_{m_1}^1, \ldots, b_{m_\gamma}^1 \right\rangle, \ldots \left\langle w_{n_1}^1, \ldots, w_{n_\lambda}^1 \right\rangle \right]\!\right], \tag{18}$$

where $a_{l_i}^1 \in \overline{a^1}, a_{l_i}^1 \in \overline{a^2}, i \in \langle 1, \ldots, \alpha \rangle; b_{m_k}^1 \in \overline{b^1}, b_{m_k}^1 \in \overline{b^2}, k \in \langle 1, \ldots, \gamma \rangle; w_{n_u}^1 \in \overline{w^1}, w_{n_u}^1 \in \overline{w^2}, u \in \langle 1, \ldots, \lambda \rangle$.

If aggregates are incompatible ($A_3 \stackrel{\circ}{=} A_4$), their *exclusive intersection* is the null-aggregate, and if aggregates are quasi-compatible ($A_5 \doteq A_6$), their *exclusive intersection* is an aggregate which includes components of the first aggregate (A_5) common for both aggregates only in tuples of common sets:

$$A_5 \neg A_6 = \left[\!\left[M_1, \ldots, M_x \middle| \left\langle a^1_{l_1}, \ldots, a^1_{l_\alpha} \right\rangle, \ldots, \left\langle f^1_{t_1}, \ldots, f^1_{t_\rho} \right\rangle, \right]\!\right] \tag{19}$$

where $a^1_{l_i} \in \overline{a^1}, a^1_{l_i} \in \overline{a^2}, i \in 1, \ldots, \alpha; f^1_{t_e} \in \overline{f^1}, f^1_{t_e} \in \overline{f^2}, e \in 1, \ldots, \rho.$

It is important that in contrast with logical operations on sets, logical operations on aggregates in ASA are non-commutative because the sequence order of elements and tuples is considered.

Ordering operations include *sets ordering*, *ascending sorting*, *descending sorting*, *singling*, *extraction*, and *insertion* [24].

A *sets ordering* operation reorders an aggregate according to a template aggregate. The template aggregate can be arbitrary, undefined or empty. If the aggregate A is defined as:

$$A = \left[\!\left[M_3, M_1, M_2, \ldots, M_N \middle| \left\langle a^3_{i_3} \right\rangle^{n_3}_{i_3=1}, \left\langle a^1_{i_1} \right\rangle^{n_1}_{i_1=1}, \left\langle a^2_{i_2} \right\rangle^{n_2}_{i_2=1}, \ldots, \left\langle a^N_{i_N} \right\rangle^{n_N}_{i_N=1} \right]\!\right] \tag{20}$$

and the template aggregate is defined as $A_t = [\![M_1, M_2, M_3, \ldots, M_N|\langle \_ \rangle]\!]$, then, the result of *sets ordering* operation on the aggregates A and A_{tem} is the aggregate defined as:

$$A \vDash A_t = \left[\!\left[M_1, M_2, M_3, \ldots, M_N \middle| \left\langle a^1_{i_1} \right\rangle^{n_1}_{i_1=1}, \left\langle a^2_{i_2} \right\rangle^{n_2}_{i_2=1}, \left\langle a^3_{i_3} \right\rangle^{n_3}_{i_3=1}, \ldots, \left\langle a^N_{i_N} \right\rangle^{n_N}_{i_N=1} \right]\!\right] \tag{21}$$

Operations *ascending sorting* and *descending sorting* enable reordering of all tuples according to new elements order (ascending or descending) of a certain tuple, which is called a *primary tuple*, among all tuples of the aggregate. For example, if $A_1 = \left[\!\left[M_j | \langle a^j_i \rangle^{n_j}_{i=1} \right]\!\right]^N_{j=1}$ and $\exists k$ such as $1 < k < N, k \neq 2$ and $n_1 > n_k > n_N, n_2 = n_k$, then the result of *ascending sorting* operation of A_1 according elements of tuple $\bar{a}^k$ is the aggregate such as:

$$\begin{aligned} A_1 \uparrow \bar{a}^k = [\![M_1, M_2, \ldots, M_k, \ldots, M_N | \langle a^1_\alpha, a^1_\beta, \ldots, a^1_\nu, \ldots, a^1_\omega, a^1_{n_k+1}, \ldots, a^1_{n_1} \rangle, \\ \langle a^2_\alpha, a^2_\beta, \ldots, a^2_\nu, \ldots, a^2_\omega \rangle, \ldots, \langle a^k_\alpha, a^k_\beta, \ldots, a^k_\nu, \ldots, a^k_\omega \rangle, \ldots \langle a^N_\alpha, a^N_\beta, \ldots, a^N_\nu \rangle]\!] \end{aligned} \tag{22}$$

where $1 \leq m \leq n$, $a^k_\alpha < a^k_\beta < \ldots < a^k_\nu < \ldots < a^k_\omega$, $a^j_m \in \langle a^j_i \rangle^{n_j}_{i=1}$, $j = 1 \ldots N$, $m \in [\alpha, \beta, \ldots, \nu, \ldots, \omega]$, $1 \leq m \leq n$, and $n = n_k$ if $n_j \geq n_k$ or $n = n_j$ if $n_j < n_k$.

A *singling* operation removes repetitions of the same value in a tuple. For example, if aggregate A_1 is defined as:

$$A_1 = [\![M_1,\ldots,M_k,\ldots,M_N|\langle a_1^1,\ldots,a_m^1,\ldots,a_{m+p}^1,a_{m+p+1}^1,\ldots,a_{n_1}^1\rangle,\ldots,\\ \langle a_1^k,\ldots,a_m^k,\ldots,a_{m+p}^k,a_{m+p+1}^k,\ldots,a_{n_k}^k\rangle,\ldots,\\ \langle a_1^N,\ldots,a_m^N,\ldots,a_{m+p}^N,a_{m+p+1}^N,\ldots,a_{n_N}^N\rangle]\!] \tag{23}$$

where $1 \le k \le N$, and let $\exists m_l, \forall l$ such as $a_m^k = a_{m+1}^k = \ldots = a_{m+p}^k$, $1 \le m \le (n_k - p), 1 \le p \le n_k$, then the result of *singling* operation on the aggregate A_1 by the tuple $\bar{a}^k$ is:

$$A_1 \parallel \bar{a}^k = [\![M_1,\ldots,M_k,\ldots,M_N|\langle a_1^1,\ldots,a_m^1,a_{m+p+1}^1,\ldots,a_{n_1}^1\rangle,\ldots,\\ \langle a_1^k,\ldots,a_m^k,a_{m+p+1}^k,\ldots,a_{n_k}^k\rangle,\ldots\langle a_1^N,\ldots,a_m^N,a_{m+p+1}^N,\ldots,a_{n_N}^N\rangle]\!] \tag{24}$$

An *extraction* operation allows us to remove an element from the tuple and an *insertion* operation enables adding a new element in the tuple. For example, if the aggregate A_1 is defined as:

$$A_1 = [\![M_1,\ldots,M_k,\ldots,M_N|\langle a_{i_1}^1\rangle_{i_1=1}^{n_1},\ldots,\langle a_1^k,\ldots,a_{m-1}^k,a_m^k,a_{m+1}^k,\ldots,a_{n_k}^k\rangle,\\ \ldots,\langle a_{i_N}^N\rangle_{i_N=1}^{n_N}]\!] \tag{25}$$

where $1 \le m \le N$ and $\exists d$ such as $d \in M_k$, $1 \le k \le N$, then the result of *extraction* of the element a_m^k from the aggregate A_1 is an aggregate such as:

$$A_1 \ltimes a_m^k = [\![M_1,\ldots,M_k,\ldots,M_N|\langle a_{i_1}^1\rangle_{i_1=1}^{n_1},\ldots,\langle a_1^k,\ldots,a_{m-1}^k,a_{m+1}^k,\ldots,a_{n_k}^k\rangle,\\ \ldots,\langle a_{i_N}^N\rangle_{i_N=1}^{n_N}]\!] \tag{26}$$

The result of *insertion* of d to the aggregate A_1 is an aggregate such as:

$$A_1 \rtimes \left(d \prec a_m^k\right) = A_1 \rtimes \left(d \succ a_{m-1}^k\right) = [\![M_1,\ldots,M_k,\ldots,M_N|\langle a_{i_1}^1\rangle_{i_1=1}^{n_1},\ldots,\\ \langle a_1^k,\ldots,a_{m-1}^k,d,a_m^k,a_{m+1}^k,\ldots,a_{n_k}^k\rangle,\ldots,\langle a_{i_N}^N\rangle_{i_N=1}^{n_N}]\!] \tag{27}$$

Thus, the ordering operations in the ASA enable reordering of both tuples in the aggregate and elements in tuples.

Relations in ASA include relations between tuple elements, relations between tuples, and relations between aggregates [29].

Relations between tuple elements are *is greater*; *is less*; *is equal*; *proceeds*; *succeeds*. The first three relations are based on elements value and the last two relations concern elements position in a tuple. Naturally, elements must belong to the same tuple.

Relations between tuples includes arithmetical relations, frequency relations, and temporal relations.

Arithmetical comparison is elementwise and includes the following relations: *is strictly greater*; *is majority-vote greater*; *is strictly less*; *is majority-vote less*; *is strictly equal*; *is majority-vote equal*.

Frequency relations are *is thicker*; *is rarer*; *is equally frequent*. If there are two time value tuples $\bar{t}^1 = \langle t_i^1 \rangle_{i=1}^{n_1}$ and $\bar{t}^2 = \langle t_i^2 \rangle_{i=1}^{n_2}$, then the tuple $\bar{t}^1$ can be *thicker* than the tuple $\bar{t}^2$ ($|\bar{t}^1| > |\bar{t}^2|$); $\bar{t}^1$ can be *rarer* than $\bar{t}^2$ ($|\bar{t}^1| < |\bar{t}^2|$), or they can be *equally frequent* ($|\bar{t}^1| = |\bar{t}^2|$).

Temporal relations are *coincides with*; *is before*; *is after*; *meets*; *is met by*; *overlaps*; *is overlapped by*; *during*; *contains*; *starts*; *is started by*; *finishes*; *is finished by*. The temporal relations between tuples are possible only if all elements of these tuples are unique discrete values $\bar{t} = \langle t_i \rangle_{i=1}^{n}$ such as that either $t_i < t_{i+1}$ or $t_i > t_{i+1}$ is true for all pairs (t_i, t_{i+1}), $\forall i \in [1...n-1]$. Then tuple $\bar{t}^1$ can *coincides with* $\bar{t}^2$ ($t_1^1 = t_1^2$, $t_{n_1}^1 = t_{n_2}^2$ and $n_1 = n_2$); $\bar{t}^1$ can be *before* $\bar{t}^2$ ($t_{n_1}^1 < t_1^2$); $\bar{t}^1$ can be *after* $\bar{t}^2$ ($t_1^1 > t_{n_2}^2$), etc. The temporal relations between tuples are based on Allen's interval-based temporal logic [30, 31].

Relations between tuples of aggregates are identical to relations between single tuples. However, relations between tuples can be applied only to compatible and quasi-compatible aggregates. Hiddenly compatible aggregates must be first transformed to compatible ones [23, 24].

4 Multimodal Data Representation and Processing

In this paper we consider two use cases for multimodal data representation and processing: an online laboratory and an immersive environment. In both cases we suppose that there is a real object (process, phenomenon, etc.) of study. The investigation of this object is conducted by registration of its characteristics by using of a set of digital sensors; these characteristics can be also generated. Since the characteristics reflect different sides of the object's nature and/or behaviour, we obtain sequences of multimodal data (e.g. temperature, humidity, speed, etc.).

4.1 Online Laboratory

An online laboratory is a hardware-software system which allows students to use remote equipment for making experiments for learning purposes. The hardware part of the online laboratory consists of digital sensors which measure the object's characteristics. The software part of the laboratory includes software tools for data processing, analysis, storing, transferring, visualization, etc.

A general architecture of the online laboratory based on the use of multimodal data is shown on Fig. 1. As it was mentioned above, we receive sequences of real-time data from digital sensors. Each of these N sensors might generate data in its specific format but it must always provide us with a pair "object's characteristic value"–"time value". Thus, in general case, we obtain N pairs of sequences ($\bar{t}^n$, $\bar{a}^n$) such as $\bar{t}^n = \langle t_i^n \rangle_{i=1}^{L_n}, t_i^n \in T, \bar{a}^n = \langle a_i^n \rangle_{i=1}^{L_n}, a_i^n \in \mathbb{R}$, $0 < L_n \leq d_{exp}$, where $\bar{t}^n$ is a sequence of values defining time moments t_i^n when values a_i^n, which reflect a certain characteristic of the object, are measured, T is a set of time values, d_{exp} is a duration of the experiment.

To obtain a complex time-wise description of the object's behaviour, we need to compose its multi-image according to (2):

$$I = \left[\left[T, M_1, \ldots, M_N | \langle t_1, \ldots, t_\tau \rangle, \left\langle s_1^1, \ldots, s_{n_1}^1 \right\rangle, \ldots, \left\langle s_1^N, \ldots, s_{n_N}^N \right\rangle \right]\right]$$

where s_i^j is a value obtained from $\bar{a}^n$ as a result of data consolidation.

Thus, to ensure correct interpretation of the data sequences received from the sensors, at first, we need to synchronize them before composing a multi-image of the object. The synchronization has especial importance when the remote connection with sensors is not stable because of high-loaded network conditions.

If there are two concurrent measurements presented as tuples $\bar{a}^1 = \langle a_i^1 \rangle_{i=1}^{n_1}$ and $\bar{a}^2 = \langle a_i^2 \rangle_{i=1}^{n_2}$ which relate to the same object of study and registered in discrete moments of time defined by tuples $\bar{t}^1 = \langle t_i^1 \rangle_{i=1}^{n_1}$ and $\bar{t}^2 = \langle t_i^2 \rangle_{i=1}^{n_2}$, then in terms of ASA these tuples are connected by temporal relation *coincides*. Let us consider mathematical models of data synchronization [29] for the case when $\bar{t}^1$ *is equally frequent* to $\bar{t}^2$, i.e. $n_1 = n_2$.

For example, if values a_i^1 and a_i^2 are measured simultaneously ($t_i^1 = t_i^2$) but value a_{i+1}^1 is measured later than value a_{i+1}^2 is measured ($t_{i+1}^1 > t_{i+1}^2$) and we know that $i \in [2\ldots(n_1 - 2)]$ and $i\ mod\ 2 = 0$, then the multi-image I is to be composed from values defined as follows:

$$\begin{array}{lllll} t_1 = t_1^1 & t_{j-1} = t_i^1 & t_j = t_{i+1}^2 & t_{j+1} = t_{i+1}^1 & t_n = t_{n_1}^1 \\ s_1^1 = a_1^1 & s_{j-1}^1 = a_i^1 & s_j^1 = \emptyset & s_{j+1}^1 = a_{i+1}^1 & s_n^1 = a_{n_1}^1 \\ s_1^2 = a_1^2 & s_{j-1}^2 = a_i^2 & s_j^2 = a_{i+1}^2 & s_{j+1}^2 = \emptyset & s_n^2 = a_{n_2}^2 \end{array} \quad (28)$$

where $j = \frac{3i}{2}$ and $n = \frac{3n_1}{2} - 1$.

If $t_i^1 = t_i^2$ and $t_{i+1}^1 < t_{i+1}^2$, it means that value a_{i+1}^1 is measured earlier than value a_{i+1}^2 is measured, then the tuple values of the multi-image I are the following:

$$\begin{array}{lllll} t_1 = t_1^1 & t_{j-1} = t_i^1 & t_j = t_{i+1}^1 & t_{j+1} = t_{i+1}^2 & t_n = t_{n_1}^1 \\ s_1^1 = a_1^1 & s_{j-1}^1 = a_i^1 & s_j^1 = a_{i+1}^1 & s_{j+1}^1 = \emptyset & s_n^1 = a_{n_1}^1 \\ s_1^2 = a_1^2 & s_{j-1}^2 = a_i^2 & s_j^2 = \emptyset & s_{j+1}^2 = a_{i+1}^2 & s_n^2 = a_{n_2}^2 \end{array} \quad (29)$$

where $j = \frac{3i}{2}$ and $n = \frac{3n_1}{2} - 1$.

If the measurement of value a_i^1 precedes the measurement of a_i^2 ($t_i^1 < t_i^2$) but values a_{i+1}^1 and a_{i+1}^2 are measured are measured simultaneously ($t_{i+1}^1 = t_{i+1}^2$), then the tuple values of the multi-image I can be defined as:

$$\begin{array}{lllll} t_1 = t_1^1 & t_{j-1} = t_i^1 & t_j = t_i^2 & t_{j+1} = t_{i+1}^1 & t_n = t_{n_1}^1 \\ s_1^1 = a_1^1 & s_{j-1}^1 = a_i^1 & s_j^1 = \emptyset & s_{j+1}^1 = a_{i+1}^1 & s_n^1 = a_{n_1}^1 \\ s_1^2 = a_1^2 & s_{j-1}^2 = \emptyset & s_j^2 = a_i^2 & s_{j+1}^2 = a_{i+1}^2 & s_n^2 = a_{n_2}^2 \end{array} \quad (30)$$

where $j = \frac{3i}{2}$ and $n = \frac{3n_1}{2} - 1$.

If the measurement of value a_i^1 precedes the measurement of a_i^2 ($t_i^1 < t_i^2$) and the measurement of value a_{i+1}^1 succeeds the measurement of a_{i+1}^2 ($t_{i+1}^1 > t_{i+1}^2$), then the tuple values of the multi-image I are the following:

$$\begin{array}{llllll} t_1 = t_1^1 & t_{j-2} = t_i^1 & t_{j-1} = t_i^2 & t_j = t_{i+1}^2 & t_{j+1} = t_{i+1}^1 & t_n = t_{n_1}^1 \\ s_1^1 = a_1^1 & s_{j-2}^1 = a_i^1 & s_{j-1}^1 = \emptyset & s_j^1 = \emptyset & s_{j+1}^1 = a_{i+1}^1 & s_n^1 = a_{n_1}^1 \\ s_1^2 = a_1^2 & s_{j-2}^2 = \emptyset & s_{j-1}^2 = \emptyset & s_j^2 = a_i^2 & s_{j+1}^2 = a_{i+1}^2 & s_n^2 = a_{n_2}^2 \end{array} \quad (31)$$

where $j = 2i$, $n = 2(n_1 - 1)$.

In the case, when $t_i^1 < t_i^2$, $t_{i+1}^1 < t_{i+1}^2$ and $t_{i+1}^1 > t_i^2$, the tuple values of the multi-image I are:

$$\begin{array}{llllll} t_1 = t_1^1 & t_{j-2} = t_i^1 & t_{j-1} = t_i^2 & t_j = t_{i+1}^1 & t_{j+1} = t_{i+1}^2 & t_n = t_{n_1}^1 \\ s_1^1 = a_1^1 & s_{j-2}^1 = a_i^1 & s_{j-1}^1 = \emptyset & s_j^1 = a_{i+1}^1 & s_{j+1}^1 = \emptyset & s_n^1 = a_{n_1}^1 \\ s_1^2 = a_1^2 & s_{j-2}^2 = \emptyset & s_{j-1}^2 = a_i^2 & s_j^2 = \emptyset & s_{j+1}^2 = a_{i+1}^2 & s_n^2 = a_{n_2}^2 \end{array} \quad (32)$$

where $j = 2i$, $n = 2(n_1 - 1)$.

The next case is when $t_i^1 < t_i^2$, $t_{i+1}^1 < t_{i+1}^2$ and $t_{i+1}^1 = t_i^2$; then the tuple values of the multi-image I can be obtained as:

$$\begin{array}{lllll} t_1 = t_1^1 & t_{j-1} = t_i^1 & t_j = t_i^2 & t_{j+1} = t_{i+1}^2 & t_n = t_{n_1}^1 \\ s_1^1 = a_1^1 & s_{j-1}^1 = a_i^1 & s_j^1 = a_{i+1}^1 & s_{j+1}^1 = \emptyset & s_n^1 = a_{n_1}^1 \\ s_1^2 = a_1^2 & s_{j-1}^2 = \emptyset & s_j^2 = a_i^2 & s_{j+1}^2 = a_{i+1}^2 & s_n^2 = a_{n_2}^2 \end{array} \quad (33)$$

where $j = \frac{3i}{2}$ and $n = \frac{3n_1}{2} - 1$.

Another possible case is when the measurement of value a_{i+1}^1 precedes the measurement of a_i^2 ($t_{i+1}^1 < t_i^2$); then the tuple values of the multi-image I are the following:

$$\begin{array}{llllll} t_1 = t_1^1 & t_{j-2} = t_i^1 & t_{j-1} = t_{i+1}^1 & t_j = t_i^2 & t_{j+1} = t_{i+1}^2 & t_n = t_{n_1}^1 \\ s_1^1 = a_1^1 & s_{j-2}^1 = a_i^1 & s_{j-1}^1 = a_{i+1}^1 & s_j^1 = \emptyset & s_{j+1}^1 = \emptyset & s_n^1 = a_{n_1}^1 \\ s_1^2 = a_1^2 & s_{j-2}^2 = \emptyset & s_{j-1}^2 = \emptyset & s_j^2 = a_i^2 & s_{j+1}^2 = a_{i+1}^2 & s_n^2 = a_{n_2}^2 \end{array} \quad (34)$$

where $j = 2i$, $n = 2(n_1 - 1)$.

If the measurement of value a_i^1 succeeds the measurement of a_i^2 ($t_i^1 > t_i^2$) and the measurement of value a_{i+1}^1 precedes the measurement of a_{i+1}^2 ($t_{i+1}^1 < t_{i+1}^2$), then the tuple values of the multi-image I are defined as follows:

$$\begin{array}{llllll} t_1 = t_1^1 & t_{j-2} = t_i^2 & t_{j-1} = t_i^1 & t_j = t_{i+1}^1 & t_{j+1} = t_{i+1}^2 & t_n = t_{n_1}^1 \\ s_1^1 = a_1^1 & s_{j-2}^1 = \emptyset & s_{j-1}^1 = a_i^1 & s_j^1 = a_{i+1}^1 & s_{j+1}^1 = \emptyset & s_n^1 = a_{n_1}^1 \\ s_1^2 = a_1^2 & s_{j-2}^2 = a_i^2 & s_{j-1}^2 = \emptyset & s_j^2 = \emptyset & s_{j+1}^2 = a_{i+1}^2 & s_n^2 = a_{n_2}^2 \end{array} \quad (35)$$

where $j = 2i$, $n = 2(n_1 - 1)$.

In the case when measurement of value a_i^1 succeeds the measurement of a_i^2 ($t_i^1 > t_i^2$) but values a_{i+1}^1 and a_{i+1}^2 are measured are measured simultaneously ($t_{i+1}^1 = t_{i+1}^2$), the tuple values of the multi-image I can be defined as follows:

$$\begin{array}{lllll} t_1 = t_1^1 & t_{j-1} = t_i^2 & t_j = t_i^1 & t_{j+1} = t_{i+1}^1 & t_n = t_{n_1}^1 \\ s_1^1 = a_1^1 & s_{j-1}^1 = \emptyset & s_j^1 = a_i^1 & s_{j+1}^1 = a_{i+1}^1 & s_n^1 = a_{n_1}^1 \\ s_1^2 = a_1^2 & s_{j-1}^2 = a_i^2 & s_j^2 = \emptyset & s_{j+1}^2 = a_{i+1}^2 & s_n^2 = a_{n_2}^2 \end{array} \quad (36)$$

where $j = \frac{3i}{2}$ and $n = \frac{3n_1}{2} - 1$.

If $t_i^1 > t_i^2$, $t_{i+1}^1 > t_{i+1}^2$ and $t_i^1 < t_{i+1}^2$ then the tuple values of the multi-image I can be obtained in the following way:

$$\begin{array}{llllll} t_1 = t_1^1 & t_{j-2} = t_i^2 & t_{j-1} = t_i^1 & t_j = t_{i+1}^2 & t_{j+1} = t_{i+1}^1 & t_n = t_{n_1}^1 \\ s_1^1 = a_1^1 & s_{j-2}^1 = \emptyset & s_{j-1}^1 = a_i^1 & s_j^1 = \emptyset & s_{j+1}^1 = a_{i+1}^1 & s_n^1 = a_{n_1}^1 \\ s_1^2 = a_1^2 & s_{j-2}^2 = a_i^2 & s_{j-1}^2 = \emptyset & s_j^2 = a_{i+1}^2 & s_{j+1}^2 = \emptyset & s_n^2 = a_{n_2}^2 \end{array} \quad (37)$$

where $j = 2i$, $n = 2(n_1 - 1)$.

If values a_i^1 and a_{i+1}^2 are measured simultaneously ($t_i^1 = t_{i+1}^2$), then the tuple values of the multi-image I are:

$$\begin{array}{lllll} t_1 = t_1^1 & t_{j-1} = t_i^2 & t_j = t_i^1 & t_{j+1} = t_{i+1}^1 & t_n = t_{n_1}^1 \\ s_1^1 = a_1^1 & s_{j-1}^1 = \emptyset & s_j^1 = a_i^1 & s_{j+1}^1 = a_{i+1}^1 & s_n^1 = a_{n_1}^1 \\ s_1^2 = a_1^2 & s_{j-1}^2 = a_i^2 & s_j^2 = a_{i+1}^2 & s_{j+1}^2 = \emptyset & s_n^2 = a_{n_2}^2 \end{array} \quad (38)$$

where $j = \frac{3i}{2}$ and $n = \frac{3n_1}{2} - 1$.

Finally, in the case when measurement of value a_i^1 succeeds the measurement of a_{i+1}^2 ($t_i^1 > t_{i+1}^2$), the tuple values of the multi-image I are defined as:

$$\begin{array}{llllll} t_1 = t_1^1 & t_{j-2} = t_i^2 & t_{j-1} = t_{i+1}^2 & t_j = t_i^1 & t_{j+1} = t_{i+1}^1 & t_n = t_{n_1}^1 \\ s_1^1 = a_1^1 & s_{j-2}^1 = \emptyset & s_{j-1}^1 = \emptyset & s_j^1 = a_i^1 & s_{j+1}^1 = a_{i+1}^1 & s_n^1 = a_{n_1}^1 \\ s_1^2 = a_1^2 & s_{j-2}^2 = a_i^2 & s_{j-1}^2 = a_{i+1}^2 & s_j^2 = \emptyset & s_{j+1}^2 = \emptyset & s_n^2 = a_{n_2}^2 \end{array} \quad (39)$$

where $j = 2i$, $n = 2(n_1 - 1)$.

All other temporal relations either do not require precise synchronization, such as relations *after* and *before*, or can be transformed into a combination of relations *coincides*, *after* and *before*.

The synchronization procedure is fulfilled by the Data Synchronization Module (Fig. 1).

After synchronization the data are ready for their aggregation. The purpose of aggregation is to obtain one complex description of the object of study. In fact, this description is a Digital Twin [26] of the real object. The aggregation procedure is based on the use of both logical and ordering operations of ASA. This procedure is fulfilled by the Data Aggregation Module and is resulted in obtaining the multi-image $I = \left[\left[T, \mathbb{R}, \ldots, \mathbb{R} | \langle t_i \rangle_{i=1}^{L_{max}}, \langle s_i^1 \rangle_{i=1}^{L_1}, \ldots, \langle s_i^n \rangle_{i=1}^{L_n}\right]\right]$, where L_{max} is the total duration of experimental session.

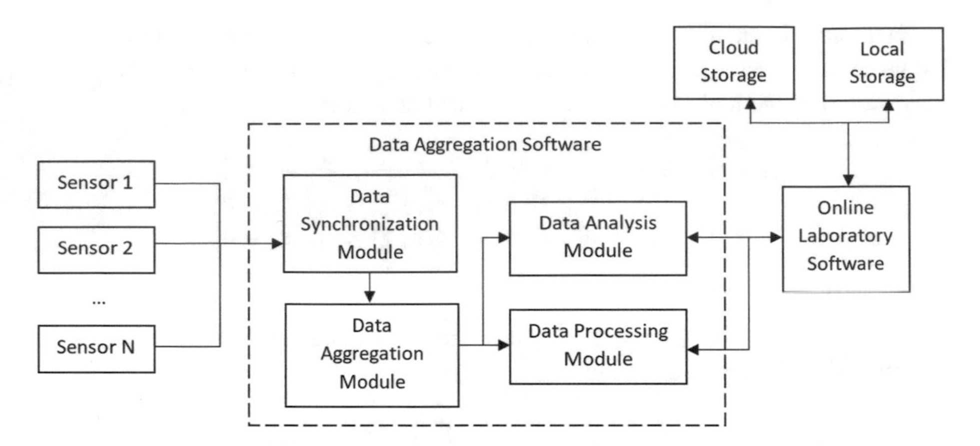

Fig. 1. A general architecture of an online laboratory based on multimodal data processing

When the multi-image is formed, we can do any further processing and/or analysis of the information we have collected about the object of study. These procedures are fulfilled by the Data Analysis Module and the Data Processing Module accordingly. These modules realize algorithms of data handling accordingly to the purpose of the online laboratory, nature of experiments, available equipment, etc. The input data set of these modules includes the aggregated data (multi-images) to be obtained from the Data Aggregation Module and requests on data processing and/or analysis to be obtained from the Online Laboratory Software. These requests are formed according to the online laboratory user's actions, i.e. experiments which a user fulfils while working with the online laboratory tools.

The Online Laboratory Software can require external database organized as local and/or cloud storages for keeping the user's experimental results and other data.

4.2 Immersive Environment

An immersive environment is a hardware-software system which enables simulation of a user's presence in an artificially generated scene (virtual world) where the user can supervise and/or interact with virtual objects (processes, phenomenon).

The hardware part of the immersive environment consists of digital sensors which monitor the use's actions in the immersive environment as well as actuators which enable 3D visualization, spatial sound reproduction, interaction of the user with virtual objects, etc. The software part of the immersive environment includes software tools for data processing, visualization, transferring, and storing.

Let us discuss the difference in the architecture of the immersive environment (Fig. 2) from the architecture of the online laboratory (Fig. 1).

In contrast to the online laboratory, in the immersive environment data for visualization and rendering can be received from different sources such as sensors, cloud and local storages. Thus, their synchronization is even more important task as in many cases these data sequences are linked to different time lines which can be both absolute and relative. The procedure of data synchronization in the immersive environment application is also based on operations and relations of ASA.

The data aggregation is fulfilled in a similar way as it is supposed to be done in the online laboratory. As a result of data aggregation in the immersive environment, we obtain a multi-image of the virtual reality scene.

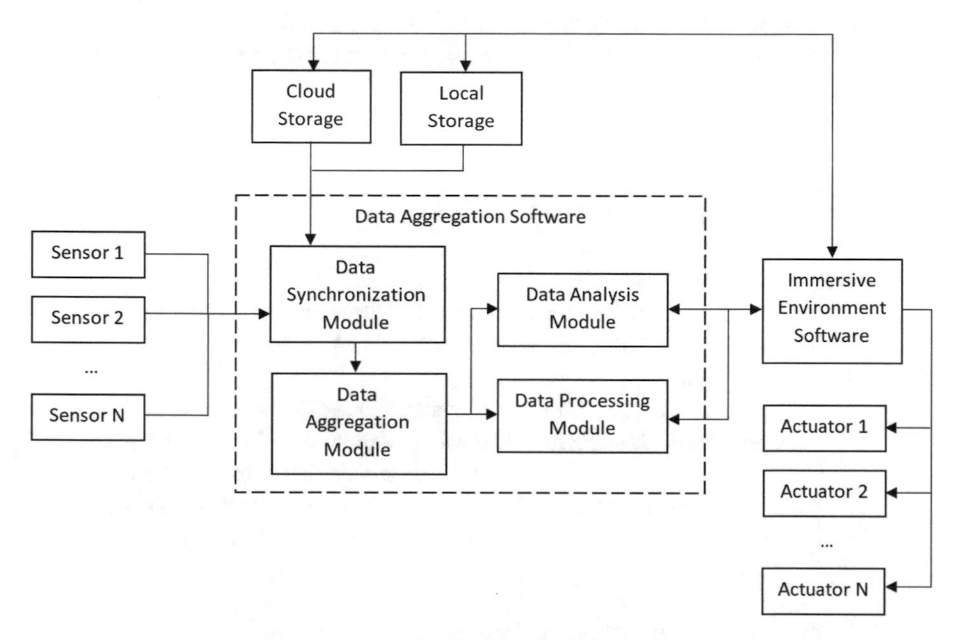

Fig. 2. A general architecture of an immersive environment based on multimodal data processing

Such immersive environment can be developed with the use of a domain-specific program language, e.g. ASAMPL. The general structure of a program in ASAMPL can be seen in [25]. The following fragment of the program in ASAMPL shows data processing principle in the immersive environment application for astronomy studying. In this example we suppose that there is 3D visualization equipment for planets demonstration as well as 2D monitors for visualization of graphs on spectroscopy of planetary atmospheres. Visual data of both types (video streams and graphs) are received from external online resources.

```
Program Astronomy {
  Libraries { ... }
  Handlers { ... }
  Renderers { ... }
  Sources {
    3DVisualStream1 Is 'https://nasa.edu.net/mars';
    3DVisualStream2 Is 'https://nasa.edu.net/moon';
    SpectrumData1 Is 'ftp://191.108.1.2320';

SpectrumData2 Is 'ftp://191.108.1.2319';
    SceneFile Is 'D:\Astronomy\lesson1.agg'; ... }
  Sets { ... }
  Elements {
    Duration1 is Time;
    Duration2 is Time;
    StartTime = 10:05; ... }
  Tuples { ... }
  Aggregates {
    Object1 = [3DVisualObject1, SpectrumDate1];
    Object2 = [3DVisualObject2, SpectrumDate2];
  Actions { ...
    Timeline StartTime : (StartTime+Duration1) {
      Download 3DVisualObject1 From 3DVisualStream1
                         With default.3DVideoLib;
      Download SpectrumObject1 From SpectrumData1
                           With default.VisualLib;}
        ...

    Timeline (StartTime+Duration1) :
(StartTime+Duration1+Duration2) {
      Download 3DVisualObject2 From 3DVisualStream2
                         With default.3DVideoLib;
      Download SpectrumObject2 From SpectrumData2
                           With default.VisualLib;}
        ...

    Scene is [Object1, Object2];
    Upload Scene To SceneFile With default.all;
    Render Scene With [VisualRen, GraphDataRen];} }
```

The advantage of ASAMPL for development of such applications consists in special focus on time-wise data representation and processing. However, general purpose program languages can also be effective for the cases discussed in this paper.

5 Conclusion

The research presented in this paper is focused on the mathematical background which is supposed to be used for development of educational software applications such as immersive environments and online laboratories. The main novelty of this research is that we propose a complex multimodal representation of an object of study in time domain. It gives new opportunities for data analysis, behaviour modelling and other tasks which can be a subject of study in engineering education. The proposed approach can be realized by using domain-specific programming language ASAMPL.

References

1. Sáenz, J., Chacón, J., de la Torre, L., Visioli, A., Dormido, S.: Open and low-cost virtual and remote labs on control engineering. IEEE Access (2015)
2. Sancristobal, E., Martín, S. Gil, R., Orduña, P., Tawfik, M., Pesquera, A., Diaz, G., Colmenar, A., García-Zubia, J., Castro, M.: State of art, initiatives and new challenges for virtual and remote labs. In: Proceedings of 12th IEEE International Conference on Advanced Learning Technologies, pp. 714–715 (2012)
3. Kalúz, M., Cirka, L., Valo, R., Fikar, M.: ArPi lab: a low-cost remote laboratory for control education. In: Proceedings of the 19th World Congress the International Federation of Automatic Control, Cape Town, South Africa (2014)
4. Oyediran, S.O., Ayodele, K.P., Akinwale, O.B., Kehinde, L.O.: Development of an operational amplifier iLab using an android-based mobile platform: work in progress. In: Proceedings of 120th ASEE Annual Conference & Exposition (2013)
5. Nedic, Z., Machotka, J., Najhlski, A.: Remote laboratories versus virtual and real laboratories. In: Proceedings of the 33rd ASEE/IEEE Frontiers in Education Conference (2003)
6. Andújar, J.M., et al.: Augmented reality for the improvement of remote laboratories: an augmented remote laboratory. IEEE Trans. Educ. **54**(3), 492–500 (2011)
7. Tawfik, M., et al.: Virtual Instrument Systems in Reality (VISIR) for remote wiring and measurement of electronic circuits on breadboard. IEEE Trans. Learn.Technol. **6**(1), 60–72 (2013)
8. Virtual Labs. Computer Science and Engineering. http://vlab.co.in/ba_labs_all.php?id=2
9. Learn Genetics. Virtual Labs. http://learn.genetics.utah.edu/content/labs/
10. BioInteractive. Virtual Labs. http://www.hhmi.org/biointeractive/explore-virtual-labs
11. Huawei Developer. Remote Lab. http://developer.huawei.com/ict/en/remotelab
12. Internet Remote Laboratory. http://remote-lab.fyzika.net
13. Gravier, C., et al.: State of the art about remote laboratories paradigms - foundations of ongoing mutations. Int. J. Online Eng. **4**(1), 1–9 (2008)
14. Valera, A., Díez, J.L., Vallés, M., Albertos, P.: Virtual and remote control laboratory development. IEEE Control Systems Magazine, pp. 35–39 (2005)
15. Esquembre, F.: Facilitating the creation of virtual and remote laboratories for science and engineering education. ScienceDirect, IFAC-PapersOnLine **48–29**, 049–058 (2015)
16. de Jong, T., Linn, M.C., Zacharia, Z.C.: Physical and virtual laboratories in science and engineering education. Science **340**, 305–308 (2013)
17. Potkonjak, V., et al.: Virtual laboratories for education in science, technology, and engineering: a review. Comput. Educ. **95**, 309–327 (2016)

18. Parkhomenko, A., et al.: Development and application of remote laboratory for embedded systems design. Int. J. Online Eng. **11**(3), 69–73 (2015)
19. Wenshan, H., Liu, G.-P., Zhou, H.: Web-based 3-D control laboratory for remote real-time experimentation. IEEE Trans. Industr. Electron. **60**(10), 4673–4682 (2013)
20. Salzmann, C., Govaerts, S., Halimi, W., Gillet, D.: The smart device specification for remote labs. Int. J. Online Eng. **11**, 20 (2015)
21. Chen, X., Song, G., Zhang, Y.: Virtual and remote laboratory development: a review. In: Proceeding of Earth and Space 2010: Engineering, Science, Construction, and Operations in Challenging Environments, pp. 3843–3852 (2010)
22. Heradio, R., et al.: Virtual and remote labs in education: a bibliometric analysis. Comput. Educ. **98**, 14–38 (2016)
23. Dychka, I., Sulema, Ye.: Logical operations in algebraic system of aggregates for multimodal data representation and processing. In: Research Bulletin of the National Technical University of Ukraine "Kyiv Polytechnic Institute", vol. 6, pp. 44–52 (2018)
24. Dychka, I., Sulema, Ye.: Ordering operations in algebraic system of aggregates for multi-image data processing. In: Research Bulletin of the National Technical University of Ukraine Kyiv Polytechnic Institute, vol. 1 (2019)
25. Sulema, Y.: ASAMPL: programming language for mulsemedia data processing based on algebraic system of aggregates. In: Advances in Intelligent Systems and Computing, Springer, vol. 725, pp. 431–442 (2018). https://doi.org/10.1007/978-3-319-75175-7_43
26. Maltsev, A.I.: Algebraic systems. Nauka, 392 p. (1970). (in Russian)
27. Fraenkel, A.A., et al.: Foundations of Set Theory. Elsevier, 415 p. (1973)
28. What Is Digital Twin Technology and Why Is It So Important? https://www.forbes.com/sites/bernardmarr/2017/03/06/what-is-digital-twin-technology-and-why-is-it-so-important/#573fc3c2e2a7
29. Sulema, Y., Kerre, E.: Multimodal Data Representation and Processing Based on Algebraic System of Aggregates, 37 p. (2019), preprint
30. Allen, J.F.: Maintaining knowledge about temporal intervals. Commun. ACM, 832–843 (1983). https://doi.org/10.1145/182.358434
31. James, F., Allen, P., Hayes, J.: Moments and points in an interval-based temporal logic. Comput. Intell. **5**(3), 225–238 (1989). https://doi.org/10.1111/j.1467-8640.1989.tb00329.x

Virtual Control Units in Remote Labs

Heinz-Dietrich Wuttke(✉), Karsten Henke, and René Hutschenreuter

TU Ilmenau, 98693 Ilmenau, Germany
Heinz-Dietrich.Wuttke@tu-ilmenau.de

Abstract. Enforced by current results of research in the field of the Internet of Things, working with virtual models has taken a significant step forward. To use such technologies in interactive collaborative and blended learning scenarios, virtual, remote and hybrid online labs has a long tradition in the engineering education process. In virtual or hybrid labs, parts of the experiments take place not on real, but on virtual instruments. In the paper, we present virtual design tools, used in a hybrid online lab that supports a course in designing combinational and sequential logic for control systems. We describe a new integrated hardware-oriented design tool called BEAST (Block-diagram Editing And Simulation Tool) supporting a circuit-oriented design process and works as virtual control unit in a hybrid online lab. We also discuss in which learning scenarios virtual control units have proven themselves.

Keywords: Virtual devices · Problem based learning · Digital design · Remote lab · Learning scenario · Collaborative lab-work

1 Introduction

Labs are divided into remote, virtual and hybrid [1]. Remote Labs allow remotely controlled experiments. Virtual labs work in virtual artificial worlds. Hybrid labs combine both approaches by allowing work with simulations in addition access to remote lab equipment. The discussed lab flexibly implements all these possibilities. This makes it possible to carry out all experiments entirely virtually, either entirely realistically or in a combination of both. In the experiments, the students have the task of designing a control algorithm for a given movement of a control object for a choice of ECUs. Microcontrollers, FPGAs, PLCs are available as real control units in the remote lab. As control objects, there are physical systems (electromechanical models of an elevator, a production cell, a pump station and further objects). To teach the design of control algorithms for these control units, different approaches are possible:

- a more abstract approach on the basis of Finite State Machines.
- a more theoretical oriented approach on the basis of Boolean equations and
- a hardware oriented approach on the basis of logic gates.

Our long-term teaching experiences show, that students often learn design steps of these approaches separately without seeing the equivalences between them. We have actually finished a tool-chain for demonstrating the connections between the three approaches. Thus, students can immediately validate the results in our remote lab by applying the design in remote experiments.

M. E. Auer and D. May (Eds.): REV 2020, AISC 1231, pp. 223–231, 2021.
https://doi.org/10.1007/978-3-030-52575-0_18

2 The Abstract FSM-Based Approach

To control a physical system, it has to be equipped with sensors, characterizing its actual positions and actuators to fulfil a required movement. Using these sensors and actuators, students have to design a control algorithm to stimulate the physical system (called the control object) to realize the required behavior and implement it in hardware or software on a control unit.

For a systematic and probable design of a control algorithm, the students learn formal methods based on automaton theory, namely the approach with Finite State Machines (FSM), which are also a basic part of the state diagrams in the UML-specification [2, 3].

To minimize the sources of errors it is beneficial to test the logic of the control algorithm before implementing it on a real control unit. For that purpose, the GOLDi-lab provides a virtual control unit. This control unit interprets Boolean equations for the next state function and the output function of a finite state machine.

To get these equations, a graphical tool named “graphical interactive FSM tool” (GIFT) supports the students in the design process. It shows in Fig. 1 e.g. contradictions and incomplete specifications.

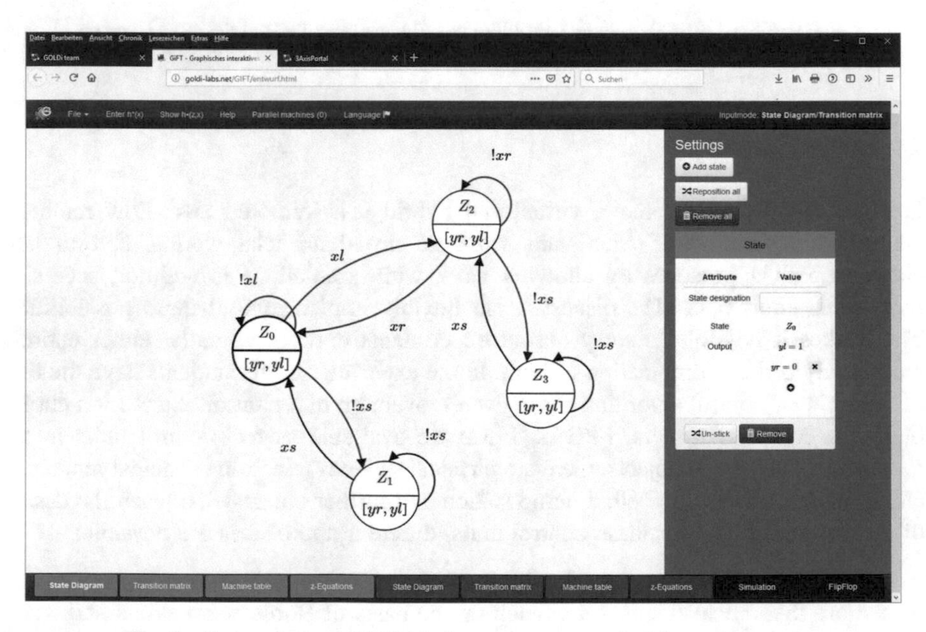

Fig. 1. Screenshot of GIFT (design errors are marked by red arrows)

An export function of the GIFT- system allows a direct test of the control algorithm on the control object. This export function automatically produces in the background the equations, necessary for the virtual control unit of the GOLDi-lab.

Equations, derived from an erroneous design like in Fig. 1 of course produce an erroneous behavior of the control object. However, even such errors are useful for teaching purposes and discussions with the students (see Sect. 5). For a deeper insight of the tool, please refer to [4].

3 The Approach Based on Boolean Equations

The GIFT-system derives automatically Boolean equations for the next state function and the output function of the abstract graphical FSM-design. The interpreter of the GOLDi-Lab is able to process these equations and therefor functioning as a virtual control unit. Figure 2 shows a screenshot of the experiments user interface with the imported equations from the GIFT- system for the output function of the FSM. The next state function is available by pressing the button “Machines”.

The design of the control algorithm is also possible by just editing the Boolean equations and optional importing them to the GIFT-system to visualize the corresponding automaton graph for a better understanding.

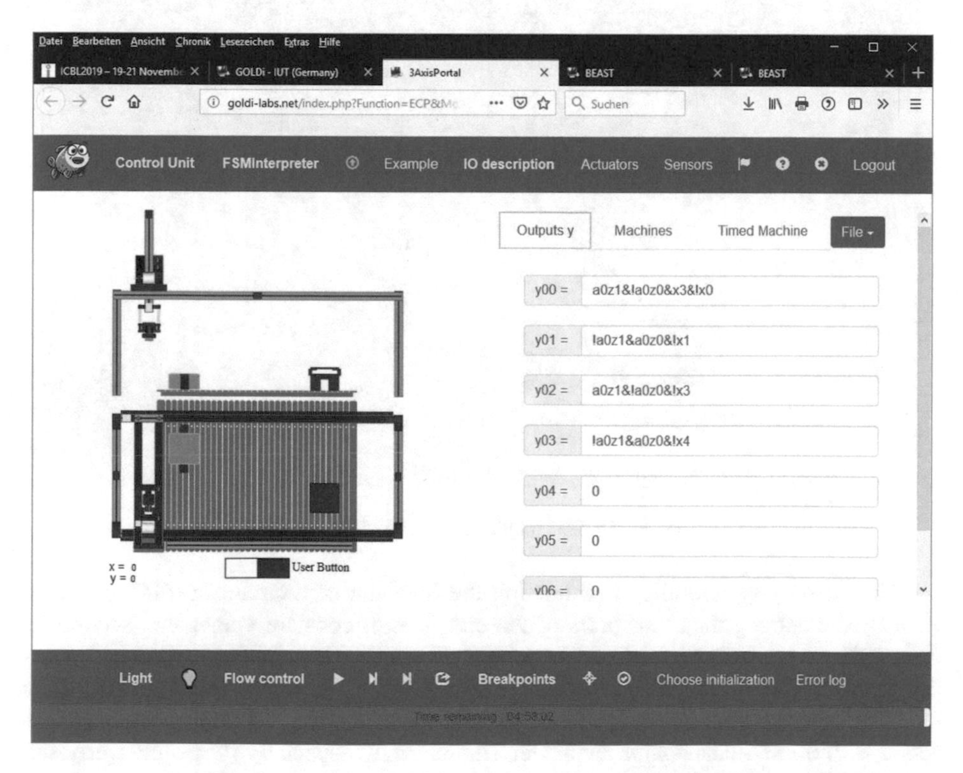

Fig. 2. Imported Boolean equations from the tool-chain

4 The Hardware-Oriented Approach

As a third design method, we provide the design based on hardware circuits. In the hardware-oriented approach, students design a control algorithm supported by the BEAST-system [5]. BEAST stands for "Block-diagram Editing And Simulation Tool" and allows the design of combinational and sequential circuits, based on elementary gate structures and more complex circuits like arithmetic operation networks, decoders counters and so on from a library. It is possible to extend the library by own designed components.

Figure 3 shows the design panel of the BEAST-system with a sequential circuit design of a FSM. Each design is organized as a project with its own unique name and can be saved locally and extended at any time.

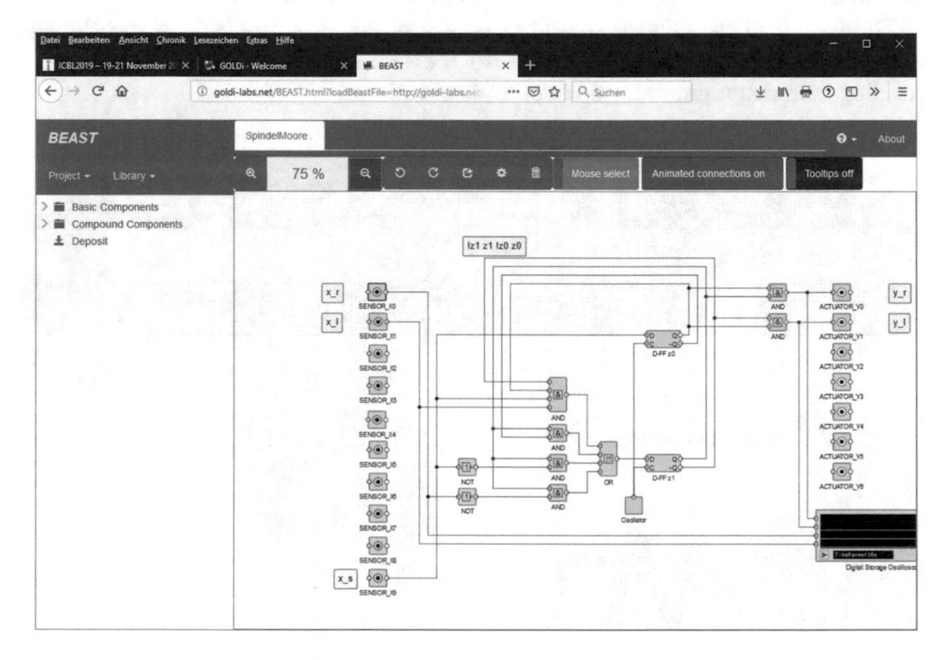

Fig. 3. Design panel of the BEAST-system

The BEAST-system allows simulating the behavior of a circuit in different ways: First by connecting the input ports of the circuit with constant values and second by connecting them with oscillators that produce changing input values. Active connections (logical "1") are colored red and inactive black. To visualize sequential behavior on the output or on any other point of interest in the circuit, a digital storage oscilloscope with up to 4 channels is available. This is useful especially for testing purposes.

A third variant to simulate the circuits' behavior is connecting them with the GOLDi-lab. Therefor the user has to open the BEAST-system after choosing an experiment configuration, defining BEAST as the virtual control unit and one of the virtual or real physical systems as the control object.

Figure 4 shows a screenshot of an experiment after choosing BEAST as virtual control unit. The control object and its digital twin is presented.

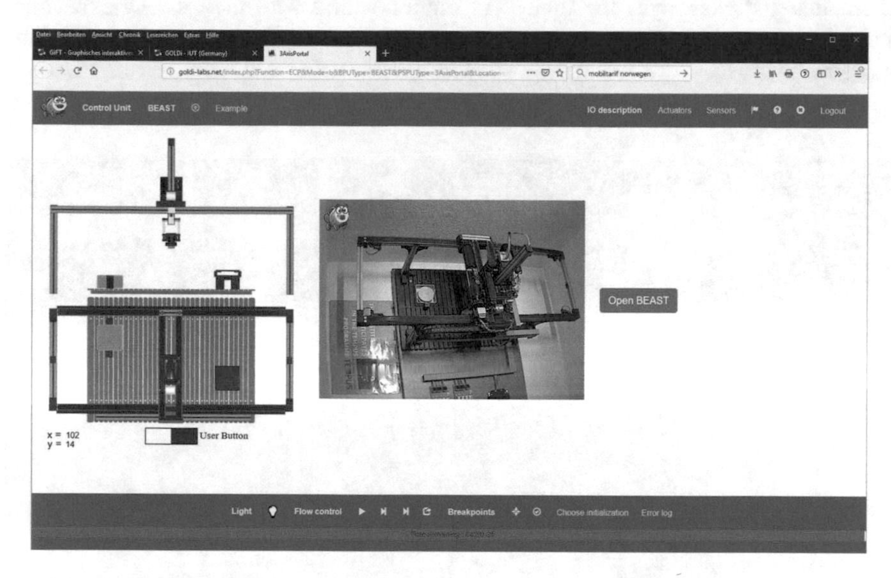

Fig. 4. Screenshot of an experiment after choosing BEAST as virtual control unit

By clicking the BEAST-button the user interface switches to the BEAST interface, presenting all in ports (sensors of the control object) and out ports (actuators of the control object) in a new project, see Fig. 5.

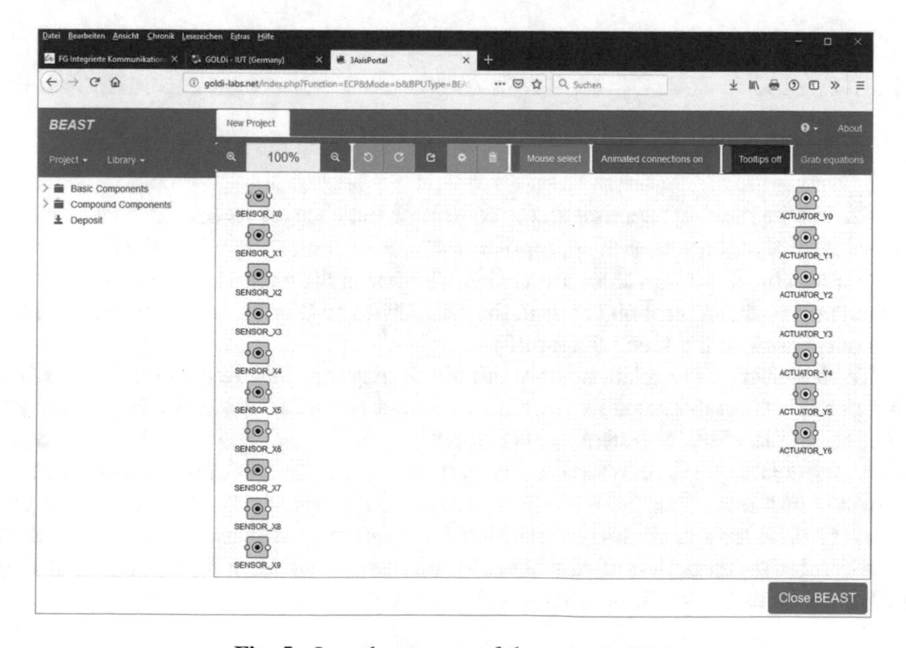

Fig. 5. In and out ports of the control object

Between the in and out ports the circuit of the control algorithm will be placed like shown in Fig. 6.

The inherent simulator of the BEAST-system simulates the behavior of the block schematics and exchanges the input and output values with the GOLDi-experiment. That way it acts like a virtual control unit and controls the physical system in the lab or its digital twin [6].

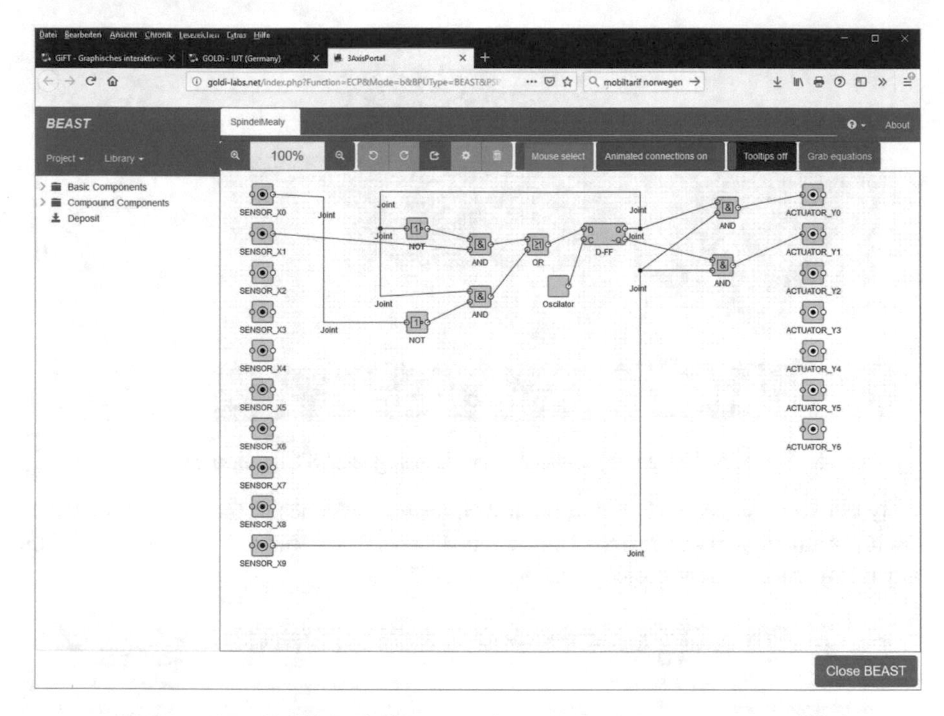

Fig. 6. Animated block diagram of the FSM for the control algorithm

During the running experiment, the control object changes its sensor values and the virtual control unit reacts with appropriate changes of control signals to the actuators of the control object. These values are visible whether in the experiment view (Fig. 4) by movements of the control object or in the BEAST view (Fig. 6) by changing colors of the connections in the block diagram.

To demonstrate the relations between block diagrams and Boolean equations it is also possible to grab equations from a developed circuit as shown in Fig. 7. On the other hand, the BEAST-system is able to derive block diagrams from Boolean equations. A special input function to the library allows the definition of new components as Boolean equations. With these features the BEAST-system is an ideal additional tool for the GOLDi-lab and allows new problem based learning scenarios and a contribution to the further digitalization of educational processes. Some learning scenarios, dealing with the new features will be discussed in the next section.

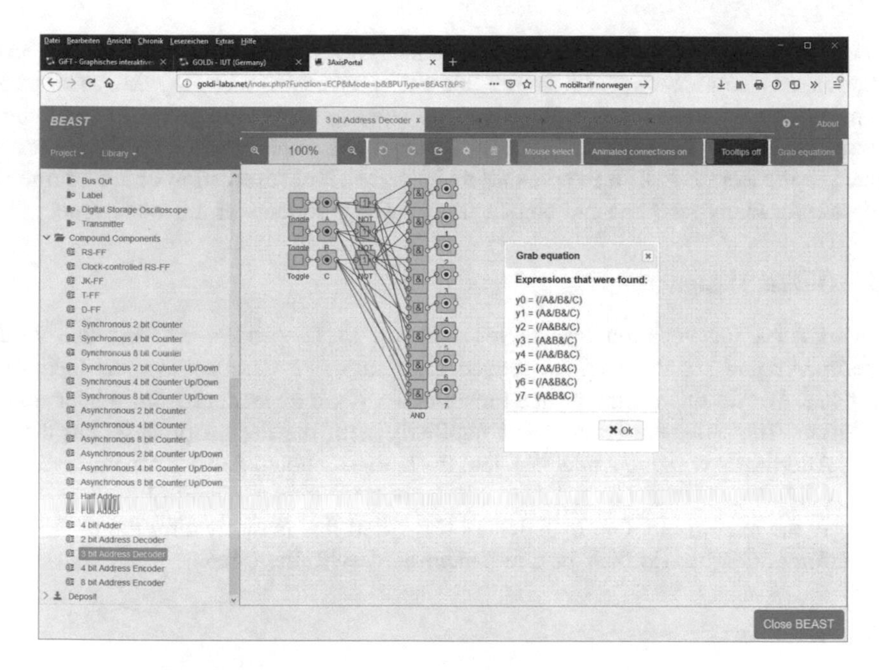

Fig. 7. Boolean equations grabbed from a block diagram

5 Learning Scenarios

The use of the tool-chain allows us teaching in the classroom, where we enforce students using the virtual control objects during the design phase for exploring side effects. They can occur, on wrong designs. The virtual control object behaves in the same way as the real objects. So once the algorithm works in the simulation, it will work in the real environment as well.

5.1 Lecture Hall

In the lecture hall, the design process of an FSM-based control algorithms can be demonstrated. We use this mode to explain the concepts of Boolean constants, variables and expressions as well as the concept of finite state machines in a first year basic course in computer science.

We can insert constant Boolean values (e.g. set the actuator y_7 to logical “1”) or variables or expressions (e.g. $x_1\&!x_2$, if sensor x_1 is true and x_2 is not true) to the virtual control unit and demonstrate immediately its effects on the experiment. For first year students it is very motivating to see the application of what they are enforced to learn.

5.2 Reflection/Flipped Class Room

For flipped classroom scenarios as well as for self-studies **virtual experiments** virtual control units and objects are beneficially. These kinds of experiments run offline on the browser once they were configured and started. Therefor the students can make their experiments also independent from the Internet-connectivity and prepare questions for the discussion with the coach in the seminars. They can try different variants of control algorithms and explore the differences or repeat the experiments that were demonstrated in the lecture hall. It requires no maintenance and reservation of lab equipment and therefor many students can benefit from the experiments at the same time.

5.3 Guided Design

In workshops for vocational education with up to 15 people, we first use virtual experiments and let the learners retrace each step, we demonstrate, on their own computer. We do this on the abstract level of FSMs, independent from a later implementation. That way the learners get familiarly with the design process and the lab work. Afterwards we give a new task that the learners should solve by themselves. We discuss different solutions on the desk and the best solution can control the real control object in the remote lab. Our experiences show that this is motivating the participants and enforces them to do their best in concurrency with the others.

5.4 Lab Exercises

To replace real hands on experiments by using remote labs a booking system allows reserving a dedicated configuration for a defined time slot. The architecture of the lab enables an easy extension of control objects and units because they are connected to a LAN and their presence is scanned permanently. In case, a new device is connected to the LAN, it is immediately accessible for experiments. For example, if there are three instances of the same control object installed, for the user this is transparent. He/she will not see which concrete device is connected to the experiment. Only if all devices are occupied, the user has to look for another time slot in the booking system.

6 Conclusion

The paper describes the tools and usage of the tool-chain for designing digital control algorithms following different approaches in a remote laboratory. Main advances of the concept are the possibility to demonstrate the equivalence of these approaches during a lecture by running experiments with control algorithms on virtual control units and real control objects in the remote lab. Thus, students get a better insight of a systematic approach to the design process of digital systems in hard- and software.

Acknowledgement. The authors thank the students of the GOLDi Team (see: http://goldi-labs.net/index.php?Site=3), namely Johannes Nau, for their implementation work during the last semester.

References

1. Ayodele, K.P., Kehinde, L.O., Komolafe, O.A.: Hybrid online labs. Int. J. Online Eng. (iJOE) **8**(4), 42–49 (2012)
2. O.M.G. (OMG®). Unified Modeling Language Specification Version 2.5.1. https://www.omg.org/spec/UML/2.5.1/PDF. Accessed June 2019
3. Sparx Systems Pty Ltd., UML 2 Tutorial - State Machine Diagram. https://sparxsystems.com/resources/tutorials/uml2/state-diagram.html. Accessed June 2019
4. Henke, K., Fäth, T., Hutschenreuter, R., Wuttke, H.-D.: Gift – an integrated development and training system for finite state machine based approaches. Int. J. Online Eng. (iJOE) **13**(08), 147–162 (2017)
5. TU Ilmenau/IKS, “BEAST Manual,” TU Ilmenau, 25 April 2017. http://goldi-labs.net/BEAST/doc/beast_manual_english.pdf. Accessed 28 May 2019
6. Wuttke, H.-D., Henke, K., Hutschenreuter, R.: Digital twins in remote labs. In: Auer, M.E., Ram, B.K. (eds.) REV2019 2019. LNNS, vol. 80, pp. 289–297. Springer, Cham (2020). https://doi.org/10.1007/978-3-030-23162-0_26

Augmented Reality

Augmented Reality Application for the Mobile Measurement of Strain Distributions

Oleksandr Mogylenko(✉), Alessandro Selvaggio, Siddharth Upadhya, Joshua Grodotzki, and A. Erman Tekkaya

Institute of Forming Technology and Lightweight Components, Dortmund, Germany
{oelksandr.mogylenko, alessandro.selvaggio, siddharth.upadhya, joshua.grodotzki, erman.tekkaya}@iul.tu-dortmund.de

Abstract. In mechanical engineering studies hands-on laboratories are an integral part of the students' education. In manufacturing related fields, material characterization labs are often used to enable students to foster their understanding of different materials. To enhance the laboratory experience and to educate about specific aspects of the uniaxial tensile test, an Augmented Reality (AR) application has been developed. With this applications, it is possible to visualize the inhomogeneous strain field that arises during the experiment on the surface of the specimen. The technical components and structure of the implementation are described in this paper. The usability of several algorithms, technical and software implementation is discussed and evaluated.

Keywords: Augmented reality · Mobile application · Process visualization · Engineering education

1 Introduction

Students of manufacturing related fields often face the task to design and calculate manufacturing processes during their studies. To successfully accomplish such tasks, a fundamental understanding of the material behavior is necessary since the properties of the material being used directly influence the manufacturing process as well as the final product properties. Especially in metal forming processes, various material properties, such as elastic and plastic behavior need to be taken into account during such calculations or simulations. A common approach to determine those material properties is the uniaxial tensile test, among other standardized material tests.

During such an experiment, the elongation of the specimen is tracked as well as the forces currently acting on the material. This values are commonly related to one another in order to determine the elastic and plastic properties of the testes material. Such tensile tests are frequently conducted during hands-on laboratories in numerous manufacturing related subjects.

A common assumption made during these tests, is that only the travel of the crosshead is sufficient to gather data for the elongation of the specimen, cf. Fig. 1.

M. E. Auer and D. May (Eds.): REV 2020, AISC 1231, pp. 235–245, 2021.
https://doi.org/10.1007/978-3-030-52575-0_19

Fig. 1. Overview of the tensile test experiment. The mounting system for the mobile device on the holder is also shown.

For small, soft material this will yield accurate results. But for thicker and stronger materials, the acting forces will bend the cross-head of the machine, therefore inducing an unknown error to the measured elongation. As an alternative, tactile or optical measurement systems can be employed which directly measure the elongation on the specimen from which the strain can be calculated [1]. Optical measurement systems are often referred to as DIC (Digital Image Correlation) [2]. A crucial difference between tactile and optical systems is that with optical systems, one can measure not just a simple uniaxial elongation but the non-uniform, two-dimensional strain field on the surface of the specimen. Compared to tactile systems, optical systems need to be thoroughly calibrated and the specimens have to be carefully prepared for the experiment. Therefore, conventional optical systems require a lot of experience from the operator who is usually not part of the educational setting. There, the students should focus on understanding the outcome of the experiment and how to analyze and interpret the data.

In order to incorporate the advantages of an optical strain distribution measurement system into a hands-on laboratory without the need for complicated soft- and hardware calibrations, an augmented reality application has been developed which can measure a two-dimensional strain distribution during a tensile test. This application can run on most of the commonly used mobile devices.

AR based applications are already present in various engineering disciplines such as plant learning [3], where students can virtually investigate the planting process. In the field of electrical engineering students can learn the behavior of electrical circuits of a simple transformer via their smartphone [4]. In the field of mechanical engineering a teaching assistance system based on AR technologies was introduced, which improved the efficiency of the students [5, 6]. The study of AR applications with 3D and 2D modeling have shown that there are many possibilities to apply and use such applications for the teaching process [7]. One study compared two groups of students, one who used AR tools during their classes while the other group used only traditional tools. The group using AR tools showed relatively improved results, thus proving the beneficial effects of using AR in engineering education [8, 9]. The strain measurement application being developed will act as a supplement to the already existing hands-on lab and act as a learning tool to enable the students to better understand the process.

2 Technical Components

A Zwick Z250 tensile machine is used to perform the tensile test as stated above. The machine settings are set up for flat tensile specimens of the size 120 mm × 20 mm (L × W). For the first version of the application, the students' device need to be in a fixed spatial position related to the specimen. To this end, a device holder was mounted to the tensile test machine which would not move relative to the specimen during experimentation, cf. Fig. 1.

The application itself is designed to run on all mobile devices with integrated cameras. For testing purposes, an iPhone 6S with a 12 Megapixels integrated camera was used for the following tests.

2.1 Preparation and Implementation of the Measurement

Similar to every common optical measurement system using cameras, a pattern is applied to the specimen in order to be able to measure the deformation and the resulting strain distribution. An example of the specimen with the applied pattern over the forming zone is shown in Fig. 2. Deformation of the specimen should occur only in the forming zone which can be assumed as given in these standardized tests since the remaining parts of the specimen are fixed between the clamping jaws, see Fig. 1. The pattern represents a 2D array of 3 × 23 circles. During the test, the change of circles' center position is monitored using multiple images, analyzed and are transformed to calculate the strain distribution.

At the end of the experiment the students are able to follow the changes of strain distribution in order to understand the different mechanisms acting on the material causing a non-uniform distribution of the strains. Strain distribution and stress distribution visualization are beneficial to understand the phenomena such as Lüders bands, local necking and damage accumulation. Lüders bands are localized bands of plastic deformation [10]. Local deformation of the small region on the specimen due to high strain is called local necking [11].

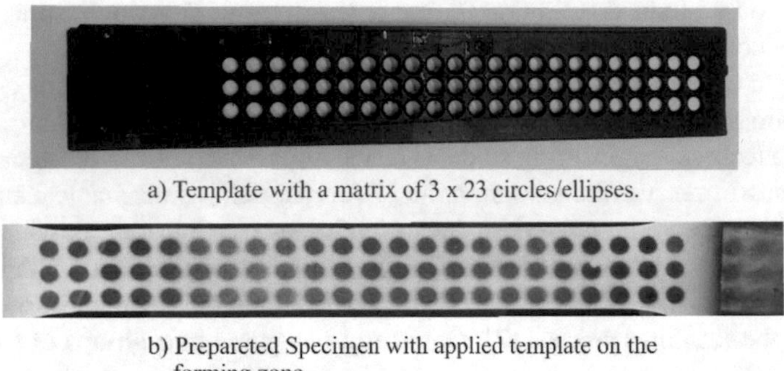

a) Template with a matrix of 3 x 23 circles/ellipses.

b) Prepareted Specimen with applied template on the forming zone.

Fig. 2. Specimen preparation. a) template with pattern and b) prepared specimen.

3 Mobile Application

As opposed to the conventional strain measurement systems, in the proposed AR-based approach, the actual distribution is measured live, during the experiment in two dimensions using a mobile device utilizing the built-in digital camera(s). The application tracks multiple target points on the specimen surface and calculates the strain based on their relative movements. The calculated strains are used to generate a color gradient strain field, which is overlaid on the real camera stream thus realizing an AR strain visualization program. The application will be published as an open educational resource (OER) on a freely accessible server. After downloading of the app, it needs to be installed on the mobile device. After the installation measurements can be performed without additional difficulties/setup. A guideline to manufacture the template for spraying the required pattern will be distributed along with the application.

3.1 Application Structure

In order to realize and execute the AR-based visualization of the strain distribution several steps need to be performed. Figure 3 shows an overview of those steps leading to an optical strain measurement using the proposed application.

In step 1, the images are captured using the integrated camera of the mobile device. The frame rate can be adapted and changed based on the temporal resolution that is requested. This frequency should be chosen in accordance with the speed at which the specimen is pulled apart. Digital reduction of the colors is applied in step 2 (Grayscaling). This step reduces the effort on further calculation due to limited hardware resources. Moreover, the color of the image does not provide any additional information for the calculation or evaluation of the strains. In order to reduce the noise and reduce irrelevant characteristics of the image, an additional image processing takes place in step 3. Here, the patterns on the specimen are additionally detected. These patterns are analyzed and the calculation of the strain distribution based on these analyzed data is executed in step 4. Afterwards the analyzed and calculated data is visualized and can be saved (steps 5 and 6). All described steps can be performed on the mobile device.

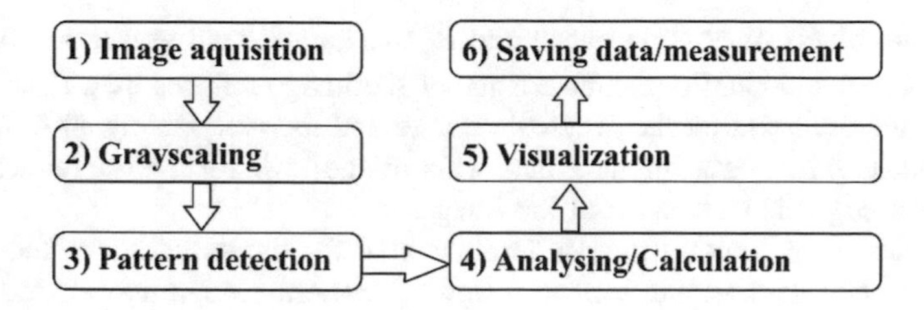

Fig. 3. Structure of the application for measurement of strain distribution.

3.2 Image Processing

Limited hardware resources need to be considered during the implementation and evaluation of the image processing algorithms. All steps, such as pattern recognition and image processing need to be applied and executed on the mobile device. In Fig. 4 all stages of the image processing are shown and described in what follows.

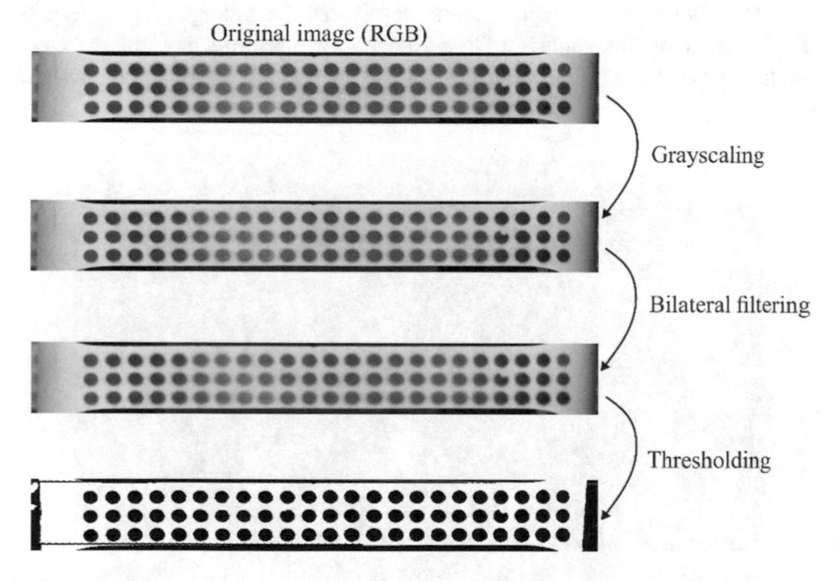

Fig. 4. Images of the processing routine. a) original image, b) grayscale image, c) bilateral filter and d) threshold value of the image.

As previously mentioned the color characteristics are of no benefit for the later image processing. Grayscaling is therefore applied using the following relation (1)

$$P_{ij} = \begin{pmatrix} Pij^r \\ Pij^g \\ Pij^b \end{pmatrix},\ \mathrm{P}'_{ij} = \left(0{,}299 \cdot P^r_{ij}\right) + \left(0{,}587 \cdot P^g_{ij}\right) + \left(0{,}114 \cdot P^b_{ij}\right) \quad (1)$$

where P_{ij} are pixels of original image and P'_{ij} are transformed grayscale pixels. With this approach, the amount of characteristics of the image reduces by a factor of three.

To reduce the noise on the grayscale image and increase the signal to noise ratio (SNR) a bilateral filter was implemented. This filter allows for a noise reduction while retaining the edge characteristics of the image.

In the next step, another filter is implemented for the reduction of the irrelevant information and characteristic for the following pattern recognition. Thresholding is applied according to (2) which represents a binary method of filtering, depending on the threshold parameters.

$$g(p_{ij}) = \begin{cases} f_{dist}(x, y), & p_{ij} > T \\ 0, & p_{ij} \leq T \end{cases} \tag{2}$$

$g(p_{ij})$ is a threshold value of the image, p_{ij} are the grayscale values from the previous step. $f_{dist}(x, y)$ is the function which represents values in the range of $[0, \ldots, 255]$. The values between 0 and 255 corresponds digitally by 8 bits, which represents the chosen color depth. Color depth can be changed upon request, but generally depends on the quality of the image. Furthermore, the function $f_{dist}(x, y)$ separates the background points from the points that contain the main characteristics of the image.

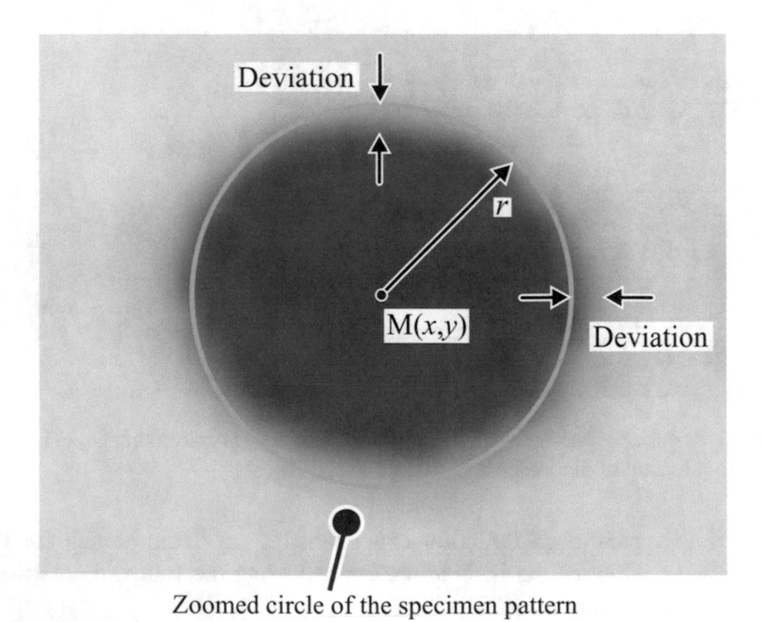

Fig. 5. Parameters of circle Hough-transformation. Middle point M(*x*, *y*) and the radius of the circle *r*.

3.3 Optical Detection Algorithms

In the following, different algorithms for the optical detection of the deformed circles are summarized and tested. Due to mechanical clamping of the flat tensile specimens and manually mounting of the mobile devices a circle could look like an ellipse due to a resulting angle between the normal axis of the specimen surface and the camera. In this case the measurement based on simple circles cannot be used for the evaluation. To avoid this restriction, the Hough-Transformation algorithm (HT) was used. With the usage of the Hough-Transformation it is possible to detect a line on the image which can be used to detect even more complex shapes as circles and other self-defined shapes.

In order to detect the circles of the used pattern, different algorithms and methods were tested. The first method was the standard HT to detect circles. With this approach circles with three parameters (angle, major and minor radius) could be detected as shown in Fig. 5. During the tests, it was observed that a circle applied on the specimen is deformed to an ellipse. The standard Hough-circle-transformation was not able to detect the resulting ellipses. For this reason, an extension of the HT was implemented which can be used to detect ellipses as shown in Fig. 6. The need for more parameters leads to the need of a higher resources effort, which represents a big disadvantage during the usage on mobile devices.

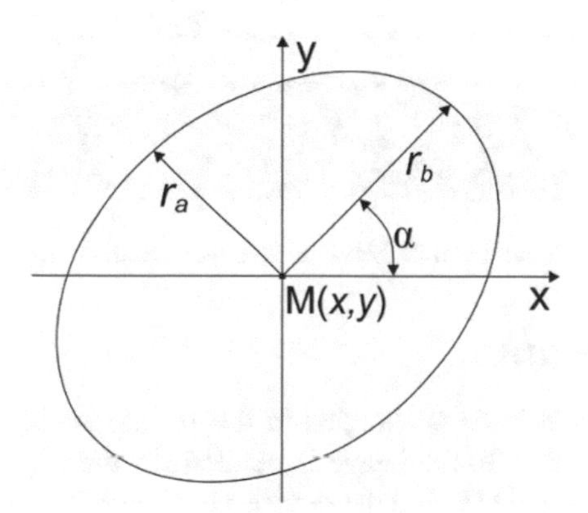

Fig. 6. Parameters of the ellipse Hough-transformation. Middle point M(x, y) shortest radius r_a and the longest radius r_b and α rotation angle along the length of the specimen.

Another method is based on the optimization of the least-square distance between the measured and detected edges (see Fig. 7). The direct least square (DLQ) method approximates the detected edge points to an ellipse. Equation (3) represents the approximation with characteristic parameters from Fig. 7.

$$\frac{(x - x_0)^2}{r_a^2} + \frac{(y - y_0)^2}{r_b^2} = 1 \tag{3}$$

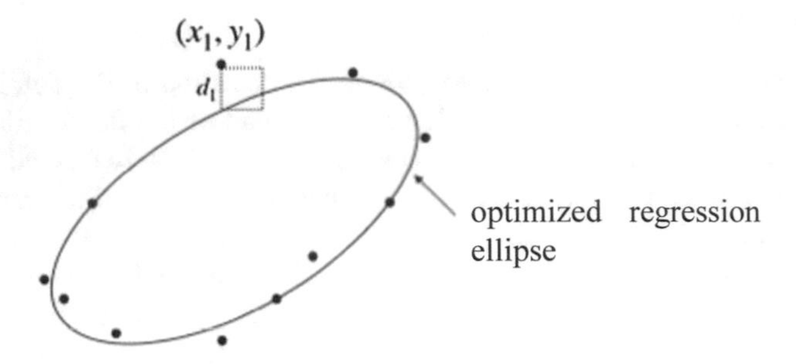

Fig. 7. Optimized regression ellipse $E_o = (x', y', r'_a, r'_b, \alpha')$ der Punktwolke D

In Fig. 8 the detected ellipses are shown with the green border. Here the center points of all ellipse are known and can be used to calculate the local strain distribution between all center points. To be able to determine the distances between the center points a mapping of the points is done between all consecutive images.

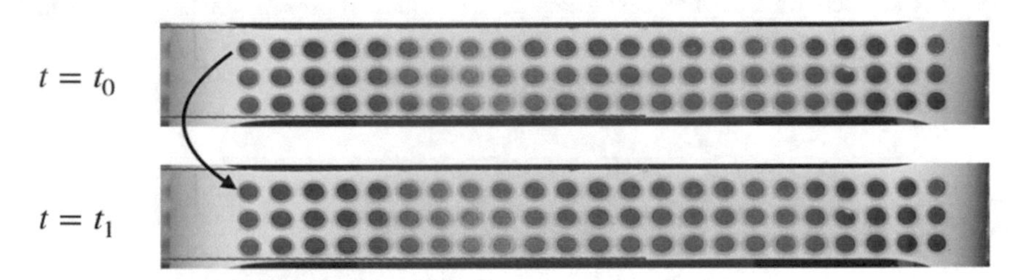

Fig. 8. Mapping of the center points between two consecutive images t = 0 and t = 0 + 1

3.4 Technical Evaluation

Finally, the prepared specimen is clamped in the tensile testing machine as shown in Fig. 9 left. The virtual line on the image is needed for the alignment of the specimen regarding to the vertical axis in case the device is not parallel to the specimen's center axis. The left part of Fig. 9 shows the specimen with a minor angle between the orientation of the specimen and the vertical axis. This angle has a negative effect on the following calculation and needs to be normalized. To this end, a new coordinate system is created with the same orientation of the clamped specimen based on the detected normalization line shown in Fig. 9 right.

After the normalization step, the part of interest, namely the pattern on the specimen, is separated from the image. In this part of the image the detection of the ellipses is performed as described before. The empirical tests have shown that the described algorithm DLQ can be used for accurate calculations of the strain distribution.

In order to evaluate the accuracy, the mean deviation between the measured values from the Z250 testing machine and the calculated values are considered. The value of the testing machine is the distance between the top and bottom clamping jaw.

Equation (4) calculates the mean deviation of the measured and calculated values, with $d_{z250}(i)$ strain values of testing machine Z250, $d_r(i)$ calculated strain values, d_m mean deviation and b total number of measured values.

$$d_m = \frac{\sum_{i=1}^{b} ||d_r(i) - d_{z250}(i)||}{b} \tag{4}$$

The deviation between the data from the Z250 machine and the calculated values is less than 0.5%. Here, the movement of the cross head of the Z250 is given with 14.67 mm while the measured displacement with DLQ is 17.74 mm.

Additionally, the timing factors were evaluated in order to realize a live visualization of the results on a mobile device. The time to calculate on image was measured with 59 ms on a standard Desktop-PC, so that the expected calculation time on a mobile device will be a little bit higher.

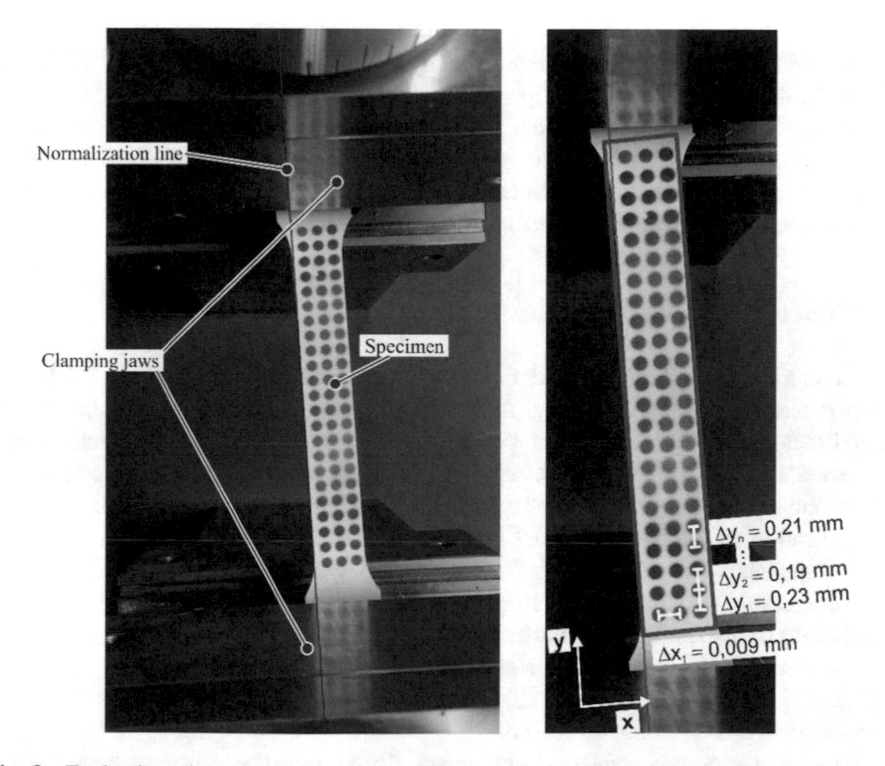

Fig. 9. Evaluation of specimen. Clamped specimen with virtual normalization line on the image (left). Users view of the local strain distribution on the mobile device (right).

4 Use in Hands-On Laboratories

The main motivation behind the development of this AR-application was to enrich the students' experience in classical hands-on material characterization laboratories. In mechanical engineering studies, typically students at the master's level will conduct such experiments hands-on because the number of participants is significantly smaller compared to the bachelor studies where remote laboratories are better suited. In the past, students had to use either no optical strain distribution measurement or rely on more experienced operators who can calibrate and use conventional DIC systems during tensile tests. Based on the bring-your-own-device approach, the students can now directly use their phones during the test to observe the strain distribution, at best, in real-time. Further, the preparation of the specimens is simplified compared to conventional DIC system, which require a complex stochastic speckle pattern. The students can now easily prepare their own specimens. It has to be mentioned, that the overall workflow is improved due to this easy-to-use approach, but at the cost of resolution. Conventional post-processing DIC systems deliver much higher local resolutions of the strain distribution compared to the application presented in this paper. However, during teaching, this aspect is of lesser importance compared to research applications. In general, understanding that the strain distribution is highly non-uniform is the most important benefit of this application. This is an important improvement for the current hands-on laboratories where usually only the cross-head movement is taken into consideration and a homogeneous strain field is assumed.

5 Conclusion and Outlook

First empirical tests have proved the functionality and usability of the detection algorithms regarding the accuracy, timing and hardware restrictions. It can be concluded that students will be able to perform 2D non-invasive strain distribution measurements using their own devices. Additionally, no complicated specimen preparation nor system calibration is required. These aspects render the application well suited for its application in hands-on material characterization laboratories enhancing the students' experience and assisting the further learning process.

The current version of the mobile app does not allow for live measurements and calculations of the strain distribution during the tensile experiment. In its current status, the app is still well suited to learn about localization phenomena as well as comparing engineering and true strains. Overall, performing the optical strain analysis in a post-processing step is still state of the art.

Based on the current stage of development, further improvements can be applied to this application. First, the real-time capabilities will be improved so that the students directly observe the strain field evolution during the experiment. On a similar line, the need for a spatially fixed holder should be overcome so that the user can simply hold his or her device in front of the specimen. If the plastic behavior of the material can be assumed to be known, the stresses could be calculated and visualized based on the information gathered from the strain distribution.

The final graphical user interface (GUI) is currently under development. Using the GUI, the students will be able to vary certain parameters such as the frame rate of the acquired images, kind of specimens they are inspecting etc. They would also be able to select which data must be overlaid on the real specimen in the AR mode, apart from being able to take static screenshots of the process being observed.

Acknowledgement. The work was done as part of the "ELLI2 – Excellent Teaching and Learning in Engineering Science" and the authors are grateful to the German Federal Ministry of Education and Research for funding the work (project no: 01PL16082C).

References

1. Oprina, A., Simion, E., Simion, G.: Computer vision principles. A contrario method. In: 2009 International Symposium on Signals, Circuits and Systems, Iasi, pp. 1–4 (2009)
2. Sutton, M.A., Orteu, J.-J., Schreier, H.: Image Correlation for Shape, Motion and Deformation Measurements: Basic Concepts, Theory and Applications, no. 364. Springer, Boston (2009)
3. Zhao, G., Zhang, Q., Chu, J., Li, Y., Liu, S., Lin, L.: Augmented reality application for plant learning. In: IEEE 9th International Conference on Software Engineering and Service Science (ICSESS), Beijing, China, pp. 1108–1111 (2018)
4. Rigenkov, N.S., Tulsky, V.N., Borisova, S.V.: Application of augmented reality technology in the study of electrical engineering. In: IV International Conference on Information Technologies in Engineering Education (Inforino), Moscow, pp. 1–4 (2018)
5. Pan, X., Sun, X., Wang, H., Gao, S., Wang, N., Lin, Z.: Application of an assistant teaching system based on mobile augmented reality (AR) for course design of mechanical manufacturing process. In: IEEE 9th International Conference on Engineering Education (ICEED), Kanazawa, pp. 192–196 (2017)
6. Fiorentino, M., Monno, G., Uva, A.: Interactive "touch and see" FEM simulation using augmented reality. Int. J. Eng. Educ. **25**, 1124–1128 (2009)
7. Chatzidimitris, T., Kavakli, E., Economou, M., Gavalas, D.: Mobile Augmented Reality edutainment applications for cultural institutions. In: IISA 2013, Piraeus, pp. 1–4 (2013)
8. Martin-Gutierrez, J., Contero, M.: Mixed reality for learning standard mechanical elements. In: 2011 IEEE 11th International Conference on Advanced Learning Technologies, Athens, GA, pp. 372–374 (2011)
9. Figueiredo, M.J.G., Cardoso, P.J.S., Gonçalves, C.D.F., Rodrigues, J.M.F.: Augmented reality and holograms for the visualization of mechanical engineering parts. In: 18th International Conference on Information Visualisation, Paris, pp. 368–373 (2014)
10. Hertzberg, R.W., Vinci, R.P., Hertzberg, J.L.: Deformation and Fracture Mechanics of Engineering Materials, 5th edn. Wiley, Hoboken (2012)
11. Bridgman, P.: Studies in Large Plastic Flow and Fracture. With Special Emphasis on the Effects of Hydrostatic Pressure. Harvard University Press, Cambridge (2014). Accessed 21 Nov 2019
12. Dzikus, P.: Optische Identifikation von Bewegungsfelder für Formveränderungen. TU Dortmund, Dortmund, Germany (2019)

Basic Requirements to Designing Collaborative Augmented Reality

Status Quo and First Insights to a User-Centered Didactic Concept

Nina Schiffeler(✉), Valcric Varncy, Esther Borowski, and Ingrid Isenhardt

IMA – RWTH Aachen University, Dennewartstr. 27, 52068 Aachen, Germany
{nina.schiffeler,valerie.varney,esther.borowski, isenhardt.office}@ima-ifu.rwth-aachen.de

Abstract. Augmented Reality (AR) has become a ubiquitous technology in daily working, learning, and spare life. However, AR still lacks research in terms of productivity and design for being effective on users and the situations it is used in. Particularly in collaborative and also educational settings, the technology has to be adapted in order to be effective for all users concerned. Moreover, the design of the scenario the technology is realised in has to be adapted to the needs and requirements of a productive digital learning environment. In this context, the question still needs investigation how these applications affected e.g. the learning process. As a basis, this paper examines the state of the art of the design process in terms of developing collaborative AR applications and their respective use cases. Besides technical requirements, this paper also focuses on user-centered criteria in terms of usability and effectiveness for collaborative processes in order to lay the groundwork for designing a didactical concept. In order to develop a list of requirements of design criteria for both the didactic scenario and the collaborative AR application, a literature-based requirements analysis has been conducted. Thus, a set of criteria for designing collaborative AR has been deduced. This set is categorised in order to give a detailed description for developing AR applications for collaborative settings on different levels and constraints, e.g. regarding didactic methods, implementation process, usability and user experience design, and necessary resources (for both developing and using collaborative AR). Also, the catalogue presented in this paper will not be finalised but iteratively adjusted to the findings of the empirical study.

Keywords: Augmented Reality · Collaboration · Requirements · Design · Scenario

1 Context

1.1 Augmented Reality (AR)

According to the Gartner Hype Cycle for Emerging Technologies [1], Augmented Reality (AR) has become a productive and, thus, not anymore emerging technology.

M. E. Auer and D. May (Eds.): REV 2020, AISC 1231, pp. 246–259, 2021.
https://doi.org/10.1007/978-3-030-52575-0_20

Concerning productivity and (everyday) use, the technology, however, still lacks evaluation and research on its effectiveness and effects on users and the situations it is used in. With respect to teaching and learning contexts, for instance, AR is used increasingly, but needs proper investigation of its effects on the learning process and outcomes.

AR technology augments, i.e. enhances, the users' perception of reality by supplementing the real world with virtual objects or additional information (e.g. text, images, assets etc.) [2]). Virtual objects and information are mapped onto the real environment. The main functions of AR are 1) embedding virtual objects into reality, 2) interaction in real-time, and 3) correct three-dimensional alignment of the virtual objects in the real world. To use AR applications, mobile devices (e.g. smartphones, tablets or AR-specific glasses) are used that have a rear camera or a flexible camera attached – in terms of laptops, for instance (e.g. all current smartphones and tablets). When scanning the environment (i.e. object-based or markerless tracking) or a particular AR marker (i.e. marker tracking as a visual trigger for the AR application), virtual objects can be placed into the environment by appearing on the device screen at the respective point in the environment. On the reality-virtuality continuum [13], AR tends more towards the reality pole since it necessarily needs a real environment to work while the part of the virtual augmentation is rather low – e.g. in comparison to Virtual Reality (VR) (cf. Fig. 1).

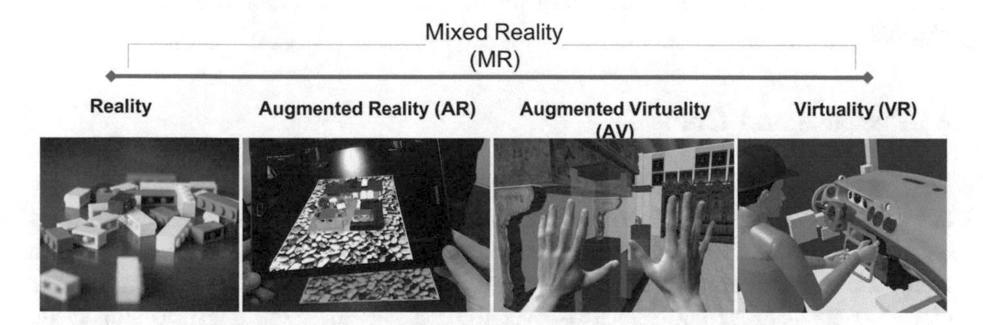

Fig. 1. Reality-virtuality continuum based on Milgram/Kishino 1994 (own depiction)

The main focus of (most) current AR applications is put on demonstrative purposes. Also, interaction is realised in terms of one person being able to interact with the technology. Due to the increased number of collaborative elements e.g. resulting from the "shift from teaching to learning" [3] in the context of the Bologna reform in higher education, suitable and particularly effective didactical tools and methods need to be taken into account. Collaborative settings can, however, also be realised by AR applications. It works via the creation of (Internet-based) network connections between different devices. This way, users – such as students, for instance – can conduct digitally supported group work, e.g. by role-plays, that is even fostered by the collaborative nature of a specifically developed AR application or scenario [4, 5].

1.2 Collaboration

In the context of people working or learning together, there are five stages of interaction: networking, cooperation, coordination, coalition, and collaboration [6–8]. In these stages, the level of communication, sharing of ideas and information, and decision-making increases. Collaboration is assumed to be the closest form of interaction due to the necessity of direct contact and coordination (e.g. to be carried out by AR instructions or simulations) [9]. It is characterised by emergence and the exploitation of synergies [10, 11] define the occurrence of collaboration as "when a group of autonomous stakeholders of a problem domain engage in an interactive process, using shared rules, norms, and structures, to act or decide on issues related to that domain". In conclusion and as a definition for this study, collaboration needs 1) different individuals working on a common task that combine their competences, 2) distribute subtasks onto these people, and 3) creating a dependent relationship for finishing the common task.

Education is also increasingly pervaded by a collaborative and social interactive learning. This style of learning is characterised by knowledge discovery as well as the development of competences via group- and team-based teaching and learning formats [12]. Understanding the competences, capabilities, and tasks of the other team members is an important foundation for teamwork and collaboration in particular [46]. In order to work efficiently in a collaborative team, thus, the collaboration needs specific instructions and supportive means of the teamwork e.g. to be realised by digital means like AR.

2 Purpose and Goal

Particularly when being used in collaborative settings, not only the technology AR has to be adapted to respective (learning) scenarios and settings it is used in to be effective but also the scenario surrounding it, i.e. the didactical environment, in order to become a productive digital learning environment. Although there have been a various number of (pilot) applications and scenarios over the past decades, the question still needs investigation how these applications affected e.g. the learning process. As a basis, this paper examines the state of the art of the design process in terms of developing collaborative AR applications and their respective use cases. Besides technical requirements, this paper also focuses on the methodological, didactical criteria for collaborative AR as well as the user-centered criteria in terms of usability and effectiveness for collaborative processes. The paper displays a state-of-the-art analysis on basic requirements on designing AR applications for collaborative purposes. It is, thus, a first step in a series of (empirical and literature-based) studies on developing a didactical concept for collaborative AR.

3 Approach/Method

In order to develop a list of requirements of design criteria for both the didactic scenario and the collaborative AR application, a literature-based requirements analysis was conducted. By means of researching (pilot) studies, journal articles, and existing applications, a set of design criteria has been derived. The analysis was conducted with 29 research papers and journal articles (cf. references [14–43]) after reducing the initial (online) search for resources based on relevance to the topic. The initial search was conducted in both English and German to find sources in the respective languages and was realised by entering different combinations of terms from the field of collaborative AR. These combinations included (left: English terms; right: German terms):

- Augmented Reality + collaborative/Augmented Reality + kollaborativ
- Augmented Reality + collaboration/Augmented Reality + Kollaboration
- Augmented Reality + interaction/Augmented Reality + Interaktion
- Augmented Reality + communication/Augmented Reality + Kommunikation
- Augmented Reality + definition/Augmented Reality + Definition
- Augmented Reality + group/Augmented Reality + Gruppe
- Augmented Reality + team/Augmented Reality + Team
- Augmented Reality + roleplay/Augmented Reality + Rollenspiel

With the help of these term combinations, the search brought 144 results. The results were manually filtered with regards to their relevance to the topic of describing collaborative AR from the human perspective. The human perspective, in this paper, means deducing criteria necessary for a good user experience, focusing on the people to be using the technology in collaborative settings. This focus aims at forming the basis for investigating the impact and effects of AR on collaborative settings as well as the teams conducting the collaboration while using the technology. Hence, 114 search results (i.e. conference and journal papers, book chapters, and books) were excluded from the further analysis. Reasons for the exclusions were:

- excerpt from patent with merely technical information on the product without further explanatory texts or information
- description of Mixed Reality without further specification on Augmented Reality (too broad definitions of the technology with no distinction between different forms of Mixed Reality)
- too vague examples of applications, i.e. not possible to decide from the text which technology is described (VR or AR)
- lack of focus on academic or vocational usage, i.e. focus on mere entertainment

The remaining 30 literature sources included conference papers, journal articles, book chapters, and books focusing on the use of AR in collaborative or interactive settings. These sources were analysed in terms of their contents by means of a qualitative content analysis according to Mayring [44]. The aim of the qualitative content analysis was on the one hand to identify criteria for designing collaborative AR applications and their respective usage scenarios. On the other hand the analysis aimed at categorising these design criteria in order to form a clearly structured criteria catalogue for describing the basic requirements when developing and implementing AR applications and scenarios for collaborative purposes.

On the one hand, a requirement catalogue lists those design criteria (see Sect. 4) for handing guidelines and recommendations for action to software developers and those, responsible for the technical realisation of AR. On the other hand, the recommendations are addressed to those responsible for developing concepts on the conduct of collaborative AR into learning or working processes as well as for integrating these concepts into scenarios and processes. This catalogue of design criteria and recommendations serves as a basis for the investigation of the use of AR in collaborative settings, i.e. what implications and possible impacts it has on collaborative learning and working processes. With the state of the art analysis of design criteria and that first set of recommendations, a specific collaborative AR application as well as an associated (didactic) scenario are developed and tested with groups of university students and teachers in order to investigate the effects the use of this technology has on the collaborative process, e.g. in terms of learning outcomes, communicational behaviour, and team spirit.

4 Outcomes and Discussion

By means of the qualitative content analysis of the literature identified by an online search, a set of criteria for designing collaborative AR has been deduced. This set is categorised in order to give a detailed description for developing AR applications for collaborative settings on different levels and constraints, e.g. regarding didactic methods, implementation process, usability and user experience design, and necessary resources (for both developing and using collaborative AR). These categories comprise technical requirements, design of the human perspective (i.e. usability, user experience, effects on the user etc.), including didactic-methodological criteria. Hence, the criteria catalogue can serve as a specification book for realising one's own collaborative AR projects and applications.

Technical Requirements

The first category, "technical requirements", resembles a rather ubiquitous category, since almost all 30 literature sources comprise a description of the necessary technical requirements to realise AR (cf. references [14–29, 31–40, 47]). Besides the realisation of AR in general, this category also describes the way AR is supposed to be designed for using it in collaborative settings from the technological point of view. In general, a mobile device is necessary for AR. Depending on the scenario of the respective AR application, a different device is recommended. While most scenarios build up on smartphones, there are also recommendations for tablets or Head-Mounted Displays (HMD) like e.g. the Microsoft HoloLens. The advantages of smartphones against tablets are various: they are lightweight, suitable for use in one hand, intuitive due to their integration in daily life and, thus, everyday use, usually have an integrated rear camera, and can easily be stabilised by putting the arm the user is holding it with to a stable ground (e.g. leaning the arm onto a table or even surface). The tablets, in comparison, often are harder to handle in terms of AR: since the size and weight of the tablets (usually) exceeds the human's capacity to hold it with one hand in a stable manner, the use in interactive scenarios is limited or difficult. However, the screen of a tablet is larger than the one of smartphones, being an advantage when displaying highly detailed AR applications.

For collaborative AR in particular, the integration of unexpected events in the application (e.g. pop-up information or animations of virtual objects) are highly motivating for the user since the interaction with the application and also real world is enhanced [15] and forms the basis for further communication and interaction with other users as well. It triggers the user's interest in exploring and researching the real world and sharing the insights with others – fostering the collaboration among users. In order to use AR in collaborative educational or vocational settings, a major criterion for purchasing the technology (or rather the respective devices to use it) is the scalability of the technology. Scalability includes on the one hand cost efficiency, on the other hand the possibility to scale it to groups, especially when using it in collaborative settings in which each user needs a device to participate in the collaboration. Smartphones and tablets meet the requirement of cost efficiency since low-budget devices are available on the market that also meet the minimum technical requirements (e.g. Wifi connection, >4 GB RAM, most current iOS or Android version can be/is installed, rear camera). In terms of the visual collaboration enabled by AR, latency is also a major technical criterion to be met when developing a collaborative AR application [47]. With a low latency, the collaboration can be supported best since the visual information match the verbal or nonverbal communication between the users collaborating as it is synchronous.

When multiple users collaborative with the help of an AR application, "to mitigate ambiguous configurations when used in the collaborative mode, each [real-world] anchor [i.e. AR marker] is registered with a server to ensure that only uniquely recognizable anchors are simultaneously active at a particular location (…) [to] span multiple sites, by associating a portal with an anchor at each site. Using the location of their corresponding AR device as a proxy for their position, AR renditions of the other participating users are provided" [38]. Thus, the collaboration needs to be supported by visual anchors in order to know where the other participants of the collaborating team are located, what they are doing, and where their virtual objects are moving (from).

As a conclusion to this category, Table 1 sums up the technical requirements mentioned in accordance with the most common AR devices. Depending on which scenario a collaborative AR application is to be developed for, this first draft of systematisation can be used for selecting the most suitable device for the respective concept and situation. As has been mentioned above, it is best for detailed scenarios, for instance, to have a large screen rather than a small one. Moreover, it is to be noted that the literature considered for this analysis shows a high accordance over all sources under investigation of this paper in terms of necessary and useful criteria for collaborative AR. This means, thus, that the sources analysed do not show contrasting, but rather supplementary results.

User-Centered Requirements

When integrating AR into collaborative settings, the mere technical development is not sufficient to be an effective means for collaborating. In terms of the scenario the

Table 1. Match of technical requirements to AR devices

	smartphone	tablet	HMD
device weight	low/small	medium	high/large
intuitive controls	high/large	high/large	low/small
stability (holding)	medium	low/small	medium
perceived easiness of usage	high/large	high/large	low/small
screen size	low/small	high/large	low/small
average price	medium	medium	high/large

Legend: ■ low/small ■ medium ■ high/large

collaborative AR is integrated in, the qualitative literature analysis shows a recommendation of using an authentic scenario for introducing AR as a working means in collaboration [14]. This authentic scenario is assumed to be helpful for identifying with the situation and accepting the (new) technology in this context. Depending on the task for the collaboration, these situations can be e.g. industrial settings, university contexts, or daily life scenarios (e.g. at home, in the supermarket etc.). In general, collaborative settings are suggested to be of a problem-solving or planning nature [16, 23]. An example for a collaborative planning process is the furnishing of an apartment for a flat-sharing community in a daily life context or the interiour design of a factory hall in an industrial context.

Especially for first-time users of AR a tutorial of the respective collaborative AR application is useful to increase its acceptance as is true for new technology in general [48]. This criterion is of major importance in collaborative settings in order to form a common ground for all people collaborating. Since all team members are necessary to come to a solution in collaboration, all users need to be introduced to the working means, i.e. the collaborative AR application. Otherwise, the process of collaboration becomes harder to realise since not all team members can participate equally well or easily. This tutorial can either be realised by “guidance of a [e.g.] remote expert helper” [29] or a tutorial inside the app itself.

From the didactical point of view, it is also necessary to include feedback [14]. This can either be realised by the collaborative AR app itself or by providing an option for giving feedback by the other users. Feedback by the app can take various forms: in terms of visual feedback by showing warning signs or red marks, the app can provide feedback on the position of virtual objects, since “sketch cues [as visual feedback method] improved the task completion time” [29]. Particularly in planning scenarios, in which objects need to be placed by multiple users, it is helpful to give feedback on whether the objects can be placed e.g. very close to each other or on top of each other (cf. Fig. 2).

Fig. 2. Screenshot of the first prototype of a collaborative AR application

Moreover, it is motivating for the user to receive positive feedback on the moves and procedural steps in the collaboration that he or she made. Thus, it is suggested that a collaborative AR app also focuses on the individual user by displaying hints and positive feedback for the respective user by e.g. showing "good position" when he or she placed a virtual object in a position without blocking other ones. Feedback by other users, in comparison, can also be realised by integrating chats or "various interaction methods to enrich collaboration, including gestures, head gaze, and eye gaze [or cursor] input, and provides virtual cues to improve awareness of the (…) collaborator" [28]. However, collaboration is best supported by face-to-face feedback among the team members [16], so it is recommended that the scenario collaborative AR is integrated into provides opportunities for direct verbal and face-to-face interaction and communication by the team members. This can, for instance, be realised by the distribution of responsibilities or tasks prior to using the AR application during the collaborative process.

In order to accept AR as a means for collaboration or supporting collaborative processes, the respective application needs to foster the interaction and communication among the team members. For this support, the application needs to provide an added value to the collaboration in terms of adding to the competences and resources that the individual team members bring into the collaboration [45]. If the application does not have a clearly recognisable added value, it does not support the collaboration. An added value can, for instance, be realised by providing additional resources (e.g. virtual objects for building an authentic 3D model) or serving as a project management tool (e.g. distributing tasks and roles, tracking procedural steps, and documenting the process of collaboration). In the scenario example of furnishing an apartment, the added value of a collaborative AR application is e.g. a catalogue with furniture as virtual objects which the apartment can be designed with.

For improving the acceptance of collaborative AR further, the application needs to be easy to understand and easy to use as well as robust in daily use [16]. The criterion of robustness can be defined in a technical or human manner: in terms of the technical development, robustness aims at low latency and minimising crashes of the app, even if several people are using it for collaborative purposes (i.e. several people are giving input into the application, resulting in an increase in computing power necessary). In terms of the human perspective on robustness, however, the users are not supposed to be able to cause crashes of the AR application. By, for instance, inserting multiple virtual objects or clicking multiple buttons at the same time, the app is supposed to stay stable and still be able to use for the participants of the collaboration, without having to restart the application and start the collaboration from scratch.

Moreover, it is of importance for users to be able to easily pass on information to the other participants of the collaboration (cf. ibid.). As collaboration is about exchange of competences and including several participants simultaneously in order to find a common solution to a problem or a task (see Sect. 1.2), the requirement is posed towards such a technological support system like an AR application to provide information exchange as well. The information exchange can, for instance, work by means of using virtual objects for visual communication or as a starting point for verbal communication about the collaboration process. In this context, HMDs can resemble a productive means for collaborating with AR since – as an AR device for hands-free usage – idle periods and external disruptive factors can be reduced [24], enhancing interaction between the collaborating team members.

Requirements such as age specifications, for instance, have hardly or not at all been investigated. However, designing AR for collaborative purposes is also assumed to include criteria on the target audience in more detail: integrating AR in collaborative scenarios might also depend on the individual participants' learning type, their age, their prior experiences with similar technologies and devices, the technical affinity in general, or the aim of the scenario in terms of which (learning) objectives the collaboration is supposed to cover. These criteria need to be also considered when developing an extensive didactic concept on collaborative AR. While some individuals, for instance, are haptic in terms of learning, understanding and working, others are oriented towards auditive, visual or textual information and means. For textually-oriented individuals, thus, such a visual technology as AR might not be suitable to support collaboration while haptically- or visually-oriented individuals appreciate the support by this technology.

First Insights to a Didactic Concept

Since the topic of collaborative AR mostly lacks empirical investigation in broad terms, there are even more criteria and requirements to meet when developing designing collaborative AR scenarios. However, the results from the literature analysis on AR requirements for collaborative purposes form a basis already for finding first insights into developing a didactic concept for integration AR in collaborative scenarios.

As far as the scenario is concerned, it is suggested from the literature analysis to choose an authentic setting, such as the furnishing of an apartment. This scenario is possible to accomplish with common sense and without prior knowledge on a specific topic since it is a basic, everyday example. Thus, it is assumed that the users do not have a high cognitive task load in order to be simultaneously able to concern themselves with the technology AR. It is, thus, considered to be cognitively exhausting if both scenario and technology as supportive means for the collaboration, i.e. AR, are (hardly) unknown or difficult to perceive by the collaborators. In order to realise this sample apartment scenario as a problem-solving process, specific requirements are suggested to give to the collaborators that they need to meet during their teamwork.

In this context, AR can support the collaboration due to its visual nature by displaying virtual objects into the real world, such as furniture, walls etc. in the suggested apartment scenario. Moreover, in collaborative settings need to enable the collaborators to directly interact with each other, which is why it is suggested to conduct the collaboration with all participants together in one physical room. This way, direct interpersonal communication, i.e. face-to-face communication, is enabled. Also, the AR application is supposed to even support this communication and interaction by providing all pieces of information necessary for all collaborators equally. It is, thus, suggested to enable an Internet connection among the participating devices in order to show e.g. movements of virtual objects for all collaborators simultaneously so that every person can comprehend and see what the others work on during the collaboration.

In terms of the device used for collaborative AR, it is suggested to use tablets with the apartment scenario. Since furniture has different sizes and characteristics, it is a scenario with a high degree of details to be displayed to the users and needs a rather large screen. Thus, tablets are considered suitable since the screens of smartphones or current HMDs are too small to show the necessary degree of details (cf. Table 1). However, the investigation of further criteria and requirements is necessary to build an extensive didactic concept for collaborative AR.

5 Conclusion

This paper focuses on the investigation of AR as a means for collaborative learning and working processes by analysing literature resources on the topic and deriving design criteria for AR applications and respective scenarios. It shows that AR is not suitable for all collaborative processes. For those cases identified in which collaborative AR is a meaningful way of working together in a team, a catalogue of design criteria and requirements for both the technical and didactical development and implementation has been set up with the aim of evaluating these criteria in future empirical studies on the investigation of the effects of using AR in collaborative (learning) processes. Also, the catalogue presented in this paper will not be finalised but iteratively adjusted to the findings of the empirical study.

As has been shown, the most suitable scenarios for using collaborative AR in are problem-solving and planning processes. In these, all team members included are necessary to solve the given task since different competences and perspectives are essential. Moreover, AR as a means for working together has the potential to enhance the collaboration since it is a visual technology that can display what is talked about and conceptualised during the collaborative process. In terms of the specific design criteria, it is suggested to use an authentic, well-known scenario in order to identify with the setting for using collaborative AR in order to enhance the acceptance of the technology.

As a conclusion, the design criteria identified serve as a basis for developing a collaborative AR application and a respective scenario the application is to be used in. The example of furnishing an apartment (see Sect. 4) has been chosen as an authentic, daily life scenario and also planning process. Moreover, the problem-solving aspect useful for collaboration is realised by restricting resources and providing conflicting information (e.g. only 3 rooms for 4 persons to share the apartment including kitchen, bathroom, and hallway; too few furniture to choose from etc.) by means of the collaborative AR application.

6 Outlook

On the basis of the catalogue of design criteria and the first basic insights into a didactic concept, a sample collaborative AR application is developed and investigated in terms of its usefulness for collaboration. The literature analysis provided in this paper is, thus, part of a larger research design including surveys on the effects and impact of the

Fig. 3. Team collaborating with tablets and AR application

technology AR on collaboration and the people working together (e.g. see Fig. 3). The next step in this research design is, thus, the test of the prototype of a collaborative AR application specifically developed on the basis of the design criteria from the literature analysis. The test includes the investigation of the usability and user experience as well as the impact of the technology on collaboration.

References

1. Gartner: 5 Trends Appear on the Gartner Hype Cycle for Emerging Technologies 2019 (2019). https://www.gartner.com/smarterwithgartner/5-trends-appear-on-the-gartner-hype-cycle-for-emerging-technologies-2019
2. Schmalstieg, D., Hollerer, T.: Augmented Reality: Principles and Practice. Addison-Wesley Professional, Boston (2016)
3. Barr, R.B., Tagg, J.: From teaching to learning: a new paradigm for undergraduate education. Change Mag. High. Learn. 27(6), 13–26 (1995)
4. FitzGerald, E., Adams, A., Ferguson, R., Gaved, M., Mor, Y., Thomas, R.: Augmented reality and mobile learning: the state of the art. In: 11th World Conference on Mobile and Contextual Learning (mLearn 2012), 16–18 October 2012, Helsinki, Finland (2012)
5. Radu, J.: Augmented reality in education: a meta-review and cross-media analysis. Pers. Ubiquituous Comput. **18**(6), 1533–1543 (2014)
6. Roschelle, J., Teasley, S.D.: The construction of shared knowledge in collaborative problem solving. In: O'Malley, C. (ed.) Computer Supported Collaborative Learning, pp. 69–97. Springer, Berlin (1995)
7. Hogue, T.: Community-based collaboration: community wellness multiplied. Oregon Center for Community Leadership, Oregon State University (1993)
8. Borden, L.M., Perkins, D.F.: Assessing your collaboration: a self-evaluation tool. J. Extension **37**(2), 67–72 (1999)
9. Schmidtler, J., Knott, V., Hölzel, C., Bengler, K., Schlick, C.M., Bützler, J.: Human centered assistance applications for the working environment of the future. OER **12**(3), 83–95 (2015)
10. Omnasch, L., Maier, X., Jürgensohn, T.: 'Mensch-Roboter-Kollaboration – Eine Taxonomie für alle Anwendungsfälle. Dortmund (2016)
11. Wood, D.J., Gray, B.: Toward a comprehensive theory of collaboration. J. Appl. Behav. Sci. **27**(2), 139–162 (1991)
12. Beckers, K.: Kommunikation und Kommunizierbarkeit von Wissen – Prinzipien und Strategien kooperativer Wissenskonstruktion. Zeitschrift für Rezensionen zur germanistischen Sprachwissenschaft, vol. 5(2) (2013)
13. Milgram, P., Kishino, F.: A taxonomy of mixed reality visual displays. IEICE (Institute of Electronics, Information and Communication Engineers) Transactions on Information and Systems, December 1994
14. Zobel, B., Berkemeier, L., Werning, S., Thomas, O.: Augmented reality am arbeitsplatz der zukunft: ein usability-framework für smart glasses. In: Mayr, H.C., Pinzger, M. (eds.) Informatik 2016, pp. 1727–1740. Bonn, Gesellschaft für Informatik e.V. (2016)
15. Söbke, H., Montag, M., Zander, S.: Von der AR-App zur Lernerfahrung: Entwurf eines formalen Rahmens zum Einsatz von Augmented Reality als Lehrwerkzeug. In: Proceedings der Pre-Conference-Workshops der 15. E-Learning Fachtagung Informatik DelFI 2017, Chemnitz, Germany (2017)

16. Zobel, B., Werning, S., Berkemeier, L., Thomas, O.: Augmented- und virtual-reality-technologien zur digitalisierung der aus- und weiterbildung – überblick, klassifikation und vergleich. In: Thomas, O., Metzger, D. and Niegemann, H. (eds.) Digitalisierung in der Aus- und Weiterbildung, pp. 20–34 (2018)
17. Zabel, C., Heisenberg, G.: Virtual-, Mixed- und Augmented Reality in NRW – Potenziale und Bedarfe der nordrhein-westfälischen VR-, MR- AR-Branche (2017)
18. Alptekin, M.: Möglichkeiten und Grenzen von Virtual- und Augmented Reality im Laborpraktikum. Tagungsband der 12. Regionaltagung, Technische Universität Ilmenau 2017 (2018)
19. Daniel, C., Schulte, S., Petersen, M.: Virtuelles Schweißen – Digitale Lernmöglichkeiten und didaktische Potenziale einer Simulation für Lehrende und Lernende. Tagungsband der 12. Regionaltagung, Technische Universität Ilmenau 2017 (2018)
20. Ermel, D., Kirstein, J., Haase, S., Saul, C., Großmann, H.: Elixier – Didaktisch-technologische Konzeption einer Mixed-Reality-Experimentierumgebung. Tagungsband der 12. Regionaltagung, Technische Universität Ilmenau 2017 (2018)
21. Huntemann, N., Krömker, H.: Patterns für die Entwicklung von interaktiven 3D-Modellen. Tagungsband der 12. Regionaltagung, Technische Universität Ilmenau 2017 (2018)
22. Nenner, C., Bergert, A.: Digital. International. Interdisziplinär. Neue Lehr-Lernformate für die Ingenieurausbildung. Tagungsband der 12. Regionaltagung, Technische Universität Ilmenau 2017 (2018)
23. Stefan, D.P., Pfandler, M., Wucherer, P., Habert, S., Fürmetz, J., Weidert, S., Euler, E., Eck, U., Lazarovici, M., Weigl, M., Navab, N.: Teamtraining und assessment im mixed-reality-basierten simulierten OP. Der Unfallchirurg **121**, 271–277 (2018)
24. Niegemann, L., Niegemann, H.: Potenziale und Hemmnisse von AR- und VR-Medien zur Unterstützung der Aus- und Weiterbildung im technischen Service. In: Thomas, O., Metzger, D., Niegemann, H. (eds.) Digitalisierung in der Aus- und Weiterbildung, pp. 20–34 (2018)
25. Choi, S.H., Kim, M., Lee, J.Y.: Situation-dependent remote AR collaborations: Image-based collaboration using a 3D perspective map and live video-based collaboration with a synchronized VR mode. Comput. Ind. **101**, 51–66 (2018)
26. Abramovici, M., Wolf, M., Adwernat, S., Neges, M.: Context-aware maintenance support for augmented reality assistance and synchronous multi-user collaboration. Procedia CIRP **59**, 18–22 (2017)
27. Elvezio, C., Sukan, M., Oda, O., Feiner, S., Tversky, B.: Remote collaboration in AR and VR using virtual replicas. In: ACM SIGGRAPH 2017 VR Village (2017)
28. Piumsomboon, T., Lee, Y., Lee, G., Billinghurst, M.: CoVAR: a collaborative virtual and augmented reality system for remote collaboration. In: SIGGRAPH Asia 2017 Emerging Technologies, Bangkok, Thailand (2017)
29. Huang, W., Kim, S., Billinghurst, M., Alem, L.: Sharing hand gesture and sketch cues in remote collaboration. J. Vis. Commun. Image Represent. **58**, 428–438 (2019)
30. Piumsomboon, T., Day, A., Ens, B., Lee, Y., Lee, G., Billinghurst, M.: Exploring enhancements for remote mixed reality collaboration. In: SIGGRAPH Asia 2017 Mobile Graphics & Interactive Applications (2017)
31. Zenati, N., Benbelkacem, S., Belhocine, M., Bellarbi, A.: A new AR interaction for collaborative E-maintenance system. IFAC Proceedings **46**(9), 619–624 (2013)
32. Piumsomboon, T., Lee, G.A., Hart, J.D., Ens, B., Lindeman, R.W., Thomas, B.H., Billinghurst, M.: Mini-me: an adaptive avatar for mixed reality remote collaboration. In: Proceedings of the 2018 CHI Conference on Human Factors in Computing Systems (2018)

33. Cheng, K.H., Tsai, C.C.: The interaction of child–parent shared reading with an augmented reality (AR) picture book and parents' conceptions of AR learning. Br. J. Edu. Technol. **47** (1), 203–222 (2016)
34. Hilliges, O., Kim, D., Izadi, S., Molyneaux, D., Hodges, S.E., Butler, D.A.: U.S. Patent No. 9,529,424. Washington, DC: U.S. Patent and Trademark Office (2016)
35. Bennett, D., Mount, B.J., Scavezze, M.J., McCulloch, D.J., Ambrus, A.J., Steed, J.T., Geisner, K.A.: U.S. Patent No. 9,292,085. Washington, DC: U.S. Patent and Trademark Office (2016)
36. Haring, T.M.: U.S. Patent Application No. 14/882,474 (2016)
37. Barzuza, T., Wiener, Y., Modai, O.: U.S. Patent No. 9,959,676. Washington, DC: U.S. Patent and Trademark Office (2018)
38. Weisman, J.K., Redmann, W.G.: U.S. Patent No. 9,779,548. Washington, DC: U.S. Patent and Trademark Office (2017)
39. Liu, R., Salisbury, J.P., Vahabzadeh, A., Sahin, N.T.: Feasibility of an autism-focused augmented reality smartglasses system for social communication and behavioral coaching. Front. Pediatr. **5**, 145 (2017)
40. Müller, J., Rädle, R., Reiterer, H.: Virtual objects as spatial cues in collaborative mixed reality environments: how they shape communication behavior and user task load. In: Proceedings of the 2016 CHI Conference on Human Factors in Computing Systems, pp. 1245–1249 (2016)
41. Akçayır, M., Akçayır, G.: Advantages and challenges associated with augmented reality for education: a systematic review of the literature. Educ. Res. Rev. **20**, 1–11 (2017)
42. Hintz, A.J.: Erfolgreiche Mitarbeiterführung durch soziale Kompetenz. Springer Fachmedien, Wiesbaden (2016)
43. Hissnauer, W. Arbeit im Team. Mainz: ILF
44. Mayring, P.: Qualitative inhaltsanalyse. In: Flick, U., Kardoff, E.V., Keupp, H., Rosenstiel, L.v., Wolff, S. (eds.) Handbuch qualitative Forschung in der Psychologie. München: Beltz, pp. 601–613 (2010)
45. Bainbridge, L., Nasmith, L., Orchard, C., Wood, V.: Competencies for interprofessional collaboration. J. Phys. Therapy Educ. **24**(1), 6–11 (2010)
46. San Martín-Rodríguez, L., Beaulieu, M.-D., D'Amour, D., Ferrada-Videla, M.: The determinants of successful collaboration: a review of theoretical and empirical studies. J. Interprof. Care **19**(1), 132–147 (2009)
47. Lincoln, P., Blate, A., Singh, M., Whitted, T., State, A., Lastra, A., Fuchs, H.: From motion to photons in 80 microseconds: towards minimal latency for virtual and augmented reality. IEEE Trans. Visual Comput. Graph. **22**(4), 1367–1376 (2016)
48. Janßen, U.D.: Einfluss von Persönlichkeitseigenschaften und immersiven Benutzerschnittstellen auf User Experience und Leistung. Aachen: apprimus (2018)

Augmented Reality Production Monitoring Control Room

Cristian-Ovidiu Ivascu[3](✉), Doru Ursutiu[1], and Cornel Samoila[2]

[1] AOSR, Transilvania University of Brasov, 500036 Braşov, Romania
[2] ASTR, Transilvania University of Brasov, 500036 Braşov, Romania
[3] Transilvania University of Brasov, 500036 Braşov, Romania
cristian.ivascu@unitbv.ro

Abstract. The idea of visiting fully virtual worlds separate from our own has long captured the public's imagination and taken precedence over the "augmenting" of our existing world. On the other hand, AR has long held many practical applications in enterprise environments, such as industrial manufacturing trough tools used for monitoring production, training for personnel and remote maintenance.

Keywords: Augmented reality · Mixed reality · Control room · Production line · Productivity · LabVIEW · OPC

1 Introduction

In the last few years, several approaches were researched for using AR technology in production engineering. Assembly and maintenance scenarios are typical fields of application for it. AR approaches were introduced which support workers with dissolved virtual instructions for assembling or repairing a machine tool. Such systems show animations and graphical hints (e.g. Virtual arrows). Users of the system can see where a part has to be applied and which tools are needed. Another benefit is that all instructions are shown in the right order. Therefore, workers can also maintain machine tools which are unfamiliar. No time consuming teaching is necessary because the required knowledge is stored in databases. [1] For more difficult damages, predefined repairing instructions may not be sufficient due to the fact that the cause of the damage is not clear. In this case expert knowledge is necessary. A remote support system can solve this issue. [2] describes an application tool which addresses this field of research.

The authors suggest an approach how a maintenance expert sitting in his office accomplishes technical support to a maintenance operator on-site. The expert has a graphical interface with several elements for creating AR based instructions (e.g. highlighting components or marking dangerous zones).

In contrast to the maintenance scenario, [3] present an approach for assembling only with AR components instead of real objects. This "AR aided interactive manual assembly design" enables to simulate a whole assembly process without the need of auxiliary CAD information.

M. E. Auer and D. May (Eds.): REV 2020, AISC 1231, pp. 260–273, 2021.
https://doi.org/10.1007/978-3-030-52575-0_21

AR assistance systems are another field of application for planning and extending assembly lines [4–6]. Beside the scenario, the technique behind is also worth investigating. Different AR frameworks were introduced to support the development of AR applications. Distributed Wearable Augmented Reality Framework (DWARF) (see [7]) is a very well-engineered and well-developed framework. It is built on a service oriented, distributed architecture and relies on the Common Object Request Broker Architecture (CORBA) [8]. This leads to a high modularity of DWARF. However, the initial training effort is very high. Performance may also be an issue for mobile devices with moderate hardware equipment. Another approach is Augmented Presentation and Interaction Language (APRIL), an authoring language to describe AR content [9]. The advantage of such a platform is also the disadvantage. The easy usable scripting language (in contrast to a programming language) is restricted to the possibilities of the authoring language. Summarizing, the presented approaches concentrate on assembly scenarios and assembly line planning. Monitoring process values with AR support was not in focus of work, yet. Available frameworks are not easy to use or limited in their possibilities.

2 Augmented Reality Framework

Programming AR applications presumes a deep knowledge of the technology behind. However, the scenario and the content are much more important for application developers than understanding the detailed concepts of AR programming. A solution proposed by [13] for this fact is the introduction of an AR framework which defines clear interfaces and hides the complexity of AR technology (see Fig. 1).

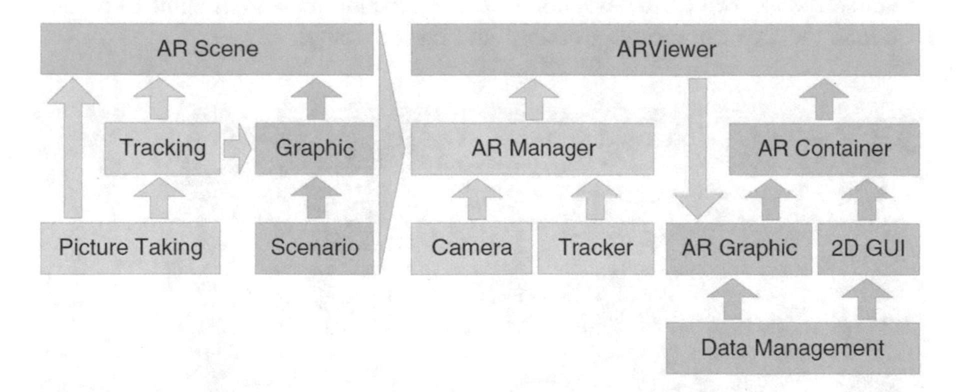

Fig. 1. Structure of the augmented reality framework.

Therefore, the goal of such a framework is to simplify and to speed up the development of AR applications.

With regard to the required components of an AR application, it was proven as an advantage to separate the AR framework into three areas:

- picture taking
- tracking
- graphic

An AR application needs a camera picture of the real world where virtual objects dissolve. This is the area of picture taking. After this, a tracking must be performed to recognize objects (e.g. markers) and to calculate the position and orientation of the virtual camera. Both parts are isolated from the user scenario and represent the basis of the AR framework.

In contrast, the graphic part is dependent from the scenario to visualize the received data. It has to be in full charge of the application programmer. Therefore, the graphic has to be mounted from outside through an interface. The strict separation in the presented three areas leads to a high cohesion resulting in a high usability and maintainability of the software design.

3 AR Monitoring for an Assembly Line

The entire process of an assembly line consists of several different steps. Many machine tools can be chained resulting in a complex system. From these machine tools it is possible to collect and evaluate a huge amount of process data (see Fig. 2). The obtained data is the basis for the visualization of important process information and applications like monitoring as well as diagnosis and control purposes for machines or the whole assembly line. Especially for large systems (like an automotive factory) it is a great benefit (e.g. to locate problems) to obtain relevant process information locally in an easy manner. To achieve this as quickly as possible and directly on-site would be a big advantage to keep the interruption of the production process as short as possible. Therefore, the user can receive warnings and react quickly.

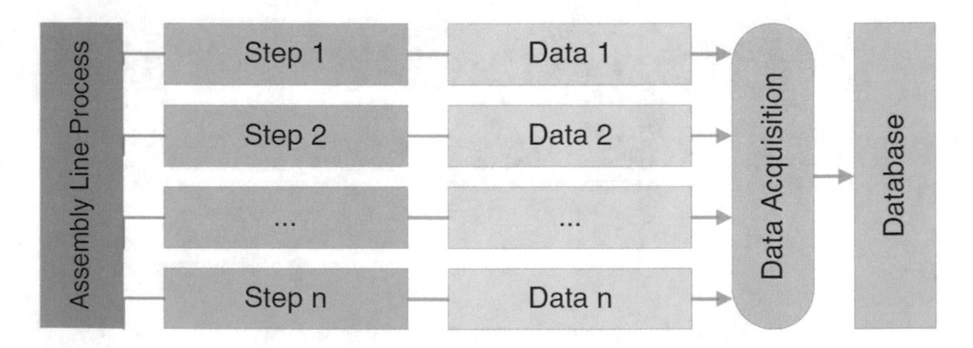

Fig. 2. Assembly line process data.

To show process-relevant information to the user, process values (e.g. energy and power of a machine or duration of a process step) can be stored and analyzed. Thus, the user does not need an extensive knowledge of the used machines and their control unit. Furthermore, the production process is not disturbed. The gathered data can be displayed in real time to the user or evaluated at a subsequent date. Therefore, the process data is stored in a database and can be retrieved by different front-end applications, e.g. for desktop or mobile usage, mobile headsets like Microsoft HoloLens.

4 Monitoring Concept

The concept consists of four clearly separated parts. Starting point is the real process which produces process data associated with work pieces, machines or the whole assembly line. The process data can be used in different software applications and for different purposes for the user, e.g. a web application to summarize the data of an assembly line or an augmented reality application to show selected data in real time (see Fig. 3).

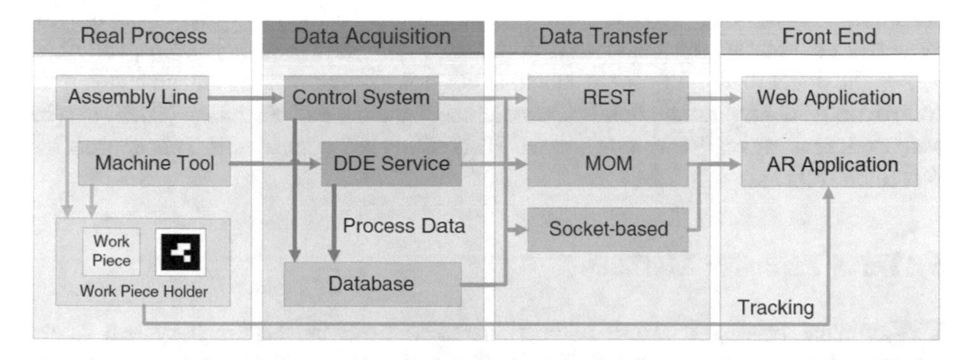

Fig. 3. AR monitoring concept.

Based on the real process it is necessary to provide data acquisition (part 2) and data transfer methods (part 3) to gather process information and to distribute them to the system. Some applications need the current data in real time and others also need the past values for analyzing and summarizing later. For this purpose, a database is used to store all the process data.

Last part of the concept is the front end which is in charge of presenting the acquired data to the user. The actual form depends on the scenario. Possible front ends are web applications or stand-alone programs like the presented AR application.

5 Process Data Acquisition

The single process values of a process step, e.g. energy consumption or power, production volumes, part numbers, times at each assembly station, defects at each station, good parts produced by station, scrap parts by station, can be gathered and stored in a database. By saving an ID with each dataset, the process data can be referenced to a particular work piece.

It is also possible to send the process data to the front-end application directly without saving it, if the data is not needed for further evaluations. The way of acquiring data from a specific system depends on the control vendor and the provided interfaces. One possible approach is to gather the necessary data from the programmable logic controller (PLC) of the control system.

A data management framework based on this was introduced in [10]. With this flexible system the data can be gathered from different sources and with different formats. By using this framework, it is not necessary to create special software for each PLC, however a program change at the PLC is still necessary.

The process data acquired from the PLC is stored in one consistent database. Therefore, an adaptive data model for semi-structured data is defined to save all necessary values efficiently and completely, also for the description of the specific acquisition task. This includes also a meta model to define the complex data structures based on the model of the plant. With this database the front-end application can access a uniform data pool to get the relevant information. Another approach, especially for NC control units, is the direct readout of data, e.g. the position or speed of the moving axes, from the human machine interface (HMI). A software tool was developed, which allows the direct read-out from the machine control without making any changes of the control units' configuration. In this approach the dynamic data exchange (DDE) service was used, running on the windows-based HMI of the NC control unit. This approach is described in [11].

6 Data Transfer Methods

The gathered process data has to be transferred from the back-end system of the assembly line to the front-end application for the user. For different application and usages of the data various data transfer methods are proposed.

One possibility to transfer the process data from the assembly line database to the AR application is a OPC Server-Client solution, Open Platform Communications (OPC) is a series of standards and specifications for industrial telecommunication.

OPC specifies the communication of real-time plant data between control devices from different manufacturers. OPC was designed to provide a common bridge for Windows-based software applications and process control hardware. Standards define consistent methods of accessing field data from plant floor devices. This method remains the same regardless of the type and source of data. An OPC Server for one hardware device provides the same methods for an OPC Client to access its data as any and every other OPC Server for that same and any other hardware device.

The aim was to reduce the amount of duplicated effort required from hardware manufacturers and their software partners, and from the SCADA (Supervisory Control And Data Acquisition) and other HMI (Human-Machine Interface) producers in order to interface the two.

Once a hardware manufacturer had developed their OPC Server for the new hardware device, their work was done with regards to allowing any 'top end' to access their device, and once the SCADA producer had developed their OPC Client, their work was done with regards to allowing access to any hardware, existing or yet to be created, with an OPC compliant server.

OPC servers provide a method for many different software packages (as long as it is an OPC Client) to access data from a process control device, such as a PLC or DCS. Traditionally, any time a package needed access to data from a device, a custom interface or driver had to be written. The purpose of OPC is to define a common interface that is written once and then reused by any business, SCADA, HMI, or custom software packages.

Next Generation OPC – UA Unified Architecture is a machine to machine communication protocol for industrial automation developed by the OPC Foundation. Distinguishing characteristics are:

- Focus on communicating with industrial equipment and systems for data collection and control
- Open - freely available and implementable without restrictions or fees
- Cross-platform - not tied to one operating system or programming language
- Service-oriented architecture (SOA)
- Robust security
- Integral information model, which is the foundation of the infrastructure necessary for information integration where vendors and organizations can model their complex data into an OPC UA namespace take advantage of the rich service-oriented architecture of OPC UA (Fig. 4).

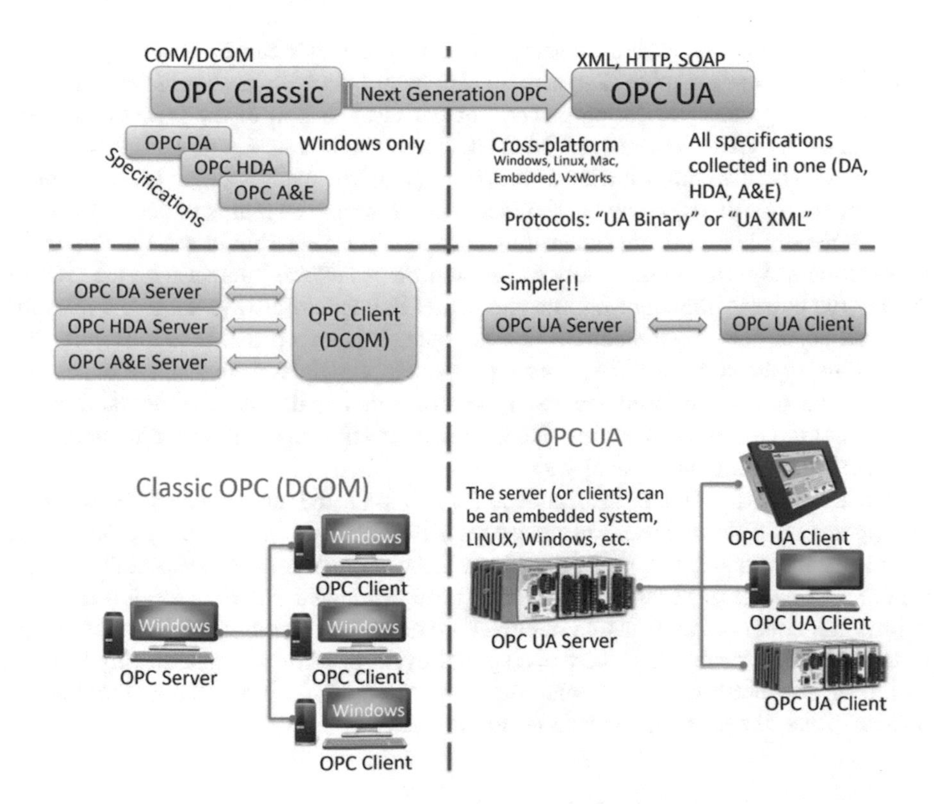

Fig. 4. Classic OPC vs OPC UA

NI LabVIEW software can communicate with any programmable logic controller (PLC) in a variety of ways. OLE for Process Control (OPC) defines the standard for communicating real-time plant data between control devices and human machine interfaces (HMIs). OPC Servers are available for virtually all PLCs and programmable automation controllers (PACs). In this tutorial, learn how to use LabVIEW to communicate with a networked PLC using OPC (Fig. 5).

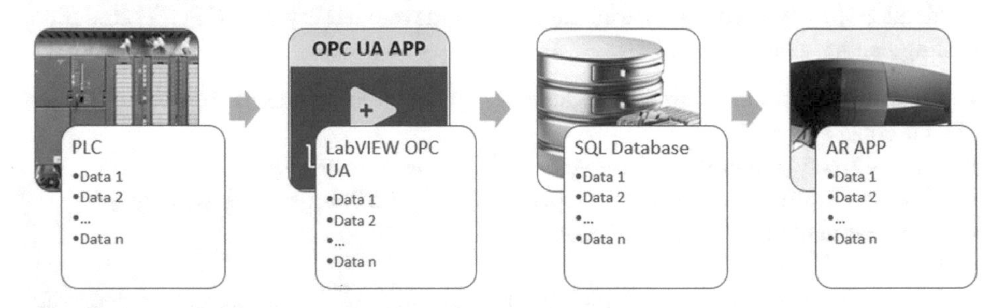

Fig. 5. Process data visualization of an assembly line.

7 Front End Concepts

As mobile devices AR glasses, smart phones or tablet PCs can be used. E.g. for the visualization of the assembly line scenario Microsoft HoloLens is chosen because of its integrated design and processing power. For the visualization of the process data, 3D graphics or 3D GUI elements can be used. For a single value, e.g. energy flow, 3D objects can visualize the behavior of the value [12]. For many various process values 3D elements give an overview of the process to the user. The aim is to design the GUI as user-friendly and problem oriented as possible. The user should see the important information and errors immediately and should be capable to interact with the GUI to display the current important information. The GUI should always show the relevant data and adjust itself to the current process state. It is also possible to customize the GUI of the application for different user groups. For example, the executive board of a company needs other information about the production as the machine operator or the designer of the machine. Also, for visitors it is interesting to see information about the processes, but in a more general way.

An important function of the front-end application is the interaction of the user with the application. Thus, the user can choose the different views as well as the process data and charts to be displayed. User friendliness is an important aspect, especially for the handling using the gestures. With the used neatly arranged 3D elements the user can retrieve the important information intuitively. Another possibility for interaction is e.g. audio and video feedback. If there is a special event or error on the assembly line the user could be notified by a special audio or video signal highlighting the area of interest. Thus, the user can react faster to the event.

8 The Potential of AR Techniques for Experiencing Space

Microsoft HoloLens is capable of tracking head movements, they make it possible to create an impression of permanent presence of holographic geospatial objects. Even if the user walks around in a defined area, commonly indoor area, holograms remain. and adopt to the user location and viewing perspective. This permanent and adaptable holographic projection may lead to visualization approaches that bring additional advantages for the cognitive processing of the geospatial area experienced.

Empirical research of cartography, spatial cognition and experimental psychology has recently led to some recommendations for the construction of user cognition-oriented cartographic media. However, to take full advantage of the possibilities of AR for geospatial applications, technical limitations of the current available AR devices must be faced. These include the precise placement and stability of holograms in the three-dimensional space, a crucial quality criterion for AR applications [13] Having found stable solutions that guarantee a high spatial precision, the AR devices can become valuable methodological tools in geospatial experiments focused on fundamental questions of spatial cognition in 3D environments. In user studies, they could be used to project holographic objects in the environment.

Moreover, AR devices could assist experimental investigators to arrange the spatial layout of movable real-world objects used in their study. The projection of "virtual place markers" can increase the precision of identical spatial object arrangements, which—from a methodological perspective—increases the comparability of acquired user data (between participants). Moreover, projected "virtual markers" can support the analysis of user tasks, such as the identification/measurement of distortion errors, for example, in location memory tasks. To exploit the possibilities of AR systems for geospatial user experiments, it is necessary to create technical methods to establish controlled procedures and to standardize the placement of holographic objects in a real 3D setting.

9 Hologram Placement and Display

In many VR and monitor-based 3D applications, positions of 3D objects are 'hard coded', i.e., their positions are predefined and cannot be changed. The advantage is that all participants see exactly the same arrangement of visual stimuli, which is often a highly relevant precondition for geospatial experiments. However, in AR-based applications, several technical limitations demand a more flexible approach for the placement of visual stimuli.

The Microsoft HoloLens uses an inside-out tracking system, i.e., the sensors used to track the position and rotation of the head are mounted to the HMD (Head Mounted Device). Six cameras integrated in the headset scan the environment [14]. The camera recordings are then used to build a spatial representation of all perceived objects [15]. As this representation can be exported as a 3D model, the HoloLens can also be used to visualize 3D space (see Fig. 6).

Head movements are registered by matching objects recorded by the cameras in real-time to objects already represented in the spatial model. Calculating the relative

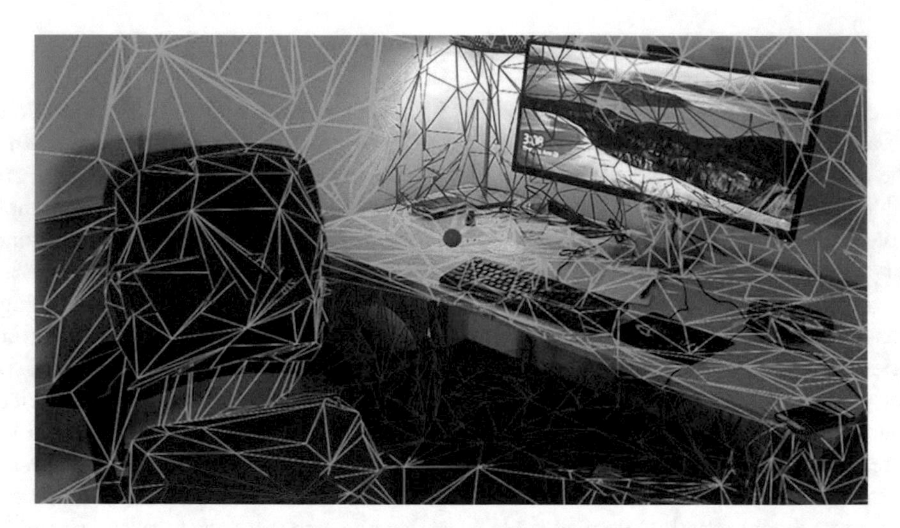

Fig. 6. HoloLens spatial mapping

position towards these objects then allows to triangulate the current head position and rotation inside the 3D space. The advantage of pure inside-out tracking is that no additional hardware is required. Given that the Microsoft HoloLens is a standalone device [14] people can walk freely and use the device seamlessly in different rooms or even floors. However, inside-out tracking also has some serious disadvantages concerning the placement of static holograms. First, image analysis, the precondition for image-based tracking, requires a lot of processing power [16]. As the processing power of a standalone device is naturally limited, this leads to only moderate tracking accuracy and occasional tracking lags. Second, the cameras need to identify at least some reference objects within the spatial range of the cameras, which according to our experience is approximately 5 m. Therefore, tracking lags regularly occur in large empty spaces, especially outdoors. Additionally, poor lighting conditions may negatively affect the capability to identify referenced objects [17]. The mentioned tracking lags often lead to a distorted or shifted internal coordinate system. In these occasions, the positions of all 'hard-coded' holograms relative to real world objects are also distorted or shifted and need to be readjusted. A third limitation of the Microsoft HoloLens concerning the placement of static holograms is rather code-based than tracking-based. Each time an application is started, the internal coordinate system is set relative to the position of the headset. None of these experiences particularly make wearing something on your face pleasant or natural (and that will always be a problem), but HoloLens is among the best of the bunch.

10 Implementing Production Monitoring Control Room

The implementation can be described as follows, the live data from the production process is read out from the PLC which is connected to the ethernet network, using LabVIEW and OPC (see Fig. 7) and then saved to a Microsoft SQL Database (see

Fig. 8) from which the AR Application is reading the data (Fig. 9) an the overlays it to the virtual plant seen through Microsoft HoloLens combined with the ERP Data from the production system, AR application is built using Unity 2018.4 and deployed & customized to the HoloLens Device using Visual Studio (Figs. 10, 11 and 12).

Parameters monitored for a production line/production station integrated with the AR application:

- Date
- Time
- Plant location
- Production hall
- Production line
- Production station/production step
- Current consumption
- Voltage measurement
- Compressed air pressure
- Compressed air consumption
- Produced part number
- Good part production
- Scraped parts production
- Poka-yoke number
- Poka-yoke state
- Crimping force measurement
- Crimping tool dimension
- Crimped part dimension
- Production operator on duty
- Raw material supplier
- Raw material part number
- Raw material patch
- Customer
- Customer Order

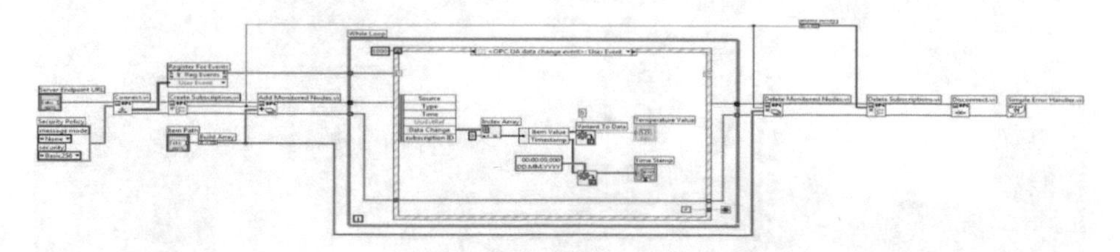

Fig. 7. OPC Data read out in example LabVIEW

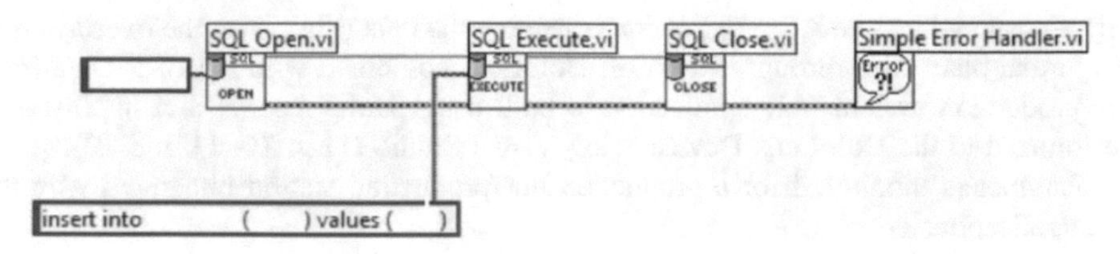

Fig. 8. SQL LabVIEW write to database example

```
using System.Data;
using System.Data.SqlClient;

private string _conString = @"Data Source = 127.0.0.1;
     user id = Username;
     password = Password;
     Initial Catalog = DatabaseName;";
 public string SQLQuery ( string _query ) {
     using (SqlConnection dbCon = new SqlConnection(_conString)) {
          SqlCommand cmd = new SqlCommand(_query, dbCon);
          try {
               dbCon.Open();
               string _returnQuery = (string) cmd.ExecuteScalar();
               return _returnQuery;
          }
          catch (SqlException _exception) {
               Debug.LogWarning(_exception.ToString());
               return null;
          }
     }
 }
SQLQuery("SELECT TOP 1 ProductionOutput FROM M223_ProductionLine WHERE ProductionStation = 'ST20'");
```

Fig. 9. C# Unity3D SQL read data example simplified

Fig. 10. General plant view

Fig. 11. Unity3D production line detail view

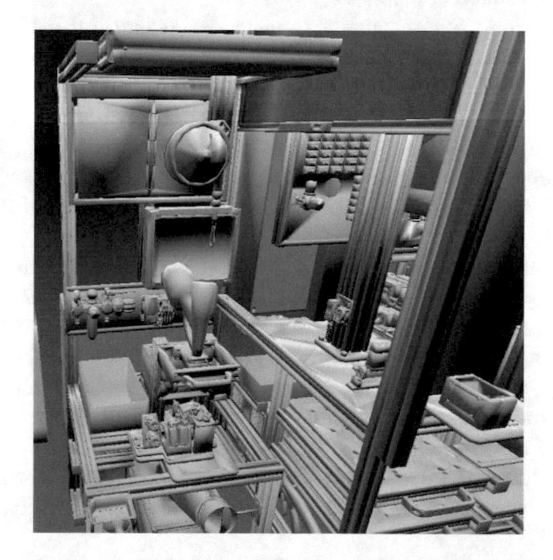

Fig. 12. Unity3D production station detail view

11 Conclusion

In this paper we have presented the concepts of AR Framework, AR Space perception ad hologram placement and space mapping for further going to the integration of AR applications in production monitoring techniques using LabVIEW as method of data transfer from PLC to Microsoft SQL data base of production parameters and then overlaying them on to a layout from production plant layout to production line and production station, monitoring several production and process parameters integrated in Unity3D as solution for developing applications for Microsoft HoloLens and Visual Studio for deploying the application to the HoloLens Device.

References

1. Haberland, U., Brecher, C., Possel-Dölken, F.: Advanced augmented reality based service technologies for production systems. In: Proceedings of the International Conference on Smart Machining Systems (2007)
2. Azpiazu, J., Siltanen, S., Multanen, P., Mäkiranta, A., Barrena, N., Díez, A., Agirre, J., Smith, T.: Remote support for maintenance tasks by the use of augmented reality: the ManuVAR project. In: CARVI 2011: IX Congress on Virtual Reality, Alava, Spain, 11–12 November 2011
3. Wang, Z.B., Ong, S.K., Nee, A.Y.C.: Augmented reality aided interactive manual assembly design. In: Int. J. Adv. Manuf. Technol. (2013). https://doi.org/10.1007/s00170-013-5091-x
4. Ong, S.K., Nee, A.Y.C.: Use of augmented reality in design and manufacturing. In: VAR^2 - Extend Reality, Chemnitz, pp. 35–45 (2013)
5. Reinhart, G., Patron, C.: Integrating augmented reality in the assembly domain – fundamentals, benefits and applications. CIRP Ann. Manuf. Technol. **1**(52), 5–8 (2003)
6. Doil, F., Schreiber, W., Alt, T., Patron, C.: Augmented reality for manufacturing planning. In: Proceedings of the Workshop on Virtual Environments, pp. 71–76. ACM (2003)
7. Bauer, M., Bruegge, B., Klinker, G., MacWilliams, A., Reicher, T., Riß, S., Sandor, C., Wagner, M.: Design of a component-based augmented reality framework. In: Proceedings of the Second IEEE and ACM International Symposium on Augmented Reality (ISAR 2001), pp. 45–54 (2001)
8. Mowbray, T.J., Ruh, W.A.: Inside Corba. Addison Wesley Longman, Amsterdam (1997)
9. Ledermann, F., Schmalstieg, D.: APRIL: a high-level framework for creating augmented reality presentations. In: Proceedings of the IEEE Virtual Reality (2005)
10. Langer, T., Neugebauer, R., Wenzel, K.: Konzepte zur flexiblen und generischen Datenerfassung, -verwaltung und - auswertung bei Maschinen und Anlagen – Teil 2: Flexible Erfassung und Speicherung von Prozessdaten. In: ZWF Zeitschrift für wirtschaftlichen Fabrikbetrieb, vol. 3, pp. 151–159 (2008)
11. Neugebauer, R., Klimant, P., Witt, M.: Realistic machine simulation with virtual reality. In: Procedia CIRP, Volume 3, 45th CIRP Conference on Manufacturing Systems 2012, pp. 103–108 (2012)
12. Neugebauer, R., Wittstock, V., Meyer, A., Glänzel, J., Pätzold, M., Schumann, M.: VR tools for the development of energy-efficient products. CIRP J. Manuf. Sci. Technol. **4**(2), 208–215 (2011). https://doi.org/10.1016/j.cirpj.2011.06.019. Energy-Efficient Product and Process Innovations in Production Engineering
13. Kollatsch, C., Schumann, M., Klimant, P.: Mobile augmented reality based monitoring of assembly lines. In: 5th CATS 2014 - CIRP Conference on Assembly Systems and Technologies Procedia CIRP, vol. 23, pp. 246–251 (2014)
14. Evans. G., Miller, J., Pena, M.I., MacAllister, A., Winer, E.H.: Evaluating the microsoft HoloLens through an augmented reality assembly application. In: Sanders-Reed, J.N., Arthur, J.J. (eds.) Degraded Environments: Sensing, Processing, and Display. SPIE, Bellingham, Washington: Proceedings of SPIE 10197. https://doi.org/10.1117/12.2262626
15. Liu, Y., Dong, H., Zhang, L., Saddik, A.E.: Technical evaluation of HoloLens for multimedia: a first look. IEEE Multimed. **25**(4), 8–18 (2018). https://doi.org/10.1109/mmul.2018.2873473

16. Liu, F., Shu, P., Jin, H., Ding, L., Yu, J., Niu, D., Li, B.: Gearing resourcepoor mobile devices with powerful clouds: architectures, challenges, and applications. IEEE Wirel. Commun. **20**(3), 14–22 (2013). https://doi.org/10.1109/MWC.2013.6549279
17. Loesch, B., Christen, M., Wüest, R., Nebiker, S.: Geospatial augmented reality—Lösungsansätze mit natürlichen Markern für die Kartographie und die Geoinformationsvisualisierung im Außenraum. In: Kersten, T.P. (ed.) Publikationen der Deutschen Gesellschaft für Photogrammetrie, Fernerkundung und Geoinformation e.V. DGPF, Münster, pp. 89–97 (2015)

Usability Study for an Augmented Reality Content Management System

Jan Luca Siewert[1], Mario Wolf[1](✉), Bianca Böhm[2], and Sigurd Thienhaus[3]

[1] Digital Engineering, Ruhr-University Bochum, Bochum, Germany
{jan.siewert,mario.wolf}@rub.de
[2] Excellent Teaching and Learning in Engineering, Ruhr-University Bochum, Bochum, Germany
bianca.boehm@rub.de
[3] Materials Discovery and Interfaces, Ruhr-University Bochum, Bochum, Germany
sigurd.thienhaus@rub.de

Abstract. With the propagation of digitalization in engineering education, tools like Augmented Reality are increasingly focused by educators. However, creating content to use in education is a time-consuming task that requires a basic understanding of 3D modeling and programming. To solve those issues the authors developed an Augmented Reality Content Management System for use in laboratory experiments. In this paper the authors describe and discuss a conducted usability study for the developed system. Students from different fields used the system to conduct a previously unknown experiment. Afterwards, a questionnaire based on the NASA Task Load Index was filled out to judge not only the perceived stress level but also the usability and usefulness of the system so that future development efforts can be based on the results.

Keywords: Augmented reality · Content management system · Engineering training

1 Introduction

With the propagation of digitalization in engineering education, visualization methods like Virtual and Augmented Reality (VR, AR) are increasingly focused by educators. The process of manually creating content for Mixed Reality (MR) applications is however still time consuming and requires special knowledge in areas of 3D modeling, programming, choice of device and software platform.

The authors developed an Augmented Reality Content Management System (ARCMS) to assist educators with creating AR manuals in the context of laboratory experiments, which allows for an easy-to-use and simple-to-learn creation of AR based step-by-step instructions.

Educators use a web-interface to create content for AR manuals. They define individual steps to be performed by the students. Each step can contain text and multimedia elements as well as a hint that is displayed in AR.

M. E. Auer and D. May (Eds.): REV 2020, AISC 1231, pp. 274–287, 2021.
https://doi.org/10.1007/978-3-030-52575-0_22

The system was implemented and is used in different departments in the authors' institution under experimental conditions to further validate the findings presented in the paper at hand. Educators pointed out the ease-of-use in creating content for their laboratory experiments. However, no formal study from the students' point of view has been conducted before the paper at hand.

1.1 Augmented Reality

Azuma defines AR as a context-aware overlay of virtual information on a fixed position of the real world. Augmented Reality is defined by three key characteristics: a combination of the real and virtual, real-time interaction and 3D registered, spatial context, so that virtual content appears at a fixed position independent of the user's position [3]. This stands in contrast to VR where the user immerses himself in a fully virtual environment. However, a clear difference between those technologies is not always obvious. Milgram et al. [14] place both technologies on their Mixed Reality (MR) spectrum where the real world dominates the user experience in AR applications and the virtual world is dominant in VR.

To allow for a spatially fixed visualization of virtual content the system needs to track the position of the user. The most common way to access AR content today is via smart devices like smartphones and tablets [1]. Those devices offer high-resolution cameras and displays as well as a variety of sensors, which assist in calculating the exact orientation and position of the device in the surrounding space. Furthermore, they provide high processing power and come in compact form factors. While other AR systems like Head-Mounted-Displays and Smart Glasses exist, the hardware is still in its infancy and not as reliable and proven as in smart devices [6].

1.2 Augmented Reality in Engineering Education

Common industrial use-cases for AR include MRO (maintenance, repair and operation) processes and marketing. AR in education however becomes an increasingly focused topic [1, 4]. Research on AR in education has been undertaken in various educational levels, including higher education at universities as well as primary and secondary education, and over a variety of fields like engineering, language teaching, mathematics [4], medicine [9] or arts [8].

Systematic reviews of literature confirm the various benefits AR brings to education. On the one hand it supports the learning process by enhancing motivation [1] and satisfaction [4]. The immersion provided by AR as well as the integration of interactive multimedia content and the personalized learning possibilities are further benefits of AR in education [4]. On the other hand, it should be noted that while most reviews of literature show positive effects on students learning, negative effects of AR are rarely mentioned and mostly focus on cognitive overload [15] or technical problems [1].

AR is used for a wide range of use cases in engineering education. Lin et al. [11] show that using AR to teach basic physical concepts like elastic collisions is notably improved over 2D animations. Akçayir et al. [2] use AR in laboratory environments to improve laboratory skills of first-year university students. They show that AR makes it easier and more comfortable to complete the experiments, which in turn allows students to complete the experiment in less time. Martín-Gutiérrez et al. [13] present different use-cases for AR in electrical engineering education that present animations and 3D models on predefined markers to provide step-by-step instructions, help with understanding electrical plan reading or teach fundamental physical concepts like magnetic fields. However, they developed an individual, specialized AR application for each use-case.

This underlines that the creation of content for AR is still a major challenge before AR can become more widespread in education [17]. Kerawalla et al. [10] identify the educator's flexibility in creating custom AR experience as a primary requirement for using AR in education. To this day, it is still time consuming and requires a very special skillset. Content is often created using 3D modelling tools, which offer a lot of advanced functionality, but also have a steep learning curve [5]. Instead of directly modelling content for AR, 3D models can be obtained by scanning real-world objects using 3D scanners or video cameras [5, 13]. While also time-consuming, this process requires additional hardware as well.

To provide an easier solution for content management in AR, Cubillo et al. [7] propose a client-server-architecture that allows educators to create multimedia content in a web application. The data is then associated with a marker that students can scan with their mobile device to view the created content. The markers are then incorporated into existing educational material.

However, this solution is neither suitable for step-by-step instructions nor does it take advantage of new, advanced AR functionality, like spatial awareness and simultaneous localization and mapping (SLAM). SLAM allows free movement even when no image marker is visible to the camera. The authors therefore propose a different approach for an ARCMS.

2 Aims and Requirements

To make the creation of content for AR manuals easier, the authors propose an intuitively usable ARCMS. In a typical use case three actors can be identified. An educator creates and prepares the step-wise instructions. He is interested in using AR but does not necessarily have any prior knowledge or experience in using AR or creating content for AR. Students are performing laboratory experiments. They might not have any prior knowledge with the experiments and are getting their knowledge from paper-based manuals or from an oral introduction by the educator. AR based manuals can help with providing structured information overlaid on the real world that improves orientation as well as the motivation of the students. To achieve this goal, a manual is divided into discrete steps that have to be completed in sequence. Each step describes a singular task and accompanying descriptions, hints and multimedia content that help completing it.

An ARCMS consists of three parts. All information needs to be stored in a central datastore. A web-application is used by the educator to create the manual and save them in the central datastore. Finally, a tablet-based AR application is used by the educator to spatially register the steps' hints at the real-world equipment. After that, the students can view the prepared manuals.

Educators must be able to select from a given set of hints, like different types of arrows, that they can position freely in AR. Instructions should contain texts, as well as multimedia elements like images. Furthermore, it should be possible to add safety or warning symbols to each individual step. Because a reliable wireless network connection cannot be always guaranteed, the system needs to provide a mechanism for storing and synchronizing all data necessary to the tablet application. Students must be able to move freely around the machinery with the AR application to perform the various tasks. While an image-based marker can be used as a reference point for initialization, it should not be necessary to have this marker visible at all times, as this would severely hinder the students' mobility.

3 Concept

To create and use an instruction, three steps are necessary. First, the content is created in a web-application by an educator. A step can contain an AR hint, like an arrow, that then needs to get a position assigned at a real-world machinery in the second step. After each step has an assigned position, the gathered data is saved in the data-store and can be viewed by students on other smart devices. This three-step process is shown in Fig. 1.

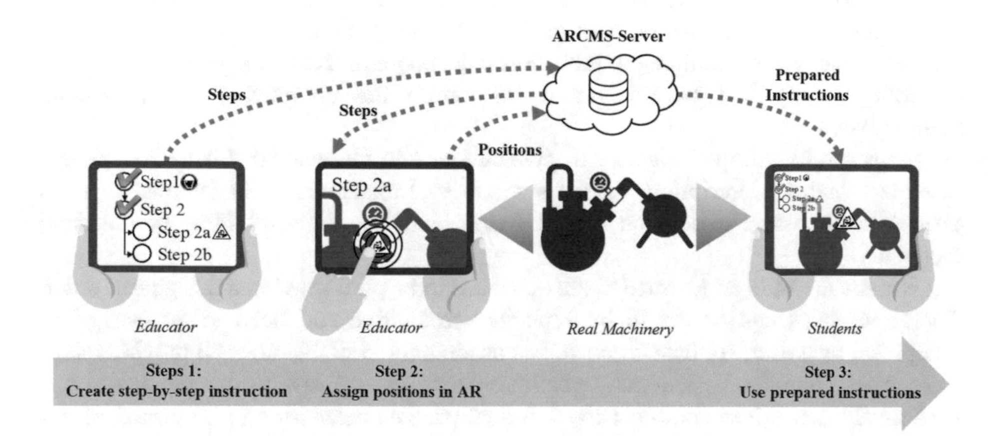

Fig. 1. Creating and using an instruction takes three steps. Creating the step-by-step instruction, assigning a position for each step and then using the instruction.

The three subsystems of the ARCMS are implemented in two components. A webserver provides the central datastore as well as the interface for the web application. It also provides an Application Programing Interface (API) built based on the Representational State Transfer (REST) paradigm. The second part of the system, the tablet-based AR application, uses the API to query and update data in the datastore. This is shown in Fig. 2.

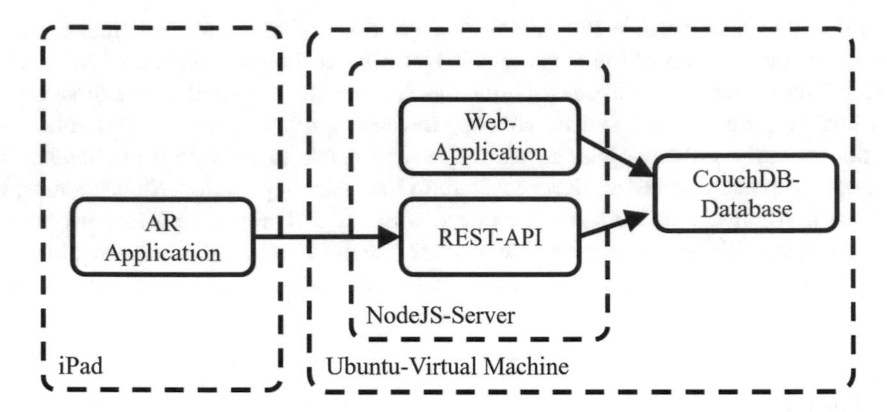

Fig. 2. Implementation of the system in its two parts, the webserver and the tablet application.

A manual contains general metadata, like a unique identifier, a revision, a title and a description, and as the central element of the approach list of tasks to perform. The actual content is structured in a list of steps and offers additional descriptions on each step. Those descriptions can not only contain short texts but also show additional images or documents. They can furthermore attract special attention by showing hazard symbols and accompanying safety instructions. For each step within an instruction, a visual hint is presented to the user in the AR overlay to assist with the orientation around the laboratory equipment. Educators can choose hints from a given set of icons, including arrows and hazard-symbols to reduce the complexity when creating instructions.

Steps can be grouped together to give structure to the manual. Like steps, groups provide a textual description, hazard symbols and attachments. In contrast to septs groups do not store a position or AR based hint, but show all hints of the child elements.

Usually, tasks in an instruction are supposed to be performed in a designated order. Therefore, steps can be set to be explicitly prohibitive and need to be marked as completed by a user, so that subsequent steps are only available after all previous steps have been completed. Because sometimes steps should always be available, e.g. because the step gives some general information, this behavior can be turned off for each individual step.

After the content has been defined in the web application, each step needs to get a position defined at which the AR hint can be shown. This is done in the tablet application. An image marker is printed out and attached close to the machinery to provide a frame of reference for spatial initialization. The application detects the marker and stores all positions and orientations relative to its detected position. After the marker has been detected, the educator can then select steps and assign a position to them by tapping on the desired location in the tablets' camera image. The spatial position in three-dimensional space is position calculated relative to the marker and the hint is placed at the appropriate position at the real-world machinery. The educator can relocate the position for each step's hint later without changing any other data of the manual. Once placed, each AR hint can be scaled and rotated with the usual smart device gestures. The manual can be accessed by students only after each step has a position assigned.

The student must detect/scan the marker to start using a manual. Afterwards, a list of all steps in the manual is presented to him. When selecting a step, it's description, attached safety and hazard symbols and images are shown in the lower-right corner of the tablet's display. The primary focus on the screen is the camera image with the relevant overlaid AR hints. When a single step is selected by the student, the chosen hint is presented at the position associated with the step. For a group, the application shows all hints from all children of the group. Then, the user can tap an AR hint to see the information the educators have provided for the step. In combination with steps that are available even if previous steps have not been marked as completed, this allows for a flexible system that can be used to provide informational content rather than just providing actionable steps.

4 Prototypical Implementation

All subsystems have been implemented prototypically. A server is responsible for both the datastore with a database built on the CouchDB NoSQL database management system and a filesystem for storing attachment files. It also runs an Express webserver on top of NodeJS serving both the web application as well as the REST-API.

CouchDB was chosen as a document-based database because of its native ability to store JSON-Data as well as it's REST-API, which makes it easy to integrate into NodeJS applications. Furthermore, CouchDB provides an identification and revisioning system that can be utilized for the synchronization of data between the sever and tablet clients. Both manuals as well as user data are stored in a single database. The webserver was implemented using the Express framework for NodeJS. The web application allows users to register and log into the application where they have the option to create a new manual or edit an existing one. When editing a manual, the user can add, remove and reorder steps using a drag-and-drop interface. The user can enter text for titles and descriptions, select AR hints and safety symbols from prepopulated dropdown-menus and upload files. When the user saves the manual from the web application, CouchDB automatically increases the revision number to avoid synchronization conflicts.

Because each step needs a defined position, adding new steps requires the educator to assign a position using the tablet application after the manual was edited. However, the position is saved if the educator only edits or reorders existing steps. Small changes and improvement can therefore be made without much redundant work. The web-interface of the prototype is shown in Fig. 3.

The tablet application is used by educators to assign positions to each step as well as by students to view manuals. It is built for iOS on top of Apples ARKit AR

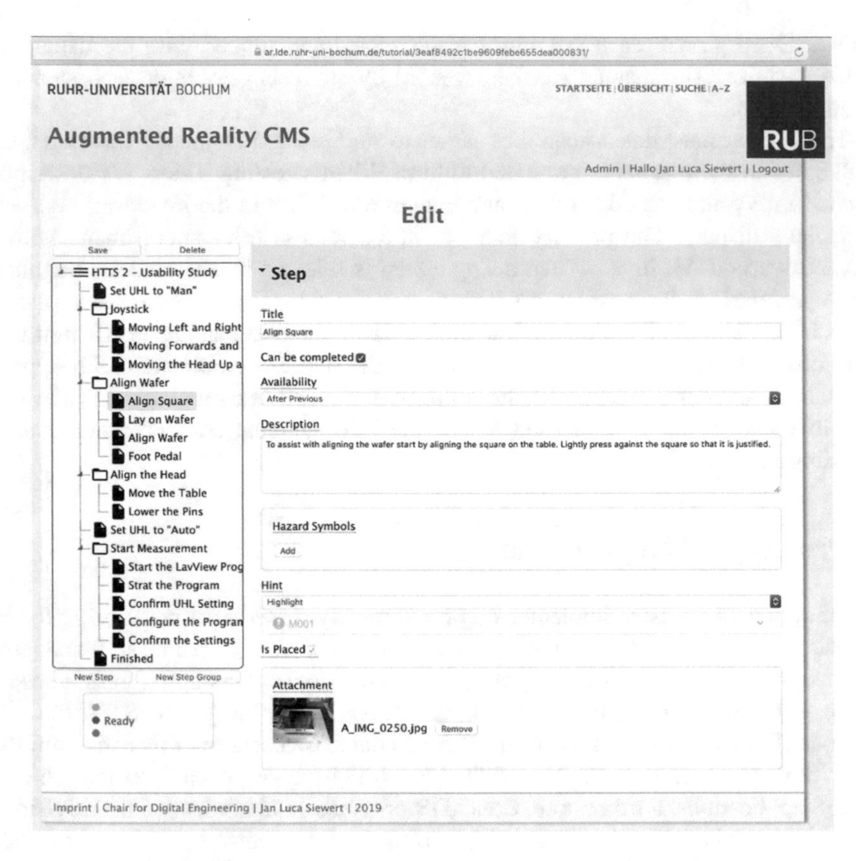

Fig. 3. The web-interface used to create and edit instructions

framework and runs on both iPad and iPhone, although the use of tablets is recommended because of the increased screen size. To use the application, the user first needs to register using the web application. Then, the same credentials can be used to log into the tablet application as well. Educators can select from a list of manuals they have created. After selecting a manual, the application requires them to point the devices camera to the marker. The application is initialized once the marker is recognized. The educator then selects each step from a sidebar and taps on the screen to assign a position to the step. After each position is defined the manual is saved into the database. The educator can then choose to publish the manual to make it available students.

Students only get to see a list of published manuals, which have positions assigned to each step and are therefore marked as published. When a marker was detected students can select each step that is available to them. In addition to the AR hint that is shown at the saved position relative to the marker, the additional information is shown to them. This includes all texts, safety symbols as well as attached images associated with the step. When a step can be marked as completed, a button is also shown that creates a checkmark next to the step and advances the manual to the next step. The tablet application is shown in Fig. 4 as seen by students.

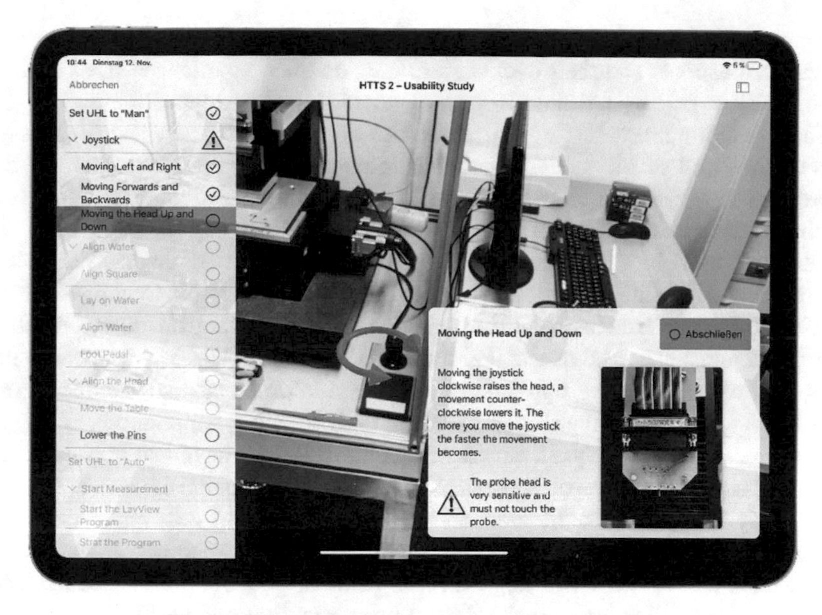

Fig. 4. The tablet application as seen by students.

5 Usability Study

An interactive AR based instruction was created using the stated ARCMS to conduct a usability study. This instruction describes a real-world experiment on actual equipment from the field of material science, namely a conductivity measurement of a thin film materials library which was sputter-deposited on a silicon wafer [12, 16]. In the experiment the participants have to perform a variety of tasks. This includes turning on and setting up measurement equipment, aligning a silicon wafer under the probe head and maneuvering the head to a defined starting position using a joystick control. Furthermore, the participants have to set up the measurement software using a PC. Participants get a short, around 1-minute-long introduction in the purpose of the experiment as well as the general usage of the AR tablet application. However, no additional information is given on what steps the participants are asked to perform. Participants are asked not to request additional help or information from supervisors during the experiment. Instead, they should perform the steps to the best of their ability using only the information provided to them by the AR application.

The instruction incorporates different features of the system and shows detailed descriptions for each step, images and hazard symbols as well as visual hints. The participants of the study are students from different fields that have not previously conducted this exact experiment. Supervisors measure the time needed and the number of handling errors whilst participants perform the experiment. After finishing the experiment, students evaluate their experience while using the system and the instruction with a provided questionnaire. Students will judge different aspects of the system, whether the system helped them performing the task and how well they felt supported with the orientation at the complex instruments, as well as rating several motivational aspects and their own success rate, as they experienced it, of solving an unknown lab experiment while using an augmented reality application.

The questionnaire asks for general metrics about the students and their usage of smart devices and social media to assess their technical knowledge and overall digital competence concerning tablets prior to the task. Furthermore, students evaluate their perceived workload while performing the experiment using a Task Load Index designed and used by the NASA and adapted into the German language and used by leading research facilities such as the German Federal Institute for Occupational Safety and Health (BAuA, Bundesanstalt für Arbeitsschutz und Arbeitsmedizin). Finally, the participants judge the ease of use and functionality of the ARCMS itself. The participants also were given the opportunity to write additional textual comments on the instruction and the used application based on their learning experiences.

The results will be analyzed with regard of the student's technical prior knowledge, their actual and their perceived success rate while working on the experiment. The goal of this study is to evaluate the motivational aspects of using AR applications in laboratory settings as well as to investigate the differences between actual and perceived learning experiences while working with AR. Another goal is to further understand the usability of the application to make it even more intuitive and usable for novices and experts alike.

5.1 Outcomes

16 participants, all students at the Ruhr-University Bochum and between 21 and 31 years old, (62.5% male and 37.5% female) took part in the study. 60.0% were students in engineering, 33.3% studied in humanities, 6.7% said they studied something else, and one participant did not give his or her subject of study. Since it was suspected that engineers would be more used to the overall experimental laboratory setup and the use of smart devices for learning, students of other disciplines were asked to take part in the experimental study for comparison. After finishing the experiment, they were asked to assess their use of smart devices using a Likert scale asking for the frequencies of certain uses. Then they had to assess their experiences while working on the experiment using a Likert scale consisting of five characteristics: Low, rather low, medium, rather high, high.

Table 1 shows the accumulated frequency of positive answers (high or very high) to the questions asked concerning the prior technical knowledge and overall use of tablets and smart devices.

The numbers show how often and for what purpose smart devices are used in the test group. Overall, 93.8% of all participants use smart devices often or very often. Smart devices are most used for communication (93.8%) or social media purposes (60.0%). 55.3% of the participants also use their smart devices for learning, but only 8.3% have used AR applications in the past.

The students were asked to validate the physical and psychological work load they experienced while using the AR application to solve the laboratory experiment. They were given a scale reaching from low to high (Table 2).

Table 1. Prior knowledge and use of smart devices (n = 16).

Question		Cumulated frequency: often or very often
1	How often do you overall use smart devices?	93.8%
2	How often do you use smart devices for social media purposes?	60.0%
3	How often do you use smart devices for communication?	93.8%
4	How often do you use smart devices for learning?	55.3%
5	How often have you used AR-applications?	8.3%

Table 2. NASA instrument for assessing mental and physical load (n = 16).

Question		Most frequently answered
1	How do you asses your mental load during the use of the application?	Rather low, medium and rather high
2	How do you asses your physical activity during the use of the application?	Rather low and medium
3	How do you asses the pressure of time during the use of the application?	Low and rather low
4	How do you overall asses your strain while fulfilling the given task?	Low, rather low and medium
5	How do you asses your frustration while working on the given task?	Low and rather low
6	How do you asses your own performance?	Good, rather good and medium

Next, the students were asked to validate the physical and psychological work load they experienced while using the AR application to solve the laboratory experiment. They were given a scale reaching from low to high (Table 2).

Concerning mental activity and cognitive workload most students (86.6%) said that it was somewhere between rather high and rather low, with the median being right in the middle of the scale. This means that most students assessed their cognitive load as neither too much nor too low. 75.0% said that their physical activity while using the AR application was rather low or medium. The remaining participants assessed their physical activity as low or rather high. None of the participants said that they felt their physical activity was too high, none of them found it straining to hold a tablet and navigate through the application. When asked to assess the perceived time pressure while working on the experiment 75.0% said it was low or rather low. While the students did not know about any time limit, they did know that the time needed for the experiment was measured. 18.8% found the time pressure to be medium. One participant found it rather high. Asked for their overall strain while fulfilling the given task 56.3% said it was low or rather low, 25.0% said it was medium. Two students found it rather high and one said the experienced strain was high. Assessing their frustration while working on the given task 87.5% said it was low or rather low. One student ticked medium and one student ticked rather high. Finally, they were asked to assess their own performance during the experiment. Despite the fact that some found it straining or difficult to finish the experiment, the majority felt they completed the task well (25.0%) or rather well (43.8%). The rest (31.3%) thought they did neither too good nor too bad.

Table 3 shows the accumulated frequency of positive answers (high or very high) to the questions asked concerning the ease of use and learning experience using the AR application. While most participants found the information given in the application helpful and useful, 68.8% of them said that it helped with their understanding of the lab experiment itself. It should be noted, however, that the particular instruction focused heavily on the actual instructions and did not offer any additional background information about the experiment.

Table 3. Ease of use and learning experience (n = 16).

Question		Cumulated frequency
1	The AR application gave me enough information to successfully finish the given task	80.1%
2	The information given in the AR application were easy to understand	80.3%
3	The AR application helped me to better solve the given task	93.8%
4	The AR application was easy to understand	86.7%
5	The AR application was easy to operate	93.8%
6	The AR app helped with my understanding of the lab experiment	68.8%
7	I would like to see more of such learning apps in education	93.8%

This underlines that the quality of the actual content heavily influences the perceived usefulness. The difference between the positive answers to this question and the significantly higher numbers in other questions might also be explained by the fact that not all participants had a background in engineering and may have lacked certain basic knowledge that helped the other students to understand the experiment at hand better.

5.2 Conclusion

Most students needed 10 to 13 min to solve the hitherto completely unknown laboratory experiment using the AR-application. One student was slightly faster and finished in under 8 min, one was slightly slower and finished it in 15 min. Most students (68.8%) made only two, three or four mistakes that could be easily corrected. Three participants made even fewer mistakes, two made five or six mistakes. All students, independent of their experience and study subject, were able to finish the experiment quickly and successfully.

The numbers show that smart devices, although not frequently used for AR applications, are already important for students when it comes to learning and are overall very widely and frequently used. Therefore, tablets work as an intuitive instrument and medium for teaching university students.

Interestingly, three questions in the questionnaire showed a wide variance of answers:

- How do you asses your mental load during the use of the application?
- How do you asses the pressure of time during the use of the application?
- How do you overall asses your strain while fulfilling the given task?

The rather high deviation in the participants' answers can be explained because some participants had already done lab experiments in the past and some didn't. Secondly, some participants had no engineering background and had not used AR applications, meaning they were not familiar with working in a lab, using an application and were, overall, not familiar with the science behind the experiment. Therefore, these students assessed their mental load higher, the time pressure bigger and the strain equally higher than the rest of the group. These outcomes might indicate that this particular instruction might not be enough for novice learners, since those still required additional help and support to finish the task.

On the other hand, even those learners who had to work harder to navigate through the experiment were able to finish it successfully in the end and were overall pleased with their own performance, meaning that the developed application fosters all students' motivation, independent of their academic background and learning experience.

Overall, most participants felt that the given information was enough to help them with the experiment, that the AR application was helpful while solving the given task, and that it was both easy to understand and easy to operate. 93.8% said that they would like to see more such learning applications in education.

To sum it up, this first study showed that students can work very well with the application, that it supports their learning and even enables them to do lab experiments independently from scientific staff, even when they have never done a lab experiment in the past. The outcomes also show that the developed AR instruction has a positive effect on student's motivation.

6 Summary

All in all, this first study regarding the usability of the designed system shows promising results. The authors developed the instruction used for the study in about one hour, which underlines how easy and fast content creation becomes with the ARCMS tool. The student participants were satisfied with the usability of the application. However, the mixed results in the satisfaction regarding the actual content shows the importance of the actual instruction independently from the used framework. However, the authors observed a boost in motivation not only for students working with the instructions but also for educators who prepare such experiments. The novelty of AR combined with the easy-to-use ARCMS can help educators in creating better and more structured instructions for their experiments.

The authors plan further development of the system to make it available to a broader range of users. This includes other use-cases for education in engineering, education in medicine as well as in sheltered workshops. Furthermore, the authors plan to test industrial applications for AR based step-by-step instructions in MRO (Maintenance, Repair and Operations) processes. Future studies should further investigate the correlation between the use of AR applications and students' motivation. Also, when other scientific fields of study start adopting AR based instructions, that different groups of participants and different kinds of students can be investigated and compared to one another.

References

1. Akçayır, M., Akçayır, G.: Advantages and challenges associated with augmented reality for education: a systematic review of the literature. Educ. Res. Rev. **20**, 1–11 (2017). https://doi.org/10.1016/j.edurev.2016.11.002
2. Akçayır, M., Akçayır, G., Pektaş, H.M., et al.: Augmented reality in science laboratories: the effects of augmented reality on university students' laboratory skills and attitudes toward science laboratories. Comput. Hum. Behav. **57**, 334–342 (2016). https://doi.org/10.1016/j.chb.2015.12.054
3. Azuma, R.T.: A survey of augmented reality. Presence Teleoperators Virtual Environ. **6** (4), 355–385 (1997)
4. Cabero, J., Barroso, J.: The educational possibilities of augmented reality. New Approaches Educ. Res. **5**(1), 44–50 (2016). https://doi.org/10.7821/naer.2016.1.140
5. Camba, J.D., Leon de, A.B., La Torre de, J., et al.: Application of low-cost 3D scanning technologies to the development of educational augmented reality content. In: 2016 IEEE Frontiers in Education Conference (FIE), pp. 1–6. IEEE (2016)
6. Chatzopoulos, D., Bermejo, C., Huang, Z., et al.: Mobile augmented reality survey: from where we are to where we go. IEEE Access **5**, 6917–6950 (2017). https://doi.org/10.1109/access.2017.2698164
7. Cubillo, J., Martin, S., Castro, M., et al.: Preparing augmented reality learning content should be easy: UNED ARLE-an authoring tool for augmented reality learning environments. Comput. Appl. Eng. Educ. **23**(5), 778–789 (2015). https://doi.org/10.1002/cae.21650
8. Di Serio, Á., Ibáñez, M.B., Kloos, C.D.: Impact of an augmented reality system on students' motivation for a visual art course. Comput. Educ. **68**, 586–596 (2013). https://doi.org/10.1016/j.compedu.2012.03.002
9. Jamali, S.S., Shiratuddin, M.F., Wong, K.W., et al.: Utilising mobile-augmented reality for learning human anatomy. Procedia – Soc. Behav. Sci. **197**, 659–668 (2015). https://doi.org/10.1016/j.sbspro.2015.07.054
10. Kerawalla, L., Luckin, R., Seljeflot, S., et al.: "Making it real": exploring the potential of augmented reality for teaching primary school science. Virtual Reality **10**(3–4), 163–174 (2006). https://doi.org/10.1007/s10055-006-0036-4
11. Lin, T.-J., Duh, H.B.-L., Li, N., et al.: An investigation of learners' collaborative knowledge construction performances and behavior patterns in an augmented reality simulation system. Comput. Educ. **68**, 314–321 (2013). https://doi.org/10.1016/j.compedu.2013.05.011
12. Ludwig, A.: Discovery of new materials using combinatorial synthesis and high-throughput characterization of thin-film materials libraries combined with computational methods. npj Comput. Mater. **5**(1), 121 (2019). https://doi.org/10.1038/s41524-019-0205-0

13. Martín-Gutiérrez, J., Fabiani, P., Benesova, W., et al.: Augmented reality to promote collaborative and autonomous learning in higher education. Comput. Hum. Behav. **51**, 752–761 (2015). https://doi.org/10.1016/j.chb.2014.11.093
14. Milgram, P., Takemura, H., Utsumi, A., et al.: Augmented reality: a class of displays on the reality-virtuality continuum. In: Das, H. (ed.) Telemanipulator and Telepresence Technologies, pp 282–292. SPIE (1995)
15. Sommerauer, P., Müller, O.: Augmented reality for teaching and learning - a literature review on theoretical and empirical foundations. In: 26th European Conference on Information Systems: Beyond Digitization - Facets of Socio-Technical Change, ECIS 2018 (2018)
16. Thienhaus, S., Hamann, S., Ludwig, A.: Modular high-throughput test stand for versatile screening of thin-film materials libraries. Sci. Technol. Adv. Mater. **12**(5), 54206 (2011). https://doi.org/10.1088/1468-6996/12/5/054206
17. Wang, M., Callaghan, V., Bernhardt, J., et al.: Augmented reality in education and training: pedagogical approaches and illustrative case studies. J Ambient Intell. Hum. Comput. **9**(5), 1391–1402 (2018). https://doi.org/10.1007/s12652-017-0547-8

Voluminis: An Augmented Reality Mobile System in Geometry Affording Competence to Evaluating Math Comprehension

Juan Deyby Carlos-Chullo(✉), Marielena Vilca-Quispe, and Eveling Castro-Gutierrez

Universidad Nacional de San Agustin de Arequipa, Arequipa, Peru
{jcarlosc,mvilcaquispe,ecastro}@unsa.edu.pe

Abstract. The spatial competence for learning three-dimensional objects in the geometry course makes Augmented Reality a perfect ally, due to its popularity among children allowing them to experience more realistic learning. Also, the use of augmented reality allows to improve the interactive and spatial skills with three-dimensional objects. A mobile interactive augmented reality system with 3D models was developed for the teaching of geometry, allowing the teacher to obtain real-time results of the student's spatial and mathematical competence in a real-world environment. In the mobile system, a playful environment capable of visualizing geometric figures (Cube, Rectangular Prism, Triangular Prism, Pyramid, Cone, Cylinder and Sphere), and formulas to calculate its volume are proposed. After learning the formulas, the student must continue with the game and obtain the highest score without visualizing the formulas in such a way that in a competitive environment they can experience more realistic and beneficial learning to improve their spatial competence. The students evaluated are in the primary level (sixth grade) of two classrooms in the city of Arequipa, Peru, enrolled in a one-week course (4 h a week) entitled "Mathematical Logic", and divided into a experimental group (they used Voluminis) and a control group (traditional methodology). A performance test (post-test) and a satisfaction questionnaire were used. In addition, a pre-test to both groups to determine the same level of knowledge about spatial geometry. The results revealed a positive impact on performance, greater academic motivation of students and above all increased competition among students in the Voluminis game. Thanks to the results in real time the professor observed the learning difficulties of the students and obtained the immediate qualifications of the space competition. Voluminis can be part of the qualified evaluation of the course, because it provides the evolution of each student in the area of geometry. It also helps to strengthen space skills thanks to the competition activities that occur in the game.

Keywords: Mobile Augmented Reality · Mathematics learning · Solid geometry · System learning

M. E. Auer and D. May (Eds.): REV 2020, AISC 1231, pp. 288–299, 2021.
https://doi.org/10.1007/978-3-030-52575-0_23

1 Introduction

Augmented Reality (AR) allows you to show the user virtual objects overlaid in the real environment [7]. Besides, AR offers a different way of interacting with information, in such a way that it improves the user experience during learning. Therefore, it is applied in subjects such as physics, chemistry, geography and mathematics [12]. Also, Mobile Augmented Reality (MAR) appears as an attractive technology that provides entertainment, space capacity development, and portability.

The use of mobile devices in AR facilitates teaching and learning, as they are widely used among young people for their novelty [1]. In recent years, AR-based mobile games such as Pokemon Go, Jurassic World Alive or Harry Potter: Wizards Unite are popular with children and young people. That is why, using AR, students will obtain content more easily from AR applications, than in a traditional classroom [13]. Moreover, the activities used in mobile augmented reality (MAR) applications can be reviewed anywhere else. MAR provides the opportunity to build independent experiences based on prior knowledge of the respective subject [1].

In this study, two applications were designed and developed: a MAR video game for primary school students, whose purpose is to help explain and practice spatial geometry exercises (such as the volume of an object); Contribute to learning, understanding geometric figures in three dimensions and mainly competition. And an application for the mentor, to monitor student results in real-time; In addition to analyzing the state in which it is. With the help of the augmented reality mobile application, the participants learned the concepts of calculating the volume of solid geometries such as a cube or a prism. Students can also interact with the geometrical figures through the mobile device screen, object scaling is also implemented. The objective is to provide students with a complementary tool and test the effectiveness and influence of this video game in a traditional geometry teaching classroom.

2 Literature Review

2.1 Geometric Theory

Geometry is an important topic of mathematics that serves as a major base for exploring space and form, in turn, is important for academic success and professional, PISA 2021 (The Programme for International Student Assessment)[1].

In [6], the use of technology in the teaching of geometry in children is proposed for the best benefit of curiosity, even in other projected investigations the use of AR in geometry, in the way that you can project and interact with 3D objects. For example, [10] introduces geometry with the use of puzzles for 6-year-olds. On the other hand, [3] proposes that the use of AR helps students visualize the design and construction of edits including geometric information. Thus it can be affirmed that geometry is very important in personal development.

[1] https://pisa.e-wd.org/.

2.2 Gamification

Gamification helps make a system enjoyable, and also increases interaction. In many serious games, there are several examples of gamification such as point system, collection systems, time and track systems [17]. For that reason, gamification is being used in teaching methods, to increase motivation, teamwork, self-knowledge, and retention of knowledge. In turn, it provides information to the teacher. In addition to the new tools, they assume an important commitment with the students, who increasingly acquire greater knowledge about technology, especially video games [15].

2.3 Augmented Reality and Education

Virtual reality and augmented reality are very popular due to their high quality and accessibility [11]. AR games incorporate students into the learning process, so that students finish all the objectives of the game, even if they suffer from any difficulty such as Attention-deficit/hyperactivity disorder (ADHD) [14]. Similarly, in [2] where they found that students with difficulties learning math (Dyscalculia), showed interest, motivation, and improvements in mathematical reasoning. On the other hand, it is necessary to have a multimodal presence (visual, auditory and tactile, among others) in MAR to recover accurate data and significant information effectively [8].

In a nutshell, MAR video games have a positive effect on students in the math course. As for spatial geometry, the need for a good spatial capacity in students, allows MAR to be an appropriate technology in the course of mathematics since it helps in the understanding of three-dimensional objects.

According to related research, AR-based learning can increase students' motivation to learn [4,5,9,16].

It is also able to improve student performance with the right resources and at the right time [9].

3 Designing and Developing a MAR System

This section describes the design and development of the Voluminis system, which is a set of applications, which are installed in the Android 7.0 or higher operating system to meet the needs of the ARCore[2]. The applications were developed in the Unity[3] game engine with the programming language C#. In addition, a real-time database of Firebase[4] was used. The core of the system is mainly using ARCore to detect and recognize surfaces and lift interactive 3D objects on the plane. This system consists of two main modules (see Fig. 1), which includes the administrator or teacher module and the student module. It should be noted that privacy and data protection are guaranteed. These modules are described in detail in the next section.

[2] https://developers.google.com/ar.
[3] https://unity.com/.
[4] https://firebase.google.com/.

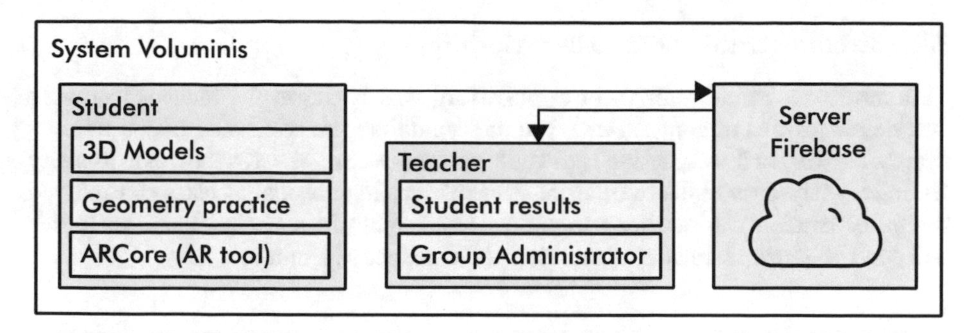

Fig. 1. System architecture

3.1 Student Module

This module represents a game that provides students with a series of exercises, as shown in Fig. 2. The session begins with an account and a group code that will allow us to store the score of each of the students and will encourage competition among the students. The number of exercises of each geometric figure is shown in Table 1. For each successful exercise, the student gets 10 points, for each wrong answer they lose 1 point and if they want to remember and get the formula to solve the exercise, they lose 5 points. The student who hits all the exercises without losing points gets a maximum of 100 points. Once the game is over, the score restarts and starts again.

Figure 3 shows one of the exercises, each exercise has three alternatives of which one is correct and the other two are incorrect. Measurements of geometric figures and alternative solutions of exercise are randomly changed for each attempt made.

Table 1. Number of exercises per spatial geometric figure

Object used	Number of exercises
Cube	1
Rectangular prism	2
Triangular prism	2
Pyramid	2
Cylinder	1
Cone	1
Sphere	1

3.2 Administrator or Teacher Module

This module is an administrator application, which allows the teacher to create and delete groups of games. Also, you can visualize the results of each student's exercise with their respective real-time attempts. So, the teacher can identify students with fewer skills to reinforce the subject in subsequent classes. Finished the game session the teacher can export the results in a spreadsheet, with this will see the score of each student with their respective name.

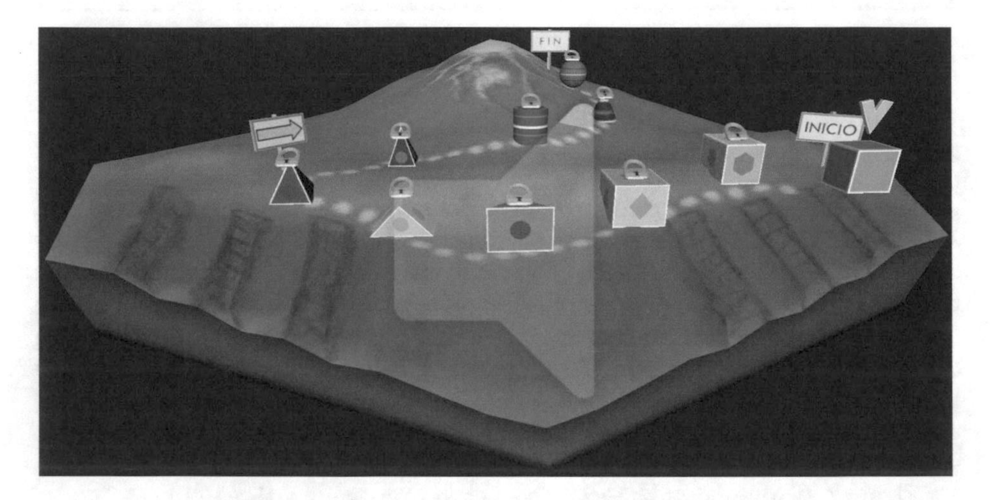

Fig. 2. Game levels: The 10 levels of the game on a journey from the countryside to the top of the volcano (Misti, Arequipa, Peru), from the easiest figure (cube) to the hardest figure (sphere).

An example of the list of names and scores of the participants with the status of their achievement is shown in Fig. 4. Where, gold means that it was perfect (10 points), red that was achieved (0 to 9 points) and lead that failed to learn.

4 Method

4.1 Research Design

As a hypothesis, it is established that by integrating applications of augmented reality on spatial geometry, a significant contribution will be made to the acquisition of mathematical skills on spatial geometry in elementary students.

The study is carried out through an experimental type design with different modalities, pre-test and post-test measures in the two groups (experimental and control), as seen in Fig. 5. Students are divided into two groups: experimental group (E.G.), whose members will use the application; and the control group (C.G.), formed by the subjects that will use traditional activities (practice exercises with sheet and pencil). Both groups were randomly selected. Following the methodological criteria of this type of research design, measures of each individual are collected, before and after the intervention.

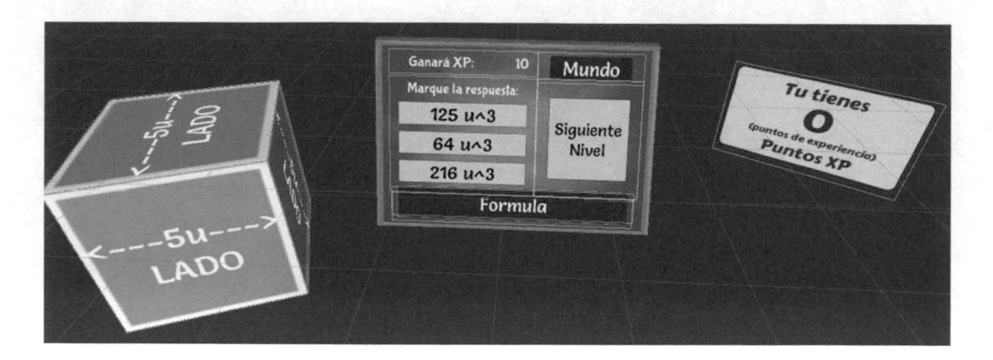

Fig. 3. Game exercise: geometric figure, exercise chart with points to be earned, alternatives, next level button, formula button, return to main menu button (World) and total points earned.

4.2 Variables

The training program about solving exercises spatial geometry is the independent variable. The dependent variable is defined as the mathematical capabilities of students on spatial geometry.

4.3 Participants

In this study involved students from the Educational Institution Señor de Huanca CIRCA (Arequipa-Peru), the sixth grade of primary school was intervened, the ages of the students range between 10 and 12 years. Before carrying out the study, students acquired prior knowledge of basic geometry with traditional methodology, this class was taught by the teacher in charge. A sample was taken randomly (n = 11, of which 5 are boys and 6 are girls). After the explanation of the intervention, this first group known experimental group identified the application during a regular math class session, while the other group (n = 10, of which 5 are boys and 5 are girls), using the traditional methodology of the students in a practice with exercises during the same time this group was called: control group. Girls represent 54% of the subjects of the experimental group and 50% of the control group.

4.4 Procedure

Figure 5 shows the experimental procedure. The research was structured based on three stages: The first stage involves the measurement of the dependent variable (pre-test) with a duration of 10 min; in the second stage, the intervention is developed with the application of the game and a duration of 25 min; and in the third stage, the evaluation test (post-test) is repeated with a duration of 10 min.

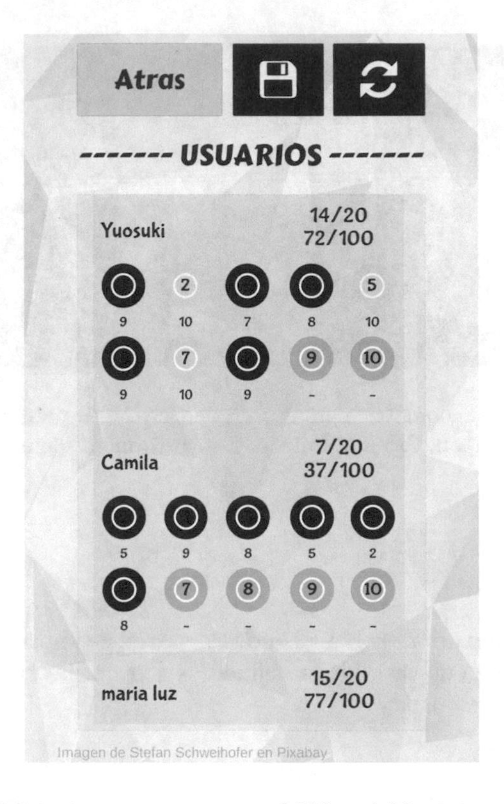

Fig. 4. Game administrator: names, scores of 100 and 20, state and score each level.

The intervention consisted of the resolution of the first 7 exercises of the application with the children of the experimental group, see Fig. 6, considering that the participants can restart the game and achieve better scores than the previous attempt, the only one Limiting is time. On the other hand, the control group focused on the resolution of 14 exercises (2 cubes, 4 rectangular prisms, 4 triangular prisms, and 4 pyramids). The application of the evaluation tests was carried out individually pre-test and post-test, to demonstrate the influence of different learning modalities on academic performance.

The pre-test consisted of 7 questions, each of which had an exercise with different measures and with different geometric figures, without placing the formulas.

4.5 Analysis of Data

To verify the influence of the Voluminis system on the acquisition of mathematical skills on spatial geometry in schoolchildren, the analysis of the results obtained from the pre-test and post-test was carried out.

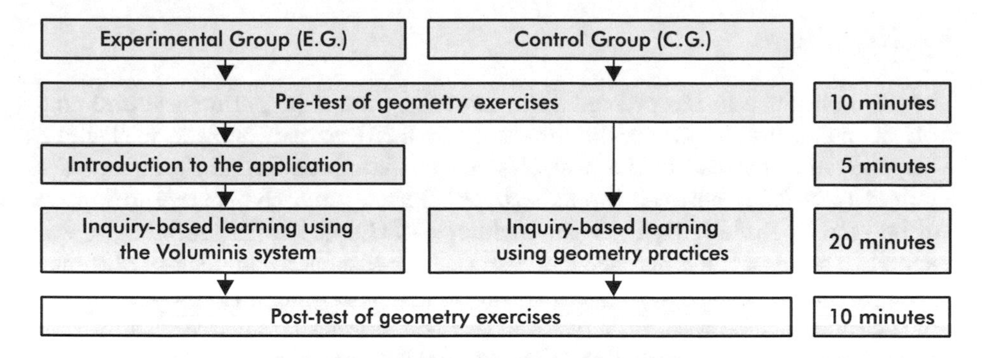

Fig. 5. Experimental procedure

(a) (b)

Fig. 6. Students using the application (a) Outside the classroom (b) Inside the classroom

First, two normality studies were carried out using the pre-test and post-test, to compare both groups. For this study, the Shapiro-Wilk normality test was used. The use of this test is recommended when the study is performed on a sample of fewer than 30 individuals, as in this case. In the statistical analyses that are carried out, <0.05 is established as a critical value. Subsequently, the normality study of the pre-test and post-test sample based on the differences in scores was carried out, the Shapiro-Wilk normality test was also used. Similarly, this data does not follow a normal distribution.

From the above, we can deduce that we must use non-parametric contrast tests such as Mann-Whitney U and Wilcoxon W.

5 Results

The data obtained in the pre-test show that the groups (experimental and control) do not show significant differences ($p > 0.05$), as can be seen in Table 2. Also, the data obtained in the post-test shows that there are no significant differences ($p > 0.05$) between the experimental group and the control group, as can be seen in Table 3. To know the incidence of the dependent variable, it was decided to analyze the difference between the pre-test scores and those obtained in the post-test, for the two experimental and control groups (Table 4).

The difference between post-test and pre-test shows that there are significant differences ($p < 0.05$), with this it is interpreted that the score (post-test) of students after traditional practices and the Voluminis system differ from the initial score (pre-test).

Some anecdotes that arose when applying Voluminis were the following: At the stage of the introduction to publicize how the Voluminis application was going to be done, some of the students were already interacting with the exercises of the game, being something intuitive for them. On the other hand, during the session with Voluminis some students used paper and pencil to solve the exercises. Regarding the aspect of competing with their peers, they wondered about the score they were getting to improve. Additionally, somebody participants were trying to pass the exercises randomly marking the answer. Also, some had difficulties in recognizing the surfaces required by the ARCore tool to lift the game.

At the end of the session, the students showed interest in the game, they are motivated, wished they had more time with the application and asked if more applications had been made to download them in their homes and that they could practice the knowledge acquired in the classroom.

Table 2. Differences in the pre-test between experimental and control groups (Mann-Whitney test)

	Points
U de Mann-Whitney	44,000
W de Wilcoxon	99,000
Z	–,976
Sig. significance (bilateral)	,329

Table 3. Differences in the post-test between experimental and control groups (Mann-Whitney test)

	Points
U de Mann-Whitney	32,000
W de Wilcoxon	98,000
Z	−1,948
Sig. significance (bilateral)	,051

Table 4. Analysis of the differences between post-test and pre-test (Wilcoxon test)

	PostTest - PreTest
Z	−3,366
Asymptotic significance (bilateral)	,001

6 Conclusion and Future Works

The results of the control group and the experimental group showed similar behavior. It is suggested that this way of acting is since the two methodologies manage to impart knowledge, and the student manages to capture it, however, the methodology with the use of technology leaves more motivated the student, wanting to continue learning and gaining more knowledge.

This study demonstrates that it is possible to increase math skills in geometry, being able to test the hypothesis. Considering that it had similar results to the traditional methodology.

Thanks to the use of MAR, great interest and acceptance were obtained by the students during the game. The way of transmitting the concept of three-dimensional space, in comparison with two-dimensional drawings, promoted its acceptance. About the use of augmented reality, it can help students to have more interest in mathematics. About the competition, it can be an essential motivator in a group of participants in the use of educational games.

In the future, it could be integrated with more interaction methods such as rotation and animations. On the other hand, to increase the interaction time in classes and at home, it should be considered that the number of students who have a mobile device with sufficient characteristics to exercise MAR applications is low. Finally, you could introduce composite geometric models such as a sphere and a cone, achieving a figure in the shape of an ice cream ball on top of an inverted cone.

Acknowledgements. Thanks to the "Research Center, Transfer of Technologies and Software Development R + D + i" - CiTeSoft EC-0003-2017-UNSA, for their collaboration in the use their equipment and facilities, for the development of this research work.

References

1. Abu Bakar, J.A., Gopalan, V., Zulkifli, A.N., Alwi, A.: Design and development of mobile augmented reality for physics experiment. In: Abdullah, N., Wan Adnan, W.A., Foth, M. (eds.) User Science and Engineering, pp. 47–58. Springer, Singapore (2018)
2. Avila-Pesantez, D.F., Vaca-Cardenas, L.A., Delgadillo Avila, R., Padilla Padilla, N., Rivera, L.A.: Design of an augmented reality serious game for children with dyscalculia: a case study. In: Botto-Tobar, M., Pizarro, G., Zúñiga-Prieto, M., D'Armas, M., Zúñiga Sánchez, M. (eds.) Technology Trends, pp. 165–175. Springer, Cham (2019)
3. Ayer, S.K., Messner, J.I., Anumba, C.J.: Augmented reality gaming in sustainable design education. J. Archit. Eng. **22**(1), 04015012 (2016). https://doi.org/10.1061/(asce)ae.1943-5568.0000195
4. Garzón, J., Pavón, J., Baldiris, S.: Systematic review and meta-analysis of augmented reality in educational settings. Virtual Reality **23**(4), 447–459 (2019). https://doi.org/10.1007/s10055-019-00379-9
5. Hsiao, H.S., Chang, C.S., Lin, C.Y., Wang, Y.Z.: Weather observers: a manipulative augmented reality system for weather simulations at home, in the classroom, and at a museum. Interact. Learn. Environ. **24**(1), 205–223 (2013). https://doi.org/10.1080/10494820.2013.834829
6. Jung, M., Conderman, G.: Early geometry instruction for young children. Kappa Delta Pi Rec. **53**(3), 126–130 (2017). https://doi.org/10.1080/00228958.2017.1334478
7. Martín-Gutiérrez, J., Saorín, J.L., Contero, M., Alcañiz, M., Pérez-López, D.C., Ortega, M.: Design and validation of an augmented book for spatial abilities development in engineering students. Comput. Graph. **34**(1), 77–91 (2010). https://doi.org/10.1016/j.cag.2009.11.003
8. Muhamad Nazri, N.I.A., Awang Rambli, D.R., Irshad, S.: Exploratory study on multimodal information presentation for mobile AR application. In: Abdullah, N., Wan Adnan, W.A., Foth, M. (eds.) User Science and Engineering, pp. 358–369. Springer, Singapore (2018)
9. Platonov, J., Heibel, H., Meier, P., Grollmann, B.: A mobile markerless AR system for maintenance and repair. In: 2006 IEEE/ACM International Symposium on Mixed and Augmented Reality. IEEE, October 2006. https://doi.org/10.1109/ismar.2006.297800
10. Radu, I., Doherty, E., DiQuollo, K., McCarthy, B., Tiu, M.: Cyberchase shape quest: pushing geometry education boundaries with augmented reality. In: Proceedings of the 14th International Conference on Interaction Design and Children, IDC 2015, pp. 430–433. ACM, New York (2015). http://doi.acm.org/10.1145/2771839.2771871
11. Ranade, S., Zhang, M., Al-Sada, M., Urbani, J., Nakajima, T.: Clash tanks: an investigation of virtual and augmented reality gaming experience. In: 2017 Tenth International Conference on Mobile Computing and Ubiquitous Network (ICMU). IEEE, October 2017. https://doi.org/10.23919/icmu.2017.8330112
12. Satpute, T., Pingale, S., Chavan, V.: Augmented reality in e-learning review of prototype designs for usability evaluation. In: 2015 International Conference on Communication, Information & Computing Technology (ICCICT). IEEE, January 2015. https://doi.org/10.1109/iccict.2015.7045712

13. Tezer, M.: The effect of answer based computer assisted geometry course on students success level and attitudes. Qual. Quant. **52**(5), 2321–2329 (2017). https://doi.org/10.1007/s11135-017-0666-5
14. Tobar-Munoz, H., Fabregat, R., Baldiris, S.: Using a videogame with augmented reality for an inclusive logical skills learning session. In: 2014 International Symposium on Computers in Education (SIIE). IEEE, November 2014. https://doi.org/10.1109/siie.2014.7017728
15. Toukoumidis, A.: Gamificación en Iberoameérica: experiencias desde la comunicación y la educación. Abya Yala Cuenca, Ecuador Universidad Politécnica Salesiana, Quito, Ecuador (2018)
16. Wei, X., Weng, D., Liu, Y., Wang, Y.: Teaching based on augmented reality for a technical creative design course. Comput. Educ. **81**, 221–234 (2015). https://doi.org/10.1016/j.compedu.2014.10.017
17. Zhenming, B., Mayu, U., Mamoru, E., Tatami, Y.: Development of an English words learning system utilizes 3D markers with augmented reality technology. In: 2017 IEEE 6th Global Conference on Consumer Electronics (GCCE). IEEE, October 2017. https://doi.org/10.1109/gcce.2017.8229353

Objective Construction of Ground Truth Images

Mark Smith[1], Ananda Maiti[2(✉)], Andrew Maxwell[1], and Alexander A. Kist[1]

[1] School of Mechanical and Electrical Engineering, University of Southern Queensland, Toowoomba, Australia
{mark.smith, andrew.maxwell}@usq.edu.au, kist@ieee.org
[2] Discipline of ICT, University of Tasmania, Newnham, Australia
anandamaiti@live.com

Abstract. In the context of virtual and augmented reality, computer vision plays a pivotal role. To benchmark performance, evaluation of computer vision models, such as edge detection is essential. Traditionally this has relied on subjective analysis of the resultant images. Alternatively, models have been assessed against ground truth images. However, ground truth images are highly subjective, relying on human judging to determine the appropriate location of features. Literature complains about the lack of objective quantitative measures for model evaluation, yet no solution has been presented. Ground truth is the objective verification of properties of an image. In the context of this paper it is a data set that includes an accurate and complete representation of the edges. The subjective nature of creating ground truth images has meant that true image analysis model evaluation has been limited. Reducing the level of subjective decisions can improve the confidence level when measuring the performance of computer vision image analysis models. This work describes a new method to improve ground truth image confidence through an automated computer vision feature detection model voting system.

Keywords: Computer vision · Edge detection · Ground truth

1 Introduction

In the context of cyber-physical systems, augmented reality and remote access laboratories, computer vision and image processing provide important system building blocks. To judge the effectiveness of Computer Vision (CV) models, both the speed of algorithms and the accuracy of detection have to be evaluated. The former can be achieved with traditional benchmarking methods, the latter has largely relied on the subject examination of the resultant spatial data. The appearance, as analyzed by the researcher or a panel, is observed and rated as a measure of the model's effectiveness [1]. Attempts to improve the scientific rigor still required human assessment of image pre and post processing [2].

Characteristics of the proposed models consist of the computational costs and feature detection ratios. While traits such as model speed and resource utilization can

M. E. Auer and D. May (Eds.): REV 2020, AISC 1231, pp. 300–312, 2021.
https://doi.org/10.1007/978-3-030-52575-0_24

be objectively compared to existing models, evaluation of the effectiveness of the models feature detection is based solely on subjective decisions. Both qualitative and quantitative measures are required for performance testing. Image analysis performances generally rely on quantitative measures. Literature complains about the lack of objective quantitative measures for model evaluation, yet no universal solution has been proposed. Often, ground truth images are used to test the accuracy and performance of image processing. The subjective nature of ground truth images has meant that true image analysis model evaluation has been limited. Ground truth testing is the only methodical means of verifying the effectiveness of a CV model, yet many researchers are loathed to employ the method because of the subjective nature.

Computer Vision edge detection models struggle to work with digital data sets that consist of noise and interference from inconsequential artefacts. Artefacts that are of no interest to the analysis of the image contribute to spurious points or edges. The clean and tight lines we view on an image are an illusion. The left image of Fig. 1 demonstrates a simple series of lines and circles in which the edges seem apparent. However, zooming into the highlighted area reveals that the edges quickly blur together. Locating an actual edge from the blended edges becomes difficult, as shown in the right image of Fig. 1.

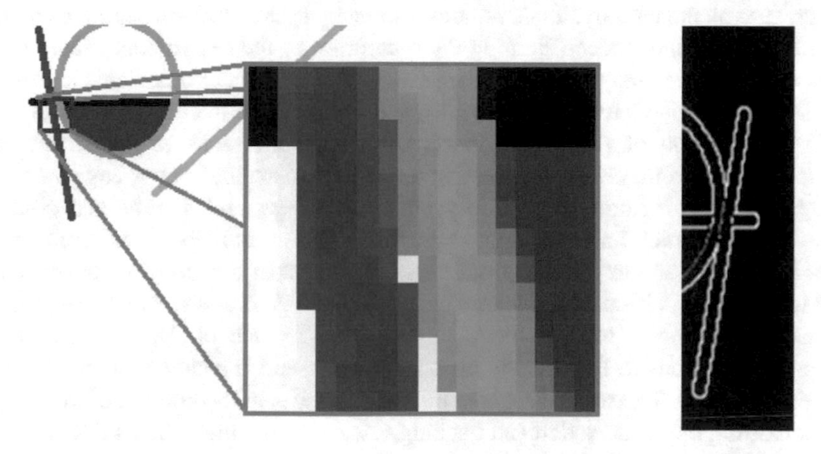

Fig. 1. Edge detection left (Color): edge and color blur right (Black/White): edge detection result

The level of edge detection becomes difficult to ascertain when comparing CV models which have seemingly similar results. However, there are also certain extremes of edge detection. An original image is shown in Fig. 2. The right image of Fig. 3 appears to have found all edges we would subjectively consider. However, there is also an obvious over processing. Subjectively, which edge detector is the most effective?

This works outlines a method to overcome this limitation by building hybrid ground truth images. Sequentially applying edge detection models to the original image and accumulating the hits to each pixel creates a depth image representing the most common potential edges. Thinning of the data is achieved through the Gaussian distribution of the pixel hits. The most common pixels are saved leaving a partially completed ground truth image. Fine-tuning still requires human intervention, but the majority of the decision-making process has been automated.

This paper is structured as follows. Section 2 provides a brief overview of the prevalent ground truth testing and methodology. Section 3 explains the purpose of ground truth testing, and the current state of the field, while Sect. 4 defines the proposed model, and demonstrates the effectiveness. Section 5 describes the proposed model's application.

2 Augmented Reality for Remote Access Laboratories

The use of Augmented Reality for Remote Access Laboratories is an amalgamation of technologies. Many current RAL systems have added networking functionality to existing equipment to achieve a remote capability with limited regard to the full advantages the technology may provide [3]. These RAL systems provide many benefits such as: practice sessions in which the student rehearses in-situ laboratory demonstrations and access to resources anywhere at any time. However, the systems still fall short in providing efficient feedback to students [4]. A level of transparency for the technology, such as the current simple remote laboratory system infrastructure, may also be distracting [5] to the student, limiting their attention to the experiment. Some Remote Access Laboratories may also limit the student social interactions, achieved from in-class sessions, which have been shown to be detrimental to learning outcomes [6].

The application of AR within the RAL framework seeks to break free of the traditional didactic processes by creating an environment that better engages the student. Supplying an immersive sensory environment generates improved contextual situations from which understanding is derived [7]. The immersive environment promotes knowledge and experience from the engagement of the students rather than just static learning [8]. Vision-based AR relies heavily on CV models, which are processor-intensive and complex to implement. Expanding the reach of AR for RAL requires unique CV solutions to the specific object detection and tracking problem within the RAL environment. The work discussed in this paper sits in the context of investigating new or existing object detection and tracking CV models, which allows AR for RAL to appropriately immerse the user into the remote environment.

Evaluation of current CV models is necessary to ascertain their suitability to support AR within the RAL framework. However, not every CV model, in isolation, is relevant to the identification of objects with a video stream. For example, a Gaussian filter provides low pass filtering, removing high-frequency noise from an image, and as such cannot be tested for suitability by itself. Additionally, the nature of the tests and determination of fitness is of a complex nature, requiring careful consideration. Fifty existing CV models have been interpreted and constructed, in the course of this research, to validate previous research and to then apply them to the AR RAL object detection problem. Two testing pathways have been selected to match the research question; the object detection testing, and the object tracking testing pathways. Ground truth images have been used for object detection testing. The images have been selected to test the CV models' abilities to detect objects with solid colours, geometric edges, varying backgrounds, multiple colours, similar objects and varying patterns. Some previous CV models validate their functionality through synthetic images, such as the SUSAN Feature Detector [9], but real images have been chosen, for this research, to validate hybrid models against real-world data. Three types of tests are performed, based on the hybrid model under consideration. Segmentation, edge detection and feature point detection are measured and calculated, counting the number correct, incorrect or missing for each ground truth image. Tracking tests are applied using the hybrid models, against real RAL experiments, where points or items within the video stream are identified and tracked. The number of frames successfully tracked is counted and compared along with the frames where the models lost track or misidentified the object. The novel method introduced in this paper improves the generation of ground truth images which, in turn, are required to evaluate the performance of the algorithms.

Fig. 2. Original image

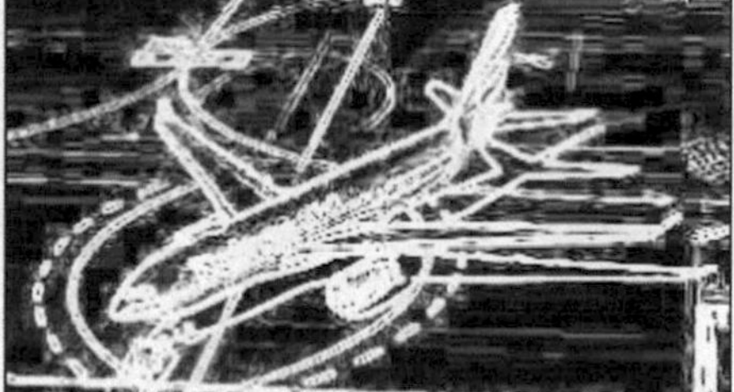

Fig. 3. Variations of true edge

From the diverse range of CV image edge detection and feature point detection models available, objective evaluation of their effectiveness is atypical. Meaningful performance measures have relied on signal-based approaches [2] which assess the local signal for false positives and false negatives, or the results from ground truth images [10]. Testing CV detection models against synthetic images had been able to provide solid measurement data [11, 12], but the effectiveness of the models on real-world environments is a far more difficult proposition. Unfortunately, these methods rely on subjective decisions by one or many humans and do not produce consistent results. The Berkeley Segmentation Data Set (BSDS300 & BSDS500) [13] constructs hand-drawn contour and segmentation boundaries in an attempt to train CV image analysis. For each sample image, there are a series of hand-drawn interpretations, and reviewing each interpretation, it becomes apparent, the difficultly in limiting the subjective nature of ground truths.

3 Ground Truth Purpose

Understanding and interpreting two-dimensional spatial digital data sets, commonly known as digital images, is non-trivial. Computer vision image analysis has produced many models to discover features of an image in an attempt to build knowledge. When deciding on the effectiveness of newly created models or attempting to determine the capabilities of an existing model, researchers must rely on either a small cohort of judges or create ground truth images to baseline their results.

Extracting the results of a CV image analysis model, such as an edge detector, involves comparing each pixel from the CV analysis image with the ground truth image. Within the ground truth image, pixels are allocated a membership to one of three possible classifications.

- *Key Point*: This is a pixel that indicates an edge or feature point that the model under review must detect.
- *No Point*: This is a pixel that is not an edge or feature point, and the model under review must not detect.
- *Don't Care*: Represents pixels that are not relevant to the detection process. In most situations this represents regions or features that are unimportant to the goals of the model.

Pixels classified as *Don't Care* are primarily selected in regions of the image which are not analysed by the CV model. For example, a CV system monitoring road traffic does not care about edge detection of the nearby trees. While the CV models may detect the trees to varying degrees, the application of the model means that the region will not be considered as part of the effectiveness score.

For the purposes of this document, the computer vision image analysis models under discussion are edge detection models.

Comparing the CV analysis model image against the ground truth image, pixel by pixel produces a measure of the model's effectiveness. True positive (TP) values represent a match between a model edge pixel and the ground truth key point pixel, while true negative (TN) values are non-edge pixels matched to no point pixels. Error detection is represented by false positive (FP) and false negative (FN) results which occur through mismatches in the model edge pixels and the ground truth points. Assessing the performance, the Receiver Operating Characteristic (ROC) curves [10] can then be used, or other evaluation methods.

Utilizing a properly formed ground truth image, a quantitative assessment of a CV edge or feature point detector can be formed.

4 Proposed Model

While the importance of ground truth images for CV analysis model assessment is argued, the subjective nature of the ground truth construction has become the primary difficulty for its support. Producing consistent ground truth images requires an automated method to determine pixel membership and minimise the subjective nature of the process.

The proposed model creates a three-dimensional ground truth object representing the current image, shown in Eq. 1.

$$GT = M \times N \times d \tag{1}$$

Where the image is of size $M \times N$ and d is the number of edge detection models engaged, producing the depth. Multiple models are applied to the test image with the edge map a new layer within the ground truth object. The accumulation of edge maps produces a ground truth image containing all of the desired features of the image under test.

The proposed model engages known CV models to create multiple feature maps (edge maps). The edge maps build upon each consecutive map.

For these works, fifteen different edge detection methods are utilised and listed below:

- Laplacian 3 × 3 [14]
- Gradient Derivative
- Gradient Edge (First Derivative)
- Gradient Edge (Second Derivative)
- Laplacian 5 × 5 [14]
- Laplacian of Gaussian [15]
- Sobel 3 × 3 [16]
- Sobel (Absolute)
- Prewitt 3 × 3 [17]
- Kirsch 3 × 3 [18]
- Canny [19]
- Homogeneity
- Compass Sobel
- Compass Prewitt
- Compass Kirsch

Pixels assigned to edge maps for each CV model, increment the corresponding pixel votes' within the ground truth image.

Pixels with a higher accumulation of votes could be assumed to have a higher probability of being a valid edge or interest point. Assuming a single-sided Gaussian distribution of pixel votes, thinning the ground truth image to return image detail is possible. Ignoring pixel votes that fall outside a specific standard deviation returns image detail. Through the selection of a nominal standard deviation filter, edge thinning of the accumulated ground truth image exposes the features/edges most likely to be valid. Further improvement of the ground truth image may occur manually, depending upon the requirements of the work.

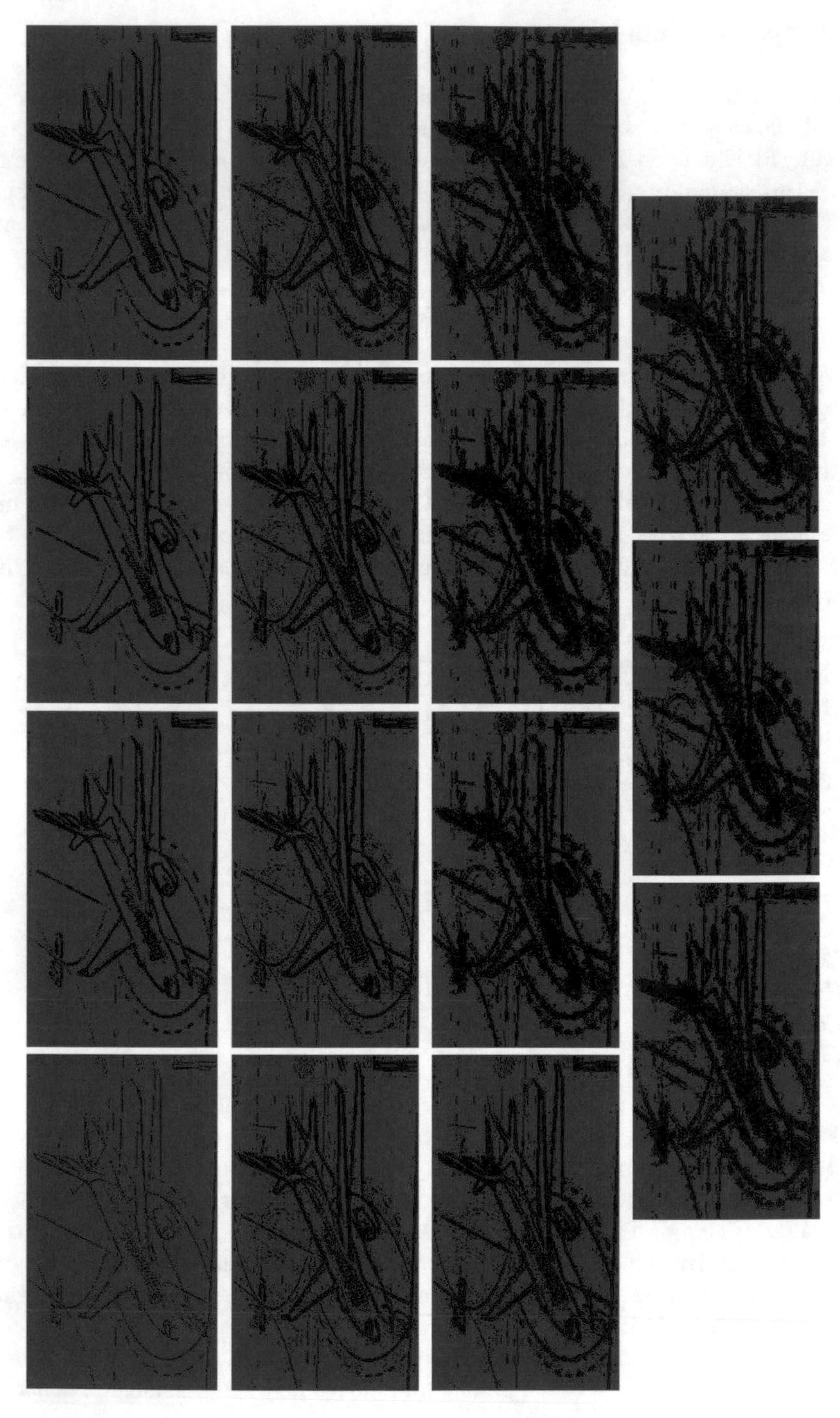

Fig. 4. Composite ground truth image building

5 Model Application

This section demonstrates how the model operates using the original sample image shown in Fig. 2. Figure 4 demonstrates the accumulation of edge maps to produces a ground truth image that captures relevant features. The increased accumulated strength of edges is clearly visible in the images. Additionally, the clarity of the ground truth image degrades as noise from each CV model builds up.

This is apparent from the minimal detail remaining in the first image (top-left) to the *overexposed* last image (bottom-right) of Fig. 4. However, accumulating pixel votes uncovers the concentration of commonly detected edge pixels.

If the distribution of pixel density is considered to be a single-sided Gaussian distribution, noise can be removed. Figure 5 shows the distribution of votes after fifteen edge detection models have been applied to the ground truth object from the original image shown in Fig. 2.

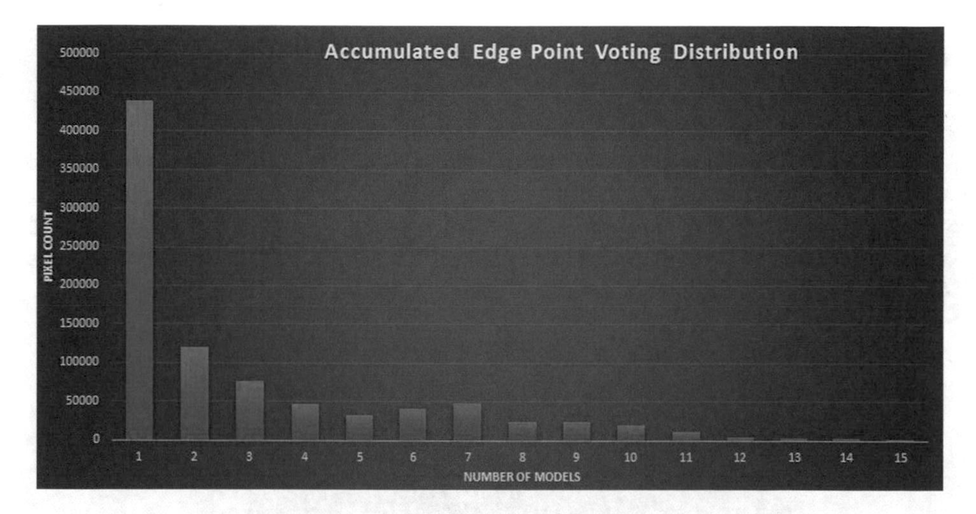

Fig. 5. Edge Detector voting distribution

Using a standard deviation of one ($\sigma = 1$), Fig. 6 demonstrates the results of the composite ground truth image building and thinning. Visually comparing Fig. 2 with 6 shows the accuracy in determining probable ground truth edges. Overlaying the two images produces a very high level of correlation. Details from the previously over-exposed image feature return and much of the noise is now removed.

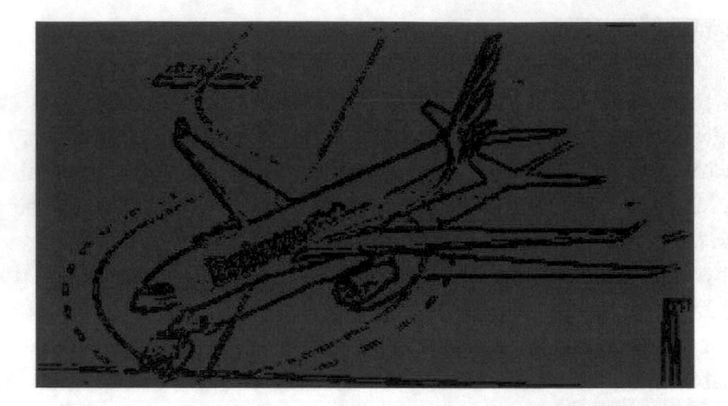

Fig. 6. Edge thinning of composite ground truth image

Figure 7 demonstrates some additional subjective work on the intermediate composite, statistically filtered image. Some lines have been completed, and regions have been marked as *Don't Care* (in white) as, for the current tests, these regions were not important in object detection of an airplane.

Outcomes from computer vision image analysis, based on ground truth performance measures, are solely a factor of the ground truth image.

Objectively, the application of known, robust edge detection models to an image will create some level of confidence in the location of valid edges. Subjectively, comparing Fig. 2 to Fig. 6 seems to indicate successful detection and localization of valid edges. The proposed model has minimized the subjective nature of ground truth construction by building a foundation of projected valid edges.

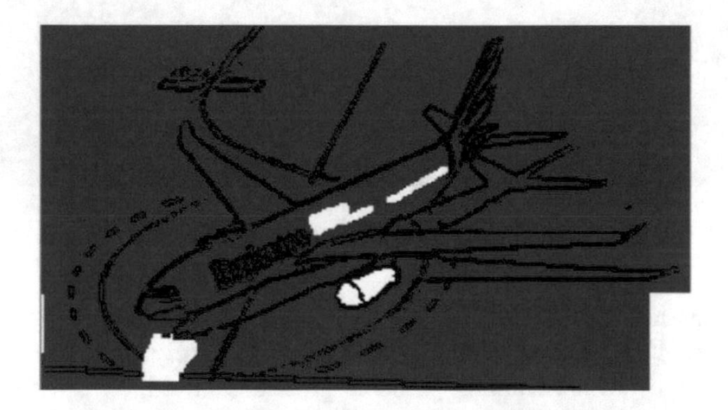

Fig. 7. Completed ground truth image

For the image in Fig. 2, there are no pixels within the image which were detected as an edge for all of the fifteen edge detection models tested. As shown in the sequence in 8, the first pixels appear for 13 out of the 15 models. Image substance only starts to appear when nine common edge detectors have voted consistently. Selection of an appropriate level of detail is another level of subjectiveness when building a ground truth image. However, as can be seen in the sequential images of 8, independent CV image analysis models still improve the subjective decision outcomes (Fig. 8).

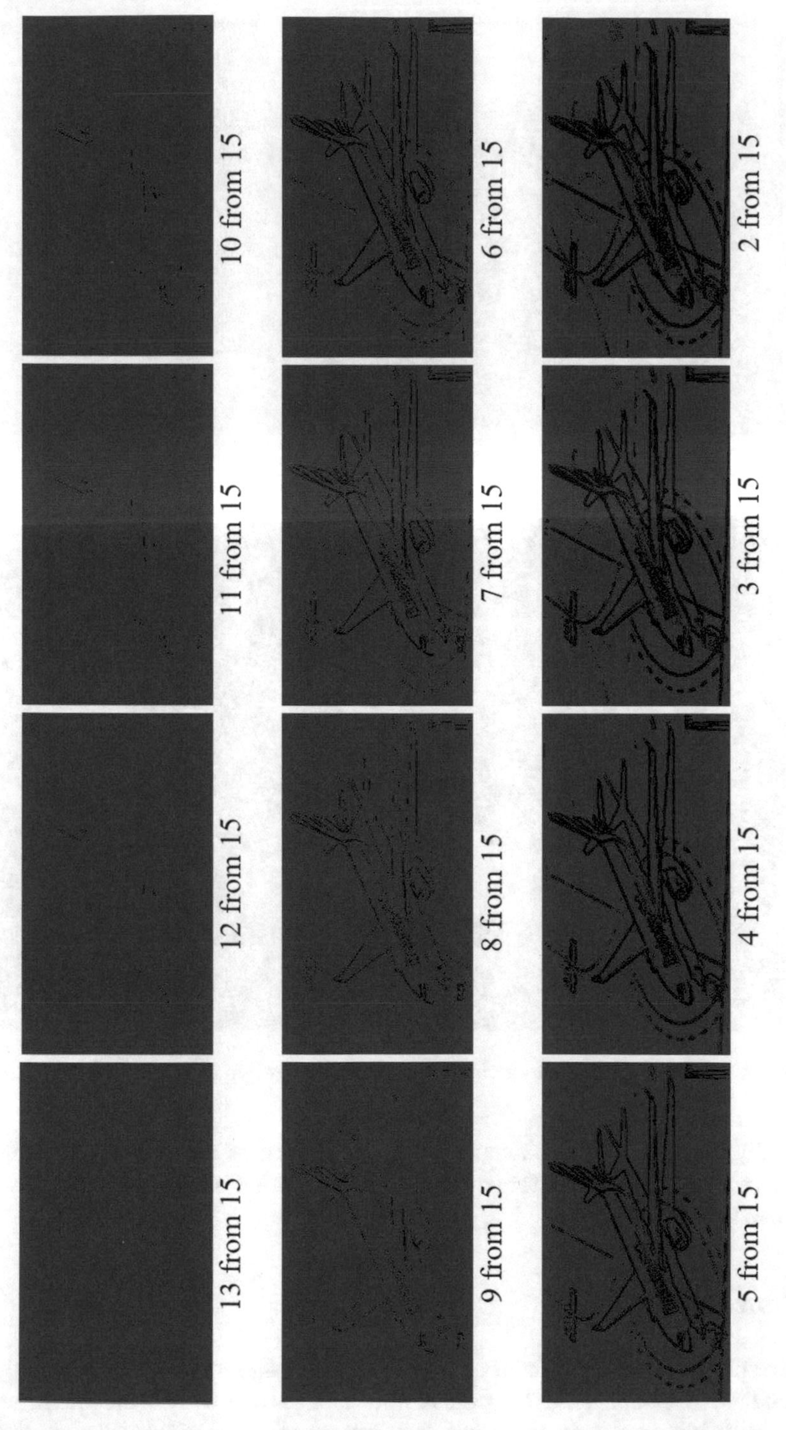

Fig. 8. Edge detection - pixels associated with the number of edge detection models

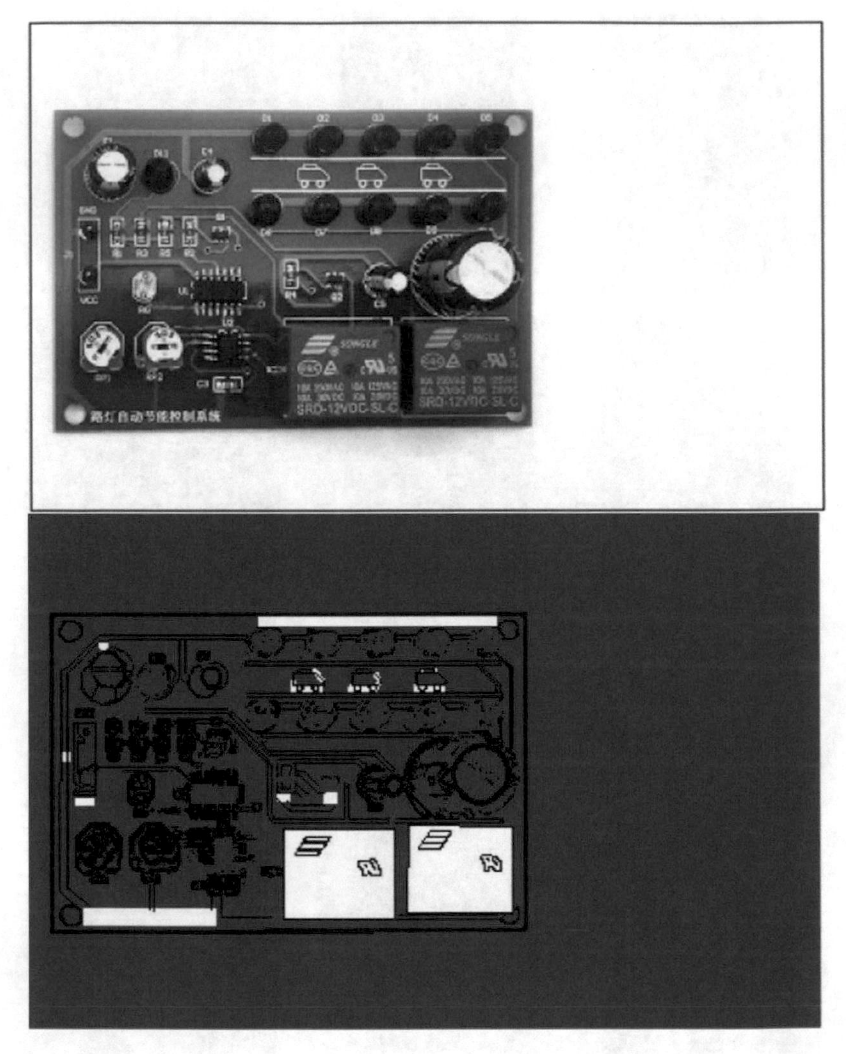

Fig. 9. Model application to a circuit board image – original and ground truth image

Finally, Fig. 9 shows an example of the original image and the ground truth image created using the algorithm for a circuit board as often used for electronics remote experiments.

6 Conclusions

Effective performance analysis of computer vision image processing has mostly relied on quantitative examination of ground truth images. However, the construction of ground truth images requires a human assessment of the original image to determine the appropriate edges. The subjective nature of ground truth construction leads to many wary of the true effectiveness of CV image feature detection models. Construction of ground truth images on a firmer foundation improves the acceptance of model testing.

Developing a ground truth method based upon the measurements of standard edge detection models limits the subjective nature of ground truth image building, but the validation of such a model is still subjective. The proposed model is asking to accept the results of objective ground truth image building, even when the assessment of such a result is still subjective. However, the proposed model achieves some level of automation and provides a means to objectively and subjectively determine valid ground truths.

Ground truth construction utilizing this method still requires a human element, but the task is no longer fully subjective. Incorporating a weighting scheme which promotes weighted voting by robust models may provide improved results. Further research for building completely quantifiable ground truth images are not foreseeable, but improvements to this method by automating the edge thinning techniques. With access to reliable ground truth image files, testing edge detection models become possible with binary performance classifiers to score measures such as accuracies, precision, and sensitivity.

References

1. van Vliet, L.J., Young, I.T., Beckers, G.L.: A nonlinear laplace operator as edge detector in noisy images. Comput. Vis. Graph. Image Process. **45**(2), 167–195 (1989)
2. Heath, M., Sarkar, S., Sanocki, T., Bowyer, K.: Comparison of edge detectors: a methodology and initial study. In: 1996 IEEE Computer Society Conference on Computer Vision and Pattern Recognition, Proceedings CVPR 1996, pp. 143–148. IEEE (1996)
3. Jeschke, S., Thomsen, C., Richter, T., Scheel, H.: On remote and virtual experiments in elearning in statistical mechanics and thermodynamics. J. Softw. **2**(6), 76–85 (2007)
4. Kist, A.A., Maxwell, A.D., Gibbings, P.: Expanding the concept of remote access laboratories. In: ASEE 2012 Annual Conference & Exposition: Spurring Big Ideas in Education, San Antonio, TX, USA. American Society for Engineering Education (2012)
5. Ijsselsteijn, W.A., de Ridder, H., Freeman, J., Avons, S.E.: Presence: concept, determinants and measurement. Hum. Vis. Electron. Imaging **3959**, 520–529 (2000)
6. Lindsay, E.D., Naidu, S., Good, M.C.: A different kind of difference: theoretical implications of using technology to overcome separation in remote laboratories. Int. J. Eng. Educ. **23**(4), 72 (2007)
7. Barak, M.: Instructional principles for fostering learning with ICT: teachers' perspectives as learners and instructors. Educ. Inf. Technol. **11**(2), 121–135 (2006)
8. Bowtell, L.A., Moloney, C., Kist, A.A., Parker, V., Maxwell, A.D., Reedy, N.: Enhancing nursing education with remote access laboratories. Int. J. Online Eng. (iJOE) **8**(S4), 52–59 (2012)
9. Smith, S.M., Brady, J.M.: SUSAN—a new approach to low level image processing. Int. J. Comput. Vision **23**(1), 45–78 (1997)
10. Bowyer, K., Kranenburg, C., Dougherty, S.: Edge detector evaluation using empirical ROC curves. In: IEEE Computer Society Conference on Computer Vision and Pattern Recognition, vol. 1, pp. 354–359. IEEE (1999)
11. Fram, J.R., Deutsch, E.S.: On the quantitative evaluation of edge detection schemes and their comparison with human performance. IEEE Trans. Comput. **100**(6), 616–628 (1975)

12. Peli, T., Malah, D.: A study of edge detection algorithms. Comput. Graph. Image Process. **20** (1), 1–21 (1982)
13. Arbelaez, P., Maire, M., Fowlkes, C., Malik, J.: Contour detection and hierarchical image segmentation. IEEE Trans. Pattern Anal. Mach. Intell. **33**(5), 898–916 (2011)
14. Torre, V., Poggio, T.A.: On edge detection. IEEE Trans. Pattern Anal. Mach. Intell. **2**, 147–163 (1986)
15. Marr, D., Hildreth, E.: Theory of edge detection. Proc. Roy. Soc. Lond. B Biol. Sci. **207** (1167), 187–217 (1980)
16. Sobel, I.: An Isotropic 3 × 3 Image Gradient Operator, Presentation at Stanford Artificial Intelligence Project (SAIL). Stanford, California (1968
17. Prewitt, J.M.: Object enhancement and extraction. Picture Process. Psychopictorics **10**(1), 15–19 (1970)
18. Kirsch, R.A.: Computer determination of the constituent structure of biological images. Comput. Biomed. Res. **4**(3), 315–328 (1971)
19. Canny, J.: A computational approach to edge detection. IEEE Trans. Pattern Anal. Mach. Intell. **8**(6), 679–698 (1986)

Work-in-Progress: Using Augmented Reality Mobile App to Improve Student's Skills in Using Breadboard in an Introduction to Electrical Engineering Course

Adel Al Weshah[1(✉)], Ruba Alamad[2], and Dominik May[1]

[1] University of Georgia, Athens, GA 30602, USA
adel.alweshah@uga.edu
[2] The University of Akron, Akron, OH 44325, USA
raa88@zips.uakron.edu

Abstract. The enhancement of first-year electrical engineering students' skills in building circuits using the breadboard is essential. To overcome difficulties in dealing with internal breadboard connections, Augmented Reality (AR) technology used to build a 3D application. AR could improve students' ability with respect to efficient design and build the electrical circuits in the breadboard due to the support for viewing the internal breadboard connections and learning step by step how to build circuits.

This paper will explore the integration of AR into the Introduction to Electrical Engineering course in order to provide an interactive learning experience, in which abstract thinking is supported, and concept visualization, exploration, and evaluation are facilitated during the various stages of the circuits building steps. Such an AR app will permit the students to visualize virtual representations of the circuit on the breadboard in real-world settings by mixing the 3D circuit element models with a camera view of the students' smartphone and 3D animations for circuit-building procedures.

Keywords: Augmented reality · Education · Breadboard · Electrical engineering · 3D visualization

1 Introduction

Augmented Reality (AR) is an interactive technology that superimposes a computer generated image or virtual object on a user's view of the real world, thus providing a composite view [7]. The essence of AR is that a device recognizes objects in the real world and renders the computer graphics registered to the same 3D space, providing the illusion that the virtual objects are in the same physical space with the user. Such technology is a big opportunity for a wide

M. E. Auer and D. May (Eds.): REV 2020, AISC 1231, pp. 313–319, 2021.
https://doi.org/10.1007/978-3-030-52575-0_25

variety of applications like medical training, repair and maintenance, classroom education and much more [7,8] and [11].

Over the past years, many AR applications for various education uses have been developed. A study was reported in 2006 that AR would be a promising technology in the long term and would have a significant impact in the coming years [5]. And, as expected, AR is getting more focus and attention, especially with the release of products like Google Glass [3]. It also documented that the number of educational AR studies has significantly increased after 2008, which indicates that more research interest has been focused on it [1].

In [4], an AR system is developed to teach the Ecosystems. The study made comparisons between the AR system and the traditional face-to-face teaching method. The study concluded that the students in the traditional-teaching group had significantly lower scores, while the students involved in the AR system had the highest ratings.

The environment AMIRE has designed and implemented a multimedia content using the AR technology. It provided AR tools that can be applied to the majority of engineering education, including Mechanical Engineering, Electronics and Telecommunications Engineering [10].

In [6], a project achieves a connection between the theoretical explanations and the laboratory practices using AR in the course of the electrical machines. The study reported a positive experience for students; they found the tool to be useful, user-friendly, and in accordance with the goal.

The main purpose of this study is to enhance students' understanding of Introduction to Electrical Engineering lab experiments with additional digital information intended to improve the needed skills to build circuits using the breadboard. The application was developed for the iOS and the Android platform, and its first version has already been tested with one section of Introduction to Electrical Engineering course.

2 Introduction to Electrical Engineering Course

Understanding and developing electric circuits is a fundamental building block in Introduction to Electrical Engineering course in the College of Engineering at the University of Georgia. Applications of electric circuits span over the entire electrical engineering curriculum via instruments and applications. Successful learning in electric circuits indicates a guaranteed completion of electrical engineering. Thus, this hurdle requires tools that will enable student success.

One of the difficulties for students is the inability to understand concepts such as voltage, current, and resistance. They are abstract and invisible. Though the mechanical analogies can be taught as the course progresses, the students struggle to understand more complex circuits and often fail to realize whether the answers to problems are within the physical regime. Moreover, carrying these concepts to a real circuit laboratory constitutes a more significant challenge. AR app has been introduced to one section of Introduction to Electrical Engineering course with 23 students, who used the application in conjunction with existing physical hardware.

The use of the AR app in Introduction to Electrical Engineering course lab experiments opened the opportunity to the students to visualize some of the mentioned abstract quantities and added more meaning to lab topics that students may not experience in the real world. For example, students would typically not be able to see the internal connection of the breadboard, flow of electric currents in the circuit, and charging and discharging of a capacitor. Interacting with augmented simulated circuits created by the AR application, students will develop adequate problem-solving skills and transferrable solving methods to apply what they learned in the classroom to different settings.

3 The Augmented Reality Application

AR has the potential to enhance student motivation, involvement, and excitement due to the active role; it allows the students to encompass [9]. Also, it enhances engagement and learning. AR experiences could provide positive impacts to the cognitive process if it is well designed [2]. Another advantage of AR is students' ability to learn the concepts and environment around them without being removed from the lab.

In particular, AR was deployed to help with historically challenging concepts. First, the AR app was used as a visualizing tool for the breadboard internal connection. Students often confuse the columns and rows on a breadboard that are connected. This application allows for the successful building skill of the circuits on the breadboard without the potential stigma of having to ask for help. Secondly, the AR application was used to develop skills in reading the pinout diagram for the integrated circuit (IC) that contains four NAND gates. ICs are active elements, meaning they require external power. This added wrinkle increases the potential of poor implementation that can lead to damage to the IC. Using the AR application, no ICs were destroyed; rather, students could visualize the internal functionally of each IC pin.

For this study, we used AR app that was designed to work on students' smartphones or tablet computers. A printed target (marker) was stuck to the breadboard that served as the scanned image for the digital objects displayed as augmented reality objects (see Fig. 1a). This means that the application displays additional digital information to the student after scanned the marker. The student just needs to open the application and scan the marker on the breadboard that is in the application's database. As soon as the application recognizes the marker, the additional information is displayed automatically (see Fig. 1b).

The application functions are as follows. First, the internal breadboard connections: The student can check the breadboard connection. In that case, a 3D breadboard model shows internal breadboard connections on the smartphone screen. Getting closer to it, the model becomes larger. In such a way, the student has a better view of the internal breadboard connections compared to a just figure. Figure 2 shows an example of a 3D breadboard model that appears on the student's smartphone using the developed AR application.

(a)

(b)

Fig. 1. AR mobile application: **a** breadboard marker; **b** app main menu

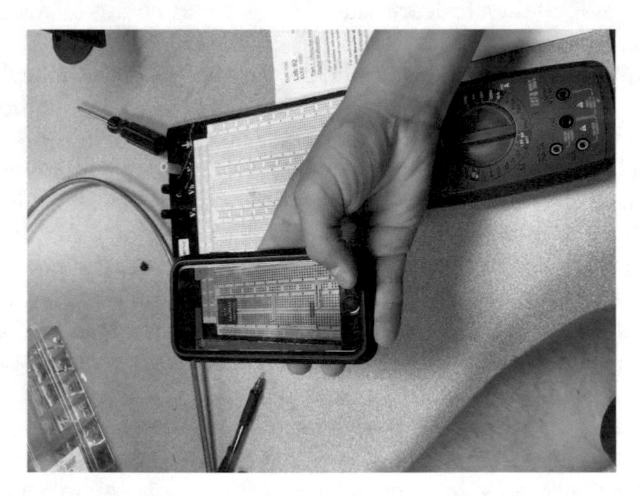

Fig. 2. 3D breadboard model with internal connections

Second, experiments mode: By selecting on the one of the two lab experiments on the AR app main menu (LED Lad or Logic Lab), a student will have access to the lab handout and step by step 3D animation instructions on how to build the lab circuit on the breadboard. In Fig. 3, the 3D animations for Logic lab experiment are shown. The 3D animations help the student to understand how to build the circuit on the breadboard. The application is developed with Unity and Vuforia. The mentioned software is chosen due to its user-friendly interface.

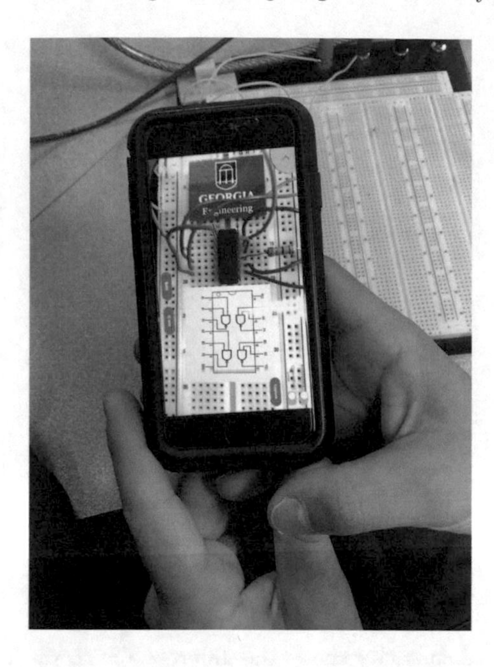

Fig. 3. 3D animations for the Logic lab experiment

4 Discussion

The AR developed application has greatly impacted the students during the Introduction to Electrical Engineering course. The application is fast and recognizes the marker easily. The usage of this application has to be seen as a pilot study, and the main goal at this point is to get a good sense on the effect of the app on a student's ability to learn the concepts.

To test the impact of the app, we performed a pilot study. Our study contains two groups. In the first group, the student taught the lab using AR app, while in the second group, the student taught the lab in a traditional way. After the lab, both groups were asked to answer one exam question about the internal connection of the breadboard. After we analyzed the results of the students in both groups, we found that the average for the section with AR application is 79.16% (SD 37.8), while the average for the section without AR application is 52.52% (SD 43.3). The following figure visualizes the difference in achieving scores when the AR app has been used and when the AR app has not been used (Fig. 4).

In addition, analysis of variance (ANOVA) was performed using JMP to check if the difference in grades between the two groups is significant and are due to the use of AR app. The result implies that the difference is significant, and the P-value is 0.0191. According to our results, the AR application is good additional material for the course lab experiments, and it led to positive impressions on the tested students. However, a few comments are stated related to further

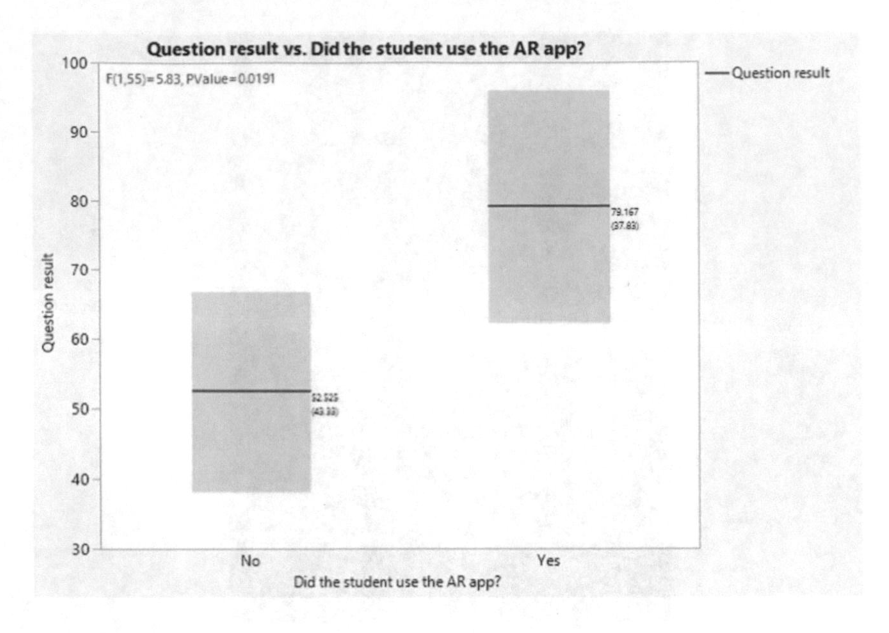

Fig. 4. A statistical summary shows the difference in scores in the two groups.

improvement of the application such as create a 3D animation for the flow of current in the circuit, develop the application for the rest of the lab experiments, and give some information about the circuit elements in the form of the textual explanations. These comments will be considered during the future development of this application.

5 Conclusion

A continuous trend in education with the support of technological developments is to use new methods and technologies; one of them is Augmented Reality (AR), which intends to make teaching and learning processes more hands-on and more tangible for abstract concepts. In this paper, an AR application developed for smartphones and tablet computers for the iOS and Android platforms. The designed lab experiments are augmented with digital information in order to help the students to build the circuit in the breadboard. Two forms of digital information are used for that purpose: 3D models and animations. So far, the AR application left a good impression on students; it also proved to significantly improve the students' grades in breadboard connection question. More experiments will be designed, more features will be added, and more analyses will be done to ensure the good impact of the application will be performed in the next stage of our project.

References

1. Akçayır, M., Akçayır, G.: Advantages and challenges associated with augmented reality for education: a systematic review of the literature. Educ. Res. Rev. **20**, 1–11 (2017)
2. Baran, B., Yecan, E., Kaptan, B., Paşayiğit, O.: Using augmented reality to teach fifth grade students about electrical circuits. Educ. Inf. Technol., 1–15 (2019)
3. Golparvar-Fard, M., Peña-Mora, F.A., Savarese, S.: Four-dimensional augmented reality models for interactive visualization and automated construction progress monitoring. US Patent 9,070,216, 30 June 2015
4. Hsiao, K.F., Chen, N.S., Huang, S.Y.: Learning while exercising for science education in augmented reality among adolescents. Interact. Learn. Environ. **20**(4), 331–349 (2012)
5. Johnson, L., Levine, A., Smith, R., Stone, S.: The 2010 horizon report. The New Media Consortium, Austin, Texas (2010)
6. Martín-Gutiérrez, J., Fabiani, P., Benesova, W., Meneses, M.D., Mora, C.E.: Augmented reality to promote collaborative and autonomous learning in higher education. Comput. Hum. Behav. **51**, 752–761 (2015)
7. Nee, A.Y., Ong, S., Chryssolouris, G., Mourtzis, D.: Augmented reality applications in design and manufacturing. CIRP Ann. **61**(2), 657–679 (2012)
8. Nee, A.Y., Ong, S.K.: Virtual and augmented reality applications in manufacturing. IFAC Proc. Vol. **46**(9), 15–26 (2013)
9. Nincarean, D., Alia, M.B., Halim, N.D.A., Rahman, M.H.A.: Mobile augmented reality: the potential for education. Procedia Soc. Behav. Sci. **103**, 657–664 (2013)
10. Olabe, M., Basogain, X., Espinosa, K., Rouèche, C., Olabe, J.: Engineering multimedia contents with authoring tools of augmented reality. In: International Technology, Education and Development Conference (INTED 2007), p. 5. Citeseer (2007)
11. Yovcheva, Z., Buhalis, D., Gatzidis, C.: Smartphone augmented reality applications for tourism. E-Rev. Tour. Res. (ERTR) **10**(2), 63–66 (2012)

Model to Evaluate Users' Engagement to Augmented Reality Using xAPI

Maria Paula Corrêa Angeloni[1(✉)], Hamadou Saliah-Hassane[2], Juarez Bento da Silva[1], and João Bosco da Mota Alves[1]

[1] Federal University of Santa Catarina, Araranguá, Brazil
maria.paula@posgrad.ufsc.br, juarez.b.silva@ieee.org, joao.bosco.mota.alves@ufsc.br
[2] TELUQ University, Montreal, Canada
hamadou.saliah-hassane@teluq.ca

Abstract. This paper proposes a way of using Augmented Reality to help students accomplish remote laboratory experiments and check their engagement while their activities are carried on. For this end, we used Experience API (xAPI) to track student's performances. First, students should access a remote laboratory. In this case, the Remote Microscope Experiment was accessed through a computer on RExLab's website and a smartphone was used to interact with augmented elements inserted in a specific context. The users' interactions with the elements inserted through the Augmania™ platform are withheld through xAPI and shown in the Learning Record Store of this same application. This made possible to acquire and analyze the students' performance and engagement through data visualization dashboards and charts. RExLab (Remote Experimentation Laboratory) is physically located at Araranguá, Brasil.

Keywords: Augmented · Reality · Online laboratory · Remote laboratory · xAPI

1 Introduction

An educational laboratory is a set of hardware or software (or both) components that are integrated to perform scientific experiments in an educational context [1]. An online laboratory is a laboratory that is accessible via computer networks, such as the Internet [1]. It can either be virtual, remote, or hybrid of the two. Remote labs are virtual laboratories with hardware in the networked loop. They are embedded systems. Online laboratories, and specifically remote labs present advantages that the physical ones do not: the students can access and perform laboratory works wherever they are – as long as they have Internet connection – and whenever they want.

Since the students can access an online laboratory from any location or at any time, they might not have an available tutor to answer their questions or guide them. Augmented reality (AR) technology can be used as a helping tool to augment in context the information related to their lab tasks to carry on. This will facilitate the users' comprehension and enhance their performance. Augmented Reality consists in adding virtual elements in a real scene [4]. Nowadays, it is possible to do that through a

M. E. Auer and D. May (Eds.): REV 2020, AISC 1231, pp. 320–329, 2021.
https://doi.org/10.1007/978-3-030-52575-0_26

smartphone camera: whereas a regular camera would mirror the real images of the world, Augmented Reality adds new items to it. Different elements can be augmented through AR to help the users' understanding of chores: texts, additional images, videos and audios.

The final result is an online laboratory shows the final results after an activity is completed, but it usually does not display the users' efforts for it to be done. The same goes to AR: it usually exhibits augmented items inserted in the real world through gadgets and tends to not record how the users interact with these items. The xAPI can check the interactions the students have engaged with during an online activity, so in this paper AR elements were inserted in a remote RExLab experiment and the interaction the users had with the augmented information were caught through xAPI. Then, the retained information was stored and displayed on a Learning Record Store, where the responsible tutor was able to analyze the students' engagement during the accomplishment of the experiment.

2 Augmented Reality

Augmented Reality has customized people's experience in public spaces [5]. Even though many areas can make good use of AR, education will be the most impacted area because the coexistence of real and virtual environments allow the students to visualize abstract concepts and from there develop important skills that could not be worked on in other technological learning environments [6].

Augmented Reality can be considered a technique that improves the users' sensorial perception regarding the real world by adding virtual elements dynamically on physical elements [7]. Unlike Virtual Reality, Augmented Reality allows the users to interact with the real world, and not with a completely digital world [8]. AR alters the perception of the real world, meanwhile VR replaces the real environment completely for a made-up one [9].

3 Online Laboratories

Online laboratories can be virtual, remote or hybrid – the last one is a combination of the first two. Virtual laboratories provide access to simulations, as it presents only simulated components. For that reason, it usually shows less realism but at the same time, it requires no physical space, is less expensive to maintain and highly scalable. Besides, it can "adapt reality to fit the teaching needs, such as by simplifying it or by displaying unobservable phenomena" [10].

Remote laboratories provide real components for the students to practice what they have learned in the classroom [10]. A remote laboratory is a system promoted on a hardware and software structure able to collect data through the Internet [11]. Hence, a remote laboratory makes it possible to perform real experiments that can be observed in real time through the computer, which exhibits images that are captured by cameras that are present in the venue that the experiment is physically located at.

Remote laboratories are able to provide real information in near real time, allowing authentic delays and other unexpected events, like measurement inaccuracies, that way contributing for the students to learn in practice about the complexities of science [10].

Remote laboratories are online systems that allow users to control devices through the Internet [12] and are very used in places with no conditions of presenting a physical one for their students to perform determined activities inside the academic space. Another advantage regarding the use of remote laboratories, besides the variety of activities, is the fact that the students can perform their tasks in and outside of the classrooms [2]. Such systems provide many benefits by overcoming time and space issues, allowing more students to access the experiments, increasing then the utilization of the equipment [12].

Hybrids laboratories are a combination of virtual and remote laboratories – laboratories that mix virtual and real components [10].

4 Experience API

The development of Experience API (also known as xAPI) was created by ADL (Advanced Distributed Learning Initiative) with the intention to design a standard to "ensure data interoperability between learning management systems (LMSs) and learning objects" [13]. xAPI serves to track learning activities, and not the learning itself, as the latter occurs in one's brain and nervous system. Even though the intention of participating in an activity is to learn, it does not happen simply by that action [14].

xAPI stood out amongst other proposals for being a pattern with a simple data model and also because of the number of providers that adopted it, such as Moodle, Blackboard, Sakai, and others [15].

A generic architecture for xAPI can be defined in three components:

1. Learning Record Provider (LRP) [16]: also known as Activity Provider (AP), the LRP is a system where learning experiences happen, such as learning portals. It is responsible for creating data in xAPI format and sending it to the LRS. Multiple LRPs can send information to a LRS at the same time [13].
2. Learning Record Store (LRS) [13, 16]: the LRS consists in a database that checks if the entry corresponds to the xAPI specification and stores all valid information for the LRC to analyze or for administrators that want to verify that data [13].
3. Learning Record Consumer (LRC) [16]: also known as Activity Consumer (AC), it is a system similar to the LRP and it can be considered a Learning Management System (LMS) that monitors a complete learning experience, as these exercises appear in the LRS [13].

xAPI is meant to track learning experiences, and those might happen in any interaction that occurs in a learning environment, so to be possible to store and analyze learning experiences, a data model was specified to call statements in the following format: <actor> + <verb> + <object> [16].

The actors, verbs and objects must be defined in the LRS as xAPI creates a pattern in learning reports by using sentences that follow the same structure. The resulting statement in an activity is stored in the LRS, and to avoid information being modified after it has been sent to the LRS, the Statement API becomes immutable, therefore "it is not possible to edit or delete a statement after it has been sent, only be invalidated by means of a statement with the verb 'voided'" [13] (Fig. 1).

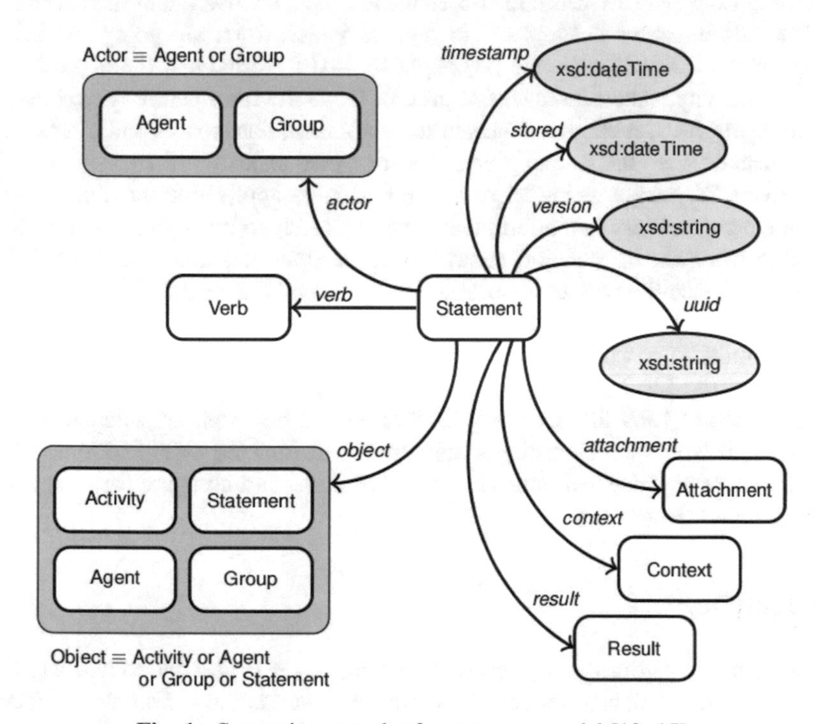

Fig. 1. Semantic network of a statement model [13, 15]

5 Learning Record Store

The LRSs are servers that store and enables access to the records generated during the performances in the learning experiences. These activities are tracked and sent to the LRS by an LRP. Then, the LRS can give access to the saved records to an LRC [16]. Figure 2 shows this architecture, showing that an LRS can give access to different LRCs.

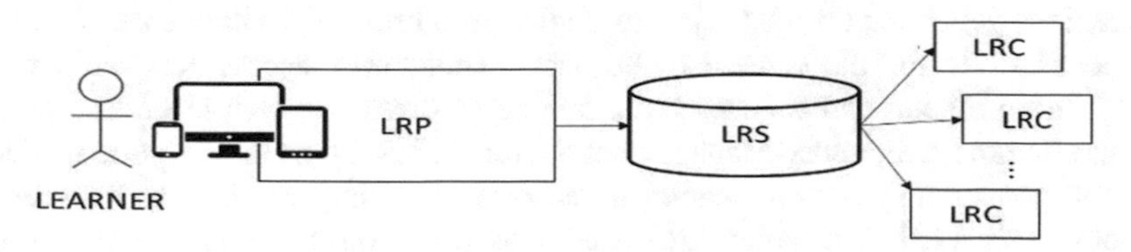

Fig. 2. xAPI architecture overview [16]

Learning Locker, developed by H2T Labs™, is a free and open source learning record store that was installed on the RExLab servers in a related work as stated in Ref. [13] and it served to store the performed interactions on the activities. The architecture of Learning Locker is divided into two elements: the LRS itself, and the xAPI service [13]. The administrator is supposed to register which users are going to play roles (actors), which are actions should be caught by xAPI (verbs) and the objects of each sentences (activity, statement, agent or group). Once the information is exported from the LRP to the LRS, it can be displayed in customized dashboards and charts, so that the LRC can analyze the learner's engagement to the tasks.

Augmania™, besides being an Augmented Reality application that allows users to augment elements, is also a platform that supports xAPI, so it has its own LRS where it is possible to check information about the interactions the users had with the augmented items once the project is uploaded and shared with others.

In this paper, Augmania™ functionalities were applied to RExLab remote activities so that it would be possible to check the interaction the users had with the augmented elements. For the LRC to verify the actions the user had towards the remote activity itself, an external LRS like Learning Locker should be used, as stated in [13]. For future work, it would be interesting to analyze the actions the users had to accomplish both the remote exercise and interact augmented items and compare their engagement regarding each feature.

6 Augmania™

The Augmania™ platform is a WebAR tool that makes it possible to add augmented elements – such as 3D objects, images, video, texts and audio – from the moment the user reaches a certain location, captures a determined image or scans a QR code. When someone uploads an experience created on Augmania™, an URL is generated and it can be accessed through any smartphone browser, therefore there is no need of downloading a specific app to augment elements, like most AR platforms.

The Augmania™ experiences support xAPI and work as their own Learning Record Store, so the application presents real-time analytics access, where the project creator can see what others have done to the augmented items they inserted in the shared content, how many times it has been accessed, their location, which devices and browsers they used, etc., while they were performing the activity.

Figure 3 shows the proposed architecture implemented in this paper to use Augmented Reality for supporting online laboratory activities: the students access the URL Augmania™ generated for the project through a web browser on their smartphones and give permissions for the camera to be used. Then, they access RexLab's website through the computer and connect to the remote laboratory, which receives a response from the Remote Laboratory Management System (RLMS), as this last one works as a mediator, considering communication usually does not happen directly between user and laboratory [13]. The students are able to carry on the remote activity while engaging with the AR elements. Augmania™ then exports the xAPI statements performed to its own LRS and displays the information in charts and dashboards designed by the own application.

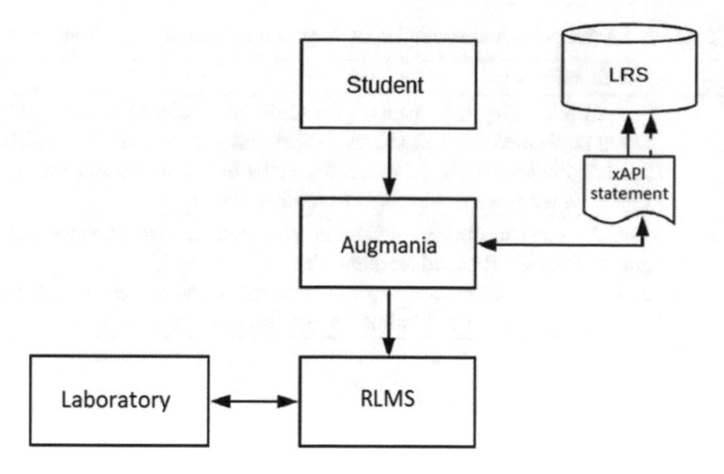

Fig. 3. Architecture overview of the AR features applied to online laboratory activities

7 Application Scenarios

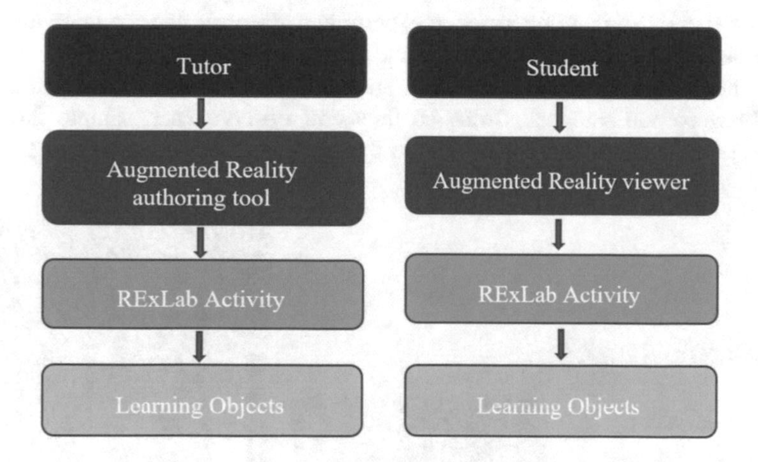

Fig. 4. Tutor and students' steps to use the proposed model

Figure 4 exhibits two different views of the proposed model: the first one shows the tutors' side, as they are responsible for inserting the Augmented Reality elements in the context, and the second one shows the students' side, where the AR application is used to view the inserted augmented items.

The proposed model can be applied to different scenarios: Table 1 presents some examples in which the Augmented Reality can help its users suggesting new or additional information.

Table 1. Actor roles and examples of Augmented Reality functionatilies

Role	AR functionality
Student	1. Add interesting information (to online laboratories) about the activity being performed through videos, additional images and 3D models 2. Help the students get through the tasks they have to accomplish through tips or video lessons, circuits schematics, etc.
Tutor/teacher	1. Help tutors to observe synchronously student performances, and to edit and post new AR based learning objects 2. Enable tutor to request captured images of the student smartphone screen sent by email or shared online during the lab work

7.1 Remote Microscope Laboratory

The proposed model was applied to Remote Microscope, a remote activity developed by RExLab, from Federal University of Santa Catarina/Brazil. While users were performing tasks, they access an URL generated through Augmania™ on their smartphones and used their cameras to augment elements and interact with them. The remote laboratory presents various experiments in the area of Physics, Biology and Robotics.

In this specific remote microscope experiment, the user can navigate through six different samples: three of them are manioc samples – two leaves and one root – and three of them are tomato samples – leaf, stalk and root. The users use the arrows to switch between the samples. Doing so they can observe each sample through the microscope lenses and read about them in the presented text Figs. 5a and 5b.

Fig. 5 (a) Not augmented screenshot of tomato root as seen on RExLab's website. (b) Augmented screenshot showing 3D root when the scene is captured by the smartphone camera

Through Augmania™ platform, each one of the microscope samples is targeted as a trigger, then the project was uploaded to Augmania™ and an URL was generated. That way, when the user accessed that URL through any web browser on a smartphone, different elements would be augmented once they pointed their cameras to each one of the triggers.

One of the images on the activity shows all the samples from above. When the users captured that image, a sentence appeared on their smartphones screens and read "select which sample you would like to observe through the arrows". Also, each one of the samples displayed some kind of augmented information: three of them show YouTube videos about the benefits of the sample in question, two of them show 3D models of the sample and one of them showed an image of the sample in its popular form – as bought in the market – and a text with the different names this sample is known for.

Although it is not possible to measure someone's learning this way, the approach enables to notice the effort a student makes to comprehend a particular situation: just participating in an activity will not make you someone understand something, he/she would have to engage to the proposed tasks [14]. The xAPI is able to catch and store meaningful kind of learning activity data.

Once the users accessed the URL and used Augmented Reality throughout their performance in the task, Augmania™ recorded their engagement and stored it in its LRS, so it was displayed for the project creator to analyze their interactions through dashboards and charts. The data visualization can be filtered by date, location, devices, browsers, etc. Figure 6 shows the first data dashboard exhibited.

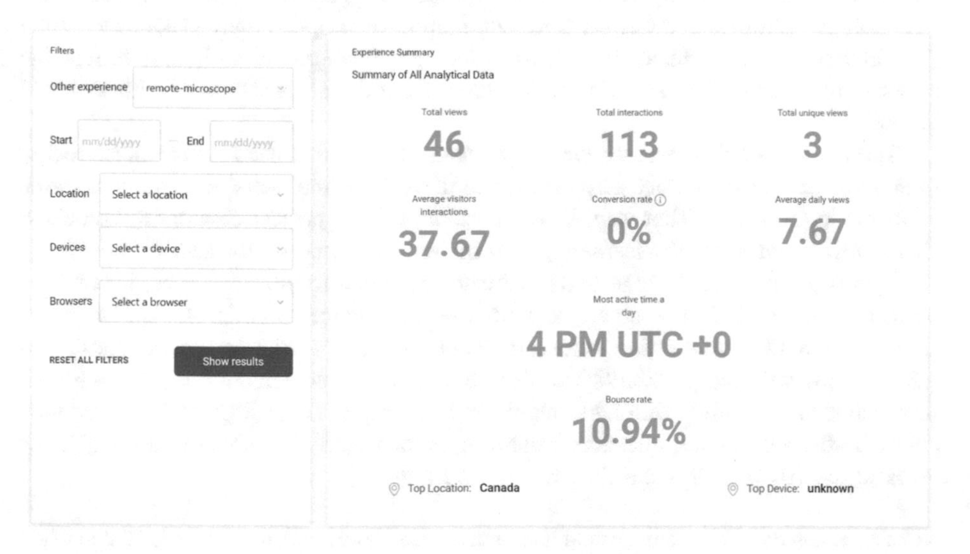

Fig. 6. xAPI results exhibited in Augmania

8 Future Work

For future studies, it would be interesting to add to the application a feature that would allow the user to share his cellphone screen live, so the tutor would be able to see the students' performance on real-time and help them accomplish tasks by giving advices or clarifying any questions.

Another idea taken from the users' interaction during the remote activity would be for him/her to take a screenshot during the exercise and export that image to a storage where the metadata would be extracted and then the information would be transformed and sent to learning analytics ready tools.

Also, the authors would like to create yet another way to try and measure the students' learning: ask questions after the performance of the remote laboratory work and depending if the students get their answer right or wrong, adapt and augment different elements for each one of them.

One more approach the authors would like to study in future work is regarding the user data privacy: it would be interesting to make some research work about the safety risks and security issues that might occur through Augmented Reality applications, as these are features that ask permissions to have access to people's camera phones.

9 Conclusion

As the Information and Communication Technologies get more present each and every day of our lives, it is important to research the best possible ways to utilize them. Technology promotes many advantages and help to accomplish daily tasks faster. As it gets inserted in the education context, education changes as well and brings the opportunity to transform knowledge measures from data that would not be withheld before.

The purpose of this research was to propose a model that used Augmented Reality to support remote laboratory activities while it recorded the users' engagement to the elements in the scene. That way, it was possible for the project creators to somehow check how much effort the learners put in the exercise they performed.

This paper presented suggestions that can be applied to different scenarios and the performed research is not complete with the termination of this work. For future studies, it would be interesting to research about the safety risks and security issues in AR, working with the possibilities to share the cellphone screens of the users – live or screenshotted, showing different augmented elements according to the students' answers after responding questions regarding certain activities, creating a protype for users to use AR to help them fix engines, and more.

Acknowledgment. The authors would like to thank the Global Affairs Canada for the mobility fellowship granted by the Emerging Leaders in the Americas Program (ELAP), which made possible for this project to be worked on in collaboration with RExLab from Federal University of Santa Catarina/Brazil and L@d at TELUQ University/Canada.

References

1. IEEE Std 1876™-2019: IEEE Standard for Networked Smart Learning Objects for Online Laboratories
2. Barbosa, T.L.M., Coan, A.L.: Laboratórios remotos e as suas contribuições para os estudos na área de programação. In: LACCEI International Multi-conference for Engineering, Education, and Technology: "Industry, Innovation, and Infrastructure for Sustainable Cities and Communities", Campinas, vol. 17 (2019)
3. Rivera, L.F.Z., Larrondo-Petrie, M.M.: Models of remote laboratories and collaborative roles for learning environments. In: 13th International Conference on Remote Engineering and Virtual Instrumentation (REV). IEEE (2016)
4. Azuma, R.T.: A Survey of Augmented Reality. Presence Teleoperators Virtual Environ. **6**(4), 355–385 (1997). MIT Press – Journals
5. Potts, R., Yee, L.: Pokémon Go-ing or staying: exploring the effect of age and gender on augmented reality game player experiences in public spaces. J. Urban Des. **24**(6), 878–895 (2019)
6. Silva, M.M.O., Teixeira, J.M.X.N., Cavalcante, P.S., Teichrieb, V.: Perspectives on how to evaluate augmented reality technology tools for education: a systematic review. J. Braz. Comput. Soc. **25**(1), 2019 (2019). Springer Nature
7. Fidan, M., Tuncel, M.: Integrating augmented reality into problem based learning: the effects on learning achievement and attitude in physics education. Comput. Educ. **142**, 2019 (2019)
8. Cai, S., Wang, X., Chiang, F.: A case study of Augmented Reality simulation system application in a chemistry course. Comput. Hum. Behav. **37**, 31–40 (2014)
9. Yun, M., Qimeng, N., Fang, W., Ying, L., Haiyang, J.: Application of augmented reality technology in industrial design. IOP Conf. Ser. Mater. Sci. Eng. **573**, 2019 (2019)
10. Rodriguez-Gil, L., García-Zubia, J., Orduña, P., López-de-Ipiña, D.: Towards new multiplatform hybrid online laboratory models. IEEE Trans. Learn. Technol. **10**(3), 318–330 (2017)
11. Vilela, D.C., Germano, J.S.E., Monteiro, M.A.A., Carvalho, S.J.: Estudo comparativo de um experimento de eletrodinâmica: Laboratório Tradicional x Laboratório Remoto. Revista Brasileira de Ensino de Física **41**(4), 1–8 (2019)
12. Maiti, A., Zutin, D.G., Wuttke, H., Henke, K., Maxwell, A.D., Kist, A.A.: A framework for analyzing and evaluating architectures and control strategies in distributed remote laboratories. IEEE Trans. Learn. Technol. **11**(4), 441–455 (2018)
13. Simão, J.P.S., Carlos, L.M., Saliah-Hassane, H., Silva, J.B., Alves, J.B.M.: Model for recording learning experience data from remote laboratories using xAPI. In: XIII Latin American Conference on Learning Technologies (LACLO). IEEE (2018)
14. Woodwill, G.: Understanding the experience API. In: Mastering Mobile Learning: Tips and Techniques for Success, vol. 43, no. 1, pp. 277–280 (2015)
15. Vidal, J.C., Rabelo, T., Lama, M.: Semantic description of the experience API specification. In: 2015 IEEE 15th International Conference on Advanced Learning Technologies (2015)
16. Barone Rodrigues, A., Dias, D.R.C., Martins, V.F., Bressan, P.A., de Paiva Guimarães, M.: WebAR: a web-augmented reality-based authoring tool with experience API support for educational applications. In: Antona, M., Stephanidis, C. (eds.) UAHCI 2017. LNCS, vol. 10278, pp. 118–128. Springer, Cham (2017). https://doi.org/10.1007/978-3-319-58703-5_9

A Case Study of AR Technology and Engineering Students: Is There a Gender Gap?

Diana Urbano[1(✉)], Paulo Menezes[2], Maria de Fátima Chouzal[1], and Maria Teresa Restivo[1]

[1] LAETA-INEGI, University of Porto, Porto, Portugal
{urbano, fchouzal, trestivo}@fe.up.pt
[2] ISR, University of Coimbra, Coimbra, Portugal
pm@deec.uc.pt

Abstract. Determining the factors influencing students' intention to use Augmented Reality (AR) allows a deeper understanding on how students react to the use of such technologies in their training as engineers. This study aims to identify the emotional and cognitive factors that influence the students' intention of using AR in their future professional life and to access possible gender differences. A group of about 150 undergraduate students from an Engineering and Industrial Management program had the opportunity to explore AR applications related to contents addressed in Sensors and Actuators course. A survey was designed and used with those students. Principal component analysis resulted in three components named interest, ease of use and attitude. Logistic regression analysis was conducted with these three components together with gender, as predictors of intention of using AR in later professional life. Attitude turned out to be the strongest predictor. This analysis has also shown that gender has no significant effect.

Keywords: Augmented reality · Motivation · Intention to use · Gender gap

1 Introduction

Educators always face the complicated question of which is the best way to motivate and communicate knowledge to students. Although as part of the learning process the students should develop the ability to understand abstract definitions and relationships, it is always necessary at some given point to help them establish connection with real world scenarios, situations, devices or objects.

Augmented reality (AR) offers a clear opportunity to bring students to "try" the system, to "run" an experiment, to run experiment enriched by additional details and thus validate some learnt principles [1, 2]. AR may be used in a safer way as real world obstacles will still be seen by users, and virtual devices, system, or experiment elements may be ecologically integrated in lab, home, or factory environments where the training will take place [3–6].

M. E. Auer and D. May (Eds.): REV 2020, AISC 1231, pp. 330–337, 2021.
https://doi.org/10.1007/978-3-030-52575-0_27

Most educators would agree that students should have the opportunity to engage in activities AR or virtual reality (VR) since they would mostly likely impact on motivation, thus facilitating the learning process. To establish to which extent using such technologies are in fact pedagogically relevant, it is desirable to get the student's opinion whenever they have the opportunity to use them. Some studies support the idea that using AR increases motivation and enhances academic performance [7–12]. As far as technology usage, AR acceptance and AR contribution to improve certain skills is concerned, results for gender gaps are still contradictory [13–15].

Carrying on earlier studies, this work is currently trying to answer to the main goals of the present study:

- To investigate the emotional and cognitive factors that influence the students' intention of using AR in their future professional life;
- To access possible gender differences.

The answers to a survey conducted in a second year course of an Industrial and Management Engineering degree were analyzed using descriptive statistics; principal component analysis and logistic regression [16]. Principal component analysis is used to reduce the data and logistic regression analysis (LRA) to determine which of the components predict intention to use AR in the future. The analysis was conducted using SPSS software package [17].

2 Methodology

A group of 150 students from Engineering and Industrial Management degree participated in an activity where they explored different AR applications related to contents addressed in their undergraduate courses related with sensors and actuators.

These students were briefly introduced to the fundaments of Marker-Based AR technologies and then they were asked to explore two distinct AR apps [18–20]. Finally, with informed consent, they answered a questionnaire based on Likert scales from 1 (strongly disagree) to 5 (strongly agree), about their opinion on the performed activity and their perception of the interest of using augmented reality technology.

3 Results and Discussion

A total of 150 students participated in the activity. However, seven of the cases (4,6%) were not considered due missing results. Table 1 shows the characterization of the N = 143 sample with average ages values for female and male and respective familiarity with AR.

Table 1. Sample description

Sex	N° of students	Age (Mean/SD)	Familiarity (%)
Female	62	19.5/1.26	18.0
Male	81	19.6/1.52	43.2

The reported familiarity with the AR was 37.1%, being male students more familiar than female ones.

Table 2. Survey questions

Question type	Question content
INT-1	I liked performing activities with AR
INT-2	I would describe the activities with the AR as very interesting
INT-3	The use of AR app attracted my attention
EU-1	For me these AR apps are easy to use
EU-2	Learning how to use these AR tools was easy
EU-3	I quickly familiarized with using these AR tools
ATT-1	As engineering student I think this lab session using AR examples was important
ATT-2	I consider the use of AR resources to be important in the framework of our course curriculum
ATT-3	It should be a good idea using AR to learn other course contents
ATT-4	It should be a good idea using AR resources in other courses
IU	I intend to use AR technologies in my professional future

The survey used in this study is shown in Table 2. The items selection was based on many previous studies that investigate technology acceptance in education [21, 22].

Principal component analysis (PCA) was applied with the first 10 items of the table and three components were extracted. They were named Interest (items INT-1, INT-2 and INT-3), [23]; Ease of Use (items EU-1, EU-2 and EU-3), [24, 25]; and Attitude (ATT-1, ATT-2, ATT-3 and ATT-4), [26–28], described as:

- Interest (INT) – interest created by experiencing the AR apps.
- Ease of use (EU) – students' opinion on how easily they have learned to use the application and how to handle it;
- Attitude (ATT) – cognitive and affective response of the students to the use of the AR tools. The cognitive part is included in items ATT1 and ATT2 that measure how valuable students consider to be experiencing "experiments" with the AR. Items ATT3 and ATT4 are related to the affective component of attitude.

IU, the intention to use AR in the professional future, is measured by one single item, and it will be the dependent variable in the regression analysis.

According to the Kaiser-Olin Mayer Test (0.793), and the Bartlett's test of specificity ($p < 0.001$), the sample is adequate to perform PCA. All the factor loadings have values above 0.5 and the total variance extracted exceeds 60%, justifying the grouping of the items in three components [16]. The correlations between the components are below 0.2, indicating distinctive measures.

Table 3 depicts the results obtained for the averages, standard deviations, medians and modes of all the items of the survey, for both female (N1 = 61) and male (N2 = 81) students, as well as the Cronbach α value, this being used to assess internal consistency or reliability of the instrument. If the items that group in one component are reasonably correlated, the value of α will be closer to one, but $\alpha \geq 0.7$ is considered to indicate good internal consistency [29].

Table 3. Summary of descriptive statistics

Item label	Mean (SD)		Median/Mode		Cronbach α
	Female	Male	Female	Male	
INT-1	4.48 (0.57)	4.54 (0.57)	5/5	5/5	
INT-2	4.34 (0.65)	4.12 (0.71)	4/4	4/4	0.71
INT-3	4.48 (0.57)	4.28 (0.66)	5/5	4/4	
EU-1	4.23 (0.66)	4.15 (0.67)	4/4	4/4	
EU-2	4.03 (0.70)	4.17 (0.57)	4/4	4/4	0.70
EU-3	4.02 (0.64)	4.02 (0.47)	4/4	4/4	
ATT-1	4.40 (0.66)	4.41 (0.72)	4/5	5/5	
ATT-2	4.15 (0.74)	4.14 (0.80)	4/4	4/4	0.79
ATT-3	4.52 (0.62)	4.46 (0.67)	5/5	5/5	
ATT-4	4.48 (0.59)	4.51 (0.51)	5/5	5/5	
IU-1	3.76 (0.80)	3.74 (0.70)	4/3	4/4	–

A non-parametric Mann-Whitney test was conducted to compare female and male values, in every individual item result as well as in the three components' scores. No statistically significant results were found, indicating similar observation in both groups of students.

Since intention to use AR technology in the future is only one item in the survey, it was transformed in a binary variable so that it can be considered the dependent variable in the logistic regression approach [16]. Results of 4 and 5, above the middle point in the Likert scale, 3, are replaced by 1, indicating intention to use AR. Results equal to 3 are replaced by 0 and indicate indifference to using AR in the future. Results of 2 and 1 would indicate no intention. In the present case none of these last cases appeared.

In order to determine which of the independent variable has effect on the dependent variable IU and to provide a classification table to assign group association, the scores of three components obtained from PCA, together with gender, were used as predictors of the dependent variable IU in LRA. In other words, the LRA allows identifying the likelihood of a student in showing intention to use AR in the future and which of the variables interest (INT), ease of use (EU), attitude (ATT) and gender have a significant effect on that intention.

Table 4 depicts the number of female and male students distributed in the two categories of IU, as well as the means and standard deviations of the variables EU, ATT and INT.

Table 4. Means of the predictors for female and male students for both IU categories

	Total		INT		EU		ATT	
			Mean (SD)		Mean (SD)		Mean (SD)	
	F	M	F	M	F	M	F	M
Positive (1)	33	48	4.59 (0.43)	4.42 (0.50)	4.12 (0.53)	4.15 (0.46)	4.51 (0.48)	4.58 (0.48)
Indifferent (0)	29	33	4.26 (0.50)	4.16 (0.47)	4.06 (0.56)	4.07 (0.42)	4.24 (0.49)	4.09 (0.55)

Using the total sample, a t-test was conducted to evaluate differences of INT, EU and ATT mean values, between categories "1" and "0". Statistical significance were found for INT and ATT. These results are still valid for both female and male samples.

The next step will be analyzing the effect of the INT, EU, ATT and gender variables on the dependent variable IU, using LRA. The output of applying LRA is a logit function:

$$ln\left(\frac{p}{1-p}\right) = B_0 + \sum_{n=1}^{4}(B_i \mathrm{Y_i}) \tag{1}$$

where p is the probability of an event outcome ($0 \leq p \leq 1$), B_0 is a constant, Y_i the corresponding independent variables (INT, EU, ATT, Gender) and B_i the corresponding unstandardized regression weights. Table 5 depicts these model estimates as well as the values of the Wald χ^2 test and the corresponding p indicating that INT and ATT are the only significant predictors of IU. Additionally, ATT has the strongest effect on IU, followed by INT, as suggested by the corresponding B values. The model fits the data well according to the Hosmer–Lemeshow test ($\chi2 = 10.347$ (8) and $p = 0.241$) [30].

Table 5. Summary of results of the LRA analysis

	B*i*	Wald χ^2 test (df)	*p*
B_0	0.455	3.389 (1)	0.066
INT	0.452	4.750 (1)	0.029
EU	−0.076	0.151 (1)	0.697
ATT	0.644	9.546 (1)	0.002
Gender	0.406	1.171 (1)	0.279

Table 6. Observed and predicted frequencies for Intention to Use with the cutoff of 0.50

		Predicted IU		Correct percentage (%)
		0 (F/M)	1 (F/M)	(F/M)
Observed IU	0 (F/M)	32 (16/16)	30 (13/17)	51.6 (46.8/40.7)
	1 (F/M)	16 (11/5)	65 (22/43)	80.2 (66.7/89.6)
Global percentage				67.8 (61.2/72.8)

To build the classification Table 6 for dependent variable IU, probabilities lower or equal to 0.5 are associated with IU = 0 and bigger than 0.5 with IU = 1.

The rate of true positives (sensitivity) is 65/(65 + 16) = 80.2% and the rate of true negatives (specificity) is 32/(30 + 32) = 51.6%, with a global percentage of correct prediction (accuracy) of 67.8%. Table 6 also shows those percentages for female and male students.

4 Conclusions

This work aimed at analyzing engineering students' reaction when using AR tools to explore some practical aspects related with measurement. No gender differences were found in the means of any of the items of a survey answered by a student sample of 143.

Applying PCA to the survey items three components were extracted: interest (INT), ease of use (EU) and attitude (ATT). Again, no statistical significance differences were found in variables INT, EU and ATT between female and male students.

IU was transformed in a binary variable, and a t-test was used to compare variables mean values. Only statistical significance in the mean values of INT and ATT were found, between the groups with IU = 1 and IU = 0.

The variables interest, ease of use and attitude, together with gender were used as predictors of intention to use in a logistic regression analysis. This analysis allowed identifying the likelihood of a student in showing intention to use AR in the future and which of the variables have a significant effect on that intention. The result identified interest and attitude as significant predictors.

The model classifies correctly 67.8% of intention of use AR, with good sensitivity (80.2%) and a specificity of 51.6%.

One major conclusion of this study is that students react very positively when given the opportunity to using AR examples and that their attitude and interest affects the intention to use AR in their future professions.

The only difference in gender was found in the familiarity with AR technology with male students reporting being more familiar with the technology than female students. It is interesting that although males have shown to be more familiar with this technology, it has been concluded in this analytical study that both genders seem to have a similar reaction to the use of the technology. These results encourage the authors to continuing using and testing the effects of AR technology in engineering learning activities.

References

1. Fuhrt, B.: Handbook of Augmented Reality. Springer, New York (2011). https://doi.org/10.1007/978-1-4614-0064-6
2. Lee, K.: Augmented reality in education and training. Techtrends Tech Trends **56**(2), 13–21 (2012). https://doi.org/10.1007/s11528-012-0559-3
3. Rodrigues, J., Andrade, T., Abreu, P., Restivo, M.T.: Adding augmented reality to laboratory experimentation. In: 2017 4th Experiment@International Conference (exp.at 2017), Faro, 2017, pp. 135–136 (2017)
4. Maiti, A., Smith, M., Maxwell, A.D., Kist, A.A.: Augmented reality and natural user interface applications for remote laboratories. In: Cyber-Physical Laboratories in Engineering and Science Education, pp. 79–109. Springer International Publishing, Cham (2018)
5. Dengzhe, M., Gausemeier, J., Fan, X., Grafe, M.: Virtual reality & augmented reality in industry. In: The 2nd Sino-German Workshop. Shanghai and Springer, Berlin (2011)
6. De Pace, F., Manuri, F., Sanna, A.: Augmented reality in industry 4.0. Am. J. Comput. Sci. Inform. Technol. **6**(1), 17 (2018). https://doi.org/10.21767/2349-3917.100017
7. Wang, Y., Anne, A., Ropp, T.: Applying the technology acceptance model to understand aviation students' perceptions toward augmented reality maintenance training instruction. Int. J. Aviat. Aeronaut. Aerospace **3**(4) (2016). https://doi.org/10.15394/ijaaa.2016.1144
8. Wild, F., Klemke, R., Lefrere, P., Fominykh, M., Kuula, T.: Technology acceptance of augmented reality and wearable technologies. J. Univ. Comput. Sci. **24**(2), 192–219 (2018)
9. Almenara, J.C., Fernandez-Batanero, J.M., Barroso-Osuna, J.: Adoption of augmented reality technology by university students. Heliyon **5**, e01597 (2019)
10. Martin-Gutiérrez, J., Fernandéz, M.D.M.: Applying augmented reality in engineering education to improve academic performance & student motivation. Int. J. Eng. Educ. **30**, 625–635 (2014)
11. Yip, J., Wong, S.-H., Yick, K.-L., Chan, K., Wong, K.-H.: Improving quality of teaching and learning in classes by using augmented reality video. Comput. Educ. **128**, 88–121 (2019)
12. Yena, J., Tsaib, C., Wu, M.: Augmented reality in the higher education: Students' science concept learning and academic achievement in astronomy. Procedia Soc. Behav. Sci. **103**, 165–173 (2013)
13. Goswami, A., Dutta, S.: Gender differences in technology usage—a literature review. Open J. Bus. Manage. **4**, 51–59 (2016). https://doi.org/10.4236/ojbm.2016.41006
14. Dirin, A., Alamäki, A., Suomala, J.: Gender differences in perceptions of conventional video, virtual reality and augmented reality. iJIM **13**(6) (2019)
15. Hou, Lei, Wang, Xiangyu: A study on the benefits of augmented reality in retaining working memory in assembly tasks: a focus on differences in gender. Autom. Constr. **32**, 38–45 (2013)
16. Everitt, B.S., Dunn, G.: Applied Multivariate Data Analysis, 2nd edn., 354 pages. Wiley (2010). ISBN: 978-0-470-71117-0
17. IBM Corp. Released 2015. IBM SPSS Statistics for Windows, Version 23.0. IBM Corp., Armonk, NY (2015)
18. Restivo, M.T., Chouzal, M.F., Rodrigues, J., Menezes, P., Patrão, B., Lopes, J.B.: Augmented reality in electrical fundamentals. Int. J. Online Eng. (iJOE) **10**(6), 68–72 (2014). https://doi.org/10.3991/ijoe.v10i6.4030
19. Menezes, P., Chouzal, F., Urbano, D., Restivo, T.: Augmented reality in engineering. In: Auer, M., Guralnick, D., Uhomoibhi, J. (eds.) Advances in Intelligent Systems and Computing, vol. 545, Springer, Cham (2017)

20. Restivo, M.T., Rodrigues, J., Chouzal, M.F., Menezes, P., Almacinha, J.: Online systems for training the evaluation of deviations of geometrical characteristics. Int. J. Online Biomed. Eng. (iJOE) **9** (2013)
21. Park, S.Y.: An analysis of the technology acceptance model in understanding university students' behavioral intention to use e-learning. J. Educ. Technol. Soc. **12**(3), 150–162 (2009)
22. Teo, T.: Modelling technology acceptance in education: a study of pre-service teachers. Comput. Educ. **52**(2), 302–312 (2009)
23. Lee, M.K.O., Cheung, C.M.K., Chen, Z.: Acceptance of internet-based learning medium: the role of extrinsic and intrinsic motivation. Inf. Manage. **42**(8), 1095–1104 (2005). https://doi.org/10.1016/j.im.2003.10.007. ISSN: 0378-7206
24. Davis, F.D.: Perceived usefulness, perceived ease of use, and user acceptance of information technology. MIS Q. **13**, 319–339 (1989)
25. Venkatesh, V., Morris, M., Davis, G., Davis, F.: User acceptance of information technology: towards a unified view. MIS Q. **27**(3), 479–501 (2003)
26. Gawronsk, B.: Editorial: attitudes can be measured! But what is an attitude? Soc. Cogn. **25**(5), 573–581 (2007)
27. Jain, V.: 3D model of attitude. Int. J. Adv. Res. Manage. Soc. Sci. **3**(3) (2014)
28. Yang, H., Yoo, Y.: It's all about Attitude: revisiting the technology acceptance model. Decis. Supp. Syst. **38**(1), 19–31 (2004). https://doi.org/10.1016/S0167-9236(03)00062-9. ISSN: 0167-9236
29. Tavakol, M., Dennick, V.: Making sense of Cronbach's alpha. Int. J. Med. Educ. **2**, 53–55 (2011)
30. Hosmer Jr., D.W., Lemeshow, S., Sturdivant, R.X.: Applied Logistic Regression, 3rd edn. Wiley, New Jersey (2013)

Towards a Remote Warehouse Management System

Ioannis Stamelos[1], Charalampos Avratoglou[2], Panayotis Tzinis[1], George Kakarontzas[3], Alexander Chatzigeorgiou[4], Apostolos Ampatzoglou[1], Dimitris Folinas[5], Iakovos Stratigakis[1], Lampros Karavidas[1], Christina Volioti[1], Theodoros Amanatidis[4], Anastasia Deliga[1], Charalampos Dimitrakopoulos[2], and Thrasyvoulos Tsiatsos[1](✉)

[1] Aristotle University of Thessaloniki, Thessaloniki, Greece
{stamelos, apamp, iakovosds, karavidas, adeliga, tsiatsos}@csd.auth.gr, panayotis.tzinis@gmail.com, chvolioti@gmail.com

[2] Entersoft, Athens, Greece
{avr, xdi}@entersoft.gr

[3] University of Thessaly, Volos, Greece
gkakaron@teilar.gr

[4] University of Macedonia, Thessaloniki, Greece
achat@uom.gr, tamanatidis@uom.edu.gr

[5] International Hellenic University, Thessaloniki, Greece
dfolinas@gmail.com

Abstract. This paper describes the software architecture and the required hardware components of a Remote Warehouse Management System using Augmented Reality, designed in the context of the project called WMS & AR project. The system architecture will augment the capabilities of an existing WMS system [1]. The main services that the extended system will provide are (a) automatic recording of the layout of the warehouse; (b) automatic measurement of goods; (c) exception handling. Moreover, the paper discusses the architectural requirements of the system, presents the proposed software architecture and refers to the future steps.

Keywords: Warehouse Management System · System architecture · Augmented Reality

1 Introduction

A warehouse management system (WMS) is software and processes that help the organizations with the warehouse operations that should be conducted so as the warehouse to be fully functional. These operations start from the moment that materials or goods enter a warehouse until the move out. In addition to the basic core functions, other ones more extensive are integrates to WMS to check the system states in order to have data to optimize the organization's strategy.

M. E. Auer and D. May (Eds.): REV 2020, AISC 1231, pp. 338–348, 2021.
https://doi.org/10.1007/978-3-030-52575-0_28

Warehouse management systems come in a variety of types. It can more or less complex and it should be adjustable to the size of the business. Some organizations build the own WMS from scratch. However, it is mostly seen businesses to implement a WMS from an established vendor.

A WMS system typically integrates the following subsystems

- An integration with an ERP, to accept requests and external events that affect the Warehouse
- The main WMS UI for back-office users to issue and plan WH actions, monitor progress, measure performance of operations etc.
- A set of mobile (typically RF) devices, held by WH operators, to issue and monitor proper W/H job order execution
- A set of touch stations to support assembly, packing or unpacking operations
- Several automation devices such as beacons, lights etc. used to automate WH processes.

In summary, it can be said that a warehouse management system (WMS) is a complex software structure with many integrated functionalities. It usually receives order from the overlying host system and manages them with the aim to be as optimized as possible in order to store as few goods as possible, using few resources and personnel for storage and delivery to the end customer.

2 Short Presentation of WMS and AR Project

The WMS+AR system aims to exploit the benefits of AR for a Warehouse Management System. The AR subsystem is envisaged to undertake the responsibility of identifying and rendering the Warehouse and its contents and provide a means to map and validate the above against the WMS logical information.

Thus, the AR system shall link to the logical entities of the WMS, such as Warehouses, Storage locations, and packages. It shall monitor the Warehouse contents, and it shall also provide near real-time rendering based on the identified 3D Model of the Warehouse at the Storage Location level and the photo images taken during the operation.

The AR system is envisaged to be used as a straightforward and easy to use means to automate stock counting tasks in a relatively low cost compared to other methods. Even though there are several methods to accomplish this, they either involve manual counting or infrastructure which has relatively high cost of ownership & operations such as RF IDs etc.

The AR system is expected to run continuously to pinpoint exceptions or discrepancies from the logical information maintained by the WMS.

The WMS user experience is expected to be substantially enhanced by 3D rending of the Warehouse while the user is expected to have increased control due to contextual image rendering, at the Storage location level, and the ability for the user to issue stock counting actions, either ad-hoc or automatically, based on predefined tactics, such as follow-the-picker, which shall revalidate the contents of the storage locations after a change is registered in the WMS.

Additional benefits provided by the AR system will be to initially map the physical coordinates to the Warehouse Storage locations and using AR data to inform or navigate users.

3 Technical Issues of the WMS and AR System

This section refers briefly to the basic technical issues of the system. The proposed system should do at least three basic functions:

- The automatic recording of the layout of the warehouse.
- The automatic measuring of the goods that are always in the warehouse.
- Management of exceptions, where the term "exception" refers to any discrepancy between the automatic measurements and the stored measurements in the Warehouse Management System (WMS) database.

These issues are presented in the following paragraphs

3.1 Automatic Recording of the Layout of the Warehouse

The layout of the warehouse must be recorded automatically and can be done either in a warehouse that is completely empty but also in warehouses with merchandise. This implies three basic technical capabilities:

- The capability of navigating the warehouse in some way. The warehouse can be navigated: by:
  - An employee who holds a camera or mobile phone with suitable software.
  - A robot that moves through the corridors of the warehouse and carries a camera or mobile with suitable software on a telescopic arm.
  - A drone which moves to the premises of the warehouse and carries a camera or mobile phone with suitable software.
  - A camera attached to a fixed-orbit mechanism.
- The capability to identify the storage locations of the warehouse given the storage system. Possible storage systems are (a) Back-to-back (b) Stack and (c) Drive-in (see Fig. 1). In this project we will focus on warehouses with back-to-back systems. The identification can be carried out by a trained neural network depending on the type of warehouse and can be done
  - On the device from (1) with the appropriate software, or
  - In a middleware server on which Video recorded on the device (1) is transmitted.
- The capability to measure the dimensions of storage spaces (length, width, height). This feature involves cameras suitable for capturing accurately these dimensions (e.g. special telephones or cameras).

Fig. 1. Storage systems

Finally, in Entersoft's WMS system there should be a spatial representation as shown in Fig. 2:

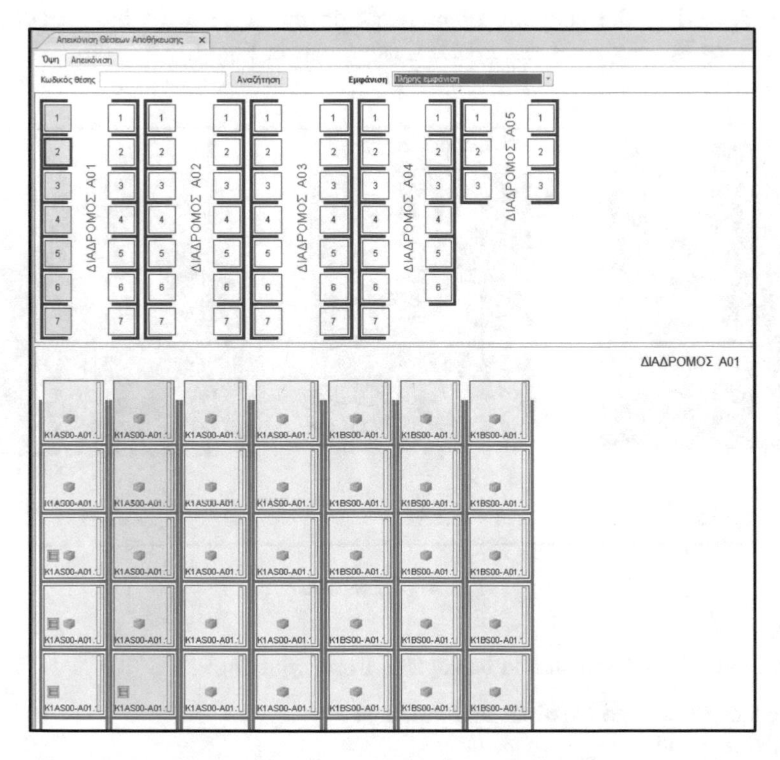

Fig. 2. Warehouse layout in Entersoft WMS

3.2 Automatic Measurement of Goods

Every kind of merchandise in the warehouse has various characteristics, which are stored in Entersoft WMS, as Code, Description and Barcode (see Fig. 3).

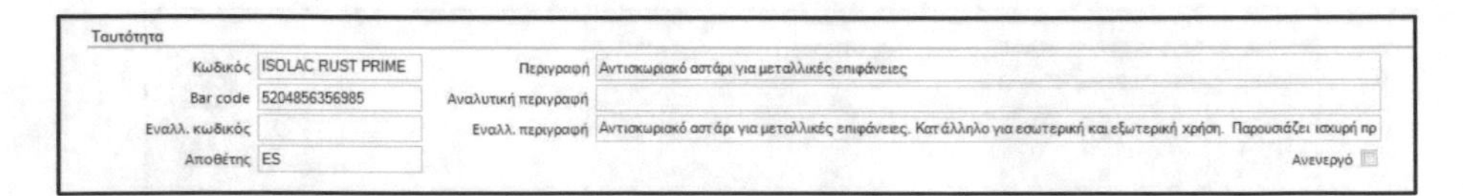

Fig. 3. Merchandise kind description in Entersoft WMS

The system should recognize the barcode of each product, which is packaged in the warehouse and should automatically count the goods. The measurement should also take into account the characteristics of the packaging:

- Measurement unit: which can be KIB (crates), KIΛ (kg) or TEM (pieces), while there may be a basic unit of measurement of the packaging (e.g., TEM) and secondary unit of measurement (for example, pieces have weight).
- Volumetric characteristics: Height, length, width (in centimeters), area and volume
- Weight: net and gross

Also related to the counting of products, is the concept of the container. The container may be sacks, barrels or boxes, as shown in Fig. 4.

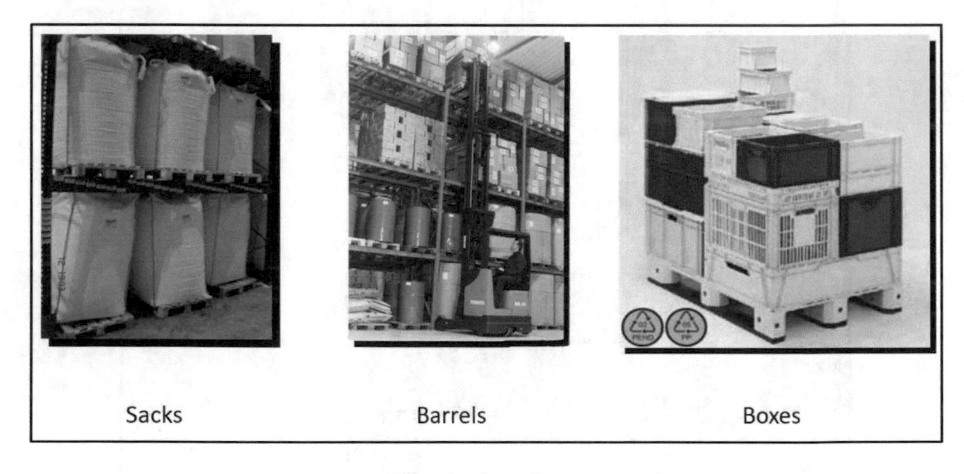

Fig. 4. Containers

Therefore, the system should be capable in recognizing:

- Barcodes of storage locations and products
- Containers and
- Packagings

Considering the above, the system should be capable to measure the goods in the warehouse.

3.3 Exception Handling

The term "exception" refers to differences between the observed state in the warehouse by the devices and the recorded state in the WMS database.

The exceptions that will be handled by the system include at least the following:

- Quantity of stock,
- Location of stock,
- Constraints of placement regarding the various stored items.
- And possibly other conditions,

The handling of the exceptions will begin when there is a discrepancy between the state of the warehouse recorded in the WMS database and the state observed from the WMS&AR system.

In the proposed architecture several software components will provide the role of "middleware" between the WMS devices and the WMS system. These components will be executed in the WMS&AR Services server in the warehouse premises They will be responsible for (a) recording the observed values and sending them back to the WMS and (b) receiving actions from the WMS (e.g. an action would be a request that an employee should move to a specific WMS location for handling a possible exception etc.)

4 Architectural Requirements

4.1 Stakeholder Concerns

Both Entersoft and the users of the system (i.e. companies that use the WMS system) require an economically efficient solution that improves the current status concerning the accuracy of the reported state of the warehouse. In addition, a solution that can be integrated easily with the current WMS system is required.

4.2 Constraints

The proposed solution should be economically not very expensive and no other special hardware or software should be required. The solution should be usable in both empty and filled warehouses (e.g. we do not assume that the warehouse is currently empty for the first automated layout recognition of the warehouse by the system). Moreover, the proposed architecture of the WMS&AR Services middleware should be interoperable with the Warehouse Management System.

4.3 Non-functional Requirements

The most important non-functional requirements are the following:

1. Energy savings and recharging of the motion device (e.g. drone). The motion device should be capable of some autonomy before recharging and it should be possible to work efficiently. To this end, our solution is the handheld device to be able to carry the same actions as the motion device. Hence, when the motion device is charging

on a docking point in the warehouse, an urgent request can be handled by the handheld device. Lastly, the motion device will carry out certain actions in periodic intervals and others non-periodic when available and told to do so by the Manager component of the WMS & AR Services middleware.

2. Speed of calculations for the volume measurements, barcode recognition and other possibly intensive algorithms. These measurements in the proposed architecture are carried out on the WMS&AR Services middleware server, which will be located in the warehouse premises in order to not be carried out by the devices. There might also be some exceptions to this, such as real – time recognition of objects and real – time object measuring.
3. Energy efficiency for the motion and handheld devices. The motion devices should be used in such manner so that energy is preserved. In the proposed architecture the majority of the intensive calculations will be carried out in the WMS&AR Services middleware server so that energy of these devices in not unnecessarily wasted.

4.4 Project Risks

The major risk of the project is that the WMS&AR Services middleware will not be able to provide accurate measurements of the warehouse for the purposes of identifying exceptions. To this end some intensive prototyping already has been done and we have sufficient evidence that the proposed architecture will be able to address the risk.

5 Proposed Software Architecture

This section describes the software architecture and the required hardware components of the WMS & AR (Warehouse Management System and Augmented Reality). It follows largely the software documentation template proposed by Gorton in [8]. In Fig. 5 the high-level overview of the architecture is depicted. The system under development (SuD) is communicating with the existing WMS.

The main concern of this document is the Augmented Reality Warehouse Management Services (AR WMS Services) components, which comprise components that are executed in the premises of the warehouse. The reason that this document will not be concerned further with the current WMS, it that the WMS software architecture will not change but only additions to the existing WMS system will be made for integrating the new functionality of the AR WMS Services components.

The main components of the AR WMS Services are the following:

1. **High Level - AR Actions API:** This component will provide an API that will allow calls from the WMS server to give instructions to the AR WMS Services subsystem. Actions can include instructions for fetching various pieces of information (e.g. volume of stock on a shelf etc.), but also sets of instructions (e.g. a set of measurements).
2. **Visualization Server:** This server will provide a visualization of the warehouse and specific shelves at a given time which will be accessed by the WMS Windows client. It provides access to the various .obj model files stored by the Visual Processor component at a specific storage location.

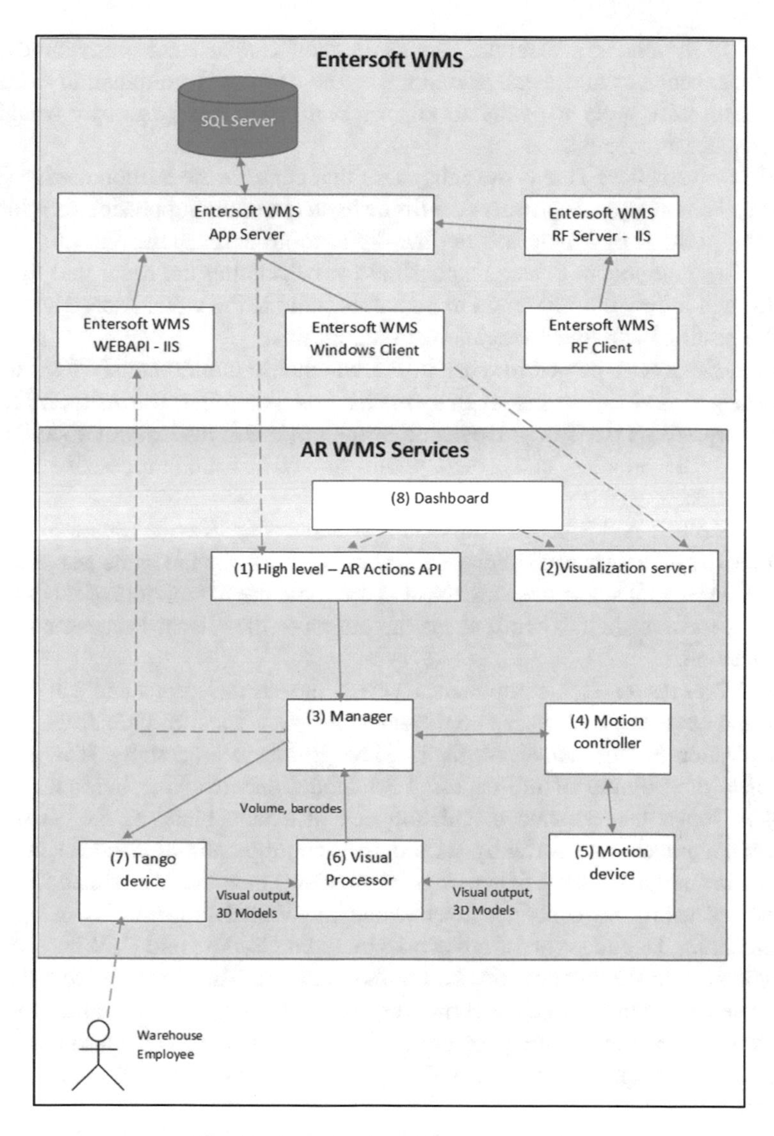

Fig. 5. High-level architecture overview

3. **Manager:** This will be the core component of the AR WMS Services system coordinating all the other components. Its purpose is to forward messages between the components. For example, it will forward incoming actions to the motion controller and also to the Tango device depending on the system configuration. It will also schedule incoming actions via a queue to keep track of the pending, currently operating, completed and failed actions. This component will be also responsible for marshalling and transferring the shelf measurements (barcodes, volumes etc.) to the WMS using the existing WMS WebAPI. Notice that this component is also responsible for making micro-corrections in the data sent back to

the WMS system. For example, the same shelf can be read many times by the warehouse devices and small variations in the computed volume can occur. This subsystem will apply a policy to correct such variations (e.g. by averaging the various measurements).

4. **Motion Controller:** This is the subsystem that controls the motion device (e.g. the Drone). For example, it instructs the drone to start a scanning session or return to its docking position or it may instruct the drone to go and capture visually a specific shelf. This component is also responsible for calculating the route that the motion device will follow in order to go to a specific point in the warehouse. The route will be the result of employed shortest-path algorithms.
5. **Motion Device:** A device moving in the warehouse autonomously (e.g. a Drone) according to the instructions given by the Motion Controller. Basically, it fulfills the same purposes as the Tango Device: it scans the warehouse and/or specific selves, generates 3D models and sends them to the Visual Processor for further computation.
6. **Tango Device:** A device that runs Google Tango (e.g. Tango Smartphone) and is operated by a Warehouse Employee. Basically, it fulfills the same purposes as the Motion Device: it scans the warehouse and returns the 3D models of the warehouse and/or a specific shelf. Then it sends the output to the Visual Processor for further computation.
7. **Visual Processor:** This component receives the visual output and the 3d models from the devices. It saves the .obj model files at a specific path from which the Visualization Server can access them. Through image processing, it is possible to calculate the volume of the captured 3d model and read the barcodes from the image. Notice that volume calculation can also take place at the Visualization Server component and currently we are assessing the optimal place for this service from a technology appropriateness point of view since the Visualization Server is developed using the Unity framework and the Visual Processor is using various programming languages and frameworks including Python and C. When completed the volume and the barcodes are sent to the Manager. The above process could also be implemented in the devices; however, this is an energy-consuming process that would overload the resources of the devices. This is the main reason why it was decided to be implemented in the WMS&AR Services middleware and not in the devices.
8. **Dashboard**: The Dashboard is going to communicate with (1) the High-Level AR Actions API (ordering of actions, listing of actions, cancellation of actions, listing of WMS-AR exceptions) and (2) the Visualization Server, via a REST API over HTTP(S).

6 Conclusions and Next Steps

In conclusion we are confident that the proposed architecture is solid enough and ready to be a reality. Thinking towards this direction our next steps are to find hardware and software that fit our needs and acquire the most appropriate one. A state of the art analysis was conducted to figure out our options.

Considering what was analyzed previously and the needs of a WMS AR system, it is obvious that our system should be able to record the layout of the warehouse. To do so the following solutions were found to be the most suitable one.

- **Matterport Pro2 3D Camera:** As found in the company's official site [2], the camera has a built-in Wi-Fi sensor in order to be paired with an iOS device, using it as a controller. With an intuitive interface the user can produce a 3D model of a space. The camera is equipped with an infrared 3D sensor and is 99% accurate. However, it must stand on a tripod to scan an area and must be paired with an iOS device so as to be controlled and return the data through the Capture app.
- **Structure sensor by Occipital:** A sensor that can be attached to the camera of an iOS device [3] and produce a 3D map or even a model of an object or an area. It is a light (95 g) sensor and can be used for 3–4 h straight of active sensing or can be even charged throughout its use. The camera itself uses an infrared structured light projector. The operating temperature is between 0 and 35 °C.
- **Google Tango supported device:** Combined with a motion tracking camera, a 3D depth sensor, an accelerometer, a barometer, a gyroscope and a GPS, a Tango – compatible device, the ability not only to detect their position relative to their world but also construct a 3D map of their environment.

Moreover, to achieve the automated measurement of the commodities inside the warehouse an autonomous or semi-autonomous machine should be used to go through the corridors of the warehouse with the ability to find a specific location. By thinking towards that direction mobile robots either on the ground or drones match our needs.

A mobile robot is a robot that is capable of locomotion and is not fixed to one physical area. Mobile robots can be "autonomous" (AMR - autonomous mobile robot) or depend on direction gadgets (AGV - autonomous guided vehicle). Some of our solutions are the following:

- **KMR iiwa by KUKA:** A location – independent and highly flexible lightweight robot [4]. Equipped with seven special joint torque sensors on each axis that make it exceedingly sensitive to its environment. It screens its environment and responds quickly if an individual or item is standing out. It can move omnidirectionally and its positioning accuracy is up to ±5 mm. Lastly it is independent as the vehicle and robot are supplied directly with power from Li-ion batteries.
- **Gapter Drone:** Gapter is an unprecedented drone [5] that supports Robot Operating System (ROS), the MAVLink protocol, and the mavros wrapper layer between ROS and MAVLink. Gapter can work both indoor and outdoor with its inbuilt sensors like GPS and optical flow sensors and allows can be extended by add new modules or sensors. The Ordroid XU4 single-board computer integrated to provide the best performance possible for the autopilot and combined with a Rplidar laser scan it can build a map of the space and navigate within it.

In the case that an AGV is used some directional gadgets should be used in order to make it aware of its surroundings and make it possible for it to move with no collisions.

- **Dragonfly:** Dragonfly is a visual 3D positioning/location system [6] based on Visual SLAM. Visual SLAM, otherwise called vSLAM, is an innovation ready to fabricate a map of an obscure domain and guide, at the same time utilizing the mostly manufactured map, utilizing just computer vision. The main sensor required is a camera that must be mounted on board of the gadget. No other outside sensors are required.
- **R2000 2-D-LiDAR:** R2000 2-D-LiDAR is a photoelectric sensor [7] that monitors the AGV's surroundings with a 360° angle and provides measurements to the vehicle controller. Employees' legs can be filtered out of the high-resolution data and utilized as a source of perspective point to distinguish their course of movement and follow them closely.

Trying to fulfill the needs and restrictions of our project we will proceed into the making of our architecture by using a Google Tango-compatible device and a Gapter Drone. To start with, both seem economically not so expensive as the other "contestants". In addition, the Tango-compatible device is preferred to the other options as it can be carried around without a tripod and has no room temperature limitations. Lastly, the Gapter drone is so powerful that can navigate on its own and versatile enough to operate indoors, such as warehouse.

Acknowledgment. This research has been co-financed by the European Union and Greek national funds through the Operational Program Competitiveness, Entrepreneurship and Innovation, under the call RESEARCH – CREATE – INNOVATE (project code: T1EDK-03502).

References

1. Entersoft WMS product page. https://www.entersoft.gr/products/wms/. Accessed 04 June 2019
2. Matterport Pro2 Camera page. https://matterport.com/pro2-3d-camera/. Accessed 04 June 2019
3. Structure sensor page. https://structure.io/structure-sensor. Accessed 04 June 2019
4. KMR iiwa page. https://www.kuka.com/en-my/products/mobility/mobile-robots/kmr-iiwa. Accessed 04 June 2019
5. Gapter overview page. http://edu.gaitech.hk/gapter/overview-and-features.html. Accessed 04 June 2019
6. Dragonfly. https://www.dragonflycv.com/. Accessed 04 June 2019
7. Pepperl + Fuchs Sensors for AVG Vehicles page. https://www.pepperl-fuchs.com/global/en/32763.htm. Accessed 04 June 2019
8. Ian, G.: Essential Software Architecture, 2nd edn. Springer, Heidelberg (2011)

Towards a Flipped Optical Laboratory

Thrasyvoulos Tsiatsos[1](✉), Nikolaos Politopoulos[1], Panagiotis Stylianidis[1], Nikolaos Pleros[1], Lampros Karavidas[1], Kyriaki Karypidou[1], Andreas Pester[2], and Thomas Klinger[2]

[1] Aristotle University of Thessaloniki, Thessaloniki, Greece
{tsiatsos,npolitop,pastylia,npleros,karavidas, karypidk}@csd.auth.gr

[2] Carinthia University of Applied Sciences, Klagenfurt, Austria
{A.Pester,t.klinger}@fh-kaernten.at

Abstract. This paper presents the main steps for creating an innovative educational Optical Laboratory, based on the flipped classroom educational model. This laboratory called FlipOL: Flipped Optical Laboratory. FlipOL intends to support hands-on training and skill enhancement on photonics by offering hands-on educational activities, broadening the audience and extending existing facilities expediting access to photonic components, measurement equipment of Photonic Integrated Circuits (PICs) and support services. Doing so, FlipOL aims to introduce, adapt and exploit the rich capabilities of current ICT distance learning toolkit towards negating the barriers associated with geographical location, localization and cost in outreach and learning activities. It will establish a single one-stop-shop Virtual Space for educational material, bringing the community of photonics in contact with well-established principles from the industrial and educational sector and shaping a portal to a rich on-line remote access portfolio. FlipOL specific objectives are: to establish a Photonics Community of Practice and the FlipOL Virtual Space in order to form a portal to support a broad range of training actions in Photonics and PICs, to extend and utilise existing facilities in order to facilitate access to photonic components and related equipment through creating (i) an online photonics design environment, (ii) a remote lab infrastructure, and (iii) an augmented reality on-site lab app, to deploy a Remote MakerLab that exposes the user to all integrated photonic technology processing cycles from designing, fabricating and measuring his/her own device, to organize a Lifelong Learning Program in Photonics based on Massive Online Open Courses (MOOCs).

Keywords: Photonics · Augmented reality · MOOCs

1 Introduction

Photonics has definitively played a key role in the revolutions witnessed in our daily lives during the last 20 years. Photonic technology is everywhere, spanning from communications and health, to materials processing in production, to lighting and photovoltaics and to everyday products like DVD players and mobile phones. Its wide penetration across several socio-economic application areas has been naturally reflected

M. E. Auer and D. May (Eds.): REV 2020, AISC 1231, pp. 349–356, 2021.
https://doi.org/10.1007/978-3-030-52575-0_29

also in the financial and employment landscape, with the global market size in photonics reaching now more than 600€ billion [1]. The Europe photonics industry alone employs almost 290,000 employees, having an indirect impact to around 30 million jobs since almost 20-30% of the economy and 10% of the workforce is depended on photonics.

These figures are simply expected to grow even more in the future, as big revolutions are still expected to come by photonics in several technological sectors like computing, bio-imaging, neuro-photonics etc. Retaining this pathway can be, however, realized only by ensuring that an increased number of people forming the workforce of tomorrow get interested in photonics, bringing with them an enhanced innovation capacity that goes beyond today's standards.

Raising interest and enhancing skills and innovation capacity in the field of photonics is, however, certainly not an easy task to perform, with geographical location and cost forming the most critical barriers. Photonic equipment facilities are usually concentrated in localized areas in order to promote their cost-efficient utilization, requiring also highly experienced personnel for their operation. At the same time, photonic components are usually high-cost items that can't be easily afforded by individuals or non-expert communities. These constraints have forced awareness raising as well as training activities in photonics to primarily focus on information-conveying actions, limiting their potential to expose people to real photonic settings and hands on experience. Within the same frame, access to photonic components has been restrictive for a wide audience and has been mainly facilitated for groups being already experts in the field.

FlipOL intends to support hands-on training and skill enhancement on photonics by offering hands-on educational activities, broadening the audience and extending existing facilities expediting access to photonic components, measurement equipment of photonic integrated circuits (PICs) and support services.

Doing so, FlipOL aims to introduce, adapt and exploit the rich capabilities of current ICT distance learning toolkit towards negating the barriers associated with geographical location, localization and cost in outreach and learning activities. It aims to be the first project that will establish a single one-stop-shop Virtual Space for educational material, bringing the community of photonics in contact with well-established principles from the industrial and educational sector and shaping a portal to a rich on-line remote access portfolio.

2 Aims and Objectives

In order to realize FlipOL, specific objectives must be accomplished. First and foremost, the establishment of a Photonics Community of Practice and the FlipOL Virtual Space in order to form a portal to support a broad range of training actions in Photonics and PICs. Second of all, the extension and utilization of existing facilities in order to facilitate access to photonic components and related equipment through creating (i) an online photonics design environment, (ii) a remote lab infrastructure, and (iii) an augmented reality on-site lab app. In addition, the deployment of a Remote MakerLab that exposes the user to all integrated photonic technology processing cycles from designing, fabricating and measuring his/her own device. Lastly, the organization of a Lifelong Learning Program in Photonics based on Massive Online Open Courses (MOOCs). FlipOL will implement a wide range of activities such as:

- Boosting innovation in Photonics higher education, business and in the broader socio-economic environment through: Learning Technologies like Augmented Reality, Remote Labs, MOOCs.
- Stimulating the flow and exchange of knowledge between higher education and enterprises in the area of Photonics and PICs.

The primary target groups of the project are the following:

- University students at undergraduate and postgraduate level, aiming to communicate the importance of photonics and its interaction with a broad range of scientific disciplines and to train and enhance the skills of tomorrow's workforce. In addition, remote training will reduce or even avoid the costs related to travelling.
- University Professors, aiming at broadening their knowledge on cutting-edge applications of photonics and at expediting the interest and knowledge transfer to their students locally and abroad.
- Technicians and Young professionals, intending to enhance their skills, improving their innovation capacity and offering them a holistic view of the photonics field and its underlying processes breaking the barriers usually enforced by specialization and geographical limitations. This is expected to strengthen the cooperation between education/training and industry helping to better match the needs of industry into the university educational programs.

According to the above many organizations could benefit from FlipOL:

- Universities/Research & Education Institutes that want to include photonics education in their programs of studies or to start making research in the area while attracting international students or second via e-learning platforms. Remote study and training programs will be enabled offering a broad theoretical and experimental knowledge base to an international environment.
- Companies/SMEs/businesses with the same or similar expertise of the companies that are participating in the project (namely VPI, VLC, PHX) from the sectors of photonic component sand circuit design that would like to train their clients as well as work force while offering attractive free trial versions of the remote training to their customers worldwide.
- Wider public for having better awareness about photonics while witnessing the true possibilities opened by photonics to everyday life applications.

All the above categories of end users still phase challenges associated to fragmented educational, training and innovation resources while inherent barriers still impede hands-on practice and centralized knowledge base mainly due to geographical limitations and the high cost of state-of-the-art equipment, facilities and services associated to photonics eco-system. Target groups and organizations will benefit from FlipOL by exploiting for the first-time modern e-learning practices, remote hands-on experiments and inspiring augmented reality technology in a unified, interactive education and training platform to students and professionals with cross-disciplinary expertise and form different country of presence, boosting innovation and interdisciplinary synergy in photonics and their applications.

3 Implementation

The project will have three main outcomes:

- To create a short **Lifelong Learning Program (LLP)** supporting photonics trainings
- To implement **digital services and educational content** to support and offer LLP from distance
- To **evaluate** the above outcomes

Therefore, the project methodology will follow 3 directions based on well-known methodologies in each outcome.

1. The **1st direction** concerns the creation of an LLP for supporting dual career of athletes. To this direction the project will follow the **EADTU – Business model for lifelong learning** [3].
2. The **2nd direction** concerns the instructional design and the development of digital services. To this direction the project will follow the **ADDIE** [4, 5] model, which is a framework that lists generic processes that instructional designers and training developers use.
3. The **3rd direction** concerns the educational evaluation where the project will follow **Kirkpatrick's training evaluation model** [6].

3.1 FlipOL Ecosystem

The FlipOL ecosystem will focus on the support of **distant, hands on training on photonics,** involving:

- **Massive Open Online Courses (MOOCs)**, being online photonics courses aiming at unlimited participation and open access via the web, offering both traditional course materials such as filmed lectures, readings, and problem sets but also interactive user forums to support community interactions.
- **Mobile and Augmented Reality Tools.** The three essential properties of Augmented Reality (AR) are the combination of virtual and real objects in a real environment; a system that aligns/registers virtual and real objects with each other; and that runs interactively in real time. FlipOL will take advantage of this technology for visualizing the invisible processes taking place in photonic components and applications as well as for enabling the cost-efficient and as such massive production of 2D/3D cardboard-based educational kits directed for hands-on-training of students.
- **Remote Labs**, exploiting the use of telecommunications to remotely conduct real photonic experiments, at the physical location of the operating technology, whilst the trainee will simply utilize his/her own computer from a separate geographical location.

- **Community of Practices.** FlipOL aims to establish a Photonics CoP increasing the interaction between users interested in photonics, raising the awareness of photonics workforce about other practices in photonics and enhancing the innovation capacity of the CoP participants.

The FlipOL high level architecture for supporting hands on and Awareness/ eLearning experience of students, professors, technicians and young professionals is depicted in the following figure (Fig. 1).

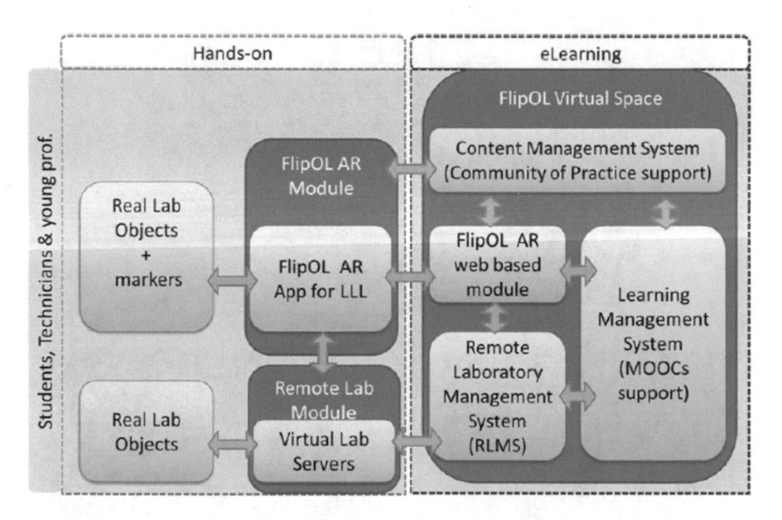

Fig. 1. FlipOL high – level architecture

3.2 Mobile and Augmented Reality Tools

The main aim of this task is to create an on-site Augmented Reality (AR) application in order to enhance hands-on training in photonics, implementing:

- **FlipOL on-site AR App:** Using this app user will scan an AR marker positioned on the lab equipment, allowing them to access training and educational material and interact with a 3D virtual model objected in FlipOL AR platform.
- **FlipOL AR web-based module:** for collecting the objects and sharing them with other users. The FlipOL AR Platform will interoperate with FlipOL CoP and MOOCs, in order to recommend to the user educational content like videos, animations, photos, papers, etc.

FlipOL AR App for LLL will **extend** the experience with **real photonics lab** objects, so that the reality of photonic components, devices and measurement equipment that is often perceived by trainees as a box with unknown content will have its inner-running processes visualized, correlating the box-inside with elementary functions and fundamental theory principles. For example, an AR App for lasers could easily visualize the laser cavity and electron-hole recombination processes taking place in the cavity, helping the trainee to visualize the roundtrip travel of photons and to correlate different laser parameters with the outgoing light. On the same line, an AR app for optical fibers could help the visualized perception of light propagation and Total Internal Reflection associating the geometrical parameters and refractive index values with the properties of propagating light. It should be noted that industries and universities will have access to this service through the **FlipOL Community of Practice** in order to create, upload and share their objects by extending the access to them (Fig. 2).

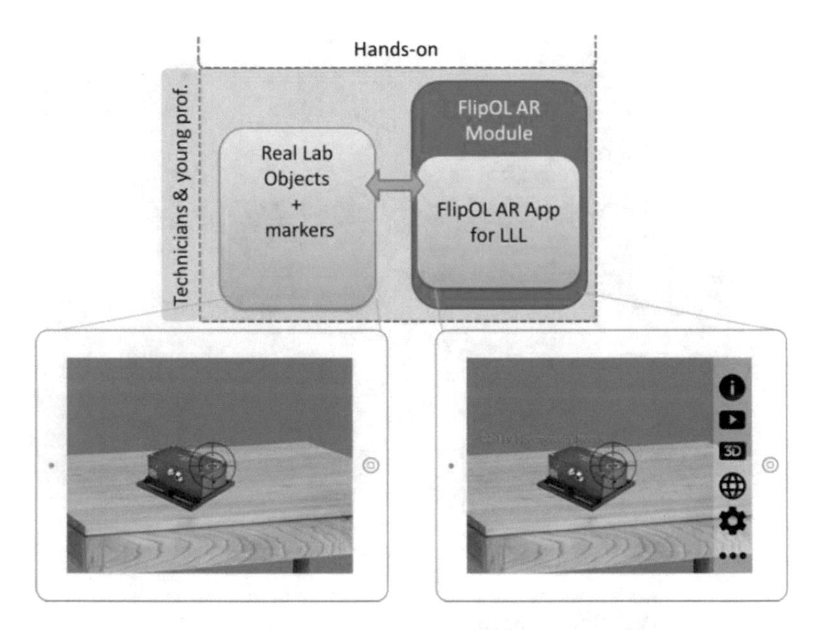

Fig. 2. FlipOL AR for technicians and young professionals

3.3 Massive Open Online Course (MOOC) Design

This activity will select and install a Learning Management System (LMS) in order to support asynchronous learning services. Main criteria for the selection of the platform will be:

- The Mode of delivery (e.g. synchronous, asynchronous, AR, Remote Labs, etc.) of the lessons.
- The cost of usage, installation and support of the platform.
- The exploitation of current infrastructure provided in the participating organizations (for example AUTH could host a "Moodle" LMS platform)

Furthermore, the whole courses will be embedded to a MOOC Open Education Scoreboard28 (or similar gateway) and it will be available via MOOC search services.

The outcome of this activity will be to create the infrastructure for a European Massive Open Online Course (MOOC) integrating educational content for supporting photonics education, in accordance with the FlipOL's Curriculum.

3.4 Remote Lab Services

The expected output will be a remote lab, including real experiments on the CUAS cloud laboratory infrastructure named Experiment Dispatcher. The Experiment Dispatcher abstracts the setup of a laboratory server, allowing the lab owner to deploy several instances of virtual lab servers. It additionally implements the API of some well-known RLMS enabling the Experiment Dispatcher to act as a federated node for these systems by serving up experiment execution requests.

The experiment engines comprise the laboratory specific parts of the architecture. An engine dequeuers experiments, executes them by interacting with the laboratory equipment and sends the experiment results to the Experiment Dispatcher for later retrieval by the RLMS or client application. An experiment engine runs on the lab owner's side since it must interface with the laboratory hardware. This implementation allows the Experiment Engines to reside anywhere on the Internet, even in a private network (Fig. 3).

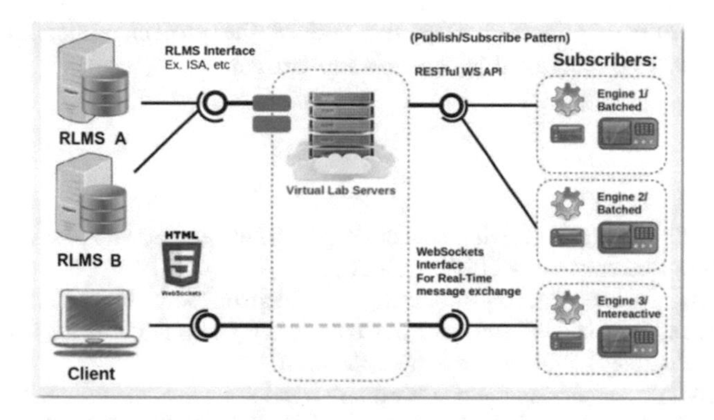

Fig. 3. FlipOL remote lab module

The core of the online lab will be an automated probe station where the trainee will evaluate integrated photonic structures (PICs). The probe station will comprise a Tunable Laser, two XYZ axis piezoelectric actuators for fiber alignment to the chip and a power meter, all connected through LabView and controlled via a web-based GUI (compatible with smart phones and tablets) or client application of the laboratory. As soon as a link is established by the trainee, between the input/output ports, the online laboratory user will be able to measure various parameters including among others; insertions losses of passive components, the frequency response, the transmission and reflectance properties vs. the launched optical power etc., having the opportunity to confirm the results with the theoretical expectations.

4 Outcomes and Future Work

FlipOL aims to fight the fact that photonics technology, educational, training and innovation resources remain fragmented by introducing for the first-time modern e-learning practices, remote hands-on experiments and inspiring augmented reality technology in a unified, interactive education and training platform to students and professionals with cross-disciplinary expertise and form different country of presence. The platform will unite currently fragmented photonics research and education resources across Europe in a single yet multifunctional ecosystem that will boost awareness and active involvement of future workforce on all levels of photonics science and its capabilities. In this context, FlipOL will go far beyond the state of the art, by simultaneously exploiting modern AR technologies, MOOCs, user-instructor interaction space and remote lab facilities to con-currently provide interactive education and hands-on training, shortening the learning curve of photonics technology in an engaging experience while removing access barriers related to experimental facility cost and geographical location. Finally, the operation model of FlipOL will be open to contributions from other parties after the end of the project, under the same OER strategy, that will be embodied in the existing program.

References

1. Towards 2020 – Photonics driving economic growth in Europe, Photonics21 PPP Multiannual Strategic Roadmap 2014–2020, April 2013
2. A Photonics Private Public Partnership in Horizon2020, Photonics21. http://www.photonics21.org/download/Photonics21_Association/A_Photonics_Private_Public_Partnership_Photonics_PPP_proposal_final-final.pdf
3. Bang, J.: EADTU, organising lifelong learning - a report on university strategies and business models for lifelong learning in higher education (2010). http://lll-portal.eadtu.eu/images/files/Manual_Organising_EADTU%2024-09-2010.pdf
4. U.S. Air Force. (Instructional System Development (ISD). AFM 50-2. U.S. Government Printing Office, Washington, DC (1970)
5. http://www.nwlink.com/~donclark/history_isd/addie.html
6. Kirkpatrick, D.L.: Techniques for evaluating training programs. J. ASTD **11**, 1–13 (1959)

Virtual and Mixed Reality

Enhancing Engineering Education by Virtual Laboratories

A Comparison Between Two Different Approaches

Diana Keddi and Sulamith Frerich(✉)

Ruhr-Universität Bochum, Bochum, Germany
{keddi, frerich}@vvp.rub.de

Abstract. The aim of this contribution is to compare two different settings of two virtual laboratories. Both of them are situated in the context of chemical engineering. One of them is used as online preparation for international students, while the other is implemented in lectures and seminars as demonstrating unit of subjects related to porous materials. While the online preparation for international students has already been at use, the demonstration unit is still work in progress. The students benefit from this kind of digital preparation to a high degree. Theoretical knowledge is available on an individual level, and they can choose time and place when to attend the courses. Many students mastered their course, understood the underlying concepts, and also exceeded usual expectations with their final reports. Regarding their comments, the implemented visualizations were highly appreciated, and the students also rated the set-ups as affirmative. However, a reasonable amount of participants complained about the absence of a real person in charge throughout the experiment, as they have experienced it in hands-on laboratories on site. Although it was found that virtual laboratories are an appropriate way to explain scientific topics, it can be observed that the actual implementation is still facing some issues. This contribution gives an overview of experiences made and discusses the potential for future applications of virtual laboratories in engineering education.

Keywords: Virtual laboratory · Online-learning · Visualization

1 Introduction

Recently, it was shown that student exchange programs in engineering sciences between Germany and USA benefit a lot from "Summer School" formats. They are targeting the summer break in US curricula to offer US-American students a stay abroad at a German host university. The exchange program is usually complemented by offering German students a stay at the respective partner university. Hence, the obstacle of a mismatch between the terms of an academic year in both countries can be overcome. The project ELLI (Excellent Teaching and Learning in Engineering Sciences), funded by the German Government as part of the Teaching Quality Pact, supported such exchange programs at the Faculty of Mechanical Engineering at Ruhr-University Bochum (RUB) since 2014 [1].

M. E. Auer and D. May (Eds.): REV 2020, AISC 1231, pp. 359–365, 2021.
https://doi.org/10.1007/978-3-030-52575-0_30

In 2015, a new student exchange program between RUB and Virginia Tech (Virginia Polytechnic Institute and State University) was launched, with increasing numbers of participants every year. It addresses one of the major learning objectives in engineering education: Students are given the opportunity to experience unit operation labs in Germany, where they are working on problem solving tasks, gaining practical experience and using theoretical knowledge to target practical applications. Although first evaluations did show that US students highly appreciated these opportunities, the short duration of the above mentioned summer school format of only 8 weeks in total caused feelings of being non-integrated into German routines and campus life. Therefore, an online tool for preparing US students for their upcoming stay at RUB was set up in 2016. The so-called "VTprep" class using moodle as learning management system comprises both curricular and extra-curricular content and offers incoming US students and their future RUB supervisors and fellow students to get in touch prior to their arrival in Germany [2].

Apart from offering practical experience to international students, RUB study programs send their own engineering students to participate in unit operation laboratories as well. However, laboratory capacity is limited, due to time and safety issues. Thus, the availability of laboratory equipment is limited, too. Therefore, the same project ELLI as mentioned above is also interested in enhancing the opportunities for regular students of participating in these laboratories as well.

Virtual laboratory setups use simulations, modelling, and visualization to create realistic scenarios, explaining scientific phenomena and demonstrating varying plant operation states. They can be used to show logical causalities between operating parameters and visualize effects which are usually not to be seen in a hands-on experiment on site [3]. In addition, they can be carried out with greater flexibility: There are no time restrictions due to online availability, and they can be assessed from virtually anywhere [4].

This contribution is bringing both perspectives together. There are two virtual laboratories which have been implemented separately, "Silo Design" and "Foam Analysis". Their identical aim is preparing engineering students for their tasks in real laboratories on site, being situated both in chemical engineering. Therefore, both are providing theoretical knowledge about their topics, respectively, by addressing issues, critical parameters and selected methods the participating students need to know in the future. However, their individual set-up has been different, due to differing resources and methods.

2 The Setups of Both Virtual Laboratories

2.1 Silo Design – An Avatar in an Artificial Environment

The virtual laboratory "Silo Design" has been developed in cooperation with Labster, a Danish company working on developing fully interactive laboratory simulations. They are using mathematical algorithms that support open-end investigations, mainly focusing on laboratory experiments taking place in natural sciences and medicine [5].

Over a time period of about 1.5 years in total, the set-up was undertaken. Based on pictures and photos from the real lab environment at RUB, an artificial environment was created. Referring to learning objectives and anticipated learning outcomes, several activities and interactions were defined that should be possible to be carried out by future participants. Still, the process parameters and related results determining the actual experiment to be run by an avatar marked the central part of this virtual laboratory.

Topics of this virtual lab include the actual design of a bulk material silo, a characterization of the bulk material with a shear cell, and the design of a feed pump. Any standard PC or Mac can be used, since the virtual lab is provided by using an internet browser. Several iterations for optimization were run, based on video streaming test runs and self-assessed trials, prior to its initial operation. A language selection has also been implemented, enabling both English and German students to pursue actions in their native tongue. Figure 1 shows images of the real silo site at RUB as well as pictures taken from the virtual lab "Silo Design".

Fig. 1. Real silo site (left) and pictures of the virtual lab "Silo Design" (right, top and bottom)

2.2 Foam Analysis – A Modular Experimental Setup

The virtual laboratory "Foam Analysis" was designed as a modular experimental set-up. In contrast to the artificial environment created for the virtual lab "Silo Design", all components here were put together as individual topics, using the tools provided by the Learning Management System Moodle.

Since the tasks of this virtual lab are targeting polymeric foams, generated by a tandem extrusion plant using high-pressure technology, the participants need to know how to operate extruders, use compressed carbon dioxide, and analyze the generated samples about their pore size distribution, morphology and mechanical strength.

Figure 2 shows a screenshot of one task referring to the extruder used in this set-up, including both a picture of the real experimental site and the schematic figure of the tandem extrusion plant. The whole set-up was also including a language selection, to run the virtual in English or German, too.

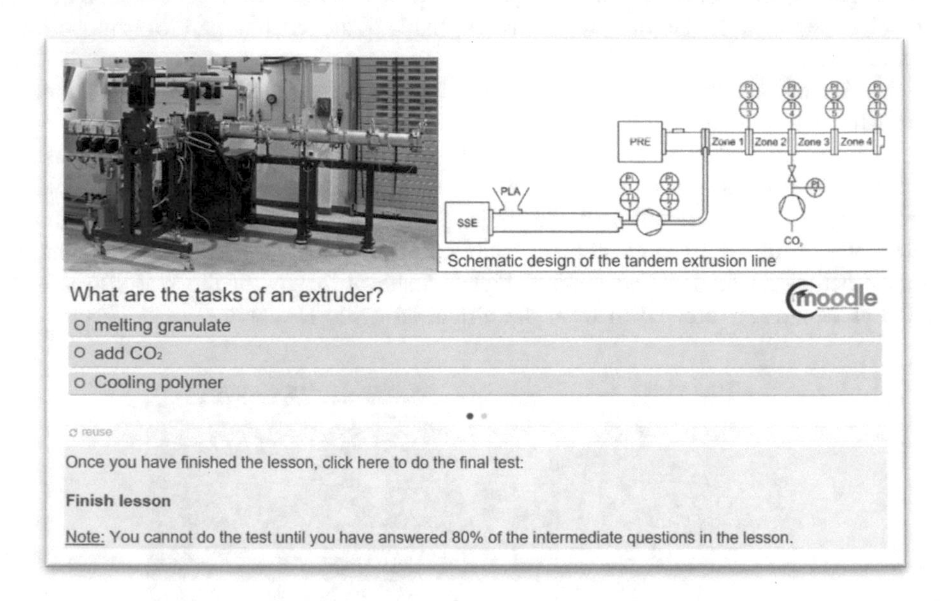

Fig. 2. Screenshot of virtual lab "Foam Analysis"

3 Embedding of Virtual Laboratories

As mentioned above, both virtual laboratories are supposed to prepare engineering students for their tasks in real laboratories on site. Therefore, they are supposed to enhance the individual student preparation prior to the hands-on experience, to provide theoretical information if needed, and explain major tasks in advance. The most important aspect of undertaking real experiments is about safety precautions. Participating students have to be aware of hazardous substances, dangerous process conditions and required actions in case of emergencies. Usually, access to lab sites is only permitted after a safety instructions survey has been passed. Thus, the same concept has been implemented in both virtual laboratories. Prior to the start of the actual experiment, the students have to go through safety instructions. The respective theoretical input is tested by additional quizzes and tests, and the virtual laboratory lists a score of the achievements already accomplished, in order to keep a high motivational level. The current status of the students within the virtual lab is individually documented and saved. Being at the interim state of having completed the virtual laboratory and looking forward to the real test site, the students are interviewed, as it is shown in Fig. 3.

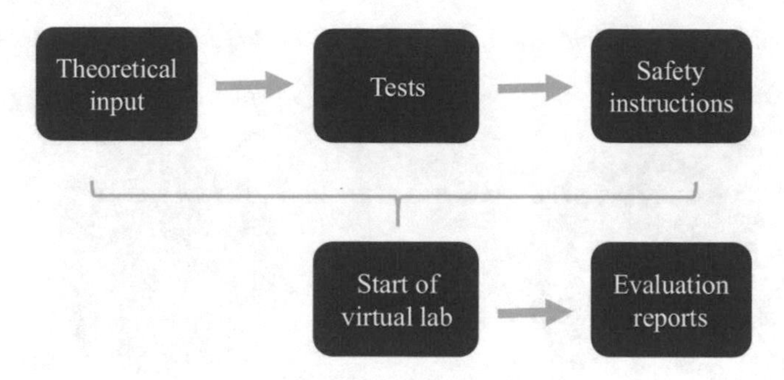

Fig. 3. Embedding of the virtual laboratory

The overall evaluation is based on the students' experiences in the virtual laboratory and addresses their initial expectations as well. Their opinion in feeling prepared for the real laboratory test is also recorded. Once the students have carried out the real laboratory experiment as well, the acquired knowledge and experiences of the combined experience of both the virtual and the real laboratory is evaluated. At the same time, the supervisors who guided the students through the real experiment are questioned about their experiences as well. In this way, a comprehensive examination of the implementation of the virtual laboratories in engineering education possible.

While the virtual lab "Silo Design" has been used for preparing US-American students for their real lab experiments at RUB, the virtual lab "Foam Analysis" has been incorporated into the regular RUB lecture "Heat and Mass Transfer". Thus, the number of participants varies from 14 to 70 students.

4 Evaluation

The survey of both virtual laboratories showed that about 85% of the participants already gained experiences with real laboratory experiments beforehand. This was to be expected, as most of the students were already in their third year of studying and should therefore have had the opportunity to be part of practical laboratory experiments. However, only 50% of them have already carried out a virtual laboratory. It was assumed that these were the ones able to easily master the virtual laboratories shown in this contribution.

Being interviewed about their expectations regarding the virtual lab they were about to conduct, they expected to get a first grasp on the tasks lying ahead. In addition, the students mentioned that they hoped to memorize the theoretical knowledge better. However, the acquisition of practical skills, which are usually part of a real laboratory experiment, is viewed critically. The majority of students assumes that they can hardly acquire these skills using the virtual laboratory only. However, most of them consider a virtual laboratory to be helpful to very helpful. Figure 4 shows two aspects of the final evaluation of both the virtual lab and the combination of virtual and real lab, too.

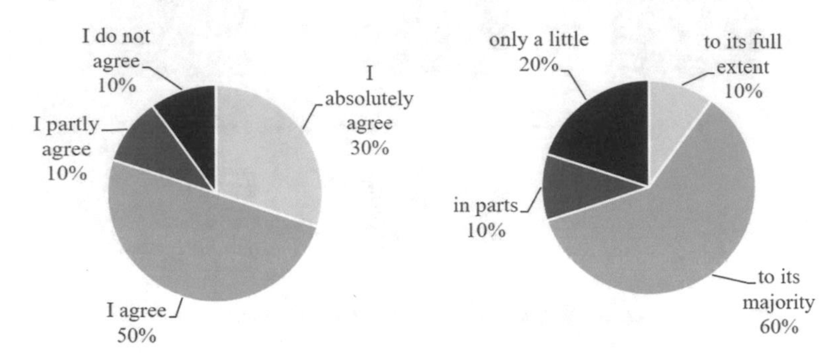

Fig. 4. Evaluation of the virtual laboratory

About 90% of the students agreed that the design of the virtual lab and their expectations about its outcome were fulfilled. The students particularly appreciated various visualizations, because they would not have been possible in an exclusively real experiment. At least 70% felt well prepared for the real laboratory experiment at this point, even if criticism is expressed on various occasions about lacking of some references to real life applications or literature sources as too extensive to read.

Once the evaluation results are comparing both virtual labs with each other, it is found that although moving an avatar has been highly motivational, the progression within "Silo Lab" was perceived as quite time-consuming. Regarding "Foam Analysis", the modular set-up was considered as quite favorable. Still, the students stated that the level of immersion could have been higher.

It is striking to see that about 50% of the students specified in their answers that they lacked a real contact person or supervisor during the virtual experiment, as it is usually the case during a real lab experiment. Apparently, the way of conducting the virtual has not been clear to everyone from the beginning. However, it is not clear from the data whether this correlates with the above-mentioned 50% of students who have not yet carried out a virtual laboratory test.

After having carried out the real laboratory, the final survey of the students revealed an almost consistently positive picture. Even though their individual statements differed slightly from each other, the students agreed to have rated the combination of virtual and real experiment about one topic, respectively, to be helpful or even very helpful. In particular, they mentioned that the preparation of the experiments by using the virtual laboratory as auxiliary helped them to understand content and applications better. At the same time, though, they were in favor of retaining an additional script and the extensive literature, too, to better understand the underlying theoretical content.

This view is shared by the supervisors who were in charge of conducting the real laboratory experiments: It became apparent throughout these experiments that the students had to look up frequently the theoretical basics to be able to create the report about the results and their analysis successfully. Overall, the supervisors noted enthusiastic, intrinsically motivated students, who participated vividly by questioning the contents individually. In comparison to previous participants, the students of this round seemed to be particularly active, according to the supervisors. However, it cannot be proven that this is due to the new digital preparation and not to the student group itself. In any case, the supervisors were quite satisfied with the changes made to the digital preparations of the labs.

5 Summary and Outlook

The aim of this article was to show that a virtual laboratory experiment as part of an online service to prepare students for their summer school is a good complement to real-life experiments. This is also in line with earlier studies, see ref [6, 7]. The majority of the students found the virtual laboratory experiment to be useful in order to prepare the content for the real laboratory experiment. For the execution and evaluation of the real laboratory test, the students also see the necessity of a detailed script. The lack of direct contact with lecturers and the inability to train manual skills are negative aspects of a virtual laboratory. However, the improved availability of the virtual supplementary laboratory is to be extended to further laboratory experiments in the future. Since the modular concept are worked out well, there is no necessity for creating an artificial environment by all means.

Acknowledgement. This work was supported by the German Federal Ministry of Education and Research under Grant 01PL16082B in the Teaching Quality Pact.

References

1. Strenger, N., Petermann, M., Frerich, S.: Student exchange programs in engineering sciences between USA and Germany. In: Engineering Education Conference (EDUCON), pp. 1038–1041, Istanbul, Turkey (2014)
2. Strenger, N., May, D., Ortelt, T.R., Kruse, D., Frerich, S., Tekkaya, A.E.: Internationalization and digitalization in engineering education. In: 3rd International Conference on Higher Education Advances (HEAD), pp. 558–565, Valencia, Spanien (2017)
3. Strenger, N., Frerich, S.: How to design digitalized laboratories? Lessons learned from implementing virtual and remote labs. In: IEEE EDUCON (2019, submitted)
4. Kruse, D., Frerich, S., Petermann, M., Kilzer, A.: Virtual labs and remote labs: practical experience for everyone. In: IEEE EDUCON (2014). https://doi.org/10.1109/educon.2014.6826109
5. Labster (2019). https://www.labster.com. Accessed on 23 Oct 2019
6. Ulbrich, N.: VTprep – digitales Begleitprogramm einer Summer School. In: Ittel, A., Meyer do Nascimento Pereira, A., (eds.) Internationalisierung der Curricula in den MINT-Fächern, pp. 240–247. wbv Media, Bielefeld (2018)
7. Kruse, D., Kuska, R., Frerich, S., May, D., Ortelt, T.R., Tekkaya, A.E.: More than "Did you read the script?" - different approaches for preparing students for meaningful experimentation processes in remote and virtual laboratories. In: 14th International Conference on Remote Engineering and Virtual Instrumentation (REV), pp. 160–169 (2017). https://doi.org/10.1007/978-3-319-64352-6_16

The 3D Motion Capturing Process of Greek Traditional Dance "Syrtos in Three" and a Proposed Method for the Sex Identification of Performing Dancers (Terpsichore Project)

Styliani Douka[1], Efthymios Ziagkas[1](✉), Vasiliki Zilidou[1], Andreas Loukovitis[1], and Thrasyvoulos Tsiatsos[2]

[1] Faculty of Physical Education and Sport Science, Aristotle University of Thessaloniki, Thessaloniki, Greece
eziagkas@phed.auth.gr

[2] School of Informatics, Aristotle University of Thessaloniki, Thessaloniki, Greece

Abstract. The UNESCO considers as part of the cultural heritage of a place, the intangible cultural heritage (ICH) which is a practice, an expression or representation and knowledge or skill, as well as instruments, objects and cultural places. The cultural heritage includes the traditions or living expressions inherited from the ancestors and passed on to the descendants, not only the monuments or collections of objects. Such features are oral traditions, social practices, arts, festive events and generally, knowledge and practices concerning nature and the universe or the knowledge and skills to produce traditional crafts. Terpsichore project targets at integrating the latest innovative results of photogrammetry, semantic technologies, computer vision and time evolved modelling, along with traditional choreography and narrative. The study, design, research, education, analysis, implementation and validation of an innovative framework for accessible digitization, modelling, archiving, online preservation and presentation of ICH content related folk dances is reflected in the project. The Department of Physical Education and Sport Sciences of the Aristotle University was involved in the project in the performance of six traditional Greek dances and the three-dimensional capturing. Specifically, in the present work with the process of 3D recording of the movement of the Greek traditional dance "Syrtos in three", the method for identifying the dancers' gender will be proposed. The results showed that using 3D motion capturing we may recognize the sex of the dancer through the trajectories of toes markers and the kinematic data of joints angles. Regarding kinematic data founded differences at the means of angles of the hip, the knee and the ankle joints. Male dancers showed narrower angles in all mentioned joints of all six steps than the angles of female dancers during the implementation of the Greek traditional dance "Syrtos in three steps".

Keywords: Traditional dances · Movement analysis · Intangible cultural heritage · Capturing · Vicon

M. E. Auer and D. May (Eds.): REV 2020, AISC 1231, pp. 366–374, 2021.
https://doi.org/10.1007/978-3-030-52575-0_31

1 Introduction

The UNESCO considers as part of the cultural heritage of a place, the intangible cultural heritage (ICH) which is a practice, an expression or representation and knowledge or skill, as well as instruments, objects and cultural places [1]. The cultural heritage does not end in monuments or collections of objects. It also includes the traditions or living expressions inherited from the ancestors and passed on to the descendants. Such features are oral traditions, social practices, arts, festive events and generally, knowledge and practices concerning nature and the universe or the knowledge and skills to produce traditional crafts. An understanding of the intangible cultural heritage of different communities helps intercultural dialogue and encourages mutual respect for other kinds of lifestyle. The critical thing in intangible cultural heritage is the variety of knowledge and skills that are passed on from one generation to the next and not the cultural event itself. The economic and social value of this transit of knowledge considered as necessary such as for both of minority groups, the dominant social groups within a state and for developing and developed countries [2, 3].

The UNESCO in 2001, conducted a survey among state and NGOs trying to agree on a definition. The Convention on the Safeguarding of the Intangible Cultural Heritage was established in 2003 for the protection and promotion thereof. The intercultural dialogue between people, different cultures and countries provide a multilevel scheme through the digitalization technology-interface intelligence-emotional intelligence according to the field of traditional dances. Performance arts, cultural diversity it's a privilege especially to the young people to explore and to discover all of their capacity in order to understand emotionally and spirituality all the benefits of the interaction with the culture sector.

As consider the main structure of ICH content, includes a combination of interactive and powerful strength of the participant's nature, combined with the emotional environment so that, he or she can express thoughts and feelings most effectively. What really counts is the personality and the style of each performer-dancer. Before the presents of its program, we have to mention the contribution of the i-Treasures project and the effective way of a specific platform which has given access directly to essential resources, so that, the transmission of rare know-how from Living Human Treasures to apprentices was quite successful [4]. We also have to mention the importance of the RePlay project which provided most understandably the accurate knowledge to the traditional sports [5, 12].

The primary purpose of the Terpsichore project is emphasised to educate, analyse, create, research, validate on a new design gestalt model via digital environment under of the financial umbrella of European Union's Horizon 2020 in order to transmit the innovative approach to the traditional dances, in a wide range of users (dance teachers professional dancers, creative industries and general public).

As consider the primary purpose of digging dipper into the meaning of the Terpsichore project is that we have managed with tools of the 3D virtual content, 3D modelling and reconstruction, 3D motion capturing technology, computer vision and learning, video capturing, virtual and augmented reality, computer graphics and data aggregation for metadata extraction in a low cost digitisation to aggregate the knowledge in combination with the technology to collect the needed data [12].

According to the European studies, it is proved the full range of internet users free of charge info in traditional topics sites. The difference of ICH content and the general representation is that it is presented in variable sectors such as education programmes, arts, media, tourism sector, science and leisure settings. Cultural Heritage is protected efficiently through the use of digital technology because it improves the digital era. More than less, the contribution of the traditional literature form is achieved through ICH to be modificative in a digital transformation type (database). So that, create enriched virtual surrogates as previous research have been proposed [6]. To improve all the above-mentioned significant accomplishments for a better understanding, presentation, protection and re-use of the Cultural Heritage (CH) we've been supported by the tool of the Digital Library of EUROPEANA. We have established re-use of ICH in several fields such as, traditional music, fashion, tradition and handcrafting items to analyse all the impact of the digitalization technology through multimedia metadata and ontologies using 3D modelling for a more accurate e-documentation.

Menier et al. in 2006, in order to identify human body motions from a 3D perspective, he has used 3D human skeletal models which have imitate the kinematic cycle of joints representing the human body posture. 3D modelling defined by three primary lines: a) a skeleton origin, b) separation of space into subspaces and c) mesh reconstruction - everything of this, based on computational geometry techniques [7]. The 3D typification for visualizing the modelled content, e.g. traditional dance, constitute for the computer science a big challenge. In order to handle the high-cost expense of the detailed animated characters, it is suggested to use methods based on high definition approach. The use of a textured polygon has improved the closest simulation to the human kinetic scheme, so that, the movements and the sense of the human body could be visualized in the most efficient way [8]. Despite the memory procedure, there is also another efficient method, which is related to the typification – projection. The animated character - avatar is categorized in particular parts and sub-parts of kinetical networks and statically fields and also false networks [9]. A three-dimensional recording method was described, which combined high-quality dynamic grid movement with the high performance provided by static meshes and impostors [10]. It is crucial to mention the use of a shorted mapping approach for the coding of details in every 3D animated model – avatar with minimum geometrical claims [11].

Terpsichore project (http://terpsichore-project.eu/), targets at integrating the latest innovative results of photogrammetry, semantic technologies, computer vision and time evolved modelling, along with traditional choreography and narrative. The study, design, research, education, analysis, implementation and validation of an innovative framework for accessible digitization, modelling, archiving, online preservation and presentation of ICH content related folk dances is reflected in the project. One of the significant outcomes of this project is anticipated to be a Web-based cultural server-viewer that will permit user's interaction, visualization, interfacing with existing cultural libraries (EUROPEANA) and enrichment functions to provide virtual surrogates and media application scenarios releasing the inevitable economical conflict of ICH. The final product will support a range of services such as virtual/augmented reality, interactive maps, social media interactions, presentation and learning of traditional European dances with a substantial impact on European society, culture and tourism [12].

In Greece, dance used to be one of the most amusing way for the communities. Everyone who has participated or has danced Greek dances in both cases enjoys the feeling that dance provoke. There are many different types of dances, approximately over than 10.000 all over Greece [13]. It is quite important to emphasize to the strong relation among the dancer and the music because of their origin-roots.

Regular physical activity is considered one of the most important factors in maintaining good health. Dance is considered an activity that involves coordinating movements combined with music. More specifically, is a procedure which activates main brain areas, because it is constantly necessary to learn and to remember new steps. Dancing as a motor-kinetic skill requires the coordination of body movements with rhythmic stimuli, developing adaptability of movement [14].

"Syrtos in three steps", is the most basic and widespread traditional dance in Greece. It is the dance that everyone first learned at an early age because it has a slow tempo and simple steps. Men and women participate and dance in a circle. The handle is from the palms with the hands up and the elbows bent. The rhythm is 3/4 and is completed in six musical meters. In Epirus, area in Greece, dance has been combined with specific popular songs such as "Paidia tis Samarinas", "Yianni mou to mantili sou" and "Kontoula Lemonia".

During the learning process of dance "Syrtos in three steps", dancers achieve steps of 2/4-time value in the rhythm of "Syrtos in three steps" music, is directed to the cycle. Then they accomplish 4/4 cross-steps to both directions, right and left. Specifically, for the cross-steps they do a lateral step on the right foot to the right side, 2/4 of time value. Continuously, they cross the left foot forward in front of and over the right foot, 1/4 of the time value (only men is allowed to lift up) and then they pressing the left foot on the left and they are crossing the right foot in front of and above the left foot, 1/4 of the time value (only men is allowed to lift up). When observing 3D motion capturing data macroscopically, an expertised in Greek traditional dances, observer is able to identify the gender of each dancer by "analysing" qualitative kinematic data. In contrast for informatics there is the need of quantitative data in order to identify the gender differences in the performance of the Greek traditional dance "Syrtos in three steps".

In a previous study, we have presented a new methodology using 3D Motion Capturing System and markers in order to improve the identification of traditional Greek dances based on the dance rhythm pattern analysis by calculating the number of frames between key-events [5]. Our main purpose in this work, is to analyse the specific dance "Syrtos in three steps", to present the differentiation of the gender among female and male dancers by using 3D Motion Capturing technology.

2 Methods

2.1 Participants

The recordings included four professional dancers of traditional Greek dances, two men and two women who were called to perform, one by one, the steps of "Syrtos in three" guided at first by counting the rhythm orally and then, by the proper music for each dance. All performances and captures took place in the Laboratory of Motor Behaviour and Adapted Physical Activity at the Department of Physical education and Sport Sciences at the Aristotle University of Thessaloniki. All four participants were chosen because of their expertise in the performance and didactics in Greek traditional dances and their previous experience in human motion capturing processes. The procedure lasted 3 days. The 3D motion capturing for each participant lasted one day, from 9 a.m. to 1 p.m.

2.2 Instruments

The place where the measurements were carried out was the Laboratory of Motor Behaviour and Adapted Physical Activity at the Department of Physical Education and Sport Sciences of the Aristotle University of Thessaloniki. We performed the Greek traditional dance capturing using the VICON system (Nexus Vicon, Oxford, UK). This system consists of 10 high precision and sampling camcorders to record human motion (Bonita 3, Nexus Vicon, Oxford, UK).

2.3 Procedure

We captured dancing trials from four different dancers performing several parts of traditional Greek dance “Syrtos in three”. Dancers performed and we captured 9 different trials of the traditional Greek dance “Syrtos in three” (one trial for each choreographic step and three trials for the whole cycle of all choreographic steps). The music which guided the steps of each dancer was the traditional Greek song “Kontoula lemonia” with the rhythm pattern of 4/4. After collecting anthropometric measurements of the dancers, we calibrated the Vicon system capturing area and started capturing. The capturing procedure lasted about an hour for each dancer. The capturing frequency of the Vicon system was set to 100 Hz. For the 3D video capture, we used the PIG (plug-in gait full body) reflective market placement using a total of 36 reflective markers placed on specific anatomical points of the body of dancers. Before capturing each dancer trials, we performed a static subject calibration for each dancer. After the 3D video capturing, we labelled all reflective markers based on the full-body plug-in gait model and exported data concerning, segments, joints and trajectories of each market. After the post capturing data processing, we exported a 3D format type file and a txt format type file containing all those variables for further analysis.

3 Results

For descriptive data analysis, we took into account the fourth and the sixth choreographic step of “Syrtos at three”. The maximum displacement of the toe reflective marker shows the differences between male and female dancer (Table 1). We found that the toes markers on both feet of men dancers had different maximum trajectories at the z-axis than markers displacement of female dancers at the fourth (4th) and the sixth (6th) step of this dance.

Table 1. A visual representation of the differences between the fourth and the sixth choreographic step of "Syrtos in three" dance among male and female dancers using 3D motion capture technology.

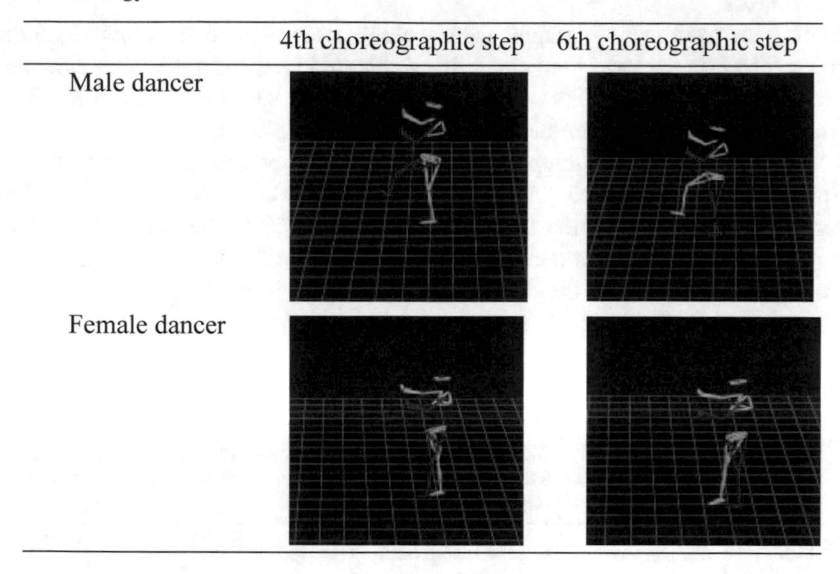

	4th choreographic step	6th choreographic step
Male dancer		
Female dancer		

More specifically, as regards toes marker trajectory on the left foot movement during the 4th choreographic step, male dancers showed a maximum displacement on the z-axis of 448 mm while female dancers showed a maximum displacement on the z-axis of 72 mm. Concerning toes marker trajectory on the right foot movement during the 6th choreographic step male dancers showed a maximum displacement on the z-axis of 441 mm while female dancers showed a maximum displacement on the z-axis of 71 mm (Fig. 1).

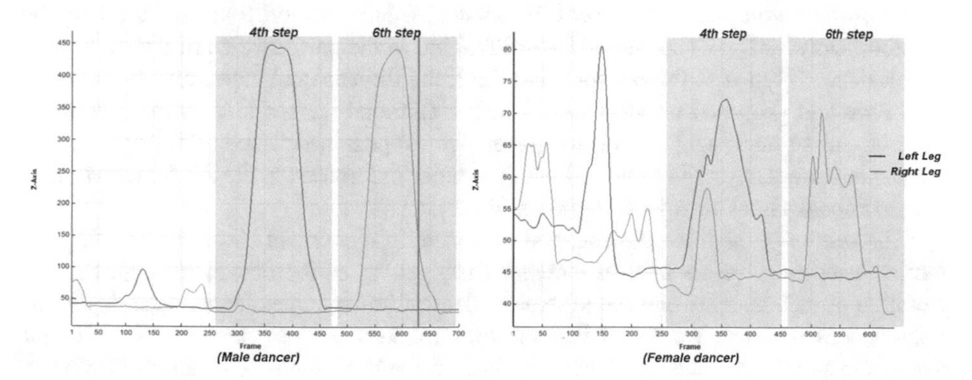

Fig. 1. Left and right toe reflective markers displacement on the z-axis among male and female dancers during the fourth and the sixth choreographic step

From a kinematic respective, descriptive statistics data intdicate that male dancers showed narrower angles on all lower limb joints (hip, knee and ankle) on both feet than female dancers.

During the fourth choreographic step, male dancers showed at means 59,25 on the hip joint, 50,97 on the knee joint and 1,46 on the ankle joint. On the other side, female dancers, during the fourth choreographic step showed 18,06 on the hip joint, 2,45 on the knee joint and −19,17 on the ankle joint (ankle extension).

During the sixth choreographic step, male dancers showed at means 58,03 on the hip joint, 49,39 on the knee joint and 2,22 on the ankle joint. In contrast, female dancers, during the sixth choreographic step showed 11,59 on the hip joint, 2,50 on the knee joint and −35,21on the ankle joint (ankle extension). Data concerning the joint angles of the lower limb of the dancer are presented in Table 2.

Table 2. Kinematic data (joint angles) concerning the mean joint angles of each lower limb among male and female dancers during the fourth and the sixth choreographic step.

Maximum joint angles at the x-axis	4th choreographic step		6th choreographic step	
	Male dancers	Female dancers	Male dancers	Female dancers
Right hip angle	–	–	58,03°	11,59°
Right knee angle	–	–	49,39°	2,50°
Right ankle angle	–	–	2,22°	−35,21°
Left hip angle	59,25°	18,06°	–	–
Left knee angle	50,97°	2,45	–	–
Left ankle angle	1,46°	−19,17°	–	–

4 Discussion

In the present work, the Department of Physical Education and Sport Sciences of the Aristotle University of Thessaloniki was involved in the performance of the traditional Greek dance “Syrtos at three steps” and the three-dimensional capturing. In previous work, we had proposed a method for Greek traditional dances identification through rhythm pattern analyses [15]. In this paper, we propose a new method, based on 3D motion capture analyses in order to identify gender differences in the implementation of the traditional Greek dance of “Syrtos at three steps”.

The results of the three-dimensional capturing of movement showed that using 3D motion capturing we are able to recognize the gender of the dancer through the trajectories of toes markers and the kinematic data of joints angles. More specifically, as regards toes markers, we found that the toes markers of men dancers had a longer displacement at the z-axis than markers displacement of female dancers at the fourth and the sixth step of this dance. Regarding kinematic data, we found differences at the means of angles of the hip, the knee and the ankle joints. Male dancers showed narrower angles in all mentioned joints of all six steps than the angles of female dancers during the implementation of the Greek traditional dance “Syrtos in three steps”.

The findings of the present work, in addition to our previous work [15] offers new methodologies using 3Dmotion capturing technology to collect quantitative data in order to improve algorithms for the archiving and the categorisation of traditional Greek dances.

5 Conclusion

Our findings indicate that the proposed method is capable in order to identify the gender of performing dancers through 3D motion kinematic data, which is an essential part for digitising and archiving of traditional Greek dances. Finally, the findings of our work were distributed to the program partners in order to further analyse them for the fulfilment of the project objectives. With the completion of the Terpsichore project, a set of services will be created including virtual and augmented reality applications, interactive maps, presentation and learning of traditional European dances with a significant impact on European society culture and tourism.

Acknowledgements. This work has been supported by the H2020-MSCARISE project "Transforming Intangible Folkloric Performing Arts into Tangible Choreographic Digital Objects (Terpsichore)" funded by the European Commission under grant agreement no 691218. The authors would like to help all partners for their contribution and collaboration.

References

1. Sullivan, A.M.: Cultural heritage & new media: a future for the past. John Marshall Rev. Intellect. Property Law **15**, 604–646 (2016)
2. https://ich.unesco.org (assessed on 11/11/2019)
3. Kyriakaki, G., et al.: 4D reconstruction of tangible cultural heritage objects from web-retrieved images. Int. J. Herit. Digit. Era **3**, 431–452 (2014). https://doi.org/10.1260/2047-4970.3.2.431
4. Dimitropoulos, K., Barmpoutis, P., Kitsikidis, A., Grammalidis, N.: Extracting dynamics from multidimensional time-evolving data using a bag of higher-order Linear Dynamical Systems. In: 11th International Conference on Computer Vision Theory and Applications, VISAPP 2016, Rome, Italy (2016)
5. Linaza, M., Moran, K., O'Connor, N.E.: Traditional sports and games: a new opportunity for personalized access to cultural heritage. In: 6th International Workshop on Personalized Access to Cultural Heritage, PATCH 2013, Rome, Italy (2013)
6. Li, R., Luo, T., Zha, H.: 3D digitization and its applications in cultural heritage. In: Ioannides, M., Fellner, D., Georgopoulos, A., Hadjimitsis, D.G. (eds.) EuroMed 2010. LNCS, vol. 6436, pp. 381–388. Springer, Heidelberg (2010). https://doi.org/10.1007/978-3-642-16873-4_29
7. Menier, C., Boyer, E., Raffin, B.: 3D skeleton-based body pose recovery. In: Proceedings of the 3rd International Symposium on 3D Data Processing, Visualization, and Transmission, 3DPVT'06, Washington, DC, USA, 2006, pp. 389–396 (2006)
8. Tecchia, F., Loscos, C., Chrysanthou, Y.: Image-based crowd rendering. IEEE Comput. Graph. Appl. **22**, 36–43 (2002)

9. Kavan, L., Dobbyn, S., Collins, S., Zára, J., O'Sullivan, C.: Polypostors: 2D polygonal impostors for 3D crowds. In: Proceedings of the 2008 Symposium on Interactive 3D Graphics and Games, New York, NY, USA, pp. 149–155 (2008)
10. Pettré, J., Ciechomski, P., Maïm, J., Yersin, B., Laumond, J., Thalmann, D.: Real-time navigating crowds: scalable simulation and rendering. Comput. Anim. Virtual Worlds **17**, 445–455 (2006)
11. Andújar, C., et al.: Omni-directional relief impostors. Comput. Graph. Forum **26**, 553–560 (2007). https://doi.org/10.1111/j.1467-8659.2007.01078.x
12. Doulamis, A., Voulodimos, A., Doulamis, N., Soile, S., Lampropoulos, A.: Transforming Intangible Folkloric Performing Arts into Tangible Choreographic Digital Objects: The Terpsichore Approach, pp. 451–460 (2017). https://doi.org/10.5220/0006347304510460
13. https://ithaca-culture.co.uk/en. Assessed 30 Sep 2019
14. Douka, S., Zilidou, V.I., Lilou, O., Manou, V.: Traditional dance improves the physical fitness and well-being of the elderly. Front. Aging Neurosci. **11**, 75 (2019). https://doi.org/10.3389/fnagi.2019.00075
15. Ziagkas, E., et al.: Greek Traditional Dances 3D Motion Capturing and a Proposed Method for Identification through Rhythm Pattern Analyses (Terpsichore Project) (2019, in press)

Comparing Virtual Reality SDK Potentials for Engineering Education

Pascalis Trentsios, Mario Wolf(✉), and Detlef Gerhard

Digital Engineering, Ruhr-University, Bochum, Germany
{pascalis.trentsios,mario.wolf,detlef.gerhard}@rub.de

Abstract. The paper at hand aims to help educators to grasp the potential and difficulties related to virtual reality hardware and software development kits when freshly starting off in the mixed reality realm. Therefore, it will present and compare major VR hardware devices and SDK's potentials while considering individual features and advantages. As a common ground to judge the capabilities, a virtual test environment is created featuring common interactions needed in engineering education scenarios when interacting with virtual representation of technical equipment. The final verdict contains two main parts, one of which is the actual judgement of current VR SDK capabilities. The second part is a taxonomy with an overview over selection criteria and solution spaces when dealing with hardware choices.

Keywords: Virtual reality · Virtual training · Software development · Engineering education

1 Introduction

As the process of digitalization continues, more and more educators try to establish offerings with visualization in Mixed Reality (MR), for the most part in either Virtual Reality (VR) or Augmented Reality (AR). The market for VR applications and hardware is mostly driven by the consumer-oriented gaming industry. Each major manufacturer of VR hardware offers individual software development kits (SDK) for several major game engines to create individual VR applications. In the context of education, typical use cases are serious games or scenario-based learning experiences [1]. Combined with the multitude of available hardware devices, educators need to compare and evaluate a lot of specific capabilities to find a suitable match of VR hardware and SDK for the educational content they want to provide. The use of VR educational applications can be beneficial to motivate high school graduates and engineering students [2].

The motivation for the paper at hand is to share year-long experience with the complex world of VR development in a structured fashion for the benefit of the engineering education community. While the authors do not wish to create obligatory best practices for every option available, the created taxonomy, criteria analysis and the reflection based on completed MR projects may help current and future researcher on their way into the Virtual Reality.

M. E. Auer and D. May (Eds.): REV 2020, AISC 1231, pp. 375–392, 2021.
https://doi.org/10.1007/978-3-030-52575-0_32

MR applications in the context of engineering education can be very heterogeneous, yet there are common basic functionalities that make use of native SDK functions. Typical examples are interactions with objects in the virtual world such as pushing or grabbing, interface interactions or more specific interactions like pulling a lever (linear motion) or opening or closing a valve via a valve wheel (rotational motion).

2 Related and Previous Work

2.1 The Reality-Virtuality-Continuum

Virtual Reality can be described as a simulated, virtual environment that can mimic the real world [3] and offer its user full immersion [4]. When the reality is augmented with virtual components the resulting visualization is called Augmented Reality (AR). Respectively, when Virtual Reality is augmented with real components the resulting hybrid is called Augmented Virtuality (AV). An absolute determination between the hybrids in the intersection between AR and AV is not always possible, therefore this field is also referred to as Mixed Reality (MR). Figure 1 shows the Reality-Virtuality Continuum as presented by Milgram [5]. VR systems such as head mounted displays (HMD) are used to experience VR environments. Since HMD cover the user's actual sight with the view into the virtual environment, simultaneously the HMD transfers the user's movement into the VR environment. A HMD offers a greater immersion than a Cave Automatic Virtual Environment (CAVE) or a Powerwall [6].

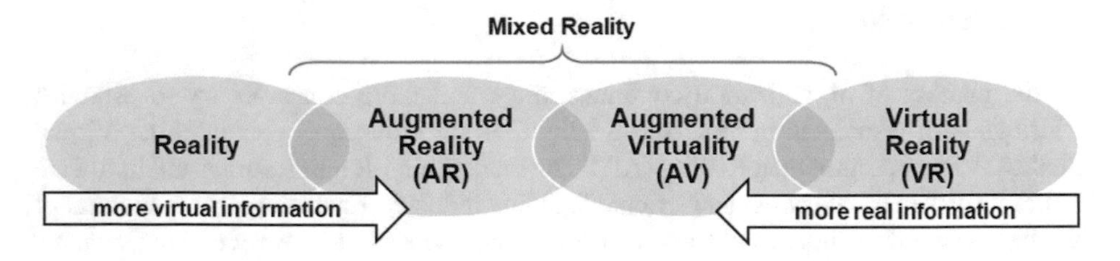

Fig. 1. Reality-Virtuality-Continuum (cp. Milgram [1])

2.2 User Tracking for VR

As expressed before the HMD transfers the user's movement into the VR environment. Movement can be described with six degrees of freedom (DoF), three of which are translational and three rotational. These degrees of freedom must be tracked by the HMD, as well as by other input devices such as hand-hold controllers. The two major tracking methods that have become established are the Outside-In (Fig. 2) and the Inside-Out (Fig. 3) tracking.

The Outside-In tracking method uses fixed base stations to track the HMD and the controllers. It is therefore not mobile at must be kept calibrated in the specific environment before use.

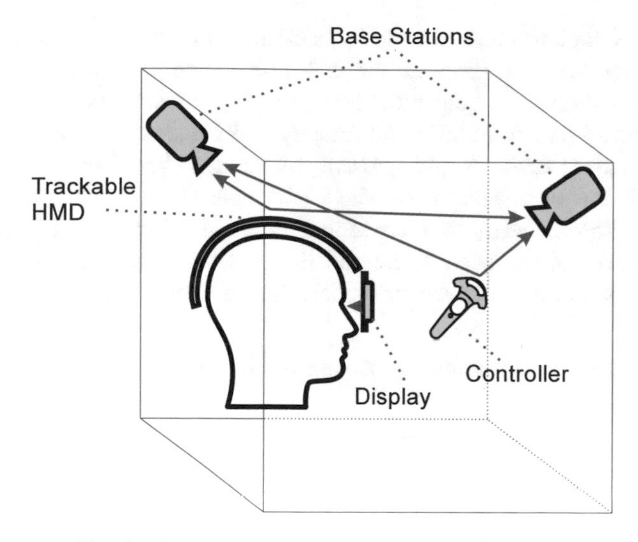

Fig. 2. Outside-In tracking method (cp. Wolf [7])

The Inside-Out tracking method uses the HMD itself to scan the surroundings and derives the user's position and movement based on the spatial movement to the surroundings. The controller's positions are tracked by the HMD relatively to the HMD's position.

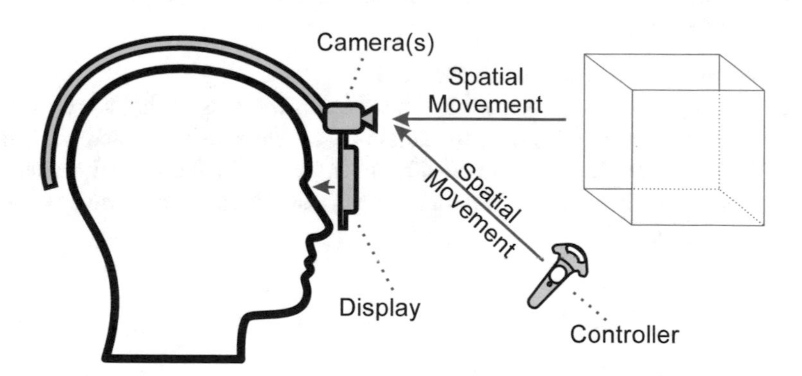

Fig. 3. Inside-out tracking (cp. Wolf [7])

2.3 General Problems When Dealing with VR

Motion sickness [8] can occur while using a VR system through different causes. One cause can be the display resolution and the so-called screen-door effect (SDE). SDE means that individual pixels are visible. The SDE appears when the display resolution is relatively low and the focal optics that are used to stretch and increase the field of view, magnify the screen. Higher resolution or lens optimizations can decrease the

SDE. Another common cause for motion sickness is a too low framerate, caused by either insufficient optimization of the VR application or weak hardware. Lower framerates are perceived as unnatural stuttering of the vision and can therefore lead to motion sickness. Apart from technical reasons, motion sickness can also appear from the displayed virtual scene. Rapid (passive) movement should be avoided.

In general, the user acceptance should be considered while building a virtual environment. Which means that ergonomics should be considered as well as the emotional comfort of the user. Therefore, the presence of avatar models, as well as movement animation of those avatar models can be beneficial.

2.4 Background Information on Movement in VR

Tracking-Based (Roomscale) Movement
The tracking-based movement tracks the actual movement of the user and synchs it with the virtual scenery. It is one of the most comfortable movement methods [9], since the movement is processed by real life movement. The movement distance is constrained to the tracking size of the VR systems. For a spacious VR scene an equal sized real-life space would be needed to reach every location, if only a tracking-based movement is used.

Teleport/Teleportation-Based Movement
Teleport is a technique to move the user in a virtual scenery. The teleport instantly changes the user's position without any movement in between the start point and the point of teleportation. Thus, the teleport doesn't interfere in a negative way with the user's sense of equilibrium.

Controller-Based (Input-Based) Movement
Controller- or input-based movement is considered a movement that starts if the user presses a certain button or joystick to initiate an accelerated movement in a desired location. The users don't move in real life but is moved in the virtual scenery. This movement technique can lead to motion sickness since it can cause conflicts with the user's sense of equilibrium.

3 Approach

The first step to evaluate all relevant SDK/hardware combinations is to create an instrument based on expected behavior in an engineering education environment. This instrument should have the form of an actual VR environment with several different virtual test stations and be able to test a multitude of abstract actions for individual evaluation. The test stations are to be based on a catalogue of crucial functionalities (Subsect. 3.2), which was designed based on the authors' past projects. The SDK/hardware combinations are evaluated based on the ability to facilitate the abstract functions needed in different engineering education scenarios.

The second step is to implement the VR environment in the preferred 3D gaming engine and to import all available VR SDKs. As those SDKs are mutually exclusive for the most part, the created VR environment will be copied to each individual test instance. Due to the extensive experience with the 3D engine, the virtual environment was developed in Unity, which is the most widely used 3D engine on the market, and then copied as a new project to ensure that each implementation of each SDK had the same base and was conflict-free with other software components.

The third step is to test the applicability of all relevant combinations. To ensure this compatibility test will benefit less programming inclined educators, the authors will focus on "mostly out of the box"-functionalities and offer a simple three-step scale to judge whether a certain functionality is obtainable with no to little coding, extensive coding or not at all. The final step is to formulate comprehensible advice, lessons learned and best practices for the creation of VR applications in engineering scenarios.

3.1 Current Major VR SDKs

This sub-chapter addresses current major VR SDKs, their supported hardware devices, the possibility to add avatar models, the availability of a VR simulator that can emulate VR devices with keyboard and mouse inputs for debugging purposes, included example scenes covering the SDK's functions, and miscellaneous special features. Some VR system manufacturers provide proprietary VR SDKs that only support their own VR systems. However, some manufacturers offer additional functionality and open up their SDKs even for third-party VR hardware systems.

It should be noted that as of the time of writing, Google announced that they will no longer support their VR systems, therefore this paper will not address Google's Daydream platform.

SteamVR Plugin

The SteamVR SDK supports a multitude of different devices, like Vive Cosmos, Valve Index, Vive and Vive Pro, as well as other VR systems like the Oculus Rift S, Quest and the Windows MR headsets. While SteamVR works without the need of major adaptations, developers should still consider the different hardware specifications of the mentioned VR systems.

SteamVR includes different 3D-hand-models, that are rigged and animated to mimic the way the user interacts with the controller buttons. In combination with the index controllers, continuous animation of individual finger movement is provided.

A VR simulator is not implemented, a physical VR system needs to be connected for testing and debugging.

The SteamVR Plugin offers an extensive example scene and multiple smaller scenes, in which every major function of SteamVR can be explored.

As of writing the paper at hand, the current version of SteamVR is 2.5.0 (SDK 1.8.19) [10].

Oculus Integration

The Oculus Integration SDK supports the VR systems Oculus Rift S, Oculus Quest and Oculus GO, developers should still consider the different hardware specifications of the mentioned VR systems.

The Oculus Integration SDK offers multiple avatar models. It provides 3D avatar head, body and hand models. All the avatar models are customizable. The 3D–hand–

models will be animated according to the way the controller buttons are touched and pressed by the user. The lip and mouth movement of the user is simulated by the avatar according to microphone input. Additionally, there are some features that aim to improve the natural behavior of the avatar, such as eye movement, eye blink modelling and expression modelling.

A VR simulator is not implemented, a physical VR system needs to be connected for testing and debugging.

The Oculus Integration offers multiple minimalistic example scenes, in which some functions of the SDK can be tested.

As of writing the paper at hand, the current version of Oculus Integration is 1.41 [11].

Microsoft Mixed Reality Toolkit (MRTK)

The Windows Mixed Reality Toolkit supports AR devices such as the HoloLens and the HoloLens 2 as well as the immersive VR systems such as the Windows Mixed Reality headsets. There are multiple Windows MR headsets distributed by different manufacturers. They have mostly the same hardware specifications and are functionally identical.

The MRTK provides a VR simulator with extensive functionalities that simulates a VR system with keyboard and mouse inputs. Developers can thereby use the VR simulator to test functions in their scene without the need to connect a real VR system.

The MRTK provides avatar hand models, that are animated according to controller inputs by the user.

The SDK provides text-to-speech (TTS) and speech-to-text (speech-recognition/SR) [12].

As of writing the paper at hand, the current version of Microsoft Mixed Reality Toolkit is v2.1.0 [13].

Virtual Reality Tool Kit (VRTK)

VRTK supports all of the mentioned VR devices as a middleware between the proprietary hardware and the generic game engine. Therefore, any virtual scene built with the VRTK can be deployed to various VR devices without making adaptations to the project.

The VRTK provides a VR simulator with extensive functionalities that simulates a VR system with keyboard and mouse inputs. Developers can thereby use the VR simulator to test functions in their scene without the need to connect a real VR system.

The VRTK offers an extensively example scene and multiple smaller example scenes, in which every major function of the VRTK can be tested.

VRTK is an open source middleware project that supports development with all major VR SDK like the SteamVR Plugin, the Oculus Integration and the Microsoft Mixed Reality Toolkit (MRTK) SDKs. Therefore, updates of those SDKs can lead to problems with the middleware VRTK.

As of writing the paper at hand, the current version of the VRTK 3.3.0 with version 4 in beta status. The version 4 beta provides an interaction system based on an event system, which is completely different to the one in version 3.3.0. It is therefore likely that the VRTK will be completely changed once version 4 will be released officially [14].

3.2 Criteria Catalogue for the VR SDK Testing

The criteria catalogue includes crucial functionalities needed in VR engineering education environments, which must be provided by each VR SDK/hardware combination in order to accomplish movement, interaction, user comfort and user guidance.

Movement

- Tracking-based (roomscale) movement
- Teleportation-based movement
- Continuous input-based movement

Interaction

- Interaction with virtual 3D-objects
- Interaction with a UI-Elements
- Interaction through speech recognition

User Comfort

- Motion sickness prevention
- Immersion through feel of presence
- Ergonomic scene

User Guidance

- Speech commands directed at the user
- Visual commands directed at the user
- Highlighting

4 Testing VR SDKs for Engineering Education

4.1 Hardware Setup

The VR system presented in the following Tables 1 and 2 were used to test the VR SDKs. The VR systems are categorized by tracking technique which can be divided into Outside-In and Inside-Out tracking (cp. Sect. 2). While the testing was done with a representative VR device for the regarding SDK (Vive Devices used to test SteamVR, Oculus Devices used to test Oculus Integration and so on, the third-party SDK VRTK was tested using Vive devices as well as Oculus devices)

Outside-In Tracking VR Systems
VR systems using this tracking technique offer a high tracking quality and stability. The use of base stations and a room setup are mandatory. Room interior and light don't considerably influence the tracking stability. From the VR systems the authors used in this paper only the Steam devices use the Outside-In tracking technique. All these devices, the HMD's as well as the controllers, use the same outside in tracking-technique. The tracking is based on infrared laser triangulation of multiple infrared sensors that are integrated in the devices. Additionally, IMU (inertial measurement unit) are used. The trackable space is limited by the room setup. With the SteamVR 2.0

Base Station stations a trackable space of 5 m × 5 m, 25 m^2 is supported, when two Lighthouse base stations are in use. Using four Lighthouse base-stations increases the room scale to 10 m × 10 m, 100 m^2. There is a possibility to add multiple trackable devices, using the Vive Trackers. Self–made controllers can be added, or a full body tracking can be realized [15]. Since no Cameras are used for the tracking, these VR systems can be used in places with a strict privacy policy. A comparison of Outside-In tracking VR systems is displayed in Table 1.

Table 1. Outside-In tracking VR systems comparison

Specifications	HTC vive	HTC vive pro	Valve index
Display resolution (per eye)	1080 × 1200 pixel	1440 × 1600 pixel	1440 × 1600 pixel
Display resolution (total)	2160 × 1200 pixel	2880 × 1600 pixel	2880 × 1600 pixel
Framerate	90 Hz	90 Hz	80/90/120/144 Hz
Field of View	110°	110°	110°, mechanically adjustable
Audio	headphones not integrated, integrated 3,5 mm headphone jack	integrated headphones, headphone jack not integrated	integrated headphones, integrated 3,5 mm headphone jack
Degrees of Freedom	Six (three rotational + three translational)	Six (three rotational + three translational)	Six (three rotational + three translational)
Costs	Ca. 600 $ complete VR-System with SteamVR 1.0 Base Station, HMD and two Vive controllers	Ca. 1.399 $ complete VR-System with SteamVR 2.0 Base Station, HMD and Vive two controllers	Ca. 1100 $ complete VR-System with SteamVR 2.0 Base Station, HMD and two Index controllers
Features	USB 3.0 port		USB 3.0 port and free space for modifications

Inside-Out Tracking VR Systems

The Oculus Quest, the Oculus Rift S and the Windows MR headset use a camera–based tracking method. Cameras observe the surroundings and calculate the users position based on movement-based changes. The Controllers however do not use this technique. They are tracked by the headset via camera recognition additionally they use IMU. The relative tracking of these HMDs and controllers is good and precious. However, problems that the authors often experienced with the camera–based tracking is imprecise repeatability of the room tracking. Especially when it comes to the tracking of the ground position, which tends to shift. Since this tracking method is based on

Table 2. Inside-Out tracking VR systems comparison

Specifications	Oculus rift S	Oculus quest	Windows MR headset	Vive cosmos
Display resolution (per eye)	1280 × 1440 pixel	1440 × 1600 pixel	1440 × 1600 pixel	1440 × 1700 pixel
Display resolution (total)	2560 × 1440 pixel	2880 × 1600 pixel	2880 × 1600 pixel	2880 × 1700 pixel
Framerate	80 Hz	72 Hz	60 Hz, 90 Hz	90 Hz
Field of View	Undisclosed	Undisclosed	110°	110°
Audio	integrated headphones, integrated 3,5 mm headphone jack	integrated headphones, integrated 3,5 mm headphone jack	integrated headphones, integrated 3,5 mm headphone jack	integrated headphones, integrated 3,5 mm headphone jack
Degrees of Freedom	Six (three rotational + three translational)	Six (three rotational + three translational)	Six (three rotational + three translational)	Six (three rotational + three translational)
Costs	Ca. 450 $ complete VR-System with HMD and two controllers	Ca. 450 $ complete VR-System with HMD and two controllers	Ca. 299 $ complete VR-System with HMD and two controllers	Ca. 699 $ complete VR-System with HMD and two controllers

cameras it is highly dependable on the surroundings. Characteristic features in those surroundings enable the tracking. A plain monochrome room with no interior and bad lighting is highly inconvenient for this tracking method.

While the Rift S and Windows MR headset use the Hardware of the PC on which they are running on, the limitations of the provided content depend on the connected PC's hardware. The Oculus Quest is a standalone VR headset, meaning that it runs applications autonomously without a connected PC. A comparison of Inside-Out tracking VR systems is displayed in Table 2.

Miscellaneous

The most simplistic VR Headsets, like Google Cardboard, are based on inserting a smartphone into a "non-smart" carrier to use the IMU of said smartphone for purely rotational tracking. Samsung's GearVR is mechanically similar but offers an external IMU with improved performance. Typically, if there is a controller available in this kind of VR device, the controller will also be tracked rotationally.

4.2 Virtual Test Scene Setup

The purpose of the Virtual Test Scene Setup is to establish a tool, with which the authors can judge VR SDK/hardware combinations. The authors therefore created a virtual, engineering themed environment inside the Unity engine [16], that consist of a workshop with workbenches covering different tools, multiple storage racks, hydraulic systems and a floating UI. The Virtual Test Scene Setup is shown inside the Unity engine from an above view in Fig. 4.

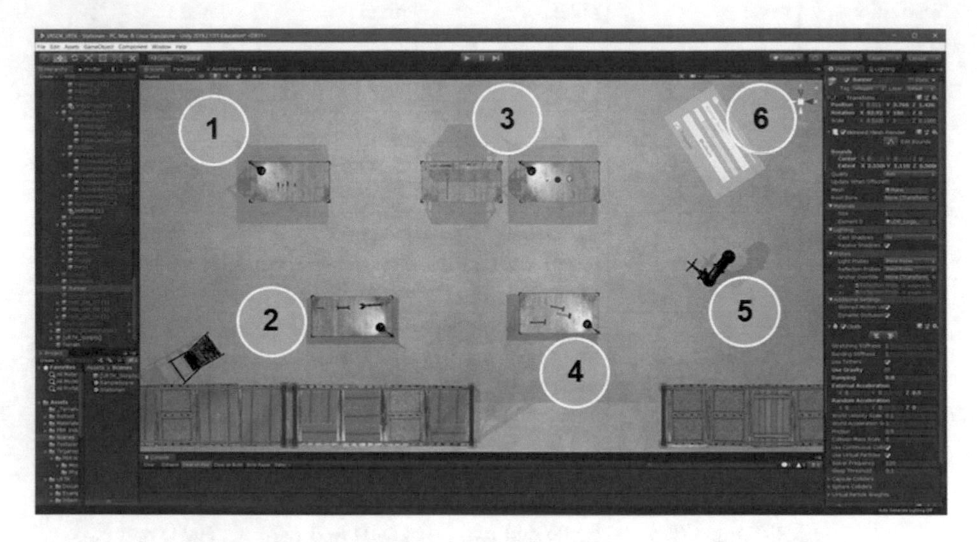

Fig. 4. Screenshot of the virtual test scene setup (with position numbers)

The Virtual Test Scene Setup contains multiple fictional tasks, that are based on real-world use cases needed in the context of engineering education and represent multiple combinations of functions mentioned in the criteria catalogue from Subsect. 3.2. These real-world engineering use cases are extrapolated into abstract tasks, with which the authors can evaluate the challenges for the individual SDKs. Even though the whole setup acts as the testing tool, there are special testing stations that cover specific testing purposes. These stations are described in Table 3, their position in the Virtual Test Scene Setup is marked in Fig. 4.

One of the testing stations is shown in Fig. 5, where the user must pick a highlighted cardboard box (1) out of a storage rack. The box however is out of the user's range and to reach it the user must climb a ladder (2). The SDKs must provide a way to climb the ladder, so that the box can be reached by the user. The box must be maintain picked by the user when returning to the former position in the workshop.

Another testing station is shown in Fig. 6, where the user must pick a rasp (1) laying on the workbench and place it on a tool wall (2). The SDKs must provide the ability to pick the rasp and to locate the desired point. Preferable the desired point should be highlighted.

Table 3. Virtual test station descriptions

Nr.	Real-world example	Abstract task	Challenge
1	Using the right size and type of a tool to manipulate an object. Specific screwdriver for each screw type	Use a specific object to manipulate another object. lock-and-key model	The SDK must match individual virtual objects with each other
2	The workshop is too spacy to be covered merely using roomscale tracking. Additionally, there are height differences that must be bridged	Move to a position that is not reachable with roomscale tracking	The SDK must provide a way to reach desired positions in the scene, while maintaining the user's comfort
3	Some objects must be put in certain, fixed places. A bottle must be closed with the fitting tap. A rasp must be hung on a tool wall	Pick an object and snap it to a certain position	The SDK must provide an interaction between the handheld controllers and the virtual object. The SDK must be able to recognize or detect when the object is close to the snap position. Furthermore, it must detect when the object is released in the snap position at should then place it recedingly
4	Making an object fit or looking at it with magnification	Manipulate a picked object. Temporarily re–scale an object	The SDKs must provide an option to manipulate the geometry of a virtual object
5	Real World physics. Turning a valve Wheel or pulling a lever. Objects should keep their momentum when released	Emulate physical properties, like friction, dampening, gravity etc. Combine objects with hinges and joints. Constrain certain degrees of freedom	Unity provides physical simulation of virtual objects. However, thc SDK must be able to use this option and to co-function with Unity's build in physics simulator
6	UI panel that contain multiple UI elements. Each controlling different things in a different manner	Manipulate various UI elements like sliders, buttons, triggers, etc.	While interacting with an UI the SDKs must be able to interpret which UI element the user is referring to. Since UI elements can be touched directly, or pointed at with the handheld controllers, or the HMD itself. Speech commands are also possible

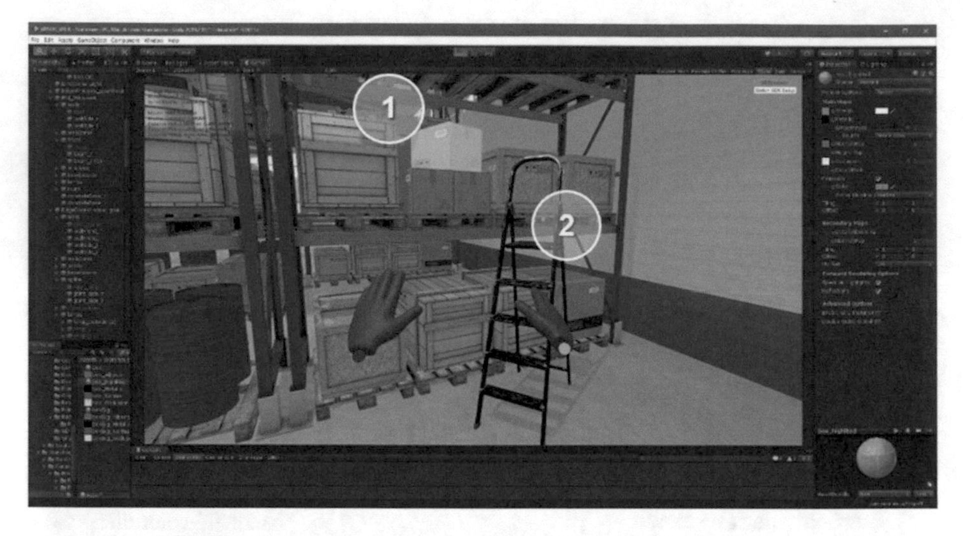

Fig. 5. Station to move to a specific position to pick an object

Fig. 6. Station to place an object to a certain position

4.3 VR SDK Comparison Results

Every VR SDK combination was tested with the Virtual Test Scene Setup and compared with each other. The comparison of the VR SDKs is summed up in Table 4. In total 24 individual features were tested. The VR SDKs were rated based on those features. There are three different rating types for each functionality: out of the box included (+++), included but minimal coding/effort is necessary (++), included but major coding/effort is required (+).

Table 4. VR SDK comparison

No.	Feature	SteamVR	Oculus integration	MRTK	VRTK
1	Teleport to a free point	+++	+++	++	+++
2	Teleport to a set point	+++	+	+	+++
3	Pick an object	+++	+++	+++	+++
4	Snap an object to a certain place	+	+	+	+++
5	Move an object (translation)	+++	+++	+++	+++
6	Move an object (rotation)	+++	+++	+++	+++
7	Constrain object movement	+	+	+	+++
8	Manipulate a picked object	+++	+++	+++	+++
9	Move the player (translation)	+++	+++	+++	+++
10	Move the player (rotation)	+++	+++	+++	+++
11	Possibility to add trackable devices	+++	+	Spatial awareness	++
12	Number of trackable devices	Two controllers. Additional vive trackers	Two controllers	Two controllers. Hands with HoloLens 2	Depends on the used VR system
13	Bungee-motion “fishing rod”	+	+	+++	+
14	Motion sickness prevention	+	+	+	+++
15	Avatar model:	++	+++	++	++
16	Avatar head	+	+++	+	+
17	Avatar Mouth/Lips	+	+++	+	+

(*continued*)

Table 4. (*continued*)

No.	Feature	SteamVR	Oculus integration	MRTK	VRTK
18	Avatar hands	+++	+++	+++	+++
19	Avatar arms	+	+	+	+
20	Avatar body	+	+++	+	+
21	Avatar legs	+	+	+	+
22	Interactable UI elements	+++	+++	+++	Unity UI/3D UI elements
23	Text to speech (TTS)	+	+	+++	+
24	Speech to text (SR)	+	+	+++	+

5 Conclusion

5.1 Lessons Learned

Before conducting the study at hand, the usual recipe for a VR implementation at the authors chair was to use Unity as a game engine, the SteamVR SDK on HTC Vive with the accompanying controllers. The obvious downsides of this VR system are the high costs of a setup that is immobile and therefore inflexible, with a necessary space of around 25 m^2. The upsides are high quality tracking with a calibrated, fixed ground height and the possibility to add trackable objects, which makes it the sole out–of–the–box solution for augmented virtuality applications.

The authors now tend to use Oculus Rift S and Quest with the Oculus SDK whenever a more flexible, easier to transport and less prone to fail/recalibration from stuff/staff between headset and base station solution is needed. The combination also offers simultaneous development of stationary and mobile platform applications.

The VR system choice remains a difficult task for each educator. Due to the very short lifecycles of consumer hardware and the ever-rising demand in computing power for improved graphics, a singular investment won't last forever. The authors generally expect a span of two years for their VR equipment to count as "new" and around four years as "acceptable still". As opulent graphics are not at the core of education applications, middle-class gaming PCs should outlast two generations of VR headsets.

SteamVR

The Vive devices in combination with SteamVR should be used by users that have high needs when it comes to tracking precision and robustness. Since controllers that are tracked by cameras are more prone to be covert by the user and thereby inhibit the tracking of the controller. SteamVR is also recommendable for users that need to track custom objects. For this purpose, the Vive trackers can be used, they are sold separately to the VR systems. Besides the tracking, the Vive tracker can also provide any input (button presses) and outputs (vibration) of which the Vive controllers are also capable of. The major downside of the Vive devices is the pricing, since they are by far the costly devices compared to VR systems from other manufacturers.

The Vive Cosmos is a soon to be hybrid device, using camera based Inside-Out tracking as well as Valves lighthouse-based Outside-In tracking.

MRTK
Inexpensive devices compared to other manufacturers. The MRTK offers some interesting features, like the TTS and SR functions, that other SDK do not yet have. MRTK should also be considered by developers that want to work with the soon to be released HoloLens 2, since it supports the HoloLens as well.

Oculus Integration
Oculus offers one of the best cost effectiveness with their devices. some interesting features, like the extensively customizable Avatars, that other SDK do not yet have. Oculus announced a PC-connected-mode via USB C connection for the Oculus Quest, which is currently a standalone VR system. This would enhance the possibilities of the Oculus Quest.

VRTK
The Virtual Reality Tool Kit offers functionalities when it comes to interaction with the virtual scene. While the other SDK still offer the same core functionalities the VRTK achieves those with various technical approaches. VRTK provides plenty of in scene mechanics between individual objects. For developers that need certain object mechanics but don't want to implement them themselves, VRTK should be considered. It also provides a physical 3D-UI. Thus, making the VRTK a substantial SDK. Furthermore, VRTK isn't bound to a specific hardware.

5.2 Resulting Selection Criteria Taxonomy

The authors summed up all the crucial criteria and functions in the taxonomy found in Table 5. The main criteria the authors formulated are costs, visual fidelity, audio fidelity, tracking, room requirements, expandability. Each criterion comes with individual characteristics going from low, to mid, to high class. A developer new to VR should consider what criteria is needed for the project. Regarding the consideration, the developer can together with a VR expert choose the fitting VR system and VR SDK combination. Even though the SDK's presented offer a huge variety of functionalities, developers should consider to manually code specific functions to fulfill their individual needs.

As a result of the in-depth analysis, the afore mentioned focus on a single implementation strategy now shifts based on the use case. Each SDK/hardware combination has different strengths and weaknesses, but one must account for the complexness of implementing a previously unknown SDK.

Specific Example for Using the Taxonomy
The usage of the developed taxonomy becomes clearer when looking at an example. The authors plan to introduce VR hardware on a broader scale in engineering education, specifically in the realm of 3D computer-aided design and the further use of the generated 3D models. The criteria characteristics for this use case are stated in Table 6.

Table 5. Selection criteria with solution spaces

Criteria	Characteristics		
Costs	Low	Mid	High
Visual fidelity	Low	Mid	High
Audio fidelity	No	Stereo	Spatial
Tracking	Rotational	Rotational + Translational	Rotational + Translational + Roomscale
Room requirements	Low	Mid	High
Expandability	Software only		Additional hardware

Table 6. Criteria selection for the given example

Criteria	Characteristics		
Costs	***Low***	Mid	High
Visual fidelity	Low	***Mid***	High
Audio fidelity	***No***	Stereo	Spatial
Tracking	Rotational	***Rotational + Translational***	Rotational + Translational + Roomscale
Room requirements	***Low***	Mid	High
Expandability	***Software only***		Additional hardware

Low costs, reasonable visual fidelity and good tracking capabilities with minimal space requirements are the criteria for the authors' previously described use case.

The result of filling out the taxonomy can be reflected with the presented findings, which will inevitably need updating every few hardware or software release cycles (at best once per year). The taxonomy itself will only need updating once major new possibilities are presented by hardware and software manufacturers.

A suitable VR system for this use case would be an Oculus Quest in combination with the Oculus Integration SDK. For a low-cost criteria characteristic, the Oculus Quest is a good fit since it is reasonable priced (Table 2) and as a standalone VR system no additional PC is needed. For example, an Oculus Rift S or a Windows MR Headset, which are similar to the Oculus Quest, would still need a connected PC which would raise the costs by roughly 1000-1500€. The visual fidelity of the Oculus Quest is limited by the built–in hardware of the HMD, which makes it a fitting option for mid–range visual fidelity applications. No significant audio fidelity is needed for the use case, however even if it was the Oculus Quest and the Oculus Integration SDK would still be capable enough and the HMD offering build-in speakers. The Oculus Quest provides Inside-Out roomscale tracking (Subsect. 2.2) making it a fitting choice for the authors use-case were a rotational and translational tracking is required, with a precise roomscale tracking not intended. The room requirements for the Oculus Quest are low, since it uses Inside-Out roomscale tracking (Subsect. 2.2) and is a standalone device. However, the room should still offer proper lighting and enough feature points in the interior (Subsect. 4.1). While there is only software expendability needed for the mentioned use-case, the Oculus Quest fits this criteria characteristic as well. For example, if trackable Hardware expandability would have been necessary, the HTC Vive VR system (Table 1) with Vive trackers using the Valve Lighthouse Outside-In tracking technique (Subsect. 4.1) would be preferable. The combination of the Oculus Quest and the Oculus Integration SDK covers all stated criteria characteristics marked in Table 6.

Even though an estimation of a suitable VR system can be done with the criteria catalogue alone, the final decision should be based on deeper knowledge or information, like the our findings presented above or in direct cooperation with a VR expert.

References

1. Pruna, E., Rosero, M., Pogo, R., Escobar, I., Acosta, J.: Virtual Reality as a Tool for the Cascade Control Learning, pp. 243–251. Springer, Heidelberg (2018)
2. Luthon, F., Larroque, B., Khattar, F., Dornaika, F.: Use of gaming and computer vision to drive student motivation in remote learning lab activities. In: 10th annual International Conference of Education, Research and Innovation, ICERI 2017 (2017)
3. De Paolis, L.T., Bourdot, P.: Augmented Reality, Virtual Reality, and Computer Graphics, vol. 10850. Springer, Cham (2018). 978-3-319-95269-7
4. Nilsson, N.C., Nordahl, R., Serafin, S.: Immersion revisited: a review of existing definitions of immersion and their relation to different theories of presence. Hum. Technol. **12**(2), 108–134 (2016)
5. Milgram, P., Takemura, H., Utsumi, A., Kishino, F.: Augmented reality: a class of displays on the reality-virtuality continuum. In: Proceedings of SPIE 2351, Telemanipulator and Telepresence Technologies, pp. 282–292 (1994)
6. Shu, Y., Huang, Y.-Z., Chang, S.-H., Chen, M.-Y.: Do virtual reality head-mounted displays make a difference? a comparison of presence and self-efficacy between head-mounted displays and desktop computer-facilitated virtual environments. Virtual Reality **43**(3), 555 (2018)
7. Wolf, M., Teizer, J., König, M.: Mixed Reality Anwendungen und ihr Einsatz in der Aus- und Weiterbildung kapitalintensiver Industrien. Bauingenieur **93**, 73–82 (2018)
8. McCauley, M.E., Sharkey, T.J.: Cybersickness: perception of self-motion in virtual environments. Presence Teleoper. Virtual Environ. **1**(3), 311–318 (1992). https://doi.org/10.1162/pres.1992.1.3.311
9. Pirker, J., Lesjak, I., Parger, M., Gütl, C.: An educational physics laboratory in mobile versus room scale virtual reality - a comparative study. In: Auer, M.E., Zutin, D.G. (eds.) Online Engineering & Internet of Things, vol. 22, pp. 1029–1043. Springer, Heidelberg (2018). https://doi.org/10.1007/978-3-319-64352-6_95
10. Valve Corporation: SteamVR Unity Plugin| SteamVR Unity Plugin (2019). https://valvesoftware.github.io/steamvr_unity_plugin/index.html. Accessed 17 June 2019, Checked 9 Nov 2019
11. Facebook Technologies, LLC.: Unity Integration| Developer Center| Oculus. https://developer.oculus.com/downloads/package/unity-integration/. Accessed 9 Nov 2019

12. Callaghan, M.J., Putinelu, V.B., Ball, J., Salillas, J.C., Vannier, T., Eguíluz, A.G., McShane, N.: Practical use of virtual assistants and voice user interfaces in engineering laboratories. In: Auer, M.E., Zutin, D.G. (eds.) Online Engineering & Internet of Things, vol. 22, pp. 660–671. Springer, Heidelberg (2018). https://doi.org/10.1007/978-3-319-64352-6_62
13. Windows: What is the Mixed Reality Toolkit| Mixed Reality Toolkit Documentation (2019). https://microsoft.github.io/MixedRealityToolkit-Unity/README.html. Accessed 9 Nov 2019
14. Extend Reality Ltd.: VRTK - Virtual Reality Toolkit (2019). https://www.vrtk.io/. Accessed 31 Mar 2019, Checked 9 Nov 2019
15. Caserman, P., Garcia-Agundez, A., Konrad, R., Göbel, S., Steinmetz, R.: Real-time body tracking in virtual reality using a Vive tracker. Virtual Reality **23**(2), 155–168 (2018). https://doi.org/10.1007/s10055-018-0374-z
16. Unity Technologies: Unity - Manual: Unity User Manual (2019). https://docs.unity3d.com/Manual/index.html. Accessed 11 Aug 2019, Checked 11 Sept 2019

Improvement in Quality of Virtual Laboratory Experiment Designs by Using the Online SDVIcE Tool

Anita S. Diwakar(✉)

Indian Institute of Technology, Bombay, India
anitasd2008@gmail.com

Abstract. The major problem engineering instructors have when they wish to effectively integrate any educational technology such as virtual laboratories is the lack of comprehensive guidelines. This paper discusses the design and development of the experiment design guidelines, which enable engineering instructors to design effective virtual laboratory experiments. The instructors follow a step-by-step process for the scientific design of various virtual laboratory experiments and the online SDVIcE tool provides scaffolds with description of various aspects of the experiment design. The effectiveness of the SDVIcE tool is established with the help of a quasi-experimental study carried out with 39 UG engineering students. The students were divided into two groups. The control group students perform the experiment in virtual lab using the traditional experiment design and the students from the experiment group perform experiments designed using the SDVIcE tool. The results of the study indicate that the laboratory learning outcomes and skills developed are improved when they perform experiment in the virtual laboratory with experiment designs carried out using the SDVIcE tool. So it can be claimed that the SDVIcE tool is effective in improving the quality of the virtual laboratory experiment designs.

Keywords: Virtual laboratories · Quality · Experiment designs

1 Introduction

1.1 Problems in Current Engineering Laboratory Practices

Laboratory work is an integral part of science and engineering education [1]. The laboratory is the place where various theories are tested and relation between theoretical knowledge and physical materials is established. The Engineering laboratory instruction has reached a crisis level due to inadequate instructional resources and the desirable learning outcomes are not being achieved [14]. There is a lack of challenge and initiative provided to the students in performing experiments. There is a need of improving the quality of experiment designs.

1.2 Effective Laboratory Experiment Designs

Instructors play a critical role in designing effective laboratory experiences. Improving instructors' capacity to design effective laboratory experiments is critical to advancing

M. E. Auer and D. May (Eds.): REV 2020, AISC 1231, pp. 393–410, 2021.
https://doi.org/10.1007/978-3-030-52575-0_33

the educational goals of these experiences. This can be achieved by developing more comprehensive systems of support for Instructors. The Instructors can achieve their laboratory goals if they design student centered effective experiments, based on scientifically proven instructional strategies and exploiting the features of virtual labs [12]. The instructors find the learning design for effective virtual laboratory experiments difficult due to a number of factors such as lack of suitable training and non-availability of appropriate guidelines. Hence the aim of this research is to design and develop comprehensive guidelines for engineering instructors so as to facilitate the process of effective learning designs for virtual laboratory experiments. The guidelines have been developed for engineering instructors so as to facilitate the design of effective experiments for using virtual laboratories and converted to the online version in the form of the SDVIcE tool (Scientific Design of Virtual Laboratory Experiments) to increase accessibility.

2 The Characteristics of the Online SDVIcE Tool

The online SDVIcE tool takes the user through the various steps in the scientific experiment design process.

2.1 Step by Step Experiment Design Process

The user is taken through the details of each and every step of the experiment design process and a text box is provided adjacent to the content in which the user can type the content for each step as follows:

1. Broad Goal of the experiment [1, 2, 12]
2. Learning objectives of the experiment [1, 3, 12]
3. Instructional Strategy [4, 14]
4. For - Expository Instructional Strategy [3, 7, 13] – Design tasks for following phases
   a. Conception, planning and design of experiment
   b. Execution of experiment
   c. Analysis and interpretation
   d. Applications
5. For - Discovery Instructional Strategy [3, 5, 16] – Design tasks for following phases
   a. Initiation Phase
   b. Exploration Phase
   c. Experimentation Phase
   d. Presentation Phase
6. For – Well-Structured Problem Solving Instructional Strategy [7, 10] – Design tasks for following phases
   a. Step 1: Review Prerequisite Component Concepts, Rules, and Principles
   b. Step 2: Present Conceptual or Causal Model of Problem Domain
   c. Step 3: Model Problem Solving
   d. Step 4: Present Practice Problems

   e. Step 5: Support the Search for Solutions Step 6: Reflect on Problem State and Problem Solution
7. For – Problem-Based Instructional Strategy [7, 11] – Design tasks for following phases
   a. Formulate learning objectives
   b. Phase 1: Problem Definition Phase
   c. Phase 2: Research Phase
   d. Phase 3: Proposed Solution Phase
   e. Phase 4: Implementation Phase
   f. Phase 5: Desired results
8. Design assessment [9, 15]
   a. Properties of assessment
   b. Measurement metric
   c. Method
   d. Instruments used
9. Use features of Virtual laboratories [6, 8, 16]

The figures illustrate the screen shots of the SDVIcE tool for each step of the experiment design. After the user has completed all the steps and the experiment design is finalized they can click on the submit button and a pdf document is generated. This is the final experiment design that can be given to the students. The students can perform the experiment as per the design and submit their solution online to the instructor (Figs. 1, 2, 3, 4, 5, 6 and 7).

Fig. 1. Landing page of SDVIcE tool

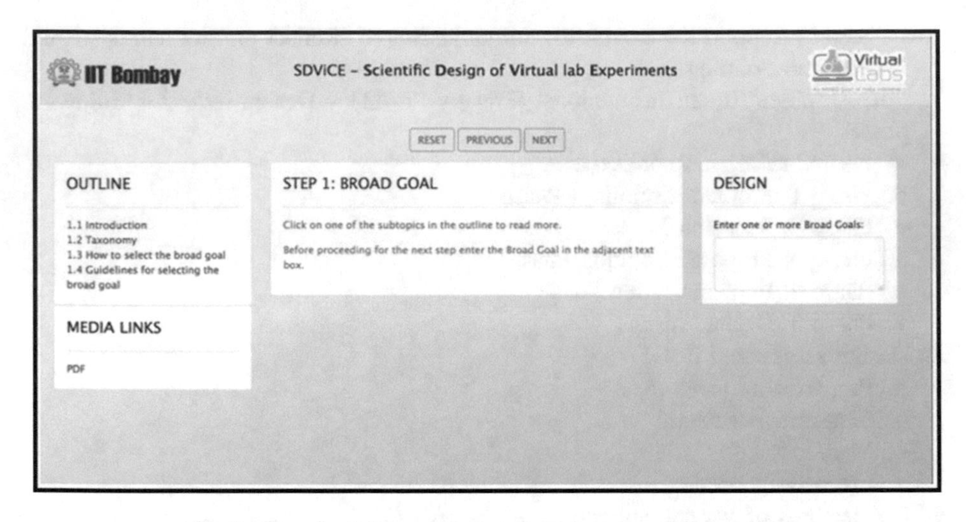

Fig. 2. Step 1 of the experiment design process – Broad Goal

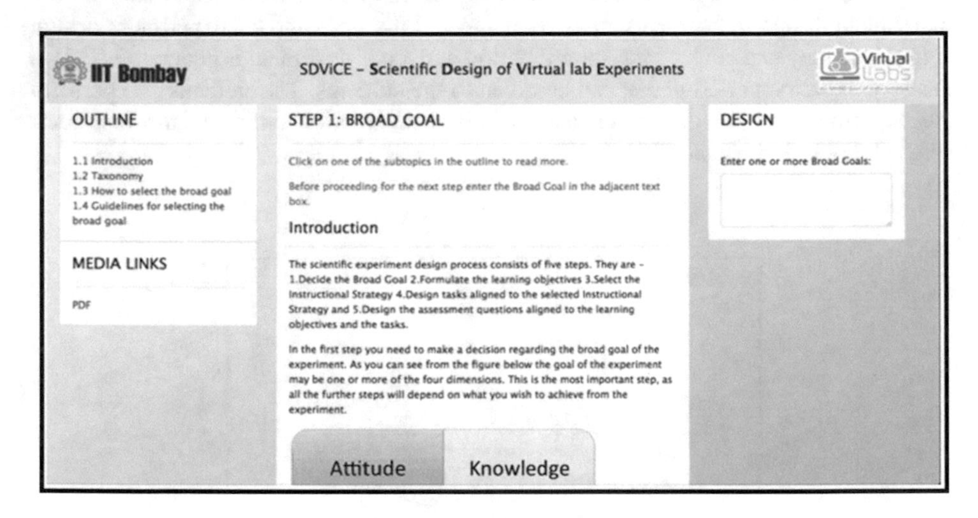

Fig. 3. Step 1 of the experiment design process – Broad Goal

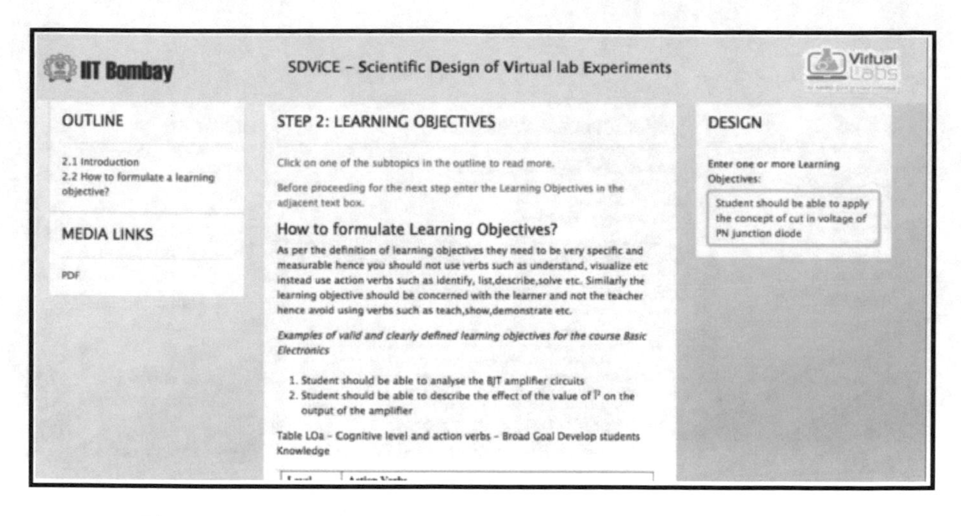

Fig. 4. Step II of the experiment design process– Learning objectives

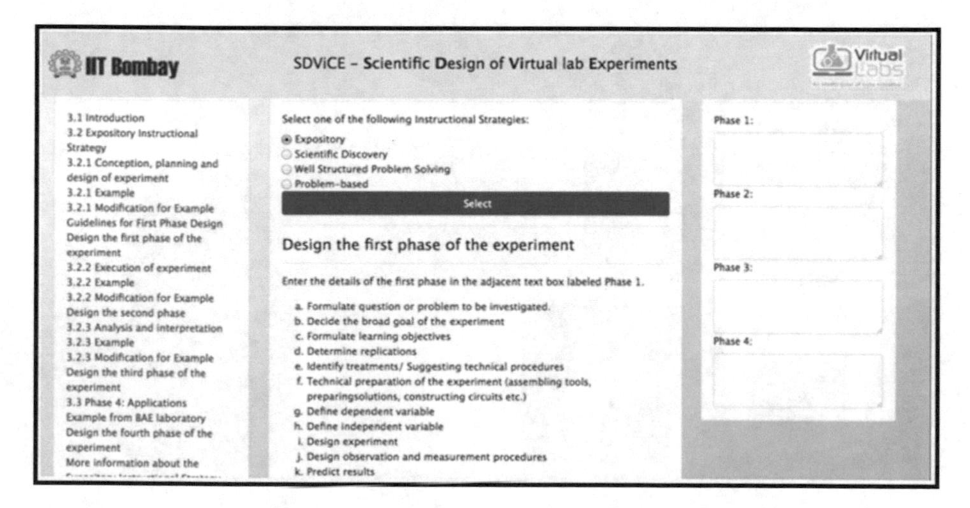

Fig. 5. Expository Instructional Strategy

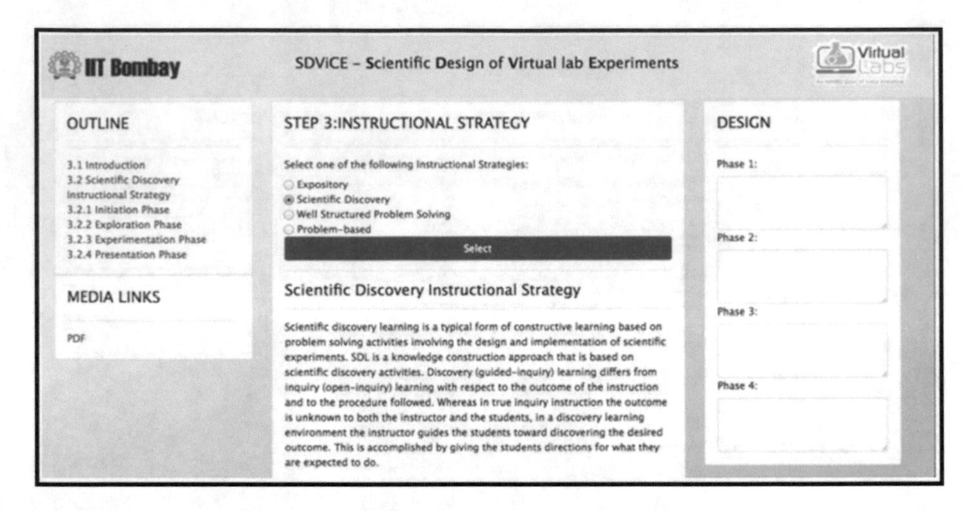

Fig. 6. Scientific Discovery Instructional Strategy

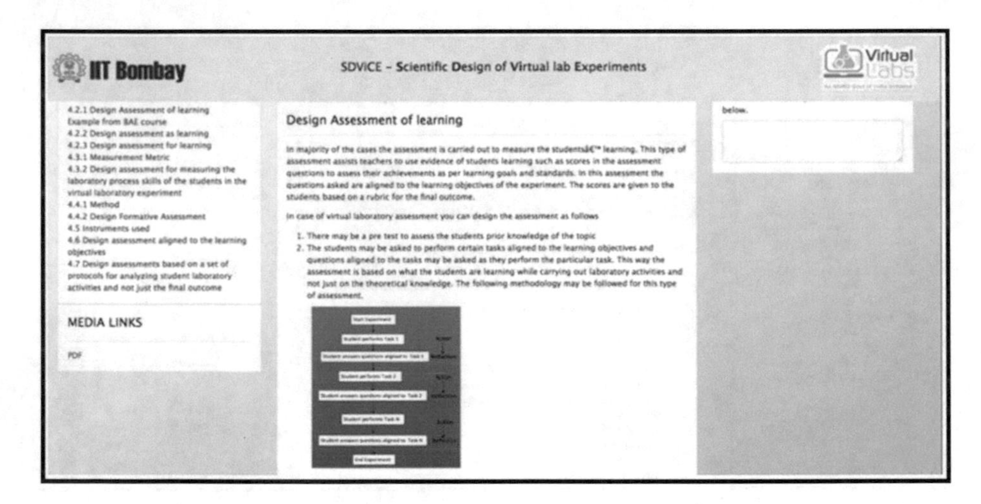

Fig. 7. Assessment design

Feature 1: A library of a variety of components and types of instruments are available for the students. This provides them an opportunity to play around with them. This is one of the most important features as although there are a variety of components and equipment in physical lab too there are constraints and we do not allow the students to play around. This helps in the development of practical skills (Fig. 8).

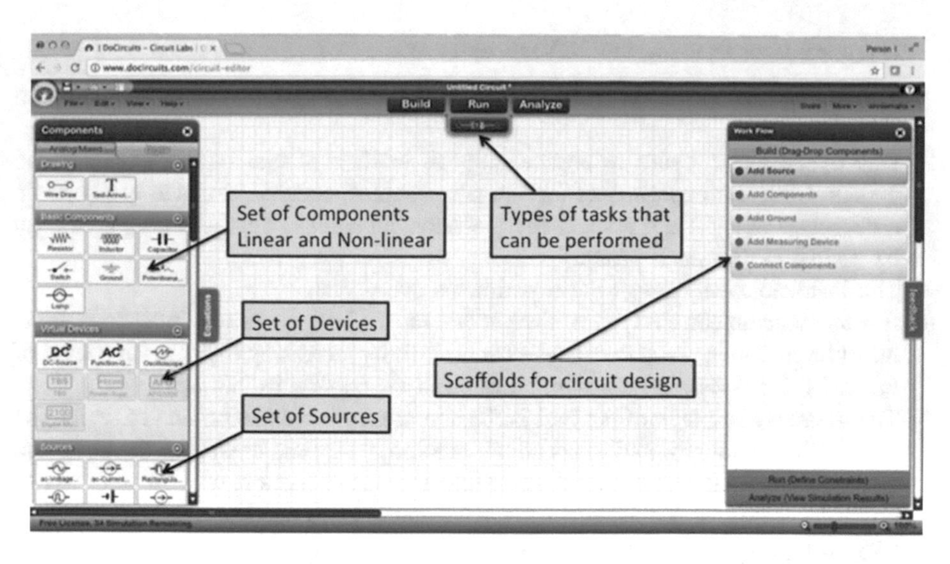

Fig. 8. Virtual laboratory feature 1

Guideline: This feature can be used to achieve the following set of learning objectives. You can design the tasks aligned to the learning objectives and the above feature makes it possible to perform the given tasks (Table 1).

Table 1. Guidelines for Feature 1 usage in experiment design

Vlab tasks	Learning objectives	Broad Goal achievable
Select particular components and equipment from the comprehensive set available	Student should be able to identify the components required as per the circuit diagram given	This feature is necessary to achieve all the Broad Goals of the laboratory experiment as discussed in Section…
Select particular components and equipment from the comprehensive set available	Student should be able to apply their knowledge of the components and equipment and select the most suitable for the particular experiment	
Select particular components and equipment from the comprehensive set available	Student should be able to analyze the functions of the components and equipment and select the most suitable for the particular experiment	
Select particular components and equipment from the comprehensive set available	Student should be able to evaluate the functions of the components and equipment and select the most suitable for the particular experiment	

The advantages of using the SDVIcE tool are

1. The faculty can design virtual laboratory experiments in their course with minimum efforts.
2. One of the major issues in engineering lab work is plagiarism. The faculty can generate different experiments for a small group of students. So each group has a different set of tasks and assessment questions but similar. This way plagiarism can be limited to that small group.
3. The instructors can change the experiments every year.
4. Another issue in current lab practices is that most of the assessment questions are at recall level. The instructors can take care of this by increasing the cognitive level of the tasks and assessment questions from first experiment to the last.
5. The greatest advantage of the tool is the comprehensive database due to which the instructors are not required to spend too much of their time in writing the tasks and questions for BAE course. At the same time if some instructors wish to add tasks or questions there is a provision for the same and it gets added to the database. This way each year the database gets automatically updated.
6. This availability of online content for virtual laboratory experiments can lead to some level of standardization and improve the overall quality of the students' laboratory learning outcomes.

3 Effectiveness of the Experiment Design Guidelines with Respect to Students' Learning Outcomes in Virtual Laboratory

A longitudinal mixed method study spanning over a period of one semester was carried out in order to evaluate the impact of the lab experiments designed using the Experiment design guidelines. The sample for the study was 39 UG second year engineering students from a self-financed engineering institute affiliated to Mumbai University, India.

RQ: What is the impact of experiments designed using the experiment design guidelines for virtual laboratories on the students' laboratory learning outcomes?

In order to measure the impact of the guidelines when students perform experiments using virtual labs following steps were carried out.

- Selection of the course
- Selection of experiments to be performed in virtual labs
- Selection of the Virtual lab
- Identification of learning objectives for the experiments as per the guidelines
- Design of tasks aligned to the learning objectives as per the guidelines
- Design of assessment questions aligned to tasks and learning objectives as per the guidelines
- Design of pre test, post test and learning outcomes test for all the experiments

The researcher in collaboration with one engineering faculty designed and developed five experiments to be performed using the virtual laboratory for the Basic and Advanced Electronics course.

Research Design: The research design used to study the impact of the experiments designed using the Experiment design guidelines on the students learning was a control group experimental group mixed method study. The experimental group consisted of the students taking the virtual lab audit course with experiments designed using the Experiment design guidelines. The control group consisted of students who took the virtual lab audit course but were given the lab manual given in the traditional labs and not designed using the Experiment design guidelines. The students gave two tests – onc pre-test at the beginning of the each experiment and a post-test at the end of each experiment. While performing the various experiments they also gave the learning outcome test (LOT). The entire study conducted over a period of one semester from July 2016 to December 2016 was carried out in authentic settings.

Sample: The 39 participants of the study were second year engineering students from Electronics and Electronics and Telecommunication branch from a self-financed engineering educational institute. The study was conducted in real settings where the students were undertaking the audit course on Virtual laboratories. As part of their curriculum requirements the students are required to complete two audit courses along with other credit courses. Thus the 39 students who chose to opt for the virtual laboratories course are the sample for the study with 20 students in the experimental group and 19 in control group. Hence the sampling was convenience sampling. The students were informed about the research study by the faculty and they volunteered to participate. The scores of the students in the various tests conducted as part of the study were considered for the certification of the course and not for the final grades allocated for the semester results.

Implementation: The experiment was conducted with the following procedure.

1. At the beginning of the lab course the students from both groups were given a pre-test based on the topics covered in the class.
2. The students from both the groups were taught the topics in the class by the faculty and they performed five experiments using the traditional lab with breadboard, components, wires and equipment spanning over a period of one semester.
3. All the students were appraised about virtual labs and the researcher gave a demo of one experiment.
4. These students worked with virtual labs for nearly two hours and performed the experiments on the same topic.
5. The control group students worked with the Basic Electronics virtual lab with the lab manual having the traditional cookbook approach.
6. The experimental group students worked with the Basic Electronics virtual lab with the experiments designed as per the guidelines.
7. All the students worked with the Basic Electronics lab for the same duration of time and answered the questions after completion of allocated tasks.
8. The entire learning process of a few students is captured using the screen capture software.

9. The same procedure is repeated for all the remaining four experiments over the complete period of one semester.
10. Once the students have completed all the five experiments they were given a Survey Questionnaire.
11. A few students belonging to each group were also interviewed.
12. After the students completed the performance of experiments along with the completion of test questions in the online material it is submitted to the instructor. In the study presented the researcher in collaboration with one faculty graded the test answers.

Data Collected: The following data was gathered for analysis at the completion of the study.

1. The students scores in the pre-test
2. The students scores in the LOT (Learning Outcome test) for each experiment
3. The students scores in the post-test
4. The screen capture videos of a few students
5. Student responses to the Survey questionnaire
6. Responses to the open ended questions in the interviews of a few sample students

Instruments Used: The mixed method study was carried out where the following instruments were used

1. Experiment designs for each experiment as per the VLEDG
2. Pre-test question paper
3. Learning outcome test questions
4. Post-test question paper
5. Survey questionnaire
6. Rubrics to measure the various skills

1. Experiment designs for each experiment as per the VLEDG
   The five experiments given to the experimental group were designed using the VLEDG. The topics these experiments covered were as follows:
   (i) V-I Characteristics of PN Junction Diode
   (ii) PN Junction Diode as Clipper
   (iii) PN Junction Diode as Clamper
   (iv) Common emitter characteristics of BJT
   (v) CE Amplifier using BJT
   The instructional strategies used for each of the experiments is as follows:
   (i) Experiment 1 – Expository with various difficulty levels.
   (ii) Experiment 2 – Expository with active learning methods
   (iii) Experiment 3 – Discovery
   (iv) Experiment 4 – Well-Structured Problem Solving
   (v) Experiment 5 – Problem-based

The complete experiment designs and sample answers submitted by students are given in Appendix…

2. Pre-test question paper
   The pre-test was conducted in order to find out the equivalence between the two groups control and experimental. The questions were based on the participants' prior knowledge about the particular topics in Basic Electronics on which the experiments were based.
3. Learning outcome test questions
   This is an online test which the students give while they are performing the experiment and after the completion of various tasks. After the end of each task performed in the virtual lab the students reflect on their learning by answering to the questions. Each question is aligned to the task and also to the learning objectives. This has questions similar to the post-test such as
   a. Multiple Choice Questions
   b. Numerical Problems
   c. Open Ended questions

   These questions test the knowledge achieved by the students to find out if the learning objectives selected for the research are achieved. The marks obtained by the student in each task correspond to the marks obtained for the various learning objectives. So if a student scores well in a task it indicates that the student has performed well in the learning objectives targeted by the task. This test ensures that the students' assessment is authentic and the questions asked assess their knowledge and skills as per the target learning objectives. It can be inferred that the Broad Goals are achieved if the target learning objectives are met.
4 Post-test question paper
   A few sample questions are given at the end after the reference section. Each of the questions in this test is designed by a subject expert and then validated by two other subject experts and an educational technology expert.
   In this test all students had to perform two experiments based on the topics covered in the Basic Electronics theory course but on which they had not performed experiments using the virtual laboratories. They had to answer the questions after performing each task. The questions were similar to the learning outcome test questions aligned to the learning objectives.
5 Survey questionnaire
   The Survey questionnaire is administered after the students from both the groups completed performing all the experiments in the course and appeared for the post-test. The questionnaire consisted of five point likert scale questions and two open ended questions. The following Table 2 gives the structure of the survey questionnaire.

Table 2. Structure of survey questionnaire

S.No	Section Title	Number of questions	Type of question	
			Four point Likert Scale	Open-ended
1	General information	4		4
2	Virtual lab helpful in improving understanding of concepts	2	1	1
3	Attractiveness	3	2	1
4	Virtual lab helpful in Data analysis	2	1	1
5	Useful Vlab features	4	4	4
6	Virtual lab helpful in developing practical skills	2	1	1
7	Virtual lab helpful in problem solving	2	1	1

The questions were related to the features of the virtual lab, which they considered useful in achieving the desired learning outcomes, and also helped them in solving problems. The analysis of the survey data is given in the next section.

6 Rubrics to measure the various skills
In the learning outcome test and the post-test there are a number of questions aligned to the learning objectives. The score of the students for a particular learning objective is calculated as the sum total of the scores obtained for the particular questions in the experiment. The learning objectives are aligned to the Broad Goals that are knowledge at different cognitive levels, various skills and cognitive ability. The attainment of the learning outcomes is measured based on the scores of students in the LOT. The attainment of the skills and cognitive ability is measured by means of rubrics.

Data Analysis: The data analysis techniques used in the study are

1. Comparison of means of students' scores of the two groups in the pre-test.
2. Comparison of means of students' scores of the two groups in the post-test.
3. Analysis of the scores of experimental group students in the LOT.
4. Comparison of mean time spent by students in
   a. Each lab experiment to arrive at the results and answer the questions
   b. Using various features

The data obtained from the answers to open-ended questions and recorded videos when students are performing the experiment is analyzed using qualitative methods.

Comparison of Scores of Students from the Two Groups in the Pre-Test
The following Table 3 gives the statistical analysis data of the experiment. As there are two groups – control group and experimental group in the research design, initially an independent sample t test is conducted on the pre-test scores of the two groups.

Table 3. Results of independent samples t-test – Pre test

Independent samples t-test	Levene's test for equality of variances		t-test for equality of means					
	F	Sig	t df	Sig. (2-tailed)	Mean difference	Std. error difference	95% Confidence interval of the difference	
							Lower	Upper
Equal variances assumed	11.236	0.002	1.252 28	0.221	0.23864	0.19057	−0.15174	0.62901
Equal variances not assumed			1.462 17.46	0.162	0.23864	0.16323	−0.10506	0.58233

As seen from the results the Sig values is 0.002 which is less than 0.005 and hence the equal variances are not assumed and the row number two is used to find if the two groups are different. As seen the Sig(2-tailed) value is 0.162 which is greater than 0.005 it can be concluded that the two groups are have not scored significantly different in the pre-test. This indicates that the two groups are equivalent before they used the virtual labs and performed the various experiments during the audit course.

The same test is then carried out on the post-test scores.

Comparison of Scores of Students from the Two Groups in the Post-test
From this Table 4 the Sig value is 0.002, which is much less than 0.005, and so the equal variances are not assumed. The second row values are to be used for the analysis. As seen the value of Sig (2-tailed) is 0.000, which is much less than 0.005, and hence it can be inferred that the scores of the two groups in the posttest are significantly different. This indicates that the students who carried out experiments designed using the Experiment design guidelines performed better than the students who were given

Table 4. Results of Independent samples t-test- Post test

Independent samples t-test	Levene's test for equality of variances		t-test for equality of means					
	F	Sig	t df	Sig. (2-tailed)	Mean difference	Std. error difference	95% Confidence interval of the difference	
							Lower	Upper
Equal variances assumed	9.728	0.002	4.533 28	0.000	0.25435	0.05612	−0.36479	0.14392
Equal variances not assumed			4.533 17.46	0.000	0.25435	0.05588	−0.36433	0.14438

Table 5. Analysis of survey questionnaire data

S.No	Virtual lab helped in	Experimental group N = 20			Control group N = 19		
		Percent students agree	Percent students neutral	Percent students disagree	Percent students agree	Percent students neutral	Percent students disagree
1	Improving understanding	96	4	–	73	8	19
2	Data analysis	90	4	6	64	4	32
3	Problem solving	94	4	2	70	5	25
4	Developing practical skills	91	4	5	85	2	13
5	Attractiveness	90	7	3	89	6	5

the experiments similar to the traditional methods without using the Experiment design guidelines (Table 5).

The response of the students to the survey questionnaire was more positive from the experimental group than the control group. The response was similar for the two groups on the metric of attractiveness. All the students responded that they felt motivated and enjoyed performing experiments using virtual labs. There were a few students from the control group who perceived that the lab work did not help them in understanding concepts, data analysis, problem solving and developing practical skills. Whereas most of the students from the experimental group perceived that the virtual labs helped them in understanding concepts, data analysis, problem solving and developing practical skills.

Analysis of Qualitative Data – Recorded videos

In order to get better insights into the students learning process the activity of the students while they performed experiment was recorded using CAM studio software. Then a detailed analysis of videos from students of each group was carried out.

The videos were analyzed to find out the time spent by the student for each task and using each feature of the lab by the free and open source software named Tracker. The analysis of the video of three students randomly selected from the control group and three from experimental group who performed the experiment using virtual labs was carried out.

The following Figs. 9 and 10 illustrate a sample interaction of two students – one from control group and the other from experimental group.

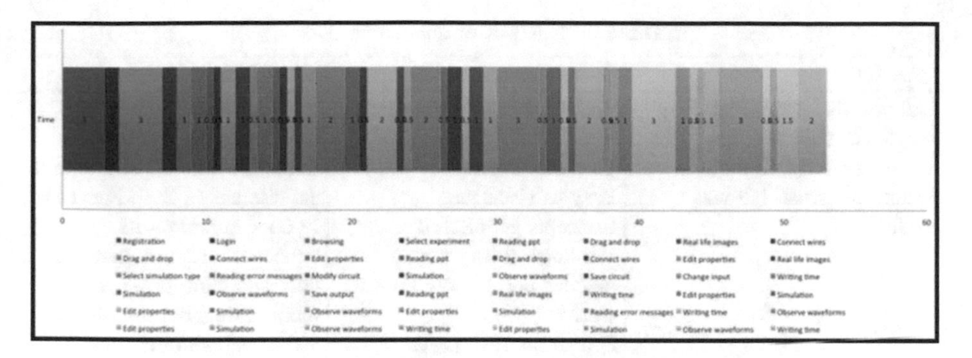

Fig. 9. Interactivity - Experimental group

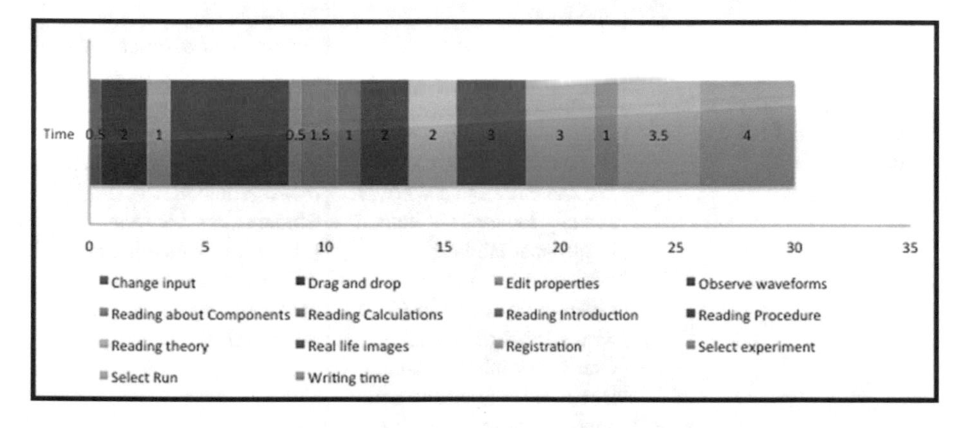

Fig. 10. Interactivity – Control group

The control group student completes the given experiment in 46 min while the student from experimental group needs 81 min to complete the same experiment. The reason for this difference in the time is that the experimental group students reflect on their tasks as per the assessment questions asked after each task. The experiment design incorporates the formative assessment methodology as per the experiment design guidelines. The control group students complete all the tasks and then answer a few questions, which are as per the traditional lab experiment design. It can be observed that the students from experimental group have lot of interactions with the virtual laboratory components as these have been incorporated in the experiment design.

Analysis of Qualitative Data – Open-Ended Questions and Semi-Structured Individual Interviews

The analysis of the responses of the students to open ended questions and interview questions was carried out by content analysis method. All the answers were written verbose in a document. Then each response was given tags based on the content. The tags were then classified into different categories. These categories were related to the specific feature in the virtual lab, which the students found useful in conceptual understanding, development of technical skills, data analysis and problem solving. The following Table 6 gives the results of the analysis.

Table 6. Analysis of qualitative data

Question	Student responses	Virtual lab features
Did you find virtual lab helpful in understanding concepts? Why? Which feature of virtual lab was useful?	Performance over and over again User convenient Easy to visualize Concepts get cleared We know theory but in practice don't know how it works Bridge between practical and theoretical concepts Conceptual understanding improved	Interactive exploration of unobservable phenomena Offered students more time to experience an experiment and to concentrate on its conceptual aspects than the corresponding physical laboratories, because the virtual laboratories allowed faster manipulation of the materials involved in the experiments
Did you find virtual lab helpful in developing practical skills? Why? Which feature of virtual lab was useful?	Objective was met. Various parameters could be varied Many components are present on the screen We can click and see how it actually looks. This helped in physical lab work We could see and use different functions in CRO without fear of damage One varies other constant Demo video provided made concepts clear	A library of a variety of components and types of instruments Full function simulations of the instruments to take the measurements Students can vary the properties of components All parameters may be modified, which cannot be done with the real system
Did you find virtual lab helpful in data analysis? Why? Which feature of virtual lab was useful?	Practical knowledge more important Virtual lab is easy to use Put numerical values in formula and you get the output Feature - output - entire visualization of output Get the output immediately so we can vary values again and again and find the output Time is not wasted in drawing graph	Facilitates recording measurements and plotting of data Students can perform more experiments and thus gather more information in the same amount of time it would take to do the physical experiment

(*continued*)

Table 6. (*continued*)

Question	Student responses	Virtual lab features
Did you find virtual lab helpful in problem solving? Why? Which feature of virtual lab was useful?	Calculate the values Various parameters could be varied Help given in the form of what to do next step Questions are asked as we proceed so it helps Hints were given if error Online help provided	Detailed help can be invoked by choosing either the Demo or Problem Assistance selection Provide the guidelines for selection of parameters and the testing of the selected values If students do not complete correctly, an explanation is provided in the form of a concise informative message

4 Conclusion and Discussion

Through this study the answer to the following research question was obtained.

In order to evaluate the students' learning Donald Kirkpatrick's Learning Evaluation was referred. Kirkpatrick's has defined 4 levels of evaluation:

- Reaction - what participants thought and felt about the training (satisfaction; "smile sheets")
- Learning - the resulting increase in knowledge and/or skills, and change in attitudes. This evaluation occurs during the training in the form of either a knowledge demonstration or test.
- Behavior - transfer of knowledge, skills, and/or attitudes from classroom to the job (change in job behavior due to training program). This evaluation occurs 3–6 months post training while the trainee is performing the job. Evaluation usually occurs through observation.
- Results - the final results that occurred because of attendance and participation in a training program (can be monetary, performance-based, etc.)

The learning of students on was measured for the two dimensions – Reaction and Learning. On both these dimensions the virtual laboratory experiment designed as per the guidelines receive a higher score than the experiment design with traditional methods.

So it can be inferred that if the guidelines are properly implemented and the experiments are designed using the virtual laboratory experiment design guidelines developed with scientific methodology the students' laboratory learning outcomes can be improved. Thus the experiments designed using the Experiment design guidelines have a positive influence on the students' laboratory learning outcomes.

5 Limitations

The study was conducted in authentic settings but the sample size was very small. The triangulation methods have therefore been used to ensure the reliability of the results obtained. The researcher along with a faculty member conducted the study and hence the experiment design was validated for internal and external factors. In order to establish the results with more confidence many more such studies need to be conducted by other faculties from the Electronics domain and other engineering domains implementing the experiments after careful design incorporating the guidelines and taking into account other confounding variables.

Acknowledgements. We would like to thank all the engineering students for participating in the study and providing their valuable feedback to our survey questionnaire.

References

1. Feisel, L.D., Peterson, G.D.: A colloquy on learning objectives for engineering education laboratories (2002)
2. Draper, S.W., Brown, M.I., Henderson, F.P., McAteer, E.: Integrative evaluation: an emerging role for classroom studies of CAL. Comput. Educ. **26**(1–3), 17–32 (1996)
3. Watson, T.J.: The Role of the laboratory in applied science. J. Eng. Educ. (2005)
4. Gomes, L., García-zubía, J.: Advances on remote laboratories and e-learning experiences (n. d.)
5. Abdulwahed, M., Nagy, Z.K.: Computers & education the TriLab, a novel ICT based triple access mode laboratory education model. Comput. Educ. **56**(1), 262–274 (2011)
6. De Jong, T., Linn, M.C., Zacharia, Z.C.: Physical and virtual laboratories in science and engineering education. Science **340**(6130), 305–308 (2013)
7. Jona, K., Adsit, J.: Goals, Guidelines, and Standards for Student Scientific Investigations, June 2008
8. Ercil, N.: Remote RF Laboratory Requirements : Engineers ' and Technicians ' Perspective, pp. 80–95, October 2007
9. Pati, B., Misra, S., Mohanty, A.: A Model for Evaluating the Effectiveness of Software Engineering Virtual Labs (2011)
10. Hofstein, A., Lunetta, V.N.: The laboratory in science education: foundations for the twenty-first century. Sci. Educ. **88**(1), 28–54 (2004)
11. Hofstein, A.: The laboratory in chemistry education: thirty years of experience with developments, implementation, and research laboratory activities have long had a distinctive and central role in the science to quote from Ira Ramsen (1846–1927), who wrote his me. Chem. Educ. Res. Pract. **5**(3), 247–264 (2004)
12. Harris, J., Koehler, M., Koehler, M.J., Mishra, P.: What Is Technological Pedagogical Content Knowledge?, vol. 9, pp. 60–70 (2009)
13. Nachmias, R., Ram, J., Segev, L., Mioduser, D.: A Campus-Wide Project of Web based Academic Instruction : Research Framework and Preliminary Results. Pedagogies (2000)
14. Chin, S.T.S., Williams, J.B.: A theoretical framework for effective online course design. MERLOT J. Online Learn. Teach. **2**(1), 12–21 (2006)
15. Scalise, K., Timms, M., Moorjani, A., Clark, L., Holtermann, K., Irvin, P.S.: Student learning in science simulations: design features that promote learning gains. J. Res. Sci. Teach. **48**(9), 1050–1078 (2011)

Collaborative Virtual Reality Training Experience for Engineering Land Surveying

Anton Franzluebbers, Alexander James Tuttle, Kyle Johnsen(✉), Stephan Durham, and Robert Baffour

University of Georgia, Athens, GA 30602, USA
kjohnsen@uga.edu

Abstract. Training students to perform accurate and efficient land surveys through hands-on laboratory experience takes time, space, and expensive equipment. To alleviate demands on these resources, we designed and implemented an immersive virtual reality training simulator that closely replicates both the equipment and social experience of land surveying. Our pilot study of 85 undergraduate engineering students shows high levels of student enthusiasm for the approach, which, by virtue of the recent affordability of virtual reality equipment, is now feasible for a multitude of use cases throughout the engineering curriculum.

Keywords: Virtual reality · Remote laboratory · Training · Education

1 Introduction

Land surveys are integral parts of a multitude of engineering projects, which rely upon accurate topographic measurements for analysis, design, and construction processes. In practice, land surveying involves the use of a suite of sophisticated measurement technologies, such as the "total station", a composite surveying device used to perform many measurement and computation tasks. It also requires teamwork to handle equipment and perform surveys efficiently. As such, hands-on training experiences currently require significant investment in the requisite technology alongside appropriate space and time management to allow for student teams to practice. This may be practical for a dedicated technical training program, but is burdensome for a undergraduate engineering program, where students may only have a single semester course that introduces basic principles, but are later expected to perform land surveys and work with land surveyors. Under such constraints, students may have few opportunities to gain hands-on practice and feedback, little or no access after the course, and hence difficulties learning, retaining, and using land surveying skills.

Much like other training situations that feature expensive equipment (e.g. pilot training), land surveying simulators have been proposed as a practice tool

M. E. Auer and D. May (Eds.): REV 2020, AISC 1231, pp. 411–426, 2021.
https://doi.org/10.1007/978-3-030-52575-0_34

that greatly alleviates resource constraints [6,10]. Both realistic environments and surveying equipment functionality may be simulated effectively. However, their validity has not received much attention. In terms of face validity, an issue with extant simulators has been the use of a desktop/laptop based interface. These interfaces do not replicate the perceptual experience of surveying. To view the virtual environment, a joystick or mouse is used, rather than head motion. A monoscopic view of the environment is presented, reducing depth perception. Interaction with the virtual equipment involves mouse clicks and button presses, reducing the likelihood skills would transfer to real equipment. Furthermore, though multi-user experiences are possible, the interface limits natural, nonverbal communication from being employed.

A oft-proposed solution to the above problems involves the use of more "immersive" virtual reality technologies that more closely replicate real-world training. This would be akin to a full-motion flight simulator; however, until very recently (c.a. 2016), such technology was more expensive and difficult to deploy than real surveying equipment, making it entirely impractical to consider. Both of these issues have been largely addressed. Using affordable technology (as of this writing, $400 USD), this work demonstrates that it is now possible to simulate a first-person, immersive experience in a virtual world to a sufficient level of visual, and interaction fidelity. The common technology consists of a head-mounted display and two controllers, both of which have their position and rotation accurately tracked with respect to the physical environment. Notably, this can now be achieved without any world-mounted hardware, and without exposed wires, or the requirement of a separate computer, making the technology entirely self-contained and portable, overcoming many previous barriers to adoption (See Sect. 3).

This paper explores the readiness of contemporary virtual reality technology for land surveying training through a pilot study. The study, described in detail in Sect. 4, involved 85 students enrolled in an introductory engineering course. Participants learned to operate a total station device to take accurate measurements. Afterwards, they provided feedback. Results strongly support continued integration of virtual reality technology into land surveying training, as well as integration to further courses that may now provide training where it was difficult or impossible to employ before.

2 Related Work

Kuo et al first demonstrated the feasibility of Using interactive 3D graphics technology to provide a simulated surveying experience [6]. The tool, SimuSurvey, featured an interface with several control panels that included an exocentric 3D view of the environment and equipment, which could be controlled through a set of associated widgets. A preliminary evaluation showcased the efficiency and effectiveness of SimuSurvey in teaching and learning relative to outdoor laboratories, with high levels of student and instructor enthusiasm for the approach. However, they noted that the interface lacked authenticity with respect to the real world and that use of real instruments was still necessary for skills training. A follow-up study vastly improved upon this interface through human-centered design practices, optimizing the training tool for the presented tasks, yet the lack of realism in the interface was still noted as a major barrier for users [7]. Further work included the simulation of systematic errors in the manufacturing and calibration of the device [9].

Continued work from this research group developed SimuSurveyX, a serious-game built for the Microsoft XBox gaming console (also compatible with Microsoft Windows systems), releasing it for public use. This system was a notable departure from Windows-Icon-Mouse-Pointer (WIMP) interface used in SimuSurvey to first-person, exploration-based simulation using a game controller. Rather than using abstract widgets to control virtual devices and visualize measurements, these functions were mapped directly to devices and users could more directly perform many setup and control functions. Graphical realism was also greatly improved. Though details on how this system has been used subsequently are sparse, at least one research project has used it to improve an online surveying course [3].

The work presented here extends the approach of SimuSurveyX to focus on interaction fidelity, to not only integrate the visual and functional experience of surveying, but to also allow users to control the surveying instruments in naturalistic ways through motion control. Further, the system allows for multiple simultaneous users to both operate equipment together and to communicate with each other through voice and avatars.

3 System Description

3.1 Design

Rather than build a dedicated surveying simulation application, as has been previously done, this work instead focused on developing a new education and collaboration platform that could be used for surveying. The major difference between these two approaches is that the interface for a dedicated simulator can be tailored to surveying in all aspects, while a platform must consider how an interface can be used for many potential applications. For example, two prior efforts in using immersive virtual reality interfaces for engineering education have focused on engineering statics as the target application [8,11]. Integrating such applications together with surveying could provide a stronger justification for using virtual reality throughout an engineering curriculum, such that students would not have to relearn the interface for each application and could navigate between applications from the same program.

The key to this approach is the concept of a reality-based interface [5]. In a reality-based interface, the real-world is the basis for design, both in terms of the simulated world and the user interface. Instead of the objective being to make the system easier to use to accomplish tasks, the goal is to require mimic how those tasks would be performed in the real world. This has two primary advantages. The first is that the simulations can be self-contained, insofar as they are in the real world. For example, a virtual total station that directly mimicked a real total station would use the same interface as the real one, such that its functions were activated by buttons, and it could be adjusted by rotating components and knobs. Similarly, activating those buttons and turning the knobs would be accomplished through simulated physical collisions. This has the logistical advantage of allowing many simulations to co-exist without requiring adaptation of the user interface to support them.

Enabling this approach, physically immersive virtual reality interfaces simulate the perceptual experience of first-person, hands-on interaction, which further support the phenomenon of mental immersion (presence) in a virtual world [2]. Modern systems have made great strides in supporting reality-based interaction at low cost. The system used in the current work, the Oculus Quest (approximately $400 USD), provides a 1440×1600 pixel resolution display per eye at 72 Hz frame rate, inside-looking out display position and rotation tracking (using pure computer vision from 4 integrated cameras, without pre-defined reference points), two hand-held and tracked (from the same headset cameras), and a built in Android-based computing platform for simulation and rendering, without any external wires. Though some compromises must still be made, primarily due to the lack of high-fidelity haptic (touch) feedback for controls and insufficient resolution of the displays for fine-detail, the overall usability was determined to be high enough to support the current applications.

3.2 Platform Features

The platform, built using Unity3D, goes by the name ENGREDUVR (pronounced engr-ed-oov-er, short for ENGineeRing EDUcation in Virtual Reality). It is, first and foremost, built for collaborative learning. To this end, it focuses on two key team-communication supports - avatars that can communicate non-verbally through gestures and drawing surfaces that can be used to illustrate and take notes. It also supports verbal communication over a network, though in collocated situations, this is turned off in favor of speaking to each other directly.

The avatars, though customizable through the Oculus Avatars platform feature, are intentionally set as generic. However, to enable rapid identification of each person at a distance, one of several colors may be chosen by each user, which color the visualized headset and hands, leaving the rest of the body as white. As only the head and hands have tracked references, a full virtual body is not used (common in most virtual reality games, as simulating a realist virtual body without it being tracked has little value for non-verbal communication). In addition, the user may select a name that is displayed above their avatar's head. This interface is shown in Fig. 1.

The movements of avatars and all virtual objects are synchronized across a network, using a networking platform called Photon Unity Network. Photon enables network synchronization across a wide variety of network topologies, and is cross platform. Each device connects to a public Photon server, which then synchronizes messages across devices. For ENGREDUVR, this allowed support

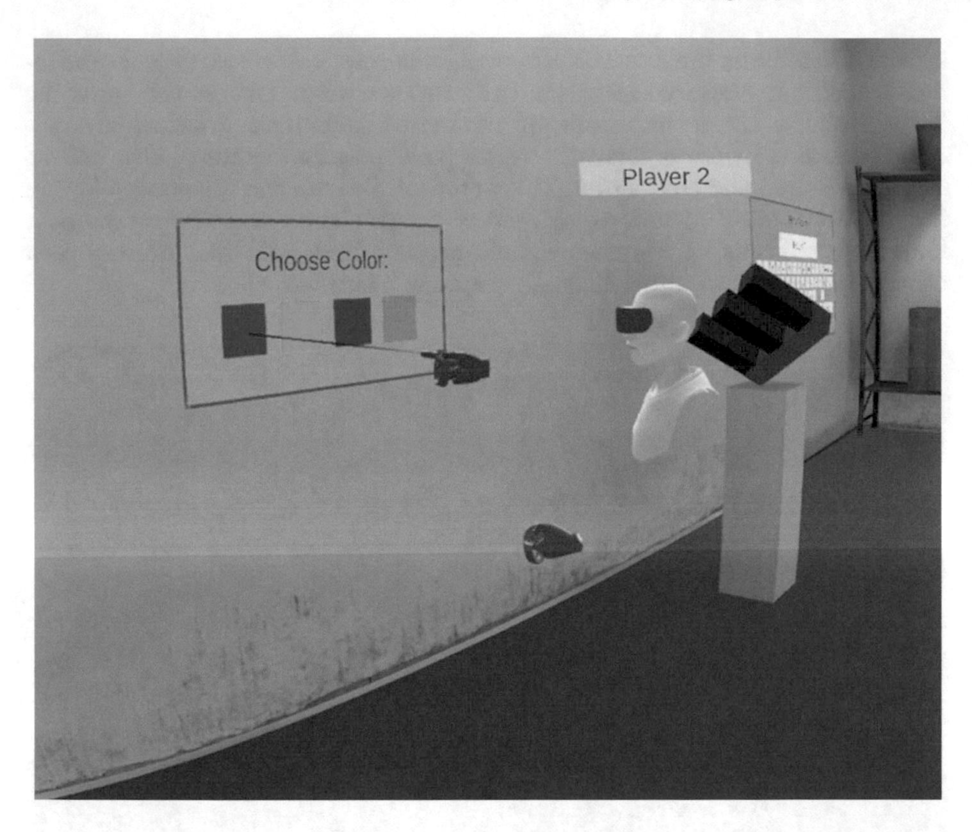

Fig. 1. A user chooses a coloring for their avatar. Afterwards, they may customize their name using the virtual keyboard in the background.

for heterogeneous hardware platforms (as was used in the pilot study, described below), and for the use across large distances without significant regard for how a device is connected to the internet (for example, behind network address translation and firewalls). To control for network bandwidth, avatars are synchronized at a slow update rate (10 Hz), with interpolation used to smooth visual movement.

A mechanism was introduced to allow for collaborative note taking and design. This idea has been explored previously for collaborative problem solving [11], though in that case required the user to have an electronic drawing tablet. As this approach required the user to hold a physical device, inhibiting manipulation of the instruments, we opted to build virtual whiteboards (see Fig. 2) that could be drawn to using only the tracking controllers (as is commonly done in virtual reality games). Though lower accuracy, precise diagrams was not strictly necessary for surveying. More writing-centric applications for the ENGREDUVR platform would need to consider the integration of higher accuracy writing instruments, which is possible and complementary to the ability to write and draw using only the controllers. As with avatars, diagrams are synchronized across the network. However, unlike avatars, these diagrams introduce significant synchronization burden when new users join, as they must be "caught up" to the current state. To minimize this burden, drawings are represented as a sequence of line trajectories (as opposed to bitmaps) that can be efficiently stored and transferred. This approach also has the benefit of allowing for efficient "undo" operations, and enforcing a strict ordering for overlapping lines between users. If two users simultaneously draw lines that overlap each other, after synchronization, they see the same image.

Fig. 2. A user writes measurements on a virtual whiteboard.

Though most of the interface is reality-based, i.e. it matches that of real-world surveying, including devices and terrain, there are several notable features that must be introduced to users that address the practical limitations of virtual reality.

First, assuming the real world environment in which the virtual reality system is used is significantly smaller than the virtual environment, a travel technique must be introduced. The choice for many extant virtual reality systems is a technique called "raycast teleportation". Using either of the tracked controllers, the user indicates their intent to teleport by pushing up on a thumbstick. This activates a parabolic-shape indicator line that originates from the tracked controller and extends a short distance outward until the line intersects a surface that can be teleported to, as seen in Fig. 3. The intersection point is where the user will be teleported to when they release the thumbstick. This enables rapid exploration of the virtual environment without the need to physically walk (though walking is still possible, within the physical space, and does control virtual position). To turn, users may simply turn their physical bodies, or they may tap left or right on the controller thumbstick to "snap-turn" in discrete, 45° intervals, with discrete turning preferred over continuous turning to mitigate cybersickness associated with the latter.

Fig. 3. A user teleports to join their partners. The black line is the arc-raycast technique.

Next, a mechanism was introduced to manipulate the movable components of virtual devices, which lack the tangibility of real devices. When a movable virtual device component is intersected by the user's hand-avatar, it highlights, indicating that it can be manipulated. Upon pulling the index-finger trigger, the movement of the component becomes mapped to the movement of the controller, with constraints set within the component that was grabbed. Reusable code allows the exact specification for different types of components. For example, small knobs may be rotated by twisting the controller after grabbing. Larger joints, such as the total station body, are turned by moving the controller in an arc. This is only semi-intuitive for new virtual reality users at first, but is quickly learned by practice. Two issues occurred due to hardware limitations. First, the resolution of the display was insufficient for seeing small text (e.g. button labels) at normal distances (text could still be seen by getting very close). To overcome this, when the user brings their hand towards a virtual button, the hand morphs into a small magnifying glass that makes text significantly more visible. Second, the computer vision tracking of the controllers fails when the controllers are very close to the head. This would naturally occur when adjusting knobs on the total station while peering through the optics. As this mechanic was also difficult because of the lack of tactile feedback, we chose to avoid having the user look through the optics, instead creating a virtual viewfinder that allowed the user's head to remain farther away from the total station while manipulating.

Finally, several in-world menus are introduced for simplicity in controlling the less educationally relevant, but still functionally important aspects of the simulation such as avatar choices and navigating between surveying scenes. These were used by pointing a virtual laser line at a menu item and pulling the trigger. The laser line could also be used as a communication device, since it's visibility was shared between users.

3.3 Surveying-Specific Features

Arbitrary virtual landscapes can be surveyed, with some limitations on scene complexity due to the need for a high frame rate that mitigates cyber sickness. As the landscape is virtual, the exact values of all surface points are known, which allows for rapid specification of assignments with known solutions (e.g. determine the Cartesian position of each landmark). Beyond this, two devices were simulated, a total station and a prism-rod.

The total station was the most complex device to simulate. From a simulation perspective, it encompassed two distinct components, the station itself and a tripod that it could be attached to. Once attached, the station must be manually leveled through a manipulating a sequence of translational and rotational joints. The tripod could be moved to an arbitrary location in the virtual world by grabbing and teleporting. When grabbed, the legs automatically extended such that the feet hit the ground, if possible. Once positioned, the legs could be independently adjusted in length by grabbing and translating. Like a real surveying tripod, a 2D circular bubble level is used to determine the rough orientation of the station platform. Once the tripod is leveled, the station is finely leveled throughout its 360° range by adjusting three foot screws in a systematic fashion. Once leveled and positioned, the station can be used to take measurements through course (directly rotating the device and adjusting the pitch) controls, and fine grain controls (rotation knobs). Finally, the device control panel can be used in a nearly identical manner as a real total station to take measurements. The prism rod operates similarly, with a circular level bubble and automatically adjusting the length of the leg, while turning the prism to face the station. These features are depicted in Fig. 4.

Total Station and tripod.

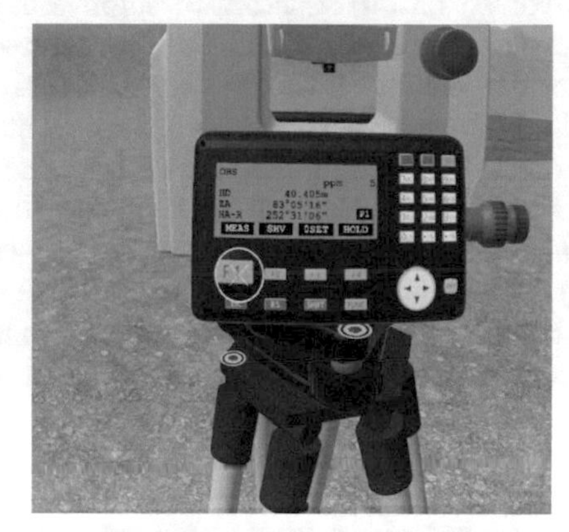

Total station control panel.

Course rotation of the station.

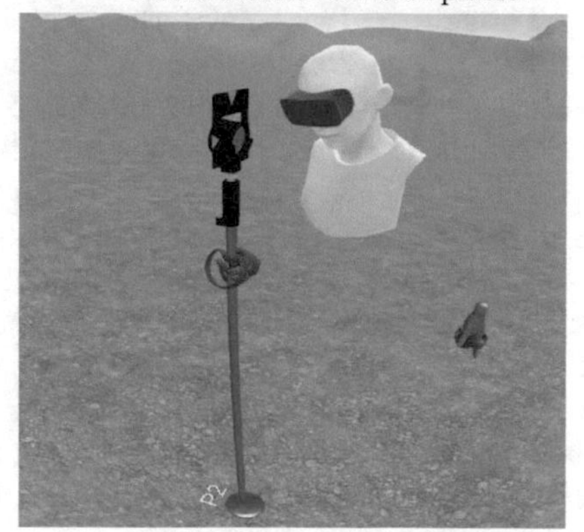

Placing the prism rod.

Fig. 4. Views of the surveying instrument interfaces.

4 Study

After Institution Review Board approval, a pilot study was conducted to determine the usability of the immersive virtual reality surveying system from Sect. 3 and to obtain student feedback on the approach, with the intent on using this feedback to prepare for integration into land surveying coursework. In addition, a single independent variable was investigated, whether students performed a test survey in pairs or individually, to mimic how students may use system in practice.

4.1 Population and Environment

Given that teaching surveying skills was not the intent of this study, we instead involved an introduction to engineering course to evaluate the technology. The instructor (a coauthor) for this course, who also teaches the land surveying course, created a course assignment that required students to sign up for time slots in pairs of two, where they would visit the laboratory and go through a practice exercise in land surveying. The purpose of the assignment was unrelated to land surveying, but instead was to evaluate the experience of using virtual reality technology in education. This assignment was worth 5% of their course grade, and could be performed without agreeing to participate in the study. To protect students from potential coercion, the instructor was not involved in conducting the experiment and was not made aware of the identity of participants, although all eligible (4 students could not participate due to being under 18 years of age) students chose to participate in the study. In total, 85 eligible participants were recruited from the class of 89 students, with the study being conducted over 3 weeks due to the need to schedule 30 min session in pairs.

Though the technology was capable of working in any indoor space with an internet connection (or without, if there is only one user), to control for external factors, the study was conducted in a dedicated virtual reality laboratory that had ample space for multiple collocated participants to spread out (See Fig. 3, lower left). As discussed earlier, by collocating participants, they could speak to each other directly, without using microphones.

The virtual environment used for the study was designed to look like a quarry or building site, and is shown in Fig. 5. This simple environment does not stress the virtual reality hardware, containing a low number of polygons to render, but provides context for significant elevation changes in a small area. Within the environment, participants had access to a single total station and tripod. Two prism rods were available. This matches a realistic surveying scenario, where one person would operate the tripod, while others would hold prism rods at designated locations.

4.2 Measures

As the study was mostly about obtaining user feedback, surveys were the primary instruments, each administered electronically using a Google Form at the experiment facility. A background survey gathered variables of interest, such as age and gender, vision correction glasses use, experience with VR, gaming, and surveying, and for dyads, how well they knew their partner. Three common post-experience surveys were also used. The NASA Task Load Index (TLX) assesses task difficulty along several dimensions [4]. The Steed-Usoh-Slater presence questionnaire [12] and Bailenson co-presence questionnaire [1] measure their eponymous constructs. Lastly, a feedback survey was administered that asked students to rate their opinion on usefulness of the tool for practice, their perceived likelihood to use it independently for study, and their perceived performance in the task.

Fig. 5. The virtual environment used for the pilot study. One total station and tripod are available, with two prism rods.

Finally, there were two questions for open-ended feedback on what difficulties they experienced and for additional comments about the experience.

In addition to survey measures, all writing done using the whiteboards was logged, as were a detailed log of all participant movements (head and hands) and actions taken (e.g. buttons pressed).

4.3 Procedure and Task

The majority of participants signed up and arrived in groups of two, however a small portion of the participants arrived alone. This formed three study groups, labelled as GROUP (N = 48, 24 dyads) - corresponding to dyads who both trained and were evaluated as a group, INDIVIDUAL (N = 33, 15 complete dyads, 3 single datapoint due to elimination of underage participant) - corresponding to dyads who trained as a group, but were evaluated individually, and ALONE (N = 4) - corresponding to those who were trained individually and evaluated individually. In all cases, the experimenter training the participants was the same (1st author). In the INDIVIDUAL case, a second experimenter (2nd author) was involved in the evaluation.

After completing a brief background and demographics survey, the experimenter explained how to wear the headset and hold the controllers. Two separate, non-overlapping areas were designated for each participant in the same room, so as to avoid physical collisions. Once both participants were wearing the headsets, the experimenter continued the instruction of the system from within VR using a third VR system. Participants were walked through the teleportation system, entering their name and choosing a color so that they could more easily identify each other. Following this, they were trained on using the virtual surveying equipment, first in a staging environment (called the "hub" world), and then to perform total station setup and take measurements in the quarry scene. Afterwards, each study group performed the setup and measurement without experimenter assistance (unless requested). For the INDIVIDUAL group, they were taken to separate (but identical) quarry scenes, where the second experimenter was involved in case help was needed. After completing their task, they filled out the post-experience surveys and were sent the results via email so that they could write their report for the class assignments.

4.4 Results

Background Survey. Results from the background survey were consistent with recruitment from a freshman-level undergraduate introduction to engineering class. A majority of the 85 participants were male (24 Female, 61 Male) with an average age of 19.1 years. Considering dyads, a majority were same gender (32 out of 39), likely the result of coordination in the sign-up process. As seen in Fig. 6, surveying and VR experience were low for both genders, with slightly more experience indicated by male participants (M = 1.97, SD = 1.18) than female participants (M = 1.46, SD = 0.66). Game playing experience was significantly ($p < .05$) higher for male participants (M = 3.38, SD = 1.98) than female participants (M = 1.92, SD = 1.18). Male participants indicated a slightly

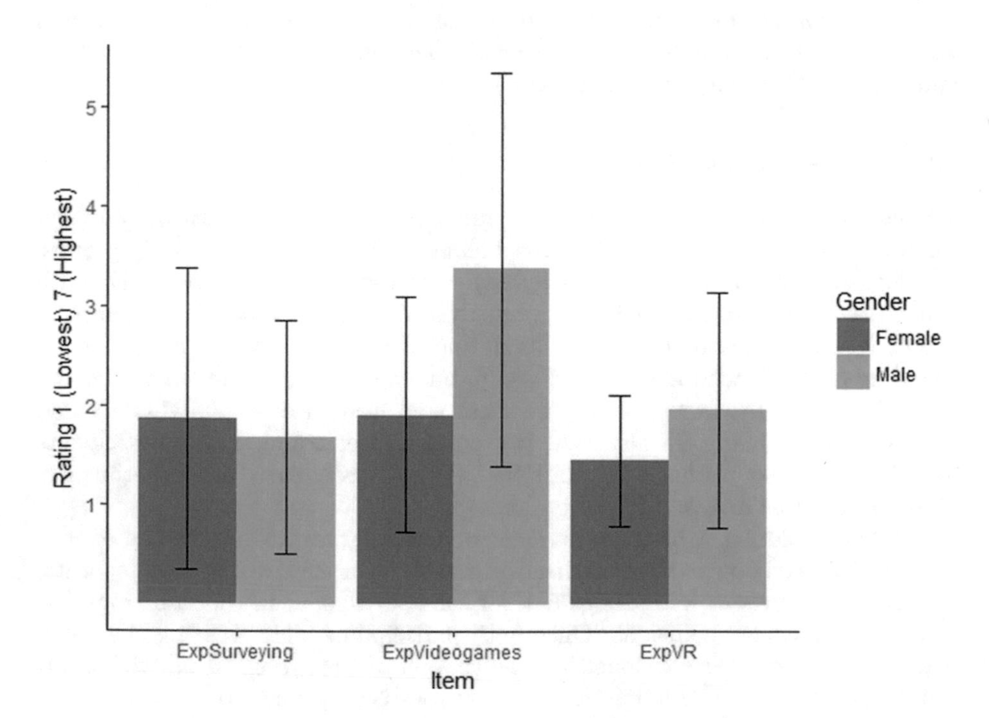

Fig. 6. Background survey results

higher rating of knowing their partners (M = 3.70, SD = 1.80) than female participants (M = 3.09, SD = 1.31).

Presence and Copresence. The preferred technique was used to calculate presence as the number of answers on the questionnaire of 6 or 7 (indicating a high level of agreement or above, maximum 7). Despite controlling for game playing experience, it was found that presence was significantly higher (p = 0.018) for male participants (M = 2.57, SD = 1.77) than female participants (M = 2.00, SD = 1.56). For copresence, the same technique was used (maximum 3). As expected, those in the GROUP condition rated copresence significantly ($p < .001$) higher (M = 2.3, SD = .99) than the INDIVIDUAL condition (M = 1.55, SD = 1.15). The SINGLE condition (N = 4) was surprisingly highest for this category (M = 2.5, SD = .58), though the sample size was small, and acknowledging the fact that in all cases, the users were in the virtual reality world with at least one experimenter.

Usability and Difficulty. No significant differences were found for gender or study condition for any of the items on the usefulness survey. However, these ratings (shown in Fig. 7 were quite high, suggesting strong student enthusiasm for the system. Participants generally thought the system was easy to use (M = 5.80, SD = 1.01), and they thought it would be useful as a practice (M = 6.35, SD = 1.02) and study (M = 5.74, SD = 1.44) tool. However, as shown

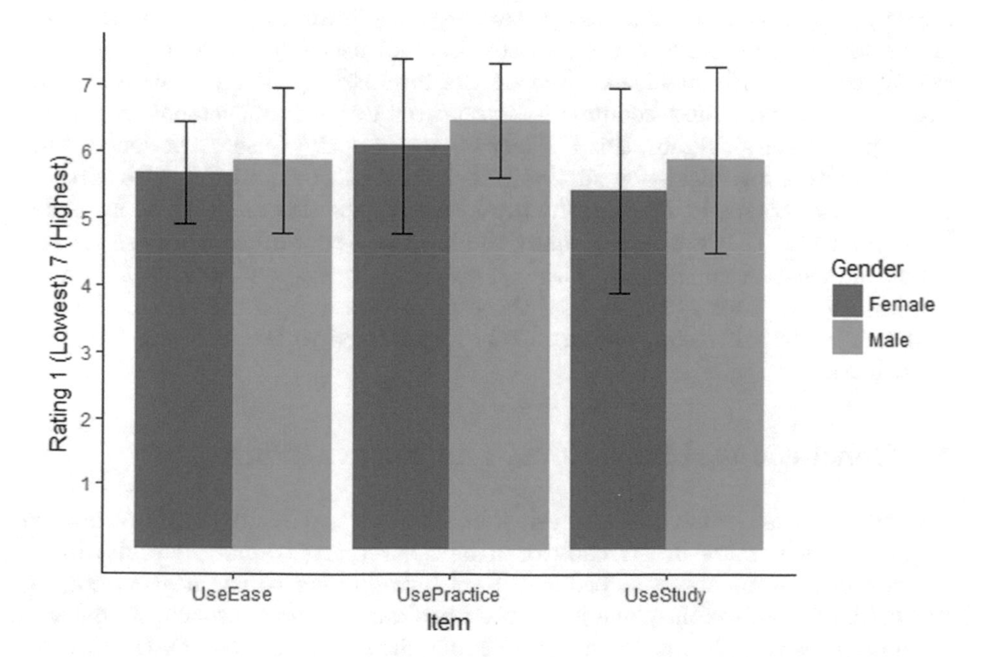

Fig. 7. Usefulness survey results

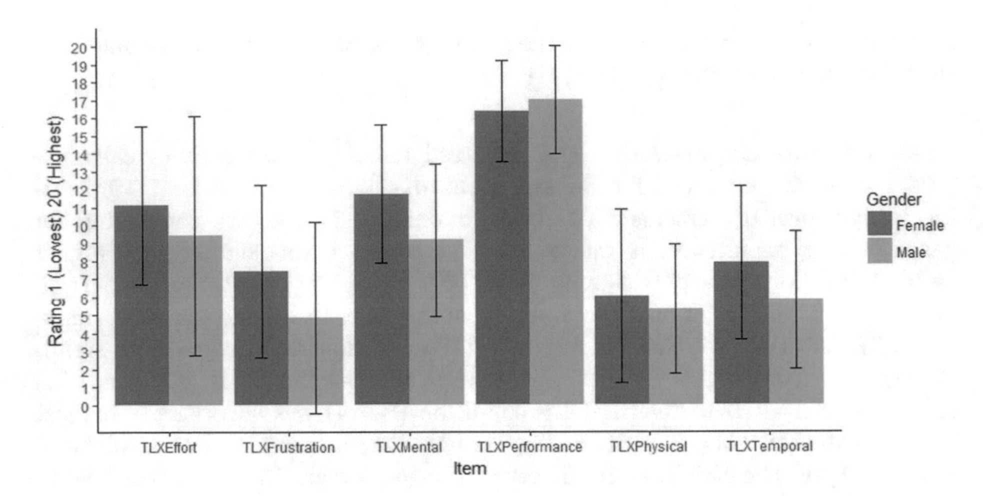

Fig. 8. NASA TLX results

in Fig. 8, the NASA TLX instrument showed signs of higher task load along the mental ($p < .01$), temporal ($p = .07$), and frustration ($p < .05$) dimensions for female participants.

Comments on usability were also analyzed by determining the amount of overlap in user issues reported about various system components. Of this, the most common issue reported was a difficulty in reading text (reported by 24 participants). As discussed in Sect. 3, text legibility is directly related to the size of the text and distance to the headset. Lack of user experience with virtual reality, coupled with small text used on the total station, likely contributed to this issue. The next most common issue reported was with movement (reported by 12 participants). Again, this is likely related to lack of user experience, and overall, quite a low incidence relative to the amount of experience with virtual reality). Encouragingly, leveling the total station was also reported as an issue by 12 students. This was intentionally the hardest part, and so it was seen as a validation measure of the task. Other notable issues included writing (11) and turning objects/knobs (9). Both of these issues are due to the lack of tactile feedback in virtual reality, and are likely also related to lack of virtual reality experience.

5 Conclusions, Limitations, and Future Work

The current work evaluated a novel, immersive virtual reality application for engineering education in the context of land surveying training. Results from a study of 85 students in a freshman-level introduction to engineering course suggest both high usability and student enthusiasm for the approach. A concern remains, however, that there will be gender disparities related to the use of virtual reality systems. The current study showed small, but significant biases favoring male participants on several key (self reported) factors including task load and presence, even after controlling for video game playing experience. Further studies are necessary to determine if these effects would persist with consistent exposure and training in virtual reality and in objective measures such as task performance, which was difficult to measure with the current group.

Related to this, a clear limitation of the current work is that it was not used to teach students as part of a land surveying course, and thus lacks ecological validity. However, this was necessary, as integrating into a course would require more infrastructure. While the cost is not a major issue relative to the cost of surveying equipment, there are still logistical issues to deal with in terms of user training and improvements in usability before course integration becomes viable. A pilot study on course integration is underway, which aims to determine the value that experienced surveying students perceive, and the extent to which their prior experience allows them to perform virtual surveys on the mimicked equipment. This is also being performed with a more authentic task that asks users to determine the volume of material that has eroded from a hill into a nearby river.

The ENGREDUVR platform is also evolving into a course-ready tool. Key improvements have been made that allow students to go through an automated tutorial, and to form *ad hoc* groups by entering a "group code" that can be shared through ordinary communication channels. This will enable a virtual reality laboratory to run without much human assistance, or for students to use their own virtual reality devices. As this happens, more opportunities will be found to leverage the equipment throughout a curriculum, with potentially compounding benefits, both logistical and educational.

References

1. Bailenson, J.N., Swinth, K., Hoyt, C., Persky, S., Dimov, A., Blascovich, J.: The independent and interactive effects of embodied-agent appearance and behavior on self-report, cognitive, and behavioral markers of copresence in immersive virtual environments. Presence: Teleoperators Virtual Environ. **14**(4), 379–393 (2005)
2. Cummings, J.J., Bailenson, J.N.: How immersive is enough? a meta-analysis of the effect of immersive technology on user presence. Media Psychol. **19**(2), 272–309 (2016)
3. Gilliéron, P.-Y., Vincent, G., Merminod, B.: Blending a moocs with interactive teaching. Technical report (2015)
4. Hart, S.G., Staveland, L.E.: Development of nasa-tlx (task load index): results of empirical and theoretical research. In: Advances in Psychology, vol. 52, pp. 139–183. Elsevier (1988)
5. Jacob, R.J.K., Girouard, A., Hirshfield, L.M., Horn, M.S., Shaer, O., Solovey, E.T., Zigelbaum, J.: Reality-based interaction: a framework for post-wimp interfaces. In: Proceedings of the SIGCHI Conference on Human Factors in Computing Systems, pp. 201–210. ACM (2008)
6. Kuo, H.-L., Kang, S.-C., Lu, C.-C., Hsieh, S.-H., Lin, Y.-H.: Feasibility study: using a virtual surveying instrument in surveyor training. In: Proceedings of the 2007 International Conference on Engineering Education (ICEE 2007) (2007)

7. Cho-Chien, L., Kang, S.-C., Hsieh, S.-H., Shiu, R.-S.: Improvement of a computer-based surveyor-training tool using a user-centered approach. Adv. Eng. Inform. **23**(1), 81–92 (2009)
8. Melatti, M., Johnsen, K.: Virtual reality mediated instruction and learning. In: 2017 IEEE Virtual Reality Workshop on K-12 Embodied Learning Through Virtual & Augmented Reality (KELVAR), pp. 1–6. IEEE (2017)
9. Shiu, R.-S., Kang, S.-C., Han, J.-Y., Hsieh, S.-H.: Modeling systematic errors for the angle measurement in a virtual surveying instrument. J. Surveying Eng. **137**(3), 81–90 (2011)
10. Stuart, D.G.: The development of a teaching tool using Sketchup to enhance surveying competence at the Durban University of Technology. Ph.D. thesis (2015)
11. Tuttle, A.J., Savadatti, S., Johnsen, K.: Facilitating collaborative engineering analysis problem solving in immersive virtual reality. In: 2019 ASEE Annual Conference & Exposition, Tampa, Florida, June 2019. ASEE Conferences. https://peer.asee.org/32830
12. Usoh, M., Catena, E., Arman, S., Slater, M.: Using presence questionnaires in reality. Presence: Teleoperators Virtual Environ. **9**(5), 497–503 (2000)

Online and Biomedical Engineering

Work in Progress: Pilot Study for the Effect of a Simulated Laboratories on the Motivation of Biological Engineering Students

Ryan Devine and Dominik May(✉)

University of Georgia, Athens, Georgia
{rdevine4,Dominik.May}@uga.edu

Abstract. The main challenge for biological engineering educators is that BE is a relatively new and broad discipline that integrates a diverse array of knowledge from the basic sciences and engineering sciences towards application in the biological and medicinal fields. Due to variety of possible career outcomes and limited department resources (money, laboratory space, etc.), it is not possible to create a single undergraduate curriculum that will cover all of the technical skills spanning the entire biological engineering field in a hands-on setting. As laboratory sections have been shown to provide a variety of educational benefits, it is imperative that university engineering departments seek alternative methods to deliver real-life application of classroom concepts.

As such, interest in the development and usage of simulated lab sections has risen. While these lab experiences offer economic benefits to educational institutions and are more convenient for students to access, the exact educational outcomes of simulated labs, especially when compared to traditional hands-on labs, is still unclear. Due to the previously stated challenges, the biological engineering education community could benefit greatly by the implementation of simulated labs; however, there is a limited amount of literature on simulated labs in the context of biological engineering.

Therefore, the purpose of this study is to provide a case study on how the implementation of a commercially available simulated lab alters the motivation of students in a BE course. Data collection will help us answer three research questions: 1) How well does the simulated lab intervention work? 2) How do BE students experience disciplinary-specific simulated labs? 3) How do those experiences inform us on the student motivation? By utilizing the MUSIC® Model of Academic Motivation, we aim to pin down the social realities surrounding simulated BE labs, clarify the unique motivations of BE students, and provide vital information for the national discourse of proper BE curricula development.

Keywords: Simulated laboratory · Biological · Biomedical · Online education

1 Introduction

Laboratory sections are at the core of undergraduate STEM education as they grant students the ability to observe how the physical world compares to the concepts taught in the classroom. In the context of engineering (as an applied science field), focus on

M. E. Auer and D. May (Eds.): REV 2020, AISC 1231, pp. 429–436, 2021.
https://doi.org/10.1007/978-3-030-52575-0_35

the application of concepts in an educational setting is especially crucial towards proper career development. For any given lab section, students must learn a measurement procedure in order to conduct an experiment and generate data. The data attained from lab assignments can remarkably improve students' understanding of classroom concepts by allowing students to observe the strengths and weakness of various scientific theories.

While theoretical analysis can be used to closely predict the behavior of physical phenomena, each theory contains simplifications used to achieve a prediction. For example, while a homework problem would include the assumption to "ignore the effects of resistance" in calculating an electric potential difference, in an electrical circuits lab the data generated is affected by such forces and students must account for them accordingly. By seeing firsthand the difference between theory and application, students are then able to develop their scientific writing skills by communicating how the data generated does or does not differ from theoretical calculations.

2 Literature Review

Compared to traditional engineering (TE) disciplines (such as civil, mechanical or electrical), biological engineering (BE) students have been found to have different motivations for entering the engineering field; therefore, it is paramount that the BE engineering education community capitalizes on these differences in order to address the systemically lackluster engineering student retention rate [1]. BE students are largely driven to the field for the opportunity to benefit society, which differs compared to TE majors who cited their love of designing and building [2, 3]. These unique motivational differences of BE students compared to TE students warrants further study, as previous motivational studies in a TE setting may not be applicable to BE students.

The best way to take advantage of the unique motivation of BE students is to engage them in real-world issues and application in the early years of university study, as well as improving the cooperation between the variety of departments used in BE curricula [4]. This notion aligns with previous literature stating how the educational benefits of laboratory sections have been a critical component towards growing the STEM field [5, 6]. Lab sections allow for student involvement in real-world investigations and interaction with lab equipment, which has been recognized to improve educational outcomes [7, 8]. Additionally, labs aid in students' development of communication and team-work skills due to the tradition of team experimentation, which is of utmost importance to develop in an interdisciplinary field such as BE [9].

However, taking advantage of the unique motivations of BEs has proven difficult due to the report of BE students struggling more than TE students in their introductory courses [2]. The main challenge lies in the fact that BE is a relatively new and broad discipline that integrates a variety of knowledge from the basic sciences and engineering sciences towards application in the biological and medicinal fields. For example, BE students face pedological conflict between their basic science classes, which tend to be memorization-based, and engineering classes, which tend to focus on problem solving skills [10]. These basic science classes are required prerequisites for BE courses, which are generally upper-class courses; so while every first year BE student is likely to have taken a biology lab, the motivational desires or problem solving needs of BE students are not addressed in these early lab sections. Additionally, BE educators face a variety of logistical challenges towards implementing hands-on BE labs, such as limited campus space and expensive equipment that are not shared with other engineering disciplines (cell culturing hoods, incubators, etc.) [11].

With advances in computer science, a variety of alternative labs have been developed by tech companies and educational institutions to address the logistical challenges of hands-on labs [12]. These labs are mainly divided into two categories: remote labs and simulated labs. Remote labs involve students conducting experiments with real equipment that is accessed via computer internet; whereas simulated labs are completely virtual and are accessed via computer or VR software. Both types of alternative labs have been found to cost less and require less setup time in an educational setting, which offers a solution to some of the challenges of hands-on labs [13].

It is difficult to directly compare lab sections as each type has a different educational objective, as hands-on labs are emphasized for design skills whereas simulated/remote labs for conceptual understanding [14]. However, in a study comparing all three lab types and recorded both design and conceptual outcomes, it was found that the attitudes toward the lab assignment is highly influenced by the convenience of the lab type [15]. Students liked that the remote and simulated lab required less time on setup and teardown, which aligns with previous literature [13]; although, this convenience seemed to impact students motivation towards the rest of the assignment, as they also tended to spend less time analyzing data and communicating the results. In the end, there was little difference in conceptual outcomes between the three lab types, which suggests that alternative labs can be as effective as hands-on labs.

When solely comparing alternative labs, remote labs have shown a motivational benefit over simulated labs by granting students the ability to interact with real scientific equipment; [16] however, in the context of BE, remote labs are not currently feasible due to technological restraints. While it is currently possible to develop a remote laboratory that takes digital measurements from a controlled environment (Ex. electrical circuitry, material mechanics, etc.), taking measurements from living biological systems is a far more complicated process that requires physical interaction with the system and a variety of equipment [12]. Previous literature research has suggested that the motivational shortcomings of simulated labs can be overcome by the level of peer collaboration and student-to-instructor discussion involved in the assignment, [17] and that the cooperative benefits of labs can still be achieved through online collaboration [7, 18]. Therefore, it is our belief that simulated labs are the most feasible and cost-effective alternative lab available for BE educators; however, it is of utmost importance to ensure that BE simulated labs are able to properly address BE student motivations before widespread application into BE curricula.

3 Research Design and Plan

The research project will be conducted within the BE department at the University of Georgia in coordination with the college's Engineering Education Transformation Institute. The researchers will look at answering two 2 key questions:

1) *Educational Outcome Question* – How well does the simulated lab intervention work?
2) *Knowledge-Generating Question* – How do bioengineering students experience disciplinary-specific simulated labs? And how do those experiences inform us on their motivation?

To answer these questions, the lab intervention was placed within a senior-level undergraduate Tissue Engineering course that does not currently have any type of lab section. As a field of, Tissue Engineering is over 30 years old; however, Tissue Engineering was only first defined and established into curriculum in the late 90s [19]. BE student motivation in the context of a Biomaterials course has been previously investigated, and the results suggest that student motivation towards Biomaterials is indicative of the motivation of BE students overall [3]. Since Biomaterials and Tissue Engineering are closely related fields, it can be inferred that the Tissue Engineering course would then be a great representation of the current challenges facing BE educators due to it covering a recently developed technical field and the current lack of laboratory resources available to the department.

A senior-level class was chosen for this pilot study in order to eliminate any conceptual gaps that students may have with the covered material. Further study will look at the implementation of simulated labs with first-year BE students, as they will be the most likely to gain motivation benefits by viewing firsthand the content at the end their major. It is important to note at the University of Georgia there is no biomedical engineering major, although there is a biomedical area of emphasis for BE majors. The Tissue Engineering course is required for BE students wanting to graduate with a biomedical emphasis; however, any BE or biochemical engineering students with a biological emphasis may take it as an elective course. The course enrollment had 45 students in total with 26 being BE majors and 19 being biochemical engineering majors.

The chosen lab intervention is a simulated Tissue Engineering lab developed by Labster™. In the Tissue Engineering lab, students are prompted to help treat an injured soccer player by developing a scaffold to help regenerate the soccer player's cartilage. Students will accomplish this by relying on their background of chemistry and material science; therefore, we believe the Labster™ Tissue Engineering activity is a perfect fit for the class as it covers material relevant to both BE and biochemical engineering students.

4 Theoretical Framework

It has been proven difficult to conduct social research in the engineering field due to how unfamiliar the engineering community is with social research. In fact, a majority of previous studies looking at motivation in an engineering education context has been found to inconsistently apply any theoretical framework at all, and on the occasion that a framework was applied, only a limited number of traditional theories were used [20].

Therefore, for this study Jones' MUSIC® Model of Academic Motivation was utilized as it is a non-traditional model of motivation, is backed by variety of resources to ensure proper framework, and made specifically to assess specific elements of a course [21]. The power behind the MUSIC® Model of Academic Motivation is derived from the five separate motivation theories used to create it: 1) Student e**M**powerment through the ability "control" one's learning 2) Course **U**sefulness towards student short- or long-term goals 3) Student understanding of what it takes to **S**ucceed in the course 4) Student **I**nterest in the course content 5) Belief that the academic structure around students truly **C**ares for their well-being and future. By combining separate motivational theories, the MUSIC® Model of Academic Motivation will be able to paint a wholistic picture of the social realities surrounding the lab intervention, and be able to pinpoint what aspects of motivation can best be improved in future BE simulated lab software. Regarding this project, it is hypothesized that a simulated lab will not show significance in the empowerment category of motivation due to the limited control a student has over a guided simulated lab; however, the other four pillars of motivation are variables capable of delivering valuable data.

5 Data Collection and Analysis

Quantitative and qualitative data will be collected from a variety of sources. Post-intervention surveys will utilize Jones' MUSIC® Inventory, which was also developed by Jones to measure the extent to which college students perceive the presence of each of MUSIC model components in a college course [22]. The survey compromises 26 questions which students will rank from 1–6 with 1 being "Strongly Disagree" and 6 being "Strongly Agree." Therefore, the higher the score, the more that individual motivation theory resonated with students. In addition to the numerical responses, 6 short answer questions were added in order to get qualitative data on student's experience with the simulated lab. These questions were used to get a more in-depth understanding of the numerical data and to select 5 students to undergo narrative analysis interviews. The purpose of the narrative analysis interviews is to attain a complete understanding of student experience with the lab section.

The data analysis will follow a case study [23] and narrative analysis [24, 25] approach to develop a theory on how the implementation of simulated labs affects the motivation of BE students. The case study approach to analysis will allow for accurate assessment of the efficacy of the lab intervention and identify what factors of motivation are at play during the lab intervention; whereas the narrative analysis approach will be used to contextualize student experience with the lab intervention. This combined approach will ensure that no assumptions are made in regards to the data collected, and that the social realities surround BE simulated labs are accurately described.

6 Preliminary Data

Of the 45 students enrolled in the course, 43 participated in the study by completing the activity and survey. The activity was completed outside of the classroom as a distance learning activity, so students had no direct help from the instructors. Additionally, the activity was completed by students in an average of 32 min. Considering some of the steps in the simulated experiment required "20–30 min" of incubation (which was fast-forwarded), we can assume that the experiment in a hands-on lab would take more time, which aligns with the previous reported expedient/convenience benefit of simulated labs [13].

After completion of the activity, students were directed to fill out the MUSIC® Inventory, which was administered through an online survey. Out of the 43 students, 1 student was omitted due to answering every question with a "Strongly Agree" and responding to the short answer section with minimal detail. The results of the numerical potion of the survey are presented in Table 1.

Table 1. Survey numerical data

	Empowerment	Usefulness	Success	Interest	Caring
Avg.	4.38	4.44	4.89	4.77	5.21
Std.	0.58	0.89	0.46	0.94	0.52

Overall, all 5 pillars of the MUSIC® theory exhibited a positive effect on students as the answers averaged out from "Agree" to "Strongly Agree". As hypothesized, the empowerment category was the lowest of the 5 ranked categories; however, students still generally agreed that they had power over their learning. This was somewhat of a surprise as, given that it's a guided simulated laboratory, it was expected that students would generally disagree with the empowerment questions. It was revealed in the qualitative short answers that students had multiple options throughout every step of the experiment, which included wrong answers that would derail the experiment. This sense of choice and the need to critically think, rather than just simply clicking through an animation, resonated with students sense of empowerment.

The survey data was collected the week that "Work in Progress" submission were due, so further analysis has not yet been conducted; however, it should be noted that a quick two-sample t-test revealed that the Caring pillar has a statistically significant difference than all 4 other pillars. Additionally, the Success pillar was statistically different from both the Empowerment and Usefulness pillars, which were the two lowest rated pillars. Further analysis behind why Caring and Success were rated so well compared to the other pillars will be conducted in the future.

The short answer portion of the survey was used to select 5 students for narrative analysis interviews. Students who gave unique and elaborate answers were chosen for the interviews as it was expected that the most insight would be gained from these students. The selected 5 students were interviewed for roughly 30 min on their interaction with the laboratory based off of the 5 MUSIC® theory pillars and were prompted to walk through their experience. These interviews are currently being transcribed and will allow for more in-depth analysis of the social realties surrounding students' experiences.

7 Research Outcomes

As illustrated above, the primary focus of this project is to establish the efficacy of simulated labs in the context of BE. The goals of this study are to serve as a pilot study for further simulated lab implementation at the University of Georgia College of Engineering, to further the available literature on BE student motivation, to provide information to technology companies on how to better develop BE simulation software, and to drive the national discourse on BE education.

References

1. Geisinger, B.N., Raman, D.R.: Why they leave: understanding student attrition from engineering majors. Int. J. Eng. Educ. **29**(4), 914 (2013)
2. Benson, L., Kirn, A., Faber, C.J.: CAREER: student motivation and learning in engineering. In: 2014 ASEE Annual Conference & Exposition, pp. 24.261. 1–24.261.9 (2014)
3. Keshwani, J.R., Curtis, E.: Motivating undergraduate engineering students through real-world applications of biological materials. Trans. ASABE **60**(5), 1421–1427 (2017)
4. Alpay, E., Ahearn, A.L., Graham, R.H., Bull, A.M.J.: Student enthusiasm for engineering: charting changes in student aspirations and motivation. Eur. J. Eng. Educ. **33**(5–6), 573–585 (2008)
5. Hofstein, A., Lunetta, V.N.: The laboratory in science education: Foundations for the twenty-first century. Sci. Educ. **88**(1), 28–54 (2004)
6. White, R.T.: The link between the laboratory and learning. Int. J. Sci. Educ. **18**(7), 761–774 (1996)
7. Huang, I., Hwang, G.-J., Yang, I.-J.: Optimization of a cooperative programming learning system by using a Constructivist approach. In: Proceedings of the 18th International Conference on Computers in Education (ICCE-2010). Asia-Pacific Society for Computers in Education, Putrajaya (2010)
8. Gagnon, G.W., Collay, M.: Constructivist Learning Design: Key Questions for Teaching to Standards. Corwin Press, Thousand Oaks (2005)
9. Johnson, D.W., Johnson, R.T.: An educational psychology success story: social interdependence theory and cooperative learning. Educ. Res. **38**(5), 365–379 (2009)
10. Harris, T.R.: Seeking improvement in bioengineering education: academic and organizational concerns. In: Proceedings of the Second Joint 24th Annual Conference and the Annual Fall Meeting of the Biomedical Engineering Society, Engineering in Medicine and Biology, vol. 3, pp. 2648–2649 (2002)
11. Perreault, E.J., Litt, M., Saterbak, A.: Educational methods and best practices in BME Laboratories[1]. Ann. Biomed. Eng. **34**(2), 209–216 (2006)
12. Potkonjak, V., Gardner, M., Callaghan, V., Mattila, P., Guetl, C., Petrović, V.M., Jovanović, K.: Virtual laboratories for education in science, technology, and engineering: a review. Comput. Educ. **95**, 309–327 (2016)
13. Scanlon, E., Colwell, C., Cooper, M., Di Paolo, T.: Remote experiments, re-versioning and re-thinking science learning. Comput. Educ. **43**(1–2), 153–163 (2004)
14. Ma, J., Nickerson, J.V.: Hands-on, simulated, and remote laboratories: a comparative literature review. ACM Comput. Surv. **38**(3), 7 (2006)

15. Corter, J.E., Esche, S.K., Chassapis, C., Ma, J., Nickerson, J.V.: Process and learning outcomes from remotely-operated, simulated, and hands-on student laboratories. Comput. Educ. **57**(3), 2054–2067 (2011)
16. Sauter, M., Uttal, D.H., Rapp, D.N., Downing, M., Jona, K.: Getting real: the authenticity of remote labs and simulations for science learning. Distance Educ. **34**(1), 37–47 (2013)
17. Chou, S.-W., Min, H.-T.: The impact of media on collaborative learning in virtual settings: the perspective of social construction. Comput. Educ. **52**(2), 417–431 (2009)
18. Looi, C.-K., Chen, W., Ng, F.-K.: Collaborative activities enabled by GroupScribbles (GS): an exploratory study of learning effectiveness. Comput. Educ. **54**(1), 14–26 (2010)
19. Reyes, J., Lysaght, M.K.: The growth of tissue engineering. Tissue Eng. **7**(5), 485–493 (2001)
20. Brown, P.R., McCord, R.E., Matusovich, H.M., Kajfez, R.L.: The use of motivation theory in engineering education research: a systematic review of literature. Eur. J. Eng. Educ. **40**(2), 186–205 (2015)
21. Jones, B.D.: Motivating students to engage in learning: the MUSIC model of academic motivation. Int. J. Teach. Learn. High. Educ. **21**(2), 272–285 (2009)
22. Jones, B.: User guide for assessing the components of the MUSIC® Model of Motivation (2017)
23. Seawright, J., Gerring, J.: Case selection techniques in case study research: a menu of qualitative and quantitative options. Polit. Res. Q. **61**(2), 294–308 (2008)
24. Rogan, A.I., de Kock, D.M.: Chronicles from the classroom: making sense of the methodology and methods of narrative analysis. Qual. Inq. **11**(4), 628–649 (2005)
25. Smith, C.P.: Content analysis and narrative analysis. In: Handbook of Research Methods in Social and Personality Psychology, pp. 313–335. Cambridge University Press, New York (2000)

"Hidden" Integration of Industrial Design-Tools in E-Learning Environments

Karsten Henke(✉), Johannes Nau, René Hutschenreuter, Robert-Niklas Bock, and Heinz-Dietrich Wuttke

Ilmenau University of Technology, Ilmenau, Germany
{karsten.henke, johannes.nau, rene.hutschenreuter, robert-niklas.bock, dieter.wuttke}@tu-ilmenau.de

Abstract. One of the problems that has emerged during the intensive usage of the GOLDi remote lab over the years is the usage of external third-party development tools, necessary for the software- or hardware-oriented design. The installation and setup (e.g. of Atmel Studio or Intel Quartus Prime) on a private PC or laptop is quite complicated, requires some expert knowledge to use these tools safely and effectively, and is not always platform-independent. Above this, these IDE occupy several gigabytes in memory size and sometimes take hours to install. These very powerful tools, of which only a fraction of the offered functionality is needed to solve the given educational task, generate a large overhead. In addition, high school and university students may have only little or no knowledge of microcontrollers, FPGA, compiler linker toolchain, hardware-related languages, etc.

That is why we developed a new tool as an integral part of the GOLDi remote lab infrastructure – called **WIDE** (WEB IDE) – which will be described in this paper. WIDE supports all the design flows with the only requirement of an Internet browser. This means WIDE is running in standalone mode to write and compile code in the specific language or directly inside the Experiment Control Panel (ECP) of the student's Web browser.

1 Introduction

Our hybrid interactive online lab GOLDi (Grid of Online Lab Devices Ilmenau [1]), described in several papers, is a cloud concept to offer a universal remote lab platform [2–5]. Actually, 10 GOLDi lab instances are running at 10 partner universities located in Armenia, Georgia, Ukraine, and Australia as a cloud system. With our remote lab infrastructure, we support a unique design process of digital control systems - independent of the selected specification technique. According to the given task, this process usually consists of the conceptual formulation and the design of the control algorithm by using different development systems to achieve a validated control finally. This can be uploaded via the Experiment Control Panel (ECP), running inside the student's Web browser to the selected control unit in the remote lab, which will be programmed automatically in the background in the remote lab. Then the student can execute his experiment remotely. An actual decisive disadvantage is that the student has to use additional third-party tools on his computer (see Fig. 1).

M. E. Auer and D. May (Eds.): REV 2020, AISC 1231, pp. 437–455, 2021.
https://doi.org/10.1007/978-3-030-52575-0_36

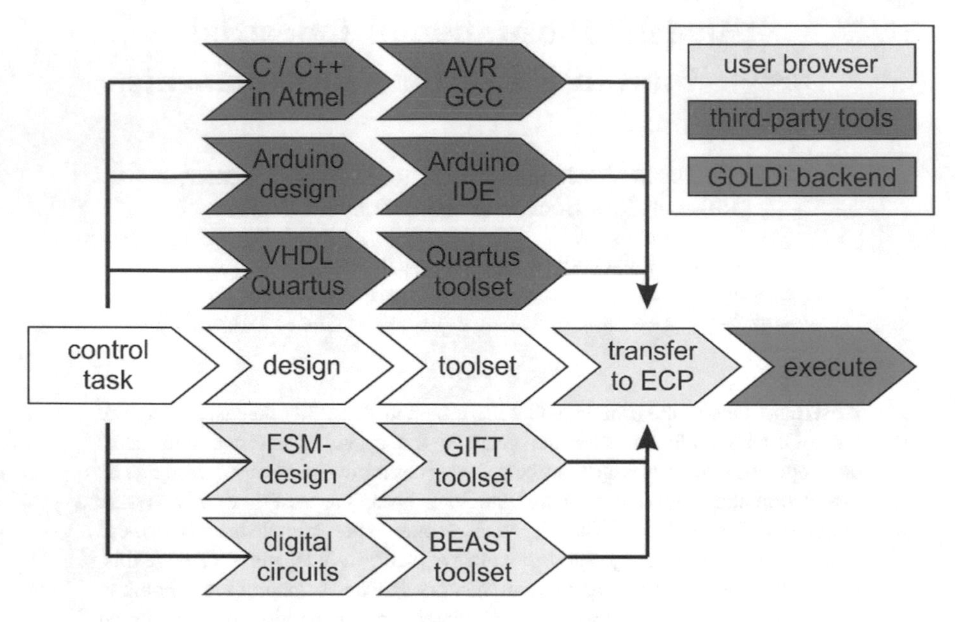

Fig. 1. Unique design flow of the GOLDi remote lab

The various design flows and the design tools required for them will be explained in more detail in Sect. 2. Section 3 describes our vision of a future design flow and Sects. 4 and 5 explain the architecture of the WIDE-system. In Sect. 6, the users` point of view is discussed and Sect. 7 summarizes the paper.

2 Currently Used Design Flows

Based on the flexible grid structure of the GOLDi system an experiment consists of two components: on the one hand, there are various control units (e.g. FSM-interpreter, microcontroller, FPGA). On the other hand, there are physical systems (e.g. elevator, 3-axis model or warehouse) as control objects. Physical systems are available as a real electromechanical hardware model and/or as a simulated virtual model [6].

Depending on the desired control unit, we support the specification techniques, described in the following.

2.1 Software-Oriented Design Flow

For a software-oriented implementation, students can implement their control algorithm directly for a microcontroller. We support the software design for microcontrollers (Atmel, Arduino). Therefore, it is necessary to use external but free available development tools (e.g. Atmel Studio [7], Arduino IDE [8]) to enter the C/C++ or Arduino code (see Fig. 2) and for compiling the design. The generated firmware file must be uploaded via the ECP to the GOLDi system, which will program the microcontroller automatically in the background (Fig. 3).

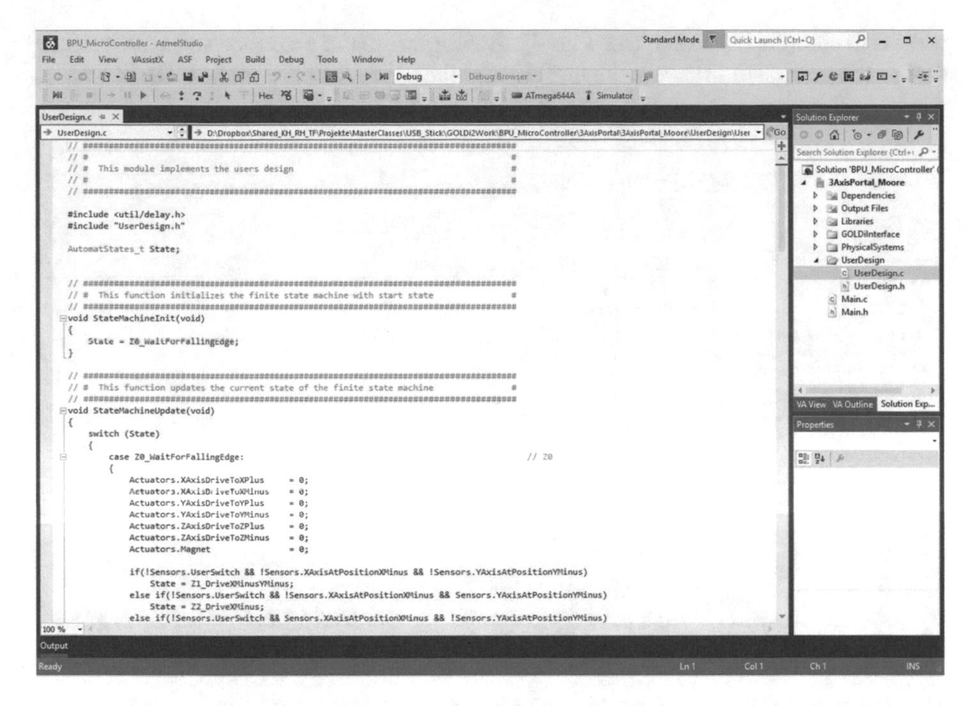

Fig. 2. Usage of third-party tool Atmel Studio for a software-oriented design

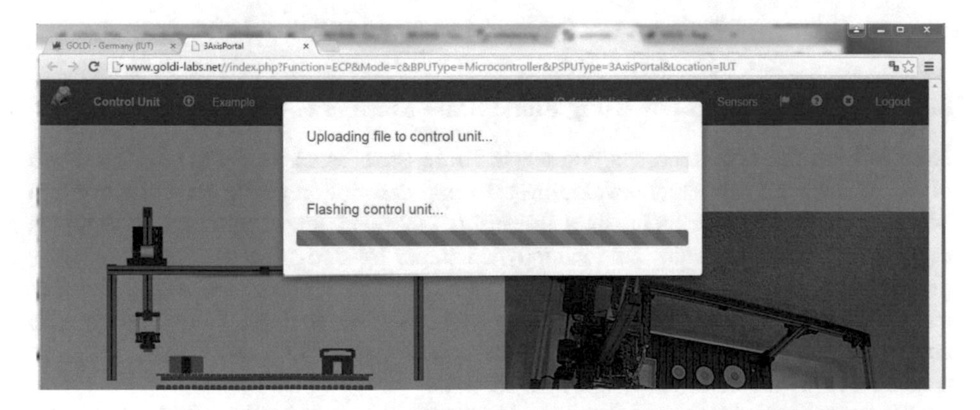

Fig. 3. Programming the microcontroller in the remote lab via the ECP

2.2 Hardware-Oriented Design Flow

If the student prefers a more hardware-oriented design using an FPGA, he can use our hardware-orientated design flow by specifying his task with hardware-description languages like VHDL.

For this design, the student again has to use huge development tools (e.g. Quartus Prime from Intel [9]) to enter his textual, block diagram, or schematic description as well as to compile his design (see Fig. 4). Analogous to the described software-oriented design the generated firmware file must be uploaded via the ECP to the GOLDi system, which will program the FPGA automatically in the background.

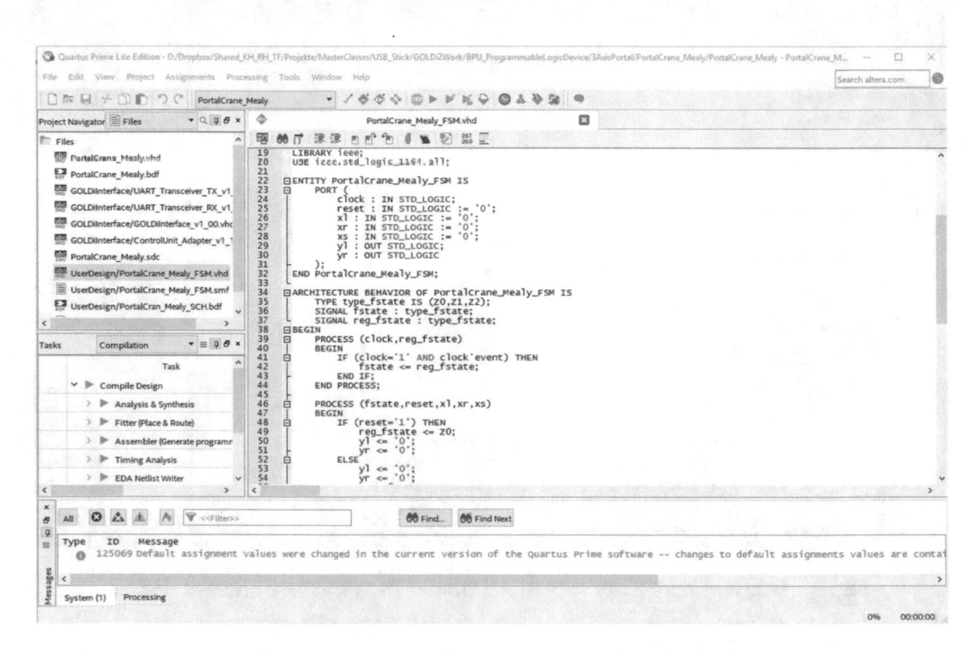

Fig. 4. Usage of third-party tool Quartus Prime for a VHDL-based hardware-design

2.3 Abstract Design Flow Using Finite State Machines

Especially in the first semesters, we teach Finite State Machine (FSM) design. Such kinds of tasks are solved by developing a formal description of the control algorithm based on an automaton graph or a transition table and the corresponding Boolean equations. Students can do this manually on paper by designing such an automaton graph. From this, they can derive the next state and output equations, which they can enter directly into the ECP or to use these equations as the basis for their C program in case of a software-oriented design with microcontrollers or for their VHDL design in case of a hardware-oriented design with FPGAs.

To further simplify the task, with the GIFT (Graphical Interactive FSM toolset, [5]) system we offer the possibility to enter and test this design directly within the GOLDi environment. Figure 5 shows two possibilities of corresponding automaton graphs by using the graph editor of the GIFT system for the 3-Axis Portal example (described in Sect. 6):

- a Mealy automaton graph with two states (left) as well as
- a Moore automaton graph with four states (right).

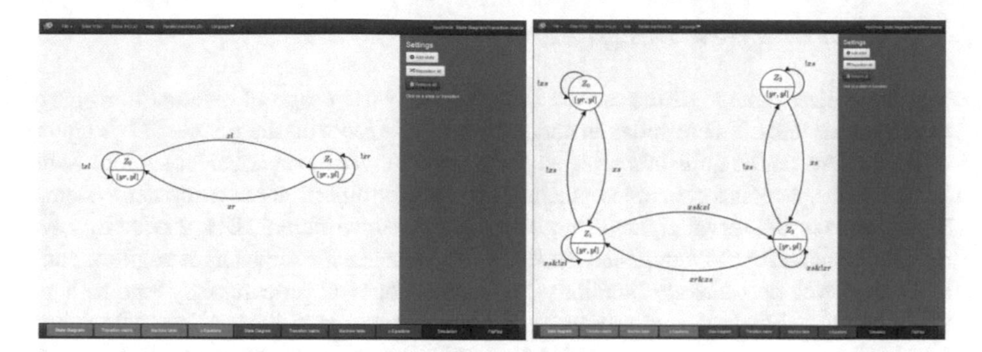

Fig. 5. GIFT as integral part of GOLDi for an FSM design

For a software- or hardware-oriented design, GIFT can generate the next state and output functions (minimized or not minimized) to use them in the third-party tools (e.g. Atmel Studio or Quartus Prime) afterward.

For control of the hardware models via FSM, the student can proceed to the simulation process using graphical controls. This leads to appropriate state transitions caused by the changing set of input variables. Via waveform simulation temporal sequence of input and output variables and the internal state variables (coding the states) of partial automata can be shown. The design can be exported to the ECP to control the experiment afterward. Because the design is typically made with user-defined variables (e.g. xl, yr) the student must adapt them to the real sensor and actuator interface notations (e.g. xr must be replaced by x00 of the 3-Axis Portal sensor signals, which means "x-Axis: Crane position right"). This can be done interactively for each input and output variable within the ECP (see Fig. 6).

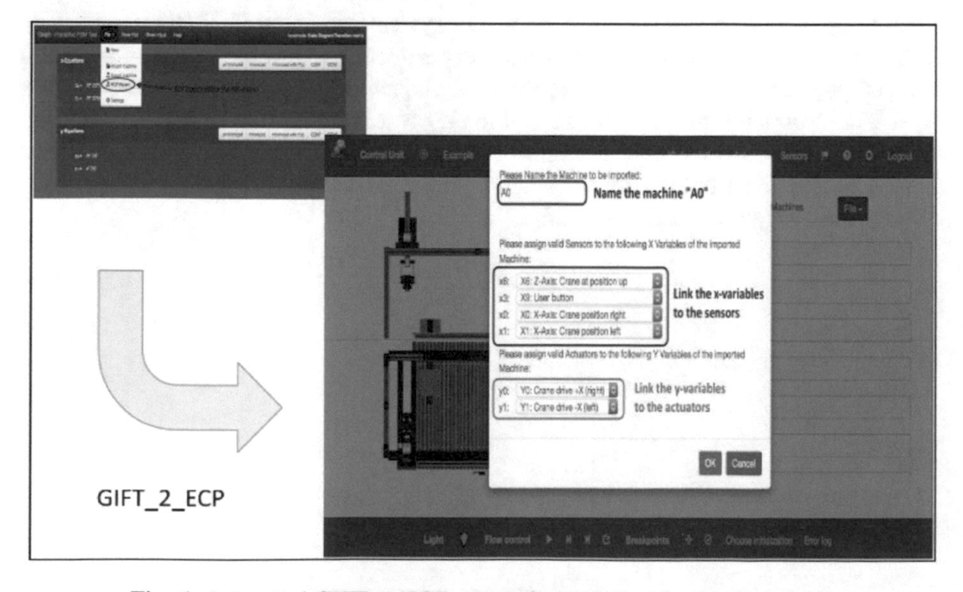

Fig. 6. Integrated GIFT to ECP export functionality, for details, see [5]

3 Vision of a New Design Flow

The main disadvantage of the current design flow is the usage of external third-party development tools. The installation and setup of these tools on the private PC or laptop of the student can be quite complicated for beginners. We also encounter students with limited disk space, incompatible operating systems, or inperformant computer systems. They had to install serval gigabyte large tools, to run the different IDEs needed to solve our design tasks for the remote lab control units. Because the tools target engineers and software developers already familiar with the development process, they tend to have many additional features and settings, a beginner cannot intuitively grasp. Altogether, this leads to a very high entry barrier for novice students. As we strive to keep this barrier as low as possible and get the students programming and designing in minutes rather than in hours it is clear that the currently used design flow needs to be renewed.

Our vision for a new design flow, therefore, encompasses two aspects. The first is a simple, clean, and intuitively user interface. This idea has already proven to be valuable in the Arduino Project [8]. The second aspect is the reduction of software the user has to install. In fact, we visualize and develop a completely web-based system. This way we can reduce the effort for a student from searching for the right software, downloading and installing a software package to just visiting our website. This approach is also implemented e.g. by [10] but is only available for a hardware-oriented approach in VHDL for FPGAs. We want to support all various design flows, described in Sect. 2.

Figure 7 shows an updated version of the design flow as we envision it. Most obviously is that there are no third-party tools, as we moved them from the user computers to our cloud-based backend. Less obvious is that the user will now (when he chooses to make a design for an actual control unit in contrast to an abstract machine like an FSM) be presented with one unified tool we call WIDE (Web IDE). This will lower the barrier from switching programming languages, as the user environment will be the same except the used language during the design process.

An important aspect when using the WIDE system within the GOLDi infrastructure is that WIDE must be assigned to a specific experiment, i.e. a combination of a control unit and physical system. This can be, for example, an FPGA control unit of an elevator in VHDL. Normally this assignment is done by reserving an experiment and starting the ECP. Up to now, the ECP has offered the possibility to upload the firmware files created offline using third-party tools and to program the corresponding control unit. This process only takes a few seconds.

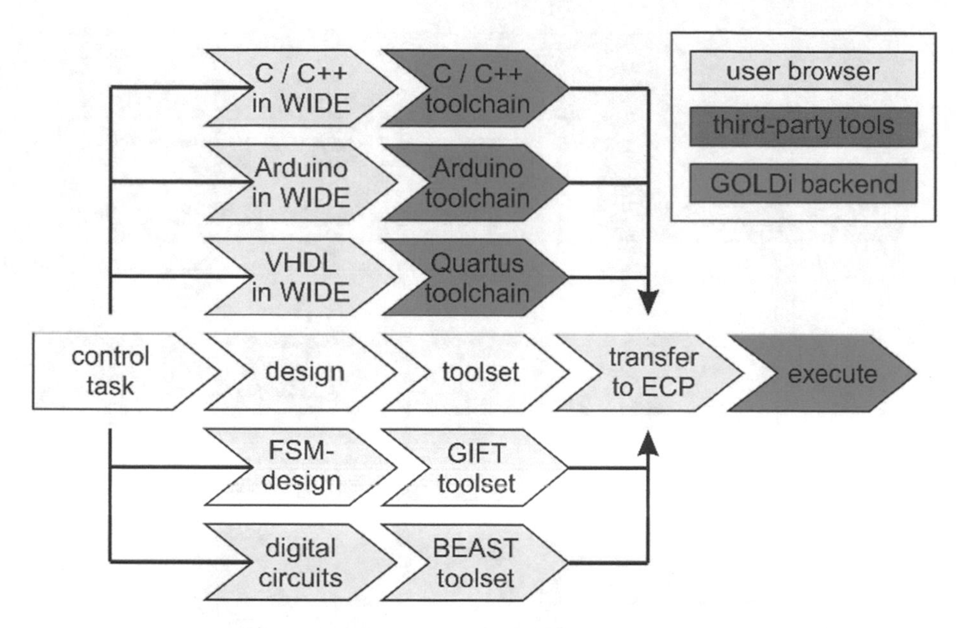

Fig. 7. New unique design flow

If we would integrate WIDE in a similar way, this would have a negative impact on the availability of our remote lab. This is because the reservation of the experiment will block one physical system and one control unit. This was no problem with the design flow used so far, where the design had already been created offline. However, since the user will now develop his solution in our WIDE system, we need to make sure that he will not occupy any physical resources longer than needed. To do this, we implement two working modes for the WIDE system:

1. The first is a standalone mode, where the user can work on his design independent from any experiment reservation. Of course, it is necessary for the user to choose the right programming language and experiment setup prior to developing in this mode.
2. In the second working mode, WIDE is coupled and integrated into the ECP. In this mode, we allow the user to upload his design directly to the chosen control unit and test it with the physical system. This mode is in contrast to the standalone mode time limited to a few minutes to encourage the user to develop his solution using the standalone version.

Figure 8 gives an overview of the WIDE system architecture, which implements the new design flows. In the following two sections, we will describe this architecture in detail.

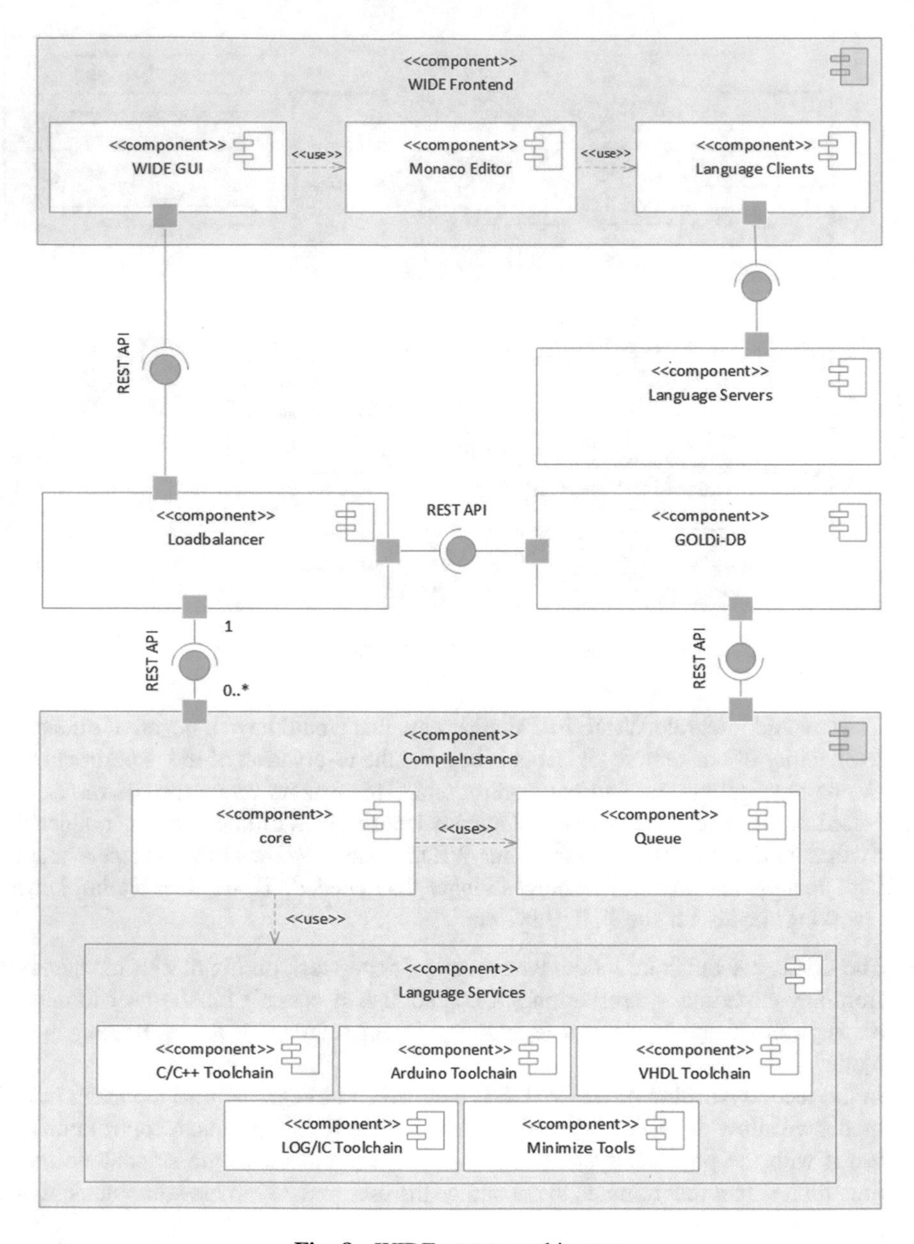

Fig. 8. WIDE system architecture

4 WIDE Backend System Architecture

The WIDE backend system offers services that previously needed to be executed at the user's computer. These services include mainly the compilation of source code from different languages to firmware files. Furthermore, tasks with high performance or complexity that cannot be realized in the web client will be implemented here. Such tasks are the minimization of logic equations and the syntax parsing of source code as provided by many popular IDEs (e.g. Visual Studio [11]).

Because of the different requirements of the toolchains and their usually high demand for system resources our system architecture as seen in Fig. 8 allows for multiple so-called compile instances. All provided services except the syntax parsing and checking are implemented as part of such a compile instance. To ensure a distributed load between all compilation instances and to ensure that each user request is sent to a compilation instance that can process it, i.e. to an instance that has the necessary toolchains installed, we use a single load balancer before all compilation instances. The syntax checking is not integrated into the compile instance, because with the Language Server Protocol (LSP) [12] we use an already tested and well-distributed solution.

4.1 Compile Instance

The compile instance is implemented as a Node.JS [13] application. This decision was made because the GOLDi remote Lab already uses Node.JS and especially JavaScript for the implementation of other services. To reduce the maintenance effort, we strive to unify the used technologies. The compile instance is developed in a version control system. The same system is used to distribute the software together with additional files like hardware templates (see below). The software itself checks in regular intervals if the repository is up to date and updates itself through the version control client.

As seen in Fig. 8 the compile instance consists of three parts. The main abstraction we did was to consider everything as a toolchain that allows for compilation, reports errors, and other information during this time. An arbitrary number of such toolchains are organized in the "Language Services" component. Based on this, abstraction the "core" component is used to receive any request from the frontend and schedule the request to the proper toolchain. When a compilation process is in progress, the compile instance will not start another compilation, even if it is another language. This prevents race condition and deadlocks between tools that are used potentially in multiple toolchains (e.g., both the C-Toolchain and the Arduino-toolchain are using GCC). This, however, means that the "core" component needs to keep track of open requests. For this, it uses the "Queue" component. While a request is kept in the queue, it is assigned an estimated waiting time, based on the current compilation in progress and the tasks waiting in front of the considered request. This waiting time is updated and communicated to the client constantly. The estimated time to completion for all tasks in the system is transmitted in regular intervals to the load balancer to ensure an equal workload between all compile instances.

Language-Specific Toolchains. Most toolchains are executed similarly: The first step is to unpack the files that are part of the compile request to a temporary directory on the file system. These files will then be integrated with a hardware template that is specified as part of the request. Most of the time the files are extended by hardware-specific software libraries and project files. The next step is to compile the generated project, which is of course highly dependent on the used language. During this process, all the output messages from the vendor toolset will be captured and transmitted to the client to be analyze for potential errors. When the compilation is successful the toolchain may, dependent on the original request, upload the firmware file to the GOLDi-Database via a REST-Interface – to allow for a device programming issued by the GOLDi-Infrastructure itself. Because many steps are very common between multiple languages, we use an independent helper class to reduce redundant implementations.

Currently, we already support or implement the following toolchains in WIDE:

- **C-Toolchain:** The toolchain from Microchip using AVR-GCC to compile C and C ++ code to firmware for our Microchip AVR microcontrollers.
- **Arduino-Toolchain:** Extension of the C-Toolchain with added support for the Arduino software library as well as some simplifications originally made by the Arduino IDE. With this toolchain, we support Arduino compatible code in our remote lab.
- **VHDL-Toolchain:** A Toolchain to synthesize VHDL code using Intel Quartus Prime to a netlist file for usage in Intel MAX V CPLDs.
- **LOG/IC-Toolchain:** This toolchain takes our proprietary logic description Language LOG/IC and translates it to VHDL code, which then is synthesized using Intel Quartus Prime. Therefore, it can be seen as an extension of the VHDL-Toolchain.

Hardware Specific Templates. As already mentioned, toolchains can support multiple Templates. The templates are included in our version control repository and therefore will be updated in the same procedure as the main application. These templates solve four problems:

1. The first one is that sometimes some additional files besides the actual source code are necessary to compile. These can include project files, where the project structure, used libraries, and other configurations are defined or additional code that is necessary to communicate with our laboratory. For example, in our laboratory, we use a UART interface to communicate the state of different sensors to and the desired state of the actuators back from the CPLD that runs the user design. The dedicated pins for this interface as well as the hardware, i.e. the chip, the design will be executed with, need to be defined in a configuration file. The usage of templates eliminates the overhead of defining such files and allows the students to concentrate on the actual design.
2. The second problem we solve with using templates is that some hardware can be misconfigured by some project settings to harmful conditions that will not allow the programming of a new design or physically damage the hardware by causing a short circuit. The usage of templates allows us to fix the project parameters to safe values and protect us from harmful usages of our laboratory.
3. The usage of templates also allows for the transparent adaption to new or different hardware. For example, we use different templates to interface different hardware revisions of our control boards, in which we have different means of communicating with the physical model (e.g., we have hardware that communicates using the UART interface and hardware that communicates directly by using multiple GPIOs).

4. Besides adapting to the selected control hardware with the help of templates, we also can adapt to the model the user will control. This will allow us to predefine macros, functions, and structures for the user to access the model intuitively. This can range from renaming the pin names of the microcontroller to self-explanatory ones to implementing whole subroutines for simple control tasks involving multiple sensors and actuators.

Integrated Logic Minimization. We want to offer the students an additional feature while they design logic equations in the LOG/IC language: They should be able to select an equation in the code editor and let the system minimize that. As the problem is NP-complete [14], this task requires a large amount of memory and computational power as the input equation gets larger. Client code in JavaScript inherently runs slower compared to a hardware optimized application in C++, especially if you consider that the client could be a mobile device. Therefore, we decided to offload the minimization from the client to our servers.

The minimization integrates well in the architecture described above because you can see the logic equation as a kind of language that is compiled to a minimized equation. Consequently, we integrate the minimization as another toolchain besides the four other toolchains described above.

4.2 Load Balancer

The load balancer is a Node.JS application developed for two purposes. The first purpose is to forward a request from the client to the right compile instance, i.e. the instance that has the right toolchain installed to handle the request. The second purpose is to keep the workload evenly distributed between the compile instances.

To archive this, the load balancer will query the GOLDi-Database in regular intervals to get a list of available compile instances. From each compile instance, it will constantly retrieve the following information:

- current utilization, i.e. if there is a task already running
- queued tasks
- supported toolchains

The routing of an incoming request is based on this information and the request itself, which contains e.g. the inquired toolchain. In case of an erroneous condition, the load balancer will transparently forward the request to another instance. In case no compile instance is available, it will inform the client about the error. Such errors could be timeouts or connection failures.

4.3 Language Server

It is hard to implement a full syntax check and auto-completion. This is because certain programming languages themselves have a very complex syntax. Furthermore, as described above with the usage of templates, the client is not in possession of the full source code including all the libraries. Despite this, we want to support our students in the best possible way and strive to offer this kind of convenience that is already offered by most offline IDEs today.

Luckily, in the last years, there has been an effort to standardize an interface that allows extending text editors by this functionality. This interface is defined by the Language Server Protocol (LSP) created by Microsoft [12]. By utilizing this protocol, we can use already available language servers like clangd [15] for C/C++ code or implement our own server to provide support for new or proprietary languages like LOG/IC.

The various language servers allow us to analyze and transmit information regarding the code completion, hover information (e.g. the type of a variable), location of definition, references, and precompiling warnings and errors. All this information will be transmitted to the web client via LSP, which will display it accordingly, e.g. like shown in Fig. 9 for detected errors.

```
22      while(Sensors.PositionX != x)
23      {
24          if(Sensors.PositionX < x)        expected ';' before numeric constant
25          {                                Quick Fix...   Peek Problem
26              Actuators.XAxisDriveToXPlus    1;
27              Actuators.XAxisDriveToXMinus = 0;
28              Actuators.YAxisDriveToYPlus  = 0;
29              Actuators.YAxisDriveToYMinus = 0;
30              Actuators.ZAxisDriveToZPlus  = 0;
31              Actuators.ZAxisDriveToZMinus = 0;
32              GOLDiInterfaceSendData();
33              while(Sensors.PositionX < x)
34                  GOLDiInterfaceSendData();
35              Actuators.XAxisDriveToXPlus    0;
36              Actuators.XAxisDriveToXMinus = 0;
37              Actuators.YAxisDriveToYPlus  = 0;
38              Actuators.YAxisDriveToYMinus = 0;

Console
3AxisPortal_Plotter/UserDesign/UserDesign.c: In function 'DriveToPositionX':
3AxisPortal_Plotter/UserDesign/UserDesign.c:26:4: warning: statement with no effect [-Wunused-value]
  Actuators.XAxisDriveToXPlus   1;
  ^
3AxisPortal_Plotter/UserDesign/UserDesign.c:26:34: error: expected ';' before numeric constant
  Actuators.XAxisDriveToXPlus   1;
                                ^
3AxisPortal_Plotter/UserDesign/UserDesign.c:35:4: warning: statement with no effect [-Wunused-value]
  Actuators.XAxisDriveToXPlus   0;
```

Fig. 9. Example of a syntax error shown in WIDE

5 WIDE Frontend Architecture

The basic architecture of the frontend can be seen in Fig. 8. It has three main components WIDE GUI, Monaco editor, and Language Clients. The application was developed as an HTML5 single page application (SPA) using modern technologies like web components [16]. As such, the whole GUI is structured accordingly to its graphical representation shown in Fig. 10. All components interact with each other using the browser event system. The Monaco editor [17] is embedded in a custom HTML element but is actually an independent third-party component. As described in the last section, we use the LSP to extend the web client with source code aware suggestions and information. The implementation of the client for the language servers provided by the backend is encapsulated in the Languages Client component. Actually, we utilize a separate language client for every language we support. This allows us to display any information in the most sensible way.

The GUI (see Fig. 10) can be divided into 3 regions: At the top, we have a toolbar concerning project management. The left side will show an interactive file tree. The right and the biggest surface is reserved for Monaco code editor. In the following, we will discuss these regions in more detail.

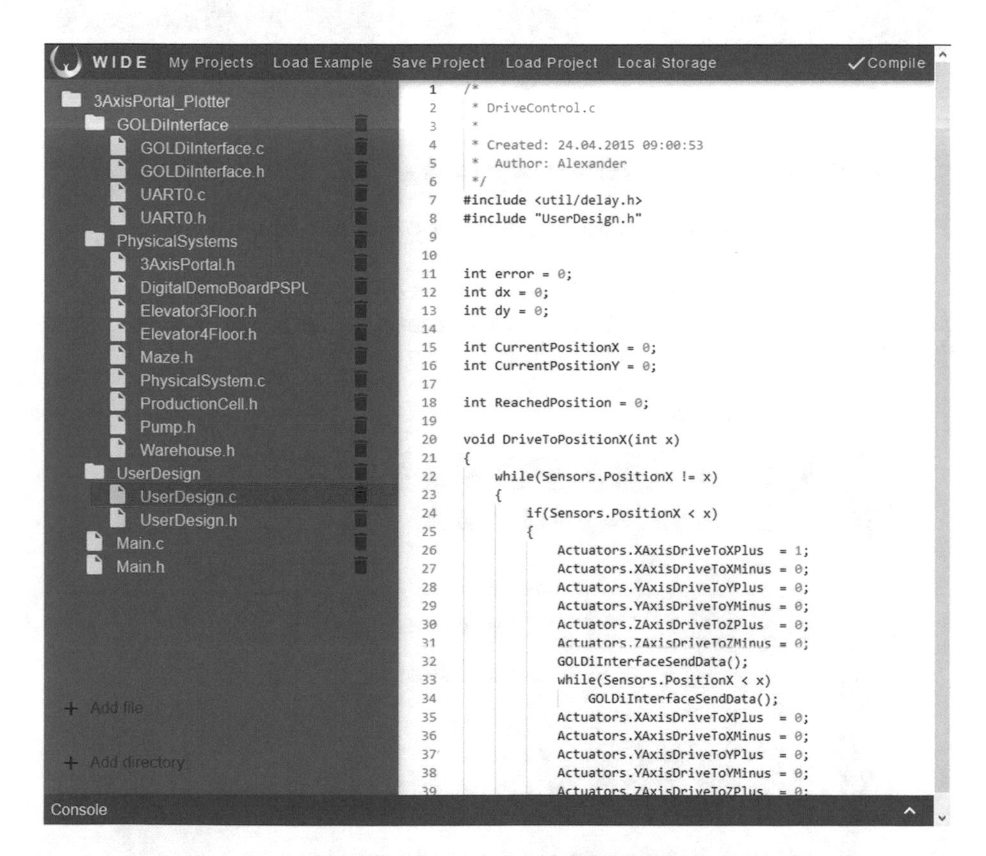

Fig. 10. Overview of the WIDE GUI

5.1 Project Management

The button "My Projects" allows the user to save and switch between multiple designs like shown in Fig. 11(a). Therefore WIDE uses the Local Storage of the web browser. When the user chooses to create a new project, he will be asked to specify the desired control unit, language, and physical system as shown in Fig. 11(b). The next button

"Load Example" will show a list of available examples and templates (Fig. 11(c)). These show how to interact with the physical system. The two buttons "Save Project" and "Load Project" enable the user to either download the current project as a proprietary ".wide" file format or as a zip container respectively upload them back to the WIDE system.

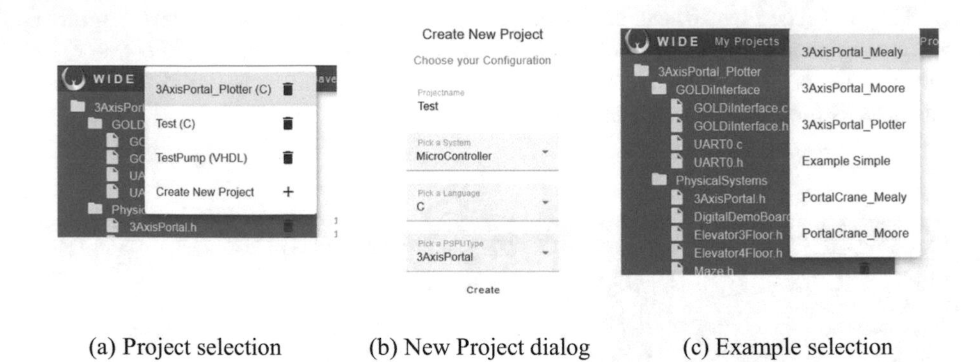

(a) Project selection (b) New Project dialog (c) Example selection

Fig. 11. Various WIDE project management tools

The button "Local Storage" will open a modal dialog, listing all projects the user has created and allows to download or delete them from the local storage as seen in Fig. 12.

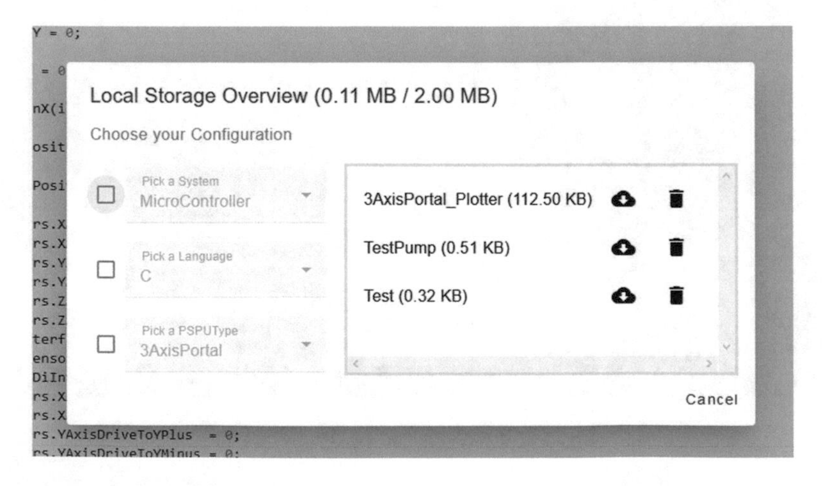

Fig. 12. Overview of the local storage management of WIDE

5.2 File Management

The file tree on the left side of the GUI (Fig. 10) enables standard file operations like the creation, renaming, and deletion of files and directories. To manage files between different directories, we support an intuitive drag and drop based file movement the user already knows from various other file managing application like windows explorer. In addition to this great flexibility, we also monitor for numerous errors the user could make while managing his project structure. The most common one is the renaming of files to a nonconforming file extension. In case we detect such mishandling, we will prevent e.g. the renaming and inform the user about the situation. In addition, files that have warnings or errors connected to them will be marked in a different color.

5.3 Code Editor

For the implementation of the code-editing component in our GUI, we chose to utilize the Monaco text editor used in the popular Visual Studio Code IDE. The text editor is written in JavaScript. It has great support for various partly quite extensive features, e.g. line numbering, search and replace functionality, multiline editing, auto completion, hover information, shortcuts, and syntax highlighting. Some of these features are shown in Fig. 13.

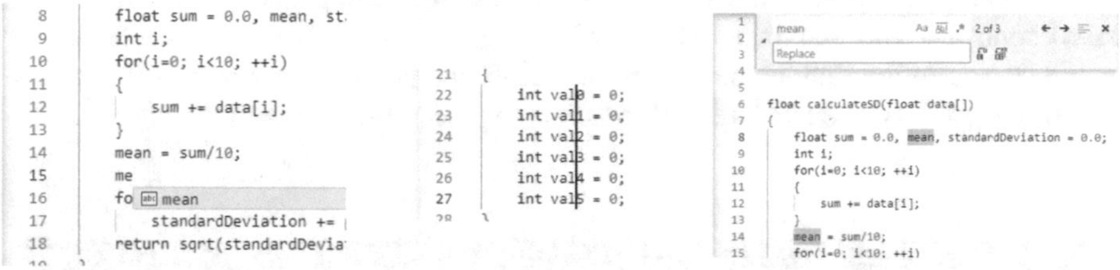

(a) Autocompletion (b) Multiline editing (c) Search and replace

Fig. 13. Various features supported by the Monaco code Editor

In our editing component, we wrap the Monaco code editor, retaining all the features above, configure it for our needs, do some graphical styling, and extend its functionality. Furthermore, we support the editing of multiple project-files. In connection with this, we have extended the component in a way that based on the file extension the right programming language and preset for the text editor is applied automatically. One of the bigger extensions is the support of the Language Server Protocol as described in Sect. 4.3 because it tightly integrates into the features, the Monaco editor already implements on the one hand and interfaces with a completely separated system for source code analytics.

With this approach, we achieve a state of the art code editing experience for the student while providing a clean and easy to understand user interface for beginners in programming and hardware design not comfortable with the cluttered user interface of traditional design tools and IDEs.

6 User Experience

As an example, to introduce Finite State Machines for students in the first semester, we will discuss a design task by using the electro-mechanical model “3-Axis portal” from the GOLDi remote lab infrastructure:

“On one spindle of a 3-Axis Portal crane, a tool carriage can be moved to the right and to the left. Limit switches provide input information on the left end position (x_l) as well as the right end position (x_r) of the tool carriage (x_r, x_l).

The motion can be controlled via the output variables (y_l, y_r) between

- *motion to the left (y_l= 1, y_r= 0),*
- *motion to the right (y_l= 0, y_r= 1) and*
- *stop (y_l= y_r= 0).*
- *An additional input variable x_s signalises stop motion (x_s= 0) or*
- *movement (x_s= 1) to the left or right.*

After a possible break, the movement in the original direction should be continued.”

To realize the given task, the student can choose his preferred design flow. We assume that he has chosen a software-orientated design using C for a microcontroller-based control of the 3-Axis portal. After logging into our GOLDi website the student would first open the standalone version of WIDE and create a new project for the task. To do this, he has to enter basic information for the project as seen in Fig. 14.

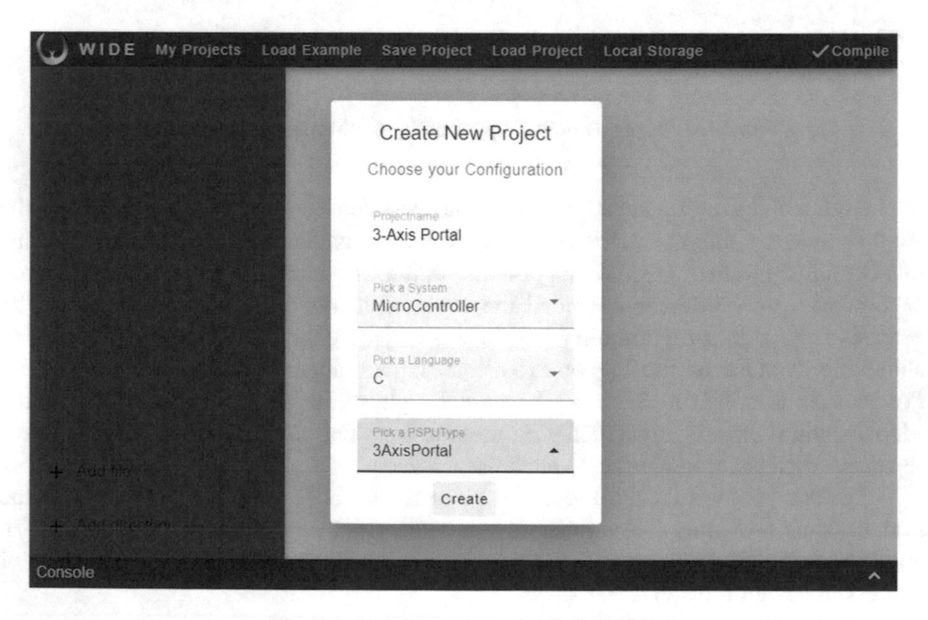

Fig. 14. Configuration of a new project

After the project setup, he has to create a new "main.c" file and can now start programming his solution for the task, shown in Fig. 15. At any time during the development, the student can try to compile his design in our system to check for possible compilation errors.

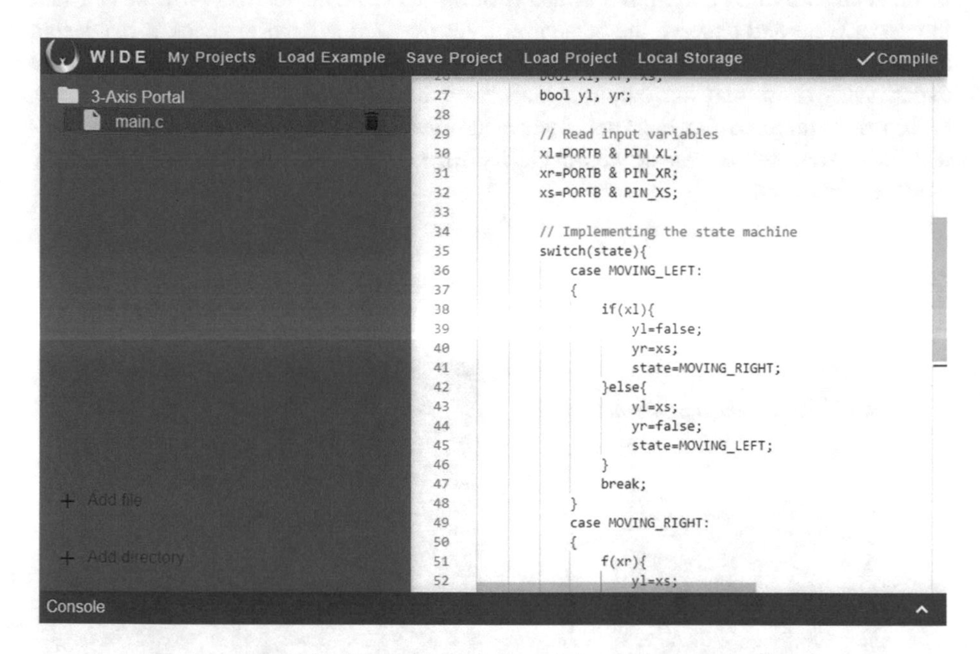

Fig. 15. Editing the solution of the student's task

Once the student is satisfied with his design and wants to try it on the real hardware he reserves the experiment (in this case consisting of the microcontroller and the 3-Axis portal) in our booking system (see Fig. 16).

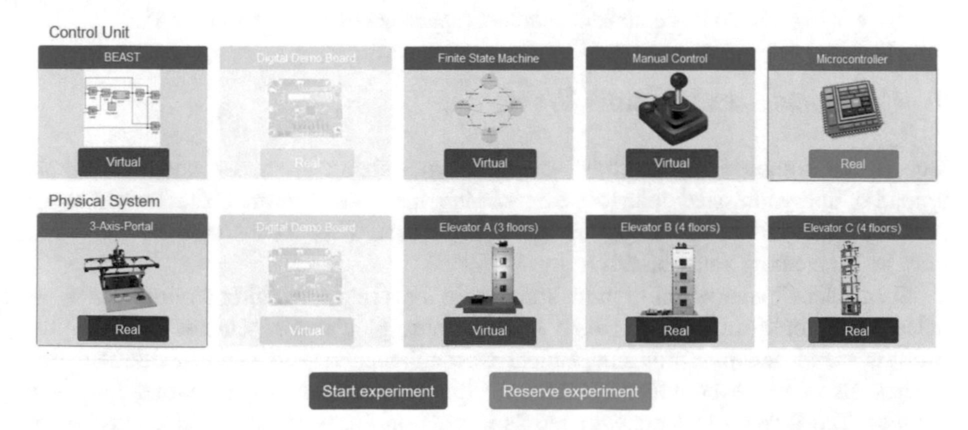

Fig. 16. Experiment configuration: microcontroller – real 3-axis portal

Usually, he will get access to the system within seconds or minutes. Once he has control over the experiment, he can start the WIDE version that is integrated within the ECP. The project created with the standalone version of WIDE already exists and can now be uploaded to the real microcontroller. Therefore the integrated WIDE version features an additional button; not available in the stand-alone version. Now he can start the experiment and observe the behavior of the physical system to check if his design fulfills the given task (see Fig. 17). In case there is a simple error in the design (e.g. the experiments show that the output signals must be swapped), the student can fix his design right there within ECP and upload the reworked design as well. All the modifications made will be saved automatically for later use e.g. in hands-on laboratory work in classroom.

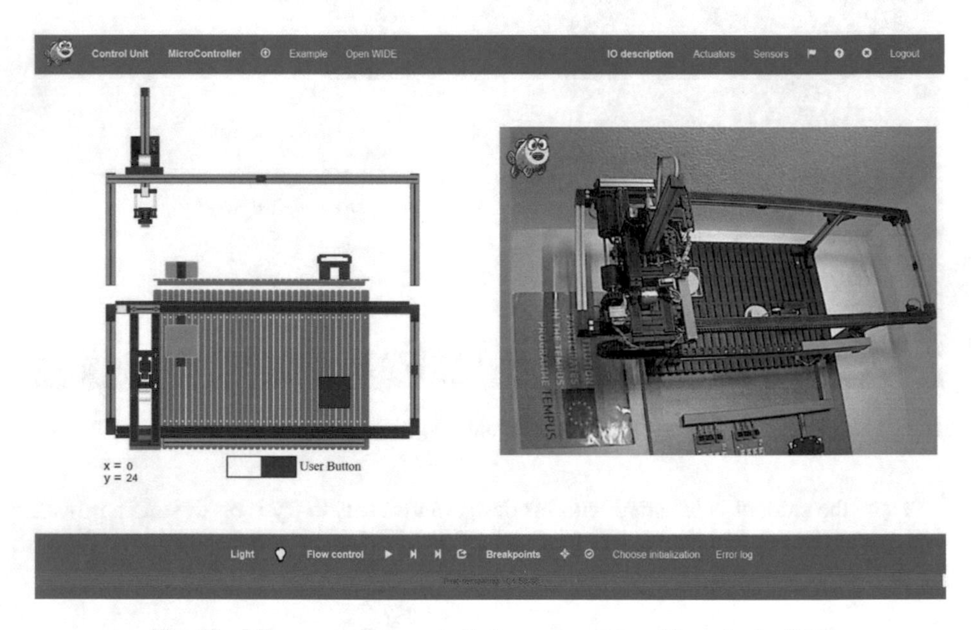

Fig. 17. Microcontroller controlled experiment handling via the ECP

7 Conclusion and Future Work

Due to the complete “abstraction” of build/synthesizing processes and the manual upload of the synthesized code to the remote lab, there are no more obstacles due to the required “additional knowledge” of third-party tools. This means that the focus can now be on creating functionality.

By “hidden” integration of industrial design tools into e-learning environments, we achieve a simplification from huge and complex platform-dependent IDEs to just browser-based and platform-independent web editors. Entered code in different languages also can be compiled/synthesized and programmed to the target hardware devices. The focus for students or pupils is now on the pure design of control algorithms in the preferred specification language.

Acknowledgment. The authors would like to acknowledge the work of Pierre Helbing for the Arduino integration and templates, Jan Höpfner for the development of the Language Server, Sven Mollenhauer for the WIDE backend implementation, Luisa Bortz, Ziyang Song, Eric Winzer for the software project WIDE as well as Florian Köhler, Patrick Langer, Söhnke-Benedikt Fischedick for their contribution to the ELWS project.

References

1. TU. Ilmenau, GOLDi-labs cloud Website, 12 Dec 2019. http://goldi-labs.net
2. Henke, K.: Fields of applications for hybrid online labs. Int. J. Online Eng. (iJOE) **9**, S3 (2013)
3. Henke, K., Vietzke, T., Hutschenreuter, R., Wuttke, H.-D.: GOLDi — Grid of online lab devices Ilmenau, pp. 283–284. IEEE, Piscataway (2016)
4. Henke, K., Vietzke, T., Hutschenreuter, R., Wuttke, H.-D.: The remote lab cloud "GOLDi-labs.net", pp. 37–42. IEEE, Piscataway (2016)
5. Henke, K., Fäth, T., Hutschenreuter, R., Wuttke, H.-D.: GIFT - an integrated development and training system for finite state machine based approaches. in online engineering & Internet of Things, vol. 22, pp. 743–757. Springer, Cham (2018)
6. Wuttke, H.-D., Henke, K., Hutschenreuter, R.: Digital twins in remote labs. In: Proceedings of the REV Conference 2019. Springer (2019). p. in print
7. Atmel Studio, 13 12 2019. https://www.microchip.com/mplab/avr-support/atmel-studio-7
8. Arduino, 13 December 2019. https://www.arduino.cc
9. Quartus Prime, 14 December 2019. https://www.intel.com/content/www/us/en/software/programmable/quartus-prime
10. Angulo, I., Garcia-Zubia, J., Orduna, P., Rodriguez-Gil, L., Villar, A.: Integral Remote laboratory for Programmable Logic, pp. 253–255. IEEE (2019)
11. Visual Studio, 14 December 2019. https://visualstudio.microsoft.com/de/
12. Language Server Protocol, 14 December 2019. https://microsoft.github.io/language-server-protocol/
13. Node.JS, 13 December 2019. https://nodejs.org/en/
14. Hemaspaandra, E., Wechsung, G.: The minimization problem for boolean formulas. SIAM J. Comput. **31**(6), 1948–1958 (2002)
15. Clangd, 13 November 2019. https://reviews.llvm.org/diffusion/L/browse/clang-tools-extra/trunk/clangd/
16. W3C, Web Components, 15 October 2019. https://github.com/w3c/webcomponents
17. Monaco Editor, 14 November 2019. https://microsoft.github.io/monaco-editor/

Remote Lab to Illustrate the Influence of Process Parameters on Product Properties in Additive Manufacturing

Siddharth Upadhya(✉), Joshua Grodotzki, Alessandro Selvaggio, Oleksandr Mogylenko, and A. Erman Tekkaya

Institute of Forming Technology and Lightweight Components, Dortmund, Germany
{Siddharth.Upadhya,Joshua.Grodotzki,Alessandro.Selvaggio,Oleksandr.Mogylenko,Erman.Tekkaya}@iul.tu-dortmund.de

Abstract. Additive manufacturing, which enables the production of highly complex components that were previously next to impossible to manufacture, has evolved from a tool for rapid prototyping to an integral part of many production lines in the metal manufacturing industry. Therefore, it is critical that today's students, the manufacturing engineers of tomorrow, get a fundamental understanding of the process and the influence of the various process parameters on the process and the final product properties. The developed remote lab offers the students an opportunity to vary different process parameters and characterize the performance of specimens manufactured under different conditions with help of the uniaxial tensile test and to quantitatively analyze the interplay of different process parameters on the final product. From the educator's point of view, the remote lab will allow for a higher throughput of students in the field of additive manufacturing without compromising on the machine and user safety, as well as the effectiveness of the lab.

Keywords: Engineering education · Additive manufacturing · Remote labs

1 Introduction

The advent of additive manufacturing (AM) has changed the way products are designed as well as manufactured. The manufacture of hitherto impossible designs, such as bionic designs or topology optimized structures, have been made possible and these parts are finding a place everywhere from customizable shoes to new age aeroplane interiors. Though the possibilities that AM offers is vast, it is still a relatively unknown territory compared to the other manufacturing processes, such as casting, machining and forming. It is thus essential that the students of today, the engineering workforce of tomorrow, get first-hand exposure to AM. Here, understanding the numerous process parameters, of which there are many more to consider compared to conventional manufacturing methods, is a crucial aspect of their education.

M. E. Auer and D. May (Eds.): REV 2020, AISC 1231, pp. 456–464, 2021.
https://doi.org/10.1007/978-3-030-52575-0_37

1.1 Additive Manufacturing

Additive manufacturing is a layer by layer manufacturing method where a part is built up gradually by the addition of material. This allows the manufacture of complex parts without the need for specific tools [1]. Raw material and the 3D CAD data of the part is fed as inputs to the machine and a part ready for finishing is obtained as the output.

The benefits of AM are plenty. It allows the creation of lighter, more complex designs that are too difficult or even impossible to build using traditional dies, moulds, milling and machining. AM also excels at rapid prototyping. Since the digital-to-digital process eliminates traditional intermediate steps, it is possible to make alterations on the run. When compared to the relative tedium of traditional prototyping, AM offers a more dynamic, design-driven process [2]. AM is also an integral part of the "Batch Size 1" philosophy of future manufacturing where each product will be customized as per user requirements.

These and other potential benefits have led to accelerated investment in the validation of current AM technologies and the rapid exploration of new concepts. As one point of evidence, the worldwide revenue from AM products and services increased by 35.2% from 2013 to 2014, now totalling $4.1 billion [3]. Many research centers have been and are being established to explore opportunities for innovation and adaptation of AM [4].

1.2 Motivation and Aim of a Remote SLM Laboratory

The importance of remote labs has been long understood in the engineering community and so the Institute of Forming Technology and Lightweight Components (IUL) at the TU Dortmund University has developed in collaboration with other partner universities more than 10 remote laboratories under the "ELLI- Excellent Teaching and Learning in Engineering Education" banner with the goal of improving engineering education. The capstone of this effort is the remote material characterization lab which allows the students to remotely perform various material characterization tests [5].

At the IUL, a special lecture on AM has been offered for the past four years and the interest and involvement shown by the students has been on a constant upward trajectory. The students already have access to FDM 3D printing machines for their student projects as part of their coursework. But, it is equally critical to understand the selective laser melting (SLM) printing of metals, which is much more complex and can have as many as 200 influencing process parameters [6]. Naturally, some are more influential and critical than others and the students must gain a thorough understanding of at least the critical parameters.

A clear trend is being observed where more and more additively manufactured parts are being subsequently formed which puts even more significance in proper quantification and understanding of the product characteristics as their behavior is completely different compared to conventional formed parts.

It is clear that AM will gain more importance among the other, traditional manufacturing methods in the future and in-depth understanding of this subject will be a vital asset to any engineer. This served as the motivation to develop a lab that will enable the students to understand the process, the various process parameters and the influence and interplay of these process parameters on the final product characteristics.

The uniaxial tensile test serves as the most basic and fundamental test to characterize the performance of a material or, in this case, of a product. Additionally, the tensile specimens are ideally suited to be additively manufactured as well since they are small in size, lightweight and multiple specimens can be printed at once, reducing the overall time to manufacture which is typically a drawback of AM.

Being in the unique position of having a remote tensile test at the institute, gives the perfect opportunity to offer the entire lab – SLM production of the specimens and subsequent material characterization – in a remote manner. Hence, the remote SLM lab will be complemented by the remote tensile lab to aid in the characterization of the printed products.

2 Selective Laser Melting

Amongst the various different types of AM processes, SLM is one of the most prevalently used processes for metal 3D printing in the industry owing to its ability to produce complex parts without compromising on the quality [7].

2.1 Pre-processing

As in every AM process, pre-processing is an essential stage in the manufacture of a part by SLM. First, a 3D STL model of the part to be printed is generated. After this, the 3D file is imported into the slicing software and here careful consideration must be given to the positioning, arrangement and orientation of the part in the build chamber as this has a significant influence on the process cost, stability and final product characteristics. Apart from the spatial parameters, various process parameters have to be set as well. These parameters, which the students over the course of the lab will vary and understand the influence of each, will be discussed in detail in the upcoming chapters. Once all the parameters have been set and the part model is sliced into layers, the actual manufacture can begin.

2.2 Manufacturing

This stage consists of three steps as seen in Fig. 1:

- Powder deposition: A thin layer of metal powder is uniformly deposited over the build plate with the help of a wiper.
- Melting: The laser beam, which is deflected with the help of scanning mirrors, scans the powder bed according to the layer cross-section of the part. The particles exposed to the laser melt and fuses together on solidifying.
- Lowering: On the completion of one layer, the build plate is lowered by an amount corresponding to the layer thickness.

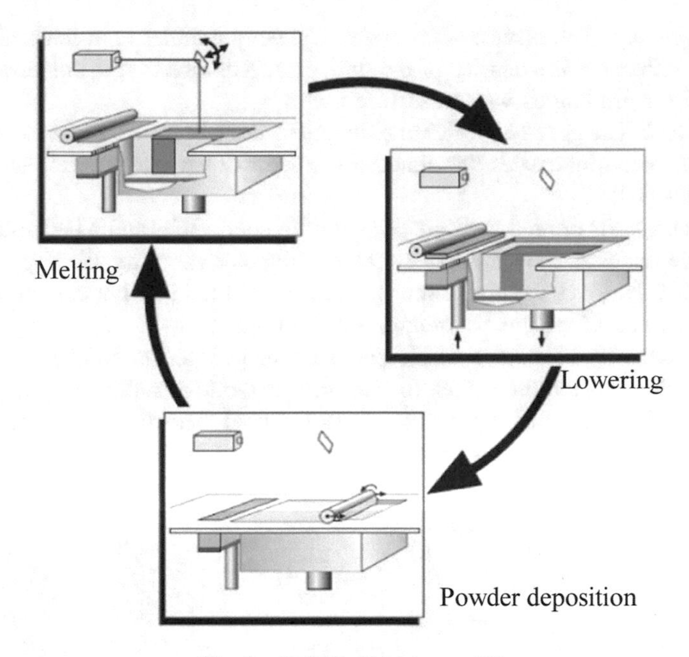

Fig. 1. SLM build process [8]

The above three steps are repeated layer by layer until the entire part is completed.

3 Process Parameters

There are a number of different parameters that can be varied while setting up the print job. The influence of certain parameters on the final product property is negligible, while certain others are critical to the proper running of the machine and must not be changed under normal circumstances. Therefore, only those parameters which the students will eventually be able to control are discussed below:

- *Layer thickness:* This defines how thick each individual layer will be. The lower the value, the smaller the thickness of each individual layer and thus the surface finish and geometrical reproduction will be better. But on the other hand, owing to the higher number of layers needed to be printed, the time taken for printing increases significantly. Thus, a layer thickness should be chosen that gives the best balance between print quality and print time. Conventionally in SLM printing, the thickness is in the range of 20–100 μm.
- *Laser Power:* This defines the intensity of the laser beam for a given beam diameter. The intensity of the beam influences the density of the part and thus has a significant influence on the part's ability to undergo deformation before failure. It also influences the geometrical accuracy of the final part due to thermal expansion [9].

- *Hatch Spacing:* The distance between two nearby parallel scan lines. This parameter too influences the density of the final part. Additionally, it influences the time required for printing as well as surface quality.
- *Scan Speed:* The speed at which the deflected laser beam scans across each layer. The scan speed influences the dimensional error of the part as well as the surface roughness [10].
- *Infill Density*: 3D printed parts are printed with solid shells and a lattice-like infill so as to save material, time and money. This parameter varies the density of this infill. The higher the infill, the higher the strength of the final part but conversely the cost of the product, as well as its weight, will be higher.
- *Orientation:* The orientation of the part with respect to the build plate, as seen in Fig. 2, influences the time taken to print the part, due to differences in print height as well as the number of support structures required. The dimensional accuracy and the residual thermal stresses are also significantly influenced by this parameter.

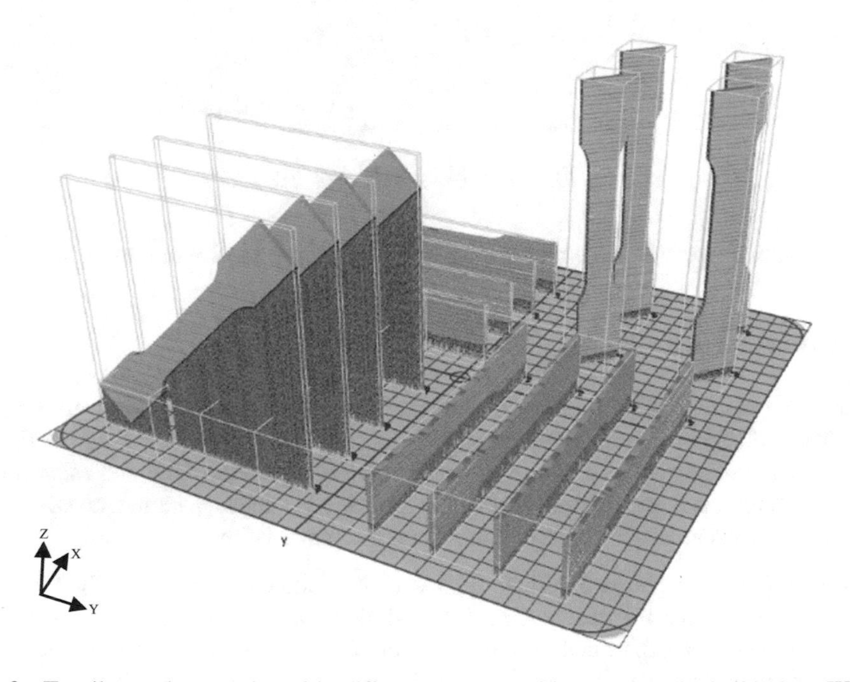

Fig. 2. Tensile specimens oriented in different manners with respect to the build plate. Wiper movement along Y-direction.

The above-mentioned parameters do not work in isolation but have interdependent effects as well. For example, increasing the layer thickness without increasing the laser power might lead to insufficient energy being deposited into each layer and thus lead to a weaker final product. Hence, it is critical to understand the influence of each parameter as well as how they are dependent on each other.

4 Remote SLM Lab

The benefits of a remote lab, which allows students time- and location-independent access to labs, while allowing them to learn and explore at their own pace and convenience are well established [11]. In the case of the SLM lab, a remote version of it makes a really strong case compared to a hands-on lab. This is due to certain unique challenges or restrictions that are associated with this manufacturing process and which a remote access lab is best suited to overcome them. Nevertheless, due to the basic nature of the process and set up of the machine, running it in a completely remote modus is impossible and human intervention is a must during certain stages, as will be explained in the following.

4.1 Challenges

There are safety and cost-related challenges in offering the SLM lab in a hands-on manner. The particles of the metal powder used in the SLM process are in the nanometre range, smaller than the pores on the human skin leading to skin penetration on contact and need to be handled with care. Certain metallic powders like Magnesium, Aluminium are easily combustible while certain others like Nickel have shown to be carcinogenic [12]. Thus, the ever-updating safety regulations mandate the use of full-body hazmat suits with a face mask during specimen removal to prevent any bodily contact or inhalation of these tiny metal particles. Providing each student with this safety equipment would not be feasible, especially in the case of large student groups as typically seen in German bachelor engineering courses. Similarly, the university safety regulations mandate that every user must undergo safety training plus special machine-specific training. This would lead to huge time and cost-related issues as well. All these challenges can be overcome by offering the lab in a remote fashion which will ensure the effectiveness of the lab without compromising on machine and user safety. Furthermore, the actual manufacturing process is a remote like operation anyways with no human interaction once the print job has started. The remote lab, therefore, does not take away any part of the user experience related to the actual process. However, loading of the sliced layer file, starting the print job and removal of the specimens still need to be done manually by an experienced and certified user since the machine can be easily damaged.

4.2 Workflow

The pre-processing software of the machine has been simplified and recreated as a web page that the students can access as seen in Fig. 3. The webpage gives the students a graphical view of the part, in this case, a standard A_{50} tensile test specimen, with respect to the build plate. They can vary the orientation of the part in the build space with the slider buttons.

Similarly, with the help of different slider buttons, they can set the values for the different process parameters mentioned in Sect. 3. Once all the parameters have been set, the job can be submitted. A data file will be generated with all the different parameters and sent over to the lab admin.

The lab admin will check all the data files and batch them according to the parameters. Since printing just a couple of specimens during a print run would be a waste of resources and time, files from different students with same or similar parameters will be batched together and printed all at once, in case they have chosen the same layer height.

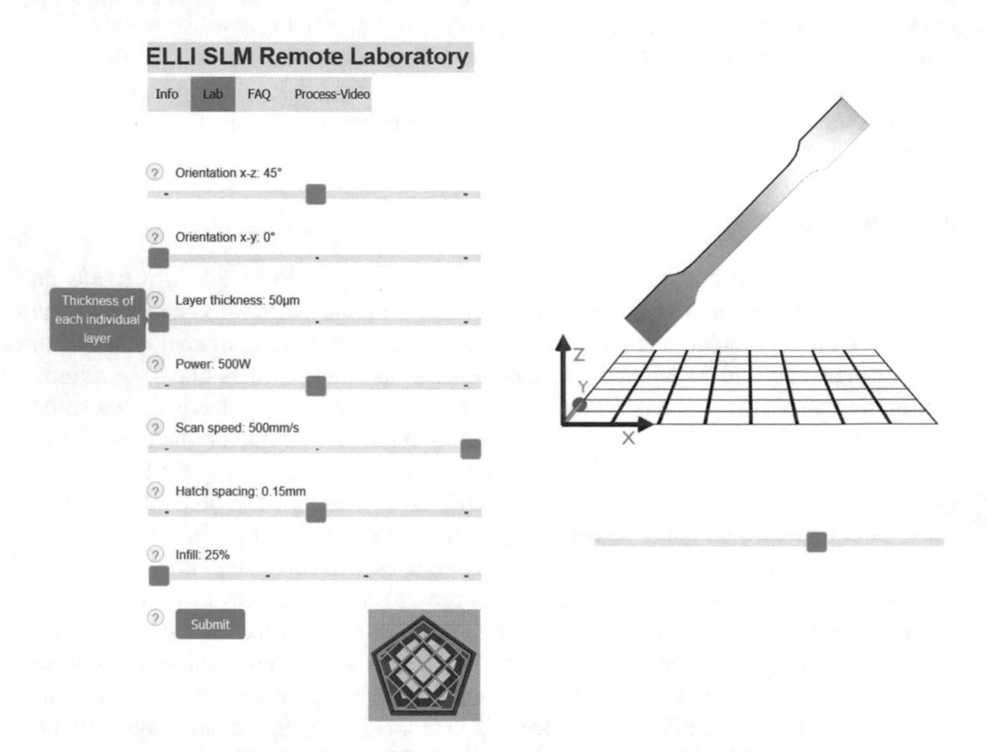

Fig. 3. A simplified web page to set up the print job

The actual printing process takes several hours to days depending upon the chosen parameters. Thus, a live stream of the print will be shown to the students so that they get a feel of the actual printing process. On top, the video is recorded and will be made accessible to the students after the print job is finished so that they can also watched at an accelerated playback speed. A screenshot of the livestream is seen in Fig. 4 on the left.

The printed specimens, after post-processing to remove support structures and cleaning, will be transferred to the remote material characterization lab. The remote material characterization lab enables the students to conduct, amongst others, remote tensile tests.

The students will conduct remote tensile tests of their printed specimens and based on the force-displacement test data, they can compare the performance of the different specimens and quantify the influence of the different process parameters on the product properties. For example, based on the orientation of the part with respect to the build plate, the final tensile test results can have a significant difference as seen in Fig. 4 on the right.

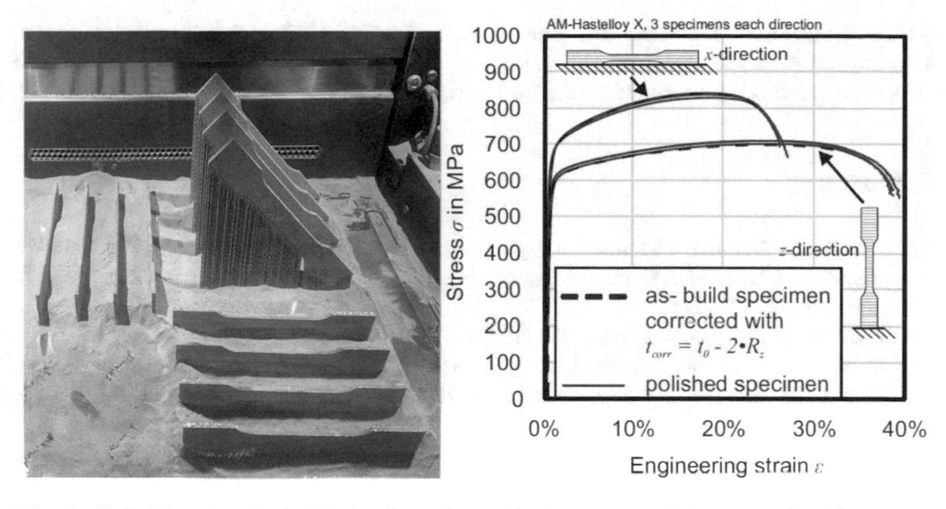

Fig. 4. (left) View into the build chamber after preliminary removal of the powder. The camera position is outside the chamber, since there is no power supply inside the build chamber (right) Stress-Strain diagrams for parts oriented along x- and z- direction [13].

Additionally, with the tensile test data the students can also generate flow curves that they can subsequently use in numerical simulations. These simulations are used to predict e.g. limits of a subsequent forming operation. Those limits can only be calculated correctly, if the material data provided to the software depicts the actual material behavior accurately. Modelling the complex behavior of additively manufactured parts is also an ongoing subject of current research.

5 Conclusion and Outlook

Once the initial trial runs have been completed, the SLM remote laboratory will be offered to the students as part of an advanced manufacturing lab. Through this lab, the students get a deeper understanding of modern additive manufacturing processes and shall be able to quantitatively analyse the interplay of different process parameters on the final product. They will also understand the possibilities as well as limits of the process. From the educator's point of view, the remote lab will allow for a higher throughput of students without compromising on the machine and user safety, as well as the effectiveness of the lab. Initially, the product characteristics will be evaluated only using the tensile test data. But considering the complex stress states usually observed in conventional forming processes such as deep drawing and the pronounced tensile-compressive asymmetry shown by additively manufactured parts, it is essential to test the parts under other loading conditions as well. To this extent, the remote SLM lab will be extended to make use of the already developed remote compression test [14] as well as remote cup drawing labs [15].

Acknowledgement. The work was done as part of the "ELLI2 – Excellent Teaching and Learning in Engineering Science" and the authors are grateful to the German Federal Ministry of Education and Research for funding the work (project no: 01PL16082C).

References

1. VDI 3405. Additive manufacturing processes, rapid manufacturing. Basics, definitions, processes. VDI – Handbuch Produktionstechnik und Fertigungsverfahren, Teil 2: Fertigungsverfahren (2014)
2. What is additive manufacturing? GE Additive 2019
3. Wohlers, T.T., Caffrey, T.: Wohlers Report 2015: 3D Printing and Additive Manufacturing State of the Industry Annual Worldwide Progress Report. Wohlers Associates, Fort Collins (2015)
4. Go, J., Hart, A.J.: A framework for teaching the fundamentals of additive manufacturing and enabling rapid innovation. Add. Manuf. **10**, 76–87 (2016)
5. Ortelt, T.R., Sadiki, A., Pleul, C., Becker, C., Chatti, S., Tekkaya, A.E.: Development of a tele-operative testing cell as a remote lab for material characterization. In: 2014 International Conference on Interactive Collaborative Learning (ICL), Dubai, pp. 977–982 (2014)
6. Rehme, O.: Cellular Design for Laser Freeform Fabrication. Cuvillier, Göttingen (2010)
7. VDI 3405 Part 3. Additive manufacturing process, rapid manufacturing. Design rules for part production using laser sintering and laser beam melting. VDI – Handbuch Produktionstechnik und Fertigungsverfahren, Band 2: Fertigungsverfahren (2015)
8. Gebhardt, A.: Rapid Prototyping, p. 392. Hanser Publishers, Munich (2003)
9. Liverani, E., Toschi, S., Ceschini, L., Fortunato, A.: Effect of selective laser melting (SLM) process parameters on microstructure and mechanical properties of 316L austenitic stainless steel. J. Mater. Process. Technol. **249**, 255–263 (2017)
10. Delgado, J., Ciurana, J., Rodríguez, C.A.: Influence of process parameters on part quality and mechanical properties for DMLS and SLM with iron-based materials. Int. J. Adv. Manuf. Technol. **60**, 601–610 (2012)
11. Frerich, S., Meisen, T., Frerich, S., Richert, A., Petermann, M., Jeschke, S., Wilkesmann, U., Tekkaya, A.E. (eds.): Engineering Education 4.0 – Excellent Teaching and Learning in Engineering Sciences. Springer International Publishing AG, Cham (2016)
12. Oller, A., Bates, H.: Inhalation carcinogenicity study with nickel metal powder in Wistar rats. Toxicol. Appl. Pharmacol. **233**, 262–275 (2008)
13. Rosenthal, S., Platt, S., Hölker-Jäger, R., Gies, S., Kleszczynski, S., Tekkaya, A.E., Witt, G.: Forming properties of additively manufactured monolithic Hastelloy X sheets. Mater. Sci. Eng. A **753**, 300–316 (2019). ISSN: 0921-5093
14. Selvaggio, A., Upadhya, S., Grodotzki, J., Tekkaya, A.E.: Development of a remote compression test lab for engineering education. In: 2019 16th International Conference on Remote Engineering and Virtual Instrumentation, pp. 389–399 (2019)
15. Selvaggio, A., Sadiki, A., Ortelt, T.R., Meya, R., Becker, C., Chatti, S., Tekkaya, A.E.: Development of a cupping test in remote laboratories for engineering education. In: 2016 13th International Conference on Remote Engineering and Virtual Instrumentation (2016)

Work-in-Progress: Bio Radar for Remotely Vital Sign Monitoring

Fiza Alina, Nabeel Khan, Syed Ubaidullah Bukhari, Syed Usman Amin, Siraj Anis, and Sarmad Ahmed Shaikh(✉)

Microwave and Antenna Research Group, Department of Avionics Engineering, PAF- Karachi Institute of Economics and Technology (KIET), Karachi, Pakistan
{Falinall.ak,Nabeelkhl997}@gmail.com, Slimshadyy99@yahoo.com, {usman.amin,siraj,sarmad}@pafkiet.edu.pk

Abstract. Contactless detection of heart and breathing rate is very impressive and interesting technology in biomedical field. This technology can play a vital role in life saving in different scenarios. For example, the burnt patients mostly suffer significantly with pain when ECG machine electrodes are hanged with finger and/or with chest. Thus, in this paper, we propose a contactless bio sensing apparatus, i.e., ultra-sonic bio radar, to detect the heart and breathing rates remotely. We aim to obtain these vital signs of human without any physical contact of the device with a human body by detecting the Doppler frequency shift in the received electromagnetic signal. Furthermore, the proposed approach can be also used for life detection buried under the debris in any disaster area. Since the work is still in progress, the results obtained so far indicate that the proposed method has capability to find the desired vital signs.

Keywords: Breathing rate · Doppler frequency shift · Fast Fourier Transform (FFT) · Heartbeat rate · Printed Circuit Board (PCB) · Non-contact vital sign · Ultra sonic sensor · Vital sign monitoring

1 Introduction

The lack of technology specifically in cardiovascular health section has resulted in a void of effective treatment of patients. One such well-known traditional technology is ECG machine used for vital sign monitoring i.e., breathing and heartbeat rates of patient are measured using in-contact body equipment. One disadvantage of using such equipment is that it requires direct contact with skin which in some cases including burnt patient is not possible. From decades the equipment used for patient monitoring is in-contact technology-based machines with bundles of wires. Such technology limits patient movements depending on cable's length. Also gives a rise to irritation factor in patients for being bound to cables. Moreover, in disastrous areas, such machines cannot be used to detect the vital sign monitoring of the people buried under debris. Thus, there is a need of a contactless device which could replace these traditional heavy and unmovable machines.

M. E. Auer and D. May (Eds.): REV 2020, AISC 1231, pp. 465–471, 2021.
https://doi.org/10.1007/978-3-030-52575-0_38

The basic electromagnetic theory was given by James Clerk Maxwell. Electromagnetic waves can be used in several ways. There are many new inventions, discoveries and applications that took place in many fields such as radio detection and ranging (RADAR). Generally, we use radar to determine the range, altitude, direction, location, speed and other features by utilizing the Doppler frequency shift phenomenon.

Christian Andreas described the theory of Doppler Effect [1]. For respiration monitoring, first non-invasive Doppler radar was suggested in 1975. After that in 1979 X-band Doppler transceiver was demonstrated [1–3]. There are several researches that were carried out in which the system basically adopts the concept of Doppler radar technique which potentially can be applied in several fields of monitoring such as in health monitoring, long range life sign detection, heart rate observation.

The benefit of non-invasive heartbeat monitoring is that treatment of patient is easy, time saving and also provides a solution for health monitoring of burnt patients. The straps and gel are not required to this non-contact heartbeat detection system.

For motion detection the Doppler shifted signal is used by Doppler radar. Generally, displacement of chest due to respiration and variation of heartbeat is measured between 4 mm to 12 mm and 0.2 mm to 0.5 mm, respectively.

Therefore, in this paper a contactless remotely operated ultra-sound bio radar for vital sign monitoring is proposed. This bio radar provides an advantage of non-invasive measurement of heart rate and breathing rate. The main aim of this technology is to detect vital signs for health care emergency and provide rescue and security.

2 Proposed Methodology

Aim to detect the vital signs is to find the doppler frequency shift, introduced by movement of heart/body (due to respiration cycle), from the received reflected electromagnetic (EM) signal of bio radar. In this paper, EM wave pulses of short duration are transmitted by transmitter having an operating frequency (40 kHz). The EM pulses which are reflected from human body are varied in frequency (due to heartbeat and breathing frequency) and received in a receiver module. The change in frequency is Doppler shift which is directly proportional to the heart/breathing frequency. The received signal is then amplified through LM741 ICs and then acquired using Arduino after that acquired signal is sent on PC on which MATLAB tool is used to apply filters (signal processing) and to remove other frequency components except the Doppler shifted frequency component which gives the information about the heart and breathing frequency. The block diagram of the proposed approach is shown in Fig. 1.

Procedure:

1. Designing a carrier transmitting circuit that works at a frequency of 40 kHz Also a receiving circuit which will amplify the received signal level to readable level.
2. Sending the output of the receiver circuit to MATLAB in real time. In order to do that, Arduino Due is used which has a typical sampling rate of 100 kHz which satisfy the Nyquist sampling criteria i.e. greater than double the max signal frequency but the time domain signal plot is not good at this sampling rate; In order to

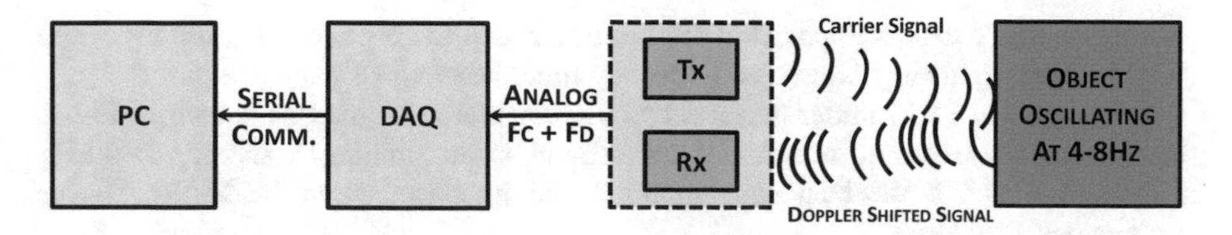

Fig. 1. The proposed approach to detect the Doppler shift.

get better results sampling rate of Arduino Due was increased to acquire a proper signal of 40 kHz. Increasing sampling rate also decrease the quantization noise of our data acquisition unit.

3. Designing a code that can increase the sampling rate of Arduino Due to 666.66 kHz. This can be achieved by coding the controller at system level because normal architectural level code only provides the sampling rate of 100 kHz.
4. Serially communicating between Arduino Due and MATLAB to acquire the received Doppler shifted signal on MATLAB that later can be used to extract information and can be visualized by plotting its FFT and time domain plots.
   To remove any garbage values on the receiving end (on MATLAB) the data must be in between the minimum and maximum range of Arduino Due ADC limit which is 0 to 4096 because of its 12-bit ADC. Therefore, we discard all those received values which are greater or less than this range. There is also a chance of receiving null values which also needs to be removed for which the designed code can run without any error in continuity. Such conditions are driven by using a suitable code that can eliminate any such error that arises due to serial communication.
5. Future aim is to design a code that generates the fast Fourier transform of the received signal that provides us a frequency domain of our signal so that the operating frequency and Doppler shifted frequency can be observed Realtime.
6. Design a filter which is capable of eliminating carrier and all unwanted noise signals while only pass the desired Doppler shifted frequency.

2.1 Digital Signal Processing

For analyzing, modifying and synthesizing signals such as sound, images and biological measurements Signal processing technique is used. Digital signal processing is the process in which signals are modified to improve the performance and efficiency. Signal processing is applying on various mathematical and computational algorithms. Signal processing is also used to produce a higher signal with good quality.

Application of digital signal processing in medical field is to analyze the bio medical signals, patient monitoring, health care, artificial organs. It's one of the well-known examples is electrocardiogram (ECG) signal which provides the information to doctor about the condition of patient's heart rate.

2.1.1 Frequency Modulation

In this paper frequency modulation technique is applied. This technique includes varying the frequency of carrier wave. An Imposed signal also known as carrier signal is needed where data is imposed in the original signal to generate the resulting signal with variable frequency known as frequency modulated signal.

Along with, a transmitter and receiver circuits were designed as shown in Fig. 2 and 3, respectively. The transmitter ultra-sound is transmitting a signal at 40 kHz frequency which is received and amplified and interfaced with the MATLAB for further post-processing.

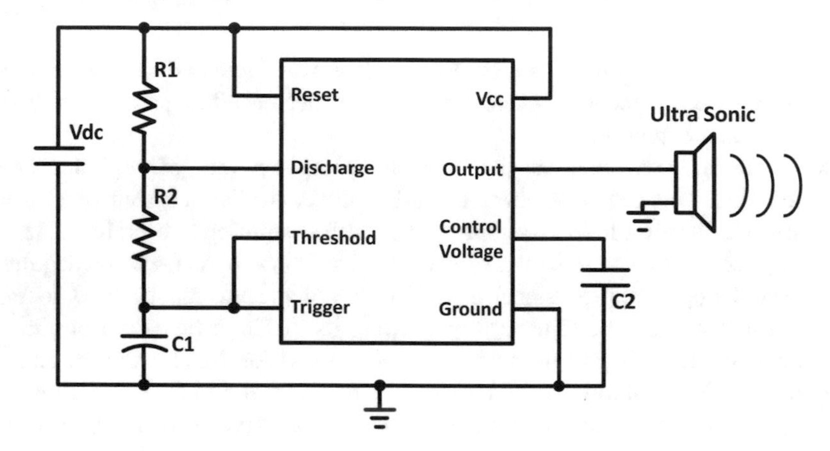

Fig. 2. Transmitter circuit for carrier pulse

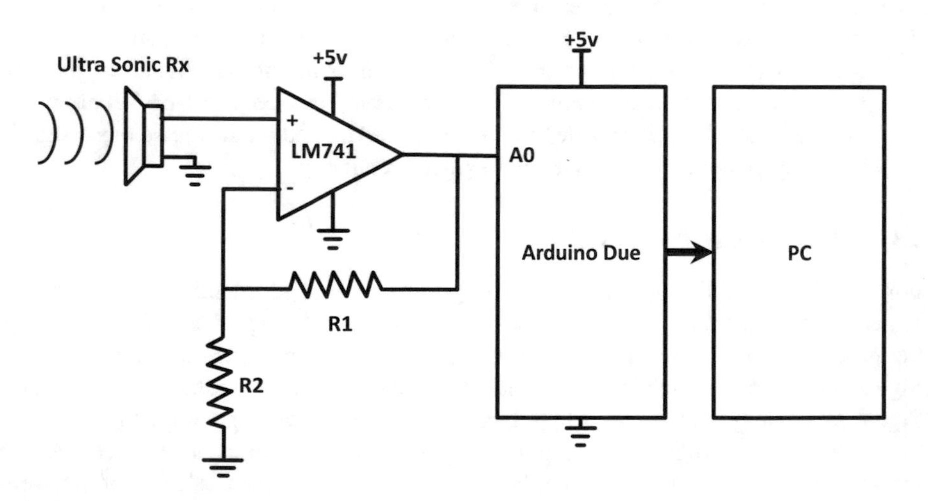

Fig. 3. Receiver circuit.

2.2 Issues Faced

A few issues were faced during the research and testing phase. The hardware limitation including the IC's which are only used by armed forces and the ones available are not easy to implement on SMD's or PCB.

3 Results

Since the work is still in progress, here are shown the transmitted and received signals graphs in Fig. 4 and 5 respectively. The received signal has been analyzed using FFT in MATLAB. A 3Vp-p Signal was transmitted, this signal is reflected back from the oscillating body which is oscillating at 4–8 Hz, this reflected signal now contains a carrier and a doppler frequency. At receiver end 5000 samples were taken using Arduino Due, those samples were sent to MATLAB via serial communication using baud rate of 115200. only 100 samples are shown in the Fig. 5a representing received signal to get thc clcar picturc of what results were obtained at the receiving end. While calculating FFT all the samples are used to get more accurate FFT of this received signal (Fig. 5b). A 4th order digital low pass filter was designed to extract the doppler shifted frequency shown in Fig. 5c.

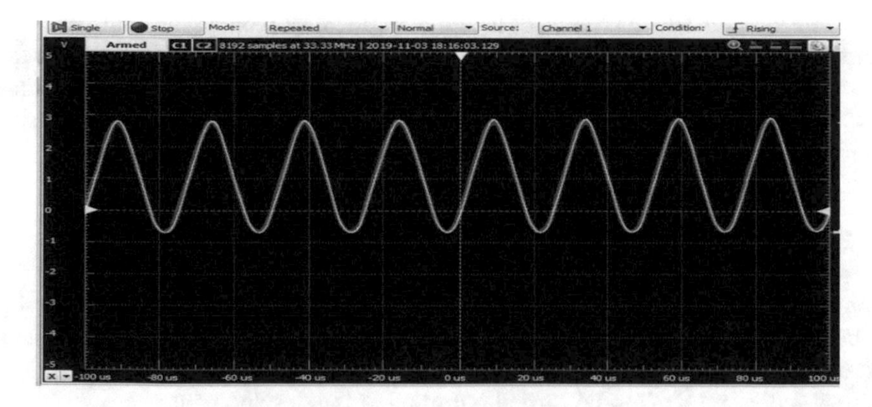

Fig. 4. Transmitted signal on oscilloscope.

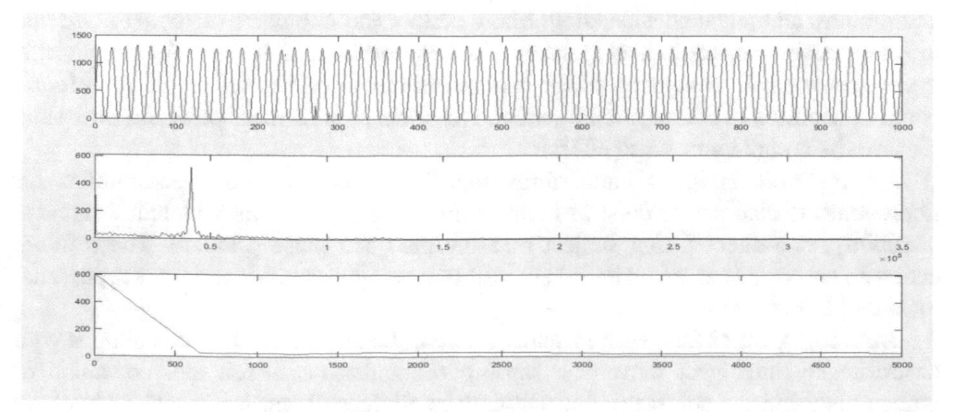

Fig. 5. Receiver output signal and its FFT.

Currently the circuit receives noise at lower frequency range as the dominant noise in this range is 1/f also known as pink noise which can be seen in Fig. 5c. However, the research is still on-going in detecting the heart and breathing rate which will be available in future publications. The desired outcome is expected approximately 1 Hz for heart frequency and around 0.2 to 0.35 Hz for breathing frequency from reflected received signal i.e., Doppler shift frequency.

4 Conclusion

Bio radar for vital sign monitoring will become a solution of many daily life problems. As the development in technology, use of radar increasing in bio medical field because of low complex, low cost, highly efficient and interesting features. And in future there are many possibilities that can allow us to replace the contact system with contactless measurement instruments.

In this paper, an attempt was made to design a system that could sense and measure the heartbeat and breathing rate without being in contact with body i.e., remotely. So far, a transmitter and receiver are designed, and interfaced with MATLAB for post processing of the received signal. However, the exact detection of heart and breathing rate through signal processing is under development and will be completed and published in future.

5 Future Recommendation

In recent years Bio Radar vital sign monitoring technology created a lot of interest [4–9], because this technology has many daily life applications. Much more efforts are required for progress in this technology. There are still many challenges though that requires further research to be carried out to enter the development phase.

Spectral algorithm is required [14–16] that can estimate the exact sinusoidal frequencies before and after the introduction of Doppler shift due to the movement of chest caused by heartbeat and respiration.

Noise is also another main factor that plays an important role in signal corruption and requires additional consideration. Many factors add into noise including noise due to body movement, equipment resistance and atmospheric noise etc. As known; this system sense small movement, which is in millimeter or centimeter range, so in future continuing efforts in this regard are required to come up with more powerful filters and remove these clutters or noise effects.

As this work is in its initial form, therefore, there are many possibilities for improvements that can be done in future. Future and improvement include heart rate variability, non-direct facing subject measurement and pulse pressure. These future achievements in this area will definitely lead to new opportunities and market potential towards bio medical field.

Another aspect of this work in future that can be unveiled is implementing it with an artificially intelligent movement sensing robot that can sense the movement of human body and can adjust itself accordingly in all three dimensions. Such technology will provide us an additional advantage of continuous monitoring of patient in a mobile environment where patient's movements are not limited because of equipment limitations as in the past where cables and hooks limited patient's movement.

One can also implement IOT in Bio Radar that can continuously provide heart rate and breathing rate on internet cloud where these values can be used for statistical analysis of a person's health.

References

1. Sengupta, D.L., Zhang, Y.: Maxwell, Hertz, the Maxwellians and the early history of electromagnetic waves. In: Antennas and Propagation Society International Symposium, vol. 1, pp. 14–17. IEEE (2001)
2. Dibner, B.: Ten founding fathers of the electrical science: X. James Clerk Maxwell: and electromagnetic forces mathematically demonstrated. Electr. Eng. **74**(1), 40–41 (1955)
3. http://www.ieeeghn.org/wiki/index.php/Heinrich_Hertz_(1857-1894)
4. Murai, K., Hayashi, Y., Stone, L.C., Inokuchi, S.: Basic study of navigator's recognition of radar target direction. IEEE Syst. Man Cybern. (SMC) **1**, 796–801 (2006)
5. Lin, J.C.: Non-invasive microwave measurement of respiration. Proc. IEEE **63**(10), 1530 (1975)
6. Lin, J.C., Kiernicki, J., Kiernicki, M., Wollschlaeger, P.B.: Microwave apexcardiography. IEEE Trans. Microw. Theor. Tech. **27**(6), 618–620 (1979)
7. Lubecke, O.B., Ong, P.W., Lubecke, V.M.: 10 GHz Doppler radar sensing of respiration and heart movement. In: Northeast Bioengineering Conference, pp. 55–56 (2002)
8. Zhuang, W., Shen, X., Bi, Q.: Ultra-wideband wireless communications. Wirel. Commun. Mob. Comput. **3**, 663–685 (2003)
9. Staderini, E.M.: UWB radars in medicine. IEEE Aerosp. Electron. Syst. Mag. **1**, 13–18 (2002)
10. McEwan, T.E.: Body monitoring and imaging apparatus and method, US Patent 5,766,208
11. Bilich, C.G.: Bio-medical sensing using ultra wideband communications and radar technology: a feasibility study. In: IEEE Pervasive Health Conference and Workshops, pp. 1–9, November 2006 (2006)
12. Ossberger, G., Buchegger, T., Schimback, E., Stelzer, A., Weigel, R.: Non-invasive respiratory movement detection and monitoring of hidden humans using ultra wideband pulse radar. In: Proceedings of the 2004 International Workshop on Ultra Wideband Systems Joint with Conference on Ultra Wideband Systems and Technologies, Piscataway, NJ, USA, pp. 395–399 (2004)
13. Lin, J., Li, C.: Wireless non-contact detection of heartbeat and respiration using low-power microwave radar sensor. In: Proceedings of the 19th Asia Pacific Microwave Conference, vol. 1, Bangkok, Thailand, December 2007, pp. 393–396 (2007)
14. Bilich, C.G.: Biomedical sensing using ultra wideband communications and radar technology: a feasibility study. In: Proceedings of the 1st International Conference on Pervasive Computing Technologies for Healthcare, Innsbruck, Austria, December 2006, pp. 1–9 (2006)
15. Immoreev, I.Y., Samkov, S.V., Tao, T.H.: Short-distance ultra-wideband radars. Theory and designing. In: International Conference on Radar Systems (RADAR 2004), Toulouse, France (2004)
16. Pavlov, S.N., Samkov, S.V.: Algorithm of signal processing in ultra-wideband radar designed for remote measuring parameters of patient's cardiac activity. In: 2nd International Workshop on Ultra-wideband and Ultrashort Impulse Signals, Sevastopol, Ukraine, September 2004, pp. 205–207 (2004)

Potential of Embedded Processors and Cloud for Remote Experimentation

Mohammed Misbah Uddin, Suresh Vakati, and Abul K. M. Azad(✉)

College of Engineering and Engineering Technology,
Northern Illinois University, Dekalb, IL, USA
aazad@niu.edu

Abstract. Traditionally remote laboratories are developed by utilizing personal computers (PC) or workstations as the main controller unit on the experiment side and a local server for database and user management. However, given the emergence of new technologies, embedded processors are becoming more powerful, faster, and resourceful. It is now possible to replace personal computers or workstation with embedded processor boards. In addition, remote laboratory Based on this scenario, this paper reports the development of several remote experimentation facilities in which embedded processors and cloud are utilized to replace PCs and local servers.

Keywords: Remote laboratories · Embedded systems · Cloud systems

1 Background

Most of the styles are intuitive. However, we invite you to read carefully the brief description below. It is vital to provide laboratory activities to maximize learning in STEM disciplines. Traditionally, students perform experiments by being present in a laboratory and working with physical systems. However, when considering the financial involvement, manageability, and accessibility, this arrangement is not always effective. With the emerging technologies in computing hardware and software, researchers and academics are leaning toward Internet accessible remote testbeds to replace, or supplement, existing laboratory experiments [1]. Remote testbeds can maximize utilization time, increase collaboration among universities and research centers, and provide access to expensive experimental resources [2].

Researchers are using various hardware and software technologies for remote testbed developments [3, 4]. Thinking broadly, a remote testbed system can be divided into a few major components: the experimental system, interface of the experiment with a local computer, graphical user interface (GUI), and server for remote access and access management. A number of software tools can be used for interfacing, GUI development and server applications. Some of these tools are LabVIEW, Matlab, and . NET. These are expensive proprietary software, and the first two have some limitations in terms of flexibility in development and browser adaptability.

With these in mind, this paper will report the design and development of experimental testbeds using Python, in which a single software can provide computer interfacing, GUI development, and remote access. Python is a scripting language

M. E. Auer and D. May (Eds.): REV 2020, AISC 1231, pp. 472–487, 2021.
https://doi.org/10.1007/978-3-030-52575-0_39

available on the market for software development particularly suited to Internet applications. As Python is an open source tool, it has a wide variety of modules, exceptions, high-level dynamic data types, classes, and interfaces to huge libraries. This paper will describe the development of two remote testbeds using Python. The testbeds are a mobile platform with self-navigation and an embedded system with remote programming capability.

The first testbed is a self-navigated mobile platform fitted with a vacuum cleaner. The mobile platform is wirelessly connected to a server for operation and control. The system is fitted with a number of sensors to implement self-navigation around obstacles and has an onboard microcontroller system for control. A GUI was developed to facilitate the interaction between the system and a remote user. Within the second testbed, an embedded processor (Arduino board) was interfaced with a number of output devices (liquid crystal display, seven segment display, light emitting diodes, and a stepper motor). Using a GUI, remote clients can run a few pre-developed programs provided within the system. They can also upload their own program to the embedded processor board. The testbed has the capability to compile an uploaded program and send the compilation report to the remote user. If needed, the remote user will then debug the program and upload again. Finally, the user can verify the program implementation through a video feedback provided within the testbed.

2 Remote Laboratory Technologies

To explore technologies involved in the development, Fig. 1 shows a conceptual structure of a remote testbed. The main components can be identified as the experimental setup, local computer/server, Internet cloud, and remote clients. The experimental system is connected with a local computer/server, which plays the role of a gateway between the experiment and the remote computer for clients. There should be some middleware that facilitates the information exchange between the local and remote computers.

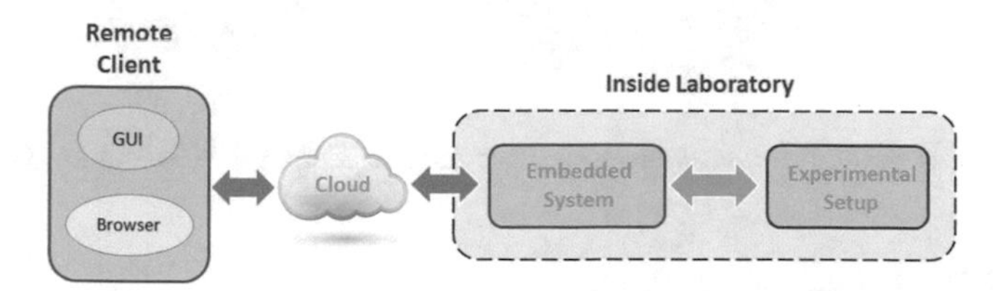

Fig. 1. Concept diagram of a remote testbed

Within a laboratory, the experimental system(s) is connected with an embedded system. The embedded system collects all the sensor data as needed by the experiment. The collected sensor data will pass through initial processing in terms of filtering, digital to analogue conversion, and other preliminary activities. In some cases, the embedded processor can perform certain local control tasks. The data will then pass to the Cloud for processing and presentation to the clients via a suitable graphical user interface (GUI). The remote client can interact with the experimental system via the GUI. In this case, the client's requests will be passed along via the cloud and the embedded system.

From a technical point of view any experimental system that can be interfaced with a computer via a transducer (sensors and actuators) is a potential candidate for a remote testbed. These can be electrical, mechanical, industrial, chemical, and even biological systems. Of course given the nature of an experiment, appropriate measures should be taken to ensure the safety and integrity of the system and the environment around it. Within the given structure, the local computer/server gathers all the system input, processes those to produce commands for the desired output, and passes them to the experiment to drive the actuators. In almost all of the cases, developers provide a GUI to facilitate a client-friendly interaction so someone with very little technical knowledge can operate the experiment. Connections between an experiment and a local computer/server can be wired or wireless depending on the technology used and the nature of the experiment. The next phase is to make the GUI available over the Internet so clients can have access to the GUI from a remote computer.

A number of software technologies are being used to develop remote laboratories. Gravier and his coworkers reviewed approximately 60 reported initiatives and identified a breakdown of the software used for remote laboratory implementation [5]. A pie chart showing the breakdown of the use of software packages is provided in Fig. 2.

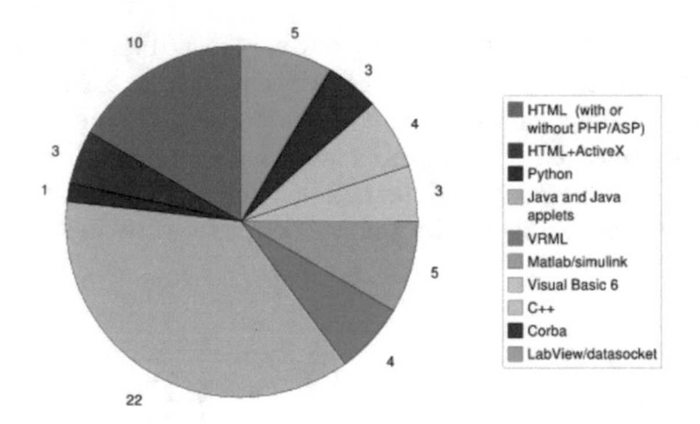

Fig. 2. Distribution for software usage for remote testbed developments (Gravier)

3 Case Studies

The systems included in this paper are the remote vacuum cleaner, remote programing of embedded systems, coupled task system, and structure monitoring. The first three are targeted to support educational activities, while the last one is for infrastructure monitoring.

3.1 Vacuum Cleaner

The first testbed is a self-navigated mobile platform fitted with a vacuum cleaner. The mobile platform is wirelessly connected to a server for operation and control. The system is fitted with a number of sensors to implement self-navigation around obstacles and has an onboard microcontroller system for control. A GUI was developed to facilitate the interaction between the system and a remote user. Within the second testbed, an embedded processor (Arduino board) was interfaced with a number of output devices (liquid crystal display, seven segment display, light emitting diodes, and a stepper motor). Using a GUI, remote clients can run a few pre-developed programs provided within the system. They can also upload their own program to the embedded processor board. The testbed has the capability to compile an uploaded program and send the compilation report to the remote user. If needed, the remote user will then debug the program and upload again. Finally, the user can verify the program implementation through a video feedback provided within the testbed.

The remote vacuum cleaner consists of a mobile platform fitted with a dc vacuum cleaner along with a number of sensors for navigation. A general block diagram of the system is shown in Fig. 3.

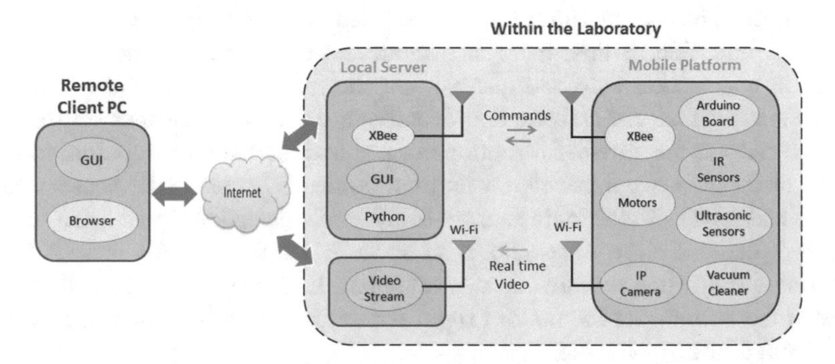

Fig. 3. System diagram for remote vacuum cleaner

The system has three main components: a mobile platform as the remote testbed system, a local server is the gateway between the testbed and remote clients and the remote client. The mobile platform consists of a drive system, sensors for navigation, an embedded processor (Arduino board) for local control and data management, an XBee for wireless communication with the local server, and an IP camera for real time video. The IP camera has its own communication route via a WiFi channel. The video is then embedded within the GUI for user monitoring. Images of completed mobile platform are shown in Fig. 4.

Designing a GUI to provide users with excellent visual composition is a vital part of remote testbed designs. The goal is to improvise and enhance the visual experience between the human eye and computer. Considering the issues for an effective GUI, HTML was used as a software tool for testbed GUI development. Along with HTML, Cascaded Style Sheets (CSS) were provided to improve the overall visual experience.

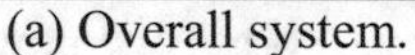

(a) Overall system. (b) Close up view of electronics.

Fig. 4. Images of the mobile platform

The GUI contains all the input and output selection buttons and variables, which can be adjusted by a client. Figure 5 shows an image of the developed GUI showing all the functionalities. There is an on/off switch provided at the top right corner of the page. Just to the right side of this, there is a mode selection switch. The system can be operated in two modes: auto and manual. On the top right corner there is a vacuum on/off switch for that can activate and deactivate the vacuum cleaner. In the manual mode one can use the arrows for motion control and the red square button to stop. In the auto mode the system travels on its own guided by a set of IR sensors for wall detection and a pair of ultrasonic sensors for obstacle detection. In addition, users can adjust the speed and allow proximity to an obstacle. On the mid left, there is an arrow to determine the distance from the nearest obstacle at any point. The distance will display at the bottom. At the bottom right, the safety status is displayed to show the danger from a nearby obstacle.

Robot On/Off
AutoMode
Vacuum On/Off
off
on
off
Speed(%):
50
50
Proximity:
Get
Allowed Proximity(cm):
30
30
Distance(cm):
Safety Status:

Fig. 5. Images of the GUI used for the remote vacuum cleaner

Figure 6 is a flowchart showing a high level view of the program execution cycle for the server system. To start the server, the Tornado Python script should be executed [6]. At the beginning, Python takes in all the modules required for operation using import method. It initiates the opening of a serial port. There will be an error message if the serial port is already being used or is not avaialble. The server communicates with the Arduino board once it opens the serial port. The server script requires all of the HTML/CSS files when a request is made from the client side. The files are loaded before starting the server, and whenever a request is made, it will transfer the files to the client's browser as a webpage. In case of any irregularity in the procedure, it will exit the loop and will display an error message.

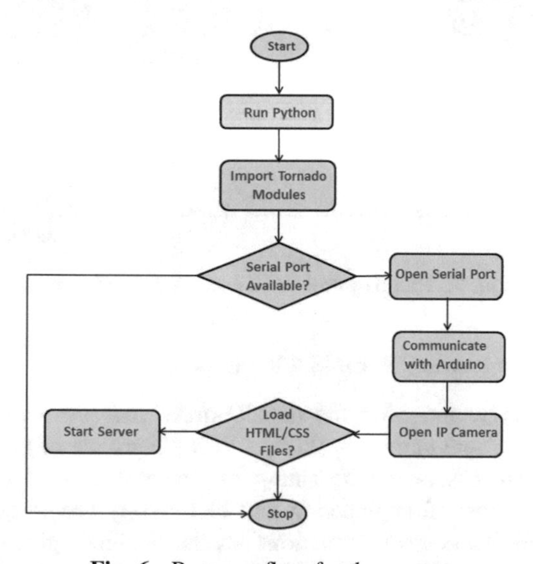

Fig. 6. Program flow for the server

When a client enters the URL address on a browser, this directs the user to an HTML page that contains all of the GUI controls of the experiment. Clients can choose different inputs from the GUI, and based on the inputs, the server responds to the user. The program dataflow between the client and the server is shown in Fig. 7. The server establishes a new connection when a user enters the URL. It waits until the user sends some input data. After receiving an input from the user, the HTML code calls the Java Script written in JQuery. JQuery assigns the input request to a variable. The assigned variable is passed to the Python server, and the Python script compares the client's input command with the pre-defined commands. If they match, Python then sends the appropriate pre-defined characters to the Arduino board. If the input command does not match the pre-defined commands, then the Python raises an error on the client's web page as "404 not found" [7, 8]

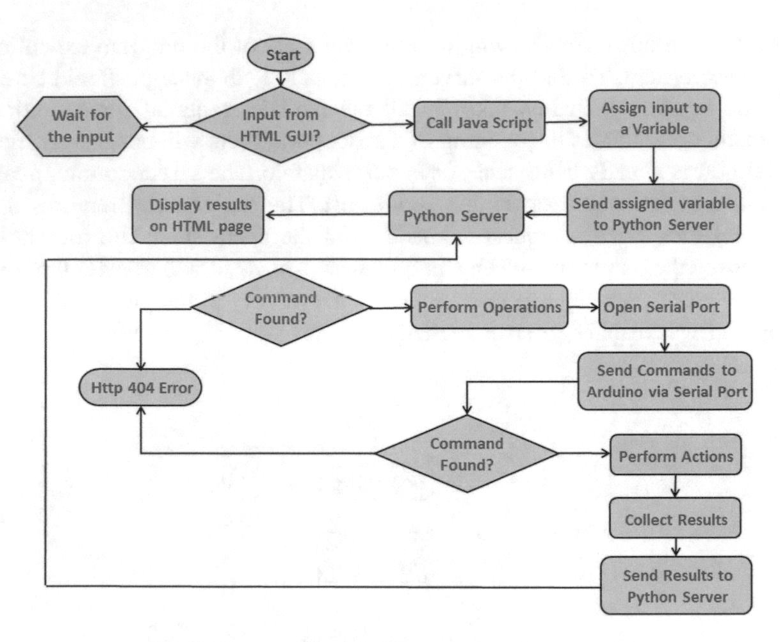

Fig. 7. Program flow for handling a client request

3.2 Remote Programming of Embedded Systems

The second case study describes the development and operation of an embedded system with remote programming capability. With the growth of the Internet of Things (IoT), embedded processors are becoming an integral part of electronic systems. Considering this, almost all engineering and technology programs offer one or two courses on embedded processors at various levels. Within a given hardware system, embedded processors can be programmed to perform various activities and clients can explore them to design and verify different ideas and concepts [8]. Remote programming capability for an embedded system opens a new horizon for clients in which they can use the facility 24/7 and enhance the learning process.

This section provides the development process and describes the features of an embedded system with remote programming capability. An Arduino board was used for this development along with a number of output devices. The output devices are an LCD, LEDs, a stepper motor, and a seven segment display. The same as in the previous test case, Python was used for interfacing, GUI development, and web services. A block diagram of the developed system is provided in Fig. 8.

The main idea of this project was to design an embedded processor system that could be programmable from remote locations. An Arduino Mega-2560 microcontroller board was used as the embedded processor [9]. All of the output devices were connected via the Arduino board. Figure 9 shows a schematic diagram of the hardware system, while Fig. 10 shows images of the completed system.

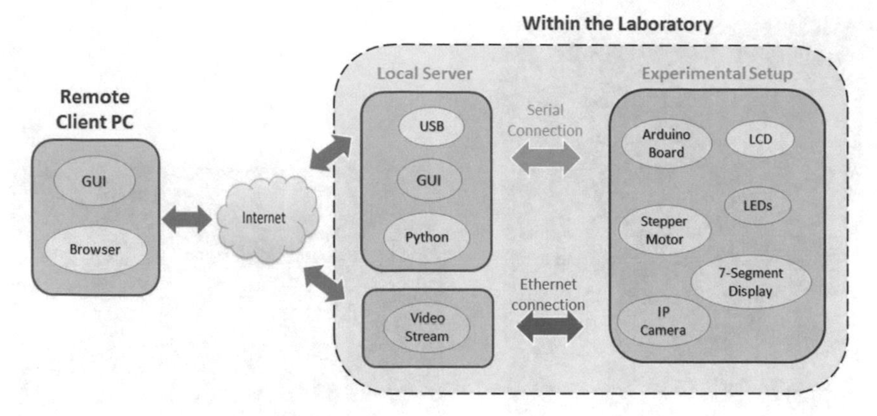

Fig. 8. Program flow for handling a client request

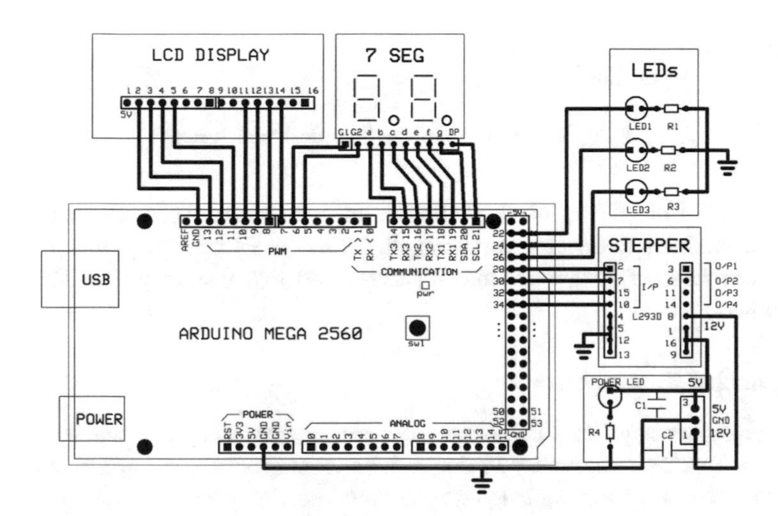

Fig. 9. Hardware connection diagram of the system

Serial communication was established between the server and experiment set-up to achieve remote programming. When a user uploads a program to the server, it handles the incoming files and stores them in its database to process and then transfer to the processor board. A Python script helps to handle the files and their names in the server. With this arrangement, the remote user does not need any plug-ins to operate the remote testbed. The clients' PC just needs an Internet connection and a latest version browser like Firefox, Opera, Chrome, or Safari.

The server is a computer that handles all incoming clients and responds to the users for their inputs. This experiment mainly deals with uploading files, so the CPU must be 2.66 GHz 128 cache or above, the RAM should be 2 GB or above, and the minimum database space should be 10 GB. The server is installed with an Ubuntu 14.10 version operating system rather than Windows, which allows easy access to the installed

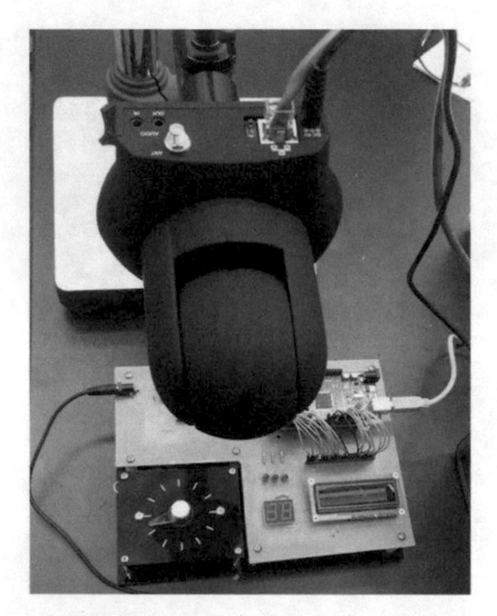

(a) IP camera is directed to the experiment.

(b) Image of the developed experiment.

Fig. 10. Pictures of the final designed system

software tools from the terminal. Python is an open source software that provides a variety of modules to develop server applications. A Python script using a Tornado module can turn any computer into a server, and this feature was utilized for this development.[7]

3.3 Coupled Tank System

In this project we have used Python as the main programming language to develop the required software on Raspberry Pi due to its advantages: As one of the most widely-used high-level programming languages, a large knowledge base is available for novice programmers to start developing their programs. Additionally, Python has a simple and easy to understand syntax that emphasizes readability and thereby reduces the cost of program maintenance. As will be presented next, it is easy to develop internet applications with the aid of open source modules and packages developed for Python.

The operating system being used on Raspberry Pi is Raspbian, a Debian Linux based operating system customized for Raspberry Pi. Each Raspberry Pi runs its own web server, a built-in Apache webserver that allows Raspberry Pi to render webpages for the client-side graphical user interfaces (GUIs) and facilitate remote interaction with hardware attached to the Raspberry Pi. Python is used to program the web server through Flask to facilitate this communication and execution procedure. Flask is one of the popular web frameworks, such as Django, that expedites development of web applications using Python by providing a solid core with basic services. Moreover, Flask supports databases, web forms, authentication, and other high-level tasks by the application of its extensions. Flask extensions are readily available to integrate with the

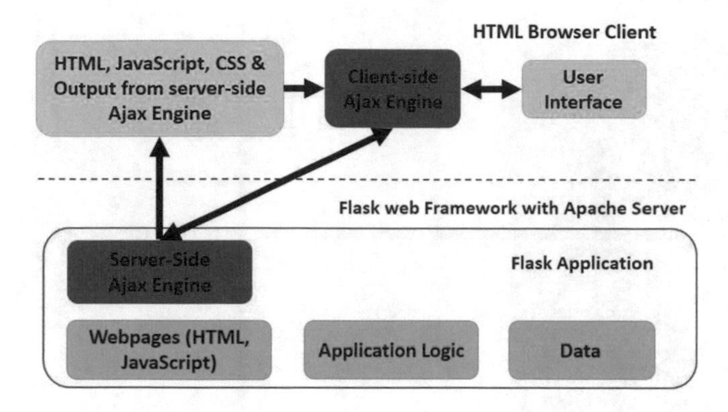

Fig. 11. Software flow diagram

core packages, and not all extensions need to be installed when developing an application. The web technologies used in this project are illustrated in Fig. 11.

Bootstrap, a front-end component library, was used to develop the GUIs on the client side. Bootstrap is an open source toolkit provided by Twitter for developing with HTML, cascading style sheets (CSS), and JavaScript. When a remote client requests a specific function using a GUI on his browser, JavaScript sends the request to the server through JSON (JavaScript Object Notation) data. The request is redirected to the Flask application, which executes the Python function mapped to return the request from the given URL.

The second module in the remote laboratory system project is a Raspberry Pi integrated with a coupled tank system, shown in Fig. 12, to control the water level in the tanks. The coupled tanks system serves as a process control experiment that provides students exercise various concepts they have learned about control theory, including transfer function representation, linearization, level control, PID, feedforward, and control parameter tuning, etc.

The system consists of a water pump with two tanks. A pressure sensor is mounted at the bottom of each tank to measure the water level. The pump is used to drive the water from the basin at the bottom. The two tanks are mounted one below the other in such a way that the outlet of Tank 1 (top) flows into Tank 2 (bottom). The outlet from Tank 2 flows directly into the basin at the bottom of the system. The outlet of each tank is configurable and can be replaced by changing the insert that screws to the bottom of the tank to adjust the rate of flow as desired. The inserts are available in three sizes: small, medium, and large. The inlets for the tanks come from the pump and are split equally through Out 1 and Out 2 orifices. Tank 1 is also provided with a drain tap so water can flow directly into the basin. Each tank is additionally equipped with a vertical scale (in centimeters) alongside the tank to visualize the water level in each tank.

The schematic diagram, shown in Fig. 13, represents how the coupled tanks are connected to the embedded system. In this configuration, an Arduino board was used as an intermediate stage between the Raspberry Pi and the control module of the coupled tanks. The pressure sensors mounted at the bottom of each tank produce analog readings that are converted into digital data by Arduino and sent to Raspberry Pi.

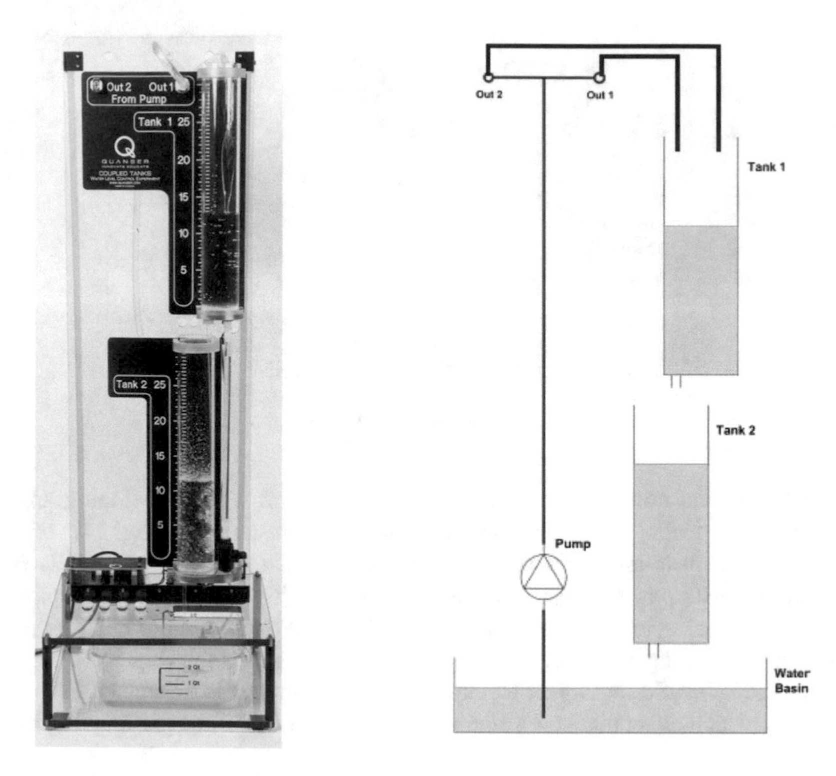

Fig. 12. Images of the coupled tank system

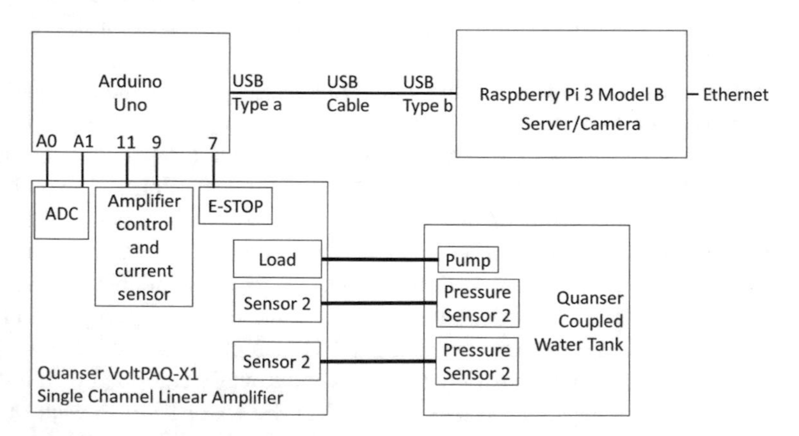

Fig. 13. Schematic diagram of the desired system

A linear amplifier, controlled by Arduino, supplies power to the pump and activates it. As an additional feature, the amount of electrical current used by the pump is sensed and reported to Raspberry Pi through the Arduino.

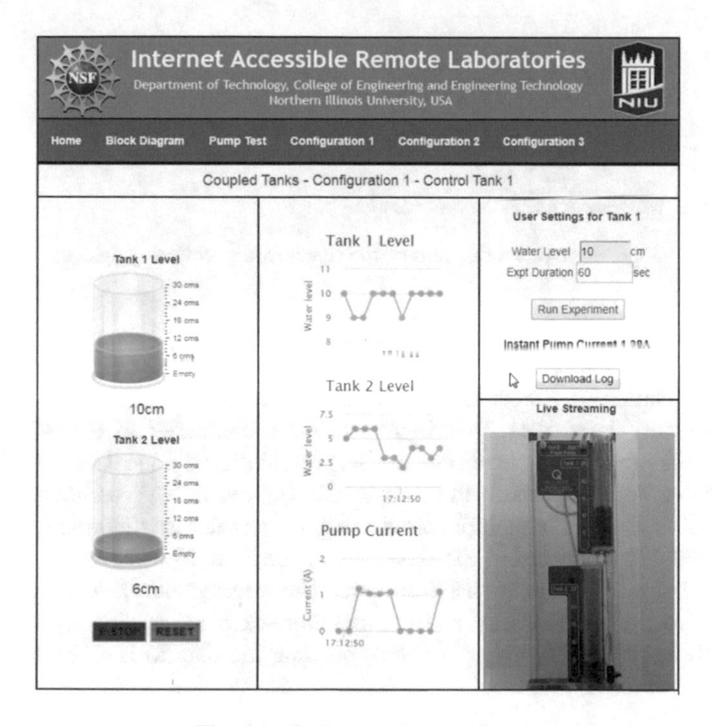

Fig. 14. GUI viewed by a client

It should be noted that in the experiment configurations presented below, the pump is operated in on/off mode. The pump can be operated with variable speed by controlling the power supplied by the amplifier, which will provide experiments to exercise concepts such as PID control (Fig. 14).

3.4 Structural Monitoring

The paper reports the design and development of an Internet of Things (IoT) enabled SHM system to provide an automated real-time diagnosis of the structural health of an infrastructure. For this study a laboratory scale suspended-bridge was used along with accelerometers mounted for data collected. Figure 15 shows the system diagram. The sensors are connected with an embedded processor system for data collection and pre-processing. Processed data are simultaneously passed to the cloud as well as a remote server for client access.

The center of the cloud integration is the NodeJS server program running Heroku cloud. This server is connected with the NVIDIA TK-1 board as well as a local web server and Datastore (Fig. 16). The Heroku cloud accepts/requests data from all the

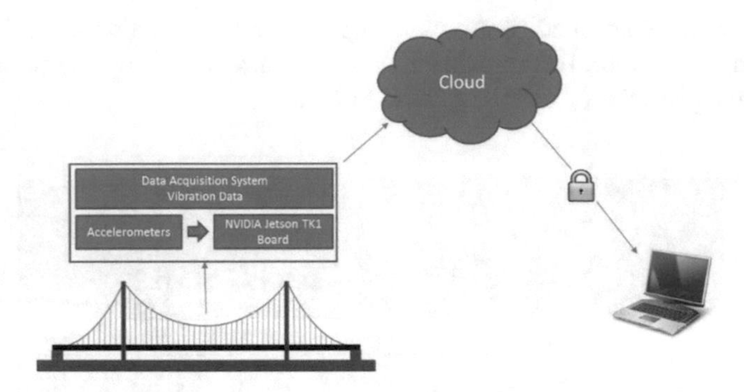

Fig. 15. System diagram of the developed monitoring system.

sensors, TK-1 board's, supporting programs, and other users [10, 11]. Sensor data are stored in Elasticsearch (Lucene) which is running in the local Datastror. Elasticsearch is a Non-SQL Datastore, which helps provides faster indexing, fuzzy searching, and analytics of very large data. It is highly scalable compared to the SQL [12]. The NodeJS server program in cloud provides Representational State Transfer (REST) end point to TK-1 board and push the data to the Datastore. This architecture makes it highly scalable to connect multiple monitoring sensors across multiple infrastructures. The collected data is pushed to Elasticsearch by the NodeJS program after receiving it from Python program running on TK-1 board. The central NodeJS server perform tasks which require high availability and computation such as serving multiple devices, clients such as web users, triggering alert, pushing the data to Datastore, and spectral analysis.

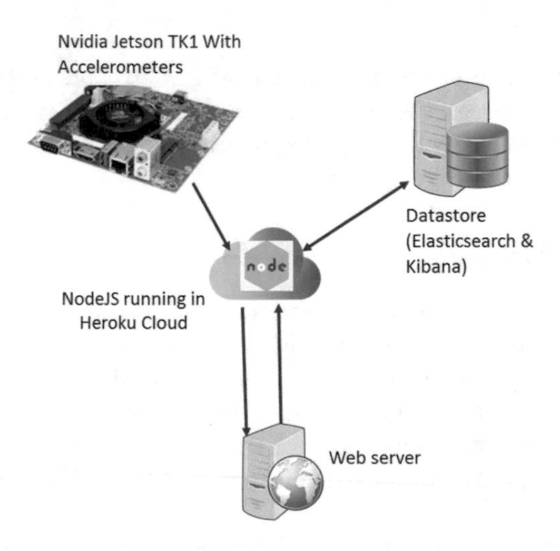

Fig. 16. System interaction using Cloud.

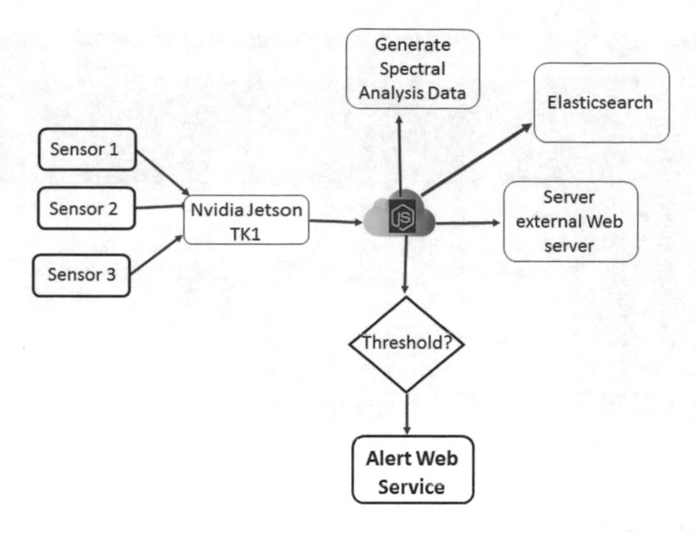

Fig. 17. Data collection threads

Considering limited storage capacity on TK-1 board, data is never stored locally but is pushed to the external server where large volume storage is possible through central NodeJS server. In addition to transferring data between the servers there is an additional service running in the web server, which trigger alerts such as e-mail, when there is critical breach in set threshold value. NodeJS in Heroku cloud automatically checks this threshold in real-time.

Data collection is done through python script running in TX-1 board. Multiple scripts run in parallel in multi-threading mechanism for each sensor (Fig. 17).

There are multiple services running in NodeJS. It fetches and pushes data from Elasticsearch, perform spectral analysis, provide service to running program to trigger threshold alert, support web application such as reading and updating threshold. Node.js is a JavaScript runtime that uses an event-driven, non-blocking I/O model that makes it lightweight and efficient. NodeJS runs on a single thread but its event driven approach makes it highly efficient with increased throughput [12].

There are three accelerometers fitted with the bridge, each providing data for three axes (x, y, and z). Data from each accelerometer will have two graphs, one is the time series and the other is the spectral analysis. Graphs are plotted using Kibana, which is an open source analytics and visualization platform designed to work with Elasticsearch [13]. Kibana graphs are embedded with the web page using 'IFrames'. Kibana also provides advanced filtering options in the graph. In addition, it can run Elasticsearch Query DSL within the graph to apply custom filter.

Figure 18 shows two graphs for one accelerometer that will be displayed within a web browser. The left-hand side graph is showing time series for one axes and the right-hand side graph is showing its spectral analysis. Within the web page the users can choose the start date for plotting. The time series graphs are dynamic and fetches new data from the Elasticsearch Datastore every 5 s. Spectral analysis is done by NodeJS server running in the cloud and pushed into Elasticsearch every minute. Each

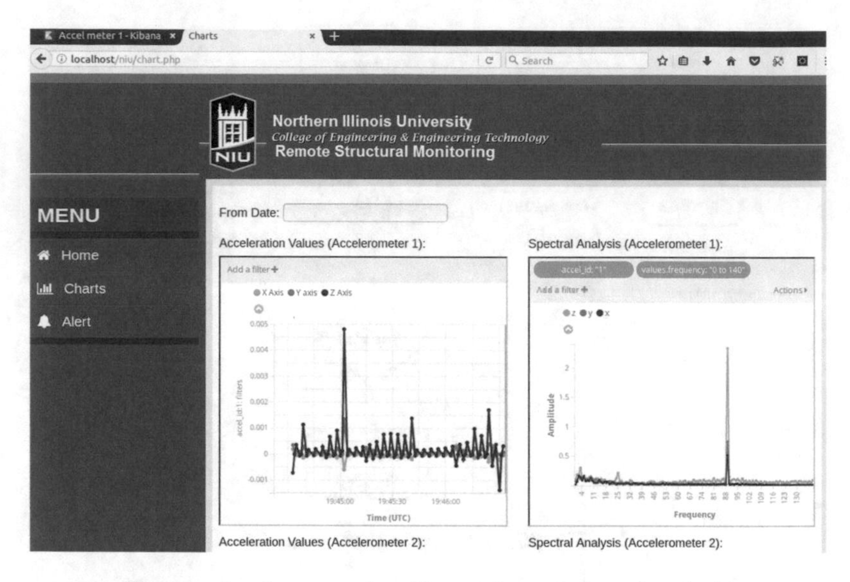

Fig. 18. Accelerometer data (time series and spectral analysis)

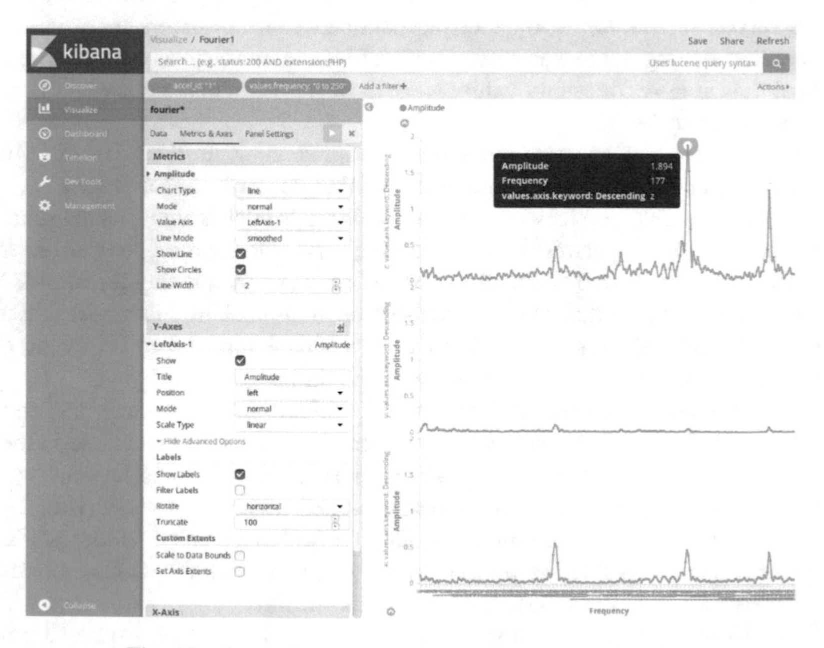

Fig. 19. Spectral graphs along with Kibana user interface.

time NodeJS takes recent 2048 data points from time-series data to perform spectral analysis and pushes them to Elasticsearch to be plotted by Kibana via the graphical user interface (GUI). Figure 19 shows spectral analysis of three axis of a single accelerometer along with Kibana user interface.

4 Conclusions

The paper describes the design, development, and implementation of a number of remote experimentation developments utilizing embedded systems as well as the Cloud. The first three used only embedded processor systems for both controlling the experiment and as the server for the complete system. When the structural monitoring system utilized an embedded processor board as initial processing and the Cloud for data analysis and user access. There are two important features of this facility. One is the use of a single software package for communicating with the experiment and the development of the GUI as well as data processing. The second is the simplicity of client access to the system; a client simply needs a browser without installing any plugins.

References

1. Budiman, R.: Utilizing skype for providing learning support for Indonesian distance. Procedia-Soc. Behav. Sci. **83**, 5–10 (2013)
2. Gomes, L., Bogosyan, S.: Current trends in remote laboratories. IEEE Trans. Ind. Electron. **56**(12), 4744–4756 (2009)
3. Chen, S., Huang, Y, Zhang, C.: Toward a real and remote wireless sensor network testbed. In: Third International Conference Wireless Algorithms, Systems, and Applications, WASA 2008, Dallas, TX, USA, 26–28 October 2008 (2008)
4. Axaopoulosa, P.J., Moutsopoulos, K.N., Theodoridis, M.P.: Engineering education using a remote laboratory through the Internet. Eur. J. Eng. Educ. **37**(1), 39–48 (2012)
5. Tripathi, P.K., Mohan, J., Gangadharan, K.V.: Design and implementation of web based remote laboratory for engineering education. Int. J. Eng. Technol. **2**(2), 270–278 (2012)
6. Gravier, J., Fayolle, B., Bayard, M.A., Lardon, J.: State of the art about remote laboratories. Int. J. Online Eng. **4**(1), 19–25 (2008)
7. Python. Retrieved from Python Software Foundation (2015). https://www.python.org/
8. Tornado. Retrieved from Tornado Stable (2015). http://www.tornadoweb.org/en/stable/
9. Boldt, E.: Python Web UI with Tornado (2013). http://robotic-controls.com/learn/python-guis/python-web-ui-tornado. Accessed 01 Apr 2015
10. Arduino Mega. Retrieved from Overview of Arduino mega (2015). http://arduino.cc/en/Main/ArduinoBoardMega
11. NodeJs, About Node.js. (2017). https://nodejs.org/en/about/
12. Heroku (2017). https://www.heroku.com/
13. Elastic Search (2017). https://www.elastic.co/
14. Elastic Kibana User Guide (2017). https://www.elastic.co/guide/en/kibana/current/introduction.html

Integration of New Technologies and Alternative Methods in Laboratory-Based Scenarios

Martin Burghardt[1], Peter Ferdinand[1], Anke Pfeiffer[2], Davide Reverberi[3(✉)], and Giovanni Romagnoli[3]

[1] Institut für Wissensmedien Koblenz-Landau, Koblenz, Germany
{burghardt, ferdinand}@uni-koblenz.de
[2] Hochschule für Technik Stuttgart, Stuttgart, Germany
Anke.Pfeiffer@hft-stuttgart.de
[3] Università di Parma, Parma, Italy
davide.reverberil@studenti.unipr.it,
giovanni.romagnoli@unipr.it

Abstract. In this study, we report a preliminary requirements analysis to recognize needs and possibilities for integrating new technologies and methods for lab-based learning in the field of Industry 4.0 and Internet of Things. To this aim, different scenarios, such as real, remote and virtual labs, are considered to be addressable within an integrated learning environment that focuses on alternative methods (i.e. Serious Games, Self-Regulated and Collaborative Learning) and new technologies (i.e. Open Badges, Mixed Reality and Learning Analytics).

To support the design of the laboratory-based learning environment, qualitative interviews were conducted with both expert lecturers and relevant students in the field of engineering, to provide complementary perspectives. These interviews were carried out to analyze the requirements, and to identify possible benefits that relevant stakeholders expect by using these teaching and learning methods and technologies. A qualitative content analysis has been started on the interviews to define which is the perception of the new technologies and teaching methods. The different points of view about technologies and methods coming from expert lecturers' and relevant students' interviews are provided.

Keywords: Lab-based learning · Self-Regulated Learning · Collaborative learning · Learning Analytics · Serious games · Mixed Reality · Open Badges

1 Introduction

It is nowadays impossible to deny the importance of laboratories and experimentation in Science, Technology, Engineering and Mathematics (STEM) education, as lab-based learning allows to achieve important pedagogical objectives, such as learning to manipulate and understanding the constraints of the surrounding physical environment, by applying theory to practice [1]. Lab-based education pursues a wide variety of objectives, such as: (i) connecting theory to what is implemented and observed in the

M. E. Auer and D. May (Eds.): REV 2020, AISC 1231, pp. 488–507, 2021.
https://doi.org/10.1007/978-3-030-52575-0_40

laboratory, thus possibly validating the theory and justifying analytic concepts; (ii) identifying differences between models and physical systems, especially by introducing real world factors usually disregarded in analytical solutions; (iii) model and simulate systems, and identifying trade-offs between simplification and accuracy; (iv) providing students with the opportunity to experience professional techniques and practice, such as writing reports, team building, and problem solving [2]. With regard to STEM education labs allow to join the practical experience to the math and physics behind the concept, by implementing the theoretical solutions in real-world problems, and by using real devices. This is particularly true for technology intensive subjects, such as Internet-of-Things (IoT) and the set of sensor systems, automation and ICT for manufacturing, often referred to as "Industry 4.0" (I4.0). Although defining those topics, as well as ascertaining their connection and interrelation, remains outside the scope of the present paper, it is undeniable that IoT and I4.0 are linked and greatly benefit from the opportunities provided by lab-based learning, especially that of applying theory to practice in the attempt to solve real-world problems, and with the support of real devices [3]. Despite the importance of lab-based learning in the above-mentioned subjects, the scientific literature often reports cases of non-satisfactory deployments of lab-based learning and ICT aided learning paradigms [4]. Amongst the possible causes of this issue, some authors report the difference between practice-based learning that is common at workplaces and the ICT-supported formal learning that is typical of university education [5]. Still, [6] suggests that new technologies will not make the teacher obsolete: ICT can promote more efficient and effective learning processes only if adequately supported, and ICT-empowered learning activities guided by a teacher have shown huge impact on learning outcomes. For these reasons, the interest of the scientific community in lab-based learning, and especially in non-traditional labs, has been increasing regularly in the last two decades [7]. Some recent researches suggest that the learning outcomes of non-traditional labs can be comparable, and sometimes even broader, than those of traditional labs [8]. In the course of the upcoming digitization of lab-based learning and teaching, this field is increasingly opening up for the use of innovative didactical methods [9]. Still, opportunities and criticalities related to the integration of new methods and technologies in lab-based learning are not well-investigated topics. The goal of this study is to perform a preliminary requirement analysis to identify needs and possibilities for the meaningful integration of new technologies and methods for lab-based learning in the field of I4.0 and IoT. We focus on alternative methods for teaching, learning and assessment that can be applied to lab-based learning, such as: (i) Serious Games; (ii) Self-Regulated Learning, and (iii) Collaborative Learning. Moreover, we considered new approaches and technologies such as: (i) Open Badges, (ii) Virtual Reality (VR), (iii) Augmented Reality (AR) and (iv) Learning Analytics. This study is based on a funded project called Open Digital Lab for you (DigiLab4U), whose goal is to integrate new methods and technologies in lab-based learning. In the course of this project, didactical methods that have already been successfully tested and implemented in other contexts are to be further developed by pedagogical sound integration of digital media. Thus, the whole range of possible scenarios, from real, remote and virtual laboratory-based learning scenarios, in the field of I4.0 and IoT should be addressable within an integrated learning environment. The remainder of the paper is organized as follows: in Sect. 2, we will briefly report an overview of innovative didactical methods, while Sect. 3 will sketch the new approaches and technologies. Section 4 describes the methods used to define the interviews' guidelines and to process data. The anticipated outcomes from the experts' and students' interviews will be reported and discussed in Sect. 5, while conclusions are presented in Sect. 6.

2 An Overview on Innovative Didactical Methods

2.1 Serious Games

Serious games (SGs) can be defined as games that are not just entertaining players, but also hold learning potential, as they foster the acquisition of skills, such as dealing with complex problems and dilemmas [10]. Although their popularity is not yet widespread, many researchers indicate that their potential is very high [11]. Amongst the advantages of SGs, we may report an holistic approach to the topic, that is experienced as a whole, the suitability of SGs to transmit characteristics of complex systems, and the active, and often enthusiastic, involvement of gamers [12]. Also, SGs are considered effective didactical methods to promote social collaboration increase motivation, improve attention and enhance technological competences, as well as encourage self-regulated learning [13]. It is also widely recognized that SGs and simulations can be incomparable tools to provide experiences that could not be made otherwise, due to safety issues, unbearable costs or time requested. On the contrary, amongst the negative factors to the use of video games in education we could report frustration, gender differences and staffing concerns, lack of instructional design models for gaming, game development costs and times, lack of understanding of educational benefits and of quality SGs [14]. However, despite these limitations, the use of SGs is increasing rapidly, as they are applied to a great number of different fields of education and training, ranging from engineering to medicine, and to several areas, from university to corporate, and even in military [15]. As Gee and Shaffer [16] suggest, SGs are likely to lead towards radical transformation of learning for twenty-first century skills. Still, to this day we notice the lack of research on SGs, especially in comparing serious game-based and nongame-based courses.

2.2 Self-regulated Learning

In German-speaking regions, the understanding of Weinert [17] is frequently used, according to which in self-directed learning the essential decisions about learning goals, strategies, assistance, location, time and control are made by the learner [18]. More recent deterministic approaches, such as that of [19], try to create a more differentiated and more precise foundation for definition by bringing together the most important common features from different explanatory approaches and thus to make a contribution to standardization. Although the question of definition has still not been finally resolved, the relevance and need for self-directed learning are beyond question.

This is because, in learning, it has a motivating effect on the learner because it addresses the three basic needs [20] of autonomy, competence and relatedness. In addition, its importance continues to grow in several areas. As a result of the change in knowledge psychology over the past few years from a behaviorist to a more cognitivist-constructivist understanding, learners are increasingly seen as independent. Teaching is seen as stimulating and supporting the learner's own activity [21]. Profound changes are also taking place outside educational institutions. Based on technical and scientific progress as well as various socio-political changes in the recent past, lifelong learning is playing an increasingly important role in education [21]. In addition, the learning environment must also be suitable for self-directed learning. Multimedia learning environments in particular are usually very well suited to supporting self-directed learning processes, as they can, for example, eliminate the place and time dependency of learning or facilitate the provision of various learning resources [22].

2.3 Collaborative Learning

In the engineering sciences, as in many higher education disciplines, collaborative learning is an important aspect to consider while designing lab lessons. It is even part of many competence goals, where it is included via teamwork skills. For example, in the "Criteria for accrediting engineering programs" it is stated under student outcomes, that students should have "[…] the ability to function effectively on a team whose members together provide leadership, create a collaborative and inclusive environment, establish goals, plan tasks, and meet objectives"[1]. Despite its importance, collaborative learning is often used as a practical learning organization to minimize workload for teachers and tutors or because the available time and space makes accessibility to laboratories for groups the only worthwhile option. Others would define it more stringently based on certain criteria to separate it from cooperative learning and other forms of learning. But most literature does not concern itself too much with more precise definitions as there can be value in taking a broader perspective. The same is true for the DigiLab4You project, where all forms of social learning are considered. But there are certain aspects which should be kept in mind while designing laboratory lessons, or any teaching method with groups for that matter. Group size, prior domain knowledge, prior experience with collaborative learning and most likely other personal factors like gender, motivation and personality play a role in a successful social learning setting [23]. While collaborative learning provides a lot of benefits in and of itself [24] it comes with its own challenges, like missing social clues while conversing, which might result in a lack of social presence [25]. Therefore, the project can research and make use of modern technologies like Mixed-Reality and Learning Analytic to support a digital collaborative learning environment and also design didactical recommendations for our target groups.

[1] https://www.abet.org/accreditation/accreditation-criteria/criteria-for-accrediting-engineering-programs-2019-2020/#GC3.

3 Outlining New Approaches and Technologies

3.1 Open Badges

Open Badges are digital artefacts which represent certain achievements of a person in a visual way. They operate as a micro-credentialing system to demonstrate and recognize accomplishments, knowledge, skills or competencies in a digital environment [26, 27]. Open Badges are digital image files that contain metadata that provides evidence of specific achievements or claims in formal or informal learning as well as in community engagement. They can represent both digital and non-digital achievements. The use of Open Badges is based on the Open Badge standard initiated by the Mozilla Foundation in 2011 and version 2 of which was published in 2018. Today Open Badges are used worldwide by organizations, including universities, to visualize and demonstrate, in particular, competences and skills acquired in formal education, but not being the subject of formal certification [28]. In the domain of higher education, [29] conducted a qualitative study on how digital badges are being used with ten individuals who lead digital badge initiatives in their education institutions. Results show that the purpose of the badge usage, their transferability (e.g. across courses or institutions) and the fit of learning objectives and assessment criteria of the badges were crucial points for the implementation of badge offers in higher education. The interviewees stated that skill-based badges, which persons acquire along described evaluation criteria, are valued higher than badges for pure participation. Moreover, badges are considered more motivating if they are used as intrinsic rather than extrinsic motivators (e.g. as pure rewards in competitions). However, to consider the value and efficacy of Open Badges in educational settings of higher education, further research is needed [29]. Especially, the questions whether digital badges are effective pedagogical tools and for what purpose they are best suited in higher education is still open. The project investigates possibilities of using Open Badges to foster individual and collaborative learning within a hybrid and partly augmented environment for lab-based learning. Especially the deployment of Badges for the support of self-directed learning (e.g. by modelling learning paths) and as motivation within serious gaming is focused. In addition, the project examines the acceptance of badges regarding the documentation and certification of competences of lab-based learning, and the potential of L.A. data as evidence for issuing Badges. A requirements analysis is currently being carried out to ensure a pedagogical sound use of badges in the project.

3.2 Mixed Reality

Mixed Reality (MR) acts as an umbrella term and describes the fusion of the real and the virtual world. Taxonomically we can differ between Augmented Reality when the real world is enriched with virtual information and Virtual Reality (VR) when we are in completely computer-generated worlds [30]. In the 90s already a lot of attention was put into the area of virtual reality. Especially in the industrial sector there were many exciting demonstrations, but they were not quite practical yet [31]. Nowadays, virtual reality applications can be found in many branches of industry and are used successfully in everyday work [32]. Therefore, the interest in using this new technology in engineering education is not surprising, since it is able to foster creative learning and prepares students with business relevant [33]. VR as a tool alone, however, is not sufficient to enable successful learning. A didactic approach has to be found that takes into account the interactivity and realism that virtual reality offers [34]. In the area of augmented reality, a similar trend can be seen, although the hardware and software are not as technologically advanced. Nevertheless, there exists may engineering-based research like [35]. It has also been shown that AR has a positive impact on learning, unlike traditional 2D desktop interfaces [36].

3.3 Learning Analytics

Due to the digitization of lab-infrastructure in the engineering education, a large amount of digital data is available that can be used to support learning and teaching processes. Research results of the past years show, that the field of Learning Analytics (LA) is increasing rapidly in the field of remote and virtual laboratories in the engineering education [37]. Siemens and Long define LA as "the measurement, collection, analysis and reporting of data about learners and their contexts, for purposes of understanding and optimizing learning and the environments in which it occurs" [38]. LA is always linked to the goal of eliciting, assessing and evaluating learner data to provide individual support and to "optimize learning processes, learning environments and educational discussion making" [39]. In laboratory-based learning processes a variety of information can be extracted, interpreted and if necessary be visualized, because the users leave a huge amount of "traces" behind, while operating in the lab-environment. Depending on the didactical objectives, entirely different data can be collected and accordingly be relevant for the interpretation. For example, Hawlitschek *et al.* analyzed cognitive and motivational differences of laboratory based learning to explain dropout factors in laboratories, while Venant *et al.* focus on students' awareness of their learning performance by using a social comparison tool that aims to engage learners in deep learning processes [40]. For the hybrid lab-environment in DigiLab4U it is planned to carry out the acquisition of LA data while students are using the provided digital tools to solve their lab-exercises. Foremost a learning management system, that can capture automatically the interaction of the user with the system will be used. Here, the data that mostly focuses on duration and frequency in using the activities and learning resources, can easily be collected and analyzed. According to Ochoa a more challenging task for laboratory-based learning is to deal with the different modalities like gestures, actions, speech or writing skills that occur during learning processes [41]. For these cases the main goal will be to extend the application of LA tools and methodologies to learning contexts that do not readily provide digital traces and to capture the required modes to understand and support the learning processes appropriately - for instance, with video-recording and an augmented reality application. Against this background the following objectives should be pursued with the implementation of LA in the DigiLab4U laboratory-based learning environment: monitoring and analysis, tutoring and coaching, assessment and feedback, adapting and awareness and reflection. To examine the advantages, boundaries and concerns of LA in order to further develop the concept behind DigiLab4U, the LA-topic was also part of the guided interviews.

4 Methods

In order to achieve the goal of this study, that is the identification of possibilities and needs in technology and method integration, we conducted interviews with teachers and learners in the course of a requirements analysis. The interviews, or surveys, have the advantage that, on the one hand, they can provide the directly affected groups of actors with views, interpretations and evaluations and, on the other hand, they can also provide their insights into action processes and knowledge from experience. The information obtained in this way should lead to a collection of a broad spectrum of types of needs and thus contribute to the achievement of the study objective. They are, however, not representative, due to the small size of the sample, and therefore not suitable for generalization.

4.1 Guidelines for Interviews

Teachers interviewed have many years of experience in the field of information technology, logistics, production and supply chain management. All eight interviews were carried out by employees of the DigiLab4U project in Stuttgart, Parma, Koblenz and Bremen, from May to July 2019. A thematically structured guideline was used, with appropriate questions, which could be omitted, supplemented or modified by the interviewers, based on the context. Questions offer as much space as possible for open answers. Students' interviews were adapted from those for experts. Indeed, guidelines was reduced to essential topics such as motivation, L.A., Open Badges, SGs, C.L., S.R. L., digital teaching-learning environments and laboratory-based learning.

4.2 Data Processing Methods

In the preparation of the evaluations, the interviews were first transcribed according to uniform rules to ensure that all important information could be included in the analysis. According to Mayring, the method of qualitative content analysis was applied. To summarize the available material according to certain topics and content areas, the procedure of the "content structuring analysis" was chosen [42], composed by the following steps: (i) the content categories and sub-categories were defined, derived from the interview-guidelines' topics, to assign text passages to them, based on sample passages; (ii) the category table resulting from step 1 was then filled by extracting and assigning text passages relevant to the objective; (iii) text passages are further paraphrased to make them more homogeneous; (iv) category system is revised accordingly; (v) text passages are finally assigned to appropriate categories.

5 Anticipated Outcomes and Discussion

The focus of the evaluation is on identifying needs, views, and possibilities in the integration of new technologies and methods. However, since the category system is currently still under revision, the associated results are preliminary.

5.1 Context and Motivation

Teachers Results

Experts agreed that the students' degree of motivation varies greatly, by type and quantity. Some learners were motivated by the contents, while others were motivated by the passing of the courses or achieving good grades. This depends on teaching and learning content and required work effort. The motivation could be read from students' commitment or simply from the regular presence and carrying of learning materials. To increase the students' motivation, communication also plays an important role. Some experts specifically approach learners and ask them about possible problems or discuss their learning progress. One expert likes to talk about his past experiences. The same goal can be reached by means of practical activities, smaller tasks, or by using multimedia elements. Rewards are not mentioned as tools: one interviewee consciously avoids the use of rewards, as they could be demoralizing for those who do not receive them.

Students Results

The main drawbacks faced by the students is the overall detached approach usually experienced at Universities. In fact, the difficulty to interact with professors and with other students is a big obstacle, particularly at the beginning of students' career. This is even more striking in mandatory courses not in line with student's interests, or because of poor material quality, frontal lessons, or when the professor's engagement in the topic is low. Here motivation can be found either in the lecturers' approach or in the utilization of alternative technologies. Seven interviewees have experience with recognitions as a tool to enhance involvement; in particular, they had experience with two different types of recognition: the first consist of a potential higher evaluation in the exam; the second consist of additional certification or personal experience (e.g. conference paper presentation) not directly related to the final exam. One doubt raised a single student is about the lack of equity of the recognition which could alter the outcome of the final exam.

Comparison and Discussion

Results shows that communication plays an important role for successful teaching and learning, especially in hybrid environments. Indeed, interviewees criticize a lack of motivation and communication, which could hinder the learning experience. There seems to be a strong case for integrating coaching and mentoring processes into laboratory-based scenarios, to involve learners right from the start and to ensure feedback for learners as well as for teachers to improve their performance. Some of the teachers already apply specific methods and approaches for lab-based learning to foster students' engagement. Such effort is appreciated by students, and feel motivated consequently.

5.2 Laboratory-Based Learning

Teachers Results

Laboratory-based learning is emphasized by all groups surveyed. An important aspect for many teachers is that students could gain their first practical experience and apply acquired knowledge under real circumstances. Also, it would give students the opportunity to experience typical problems of everyday work and learn how to solve them. In such environments, they would be forced to engage intensively with the topic themselves and would benefit from it. They would have the opportunity to test and experiment extensively and to examine the results in reality. It must be noted that interviewees of logistics/supply chain management see the most compelling issue in managing groups, given that groups can be too large, attendants and equipment can be scarce. Therefore, many participants are often denied the practical experience. Further obstacles are the often-malfunctioning technology, the cancellations of many events, caused by lack of irreplaceable experts. Accordingly, the teachers' wishes for improvement in their own laboratory environments relate primarily to the elimination of the problems mentioned. Also, they demand more personnel for a better support of their students.

Students Results

Here, the core issue was to assess the experience level of the students in laboratory-based learning. Almost all students are aware of lab-based learning. 70% of them had significant experience with it, 20% a little experience and just one student never had experience with it. A common perception is that a laboratory is a place where theoretical knowledge is put into practice. Two different kind of laboratory can be discerned, machine-based laboratory and computer-based laboratory. In machine-based laboratories, the students practice also with different software, coupled with the machines, and use it to set them up or to gather data. Instead, in computer-based laboratories, the software is used to simulate practical aspect of each topics. The software used by the students are Microsoft Excel©, Simul8©, Solid Works©, MATLAB©. Generally, the major advantage coming from lab-based learning is a higher skills' level. In fact, lab-based learning can offer students both a deeper understanding of theoretical knowledge and the development of practical skills. Nevertheless, this kind of experience expands auxiliary knowledge, like problem solving skills, resulting from the various situation faced during lab experiments. On the other hand, many criticalities have been raised. The biggest problem to face is find the correct combination of class and laboratory session. Indeed, one of the main issues raised by the students about lab-based learning is that, if too much emphasis is given to lab session, a lack of theoretical teaching can be experienced. Other problems are about implementation, which is about how the lab-based learning has been implemented in different reality. It was noticed that a balanced level of tutoring is necessary. Certainly, an absence of tutoring will inhibit the students to overcome difficult situations. On the contrary, a too detailed guidance can inhibit the development of problem-solving skills. Finally, the last issue is about available assets, which often are old or not available for everyone. Indeed, the main students' request for an optimal laboratory environment is for a higher availability of assets. Other students' inquiries are about more laboratories and more courses, as well as to have enough PCs for all the participants and equipment closer to a real industrial environment. Another wish coming from the students is the possibility to perform the experiment without pressure and have free access to the labs.

Comparison and Discussion

Both professors and students see laboratories as an important facility for experiencing the transfer of theory and practice, improving and deepening theoretical as well as technical and methodical skills, like problem solving and experimenting. From the results it can be clearly deduced that, in addition to a good infrastructure and equipment of the laboratories, appropriate supervision should be ensured. Laboratories should take up current needs – which in this case includes competences in the field IoT and I4.0 – the and support the transfer of theory and practice and to ensure that laboratory work stimulates and promotes students' technical, practical and theoretical skills in equal measure. From a teacher and learner perspective the following requirements should be met: (i) sufficient industries related equipment, (ii) regular maintenance to avoid technical problems as well as (iii) adequate financial and personal resources.

5.3 Digital Teaching and Learning Environments

Teachers Results

The use of Learning Management Systems (LMS) is considered useful and helpful by most of the experts surveyed, which is why they are frequently used. However, some of them link this circumstance to certain conditions, such as a minimum level of quality of the LMS or an added value for teaching and learning associated with its use. The basic willingness for use is, however, present in all respondents. While on the one hand there appears to be a great potential to support teaching-learning processes, administrative processes and communication procedures, on the other hand a certain reluctance remains, which seems to be mainly due to the problems of current learning management systems. For example, it is criticized that many systems would be equipped with too few functions and inadequate tools. In some cases, they would be difficult to operate and inflexible, which means that unforeseen situations or needs often result in dead ends. But there would also be many advantages to using them. This category includes not only the decoupling of space and time for learning, but also many of the possible functions of the LMS, such as communication tools, the provision of learning materials, the integration of projects or the submission of work - if necessary, with automated feedback. In the opinion of some teachers, this would give them the opportunity to learn more about students' performance and problems, so that they can then offer them better support. To this end, these experts would like to see, among other things, the integration of progress controls.

Student Results

The third significant aspect is linked to the utilization of Learning Management System (LMS), given that the new technologies that have to be integrated in lab-based learning must somehow communicate with an LMS. In fact, it's resulted that LMSs are widely spread in all the university analyzed. In fact, it has been noticed that all the interviewees need to interact almost daily with the different LMS from their universities. At the moment, the most common use-cases are (i) repository, for the download and upload of materials, and (ii) to receive communications from professors; half of them have had also some kind of deeper interactive experience: of these only a couple of cases weren't superficial experiences. The first student used the LMS even to follow Podcasts and Webcasts. The second one used it to perform tests and for scrutinies. The overall limited utilization of LMS led students to an absence of knowledge acquired via hybrid teaching: just one of the interviewees had experience with it, but not during his/her studying period. The utilization of Mixed Reality (VR/AR) is limited too, just three students out of ten had an experience with VR/AR devices and, in these cases, no further use of the tool happened. Despite this limited utilization of LMS and VR/AR, the usage of technical media is requested daily to the learners. Every student interviewed uses a laptop to take notes during class, to reorganize them or to do analysis, exercises or projects. Tablets or smartphones sometimes replace laptops to take notes or consult materials. Microsoft Office is widely used, especially the software Word©, Excel© and Power Point©, based on the different courses chosen, also other software have been used like Simul8©, MATLAB© and platforms like GitHub© and Google Docs© in order to share files among the members of the same team and work on it at the same time.

Comparison and Discussion
The interviews showed, that teachers and students appreciate the use of a learning platform. But there are some reservations on this part from the teachers' side, that can explain the results of the student interviews, namely why the actual adoption is usually limited to providing learning resources, even though the LMS provide divers functionalities and tools. On the students behalf they would wish to have the possibility of integrating digital tools they use anyway, like Google Docs© and so forth. While students are keen learning to know new technologies, that might be relevant for their future professional life, like VR or AR, these technologies play hardly any role at the moment for the teachers at the moment. Here can currently be a gap identified, that needs to be closed in a timely manner with view to developments in the course of digitalization in the engineering sciences. Consequently, for DigiLab4U a targeted acquisition of new laboratories needs to be discussed, that provide AR-/VR-technologies in their laboratory environment with the aim to foster engineering competences.

5.4 Collaborative Learning

Teachers Results
Many experts use collaborative learning in their seminars, but mainly in project work and exercises. However, especially those from the field of "Surveying and Computer Science" emphasize in this context that they did not consciously incorporate the method. For many, even from other disciplines, it is simply effective and often leads to good learning results. Furthermore, it would offer the advantage that students can learn to work together. But the evaluation also reveals some problems with collaborative learning. Thus, successful implementation would depend on a suitable environment.

Problems therefore would arise with the support of a high number of groups and participants, since this could no longer be guaranteed and would entail a considerable additional expenditure. A wrong group composition could also become problematic and lead to interpersonal difficulties that prevent collaborative learning from functioning. In addition, collaborative learning would end many times in cooperative learning.

Students Results

This kind of interaction has become fundamental with the increasing use of collaborative learning. Nine students interviewed out of ten had experience in collaborative learning; it's used in study group and in group project. It's recognized by all the relevant students that the greatest advantage coming from collaborative learning is the development of social and interaction skills that will be necessary in future works. It has emerged that students can fill each other weakness by cooperation and help one each other. Study group can help to boost the speed of study due to the deadlines that, each member, must respect. Group projects teach how to manage different members, especially if not chosen. However, especially when members are not chosen, there is a difference in the engagement. Usually some members work hard, while others work less. When groups' composition is forced it, some organizational problems can arise, such as communication, coordination, and some sabotaging behavior within the group. In order to have a better communication within the group, face-to-face communication is preferred when tasks have to be split or when final results have to be finalized. On the contrary, for fast briefing or to match and summarize the notes took during a class, it's preferred to have digital meeting on Skype© or work on the same document with Google Docs©; these tools are used especially when the students live quite far.

Comparison and Discussion

The purpose whereby the experts use the collaborative learning are totally in line with the benefits perceived by the students. Indeed, an interesting observation is that the learning goal "learning to work together", whom the teachers are looking for, if achieved through forced group is usually seen as a problem by the students. However, both the parts agree on the faster achieving of learning goals and improvement of the skills within the group. Indeed, the nature of collaborative learning allows the students to fill each other weakness. This advantage is used by the students also in study group and not only when request for project work. As well as the advantages, also disadvantages are shared by students and experts. It is a common opinion that a too heterogeneous group can create a sabotaging environment and slow down the learning process and the project development, also on motivation it can have a negative effect due to the variability in the engagement of the different group members. Despite these possible issues, the collaborative learning is widely used in laboratory work and can provide a greater improvement in learners behavior if suitable tools will be granted together with a proper group composition.

5.5 Self-directed Learning

Teachers Results

Self-directed learning is rated positively by the majority of respondents. Nevertheless, the necessary conditions for this are also pointed out. For example, it is mentioned that a meaningful use of the method would depend on course content, student skills and course form. Some students would be simply overwhelmed by this because they had never been taught the skills they needed at school and afterwards. Others, on the other hand, would lack the necessary attitude and willingness to use this method. One teacher even found that students with self-directed learning needed three times as much learning time. But respondents also put forward many arguments in favor of self-directed learning. On first hand, it would be generally important for learning to learn and also beneficial for personality development. On the second hand, this could make learning more effective and stimulating. Self-directed learning could increase the motivation to learn, because it enables one to make one's own didactic decisions and focus on one's own content. It would be especially suitable for project work.

Students Results

Just four students had experience in Self-Regulated Learning (SRL), three of them have used it to reach university goals, the other one used it for an extra-curricular language course. The common sensation, collected also from inexperienced students, is that SRL can increase personal responsibility, motivation, problem solving and organizational skills. However, various problems have been pointed out. The first one is linked to the self-determination of the objectives. In fact, students could focus their efforts on marginal aspects and, due to the non-mandatory nature of the SRL, miss some self-defined milestone. The little experience of the beginners represents a problem in SRL, if the organization and the decisions are left to the students, as they can feel disoriented and overwhelmed. Another obstacle is that, with a complete application of SRL (ranging from courses selection to milestones definition), some students could erroneously decide to skip fundamental courses in order to choose something easier. Taken to the extreme, such condition could lead to the scenario in which the goals set by the students are not in line with the knowledge requested from the professors. The students that have experience in SRL describe the interaction with teachers as two opposite behavior. There are some teachers that are too present during students' activities, preventing them from acquiring those skills deriving from the SRL; on the other side, some the professors don't interact at all with students, not even to clarify doubts or to reply to questions. The interviewee suggests an interaction where professors are available once per week or every two weeks, for clarification and to set some pillars that drive the students through their studies.

Comparison and Discussion

Due to the heterogeneous prerequisites concerning the students' ability of self-regulated learning, that were explained by the teachers, it should be taken in account to provide scaffolding in the lab-tasks to address different needs of learner. This can be organized according to different learning phases a lab-task respectively lab environment requires. In the interviews also reveal that teacher with less or mostly negative experience in guiding self-regulated learning could need methodical-didactical support to foster these processes in the lab environment. According to this, a digital lab-environment should not only deliver support for students but should also deliver design ideas for teachers for implementing lab-based learning scenarios incl. self-regulated methods and guidance in applicating these methods in lectures. Even if students describe themselves as self-regulated learner, the results show, that they are dependent on support in certain situations, especially if it comes to lab-environments when experience is still lacking. The students emphasize the need to get a good balance between instruction and self-regulated learning. Keeping this balance depending on the learning conditions and the subject determines the teacher's work in the digital/hybrid lab-environment. For laboratory-based learning this shows - what is already clear in other learning contexts - that the students expect the teacher to take more the role of a consultant and coach, who instructs, presents and explains.

5.6 Serious Games

Teachers Results

The wealth of experience on SGs differs greatly within the survey group: while some lecturers deal with the topic daily, others have little or no experience at all. Among the advantages are the playful and entertaining qualities as well as the motivating ones. Especially the idea that SGs can provide a safe and flexible environment to try out and change perspectives is appreciated by the teachers. For this reason, they would also be well suited for security briefings or technology introductions and, depending on the context of use, would be considered useful by many respondents. But there are also critical attitudes, which mainly refer to the cost-benefit ratio of the use. Many experts note that the realization of such games would be very complex and expensive. In addition, they would not be appropriate for all learning content and learning types. Among learners there would be a remarkable number of people who simply did not like games. But even those who are willing to play need to be convinced of the game and its meaningfulness. It would therefore be important to design SGs in such a way that they had the necessary closeness to reality but were still so abstract that they could still be implemented.

Students Results

The SGs are practically unused by the interviewees. In fact, just a couple of students had experience with them and only in a marginal way. Nonetheless they are perceived like a good way to have practical experience in an environment more dynamic then classical paper simulation. Through this tool, the commitment to learn is easier because it is encouraged by the pleasure of playing. Also, it results useful to develop problem solving and collaboration skills when used among a group. This can be done thanks to the treatment of difficult issues as if they were competition without the problem of excessive psychological pressure. The main challenge is to find the right combination of face-to-face class and SGs sessions, as well as the definition of how the scores are awarded. Despite a big willingness and interest to use them, they are not seen, nowadays, as a fundamental learning tool; a suggestion could be to use them like midterm exams or verification of skills.

Comparison and Discussion
For almost all the interviewees, SGs are poorly known. Indeed, just a couple of students had experience and a couple of experts deal with them daily. The flexibility of this methods is well known by both sides, which make SGs exemptional to be used in different topics. To complete their advantages, an easier and safer learning environment is provided through the pleasure of play and through the digitalization. The main difficult raised are (i) find the right balance between face-to-face class and SGs session, (ii) the cost-benefit ratio, and (iii) the non-unanimous willingness of students to use them.

5.7 Open Badges

Teachers Results
Experience with Open Badges is extremely limited among the eight teachers surveyed. Without exception, they state that they have either no or little experience in this field. Accordingly, most of them rate their knowledge of Open Badges as low. Doubts are expressed about the usefulness, feasibility and need of the tool, as well as interest and commitment. For some it is important that the certificates awarded represent a certain minimum effort and that not every little thing is rewarded. Others believe that Open Badges could increase the attractiveness of webinars, for example, because they link the tool to the documentation of previously unrecognized achievements and skills, and because this process could be partially automated and thus carried out without further bureaucracy. They could also be motivating for some learners.

Students Results
Just as for SGs, Open Badges are new for almost all the students interviewed. Indeed, only two of them had experience with them. The first student has a badge in Arduino training, but only at an introductive level, which he achieved with the university. The second student has some badges obtained on Udemy. From the interviews it results that Open Badges are perceived as a good way to prove the acquired skills; their value can be exploited especially during work interviews. They can also represent a good motivating factor due the opportunity to reach new level of certification and as a proof of the owner's willingness to deepen some topics. Due to the willingness to adopt them, some problems about Open Badges assignment dynamics have been raised. In fact, the utilization of shortcut or not totally acquired skills could lead to a value loss of the Open Badges released. According to this, has emerged a doubt about the utility of Open Badges, in fact it depends on the stakeholders that recognize their value and where and when can they be applied. In the end, too costly Open Badges, and an excessive focus of the stakeholders on them, can lead to a wrong candidate selection and to focus only on the learners that can afford them.

Comparison and Discussion

In general students and professors expressed only a little to no experience with Open Badges. Basically both, students and teachers emphasized Open Badges as a possibility to certify personal achievements/accomplishments in order to share and display Badges as verifiable record of learning and potentially teaching processes. Due to the missing knowledge about Open Badges the professors do not currently see the advantages, like identifying progress and content trajectories or signifying and credential engagement and achievements. Students see Open Badges also as a motivational factor e.g. as a reward for a task performed or as a way of teachers recognition of their own performance. Nevertheless, for teacher and learner the value of these "certificates" plays a major role and needs clarification. Who is the issuer, which standards should be achieved and certified and who is responsible for the quality of products and performances? These results show a general interest of teacher and learner in Open Badges for lab environments, also with a view to lifelong learning (e.g. relevance in work interviews). This leads to the question how the lab-experience on the platform DigiLab4U can be certified in an appropriate way in form of Open Badges, to provide meaningful certificates for the use in higher and further education soon.

5.8 Learning Analytics

Teachers Results

Experience and knowledge in this field are largely restricted. Nevertheless, many of the teachers are willing to use L.A., especially in order to identify content and organizational difficulties for the students. They hope that they will be able to offer the students better support and materials so that they will have the best possible conditions for learning. For other experts, the added value of L.A. is not apparent, for example because it would not contain useful information or only information that could be identified by observation. One interviewee also suspects that most students would not be comfortable working with such tools. Students with learning problems could be demotivated by the consequences of the use, they fear. In addition, some respondents expressed doubts about the functionality of L.A., as they often could not fulfil the expectations placed in them.

Students Results

All the students interviewed are unfamiliar with them, in fact nobody have never had experience with L.A. After a quick explanation, students said that they would like to use L.A. to understand which topics and which method of learning is the best, for each one, in order to save time and study, since the beginning, with the method that best suits them. LA can also be used to evaluate if the materials provided by the professors is enough and their quality is good. This kind of information can be used also by professors to understand if the teaching method used is correct, if the material provided is used, when it's used, and which are the students' preferred topics. However, without proper integration and interaction with LMS, without a large amount of data, LA are useless. Some students are also concerned that the results coming from LA may influence the final evaluation of the exam. There's also the risk that the students leave false data to trick the system. The students are asking to have the possibility to choose the data that have to be gathered, analyzed and then shared; in the details, they are asking for statistics about tests, possibility of comparison with other students' results, the number of downloads of a document, the preferred topics and the average time for exams preparation. Moreover, LA is not perceived like a fundamental tool for the learning process.

Comparison and Discussion
Neither for students nor for teachers does LA at the current stage play an essential role in their teaching and learning processes. On the positive side, some teachers see the possibility of supporting the learning process in a data-based way in the future, while others doubt that the data can provide them with more information than their own observations. The latter shows, that data is still not seen as possibility to visualize learning processes to students and to use these to reflect on personal achievements. Before launching LA-tools in the lab environment it will be a necessary to create an awareness of the added value that LA can offer for fostering lab-based learning, but also to name clear limits. Some students are highly positive about an opportunity to see their own learning status or possibilities to compare their learning achievements with fellow students, while other students are rather critical of LA, because they cannot assess the impact it will have on their assessment. Further they would like to have control of their data. Although there is little knowledge on the topic so far, the interview results of the teachers and students point out important facets that have to be considered in the context of LA-implementation. This includes enabling teacher to gain and interpret data, enabling students to control the learning-process based on their learning data, building an environment, that is based on trust-factors (like transparency, data privacy, ethical concerns, clear goals and learning objectives) as starting point for the acquisition to ensure a trusted framework.

6 Conclusions

Regarding the students' interviews' results, it must be stressed that the main issues is to find the right balance of teacher presence which is also the main motivational factor. Indeed, intrinsic and extrinsic student's motivation could be fostered by the teacher's attitude. Experts emphasize the importance of communication to increase motivation. Almost all students have experienced laboratory-based learning in hands-on and computer environment which is recognized, and supported by teachers, as an enhancer for both theoretical and practical knowledge acquirement. Nonetheless, students request more investments in assets, and a better exploitation of existing tools. Lab-based learning is usually coupled with collaborative learning, which have been experienced by 90% of students, mostly used for project works. Biggest difficult is group management, in particularly when group composition is forced. On the other hand, other technologies, methods and tools are less widespread. Self-directed learning, even if it can increase motivation, responsibility and problem solving skills, can be difficult to manage by beginners. Virtually no interviewee has experienced SGs. Even though they are perceived as an intriguing new way to perform simulation, they are considered optional tools. On the other hand, Open Badges have a great appealing even though they are not so prevalent, and the perceived effectiveness depends on the stakeholders' recognition. Secondly, experts are willing to adopt CL and SRL to have better learning outcomes depending on the framework conditions and context, as well as considering individual prerequisites. The same applies for SGs, Open Badges, and LAs. It must be stressed that teachers often have little knowledge and previous experience in these subject areas, and sometimes see no benefit in the use of these tools. Even if the results of the study are still preliminary, the requirements of teachers and learners collected here contain many important clues as to what is important for a meaningful inclusion of the methods and tools considered. In the future, it will therefore be necessary to complete the requirements analysis and then to initiate an iterative process (design-based research approach - DBR) with the completed information. The first step will be to develop laboratory-based learning scenarios that meet the needs of the target groups surveyed. This will raise the question of whether and how the teaching-learning methods and tools discussed here can be integrated into didactic concepts. The next step will be to implement and test the corresponding designs. Subsequently, surveys will be conducted to evaluate, for instance, how the concepts as a whole as well as the methods and tools used are received by the user groups, what advantages or disadvantages they see in them and where there is potential for improvement, in order to begin again with the first step of the DBR and optimize the laboratory-based scenarios in several runs.

Acknowledgement. This research was funded by German Ministry of Education and Research BMBF, grant numbers 16DHB2112 (HFT Stuttgart), 16DHB2116 (University of Parma) and 16DHB2115 (IWM Koblentz), and was developed within the project DigiLab4U (https://digilab4u.com/).

References

1. Feisel, L.D., Rosa, A.J.: The role of the laboratory in undergraduate engineering education. J. Eng. Educ. **94**(1), 121–130 (2005)
2. Antsaklis, P., Basar, T., DeCarlo, R., McClamroch, N.H., Spong, M., Yurkovich, S.: Report on the NSF/CSS workshop on new directions in control engineering education. IEEE Control Syst. Mag. **19**(5), 53–58 (1999)
3. Lensing, K., Friedhoff, J.: Designing a curriculum for the Internet-of-Things-Laboratory to foster creativity and a maker mindset within varying target groups. Procedia Manuf. **23** (2017), 231–236 (2018)
4. Nedic, Z., Machotka, J., Nafalski, A.: Remote laboratories versus virtual and real laboratories. In: Proceedings - Frontiers in Education Conference FIE, February 2003, vol. 1, pp. T3E1–T3E6 (2003)
5. Tvenge, N., Martinsen, K.: Integration of digital learning in industry 4.0. Procedia Manuf. **23**(2017), 261–266 (2018)
6. Mincu, M.E.: Teacher quality and school improvement: what is the role of research? Oxford Rev. Educ. **41**(2), 253–269 (2015)

7. Heradio, R., De La Torre, L., Galan, D., Cabrerizo, F.J., Herrera-Viedma, E., Dormido, S.: Virtual and remote labs in education: a bibliometric analysis. Comput. Educ. **98**, 14–38 (2016)
8. Brinson, J.R.: Learning outcome achievement in non-traditional (virtual and remote) versus traditional (hands-on) laboratories: a review of the empirical research. Comput. Educ. **87**, 218–237 (2015)
9. Orduña, P., Rodriguez-Gil, L., Garcia-Zubia, J., Angulo, I., Hernandez, U., Azcuenaga, E.: Increasing the value of remote laboratory federations through an open sharing platform: LabsLand. In: Auer, M., Zutin, D. (eds.) Online Engineering & Internet of Things, pp. 859–873. Springer, Cham (2018)
10. Hummel, H.G.K., Joosten-ten Brinke, D., Nadolski, R.J., Baartman, L.K.J.: Content validity of game-based assessment: case study of a serious game for ICT managers in training. Technol. Pedagog. Educ. **26**(2), 225–240 (2017)
11. Lewis, M.A., Maylor, H.R.: Game playing and operations management education. Int. J. Prod. Econ. **105**(1), 134–149 (2007)
12. Van Der Zee, D.J., Holkenborg, B., Robinson, S.: Conceptual modeling for simulation-based serious gaming. Decis. Support Syst. **54**(1), 33–45 (2012)
13. Annetta, L., Mangrum, J., Holmes, S., Collazo, K., Cheng, M.T.: Bridging realty to virtual reality: investigating gender effect and student engagement on learning through video game play in an elementary school classroom. Int. J. Sci. Educ. **31**(8), 1091–1113 (2009)
14. Hess, T., Gunter, G.: Serious game-based and nongame-based online courses: learning experiences and outcomes. Br. J. Educ. Technol. **44**(3), 372–385 (2013)
15. Susi, T., Johannesson, M., Backlund, P.: Serious games – an overview: Technical report, HS-IKI-TR-07-001 (2006)
16. Gee, J.P., Shaffer, D.W.: Looking where the light is bad: video games and the future of assessment. Phi Delta Kappa Int. EDge **6**(1), 3–19 (2010)
17. Weinert, F.E.: Selbstgesteuertes Lernen als Voraussetzung, Methode und Ziel des Unterrichts. Unterrichtswissenschaft **10**(2), 99–110 (1982)
18. Breuer, J.: Lernmethoden. Artikel, p. 87 (2000)
19. Dyrna, J., Riedel, J., Schulze-Achatz, S.: Wann ist Lernen mit digitalen Medien (wirklich) selbstgesteuert? Ansätze zur Ermöglichung und Förderung von Selbststeuerung in technologieunterstützten Lernprozessen. In: Communities in New Media: Research on Knowledge Communities in Science, Business, Education and Public Administration - Proceedings of 21th Conference GeNeMe, pp. 155–166 (2018)
20. Deci, E.L., Ryan, R.M.: Die Selbstbestimmungstheorie der Motivation und ihre Bedeutung für die Pädagogik. Zeitschrift für Pädagogik **39**(2), 223–238 (1993)
21. Ferdinand, P.: Selbstgesteuertes Lernen in den Naturwissenschaften - Eine Interventionsstudie zu den kognitiven und motivationalen Effekten eines Blended Learning Ansatzes. Kovac, Dr. Verlag, Hamburg (2007)
22. Sander, E., Ferdinand, P.: Empirische Befunde und pädagogische Chancen im Kontext selbstgesteuerten, experimentierenden Lernens in den Naturwissenschaften. Empirische Pädagogik **27**(1), 47–85 (2013)
23. Dillenbourg, P.: "What do you mean by collaborative leraning? In: Dillenbourg, P. (ed.) Collaborative Learning: Cognitive and Computational Approaches. Elsevier, Oxford (1999)
24. Laal, M., Ghodsi, S.M.: Benefits of collaborative learning. Procedia - Soc. Behav. Sci. **31**(2011), 486–490 (2012)
25. Tu, C.H.: On-line learning migration: from social learning theory to social presence theory in a CMC environment. J. Netw. Comput. Appl. **23**(1), 27–37 (2000)

26. Mah, D.K., Bellin-Mularski, N., Ifenthaler, D.: Foundation of Digital Badges and Micro-credentials: Demonstrating and Recognizing Knowledge and Competencies, pp. 1–530. Springer, Cham (2016)
27. Muilenburg, L., Berge, Z. (eds.): Digital Badges in Education. Routledge, New York (2016)
28. Buchem, I.: Entwurfsmuster für digitale Kompetenznachweise auf Basis von Open Badges im Kontext virtueller Mobilität (Design patterns for digital competency credentials based on open badges in the context of virtual mobility). In: Proceedings der Pre-Conference-Workshops der 16, E-Learning Fachtagung Informatik Co-located with 16th e-Learning Conference of the German Computer Society (DeLFI 2018), Frankfurt, Germany, 10 September 2018, vol. 2250 (2018)
29. Carey, K.L., Stefaniak, J.E.: An exploration of the utility of digital badging in higher education settings. Educ. Technol. Res. Dev. **66**(5), 1211–1229 (2018)
30. Milgram, P., Kishimo, F.: A taxonomy of mixed reality. IEICE Trans. Inf. Syst. **77**(12), 1321–1329 (1994)
31. Brooks, F.P.: What's real about virtual reality? IEEE Comput. Graph. Appl. **19**(6), 16–27 (1999)
32. Berg, L.P., Vance, J.M.: Industry use of virtual reality in product design and manufacturing: a survey. Virtual Reality **21**(1), 1–17 (2017)
33. Abulrub, A.-H.G., Attridge, A., Williams, M.A.: International journal of emerging technologies in learning. Int. J. Emerg. Technol. Learn. **6**(4), 4–11 (2011)
34. Vergara, D., Rubio, M., Lorenzo, M.: On the design of virtual reality learning environments in engineering. Multimodal Technol. Interact. **1**(2), 11 (2017)
35. Huang, J.M., Ong, S.K., Nee, A.Y.C.: Visualization and interaction of finite element analysis in augmented reality. CAD Comput. Aided Des. **84**, 1–14 (2017)
36. Billinghurst, M., Clark, A., Lee, G.: A survey of augmented reality. Found. Trends Hum.-Comput. Interact. **8**(2–3), 73–272 (2014)
37. Orduña, P., Almeida, A., López-De-Ipiña, D., Garcia-Zubia, J.: Learning analytics on federated remote laboratories: tips and techniques. In: IEEE Global Engineering Education Conference EDUCON, April 2014, pp. 299–305 (2014)
38. Long, P., Siemens, G.: Penetrating the fog: analytics in learning and education. Educ. Rev. **46**(5), 30–32 (2011)
39. Ifenthaler, D.: The SAGE Encyclopedia of Educational Technology. SAGE Publications Inc., Thousand Oaks (2015)
40. Broisin, J., Venant, R., Vidal, P.: Learning analytics for learner awareness in remote laboratories dedicated to computer education. In: CEUR Workshop Proceedings, April 2016, vol. 1596, pp. 31–37 (2016)
41. Ochoa, X.: Multimodal learning analytics. In: Handbook of Learning Analytics, pp. 129–141 (2017)
42. Mayring, P.: Qualitative content analysis: theoretical foundation, basic procedures and software solution (free download via Social Science Open Access Repository SSOAR). In: Forum Qual. Sozialforschung/Forum Qualitative Social Research (2014)

Process of Adapting a Local Laboratory to Connect to an Online Laboratory System Based on Standard Definitions and Open Source Technologies

Luis Felipe Zapata-Rivera[1(✉)], Maria M. Larrondo-Petrie[1,2], Başar Şahinbeyoğlu[3], and Serhan Argun[3]

[1] Latin American and Caribbean Consortium of Engineering Institutions LACCEI, Boca Raton, USA
l.f.zapatarivera@ieee.org

[2] Department of Computer and Electrical Engineering and Computer Science, Florida Atlantic University, Boca Raton, USA
petrie@fau.edu

[3] Acrome Robotics, San Francisco, USA
{basarsahinbeyoglu,serhanargun}@acrome.net

Abstract. Local educational laboratories are evolving from isolated resources into a more accessible, shareable and interoperable resources. Due to the flexibility of desktop, and mobile software development tools, developers can now easily add to the educational laboratory experiments; multimedia content such as: videos, text, animation, simulation, among others. Companies such as: Acrome, Quanser, Labs Land, among others, offer flexible educational Laboratory solutions for schools and universities in which the professor or laboratory manager can make a content selection and, in some cases, they can add their own content to the student's laboratory interface. This paper develops a methodology for extending the educational local laboratories with online capabilities that enables the interoperability and integration within a e-learning ecosystem and presents a real case of adaptation of a local laboratory into a fully functional online remote laboratory.

Keywords: Online laboratories · Educational laboratories industry · Control laboratories

1 Introduction

Online capabilities can be added to educational local laboratories without affecting current level of interactivity, enabling remote students to perform the laboratory experiments from any location through the use of internet and enhancing the students experience by the inclusion of the online integration with Learning Management Systems (LMS), Remote Laboratory Management Systems (RLMS), Open Educational Repositories (OER) Repositories, etc.

M. E. Auer and D. May (Eds.): REV 2020, AISC 1231, pp. 508–518, 2021.
https://doi.org/10.1007/978-3-030-52575-0_41

The process of adaptation of the local laboratories may require adjustments in the current local laboratory architecture. This can be done following the considerations and definitions of the recently published IEEE Standard 1876 – Networked Smart Learning Objects for Online Laboratories, which describes the architectural layers that ease the process of development and integration of educational laboratories, as well as the use of Experience API (xAPI) [2] and Learning Tool Interoperability (LTI) [1] standard that will support the information sharing and interfaces integration processes respectively.

This paper is organized as follows: Sect. 2 presents concepts related with online laboratories taxonomies, online laboratories supporting technologies and the role of the industry in the development of this type laboratories. Section 3 presented a methodology for extending educational local laboratories with online capabilities that enables the interoperability and integration within a e-learning ecosystems. Section 4 shows a real example of an adaptation of a local laboratory into a fully functional online remote laboratory. Finally conclusions and future work are presented in Sect. 5.

2 Online Laboratories and Industry Development

Online laboratories are a type of laboratories that include virtual, remote, mobile and hybrid laboratories [3]. Figure 1 presents a classification of laboratories considering the type of relations that can occur between them.

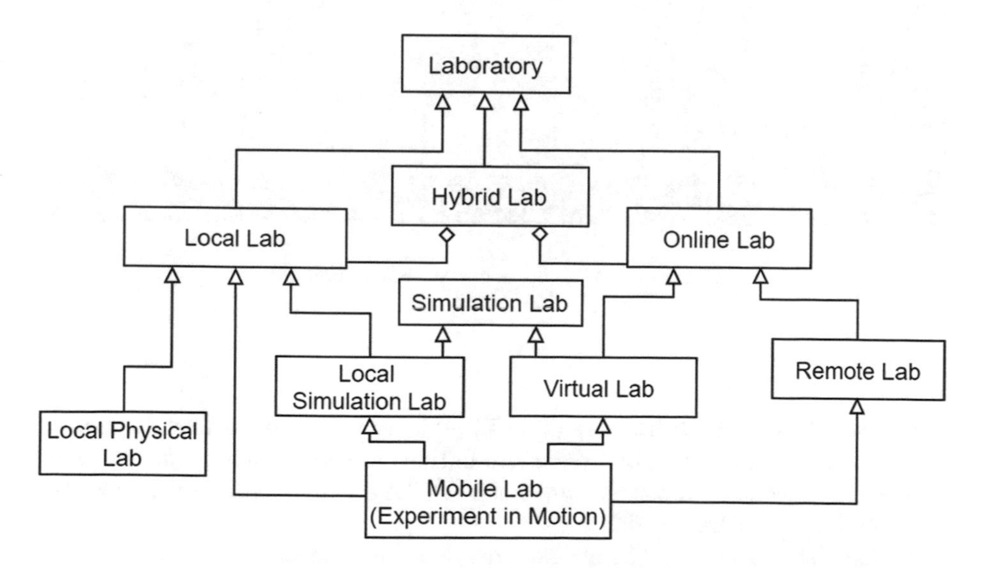

Fig. 1. Online laboratories taxonomy

The current panorama of development of educational online laboratories can be analyzed from the perspective of developments coming from academic research projects and industry efforts.

The bigger efforts on the development of the online laboratories have happened in the area of virtual laboratories (simulation based experiments). With the current development of internet and programming tools is relatively easy to develop virtual laboratories that show a very accurate and consistent results, given the users possibilities to change variable values and collect results.

This laboratories are more common in areas such as Physics, Chemistry and Biology. An example of these laboratories are:

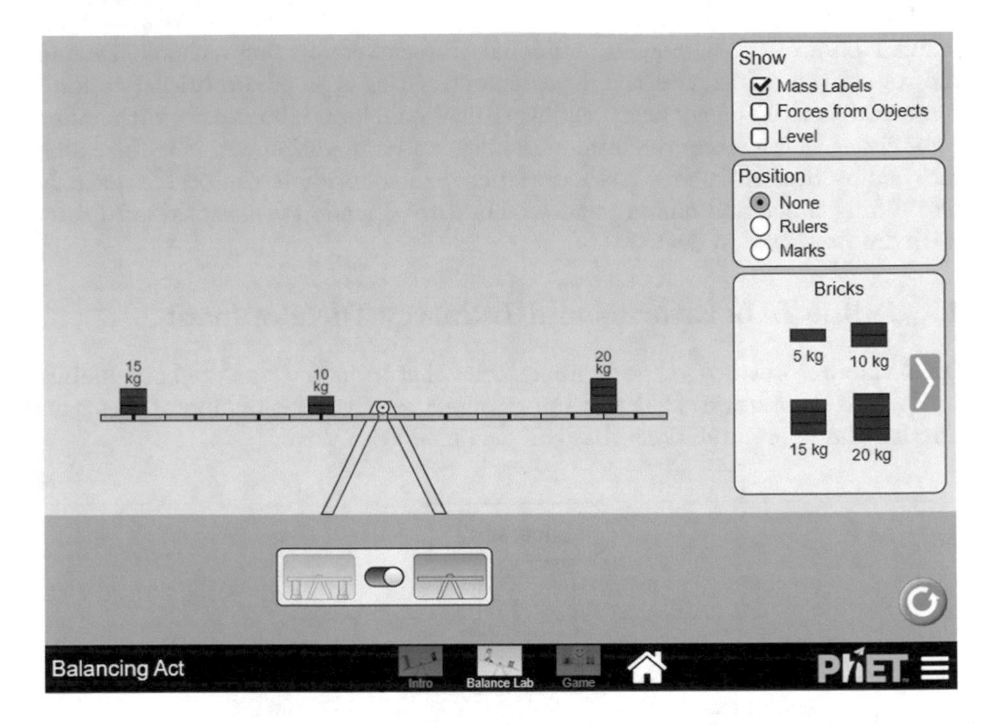

Fig. 2. PhET laboratory user interface [4]

- PhET
  Physics Education Technology (PhET) [4] interactive simulations, is an open platform with hundreds of virtual simulations in topics such as physics, biology, chemistry, earth science and math. All the PhET simulations are open educational resources (OER).
  The project was developed at the University of Colorado and funded by the Nobel prize winner Carl Wieman. Their mission is "To advance science and math literacy and education worldwide through free interactive simulations."
  Fig. 2 presents a Physics virtual laboratory about load balancing. In this laboratory the user can add different loads in different positions to understand how the torque phenomenon happens.

- Labster
  Labster [5] develops interactive virtual laboratories including simulations based on mathematical algorithms that support open-ended investigations. Their products are also developed including gamification elements such as 3D environments, storytelling and a scoring system.
  The laboratories are being used for different universities including Harvard, MIT, Stanford among others. Figure 3 presents the interface of one of the Labster laboratories about bacterial quantification by culture. This laboratory guides the student or trainee through complete realistic virtual laboratory experience.

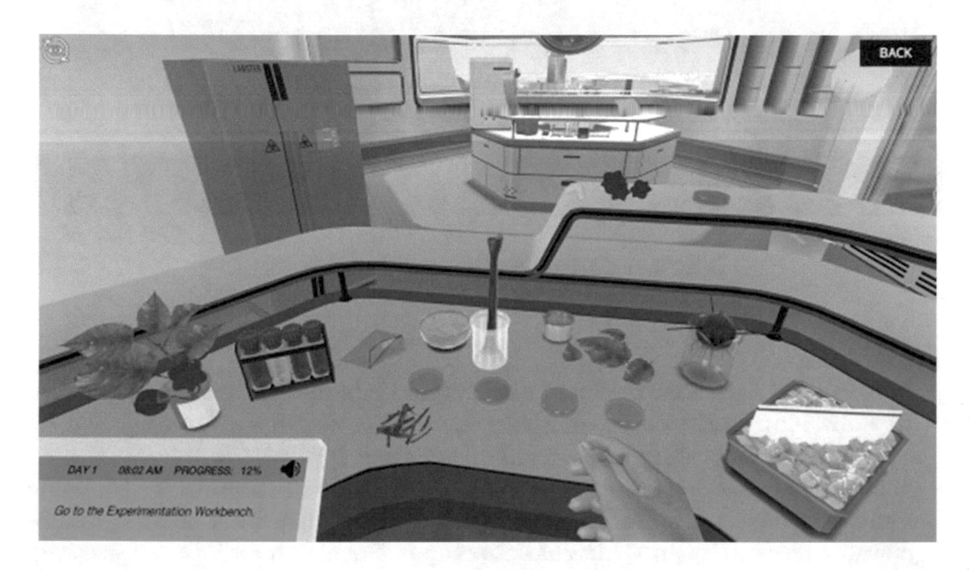

Fig. 3. Labster laboratory user interface [5]

In the Remote laboratories aspect, two projects can be mentioned as initiatives that have generated a big interest in the online laboratories community.

- VISIR
  The Virtual Instruments Systems In Reality (VISIR) [6], is a project that developed online laboratories specifically in the areas of Electrical and Electronics Engineering, including hands-on, virtual, and remote experiments.
  The VISIR experiments include: breadboards, power supplies, signal generators, meters, oscilloscopes and components (resistors, capacitors, inductors, diodes, etc.).
  All the experiments can be remotely operated through the Internet. This characteristic allows users from all over the world to use the laboratory 24/7. VISIR was developed by Blekinge Institute of Technology in Sweden (BTH). Figure 4 shows the architecture of VISIR.

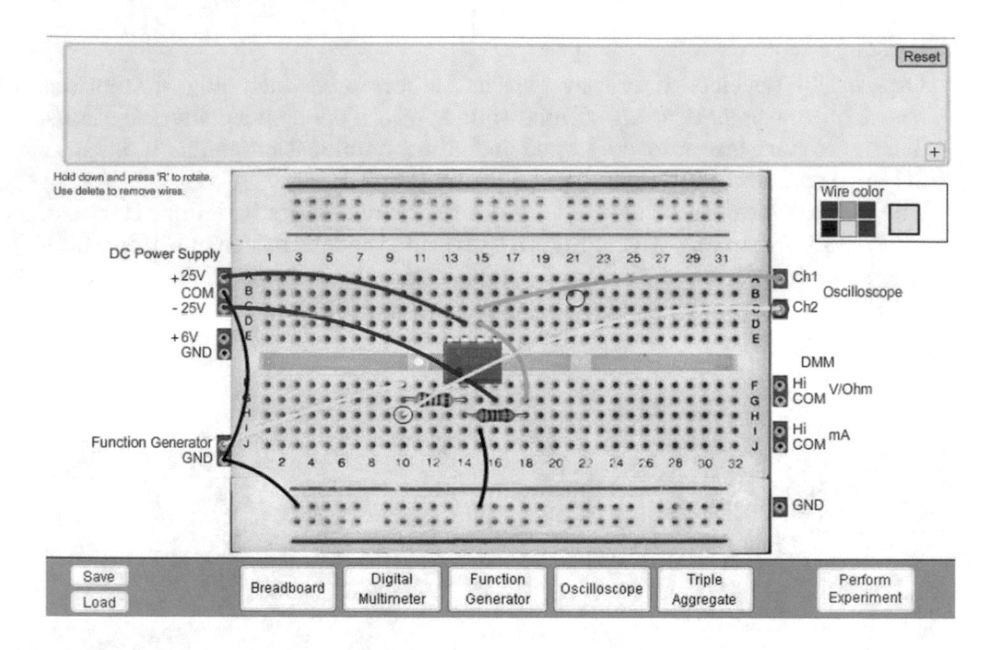

Fig. 4. VISIR laboratory user interface [6]

- LabsLand
  LabsLand is a network of distributed remote laboratories around the world. In LabsLand, institutions can share their laboratories and/or they can be consumers of laboratory experiments from other institutions.
  "LabsLand connects schools and universities with real laboratories available somewhere else on the Internet. A real laboratory can be a small arduino-powered robot in Spain, a kinematics setup in Brazil or a radioactivity testing laboratory in Australia. They are real laboratories, not simulations: the laboratories are physically there, and students from these schools and universities access them" [7]. Figure 5 presents the user interface of a remote laboratory in the topic of Robotics.

3 Methodology for the Laboratory Adaptation and Integration

The methodology includes 4 stages. First is the validation of feasibility. Second, the hardware adaptation. Third, the software level activities and finally the integration aspects that allow the deployment of the laboratory in an Online Laboratory Management System, see Fig. 6.

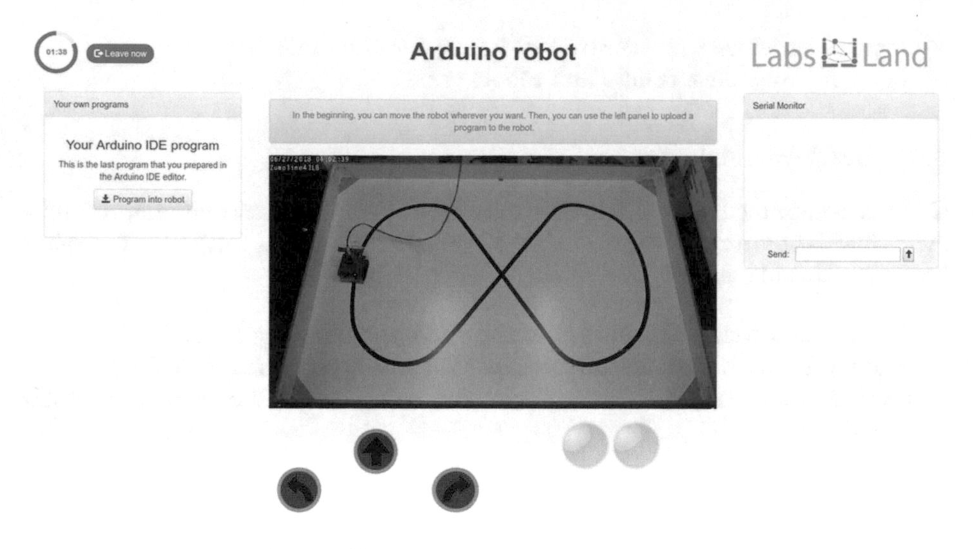

Fig. 5. LabsLand robotics laboratory [7]

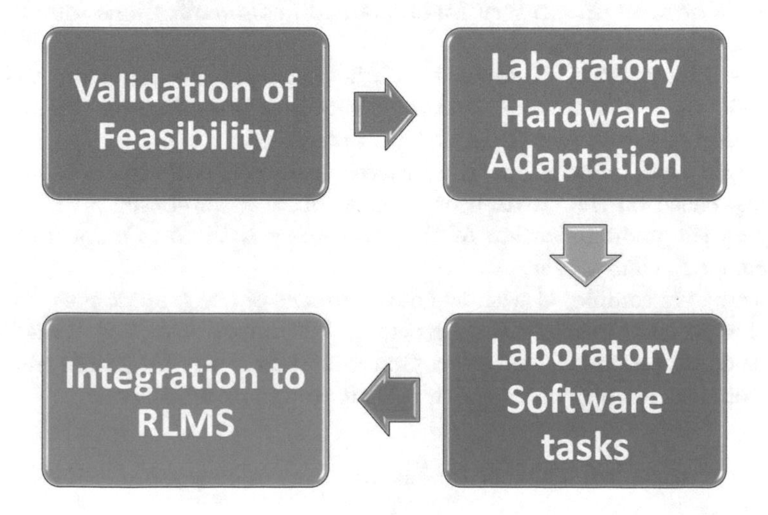

Fig. 6. Methodology

3.1 Validation of Feasibility

Before starting with the laboratory adaptation tasks, it is important to evaluate how feasible in terms of how much of the local laboratory functionality and activities can be adapted to be used through an online interface. The validation of the feasibility of this adaptation includes:

- Determine the nature of the experiment
- Experiment data capturing (Conversion of analogue variables to digital values)

- Type of variables to be controlled during the experiments
- Type of experiment results and report

3.2 Hardware Adaptation

Once is determined that the local educational laboratory can be adapted into an online laboratory, the process of adaptation continues with the following hardware considerations:

- Adding networking capabilities to the current laboratory controller
- Adding feedback mechanisms such as cameras, sensors, among others.
- Implementing hardware features for the safety and security of the online experience.

3.3 Software Level Activities

As part of the software activities, the adaptation requires the consideration of three levels of tasks: laboratory functionalities, communications interfaces and graphic user interfaces (GUI).

For the adaptation of the laboratory functionalities it is important to identify the current local laboratory station functionalities (atomic level tasks) and the identification of the services (higher level activities).

Based on that information, the process continues with the communication interfaces, based on the creation or adaptation of a web-based API of services to support the communication of the laboratory with the Online Laboratory Management System server.

This stage is completed with the development of the graphic user interfaces, that will expose to the user the controls and laboratory activities. This GUI can use or be developed based on the original local laboratory interfaces (only if the local laboratory already counts with a local computer interface).

3.4 Integration with a Online Laboratory Management System

For the deployment of the laboratory into an Online laboratory Management System (OLMS), or RLMS for the specific case of a Remote Laboratory, has to be considered the requirements of each specif platform. The Standard IEEE-1876 Networked Smart Learning Objects for Online Laboratories proposed architectures and a set of good practices for the integration and deployment of educational online laboratories.

There are multiple benefits of having the laboratory accessible through an OLMS or RLMS such as: authentication and role-based access control, possibilities for tracking and generation of learning analytics and usage reports, security controls, integration with Learning Management Systems, among others.

4 Example of an Adaptation of a Local Laboratory into a Functional Online Remote Laboratory

The implementation of this example was done by the collaboration between Acrome Robotics [8], LACCEI [9] and Florida Atlantic University [10].

For this example, the ball balancing table educational laboratory from the Acrome Robotics Company was used to perform the adaptation to an online laboratory. This ball balancing table provides an alternative to be controlled through a local computer interface to allow users to send commands to the table. Figure 7 shows the ball balancing laboratory experiment.

The ball balancing table offers a practical training and research system with 2 degrees of freedom. It uses an open architecture with extensive course-ware, suitable for practical training and research related to control systems. It is equipped with 2 servo motors with its actuator interface, resistive touch panel that detects x, y coordinates, a metallic heavy ball, and supports different controller options (Arduino, myRIO, Raspberry Pi). In our work, the Raspberry Pi option has been selected as the system's controller.

The functionalities implemented by the company for this laboratory station include:

- Calibration routines
- Customizable Control functions (P, PD, PV, PID and Fuzzy Logic examples shipped by default)
- Drawable surface to define a path that the ball will follow
- Software interface that allows setting the parameters
- Compatibility with MATLAB®/Simulink®, LabVIEW™
- Simulations in the engineering software

The curriculum that has been proposed by Acrome to teach using this laboratory includes:

- Fundamentals of Pulse Width Modulation (PWM)
- System Modeling
- Feedback In Control Systems
- Performance Measures
- Control System Design
- Control System Verification

Following the methodology proposed in Sect. 3, the networking capabilities were added by the use of Raspberry Pi3 single board computer that follows the Smart Adaptive Remote Laboratory architecture (SARL) [11].

SARL architecture complies with the definition of the architecture proposed by the IEEE-1876 Standard for Networked Smart Learning Objects for Online Laboratories [12]. SARL allows the communication of commands back and forth between the laboratory station and the SARL server. Figure 8 presents the view of the adapted local laboratory into an online laboratory.

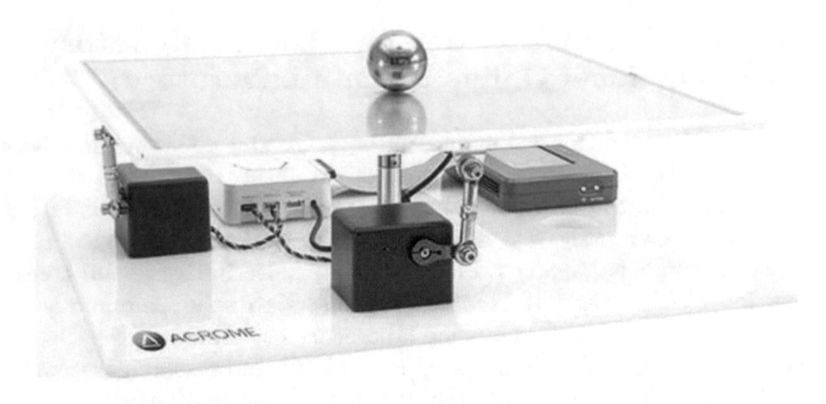

Fig. 7. Acrome Robotics ball balancing laboratory [8]

Fig. 8. Acrome Robotics ball balancing laboratory online laboratory

The feedback mechanism was implemented by the use of two Raspberry Pi cameras that were connected directly to the Raspberry Pi and its video was streamed directly to the server. This video streaming was implemented using the Motion library [13], an open source service available for Linux operating systems.

SARL architecture offers a basic level of safety and security support by controlling the remote users access through a role based access control system and an appointment based system. As well as, a set of routines for input validation and auto resetting of parameters after each user laboratory session is completed, this prevent the equipment to be damaged or misused.

The software level tasks included the re-writing of the code for the current functionalities using Python Language that is the language supported by the SARL system. The development of the web interface included the coding of the graphic user interface elements that allow the user controlling the ball balancing table from the web browser. For this, a combination of PhP and Python functions were developed.

The academic activities are loaded on the SARL data base to allow the system to propose different activities to the users. This feature, will allow in the future to have adaptive automatic activity suggestion and for the teacher the possibility to compose customized laboratory experiences.

The inclusion of the SARL xAPI statements generator, inside specific functions of the laboratory station, allows the system to collect detailed information about the users interactions. The aggregation of this data serves as input for the generation of learning analytics reports.

5 Conclusions and Future Work

As a result of this process we have a detailed process that can be followed in order to make the adaptation of a local laboratory to an online laboratory in a standard systematic way. This process complies with the architecture defined by the IEEE-1876 Standard on Networked Smart Learning Objects for Online Laboratories.

The implemented/adapted online laboratory has been integrated to the xAPI framework supported by the Smart Adaptive Remote Laboratory system, that will be hosting the access to this laboratory experiment at least during the testing process.

The development of this project has benefit both parties, the university research group as well as the company involved, the two teams have interchanged knowledge and ideas.

The next steps include the testing and validation of the adaptation results. Additionally, provide laboratory authoring mechanisms to facilitate the connection of more laboratory activities with the implemented online laboratory experiments and finally, continue with the adaptation of more laboratory stations that the company has developed in the past.

Possibilities to provide hands-on training during the online laboratory sessions are also planned to be implemented through the use of hybrid laboratory configurations. In this scenario there is also laboratory equipment components in the user side, in this way the users should implement part of the laboratory experiments using their own components and complete the experiments interacting online with the remote laboratory components.

References

1. Learning Tools Interoperability LTI, IMS Global Learning Consortium. https://www.imsglobal.org/activity/learning-toolsinteroperability. Accessed 20 Jan 2019
2. Experience API (xAPI). https://xapi.com/. Accessed 20 Jan 2019
3. Zapata Rivera, L.F., Larrondo Petrie, M.M.: Models of collaborative remote laboratories and integration with learning environments. Int. J. Online Eng. (IJOE) **12**(9), 14–21 (2016)
4. PhET Physics Education Technology, Projects Inc. https://phet.colorado.edu. Accessed 20 Jan 2019
5. Labster. https://www.labster.com/. Accessed 20 Jan 2019
6. Gustavsson, I., Zackrisson, J., Hakansson, L., Claesson, I., Lago, T.: The visir project - an open source software initiative for distributed online laboratories. In: 4th International Conference on Remote Engineering and Virtual Instrumentation, REV 2007, Porto - Portugal (2007). http://openlabs.bth.se/static/igu/Publ/Konferensbidrag/GustavssonREV2007VISIRinitiative7.pdf. Accessed 20 Jan 2019
7. Labs Land. https://labsland.com/es. Accessed 20 Jan 2019
8. Acrome Robotics. https://acrome.net/. Accessed 20 Nov 2019
9. LACCEI, the Latin American and Caribbean Consortium for Engineering Institutions. http://www.LACCEI.org/. Accessed 20 Jan 2019
10. Florida Atlantic University. https://www.fau.edu/. Accessed 20 Nov 2019
11. Zapata Rivera, L.F.: Models and implementations of online Laboratories; A definition of a standard architecture to integrate distributed remote experiments. Ph.D. Dissertation. Florida Atlantic University (May 2019)
12. IEEE 1876 Networked Smart Learning Objects for Online Laboratories Working Group. http://sites.ieee.org/sagroups-edusc. Accessed 13 Feb 2019
13. Motion, Motion detection Web Camera service. http://www.lavrsen.dk/foswiki/bin/view/Motion/WebHome. Accessed 20 Jan 2019

Development of a Testing Device for Measuring the Functional Parameters of UAV Recovery Systems

Sebastian Pop, Marius Cristian Luculescu, Luciana Cristea, Attila Laszlo Boer, and Constantin Sorin Zamfira(✉)

Transilvania University of Braşov, Eroilor 29, 500036 Braşov, CP, Romania
{pop.sebastian, lucmar, lcristea, boera, zamfira}@unitbv.ro

Abstract. In the context of the increasingly dynamic development of Unmanned Aerial Vehicles (UAVs) and the fields in which they are used, as well as by the increasingly restrictive legislative requirements, there is a need to implement a recovery system in the event of hardware or software malfunctions. This system is necessary in order to limit the damage that can occur to the onboard equipment, as well as to the goods and people on the ground. In addition to the classic electronic fail-safe systems, currently implemented at the software level in most flight algorithms used for autopilots, the last barrier to their recovery is the classic parachute recovery system.

Our goal was to test and optimize a classic parachute recovery system having two types of veils, a cruciform one (surface of 2.26 sqm) and a spherical one (surface of 2.75 sqm). Our testing device developed in order to acquire the functional parameters of the recovery systems let us measure: optimal launch speed, minimum opening speed, maximum opening speed, shock at opening, opening time, oscillation after opening.

The purpose of the experiments was to reduce the descending speed of a UAV with a mass of 3 kg up to a speed of 5 m/s.

All of the functional parameters are necessary in order to choose the optimum wing shape, the folding mode, the need to implement the folding bag, the use of the extractor parachute, as well as determining how the UAV's fuselage elements can interfere with the opening process.

Keywords: Classic parachute recovery system · Unmanned Aerial Vehicles · Sensors

1 Introduction

In the context of the increasingly dynamic development of Unmanned Aerial Vehicles (UAVs) and the fields in which they are used, as well as by the increasingly restrictive legislative requirements, there is a need to implement a recovery system in the event of hardware or software malfunctions. This system is necessary in order to limit the damage that can occur to the on-board equipment, as well as to the goods and people on the ground. In addition to the classic electronic fail-safe systems, currently

M. E. Auer and D. May (Eds.): REV 2020, AISC 1231, pp. 519–527, 2021.
https://doi.org/10.1007/978-3-030-52575-0_42

implemented at the software level in most flight algorithms used for autopilots, the last barrier to their recovery is the classic parachute recovery system.

Specialized works dealing with recovery systems consider landing parachutes that work at altitudes much higher than the flight of a UAV. Topics related to parachute flying physical model and inflation simulation analysis [1], structural modeling of parachute dynamics [2, 3], parachute recovery systems [4, 5] and parachute recovery for UAV systems [6], were studied in order to be applied to design our testing device for measuring the functional parameters of UAV recovery system.

2 Problem Formulation

Our goal was to test and optimize a classic parachute recovery system having two types of veils, a cruciform one (surface of 2.26 mp) and a spherical one (surface of 2.75 mp). Our testing device developed in order to acquire the functional parameters of the recovery systems let us measure the following quantities: optimal launch speed, minimum opening speed, maximum opening speed, shock at opening, opening time, oscillation after opening.

The purpose of the experiments was to reduce the descending speed of a UAV with a mass of 3 kg up to a speed of 5 m/s.

3 Approach

Because the UAV flight takes place at low speeds and heights, the recovery of a UAV system has more features than the classic launches with human or cargo personnel. Mounting the recovery system inside the UAV, as well as the existence of its control elements (wings, propellers, sensors, etc.), requires studying the operating mode in order not to interfere with these control elements. In this study, our team tried to find an efficient solution for testing these recovery systems. Thus, the present mobile platform has been reached, which simulates the movement of the system, and can record both images and parameters. The system is modular so that any model of recuperator can be tested. Another major advantage of the platform is that it is not destructive. The system developed by us can record two types of data correlated with each other, namely: numerical data collected using sensors, as well as video data collected with the high-speed camera. From sensors, the following parameters are collected: optimal launch speed, minimum and maximum opening speed, shock at the opening, opening time, oscillation after opening the parachute. As the recovery system opens in an extremely short time (up to 2 s), the only way to understand the phenomena that occur during the opening is to correlate the video images with the numerical data. All these parameters are necessary in order to choose the optimum wing shape, the folding mode, the need to implement the folding bag, the use of the extractor parachute, as well as determining how the UAV's fuselage elements can interfere with the opening process. Our device involved a data acquisition setup, mounted on a mobile platform. The block diagram is presented in Fig. 1.

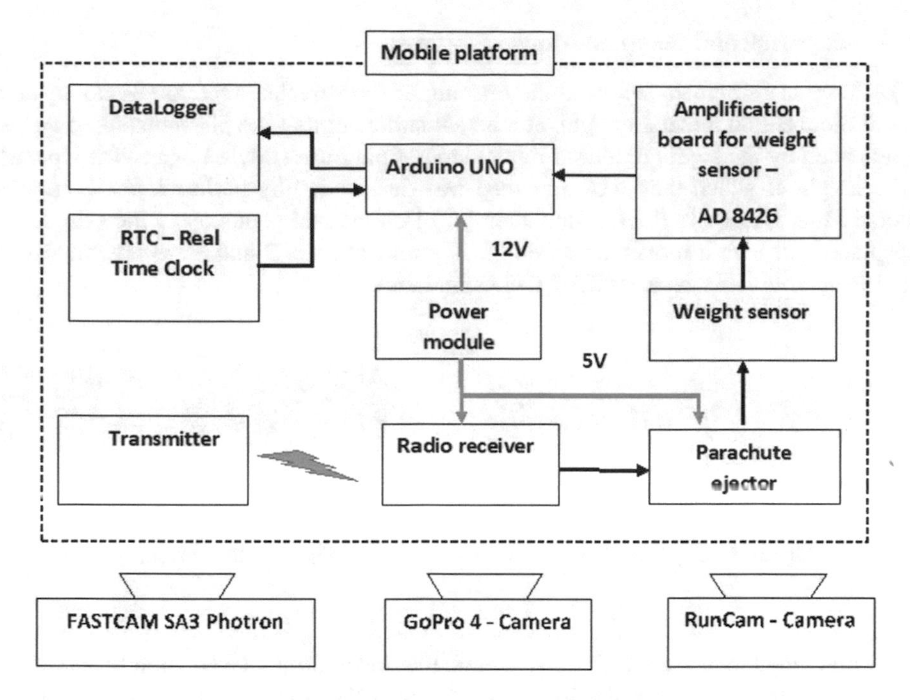

Fig. 1. Block diagram of the testing device

3.1 The Numerical Data Acquisition System

For data acquisition purposes, the system is equipped with an Arduino UNO development board, an amplification board (AD8426), a Real-Time Clock (RTC) module, and a Data Logger with an SD card module.

The recovery system was triggered by a TaranisFRsSky (TFRS) radio control system. TFRS was installed near the high-speed camera operator and the radio receiver was connected to the triggering actuator.

3.2 Video Data Acquisition System

The video data were collected through 3 cameras: a FASTCAM SA3 Photron ultra-fast recording and two sports cameras (GoPro4 and RunCam). The FASTCAM SA3 Photron camera used a fixed observation point and has been synchronized with the recovery system trigger. The camera recorded 2000 frames per second, so all the opening stages of the parachute were captured.

The two sports cameras were mounted in different positions in order to capture details of the parachute openings while the vehicle was in motion.

3.3 Experimental Setup on Mobile Platform

The force measurement sensor at the opening of the parachute and the ejector system were mounted on a metal upright, at a height sufficient that the phenomenon to not be influenced by the swirls produced by the mobile platform (Dacia Logan Pick-Up car). The height at which they were mounted was determined by performing a Computational Fluid Dynamics (CFD) simulation [7, 8] on the real geometry of the vehicle for displacement with a maximum speed of 80 km/h. In Figs. 2 and 3 are presented side and isometric views as a result of CFD simulations.

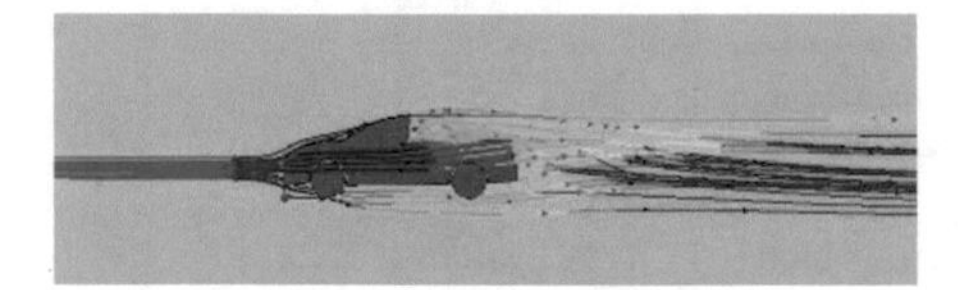

Fig. 2. CFD - side view

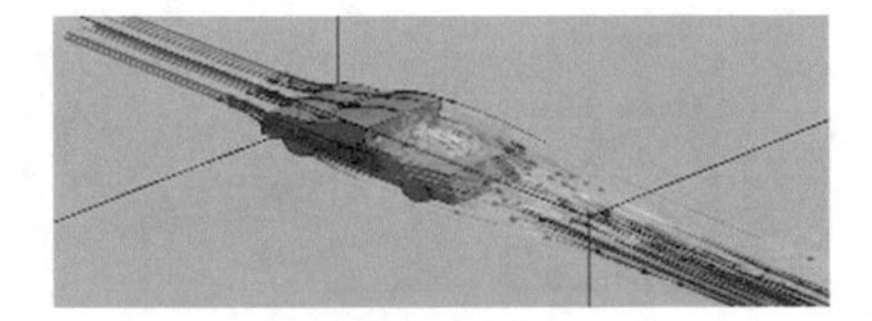

Fig. 3. CFD - isometric view

Following the data analysis, it was found that the optimum installation height of the sensor and the release mechanism is 1.6 m from the upper platform of the mobile platform (Dacia Logan Pick-Up car).

For the parachute release, a container has been designed in which the recovery system is arranged along with the elastic spring and the servomechanism that releases the retaining cap. In Figs. 4 and 5 parachute container design and practical implementation with the servomotor are presented respectively.

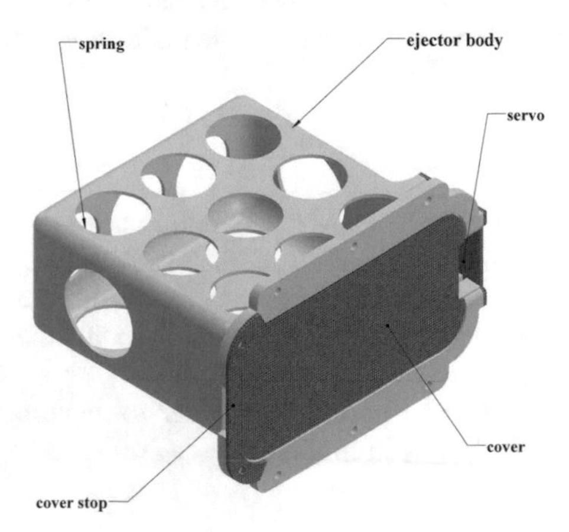

Fig. 4. The container for the parachute

Fig. 5. Practical implementation with the servomotor

Implementation of the container was made by 3D printing technology from PLA material, and the retaining cap, made of composite material, was cut using a CNC machine.

We must point out that this design is flexible because it can be scaled according to the recuperator being tested. Practical realization involves minimal labor time and the cost of materials is extremely low (cost of materials for the container ejector developed by us is approximately $ 20).

Another advantage of this type of container designed by our team is that it can be used later in order to be mounted on UAVs that carry out real missions.

4 Results

We present the data obtained with our system for a spherical parachute whose nominal diameter is 1.25 m and surface of 1.05 sqm. The study of the phenomenon involved ejecting the parachute at an average displacement speed of 6.8 m/s and launching the parachute without a folding bag or parachute extractor. The shutter release is performed in sync with the high-speed camera's shutter release when the machine enters the camera lens area.

In Fig. 6 the dimensions for one of the test parachutes are presented.

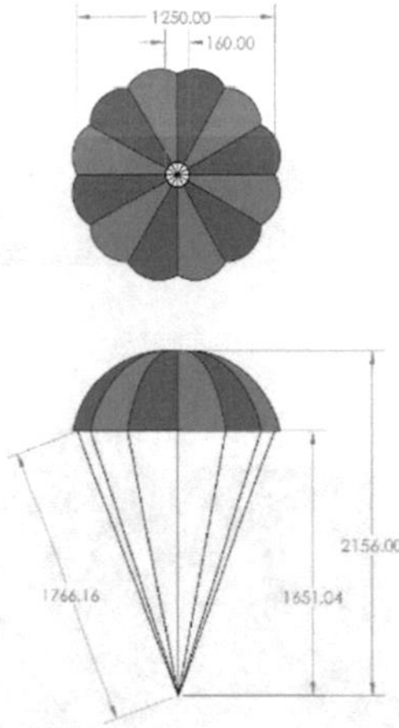

We must mention that the test device can be applied also to another type of parachutes, for example to a cruciform one.

In Fig. 7 the variation of the force that appears when the parachute is released is presented.

Figures 8, 9, 10, 11, 12, 13, 14 and 15 present the data from the data recorder synchronized with the images from the ultra-fast recording camera, for the above 8 points marked on the graph.

Fig. 6. The dimensions of one of the test parachutes

After about 1 s from parachute release, only free oscillations remain.

Following the analysis of the recorded data it was possible to determine for this type of parachute the following characteristics:

- Total opening time: 1.16 s
- Shock force at opening: 170 N
- Minimum opening speed: 10 m/s
- Maximum opening speed: 32 m/s
- For the evaluation of the optimal launch speed, more experiments must be done;
- Oscillation after opening: between 15° and 20° with respect to the longitudinal axis of the vehicle.
- A critical element determined after the launch of the recovery system is represented by the fall of the recuperator about 0.8 m below the launch level.

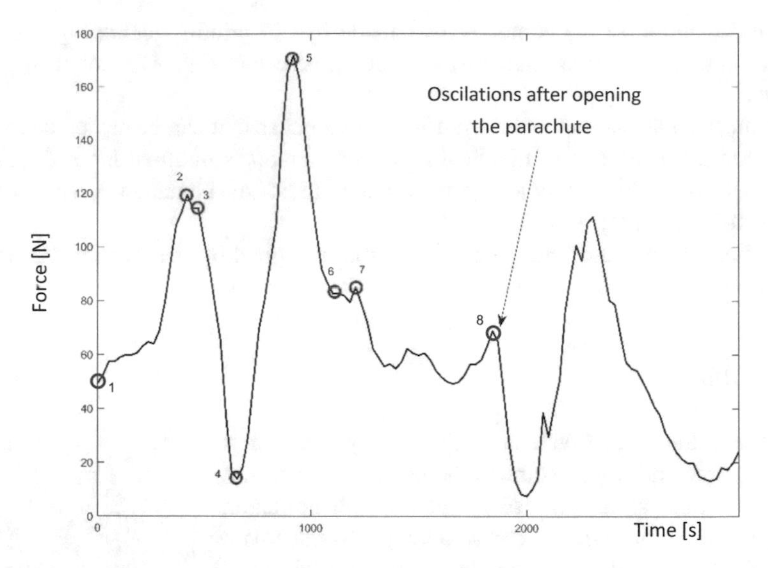

Fig. 7. Time representation of force that appears at the parachute release

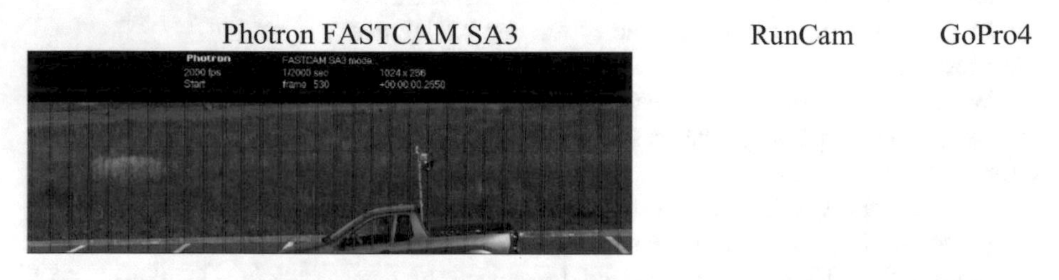

Fig. 8. Point 1: Time: 0.265 s, the command to open the ejector system cover was sent, simultaneously with the start of the data logger recorder

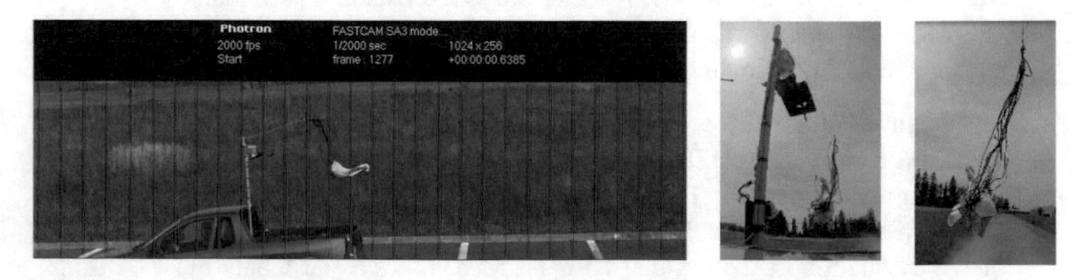

Fig. 9. Point 2: Time: 0.6385 s, the tractor cable connecting the UAV to the parachute is maximally stretched, at which point the suspensions begin the full extension process.

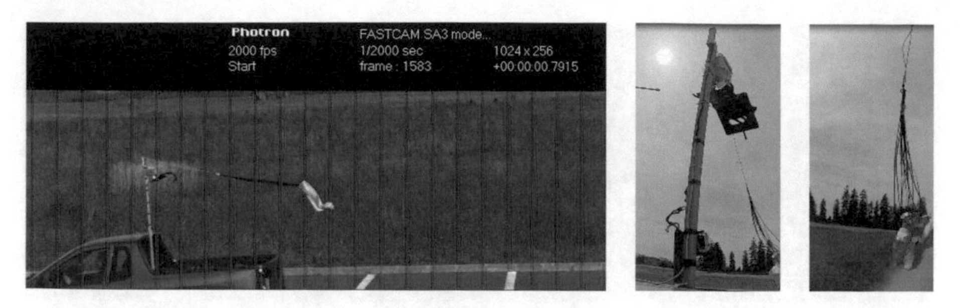

Fig. 10. Point 3: Time: 0.7195 s, tractor cable, and suspensions are fully extended, reaching the moment before the inflation process.

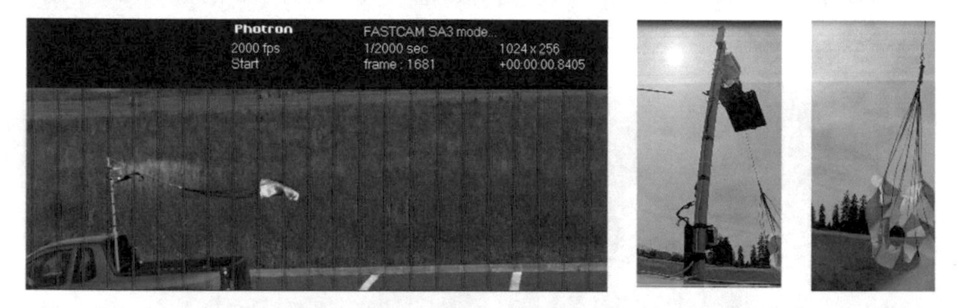

Fig. 11. Point 4: Time: 0.8405 s, starts the process of filling the veil with air

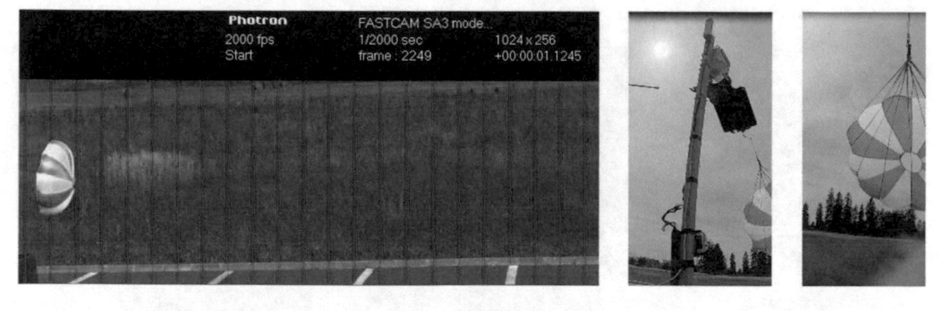

Fig. 12. Point 5: Time: 1.1245 s is the maximum point where the full opening of the wing takes place, at which point the maximum opening force is developed

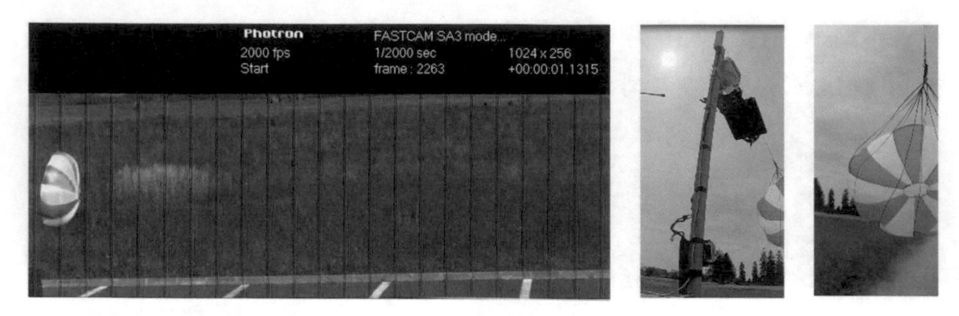

Fig. 13. Point 6: Time: 1.1315 s the process of decompressing the veil takes place

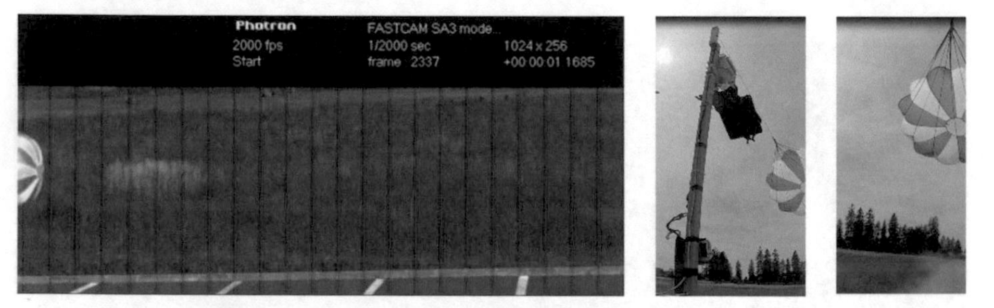

Fig. 14. Point 7: Time: 1.1685 s the opening process is complete

Fig. 15. After the final opening process, the recovery system will have more left/right oscillations with respect to the vehicle axis.

This results in the following conclusion: the use of a vertical ejection system mounted on the upper part of the fuselage, combined with the folding of the parachute without a bag, will inevitably result in the fuselage's components being hit by it, which may cause the recovery system to not open.

5 Conclusions

This launch method is recommended to be used by installing the system in the lower area, and at the same time, it is recommended to be used only as a recovery (not a salvage) solution, and only in those procedures where there is control over the evolution of the UAV.

The advantages of this test system are the following:

- Much cheaper compared to using an aerodynamic tunnel;
- The opening stages can be studied visually and these data can be analyzed in relation to the data collected from the sensors;
- The phenomenon can be studied in relation to the weather conditions (ex lateral wind);
- Modular as the studied parachute types can be easily replaced;
- The container made by the 3D printing method can be adapted to any type of size;
- Adaptable to any type of UAV recovery test.

References

1. Aircraft Engineering. Irving Air Chute – Brake Parachute Installation. Aircraft Engineering (1968)
2. Cao, Y., Xu, H.: Parachute flying physical model and inflation simulation analysis. Aircr. Eng. Aerosp. Technol. **76**(2), 215–220 (2004)
3. Accorsi, M., Leonard, J., Benney, R., Stein, K.: Structural modeling of parachute dynamics. AIAA J. **38**(1), 139–146 (2000)
4. Ewing, E., Bixby, H., Knacke, T.: Recovery system design guide. Technical Report AFFDL-TR-78-151, Air Force Flight Dynamics Laboratory, Wright-Patterson Air Force Base, OH (1978)
5. Knacke, T.W.: Parachute Recovery Systems: Design Manual. Para Publishing, Santa Barbara (1992)
6. Wyllie, T.: Parachute recovery for UAV systems. Aircr. Eng. Aerosp. Technol. **73**(6), 542–551 (2001)
7. Peterson, W.C., Strickland, J.H., Higuchi, H.: The fluid dynamics of parachute inflation. Ann. Rev. Fluid Mech. **18**, 361–387 (1996)
8. Keith Stein, R.B.: Fluid structure interactions on a Cross Parachute. Comput. Methods Appl. Mech. Eng., 673–687 (2001)

Online Creativity in Engineering Education

A Flipped Creativity Approach for the Engineers Without Borders Challenge

Tobias Haertel[1(✉)] and Claudius Terkowsky[2]

[1] TU Dortmund University, Otto-Hahn Street 6, 44227 Dortmund, Germany
tobias.haertel@tu-dortmund.de

[2] TU Dortmund University, Vogelpothsweg 78, 44227 Dortmund, Germany
Claudius.terkowsky@tu-dortmund.de

Abstract. Creativity is a key competence in times of industry 4.0 - however, it is hardly systematically promoted in engineering courses. The Engineers Without Borders Challenge is a teaching and learning format in which students have to develop creative ideas and which is therefore excellently suited to promote generic competence along the acquisition of engineering competences. However, it is difficult to allow students to go through a creative process completely and professionally with large numbers of students. This paper presents an approach that, in the sense of the flipped classroom, outsources part of the creative process to an online self-learning unit. On the one hand this meets the interests of the students and on the other hand reduces the effort for the teachers in the course.

Keywords: Online creativity · Generation of ideas · Large student groups

1 Introduction

Students in the engineering sciences are usually expected to find creative solutions to problems. However, they rarely learn to develop creative ideas, or in many teaching/learning scenarios there is no time at all for the development of ideas. Ideas are rather casually expected, students should simply have them on the side. As a consequence, they are often rather bad or do not deviate from the usual standard solutions. If, however, the generation of ideas is given the necessary space in courses, they become much more creative. This can easily be achieved in small groups by using appropriate creativity techniques - but it becomes difficult with large groups.

Against this background, the aim is to be able to professionally generate ideas even in larger study groups. Specifically, this is about a course for Engineers Without Borders Challenge. The students there have to solve a problem arising from the development cooperation of Engineers Without Borders and create a prototype as part of a product development. This course is becoming more and more popular among the students. Instead of 40 in the winter semester 2017/18, 100 students take part in the winter semester 2019/20. This means there is no way to complete the creative process with all students in the presence time. To ensure that the prototypes to be developed are

M. E. Auer and D. May (Eds.): REV 2020, AISC 1231, pp. 528–535, 2021.
https://doi.org/10.1007/978-3-030-52575-0_43

still original and do not only include the replication of known standard solutions, the course was extended by an online offer for generating ideas.

2 Engineers Without Borders Challenge

The Engineers Without Borders Challenge is a competition organized by the NGO Engineers Without Borders for students (and in some countries also for schoolchildren). Within the framework of the development cooperation of Engineers Without Borders, real problems from the work on site are identified and defined as tasks for the Challenge. Each year 3-4 different tasks are identified, which can be worked on by the students in teams of about 5 students each. This could be, for example, the more ergonomic design of a brick press, the cool storage of seeds, the uniform irrigation of agricultural land or the hygienic removal of drinking water from local cisterns. The aim is to produce a prototype that works in the best case or otherwise explains how the model works. What makes the Engineers Without Borders Challenge a particular challenge to the creativity of the students is that the solution must not cost a lot of money or consume resources and must be implemented using materials that are also available on site. These requirements are initially very daunting for the students: on the one hand, they have to find solutions to problems that generations of engineers before them have usually been looking for, and on the other hand, the conditions for the solution are very restrictive. In this dilemma, the students realize that they need creative, original ideas in order to move forward, and open themselves up to the use of otherwise rather unusual creativity techniques and the systematic passage through a creative process. In order to support the necessity of the creative process, the students do not receive any professional guidance from the teachers, but only support in questions concerning the realization of the ideas and solutions they have raised. For example, there is deliberately no positive list of materials available in the target countries, but students should think for themselves and can ask representatives of Engineers Without Borders whether certain materials can be used on site. The students have to realize the prototypes alone in their universities (in a makerspace, FabLab or using other facilities). Practical implementation usually requires a great deal of creativity, since not everything works as the students theorized.

3 Think Big! Visionary Thinking as a Scarce Commodity in Engineering Studies

If creativity plays a role in engineering studies at all, it is usually just a form of brainstorming. Normally, however, the actors involved do not have the appropriate background for the correct application of the method or the creative process as a whole and understand brainstorming simply as a collection of ideas. However, this does not go far enough. In order to promote the creativity of students in a targeted way, it is important to understand that the creative process basically consists of two phases: a divergent phase and a subsequent convergent phase [1–4]. In the divergent phase, the first step is to collect a quantitatively large number of different ideas. Quantity takes precedence here over quality in order to gain the necessary breadth. In this phase it is important not to censor an idea: No idea is bad, no idea is impossible - on the contrary: everything is possible, natural laws do not apply, resources are available indefinitely. This phase is about generating crazy ideas that should not be taken too seriously: Think big! What is the ideal solution for our problem? This is exactly what we want to achieve. In unmoderated brainstorming this is often ignored and participants quickly begin to censor themselves if an idea is not realistic [5]. The high number of ideas ensures that the human brain overcomes itself to think beyond the first obvious ideas and to come to the more distant, around the corner, more original ideas [6]. These ideas are almost never directly realizable - but they form the basis for really original ideas to prevail in the convergent phase. In this second phase of the creative process, the aim is to select from the large number of ideas those that are promising in terms of real solutions, and to further develop them accordingly. At the end of the convergent phase there are one or more solutions to a problem.

4 Encouraging the Creativity of Students with the Engineers Without Borders Challenge

As long as the Engineers Without Borders Challenge at the TU Dortmund University did not have so many participants, the creative process could be controlled by two experienced moderators.

4.1 The Old Approach

Before the students could realize their prototype for the Engineers Without Borders Challenge, they should systematically go through the creative process and develop original ideas. At the beginning of the course, one-day creativity workshops were held: 2 creativity coaches took over 8–10 students each on 3 consecutive days. All students had to assign themselves to one of the 6 groups. Within a single group, the morning was marked by the divergent phase: the students used the force-fit technique "pictures" under the guidance of the coaches. They were shown various pictures that had nothing to do with the problem of the Engineers Without Borders Challenge. The students were asked what the pictures might have to do with the problem at second glance in order to develop new associative ideas. Within a short time, the technology generated many ideas, of which the students then presented each other the best three ideas on a metaplan wall.

This process was relatively time-consuming, as the students presented and discussed a total of 24–30 ideas. Then the students were given the task of finding themselves along the discussed ideas to 2 groups with 4–5 students each. Afterwards, the afternoon was used to further elaborate the ideas chosen by the groups using convergent techniques.

4.2 Advantages and Disadvantages of the Old Approach

This approach had two advantages and two disadvantages:

- Advantage 1: The students underwent a professionally moderated whole creative process
- Advantage 2: The students formed their groups not on the basis of an external allocation or along existing friendships, but around the ideas they found exciting.
- Disadvantage 1: The students reported back that due to the small group size (8–10 students) only 2 groups could be formed, but there were many more exciting ideas and so students had to join groups with ideas they found far less interesting than other ideas.
- Disadvantage 2: The effort for the teachers was relatively high, on three days 2 teachers were needed as moderators of the creative process and could not do their other tasks during this time.

When the number of students rose to 100 due to the great success of the Engineers Without Borders Challenge, it was clear that this mode could no longer be offered realistically. The students' feedback that they liked to be guided through the creative process in larger groups also called for a better scenario. However, simply increasing the number of students in the workshops was not an alternative, as the moderators would no longer have been able to ensure the correct application of the creativity techniques.

5 The Solution: Digital Online Creativity

The feedback from the students was justified. In addition, the increased number of participating students had to be taken into account. It was therefore necessary to develop a new approach.

5.1 The New Approach

The basic idea was to outsource the more time-consuming part of the divergent phase in the sense of the flipped classroom and make it available online. The nearly 100 students were divided into two groups of about 50 students each. The force-fit technique "Images" was digitized, an online learning unit was developed with Adobe Captivate, which was then set to the moodle room of the course "Engineers without Borders Challenge" (see Fig. 1).

The technique was first explained fundamentally, in particular it was repeatedly pointed out not to censor oneself and to write down every crazy or absurd idea on a cart. Since the students initially practiced the technique on their own, the danger of looking foolish with "strange" ideas in front of the other students was initially reduced. After the description of the technique, 5 pictures, which had nothing to do with the problem, were faded in for 2 min each. The learning unit was set up in such a way that

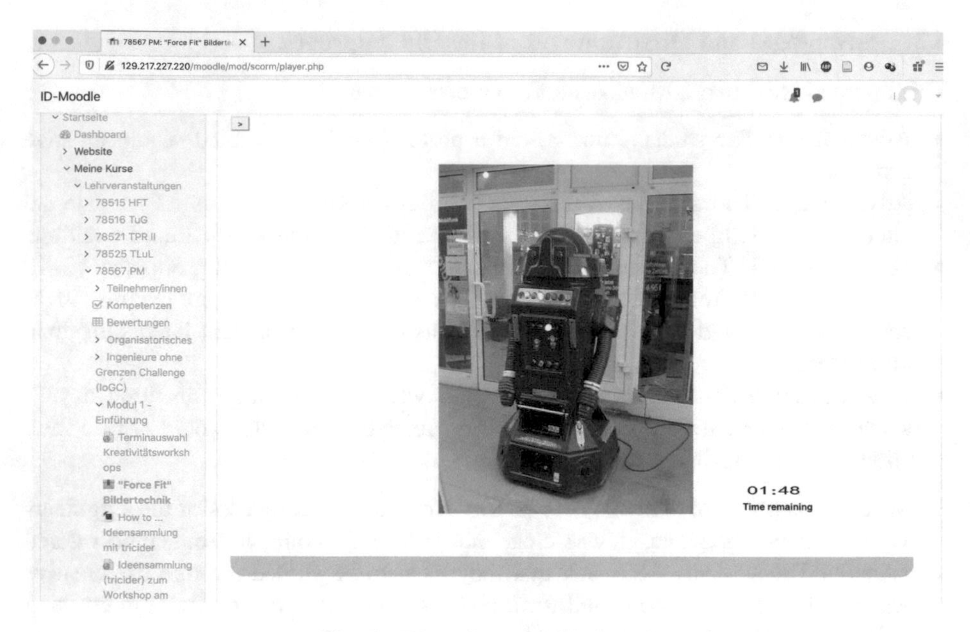

Fig. 1. Force fit method "picture"

the students could not or did not have to click on the pictures, after the confirmation of finishing the preparations the creativity technique runs automatically. Following the generation of ideas along the pictures shown, the students were asked to contribute their three best ideas to tricider (see Fig. 2).

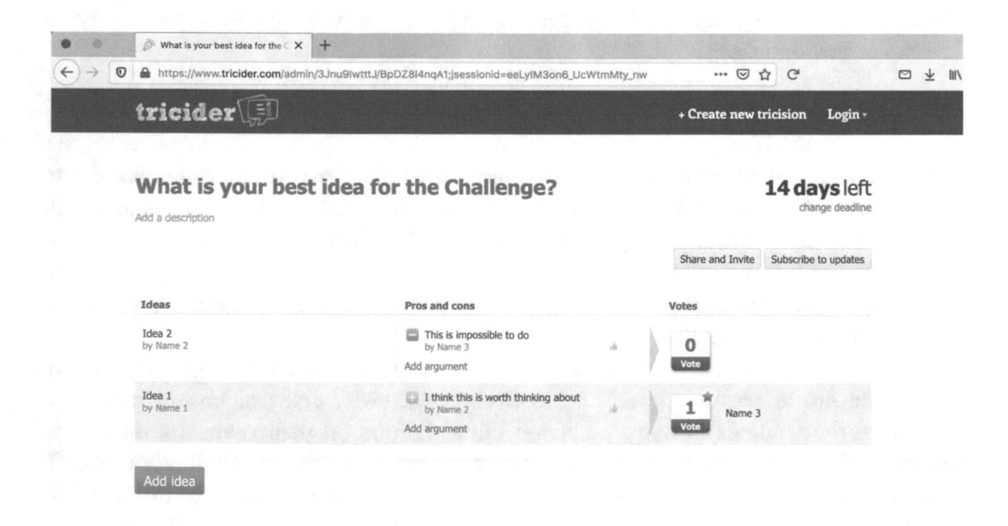

Fig. 2. Example of tricider

Tricider is a decision-making website where suggestions can be made, discussed and voted on. It can be used free of charge and without registration. A separate room in tricider has been set up for each of the two groups of students. The students had the task of submitting their three best ideas by a deadline, stating their names. After this deadline they had the task to comment and discuss the ideas until another deadline. They were also asked to vote for the three best ideas from their point of view. In this way, the 15 best ideas per group could be identified, and the students were already familiar with the ideas and had already discussed them. After this phase, the two student groups met for a one-day workshop. The 15 best ideas were briefly presented on a meta-plan wall. Then the students were asked to come together in groups of 4–5 students each along the ideas they favoured (see Fig. 3).

Fig. 3. Students finding together in small groups based on the ideas they prefer

After the group finding the convergent phase began, which small groups completed with Edwards de Bono technique of thinking hats (see Fig. 4).

Fig. 4. Convergent thinking using thinking hats

5.2 Advantages and Disadvantages of the New Approach

The new approach eliminates the disadvantages of the old approach and offers the following advantages:

- The students had much more opportunities to assign themselves to an idea. Instead of 2 groups, 8–10 groups were now formed in the workshops, which increased the chance to devote themselves to the preferred idea.
- The effort for the teaching event has decreased. Moderators were only needed on 2 instead of 3 days.
- Due to the high number of students in the workshops the dynamics increased considerably, the students had the feeling to be part of an event in teaching and could help themselves beyond group boundaries.

6 Fazit

Only through the use of digital technology was it possible to solve the problems of the valuable teaching/learning arrangement of the Engineers Without Borders Challenge. The online unit for the divergent phase of the creative process worked very well. The switch to a blended learning format has made it possible to host the event with large numbers of students, while at the same time emphasizing the importance of the Engineers Without Borders Challenge.

Acknowledgment. The authors acknowledge the financial support by the Federal Ministry of Education and Research of Germany in the framework of ELLI 2 (project number 01PL16082C).

References

1. De Bono, E.: Lateral Thinking. Viking, London (2010)
2. Haertel, T., Terkowsky, C.: The shark tank experience: how engineering students learn to become entrepreneurs. In: Proceedings of the ASEE Annual Conference and Exposition, New Orleans, Louisiana, USA (2016). S. Paper ID #16542
3. Cropley, D.: Creativity in Engineering: Novel Solutions to Complex Problems. Academic Press, An Imprint of Elsevier, London (2015)
4. Haertel, T., Terkowsky, C., Frye, S.: Kreativität in der Industrie 4.0: Drei zentrale Thesen für die Ingenieurdidaktik. In: Haertel, T., Terkowsky, C., Dany, S., Hrsg, H.S. (eds.) Hochschullehre & Industrie 4.0. Herausforderungen - Lösungen – Perspektiven, pp. 13–26. WBV Media, Bielefeld (2019)
5. Gray, D., Brown, S., Macanufo, J.: Game Storming. O'Reilly Media, Beijing (2011)
6. de Bono, E.: De Bonos neue Denkschule. MVG Verlag, Heidelberg (2005)

EOLES Course – Five Years of Remote Learning and Experimenting

André Fidalgo(✉), Paulo Ferreira, and Manuel Gericota

School of Engineering, Polytechnic of Porto, Porto, Portugal
{anf,pdf,mgg}@isep.ipp.pt

Abstract. The EOLES (Electronics and Optics e-Learning for Embedded Systems) course consists of a 3rd year Bachelor degree that relies exclusively on e-learning and remote laboratories, developed as the result of an EU funded ERASMUS+ project, involving 15 institutions from four European and three North African countries and concluded in 2015. This paper presents an overview and overall results for this initial period and a more detailed analysis of the Digital Systems Teaching Unit contents, pedagogical approach, grading methodology and results. The focus is on the unit specific characteristics and features, student and teacher experiences and the methodologies that were applied to enhance learning success. The Teaching Unit expositive material is provided as the student progresses, with progressive unlocking of content depending timeline and automatic quizzes results. Grading is divided between weekly assignments, an online exam at the end of each TU and a final exam at the end of the academic year. In short, students are allowed and encouraged to adjust their learning rhythm within the limits allowed by time restraints and evaluation criteria. The developed course was accredited as a specialization year in most partner institutions and has been running non-stop since then, mainly with students from North African institutions. Although no longer supported by an EU project, the course is a good example of sustainability as it already had 4 effective editions with successful approval rates and always with many more candidates than available vacancies.

Keywords: E-learning · Engineering teaching · Remote laboratories · Online course

1 Introduction

The degree is EOLES degree is fully delivered on-line using e-Learning 2.0 [1, 2] synchronous and asynchronous tools and fully in English language, allowing students to be part of a "virtual learning community" and empowering teamwork, even if the team members are far apart. A dedicated remote laboratory based on virtual experimentation and modelling and simulation platforms, and on remotely operated real instrumentation equipment installed in different universities was used by students to acquire essential practical skills. Degree accreditation is a major advantage of the EOLES course and with particular interest to its main target group. The degree was recognized by the educational authorities of France, Morocco and Tunisia. As a result,

M. E. Auer and D. May (Eds.): REV 2020, AISC 1231, pp. 536–545, 2021.
https://doi.org/10.1007/978-3-030-52575-0_44

all successful students receive a diploma recognized inside the European Higher Education Area (EHEA).

2 The EOLES Degree

2.1 Organization

The program was defined in cooperation with the North African Universities participating in the project, considering the priorities defined by their countries' governments. The program's focus on electronics and optics for embedded systems responds to the current tendency for integration of hardware/software into single reconfigurable platforms and to the increase on the amount of data produced and transferred requiring high-speed optical transmission, and to the need of training highly qualified professionals able to keep their countries' pace with these new technologies. The program is divided in fourteen technical units (TUs) and in three optional units. The TUs are divided in two semesters and the detailed content of each one of the TUs is available in the project website [3]. The degree runs for 31 weeks, plus 3 weeks reserved for examinations – one in the end of the first semester, another one in the end of the second semester, and a last one in the final week of the course for make-up exams. The Learning Management System (LMS) that supports TU organization, materials' access and delivery, on-line assessments, virtual and experimental lab access, tracking and reporting, forums and chats and all other course related activities was initially based on a Moodle 2.7 version platform [4], since then upgraded to version 3.5 for the 2018/19 edition.

2.2 Remote Laboratories

An effective practical innovation of the L3-EOLES degree are the remote laboratories used to perform on-line practical works. A multi-user approach is implemented allowing a group of students to work and interact in real time over the same Practical Work (PW), guaranteeing a strong collaboration among them during the training. Each hardware setup (function generator or oscilloscope, for instance) is connected to the internet. From each TU's Moodle page students have access to the related lab's webpage and to the TUs' proposed lab works. Students can change the hardware configuration in real-time and have an immediate feedback of their actions, via the virtual instrument interfaces that are deployed remotely and through a high-definition camera (or another interface).

Figure 1 shows one of those lab setups using internet-controlled instrumentation and a camera and part of the user interface. This enables students to see what is going on the real lab and how the real instruments react to their remote commands. This feedback is important for students to be sure that the interface they are seeing in their own monitor is not the visible face of a virtual world but the virtual interface of a real instrument.

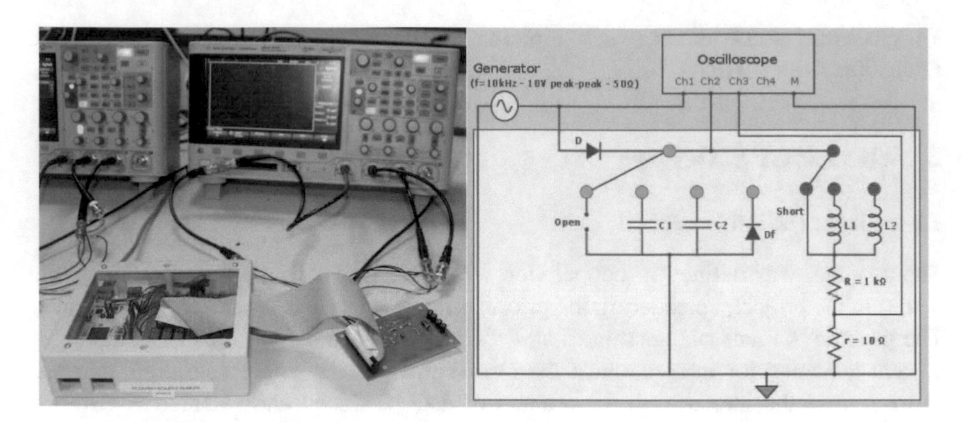

Fig. 1. Remote laboratory setup and interface

The remote laboratory is expected to have a substantial learning impact as each student or group of students have the possibility of repeating the same experiment several times and trying different configurations in a controlled and safe environment. These remote laboratories are used in subjects where the real equipment is more important and were presented and demonstrated on several technical events [5, 6]. In some TUs those online labs are replaced or complemented by simulator tools or remote access to advanced software tools.

2.3 Assessment and Grading

The EOLES degree follows the French university assessment system with some adaptations. Grading is made on a 0–20 scale, where 10 is the passing grade. Each student is required to have an average of 10.0 or more at the year's end for successful graduation, being possible to have less than 10 on any individual TU, although no grade can be below 5. At the end of the year, there is a final recourse exam, where each student can try to improve his grades on any specific TU in order to achieve passing results. Students that fail to graduate at years end, can repeat only part of the degree (where the failed to achieve a passing grade) on the following year. On each individual TU the grade is composed of three components, namely: (1) mandatory practical works or assignments; (2) an one-hour on-line exam held at the end of each TU; (3) a two-hour final exam held by the semester's end is worth 50% of the TU's final grade.

The on-line exams are designed to allow the students to consult any technical or pedagogical resource they deem necessary, therefore having a strict time limit and requiring students to be online and visible (through webcam) during the entire exam. The final exam is performed at a university room on a scheduled date, requiring the students to be physically on the same space and under staff supervision during the duration of the exam. A bonus between 0 and 2 points could be attributed at tutor's discretion to each student according to his/her level of participation in the synchronous sessions, forums and live chats. The specific weight of the grading components can be adjusted by each TU staff, varying between 20% and 35% for components 1 and 2 and

40% and 50% for component 3. The higher weight of the final exam being mostly due to the more controlled environment which provides a fairer grading.

3 Digital Electronics TU

3.1 Organization

To better illustrate the degree, we will present and discuss the pedagogical solutions implemented on TU05-Digital Electronics for Embedded Systems. The proposed framework was similar in all TUs, but some implementation adaptations were required as the subjects and difficulty levels are considerably different in some cases. In TU05 the subject is basic digital systems and operations, finite state machines, combinational logic and sequential logic. The lectures consist of a set of 21 pre-recorded asynchronous classes with a duration never exceeding 20 min, where an instructor explains the theoretical basis of a subject supported by different types of visual materials as illustrated in Fig. 2.

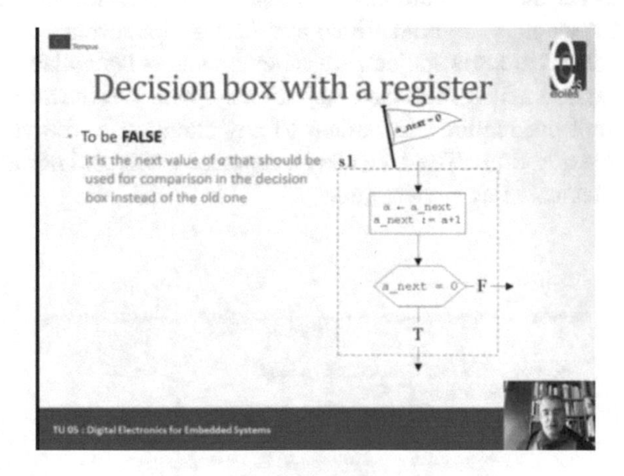

Fig. 2. Synchronous class

Most classes rely on PowerPoint slides presentations, recorded as online videos, with the teacher image superimposed and several visual aids (arrows, circles, etc.) used to illustrate key points. When required the classes also use external links and access to simulated equipment. The classes are interspersed with self-evaluation quizzes, composed of multiple-choice, fill-in-the-blanks, matching exercises. These are intended to keep students' interest and attention, breaking long expositive classes and have no weight on the TU grading. Additionally, these self-evaluation questions provide students with an immediate feedback about their degree of understanding of the subjects being taught. A Quiz example is presented in Fig. 3.

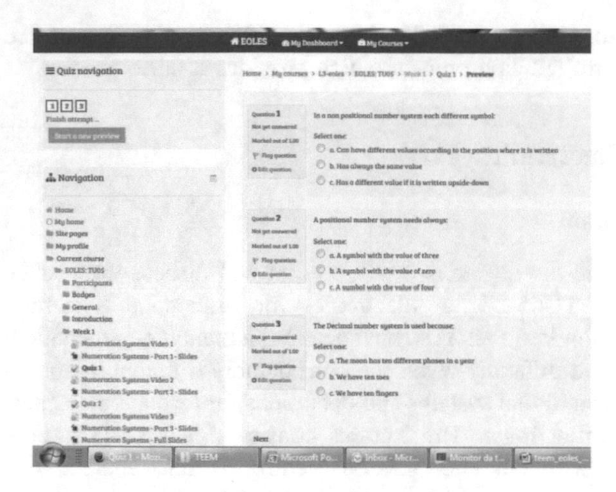

Fig. 3. Self-assessment quiz.

Figure 4 presents the initial options available to the students in week 1 of TU05. The first subject is available in both Video and PDF slides formats as well as the first automatic quiz, related to same subject. All other content is unavailable to students and is progressively unlocked as they successfully complete the quizzes. These are not particularly difficult but require the student to pay attention to the video or slides in order to answer the questions. The quiz can be repeated freely and has no effect on final grade, being implemented as a checkpoint.

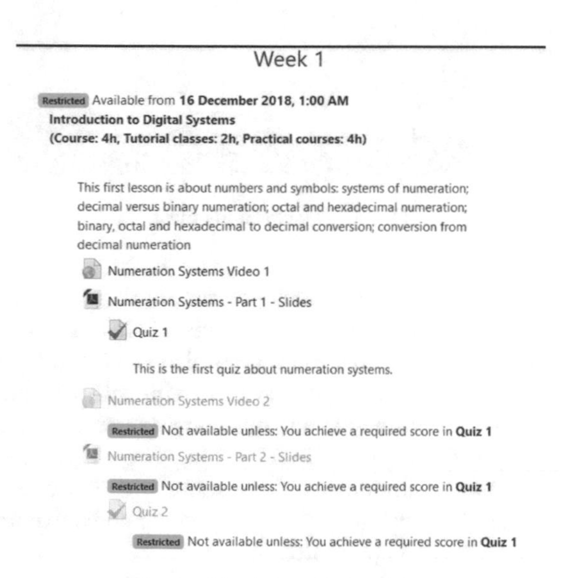

Fig. 4. Week 1 initial syllabus

Students can progress at their own pace, viewing or reviewing this visual material anytime, any number of times, without restrictions. However, the student can only proceed to the next lecture after the successful completion of the self-evaluation questions associated to the previous one. A range of other materials is also available to support the study, including companion books freely downloadable from Internet, web links to other sites containing specialized information and other complementary data, depending on the TUs subject.

Each Week, there is a specific assignment that must be completed by the students and delivered for grading (via moodle). The first week, the assignment is not graded and is based on the successful installation and tutorial completion of the required software for TU05. On subsequent weeks, each student must use the software to solve problems of increasing complexity using the concepts learned on the course. As an expositive example, Fig. 5 presents the proposed solutions of parts of two assignments. One is based on the use of Karnaugh Maps to simplify and implement a logical circuit designed to solve a real problem presented in Week 2, and the other is a State Diagram for a different problem presented in Week 4.

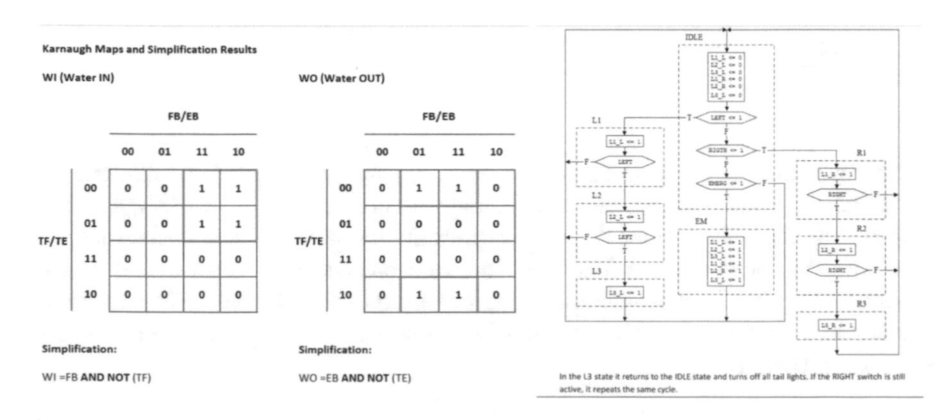

Fig. 5. Assignment solution examples

The emphasis in on the application of the learned concepts to practical and operational problems, although simplified due to the remote restrictions. In the case of TU05, all hardware is virtual (VHDL models) and simulated, with the simulation results being used as "proof" of execution. On other TUs, remote laboratories or remote access to software tools are also used to same effect. Each TU uses a different mix of remote labs, simulation and paper exercises depending on technical constraints and course preferences, but all students must do several assignments on the different courses and technologies. Additionally, TU05 relies on tutorial classes which are synchronous classes based on the use of a web conferencing tool. Their aim is to enable students to clarify any issues and ask questions related to the content of the TUs. These classes are also recorded, and the records made available to students. During the synchronous classes tutor and students are required to have their cameras on, with an example being presented in Fig. 6.

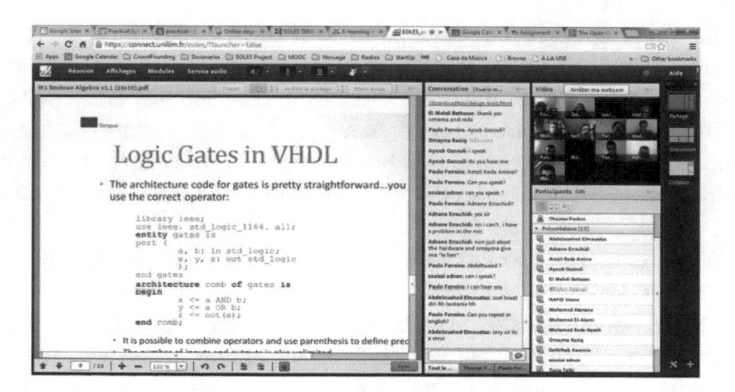

Fig. 6. Synchronous class

The aim is to have a visual feedback of the whole class making students feel part of a group and be able to interact not only with the tutor but also among each other.

4 Results

4.1 EOLES Degree

The number of applicants on all degree editions largely exceeded the expectations, albeit and the number of vacancies were highly concentrated in one of the EOLES partner countries (i.e. Morocco). Table 1 presents the number of students that were enrolled in the degree, those that finished (were present at the final exam) and the approval rates, defined as the number of approved students compared with the number of students that finished the course (attended the exams) or the number originally enrolled, respectively.

Table 1. EOLES degree results

Edition	Enrolled	Finished	Graduated	Approval	
14/15	25	21	11	52%	44%
15/16	32	26	21	81%	66%
16/17	37	34	26	76%	70%
17/18	27	22	19	86%	70%

The results are very satisfactory, proving that the issues present in the first edition were sorted and an adequate success rate was achieved. It should be noted the degree is deemed as challenging and requiring effort equivalent to a normal 3rd year degree in a French University, hence the adequacy of the presented success rates. Even so, failure to graduate is always analyzed and three main reasons were identified, namely (1) abandonment of the degree for personal reasons (usually professional); (2) inability to complete some specific TU due to lack of previous knowledge or lack of adequate effort; and (3) language problems, albeit having prior English knowledge attested by their TOEFL certificates.

4.2 Digital Systems Technical Unit

In the particular case of TU05, the results were in general good, with approval rates of more than 80% every year and average grades above 12,5. The online quizzes to get access to content are a key feature of this TU and allowed some specific conclusions. Students would sometimes require several attempts to progress, some had to review the online classes after failing a quiz and others posted their doubts in forums or sent messages to the teachers. These actions were monitored using the Moodle logging functions, and show the quizzes working as intended, promoting interactivity and the need for seeking additional information. The synchronous classes were not used to present new subjects, although several times it was necessary to clarify and repeat issues presented on recorded classes, as some doubts remained. Participation was very variable, with between 25 and 80% of students present as viewers. However, video and audio participation were limited, with a few students being responsible by the majority of questions and discussion. A very important feature was the ability to share documentation and visual aids, as several questions required the discussion and revision of available materials. The final online exam had excellent results, the grade distribution being presented in Fig. 7.

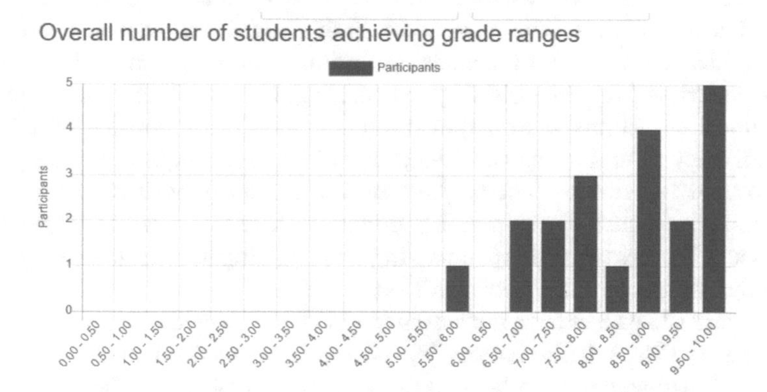

Fig. 7. Synchronous class.

The practical assignments were a mixed experience. In TU05 the initial assignments were simpler and direct, with most assignment providing the student with a good grade. In the last week the assignment was more complex and was more frequent that students would not even attempt it (with the negative effect on the final grade) and several others being obviously incomplete attempts or rushed by the students. In all assignments there was also some need to analyze the uniqueness of the delivered reports, as the online character of the course sometimes promoted the sharing of results between students. The online exams usually present much better grades that the final exams, as was to be expected, and showing that some type of attendant assessment is still required for a fairer grading.

In this TU, most grades were above the 50% mark, confirming that the TU subjects were adequately delivered. The examination results were better than assignment work results, by a considerable margin in several cases, due to somewhat tough learning curve of the high-end software being used on assignments [7]. Nevertheless, this experimental work is deemed as a vital learning resource and every year the documentation and support are improved.

5 Conclusions

EOLES is an excellent example of sustainability, providing an effective and self-sustained online degree with continuous operation and successful editions. One of its main strengths derives from the its formal accreditation on several countries that provides the main motivation for annual student applications. As all online and traditional courses, the longevity required it to provide the necessary tools and knowledge to allow students to succeed into a very specialized labor market, and that success to assist in the degree continuous evolution.

As to the effectiveness of the learning environment, several non-quantitative conclusions were possible from the first years' experience, namely: (1) Students are more used to interactive classes, preferring those as an initial approach. This solution is feasible for simpler subjects; (2) Recorded classes are a much more time efficient way to deliver complex subjects, as they allow the students to study and repeat at their own pace; (3) Interactive classes are required to clarify doubts and answer questions. The use of recorded classes is not a complete solution to most students; (4) Student participation in interactive classes is very diverse, requiring the teacher to be proactive; (5) Additional asynchronous resources (forums, emails) are often preferred by some students, namely when lacking communication skills (e.g. English language)

The overall experience of the teaching staff is extremely positive and allows for a permanent improvement of the teaching process. There are new challenges on each edition, and it is necessary to keep improving and evolving in order to keep up with the technological advances and the student's expectations.

The experience gained from the development of this degree is being used as a base of an ongoing ERASMUS+ project [8], where it is intended to develop a set of practices and methodologies to be used in the implementation of other online courses and laboratories in general, but with particular attention to engineering degrees.

References

1. Fidalgo, A., et al.: The EOLES project – engineering labs anywhere. In: Proceedings of the IEEE Global Engineering Education Conference (EDUCON 2014), April 2014, pp. 943–946 (2014)
2. Bates, T.: Understanding web 2.0 and its implications for E-learning. In: Lee, M.J.W., McLoughlin, C. (eds.) Web 2.0- Based E-learning: Applying Social Informatics for Tertiary Teaching, IGI Global, New York, pp. 21–42 (2011)

3. Course Content, EOLES project Website, September 2019. www.eoles.eu
4. Moodle, April 2019. moodle.org
5. Farah, S., Benachenhou, A., Neveux, G., Barataud, D., Andrieu, G., Fredon, T.: Flexible and real-time remote laboratory architecture based on Node.js Server. In: 3rd Experiment@ International Conference (exp.at15), Pont Delgada, Azores, Portugal, June 2015
6. Farah, S., Benachenhou, A., Neveux, G., Barataud, D.: Design of a flexible hardware interface for multiple remote electronic practical experiments of virtual laboratory. Int. J. Online Eng. **8** (2), 7–12 (2012)
7. www.xilinx.com/products/design-tools/vivado.html. April 2019
8. Gericota, M., Ferreira, P., Fidalgo, A., Andrieu, G., Dalmay, C.: The e-LIVES project: e-Engineering where and when students need. In: 2019 IEEE Global Engineering Education Conference (EDUCON), Dubai, UAE, April 2019 (2019)

e-Engineering Education: Issues and Perspectives for Higher Education Institutions

Paulo Ferreira(✉), André Fidalgo, and Manuel Gericota

School of Engineering, Polytechnic of Porto, Porto, Portugal
{pdf,anf,mgg}@isep.ipp.pt

Abstract. Higher Education Institutions, specially those offering engineering degrees face serious issues related to the maintenance and scheduling of engineering laboratories. When faced of the prospect of having also remote laboratories, their reaction is very negative. On this article is explained the history of remotes laboratories, their relation with distance education, limitations and future possibilities in e-Engineering, and Engineering Education in general. Usually, remote labs are explained in a engineering related way (how they work) forgetting their integration in an engineering degree.

e-Engineering has to be considered as the future of engineering, provoking changes in current engineering degrees, in order to give the students of engineering the right set of competences and skills.

Keywords: e-Engineering · Engineering education · Remote laboratories · Higher education

1 Introduction

Higher Education Institutions (HEIs) all over the world are facing serious issues. Too many prospective students in some cases, and too few students in other cases, due to demographic changes. In both situations the answer is reach for more students. Enroll more students, in order to satisfy market demands, or captivate more students to create market demands. And among the "market demands" there is the demand for profit [37,38] or performance based financial constraints [21].

In Engineering Degrees these problems are more polarized, because (among other Higher Education areas) engineering contributes to the social and financial development [40], but also requires expensive equipment for practical works. The equipment needs may be greater for undergraduate degrees where contact with the equipment facilitates learning [2]. In more advanced degrees the knowledge previously acquired by the student may make possible the use of simulations instead of real equipment, lowering costs. The importance of undergraduate degrees or "Engineering Technology" degrees is many times ignored or misunderstood in common education surveys [14]. These undergraduates form the majority of the staff, in a typical engineering enterprise.

M. E. Auer and D. May (Eds.): REV 2020, AISC 1231, pp. 546–553, 2021.
https://doi.org/10.1007/978-3-030-52575-0_45

Another pressing issue is the internationalization/globalization of HEIs. Today, competing globally is not an option for HEIs, it is a reality [23]. Even if the HEI has a "geographically limited" vision, other HEIs all around the world are trying to enroll the same students, and the graduates are competing in a global market [33].

Education is one of the areas where the influence of technology in social issues can be felt, so for more than a century educator have under pressure to "optimize" education, The most common views on the influences of technology on education, are either utopian [9] or dystopian [13]. Technology is presented a means to solve all education issues or as a terrible obstacle to all types of education.

Of all the HEI's, those involved in Engineering Education suffer more financial constraints due to equipment and technology related expenses. The investment in Remote Labs and e-Engineering equipment may seem foolish, or last minute effort in order to modernize an institution. We hope to show that e-Engineering is a valid and way forward for Engineering Higher Education.

2 History

In one wants to consider e-Engineering, the first step is to define engineering. The etymology of the word is related to engine, as the engineer was someone who kept the (steam) engine working. But, the root of the word is also related to ingenious, or possessing *genius* or skills (*le génie* in french). This is related to the ability to conceive or create [26].

While in England after the industrial revolution, the engineering profession was learned in contact with other engineers in the workshop, in France the engineering profession was more regulated and was taught in state regulated schools. This may seem to indicate a strong divide, with low interchanges.

This distinction is not exclusive as a strong interchange as always happened even in Middle Ages between universities [29]. One paradigmatic example of the international exchanges in engineering education, is Isambard Kingdom Brunel, who is one of the most famous British engineers, but went to France to study, besides all the formation received from his father, a french engineer [8].

Besides the obvious military roots of engineering, the first American Engineering Schools had a mix of British and French influences [18], and a strong German influence was a model for some of the first US universities [29].

In some ways, the so called *Industrial* Engineering was not seen a subject with enough importance to be studied in a traditional university, only reaching the British universities in the late 19th century reforms [1]. In France a prejudice was also felt [10], but this time between military and civil engineers, from upper schools (*École Polytéchnique*) versus more more industrial oriented schools (*École d'Arts et Métiers*) [10].

The need for equipment in engineering degrees has always motivated the search for innovative solutions like the cooperation between University and

Industry, dating at least from the 1906 [4]. Of those courses a remarkable example is a joint course in Electrical Engineering by the Massachusetts Institute of Technology and the General Electric Company [41], where on last three years of a five year degree, the students received formation on the factory during roughly half of the school year.

The proliferation of computers in the 1960s brought to the universities, the so called computer "time-shared" systems [35] and in some universities some very valid educational experiments were made, like the PLATO system [11].

An interesting timeline can be found on the name of some chapters of the work by Hamilton [20], and is reproduced in Table 1.

Table 1. Different phases of online education [20].

Name	Description
The age of automation	The technical code of online education to 1980
The age of ambivalence	Early experiments in educational computer conferencing
The age of evangelism	From online education to the virtual university
The age of openness	From critical interventions to the encoding of online education
The ambivalence of openness	MOOCs and the critical practice of online education

3 Technology and Education

If on the beginning the main idea was one of automation, viewing the online education process as something that could "optimize" the traditional teaching process, the ideas have evolved to education as a more participative and student centered process, with online conference systems as fundamental piece [19]. Further evolutions have given origin to Virtual Universities [32] and the proliferation of MOOCs (Massive Online Open Courses) [36], that have also their critics [27].

The engineering degrees, due to technical difficulties, have been absent of those discussions and the first accredited on-line engineering degree is very recent [16]. But the discussion about online engineering degrees has started some time ago, with a very good book titled "The Influence of Technology in Engineering Education" [7]. For a 25 year old book, it has aged very well, it has a very sound methodology and many of it's conclusions are still valid today.

Some of the interesting topics discussed include:

- The influence of technological advances cannot be limited to engineering education. One cannot "apply technology" to education and compare it with education "before technology" [6]. The technological impact will change society, the engineering profession including engineering roles, and technology is not something that be "added" to improve an engineering degree, but a subject to be studied and considered in it's entirety, in order to adapt engineering to technology, and technology to engineering. In reality, the introduction of

technology is not "additive" but "ecological" [34]. When "adding" technology we do not get "the old system" plus technology, but a different system.
- There is a need for a curriculum change. The recommendation for the curriculum included a reduction in total semester credits, but also an increase in active participation of the students, more and better laboratory experiences and the adoption of computing and networking ("telecomputing") tools [25].
- Besides the generic influence of technology on Engineering Education [5], two important topics considered were the changes in the engineering laboratory [3] and the contribution of distance education to Engineering Education [28]. So, the areas of remote laboratories and distance education are critical to the evolution of Engineering Education.

4 From Distance Education to Remote Labs

The issue with "Distance Education" today, is deeply related to the semantics of the word "Distance". Distance education has appeared (by physical mail) as means to eliminate physical distance, but "distances" today are more "time related" than geographical. Any teacher involved in school management knows that one of the most terrible activities he has to do is scheduling. Normal scheduling implies the synchronization "in time" of teachers, students and resources. That is the most visible part of scheduling, but classes also involve the "geographical synchronization" in a place (classroom or laboratory) of all as can be seen on Table 2.

Table 2. Education synchronisms.

	Same place	Different place
Same time	Classroom	On-line real time conferencing
Different time	Library; Mailbox	On-line bulletin boards

Modern life implies not only less free time, but also the fragmentation and randomness of that free time. One of the origins of the stress of our times, is not only the absence of "free time" but that we don't know when we will have some time, and for how long. This implies a radical change in the priorities of distance education. If the main difficulty in the past was the physical distance, solved by the use of (traditional) mail to deliver the physical educational media, the main obstacle today is the randomness of free time, that should imply different educational strategies, in order to reach the students in an effective manner. The remote labs advocates [42] considers not only distance issues, but also the big change that is providing experimentation when the students want, and not only on a fixed schedule.

One of the most debated issues in remote (and virtual) labs [17] is the quality of the graphical interface. Is the graphical interface (always virtual) an accurate representation of a physical instrument? The reasoning behind this question, is that when the student will change from remote labs with virtual instruments to "live" labs with real instruments, (s)he will be able to work with the real instruments.

This supposes that in the work life of the future engineer, all the instruments will be "physical", with a physical interface made from knobs and buttons. Almost all the electronic instruments available today (multimeters, signal generators, oscilloscopes, spectrum analyzers...) have a physical interface. But, many of them, even at a low price have also an USB or network connection, allowing the computer control of the instrument. Some of them, have only computer control, motivated not only by their final use (integration into automated laboratories) but also, for economic reasons. Today, is cheaper to place a microcontroller based network (and/or USB) interface in an instrument, than to provide that instrument with good quality physical knobs, buttons and a display. While physical knobs, switches and buttons are specified according to a maximum number of cycles, network interfaces have no maximum packet transfer limitations.

Due to greater system complexity, or system implementation requirements, an engineer has an almost constant need to automate data acquisition and collection tasks. This is facilitated by the use of network enabled instruments. So, instead of excluding remote labs from an engineering curriculum, because the instruments are nor "real", the contrary is true. In the future, the engineering students of today will work more with virtual and synthetic instruments than with physical instruments. Those instruments will be cheaper and more flexible than the traditional "physical interface" instruments.

One of the possible valid reasons for the use of Remote Labs in an engineering degree could be the use of virtual/remote instruments, in order to familiarize the students with their utilization in engineering tasks.

This is only an example of a "minor" detail (user-interface fidelity) that must be considered from different economical, technological, pedagogical and engineering perspectives.

Remote labs and e-Engineering must be seen not a extra expense for HEIs, because they are usually a low cost alternative to normal labs [24,30,39] but putting aside the financial argument, they are a worthy path to:

- Internationalization and Globalization of the offered degrees.
- Flexibility in Curriculum building, offering more optional courses to more students.
- Optimization of the use of physical laboratories.
- Introduction of pedagogical innovation to faculty members.
- Enrolling more students with physical disabilities and supporting students with temporary illnesses.
- Enrolling students from other regions, and/or with limited time availability.
- Enrolling lifelong learners in a flexible manner.
- Offering individual courses, independently of the curriculum of a particular degree.
- Saving money in equipment used (or damaged) by students in "live" laboratories.

5 Conclusion

Engineering education should also reflect the career [31] and flexibility of the future engineers [22]. The adoption of remote labs and e-engineering oriented courses, has very complex aspects that should be taken in to account. One good guide to that complexity is the e-Engineering Good Practice Guide [12] produced by the e-Lives project [15]. It covers many areas of e-Engineering, from e–Learning strategy applied to e–Engineering, legal aspects, financial models, and all the organizational, pedagogical and technological aspects of Remote Labs.

e-Engineering has to be seen not as an interesting fashion, but as the near future of engineering and that should be reflected on the education of the engineers of tomorrow.

References

1. Andrews, M.: Universities in the Age of Reform, 1800–1870. Palgrave Macmillan, London (2018)
2. Andrieu, G., Farah, S., Fredon, T., Benachenhou, A., Ankrim, M., Bouchlaghem, K., Aknin, N., Barataud, D., Gericota, M., Craemer, R.D., Cristea, M.: Overview of the first year of the L3-EOLES training. In: 2016 13th International Conference on Remote Engineering and Virtual Instrumentation (REV). IEEE, Madrid (2016). https://doi.org/10.1109/rev.2016.7444511
3. Avery, J.P.: The impact of technology in the engineering laboratory. In: Bourne, J.R., Brodersen, A.J., Dawant, M.M. (eds.) The Influence of Technology on Engineering Education, chap. 8, pp. 171–177. CRC Press, Boca Raton (1995)
4. Barbeau, J.E.: Cooperative Education in America – Its Historical Development, 1906-1971. Northeastern University, Boston, Massachusetts (1973). https://files.eric.ed.gov/fulltext/ED083913.pdf
5. Bourne, J.R., Brodersen, A.J.: The impact of computer and communication technologies on engineering education. In: Bourne, J.R., Brodersen, A.J., Dawant, M.M. (eds.) The Influence of Technology on Engineering Education, chap. 9, pp. 178–201. CRC Press, Boca Raton (1995)
6. Bourne, J.R., Brodersen, A.J.: Paradigms shifts in engineering education: Summary and conclusions. In: Bourne, J.R., Brodersen, A.J., Dawant, M.M. (eds.) The Influence of Technology on Engineering Education, chap. 11, pp. 237–248. CRC Press, Boca Raton (1995)
7. Bourne, J.R., Brodersen, A.J., Dawant, M.M. (eds.): The Infuence of Technology on Engineering Education. CRC Press, Boca Raton (1995)
8. Brunel, I.: The Life of Isambard Kingdom Brunel, Civil Engineer. Longmans, Green & Co. Ltd., London (1870). http://www.gutenberg.org/ebooks/41210
9. CWNIT, C.o.W.N.i.I.T.: Building a workforce for the information economy. National Academy Press, Washington (2001)
10. Day, C.R.: Education for the Industrial World: The École d'Arts et Métiers and the Rise of French Industrial Engineering. MIT Press, Cambridge (1987)
11. Dear, B.: A Friendly Orange Glow: The Untold Story of the PLATO system and the Dawn of Cyberculture. Pantheon Books, New York (2017)
12. e-Lives: e-Engineering Good Practice Guide (2019). https://e-lives.eu/?page_id=10241

13. Ellul, J.: Le Bluff Technologique. Hachette, Paris (1988)
14. Frase, K.G., Latanision, R.M., Pearson, G. (eds.): Engineering Technology Education in the United States. The National Academies Press, Washington (2017). https://doi.org/10.17226/23402. https://www.nap.edu/catalog/23402/engineering-technology-education-in-the-united-states
15. Gericota, M., Ferreira, P., Fidalgo, A., Andrieu, G., Al-Zoubi, A., Batarseh, M., Garbi-Zutin, D.: e-LIVES – extending e-engineering along the south and eastern Mediterranean basin. In: Smart Industry & Smart Education (REV 2018), pp. 244–251. Springer, Heidelberg (2018). https://doi.org/10.1007/978-3-319-95678-7_27
16. Gericota, M., Fidalgo, A.V., Barataud, D., Andrieu, G., Craemer, R.D., Cristea, M., Benachenhou, A., Ankrim, M., Bouchlaghem, K., Ferreira, P.: EOLES course: the first accredited on-line degree course in electronics and optics for embedded systems. In: 2015 IEEE Global Engineering Education Conference (EDUCON). IEEE, Tallinn (2015). https://doi.org/10.1109/educon.2015.7096004
17. Gomes, L., Bogosyan, S.: Current trends in remote laboratories. IEEE Trans. Ind. Electron. **56**(12), 4744–4756 (2009)
18. Grayson, L.P.: The Making of an Engineer: An Ilustrated History of Engineering Education in the United States and Canada. John Wiley & Sons Inc., New York (1993)
19. Hamilton, E., Feenberg, A.: Alternative rationalisations and ambivalent futures: a critical history of online education. In: Feenberg, A., Friesen, N. (eds.) (Re)Inventing the Internet: Critical Case Studies, chap. 3, pp. 43–70. Sense Publishers, Rotterdam (2012)
20. Hamilton, E.C.: Technology and the Politics of University Reform: The Social Shaping of Online Education. Palgrave Macmillan, Basingstoke (2016)
21. Herbst, M.: Financing public universities: the case of performance funding. No. 18 in Higher Education Dynamics. Springer, Dordrecht (2007)
22. Jarboe, K.P., Olson, S. (eds.): Adaptability of the US Engineering and Technical Workforce: Proceedings of a Workshop. The National Academies Press, Washington (2018). https://doi.org/10.17226/25016. https://www.nap.edu/catalog/25016/adaptability-of-the-us-engineering-and-technical-workforce-proceedings-of
23. Johnstone, B., d'Ambrosio, M.B., Yakoboski, P.J. (eds.): Higher Education in a Global Society. Edward Elgar, Cheltenham (2010)
24. Jona, K., Roque, R., Skolnik, J., Uttal, D., Rapp, D.: Are remote labs worth the cost? insights from a study of student perceptions of remote labs. Int. J. Online Biomed. Eng. (iJOE) **7**(2), 48–53 (2011). https://online-journals.org/index.php/i-joe/article/view/1394
25. Kulacki, F.A., Vlachos, E., Johnson, G.R.: Engineering curriculum: the contextual basis for reform in the information age. In: Bourne, J.R., Brodersen, A.J., Dawant, M.M. (eds.) The Influence of Technology on Engineering Education, chap. 2, pp. 17–35. CRC Press, Boca Raton (1995)
26. Lienhard, J.: The Engines of Our Ingenuity. Oxford University Press, Oxford (2000)
27. Littlejohn, A., Hood, N.: Reconceptualising Learning in the Digital Age: The [Un]democratising Potential of MOOCs. Springer, Singapore (2018)
28. Mattson, R.: The role of distance learning in engineering education. In: Bourne, J.R., Brodersen,A.J., Dawant, M.M. (eds.) The Influence of Technology on Engineering Education, chap. 7, pp. 118–170. CRC Press, Boca Raton (1995)
29. Moore, J.C.: A Brief History of Universities. Palgrave Macmillan, Cham (2019)

30. Morales-Menendez, R., Ramírez-Mendoza, R.A., Guevara, A.J.V.: Virtual/remote labs for automation teaching: a cost effective approach. IFAC-PapersOnLine **52**(9), 266–271 (2019). https://doi.org/10.1016/j.ifacol.2019.08.219. http://www.sciencedirect.com/science/article/pii/S2405896319305567. 12th IFAC Symposium on Advances in Control Education ACE 2019
31. N.A.P.: Understanding the Educational and Career Pathways of Engineers. The National Academies Press, Washington (2018). https://doi.org/10.17226/25284, https://www.nap.edu/catalog/25284/understanding-the-educational-and-career-pathways-of-engineers
32. Pfeffer, T.: Virtualization of Universities: Digital Media and the Organization of Higher Education Institutions. Springer, New York (2009)
33. Portnoi, L.M., Rust, V.D., Bagley, S.S. (eds.): Higher Education, Policy, and the Global Competition Phenomenon. Palgrave Macmillan, New York (2010)
34. Postman, N.: Technopoly - The Surrender of Culture to Technology. Vintage Books, New York (1992)
35. Rankin, J.L.: A People's History of Computing in the United States. Harvard University Press, Cambridge (2018)
36. Rhoads, R.A.: MOOCs, High Technology, and Higher Learning. Johns Hopkins University Press, Baltimore (2015)
37. Rondo-Brovetto, P., Saliterer, I. (eds.): The University as a Business. VS Verlag für Sozialwissenschaften, Wiesbaden (2009)
38. Ruch, R.S.: Higher Ed, inc.: The Rise of the For-Profit University. Johns Hopkins University Press, Baltimore (2001)
39. Sáenz, J., Chacón, J., De La Torre, L., Visioli, A., Dormido, S.: Open and low-cost virtual and remote labs on control engineering. IEEE Access **3**, 805–814 (2015). https://doi.org/10.1109/ACCESS.2015.2442613
40. Thomas, L., Quinn, J.: First Generation Entry into Higher Education: An International Study. McGraw-Hill, Maidenhead (2007)
41. Timbie, P.W.H.: The cooperative course in electrical engineering at the massachusetts institute of technology. Sci. LI **I**(1338), 163–165 (1920)
42. Zubía, J.G., Alves, G.R.: Using Remote Labs in Education: Two Little Ducks in Remote Experimentation. University of Deusto, Bilbao (2011)

Using Chatbot for Augmenting the Design of Virtual Lab Experiments

Prafful Javare[1], Divya Khetan[1](✉), and Anita S. Diwakar[2]

[1] K.J. Somaiya College of Engineering, Mumbai, India
{prafful.j,divya.khetan}@somaiya.edu
[2] AIEE Training and Services LLP, Thane, India
anitasd2008@gmail.com

Abstract. There is a variety of conversational chatbots that aid in education and they aid students in the process of learning. Similarly, many online tools exist to aid teaching. However, there is a scarcity of tools that help in experimental design and hence to bridge the gap, this paper introduces a conversational agent (chatbot) to augment the design of virtual lab experiments. The chatbot was implemented to answer queries and to help the teachers while using SDViCE tool. The data analysis and results of the research carried on the instructors reveals that chatbots will be useful for designing of virtual lab experiments. The research question we are trying to answer is - Whether the engineering instructors find the chatbot helpful while designing virtual laboratory experiment in the online SDVIcE tool?

Keywords: Chatbot · Virtual laboratories · Experiment designs

1 Introduction

1.1 Context

The engineering laboratories have played a major role in the development of technologies, materials and information, which have made our lives easier. In engineering education, and engineering technology in particular, laboratory courses play a central role in helping students learn challenging concepts and develop practical skills. Labs are often one of the college learning experiences that students enjoy the most (Lyle d. Feisel et al. 2005). The Engineering laboratory instruction has reached a crisis level due to inadequate instructional resources and the desirable learning outcomes are not being achieved. There is a lack of challenge and initiative provided to the students in performing experiments. There is a need of improving the quality of experiment designs. The major problem engineering instructors have when they wish to effectively integrate any educational technology such as virtual laboratories is the lack of comprehensive guidelines.

M. E. Auer and D. May (Eds.): REV 2020, AISC 1231, pp. 554–562, 2021.
https://doi.org/10.1007/978-3-030-52575-0_46

1.2 Purpose

The Virtual labs combine technology resources, reusable software environments, and automation, along with tried and true training concepts, to enable hands-on training that can be delivered to anyone, anywhere, anytime (Alan Greenberg 2004). The virtual laboratories are being used world over in order to overcome problems in traditional laboratories or to augment labwork in traditional labs. Virtual experiments provide educationally valuable features not available in physical experiments. There are many broad guidelines available for the learning outcomes to be achieved in science and engineering laboratories. Improving instructors' capacity to design effective laboratory experiments is critical to advancing the educational goals of laboratory experiences. The instructors find the learning design for effective virtual laboratory experiments difficult due to a number of factors such as lack of suitable training and non-availability of appropriate guidelines. The guidelines have been developed for engineering instructors so as to facilitate the design of effective experiments for using virtual laboratories and converted to the online version in the form of the SDVIcE tool (Scientific Design of Virtual Laboratory Experiments) to increase accessibility. We developed a semi-automatic process incorporating chatbots to help the instructors in designing quality experiments and also assess the quality of the experiment designs. If the instructors get stuck at any stage, the chatbot will help them by answering their questions like a real conversational partner, thus simplifying and augmenting process of virtual laboratory experiment design. The chatbots have been found to be useful in various applications such as teaching agents, helping clients in various FMCG products etc. The chatbot in the SDVIcE tool answers to various common queries that instructors have while designing their virtual laboratory experiment while using the online tool. These queries were gathered from the instructors while they designed experiments using the tool and then the chatbot was designed to answer these queries. A pilot study was carried out with five engineering instructors to predict whether the chatbot is helpful in their experiment design process. The research question for the pilot study was: **RQ: Do the engineering instructors find the chatbot helpful while designing virtual laboratory experiment in the online SDVIcE tool?** The results of the pilot study indicate a positive effect of the chatbot. As a future work we propose to carry studies with larger number of instructors for two purposes. One is to gather more queries so as to make the chatbot more effective and second purpose to assess the effectiveness of the SDVIcE tool along with the chatbot with a larger sample size.

2 Recent Work

After analysis of the recent work in this domain, we understood that although there are various tools using chatbot for the purpose of education and there are tools to facilitate teaching, there is no tool that integrates chatbot for designing of virtual labs.

Hence, this paper aims to bridge the gap between the above mentioned tools.

[1] describes a "chatbot" as a computer program that simulates how a computer would behave as a conversational partner, by recognizing natural language patterns and generating smart and relevant responses. Then the benefits of chatbots are highlighted

in the fields of entertainment (recommendations), industry (customer service) and education, as a learning assistant or partner. An example of Amazon's Alexa as a multipurpose chatbot is given and its importance in education is highlighted. Examples of different "skills" that aid in education such as a service that presents the user with random facts, a service that gives information about recycling and one that gives the user a plethora of audio books are given to endorse the fact that chatbots are beneficial for education.

To provide individualized support to the ever increasing number of students per lecturer, "chatbot" was used as a solution in [2]. Chatbots can help in management education by helping the manager to develop skills such as making judgments and decisions, providing and receiving feedback etc. They can also we used to provide feedback to lecturers and students. The literature survey performed gave answers to various questions such as, "In which educational settings are chatbots applied?",

"What approaches are being used to build and design a chatbot in learning settings and how does that influence CML processes and CML outcomes?". The conclusion derived from the literature survey was that the effectiveness of chatbots in education depends on individual student differences, the ways of building chatbots, and the chatbot mediated learning process quality.

[3] helps the student to interact and asks questions which they otherwise would have avoided to avoid a face to face interaction. An AI-based chatbot, named "Edubot" was initially populated with limited knowledge base. The knowledge base was then further populated based on the interactions with the students. This chatbot was helpful for interacting and answering the questions of students regarding the introduction to programming course. The survey performed revealed that the initial interest of students was due to curiosity to know the kind of responses that they could receive. Chatbots can be used to get specific answers as compared to the general answers that are otherwise available on the internet. Also, they can be very useful for students who are not confident enough to participate in regular class sessions.

[4] helps students learn basic computer science fundamentals by building their own chat bot using the open source programming tool "Chatbot". The deployed chatbots can connect to different social media messengers such as facebook messenger and reply to chats automatically. By building their own bots using the platform, the students learn different computer science concepts such as variables, conditionals, finite automata, recursion, randomness, and regular expressions, among others. Advanced concepts such as the Turing test and natural language processing can also be taught. Chatbots can be built on the platform using (pattern, effect) pairs, where the bot responds with the effect when a certain pattern is observed.

3 Implementation

The chatbot agent was developed using a tool called DialogFlow, which is a free tool for making chatbots. The user input is interpreted and classified into categories called intents and answers are formulated for each intent. Separate intents were created for each type of question (Fig. 1).

Fig. 1. The list of intents.

1. The default welcome intent answers questions such as hi, hello etc.
2. The Step1.Broad.Goal answers questions such as-
   a. What different options of broad goals do I have?
   b. What can I refer to choose my broad goals?
3. The Step2.Learning.Objective answers questions such as-
   a. How do I create my learning objective?
   b. Are there any guidelines for learning objectives?
4. The Step2.Action.verbs.create answers questions such as-
   a. Action verbs for create?
   b. Create action verbs?
5. The Step2.Action.verbs.evaluate answers questions such as-
   a. Action verbs for evaluate?
   b. Evaluate action verbs?
6. If no intent is matched i.e. the agent is unable to interpret the user input, the default fallback intent is invoked (Fig. 2, 3, 4 and 5).

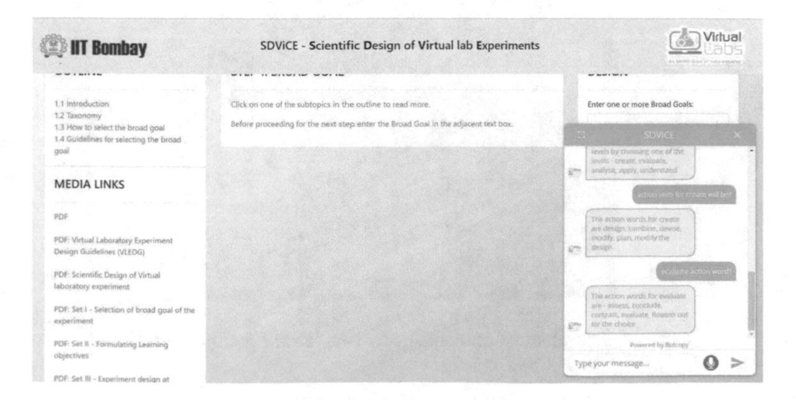

Fig. 2. The chatbot embedded on the SDViCE tool.

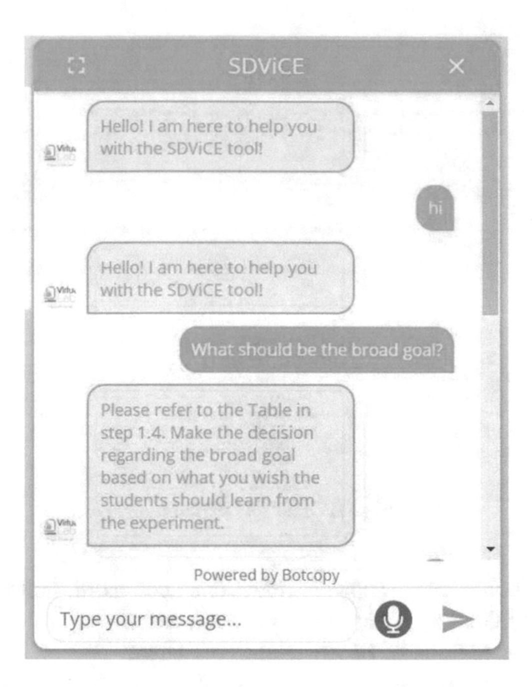

Fig. 3. The chatbot answering to the broad goal intent

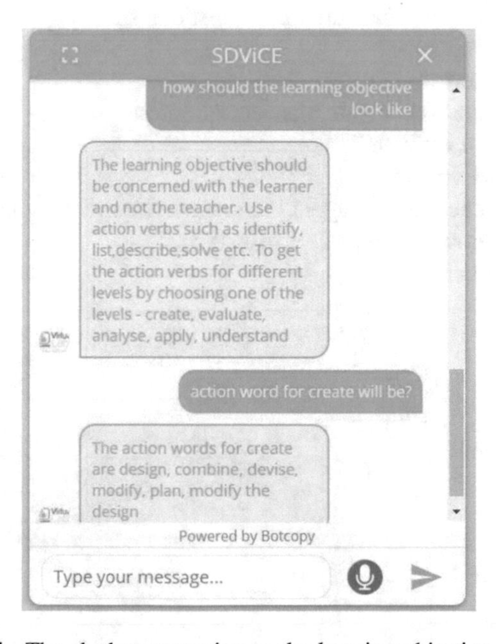

Fig. 4. The chatbot answering to the learning objective Intent

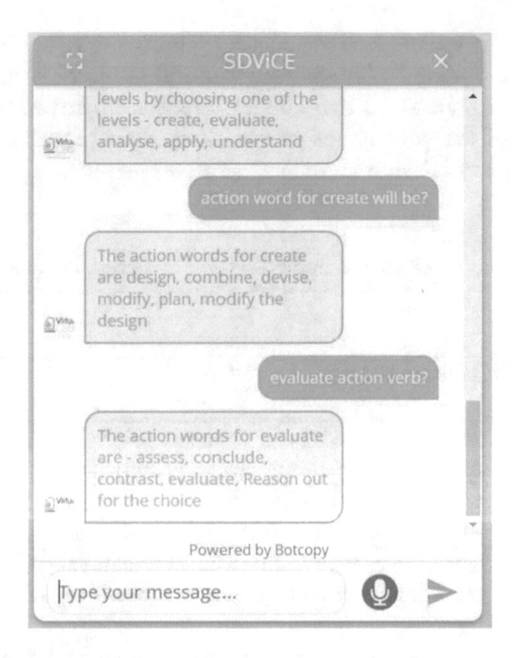

Fig. 5. The chatbot answering to the action.verb.create and action.verb.evaluate intent.

4 Research Method

In this section we discuss the survey study we carried out in order to measure the usefulness of the chatbot in experiment design guidelines. The main research questions of the study were.

RQ: Do the engineering instructors find the chatbot helpful while designing virtual laboratory experiment in the online SDVIcE tool?

4.1 Sample

In order to gather the perceptions of the instructors at diploma and degree level the researcher initially gave a 10-min presentation to the instructors describing the use of chatbots in experiment design guidelines and then the engineering instructors were asked to use the chatbot in experiment design guidelines for about an hour. After obtaining the consent the online survey questionnaire was administered and given enough time to fill up the survey. They were given enough time to fill up the survey.

The total number of participants who responded to the survey and gave their feedback was 17 undergraduate engineering instructors from Mumbai and Nagpur University.

4.2 Instrument

The usefulness survey questionnaire is given below based on the TAM survey (Davis 1989) consisted of fourteen questions with ten questions of five point likert scale format, and four with open-ended responses. There were four questions with one question each to find out the limitations, most useful sections, sections that instructors find most difficult to understand and suggestions for improvement. There were ten questions to find out the perceptions of the instructors regarding the usefulness of chatbot in the experiment design guidelines (Table 1).

Table 1. Questions in the usefulness survey instrument

Sr. No	Question	Percent agree
Q1 to Q10	(Five point Likert scale)	
Q1	I was able to use the chatbot in experiment design guidelines to decide the steps in the experiment design process	90
Q2	I was able to use the chatbot in Experiment design guidelines for selecting the broad goal for my virtual lab experiment	88
Q3	I was able to use the chatbot in Experiment design guidelines for selecting/formulating the learning objectives for my virtual lab experiment	90.4
Q4	I was able to use the chatbot in Experiment design guidelines for selecting the Instructional Strategy for my virtual lab experiment	91
Q5	I was able to use the chatbot in Experiment design guidelines for selecting/designing the tasks as per the instructional strategy and aligned to the learning objectives for my virtual lab experiment	83
Q6	I was able to use the chatbot in Experiment design guidelines for formulating the assessment questions for my virtual lab experiment	90
Q7	Using the chatbot in Experiment design guidelines helped me in asking questions at higher cognitive levels	88
Q8	Using the chatbot in Experiment design guidelines helped me in designing experiments with different difficulty levels	82
Q9	Using the chatbot in Experiment design guidelines helped me in Incorporating active learning methods in the experiment designs	91
Q10	I was able to use the chatbot in Experiment design guidelines for selecting the virtual lab as per my requirements	81
Q11 to Q14	Open-ended	
Q11	What according to you are the limitations of the chatbot in experiment design guidelines?	
Q12	Which section in the chatbot in experiment design guidelines did you find most useful?	
Q13	Which section in the chatbot in experiment design guidelines did you find most difficult to understand?	
Q14	What suggestions would you give to improve the chatbot in experiment design guidelines?	

4.3 Data Analysis Technique

By combining the responses to the two scales of strongly agree and agree and strongly disagree and disagree we carried out the analysis of responses to likert scale questions. All the 17 participants responded to the survey. The responses of the participants to the open-ended question were analysed using the thematic content analysis method.

4.4 Results

The analysis of the Likert scale data indicates that on an average 87.44 percent of the engineering instructors find the chatbot in experiment design guidelines useful for the selection, formulation and design of the various aspects of the experiment design. The percentage agreement is above 90 for the aspects – decision regarding the steps in the design and broad goals, formulation of learning objectives and assessment questions. The percentage agreement is above 80 percent for the aspects – selection of Instructional Strategy, designing tasks as per the Instructional Strategy and aligned to the learning objectives and incorporating active learning strategies.

The instructor agreement was least for the two aspects of designing experiments at various difficulty levels and selection of virtual lab.

The thematic content analysis of responses to open ended questions led to identification of the categories as follows.

1. Shortcomings or limitations of the chatbot

The instructors pointed out that the chatbot is not able to answer their queries satisfactorily for certain sections such as selection of most suitable instructional strategy. It provides standard answers to certain questions and does not vary if they have some specific requirements.

2. Most useful sections in the design with chatbot help

The chatbot was found to be most useful for designing learning objectives, tasks and assessment questions.

3. Most difficult sections in the design

The chatbot was found least useful for selection of instructional strategy and virtual lab.

5 Conclusion

We can draw the following conclusions from the results of the quantitative and qualitative analysis of the data of the usefulness study. The engineering instructors perceive that they find the chatbot in experiment design guidelines useful. They find a few sections most useful while they had difficulty in a few sections. They suggested a few modifications in the design of the chatbot in order to improve the usefulness. The limitations experienced by the instructors during experiment design will be overcome as the work progresses and more and more questions get added to the database. This is being considered as the future work. But overall the chatbot helped the instructors from even other domains than Electronics to design the experiments as their queries was easily being answered by the chatbot. Hence it can be concluded that the chatbot aided the designing of experiments and the instructors find chatbot useful.

References

1. Georgescu, A.A.: Chatbots for education–trends, benefits and challenges. In: Conference Proceedings of eLearning and Software for Education, vol. 2, no. 14. "Carol I" National Defence University Publishing House (2018)
2. Winkler, R., Söllner, M.: Unleashing the potential of chatbots in education: a state-of-the-art analysis (2018)
3. Verleger, M., Pembridge, J.: A pilot study integrating an AI-driven chatbot in an introductory programming course. In: 2018 IEEE Frontiers in Education Conference (FIE), San Jose, CA, USA (2018)
4. Benotti, L., Martínez, M.C., Schapachnik, F.: Engaging high school students using chatbots. In: Proceedings of the 2014 Conference on Innovation & Technology in Computer Science Education. ACM (2014)
5. Urban-woldron, H.: Interactive simulations for the effective learning of physics. J. Comput. Math. Sci. Teach. **28**, 163–176 (2009)
6. Wieman, C.: Comparative cognitive task analyses of experimental science and instructional laboratory courses. Phys. Teach. **53**(6), 349–351 (2015). https://doi.org/10.1119/1.4928349
7. Gomes, L., García-zubía, J.: Advances on remote laboratories and e-learning experiences (2008)

Industry 4.0

Adaptive Control System for Technological Type Control Objects

Maksym Levinskyi and Vladlen Shapo(✉)

National University Odessa Maritime Academy, Odessa, Ukraine
maxlevinskyi@gmail.com, vladlen.shapo@gmail.com

Abstract. Complications arise during automatic control of technological type control objects which are linked to their properties variations. They are caused by changes in raw material parameters, energy carriers, equipment degradation, etc. These changes have an influence on not only controlled variables but change the specifics of closed loop control system own motion. In mathematical models these changes are considered as parametric disturbances and often are represented by variable transition coefficient of the control object. The range of its change can reach and even exceed the value of ten. Common automatic control systems with constant controller parameters, in this case, loose stability and are switched to positional mode.

Corresponding mathematical apparatus is proposed. Necessary data for computer experiments are prepared. Ways of Matlab/Simulink software Design Optimization module application are determined.

Keywords: Self-tuning automatic control systems · Object gain · Controlled variable own motion component · Adaptive systems

1 Context

Technological processes, as special type of control object (CO), have specific characteristics, which differs them from other types: mobile, mechanical, electro technical, etc. [1, 2]. These include:

a) physical distribution of control channels, which reveals itself in significant delays of controlled variables reactions to control actions;
b) non-linearity of physical processes in technological unit, control valve, actuator, which reveal themselves in nonlinearities of CO model static characteristics;
c) high level of uncertainty of cause and effect relations between variables, which makes it impossible to obtain models of these relations with high enough level of adequacy and is represented as uncertain components of the parameters values;
d) a lot of independent of each other factors, which have impact on CO, but practically unavailable for measurement (characteristics of raw material and energy flows, state of the working surfaces, surfaces of heat transfer, etc.), which reveal themselves as non-controllable coordinate and parametric disturbances, which change the values of controlled variables and characteristics of control channels.

M. E. Auer and D. May (Eds.): REV 2020, AISC 1231, pp. 565–575, 2021.
https://doi.org/10.1007/978-3-030-52575-0_47

Separation of disturbances to coordinate and parametric is largely arbitrary. It is needed as a necessity for mathematical description and targeted automatic control system (ACS) analysis. To coordinate disturbances, belong such, which reveal themselves in CO variables values changes (their state coordinates) and are presented in models as determinate and stochastic components. To parametric disturbances belong such, which reveal themselves in changes of the CO model parameters. They are not only changing the values of controlled variables, but, which is very important, change the behavior of ACS own motion (motion in closed-loop), changing its stability.

As practice shows, coordinate disturbances are caused by more quickly processes, than parametric disturbances. This allows in most cases to accept the assumption about CO model parameters quasi-stationary in time intervals, whose length is sufficient for the statistical stability of coordinate disturbances characteristics estimations.

Disturbances which cause high-frequency changes of controlled variables and cannot be compensated by control action in ACS closed loop, are considered as noises. Obviously, that this noises definition includes measurement noises.

Practice shows, that parametric disturbances, and to them can be related non-linearity of control channels static characteristics, most often cause changes of CO gain k_o. Range of its change can reach and even be more than ten [3].

Behavior of gain changes $k_o(t)$ in time can be slow, for example, due to deterioration of heat transfer conditions in heat exchangers as a result of scale and mud formations, or quick, for example, when type of raw material for technological process changes.

For CO, in which gain changes in wide range, ACS with constant controller parameters is not capable of providing necessary quality indexes along with maintaining necessary stability margins. Research, carried out by Honeywell [4], shows, due to this cause at different facilities from 49% to 63% control loops work with «weak» (close to loop opening) controller parameters. To maintain compromise between quality indexes and stability margins in processes in ACS with such CO, it is necessary to utilize self-tuning of the controller parameters.

Concept of self-tuning with active identification of CO or ACS characteristics and methods which implement it (sending to control system artificial testing signals, opening the closed loop or carrying it to the stability margin) often is not acceptable for control objects of technological type. Carrying out active experiments implementing these methods increases the risk of technological regulations violations and provokes the emergency situations.

2 Purpose or Goal

Research subject in the current work – self-tuning ACS (STACS) of the technological type control object with the passive identification of its gain, which changes in wide range.

Research goal – functioning effectiveness increase of such COs by improving ACS self-tuning algorithms.

Efficient way of building self-tuning ACS for technological type CO, in which control channel gain changes in wide range, is the concept of controller self-tuning in

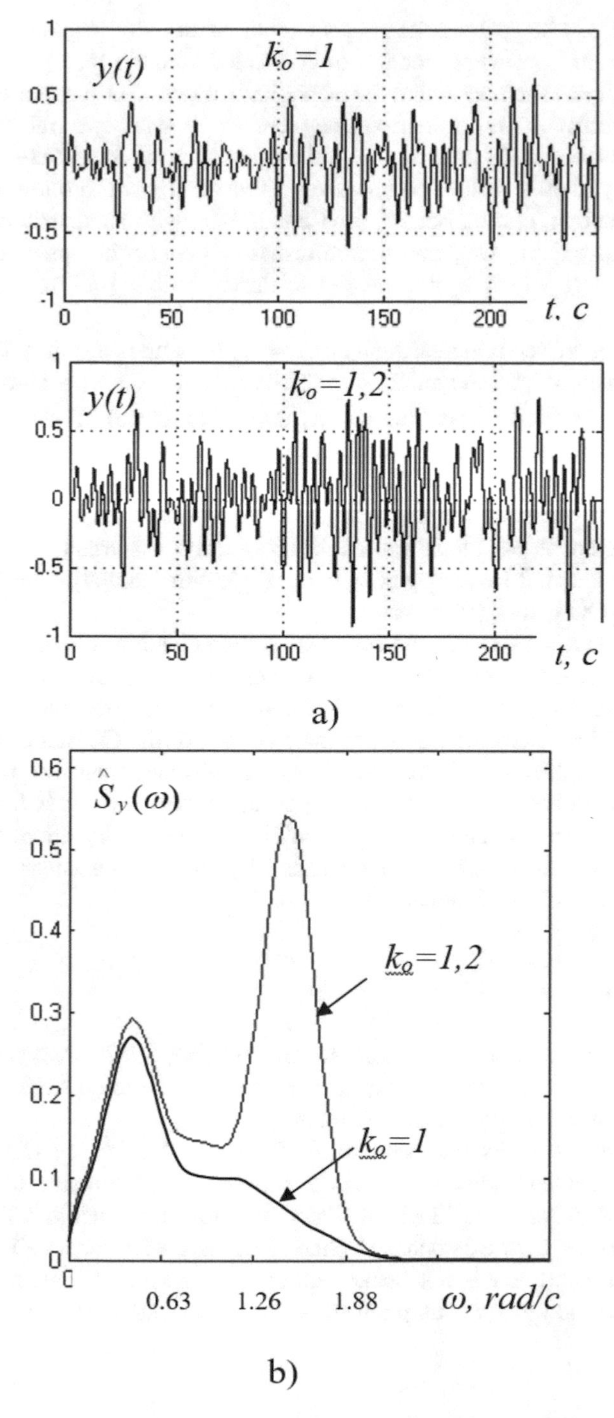

Fig. 1. Illustration of CO gain $k_o(t)$ changes impact on spectral density of the controlled variable $y(t)$: a) – time samples of $y(t)$; b) – $y(t)$ spectrum evaluations

closed-loop ACS with CO model passive identification by separating system own motion component from whole motion of controlled variable [5, 6].

This concept has such basis that for technological type CO spectral composition of parametric disturbances is lower-frequency compared with spectral composition of coordinate disturbances. In its turn, spectral composition of coordinate disturbances is lower-frequency then spectral composition of system own motion in the closed loop. ACS modeling, carried out for such conditions, with stochastic coordinate disturbances $f_k(t)$ impacting CO, revealed informative part of the spectrum of the controlled variable $y(t)$, which is efficient to use during building STACS structures (see Fig. 1).

During CO gain value increasing $k_o = 1 \Rightarrow 1,2$ changes not only dispersion, but the spectral density of $y(t)$, including due to changes of ACS own motion.

In evaluations of power spectrum density $\hat{S}_y(\omega)$ of the controlled variable $y(t)$ parts of spectrum increase in frequency range, close to resonance frequency of the ACS. If it is possible, using band pass filtration, to separate informative mid-frequency component of the $y(t)$ spectrum, which is determined by ACS own motion, from low-frequency component, which is caused by coordinate disturbances $f_k(t)$, then it is possible to evaluate mid-frequency component dispersion estimation increase, which is proportional to CO gain $k_o(t)$ increase.

Main advantage of CO parameters passive identification, during ACS controller self-tuning, is that the system functioning modes are not violated. However, STACS structure is more complicated, then ACS structure, because additional closed loop of self-tuning is added. Interaction between the feedback of the CO status and parametric feedback in the self-tuning loop determines complicated behavior of the STACS movement. This defines main disadvantage of these systems – it is hard to simultaneously provide accuracy and stability of additional self-tuning loop, which is non-linear in its nature and which composition usually include non-linear elements (multipliers, dividers, squares, switches, etc.).

3 Approach

Make specific the possibility of using information about ACS own motion $y_c(t)$ for designing STACS for CO gain changing. For this let's examine fragment of STACS block diagram, which is presented on Fig. 2.

STACS includes in its composition ACS, which consists of CO with transfer function $W^o(s)$ and controller with transfer function $W^r(s)$, tunable CO model with transfer function $W^m(s)$, two identical filters with transfer functions $W^{bf}(s)$ and two identical calculators Calc of stochastic characteristics evaluations. CO model parameters self-tuning controller is not being examined at this point (parametric controller with transfer function $W^{pr}(s)$), its point of action is presented by dashed line.

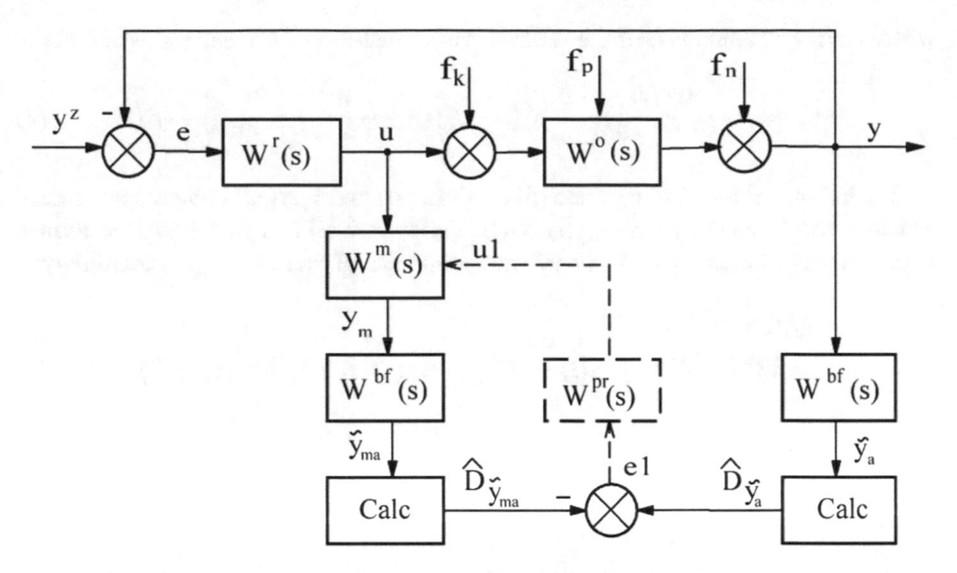

Fig. 2. Fragment of STACS block diagram

Controlled variable $y(t)$ time representation can be implemented as summation model of components with different spectral composition:

$$y(t) = \bar{y}(t) + y_l(t) + \tilde{y}(t) + y_n(t), \tag{1}$$

where $\bar{y}(t)$ – constant or slowly changing component, caused by set point $y^z(t)$ and coordinate disturbances $f_k(t)$ changes. At particular case, when $y^z(t)$ = const and $f_k(t)$ = const, then in non-static ACS $\bar{y}(t) = y^z$;

$y_l(t)$ – low-frequency component, caused mainly by influence of coordinate disturbances $f_k(t)$;

$\tilde{y}(t)$ – centered mid-frequency component caused by own motion $y_c(t)$ of ACS, in fact $\tilde{y}(t) \equiv y_c(t)$;

$y_n(t)$ – high-frequency component, caused by noises $f_n(t)$ as a rule

It is important to mention, that changes of controlled variable $y(t)$ are being formed by the influence of control variable $u(t)$ of the stabilization loop controller and disturbances $f_k(t), f_p(t), f_n(t)$, and the changes at the output of the tunable CO model $y_m(t)$ – only by the influence of control variable $u(t)$. Since parametric disturbances $f_p(t)$ change much slower then changes of $y(t)$ under the influence of $f_k(t), f_n(t)$ then it is legitimate to use hypothesis about CO parameters quasi-stationary at the interval T_{ks}. Then the following movement equations, written in operator form, are correct:

$$y(t) = W^o(s) \cdot u(t) + W^o(s) \cdot f_k(t) + f_n(t), \tag{2}$$

$$y_m(t) = W^m(s) \cdot u(t), \tag{3}$$

where control variable $u(t)$ of the controller in closed-loop ACS defines as follows:

$$u(t) = \frac{W^r(s)}{1 + W^o(s) \cdot W^r(s)} \cdot [y^z(t) - f_n(t)] - \frac{W^o(s) \cdot W^r(s)}{1 + W^o(s) \cdot W^r(s)} \cdot f_k(t) \qquad (4)$$

Filters' function – is to separate (filter) from $y(t)$ and $y_m(t)$ all components, except centered mid-frequency parts $D_{\tilde{y}_{ma}}(t)$, $\tilde{y}_m(t)$. Variables $\tilde{y}_a(t)$ and $\tilde{y}_{ma}(t)$ must be formed at the outputs of filters, which characterize properties of $\tilde{y}(t)$ and $\tilde{y}_m(t)$ accordingly:

$$\begin{aligned} \tilde{y}_a(t) &= W^{bf}(s) \cdot y(t) \\ &= W^{bf}(s) \cdot W^o(s) \cdot u(t) + W^{bf}(s) \cdot W^o(s) \cdot f_k(t) + W^{bf}(s) \cdot f_n(t) \end{aligned} \qquad (5)$$

$$\tilde{y}_{ma}(t) = W^m(s) \cdot W^{bf}(s) \cdot u(t) \qquad (6)$$

At the time domain to compare between one another immediate values of the variables $\tilde{y}_a(t)$ and $\tilde{y}_{ma}(t)$ is irrational, because:

1) model always only approximately describes dynamic properties of the control object;
2) real filters are not capable of ideal filtration of disturbances $f_k(t)$, $f_n(t)$ action consequences;
3) disturbances $f_k(t)$, $f_n(t)$ action apply points to CO and its model are different, consequently phases of the variables $\tilde{y}_a(t)$ and $\tilde{y}_{ma}(t)$ will not coincide. Solving these abovementioned problems is possible by statistical averaging of $\tilde{y}_a(t)$ and $\tilde{y}_{ma}(t)$ variables and comparing their stochastic characteristics, for example, dispersions $D_{\tilde{y}_a}$, $D_{\tilde{y}_{ma}}$.

Let's examine dependencies (5)–(6) in frequency domain. Control action $u(t)$ time change (4) is replaced by its spectral density $S_u(\omega)$ [7]:

$$\begin{aligned} S_u(\omega) = {} & \left| \frac{W^r(s)}{1 + W^o(s) \cdot W^r(s)} \right|^2 \cdot S_{y^z}(\omega) - \left| \frac{W^r(s)}{1 + W^o(s) \cdot W^r(s)} \right|^2 \cdot S_{fn}(\omega) \\ & - \left| \frac{W^o(s) \cdot W^r(s)}{1 + W^o(s) \cdot W^r(s)} \right|^2 \cdot S_{fk}(\omega) = A^2_{y^z-u}(\omega) \cdot S_{y^z}(\omega) \\ & - A^2_{fn-u}(\omega) \cdot S_{fn}(\omega) - A^2_{fk-u}(\omega) \cdot S_{fk}(\omega) \end{aligned} \qquad (7)$$

where $S_{y^z}(\omega)$ – spectral density of the set point value $y_z(t)$; $S_{fk}(\omega)$ – spectral density of coordinate disturbance $f_k(t)$, which is appointed to the CO input; $S_{fn}(\omega)$ – noise $f_n(t)$ spectral density, which is appointed to the output of the CO; $A(\omega)$ – amplitude-frequency characteristics of closed-loop ACS through according channels. Dependencies (5) and (6) take the following form:

$$\begin{aligned} S_{\tilde{y}_a}(\omega) &= |W^o(j\omega)|^2 \cdot \left|W^{bf}(j\omega)\right|^2 \cdot S_u(\omega) \\ &\quad + |W^o(j\omega)|^2 \cdot \left|W^{bf}(j\omega)\right|^2 S_{fk}(\omega) + \left|W^{bf}(j\omega)\right|^2 S_{fn}(\omega) \\ &= A_o^2(\omega) \cdot A_{bf}^2(\omega) \cdot S_u(\omega) + A_o^2(\omega) \cdot A_{bf}^2(\omega) \cdot S_{fk}(\omega) \\ &\quad + A_{bf}^2(\omega) \cdot S_{fn}(\omega) = S_{\tilde{y}_a - u}(\omega) + S_{\tilde{y}_a - fk}(\omega) + S_{\tilde{y}_a - fn}(\omega) \end{aligned} \tag{8}$$

$$S_{\tilde{y}_{ma}}(\omega) = |W^m(j\omega)|^2 \cdot \left|W^{bf}(j\omega)\right|^2 \cdot S_u(\omega) = A_m^2(\omega) \cdot A_{bf}^2(\omega) \cdot S_u(\omega) \tag{9}$$

where $S_{\tilde{y}_a}(\omega)$, $S_{\tilde{y}_{ma}}(\omega)$ – spectral densities at the outputs of according filters; $W^o(j\omega) = k_o W^{oi}(j\omega)$, $W^m(j\omega) = k_m W^{mi}(j\omega)$ – CO and its tunable model transfer functions in frequency domain; $W^{bf}(j\omega)$ – filters transfer function in frequency domain; $A(\omega)$ – according amplitude-frequency characteristics; $S_{\tilde{y}_a - u}(\omega)$, $S_{\tilde{y}_a - fk}(\omega)$, $S_{\tilde{y}_a - fn}(\omega)$ – components of spectral density $S_{\tilde{y}_a}(\omega)$.

From (8) and (9) follows, that spectral densities $S_{\tilde{y}_a}$, $S_{\tilde{y}_{ma}}$ at the filters outputs substantially differs. That means, that it is impossible to use them directly in the task of tuning the gain $k_m(t)$ of CO model to the changing, under the influence of parametric disturbances $f_p(t)$, gain $k_o(t)$ of the CO. To solve this task it is necessary to minimize influence of the component $S_{\tilde{y}_a - f_k}(\omega)$ of external coordinate disturbances $f_k(t)$ and component $S_{\tilde{y}_a - f_n}(\omega)$ of measurement noises $f_n(t)$ in spectrum of the signal $S_{\tilde{y}_a}$ (8). This function in STACS is implemented by band pass filters, which structures and parameters should be chosen in a such a way, to maximally possible bring closer spectral density at the output of filter $S_{\tilde{y}_a}$ (8) to the spectral density at the output of filter $S_{\tilde{y}_{ma}}$ (9).

To show that dispersion values $D_{\tilde{y}_a}$, $D_{\tilde{y}_{ma}}$ depend on CO gain k_o and its model gain k_m, using (8)–(9), follows from next dependencies :

$$D_{\tilde{y}_a} = \frac{1}{2\pi} \int_{-\infty}^{\infty} \begin{bmatrix} |W^o(j\omega)|^2 \cdot \left|W^{bf}(j\omega)\right|^2 \cdot S_u(\omega) \\ + |W^o(j\omega)|^2 \cdot \left|W^{bf}(j\omega)\right|^2 S_{fk}(\omega) + \left|W^{bf}(j\omega)\right|^2 S_{fn}(\omega) \end{bmatrix} d\omega \tag{10}$$

$$D_{\tilde{y}_{ma}} = \frac{1}{2\pi} \int_{-\infty}^{\infty} |W^m(j\omega)|^2 \cdot \left|W^{bf}(j\omega)\right|^2 S_u(\omega) d\omega \tag{11}$$

Gains k_o, k_m do not depend of frequency ω. If it is accomplishable to mainly suppress consequences of external coordinate disturbances $f_k(t)$ and noises $f_n(t)$ acting on control variable $y(t)$ and to make them insignificant, then in (10) components $|W^o(j\omega)|^2 \cdot \left|W^{bf}(j\omega)\right|^2 S_{fk}(\omega) \to 0$, $\left|W^{bf}(j\omega)\right|^2 S_{fn}(\omega) \to 0$. Consequently, Eqs. (10), (11) take form:

$$D_{\tilde{y}_a} \approx k_o^2 \frac{1}{2\pi} \int_{-\infty}^{\infty} |W^{oi}(j\omega)|^2 \cdot \left|W^{bf}(j\omega)\right|^2 \cdot S_u(\omega) d\omega \tag{12}$$

$$D_{\tilde{y}_{ma}} = k_m^2 \frac{1}{2\pi} \int_{-\infty}^{\infty} \left|W^{mi}(j\omega)\right|^2 \cdot \left|W^{bf}(j\omega)\right|^2 \cdot S_u(\omega)d\omega \tag{13}$$

From (12), (13) follows CO model $W^{mi}(j\omega)$ precise enough describes dynamic of the real CO $W^{oi}(j\omega)$, i.e. $W^{oi}(j\omega) \approx W^{mi}(j\omega)$, then if dispersions at the outputs of filters are approximately equal $D_{\tilde{y}_a}(t) \approx D_{\tilde{y}_{ma}}(t)$ then gains of CO and its model are also approximately equal $k_o^2 \approx k_m^2$. Exactly this fact is being used during designing of STACS.

In accordance with hypothesis about quasi-stationary of CO parameters at the time interval T_{ks}, changes if gain $k_o(t)$ take place relatively slowly. This makes it necessary and gives opportunity to calculate dispersions estimations (12), (13) at the sliding time interval $t_{oc} \leq T_{ks}$. At the real time these estimations are being calculated, as a rule, by uniform or exponential averaging of stochastic processes, in this case – $\tilde{y}_a^2(t)$ and $\tilde{y}_{ma}^2(t)$, «taking them through» according dynamic links with transfer functions $W^{oe}(s) = \frac{(1-\exp(-\tau_{oc}s))}{t_{oc}s}$ or $W^{eo}(s) = \frac{1}{0{,}5t_{oc}s+1}$ [8]:

$$\hat{D}_{\tilde{y}_a}(t, t_{oc}) = \frac{(1-\exp(-t_{oc}s))}{t_{oc}s} \cdot \tilde{y}_a^2(t),$$

$$\hat{D}_{\tilde{y}_{ma}}(t, t_{oc}) = \frac{(1-\exp(-t_{oc}s))}{t_{oc}s} \cdot \tilde{y}_{ma}^2(t) \tag{14}$$

$$\hat{D}_{\tilde{y}_a}(t, t_{oc}) = \frac{1}{0,5t_{oc}s+1} \cdot \tilde{y}_a^2(t),$$

$$\hat{D}_{\tilde{y}_{ma}}(t, t_{oc}) = \frac{1}{0,5t_{oc}s+1} \cdot \tilde{y}_{ma}^2(t) \tag{15}$$

At current work estimators Calk in STACS structure (Fig. 2) evaluate estimations using dependencies (15), because it is easier to implement them in a software.

It is important to note that difference between estimations $e_1(t, t_{oc}) = \hat{D}_{\tilde{y}_a}(t, \tau_{oc}) - \hat{D}_{\tilde{y}_{ma}}(t, t_{oc})$, likewise as dispersion difference $D_{\tilde{y}_a} - D_{\tilde{y}_{ma}}$, is proportional to the difference $k_o^2 - k_m^2$. Dispersion estimation changes $\hat{D}_{\tilde{y}_a}(t, t_{oc})$ signal about CO gain $k_o(t)$ changes that has begun.

From the abovementioned follows, that precision of self-tuning in STACS depends on quality of filters operation, their ability to separate own and forced motions of the ACS.

Additional researches show that as an error of self-tuning loop of CO model gain $k_m(t)$ towards changing gain of the CO $k_o(t)$, it is better to use not dispersions estimations difference but the difference of standard deviations estimations (SD) $\tilde{y}_a(t)$ and $\tilde{y}_{ma}(t)$. This allows to obtain the following STACS structure (Fig. 3):

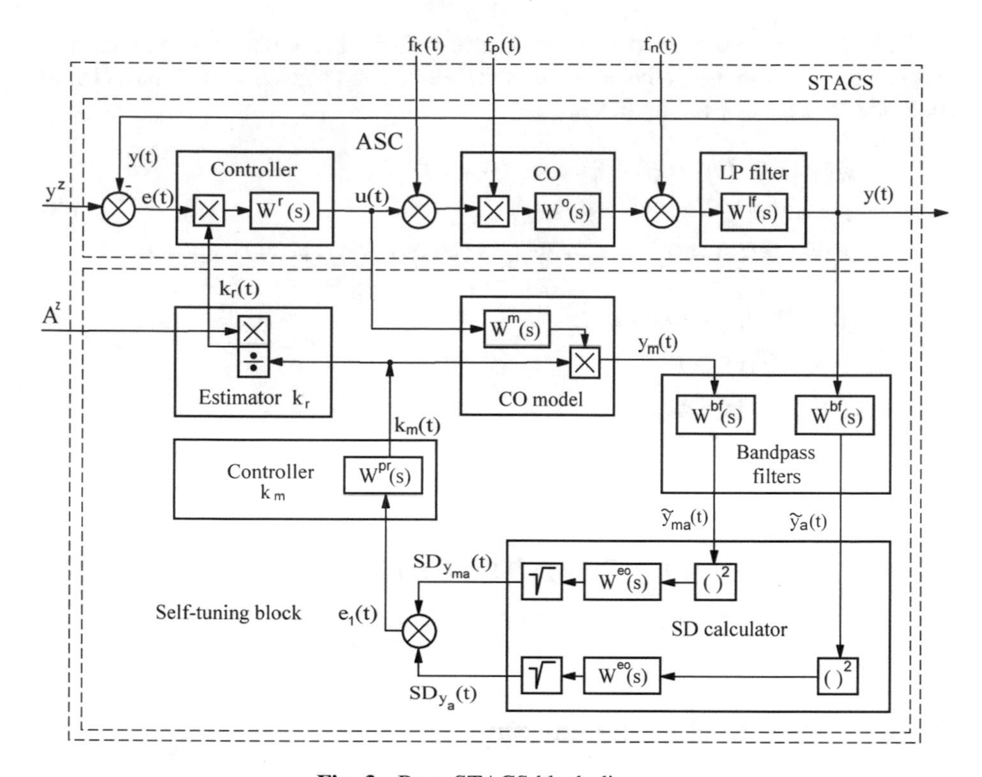

Fig. 3. Base STACS block diagram

Now let's take a look at functioning principle of the base STACS self-tuning block. Its composition involves CO model on the input of which control action $u(t)$ of the ACS controller is being put. Controlled variable $y(t)$ of the CO and its model $y_m(t)$ are coming to the inputs of the band pass filters. They are suppressing the consequences of the coordinate disturbances $f_k(t)$ and noises $f_n(t)$ impacts on $y(t)$ and $y_m(t)$. At the outputs of filters forms variable $\tilde{y}_a(t)$ – own motion component in the general movement of the CO controlled variable $y(t)$ and estimation of this component $\tilde{y}_{ma}(t)$ at the CO model output. SD estimators at the sliding time interval t_{oc} average variables $\tilde{y}_a(t)$ and $\tilde{y}_{ma}(t)$, evaluating estimations of their SD: $SD_{\tilde{y}_a}(t)$ and $SD_{\tilde{y}_{ma}}(t)$, which proportional to the current values of CO and its model gains $k_o(t)$, $k_m(t)$. SD changes estimation $SD_{\tilde{y}_a}(t)$ signals about beginning of CO gain $k_o(t)$ changes.

Self-tuning task in STACS is reduced to the stabilization task. Controller of the km receives control error – estimations difference $e_1(t) = SD_{\tilde{y}_a}(t) - SD_{\tilde{y}_{ma}}(t)$ and stabilizes estimation $SD_{\tilde{y}_{ma}}(t)$ of model on the level of CO SD estimation $SD_{\tilde{y}_a}(t)$ by changing the model gain $k_m(t)$. In fact, km controller detects at the real time changes of the CO gain $k_o(t)$ and tunes the model gain $k_m(t)$.

Current value $k_m(t) \approx k_o(t)$, from the output of the km controller comes also to the k_r calculator, which defines the value of the ACS controller gain $k_r(t)$, using dependency $k_r(t) = A^z/k_m(t)$. By that stable operation of the main ACS stabilization loop is ensured during changes of CO gain.

STACS mathematical model, in accordance with the block diagram, is represented by two dependent on each other systems of differential and algebraic equations: (16) for ACS and (17) for self-tuning block.

$$\begin{aligned} y_o(t) &= W^o(s) \cdot (u(t) + f_k(t)) \cdot k_o(t) + f_n(t) \\ y(t) &= W^{lf}(s) \cdot y_o(t), \quad y(t=0) \\ u(t) &= W^r(s) \cdot e(t) \cdot k_r(t) = W^r(s) \cdot (y^z(t) - y(t)) \cdot k_r(t), \quad u(t=0) \end{aligned} \tag{16}$$

$$\begin{aligned} y_m(t) &= (W^m(s) \cdot u(t)) \cdot k_m(t) \\ \tilde{y}_a(t) &= W^{bf}(s) \cdot y(t) \\ \tilde{y}_{ma}(t) &= W^{bf}(s) \cdot y_m(t) \\ SD_{\tilde{y}_k}(t) &= \sqrt{W^{eo}(s) \cdot \tilde{y}_a^2(t)} \\ SD_{\tilde{y}_{mk}}(t) &= \sqrt{W^{eo}(s) \cdot \tilde{y}_{ma}^2(t)} \\ k_m(t) &= W^{pr}(s) \cdot e_1(t) = W^{pr}(s) \cdot \left(SD_{\tilde{y}_k}(t) - SD_{\tilde{y}_{mk}}(t)\right), \quad k_m(t=0) \\ k_r(t) &= A^z / k_m(t) \end{aligned} \tag{17}$$

4 Actual or Anticipated Outcomes

STACS (16)–(17) includes non-linear differential equations with variable parameters (coefficients); coordinate disturbances $f_k(t)$ and noises $f_n(t)$ are stochastic processes. For solving the tasks of analysis and synthesis of such systems by using analytical methods is drastically complicated. That is why, as a main instrument for STACS research was chosen computer modeling by conducting of series of many-factors planned computer experiments in the MatLab/Simulink program package environment.

5 Conclusions/Recommendations/Summary

1. Results computer modeling of base structure STACS proves theoretical grounds about possibility of changing CO gain identification in ACS closed-loop by separating the own motion component from the whole motion of controlled variable, which arises as a consequence to coordinate disturbances impact.
2. To separate the components of own motion of CO and its model controlled variables in STACS it is reasonable to use band pass filters, particularly Butterworth filters. Choosing band pass filters of the second order is a good compromise between filtration effectiveness and their software implementation complexity.
3. STACS self-tuning loop parameters influence integral quadratic index of ACS error has extreme nature, which gives opportunity to conduct optimal parametric synthesis of this loop.

4. Further STACS effectiveness increase, including self-tuning accuracy with maintaining system stability, potentially possible by: a) lowering inertia and increasing performance of the self-tuning loop; b) self-tuning loop gain stabilization, which in STACS of base structure changes depending on values of CO gain.

References

1. Ray, W.H.: Advanced Process Control. McGraw-Hill (1981)
2. Astraom, K., Wittenmark, B.: Adaptive Control, pp. 1–580. Addison-Wesley, Massachusetts (1989)
3. Isermann, R.: Digital Control Systems. Springer Verlag, 2nd ed. (1989)
4. Li, Y., Ang, K., Chong, G.: Patents, software and hardware for PID control: an overview and analysis of the current art. IEEE Control Syst. Mag. **26**(1), 42–54 (2006)
5. Khobin, V.A., Levinskyj, M.V.: Adaptyvne keruvannya ob'yektamy texnolohichnoho typu. Odesa, Helvetyka (2019). 228 s. ISBN 978-966-717-0
6. Khobin, V.A., Levinskyj, M.V.: Patent UA № 117038 from 11.06.2018
7. Bendat, J.S., Piersol, A.G.: Random Data: Analysis and Measurement Procedures. Wiley, p. 640, 4th ed. (2010). ISBN: 0470248777, 9780470248775
8. Abraham, B., Ledolter, J.: Statistical Methods for Forecasting. Wiley, New York (1983)

Role of Education 4.0 Technologies in Driving Industry 4.0

Venkata Vivek Gowripeddi[1,2](✉), Manav Chethan Bijjahalli[3], Nikhil Janardhan[2], and Kalyan Ram Bhimavaram[4]

[1] International Institute of Information Technology, Bangalore, India
venkata.gowripeddi@iiitb.org

[2] Cymbeline Innovation Private Limited, Bangalore, India
{vivek,nikhil}@cymbelinein.com

[3] Dr Ambedkar Institute of Technology, Bangalore, India
manavbijjahalli885@gmail.com

[4] BITS - Pilani KK Birla Campus, Goa, India
kalyanram.b@gmail.com

Abstract. Industry 4.0 is the dawn of an era of data oriented and connected industrial processes. Education 4.0 is the need for way of preparing the next generation for Industry 4.0. The purpose of the paper is to identify, measure and evaluate the role of advanced and collaborative learning practices in education, collectively termed as Education 4.0 in building Engineers and Researchers of Tomorrow in fields of Data Driven Technologies. The paper analyses the three case studies of Remote Engineering in Manufacturing sector, Remote Labs in Academic Environment and transition of students from Remote Lab builders to Remote Engineering Framework builders. The key outcomes of the paper include mapping of common skills, impact of industrial work exposure and correlation between the two technologies This paper provides an insight into the various education technologies and their importance in Industry to facilitate modern industrial technologies.

Keywords: Industry 4.0 · Education 4.0 · Remote Engineering · Collaborative learning

1 Introduction

1.1 Industry 4.0

Industry 4.0 is the dawn of an era of data oriented and connected industrial processes. Education 4.0 is the need for way of preparing the next generation for Industry 4.0. As the world leaps towards Industry 4.0, Cyber Physical system and Digital Twin Technologies, it becomes more and more important for educational technologies to catch up and provide the right foundation and framework to enable this.

Digitization and the use of technologies have begun in Hungary, but there is still a great lag behind the other, more developed countries. By examining the previous period

M. E. Auer and D. May (Eds.): REV 2020, AISC 1231, pp. 576–587, 2021.
https://doi.org/10.1007/978-3-030-52575-0_48

and evaluating results, we can establish that the number of new investors will increase in the coming years as companies want to gather, store and process data from an increasing number of processes. The greatest barrier to the implementation of Industry 4.0 is the lack of a clear digital strategy in value-creating (production and logistics) processes and a lack of support from leadership [1]. The main purpose of Industry 4.0 is to achieve improvements in terms of automation and operational efficiency, as well as effectiveness [2]. Computers, automation and robots existed in previous decades, but the opportunities provided by the Internet revolutionize their use, and the opportunities they provide [3]. The emerging Industry 4.0 concept is an umbrella term for a new industrial paradigm which embraces a set of future industrial developments including Cyber-Physical Systems (CPS), the Internet of Things (IoT), the Internet of Services (IoS), Robotics, Big Data, Cloud Manufacturing and Augmented Reality [4]. The Cyber-Physical System (CPS) is a promising paradigm for the design of current and future engineered systems and is expected to make an important impact on our interactions with the real world. The idea behind CPS places the focus on the integrated system design instead of on the cyber or the physical system independently [5].

1.2 Education 4.0

The students that they have now have different preference than students that they had 10 years ago. Integrating more current technologies will make the instructors more creative in designing their lessons, thus making the learning more interesting [5]. Teaching factories 4.0 serve as an introduction for the aspiring engineers to the newly developed and implemented technologies, through workshops that call the participants to utilize these technologies as a mean that will improve the quality and the effectiveness of their tasks, potentially unlocking new capabilities. The implementation of these technologies in the teaching factories will also boost their integration in manufacturing, as the new engineers that have familiarized themselves with the true potential and the capabilities offered by Industry 4.0 will seek for opportunities to consider these technologies in their works [6]. To be able to live in a society and to be equipped with the best of his/her ability. Therefore, Education 4.0 will be more than just an education. it is not enough to define the education for 21st Century Skills only, but also to consider the learning management in the aspect of Social & Virtual Learning; that the graduates will become intellectuals and eventually that would help to build an intelligent nation [7]. Collaborative learning research emerged in the 1970s, for practical reasons (for example, scarce resources such as computers had to be shared between students) and as a reaction to psychological and education approaches centred on the individual learner [8]. CL compared with competitive and individualistic efforts, has numerous benefits and typically results in higher achievement and greater productivity, more caring, supportive, and committed relationships; and greater psychological health, social competence, and self-esteem [9].

2 Purpose

The purpose of the paper is to identify, measure and evaluate the role of advanced and collaborative learning practices in education, collectively termed as Education 4.0 in building Engineers and Researchers of Tomorrow in fields of Data Driven Technologies. The key goals of the paper are as follows:

(1) Identify common key elements in Education 4.0 and Industry 4.0.
(2) Design a framework that emphasizes on these elements.
(3) Measure the results and evaluate the framework

3 Approach

Remote Engineering as Education Technology is chosen for this study.

The approach consists of studying three case studies:

(1) Remote Engineering in Industry 4.0
(2) Remote Engineering in Education 4.0 and
(3) Relationship between Education 4.0 and Industry 4.0.

3.1 Case 1: Studies a Use Case of Remote Engineering in a Automotive Company

A Typical Industry 4.0 architecture is as follows:

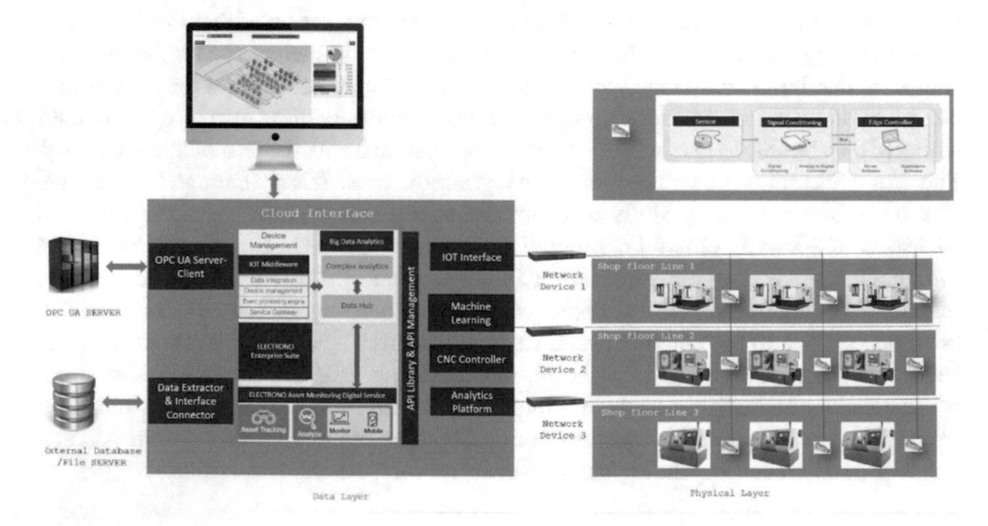

Fig. 1. Industry 4.0 architecture

As shown in Fig. 1., Industry 4.0 implementation consists of Networked Machines – CNCs and PLCs programmed to send data over network. The data is analyzed and in sights are drawn with deep domain expertise of Machine Processes (Fig. 2).

Table 1. The key Skillsets required for industry 4.0 are as follows:

Skill	Discipline
1. Networking	Information technology
2. Machine process analyst	Industrial engineering
3. CNC PLC programming	Industrial engineering
4. Data science	Statistics/Computer science
5. IoT expertise	Electronics and communication

Fig. 2. Industry 4.0 implementation screens and data

3.2 Case 2: Studies a Use Case of Remote Labs at a University

We examine the case of an Analog Remote Lab whose architecture and implementation are as follows:

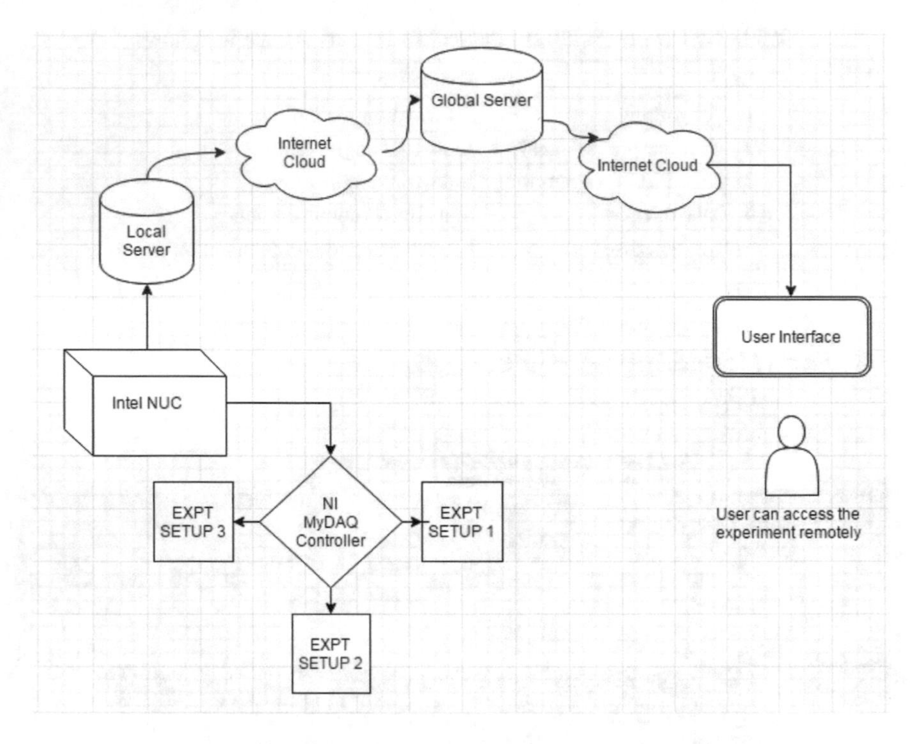

Fig. 3. Architecture of analog remote lab

A Remote lab consists of different experimental setups – Exp Setup1, Exp Setup2, viz.. controlled by Data acquisition and Signal Conditioning system. This system is connected to Local Server through host computer. Local Server along with Cloud form the top layer of network. A front end is developed to enable access to these systems from anywhere through internet (Fig. 3).

By examining Tables 1 and 2, we can understand that the following skill sets borrowed from various disciplines of engineering need to be developed.

Table 2. The key Skillsets required for Remote Labs are as follows:

Skill	Discipline
1. Data acquisition and signal processing	Electronics and communication
2. Communication technologies and protocols	Electronics and communication
3. UI/IX design	Information and technology
4. Circuit design and testing	Electronics and communication
5. Research skills	Critical thinking and writing

Table 3. Indentified necessary skills for Industry 4.0

Skill	Discipline	In Context of Industry 4.0
Networking	Information and technology	Industrial networks
IoT	Electronics and communication	Industrial wireless sensor networks, data communication protocols
Data science	Statistics/Computer science	Data analytics for machines
UI/UX	Information and technology	Production dashboards
Research skills	Critical thinking and writing	Industry 4.0 research

In the coming sections, we fill the gap of necessary skills as identified in Table 3 by suggest new subjects and their fitment into the existing curriculum.

3.3 Case 3: Map Out Current Curriculum, with Respect to Modern Day Relevance and Industry 4.0 and Suggest a Curriculum for Industry 4.0

In Case 3, The entire curriculum of Industrial Engineering Undergraduate Curriculum is analyzed according to the Credits, Skillset Developed, Potential Job Profiles and scored against the Modern Day Relevance of the subject and its significance for Industry 4.0. The Syllabus is broken down into 3 years – Year I, Year II and Year III and a score as below is assigned

$$\text{Score} = \text{Credits} * (2/3 * \text{Modern Day Relevance} + 1/3 * \text{Industry 4.0 Relevance}) \quad (1)$$

'Rubric' is a term with a variety of meanings. As the use of rubrics has increased both in research and practice, the term has come to represent divergent practices [10] (Table 4).

Table 4. Rubrics for scoring Modern day relevance

Criteria	Score		
	[1–3]/10	[4–7]/10	[8–10]/10
Applications in present day score	Very Poor - Poor	Average - Good	Very Good - Excellent
Collaborative learning score	Very Poor - Poor	Average - Good	Very Good - Excellent
Practicality score	Very Poor - Poor	Average - Good	Very Good - Excellent

Engaging with authentic scientific tools and practices such as controlling remote laboratory experiments, improve conceptual understanding, and increase motivation (Table 5).

$$\text{Modern Relevance Score} = [\text{Applications in Present Day Score} + \text{Collaborative learning Score} + \text{Practicality Score}]/3 \quad (2)$$

$$\text{Industry 4.0 Score} = [\text{Industry 4.0 Applications Score} + \text{Industry 4.0 Use Cases Score} + \text{Practicality Score}]/3 \quad (3)$$

Table 5. Rubrics for scoring Industry 4.0 relevance

Criteria	Score		
	[1–3]/10	[4–7]/10	[8–10]/10
Industry 4.0 applications score	Very Poor - Poor	Average - Good	Very Good - Excellent
Industry 4.0 use cases score	Very Poor - Poor	Average - Good	Very Good - Excellent
Practicality score	Very Poor - Poor	Average - Good	Very Good - Excellent

Year I.

Subjects which ranked the least significant in terms of Modern Day Relevance and Industry 4.0 were to be replaced by the following two subjects which focus on Industry 4.0 (Tables 6, 7, 8, 9, 10 and 11).

Table 6. The curriculum analysis and scoring for Year I

Sl no.	Subject	Credits	Skillset Developed	Potential Job Profiles	Modern Day Relevance (Out of 10)	Industry 4.0 Relevance (Out of 10)	Score
1	Computer integrated manufacturing & Lab	4	CAD/CAM based product development	CAMD designer	8	9	33.33
2	Manufacturing technology & Lab	4	Metal cutting techniques, cutting tool materials, drilling machines, milling machines & grinding machines, able to forge metals and foundry	Manufacturing engineer, consultant	8	8	32
3	Thermal & Fluids engineering	4	Heat processes and thermodynamic cycles & its applications, analysis performance of refrigeration and IC engine	Thermodynamics engineers/physicist	8	7	30.67
4	Theory of machines	4	Gear mechanics	Gear design engineer	8	7	30.67
5	Python programming & Lab	3	Able to do testing and debugging of code written in Python.	Python developer	9	10	28

(*continued*)

Table 6. (*continued*)

Sl no.	Subject	Credits	Skillset Developed	Potential Job Profiles	Modern Day Relevance (Out of 10)	Industry 4.0 Relevance (Out of 10)	Score
6	Mechanics of materials & Material testing lab	4	Analyze materials under different load conditions. Able to operate and test using Universal testing machine & hardness testing machine	Material testing engineer	7	5	25.33
7	Work study & Ergonomics lab	5	Process Planning, Time Management, Work Study, Time Study, standard time calculations	Operation and Process Planner	5	5	25
8	Computer aided machine drawing & Lab	4	Drawings in 2D & 3D, assemble different mechanical parts together, Operate modern manufacturing using NC, CNC and DNC	CAD designer	5	6	21.33

Table 7. Replacement Subjects for Year I

Year I	Industrial Internet of Things (IIoT)	Transformation of industrial processes through the integration of modern technologies such as sensors, communication, and computational processing
	IR 4.0 use cases & challenges	Opportunities, challenges brought about by Industry 4.0 and how organizations and individuals should prepare to reap the benefits

Year II.

Table 8. The curriculum analysis and scoring for Year II

Sl no.	Subject	Credits	Skillset Developed	Potential Job Profiles9	Modern Day Relevance (Out of 10)	Industry 4.0 Relevance (Out of 10)	Score
1	Facilities planning & Materials management	7	Analysis, inventory control planning, Solve facility location problems	Purchase manager, inventory planning & control supervisor, Occupancy planner, Facility & planning	8	8	56
2	Operations research & Management	7	Forecasting demand, operations decision making, Game theory analyze transportation models, use of network analyses	Operation research analyst, Operation research scientist, Data scientist operation research	7	8	51.33

(*continued*)

Table 8. (*continued*)

Sl no.	Subject	Credits	Skillset Developed	Potential Job Profiles9	Modern Day Relevance (Out of 10)	Industry 4.0 Relevance (Out of 10)	Score
3	Statistics for engineers & Lab	5	Design of experiments, estimate hypothesis & give inference to random experiments	Statistical specialist, data analyst, statistical programmer	8	7	38.33
4	Engineering economy	4	Estimation, costing, Replacement analysis, rate of return analysis	Economist, Financial advisor	9	9	36
5	Design of machine elements	4	Design springs, mechanical joints & gears	Mechanical designer	8	7	30.66
6	Marketing management	3	Develop marketing plan, advertise, promotion, brand the product, label and package	Marketing manager, consultant, social media manager, SEO specialist, Market research analyst	9	8	26
7	Advanced manufacturing process	3	Electric discharge machining operations, laser beam operations, electron beam machine operations, & Plasma arc machine	Advance manufacturing engineer, welding engineer	6	7	19
8	Mechanical measurements & Metrology, lab	4	Material properties, Caliberation techniques & methods, calculation of errors in calibration	Material Analyst, Caliberation engineer	4	6	18.66

Table 9. Replacement Subjects for Year II

Year II	Smart Logistics	foundational business skills needed in SCM settings, make decisions affecting supply chain's plan, deliver and customer management functions.
	Role of Manufacturing data & internet model	Industrial asset tracking, Inventory management Predictive maintenance

Year III.

Table 10. The curriculum analysis and scoring for Year III

Sl no.	Subject	Credits	Skillset developed	Potential job profiles	Modern day relevance (Out of 10)	Industry 4.0 Relevance (Out of 10)	Score
1	Quality assurance & Reliability & Lab	4	Control charts design analyse various failure models	Reliability engineer, Quality assurance analyst	8	8	32
2	Lean manufacturing	4	Apply the various tools and techniques of lean manufacturing	Lean manager, consultant & analyst	8	8	32
3	Simulations modelling & Analysis	4	Analysis of simulation data, output analysis	Modelling & simulation analyst, process modelling analyst	7	9	30.67
4	Financial accounting & Management	4	Preparation of financial statements, cash flow diagrams, budgeting	Accounts & finance manager, account management analyst	8	6	29.33
5	Financial accounting & Management	4	Preparation of financial statements, cash flow diagrams, budgeting	Accounts & finance manager, account management analyst	8	6	29.33
6	Supply chain management	3	Analyze supply chain, build supply chain models	Supply chain manager, analyst, sourcing & procurement manager, supply chain consultant	7	8	22
7	Organizational behavior	3	Decision making abilities	Human Resource Manager	6	6	18
8	Enterprise resource planning & Lab	1	Implementation of ERP systems	ERP consultant, analyst	6	5	5.66

Table 11. Replacement Subjects for Year III

Year III	Predictive Analytics	Extracting information from existing data sets in order to determine patterns and predict future outcomes and trends
	Cyber Physical systems	Analyze the functional behavior of CPS based on standard modelling formalisms. Implement specific software CPS using existing tools. Design CPS requirements based on operating system and hardware architecture constraints

4 Results

In Depth analysis of above three case studies is done to better understand the role of Remote Engineering in these Framework and provide certain conclusions and premises for more research to be conducted upon this.

The key outcomes of the paper include mapping the skills, experiences and learnings from three case studies and forming a framework for preparing students for Industry 4.0 challenges and requirements. The research answers five important topics as below in detail:

(1) The key skills required to be built for training students for Industry 4.0
(2) Education Technologies that provide these skills.
(3) Evaluation of Role of Remote Engineering in Education 4.0
(4) Correlation between Remote Engineering in Education 4.0 and Remote Engineering in 4.0.
(5) Industry exposure and its benefits.

5 Conclusion

This paper provides an insight into the various education technologies, mapping of common skills, impact of industrial work exposure and correlation between the two technologies. The paper analyses the three case studies of Remote Engineering in Manufacturing sector, Remote Labs in Academic Environment and transition of students from Remote Lab builders to Remote Engineering Framework builders. This paper provides an insight into the various education technologies and their importance in Industry to facilitate modern industrial technologies.

The paper also talks about the necessary modifications which is needed in the present curriculum of engineering education.

Acknowledgement. The authors would like to thank International Institute of Information Technology, Bangalore, Management of PVP welfare trust, Dr Ambedkar Institute of Technology, for supporting this research work. Authors would like to express their gratitude to Department of Industrial Engineering and Management [IEM], Dr AIT for their valuable guidance.

References

1. Nagy, J., Oláh, J., Erdei, E., Mate, D., Popp, J.: The role and impact of industry 4.0 and the internet of things on the business strategy of the value chain—the case of hungary. Sustainability **10,** 3491 (2018). https://doi.org/10.3390/su10103491
2. Slusarczyk, B. Industry 4.0—Are we ready? Pol. J. Manag. Stud. **17**, 232–248 (2018)
3. Deloitte. Industry 4.0, Challenges and Solutions for the Digital Transformation and Use of Exponential Technologies. Deloitte, Swiss, Zurich (2015)

4. Weyer, S., Schmitt, M., Ohmer, M., Gorecky, D.: Towards industry 4.0-standardization as the crucial challenge for highly modular, multi-vendor production systems. IFAC-Papers Online **48**, 579–584 (2015)
5. Gunes, V., Peter, S., Givargis, T., Vahid, F.: A survey on concepts, applications, and challenges in cyber-physical systems. KSII Trans. Internet Inf. Syst. **8**, 4242–4268 (2014). https://doi.org/10.3837/tiis.2014.12.001
6. Mourtzis, D., Vlachou, K., Dimitrakopoulos, G., Zogopoulos, V.: Cyber- physical systems and education 4.0 –the teaching factory 4.0 concept. Procedia Manuf. **23**, 129–134 (2018). https://doi.org/10.1016/j.promfg.2018.04.005
7. Ponce, O., Pagán Maldonado, N.: Educational research in the 21st century: challenges and opportunities for scientific effectiveness. Int. J. Educ. Res. Innov. **8**, 24–37 (2017)
8. Baker, M.: Collaboration in collaborative learning. Interact. Stud. **16**, 451–473 (2015). https://doi.org/10.1075/is.16.3.05bak
9. Laal, M., Ghodsi, S.: Benefits of collaborative learning. Procedia Soc. Behav. Sci. **31**(2012), 486–490 (2012). https://doi.org/10.1016/j.sbspro.2011.12.091
10. Dawson, P.: Assessment rubrics: towards clearer and more replicable design, research and practice. Assess. Eval. Higher Educ. **42**(3), 347–360 (2015). https://doi.org/10.1080/02602938.2015.1111294

Enabling Software Defined Networking for Industry 4.0 Using OpenStack

Venkata Vivek Gowripeddi(✉), Nithya Ganesan, Samar Shailendra, B. Thangaraju, and Jyotsna Bapat

International Institute of Information Technology, Bangalore, India
vivek@cymbelinein.com, nithya.ganesan@iiitb.org, {samar.shailendra, b.thangaraju, jbapat}@iiitb.ac.in

Abstract. Industry 4.0 refers to the fourth generation of advancements in Manufacturing that enables data driven processes, Predictive maintenance, Lights-Out Manufacturing among others. Industry 4.0 comprises of elements from Cyber Physical Systems, Industrial IoT, Big data, Cloud Computing, AI and ML. However, Networking at the which is at the heart of these technologies has not had a major improvement, we wish to address this in this paper. The goal of this paper is to highlight the challenges in the current networking infrastructure for Industry 4.0 and propose a new standard for networking in Industry 4.0. Software Defined Networking is a emerging technology in the field of networking which provides for a centralised Control Plane and enables Network Function Virtualisation for Industry 4.0. OpenFlow Stack is used for building the solution. The outcomes of the paper include: Provide an SDN Architecture for the above system. Different zones have to be integrated to provide the ability for Centralized Control plane without compromising confidentiality. The system should be scalable for widespread deployment across multiple plants and manufacturers. This project deals with design and development of a Software Defined Networking (SDN) architecture for Industry 4.0, which involves integration of various components of OpenStack and openFlow. In this project, we identify the limitations and gaps in the current networking architecture for Industry 4.0 and propose a solution for the same. The openstack implementation of the project is done, Different Results of the implementation and its advantages are discussed along with scope for future implementation. This addresses a key challenge in Networking for Industry 4.0.

Keywords: Industry 4.0 · Networking · Data science · SDN · OpenStack

1 Introduction

1.1 Industry 4.0

Industry 4.0 refers to the fourth generation of advancements in Manufacturing that enables data driven processes, Predictive maintenance, Lights-Out Manufacturing among others.

Industry 4.0 integrates the Internet of Things (IoTs), cloud computing, big data and artificial intelligence. To establish smart factories, various problems have to be solved.

M. E. Auer and D. May (Eds.): REV 2020, AISC 1231, pp. 588–602, 2021.
https://doi.org/10.1007/978-3-030-52575-0_49

For example, when a company has various orders, it must implement the required, but the order in which the products need to undergo the processes and steps differs from different requirements [1].

Smart industry of the future will need to process large amounts of complex information, and production processes will depend on real-time monitoring of equipment status. To perform huge amounts of data processing and complex operations, enterprises must be able to perform many calculations, transmit data at high speed and easily manage for high performance and reliability.

One of the most significant collections of technology that will contribute to Industry 4.0 and smart manufacturing is the Industrial Internet of Things (IIoT). The IIoT is a new revolution resulting from the convergence of industrial systems with advanced computing, sensors, and ubiquitous communication systems. It is a transformative event where countless industrial devices, both old and new, are beginning to use Internet Protocol (IP) communication technologies [2, 3]. The Industrial Internet of Things is a subset of what we have come to know as the Internet of Things (IoT). The IoT is an abstract idea that captures a movement that started when we began integrating computing and communication technology into many of the "things" that we use at home and work. It started with the idea of tagging and tracking "things" with low cost sensor technologies such as radio frequency identification (RFID) devices. However, the paradigm shifted as the market began delivering low-cost computing and Internet-based communication technologies, simultaneously with the rise of the ubiquitous smartphone [4]. This perfect storm of low cost computing and pervasive broadband networking has allowed the IoT to evolve. Now, the IoT includes all types of devices ranging from home appliances, light bulbs, automation systems, watches, to even our cars and trucks. Technically speaking, the IoT is a collection of physical artifacts that contain embedded systems of electrical, mechanical, computing, and communication mechanisms that enable Internet-based communication and data exchange. The Industrial IoT follows the same core definition of the IoT, but the things and goals of the Industrial IoT are usually different. Some examples of the 'things' of the Industrial IoT include devices such as sensors, actuators, robots, manufacturing devices such as milling machines, 3D-printers, and assembly line components, chemical mixing tanks, engines, healthcare devices such as insulin and infusion pumps, and even planes, trains, and automobiles. Indeed, it is a vast spectrum of devices. Another term commonly used when discussing the Industrial IoT is operational technology. Operational technology (OT) refers to the traditional hardware and software systems found within industrial environments. Some examples include programmable logic controllers (PLC), distributed control systems (DCS), and human-machine interfaces (HMI). These systems are also known as Industrial Control Systems (ICS) because they "control" the various processes that occur within an industrial environment. These traditional control systems are rapidly beginning to use Internet-based communication technologies so that they can be integrated into manufacturing organizations' information technology (IT) systems and infrastructures. This OT/IT integration movement is currently happening in large scale across numerous industries, and it provides a technological alignment with the needs of future smart manufacturing systems and Industry 4.0 [5, 6] (Fig. 1).

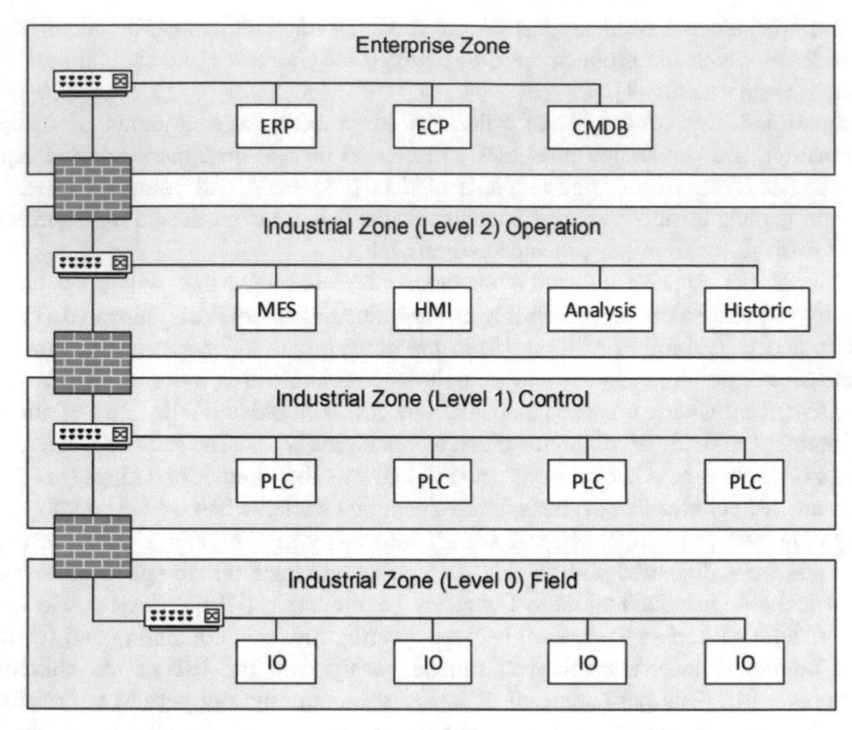

Fig. 1. Existing Industry 4.0 architecture

Here, the Level 0 refers to the Input/output Peripherals which capture the sensor data and actuate motors. Level 1 consists of Programmable Logic Controllers (PLCs) which take actions as per the algorithm written into them, they are. Level 2 is the Data Layer which stores data from the machines, displays through HMI and helps in the Analysis. Level 3 is the final layer which has Enterprise Resource Planning (ERP), which the key interface for Plant Managers/Decision makers to make decisions/manage resources. A software defined network is modelled in OpenStack framework that allows. Its performance is measured and compared against traditional networking architecture.

1.2 OpenStack

OpenStack is an open-source software platform for cloud computing mostly providing IaaS. The openstack platforms has so many services each performs different tasks. where each offers a specific service, and acts as a Software Defined Data Center (SDDC) (Fig. 2).

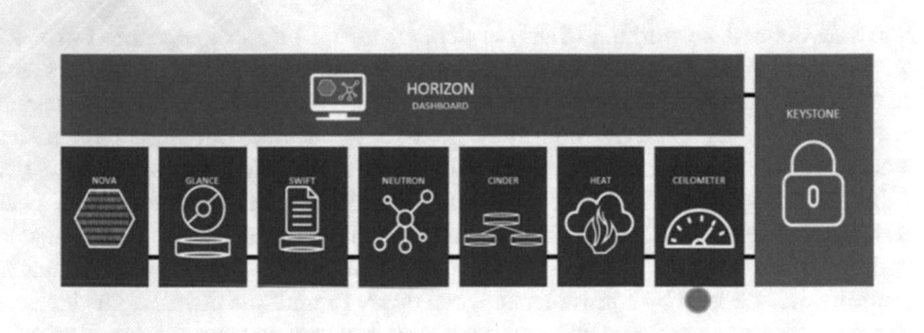

Fig. 2. Various components. Services of Openstack

It has a modular nature with different name for its components. they also have high level of integration of services within a SDDC [7, 8]. A SDDC has a data storage facility where, networking, storage, CPU and security are virtualized and delivered as a service. Each component is provisioned by API which horizon dashboard is supported. It also has architectural components which consists software defined datacenter as

- Compute
- SDN (networking)-supports load balancing as a service by NEUTRON.
- Business logic

This can be tested with openstack JUNO which has so many capabilities and features for the same, it also has keystone for user and group categories [9]. For the greater capacity needs ceilometer can be used for metering services which in turn supports separation of deployment, testing and implementation to be partial. A common antivirus and backup kept readily available. openstack also support for industry 4.0 in few relevant activities like monitoring and any vulnerable activities for the management. The installation of hot fixes and other parts of patches has to be taken into account for other process. Cinder services can be used for management of volumes and snapshots (DLCM02) for use with the Block Storage API [10].

The possible levels are "debug", "info", "audit", "warning", "error", "critical" and "trace". Horizon offers the dashboard for the end user, with limited access to the content of the log-files (LMF01). MONASCA Provides a monitoring as a service which can be implemented for SDN(INDUSTRY 4.0).

Why OpenStack for Industry 4.0?

An SDN openstack has greater capacity management functionalities than ESXi (BASIC FUNCTIONALITIES). Because ESXi is the small hypervisor for protection of log information. Here openstack has so many components each deliberately performs each function which make the system faster and faster. Also openstack follows standard mechanism for logging and other credentials which is very much useful for production services like Industry 4.0 and other IOT services.

1.3 SDN

Software-defined networking (SDN) is defined as the physical separation of the network control plane from the forwarding plane where a control plane controls several devices. To understand the ramifications of this design, one must consider the paradigm it is replacing. Particularly, non-SDN networking devices are based on a design whereby each network device is totally isolated from the other devices in its network. Although it might coordinate and work with other devices, its so-called control plane is isolated to itself and its control plane functionality cannot be modified (outside of traditional patching, upgrades, etc.) [11]. With SDN, the control plane is managed centrally, it is defined by software, and it can apply to multiple devices. The idea is that network devices have "generic" hardware that does not require vendor specific software, and the control plane functionality can be molded to fit a given design goal and can apply to multiple devices. SDN is known to be flexible, manageable, adaptive, and very cost-effective. It allows the control plane to be directly "programmable" instead of fixed software that is only "configurable".

SDN has the advantages of programmability, automation, and network control, enabling operators to establish highly scalable and flexible networks while adapting to changes in the network environment [12].

2 Current Implementation and Challenges

The current architecture of Industry 4.0 is a follow (Figs. 3 and 4):

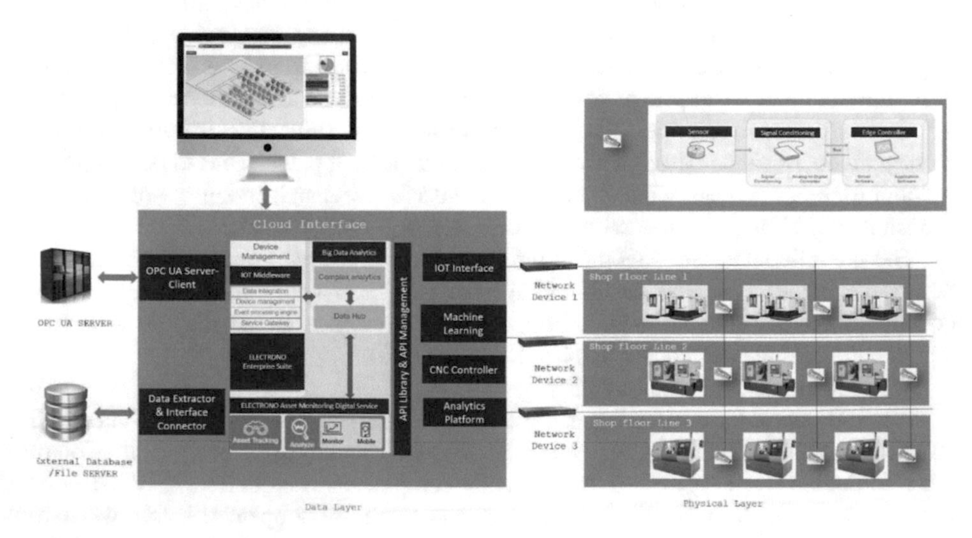

Fig. 3. Industry 4.0 architecture

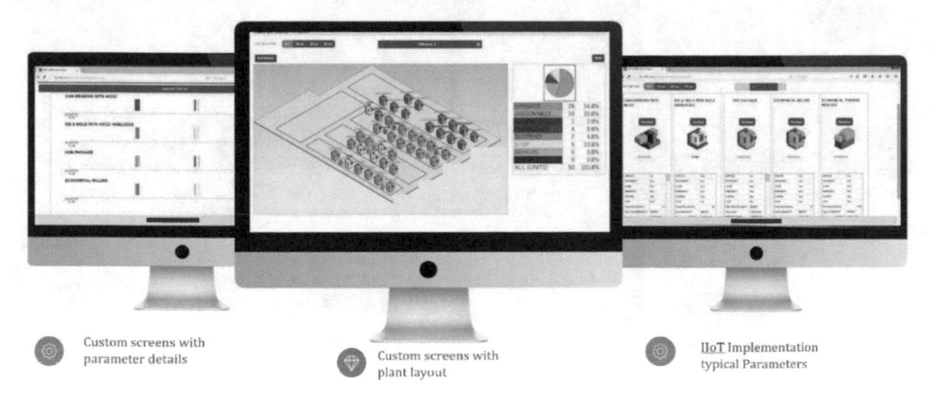

Fig. 4. Industry 4.0 implementation screens and data

In the Industry 4.0, there may be hundreds of smart devices, devices that require network connections to cooperate with each other, and to perform increasingly complex tasks without manual assistance. In addition to the proposal for an architecture that can connect the various components of Industry 4.0, from sensors to customers, there are also specific goals that would approach the requirements necessary for the operation of Industry 4.0.

- Process of Production processes have to meet with respect to production orders.
- Sometimes the process should act intelligent with different aspects as time, size and cost.
- Remote monitoring, collecting values for big data.
- Predictive maintenance and protective maintenance has to be kept high in industry 4.0. Self-configuration is a must in all scenarios.
- Energy management & security.

The goal of this paper is to highlight the challenges in the current networking infrastructure for Industry 4.0 and propose a new standard for networking in Industry 4.0. Software Defined Networking is a emerging technology in the field of networking which provides for a centralised Control Plane and enables Network Function Virtualisation for Industry 4.0 [13].

3 Implementation

3.1 Architecture

Below is a description of each component of the topology (Fig. 5):

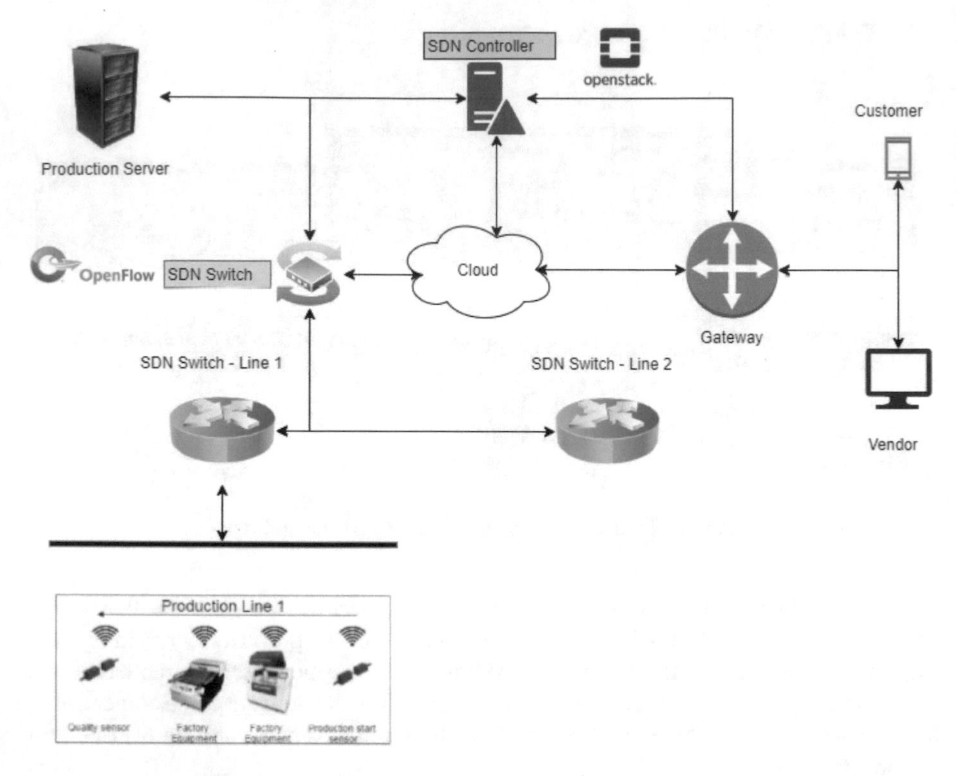

Fig. 5. Architecture of proposed SDN for Industry 4.0

1. **Production Controller:** It is the main component of the industry, it controls the entire Industry 4.0 production process. In addition, to collecting information about each product throughout the production line, the production controller makes decisions such as: load distribution, raw material request to supplier, use of idle production line, attendance of customer requests, inventory check, etc.
2. **Manufactured product:** The product is an active object in production, where each product has an Radio Frequency Identification (RFID) tag, which will be used to identify the product, from the start of production until such time that it is ready to be sent to a client.
3. **Client:** It is an agent, which might be a person or another industry, which should be able to place orders on demand and monitor the production process of your order. The manufacturing process is a response to a client request • Supplier: It is notified by Production Controller when there is a need for more raw material for production and refills it.
4. **Stock:** It is responsible for storing the raw material. When the stock is below the minimum is requested more raw material from the supplier, and during the production process provides the material to the production line.
5. **Production Start Sensor:** RFID sensor that collects product tag to begin the manufacturing process.

6. **Quality Sensor:** These Sensors are responsible for product quality control. During the manufacturing process it makes checks such as the weight and size of the product. If the product is out of the quality standard it will be discarded and the request for a new product is sent to the Production Controller.
7. **Production Equipment:** These are equipment used in the product manufacturing, which transform raw product into final product. They are also connected to the Production Controller by a switch. Eg.: laser cutting machine, laser welding system, bending machine, thermoforming machine, etc.
8. **Data Cloud:** The cloud stores all the information that is transmitted in Industry 4.0, and customers and suppliers also have access. The production control and data cloud could be implemented together, but for better organization and independence, we decided to implement them separately.
9. **SDN Controller:** It is implemented using OpenStack. SDN controller is responsible for managing the Industry 4.0 network. SDN controller exchanges information with the Production Controller to make decisions about packet forwarding, and checks networking devices status.
10. **Switch/Gateway:** It is implemented using OpenFlow protocol. In this case the SDN Switch/Gateway is managed by the SDN controller, which creates its data plan.

3.2 SDN with OPENSTACK

To achieve our main goal, we implemented an architecture of computer networks for Industry 4.0. using the paradigm of SDN to connect the components of Industry 4.0, we can mention some of the contributions:

- External communication is done with OPENSTACK integrated with SDN- gateway is provided
- Management of cloud with dynamic situation
- Speedy decision making
- Predicting failure and fault nodes

With traditional computer networks the network flow is not accepted at constant changes. when considered with IOT networks it is hard enough to predict and find a best a best architecture framework for Industry 4.0. By considering the present situation we are trying to integrate the SDN with openstack for many reasonable purpose [14, 15].

Openstack is installed on top of CentOS 7 and stein release of openstack with all the key components such Nova(Compute), Neutron(Networking) and Horizon(dashboard) along with Keystone(Authentication) are installed (Figs. 6, 7, 8, 9, 10, 11 and 12).

Not secure 192.168.1.109/dashboard/auth/login/?next=/dashboard/
IBM Knowledge Ce...
Hyperledger Comp...
Iot_Protocols
Microsoft Office Ho...
Admin console
New Tab
ICL2019 - ConfTool...
Accelerometer-GY-...
openstack.
Log in
User Name
Password
Sign In

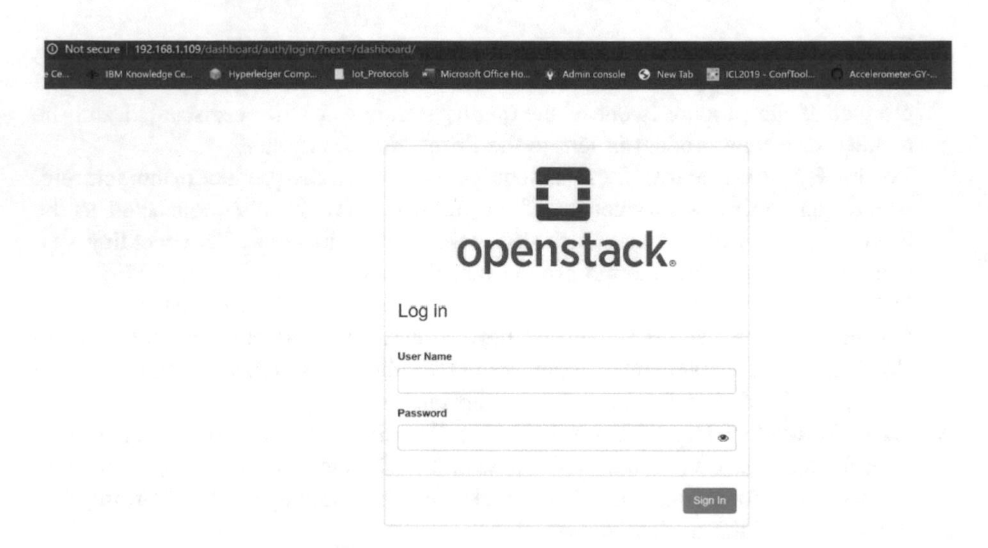

Fig. 6. Keystone authentication to OpenStack components

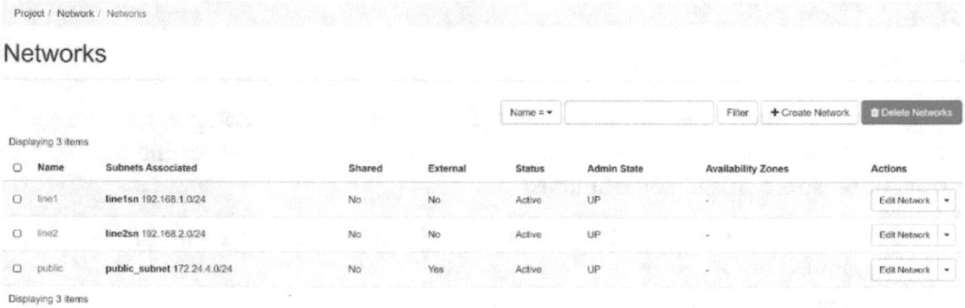

Project / Network / Networks
Networks
Name = ▾
Filter
+ Create Network
Delete Networks
Displaying 3 items
Name
Subnets Associated
Shared
External
Status
Admin State
Availability Zones
Actions
line1
line1sn 192.168.1.0/24
No
No
Active
UP
Edit Network
line2
line2sn 192.168.2.0/24
No
No
Active
UP
Edit Network
public
public_subnet 172.24.4.0/24
No
Yes
Active
UP
Edit Network
Displaying 3 items

Fig. 7. Creation of networks of different lines of production floor in Neutron

Firstly, which communicates deeply with application layer via SDN Controller (RYU OR ONOS). the openstack –SDN provided api provides best the production control over the system. It is responsible for the best decisions and management as well. Whenever the requests for decision is received it passes to the SDN controller, all this parallel process done via API. When all data is collected it passes to the datacenter which can be accessed to customers and users, with the layered architecture of openstack, it also automates the services whenever necessary, (really critical thing to consider). all this reflection is put in forward plane.

Secondly, control layer comprises of central node which is for entire network management.in the infrastructure layer we used the open flow managed by SDN controller (control plane). status of sensors and further information can be collected.

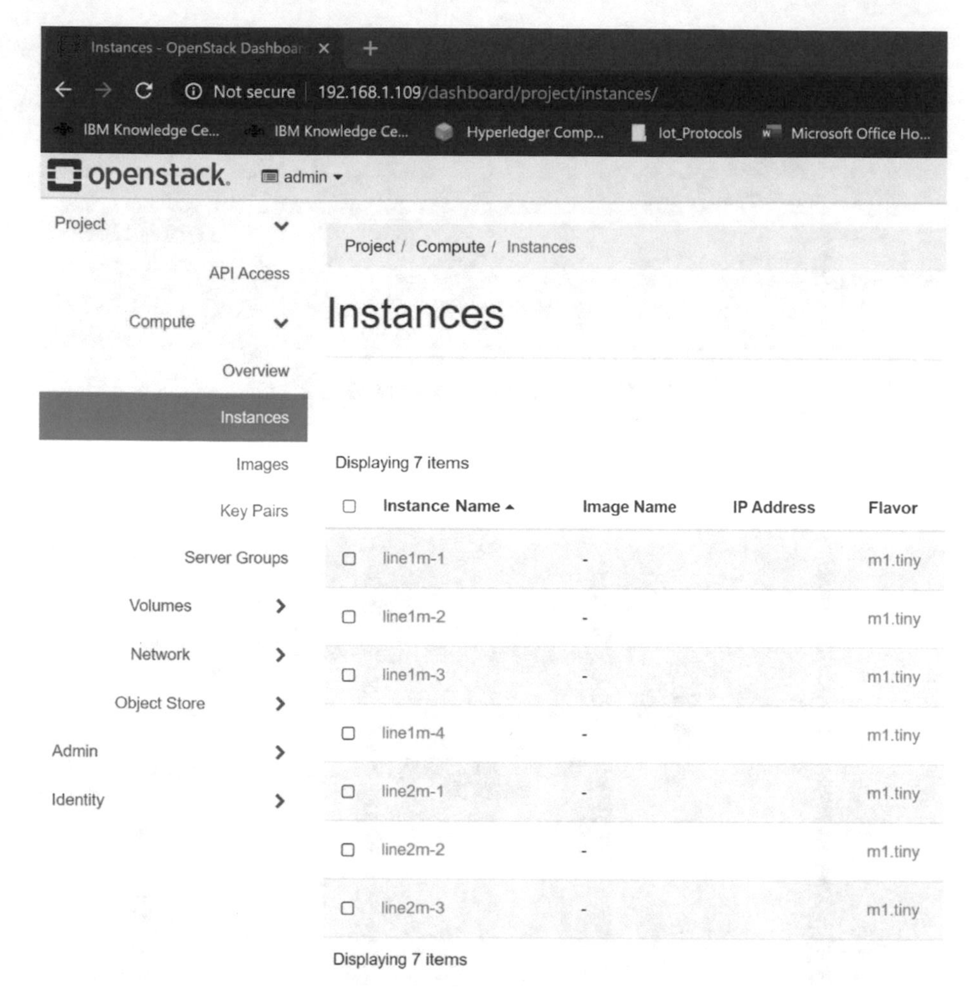

Fig. 8. Machines under different lines in Nova. Line 1 has 4 machines and Line 2 has 3 machines

Routers

Router Name = Filter + Create Router Delete Routers

Displaying 3 items

Name	Status	External Network	Admin State	Actions
line2	Active	-	UP	Set Gateway
main gateway	Active	-	UP	Set Gateway
line1	Active	-	UP	Set Gateway

Displaying 3 items

Fig. 9. Different routers/switches in the network

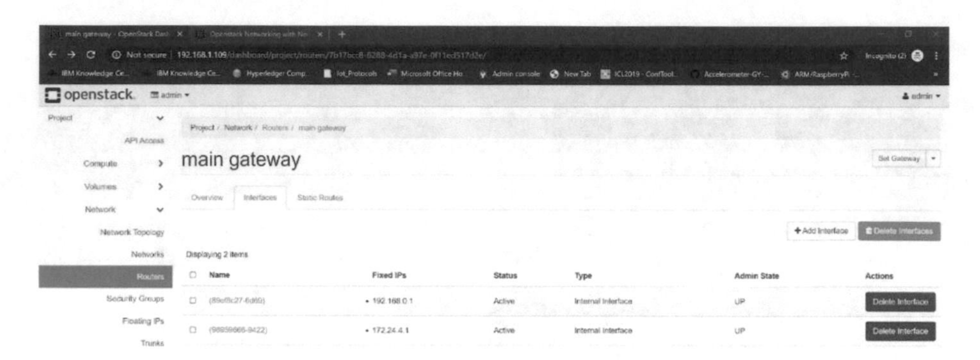

Fig. 10. Adding interfaces to switches

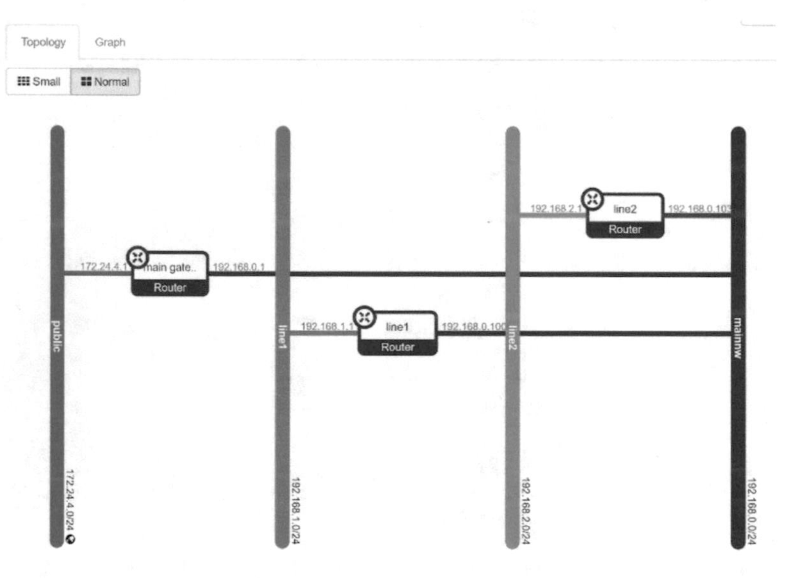

Fig. 11. Network topology

3.3 Openstack Data Cloud (ODC)

This ODC stores every information of industry 4.0 and users can have access for the same. But for better independent we implemented control and data cloud separately. This Results in flexible and effective management for better flow control of production and further resources. The integrated approach of SDN and openstack brings speed decision process in real time for production. The results shown can be implemented and can be brought so soon to industry for better convenience. The sdn working together with production controller is said to be biggest challenge for the upcoming industries.

main gateway
line2
public
mainnw
line2
line1
line1

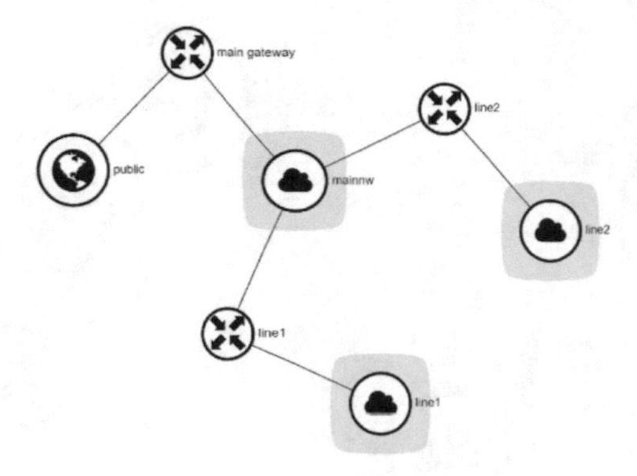

Fig. 12. Final network map of SDN architecture in OpenStack

4 Outcomes

The key outcomes of establishing a SDN Based Architecture for Industry 4.0 include:

4.1 Failover

When a Switch fail, the controller routes traffic through other switches, thus ensuring the crucial machine data is not lost due to network issues (Figs. 13 and 14).

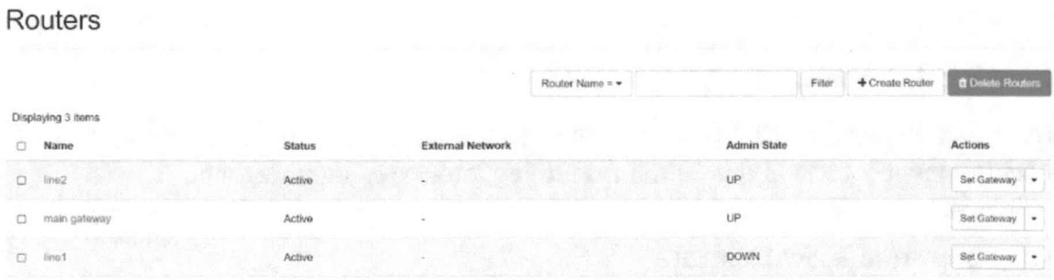

Routers
Router Name = ▾
Filter
+ Create Router
Delete Routers
Displaying 3 items
Name
Status
External Network
Admin State
Actions
line2
Active
-
UP
Set Gateway
main gateway
Active
-
UP
Set Gateway
line1
Active
-
DOWN
Set Gateway

Fig. 13. Line 1 switch is down

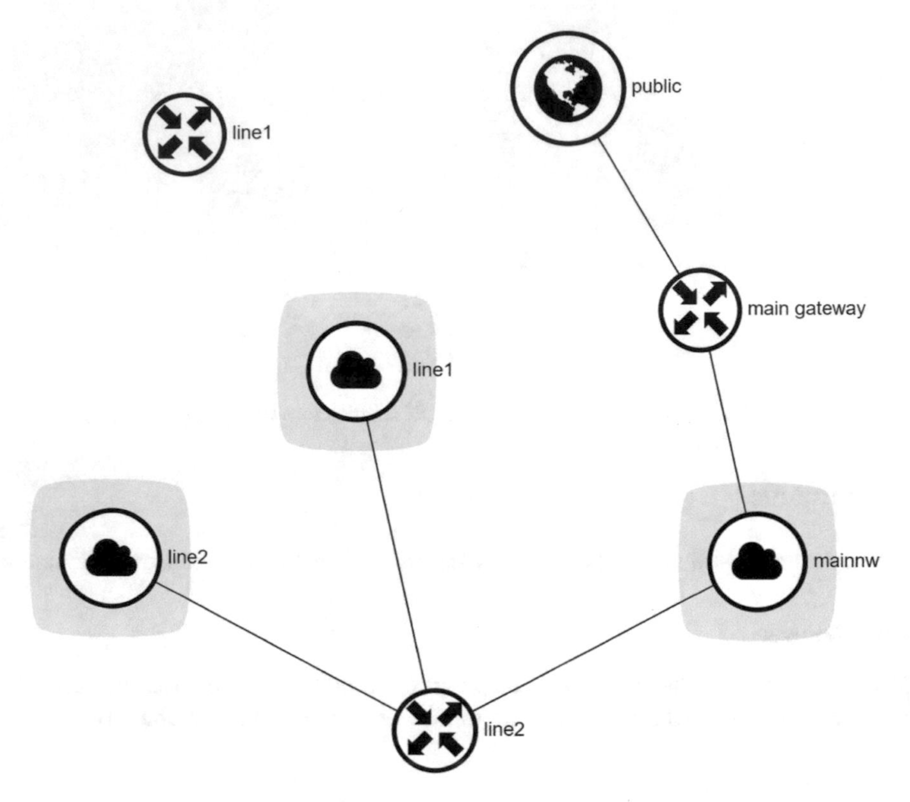

Fig. 14. Incase of failover of Line 1 switch, traffic is routed through Line 2

4.2 Load Balancing

The Load Balancing NFV can be created to manage the load of the network as the machine layouts change due to the machining processes requirements.

4.3 Improved Flow Control

Due to the centralized control plane, flow can be monitored and controlled centrally thus ensuring the reliability and efficiency.

4.4 Energy Efficiency

By Actively communicating with Switches and Hosts in the network, Controller can improve Energy Efficiency, by putting devices to sleep and waking them up when needed instead of them being always online as in a traditional network.

5 Conclusions

The paper talks about current Networking Infrastructure its shortcomings for Industry 4.0, proposes a solution of implementing solution in OpenStack, and measures the networking performances after implementation. This enables for a better networking standard for Industry 4.0.

1. Provide an SDN Architecture for the above system
2. Different zones have to be integrated to provide the ability for Centralized Control plane without compromising confidentiality.
3. The system should be scalable for widespread deployment across multiple plants and manufacturers.

Due to the steep learning curve of installing, using and debugging of components of OpenStack, mainly Failover Application was tested, in the future, other applications such as Energy Efficiency, Load Balancing NFV will be tried and tested. We wish to research and understand in depth various components of SDN and NFV and implement them in Openstack platform.

References

1. Nagy, J., Oláh, J., Erdei, E., Mate, D., Popp, J.: The role and impact of industry 4.0 and the Internet of Things on the business strategy of the value chain—the case of Hungary. Sustainability **10**, 3491 (2018). https://doi.org/10.3390/su10103491
2. Deloitte: Industry 4.0, Challenges and Solutions for the Digital Transformation and Use of Exponential Technologies. Deloitte, Swiss, Zurich (2015)
3. Gunes, V., Peter, S., Givargis, T., Vahid, F.: A survey on concepts, applications, and challenges in cyber-physical systems. KSII Trans. Internet Inf. Syst. **8**, 4242–4268 (2014). https://doi.org/10.3837/tiis.2014.12.001
4. Wan, J., Tang, S., Shu, Z., Li, D., Wang, S., Imran, M., Vasilakos, A.V.: Software-defined industrial Internet of Things in the context of industry 4.0. IEEE Sens. J., **16**(20), 7373–7380 (2016)
5. Thames, L., Schaefer, D.: Software-defined cloud manufacturing for industry 4.0. Procedia Cirp **52**, 12–17 (2016)
6. Chen, B., Wan, J., Shu, L., Li, P., Mukherjee, M., Yin, B.: Smart factory of industry 4.0: key technologies, application case, and challenges. IEEE Access **6**, 6505–6519 (2017)
7. Schluga, O., Bauer, E., Bicaku, A., Maksuti, S., Tauber, M., Wöhrer, A.: Operations security evaluation of IaaS-cloud backend for industry 4.0. In: CLOSER, pp. 392–399 (2018)
8. Ribeiro Filho, J.L., Nunes, A.C., Araújo, G., Martins, G.M., Guimarães, L.M.: A Strategy for the implementation of the Brazilian Academic Cloud (2014)
9. Kumar, R., Parashar, B.B.: Dynamic resource allocation and management using openstack. Nova **1**, 21 (2010)
10. Ahmed, S., Raja, M.Y.A.: Software defined networks for multitenant, multiplatform applications. In: 2017 14th International Conference on Smart Cities: Improving Quality of Life Using ICT & IoT (HONET-ICT), October 2017, pp. 62–67. IEEE (2017)
11. Ma, Y.W., Chen, Y.C., Chen, J.L.: SDN-enabled network virtualization for industry 4.0 based on IoTs and cloud computing. In: 2017 19th International Conference on Advanced Communication Technology (ICACT), February 2017, pp. 199–202. IEEE (2017)

12. Lins, T., Oliveira, R.A.R.: Energy efficiency in industry 4.0 using SDN. In: 2017 IEEE 15th International Conference on Industrial Informatics (INDIN), July 2017, pp. 609–614. IEEE (2017)
13. Li, X., Li, D., Wan, J., Liu, C., Imran, M.: Adaptive transmission optimization in SDN-based industrial Internet of Things with edge computing. IEEE Internet of Things Journal **5**(3), 1351–1360 (2018)
14. Pashkov, V., Shalimov, A., Smeliansky, R.: Controller failover for SDN enterprise networks. In: 2014 International Science and Technology Conference (Modern Networking Technologies) (MoNeTeC), October 2014, pp. 1–6. IEEE (2014)
15. Zhou, Y., Zhu, M., Xiao, L., Ruan, L., Duan, W., Li, D., Zhu, M.: A load balancing strategy of SDN controller based on distributed decision. In 2014 IEEE 13th International Conference on Trust, Security and Privacy in Computing and Communications, September 2014, pp. 851–856. IEEE (2014)

Using Constructive Alignment to Evaluate Industry 4.0 Competencies in Remote Laboratories for Manufacturing Technology

Claudius Terkowsky[1(✉)], Silke Frye[1], and Dominik May[2]

[1] TU Dortmund University, Dortmund, Germany
{claudius.terkowsky,silke.frye}@tu-dortmund.de
[2] University of Georgia, Athens, USA
dominik.may@uga.edu

Abstract. Digitalization, smartification and cyber-physical systems are going to reshape the working lives of future engineers. Thus, these developments will have an impact on higher education in the near future. For years, there has been a vivid discussion on the respective competencies students need to develop in order to be able to successfully face these challenges. In this article, the laboratory is presented as a cyber-physical system that can offer a setting to foster these competencies in engineering laboratory courses. We present the remote laboratory as a setting in which future-oriented teaching and learning in the light of required competencies for the 'working world 4.0' can take place. In order to identify these competencies and to analyze an existing laboratory, a qualitatively oriented content-analytical procedure is used. In a first step, based on current scientific studies, competencies expected to be shown by future engineers are named and summarized. In a second step, an exemplary existing remote laboratory is analyzed with regard to explicit and implicit learning objectives, which address the identified competency requirements. It can be shown that only a few professional and interdisciplinary competencies in the context of Industry 4.0 are being addressed so far. Based on the focused example, the close connection to engineering fundamentals on the one hand and the lack of interdisciplinarity in the observed learning scenario on the other hand can be identified as critical parameters, which limit the promotion of more future-oriented competencies. This shows possibilities for future research and development in this area.

Keywords: Laboratory education · Working 4.0 · Remote laboratories · Industry 4.0

1 Introduction

Over the past 150 years, laboratory work has become an integral part of engineering courses worldwide. Existing courses are of great importance in research oriented as well as applied degree programs. In high-industrialized countries, laboratory education is comparatively extensive and will become even more important in the future as technological change and the digital transformation of the economy create a new

M. E. Auer and D. May (Eds.): REV 2020, AISC 1231, pp. 603–613, 2021.
https://doi.org/10.1007/978-3-030-52575-0_50

working world for which engineering students not only have to be technically theoretical but also practically prepared [1].

According to the results of a recent study on "Higher Education for the Working World 4.0", the digital transformation will change jobs, products and modes of production. Working and learning with digital technologies must therefore become an integral part of the academic competency profile. However, for higher education the emerging 'working world 4.0' does not mean fundamental replacement of previous educational goals, but their extension [2]:

> *"With such an academic profile, the skills added to digital skills continue to form the basis for a scientific, career-oriented, and personality-building study. For the working world 4.0, however, the relevance to application (due to the increasing integration of academic and professional competencies) and the formation of personality (due to the new, cooperative forms of working) are becoming more important than before"* [2].

The research question of this contribution focusses on whether laboratory learning offers potential for the development of these competencies, since industrial activities can be practically implemented there. However, this requires new learning content and didactic formats related to Industry 4.0 [2]. For this, the following questions are considered:

- Which competencies does Industry 4.0 require?
- How are these competencies promoted in laboratory education thus far?
- How can the acquisition of competencies for Industry 4.0 in higher education laboratory education be fostered in the future?

In order to identify the required competencies in Industry 4.0 as well as the exemplary analysis of an existing teaching and learning laboratory, a qualitatively oriented content-analytical approach has been carried out.

2 Theory

Recent studies on the effectiveness of learning in the laboratory show that active learning in the laboratory can specifically support the development of subject-related and interdisciplinary competencies [6–8]. However, a great variety of laboratory exercises, experiment instructions, and study materials for students are still based on traditional instructive laboratory methods. Lacking what is called "constructive alignment" [3], they are designed to be less supportive for innovation and thus promote outcome based teaching and learning required for successful competency-based work in Industry 4.0 only marginally [8–10].

2.1 The 'Cyber-Physicalization' of the Laboratory in Higher Science and Engineering Education

The digital transformation has far-reaching consequences for laboratory-didactic formats:

- On the one hand, the progress of knowledge in the area of Industry 4.0 is increasingly the subject matter of research and teaching, for example in production and automation technology or logistics (keyword: Industry 4.0 at the university).
- On the other hand, tools of Industry 4.0 are increasingly permeating university research and higher education (keyword: Industry 4.0-ization of the university).

This means that in order to address the overarching objective 'dealing with digital requirements in the occupational field', both content-related and media technology-structural adjustments with regard to Industry 4.0 are indispensable. There is a large number of cyber-physical laboratory systems and laboratories that are used via laboratory portals in teaching and research (see i.e. GoLab, Labster, FED4FIRE, LabsLand, VISIR). Regarding the technological settings, [4] write the following:

> *"Today, almost all definitions of cyber-physical laboratories, ..., involve either monitoring, controlling, or twinning an object in the physical world by means of software algorithms which permit the dynamic interaction between said object and the real world, maintained through either cabled or wireless communications to computer-based resources. Also, digital twins and simulations are widely used in the online laboratory field. Of course, this implies that major advantages of cyber-physical laboratories are that they are scalable, often shared resources that are not constrained by spatial-temporal considerations"* [4].

The interaction with cyber-physical experimentation facilities, i.e. the actual 'experimentation', can take place locally or arbitrarily remote while remotely controlled as a web service. Moreover, computer-generated interactive simulations can replace the physical 'hands on' laboratories and experiments [5, 6].

As the aim of this paper is to review the laboratory as a learning environment in higher education with the background of increasing 'cyber-physicalization' in terms of its suitability for promoting relevant competencies, the following section outlines which competences are currently required for Industry 4.0.

2.2 Constructive Alignment to Design and Evaluate Competencies as Learning Outcomes

In general, the term 'competence' describes a person's capability to do something adequately, or a person's mental capacity to understand the proceedings of a trial. According to the European Quality Framework for Lifelong Learning [7].

> *"Competence is the proven ability to use knowledge, skills and personal, social and/or methodological abilities, in work or study situations and in professional and personal development. In the context of the European Qualifications Framework, competence is described in terms of responsibility and autonomy"* [7].

Our aim is to ask and answer 1. whether and to what extent an online teaching and learning laboratory can promote the needed development of competencies; 2. what is already being implemented; 3. which further developments will be needed in the future. This is done by considering the concept of "Constructive Alignment" [3]. Constructive Alignment (CA) is an evaluation and design approach for teaching and learning, which has become increasingly widespread in higher education in recent years. Belonging to the outcome-based teaching and learning approaches (OBTL), Constructive Alignment focuses not on what a teacher wants to teach, but on which outcomes a learner is intended to achieve based on the teacher's teaching activity and the resulting learner's learning activity. Teaching is then designed to engage students in learning activities that optimize their chances of achieving those outcomes, and assessment tasks are designed to enable clear judgments as to how well those outcomes have been attained [8].

In the language of CA, we analyzed the intended learning outcomes (ILOs) the laboratory's developers set themselves for an exemplary laboratory. Subsequently, we evaluated to what extend these ILOs currently match with possible competencies required for the working world 4.0. We did that by running a brief case study deploying a qualitative content analysis.

3 Method: Qualitative Content Analysis

In order to identify the required competencies in the field of Industry 4.0 and to analyze an existing teaching and learning laboratory as an example, a qualitatively oriented content analysis approach has been carried out. During this approach the analysis focuses on both being led by a predefined category framework and developing new categories at the same time. The texts are systematically summarized and analyzed based on theory and rules as 'close to the text' as possible. This is particularly suitable for investigations of debates and literature [9].

In a first step, studies are explored to clarify which competencies are needed to succeed in an Industry 4.0-based economy. From this, an overall competency grid is derived in the second step. In a third step this grid is used to analyze an existing teaching and learning remote laboratory as a brief reality check. Because of its exploratory character, this exemplary analysis has only limited validity and portability, and it does not initially claim to be fully generalizable. The developed methodology and its implementation can easily be transferred to other online laboratories, thus expanding the database in the future.

4 Results

4.1 Competency Profiles for Industry 4.0

There are several resources that provide clues at different levels of abstraction about what skills are expected in the future. In parts, these sources rely on their own studies, mostly in the form of surveys of company representatives (For a detailed discussion, see also: [10–12]).

Respectively, they are:

- a 'Higher Education Report 2020' of the Donors' Association for the Promotion of Humanities and Sciences in Germany [2]
- a position paper on 'Competencies for Industry 4.0 - Qualification requirements and solutions' of the National Academy of Science and Engineering [1]
- a study of the Fraunhofer Institute for Ergonomics and Organization [13]

- a study on 'Industry 4.0 - Qualification 2025' of the Mechanical Engineering Industry Association [14], and
- a meta-analysis of 24 studies from 2014 to 2016 [15]

The summary of the relevant competencies bases on the framework model of the German Qualifications Framework for Lifelong Learning. The term competency refers to "the ability and willingness of the individual to use knowledge and skills as well as personal, social and methodical abilities and to behave thoughtfully as well as individually and socially responsible". In addition, the grid differentiates between subject-specific and interdisciplinary technical competencies (with regard to concrete technologies and organizational structures), social competencies (with regard to social interaction structures) and self-competencies (with regard to individual personality structures). Since these are learning outcomes, i.e. intended learning outcomes, competencies are defined in this sense [16].

Based on the listed studies, the following competencies are of particular relevance to the context of Industry 4.0:

Domain-Specific and Generic Technical Competencies. In the context of domain-specific and generic technical competency, learners should be able to meet the requirements of Industry 4.0. These are:

- to act and collaborate in interdisciplinary contexts
- to flexibly adapt business processes to changeable conditions to new technologies i.e. such as additive manufacturing or augmentation
- to design IT processes in the context of production and to use IT components for human-machine interaction
- to design and to control holistic and complex production processes and networked production structures as well as to manage appropriate interfaces (including the implementation of problem-solving and optimization processes)
- to establish a connection between a digital twin and its physical reality
- to deal with large amounts of data and to use appropriate statistical skills (including recognizing the importance of algorithms and the management of sensitive data)
- to demonstrate system competency by recognizing functional elements, identifying system boundaries and making predictions about system behavior
- to initiate and to implement innovation processes
- to control the legal context of the entrepreneurial act
- to act strategically in a company-specific way, and
- to use the appropriate evaluation tools in complex decision-making situations

Social Competencies. In the context of social skills, learners should be able to meet the requirements of Industry 4.0. These are

- to communicate business fluently and to cooperate, both internally (in terms of process flows) and externally (in terms of customers and supplier relations)
- to act confidently and effectively in social (including intercultural) contexts
- to lead production units and teams with goal orientation
- to design digitally supported interaction and cooperation processes

Self-competencies. In the context of self-competencies, learners should be able to meet the requirements of Industry 4.0. These are:

- to realistically assess the value of one's own subjective knowledge of experience and incorporate it accordingly into one's own action
- to act self-determined and self-organized
- to act on the basis of one's own open mindedness and creativity
- to design and implement your own lifelong learning

4.2 Reality Check by Means of an Exemplary Case Analysis

The following section analyses the extent to which the promotion of these skills for working world 4.0 is addressed within existing laboratory concepts. As an example, an engineering laboratory being considered is assigned to the field of production or production technology. The laboratory under consideration is used in the context of engineering bachelor and master's programs. Thematically, the laboratory deals with the characterization of metallic materials in the context of forming technology. Students plan a uniaxial tensile test, carry it out and analyze material characteristics from the recorded measurements. In addition to working with the test setup on site, it is also available as a remotely accessible online laboratory [17]. Both use cases are considered below.

The first step is the selection of the material to be evaluated. By integrating into the ELLI (Excellent Teaching and Learning in Engineering, funded by the German Federal Ministry Education and Research) project, there is a sufficient number of publications devoted to the content, technical, organizational and pedagogical design of the selected laboratory. Subjects of the analysis are conference contributions, book chapters and further contributions to other scientific bodies, as well as one dissertation. In order to assess the origin of the texts, it should be noted that no article directly addressed the topic of Industry 4.0, but all of them were focused on technical subjects or higher engineering education. A total of 15 publications from the past six years were evaluated (see: [17–31]). As an analytical technique, the structured content analysis was chosen to identify the learning objectives of the laboratory explicitly and implicitly in the texts. To this end, all relevant content-bearing passages were paraphrased. In eight of the 15 examined publications, statements about learning objectives were found. A total number of 29 relevant passages were identified from these publications, and after an initial reduction, 23 objectives of the laboratory could be coded.

Nine codes belong to the general objective that students plan and carry out experiments. This is referenced 5 times to the on-site experiment and 4 times to the tele operative experiment. Six codes correspond to the objective that students gain practical experience in the use of technical equipment and laboratory equipment:

"[…] our aims are that students *get into contact with real technical equipment*, understand the greater context of research and *gain technical competencies* for their future work" [26].

Each four codes refer to the technically correct recording and evaluation of measured data and characteristic values as well as the acquisition of 'problem-solving abilities':

They "*learn how to calculate the bending moment out of the bending force* in a real process" [29].

The next step was to examine the extent to which these objectives can be assigned to the required competencies in the context of Industry 4.0. To this end, the competencies formulated in 4.1 were coded as categories. Subsequently, the identified objectives were compared with these competency categories. It was possible to assign 14 of the 23 objectives to the categories. It is striking that many of the objectives of the laboratory are in the areas of social and self-competencies. Thus, seven learning objectives can be assigned to self-organized and self-determined action. Examples are the learning objectives coding 'own planning/organization of the experiment':

"Hence, the students have *to develop a working plan* in their group when to do the experimentation and *arrange the experimentation* by booking a time slot" [26].

To next come the learning objective coding 'self-organized learning':

"So students could learn about superposition of stresses *by themselves* room [sic!] and time independent" [29].

The competence 'confident internal and external communication and cooperation' can be attributed to six objectives. Examples are the learning objective coding 'communication/presentation of procedure/results'.

"After completion of the laboratory, students should be able [...] *to present* scientifically and in line with the target audience the experiments, the *procedure* and the *results* in a laboratory report, and to be able *to defend their findings in a scientific discussion*" [17].

Next to come is the learning objective coding 'group collaboration':

"They are asked *to work on this problem in small groups* by planning and carrying out experiments using the tele-operated equipment" [20].

In total, all four categories in the field of self-competency and 3 out of 4 categories in the field of social competencies are addressed. The subject-specific and interdisciplinary competencies could only be assigned four objectives, whereby only two of the 10 categories are addressed.

One example is the learning objective code 'technical process understanding', which is assigned to the competency category 'system competence':

"After finishing the laboratory coursc, students should be able to [...] *describe the problem regarding the requirements for the production* of a set *radius with the principles of forming technology on the basis of the observed forming process and to analyze based on their previous knowledge and to develop hypotheses.*" [17], as well as the learning objective code that can be assigned to the category 'apply IT components for human-machine interaction':

"They are asked to work on this problem in small groups by planning and carrying out experiments *using the tele-operated equipment*" [20].

5 Discussion

For the laboratory, a variety of explicit and implicit learning objectives can be identified which directly address the formulated competence requirements in the context of Industry 4.0. In particular, the areas of social and self-competencies are addressed, which is due to the didactic and methodological design of the laboratory. Future engineers are expected to work in groups, and the resulting challenges in communication and interaction are a good way to promote social skills for professional situations. The laboratory follows the model of experiential learning, allows for creativity and learning from mistakes, and fosters self-organization. For example, it supports the development of students' self-competencies as required in the context of Industry 4.0. The technical design of the laboratory shows the influence of the remote laboratory. This concept brings 'something digital' into the laboratory, which is essential in terms of the working world 4.0 both in the field of subject-specific (use of IT components for human-machine interaction) as well as in the field of social competencies (digitally based interaction and cooperation processes). It can also be assumed that the ability to link the digital twin with a physical reality is promoted, even if it was not formulated as a laboratory learning objective.

However, the results also show that only a few categories in the subject-specific and interdisciplinary competencies are addressed in the context of Industry 4.0 in the laboratory under study. The unavoidable systematic approach in the sense of the 'one right way' to implement and evaluate the experiment offers hardly any possibilities for 'innovations' from the students. The limited subject-specific design hinders the promotion of interdisciplinary thinking and acting. This also implies the absence of upstream and downstream processes to map holistic structures, interfaces and complex decision-making situations.

From this, it is possible to derive recommendations with regard to competence orientation in the context of Industry 4.0.:

- It can be essential to supersede the integration of the laboratory in individual disciplines and to design or embed it in a cross disciplinary context.
- Connecting with other laboratories under a common, wider problem can initiate a more comprehensive, complex teaching-learning scenario. Here, 'digitization' as a remote laboratory opens up a multitude of possibilities.
- If the given problem is aligned less to subject-specific basic knowledge, but more to the practical context of future engineers, this can foster the development of system competency and the ability to operate in complex and networked structures.

Finally, it should be mentioned that nine coded objectives of the considered laboratory cannot be assigned to the competencies in the context of Industry 4.0. They refer in particular to the "classical academic educational goals" of engineering education [2]:

- general technical skills and practical experience in handling technical equipment
- handling and using of measured values, their evaluation and technical interpretation
- connection of theory and practice

In the context of the laboratory, these objectives do not stand in the way of the required competencies in the context of Industry 4.0. This shows that the training of engineers for a working world 4.0 does not mean that only 'new' competencies requirements are relevant. They do not replace the previous educational goals, but complement them and further develop them with a view to increasingly digitized and networked systems and processes.

6 Conclusion and Outlook

This article provides an overview of teaching and learning in the laboratory, suggests possible new requirements of a working world 4.0, and summarizes the expected competencies of future engineering in the context of Industry 4.0.

As a way to foster these competencies in the course of studies, the laboratory was introduced. The extent to which the technical, pedagogical and methodical design of the laboratory as teaching-learning setting already addresses the competencies required in the context of Industry 4.0 was analyzed using a case study as an example It could be shown that this reduced pedagogical setting's potential to account for the complexity and 'digitization' (in the form of a cyber-physical remote laboratory) of the working world 4.0 opens up a multitude of possibilities. However, the results also show that in the examined laboratory, only a few competencies from the technical and interdisciplinary technical area are addressed in the context of Industry 4.0, because the lack of interdisciplinarity significantly limits the promotion of these competencies.

For a general assessment of the potential of laboratories for preparing students for working world 4.0, further research and comparative analysis with other laboratory approaches is needed.

Acknowledgment. The presented work was done in the scope of the research project "ELLI – Excellent Teaching and Learning in Engineering Sciences", funded by the German Ministry of Education and Research (project number: 01PL16082).

References

1. Acatech (ed.): Kompetenzen für die Industrie 4.0: Qualifizierungsbedarfe und Lösungsansätze. München: Utz, Herbert (2016)
2. Stifterverband, Hochschulbildung für die Arbeitswelt 4.0: Jahresbericht 2016. Edition Stifterverband - Verwaltungsgesellschaft für Wissenschaftspflege mbH, Essen (2016)
3. Biggs, J., Tang, C.: Teaching for Quality Learning at University: What the Student Does, 4th edn. McGraw-Hill/Society for Research into Higher Education/Open University Press, Maidenhead (2011)
4. Auer, M.E., Azad, A.K.M., Edwards, A., de Jong, T. (eds.): Cyber-Physical Laboratories in Engineering and Science Education. Springer, Cham (2018). https://doi.org/10.1007/978-3-319-76935-6
5. Terkowsky, C., et al.: Developing tele-operated laboratories for manufacturing engineering education platform for e-learning and telemetric experimentation (PeTEX). Int. J. Onl. Eng. **6**(5), 60–70 (2010)
6. Terkowsky, C., Pleul, C., Jahnke, I., Tekkaya, A.E.: Tele-operated laboratories for online production engineering education - platform for e-learning and telemetric experimentation (PeTEX). Int. J. Onl. Eng. **7**(S1), 37–43 (2011)
7. European Commission: The european qualifications framework for lifelong learning (EFQ). Office for Official Publications of the European Communities, Luxembourg (2008)
8. Biggs, J.: Constructive alignment in university teaching. HERDSA Rev. High. Educ. **I**, 5–22 (2014)

9. Mayring, P.: Qualitative Inhaltsanalyse: Grundlagen und Techniken, 12th edn. Beltz, Weinheim (2015)
10. Terkowsky, C., Frye, S., May, D.: Online engineering education for manufacturing technology: is a remote experiment a suitable tool to teach competences for “Working 4.0”? Eur. J. Educ. **54**(4), 577–590 (2019)
11. Terkowsky, C., Frye, S., May, D.: Is a remote laboratory a means to develop competences for the ‘working world 4.0’? A brief tentative reality check of learning objectives. In: 2019 5th Experiment International Conference (exp.at’19), Funchal (Madeira Island), Portugal, 2019, pp. 118–122 (2019)
12. Terkowsky, C., Frye, S., May, D.: Labordidaktik: Kompetenzen für die Arbeitswelt 4.0. In: Haertel, T., Terkowsky, C., Dany, S., Heix, S. (eds.) Hochschullehre & Industrie 4.0: Herausforderungen - Lösungen - Perspektiven, 1st edn., pp. 89–103. wbv Media, Bielefeld (2019)
13. Schlund, S., Pokorni, W.: Industrie 4.0 - Wo steht die Revolution der Arbeitsgestaltung? http://publica.fraunhofer.de/dokumente/N-432393.html. Accessed 01 Sep 2018
14. Pfeiffer, S., Lee, H., Zirnig, C., Suphan, A.: Industrie 4.0 - Qualifizierung 2025. https://arbeitsmarkt.vdma.org/viewer/-/v2article/render/13668437. Accessed 01 Sep 2018
15. Hartmann, F.: Zukünftige Anforderungen an Kompetenzen im Zusammenhang mit Industrie 4.0 – Eine Bestandsaufnahme. In: Facharbeit und Digitalisierung, Verbundprojekt Prokom 4.0, pp. 19–28 (2017)
16. AK DQR: Deutscher Qualifikationsrahmen für lebenslanges Lernen. https://www.dqr.de/media/content/Der_Deutsche_Qualifikationsrahmen_fue_lebenslanges_Lernen.pdf. Accessed 01 Sep 2018
17. Pleul, C.: Das Labor als Lehr-Lern-Umgebung in der Umformtechnik: Entwicklungsstrategie und hochschuldidaktisches Modell, 1st edn. Shaker, Aachen (2016)
18. Terkowsky, C., May, D., Haertel, T., Pleul, C.: Integrating remote labs into personal learning environments - experiential learning with tele-operated experiments and e-portfolios. Int. J. Onl. Eng. **9**(1), 12–20 (2013)
19. Terkowsky, C., Haertel, T., Bielski, E., May, D.: Bringing the inquiring mind back into the labs a conceptual framework to foster the creative attitude in higher engineering education. In: 2014 IEEE Global Engineering Education Conference (EDUCON), Istanbul, pp. 930–935 (2014)
20. May, D., Terkowsky, C., Haertel, T., Pleul, C.: Using E-Portfolios to support experiential learning and open the use of tele-operated laboratories for mobile devices. In: 2012 9th International Conference on Remote Engineering and Virtual Instrumentation (REV), Bilbao, Spain, pp. 1–9 (2012)
21. Terkowsky, C., Haertel, T.: Fostering the creative attitude with remote lab learning environments: an essay on the spirit of research in engineering education. Int. J. Onl. Eng. **9**(S5), 13 (2013)
22. Haertel, T., Terkowsky, C., May, D., Pleul, C.: Entwicklung von Remote-Labs zum erfahrungsbasierten Lernen. In: Frerich, S., et al. (eds.) Engineering Education 4.0, pp. 105–112. Springer, Cham (2016). https://doi.org/10.1007/978-3-319-46916-4_9
23. May, D., Terkowsky, C., Haertel, T., Pleul, C.: The laboratory in your hand: making remote laboratories accessible through mobile devices. In: 2013 IEEE Global Engineering Education Conference (EDUCON), Berlin, pp. 335–344 (2013)
24. Ortelt, T.R., et al.: Development of a tele-operative testing cell as a remote lab for material characterization. In: 2014 International Conference on Interactive Collaborative Learning (ICL), Dubai, UAE, pp. 977–982 (2014)

25. Terkowsky, C., Haertel, T., Bielski, E., May, D.: Creativity@School: mobile learning environments involving remote labs and e-portfolios. A conceptual framework to foster the inquiring mind in secondary STEM education. In: García-Zubía, J., Dziabenko, O. (eds.) IT Innovative Practices in Secondary Schools: Remote Experiments, pp. 255–280. University of Deusto, Bilbao, Spain (2013)
26. May, D., Ortelt, T.R., Tekkaya, A.E.: Using remote laboratories for transnational online learning environments in engineering education. In: E-Learn: World Conference on E-Learning in Corporate, Government, Healthcare, and Higher Education 2015, pp. 632–637 (2015)
27. May, D., Sadiki, A., Pleul, C., Tekkaya, A.E.: Teaching and learning globally connected using live online classes for preparing international engineering students for transnational collaboration and for studying in Germany. In: 2015 12th International Conference on Remote Engineering and Virtual Instrumentation (REV), Bangkok, Thailand, pp. 118–126 (2015)
28. Sadiki. A., et al.: The challenge of specimen handling in remote laboratories for engineering education. In: 2015 12th International Conference on Remote Engineering and Virtual Instrumentation (REV), Bangkok, Thailand, 2015, pp. 180–185 (2015)
29. Meya, R., et al.: Development of a tele-operative control for the incremental tube forming process and its integration into a learning environment. In: 2016 IEEE Global Engineering Education Conference (EDUCON), Abu Dhabi, UAE, pp. 80–86 (2016)
30. Selvaggio, A., et al.: Development of a cupping test in remote laboratories for engineering education. In: 13th International Conference on Remote Engineering and Virtual Instrumentation (REV), Madrid, Spain, pp. 122–126 (2016)
31. Ortelt, T.R., et al.: Concepts of the international manufacturing remote lab (MINTReLab): combination of a MOOC and a remote lab for a manufacturing technology online course. In: 2016 IEEE Global Engineering Education Conference (EDUCON), Abu Dhabi, UAE, pp. 602–607 (2016)

Work-in-Progress: Machine Development Using Virtual Commissioning

Hasan Smajic[1(✉)] and Jean Bosco[2]

[1] University of Technology, Arts and Sciences, Cologne, Germany
hasan.smajic@th-koeln.de
[2] Department for Mechatronics, Dedan Kimathi University of Technology, Kenyatta-Nairobi, Kenya
smscp.kenya@dkut.ac.ke

Abstract. Due to the changes around industry 4.0 and digital production many work processes are changing and it has to be clarified which of these new possibilities brings added value in which situations compared to the "old" approaches. In mechanical engineering and in the process industry, it is possible to create a virtual twin of a machine or plant. With the help of this twin, the control software can be completed and tested at a much earlier stage. This leads to a reduction of effort in case of an error in the software, as bugs can be detected and eliminated at a much earlier stage of the project. A fundamental design bug in the control software during commissioning on the real machine would have a major impact on completion and associated costs. In addition, design bugs can be detected before production and a functional solution can be worked out together with the design department. In this publication, a solution concept is worked out and implemented with which a virtual commissioning of the virtual twin is carried out. As an example for the implementation of the virtual concept, the control of a high-bay warehouse was developed.

Keywords: Industry 4.0 · Mechatronic design · Virtual commissioning

1 Introduction

The strong globalization of the economy in recent decades has confronted many companies with new challenges. The increasing competition forced the companies to offer a larger product portfolio with constant high quality. At the same time, however, customers are becoming more and more demanding, which often leads to highly complex unique systems, especially in mechanical engineering. By using numerous simulation tools, many machine builders nowadays extensively test the mechanics and the process flow of production plants. Since the systems are increasingly equipped with extensive electronics and control technology, appropriate simulations have become necessary in order to be able to deliver machines error-free and on schedule (Lacour 2011, pp. 40–41). The quality requirements for the built-in automation technology also continue to rise. The safety requirements for technical systems are constantly updated and the functionalities required to comply with standards, laws and directives are largely based on automation components (VDI 3693-1 2016, p. 3). In order to remain

M. E. Auer and D. May (Eds.): REV 2020, AISC 1231, pp. 614–623, 2021.
https://doi.org/10.1007/978-3-030-52575-0_51

competitive in the future, commissioning and start-up of production must therefore be carried out faster and faster (Zäh, et al. 2006, p. 595). The first companies are now relying on extensive virtual commissioning of their machines in order to test control software before actual commissioning and to detect engineering errors as early as possible.

2 Virtual Commissioning

The core idea of virtual commissioning is to test the control program of a machine on a virtual CAD model in order to detect and eliminate errors in the automation system that have been caused by engineering. The VDI Guideline 3693 Sheet 1 literally describes this as follows: "Virtual commissioning is the "commissioning which comprises the testing of individual components and subfunctions of the automation system during development with the aid of simulation methods and models adapted to the respective task" (VDI 3693-1, 2016, p. 5). This ensures that the actual commissioning takes place with a control program of significantly higher quality (Wünsch 2007, p. 30).

The first requirement for a virtual commissioning is a comprehensive CAD model of the machine. This modelling phase always causes additional effort. However, individual assemblies can often subsequently be saved in libraries, allowing them to be reused for new models. In most software applications, the basic structure of the model can be derived directly from the CAD data. Once modeling is complete, the model is connected to a real or simulated PLC and the control program is imported. This allows a large number of processes to be tested before the actual machine is assembled. Thus, the actual commissioning runs much more smoothly and can be completed with a corresponding time saving (Fig. 1).

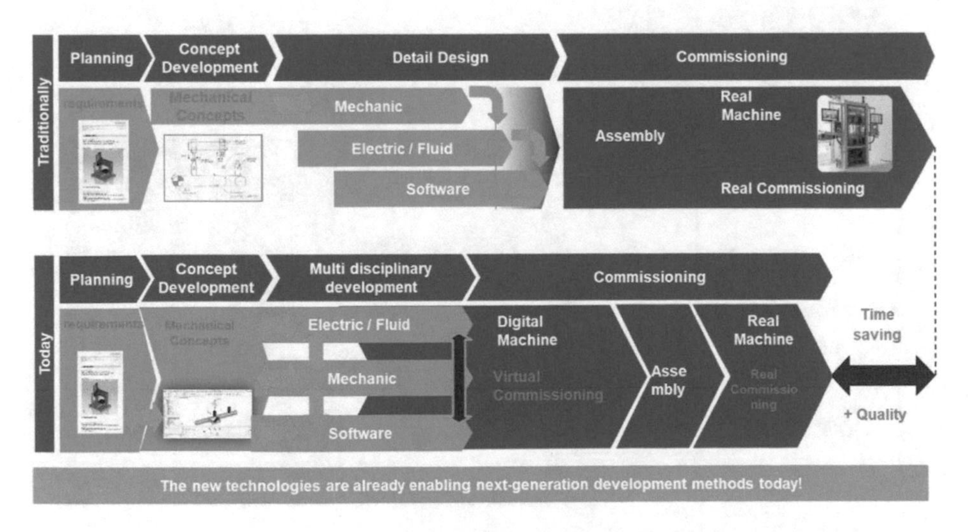

Fig. 1. Time saving by VC (Siemens AG, 2017)

The test configuration of a virtual commissioning gives a prediction of the form in which the controller is available (emulated or real) and how the control code is implemented in the simulation. Basically, there are three types of models:

- Model-in-the-Loop (MIL)
- Software-in-the-Loop or
- Hardware-in-the-Loop.

The difference between the test models is based on the execution of the control code. In the MIL simulation, the implementation does not take place in a classical control language, but in a model language, e.g. on the basis of automats. Often the control code is executed within the simulation program, so that only one software is needed for the simulation loop. SIL simulation moves one step further. The implementation takes place directly in the control language, the execution in a virtualized control, which usually belongs to the engineering tool of the automation system. In addition to the simulation program, further software is required; data is exchanged via virtual communication. HIL simulation extends even further. The control code is executed on real hardware and tested on the plant CAD model running on a simulation computer.

3 Configuration of Interfaces and Model Concept

Before the models are created, interfaces for data exchange between the PLC and the PC must be established. When the required data is available on the computer, the simulation model must access NX data and the SCADA system can read and write it (Fig. 2).

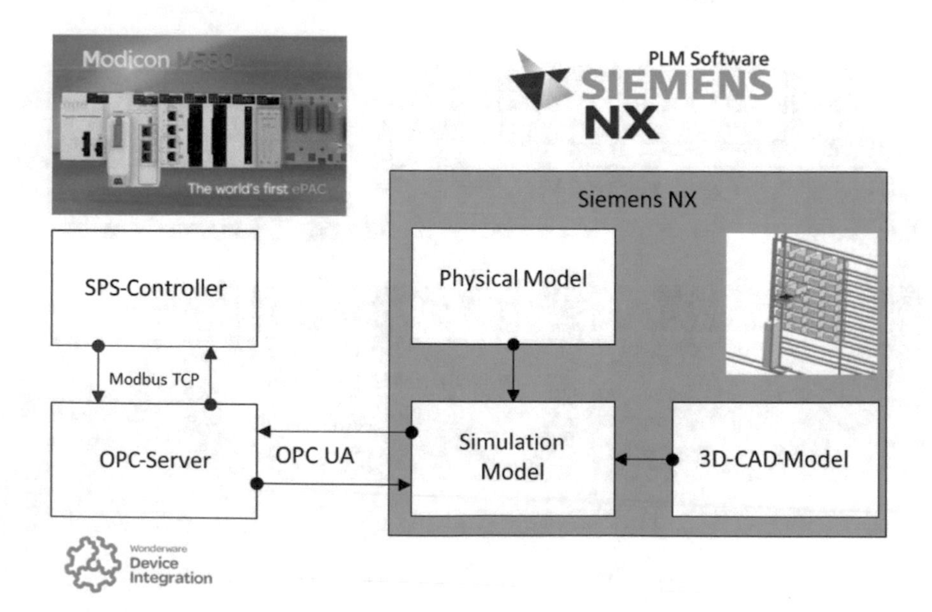

Fig. 2. Interface overview

• PLC to OPC

The Modbus TCP DaServer was used to enable data exchange between the computer and the PLC. It allows other Windows applications to access data from a Schneider Electric PLC, using Modbus TCP/IP protocol. To set up the server, the parameters relevant for the connection must be configured. Two "Device Groups" have been created. The first one for connecting to the System Platform And the second to the simulation model. This configuration is sufficient to connect variables to the SCADA system because they can be created more conveniently in the programming environment of the system platform. For communication with the simulation model, however, the variables must now be created.

To create the variables of a Modbus TCP connection, a flag area must first be made available to the variables on the PLC side. The server is then parameterized on this and as soon as another user program requests variable. It must be ensured that an offset of "1" is used for the flag words and also for the bits. In addition, the flag word plus "400,000" must to be calculated. For REAL values, an "F" for FLOAT is added after the server address. This results, for example, in a server address of "401101.1" for the BOOL PLC variable "axisXBusy" at the address "%MW1100.0".

• OPC to NX

The configuration of this interface is performed in NX. Firstly, all signals in the MCD that are to be read or written by the PLC must be defined as signals and it must be determined whether the MCD may only read the signal (configure the signal as "Input") or only write it (configure the signal as "Output"). On this way are signals connected to the corresponding MCD object (as shown in Fig. 3).

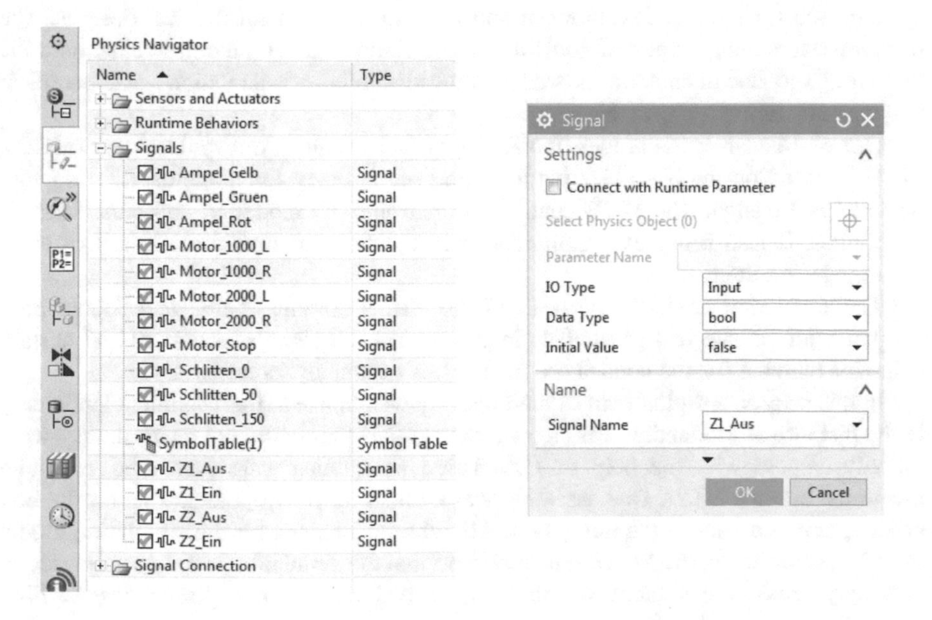

Fig. 3. Signal connection PLC to MCD

The connection between the Modbus TCP DaServer and the simulation model takes place via OPC-DA. As soon as the OPC server is running, it can be set in the MCD under SignalConnection -> Signal Mapping. The OPC server name is "Archestra.dasmbtcp". As soon as the entry is made, the variables available in the server are displayed. By selecting and confirming this option, these variables are released for signal mapping.

The last step is to connect the signals to the OPC variables. On the left the unconnected signals are displayed in the MCD and on the right the unconnected, enabled OPC variables. By creating a mapping between the OPC server and the MCD, a data exchange takes place after the start of the simulation, which has an influence on the MCD objects that are connected to the signals. This makes it possible to control the simulation model from the PLC.

- OPC to SCADA

From the system platform, the connection to the PLC is also implemented via the OPC server. But for the access to the variables a ready template is provided for the connection to an OPC-DA server, the "DDE SuiteLink Client". A separate client is registered on the server for each connection to a PLC. This client must receive the name of the server and the name of a topic on the server. The IP address and other communication parameters must be configured on the server. The Modbus address is then configured as on the Modbus TCP DaServer. After the settings have been accepted, the value in the PLC can be read or written from the entire Galaxy via the name of the variable.

4 Design of a 3D Digital Twin with NX MCD

NX with Mechatronics Concepts Designer NX is an interactive CAD/CAM/CAE system used for product development and manufacturing of mechanical designs. The program offers many important tools for different functions and requirements, both 3D modeling and documentation as well as multidisciplinary calculations and complete part manufacturing are possible.

One of the many possibilities in NX is the Mechatronics Concept Designer (MCD). MCD is based on the NX CAD platform and offers many functions used for sophisticated CAD design. The MCD enables mechatronic 3D modeling and simulation of assemblies. It also provides a comprehensive solution for multidisciplinary collaboration between them.

The simulation model is the heart of the virtual commissioning and should be a "digital twin" of the real plant. It is important that all signals, actuators and system behavior relevant for the control are mapped as accurately as necessary.

In this project, a digital twin of the high-bay warehouse at the Technical University in the Institute of Production was created. Since CAD files could not be used, they were initially created with the help of NX. Three assemblies were generated and then integrated into the MCD. One assembly was created for the high rack, one for the box holding unit and one for the travel unit. All subsequent steps for the simulation model must be performed in the MCD. It is important that the simulation is only as accurate as necessary, since the simulation can be quickly brought to its limits despite high computing power (Fig. 4).

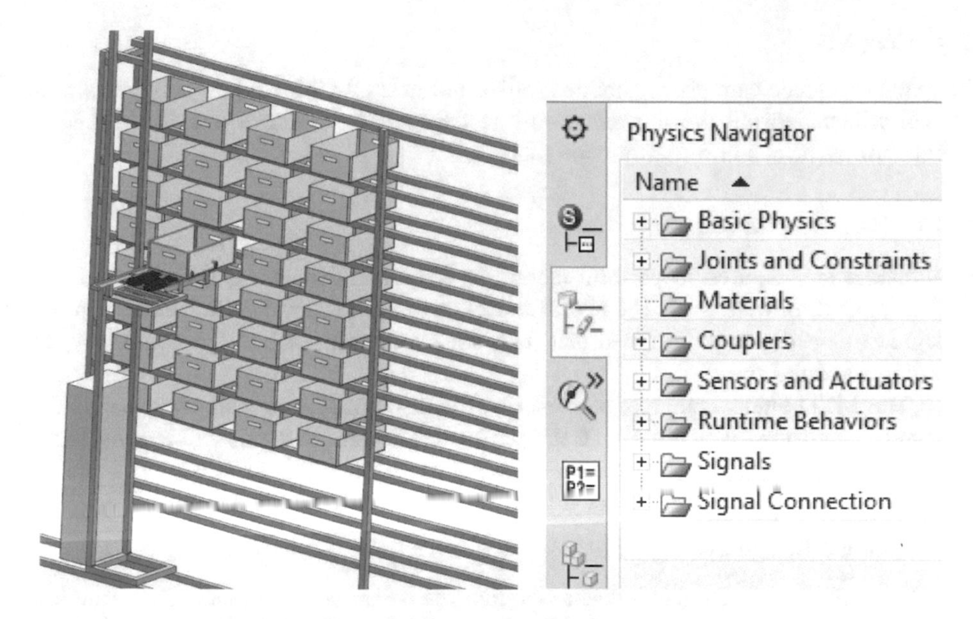

Fig. 4. NX Digital twin and MCD physical navigator

- Basic Physics

The physical properties assigned to an object are located here. An object used in the MCD must first be defined as a solid (RigidBody). Below this, mass, moments of inertia and initial velocity can be defined. Collision bodies also belong to this category. A body in the MCD only takes a collision if it is defined as a collision body, otherwise solid bodies can move against each other without influencing each other.

- Joints and Constraints

This category displays the objects that define constraints and the kinematic behavior of the objects to the system or other objects. In this project the connectors FixedJoint, SlidingJoint and HingeJoint were used. A FixedJoint binds a fixed connection with no degree of freedom between the two selected bodies, but if only the attachment is used, the object is fixed in space without degree of freedom. The SlidingJoint fixes all lines of freedom relatively between the objects except for a translational degree of freedom, which is determined by a vector. The HingeJoint also leaves only one degree of freedom free. Here it is a rotatory one, which is also given by a vector.

- Materials

Diese Kategorie zeigt Materialien an, die festen Komponenten im System zugeordnet sind. Diese Objekteinflüsse wurden in diesem Projekt nicht verwendet und werden daher nicht weiter beschrieben.

- Couplers

Couplers include cam plates, motion profiles and gears. In this project, only gears were used which transmit the movement in one degree of freedom to another degree of freedom by means of a transmission ratio.

4.1 Sensors and Actuators

Actuators are required to perform motion in degrees of freedom. SpeedControl and PositionControl were used. The SpeedControl enables the control of the movement of a degree of freedom by the speed and the PositionControl enables the control by the position and the speed.

The MCD offers a range of sensors, but it is also possible to read out actual values from the actuators, which means that apart from the light barriers, further sensors are not required. The light barriers had to be set up by a CollisionSensor. This sensor detects the collision with a collision body and can thus generate a signal state.

- Runtime Behaviors

This category can be used to influence the runtime behavior of the system. The Runtime Behavior in the MCD offers the possibility to run the own C# code in the simulation clock. The created code will run through every simulation step. For this a subset of the NX Open Namespace is used. This allows information about the objects contained in the simulation to be processed and manipulated. This makes it possible to implement necessary behavior or functionalities for a digital twin beyond the MCD library. Microsoft Visual Studio Community 2017 Version 15.4.0 was used to create the C# code in this work.

- Signals and Connection

This category includes signals and communication options that are available for external coupling.

5 PLC Programming and Simulation

The programming of the real PLC was done by Softwrae Unity Pro XL from Schneider Electric. The software allows programming in many different programming languages such as Function Block Diagram (FBD), Ladder Diagram (LD), Sequential Function Chart (SFC), Instruction List (IL) and Structured Text (ST).

For the control of the high-bay warehouse, five module templates were designed, which take control of the system in three sections. The “Axis” section controls the X and Y axes. The section “Loader” takes control of the box receiving unit and the automatic mode is taken over by the “MAIN” section.

The PLC must only be set up for a standard Modbus TCP/IP connection. Under the IP-Config tab, the IP address must be entered with the subnet mask matching the IP address of the network card of the simulation computer.

The development of the control software is the step that benefits enormously from a simulation model. This makes it much easier and more complete to test the modules and functions that have been created.

As an example for programming of a plant section the program box take_put is described below.

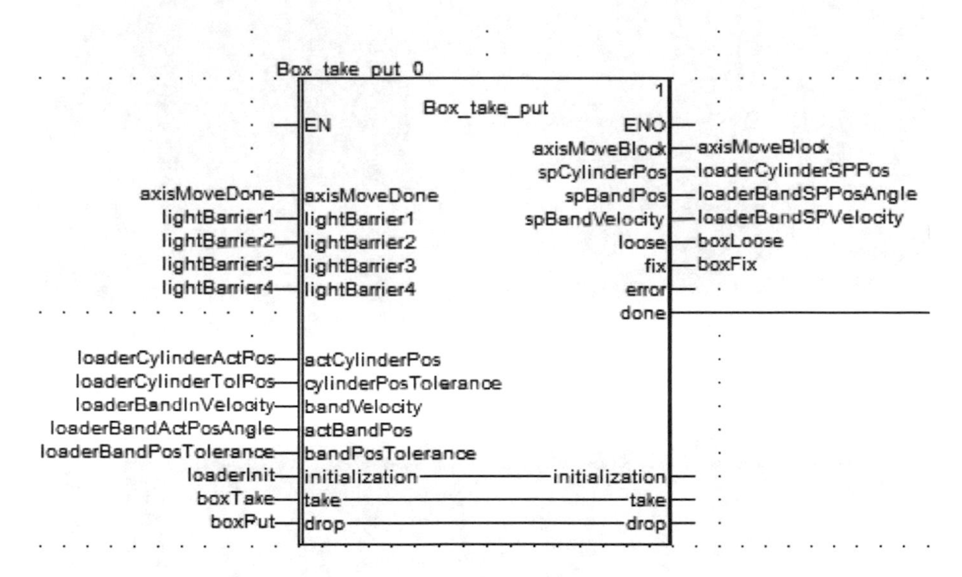

Fig. 5. NX Digital twin and MCD physical navigator

The block "Box_take_put" programmed in ST contains the step chain of the box receiving unit. The block interface has ten inputs, eight outputs and three inputs/outputs, so it is very individually adapted to the box mounting unit and cannot be used for other similar devices without adaptations. The device interface is shown in Fig. 5. Inside there is a state machine with two SwitchCase superstructures lying one inside the other, which go through the different states of picking up and placing a storage box. It uses the "Position_reached" block for a simplified evaluation of whether a position has been reached. The module would have to be connected to the sensors and actuators of the unit in the real high-bay warehouse control.

The SCADA system used here for visualization and simulation is the Wonderware System Platform. This is used to establish the connection between the operator and the PLC. The application implemented for this work could also be covered by a smaller system, but in order to allow later expansion by connecting to an order and stock management system, the decision was made to use the SCADA system. This project should be only a cornerstone in this area and should offer a comfortable possibility for the virtual commissioning.

The SCADA system of this work includes four main pages, a header and a footer. The created main pages are a title page, a page for manual operation, a page for automatic operation and a system page. In the background of the system platform is a data management system, which is created by object-oriented programming. For the implementation of this project, nine object templates were created, which are run through in different instances behind the SCADA system and thus enable the operation of the plant (Fig. 6).

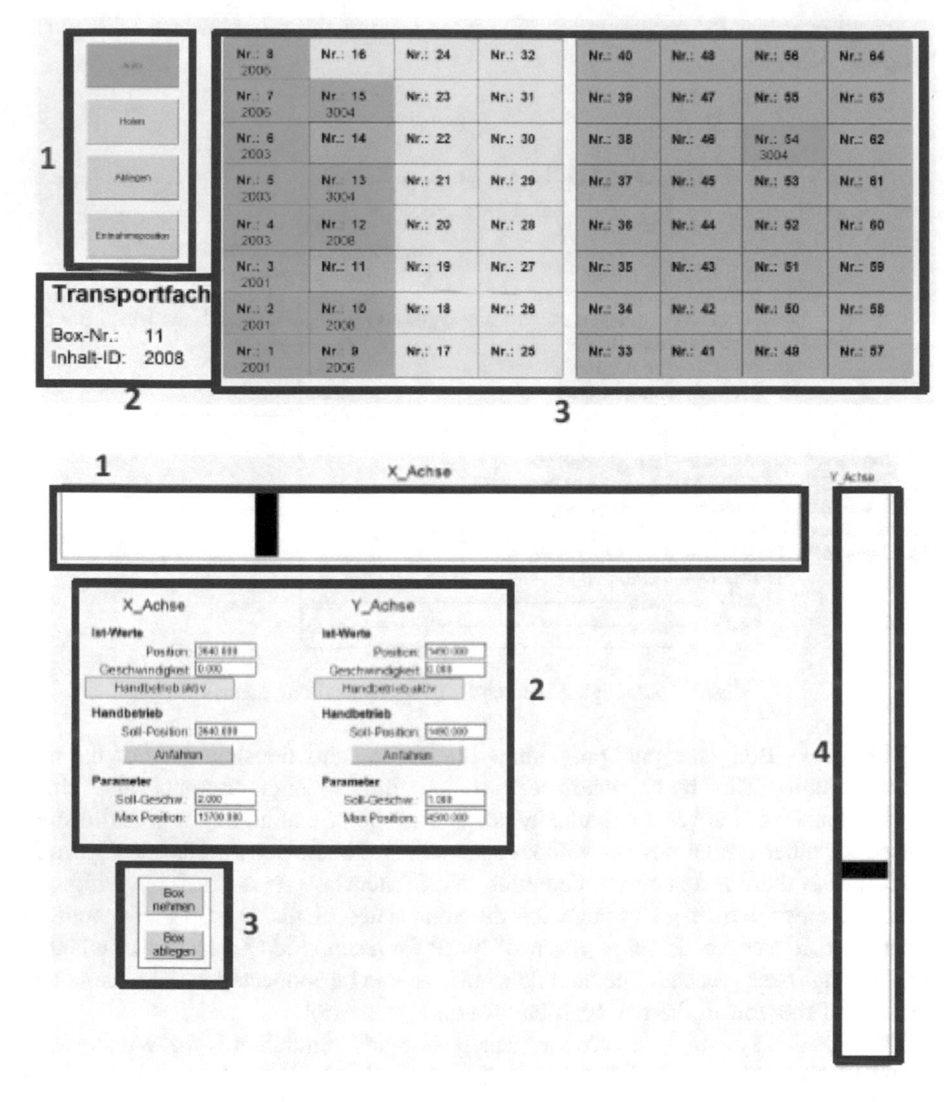

Fig. 6. Navigation and control of digital twin

6 Conclusion

The virtual commissioning provides valuable know-how for the real implementation of the high-bay warehouse and the control software can be used with small adaptations.

In addition, this project offers a great benefit for teaching as this model can be universally coupled to different control systems. Incorrect control software cannot cause any material or human damage.

The advantages of creating the control software and the function in the SCADA system were enormous. However, the design of the simulation model took up a large part of the working time. In projects, an effort and benefit analysis must always be carried out for virtual commissioning. It has to be recognized that virtual commissioning will be a very effective development tool in many situations. Complex or time-critical automation projects can thereby be designed much more efficiently.

References

Lacour, F.-F.R.: Modellbildung für die physikbasierte Virtuelle Inbetriebnahme materialfluss-intensiver Produktionsanlagen. Herbert Utz Verlag, München (2011)

VDI 3693-1: Virtuelle Inbetriebnahme - Modellarten und Glossar. Beuth Verlag, Düsseldorf (2016)

Zäh, M.F., Wünsch, G., Hensel, T., Lindworsky, A.: Feldstudie – Virtuelle Inbetriebnahme. Werkstattstechnik online **10**, 767–771 (2006)

Wünsch, G.: Methoden für die virtuelle Inbetriebnahme automatisierter Produktionssysteme. Herbert Utz Verlag, München (2007)

Osinde, N.O., Byiringiro, J.B., Gichane, M.M., Smajic, H.: Process modelling of geothermal drilling system using digital twin for real-time monitoring and control. MDPI, July 2019

Smajic, H., Bosco, J.: Simulation tools for virtual plc-testing in extrusion process. In: Hg.:DEMI 2019. University of Banja Luka, Bosnia and Herce-govina, 24–25 May 2019 (2019)

Niedersteiner, S.; Lang, J.; Pohlt, C., Schlegl, T.: Klassifikation des Arbeitsfortschritts an intelligenten Arbeitsplätzen –Support Vector Machine basierter Ansatz. In: AALE 2017 (2017)

Abé, P., Simons, S.: Virtuelle Inbetriebnahme mit NX 11 Mechatronics Concept Designer. In: AALE 2017 (2017)

Internet of Things and Smart World Applications

Measuring Noise Impact in the Cities Through a Remote and Mobile Sonometer

Unai Hernandez-Jayo(✉), Amaia Goñi, and Javier Vicente

University of Deusto and Deusto Institute of Technology (DeustoTech), Avenida de las Universidades 24, 48007 Bilbao, Spain
unai.hernandez@deusto.es
http://deusto.es, http://deustotech.eu/

Abstract. Cities are getting bigger and bigger, this means an increase in the concentration of population and therefore noise. This can pose a danger to our health, because without realizing it we could be exposing our body to noise levels much higher than permitted, but due to routine and being used to them, we do not give importance. To try to tackle this problem, the first step is to know the noise levels. To this end, cities carry out noise mapping measurement campaigns. The problem is that these campaigns are carried out every few years and the measures are implemented at specific points in the city. Nowadays, thanks to the development of small embedded systems which can integrate high capacity data processors, as well as long and short-range communications systems, it is possible to design intelligent devices, under the IoT paradigm, which provide detailed and ubiquitous information in real time about noise levels in cities.

Keywords: Remote teleoperation · Industrial Internet of Things · Smart metering

1 Introduction

The population is increasingly concentrated in urban areas. According to The World Bank Group, in 2018, an estimated 55.27 % of the world's population lived in urban settlements[1] and this trend is expected to continue. By 2030, urban areas are expected to be home to 60 % of the world's population and one in three people will live in cities with at least half a million inhabitants [1]. This reality poses new challenges for the authorities to ensure the quality of life of their inhabitants as well as the efficient use of the resources of these urban areas through better management of services, which requires significant changes in governance, decision-making and the development of specific action plans. To this end, a technological revolution is driving the change of most cities under the umbrella of the so-called Smart Cities paradigm [4,5].

[1] https://data.worldbank.org/indicator/SP.URB.TOTL.IN.ZS.

M. E. Auer and D. May (Eds.): REV 2020, AISC 1231, pp. 627–636, 2021.
https://doi.org/10.1007/978-3-030-52575-0_52

The idea of smart cities comprises a wide range of control and actuators systems aimed to improve the habitability and perception that citizens have of cities. A smart city covers many of these systems, ranging from applications that facilitate the governance of cities and encourage citizens participation to services specifically focused on improving their quality of life. Among these systems, we can highlight those using ICT to improve the environment of the city, but not only from an air quality perspective, but also to control the noise levels of the city. It is assumed by the World Health Organization that environmental noise has emerged as the leading environmental nuisance triggering one of the most common public complaints in many Member States of the European Union. The European Union tries to face the problem of environmental noise with international laws and directives (as the European Noise Directive [2]) on the assessment and management of environmental noise [3]. Noise pollution translates into both unrest and dissatisfaction of the inhabitants and high costs for public health. It is because of this negative impact on the quality of life of citizens that the control of noise pollution has been included within the paradigm of intelligent cities. For the Administrations, the need arises to monitor noise pollution in order to raise public awareness and to design and implement action plans to control and reduce it.

The objective of the previously cited European Directive is to define a common approach aimed at preventing, preventing or reducing harmful effects caused by exposure to environmental noise. To this end, noise maps must be generated and updated every five years with the aim of reporting changes in environmental conditions (mainly traffic, mobility and urban development) that may have occurred during the reference period. Updating noise maps using a standard approach requires the authorities responsible for them to collect and process new data concerning such changes. This requires a technically trained person to travel through several points making measurements. This procedure is costly in economic and time terms, and has a significant impact on the financial statements of these authorities. In addition, the result of this analysis may not be true to reality. The values taken by a device at a fixed moment do not represent the whole of the noise during the day nor can they be extended, for example, to an entire street, since the characteristics of nearby buildings or the difference in traffic can be conditioning.

In this context, the University of Deusto, together with Saitec S.A, is in the process of developing the ZARATAMAP project, whose main objective is to go one step further in solving the problem of noise monitoring in cities and provide the authorities with a low-cost tool that allows them to have dynamic noise maps of the city from which to carry out citizen awareness actions (i.e. alerts and warnings in real time) as well as having reliable information in real time on the basis of which to design more efficient action plans to combat noise pollution.

The paper is structured as follows. In Sect. 2, ZARATAMAP project is presented by its objectives and technological approach. In Sect. 2.1, its challenges are pointed out. In Sect. 3, the hardware approach proposed to address the problem of mobile acoustic monitoring is explained, and in Sect. 4 the first acoustic signal processing algorithms developed to face this challenge are described. In Sect. 5, we detail the expected outcomes of the collaborative project, and, finally, in Sect. 6 we focus on the conclusions and future work.

2 ZARATAMAP Project Description

In order to achieve the main objective of the project which has been introduced above, the solution proposed by ZARATAMAP is based on the design and development of a system that allows the identification, characterization and monitoring of noise in a dynamic way formed by (i) a set of mobile devices on board public transport vehicles capable of capturing and processing the audio signal, (ii) the communications architecture for the transmission of this information and (iii) the control centre that shows and analyzes the collected data (Fig. 1).

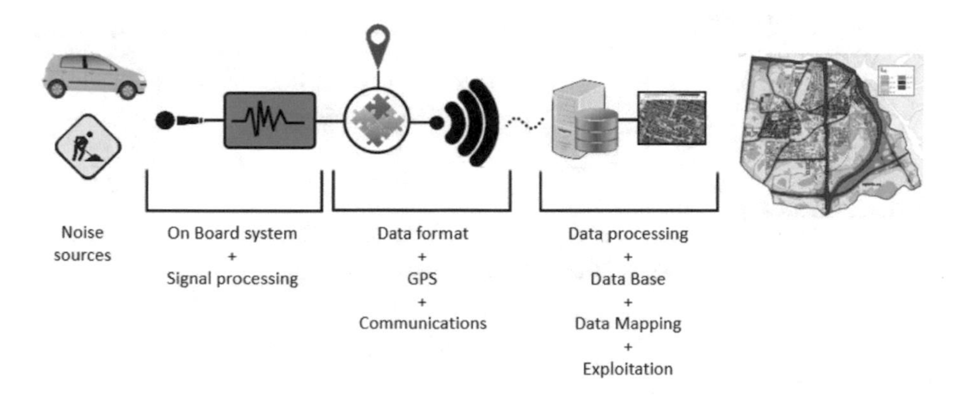

Fig. 1. ZARATAMAP concept diagram

The main advantage of this solution with respect to the alternatives currently available for noise monitoring is the use of a mobile device able to detect possible sources of noise pollution located in multiple positions at the city. The systems currently available that are capable of providing real-time data are based on the installation of sensors at fixed points in the city, under controlled conditions and with the supervision of an accredited agent. ZARATAMAP intends to add to the capture devices the possibility of being in constant movement through the target area. The advantages of a mobile sensor network are:

- Data accuracy. The placement of the sensors in public transport vehicles will allow measurements that are less distant from each other, which will give much more information and a more real representation of the noise in the streets.

– System efficiency. Having mobile receivers makes it possible to cover a wider area using the same receiver instead of having to install fixed capture stations at specific points which reduces resources and the cost required to implement the solution.

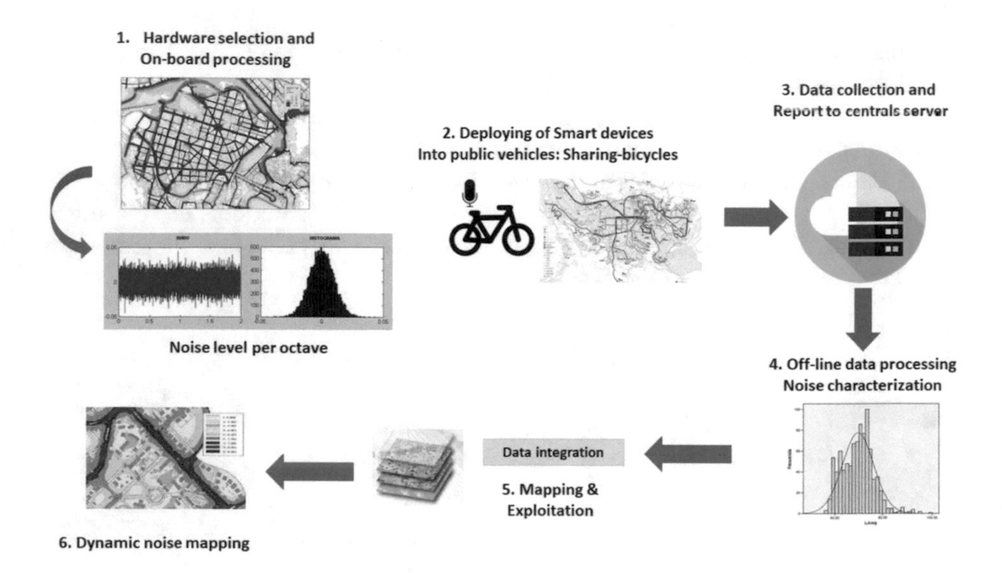

Fig. 2. ZARATAMAP technical goals

2.1 ZARATAMAP Objectives

In order to be able to develop ZARATAMAP, the project consortium has set the following technological objectives:

1. Design and develop a hardware that can be deployed in the form of a low-cost device which must be easily deployed in public transport vehicles. This smart object will allow the capture and processing of acoustic signals. It will also integrate the communication capabilities necessary for the continuous reporting the monitored and geo-positioned noise.
2. Design and codification of the required signal processing algorithms to characterise the noise. This characterization will be performed following as example how a sonometer works. Then, the output of this processing unit will be the level of noise per octave band, from 20 Hz to 20 kHz.
3. Once the noise signal is processed, the noise levels will be geo-referenced using the attached GPS into the smart device. This feature will allow to determine the noise contribution per octave in each location. This aggregate information will be sent to the central server where it will be stored.

4. Design and development of software tools for the generation and visualization of dynamic noise maps. The central server will store the data captured by the smart device in the database, simplified noise maps that are easy to interpret will be presented. To this end, ZARATAMAP will provide different levels of access to its online database, which will make it possible to provide information with different levels of complexity depending on the type of user (general public, administrations, others).
5. Characterization of noise sources: starting from pre-defined noise patterns, such as vehicles accelerating or braking, vehicle horns, working machines or others, we will try to identify for each analyzed GPS location, the noise source. Hence, the ability to implement machine learning techniques that allow the system to make these approximations will be studied. This will be of great help to the administrators of the system so that they can determine the main sources of noise in the city.
6. Design and implement a demonstration pilot, which will be decided during the first phases of the project.

Objectives 1, 2 and 3 are currently being developed, work that will be described in the following sections. In a graphical approach, project's objectives are represented at Fig. 2.

3 Hardware Platform

The first version of the hardware prototype (v1.0) developed for the remote acoustic monitoring system was developed in a previous stage of ZARATAMAP project. It was based on the following elements (Fig. 3):

- FRDM-KL25Z embedded system
- CMA-4544PF-W omnidirectional capsule microphone with auto gain control based on the MAX9814 amplier
- ESP8266 WiFi communications module
- Adafruit FONA 808, which is and all-in-one mobile communication interface plus a GPS module

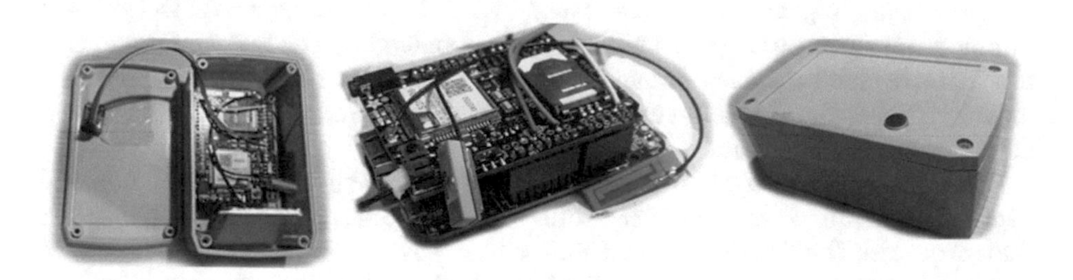

Fig. 3. ZARATAMAP prototype v1.0

But this first version had a number of deficiencies: memory capacity, real-time processing capabilities, Float Point Unit and not integration of digital processing functionalities to make easier the noise characterization. Then, in the current version (V2.0), we are using the STM32F401RE embedded system. It is based on the high-performance ARM®Cortex©-M4 32-bit RISC core operating at a frequency of up to 84 MHz. Its Cortex©-M4 core features a Floating point unit (FPU) single precision which supports all ARM single-precision data-processing instructions and data types. It also implements a full set of DSP instructions and a memory protection unit (MPU) which enhances application security.

The STM32F401xD/xE incorporates high-speed embedded memories (512 Kbytes of Flash memory, 96 Kbytes of SRAM), and an extensive range of enhanced I/Os and peripherals connected to two APB buses, two AHB buses and a 32-bit multi-AHB bus matrix. All devices offer one 12-bit ADC, a low-power RTC, six general-purpose 16-bit timers including one PWM timer for motor control and two general-purpose 32-bit timers. They also feature standard and advanced communication interfaces.

The embedded system integrates a grove sound Sensor. The main component of the module is a simple microphone, which is based on the LM386 amplifier and an electret microphone. It presents a sensitivity(1 kHz) of 52–48 dB under a working frequencies from 16 Hz to 20 kHz.

The communication capabilities will be provided by the ESP8266 self-contained Wi-Fi network solution. This module has been chosen because it provides access to 802.11 b/g/n networks and Wi-Fi Direct (P2P), which makes simplier the setup and configuration of the network connections.

4 Noise Signal Processing

At Fig. 4 the implemented block diagram in charge of performing the sampling and data processing in real time is presented.

In a first stage, the microcontroller has been set up so that the sampling and the storage of the values into the memory is done automatically without interrupting the data processing. In order to not interrupt the execution of the main threat of the program, a TIMER has been configured to generate notifications with a frequency of 60 KHz, so this event does not interrupt the execution of the processor program. Therefore, every (1/60 kHz) seconds, the ADC takes a sample from the input signal. This sample is managed by the DMA of the microprocessor, which takes care of moving sampled data to a buffer position into the memory. Then, the DMA increases the destination memory address each time a data is moved to a circular buffer.

The DMA generates two interrupts. One when the buffer is half full and another interrupt when the buffer is full. With each interrupt, a routine is called that processes the data of the half buffer that has been filled. This allows new data to be stored in the other half of the buffer while data from one half of the buffer is being processed, without the data being corrupted if the processing time of a frame is longer than the sampling period.

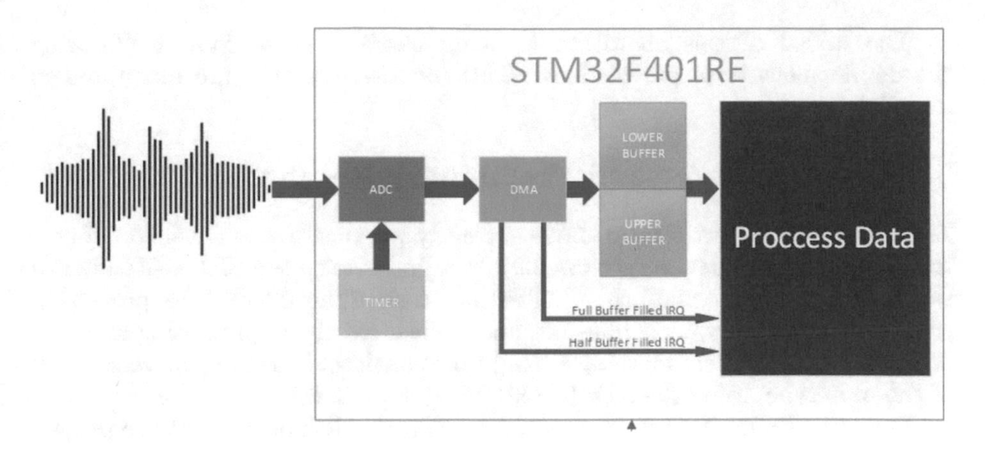

Fig. 4. ZARATAMAP on-board processing schematic

However, in order for processing to take place in real time, the processing time of a frame must be less than the product of the number of samples in a frame for the sampling period. Therefore, the calculation algorithm needs to be optimized to satisfy this restriction.

The algorithm in charge of processing the sampled data is now under development, but it is designed in four stages:

1. Anti-alising filter: is an analog low pass filter designed to limit the bandwidth of the input signal to satisfy the Nyquist–Shannon sampling theorem over the band of interest. This filter is an anti-alias filter because by attenuating the higher frequencies (greater than the Nyquist frequency), it prevents the aliasing components from being sampled.
2. A-weighting filter: it is the most common filter used of a family of curves defined in the International standard IEC 61672:2003 related to the measurement of sound pressure level. A-weighting is deployed in the proposed intelligent device in order to account for the relative loudness perceived by the human ear, as the ear is less sensitive to low audio frequencies. In a first approach, the devices is setup to work with octave bands. Then, this filter is employed by arithmetically adding a table of octave values to the measured sound pressure levels in dB.
3. Octave band filter: it is considered that the whole frequency spectrum is divided into sets of frequencies called bands, each covering band a specific range of frequencies. A band is said to be an octave in width when the upper band frequency is twice the lower band frequency. In the algorithm under development, we are working with 10 octave band filters in parallel, from 20 Hz to 20 KHz.
4. RMS calculation: the last stage consists on the calculation of the RMS value of the signal obtained in the output of each octave band filter. This RMS value will be converter, after a calibration process into a LAeq (dBA) value, obtaining the contribution of the noise signal per octave band.

The design of this algorithm is being carried out in Matlab^{TM}, using the development tools provided by STMicroelectronics that are integrated in Simulink^{TM}.

5 First Prototype and Preliminary Results

ZARATAMAP project consortium is currently working on a new version of prototype of the smart device (version 2.0). In a first version the FRDM-K64F card was used as hardware platform [4]. The main lack of this device is its processing capacity and memory, as it does not allow advanced signal processing and it is not possible to perform an octave band noise analysis. However, in version 2.0 of the prototype, using the STM32F401RE system, it will be possible.

Even so, version 1.0 has been useful to validate the approach of the project and verify that it is possible to carry out noise measurements using a connected device installed on a bicycle, as it can be shown at Fig. 5.

Measurements of LAeq obtained using version 1.0 have been compared with the levels measured using the RION NL-42 professional sonometer. As it is shown at Table 1, the readings using the prototype are quite similar to the ones provided by the sonometer, but not good enough, providing a average error equal to 3.7 dB. This could be due to two main reasons: the quality of the microphone and the precision of the signal processing performed into the FRDM-K64F-based embedded system.

As part of this first version of the prototype, also the central server has been developed. However, not all functionalities have been developed, only the database and the service that allow the representation in a map of the measurements carried out during the cycling. Figure 6 includes an screenshot of the database and different maps of the same area of the city of Bilbao with measurements carried out at different hours to appreciate the variation of the noise levels.

Fig. 5. ZARATAMAP v1.0 deployed into a bicycle

Table 1. Calibration of prototype 1.0

Sonometer (dB)	Prototype 1.0 (dB)	Error (dB)
54.4	52.4	−2
62.5	59.1	−3.4
74.0	66.2	−7.8
60.9	60.2	−0.7
60.6	57.5	−3.1
78.1	72.5	−5.6

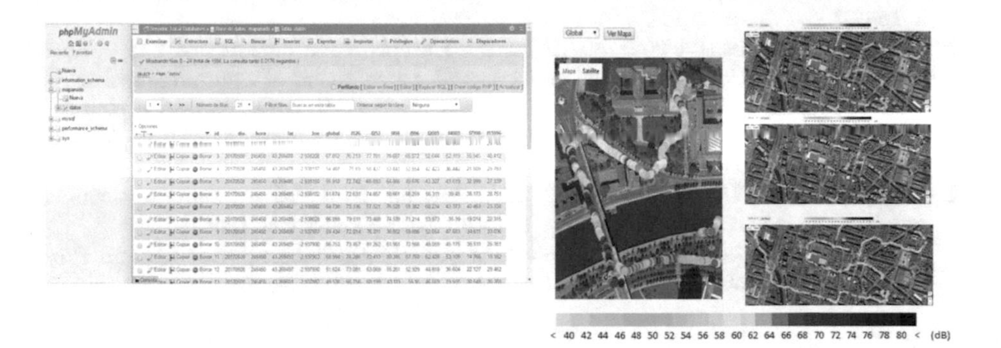

Fig. 6. ZARATAMAP v1.0 web application

6 Conclusions

The goal of ZARATAMAP project if to obtain a remote acoustic monitoring system that allows to monitor in real time city noise impact. For this purpose, a acoustic sensor (a microphone) is connected to a mobile embedded device with signal processing and data transmission capabilities.

This embedded system is designed to be installed on a the public bike sharing system, so urban noise is captured along the route travelled by riders. Therefore, the larger the number of sensors deployed, the more detailed information will be obtained to generate the noise map of the city. Finally, the collected information is sent to a central server that runs a web-based GIS application designed to display the collected noise levels in real-time.

In the current phase of the project, the algorithm in charge of carrying out the noise process is being developed. This algorithm will first be validated in the laboratory and calibrated using a professional sound level meter. Once this phase has been completed, tests will be carried out in real scenarios, using the central server developed in the first version of the prototype. This server will be adapted and improved to provide new services and utilities to end users, such as report generation.

In a more advanced phase of the project, the ability to define noise patterns will be analyzed, so that through machine learning techniques. For this purpose, a new functionality can be created that in order to characterize the monitored noise in each location and thus establish the predominant noise source in each moment.

Acknowledgment. This work is partly funded by ZARATAMAP (2018/0095) and IoTrain (RTI2018-095499-B-C33) projects.

References

1. United Nations: The World's Cities in 2016. Data Booklet (1982). ISBN 978-92-1-151549-7
2. EU Directive: Directive 2002/49/EC of the European Parliament and the Council of 25 June 2002 relating to the assessment and management of environmental noise, Official Journal of the European Communities, L 189/12, 2002, European Union
3. Night Noise Guidelines for Europe. World Health Organization 2009. http://www.euro.who.int/. Accessed Nov 2019
4. Hernandez-Jayo, U., Alsina-Pages, R.M., Angulo, I., Alias, F.: Remote acoustic monitoring system for noise sensing. In: 2017 REV Conference (2017)
5. Rabinowitz, P.: On subharmonic solutions of a Hamiltonian system. Comm. Pure Appl. Math. **33**, 609–633 (1980)

Visible Light Communication for Automotive Market Weather Conditions Simulation

Cristian-Ovidiu Ivascu(✉), Doru Ursutiu, and Cornel Samoila

Transilvania University, Brasov, Romania
cristian.ivascu@unitbv.ro

Abstract. The goal of V2V Vehicle-to-Vehicle communication is to prevent accidents by allowing vehicles in transit to send position and speed data to one another over an ad hoc mesh network. Depending upon how the technology is implemented, the vehicle's driver may simply receive a warning should there be a risk of an accident or the vehicle itself may take pre-emptive actions such as braking to slow down. V2V technology represents the next great advance in saving lives. This technology could move us from helping people survive crashes to helping them avoid crashes altogether—saving lives, saving money, and even saving fuel thanks to the widespread benefits it offers

Keywords: Vehicular networks · Visible light communication · Vehicle-to-vehicle · Vehicle-to-infrastructure · Vehicle-to-internet

1 Introduction

Li-Fi (short for light fidelity) is a technology for wireless communication between devices using light to transmit data and position. In its present state only LED lamps can be used for the transmission of visible light [1]. The term was first introduced by Harald Haas during a 2011 TEDGlobal talk in Edinburgh. In technical terms, Li-Fi is a visible light communications system that is capable of transmitting data at high speeds over the visible light spectrum, ultraviolet and infrared radiation [1].

In terms of its end use the technology is similar to Wi-Fi. The key technical difference is that Wi-Fi uses radio frequency to transmit data. Using light to transmit data allows Li-Fi to offer several advantages like working across higher bandwidth working in areas susceptible to electromagnetic interference (e.g. aircraft cabins, hospitals) and offering higher transmission speeds.

V2V communication, vehicle-to-vehicle communication, is the wireless transmission of data between motor vehicles.

Vehicular networking is one of the main enablers of Intelligent Transportation Systems (ITSs) and enables vehicles to communicate with each other and with roadside units installed along the road [18, 19]. The current ITS research activities and standardization efforts have mainly focused on radio frequency (RF) technologies for vehicle-to-vehicle (V2V), vehicle-to-infrastructure (V2I) and infrastructure-to vehicle (I2V) communications [20, 21]. RF bands have already been allocated to operate ITSs in the United States (75 MHz), Europe (70 MHz), China (20 MHz), Japan (80 MHz) and Korea (70 MHz) [22]. The impact of current V2V and V2I communications on the

M. E. Auer and D. May (Eds.): REV 2020, AISC 1231, pp. 637–651, 2021.
https://doi.org/10.1007/978-3-030-52575-0_53

RF spectrum usages is low, but this is expected to significantly increase soon with the widespread adoption of ITSs. Limited RF bands can quickly suffer from high interference levels when hundreds of vehicles located in the same vicinity try to communicate simultaneously. Furthermore, RF-based vehicular communications experience longer delays and lower packet rate because of the channel congestion. To address such issues, visible light communication (VLC) has emerged as a potential means for vehicular connectivity [23]. VLC is based on the principle of modulating light emitting diodes (LEDs) at very high speeds that are not noticeable by human eye. This lets the use of LEDs for wireless communication purposes in addition to their primary purpose of illumination. Many automotive manufacturers have started to employ LEDs due to their high resistance to vibration, improved safety performance, and long-life span.

Motivated by the looming radio frequency (RF) spectrum crisis, this paper aims at demonstrating that optical wireless communication (OWC) has now reached a state where it can demonstrate that it is a viable and matured solution to this fundamental problem. Light fidelity (Li-Fi) which is related to visible light communication (VLC) offers many key advantages, and effective solutions to the issues that have been posed in the last decade [22].

2 Goal

The goal of V2V (Vehicle-to-Vehicle) communication is to prevent accidents by allowing vehicles in transit to send position and speed data to one another over an ad hoc mesh network. Depending upon how the technology is implemented, the vehicle's driver may simply receive a warning should there be a risk of an accident or the vehicle itself may take pre-emptive actions such as braking to slow down.

V2V communication is expected to be more effective than current automotive original equipment manufacturer (OEM) embedded systems for lane departure, adaptive cruise control, blind spot detection, rear parking sonar and backup camera because V2V technology enables a ubiquitous 360-degree awareness of surrounding threats. V2V communication is part of the growing trend towards pervasive computing, a concept known as the Internet of Things (IoT).

Connected vehicles use wireless technology to connect vehicle information and location to other vehicles (V2V); to infrastructure (V2I); or to other modes, such as internet clouds, pedestrians, and bicyclists (V2X). The wireless technology typically used for connected vehicles is DSRC, dedicated short-range communications, but some functions may use cellular or other types of communication

The goal is to use VLC (Visible Light Communication) in order to communicate from one vehicle to another one and from vehicle to infrastructure or from vehicle to internet. Connected vehicles offer additional functions related to roadside devices and fleet-level information. Connected vehicles bring additional mobility and environmental benefits that cannot be achieved through automation alone.

V2V technology represents the next great advance in saving lives. This technology could move us from helping people survive crashes to helping them avoid crashes altogether—saving lives, saving money, and even saving fuel thanks to the widespread benefits it offers [2].

3 Approach

In Intelligent Transportation Systems, visible light communication (VLC) has emerged as a powerful candidate to enable wireless connectivity in vehicle-to-vehicle (V2V) and vehicle-to-infrastructure (V2I) links. While VLC has been studied intensively in the context of indoor communications, its application to vehicular networking is relatively new.

There has been a growing interest in the field of intelligent transportation systems (ITSs) in an effort to improve road safety and traffic flow and to address environmental concerns. The ITS involves the application of the advanced information processing, control technologies, sensors, and communications in an integrated approach to improve the functioning of the road transportation systems. Considerable efforts have been made in the last decade by researchers from both academia and industry to enable the cooperative ITS, which is seen as the next generation of ITSs and it is enabled by V2V and V2I communications.

We consider VLC as a complementary and/or an alternative technology to RF-based systems. An alternative communication means can turn out to be useful to offload the RF channel. VLC refers to the use of optical radiation at the visible wavelengths to transmit data in an unguided medium. Since the human eye perceives only the average intensity when light is switched on and off fast enough, then it is possible to transmit information data using LEDs without a notable effect on the light illumination level and the human eye. Recent advances in materials and solid-state technologies have enabled the development of highly efficient LEDs that are now being widely used in outdoor lighting, traffic signs, and advertising displays. Furthermore, many automotive manufacturers have started to employ LEDs due to their high resistance to vibration, improved safety performance, and long-life span. LEDs can be now found in brake lights, turn signals, and headlamps in most new vehicles. The outdoor and on-vehicle omnipresence of LEDs makes the use of VLC for V2V and V2I communications possible [3].

VLC is well positioned to address both the low latency required in safety functionalities (i.e., emergency electronic brake lights, intersection collision warning, in-vehicle signage, and platooning) and high speeds required in so-called infotainment applications (i.e., map downloads and updates, media downloads, point of interest notifications, high-speed Internet access, multiplayer gaming, and cooperative downloading). Furthermore, VLC is a cost-effective and green communication solution since the dual use of LED lighting systems on the vehicles and the roadside infrastructure is targeted. An LED-based VLC system would consume less energy compared with the RF technology, thus allowing the expansion of communication networks without the added energy requirements, potentially contributing to the global carbon emission reduction in the long run. VLC-based networks will further offer better scalability in scenarios with high vehicle density, where the RF-based vehicular communications experience longer delays and lower packet rate because of the channel congestion. The directionality of optical propagation provides an inherent advantage in vehicular VL since only a small number of neighboring vehicles, typically within the direct line of sight (LOS) of the receiver (Rx), are in the same contention domain, thereby significantly lowering collision probability and increasing scalability. However, this advantage is at the cost of reduced coverage compared with RF technologies and requires judiciously designed upper-layer protocols to handle more frequent handovers. VLC is also appealing for vehicular scenarios in which the use of a RF band is restricted or banned due to the safety regulations, e.g., industrial parks such as in oil/gas/mining industries, and military vehicle platoons, to name a few. Another attractive feature of VLC is the positioning and navigation capabilities. Although the global positioning system (GPS) is widely used today, it fails to provide a sufficient accuracy in environments where there is no LOS paths such as tunnels, indoor parking lots, and some urban canyons. For such cases, VLC-based positioning technology could be used to complement the accuracy of other localization systems, knowing that the lighting fixtures offer very high accuracy, up to of tens of centimeters, which is much more suitable for vehicle safety applications, compared with a typical positioning error of up to 10 m associated with the GPS. VLC also presents some challenges due to operation in outdoor environments such as severe weather conditions, sunlight, and ambient light. Visibility-limiting conditions such as heavy fog or snow could decrease the operation range. These degrading effects can be kept at minimum with highly sensitive RXs. Another potential concern is direct sunlight or strong ambient light, which could saturate the VLC Rx. This is usually addressed by utilizing proper optical filtering.

In this paper, we evaluate out a channel modeling study to quantify the effect of rain and fog on a V2V link with a high beam headlamp acting as the transmitter. Taking advantage of advanced ray tracing features, we first develop a path loss model for V2V link as a function of distance under different weather conditions. Then, we use this expression to determine the maximum achievable distance to ensure a given bit error rate. We further investigate the deployment of relay-assisted systems to extend transmission ranges. Numerical results are presented to corroborate our findings.

For accurately describing the advanced interaction of rays between light emitting source and a receptor within a specified environment we use Zemax® software, for a given number of rays and the number of scattering events, the received power and associated path lengths for each ray are computed these are then processed to yield the channel impulse response (CIR). In our work, we first develop a path loss model for V2V link as a function of distance under different weather types through curve fitting. Next, we use this expression to determine the maximum achievable distance to ensure a given BER. We also investigate the deployment of relay-assisted systems to further extend transmission ranges.

4 Vehicular VLC Research and Open Problems

Early research efforts on the use of LEDs for V2I communications can be traced back to a U.S. patent issued in 1997 (Patent 5 633 629), which utilizes traffic lights to optically transmit information to a vehicle. In initial performance evaluation studies [10, 11], the basic performance metrics such as the received optical power and the signal-to-noise ratio (SNR) have been derived for different road surfaces (e.g., asphalt versus cement or concrete) for a V2I communication system between a street/traffic light and a vehicle. More recent experimental works [12–14] have used off-the-shelf LEDs and PDs to investigate the throughput and the error rate performance of VLC-based V2V and V2I links. Such works have demonstrated the feasibility of the vehicular VLC.

New generation vehicles fitted with LED-based front and back lights can communicate with each other and with the roadside infrastructure through the VLC technology. While VLC has been studied intensively in the context of indoor communications [24, 25], its application to vehicular networking is relatively new [26–28]. In such an emerging topic, channel modelling is particularly important to understand the fundamental performance limits imposed by the outdoor medium. Earlier works on I2V links (i.e., from traffic light to vehicle) [29, 30] build upon the line-of-sight (LOS) channel model originally proposed for the indoor LED light sources with Lambertian pattern. However, such a model is not applicable to automotive low-beam and high-beam headlamps with asymmetrical intensity distributions. To address this, a piecewise Lambertian channel model was proposed in [31] to reflect the asymmetrical intensity distribution of scooter taillight. Measured intensity distribution patterns were further used in [32–34] to accurately reflect the asymmetrical structure of automotive lighting in V2V channel modelling. The road surface might impact the vehicular VLC system performance. The reflectance of road surface depends on its nature and physical state, and it also changes with weather conditions. In [32], Lee et al. utilized Monte Carlo ray tracing to obtain channel delay profiles for V2V, I2V and V2I links for a road surface with fixed reflectance value. In [33, 34], Luo et al. proposed a geometric V2V VLC channel model based on the measured headlamp beam patterns and a road surface reflection model. The link bit error rate (BER) performance was investigated for the clean and dirty headlamps in a wet and dry road surface. Another critical issue for vehicular VLC channels is the weather conditions which have received little attention so far in the literature. In [35, 36], Kim et al. employed a laboratory chamber and experimentally evaluated the effect of artificially generated rain and fog on the received optical signal for a red LED (that can be potentially used as a taillight). While there are some other works that quantify the effect of weather conditions on infrared LED transmission [37], these cannot be applied in a straightforward manner to vehicular VLC links at visible wavelengths.

However, in many respects, this technology is in its infancy and requires further research efforts in several areas including physical layer design, upper-layer protocols and channel modeling.

5 Physical Layer Design Issues

VLC channels are of multipath nature and exhibit frequency selectivity, which results in inter symbol interference (ISI), and leads to a reduced data rate. The conventional solution for the ISI mitigation in a single carrier system is to adopt the time-domain equalization (TDE) and use modulation formats with wide-enough pulse duration. Although nonlinear TDEs are particularly effective in handling ISI, the number of operations per signaling interval grows linearly or exponentially with the ISI span, or, equivalently, with the data rates. This results in an excessive complexity and makes

TDEs unfeasible for VLC systems where data rates of several hundreds of megabits per second are targeted. A powerful alternative to the single-carrier TDE would be the deployment of multicarrier communication techniques. The most popular form of multicarrier communications is the orthogonal frequency-division multiplexing (OFDM), which has been already adopted in various wireless RF and wireline standards and recently applied to VLC systems [8]. Different from the RF approach, in OFDM for optical systems, the dc biasing or asymmetrical clipping have been introduced to ensure the non-negativity of the intensity-modulated optical signal. Multiple-input, multiple-output (MIMO) communication, another innovative technique that was also originally proposed for RF systems, has been further applied to VLC systems in recent studies. Since a number of LEDs are used to achieve the required intensity in a typical lighting application, the MIMO techniques emerge as a natural candidate in VLC systems to boost the data rate. The combination of OFDM and MIMO is considered a powerful physical layer solution for high-speed vehicular VLC systems to support bandwidth-hungry infotainment applications. Another key component to enable connectivity in vehicular VLC networks is the concept of multi hop transmission. There are only sporadic works that have addressed multi hop VLC transmission. For example, researchers from the Disney Research Center have demonstrated toy-to-toy car communications where messages sent via VLC are passed from one toy car to another in a multi hop mode. Performance analysis of a multi hop transmission system is presented in [9] for an indoor scenario where the light signal emitted from the ceiling is relayed through a desk lamp. The current works on multi hop VLC are no longer limited to indoor environments and there is an obvious need for a thorough performance evaluation of multi hop vehicular VLC networks, autonomous vehicle platoon is an enhancement of autonomous behavior, where vehicles are organized into groups of close proximity through wireless communication. Platoon members mostly communicate with each other via the current dominant vehicular radio frequency (RF) technology, IEEE 802.11p. However, this technology leads security vulnerabilities under various attacks from adversaries. Visible light communication (VLC) has the potential to alleviate these vulnerabilities by exploiting the directivity and impermeability of light [48].

6 Upper-Layer Protocols

In addition to physical layer issues discussed previously, there are also other design considerations in the upper layers that require further attention to realize a fully functional vehicular VLC network. For example, the medium access control (MAC) protocols have been widely investigated in the literature assuming isotropic radiation of RF systems. VLC systems with their inherent directionality render conventional MAC schemes practically useless. There have been recent efforts on the design of MAC protocols taking into account this directionality feature as reported in [16]. It should be further noted that a similar line of research on the MAC and upper layers is also conducted in the context of sub terahertz communications (i.e., 60 GHz for short range wireless personal area networks, and 30–40 and 70–90 GHz for long-range wireless applications). The similarities in these works can be leveraged to some extent in MAC protocol designs for vehicular VLC networks. In contrast to RF links with the isotropic coverage, neighbor discovery and link establishments impose further challenges in vehicular VLC networks, which have not yet been addressed fully.

7 Channel Modeling Aspects

Most works on the propagation modeling and characterization of VLC channels are mainly limited to the indoor environments. For the outdoor environment, as in the case of vehicular networking, additional noise sources such as the ambient interference due to the background solar radiation and artificial light sources from cars and streetlights must be taken into consideration as mentioned earlier. Furthermore, the visibility limiting weather and environmental conditions (such as rain, snow, fog, and car fumes) and heat-induced turbulence (i.e., scintillation effects) need to be further considered. The outdoor VLC channel modeling has been investigated only in sporadic works [7], where the effects of solar irradiance and artificial light sources are modeled and incorporated in the channel model. While these initial works point out the striking difference between the indoor and outdoor VLC channels, systematic modeling and characterization of outdoor VLC channels particularly considering that the ITS environments and scenarios does not exist yet.

Channel modeling approach builds on ray tracing where numerous photons are statistically generated based on the source pattern and traced until the receiver simulating the interactions of each photon with the propagation medium. In the first step, we construct the three-dimensional model of outdoor environment in Zemax® specifying the optical characteristics of the vehicles and road surface (i.e., wavelength dependent reflectance and reflection type) as well as the refractive index, size and density of suspended water particles (e.g., fog) or hydrometeors (e.g., rain) in the propagation medium. In Zemax®, "table coating method" allows defining the wavelength-dependent reflectance of surface coating for vehicles and road. Furthermore, the "scatter fraction" parameter determines the reflection type in materials. This parameter changes between 0 and 1 such that zero indicates the purely specular reflections and unity indicates purely diffuse case. Mie scattering is used to model clear, rainy and foggy weather conditions with different visibilities [38, Chapter 3]. "Bulk scatter method" in the software allows providing the input parameters "particle index" (the refractive index of particles), "size" (the radius of the spherical particles) and "density" (the density of particles). The characteristics of various weather types are listed in Table 1.

Table 1. Characteristics of various weather types

	Particle Index	Size (μm)	Density (cm^{-3})
Clear	1.000277	10^{-4}	10^{19}
Rain	1.33	100	0.1
Fog, $V = 50$ m	1.33	10	124.6
Fog, $V = 10$ m	1.33	10	622.6

After we create the simulation environment, we use the built-in ray tracing function to determine the CIR. The nonsequential ray tracing tool generates an output file, which includes the detected power and path lengths from source to detector for each ray. We process this file and, using this information, we can express the CIR as

$$h(t) = \sum_{i=1}^{N_r} P_i \delta(t - \tau_i)$$

where P_i is the optical power of the ith ray, τ_i is the propagation time of the ith ray, $\delta(t)$ is the Dirac delta function and N_r is the number of rays received at the detector.

We consider a V2V scenario shown in Fig. 1. The coating material of vehicles is considered as black gloss paint. Following the specifications of International Commission on Illumination (CIE) [39], we assume the road type R2. This corresponds to asphalt with aggregate including a minimum of 60% gravel sized larger than 10 mm or asphalt with aggregate including a minimum of 10–15% artificial brightener aggregate. For the clear and foggy weathers, we assume mixed diffuse and specular reflections while mostly specular reflections are considered for the wet road in rainy weather [34].

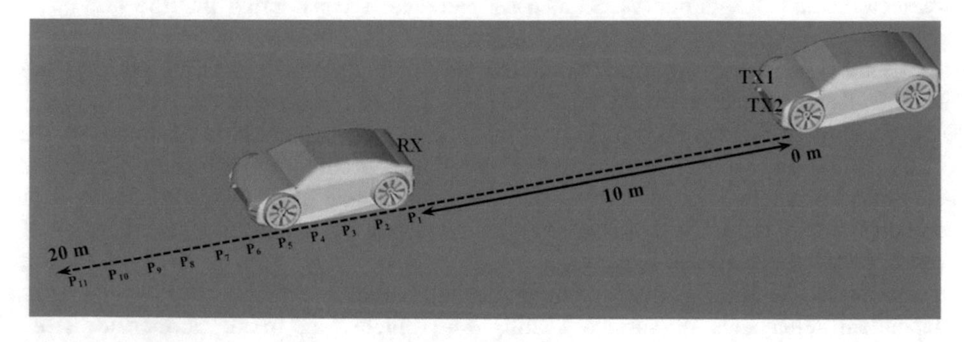

Fig. 1. V2V scenario

We use Philips Luxeon Rebel automotive white LED as the high-beam headlamp with the relative spectral power distribution shown in Fig. 2.a. Due to asymmetrical intensity distribution of luminaire [40], different cross sections indicated by C0–C180, C90–C270 and C135–C315 planes are shown in Fig. 2.b.

Two headlamps with their total power normalized to unity are placed in the front side of the first vehicle as the transmitters. One photodetector (PD) is placed in the back side of the other vehicle as the receiver (see Fig. 1). The PD has a size of 1 cm^2 and field of view (FOV) of 180. Based on the approach summarized above and for the given scenario, we obtain the CIR for the V2V link under consideration. We assume that the two vehicles are separated from each other initially at a distance of 10 m. We obtain the CIRs through all points with 1 m steps over the driving direction of the car for a range of 10 m (i.e., CIR samples were taken at P_j; $j = 1; 2; ..; 11$). As an example, Fig. 3 presents the CIRs for clear, rainy and foggy weather conditions. It is observed that the amplitude of CIR in rain decreases to 90% of that in clear weather. On the other hand, fog has much more significant impact. For V = 50 m, the CIR amplitude decreases to 45% of that in clear weather. It further decreases to 2.7% for V = 10 m.

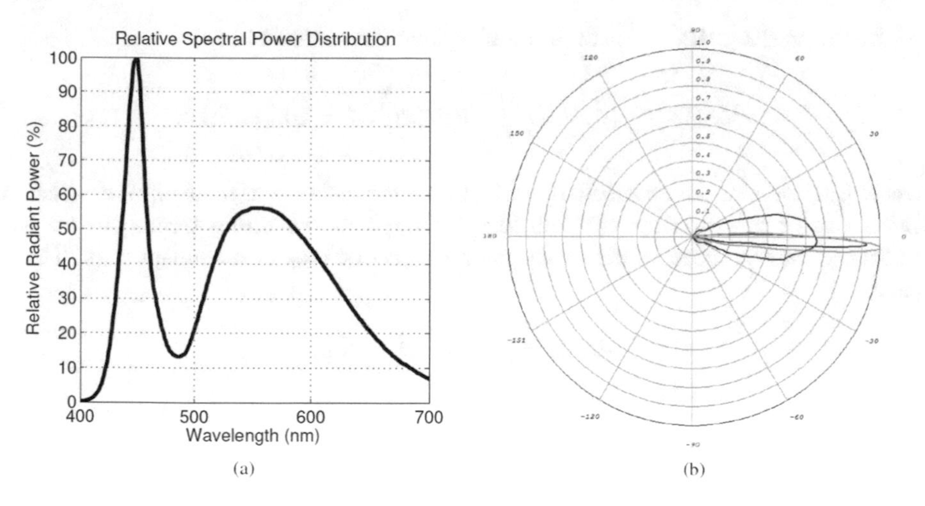

Fig. 2. (a) Relative spectral power distribution and (b) relative intensity distributions of high-beam headlamp

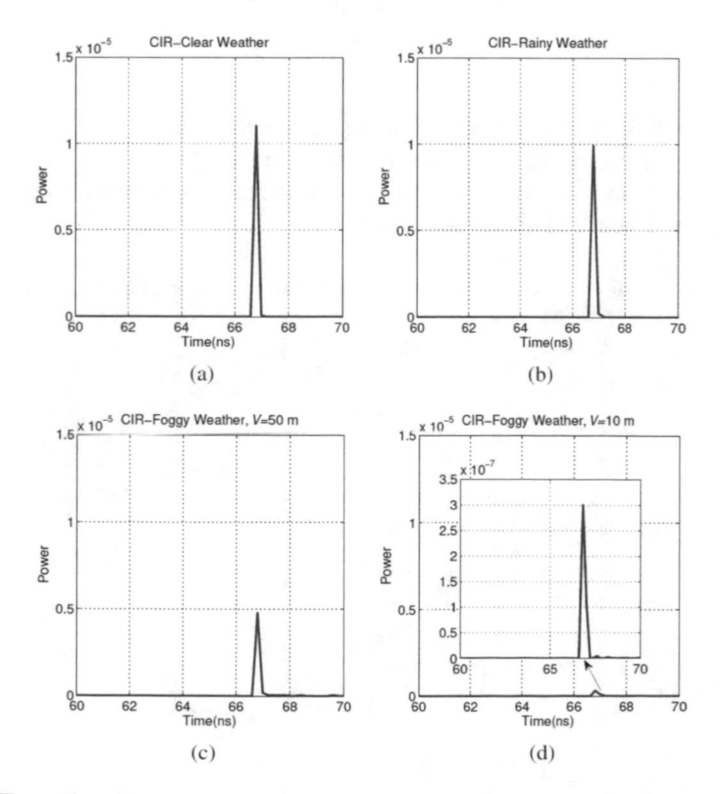

Fig. 3. CIRs at distance of 20 m for (a) clear weather (b) rainy weather (c) foggy weather with $V = 50$ m and (d) foggy weather with $V = 10$ m

Based on the CIRs, the path loss can be then calculated as [41]

$$PL_{\text{sim}}(d) = 10\log_{10}\left(\int_0^{\infty} h_d(t)dt\right), d = 10, \ldots, 20 \text{ m}$$

where $h_d(t)$ denotes the multipath optical CIR for the distance d. In an effort to obtain a closed-form expression for the path loss, we apply curve fitting techniques on our calculated values above based on the minimization of root mean square error. These yields

$$PL(d) = Ad + B, \quad d \geq 10 \text{ m}$$

Table 2. Coefficients in (3) for different weather types

	A	B
Clear	−0.44	−40.93
Rain	−0.46	−40.90
Fog, V = 50 m	−0.61	−40.46
Fog, V = 10 m	−1.20	−40.38

where the coefficients A and B are found via data fitting and depend on weather type (see Table 2). In Fig. 4, we present the path loss versus distance for different weather conditions under consideration. It is observed that the proposed closed form expression, provides a good match to simulation results.

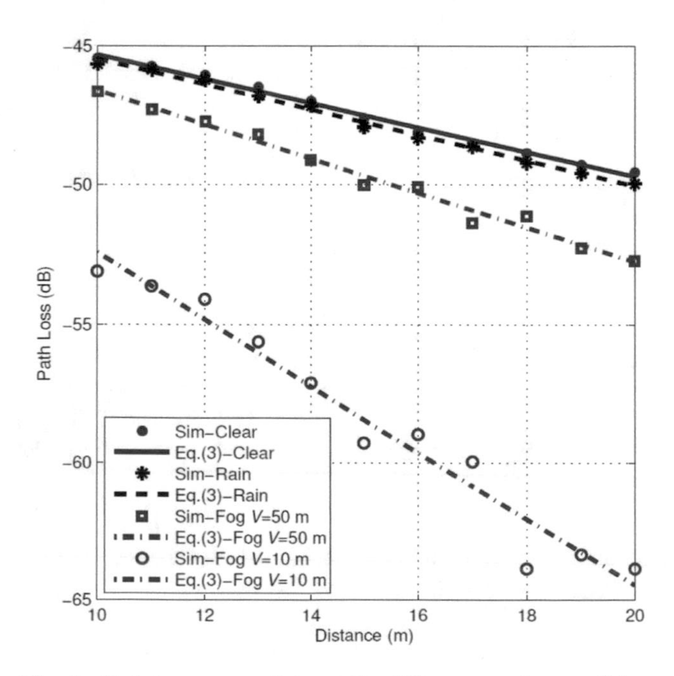

Fig. 4. Path loss versus distance for different weather conditions

8 Maximum Achievable Distance

In this section, we derive the maximum achievable link distance to achieve a specified BER. Let h denote the optical path loss measured on a linear scale (i.e., optical channel coefficient). Based on (3), h can be written as

$$h = 10^{\frac{Ad+B}{10}}$$

Let B_s and τ_d respectively denote the data rate and channel delay spread. A channel is classified as frequency-selective for $B_s\,\tau_d \geq 1$. If $B_s\,\tau_d \sim 1$, then it is classified as frequency flat channel [42]. It can be readily checked from the CIRs that the RMS delay spreads are on the order of 0.5 ns. Such vehicular VLC channels exhibit frequency selectivity only for data rates higher than 200 Megabits/sec (Mb/s). For most practical purposes in vehicular networks, data rates on the order of tens of Mb/s are sufficient. Therefore, frequency flat assumption can be justified allowing the use of single carrier systems with simple modulation techniques such as pulse amplitude modulation (PAM). For high SNR region and over AWGN channel, the BER of M-ary PAM can be approximated by [43]

$$BER \approx \frac{2(M-1)}{M\log_2(M)} Q\left(\frac{1}{M-1}\sqrt{\frac{(rhP_t)^2 T_s}{N_0}}\right)$$

where M is constellation size, r is the responsivity of photodetector, P_t is the average transmitted optical power, N0 is the noise power spectral density and Ts is the sampling interval. Let BER target denote the targeted BER value. Solving the equation for h, we have

$$h \approx \left((M-1)\sqrt{\frac{N_0}{(rP_t)^2 T_s}}\right) Q^{-1}\left(\frac{BER_{\text{target}} M\log_2(M)}{2(M-1)}\right)$$

By replacing h and solving for d, the maximum transmission distance is obtained as

$$d \approx \left(\frac{1}{A}\right)\left(10\log_{10}\left(\left((M-1)\sqrt{\frac{N_0}{(rP_t)^2 T_s}}\right) Q^{-1}\left(\frac{BER_{\text{target}} M\log_2(M)}{2(M-1)}\right)\right) - B\right)$$

9 Numerical Results

Numerical results for the maximum achievable link distance in point-to-point transmission systems under different weather types and modulation orders to achieve a specified BER. We assume $r = 0{:}28$ *A/W* [47], $P_t = 10$ dBm per each headlamp, $N_0 = 10^{-22}$ *W/Hz* and $T_s = 1$ ms. We set $\text{BER}_{\text{target}}$ as 10^{-6}. In Fig. 5, we present the BER performance versus distance for a point-to-point V2V system. In Table 3, we present the maximum achievable distances for different sizes of PAM. It is observed that the maximum distance that can be obtained for the clear weather by 2-PAM is 72.21 m. This reduces to 69.13 m, 52.85 m and 26.93 m respectively for rainy weather, foggy weather with visibility of V = 50 m and foggy weather with visibility of V = 10 m. This is a result of the fact that the light beam has more attenuation in adverse weather types. It is also observed that as modulation size is increased, the maximum distance for reliable transmission decreases.

For example, for 32-PAM, the maximum distance that can be obtained for the clear weather is 38.73 m. This reduces to 37.11 m, 28.71 m and 14.66 m respectively for rainy weather, foggy weather with visibility of V = 50 m and foggy weather with visibility of V = 10 m.

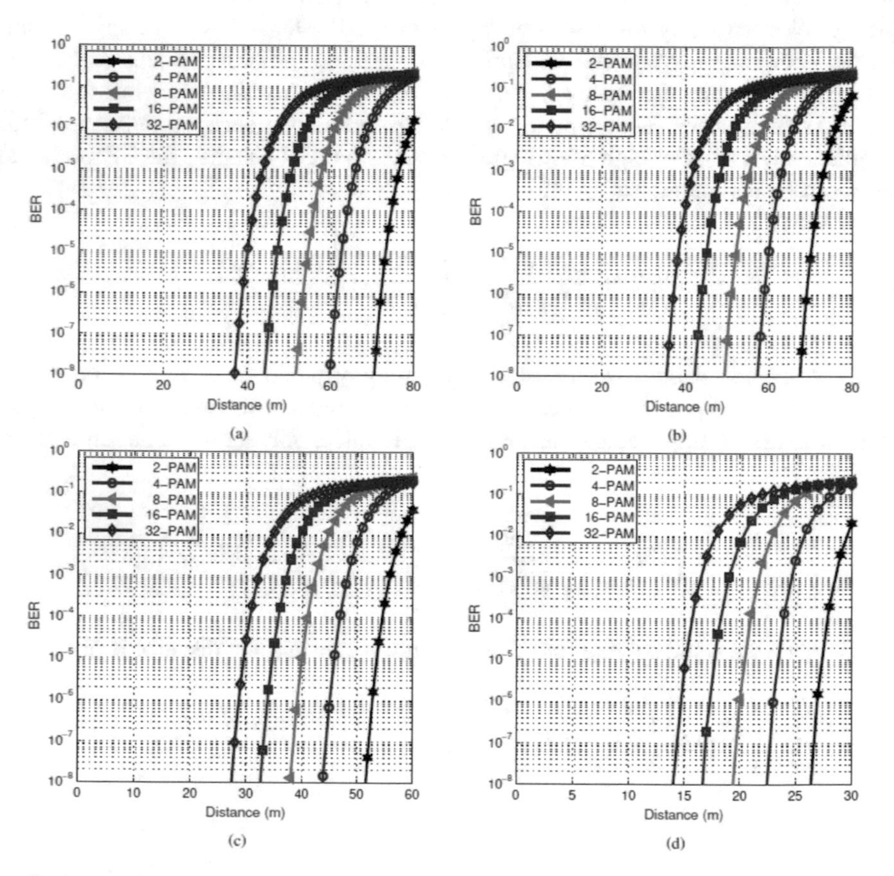

Fig. 5. BER performance of V2V transmission for (a) clear weather (b) rainy weather (c) foggy weather with V = 50 m and (d) foggy weather with V = 10 m

Table 3. Maximum achievable distance for different weather types (BER = 10^{-6})

Modulation	Maximum distance (m)			
	Clear	Rain	Fog V = 50 m	Fog V = 10 m
2-PAM	72.21	69.13	52.85	26.93
4-PAM	61.49	58.88	45.12	23.01
8-PAM	53.23	50.98	39.17	19.98
16-PAM	45.81	43.88	33.81	17.25
32-PAM	38.73	37.11	28.71	14.66

10 Conclusion

We have proposed a form pathloss expression for V2V VLC channel for different weather conditions. Building on non-sequential ray tracing, we have precisely taken into account the asymmetrical intensity distributions of automotive headlights and road reflection model. We have used the derived closed-form expression to determine the maximum achievable distance to ensure a targeted BER. Our results indicated that for a typical system configuration and assuming the deployment of 2-PAM, the maximum achievable distance in clear weather conditions is about 72 m. This reduces to around 26 m in the presence of fog.

References

1. Islim, M.S., Haas, H.: Modulation Techniques for Li-Fi. ISSN. 1673-5188, 4 February 2016
2. U.S. Transportation Secretary Anthony Foxx. http://www.nhtsa.gov/About+NHTSA/Press+Releases/NHTSA-issues-advanced-notice-of-proposed-rulemaking-on-V2V-communications
3. Uysal, M., Ghassemlooy, Z., Bekkali, A., Kadri, A., Menouar, H.: Performance study of a V2V system using a measured headlamp beam pattern model. https://doi.org/10.1109/mvt.2015.2481561
4. https://purelifi.com/lifi-products/
5. Tsonev, D., Videv, S., Haas, H.: Light Fidelity (Li-Fi): Towards All-Optical Networking
6. Morgan, Y.L.: Notes on DSRC & WAVE standards suite: its architecture, design, and characteristics. IEEE Commun. Surveys Tut. **12**(4), 504–518 (2010)
7. Lee, S.J., Kwon, J.K., Jung, S.Y., Kwon, Y.H.: Simulation modeling of visible light communication channel for automotive applications. In: Proceedings of 15th International IEEE Conference Intelligent Transportation Systems, pp. 463–468 (2012)
8. Mesleh, R., Elgala, H., Haas, H.: On the performance of different OFDM based optical wireless communication systems. J. Opt. Commun. Netw. **3**(8), 620–628 (2011)
9. Kizilirmak, R.C., Uysal, M.: Relay-assisted OFDM transmission for indoor visible light communication. In: Proceedings IEEE International Black Sea Conference Communications Networking, pp. 11–15 (2014)
10. Akanegawa, M., Tanaka, Y., Nakagawa, M.: Basic study on traffic information system using LED traffic lights. IEEE Trans. Intell. Transport. Syst. **2**(4), 197–203 (2001)

11. Kitano, S., Haruyama, S., Nakagawa, M.: LED road illumination communications system. In: Proceedings IEEE 58th Vehicular Technology Conference Fall, vol. 5, pp. 3346–3350 (2003)
12. Liu, C., Sadeghi, B., Knightly, E.: Enabling vehicular visible light communication (V2LC) networks. In: Proceedings of 8th ACM International Workshop Vehicular Inter-Networking, pp. 41–50 (2011)
13. Lourenco, N., Terra, D., Kumar, N., Alves, L.N., Aguiar, R.L.: Visible light communication system for outdoor applications. In: Proceedings of 8th International Symposium Communication Systems, Networks Digital Signal Processing, pp. 1–6 (2012)
14. Yu, S.-H., Shih, O., Tsai, H.-M., Wisitpongphan, N., Roberts, R.: Smart automotive lighting for vehicle safety. IEEE Commun. Mag. **51**(12), 50–59 (2013)
15. Takai, I., Harada, T., Andoh, M., Yasutomi, K., Kagawa, K., Kawahito, S.: Optical vehicle-to-vehicle communication system using LED transmitter and camera receiver. IEEE Photon. J. **6**(5), 1–14 (2014)
16. Tomas, B., Tsai, H.-M., Boban, M.: Simulating vehicular visible light communication: physical radio and MAC modeling. In: Proceedings of IEEE Vehicular Networking Conference, pp. 222–225 (2014)
17. Luo, P., Ghassemlooy, Z., Minh, H.L., Tang, X., Tsai, H.-M.: Undersampled phase shift ON-OFF keying for camera communication. In: Proceedings of 6th International Conference Wireless Communications Signal Processing, pp. 1–6 (2014)
18. Martinez, F.J., et al.: Emergency services in future intelligent transportation systems based on vehicular communication networks. IEEE Intell. Transp. Syst. Mag. **2**(2), 6–20 (2010)
19. Meroth, A.M., et al.: Functional safety and development process capability for intelligent transportation systems. IEEE Intell. Transp. Syst. Mag. **7**(4), 12–23 (2015)
20. Karagiannis, O., et al.: Vehicular networking: a survey and tutorial on requirements, architectures, challenges, standards and solutions. IEEE Commun. Surveys Tut. **13**(4), 584–616 (2011)
21. Zheng, K., et al.: Heterogeneous vehicular networking: a survey on architecture, challenges, and solutions. IEEE Commun. Surveys Tut. **17**(4), 2377–2396 (2015)
22. Intelligent transport systems (ITS) usage in ITU Member States: Working Document toward a Preliminary Draft New Report ITU-R M. [ITS USAGE], Annex 32 to Document 5A/469-E, June 2017
23. Uysal, M., Ghassemlooy, Z., Bekkali, A., Kadri, A., Menouar, H.: Visible light communication for vehicular networking: performance study of a V2V system using a measured headlamp beam pattern model. IEEE Veh. Technol. Mag. **10**(4), 45–53 (2015)
24. Karunatilaka, D., Zafar, F., Kalavally, V., Parthiban, R.: LED based indoor visible light communications: state of the art. IEEE Commun. Surveys Tutor. **17**(3), 1649–1678 (2015)
25. Pathak, P.H., Feng, X., Hu, P., Mohapatra, P.: Visible light communication, net-working, and sensing: a survey, potential and challenges. IEEE Commun. Surveys Tutor. **17**(4), 2047–2077 (2015)
26. Yu, S.H., et al.: Smart automotive lighting for vehicle safety. IEEE Commun. Mag. **51**(12), 50–59 (2013)
27. Cailean, A.M., Dimian, M.: Current challenges for visible light communications usage in vehicle applications: a survey. IEEE Commun. Surveys Tutor. **19**(4), 2681–2703 (2017)
28. Cailean, A.M., Dimian, M.: Impact of IEEE 802.15. 7 standard onvisible light communications usage in automotive applications. IEEE Commun. Mag. (2017). https://doi.org/10.1109/mcom.2017.1600206cm
29. Akanegawa, M., Tanaka, Y., Nakagawa, M.: Basic study on traffic information system using LED traffic lights. IEEE Trans. Intell. Transp. Syst. **2**(4), 197–203 (2001)

30. Kumar, N., Terra, D., Lourenco, N., Alves, L.N., Aguiar, R.L.: Visible light communication for intelligent transportation in road safety applications. In: Proceedings of IEEE International Conference Wireless Communications and Mobile Computing, pp. 1513–1518 (2011)
31. Viriyasitavat, W., et al.: Short paper: channel model for visible light communications using off-the-shelf scooter taillight. In: IEEE Vehicular Networking Conference (VNC), pp. 170–173 (2013)
32. Lee, S.J., Kwon, J.K., Jung, S.Y., Kwon, Y.H.: Evaluation of visible light communication channel delay profiles for automotive applications. EURASIP J. Wirel. Commun. Netw. **2012**(370), 1–8 (2012)
33. Luo, P., et al.: Fundamental analysis of a car to car visible light communication system. In: 9th International Symposium Communication Systems, Networks & Digital Signal Processing (CSNDSP), pp. 1011–1016 (2014)
34. Luo, P., et al.: Performance analysis of a car-to-car visible light communication system. Appl. Opt. **54**(7), 1696–1706 (2015)
35. Kim, Y.H., Cahyadi, W.A., Chung, Y.H.: Experimental demonstration of LED-based vehicle to vehicle communication under atmospheric turbulence. In: IEEE International Conference Information and Communication Technology Convergence (ICTC), pp. 1143–1145 (2015)
36. Kim, Y.H., Cahyadi, W.A., Chung, Y.H.: Experimental demonstration of VLC-based vehicle-to-vehicle communications under fog conditions. IEEE Photon. J. **7**(6), 1–9 (2015)
37. Schulz, D., Jungnickel, V., Das, S., Hohmann, J., Hilt, J., Hellwig, P., Paraskevopoulos, A., Freund, R.: Long-term outdoor measurements using a rate-adaptive hybrid optical wireless/60 GHz link over 100 m. In: IEEE 19th International Conference Transparent Optical Networks (ICTON), pp. 1–4 (2017)
38. Uysal, M., Capsoni, C., Ghassemlooy, Z., Boucouvalas, A., Udvary, E.: Optical Wireless Communications: An Emerging Technology. Springer (2016)
39. Stark, R.E.: Road surfaces reflectance influences lighting design. Light. Design Appl. (1986)
40. Agreement Addendum 111: Regulation No. 112 Revision 3-unece (2013)
41. Miramirkhani, F., Narmanlioglu, O., Uysal, M., Panayirci, E.: A mobile channel model for VLC and application to adaptive system design. IEEE Commun. Lett. **21**(5), 1035–1038 (2017)
42. Goldsmith, A.: Wireless Communications. Cambridge University Press, Cambridge (2004)
43. Fath, T., Haas, H.: Performance comparison of MIMO techniques for optical wireless communications in indoor environment. IEEE Trans. Commun. **61**(2), 733–742 (2013)
44. Morgado, E., Mora-Jimenez, I., Vinagre, J.J., Ramos, J., Caamano, A.J.: End-to-end average BER in multihop wireless networks over fading channels. IEEE Trans. Wirel. Commun. **9**(8), 2478–2487 (2010)
45. Florea, A., Yanikomeroglu, H.: On the optimal number of hops in infrastructure-based fixed relay networks. In: IEEE Global Telecommunications Conference (GLOBECOM 2005), pp. 3242–3247 (2005)
46. Sperling, L.H.: Introduction to Physical Polymer Science. Wiley, New York (2005)
47. Grubor, J., Randel, S., Langer, K.D., Walewski, J.W.: Broadband information broadcasting using LED-based interior lighting. J. Lightwave Technol. **26**(24), 3883–3892 (2008)
48. Ucar, S., Ergen, S., Ozkasap, O.: IEEE 802.11p and visible light hybrid communication based secure autonomous platoon. IEEE Trans. Veh. Technol. **67**(9), 8667–8681 (2018)

Usability Study of Voice-Activated Smart Home Technology

Nichole Hugo(✉), Toqeer Israr, Wutthigrai Boonsuk,
Yasmine Ben Miloud, Jerry Cloward, and Peter Ping Liu

Eastern Illinois University, Charleston, IL, USA
{nhugo, taisrar, wboonsuk, ybenmiloud,
jcloward, pliu}@eiu.edu

Abstract. One of the most recent and popular forms of technology in the consumer market is the virtual voice assistant (i.e., smart speaker) made by well-known companies such as Amazon Echo and Google Home. These devices allow users to perform voice interaction to perform a variety of tasks that can be beneficial to consumer groups such as high school students, college students, rural consumers, families, senior citizens, and disabled consumers. Thus, this study investigated the key elements of different consumer groups related to voice-activated smart home technology in order to understand how to educate and promote this technology to individual groups. In the study, there were 71 participants from six consumer groups including high school students, college students, rural consumers, consumers with disabilities, seniors, and families. Each group was surveyed prior and after a demonstrative smart home workshop to evaluate their knowledge, perceived usefulness, and preferential use of smart home technology. The results of the study showed that while the utilization of voice-activated smart home technology was not significantly different between consumer groups, the top-three benefits of using this technology were security, safety, and assurance. The major barriers that prevented consumer groups from implementing the technology was the cost of the system and concerns with user privacy.

Keywords: Smart home devices · Amazon Alexa · Google Home · Voice activated technology

1 Literature Review

An overview of smart home devices is provided by giving background information on the main two categories of smart devices (controlling and controlled smart devices), examples and features of each device used in the study and a literature review on smart home devices with different consumer groups.

1.1 Background

One of the most recent and popular forms of technology in the consumer market is the virtual voice assistant (i.e., smart speaker) made by well-known companies such as Amazon and Google. Voice assistants are software programs that can understand

M. E. Auer and D. May (Eds.): REV 2020, AISC 1231, pp. 652–666, 2021.
https://doi.org/10.1007/978-3-030-52575-0_54

human speech and respond accordingly via an artificial voice. Apple's Siri, Amazon's Alexa, Microsoft's Cortana, and Google's Assistant are the most popular voice assistants embedded in smartphones and/or dedicated home devices. These devices can take commands to perform common tasks such as playing music, streaming videos, checking the weather, making shopping lists, asking about current news, or searching the internet. The ability of these assistants can also be extended to control smart home devices (e.g., lights, thermostats, refrigerators, dishwashers, washers and dryers, garage doors, and security systems).

Each of these voice assistant software programs are enabled on smart devices manufactured by their respective companies, such as voice assistant Alexa supported by Amazon's Echo and its family of devices, and Google Assistant supported by Google's Assistant and its family of devices.

While these voice assistants are attractive to tech-savvy users, they can be beneficial to other consumer groups such as high school students, college students, families, senior citizens, and disabled consumers. For instance students whether in high school or in college can use these devices for studying, while veterans, senior citizens and people with disabilities could rely on these devices to increase convenience and make their life safer and easier, for families these devices could help a lot around the house per example for syncing everyone schedule and organize their daily life. With the variety of applications available to help with managing tasks, setting reminders and activating smart home devices to assist with safety and security, these various groups can have improved lifestyle experiences by understanding how to operate these devices.

1.2 Smart Devices

A "smart home device" is defined as any single-purpose internet-connected device intended for home use such as thermostat, outlet, blood-pressure monitor or a device that connects and controls multiple single purpose devices [1].

Smart devices can be classified into two categories: "controlling" and "controlled" smart devices. A controlling device would be a smart device with control over other smart devices, such as the Amazon Echo Dot, Google Assistant, etc. These devices are typically recipient to user's speech commands and can complete tasks such as turning off a light. A controlled device does not have a direct human-machine interface, but rather they receive their instructions from a controlling device. Examples of controlled devices would include the Philips smart bulb, smart door lock, or a smart switch.

1.3 Controlling Smart Devices

For this study, controlling smart devices manufactured by Amazon and Google were utilized. All the smart controlling devices manufactured by Amazon and Google use the voice assistants Alexa and Google Assistant respectively. These devices are in listening mode continuously and the assistant is activated by a triggering word or phrase from a user; which is set by default to "Alexa", "Okay Google", or it can be customized by a user [2]. It executes voice-operated functions while communicating through a local WiFi Internet connection with its cloud servers, or other smart home devices, to respond to the user commands [3, 4]. The same device can be used by multiple individuals and the assistant can distinguish between them via voice profile set up.

While Alexa and Google Assistant are quite similar, they are a few very important distinctions. A comparative overview between both assistants used in this study is presented in Table 1 [5].

Table 1. Google Assistant & Alexa features overview

	Google assistant	Alexa
Control smart devices	X	X
Create routines	X	X
Web browser	X	X
Customized skills		X
Voice shopping		X
Call/text		X
Access calendar	X	X
Voice profile	X	X

As mentioned, both of these are activated by certain keywords and respond back to the user intelligently (most of the time). They both have the ability to control certain compatible controlled smart devices (such as Philips smart bulb). Both can support a "routine", which is a set of commands triggered by some event or time of day. An example of a routine would be at 7 am every weekday the smart assistant would say "Good morning" and report the weather of the day.

Both of the assistants can be used as "web browsers" to perform searches on the internet, with accessibility being the difference between a traditional browser and smart assistant browser. For a traditional browser, the user would typically use a computer with a keyboard to browse. For smart assistants, the task is completed verbally. Furthermore, both assistants can be relied upon to have personal calendars to keep track of important meetings and events. With each of these voice assistants, users can have profiles in order to support multiple users.

While there are many similarities, the assistants differ in some key features. Alexa allows users to create their own "skills" using Blueprint [6, 7]. Skills are similar to apps, which can be enabled/disabled using the Alexa app or a web browser, like apps on a smartphone and/or tablet. Skills extend voice-driven Alexa capabilities to bring products and services to life. With Alexa, users can also shop using voice commands to purchase items from Amazon's shopping portal. While Google Assistant enables calling and texting people via a phone, Alexa can do the same without the need of a phone.

1.4 Amazon Products

Amazon has a variety of products which support Alexa, each with varying features. In our study Echo Dot, Echo Plus, Echo Show, and Echo Look were all used. Table 2 presents an overview of the Alexa devices from the study.

Table 2. Technical fatures of Alexa devices

	Echo dot (2nd generation)	Echo plus	Echo show	Echo look
Alexa voice service	X	X	X	X
Wi-Fi	X	X	X	X
Bluetooth	X	X	X	X
Zigbee (Built-in hub)		X		
Dolby (Room filling Speaker)		X	X	
Touch screen			X	
Camera			X	X

Amazon's most basic device is the Echo Dot. It offers its customers the possibility to access any Alexa voice service which offers preset skills such as checking the news, the weather, the traffic of a commute, access to web browser, play music, and customized skills like setting up routines, control other controlled smart home devices such as August smart lock or a Philips Hue smart bulb and play games.

Echo Plus offers the same features as Echo Dot from an assistant skills perspective. However, additionally it is an integrated hub that removes the need for an external hub to coordinate with other smart devices. It has an integrated larger speaker that improves the quality of sound for playing music.

Echo Show is an extension of an Echo Dot, with a touch screen and a camera. It offers visual support to your commands such viewing a recipe or an article from Wikipedia. It is the only device that offers the possibility to access camera feeds, and has additionally a front camera that allows the user to have video conferences with other Echo Show users or via third party compatible applications such as Skype.

Echo Look, again an extension of an Echo Dot, has a camera that can take pictures or create videos of the user's outfit. This is accomplished by both hands-free commands or through the Echo Look app. Additionally, Echo Look can provide professional feedback on the user's ensemble.

1.5 Google Products

While Google offers many products, such as Google Nest and Google Nest Hello Doorbell, we made use of Google Home Mini and Google Nest Hub. Similar to Amazon's Echo Dot, Google Home Mini supports Google Assistant which allows access to Google Assistant. This is the most basic controlling smart device by Google. In addition to all the features of a Google Home Mini, Google Nest Hub has a touch screen and a built-in Chrome cast. With the touch screens, users can use third party software to do video conferencing, such as Skype, as well as enjoy watching videos on popular streaming websites such as YouTube. Table 3 presents an overview of the involved Google devices.

Table 3. Technical features of Google Home

	Google Home mini	Google nest hub
Google voice service	X	X
Wi-Fi	X	X
Bluetooth	X	X
Microphone	X	X
Speaker	X	X
Touch Screen		X
Built-in Chromecast		X

1.6 Controlled Devices Features

The application and benefit of controlled smart home devices are numerous. In this study, a selection has been purchased to set the smart home environment in the laboratory. In this selection, there was devices for energy saving, security, entertainment, and convenience purposes that will be presented in further details in the following section.

Under smart devices useful for energy efficiency purposes, there is smart lighting offered by many companies such as Philips Hue or GE. A remote control through an app or a multi-purpose smart device such Google Assistant or Alexa allows the device to be turned on or off. Smart plugs and smart power strips give the user the opportunity to easily control the appliances that consume power anytime they are plugged in, even when not in use, known as vampire loads. Smart thermostats can help reduce heating and cooling cost just by providing the ability to be remote controlled when away from the house [8].

For devices with security purposes, there are the smart locks (e.g. August) that have ability to remote control locks and verify whether a door is open or not. In addition, smart security systems can also be used to monitor the home and sends an alert of potential trespassers when the cameras or motion sensors detect a movement.

Ring [9], per example, offers a security kit constituted of:

- A Base Station: A hub with wireless home security network. It includes a siren and backup battery power for up to 24 h
- A Keypad: Arms and disarms the ring alarm security system and mounts to walls or can be used as a portable keypad.
- Motion detector: Detects motion inside the house and mounts to room corners or flat walls.
- Contact Sensor: Notifies when doors or windows are open. Mounts to doors or window frames.

– Range Extender: Extends the signal from your Ring Base Station to other Ring Alarm components.
– Smoke & CO Listener: Sends notifications when existing smoke and carbon monoxide detectors sound their siren.
– Flood & Freeze Sensor: Detects water and low temperatures. Sends notifications to help avoid water damage.

Ring offers indoor and outdoor cameras accessible through the Ring app, giving real-time access to a camera. The camera connects to the Ring app and Echo devices to inform if someone is at the door and offers the opportunity to communicate with that person. In the Table 4, a selection of the smart home features in the manufacturer app and through Echo that were used in the experimental smart home. For devices providing entertainment, there are TV streaming technologies, such as Fire TV Stick for Alexa and Chromecast for Google Assistant that give access to favorite movies, subscription services, music, etc. Smart device assisting with improving convenience include the Behmor coffee maker, which can make coffee remotely using the Behmor app or voice assistant.

Table 4. Features of security devices

	Features			
	Smart lock	Security Kit= A set of motion detectors, sensors and sirens	Smart cameras	Smart doorbell
Brand	August	Ring	Ring	Ring
Application features	Control your locks by verify lock status; Automate unlocking	Enable/disable the monitoring of sensors (arm or disarm) Realtime access to sensors	Real time access to cameras History if status is offline or online	See who is there and answer the door from anywhere with a Video Doorbell
Echo's features	Control your locks	Enable/Disable the monitoring of sensors	Real time access to cameras if in possession of an Echo with a screen	Voice communication with the person at the door from any Echo device. In the case of an Echo with a screen, user can see them as well

1.7 Existing Studies with Various Consumer Groups

With smart technology becoming increasingly ubiquitous, it is quite essential to analyze the challenges consumers are still facing in adopting this highly popular technology. For this research, we classified our consumer groups in the following categories: high school students, university students, people living in rural areas, individuals with disabilities, seniors (elderly) and families. Research has been done with smart home technology serving users in general [10, 11] and with a focus on the elderly, healthcare patients or people with disabilities [12–20]. These studies have primarily focused on these users to assist in allowing greater independence from care facilities, maintain good health through medication reminders and the ability to call for help if in distress, and prevent social isolation while at home [21]. However, there is a research gap when it comes to analyzing a wide variety of consumer groups, outside of those mentioned.

Smart home research focusing on families, people living in rural areas, high school and university students has been rather limited. Only one academic study was found that analyzed the benefits of smart homes with families as the users. This study used dual income families to analyze how this technology could be utilized to provide families with a feeling of control in their lives, particularly focusing on family routines and communication [22]. None of the research we found investigated the evaluation of previous knowledge, training and evaluation of post training knowledge of smart home technology offered to such a wide variety of consumer groups.

2 Methodology

2.1 Participant Selection

Smart home devices offer a range of features that can be more useful for certain people than others. Per example appliance voice controls are more beneficial for people with physical disabilities, like veterans and senior citizens. Students may benefit from studying or scheduling applications. Apps to help around the house, such as setting timers or the thermostat, may be more useful for families, while online shopping tools can create conveniences for citizens in rural areas. In this study, 6 group participants were identified and contacted from senior citizens, people with disabilities, high school and students, rural residents and families to evaluate their perception of smart home device benefits. Participants were purposefully selected by researching out to organizations to ask for volunteers to complete the study. For example, a nearby high school was contacted to gain students to ask if they were interested in participating in the study.

2.2 Description

Actively engaging and educating consumers is of paramount importance in order to understand the benefits and barriers that consumers perceive related to smart home technology. This in turn could assist with understanding the limitations of the devices, so areas of improvement can be identified in order to help meet the needs of the consumer. A selection of voice activated, and other smart home devices was purchased and tested as presented in Sect. 1. The Center for Clean Energy Research and Education (CENCRE) was set up as a smart home environment for the purpose of this research.

2.3 Design

This study investigated key elements of different consumer groups related to voice-activated smart home technology through a demonstration of multiple devices, allowing potential consumers to experience and see the benefits before they invest in smart home technology. The purpose of this study was to understand the participants' perceptions of knowledge and benefits of the devices in order to determine the best methods to educate and promote this technology to individual groups. To reach a full spectrum of consumers and provide long term benefits to those consumers, a smart home setting was created in CENCERE which allowed for participants to try out multiple smart home devices like shown in Fig. 1. This project was designed to reach a full spectrum of consumers and utilized purposeful sampling to include high school students, college students, seniors, consumers with disabilities, hard-to-reach consumers (such as rural residents), and families with young children. A total of 71 participants agreed to take part in the study.

Fig. 1. High school workshop: light control station

Quantitative method was used to measure the variables associated with knowledge and attitudes towards the smart home technology through a multi-group pre-test-post-test design. The design of the project involved a pre-test questionnaire to gauge the perceived knowledge of the participants related to smart home devices, as well as their attitudes towards them. Respondents were asked to indicate their answers on a 1–5 Likert scale. Questions had been compiled based on questionnaires from previous research on opinions related to smart home technology [23]. Participants were then given a demonstration of the devices in order to better understand their capabilities.

The demonstration consisted of a small group of people working at one station and background information on the devices they would be learning about. A brief overview of the devices and the difference between them (mentioned in Sect. 1) was provided to give the participants a basic understanding of the device. The participants were then given instructions on how to control the devices and then were tasked with giving commands to the device. Once the group completed their tasks, they moved to another station with different devices. After they completed their tasks at all of the stations, a post-test questionnaire was used to measure the change in their knowledge of the devices and their attitudes towards them.

3 Results

The survey investigated how knowledgeable participants were about smart home devices, their perception on their usefulness, benefits and what keep them from buying such devices prior and after the workshop. In this section the results to each question will be presented.

3.1 How Knowledgeable Are You About Voice Controlled Smart Home Technology?

A paired-samples t test compared the mean difference of the pre-test and post-test results for all participants for knowledge of voice controlled smart home technology. Table 5 shows a statistically significant increase was found from the pre-test to the post-test for knowledge related to voice controlled smart home technology (t $(70) = -8.10$, $p < .01$) .

Table 5. Knowledge of voice controlled smart home technology

	N	Pre-test mean	Post-test mean	Mean difference	St. Dev.	t	p
Knowledge	71	2.79	4.06	1.27	1.33	−8.10	.001**

A 6 × 2 repeated measures ANOVA was calculated to examine the effects of the groups on the pre-test and post-test results. The results in Table 6 show a significant increase from the pre-test to the post-test for four of the six groups. High school students and individuals with disabilities were the only two groups that did not show a significant increase in knowledge as a result of the demonstration. However, high school students had the highest pre-test mean for knowledge, so this group did not have much room for improvement.

Table 6. Knowledge of voice controlled smart home technology by group

Group	N	Test	Mean	Mean difference	t	p
High school students	8	Pre-test	3.500	.75	−1.821	.111
		Post-test	4.250			
College students	19	Pre-test	3.158	1.26	−5.265	.000**
		Post-test	4.421			
Rural	6	Pre-test	2.167	1.83	−3.051	.028*
		Post-test	4.000			
Individuals with disabilities	12	Pre-test	2.500	.92	−1.396	.190
		Post test	3.417			
Seniors	11	Pre-test	2.364	1.64	−4.845	.001**
		Post-test	4.000			
Families	15	Pre-test	2.667	1.4	−7.359	.000*
		Post-test	4.067			

3.2 How Useful Are Voice Controlled Smart Home Technology Devices to You?

A paired-samples t test compared the mean difference of the pre-test and post-test results for all participants for the perception of the usefulness of voice controlled smart home technology. Table 7 shows a statistically significant increase from the pre-test to the post-test for usefulness related to voice controlled smart home technology ($t(70) = -4.74$, $p < .01$).

Table 7. Usefulness of voice controlled smart home technology

	N	Pre-test mean	Post-test mean	Mean difference	St. dev.	t	p
Knowledge	71	3.24	4.15	.91	1.63	−4.74	.001**

A 6 × 2 repeated measures ANOVA was calculated to examine the effects of the groups on the pre-test and post-test results. There was not a significant difference between groups for the mean difference between the pre-test and the post-test, except for families ($t(14) = -4.84$, $p < .1$) and college students ($t(18) = -4.65$, $p < .1$) Table 8 shows the mean difference for all six groups.

3.3 How Important Are the Following Benefits to You?

A paired-samples t test compared the mean difference of the pre-test and post-test results for all participants for the perception of importance of voice controlled smart home device benefits. Table 9 shows a statistically significant increase was only found

Table 8. Usefulness of voice controlled smart home technology by group

Group	N	Test	Mean	Mean difference	t	p
High school students	8	Pre-test Post-test	4.000 4.125	.13	−.284	.785
College students	19	Pre-test Post-test	3.053 4.263	1.21	−4.652	.000**
Rural	6	Pre-test Post-test	3.667 4.167	.5	−1.168	.296
Individuals with disabilities	12	Pre-test Post-test	3.000 3.833	.83	−1.034	.323
Seniors	11	Pre-test Post-test	3.182 3.818	.64	−1.208	.255
Families	15	Pre-test Post-test	3.133 4.533	1.4	−4.836	.000**

Table 9. Benefits of voice controlled smart home technology

	N	Pre-test mean	Post-test mean	Mean difference	St. Dev.	t	p
Safety	71	4.11	4.33	.22	1.26	−1.51	.135
Security	71	4.47	4.53	.06	1.32	−.36	.720
Convenience	70	2.78	3.31	.53	1.93	−2.29	.025*
Assurance	71	4.06	4.35	.29	.15	−1.91	.060
Wellbeing	71	3.8	4.01	.21	.18	−1.19	.237
Cost savingzs	71	3.96	4.01	.05	.2	−.28	.778

for Convenience from the pre-test to the post-test related to benefits of voice controlled smart home technology ($t(70) = -2.29$, $p < .05$).

The means for pre-test and post-test for benefits of voice controlled smart home technology are displayed in Table 10 by group. This shows that high school students, followed by college students, had the highest mean for the groups when looking at Convenience. Families had the lowest mean in the Convenience section, which is also the lowest mean throughout all of the benefit categories.

3.4 What Is Keeping You from Using Smart Home Technology?

A paired-samples t test compared the mean difference of the pre-test and post-test results for all participants for barriers to acquiring voice controlled smart home technology. Table 6 shows a statistically significant increase from the pre-test to the post-test for cost related to voice controlled smart home technology ($t(68) = 2.44$, $p < .05$).

Table 10. Benefits of voice controlled smart home technology by groups

Group	Safety mean (SD)	Security mean (SD)	Convenience mean (SD)	Assurance mean (SD)	Wellbeing mean (SD)	Cost savings mean (SD)
High school students	4.563 (.113)	4.813 (.091)	3.438 (.371)	4.438 (.175)	3.938 (.290)	4.437 (.147)
College students	4.150 (.121)	4.650 (.090)	3.350 (.224)	4.100 (.152)	3.875 (.166)	4.025 (.156)
Rural	3.917 (.201)	4.250 (.112)	2.833 (.459)	3.917 (.201)	4.000 (.428)	4.167 (.167)
Individuals with disabilities	4.125 (.309)	4.125 (.214)	2.833 (.310)	3.958 (.278)	4.167 (.284)	3.625 (.343)
Seniors	4.455 (.184)	4.500 (.151)	3.000 (.350)	4.182 (.216)	3.364 (.370)	3.682 (.139)
Families	4.167 (.205)	4.600 (.163)	2.733 (.243)	4.600 (.131)	4.167 (.116)	4.133 (.192)

Table 11. Barriers to using voice controlled smart home technology

	N	Pre-test mean	Post-test mean	Mean difference	St. Dev.	t	p
Cost	69	3.39	2.96	−.43	1.48	2.44	.017*
Ease of use	70	2.25	2.41	.16	1.81	.53	.6
Privacy	70	3.29	3.19	−.1	1.93	.43	.666
Lack of benefits	69	2.97	2.59	−.38	1.65	1.89	.063

The means for pre-test and post-test for barriers of using voice controlled smart home technology are displayed in Table 12 by group. This shows that seniors, followed by high school students, had the lowest mean for the barrier of Cost. This shows that these groups found the Cost of the devices to be less of a barrier after participating in the demostration. College students, followed by families, had the highest means asssociated with the barrier of Cost. This shows that these groups were not as impacted as much by the demonstration and still felt that Cost was a barrier to keep them from buying voice controlled technology.

Table 12. Barriers to using voice controlled smart home technology by groups

Group	Cost mean (SD)	Ease of use mean (SD)	Privacy mean (SD)	Lack of benefit mean (SD)
High school students	2.750 (.283)	2.000 (.134)	2.562 (.333)	2.188 (.249)
College students	3.825 (.167)	2.775 (.133)	3.850 (.171)	3.325 (.193)
Rural	3.000 (.342)	2.833 (.247)	3.500 (.342)	3.000 (.548)
Individuals with disabilities	3.042 (.317)	2.333 (.178)	3.000 (.320)	3.083 (.342)
Seniors	2.591 (.222)	2.454 (.238)	3.000 (.410)	2.272 (.264)
Families	3.233 (.128)	2.367 (.220)	3.133 (.236)	2.467 (.264)

4 Discussion

The results from this study can be used to identify areas of growth for smart home devices, so that they may be improved upon in order to better meet the needs of consumers. Understanding which areas consumers need more awareness and education on regarding these devices can assist with giving them the knowledge to be able to utilize this technology in their lives, which can allow for improvements to their overall quality of life.

Overall, the results suggest that the demonstration of voice activated technology significantly improved participants' knowledge of these devices. Before the demonstration, rural residents, senior citizens, and individuals with disabilities indicated their lack of knowledge of smart devices. These consumer groups can be targeted for more educational efforts related to these devices. After the demonstration, the results showed significant improvement related to the knowledge of several consumer groups including college students, rural residents, senior citizens, and families. Only two consumer groups, high school students and individuals with disabilities, did not showed significant improvement in their knowledge. The results for high school students are not a concern as this group already ranked the highest for knowledge of the devices in the pre-test; thus, the change in their knowledge was not likely to be significant after the demonstration. On the other hand, individuals with disabilities had little knowledge about the devices in the pre-test and only showed a slight improvement in the post-test. These results suggest that the demonstrations should be redesigned to make it more effective for this specific group.

In terms of perception regarding usefulness of voice-controlled technology, college students and families significantly perceived voice-controlled technology as more useful following the demonstration. In general, all groups considered voice-controlled technology as useful following the demonstration. Safety, security, and assurance were considered the top three benefits of voice controlled smart home technology rated among all participants. By observing specific groups, rural residents gave cost saving and wellbeing as the second and third benefits. Individuals with disabilities weighted wellbeing as the strongest benefit, followed by safety and security. This information can be very useful for developing educational materials and creating awareness of the capabilities of these devices.

The last part of this study focused on barriers which prevent consumers to adopt the voice controlled smart home technology. The results indicated that the cost of the devices and concerns regarding consumer privacy were the strongest barriers for why consumers would not purchase smart home device before the demonstration. However, there was a significant change in the mean after the smart home demonstration, which shows that more awareness regarding the devices may change the consumer perception in this regard. Concerns regarding privacy continued to persist following the demonstration, which indicates that it is an area that needs to be addressed through education and policies associated with these devices. This can be challenging to do as this information is not clear in the user agreements as to when these devices are recording or who is able to access the information. As recordings obtained from smart home devices have been used in court cases, this concern may require clarification of the laws on consumer privacy that are not simply solved by educating consumers alone.

References

1. Apthrorpe, N., Reiman, D., Feamster, N.: A smart home is no castle: privacy vulnerabilities of encrypted IoT traffic (2017)
2. Clauser, G.: What is Alexa? (2017). https://thewirecutter.com/reviews/what-is-alexa-what-is-the-amazon-echo-and-should-you-get-one/. Accessed Nov 2019
3. Dunn, J.: We put Siri, Alexa, Google Assistant, and Cortana through a marathon of tests to see who's winning the virtual assistant race – here's what we found (2016a). https://www.businessinsider.com/siri-vs-google-assistant-cortana-alexa-2016-11. Accessed 21 Nov 2019
4. Lopatovska, I., Rink, K., Raines, K., Consenza, K., Williams, H., Sorsche, P., Hirsch, D., Li, Q., Martinez, A.: Talk to me: Exploring user interactions with amazon Alexa. Jolis **51**(4), 984–997 (2018)
5. Segan, S.: Amazon Echo vs. google home: which smart speaker is best? PC reviews, Ziff Davis, 19 July 2019. https://www.pcmag.com/article/348496/google-home-vs-amazon-echo-which-one-should-rule-your-smar. Accessed 21 Nov 2019
6. Amazon.: Skill blueprint (2019). https://blueprints.amazon.com/. Accessed Nov 2019
7. Friedewald, M., Da Costa, O., Punie, Y., Alahuhta, P., Heinonen, S.: Perspectives of ambient intelligence in the home environment. Telematics Inform. **22**, 221–238 (2005)
8. Lärka, M.: Smart homes with smartphones: creating a Smart home application for smartphones. Umea University, Umea, Sweden (2015)
9. Ring.: Security system kit. Ring (2019). https://shop.ring.com/products/alarm-security-kit-10-piece. Accessed Nov 2019
10. Sanchez, A., Tercero, R.: Smart home technologies: uses and abuses. In: 2010 Ninth Mexican International Conference on Artificial Intelligence (MICAI), pp. 97–102 (2010)
11. Georgiev, A., Schlögl, S.: Smart home technology: An exploration of end user perceptions. In: Piazolo, F., Schlögl, S. (eds.) Innovative Lösungen für eine alternde Gesellschaft: Konferenzbeiträge der SMARTER LIVES 18 20.02.2018, pp. 64–78. Pabst Science Publishers, Lengerich, Germany (2018)

12. Amiribesheli, M., Benmansour, A., Bouchachia, A.: A review of smart homes in healthcare. J. Ambient Intell. Humanized Comput. **6**(4), 495–517 (2015)
13. Cheverst, K., Clarke, K., Dewsbury, G., Hemmings, T., Hughes, J., Rouncefield, M.: Design with care: technology, disability and the home. In: Inside the smart home, pp. 163–179). Springer, London (2003)
14. Courtney, K., Demeris, G., Rantz, M., Skubic, M.: Needing smart home technologies: the perspectives of older adults in continuin car retirement communities. Inf. Prim. Care **16**, 195–201 (2008)
15. Demiris, G., Hensel, B.K.: Technologies for an aging society: a systematic review of "smart home" applications. Yearb Med. Inform. **3**, 33–40 (2008)
16. Demiris, G., Rantz, M.J., Aud, M.A., Marek, K.D., Tyrer, H.W., Skubic, M., Hussam, A.A.: Older adults' attitudes towards and perceptions of 'smart home' technologies: a pilot study. Med. Inform. Internet Med. **29**(2), 87–94 (2004)
17. Dewsbury, G.: The social and psychological aspects of smart home technology within the care sector. New Technol. Hum. Serv. **14**(1/2), 9–17 (2001)
18. Lê, Q., Nguyen, H.B., Barnett, T.: Smart homes for older people: positive aging in a digital world. Future Internet **4**(2), 607–617 (2012)
19. Majumder, S., Aghayi, E., Noferesti, M., Memarzadeh-Tehran, H., Mondal, T., Pang, Z., Deen, M.J.: Smart homes for elderly healthcare—recent advances and research challenges. Sensors **17**(11), 2496 (2017)
20. McLean, A.: Ethical frontiers of ICT and older users: cultural, pragmatic and ethical issues. Ethics Inf. Technol. **13**(4), 313–326 (2011)
21. Chan, M., Campo, E., Estève, D., Fourniols, J.Y.: Smart homes—current features and future perspectives. Maturitas **64**(2), 90–97 (2009)
22. Lee, M.K., Davidoff, S., Zimmerman, J., Dey, A.K.: Smart homes, families and control. Hum. Comput. Interact. Inst. (2006)
23. Bosch.: Survey: Smart home technologies are still completely underestimated. Bosch Media Service (2016). https://www.bosch-presse.de/pressportal/de/en/survey-smart-home-technologies-are-still-completely-underestimated-58240.html. Accessed Nov 2019

Implications of IoT in 3D Printing and Remote Engineering

Using Open-Source Hardware and Software

Amogh M. Aradhya(✉)

Dr. Ambedkar Institute of Technology, Bengaluru, India
amoghmanjunath99@gmail.com

Abstract. This paper is focused on evaluating the open source hardware, software tools and add-ons for improving efficiency, decreasing time taken for the prints, predicting failures and the managerial tasks needed to be performed on an open source FDM 3D Printer in order to enable remote engineering. The experiment includes the testing of various available sensors and custom designed components which is involved in the process of 3D Printing.

Keywords: 3D printing · Internet of Things · Remote engineering

1 Introduction

1.1 About 3D Printing

Additive Manufacturing has been playing a major role in the 4th Industrial Revolution. New 3D Printing technologies are under development and the need for Rapid Prototyping is on the rise.

3D Printing is a manufacturing process which is used for rapid prototyping since it is a costly and time-consuming process. However, with the patents getting expired in recent years and introduction of open source 3D Printers has opened doors for potentially more real-time applications.

Fused Deposition Modelling is a technique in which the digital model is sliced into cross-sections by a slicing software and converted into G-Code, a raw material (usually plastic) is heated to a near-liquid state, extruded in a thin bead, and the cross sections are deposited in layers by a computer-controlled nozzle. The material fuses to the previous layer and starts to take shape.

Consumer demand for the 3D Printed parts are high but it is difficult to maintain the external conditions and predict the final model.

In this paper, we discuss an experiment with an Open source FDM (Fused Deposition Modelling) 3D Printer integrating it with the IoT and sensors. The key objective of this experiment is to enable remote monitoring and engineering with addition of sensors and testing them in real time.

M. E. Auer and D. May (Eds.): REV 2020, AISC 1231, pp. 667–673, 2021.
https://doi.org/10.1007/978-3-030-52575-0_55

2 Overview

The open source 3D printer that is used is the Ender-3 which has all its designs and software open source. Ender-3 has an accuracy of ± 0.2 mm. The software which is being used is OctoPrint. It is an Open Source Software released under the AGPL License. This software is compatible with most of the consumer 3D printers. This software is running on a Raspberry pi 3b + (an open source Single Board Computer) that connects to the 3D-Printer through the COM Port.

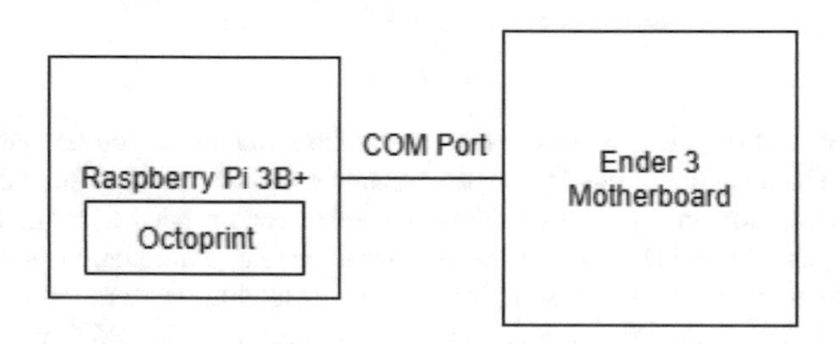

Fig. 1. Connection

3 Experimental Plan

In this section, we will have a look at the experimental setup. The 3D Printer is connected to the Raspberry Pi through the COM Port as shown in Fig. 1. List of sensors used:

1. CCS811 Air quality sensor is used to detect formaldehyde, which is one of the most toxic VOCs(Volatile Organic Compound) released during 3D Printing.
2. Custom Designed filament sensor as shown in Fig. 2 is used to detect the flow of filament and to pause the printer whenever there is breakage or shortage of the filament.
3. BMP180 Temperature sensors are placed on the motors to detect the motor temperature.
4. ACS712 Current sensor are attached to input of the Power supply to monitor the current.
5. Raspberry Pi Camera v2, a high quality 8 megapixel Sony IMX219 image sensor custom designed add-on board for Raspberry Pi is placed with an attachment to the printer as shown in Fig. 3.

All the mentioned sensors are connected to the Raspberry Pi through the GPIO pins on board.

Fig. 2. Custom designed filament sensor.

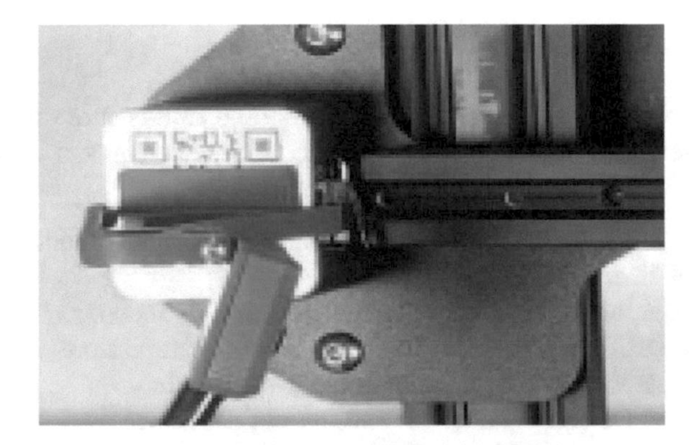

Fig. 3. Camera attachment.

4 Setup

SD card is mounted to the raspberry pi and the 3D printer is interfaced through the COM Port. All the sensors and the camera is connected to the raspberry pi. Port 80 should be open for the external traffic outside the local network. This enables the user to access the Octoprint web page from anywhere. This can be done from the Port settings page of the Router. Now, the webpage can be accessed from anywhere using http://YOURIPADDRESS:80.

The sensors were placed as required and the external factors affecting the readings were taken care of by housing them in a plastic enclosure.

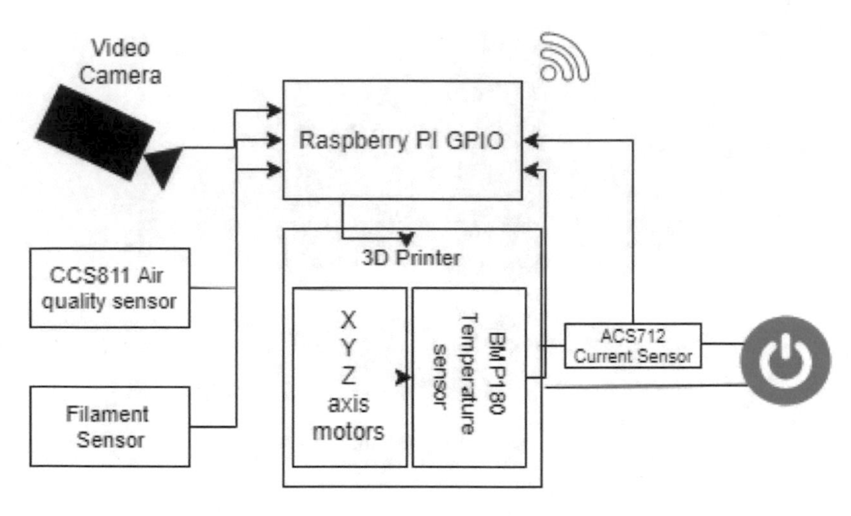

Fig. 4. Sensor setup

Based on the above idea, Fig. 4. shows a proposed architecture which detects most of the parameters required for monitoring the critical parameters in a 3D Printing process. Main Parameters which this setup monitors is:

1. Temperature (Internal): It measure the temperature from the inbuilt thermocouple of the Bed, Hot End.
2. Feed rate : The rate at which the filament is being Extruded.
3. Temperature (External): BMP180 senses the motor temperature.
4. Time: Time Elapsed and Time Remaining is predicted.
5. Filament Flow: Detects breakage or shortage of filament.
6. Air Quality: Detects and Alerts the operator of the surrounding air quality.
7. Video Monitoring: Provides real-time video stream.

All the sensors and components work insync and provides real-time data and analytics including remote control ability to the Engineer.

5 Results

The experimental setup was carried out successfully and the results were as expected. The data from sensors were also obtained on the web page as shown in Fig. 5 and inspected to ensure proper functioning of the printer. Custom events got triggered whenever a certain reading of the sensor was attained.

20 Models were printed with an average time per model being 60 min to print and the interval between each print was 100 min.

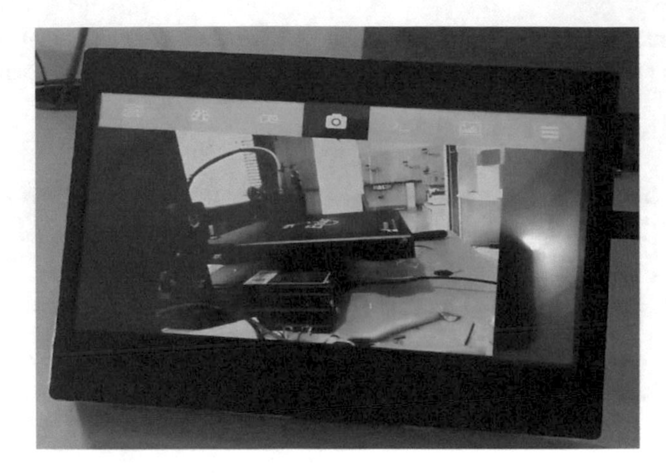

Fig. 5. Live video feed

Figure 5. Shows the live video feed at 15 frames per second which is enough to judge the quality of the print in real time and take action either to continue, pause or stop the printing process until the issue is rectified.

The most common reasons for the print to fail were Snapping of Filament, Improper Bed Adhesion and Under Extrusion. The failure rate of the prints was obtained for the following parameters and are as shown in Table 1

Table 1. Print failures

Percentage completion	Snapped filament	Bed adhesion	Under extrusion
0%–10%	2	6	0
10%–80%	0	0	5
80%–100%	3	2	1

The connection was successfully established and the ability to control was advantageous. This setup does not override the limit switches on the 3D Printer.

So, accordingly, snapped filament was detected using the filament sensor. Improper bed adhesion was fixed by monitoring the temperature of the bed during the first layer deposition by the temperature sensors and the problem with under extrusion was dealt by monitoring the camera feed.

6 Conclusion

Not only the above parameters are measured, but also other parameters like Air Quality, Motor temperature for predictive maintenance is measured in order to ensure reliability of the printer and safety of the surroundings.

Logs from the octoprint can be used by engineers to troubleshoot further problems. The log files can be found on *~/.octoprint/logs*. The final part can be 3D Scanned by optical scanning devices [6] and compared it with the original model to determine faults on the surfaces, incorrect dimensions due to shrinkage, warpage while printing.

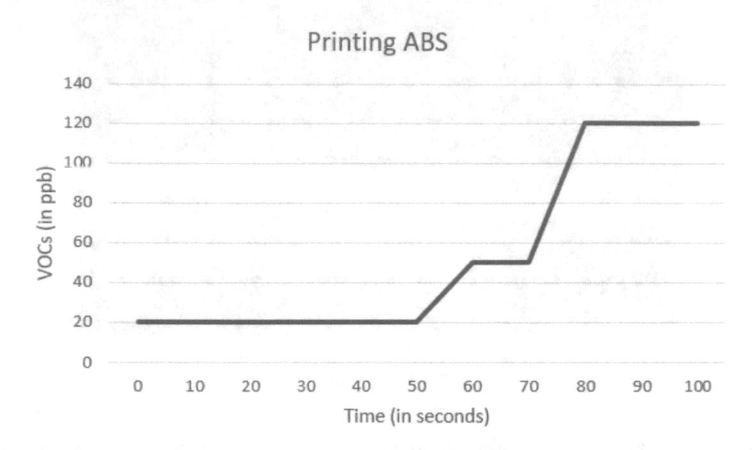

Fig. 6. Air quality of the surroundings (VOCs while printing ABS)

The Fig. 6 is a representation of the Surrounding air quality to alert the operator if there is any changes in the air quality which can also be monitored remotely.

Remote Monitoring and Remote Control is a crucial part to understand the failures or to debug and troubleshoot the faulty parts of a 3D Printer. The Remote Engineering Market was valued at 26.85bn USD in 2016 with a CAGR of 4.6%.

3D printing is really evolving very fast and even the technologies around it are improving. But in order for Industries and Companies to integrate it with their production chain, the process should be automated and it should be mass-producible. The recent studies show that the companies would be focusing on the Automation of 3D Printers until 2027. 3D Printing automation covers everything including the machine to material handling and post processing to inspection. The revenue from this market is expected to reach $11.2bn by the end of 2027 [4].

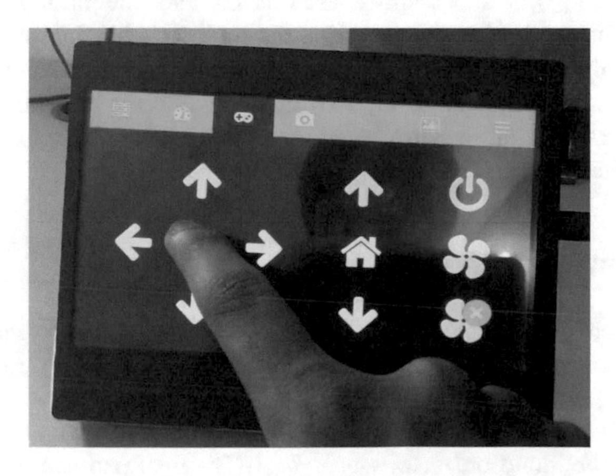

Fig. 7. Remote control dashboard

Figure 7 is a snapshot of the remote-control panel obtained from the octoprint server.

Therefore, with the help of IoT and Remote Engineering, troubleshooting of equipment and remote assistance is possible and eliminates the need for an engineer to be present full time at a place at any given point of time.

My work as of now is just an evaluation of the possible things that could be achieved and the future scope around the open source world. I would infer from my above experiment that with a robust hardware setup, scaling this to hundreds of printers is possible and an individual can monitor the all of them, initiate prints, predict failures and debug issues remotely.

The work is still in progress and it is anticipated to expand the system to cover other experiments once the work is finalized.

Acknowledgement. I would like to express my gratitude to Mr. Vivek from Cymbeline Innovations for the constructive feedback and support. I would also like to thank Prof. G. Rajendra, Prof. C. R. Mahesha for their advice and assistance for the work. I also extend my gratefulness to my institution Dr. Ambedkar Institute of Technology, Bengaluru for supporting my work and their valuable support.

References

1. Implementing a 3D Printing Service in an Academic Library Steven Pryor a a Library and Information Services, Southern Illinois University Edwardsville, Edwardsville, IL, USA
2. Lanzotti, A., Grasso, M., Staiano, G., Martorelli, M.: Department of Industrial Engineering, University of Naples Federico II, Naples, Italy
3. Yamato, Y., Fukumoto, Y., Kumazaki, H.: Proposal of real time predictive maintenance platform with 3D printer for business vehicles. In: ICSIE 2016 (2016)
4. https://www.3dnatives.com/en/3d-printed-automation-230220184/
5. Reinhart, G., Tekouo, W.: Automatic programming of robot-mounted 3D optical scanning devices to easily measure parts in high-variant assembly. CIRP Ann. Technol. **58**, 25–28 (2009)

Use of Data Mining for Root Cause Analysis of Traffic Accidents in Colombia

Hernando Vélez Sánchez(✉) and Heberto Saavedra Angulo

Universidad Distrital Francisco José de Caldas, Bogota, Colombia
{hvelezs,hsaavedraa}@udistrital.edu.co

Abstract. The high impact of traffic accidents makes it imperative to formulate public policies to reduce their occurrence. In this task, knowing the cause of accidents is of paramount importance. The use of data mining and big data adapts to the complexity of the phenomenon under study. In order to classify some possible causes of traffic accidents, we built a data model to describe the behavior and dynamic of the participant agents in the traffic accident event in Colombia. This paper presents the application of MLP and Naïve Bayes algorithms to identify the possible immediate cause and the rules decision algorithm PART for the root cause of traffic accidents. Models have been tested aiming to obtain the goodness of fit by increasing metrics like Recall, Precision, ROC and Kappa index, and minimize the RMSE.

Keywords: Traffic accidents · Big data · Neural network · Rules decision · Root cause

1 Introduction

Due to their high impact on the society, traffic accidents are considered a public health problem by the WHO [14]. Finding the root cause of traffic accidents is highly relevant to understanding the dynamic of the events related to them and formulating possible policies to prevent their occurrence. In Colombia, about 3,800 traffic accidents were recorded with fatal consequences during 2018 [15].

1.1 Related Works

Most of the literature found focuses on the application of big data, intelligent systems and data mining for the detection of traffic accidents. This section describes the most related works to root cause identification.

Jiangfeng Xi et al. [3] applied a hybrid algorithm for accident cause analysis using data mining association rules based on particle swarm optimization to analyze the correlation between attributes and accident causes. To evaluate the performance of the improved algorithm, they used the T test model and the Delphi method. This obtained ten times higher processing speeds than those of standard algorithms. The algorithm was tested in databases of more than 20,000 records with 56 attributes each.

Elfadil Abdalla [11] conducted a study to identify causes of traffic accidents using multi-class vector support machine algorithm. The author used a database of the Dubai

M. E. Auer and D. May (Eds.): REV 2020, AISC 1231, pp. 674–688, 2021.
https://doi.org/10.1007/978-3-030-52575-0_56

police in the United Arab Emirates and obtained an accuracy greater than 75% in predicting causes of traffic accidents.

Olutayo et al. [5] presented a study comparing the performance of a neuronal network algorithm and a decision tree for accident analysis on roads in Nigeria, using a database from the years 2002 and 2003. They used a multilayer perceptron with a training rate of 0.01 to minimize the absolute mean error and the mean squared error. The authors obtained values of 52.70% of hits and errors of 0.3479 MAE and 0.5004 RMSE. The decision tree algorithm performed better with 77.7% of hits and 0.1835% and 0.5029% in the mean absolute and mean squared errors, respectively.

Martin Luis [4] pointed out that, traditionally, statistical and regression analysis methods have been used to determine the relationship between accidents and their causes. These models require the formulation of hypotheses, as well as the knowledge of the relationships between dependent and independent variables.

Bigham [7] carried out a work using the algorithm of association rules with a database of the department of transport of England of 2009. The algorithm helps identifying the factors involved in the accident. It provides a basis for deepening and conducting further research on causes of accidents.

López used decision trees with a data set of traffic accidents in rural roads in the province of Granada (Spain). López took advantage of the ability of the technique to identify patterns based on data without establishing a functional relation between variables.

2 Methodology

Firstly, a model of the event (traffic accident) was built, and then mapped into a data model using an architecture of data that comprises different types of variables or attributes representing characteristics of the actors relevant to the accident and a set of values or domain of attributes. Finally, data mining algorithms were applied to obtain a classification model of the traffic accident cause.

2.1 Conceptual Model of a Traffic Accident

To carry out the conceptual model, we employed the DREAM 3.0 methodology (Driving Reliability and Error Analysis Method) developed by researchers of the Chalmers University of Technology. This is not a hierarchically organized model because the causal relationship between events may be in both senses (any factor must be antecedent or consequence). It is an interconnected network whose nodes are the different factors determinant in the accident occurrence. It consists of a set of associated events and factors, named phenotypes and genotypes. The latter contribute to the accident, but do not determine it.

The dominant, sometimes observable, situation exactly before an accident constitutes the near or proximate cause.

According to Girard [8], all accidents can be described in four stages: driving, discontinuity (continuity is interrupted by an unexpected event and the demand on the system exceeds its response capacity), emergency and collision.

There is a demand for the response of the drivers, which is related to their cognitive characteristics, observation, interpretation and planning.

The phenotypes are concerned with time, speed, distance, object, direction, and force. The genotypes are adjuvants, like human failures in interpretation, observation and planning and other more general temporary or permanent factors, such as inattention, as well as vehicle, road and organizational factors, such as maintenance, design or logistics. Among these factors there are latent failure conditions, which contribute to the accident. The blunt end and sharp end events are separated in time and space.

The theoretical point of view reflects how the contributing factors are defined in the classification scheme, and how they are related to each other.

When starting with a phenotype at the end of the chain events, the analysis can be done backwards until there are no more genotypes or significant factors. A general genotype can lead to another general one or to a specific one. The genotypes analysis allows an approximation to the root cause of the accident. The phenotype should be searched in the discontinuity phase. For each vehicle participating in the accident, only one phenotype is selected.

2.2 Data Model and Databases

Data structure is organized into three levels: factors, attributes or variables, and values of the attributes. Each attribute has its corresponding nominal, string or numerical category. [6] This data organization is shown in Fig. 1 and represents the data model corresponding to the accident model defined earlier. The structure of the model used in the analysis of the causes of accidents is largely determined by the structure of the existing databases.

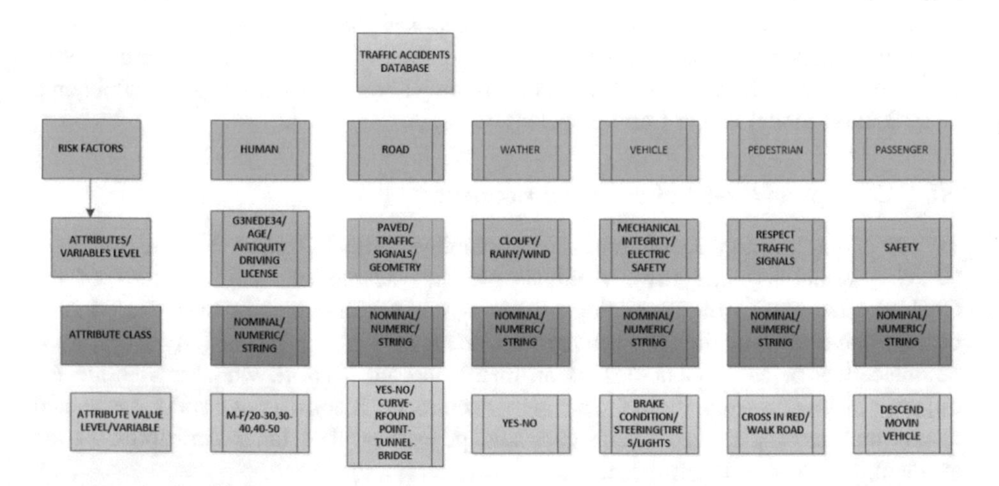

Fig. 1. Model used for the structure of the data to be managed.

Two traffic accidents databases were used. The first one has 38 fields or attributes with information related to driver, road, environment and organization, and 35000 instances. It was taken from the open data of the Colombian government [12] and registers information about accidents in all the Colombian territory during 2016. It was utilized to find the probable immediate cause of traffic accident. This data is supplied by state agencies which use it for the development of public policies aimed to reduce the rate of accidents. Attributes and values are showed in Table 1.

Table 1. Attributes for the immediate cause model database.

Attribute	Class	Values
Number of accident	Numeric	1-34232
Year	Date	Data
Month	Nominal	
Day	Numeric	
License plate	Numeric	Data
Vehicle type	Nominal	Truck, car, trailer, van, articulated bus
Gravity	Nominal	Wounded, only damage, dead
Vehicle manufacturer	Nominal	
Class name	Nominal	Shock, run over, other, occupant fall, overturn, self-injury, fire
Collision name	Nominal	Vehicle, fixed object, other, moving
Name fixed object	Nominal	Property, parked vehicle, wall, tree, pole, traffic light, railings, platform deck, signal fence, metal defense, fire hydrant, road separator, rock
Other class	Nominal	
Name other class	Nominal	Fall inside the vehicle, imprisonment, object drop on the vehicle
Address	Nominal	Data
Road type 1	Nominal	KR, C, AV, TR, AK, DG, AC
Hour occurence	Nominal	Hour
Road type design	Nominal	Intersection, Section of track, Roundabout, Bridge, Lot or property, level crossing, Overpass, Underpass, Tunnel, Trunk, Pontoon
Weather	Nominal	Normal, Rain, Wind, Rain/Rain, Normal/Normal, Rain/Normal, Normal/Rain, Wind/Normal
Type of traffic	Nominal	Normal, congested, clear
Gender	Nominal	Male, female
Cause	Nominal	Other, disobeying signal, forward closing, abrupt braking, reckless reverse, forward invading road, traffic light in red, not maintaining safety distance, transit between vehicles, speed excess, forward in prohibited area, apparent drunkenness, intoxicated crossing, disobeying signs, starting a vehicle without precautions, imperfect driving, drunkenness or drugs, turning sharply, traveling with the doors open, leaving in front of a vehicle, traveling in the opposite direction, going forward in a curve, crossing without observing, overtaking, drop off or pick up passengers in unmarked area
Ville	Nominal	Others, cross without observing, stand on the road, exit in front of vehicle, cross diagonally, cross in curve, pass red light, play on the road. Cross into a state of drunkenness
Utilization	Nominal	Brake failures, Exhaust failures, Front headlight failures, Directional failures, Other, Directional headlight failure, Steering failure, deficiency

The second database was built using the DREAM model and is the key to developing an algorithm that search for the basic or root cause of the accidents. It has 21 attributes including the immediate cause and others related to human characteristics, human behavior, psychological or physical-stress, fatigue, security involvement, risk perception and risk management, among other factors. Table 2 shows the attributes and their values in the database for the root cause model. It adapts a taxonomy developed by Arzlan and Kecesi [13] in their SHARE method for root cause analysis in maritime accidents.

Table 2. Attributes for root cause model.

1 FACTORS RELATED TO PEOPLE
Human Characteristics low learning ability, competition, low communication skills, Low learning ability, Competition, Complacency, Perception ability, risk perception, inattention, surveillance, distraction for issues other than work.
Human Behavior: culture, character.
Physical and physiological stress capacity: Sensitivity or allergy to substances, temperature, sound, etc., sensory deficiencies, Vision and hearing impairment, Other sensory deficiencies (taste, touch, balance), Temporary/Permanent Disabilities, Alcohol/drug use, Disease.
Psychological Stress-Ability: low learning ability, competition, Low communication skills.
Poor knowledge, skills and training: Improper practice, Insufficient knowledge of equipment and systems, Inadequate technical knowledge, Improper updated training, Inappropriate Initial Training, Inappropriate work environment, Lack of team training, Inadequate knowledge of vehicle operations, Lack of experience
Communication problems: Poor communication between vehicle crew members, Misunderstanding, Poor communication between vehicle crew members
Inadequate team culture; Absence of shared mental model, Lack of membership, Overreliance on the team, people or system, On trust in the supervisor, Inappropriate leadership, Conflicting relationships, Inappropriate Initial Instructions, Lack of coaching, inadequate instruction review, Lack of supervision/
Security Related Issues: Lack of safety culture, Unsafe acts, Involuntary actions, confusion, Disorder, Memory failures, Ignore, Improper attempt to save time or effort, Improper attempt to avoid discomfort, Sabotage, unsafe acts, involuntary actions, lack of safety culture,
Lack of motivation factors: improper practice, insufficient knowledge of equipment and systems, inadequate technical knowledge,
Inadequate knowledge of regulations and standards
2. FACTORS RELATED TO WORK
Management: Company management, Do not take corrective actions, Disorder in process documentation, Certification Fraud, Inadequate inspection
Risk assessment: inadequate risk assessment, Improper risk assessment process, Risk assessment process not implemented
Environmental factors; Natural environment, Normal, Heavy weather, Natural disasters, Dangerous environment, lighting, humidity, Visual environment/
Work environment: Noise, Vibration, Poor internal management, Dirty workplace, Inadequate ventilation

The root cause database was constructed from a set of 1,200 traffic accidents in Colombia, including the next cause and information from the psychosensometric tests to the drivers and stakeholders in the accident. Old and subsequent events defined in the accident model using Dream 3.0 are also included.

The database was developed based on experience, for example, considering a next cause like not maintaining a safe distance, the most probable causes associated with human factors were assigned. The accident can occur due to issues related to security, such as those mentioned above; to human characteristics, like delayed reaction or slowness, misperception of risk, neglect such as talking on the phone, inattention, among others; to human behavior such as character, culture or lack of discipline; to safety such as unsafe acts or ignoring risky situations; or to the vehicle, such as mechanical failure. It should be noted that this database can be improved to obtain greater access to data related to the subject and greater participation of experts.

2.3 Preparation of Dataset for Minable View

The preprocessing, grouping and classification stages were carried out with the Weka 3.8 program. The minable data set was obtained after cleaning and transforming the data, such as eliminating attributes that are not relevant for grouping and classification tasks (function gain ratio attribute eval in Weka), removing missing values (function remove missing values), balancing classes with balance classes function and merging similar attributes. As a result, a database with 11 attributes and 10605 instances was obtained.

2.4 Clustering

In order to obtain information on the relationships between attributes, a grouping technique is used, which is not supervised. This technique is the K-means algorithm configured to obtain five clusters. Next, the analysis of each of the obtained clusters and the respective conclusions are observed in Fig. 2.

The zero cluster in blue color is mainly characterized by accidents involving collision with another vehicle with only damage, in roundabouts, normal weather condition, in some cases presence of gaps in the road. As the accidents analyzed are within the city with a speed limit of 60 kph, high speed conditions are not expected, although possible. However, the conditions of maneuvering in the roundabouts, with simultaneous access by several vehicles, make these road sections more prone to the occurrence of road conflicts. Probably the low speed condition is related to the consequence of damage only.

In the cluster 1 (red), accidents with crash and people run over, deaths and to a lesser extent injury, with vehicles, in roundabouts and under normal environmental conditions occur in a smaller proportion and in a greater proportion in roads with potholes.

Cluster 2 (Green) shows accidents with crashes and in smaller amount overturning, with people wounded, in roundabout and in smaller amount roads and bridges, under normal atmospheric condition, and in smaller amount with wind. These are attributable to the pedestrian passing, to the other passenger or to potholes in some cases.

Cluster 3 (Aquamarine) groups accidents involving crashes and injuries and another vehicle, a fixed object in less cases, passing through a roundabout and in a smaller amount, a bridge, under normal atmospheric conditions and in a smaller amount with wind. These accidents have as a cause the other passenger and potholes in the road.

Cluster 4 (lilac) corresponds to accidents with collision, damage only, passing by roundabout, in a road or bridge in smaller quantity, under normal atmospheric condition, with wind in smaller quantity, with pedestrian stopped or passing, vehicle with failures, other passenger or potholes in the road, wet surface and posts in smaller quantity.

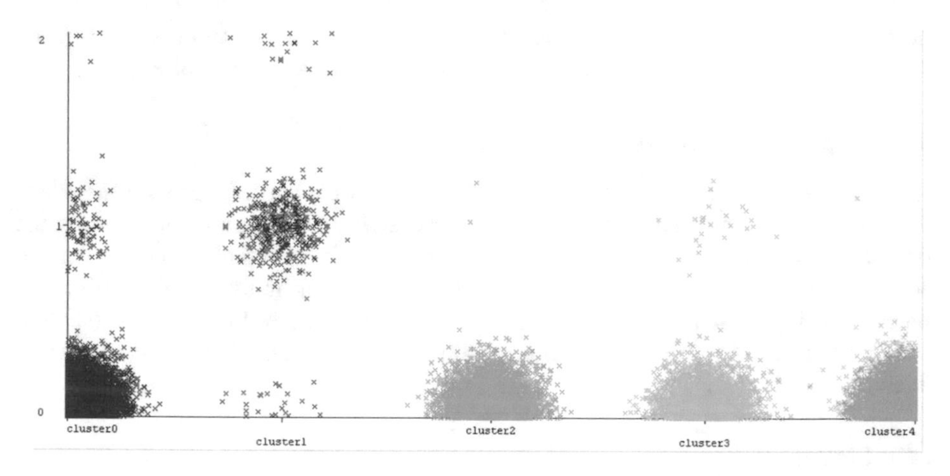

Fig. 2. Clusters from database for proximate or immediate cause.

2.5 Classification Model

Root cause model was divided into two sections; one to find the immediate cause and the other to identify the possible basic one. To do this two different learning algorithms were proposed for the immediate cause classification: artificial neural networks, Bayesian Naive and one for root cause PART, which is based on decision trees algorithm. Models are built utilizing WEKA 3.8 [2].

Proximate Cause Model

To determine the next cause of a traffic accident, the multilayer perceptron (MLP) and Bayes Naive algorithms are used and compare. These algorithms have been selected taking into account the complexity of the traffic accident, the available databases, the class of classification attributes (nominal) and the objective, that is, to classify the possible next cause as well as minimize the mean absolute error and RMSE metrics.

Multilayer Perceptron Algorithm (MLP)

With the data model outlined in Fig. 1, different levels are defined:

S: space containing the data base.

$S = [s_1, s_2, s_3, \ldots\ldots\ldots s_p]$ p: number of instances or records.

A: vector that represents the set of attributes that describe the accident. It varies between 1 and n.

X: set of values of attributes of vector A varies between 1 and n.

C: vector that contains the values of classes and allow the classification of accident cause. It varies between 1 and m.

O: vector that contains the values of vector class it varies between 1 and m.

According to [16]

Training stage

- Configure the MLP: select the number of hidden layers, learning rate and momentum.
- Scan the data set. Entries X, output C.
- Assign the weights to links between neurons of entry and hidden layers.
- Active input layer neurons

$$a_1^1 = x_i \, \mathrm{i} = 1, 2.\ldots.\mathrm{n} \tag{1}$$

- Active neurons of hidden layers.

$$f(\sum\nolimits_{j=1 j=1}^{n-1} w_{ji}^{c-1} a_j^{c-1} + u_i^c) \qquad a_j^{c-1}\, layer\, c-1\, neuron\, activation \tag{2}$$

- Active output layer neurons

$$y_i = a_i^c = f(\sum\nolimits_{j=1}^{n-1} w_{ji}^{c-1} a_{ji}^{c-1} + u_i^c) \qquad i = 1, 2.\ldots.n_c \tag{3}$$

$$Y = (y_1, y_2.\ldots\ldots.y_{nc}) \tag{4}$$

- Calculate the error between the output and the reference values.

$$e_j(n) = d_j(n) - y_j(n) \tag{5}$$

$$\varepsilon(n) = \frac{1}{2}\sqrt{e_j^2}(n) \tag{6}$$

e(n): MLP classification error for pattern n given

- Back propagate the error, reassign the weight values and calculate the new output and error. The equations used are:

$$\Delta w_{ji} = -\gamma \frac{\partial \varepsilon(n)}{\partial v_j(n)} y_i(n) \tag{7}$$

Naïve Bayes Algorihm

According to [1] steps followed to perform Naïve Bayes algorithm are;

- Scan de dataset S.
- Calculate the probability of each attribute value for X vectors using Bayes Theorem.

$$P\left(\frac{C_i}{X}\right) = [P(X/C\_i) * P(C\_i)]/P(X) \quad (8)$$

- Identify the largest P(X/Ci)P(Ci); P(Ci) = s_i/s where

  s_i: is the number of training samples of class I and
  s: is the total number of training samples.
  If P(X/Ci)*P(Ci) is maximum; assign the class Ci to the X sample.

Root Cause Model

After the approximation to the immediate cause of the accident, an analysis of the results obtained through a panel of experts is carried out. Then, a second approach is proposed, this time to the probable root cause of the accident. This information has been complemented by the classification of genotypes (antecedents or causes) and phenotypes.

These basic or root causes include two categories: those related to people and those related to the work factor. Many are oriented to the organization or company and others to individuals. As a natural person or individual driving company vehicles can intervene in accidents, the characteristics of both types are taken into account.

PART Model

It is based on the divide and rule principle. A decision tree is built using C4.5 algorithm in each iteration and the best leaf to make a decision rule is chosen.

Training stage

It follows the next steps [10];

- Scan the training data set (S)
- Check for the base case
- For each attribute a find the information gain of the division of a

$$H = -\sum\nolimits_{k=1}^{m} p_k log_2 p_k \quad (9)$$

$$p_k = \frac{n_k}{N} \quad (10)$$

H : Information entropy.
P_k: probability of state k
n_k: number of times when k happens.
N: number of samples

- Make a_best the attribute with the highest information gain.
- Create a decision node that divide a_ best node.
- Repeat in the sublists obtained by the division of a_best, and add these nodes as children nodes.

2.6 Post Processing Evaluation of Model

In order to evaluate the goodness of fit of each node, different metrics have been used, like Kappa index, RMSE, precision, ROC curve, Recall, true and false positives.

3 Actual Outcomes

3.1 Experimental Work, Analysis and Results

This section shows the work done and its results according to the methodology defined.

Neural Network Algorithm Application (MLP)
The MLP was configured according the parameters next mentioned. To control the magnitude of adjustment of the weights and the speed of convergence, the learning rate and the momentum are varied. The learning rate is 0.3, taking into account that the value should not be too small because the algorithm takes longer or too large because it can skip the optimal value you are looking for. Momentum was 0.2. The learning method was the back propagation of the error. Sigmoid function was used to activate the layers. To train the MLP cross validation with data set 1 was used. Number of hidden layers is equal to:

Number of attributes + number of classes/2 = 11 + 2/2 = 6.5 approximately 7 hidden layers.

The detailed analysis table by class shows the values of the indicators True positive rate 0.401, false positive rate 0.16, recall 0.545, accuracy 0.511, area under the ROC curve 0.744 on average and PRC area 0.311. The confusion matrix (not shown) has 1219 successes for the class not keeping a safe distance, and 225 successes for other causes of accident. Other classes have zero values because only the two more frequent classes were considered.

Naive Bayes Algorithm Analysis
This algorithm presents a quantitative probabilistic measure about how important the values of the class variable are within the problem. Among the attributes of the training set, there can be no correlations. This algorithm is based on Bayes' theorem and assumes it works well with real data and with attribute selection mechanisms to eliminate redundancy.

Table 4 shows that the number of instances correctly classified was 8107, (76.44%). There were 2498 instances classified incorrectly. The value of the Kappa indicator is 0.4817, which means that the number of hits is greater than what could be obtained randomly and, although far from 1, it is a satisfactory value for the algorithm. From the analysis of the relative absolute error, a value of 60.23% is observed which, although high, can be taken as acceptable. The ROC value (area under the R curve) is 0.525 on average. The true positive rate is on average 0.764, which is a more adequate value than expected. The false positive rate is 0.308, which is quite low, and the accuracy is 0.771, which indicates that from the total classified terms, this percentage was obtained correctly. The threshold is 0.5, the accuracy is the sum of true predictions over the sum of all the classifications not listed in the table. Metric F is 0.753 on average, which means that the relationship between the rate of true positives and the sum of the predictions, without taking into account the true negatives, is relatively high. The area under the ROC curve is 0.783, which is quite acceptable near 1. PRC area is 0.794 on average. Table 5 shows the confusion matrix. Class a – "maintain no safety distance" matches 5793 hits, followed by other class.

3.2 Root Cause Model

To perform the second part of the model, it is difficult to use classification techniques, because the root cause constitutes a characteristic of each particular accident. It is difficult to obtain information with some level of detail to perform this analysis because the root cause categories "Human factors" and "work factors" are closely related to the particular characteristics of the people involved in the accident, whether they be driver, pedestrian, passenger, maintenance mechanics, transport fleet supervisors, etc., and the organization. For these reasons, building an algorithm that makes the prediction or classification based on the root cause is not the most appropriate solution to the problem under study. This phase should be carried out using the participation of experts who, based on systematically classified information, can determine the most probable cause or causes of the accidents.

An algorithm has been constructed to provide decision rules that serve as a starting point for experts to reach conclusions consistent with the particularities of the case, following a well-defined accident investigation procedure. The model output must be validated in the accident investigation process with the one obtained by the particular accident analysis (Table 3).

Table 3. Confusion matrix for Naive Bayes algorithm.

a	b	c	d	e	f	g	h	i	j	k	l	m	n	o	p	q	r	s	t	u	v
5793	0	0	0	0	0	0	0	0	0	0	0	0	0	0	0	0	0	0	0	0	a = DO NOT KEEP SAFETY DISTANCE
0	0	0	0	0	0	0	0	0	0	0	0	0	0	0	0	0	0	0	0	0	| b = ADVANCE CLOSING
0	0	0	0	0	0	0	0	0	0	0	0	0	0	0	0	0	0	0	0	0	c = OVERSPEED
0	0	0	0	0	0	0	0	0	0	0	0	0	0	0	0	0	0	0	0	0	d = BRAKE BRUSHLY
0	0	0	0	0	0	0	0	0	0	0	0	0	0	0	0	0	0	0	0	0	e = CROSS WITHOUT OBSERVING
0	0	0	0	0	0	0	0	0	0	0	0	0	0	0	0	0	0	0	0	0	f = DISOBEY SIGNALS
0	0	0	0	0	0	0	0	0	0	0	0	0	0	0	0	0	0	0	0	0	g = CROSS IN STATE OF DRINK
1932	0	0	0	0	0	0	2314	0	0	0	0	0	0	0	0	0	0	0	0	0	h = OTHER
0	0	0	0	0	0	0	0	0	0	0	0	0	0	0	0	0	0	0	0	0	i = TRAFFIC LIGHT IN RED
0	0	0	0	0	0	0	0	0	0	0	0	0	0	0	0	0	0	0	0	0	j = ADVANCE INVESTING RAIL
0	0	0	0	0	0	0	0	0	0	0	0	0	0	0	0	0	0	0	0	0	k = TRANSIT BETWEEN VEHICLES
0	0	0	0	0	0	0	0	0	0	0	0	0	0	0	0	0	0	0	0	0	m = START WITHOUT CAUTION
0	0	0	0	0	0	0	0	0	0	0	0	0	0	0	0	0	0	0	0	0	o = ADVANCE IN PROHIBITED AREA

(*continued*)

Table 3. (*continued*)

a	b	c	d	e	f	g	h	i	j	k	l	m	n	o	p	q	r	s	t	u	v
0	0	0	0	0	0	0	0	0	0	0	0	0	0	0	0	0	0	0	0	0	p = TRANSIT THROUGH THE FOOTWEAR
0	0	0	0	0	0	0	0	0	0	0	0	0	0	0	0	0	0	0	0	0	q = IMPRUDENT REVERSE
0	0	0	0	0	0	0	0	0	0	0	0	0	0	0	0	0	0	0	0	0	r = DRUNK OR DRUG
0	0	0	0	0	0	0	0	0	0	0	0	0	0	0	0	0	0	0	0	0	s = DO NOT RESPECT PRELATION OF INTERSECTIONS OR TURNS
0	0	0	0	0	0	0	0	0	0	0	0	0	0	0	0	0	0	0	0	0	t = TRANSIT IN COUNTERWAY
0	0	0	0	0	0	0	0	0	0	0	0	0	0	0	0	0	0	0	0	0	u = BAD PARKED VEHICLE

The associated factors and variables were defined in Table 2 and the criteria for classification of genotypes and their relationship between them and with the phenotypes were defined in the DREAM 3.0 method. An arrangement has been made to establish the relationships between them and the next cause identified by the algorithm obtained in the first part of the model.

From the root cause database, a classification model was constructed using the PART technique. 27 decision rules were obtained to select the root cause. Table 5 shows the results obtained from the application of the PART technique using the second database for root cause.

Table 4. Comparison between the metrics of MLP and Naïve Bayes. Also including PART metrics.

Performance measure	Naïve Bayes	Neural networks MLP	PART
Correctly classified instances	8107 76.4451%	1444 54.47%	524 95.4463%
Incorrectly classified instances	2498 23.5549%	1207 45.53%	25 4.5537%
Kappa statistic	0.4817	0.0005	0.9463
Mean absolute error	0.0181	0.231	0.0091
Root mean squared error	0.1023	0.1253	0.0754
Relative absolute error	60.2305%	102.1792%	7.4655%
Root relative squared error	83.5118%	101.3349%	30.6183%
Total number of instances	10605	2651	549
Precision	0,771	0,401	
Recall	0,764		
F-Measure	0,753		
MCC	0,502		
ROC area	0,783	0,744	
PRC area	0,794	0,314	
Time taken to build the model: (S)			0,09

Table 5. Decisión rules from PART. Source: WEKA 3.8.

IF	THEN		
ï » ¿human characteristics = neglect AND Safety-related issues = inappropriate precautions	Slow down	22.6	1.6
fatigue = Fatigue due to lack of rest	Micro sleep	27.82	2.82
ï » ¿human characteristics = neglect AND Safety related issues = inadequate precautions	Slow down	24.76	1.76
fatigue = Fatigue due to sensory overload	Micro sleep	26.68	6.84
ï » ¿human characteristics = competence AND physical/physiological capacity = sensory deficiencies	Do not maintain a safe distance	23.61	2.61
ï » ¿human characteristics = neglect AND fatigue = Fatigue due to sensory overload	Micro sleep	26.68	2.84
ï » ¿human characteristics = competence AND physical/physiological stress capacity = sensory deficiencies	Do not maintain a safe distance	23.61	2.61
ï » ¿human characteristics = competence	Parking without security	17.45	3.45
psychological stress capacity = emotional load AND fatigue = due to lack of rest	Reckless reverse	21.0	3.0
psychological stress capacity = emotional load AND fatigue = extreme perception/concentration demand:	On the contrary	21	0
ï » ¿human characteristics = complacency	Traffic light in red	13.38/	0.38
»¿human characteristics = lack of communication AND Safety related issues = unsafe acts:	Traffic light in red	9.26	1.26
»¿human characteristics = risk perception AND fatigue = extreme perception/concentration demand AND Human behavior = character:	Do not maintain safety distance	58.0	1.0
ï » ¿human characteristics = ability to perceive AND fatigue = due to sensory overload:	On the contrary	16.0	4.0
ï » ¿human characteristics = ability to perceive:	Not respecting the priority of intersections or turns	16.59	7.59
ï » ¿human characteristics = risk perception AND fatigue = extreme perception/concentration demand AND Safety related issues = inadequate precautions: (.)	Speeding	18	0
ï » ¿human characteristics = risk perception AND physical/physiological stress capacity = sensitivity: (/	Do not maintain a safe distance	23.9	0.9

(*continued*)

Table 5. (*continued*)

IF	THEN		
Psychological stress capacity = alcohol/drug use	Not respecting priority of intersections or turns	14.55	2.55
ï » ¿human characteristics = ability to perceive AND Safety related issues = inappropriate precautions	Do not keep a safe distance	47.15	3.44
ï » ¿human characteristics = ability to perceive AND security related issues = unsafe acts AND fatigue = extreme perception/concentration demand	Speeding	20	0
»¿Human characteristics = ability to perceive AND Safety related issues = lack of safety culture:	Invade lane	59.86	21.13
physical/physiological stress capacity = AND sensory deficiencies fatigue = lack of rest:	Disobey signals	15.38	0.38
physical/physiological stress capacity = sensory deficiencies	Traffic light in red	15.0	2.0
fatigue = due to lack of rest	Overtaking invading via	17.0	6.0
fatigue = routine/monotony AND surveillance lack of motivation factors = hierarchical pressure AND Safety related issues = inappropriate precautions	Speeding	15	0
fatigue = demand for decision/extreme judgment:	Not keeping a safe distance	9	0
fatigue = routine/monotony AND surveillance Safety related issues = inappropriate precautions	Do not keep a safe distance	7	0
fatigue = due to sensory overload	Stop abruptly	9.0	4.0
Number of rules	26		

4 Conclusions/Summary

An algorithm for the identification of root cause of traffic accidents was developed and implemented using two databases: one consisting of 34,000 records and containing databases of different types of accidents in the city of Bogotá in 2016, and the other one with 1200 records was built from an assurance company database between 2009 and 2011 in different companies and areas of the country.

The first part of the model determines the next cause using MLP and NaïveBayes algorithms. Metrics show that Naïve bayes has a better performace (76.44% correctly classified instances and 0.0181 RMSE) than MLP (54.47% correctly classified instances and 0.1253 RMSE). The second part determines the possible root causes and was carried out according to the identification of the next cause and using the PART classification algorithm that provides decision rules to determine the root causes associated with the next cause determined in the first part. A performance of 95.44% hits was obtained and RMSE of 0.0091. Kappa index is 0.9463.

According to the models analyzed, the causes of accidents in Bogotá and on the roads are mainly related to the behavior of drivers in the face of traffic rules. The use of big data is a very useful tool for identifying patterns and causes of traffic accidents.

There is an opportunity to improve the root cause model with more detailed data about psychosensometric tests of stakeholders.

References

1. Janani, G., Ramya Devi, N.: Road traffic accident analysis using data mining techniques. J. Inf. Technol. Appl., 84–91 (2017)
2. Mark, H., Eibe, F.: Practical Data Mining. University of Waikato (2011)
3. Xi, J., Gao, Z., Niu, S., Ding, T., Ning, G.: A hybrid algorithm of traffic accident data mining on cause analysis. Math. Problems Eng. Procedia Soc. Behav. Sci. **160**, 607–614 (2012)
4. Luis, M., Leticia, B., Laura, B., Griselda, L.: Using data mining techniques to road safety improvement in spanish roads (2014)
5. Olutayo, V.A., Eleudire, A.A.: Traffic accident analysis using decision trees and neural networks. Inf. Technol. Comput. Sci. **02**, 22–28 (2014)
6. Shetty, P., Sachin, P.C., Kashyap, S.V., Madi, V.: Analysis of road accidents using data mining techniques, vol. 4, theme 4 (2017)
7. Bigham, B.S.: Road accident data analysis: a data mining approach. Indian J. Sci. Res. (2014)
8. Wallen Waner, H.: Dream 3.0 (Driving reliability and error analysis method) (2008)
9. Alvaro, C.: Methodological guide to obtain occupational accident patterns using data mining. Universidad de Piura, Master thesis (2013)
10. Mulay, P., Mulat, S.: What you eat matters road safety: a data mining approach. Indian J. Sci. Technol. **9**(15) (2016)
11. Mohamed, E.A.: Predicting causes of traffic road accidents using multi-class support vector machines. In: Proceeding of the 10th International Conference on Data Mining, 21–24 July 2014, pp. 37-42 (2014)
12. GOV.CO homepage. https://www.datos.gov.co/browse?tags=accidentalidad
13. Tuba, K., Ozcan, A.: SHARE technique: a novel approach to root cause analysis of ship accidents. Saf. Sci., 1–21 (2017)
14. WHO homepage. https://www.who.int/violence_injury_prevention/road_traffic/es/
15. Agencia Nacional de Seguridad Vial homepage. https://ansv.gov.co/observatorio/index.html
16. Bibing us homepage. http://bibing.us.es/proyectos/abreproy/12166/fichero/Volumen+1+-+Memoria+descriptiva+del+proyecto%252F3+-+Perceptron+multicapa.pdf

Evaluation of a Big Data System for Online Search (Case Study)

Karim Aoulad Abdelouarit(✉), Boubker Sbihi, and Noura Aknin

TIMS Research Unit, LIROSA Laboratory, Abdelmalek Essaâdi University, Tetuan, Morocco
abdelouarit.karim@gmail.com, bsbihi@hotmail.com, noura.aknin@uae.ac.ma

Abstract. As soon as they became aware of online search innovations, technologists began to design and develop tools to help users better understand this new search mode connected to massive data in constant evolution. To achieve the goal of online search, it is crucial to create a reliable environment where learners can find the information they are looking for and use it in a simpler way. In this context, we propose to design a Big Data tool that would support users in their search for information. Our main goal in this article is to provide an intelligent architecture that allows to process massive and unstructured data to provide the best result for the learner. For our study, we proposed conducting a sample survey of future system criteria that concerns the performance and the quality when using online search, through the submission of questionnaires for a set of students who represent future users of our online search platform. We have exposed the different factors that may impact the learner when using online search system, as well as the criteria for simplicity and usability of our future solution.

Keywords: Big data · E-Learning · Data structures

1 Introduction

This work is part of the project to design adaptive learning systems for teaching and online search in a massive and heterogeneous data environment generated by the Big Data phenomenon [4].

Distance learning or e-Learning is an area that is growing exponentially and relates to the management and the restitution of content and learning activities based on the use of information technology and communication. E-Learning comes in various forms and uses more or less complex technologies. Nevertheless, with the arrival of web 2.0 and web 3.0 technologies, the development of learning pedagogical systems has undergone an incomparable evolution, firstly, in order to improve the quality of learning and minimize its cost, and on the other hand, it is about giving a new form to the teaching-learning operation by offering learners an intelligent, interactive and evolving learning environment [5].

The process of adaptation is to personalize the search of data and its presentation to the learners in order to best meet their expectations and ensure their quick and efficient

M. E. Auer and D. May (Eds.): REV 2020, AISC 1231, pp. 689–701, 2021.
https://doi.org/10.1007/978-3-030-52575-0_57

results. The involved techniques are numerous, and all are based on a fictitious model to help the learner find the information he seeks in this massive and varied set of data. Considerable research efforts have been made to develop adaptation techniques, methods and tools [4].

Our main objective in this work is to propose a new solution to design an adaptive teaching system to support students in their research and scientific training. It is a solution that aims to facilitate the acquisition and evaluation of knowledge for learners by offering them learning pathways and tailored evaluations. In the same way, our work consists in facilitating to the learners the process of the information search in a massive and heterogeneous data environment. Learners can freely formulate the information search expressions they want to acquire, and the system must offer them the best result that matches their expectations [3].

In this context, it is proposed to develop a tool called "Big-Learn" based on a technique that integrates structured and unstructured data into a single layer of data to facilitate access and provide optimal search relevance with adequate and consistent results according to the expectations of the learner [4]. This solution, as shown in the figure below, is at the confluence of several domains: Cognitive Science, Artificial Intelligence, Semantic Web, Knowledge Management and Human Machine Interface (Fig. 1).

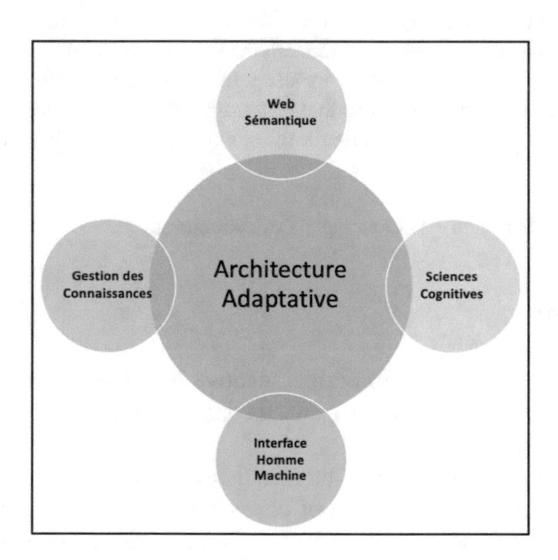

Fig. 1. Positioning of our work with various fields of research.

The following section describes the state of the art of online search and especially in the Big Data environment. We then present in paragraph 3 the case study carried out on a sample of students from Abdelmalek Essaâdi University (UAE) concerning the criteria of the future system and the indicators of performance and quality of learning when using the online search system, by submitting questionnaires to a set of students who represent future users of our search and distance learning platform. The last paragraph presents a general conclusion with a range of perspectives.

2 The Online Search in Big Data Environment

2.1 Processing Online Search

The scope and volume of information on the Web requires good search skills such as the ability to formulate relevant keywords to find the information sought. However, most users are unable to limit the topics of their search and are overwhelmed by the amount of results provided by search engines, especially when they do not have the skills or resources to access and manage this information in an intelligent way [3].

However, users of online search use keywords and very simple terms for their search, and they assume that search engines will understand their queries. Most of online search users do not adopt strategic behavior in their search, rather they expect the search engine to find the answer for them, regardless of their own strategy [2].

Indeed, search engines have developed strongly since the advent of the Big Data phenomenon. The effectiveness of the search for information, especially on the Web, would be particularly related to the use of search engine system expertise, including knowledge of online search procedures and tools, as well as strategies to be used in the search for information, as well as to quickly and correctly assess the quality of the content and the credibility of the data and information returned [4]. The significant growth of information in the Internet requires more and more efficient search tools that can distinguish relevant information from hundreds or even thousands of raw data. However, the quality of results provided by traditional search engines is not always relevant, especially when the user's query becomes more and more complex [4].

Today, information manifests itself in various forms: geolocation, mobile data, data from social networks, video and satellite imagery, customer transactions, motion data of connected objects, etc. [4]. The following figure shows the look of the process of using online search in the Big Data environment (Fig. 2).

Fig. 2. The process of the online search.

As shown in this figure, the user accesses the online search system for his particular information needs. But finally, he finds a multitude of unstructured and heterogeneous information (text, image, video, etc.) that come from different sources (Twitter, Facebook, YouTube, etc.). This situation pushes the learner to wonder about the value and reliability of data returned by online search engines [4].

This large mass of heterogeneous data has a negative impact on the user of online search and makes the search or information retrieval process difficult to achieve [1]. Thus, the use of an online information retrieval system, based on the raw data of the Big Data phenomenon, has important advantages. To do this, the need for a fictitious model to represent and process this type of data that is not necessarily textual becomes essential [4].

2.2 Using Online Search in Big Data Environment by Learner

With the emergence of Web 2.0, a new vision of the Web has been created by viewing the user as a potential producer of information and not just a consumer [6]. This radical change has significantly increased the amount of data on the Internet known as Big Data. The massive data is the largest portion of data on the Internet. This mass of data which occupies our daily life does not cease to grow and requires advanced means to capture, communicate, aggregate, store and analyze [7].

Social networks, blogs, wikis, etc., are one of the reasons for large amounts of data on the Internet. This directly impacts online search systems: where everyone starts requesting for a particular information, but since the data comes from multiple data sources, the result becomes large and rich [8].

The Big Data phenomenon has made possible the development of highly qualified online search engines. The web pages generated by search engines are based on search terms that require sophisticated algorithms and the ability to process an impressive number of requests [9]. The following figure shows the online search model in the Big Data environment (Fig. 3).

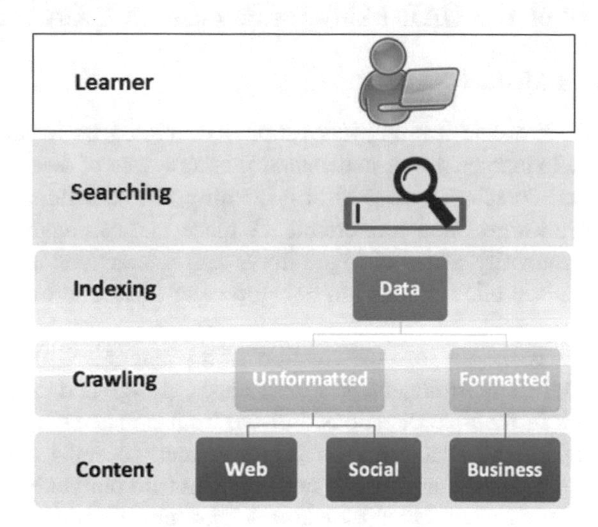

Fig. 3. Online search model in Big Data environment.

As shown in this figure, the user accesses the online search page via the web browser. He enters the keywords and validates the search form. The search engine intercepts the request of the user and begins to search the Internet for data based on the keywords entered. The collected data is referenced, indexed and classified before presenting it to the user on the results page. Analyzed data includes all Internet data such as: websites, social networks, user-generated data, and other external data sources.

However, the quality of the results provided by the search engines is not always relevant especially when it comes to composing more than one query [10]. However, online document systems, and especially search engines, have experienced a strong development since the advent of the Big Data phenomenon. The effectiveness of the search for information, especially on the Web, would be related in particular to the expertise of use of the system, which includes the knowledge of procedures and documentary tools, but also that of the heuristics and strategies to be used in documentation. to more quickly and accurately assess the quality of content and the credibility of data and information [12].

In the same context, we can mention the solution "DOCUPOLE" which was designed in 2007 to offer an online course of initiation to search documentary modeled in MOOC. Its new version, which was launched in December 2014, posted a first significant consultation on the referencing of information [11]. Similarly, we find the system "SARIOnto" which is an online information retrieval system based on domain and service ontologies, resulting from the information retrieval process and the document classification process [10]. These two systems tend towards the facilitation of the access to the documentary information without taking into account the reorganization of the heterogeneous data resulting from the search online.

3 Case Study of the UAE (Abdelmalek Essaâdi University)

3.1 Context and Methodology

Upon learning about online learning innovations, technologists began to design and develop tools to help learners better understand this new way of teaching connected to ever-changing data. To achieve the goal of e-learning, it is crucial to create a reliable environment where learners feel comfortable. A place that can aggregate content and imagine it as a community where dialogue flows and interactions and content can be simple to use. This will allow learners to develop clear ideas and evolve in their deep learning [4].

With this in mind, we propose the creation of a pedagogical platform that would support learners in their environment. Research on the design and development of this platform is at work in multiple directions, but we limit ourselves here to report some progress in pedagogy, advancements on issues related to self-learning and online search in a massive data environment. To better understand our study, we have chosen to deal with the use of online search by learners who aim to acquire information on a given course or theme for the purpose of learning and documentation. to see if there are any additional dimensions that could be added as a result of this Big Data learning study. The analysis of the results data of the online search scenario will allow us to delimit the context of our future system, as well as to better understand the design of a methodology based on the tool that will integrate the structured data mix. and unstructured in the same data layer, in order to facilitate access in addition to optimal search relevance with adequate and consistent results according to the needs of the learner [4]. The following figure shows the scenario of using the Big-Learn system illustrated by a sequence diagram (Fig 4).

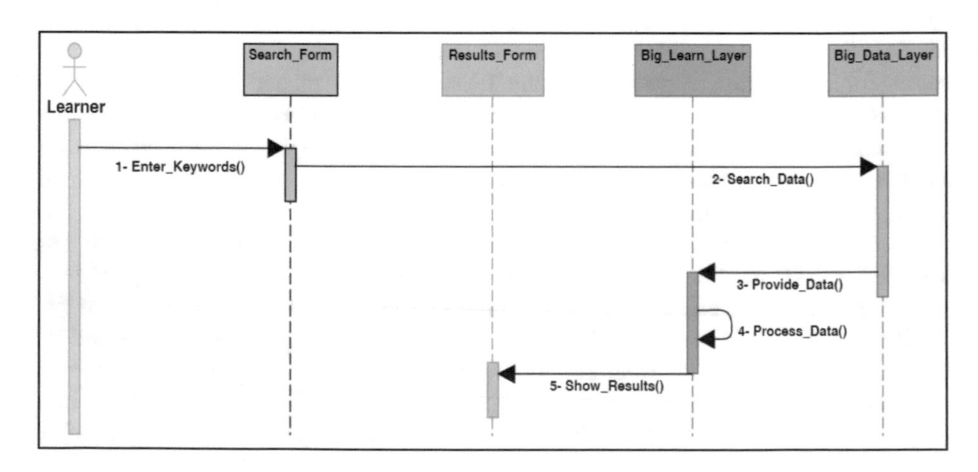

Fig. 4. General use case diagram for the Big-Learn system.

The user of the system (the learner) accesses the search interface to enter the keywords of his information request. These keywords are sent to the system for retrieval of the corresponding information from the Big Data layer. Then, the result data is processed at the Big-Learn level to structure, classify and send the processed data to the system results page, which will be displayed to the Big-Learn user.

For a better design and development of our system, we must study the circumstances of learning that takes place on online networks and distance learning. It is important to know the relevance of the learning experience of people in the online networks in which they find their information and they are likely to consume it. As part of our study, informational or learning data is defined as data collected from online open spaces where people access remotely, while communicating with others via blogs, audio-visual, wikis, as well as other sources of information and other remote communication resources. The constraints and difficulties emanating from such an environment are evident in the problems related to the study of human behavior, as well as other constraints involved, including the variability of the network and data, the power relations over the network and the size of generated content. Relevant analysis requires a mixed-method approach and leads to new ethics and questions about the confidentiality of information or data [4, 12].

For our study, we proposed conducting a sample survey of future system criteria and performance and quality of learning metrics when using online search, through the design and development of submission of questionnaires on a set of students who represent future users of our search and distance learning platform. We have described the different factors that may impact and affect the learner when using online search, as well as the criteria for simplicity and usability of our future solution. Questions were asked about the aspect of the presentation and the quality of the information in the results of the online search during documentation or distance training on a given course or theme. The classification of information also plays a dominant role in the organization and presentation of search results. The relevance of the outcome content is also in light of the expectations of the learner using online search. Thus, several elements must be studied and redefined for the design of the appropriate solution [4, 12].

3.2 Survey of Online Search Criteria

For the submission of our survey, we used the Google Forms tool for designing and completing the questionnaire. It is a tool for planning events, conducting a survey or survey, submitting questionnaires to a target population, or easily collecting information online. Forms can be created from Google Drive or an existing spreadsheet that can collect responses to form questions [4, 12].

Our target audience consists of two samples of the population, the first includes some students of the ESI (School of Information Science) of Rabat, and the second represents a group of students of the FST (Faculty of Sciences of Tetuan).

In this survey, we took on the task of identifying and defining the key factors and criteria that could impact and influence the learner's environment when using online search for learning purposes or documentation on a given topic, to enable us to derive performance and performance indicators related to search and learning in the Big Data environment to better design and achieve an adequate system that best meets the expectations of learners.

The following table summarizes the factors to be considered during this survey (Table 1).

Table 1. Factors contributing to the online search survey

Factor	Possible values
1. The type of result a learner prioritizes when searching online	a) Video (tutorial, guide, course, etc.) b) Image (diagram, graphic, illustration, etc.) c) Document (article, presentation, course, etc.) d) Web page (article, tutorial, guide, etc.) e) Audio (conference, recording, etc.)
2. The criterion of reliability of a relevant and fruitful search	a) Number of views, readings, etc. b) Publication date, most recent update. c) Element Source (author, website, blog, etc.) d) Number of sharing e) Number of comments
3. The number of results that the learner prefers per page	a) 5 b) 10 c) 20 d) 25 e) 50
4. The response time in seconds to the search request sent	a) 5 b) 10 c) 20 d) 25 e) 50

As described in this table, several factors can influence the process of using online search by the learner. Thus, the results of his request will depend on the relevance and the correct configuration of these criteria.

Students are therefore required to accurately answer the questionnaire. This questionnaire consists of four questions, each of which can have five possible answers (a), (b), (c), (d) or (e). The result of the questionnaire determines the search style and allows to propose adapted search pathways.

3.3 Results and Discussion

To find out about the search criteria that prevail at the FST and the ESI student's, we have putted an online questionnaire for the survey that we have already presented in the previous paragraph. We retained the responses of the two samples. In this subsection, we generally present the FST and the ESI student's preferences in terms of search styles.

The following figure shows the results of ESI and FST student's responses for each question in the survey (Fig. 5).

1. What type of result do you prioritize in your online search?

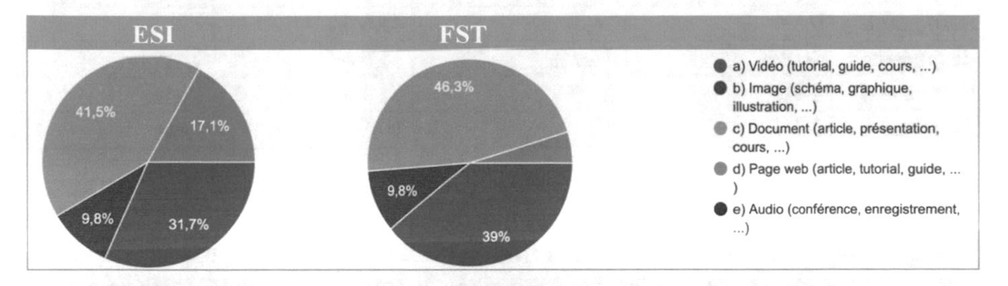

2. What is the result selection criterion do you prioritize in your online search?

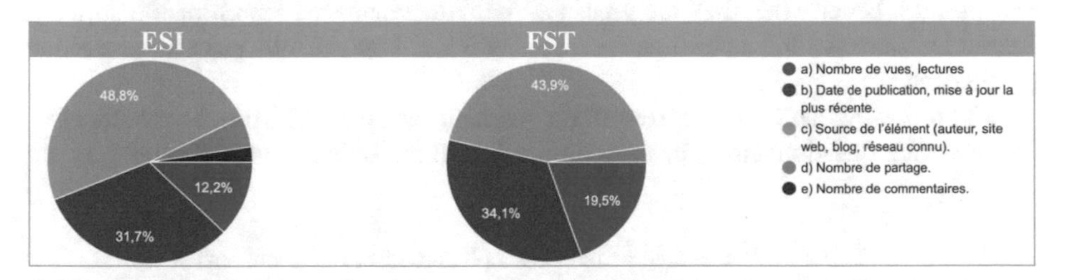

3. How many results do you prefer per page in your search results?

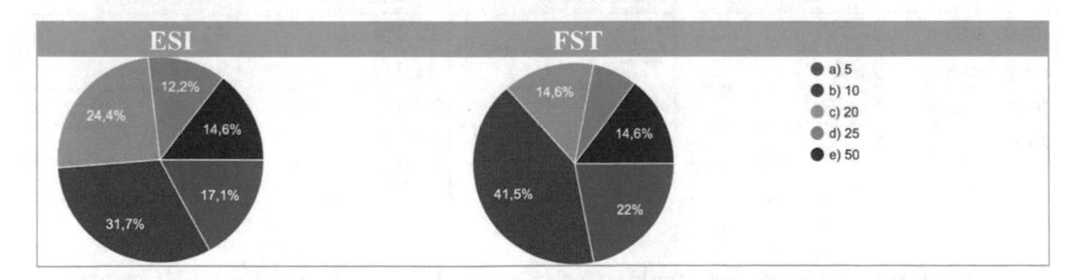

4. What is the response time to your search that you can tolerate (in seconds)?

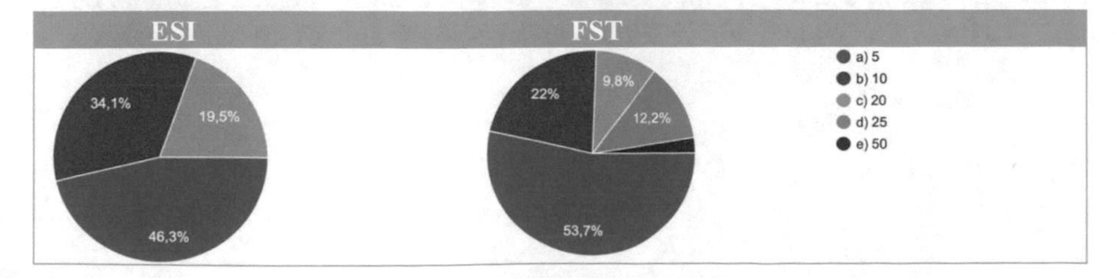

Fig. 5. Results of the online search survey for ESI and FST students.

The following table summarizes all the percentages obtained from the results of the FST student's questionnaires for each dimension of the online search (Table 2):

Table 2. Percentage of responses obtained from FST students

	Question 1 (Typology)	Question 2 (Relevance)	Question 3 (Quantity)	Question 4 (Time limit)
Response a	39%	19,5%	22%	**53,7%**
Response b	9,8%	34,1%	**41,5%**	22%
Response c	**46,3%**	**43,9%**	14,6%	9,8%
Response d	4,9%	2,5%	7,3%	12,2%
Response e	0%	0%	14,6%	2,4%

According to the preceding table, we deduce that the students of the FST are 46.3% for a typology of result in document form and 43.9% for a result depending on the source of provenance, 41.5% are for have a small amount of results on the page (10 results) and 53.7% for a search response time of no more than 5 s, so students of the FST prefer it better when they are presented with documents as search result, sorted or filtered by source within a period not exceeding 5 s and on a results page not exceeding 10 elements.

The following table summarizes all the percentages obtained from the results of the ESI student's questionnaires for each dimension of the online search (Table 3):

Table 3. Percentage of responses obtained from ESI students

	Question 1 (Typology)	Question 2 (Relevance)	Question 3 (Quantity)	Question 4 (Time limit)
Response a	31,7%	12,2%	17,1%	**46,3%**
Response b	9,8%	31,7%	**31,7%**	34,1%
Response c	**41,5%**	**48,8%**	24,4%	19,5%
Response d	17,1%	4,9%	12,2%	0%
Response e	0%	2,4%	14,6%	0%

According to the preceding table, we deduce that the students of the ESI are 41.5% for a typology of result in document form and 48.8% for a result sorted or filtered according to the source of origin, 31,7% are to have a small amount of results on the page (10 items results) and 46.3% for a search response time of no more than 5 s, so students of the ESI prefer better when they are presented with documents as search result, sorted or filtered by source within 5 s and on a results page not exceeding 10 elements.

The following table summarizes the whole percentages of responses for each question in the survey (Table 4).

Table 4. Example of the results of the replies to the questionnaire

Sample	Q1	Q2	Q3	Q4
ESI	c (82%)	c (73%)	c (65%)	b (60%)
FST	d (80%)	a (79%)	c (71%)	a (70%)

The table has four closed questions (from Q1 to Q4) with five categories (a, b, c, d and e). Each question defines a dimension for the online search model which is thus composed of 4 dimensions as shown in the following table (Table 5):

Table 5. Dimensions of the questionnaire on the criteria for online search

Dimension	Related questions
Typology	1
Amount	3
Time limit	4
Relevance	2

To locate the search criteria on a dimension, it is enough to count the number of answers on the five corresponding questions and to deduce the criterion which had the greatest number of choices in order to obtain a significant number. The results in the table above show the preferences in terms of a student's search criteria:

- The "Typology" and "Relevance" dimensions expressed respectively in questions 1 and 2 were prioritized by the two student samples from the ESI and the FST.
- The "Quantity" dimension expressed in question 3 was of moderate interest by the two student samples from the ESI and the FST.
- The "Time limit" dimension expressed in question 4 was of little interest by the two samples of ESI and FST students.

This survey will allow us to properly fit the requirements and criteria of the online search interface at the presentation layer of our solution model and which requires specific development to better meet the expectations and ergonomic needs of learners and also formatting search results returned by the integral online search system when using the Big-Learn system.

4 Conclusion and Perspectives

The aim of our paper is to propose an intelligent architecture that allows to process massive and unstructured data generated by the Big Data phenomenon in the Internet. This solution, named "Big-Learn" is based on a multiple Big Data tools, in addition to an interactive presentation layer that will facilitate the access to the data of the online search on the web so that any user can find easily the information he seeks.

The adopted method consisted initially in the study of the criteria and factors impacting the environment of the learner via the case study of using the online search by students of the UAE (Abdelmalek Essaâdi University). And this, through the creation and submission of the survey corresponding to a sample of learners using online search for their learning. This step has been followed by a detailed analysis of the results collected from this survey and that has framed the functional and technical requirements of the future solution to finally design a fictitious model for processing massive and unstructured data in the web. Thus, learners will be able to use this flexible system that can support them in their search for information or documentation needs.

As perspective of this work, we will study the possibility of integration of our solution to improve teaching and scientific research for learners of the UAE in their e-learning platform.

References

1. Padillo, F., Luna, J.M., Ventura, S.: Exhaustive search algorithms to mine subgroups on big data using apache spark. Prog. Artif. Intell. **6**(2), 1–14 (2017)
2. Leeder, C., Shah, C.: Measuring the effect of virtual librarian on student online search. J. Acad. Librariabship **42**(1), 2–7 (2016)
3. Aoulad Abdelouarit, K. Sbihi, B., Aknin, N.: Towards an approach based on hadoop to improve and organize online search results in big data environment. In: Proceedings of the International Conference on Communication, Management and Information Technology (ICCMIT 2016), April 2016
4. Aoulad Abdelouarit, K., Sbihi, B., Aknin, N.: Big-Learn: towards a tool based on big data to improve research in an e-learning environment. Int. J. Adv. Comput. Sci. Appl. (IJACSA) **6** (10), 59–63 (2015)
5. Sbihi, B., Kadiri, K., Aknin, N.: Towards a collaborative learning process based on the hybrid cloud computing and web 2.0 tools. Int. J. Eng. Technol. **1**(2) (2013)
6. Sbihi, B., Kadiri, K.: Towards a participatory E-learning 2.0 A new E-learning focused on learners and validation of the content. arXiv preprint, arXiv:1001.4738 (2010)
7. Matei, L.: Big data issues: performance scalability, availability. J. Mob. Embed. Distrib. Syst **6**(1), 1–10 (2014)
8. Gayathri, J., Saraswathi, K.: Extraction of data from streaming databases. Int. J. Comput. Trends Technol. (IJCTT) **4**(10) (2013)
9. Lakhani, A., Gupta, A., Chandrasekaran, K.: IntelliSearch: a search engine based on big data analytics integrated with crowdsourcing and category-based search. In: 2015 International Conference on Circuit, Power and Computing Technologies (ICCPCT). IEEE, pp. 1–6 (2015)

10. Soussi, R., Mustapha, N.B., Zghal, H.B., Aufaure, M.A.: Un système d'aide à la recherche d'informations en ligne basé sur les ontologies. In: CORIA 2008, pp. 483–490 (2008)
11. Blondeel, S., et al.: DOCUPOLE: un cours en ligne d'initiation à la recherche documentaire modélisé en MOOC. In: Former aux compétences informationnelles à l'heure du Web 2.0 et des discoverytools, pp. 20–28, June 2015
12. Aoulad Abdelouarit, K., Sbihi, B., Aknin, N.: Big data at the service of teaching and scientific research within the UAE. J. Educ. Vocat. Res. AMH Int. **6**(4), 24–30 (2015)

Applications and Experiences

A Methodological Proposal to Foster the Development of Team Work Mediated by ICT in the Subject Software Engineering

Ailec Granda Dihigo(✉), Dunia María Colomé Cedeño, María Teresa Pérez Pino, Marisol de la Caridad Patterson Peña, Liliana Argelia Casar Espino, and Tito Díaz Bravo

University of Informatics Sciences, La Habana, Cuba
{agranda, dcolome, mariatpp, marisol, lily, tdiaz}@uci.cu

Abstract. The purpose of this study is to show some experiences in the methodological development of Software Engineering professors at the University of Informatics Sciences by designing a set of group-work methodological and didactic actions, so that professors encourage their students to develop team work and collaborative learning with the help of ICT. The research methods employed were: analysis and synthesis, the systemic approach, the documentary analysis, interviews and scientific observations. A methodological procedure was developed based on the diagnostic analysis of the subject, taking into account the deficiencies from the previous course included in the reports. In this sense the development of teamwork and collaborative learning, using the means offered by ICT was included as a leading objective of the methodological work for the subject. The proposal was validated from the results gathered in the group interviews to professors who applied the set of actions, class observation guides, and the collaborative work developed in the virtual learning environment, as well as interviews to students to know their opinions about their professors performance. The design, development, and control of the actions integrating the methodological work contributed to the fulfillment of the research objective.

Keywords: Methodological training · Teamwork · Collaborative learning · Information and communication technologies (ICT)

1 Introduction

The scientific-technical revolution in the world promotes the development of an increasingly powerful infrastructure where new technologies are introduced, and the fundamental processes of society are associated to them. The development achieved by computer technology has led countries to be concerned about development in this field, supporting the process of training their professionals on the use of ICT [1].

In [2] it is sustained that the use of technology constitutes a genuine tool for a methodological transformation. At present, it is necessary to incorporate virtual resources in the different training actions. In [3] it is confirmed that this is the usual way to interact with young people, to access information and to build up knowledge.

M. E. Auer and D. May (Eds.): REV 2020, AISC 1231, pp. 705–717, 2021.
https://doi.org/10.1007/978-3-030-52575-0_58

Following the approaches given in [4], it is agreed that it is increasingly necessary to orient training processes in a flexible manner. In this sense, teamwork and collaborative learning play a fundamental role in the university environment. In [5] it is expressed that teamwork is that which involves several students who interact in the way designed by the teacher to achieve a common learning objective. On the other hand, in [6] it is affirmed that collaborative work, in an educational context, constitutes an interactive learning model that invites students to build together, which demands combining efforts, talents and competencies through a series of transactions that allow them to achieve the goals set in agreement.

1.1 Context

In [7] it is confirmed that Cuba is not exempt from the development achieved in the areas of ICT. Actually, a group of actions have been carried out with the objective of computerizing society. It is within this framework, that the University of Informatics Science was created in 2002, with the objective of promoting the Cuban software industry, following the strategy of developing technology in view of its three fundamental processes: Training, Production and Research. It is important to point out that these processes are linked to each other, responding to a basic principle of training: the link between study and work.

The study plan of Computer Science Engineering, a career that is studied only at the University of Informatics Sciences (UCI), has conceived the existence of several disciplines. One of them is Software Engineering and Management (IGSW for its Spanish acronym) [8]. This discipline is part of the backbone that conforms to the education and training of Engineers in Computer Science, as students develop basic skills for its application in their productive practice.

The application of modern techniques of group work, developing projects in teams and playing the different roles present in them are among the fundamental objectives of IGSW Discipline. It consists of 6 subjects: Introduction to Computer Science, Database Systems 1 and 2, Software Engineering 1 and 2 and Software Management.

The technological infrastructure of the university allows using the network for the development of the educational process of these subjects. In [1] it is stated that the development of teamwork and collaborative networking is one of the main ways to achieve the active participation of students and professors in the development of group activities, taking advantage of the possibility of socializing ideas, experiences and good practices. The Software Engineering Knowledge Body of IEEE1 [9], establishes fulfilling the need that the education and training in this discipline and its subjects incorporates students' work in the industry with the use of ICT.

The pedagogical experience presented in this paper, was carried out in the context of the methodological preparation Software Engineering teaching staff (IS for its Spanish acronym); this subject is taught to third year students of the University of Informatics Sciences (UCI). In correspondence with the Discipline, it has among its fundamental objectives, the application of modern group work techniques, as students develop team projects, playing the different existing roles within a project. The development of the different activities that are prepared for the subject are oriented towards achieving the development of individual responsibility and collectivism, as well as collaborative work, which is achieved by working in teams for the development of different learning tasks.

Currently, there are still some insufficiencies in the development of teamwork and collaborative learning in these subjects. The study showed that professors do not take advantages of all the technological resources and potentialities of ICT to support the teaching-learning process of software engineering. Although the courses in the virtual learning environment are designed with the objective of promoting collaborative work and management of the knowledge generated, professors are not sufficiently prepared to use these resources and they also lack a pedagogical preparation to fully the develop of these skills on their students.

1.2 Research Objectives

Taking into account the problem stated, this paper aims at contributing to develop the methodological training of subject Software Engineering professors at UCI, starting out with the development of a set of methodological and didactic actions that contribute to their preparation, in order to encourage the development of collaborative work, using ICT. To this end, the following components of the theoretical design are defined:

Research question: How can we contribute to the preparation of the professors of the subject Software Engineering at UCI, in order to promote the development of collaborative learning using ICT?

Objective: To elaborate a set of methodological and didactic actions related to teamwork, contributing to the preparation of UCI Software Engineering professors, in order to promote the development of collaborative learning using ICT.

Hypothesis: A set of methodological and didactic actions related to teamwork for UCI Software Engineering professors will contribute to the development of collaborative learning of students, using ICT.

Specific objectives:

1. To systematize the main theoretical and methodological foundations about teamwork and collaborative learning and their impact on the university environment.
2. To characterize the current state of the teaching-learning process of the Software Engineering subject in terms of teamwork development and collaborative learning.
3. To elaborate a set of methodological and didactic actions related to teamwork for the Software Engineering professors at UCI.
4. To evaluate the proposal of actions, through the application of different methods and techniques for obtaining information.

2 Development

The elements addressed in the introduction to this work, reveal the importance that is given to teamwork in the Career Engineering in Computer Science, and specifically in the Software Engineering Discipline and its subjects, through which students must acquire a set of tools, skills and the know-how with ICTs.

2.1 Materials and Methods Used

The units of analysis taken into consideration were: the 15 professors at UCI who belong to the subject group Software Engineering, 60 students from two third-year groups, the semester report of the subject in the academic year 2017–2018, the methodological work plan of the subject and the Discipline Software Engineering and Management, and the program of the subject and Discipline.

The scientific methods used were: Analysis-Synthesis; Systemic Approach; Documentary Analysis, Survey, Interviews and Scientific Observation. A methodological procedure was developed based on a diagnostic analysis of the Discipline and subject, taking into account the deficiencies in the previous course, collected in the mid-term reports. The degree of satisfaction with the use of ICT in these subjects was also considered, as well as the level of interaction achieved, the results achieved in the coursework which students must develop in teams, and the level of preparation of professors to promote the development of the necessary skills.

The theoretical-methodological foundations assumed in this work include what is stipulated by the resolutions of the Ministry of Higher Education [10] related to methodological work. According to what is stated in [11], the methodological work constitutes a fundamental way for the permanent improvement of teachers; so this work was also taken into account. This is based on didactic conceptions and its function is to plan, organize, regulate and control the teaching-educational process. Its essential objective is to optimize the political-ideological, scientific-theoretical and pedagogical level of the teaching staff in the different instances and levels of teaching as an indispensable factor for the quality development of the teaching-educational process.

The educational context of this research is delimited, as explained above, to UCI Software Engineering teaching staff.

2.2 Diagnostic Study

Different methods were applied as part of the diagnostic study. In order to describe the development of teamwork, the following indicators were defined:

- Number of insufficiencies identified in the methodological analyses of the IGSW Discipline Group, related to the development of teamwork in their subjects.
- Degree of satisfaction of users (students and professors) with the development of the subject Software Engineering, the support of ICT, and the skills achieved to work collaboratively and in teams.
- Level of interaction achieved with collaborative work tools, in the virtual platform they use.
- Learning outcomes achieved by students with the coursework developed in teams in the subjects of Software Engineering.
- Level of preparation of the faculty to promote the development of teamwork.

Methods Used for Diagnosis

- Documentary analysis to diagnose insufficiencies and the need to develop collaborative and team work in the Software Engineering Subject and to analyze learning achievements in the course works developed.
- Analysis of the interaction in the virtual forums of the courses of the subjects of Software Engineering, available in the EVEA UCI (Virtual learning environment)
- Questionnaires to diagnose the users' degree of satisfaction with the development of teamwork, and the professors' level of preparation to promote the development of this skill in the subjects of Software Engineering.

For data processing, the information was coded and transferred to a sheet in Microsoft Excel, using column graphics for it. Furthermore, other more advanced techniques, such as Pareto Diagrams were used. The latter were designed for specific questionnaire items, while the ones with bars in Excel were designed for the rest of the items of the instruments. As a multivariate procedure, work was done on Cluster Analysis, supported by the multivariate technique of analysis of main components. For the application of both techniques, the MiniTab 16 software was used, allowing the statistical processing of the data that was obtained.

Data Obtained from the Application of Methods and Instruments for Diagnosis Documentary Analysis

The revision of the methodological documents issued by UCI Vicerrectorship for Academic Affairs in the last two academic courses (Mid-term reports of the *IGSW* Discipline and of the Software Engineering subjects) was carried out, looking for the elements and analysis related to the development of skills for teamwork and collaborative learning.

The following are among the main inadequacies identified in the reports:

- Insufficiencies in the development of the different tasks and activities in teams, which must be carried out within the framework of the Software Engineering subjects' coursework.
- Insufficient collaborative work between students and professors, thus hindering collaborative work.
- Low levels of skills development for teamwork.
- Little use of ICT for the development of teamwork in the Software Engineering subjects.
- Professors allege not having all the necessary preparation to encourage the development of these skills in their students.

Analysis of the Interaction in the Virtual Forums of the Courses of the IGSW Subjects of the Discipline, Available in the EVEA Moodle

The analysis of the interaction achieved in various forums available in the virtual courses of the subjects of the Discipline was carried out, identifying the following behaviors:

- Low participation in the forums available in the virtual courses of the subjects of the *IGSW* Discipline.

- Most of the interventions were neither answered nor commented on.
- Low level of interaction among the participants; on many occasions participants limited themselves to introduce new topics, but not to give an opinion about those already displayed in the forums.

For the processing and analysis of the interaction in these forums, the NETDRAW (Network Visualization Program) tool was used. In all cases, it was evident that the level of access and interaction was low, and that there had been no reply to most of the posts. It was also appreciated that the interaction occurring was mainly student-teacher. Concerning the use *wikis*, it should be pointed out that the contributions were practically null, and even in some cases, this resource was not used at all.

Instruments Applied

Questionnaire I. Diagnosis of the development of teamwork in the subjects of Software Engineering:

After processing and interpreting the data collected in the questionnaire applied to the sample defined, it could be identified that most of the respondents have not been able to develop skills to be integrated into multidisciplinary teams, or to communicate effectively. This result is also confirmed with the answers related to the practical application of the different techniques to work in groups and the level of motivation to work collaboratively, taking into account the opinion and participation of the group members. The data obtained showed that the development of teamwork has not been consolidated in these subjects from a theoretical and practical point of view. Regarding the use of different resources provided by ICTs to develop more efficient teamwork, it should be noted that a significant percentage of participants have not used these resources.

Questionnaire II. Level of preparation of professors to promote the development of teamwork in the Software Engineering subjects:

The answers issued by the professors confirm the training needs in order to promote the development of teamwork. It was also possible to identify the core contents where there can be more incidences.

As a result of the diagnosis, it was possible to verify the need to develop teamwork and collaborative work on the part of the students in the Software Engineering subjects, and the necessary preparation of the faculty to promote the development of these skills.

2.3 System of Activities

The Methodological Work Plan (PTM for its Spanish acronym) of the Software Engineering Subject at UCI, constitutes the guide to develop the methodological preparation of the teaching staff and contribute to the quality of the Teaching Learning Process (PEA for its Spanish acronym) that is carried out. Its purpose is consistent with article [10], which states that:

"The methodological work is the work that, supported by Didactics, is carried out by the subjects that intervene in the educational teaching process, with the purpose of achieving optimal results in such process, hierarchizing the educational work from instruction, in order to fully satisfy the objectives formulated in the study plans".

For the planning of the methodological activities of this course, there were taken into account, the analysis of the diagnosis made, the deficiencies in the previous course included in the mid-term report and the orientations of the methodological work at University level. In this sense, the development of teamwork mediated by ICT was identified as one of its methodological lines.

Based on the identification of this line, a set of actions was designed to strengthen the didactic and pedagogical preparation of teachers so that they can contribute to their students' development of team work skills, using ICT as a means and promoting collaborative learning (Fig. 1).

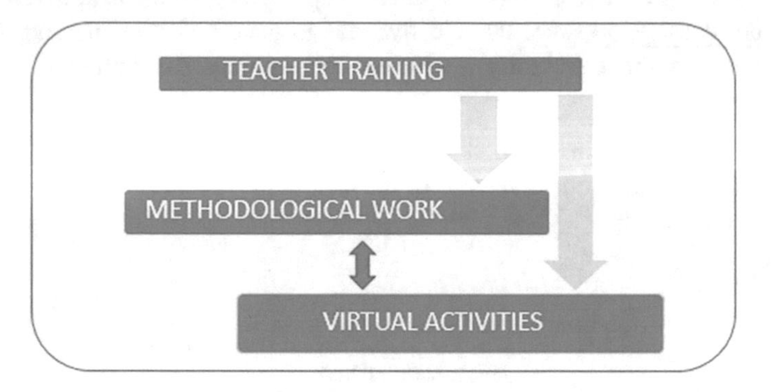

Fig. 1. General framework of the actions suggested

Methodological Work

Figure 2 displays the different planned activities as part of a methodological work cycle of the subject. This responds to the line of methodological work related to the use of ICT to promote team work and collaborative learning.

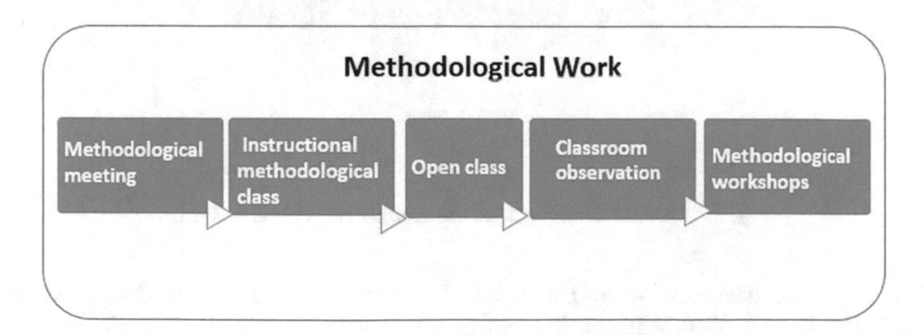

Fig. 2. General framework of the actions for the methodological work

Methodological Meeting: The objective of this activity was to describe the theory behind teamwork and collaborative learning, using ICT.

In this sense, the analysis and the debate focused on the pedagogical and didactic foundations related to teamwork.

Instructional Methodological Class: Its objective was to provide methodological guidance to the faculty on how to apply ICTs to promote the development of teamwork in a specific theme of the Software Engineering subject.
Open Class: It was suggested to consider how to promote the development of teamwork (a seminar of the subject) making the most of the different potentials offered by technology.
Classroom Observation: It is developed to verify the fulfillment of the objectives of the methodological cycle, from the practical application of the acquired knowledge.
Methodological Workshop: Its objective was to have a debate and socialization of experiences in the development of teamwork among the Software Engineering professors.

Once the structure of the methodological work conceived has been approached, the main theoretical foundations assumed for the design of virtual activities developed are conveyed (Fig. 3).

Virtual Activities

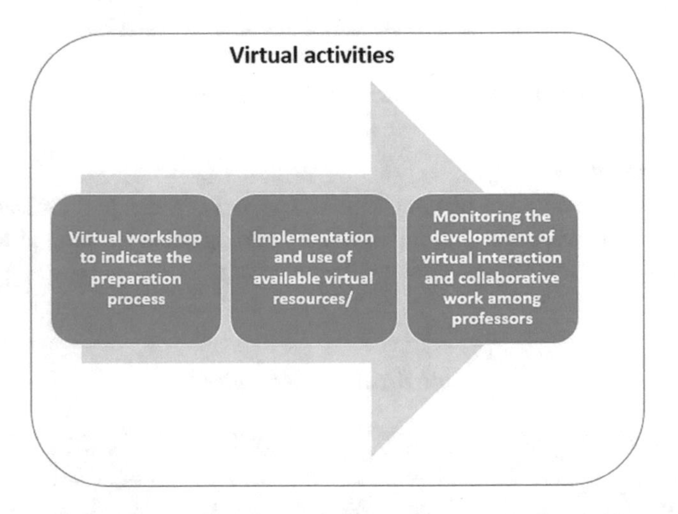

Fig. 3. General framework of the actions for virtual activities

The activities designed were based on the ADDIE instructional design model (Analysis, Design, Development, Implementation, Evaluation), which incorporates five basic steps or tasks that comprise the basis of the instructional design process. The five steps are: Analysis, Design, Development, Implementation, and Evaluation of the learning materials and the activities. The generic instructional design model is flexible, thus allowing modification and elaboration, based on the needs of the instructional situation [12].

Description of Phases

The Analysis Phase is the basis for the rests of the instructional design stages. During this phase the problem to be solved and its possible solutions were defined, outlining the educational goals and a list of tasks to be carried out for the development of the virtual activities.

The Design Phase involved the use of the results of the Analysis phase to plan a strategy for the development of instruction. During this phase, it was also identified how to achieve the educational goals determined during the analysis phase and how to broaden the educational foundations.

The Development Phase was structured upon the basis of the Analysis and the Design phases. The purpose of this stage was to generate the plans of the activities and the resources to be used in them.

The Implementation Phase referred to the actual assignment of the instruction; its purpose was the effective and valuable conveyance and use of the designed activities.

The Evaluation Phase measured the effectiveness and value of the instruction. The evaluation must be present throughout the whole instructional design process (within phases, between phases and after its implementation) Formative and Summative evaluations were developed.

Activities Developed

- Development of a virtual workshop to show the indications of the process: It was explained which were the activities designed and their objectives, as well as the resources to be used. The indications and resources for interaction were available for 15 days for professor to be able to access them and develop the activities suggested.
- Implementation and use of the virtual resources available: A work session was organized so that the professors could use the virtual resources for interaction and the development of skills. The main resources used were:
  - General Forum: It was used to present novel issues related to Software Engineering.
  - Forum for Group work: Forums were activated to discuss and have group reflections on specific Software Engineering topics.
  - Wiki: It was used so that professors would share and make it available to others, different Software Engineering topic-related materials, reflections, and links.
  - Links to the web and other folders: This resource was used in order to access Web pages and other documents related to different Software Engineering topics. All users had the possibility to add or comment existing links.
- Monitoring the development of virtual interaction and collaborative work among professors, based on their participation in different activities.

For the design, development and implementation of these activities, the work was done with a Virtual community of Software Engineering, which was the result of the doctoral thesis written by [13]. This community represents an excellent space to foster the development of group work skills. The training of the staff in the use of the different virtual resources represented an opportunity to achieve the development of these skills in the students in a more natural way. In addition, it allowed them to express their opinions on the different issues related to these activities, and improve them, both at the end of the process and during their development.

Figure 4 and 5 show the use of the different resources.

Fig. 4. Use of the forum resource

Fig. 5. Use of the wiki resource

3 Results Obtained

For the evaluation of the proposal, three indicators were defined: (1) level of acceptance, (2) level of development of collaborative work and (3) level of transfer of acquired knowledge by professors to their students. The following methods and techniques for obtaining information were applied: Iadov's test; Group interviews with professors who participated in the implementation of the action system; student interviews to learn about their assessment of their professors' performance; and observation guides used in classroom observation and observation of the collaborative work in the virtual learning environment.

Low levels of teamwork and collaborative development were observed as a result of the diagnosis made. Professors did not promote on their students the development these skills. The normative documents offer the possibility of proposing actions that contribute to work in this sense. A set of actions is designed and implemented from the integration of methodological work. A methodological work cycle was developed on the topic: Teamwork and collaborative work with the use of ICT in Software Engineering. This comprised the development of: a Methodological Meeting, an Instructive Methodological Class, the realization of classroom observations, and the development of a Methodological Workshops to close this methodological cycle. Besides, there were designed and implemented a set of virtual activities to empower the training of teachers in these topics, and achieve their preparation in promoting their students' development of team work and collaborative work in their students. The different activities were based on the use of virtual communication resources and new ways of participation and building up of information. A series of recommendations aimed at the integration of virtual modalities were offered, taking into account the orientation of the educational process.

The different methods and techniques applied demonstrate the level of acceptance, higher levels of development of collaborative work, and transfer of knowledge acquired by the professors to their students.

Concerning the first indicator, professors were satisfied (group satisfaction index ISG = 0.93, according to Iadov technique). In this sense it was suggested to develop another virtual training activity in which the pedagogical foundations of team work are addressed, in the specific didactic of Software Engineering.

The second and third indicators which reflected the results of classroom observation, showed the knowledge acquired by professors as for the theoretical-methodological foundations of team work and collaborative learning, which they were able to transmit to their students. The use of virtual learning environments and virtual communities was also promoted. In this sense, professors could assess their progress in this area and value how this influenced the way they could transmit it to their students.

What recommendations are suggested to develop the necessary teamwork skills and to develop collaborative learning? (General recommendations obtained from the result of the different methods applied).

- To design, develop and implement virtual resources from the use of instructional design models.
- To orient activities in which students must work in groups with a common and clearly defined objective.
- To encourage the request for help among the classmates and the teacher.

- To evaluate the development of skills achieved and the students' level of satisfaction with the professors' work.
- To select which digital educational resource could contribute to collaborative learning.

After processing the data obtained and as a result of the application of different methods and techniques, the level of acceptance is revealed, there are higher levels of work and collaborative learning as well as transferring of the knowledge acquired by the professors to the students.

4 Conclusion and Recommendations

The study showed that taking into account the integration of the methodological work and the implementation of virtual activities, the pedagogical and scientific training planned contributed to the professional development of UCI Software Engineering professors, promoting the development of teamwork and collaborative learning, using ICT. Furthermore, the incorporation of virtual resources in the different actions was positive.

It is recommended to continue researching on the possibilities offered by the methodological work system in order to contribute to the pedagogical and technological development of the faculty on teamwork and collaborative learning topics; it is also recommended and that this work serves as a reference for other researchers, for studies related to the need to make the most of ICT in the teaching-learning processes.

References

1. Granda, A., Gómez, Y., Santos, Y.: Las TIC en la Ingeniería de Software en la UCI. Diagnóstico de necesidades. Revista de Tecnología Educativa de la Universidad de Holguín 1 (2016)
2. García-Peñalvo, F.J., Ramírez Montoya, M.S.: Aprendizaje, Innovación y Competitividad: La Sociedad del Aprendizaje. Revista De Educación a Distancia (52) (2017). https://revistas.um.es/red/article/view/282141
3. Bartolomé Pina, A.R., García Ruiz, R., Aguaded Gómez, J.I.: Blended learning: panorama y perspectivas. RIED: revista iberoamericana de educación a distancia **21**(1), 33–56 (2018). http://dx.doi.org/10.5944/ried.21.1.18842. http://rabida.uhu.es/dspace/handle/10272/14430
4. Salinas Ibáñez, J., de Benito Crosetti, B., Pérez Garcies, A., Gisbert Cervera, M.: Blended learning, más allá de la clase presencial. RIED. Revista Iberoamericana de Educación a Distancia **21**(1), 195–213 (2018). https://doi.org/10.5944/ried.21.1.18859
5. Salinas, C.C., et al.: Técnicas de trabajo en equipo para estudiantes universitarios. Jornada Redes 2012, Universidad de Alicante (2012). https://web.ua.es/es/ice/jornadas-redes-2012/documentos/posters/246217.pdf
6. Maldonado Pérez, M.: El trabajo colaborativo en el aula universitaria. Laurus, Revista de Educación **13**(23), 263–278 (2007). http://www.redalyc.org/pdf/761/76102314.pdf
7. Granda, A., Santos, Y.: Las TIC en la enseñanza de la Ingeniería de Software en la Universidad de las Ciencias Informáticas. Pasado, presente y futuro. Edutec, Revista Electrónica de Tecnología Educativa 37 (2011)

8. Granda, A.: Diseño de Curso Virtual para apoyar el proceso de enseñanza-aprendizaje de la Disciplina de Ingeniería y Gestión de Software en la Universidad de las Ciencias informáticas. Edutec, Revista Electrónica de Tecnología Educativa, 34 Diciembre 2010
9. SWEBOK: Guide to the Software Engineering Book of Knowledge. IEEE Computer Society Press, Los Alamitos (2004)
10. Ministerio de Educación Superior: Reglamento docente metodológico. Resolución, February 2018 (2018)
11. Estrada, O., Blanco, S.M., Cancell, D.: Estrategia para la formación profesoral en el autoaprendizaje estudiantil. Opuntia Brava **10**(4) (2018). https://doi.org/10.35195/ob.v10i4.620
12. McGriff, S.J.: Modelo ADDIE. Proceso dc desarrollo de un curso. Instructional Systems, College of Education, Penn State University (2000)
13. Granda, A.: Modelo didáctico para el uso de comunidades virtuales en el proceso de enseñanza aprendizaje de la Disciplina Ingeniería y Gestión de Software en la Universidad de las Ciencias Informáticas. Tesis Doctoral, Universidad de las Islas Baleares, España (2013)

The Collaborative Learning of the Analysis and Modeling of Software with the Use of Facebook

Dunia María Colomé Cedeño(✉), Arlenys Palmero Ortega, Ailec Granda Dihigo, and Taire Faife Rodríguez

Universidad de las Ciencias Informáticas, La Habana, Cuba
{dcolome, arlenys, agranda, tfaife}@uci.cu

Abstract. Collaborative learning, by encouraging students to work together to maximize their learning, has great significance in the learning of software engineering, due to its marked teamwork character. Among the advantages of this type of learning is to promote critical thinking, by giving opportunities for its members to discuss the content of their learning. From a diagnosis made to the teaching and learning process of this discipline, in the third year of the degree in Informatics Science Engineering, were identified inadequacies in the learning of software analysis and modeling, which have caused, on occasion, errors in the modeling of the structure and behavior of a computer system. In order to minimize these difficulties, this paper presents a proposal to promote collaborative learning of software analysis and modeling, using the social network Facebook, based on a class system of the topic Analysis and software modeling of the materia Software Engineering I career Informatics Science Engineering. The results obtained show the potential of Facebook to promote the collaborative learning of Software Engineering, using as a sample a group of 30 students who actively used this social network.

Keywords: Collaborative learning · Social network · Software engineering

1 Introduction

Learning means growing in all senses, that is why the term learning associated with pedagogy acquires great relevance from the point of view of the appropriation of knowledge by people, as well as the way of transmitting and socializing them, hence learning has an individual component, but is mostly of a great collective and collaborative influence.

In this collaborative environment, the link between the knowledge we already have and the conditioning of this to a new content becomes part of a universe of discoveries that become significant for human beings, resulting in an effective mechanism to face the tasks of everyday life. In order for learning to be meaningful in a collaborative environment, it must be generally accepted that students can relate what they learn to usefulness in daily life that which commonly refers to putting into practice what is learned, taking into consideration culture, the group, the teacher, their learning style, personal and collective motivations and interests.

M. E. Auer and D. May (Eds.): REV 2020, AISC 1231, pp. 718–728, 2021.
https://doi.org/10.1007/978-3-030-52575-0_59

Collaborative learning is sometimes understood as a group of people talking or interacting at the same time, but it means more than that, it is a broad approach where groups work together to solve a situation, a case study, and above all a problem of everyday life that can be significant at an individual and collective level. This type of learning is also a process whose creative stage involves personal interests, personal experiences and knowledge acquired independently according to a common purpose.

The constructivist pedagogy on which collaborative learning is based maintains that, knowledge is not received passively, but is actively constructed by the student. The focus is not on the transmission of content, but on the construction of knowledge, which in turn is based on previous knowledge. This is why the role of the teacher changes, becoming a mediator between knowledge and the student.

At this point, reference should be made to the values that this type of learning fosters and promotes, such as solidarity, responsibility and creativity, while at the same time diminishing feelings of individuality, alienation, achieving higher levels of performance and productivity.

According to Gerlach (1994), "Collaborative learning is based on the idea that learning is a naturally social act in which the participants talk among themselves. It is through the talk that learning occurs."

There are several environments where collaborative learning acquires great relevance, but without a doubt the use of technologies has meant a look towards new ways of understanding this type of learning, through the use of tools for common creation. Some examples have come with the use of wikis, forums, academic networks and social networks, as is the case of Facebook, used mostly for entertainment purposes.

The current concern about learning through technologies and specifically social networks remains the same as when these technologies began to emerge: the pedagogical design of highly meaningful tasks for students. It can also be said that the debate becomes more complex according to the social context in which ICT learning takes place, the availability of resources and real access to the Internet. It also influences the individual technological perception and motivation on the part of the teacher towards the efficient, creative and educational use of technologies.

According to Navarro (2009): ICT not only offer a network to which individuals are added, but also act as social technologies whose improvement depends both on the diversity of their functions (social, political, cognitive, economic, etc.) and on the flexibility with which they adapt to our functional diversity (to our life cycles: from infancy to old age, our changing and oscillating motor skills or our audiovisual perception thresholds).

Bearing in mind the ever-increasing dynamics of the use of technologies as a function of learning, and in particular social networks, it becomes more imperative to promote the use of these massive networks as a function of the transmission of knowledge that does not isolate human beings, but integrates them through a motivating and increasingly participative platform.

Starting from the breadth of the concept of social network, one of the most accepted definitions is that of "a virtual space of communication between people with some common trait in which their users can contact family, friends or strangers, and share content regardless of time, space or ubiquity. Therefore, it would be an open system of social interaction, in permanent construction thanks to the continuous contributions of each user.

Orihuela (2008) defines online social networks as "web-based services that allow users to interact, share information, coordinate actions and, in general, keep in touch".

For this reason, the influence of social networks in education is important, bearing in mind that nowadays all new trends and relevant information in any socio-political, economic and also educational field, moves in social networks, being essential the integration to its use managing the didactics in an adequate and intentional way.

Modeling the structure and behavior of a system based on the analysis of its requirements, to facilitate its understanding by clients and developers, is a complex activity of software engineering. The teaching of this subject in the Software Engineering subject has deficiencies that lead to some difficulties or frequent errors in the modeling of the structure and behavior of a system. Some of these deficiencies are related to the low use of the potential of ICT to promote collaborative learning of this topic.

The objective of this paper is to instruct in the pedagogical collective on how to promote the collaborative learning of analysis and software mode, using the social network Facebook, from a system of subject classes Analysis and software mode of the subject Software Engineering. The research question is how to promote the collaborative learning of the topic Analysis and Modeling of the subject Software Engineering through the employment of the social network Facebook? The hypothesis is that the development of methodological guidelines for Software Engineering professors of the University of Informatics Sciences, which contain actions related to the use of Facebook to contribute to the collaborative learning of the topic Analysis and Software Modeling, will help to minimize student errors in modeling the structure and behavior of a system.

The development of this research was based on the analysis of semi-annual reports of formation at the University of Informatics Science; from controls to classes taught by professors of the ISW subject. As well as in the revision of reports of controls to classes of the subject ISW, made by other teachers and in the interviews to professors of ISW and students of 3rd year of the career Engineering in Informatics Sciences. In the analysis of the most common errors in the analysis and software modelling were used the following scientific methods Analysis-synthesis, Logical-Historical and Participant Observation.

The validation of the research was based on the use of the Iadov technique, through the application of a questionnaire to a sample of professors of the Software Engineering course and to a sample of students in the third year of the degree in Computer Science Engineering, which gave a level of satisfaction of 0.95%. In addition, a review of the students' activity on the Facebook social network was carried out based on their interaction with the proposed activities, in which the broad participation and exchange of knowledge between students and their teachers can be appreciated.

2 Collaborative Learning

Collaboration learning (CL) has become a twenty-first-century trend. The need in society to think and work together on issues of critical concern has increased (Austin 2000; Welch 1998) shifting the emphasis from individual efforts to group work, from independence to community (Leonard and Leonard 2001).

CL is an educational approach to teaching and learning that involves groups of learners working together to solve a problem, complete a task, or create a product. In the CL environment, the learners are challenged both socially and emotionally as they listen to different perspectives, and are required to articulate and defend their ideas. In so doing, the learners begin to create their own unique conceptual frameworks and not rely solely on an expert's or a text's framework. In a CL setting, learners have the opportunity to converse with peers, present and defend ideas, exchange diverse beliefs, question other conceptual frameworks, and are actively engaged (Srinivas 2011).

CL represents a change on the thinking that teachers should be the center of the class and students should be receptive, which doesn't mean that listening, taking notes has to disappear completely, it means putting all this into function of group discussion and active work. This type of learning is still confused today with face-to-face or face-to-face learning.

This is why design through the use of technology must take into account the personalization of content, individual experiences, and the real conditions of the group. It is not having them do the task individually and then have those who finish first help those who have not yet finished. And it is certainly not having one or a few students do all the work, while the others just put their names on the report (Klemm 1994).

It is not new that the influence of technologies is a valuable support for learning, but what happens that there is still so much debate regarding their impact on learning environments? Perhaps to some extent it is because the understanding of the main approaches to learning cannot be linked to technological means, it also influences the perception of technology as purely executive and not reflective, which limits the understanding of the important supporting role of technologies in terms of education and collaborative learning.

Collaborative learning must therefore be understood as a process of social construction of content, taking into account individual concepts and individual ownership. It is not only a question of sharing knowledge, but of pretending that the other learns from it, that he can reconstruct it, that he can retake it. Asynchronous communication in collaborative learning must contain a moment of individual thought, of individual reflection that can be consulted, analyzed and debated later.

3 Collaborative Learning Through the Use of Social Network Facebook

Some experts explain that "Computer-assisted collaborative learning experiences aim to understand learning as a social process of knowledge construction in a collaborative way. It can be defined as a teaching-learning strategy by which two or more subjects interact to build knowledge, through discussion, reflection and decision making, a process in which computer resources act as mediators. This social process results in the generation of shared knowledge, which represents the common understanding of a group regarding the content of a specific domain, (Zañartu 2011).

When the use of social networks is adequate, Muñoz Prieto et al. (2013) emphasize that:

a) They can encourage in the students the autonomy, the cooperative work and a dynamic and constant construction of diverse types of information, something fundamental in the society in which we live, allowing in addition that the student can become the mere constructor of his own knowledge.
b) To know how often students, use resources available on the Internet for academic and communication purposes.
c) Differentiate purposes and motivations of students about the use of resources available on the Internet for academic, communicational and socio-affective purposes.
d) Evaluate the level of use that students make of social networks.

Facebook is the most popular network in Cuba among young people and students, and although it is not used mainly for educational purposes nor was it created in its essence for that purpose, it is increasingly used by teachers for the development of learning tasks in a collaborative environment. Although academic networks have been created and personalized for this type of educational purpose, they do not have the same level of use, which makes teachers rethink the particular didactics of their subjects with Facebook as the ideal platform for knowledge exchange.

Meso (2010) states that one of the areas where we can develop the potential of social networks as part of education is Facebook, since it represents a collaborative space, in addition to offering a large amount of resources to illustrate applications, propose application exercises, optimize the dynamics of the class, among others, which offers the possibility of connecting students to each other in learning networks (Selwyn 2007).

If we take into account this reality, it is necessary to think about the roles occupied by both teachers and students. Evidently there is a significant change of roles that entails a change in the way of thinking of teachers, who would begin to be tutors, facilitators or organizers of online content.

Koper (2009), states that the users of a learning network, in its commitment to acquire skills, they can:

- Exchange experiences and knowledge with others.
- Working collaboratively on projects (e.g., innovation, research, jobs).
- Create working groups, communities, debates and congresses.

- Offering and receiving support to/from other users of the learning network (such as questions, observations, etc.).
- Evaluate themselves and others, search for learning resources, create and develop competency profiles.

These social networks are the most widely used by young people in Cuba, so students are much more familiar with them than with the aforementioned learning networks. In these networks students usually spend many hours a week, so teachers should take advantage of this inertia to use it in a didactic way.

4 The Teacher's Role to Promote Collaborative Learning in Social Networks

In order for networks to contribute to educational transformation, the teacher and the student build the content together, but without a doubt the design, planning and monitoring of what the students achieve on the network is mostly the responsibility of the teacher, who plays a significant role. Here the indications must be precise, it must be made clear what one wants to achieve, how to do it and how to evaluate it, what presupposes organizing in the network the group of students that will participate and the role that each one must play in the team.

However, in this type of learning situation, the teacher's effort is focused on helping the student to develop talents and competencies using new teaching schemes, which makes it a guide to the teaching-learning process. At the same time, the student becomes a more autonomous and self-sufficient being that builds its own knowledge. The teacher now has the task of helping them learn (Meso 2010).

Based on what is stated above, it is necessary that the teacher is committed to achieve much more than participation, it is necessary that the objective of the teacher is to develop skills that help the student to make effective use of social networks, as expressed Artero (2011): "Do not anchor in methods/systems that are now obsolete before the advance informative, communicative and interactive offered by the network and, even more, social networks.

For the role of students and professors of the UCI and other universities to be effective in social networks, the author considers the following elements essential:

- Student learning styles.
- Characterization of students and academic year.
- Objectives of the subjects that the students receive in the academic year
- Students' learning strategies.
- Social context in which students interact.
- Cultural background of the students.
- Integration of different subjects and curricular strategies.
- Formation of values according to the aims of the university and the model of the professional.

5 Diagnosis

As part of the diagnosis, the following activities were carried out:

- Analysis of semi-annual reports of the training at the University.
- Controls to classes taught by professors of the subject ISW 1.
- Review of reports of controls to classes of the subject ISW I, made by other teachers.
- Interviews with ISW I teachers and 3rd year students of the Computer Science Engineering career.

In addition, a diagnosis was made of the use of the social network Facebook, which resulted in the massive use of this social network by university students. Taking into account the most common errors in the analysis and modeling of software and the extensive use of the social network Facebook by university students, four activities were designed to instruct teachers on how to enhance the collaborative learning of Software Analysis and Modeling, through the aforementioned social network.

Based on the diagnosis made, it was found that 1) sometimes, group activities are oriented, but it is not taken into account in their orientation in the positive interdependence that must be achieved as a central element for collaborative learning, 2) the media are insufficient What is the facility for the students, in the way that there can be a wide exchange between the members of the group, 3) there is a low use of the potential of the information and communication technologies for the collaborative learning in the team in the subject Software engineering.

Among the main difficulties or frequent errors in the modeling of the structure and behavior of a system, are 1) use of stereotypes that do not correspond to the class diagrams of the analysis and/or to the collaboration diagrams, 2) confusion in the representations that are used to model behavior, 3) confusion between sequence and collaboration diagrams, 4) messages are not always enunciated in form of operations, 5) incorrect sense of the flow of messages in collaboration diagrams.

6 Activities to Promote Collaborative Learning Using the Social Network Facebook

The subject Software Engineering in the UCI, has a highly collaborative component, it demands the teacher greater preparation for the design of group activities, from the pedagogical conception of content.

The proposal presented is based on the creation of a group on Facebook, whose members are 22 third-year students of the degree in Computer Science Engineering. Through this group, collaborative learning has been fostered through a series of activities, some of which are presented below.

Activity No. 1
Type of class: Conference

Theme: Introduction to software Analysis and Modeling.

Objective: Characterize the modeling of the analysis for a computer system, from the approaches, models and domains of analysis.

Orientation: State at least three elements that characterize the modeling of the analysis for a computer system and show through the emoticons or answering to their peers, the approval or not of the elements they refer (Fig. 1).

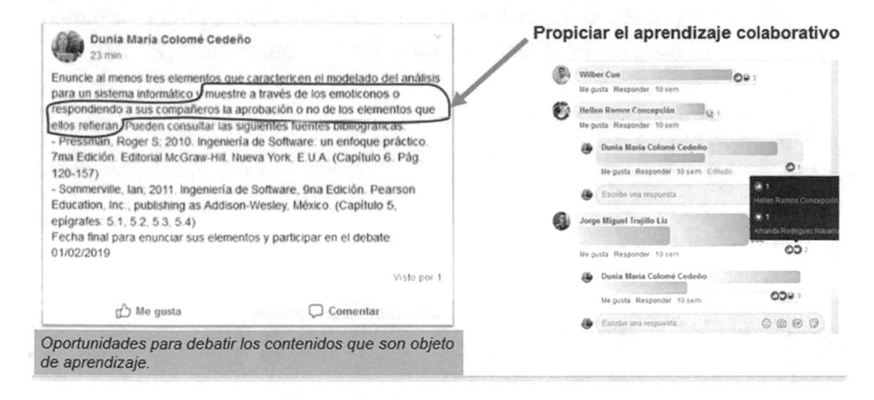

Fig. 1. Activity 1 in Facebook.

Activity No. 2

Practical class No. 9: Techniques for the analysis and modeling of the structure.

Objective: To model with a CASE tool the structure of a system using a modeling language.

Orientation: From the textual description of the use case Manage event of the case study Management of events in the ICU make the class diagram of the analysis. Publish the image of your diagram on the group page of the subject. Analyze the diagram made by your colleagues and comment on your impressions in your publication (Fig. 2).

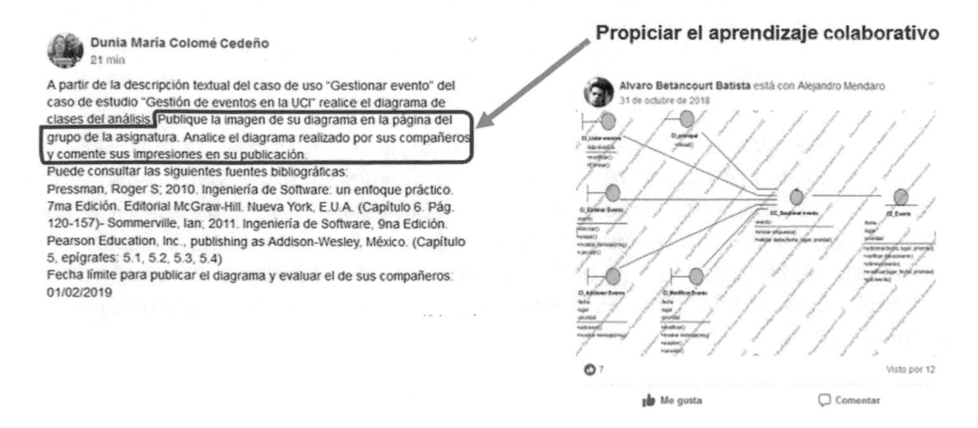

Fig. 2. Activity 2 in Facebook.

Activity No. 3

Conference No. 10: Modeling the behavior of the software structure.

Objective: To characterize the domain of the behavior analysis for a computer system, from the modeling elements of the analysis.

Orientation: In the survey published on the Facebook page of the group, entitled Representations for behavioral modeling, select the representation(s) that you consider correspond to the representations of behavioral modeling. Through the comments you can consult your colleagues or teacher if you have any questions (Fig. 3).

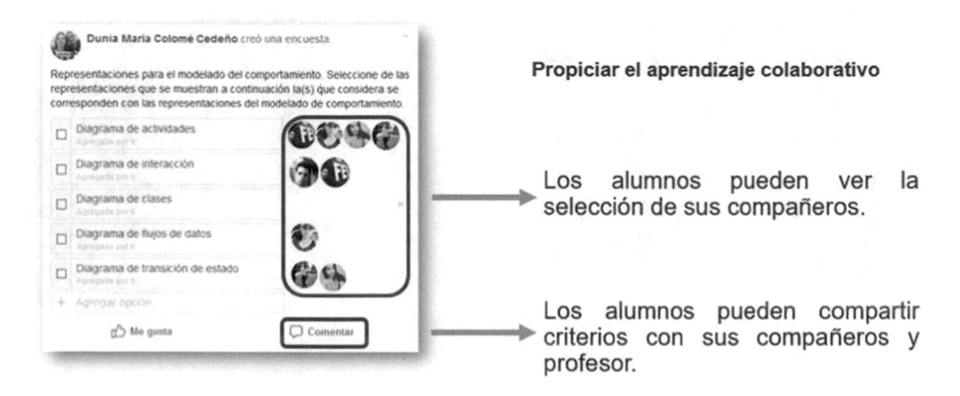

Fig. 3. Activity 3 in Facebook.

The evaluation of these activities is essential. Below are some considerations.

- All the activities must be evaluated based on the individual realization of the orientations and the interaction of each student with the rest of the group.
- Evaluations should be reflected in the record of attendance and evaluation control.
- The result of the evaluations should be discussed in the classroom.

7 Conclusions

The methodological actions enunciated constitute an initial step to promote the collaborative learning of the analysis and modeling of software using the potential of the social network Facebook. Through the analysis of pedagogical experiments it is possible to analyze the participation of students in social networks and determine the precise actions to take advantage of their potential. The evaluation of students' activity in social networks will allow teachers to evaluate the effectiveness of collaborative learning in these spaces.

8 Recommendations

It is recommended to continue the study on the use of social networks in educational contexts due to their continuous evolution.

References

Ministerio de Educación Superior. Reglamento del Trabajo Docente y Metodológico

Ortíz, T.E., Mariño, S.M.A.: La Clase Metodológica Instructiva en la Educación Superior Cubana. Revista Pedagogía Universitaria, vol. 9, no. 1 (2004)

Fuentes González, H., Pérez Martínez, L.: Fundamentos de didáctica de la educación superior. Centro de Estudios de Educación Superior (CEES) "Manuel F. Gran". Universidad de Oriente (1994)

Álvarez de Zayas, C.M.: Didáctica: la escuela en la vida. Editorial Pueblo y Educación, La Habana (1999)

Collazos, C., Guerrero, L., Vergara, A.: Aprendizaje Colaborativo: un cambio en el rol del profesor (1999)

Lara, V.R.S.: El aprendizaje cooperativo en historia: diseño de actividades y efectos cognitivos y sociales. Tesis doctoral inédita, Universidad de Murcia, España (2001)

Lucero, M.M.: Entre el trabajo colaborativo y el aprendizaje colaborativo. Revista Iberoamericana de Educación (2002)

Colás Bravo, M.P., Conde Jiménez, J., Martín Gutiérrez, Á.: Las redes sociales en la enseñanza universitaria: Aprovechamiento didáctico del capital social e intelectual. Revista Interuniversitaria de formación del profesorado (2015). ISSN 0213-8646

Gerlach, J.M.: Is this collaboration? In: Bosworth, K., Hamilton, S.J. (eds.) Collaborative Learning: Underlying Processes and Effective Techniques, New Directions for Teaching and Learning, vol. 59 (1994)

Navarro, M.: Los nuevos entornos educativos: desafíos cognitivos para una inteligencia colectiva. Comunicar **33**(17), 141–148 (2009)

Orihuela, J.L.: Internet: la hora de las redes sociales. Nueva revista **119**, 57–62 (2008)

Austin, J.E.: Principles for partnership. J. Leader to Leader **18**(Fall), 44–50 (2000)

Welch, M.: Collaboration: staying on the bandwagon. J. Teach. Educ. **49**(1), 26–38 (1998)

Leonard, P.E., Leonard, L.J.: The collaborative prescription: remedy or reverie? Int. J. Leadership Educ. **4**(4), 383–399 (2001)

Srinivas, H.: What is Collaborative Learning? The Global Development Research Center, Kobe; Japan (21 Oct 2011, last updated). Retrieved 5 Nov 2011, from: http://www.gdrc.org/kmgmt/c-learn/index.html

Klemm, W.R.: Using a formal collaborative learning paradigm for veterinary medical education. J. Vet. Med. Educ. **21**(1), 2–6 (1994)

Zañartu, L.M.: Aprendizaje colaborativo: una nueva forma de Diálogo Interpersonal y en Red) (2011). Disponible en: portal.educ.ar/debates/educacionytic/formacion.../las-tics-y-el-aprendizaje-cola.php

Muñoz Prieto, M.M., Fragueiro Barreiro, M.S., Ayuso Manso, M.J.: La importancia de las redes sociales en el ámbito educativo. Escuela Abierta **16**, 91–104 (2013)

Meso, P.M.: Gabinete de comunicación y educación (2010). Recuperado de http://www.gabinetecomunicacionyeducacion.com/files/adjuntos/Las%20redes%20sociales%20como%20herramientas%20para%20el%20aprendizaje%20colaborativo.%20presentaci%C3%B3n%20de%20un%20caso%20desde%20la%20UPV_EHU.pdf

Selwyn, N.: Web 2.0 applications as alternative environments for informal learning – a critical review (2007). Recuperado de http://www.oecd.org/dataoecd/32/3/39458556.pdf
Koper, R. (ed.): Learning Network Services For Professional Development. Springer, Heidelberg (2009). https://doi.org/10.1007/978-3-642-00978-5
Artero, B.N.: www.educaweb.com (2011). Recuperado de http://www.educaweb.com/noticia/2011/01/31/interaccion-como-eje-aprendizaje-redes-sociales-14570.html

Teaching Hardware Security: Earnings of an Introduction Proposed as an Escape Game

Florent Bruguier[1,2,3(✉)], Emmanuelle Lecointre[3], Beatrice Pradarelli[1,2], Loic Dalmasso[1,2], Pascal Benoit[1,2], and Lionel Torres[1,2]

[1] LIRMM, University of Montpellier, CNRS, 161 rue Ada, 34095 Montpellier Cedex, France
{Florent.Bruguier,Beatrice.Pradarelli,Loic.Dalmasso, Pascal.Benoit,Lionel.Torres}@lirmm.fr

[2] Pôle CNFM de Montpellier, University of Montpellier, 161 rue Ada, 34095 Montpellier Cedex, France

[3] IUT de Nîmes, University of Montpellier, 8 rue Jules Raimu, 30907 Nîmes, France

Abstract. The Internet of Things (IoT) sees the appearance of ever more connected objects. In such a context, the security aspect of these objects is more relevant than ever. That is why we have developed several courses about hardware security. Traditional security courses often start with a catalog of definitions that can sometimes be boring for students and therefore counter-productive. This study describes an escape game used as a sequence to introduce several security concepts. This serious game could be adapted according to the degree level of the students. Results show a significant improvement in acquiring skills for the students who plays to the escape game. All the contents of this course are open sourced and could be freely available on request.

Keywords: Security · Data security · Cryptography · Education · Hardware security · Security course · Serious game · Escape game

1 Introduction

Nowadays, digital systems are everywhere. They are found in a wide range of applications, including communication systems, digital instruments, and consumer products. This is even truer with the advent of the Internet of Things (IoT). They change users' habits and cater to the new needs of customers in various fields such as audio-visual, health, transport, and tourism… By 2021, around 28 billion connected devices are expected [1].

Nevertheless, this omnipresence increases the chances of user exposure to security issues. Smart fridge, smart cars, smart toys, or smart medical devices…, the number of hacked devices keeps going up [2–4]. Security breaches are easily exploitable since the users do not master technology and security aspects. For example, the use of personal information (birth date, place of residence…) as a password is still too common [5].

M. E. Auer and D. May (Eds.): REV 2020, AISC 1231, pp. 729–741, 2021.
https://doi.org/10.1007/978-3-030-52575-0_60

In this context, we have developed several courses about hardware security to educate students from high school to PhD degrees. All these courses rely on the SECNUM platform, a platform dedicated to hardware security evaluation [6]. This paper described a teaching sequence used as an introduction to different courses.

Since we address various audiences, we choose not to employ the purely transmissive method. Indeed, it keeps the audience in a passive posture and standardizes the pace of progress. In addition, we hope that this sequence has a real impact on practices through an awareness of security risks. These 6 levers of the motivation described by Turner and Paris [7] are operated in the frame of the game and the simulation. It ensures commitment and the maintenance of attention needed for deep learning. Accordingly, we have chosen to experiment gamication methods to attract attention and maintain the commitment of target audiences.

This paper describes a teaching sequence introducing the key concepts of digital security. The main contribution consists of an escape game and the associated teaching materials that are open-sourced and freely available on request. The remainder of the paper is organized as follow. In Sect. 2, an overview of related works is offered. Then, principle, objectives and pedagogical issues of escape game courses are exposed. In the following section, our escape game is deeply described. Finally, outcomes are produced.

2 Related Work and Pedagogical Issues

2.1 Related Work

Several papers present different descriptions of hardware security courses.

First, Bossuet offers a description of two different labs about FPGA security [8]. These two labs are deeply depicted. The introduction course is done in the form of a 90 min lecture. In the same way, in [9], the authors present a full course dedicated to hardware security. The paper doesn't explain in details the way the course is done. More recently, Halak offers a full course on design and evaluation of secure chip [9]. This course seems to contain everything that must contain such a design course about hardware security. The students give good feedback. The same observation is done for [10]. The authors set forth a 3-day training course for PhD students. Once again, the theoretical contributions are done in a traditional way.

To sum up, all these contributions offer terminology and definition part on a classical way.

2.2 Pedagogical Issues

We teach security since 2012 at different level from high school to PhD level in embedded systems/microelectronics. The Table 1 offers evaluation of this course since 2015, year in which we have introduced the evaluation. With an average rating better than 4 out of 5, we could conclude the course has reached the objective. Nevertheless,

Table 1. PhD course rating on 5

PhD course	2015	2016	2017	2018
Course structure	4.0	4.83	4.83	4.5
Content clarity	4.33	4.5	4.33	4.0
Used tools	4.33	4.33	4.33	4.5
Quality of materials	4.33	4.5	4.33	4.5
Instructor educational quality	4.33	4.5	4.65	4.67
Course understanding	4.5	4.5	4.5	4.67
Overall benefit of the course	4.33	4.33	4.83	4.67
Average	4.31	4.5	4.55	4.5

the students often do a redundant finding: the 3-h sequence offered to expose terminology and cryptography principles asks for many concentration efforts for the students. Indeed, their attention decreases significantly after 30 min if there is no change in their activity [11].

That's why we have decided to change this part of the course. We looked for a way to put the students in an active posture and we chose to experiment a new kind of serious game: the educational escape game [12].

3 Educational Escape Game

3.1 An Educational Vector

An escape game is a game in which a team must escape from a room in a given time. For this, it has to solve puzzles using hints hidden in the room. After beginning in the entertainment world, they are in full expansion in education world. The use of the game in class is part of what is called gamication, playfulness, or edutainment [13, 14]. As part of the latter, these are part of the category of serious games; the game becomes a learning tool.

The emotion and the positive stress generated lead the participant to the flow: a feeling of satisfaction and fullness in the realization of an activity for which all the attention is put on the task in progress [15]. All the ingredients of the video game are there to engage the participant via the storytelling or narrative frame in which they play the role of a hacker.

It's a simulated situation but with real challenges requiring periods of observation, choice and action to immediate feedback (stagnation/error-advancement/success). We aim for experiential learning through a device that allows a perceptual approach that is both abstract and concrete, and a knowledge integration that is based on both observation and action.

Thus, the sequence respects Kolb's theory of learning preferences based on constructivist theories [16]. It adapts to the different profiles and preferences of the participants since it is built in three stages. First, a phase of exchanges and reflection in small groups or among peers on the practices of each one in terms of security. An active search for information based on a questionnaire and information posters is also done. Then, the simulation of the game and finally the opportunity to reflect on the achievements during the debriefing on the experience lived: each enigma acting as a memory 'anchor' to ease the retrieval of the associated concepts of security.

Students are more receptive to the use of fun goals and to the concrete representations that are offered to them. The immersion and the pleasure of playing serve as a driving force for learning.

3.2 Importance of the Scenario

Just like any teaching sequence, the scenario is essential to any escape game. It makes it possible to put the students pursue a quest or a challenge to solve in a limited and timed time.

To create the context for the development of new skills, the assimilation of new knowledge or the application of previously acquired knowledge, students will be offered to deal with riddles, puzzles or even experiences… The objective is to propose riddles as far as possible from classical exercises to ensure success. In the same way, the non-linearity scenario will allow the appearance of the collective intelligence [17]. Aside from the initial situation, the absence or almost absence of instructions is part of the format of an escape game. It is important not to limit students' imagination and reflection to the risk that the game will lose its interest.

3.3 Teacher Posture

When designing an educational escape game, the teacher plays a leading role in planning and orchestrating the frame. Upstream, he organizes the playful experience to maximize the autonomy of the participants during the game. On the other hand, during the experience, he monitors the progress of the group and the time that elapses. He has also planned the possible dead ends in which the students could end up and has prepared 'boost' or 'help' elements for the groups strap up on a riddle.

The main difficulty will be to adapt the progress of each group to keep the sequence in the time allotted.

3.4 Debriefing

In order to ensure that students have fully integrated the concepts discussed during the game, it is important to focus on the debriefing sequence. This will allow students to put their finger on the skills and knowledge necessary to succeed but also to write them down. The teacher will take advantage of this to collect information for the improvement of the game.

4 Transposition to Hardware Security

4.1 Pedagogical Goals

The objective is to offer a game developing the skills necessary to the understanding of the security of the digital world. The terminology and definition necessary to this course have to be introduced during this teaching sequence. First, it is needful to identify them:

- We first want to make students aware of social engineering. It is a question of carrying out a psychological manipulation in order to realize a swindle. For example, "a hacker will call up, imitate someone in a position of authority or relevance, and gradually pull information out of the user" [18].
- Then, the concept of brute force attack has to be introduced. A brute force attack is the simplest method to find a secret key. It consists in testing all combinations of a secret key until the right one is found.
- The basic techniques of encryption/decryption must also be presented. We introduce the two most basic of them: substitution and transposition ciphering. These two techniques are still used in modern cryptography algorithms. For example, the Advanced Encryption Standard, AES uses this kind of encryption schemes [19].

A substitution cipher is a method of encryption. It consists in replacing a letter or a group of bits by another, according to a fixed system. For example, in Caesar cipher, a D replaces an A [20].

Transposition-based encryption relies on inverting the position of letters in a message. One example of such a system is a Scytale, a tool consisting of a cylinder with a strip of leather wound around it. The leather is used to write the message and it is in theory impossible to retrieve it without the cylinder of good diameter [21].

Depending on the level of the students, other skills/knowledge will be involved:

- Students with little knowledge of the digital world and electronics will be confronted with the principles of operation of an electrical circuit and the principle of binary coding.
- More experienced students will have the opportunity to experiment with the joy of penetration testing [22]. The principle is to measure voltages directly on a digital circuit in order to extract sensitive information.
- Differential and correlative power analysis are also of interest. Such side-channel attacks allow deducing secret keys by analyzing power consumption from multiple cryptographic operations performed by a vulnerable device [23, 24].

4.2 Implementation

The proposed teaching sequence is divided into several stages. Unlike traditional escape game, our teaching sequence includes upstream and downstream phases in addition to the real escape game phase. During the whole process, students are divided into groups of two or three. Therefore, that everyone can learn and experiment. The complete pedagogical material is duplicated for each group of students.

Upstream Phase. The upstream phase is divided into two stages.

First, an icebreaker is proposed. The students are asked to answer few quick questions. These questions about cryptography allow facilitating exchanges towards learning in order to know each other better and to turn negative emotions into positive emotions.

Then, students are led to face different skills. For this, several posters describe the concepts exposed in Subsect. 4.1 and students have to explore them to answer multiple-choice questions. This first contact with the notions to be learned allows activating the two first levels of Bloom's taxonomy: knowledge and comprehension [25].

Escape Game Phase. The game phase will be broken down as follows.

Principle. First, the teacher explains the rules of the game. Then, students are divided into binomials or trinomials. Each group has several objects enabling to open a briefcase containing the code of a safe. More details will be given in the next paragraph. All the groups are in competition since there is only one safe to open. Once opened, depending on the time remaining, the other groups could continue to search the combination of the safe.

Gamification of Knowledge. Spoiler: If you want to confront the game without knowing all the tricks, please skip the following paragraph.

At the beginning, each team is in possession of one picture frame including a picture and a briefcase as depicted at the top of the Fig. 1. Numbers in circles refer to the puzzle number shown in Table 2. The numbers are referred with a# for better comprehension.

As social engineering example, the birth date of the children on the photo is used as combination to open the briefcase (#1). Once opened, each group disposes of a "brute force" electronic card, a playing card, a "scytale" ribbon, a storage cylinder for an ultraviolet lamp, an ultraviolet lamp, a closed box, a business card, a notebook, and a pen. Using the substitution principle, each team could open the closed box using the business card address (#3). Two new objects are unlocked: two electric cables, four substitution tables. The cables power on the "brute force" card where the user has to test all the combinations to display another code (#6 and #2). The playing card and the ultraviolet lamp allow selecting one substitution table among the 4. The "scytale" ribbon and the storage cylinder illustrate the principle of transposition ciphering (#4). Finally, the codes deduced from the "brute force" electronic card and the "scytale" give the combination of the safe using the substitution card (#5). The safe is now opened.

Each skills/knowledge proposed above gives rise to an enigma proposed during the game as well as a possible associated boost that can be offered to students stuck during the game (Table 2).

In addition to the riddles above, extra puzzles are available and could be added to offer extra skills depending on the level of the students. Indeed, we propose to study two important skills of hardware security: penetration testing or pentesting and side-channel attacks. The first one could be introduced using a connected object, which is disassembled to allow voltage measurement. The last one is presented with a short video game implemented on an Arduboy. These two elements could be easily added into the briefcase.

This phase of the sequence allows enhancing the skills of the students through application and analysis: the two-second levels of Bloom's taxonomy.

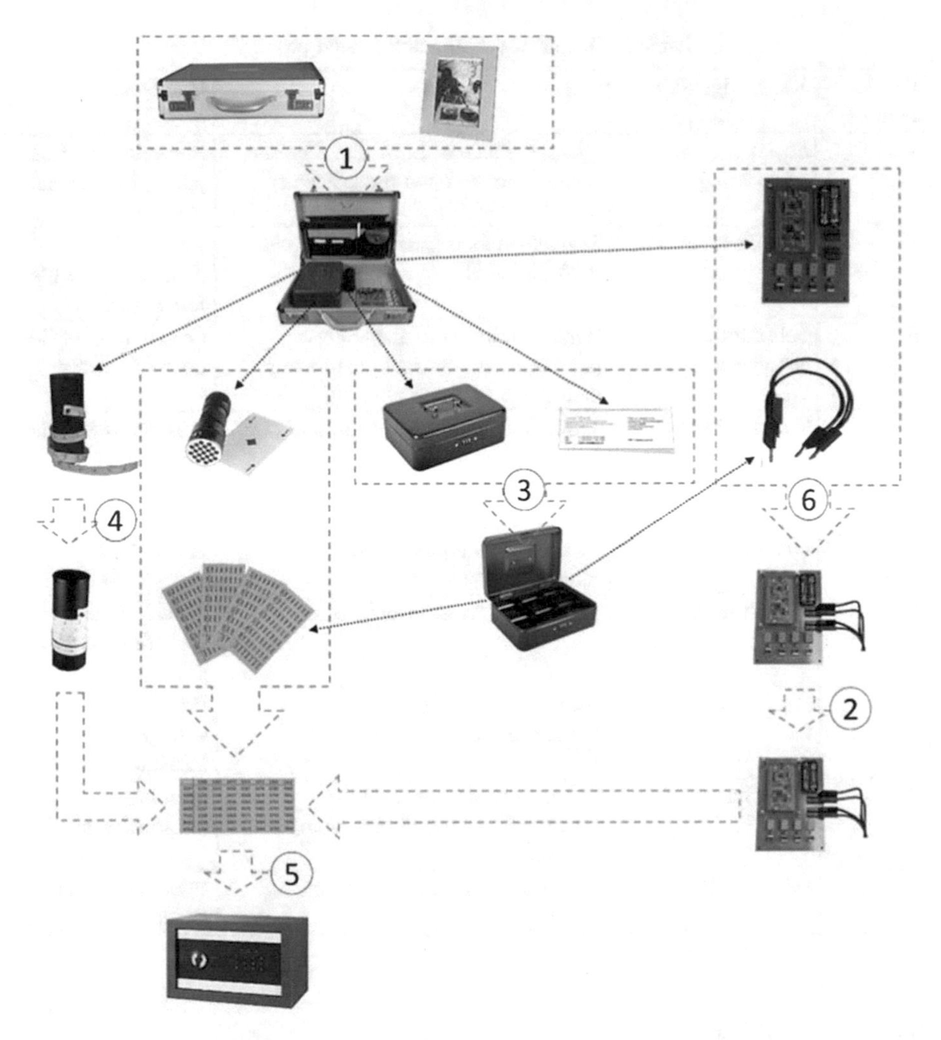

Fig. 1. Logical path between the different puzzles of the escape game. The simplest version is depicted here.

Downstream Phase. At the end of the game, the students have to fill an online multiple-choice test. This test resumes all the different skills that were addressed during the game. This allows settling them and to be sure, they are mastered. The two last levels of Bloom's taxonomy are developed here. Moreover, at the end of the sequence, a correction of the multiple choice is offered and a discussion is proposed to be sure each student has correctly understand all notions.

Table 2. Knowledge/skills and related puzzles

Puzzle number	Knowledge/Skills	Puzzle	Boost/help
1	Social engineering	Using the date of birth found on the photo frame to open the briefcase	Presentation of the principle of social engineering
2	Brute force attack	Test all combinations to find the code used on the electronic board	Recall of the definition of brute force attack
3	Substitution encryption	Replacing letters of the address provide by the business card using their position in the alphabet	Presentation of the code of Caesar
4	Transposition encryption	Winding a strip of paper around the ultraviolet lamp holder	Presentation of the principle of Scytale
5	Encryption using substitution boxes	Using the result from the electronic board and the transposition encryption as input of the substitution table	Principle of substitution boxes in the AES
6	Operation of an electronic device	Turning on the electronic board	Movie: "the seventh company" [26]
7	Binary coding	Using measured voltages on the connected object	Principle of voltage measurement
8	Pentesting	Voltage measurement on the connected object and conversion to decimal	Principle of penetration testing
9	DPA & CPA	Use of a speci c application on Arduboy	Principle of the attack

4.3 Scenarisation

In order to guarantee maximum student involvement during the phase of the game, a simple scenario has been devised. The students were summoned to the first stage of recruitment as a new security expert of the National Security Agency of their country. For this test, they have to open a safe containing state secret using all elements they find into the office of the boss of the agency. His office is almost empty except a briefcase and a photo of his son's birthday. By discovering the different enigmas presented in Table 2 as well as some additional challenges, the fastest student group will discover the combination of the safe and join the Agency.

5 Outcomes

5.1 Past Courses

A preliminary version of this game met a great success during different courses given to students from high school to PhD students. The version proposed here was used during three different courses for students from high school and bachelor degree. We chose these three courses since the students with the lowest level of study follow them. If it works for them, it will for the other too and especially for the PhD ones. Even if it is too early to draw definite conclusions, the results are very positives.

5.2 Evaluation

In order to be able to evaluate properly what students have learned, we have chosen to make a double assessment. First, before the training sequence, we ask them if they know the different skills. Finally, after the downstream phase, we evaluate them a second time. The evaluation proposed here relates to three groups. Each one is composed of 15 students. Two groups are composed of students from high school since the other one is formed by students from bachelor level.

Figure 2 depicts the number of students confident with the different skills for the three groups. The Fig. 3 shows the results of this evaluation. We can see that the progression is without appeal. Indeed, almost all students answer successfully to the evaluation. The few failures lie in the fact that students confuse substitution and transposition. This confusion shall be tackled in the next version of the upstream and downstream sessions thanks to a major focus on the differences between both concepts in the questionnaire.

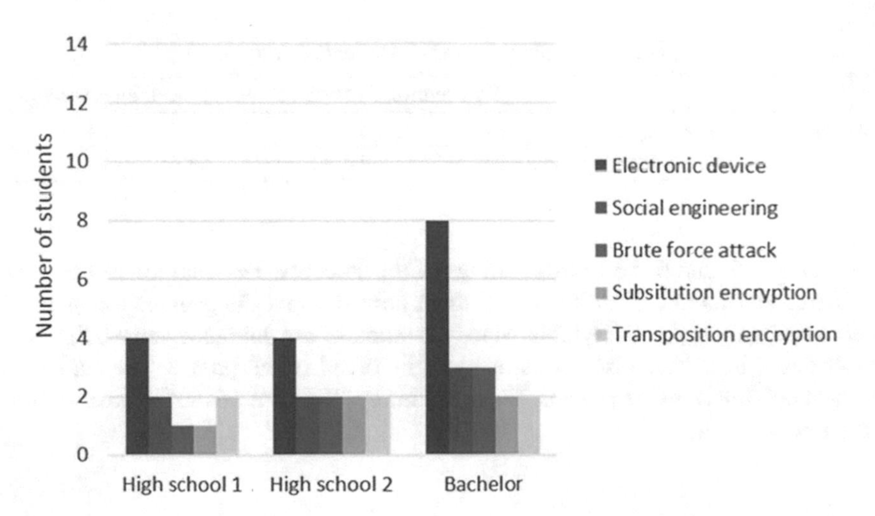

Fig. 2. Number of students thinking of knowing the different skills for each group of 15 students.

5.3 Feedback from the Students

The Table 3 depicts the average rating on different questions about the course. This sequence gave us very good student feedback. When students are asked if the escape game is a playful tool, they are unanimous and plebiscite of course the yes. The same answer is given when asked if the escape game is a good learning tool with a nuance. We still have work to convince all students of the opportunity to learn through play.

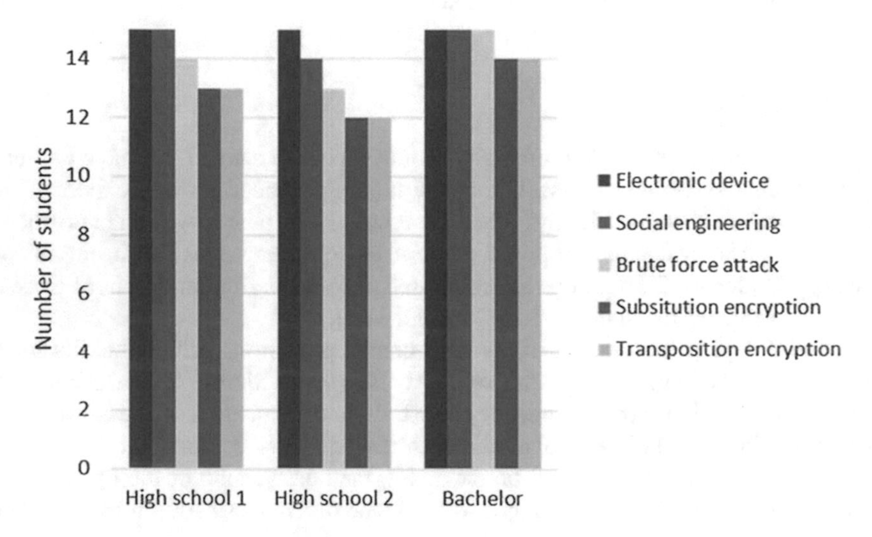

Fig. 3. Number of students having correctly answered for each skill

Table 3. Feedback from the students (rating on 5)

	High school 1	High school 2	Bachelor	Average
Is the escape game playful?	4.4	4.6	4.6	4.53
Is the escape game a learning tool?	3.8	4.33	4.6	4.04

Figures 4, 5, and 6 show other feelings of the students. The duration of the game is considered satisfactory even if some of them consider that the game is too short. The level of difficulty is also acceptable. Since the students are doing the game by groups of two or three, they were also ask is they are involved on all puzzle. The students are involved on almost all the puzzles. We are planning to tackle this in the next version of the game.

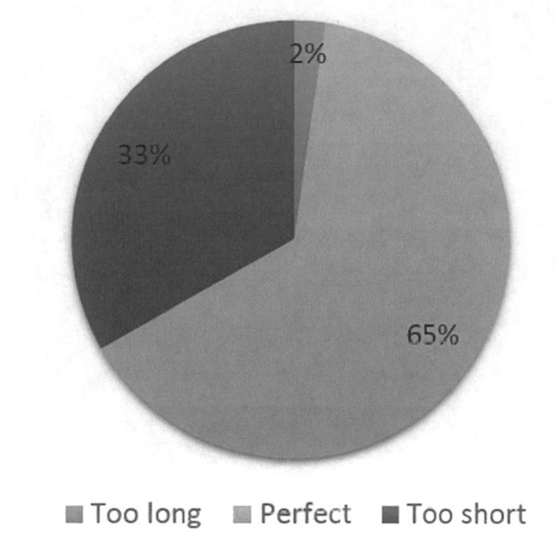

Fig. 4. How was the duration of the escape game phase?

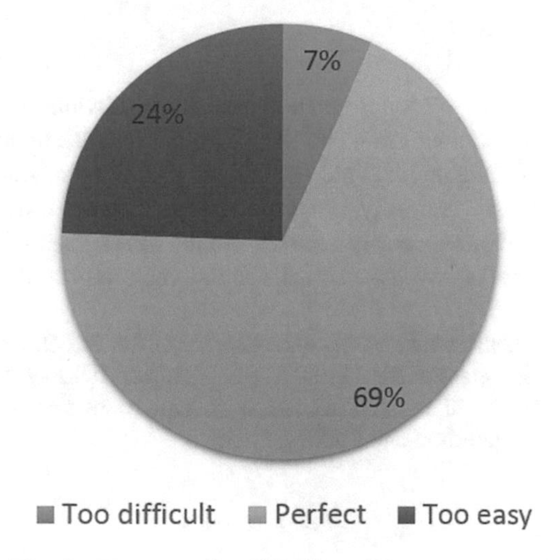

Fig. 5. How was the difficulty of the escape game?

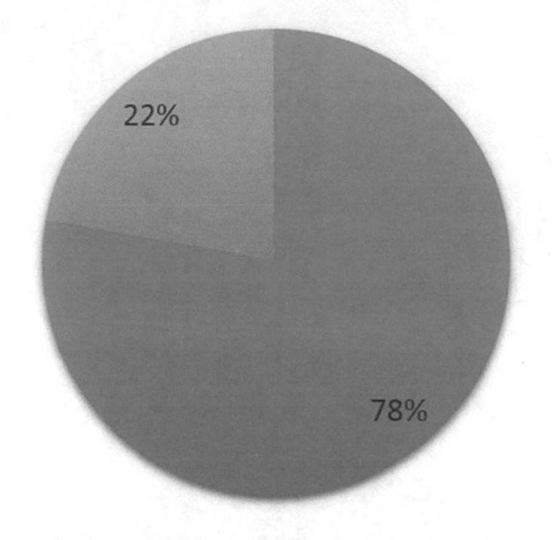

Fig. 6. Have you been involved in all the puzzles?

6 Conclusion

This paper has exposed an introduction sequence for hardware cryptography courses. In order to increase the understanding and therefore the success of students, we offer an escape game. This serious game has the advantage to make the students actor of their apprenticeship. Therefore, naturally, it provides very good outcomes and will be integrated to all the hardware security courses we are given regardless of the students' level. Moreover, this course is open-sourced. Course materials could be sent on request.

Acknowledgment. The authors acknowledge the support of the French Agence Nationale de la Recherche (ANR), under grant ANR-11-IDFI-0017 (project IDEFI-FINMINA). They also acknowledge the Occitanie and the FEDER for they support to this project as well as the University of Montpellier and the I-site MUSE.

References

1. Cellular networks for massive IoT. Technical report, Ericsson (2016)
2. Dagon, D., Martin, T., Starner, T.: Mobile phones as computing devices: the viruses are coming! IEEE Pervasive Comput. **3**(4), 11–15 (2004)
3. Wolf, M., Weimerskirch, A., Wollinger, T.: State of the art: embedding security in vehicles. EURASIP J. Embed. Syst. **2007**(1), 074706 (2007)
4. Halperin, D., Heydt-Benjamin, T.S., Ransford, B., Clark, S.S., Defend, B., Morgan, W., Fu, K., Kohno, T., Maisel, W.H.: Pacemakers and implantable cardiac debrillators: software radio attacks and zero-power defenses. In: IEEE Symposium on Security and Privacy, pp. 129–142, 2008. SP 2008. IEEE (2008)
5. Morris, R., Thompson, K.: Password security: a case history. Commun. ACM **22**(11), 594–597 (1979)

6. Bourrée, M., Bruguier, F., Barthe, L., Benoit, P., Maurine, P., Torres, L.: SECNUM: an open characterizing platform for integrated circuits. In: European Workshop on Microelectronics Education (2012)
7. Turner, J., Paris, S.G.: How literacy tasks influence children's motivation for literacy. Reading Teacher **48**(8), 662–673 (1995)
8. Bossuet, L.: Teaching FPGA security. In: 2013 International Conference on Field-Programmable Technology (FPT), pp. 306–309. IEEE (2013)
9. Halak, B.: Course on secure hardware design of silicon chips. IET Circuits Devices Syst. **11**(4), 304–309 (2017)
10. Bruguier, F., Benoit, P., Torres, L., Bossuet, L.: Hardware security: from concept to application. In: 2016 11th European Workshop on Microelectronics Education (EWME), May 2016, pp. 1–6 (2016)
11. Mackworth, N.H.: The breakdown of vigilance during prolonged visual search. Q. J. Exp. Psychol. **1**(1), 6–21 (1948)
12. Annetta, L.A.: The \i's" have it: a framework for serious educational game design. Rev. Gen. Psychol **14**(2), 105 (2010)
13. Deterding, S., Dixon, D., Khaled, R., Nacke, L.: From game design elements to gamefulness: dening gamication. In: Proceedings of the 15th International Academic MindTrek Conference: Envisioning Future Media Environments, pp. 9–15. ACM (2011)
14. Deterding, S., Sicart, M., Nacke, L., O'Hara, K., Dixon, D.: Gamication. Using game-design elements in non-gaming contexts. In: CHI 2011 Extended Abstracts on Human Factors in Computing Systems, pp. 2425–2428. ACM (2011)
15. Csikszentmihalyi, M.: Flow and the Psychology of Discovery and Invention, vol. 39. Harper-Perennial, New York (1997)
16. Kolb, D.A.: Experiential Learning: Experience as the Source of Learning and Development. FT press, New Jersey (2014)
17. Lévy, P., Bononno, R.: Collective intelligence: Mankind's emerging world in cyberspace. Perseus books, New York (1997)
18. Granger, S.: Social Engineering Fundamentals, Part i: Hacker Tactics. Security Focus, 18 December 2001
19. Daemen, J., Rijmen, V.: The Design of Rijndael: AES-the Advanced Encryption Standard. Springer Science & Business Media, Heidelberg (2013)
20. Goyal, K., Kinger, S.: Modified caesar cipher for better security enhancement. Int. J. Comput. Appl. **73**(3), 26 (2013)
21. Kelly, T.: The Spartan Scytale. The Craft of the Ancient Historian: Essays' in Honor of Chester G. Starr, pp. 141–169 (1985)
22. Engebretson, P.: The Basics of Hacking and Penetration Testing: Ethical Hacking and Penetration Testing Made Easy. Elsevier (2013)
23. Kocher, P., Jaffe, J., Jun, B.: Differential power analysis. In: Annual International Cryptology Conference, Springer, pp. 388–397 (1999)
24. Brier, E., Clavier, C., Olivier, F.: Correlation power analysis with a leakage model. In: International workshop on cryptographic hardware and embedded systems. Springer, pp. 16–29 (2004)
25. Bloom, B.: Bloom's Taxonomy of Educational Objectives. Longman (1965)
26. Mais où est donc passée la septième compagnie? (1973)

A Human-Centered Approach to Data Driven Iterative Course Improvement

Steven Moore[1(✉)], John Stamper[1], Norman Bier[1], and Mary Jean Blink[2]

[1] HCII, Carnegie Mellon University, Pittsburgh, USA
stevenjamesmoore@gmail.com, jstamper@cs.cmu.edu, nbier@cmu.edu

[2] TutorGen, Inc., Wexford, USA
mjblink@tutorgen.com

Abstract. In this paper we show how we can utilize human-guided machine learning techniques coupled with a learning science practitioner interface (DataShop) to identify potential improvements to existing educational technology. Specifically, we provide an interface for the classification of underlying Knowledge Components (KCs) to better model student learning. The configurable interface allows users to quickly and accurately identify areas of improvement based on the analysis of learning curves. We present two cases where the interface and accompanying methods have been applied in the domains of geometry and psychology to improve upon existing student models. Both cases present outcomes of better models that more closely model student learning. We reflect on how to iterate upon the educational technology used for the respective courses based on these better models and further opportunities for utilizing the system to other domains, such as computing principles.

Keywords: Learning analytics · Student model · Learning curve · Data visualization · Data-driven improvement · Educational technology

1 Introduction

The proliferation of data on students interacting with online learning environments has opened up enormous possibilities for understanding student behavior for decades [1]. It enables the construction of models on how students progress through the learning process and identify the gaps in their knowledge. Building on these student models for the purpose of tracking student learning over time has been a key area of focus in the educational technology community [2]. Cognitive tutors, such as those from Carnegie Learning, utilize student models and are adaptive to student knowledge by tracking the mastery of skills or knowledge components (KCs) [3]. The models that map KCs are generally created with the help of subject matter experts and cognitive scientists. Unfortunately, these knowledge component models (KCMs) do not always correctly model skills, which can impede student learning. When a KCM for a cognitive tutor is incorrectly modeled, it can cause incorrect problem selection and waste valuable student time on skills they have

M. E. Auer and D. May (Eds.): REV 2020, AISC 1231, pp. 742–761, 2021.
https://doi.org/10.1007/978-3-030-52575-0_61

already mastered. While it is challenging to get the models perfect, continuously iterating on the model as more data is collected can help to improve it.

Learning analytics can address this problem and presents an opportunity for continuous improvement of the models using data driven techniques [4]. In this paper, we show how we can use new user interface affordances in DataShop [5], that utilizes a novel framework for curve categorization, to assist in identifying areas of improvements in the student models of the educational technology. Using the curve categorizations as a starting points, novice users are able to make model improvements using the affordances of DataShop. We present two case studies in different educational technology systems across unique educational domains where the new DataShop features were used to improve the underlying KCM.

2 Related Work

2.1 Knowledge Components

When a student is solving a problem, there are a series of hypothesized competencies, or knowledge, that are needed to perform each step of the problem. These competencies are known as knowledge components (KCs). The KCs are fine grained representations of knowledge that includes constraints, schemas, and production rules [4]. In an educational technology system, such as a cognitive tutor, each of these problem steps has an associated action the student needs to take to solve the problem [6]. These actions can be labeled with one or more KCs to represent the required competency a student needs to successfully solve that step. Each of these problem steps also corresponds to an opportunity at which a student must demonstrate their mastery with the mapped KCs.

While it may have once been question *if* student learning could accurately be modeled through their progression of KCs, evidence has shown this to be the case for a variety of domains using knowledge tracing [7, 8]. Knowledge tracing is the practice of estimating a student's current knowledge state at a given time while they interact with an educational technology system [9]. It maintains a record of the probability that a student knows a skill or concept, based on their performance on problem steps that are mapped to KCs. This is used to inform different functions of the ed tech system, such as problem selection or advancing the student to a new content area, based on their current mastery level with the KCs. It is important to have an accurate KCM for this reason, as a poorly fit one can lead to inaccurate knowledge tracing, where over-practice or under-practice may occur [10].

2.2 Learning Curves

A learning curve is a graphical representation of the change in student performance over time. Learning curves show where students begin with their knowledge, the rate at which they learn the given KCs, and the flexibility of how the acquired skills can be used [11]. In DataShop, learning curves have the opportunities a student has with a particular KC on the x-axis and the error rate, as a percent, on the y-axis. By default, each learning curve corresponds to a single KC in DataShop and ideally, the number of students with an error, represented at each opportunity point, decreases. This decrease in students at each point occurs for a variety of reasons, such as students not attempting a problem again or skipping activities that are mapped with that KC. We know students learn more by doing, such solving problems in a course rather than just reading the text [12]. Therefore, it is expected to see a downward sloping learning curve with each additional opportunity to practice. As shown in Fig. 1, a desired learning curve is one that shows student improvement, the error rate decreasing, as the opportunity count with the specific KC increases. Additionally, these learning curves are reflective of the effectiveness of a learning system and its content, as it shows learning for the group of students using the system.

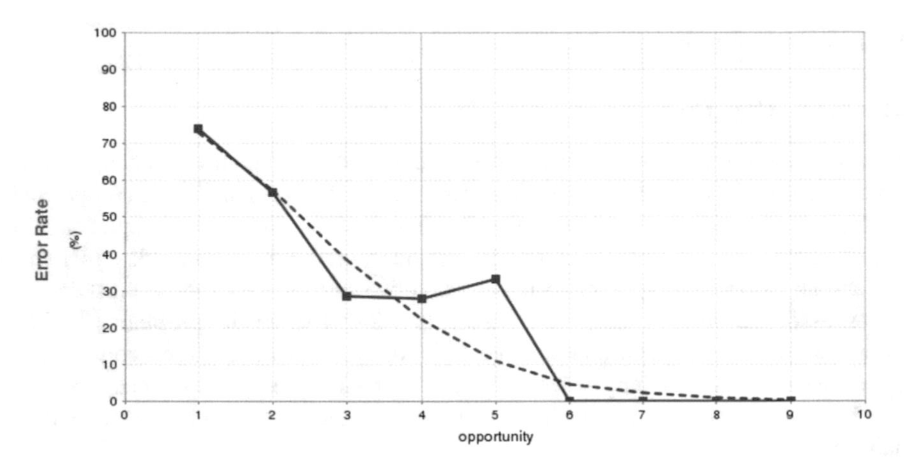

Fig. 1. A learning curve categorized as *Good*, showing a decreasing error rate per opportunity

The analysis of learning curves to provide insights into student models has been around for many years [13]. Methods involving manual human inspects of the curves to more semi-automated ones have been used to improve upon cognitive models used in intelligent tutoring systems [14]. Further analysis of learning curves and their categorizations to provide insights into courseware via crowdworkers has also been suggested [15]. Learning curves in DataShop use a specialized form of logistic regression performed on the error rate of the curve, known as the Additive Factors Model (AFM) [10]. AFM is a statistical model that makes predictions regarding student performance, in combination with item response theory that includes a growth term [16]. This model uses information about a student's prior practice opportunities on the assigned KC to predict the probability that the student will perform correctly on a given opportunity, which corresponds to a question step.

2.3 Prior DataShop Work

Previous research around DataShop has focused on improving the underlying KCMs in order to gain insights into the student learning process and provide suggestions for educational technology improvement [4, 17, 18]. Work by Stamper and Koedinger [4] showed how human-centered aspects could be combined with DataShop tools using a human-machine discovery approach to improve student models. This human aspect comes into play to make distinctions involving learner populations, sequencing, and mis-tagging of KCS that the machine learning side might miss or not take into account. For instance, teaching the same algebra course to middle vs. high school students would result in the same dataset, but it should be split amount these student populations, something the human would have knowledge of and do. The improved student model from this method was applied to two other datasets in the same domain, but collected from different sets of students, and still demonstrated improvements. Building upon that, one study utilized AI and statistical methods to discover improved models in a variety of domains using data collected from different educational technologies [18]. These resulting improved models isolated flaws in the original one, for which they demonstrated how an investigation on these flawed parts led suggested improvements for tutor design. Using these methods, a recent study was able to improve the KC model for an educational math game [19]. The authors were able to better identify parts of the game that gave the students trouble and make improvements in the form of question content and ordering.

These studies led to later work focused on an initial close-the-loop experiment, where the improved KCMs were tested inside the classroom. The results showed gains of 25% less time to master the same material and improved performance on a subset of problems using a particular skill in the course, by using the improved models in the tutor [17]. A recent study analyzed KCMs in DataShop by comparing and contrasting their metrics to measure how accurately the predictive fit of the models is to the data. They found that of the metrics, Akaike information criterion (AIC) was the best predictor of cross validation results, which is the gold-standard for model selection [20]. Another metric relevant to student models is Bayesian information criterion (BIC), while similar to AIC, denotes how well the AFM statistical model fits the data with the given KCM. In addition to these, root mean square error (RMSE) is often used, which shows how well the KCM might generalize to an independent dataset from the same educational technology, such as a cognitive tutor. Much research has focused on improving KCMs, leading us to build upon that work to find a solution for easy categorization of learning curves. These categorizations can contribute to advancing learning following a human-centered approach using data-driven iterations.

3 DataShop

3.1 Functionality

The Pittsburgh Science of Learning Center DataShop (pslcdatashop.org) is the world's largest repository of learning interaction data for research [21]. It provides a suite of tools for researchers, instructional designers, and data scientists to analyze, create, and modify educational data. Student log data from many educational technology systems are fed into DataShop, where the student interactions with the questions in the given platform can then be analyzed. The analysis features include viewing the KCs associated with the questions, viewing their problem statements, associated learning curves, and more detailed statistics, such as student accuracy and how many times the question was attempted. This collected data is fine-grained, with an average student action logged every twenty seconds. Additionally, this data is often longitudinal, spanning courses that are half a semester, a full semester, or even a yearlong.

DataShop offers statistics and categorizations on learning curves and knowledge component models. The built-in suite of tools provides a way to view the details on learning curves, including categorizations of them, to help visualize how student learning changes over time for a particular KC [22]. We focus our case studies on the use of these categorizations, showing they can be used by non-experts to make judgements about KCs to further analyze. These categories for the learning curves are driven by a set of configurable parameters found on DataShop's interface for viewing the curves. The parameters, shown in Table 1, may vary based on the data set that is being viewed. A user might modify the threshold for students to a value below ten if the class from which the data were gathered is small. Similarly, if each knowledge component is only assessed a few times, then the opportunity threshold might also be reduced. These are two parameters where knowledge of the student population from which the data was generated from would inform how a user sets these.

Both the low and high error thresholds might be tweaked depending on the domain or prior knowledge of the students. Finally, the AFM slope threshold might change based on the rate of learning students are expected to improve upon. For instance, if they are expected to reach mastery after several problems, this threshold might be lowered. If the curve does not violate any of these parameter thresholds, then it is categorized as *Good*. Ultimately the desired value of the parameters will vary based on the system that collected the data. For instance, if the tutor only contains a few questions per KC, intended to be used as review, the opportunity and error thresholds would need to be tweaked to get a better categorization of the data.

Table 1. Configurable parameters for learning curve categorization

Parameter threshold	Description	Default value
Student	The minimum number of students that have attempted the KC at each point in the curve	10.0
Opportunity	The minimum number of student attempts at a KC that must be present for a curve, if the attempts are below this value the curve is labeled too little data	3
Low error	If a point on the curve falls below this error rate value, the curve is low and flat	20.00
High error	If the last point on the curve goes above this error rate value, the curve is still high	40.00
AFM slope	If the calculated AFM slope of the curve falls below this value, the curve is labeled as no learning	0.001

3.2 Data Sources

A primary educational technology platform that feeds data into DataShop is the Open Learning Initiative (OLI). OLI is an open educational resources project, part of the Simon Initiative at Carnegie Mellon University, that allows instructors to develop and deliver online courses consisting of interactive activities. Detailed student interactions with the course materials, such as watching videos, answering a variety of traditional question types such as drag-and-drop, multiple choice, and responding to free-form question prompts are logged into DataShop. Each question in OLI is broken down into one or more problem steps, where each step corresponds to an opportunity, the x-axis for a learning curve in DataShop. For instance, if a question asks a student to set the value of three dropdown boxes, then that question has three steps. In addition to the traditional timestamps and UI element with which the student interacted, each step is assigned a set of one or more KCs required by the student to answer the question. This KC tagging of the questions in conjunction with student accuracy on the problem, time of task, and number of attempts, provides detailed insights into which concepts with which students are struggling most.

Another platform that feeds a large portion of data into DataShop is Carnegie Learning's MATHia, a cognitive tutor for algebra (carnegielearning.com). These tutors cover middle and high school math curriculum and are adaptive. Students work math problems that feature rich interactions and these interactions are logged at every step. All steps are tagged with knowledge components and associated KCMs are also exported to DataShop. With a detailed log file imported into DataShop, we can use this data to track learning over time and perform learning curves analysis. Carnegie Learning data was also featured in the 2010 KDD Cup data mining competition [5] hosted by DataShop (pslcdatashop.web.cmu.edu/KDDCup/downloads.jsp).

3.3 Curve Categorization

The categorization of learning curves feature was added to DataShop following insights from prior research on improving student models [4]. Learning curves are now automatically grouped into one of five categories, based on the parameters. This categorization is a hierarchical approach, based on the configuration variables noted for the given dataset. They provide a way for users to gain a better grouping and view of the learning curves for their data, based on the parameters the user can alter to better fit the context of the educational system from which the data was collected. The categorization of the curves first accounts for the opportunity threshold, thus categorizing any curve with fewer opportunities than that parameter as *Too Little Data*. Following this, the second parameter considered is the student threshold, which is the minimum number of students that have attempted a KC at each point in the given curve. The category formulas only use points on the curve that are at or above the student threshold, which defaults to ten. This means that a curve's final points may not be utilized in the categorization due to having too few student observations.

The first category is *Low and Flat*, which indicates that students have already mastered the target KC and do not need the additional practice. Curves in this category begin with a very low error rate and remain low as the opportunity count progresses, as shown in Fig. 2. For KCs that are grouped into this category, it is suggested that the number of assessments targeting it be reduced to avoid over-practicing [10]. A student's time is better allocated toward a set of different KCs for which they have yet to achieve mastery. When an intelligent tutoring system is the educational technology used for a dataset with curves in this category, it may also be the case that the knowledge-tracing parameters are misaligned, and the system is suggesting further practice that is redundant. It may also be the case that a different system has too many practice problems for a particular KC and that they should remove some of them in favor of other material.

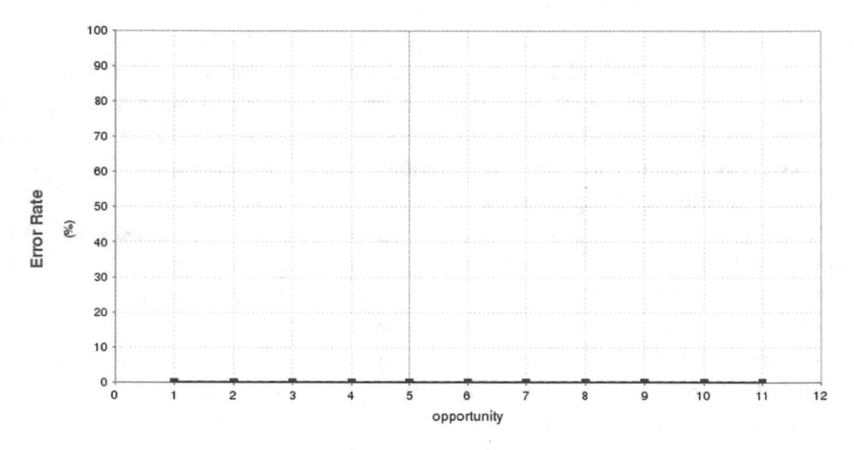

Fig. 2. *Low and Flat* learning curve showing students starting at 0% error rate and receiving up to 11 assessments mapped to this KC

Similarly, the second category, *No Learning*, is a curve representative of student learning where they do not demonstrate learning gains at a significant level. Curves in this category often begin at a moderate error rate and end around the same rate for the fitted curve, represented by the dashed line, shown in Fig. 3. These occur when the predicted learning curve's slope does not show apparent learning for the given KC. Even when the curve's final point is above the high error threshold, this category takes priority over the subsequent ones based on its use of the AFM slope in categorization. It is also important to remember these types of curves are potential cases to explore breaking down the KC into multiple KCs or disaggregated based on student subpopulations, as previous work has shown prior to the implementation of categorizations [4, 17, 22, 23].

The *Still High* category is another type of curve that is an easy area where potential improvements could be made to the KCM. Learning curves in this category have their final point, that is at or above the student threshold, above the high error threshold parameter value. This indicates that students continued to struggle with any KCs in this category, despite having sufficient opportunities. It is recommended that these curves be analyzed for another potential case of breaking down a single KC into multiple ones or providing students with additional practice opportunities. For instance, it may suggest that a better intermixing of practice be done, such as reviewing worked examples and then solving problems [24]. Figure 4 demonstrates how even when

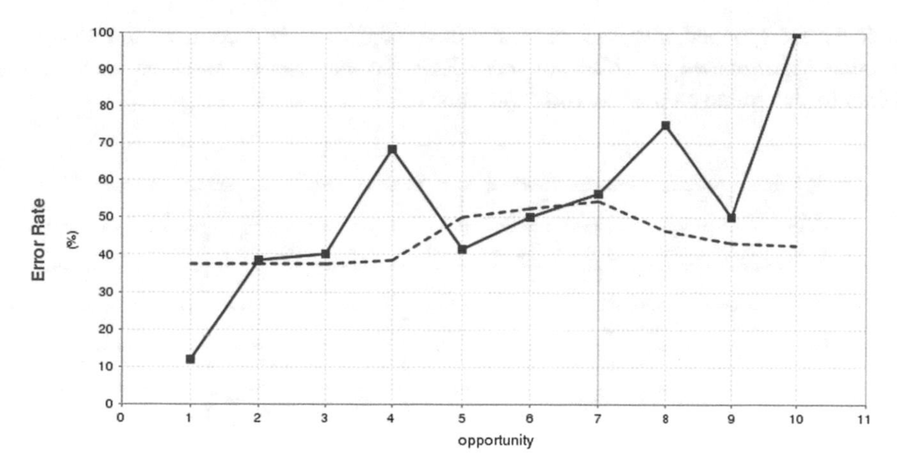

Fig. 3. *No Learning* curve where the predicted learning is below the AFM slope

students demonstrated learning, decreasing their error rate as they have opportunities, the curves end points may still fall above the set high error threshold, which was at the default value of forty.

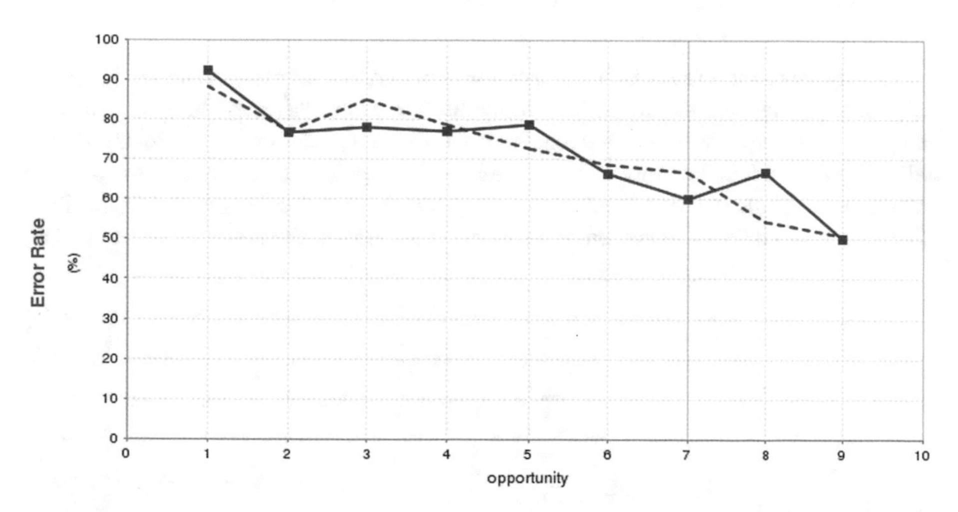

Fig. 4. *Still High* learning curve with the final point above the high error threshold (40)

When students did not have enough practice opportunities with the KCs, they are categorized as *Too Little Data,* since there are not enough opportunities for the data to be meaningful based on the configured parameters. These curves are based on the opportunity threshold and curves below the configured value are categorized as such. Even when a curve may show points above this opportunity value, the formula for generating the curves, using AFM, only includes points that meet or exceed the student threshold. Thus, by default curves with three or less opportunities of ten or more students are grouped into this category (see Fig. 5). It is recommended that more practice opportunities be added for these KCs, so they can be assessed with enough points to determine student learning progress.

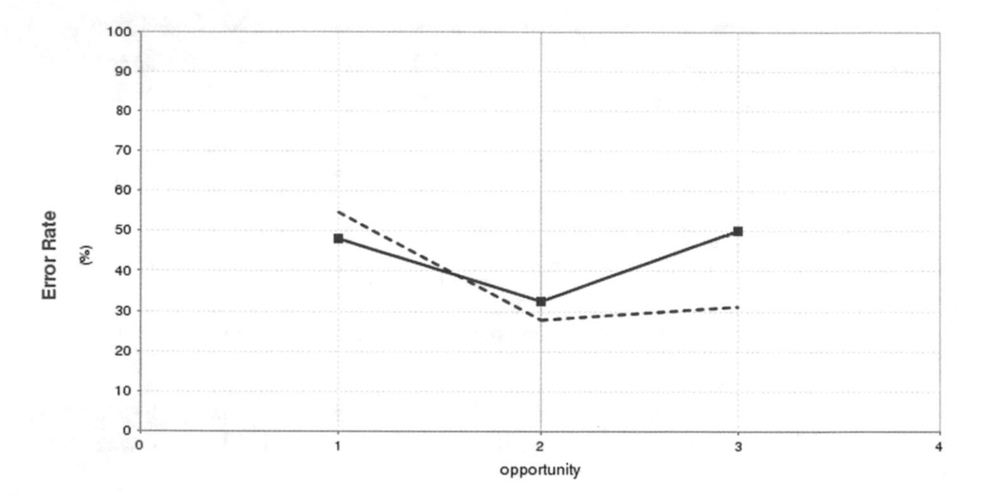

Fig. 5. Learning curve with only three points, categorized as *Too Little Data*

Finally, learning curves that did not get categorized into the aforementioned "at risk" ones are labeled as *Good*. The previous curves are "at risk" ones due to their being an opportunity for improvement in them. However, curves in the *Good* category indicate that student learning is occurring as they progress through corresponding assessments. While curves of this nature may still have room for improvement, these have an optimal balance of student improvement as opportunity count increases and are

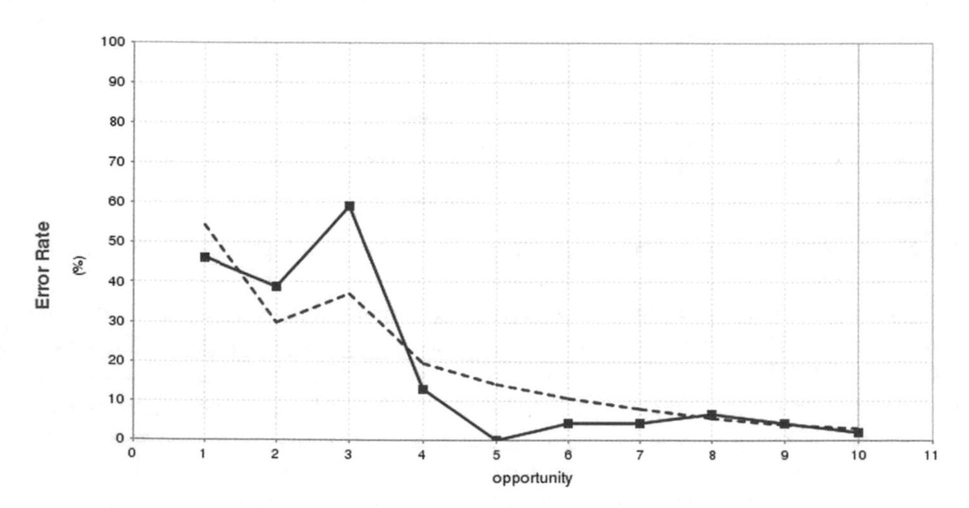

Fig. 6. Learning curve categorized as *Good* using the default parameter values

less likely to have easily identifiable areas of breakdown. A curve can still be Good even if all the points do not decrease in error rate, as demonstrated in Fig. 6.

4 Evaluation

4.1 Case Study One - Psychology

When interpreting learning curves in DataShop, one of the underlying assumptions is that the existing KC model is accurate. For this case, which models the way a student learns, we define accurate as representing valid segmentation and progression of the content. After all, it is often the case that the model in question was constructed by an expert of that particular domain who has an intuitive understanding of the content. What is important to consider, however, is that decisions regarding the KC model can be primarily driven by discipline specific standards, rather than by an iterative, data-informed process that is more broadly supported by learning sciences. When we examine the learning curves, these considerations allow us to identify candidates for human intervention to make appropriate changes so that we have a better fitting model. Additionally, by leveraging other users in the analysis and refinement process, we can hope to avoid some expert blind spot that may have been present in the generation of the original model by the domain expert.

To begin, we started our analysis of a Psychology dataset in DataShop recorded from an OLI course. The user for this case study had never used DataShop previously, but wanted to investigate a Psychology course dataset as part of their learning and participation in an educational data mining workshop. This dataset was made available to them, as it contains recent data from being used the past few years by students as part of their university class in Psychology. This data consisted of logs from 180 students who took the course during one of two semesters. When we first began analyzing the data, there were 272 total KCs. Due to the volume of this dataset, the student threshold parameter was set to 20 and the opportunity threshold was set to 5, so the data displayed in the curves would have more student attempts and a great number of assessments. Modifying these parameters changed how the learning curves for this dataset were grouped.

We were most interested in examining learning curves that have been categorized as *No Learning* or *Still High*. These curves showed plenty of room for improvement and it is common for curves with alignment issues to end up in these categories. Curves with alignment issues are ones that have problems mapped to KCs that are poor fitting and not representative of what is required to solve the problem. We first noticed a learning curve for a particular KC categorized as *No Learning*, "describe_psyhcoactive_drugs", that begins to descend normally and then spikes suddenly, shown in Fig. 7. We interpret the curve as telling us that at first, the students are learning predictably, making fewer and fewer errors. When the curve spikes, it is an indication that students have suddenly begun making errors at a much higher rate, which is confounding given the initial learning progress indicated by the beginning of the curve. To determine the cause of the increased error rate, we examined the individual corresponding items in the lesson that were mapped with this KC.

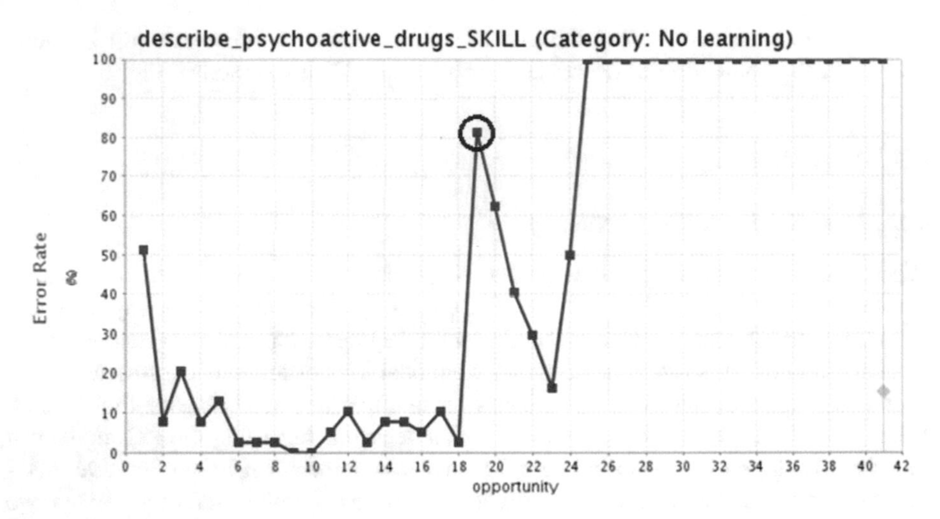

Fig. 7. Learning curve, categorized as *No Learning*, with a suspicious spike at opportunity 19

Using domain expertise, judgements were made regarding whether those opportunity items were constructed appropriately or categorized correctly. It appeared that the problems at and after the spike assessed different knowledge than prior items on the curve. The first half of the learning curve was from problems describing a particular type of psychoactive drug, while the latter part had questions about a different type of drug. As a result, we split this KC into two KCs since there were enough opportunities to provide sufficient results for each. The resulting two KCs still had enough opportunities for the given threshold parameter to be analyzed. The resulting curves, Fig. 8 and Fig. 9, show much smoother learning curves than the original one for the given KC and are now categorized as *Good*.

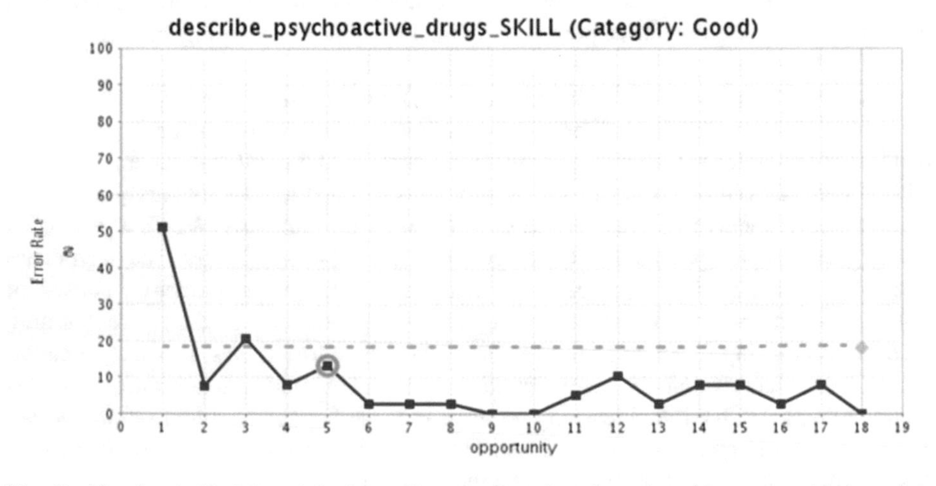

Fig. 8. The first half of the original describe_psychoactive_drugs learning curve, with the points at opportunities 19 and beyond remapped onto a different KC

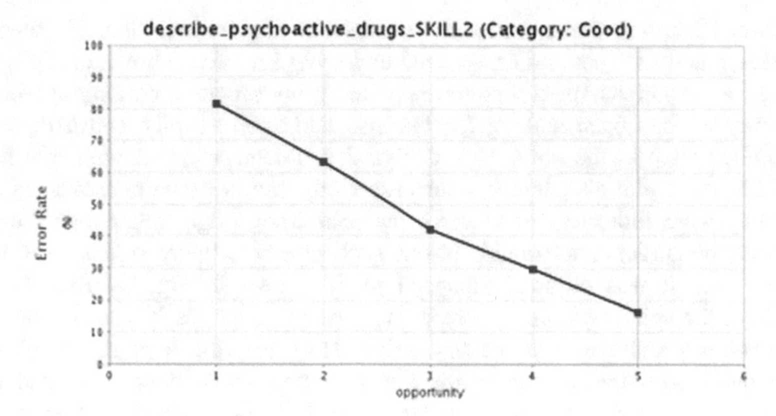

Fig. 9. The learning curve for the added KC, split from the original describe_psychoactive_drug KC

Another similar occurrence of a suspicious spike was seen when analyzing the curves categorized as *Good*. Despite this spike, students were learning this KC as the error rate was decreasing at an appropriate rate. While there were other curves to investigate, we decided to look at this one due to it have a spike like the previous one. A learning curve shown in Fig. 10, "analyze_intelligence_concepts", appeared to have a spike around the tenth and eleventh opportunities. Upon inspecting the questions in the course tagged with this KC, we noticed two of the questions were not assessing analysis of intelligence concepts. The two problems were having students identify and define, which was a better fit for an existing KC titled "describe_intelligence_concepts". Assigning those two opportunities to this different KC yielded a smoother curve and better assessed the appropriate skills. As a result, both the "analyze_intelligence_concepts" and "describe_intelligence_concepts" were improved, showing a smoother curve, as a result of this re-tagging.

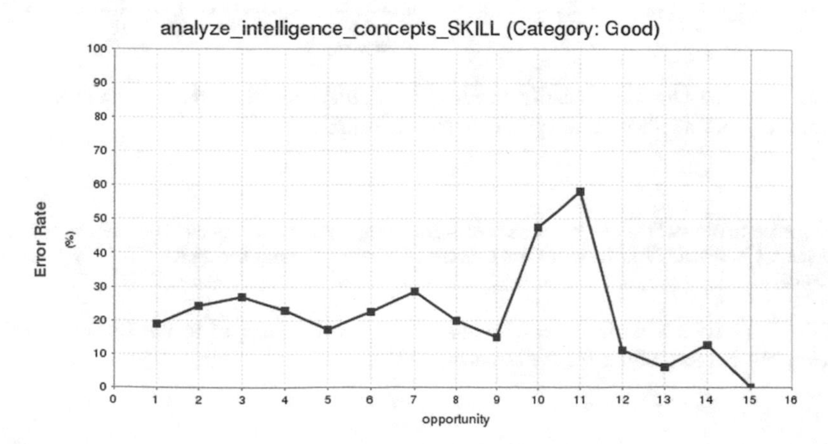

Fig. 10. A *Good* learning curve for analyze_intelligence_concepts that has an area of spiking at opportunities 10 and 11

The final learning curve we looked at was one for the KC "identify_therapies_-modalities_practices" that was categorized under *No Learning*, shown in Fig. 11. This curve had a suspicious spike, like the previously investigated learning curve, at one opportunity we had been looking for, but also had another spike consisting of three opportunities later in the curve. We analyzed problems mapped with this KC that occurred at the spike and, through our case study user's domain knowledge, determined they were indicative of research methods knowledge rather than identifying therapeutic modalities. An example of one such question, better suited for a different KC than what it was originally mapped to, is shown in Fig. 12. Remapping the problems at the spike with their more fitting research methods KC made our original curve smoother and removed the first spike. However, the later region of spiking became more pronounced, due to the fewer number of students who had done a problem at that opportunity count. Students who had not demonstrated mastery were continuing to try the problems, increasing the overall error percentage at that point. While this later spike in the curve might not warrant a remapping, it does suggest that the content might need to be improved so that students can hopefully achieve mastery before reaching this many attempts.

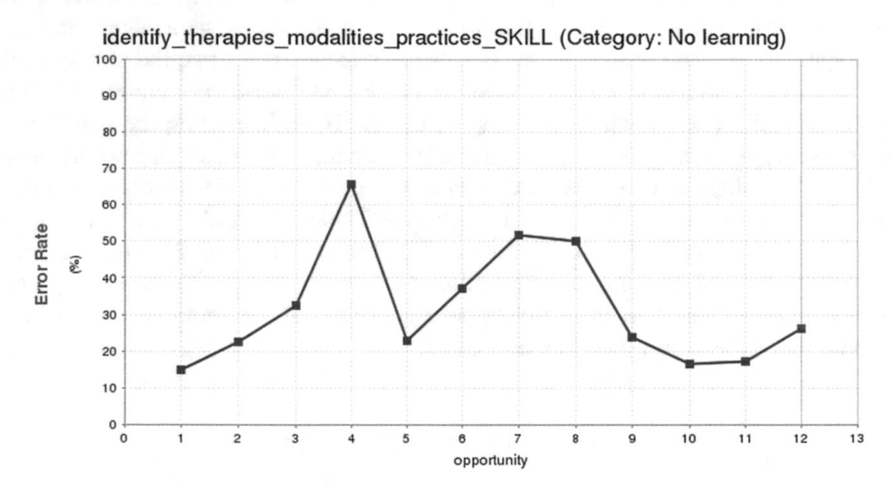

Fig. 11. Learning curve for identify_therapies_modalities_practices that has an initial spike at opportunity 4 and another set of spikes around opportunity 7

1) In an effort to monitor the effectiveness of different therapies and other medical treatments, studies are conducted. The results of such studies are published and are otherwise known as [-- ▼] (research outcomes)

Fig. 12. A question used in the course that was originally mapped to the KC for identifying therapies, but is more fitting for the research methods KC

4.2 Case Study One - Model Improvement Results

To validate the hypothesized model improvements for both cases, we performed a parallel analysis on the original student models compared to revised models with added KCs and re-tagged problem steps. The original student model for the first Psychology case, "pyschology_1-6", was created by the course's instructor, an expert in the domain. This original model consisted of 272 KCs and after the remapping and decomposition of the user in case study one, a new "psychology_1-6_model4" was developed consisting of four additional KCs for a total of 276. Utilizing AFM, we found the newer model is a better predictor of student learning when compared to the original mode, summarized in Table 2. The KC adjustment led to reducing AIC (176,705 to 176,441), BIC (183,912 to 183,728) and unstratified root mean square error (RMSE) on test set fit in cross validation (0.435319 to 0.435038). These model values support the addition of the KCs and demonstrate how the model can show improvements from just modifying a few learning curves. Not only can these modifications improve the predictive accuracy the model provides, but the analysis provides key human insights into the content that otherwise might be neglected. For instance, the analysis and improvement of these learning curves also allows users to look at the problem associated with these KCs, which might be indicative of refinement for the content, not just the KC associations.

Table 2. Knowledge Component model values for the Psychology course

Model	AIC	BIC	RMSE
psychology_1-6_model4	176,441	183,728	0.448741
psychology_1-6	176,705	183,912	0.449876

4.3 Case Study Two - Geometry

The data used in our second case study is from a Carnegie Learning cognitive tutor unit on Geometry area. This particular unit occurs later in the curriculum, and by the time students reach this unit, the skills around finding shape areas have been merged to a single skill called "Find Individual Area." Earlier units expressly break this skill into multiple skills; one skill for each shape type. By this unit, however, it is expected that students have successfully mastered these skills and now are just addressing find area as plugging in the inputs to the correct area formula for the shape given. The data largely backs up this merging of the individual area skills, but categorization in DataShop still pointed to a potential improvement of the model on this skill.

For this case study, the primary user was a Master's of Data Science student, who chose to user DataShop, for the first time, as part of a class project. When looking at the "Find Individual Area" skill using the KCM that was provided with the dataset that was used in the cognitive tutor, KTracedSkills, we see that it is categorized as *No Learning* as seen in Fig. 13. Further, visual inspection, clearly shows a spike in the error rate at opportunity 5. This led us to explore what problem steps were attempted at opportunity 5 versus the previous opportunities, which seem to show a declining curve. Sure enough, the majority of problem steps with errors were all centered on problems containing trapezoids. By retagging the "Find Individual Area" skill in all problems that address trapezoids as "Find Trapezoid Area," we were able to get a better fitting model that listed both skills in the *Good* category. This suggests that at this point in the tutor, for the group of students working, that the representation of "Find Individual Area" as 2 separate skills is a better representation of actual student knowledge. Students in the tutor were not at the same level with their skill for other shapes collectively compared to trapezoids. Making this change in the model should lead to improved learning outcomes if the model would be updated.

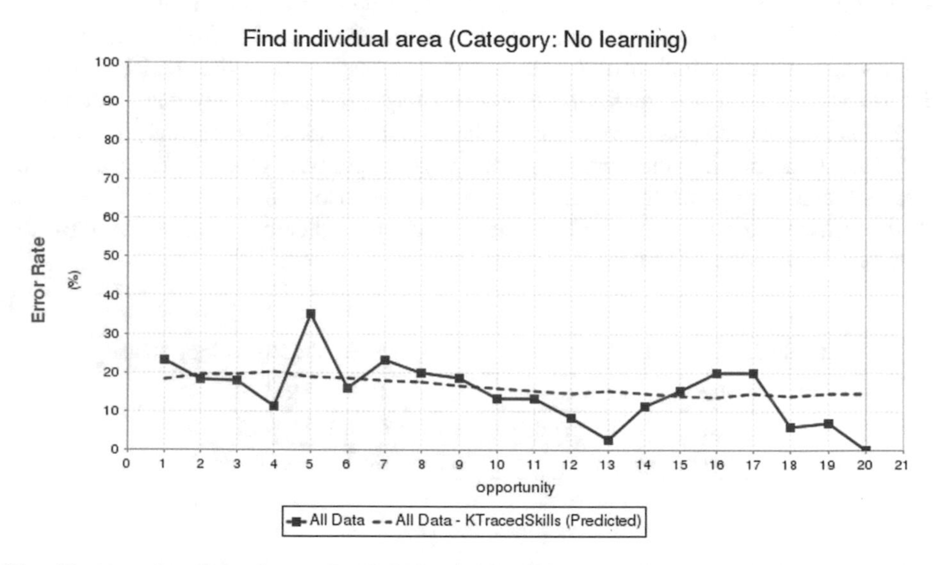

Fig. 13. Learning Curve for a KC called Find individual area that is categorized as *No Learning*. Visual inspection seems to show a potential improvement around opportunity 5, where there is a spike in error rate. Drilling down into the steps in this opportunity show that many of the errors are in problems with trapezoids, suggesting there could be different KC around these problems

4.4 Case Study Two - Model Improvements Results

The original student model for this Geometry case was one used in the cognitive tutor and was provided with the dataset, KTracedSkills. This original model contains 10 KCs and after the addition of the "Find Trapezoid Area" KC, the new model contains a total of 11. Again, using AFM we found the newer model, KTracedSkills-trap, was more predictive of the student data than the original model, summarized in Table 3. The single KC adjustment led to reducing AIC (3,409 to 3,377), BIC (4,215 to 4,196), and unstratified RMSE on test set fit in cross validation (0.304451 to 0.303349). While the improvements weren't as big as the Psychology example, there was still a noticeable difference by modifying a single KC.

Table 3. Knowledge component model values for the Geometry course

Model	AIC	BIC	RMSE
KTracedSkills-trap	3,377.25	4,196.49	0.315229
KTracedSkills	3,409.26	4,215.53	0.317503

5 Discussion

The categorization of learning curves in DataShop provides an initial grouping that allows users, regardless of expertise level, to focus their analysis of the KCs that might benefit the most from refinements. DataShop provides users the ability to filter through hundreds of KCs for their datasets, and analyze which ones are effective or which ones indicate students are not learning at an expected rate. The configurable parameters allow users to filter out learning curves that might not be as relevant for analysis, such as ones with little data, and focus on the more pertinent ones. These parameters also influence how the categorization of curves are formed, by modifying the thresholds, and allows the user a greater level of control. This presents an easy way to drill down into the learning curve to view the student error rate at each opportunity, and find anomalies, such as the high error rate spikes. We showed in our two case studies that this method of categorization to classify learning curves can accurately identify curves deserving further inspection. The analysis performed was able to determine potential issues in the model and make appropriate refinements to it.

This categorization provides a high-level view of all the curves in a manner that suggests which ones should likely be addressed first. The main analyst in the first case study regarding Psychology was a user, with a PhD in Social Psychology, who was new to DataShop. For the second case study, the main analyst was a Master's of Data Science student, also using DataShop for the first time. While these users might be familiar with general data science and statistical practices, the use of DataShop and involvement with learning curves was new to them. They were able to effectively utilize the learning curve categorizations to guide their selections into digging deeper into their analysis. These users utilized other features in DataShop as part of their investigation, but the initial grouping and visual display of all the curves served as the starting point for their analysis.

By improving, splitting, and modifying just a few learning curves, we were able to create a better fitting student model for both cases. The AIC and BIC decreased in each instance for the newer models, meaning that the AFM statistical model fits the data closer, providing a more accurate measure of student learning and progression [8, 16]. Additionally, the RMSE also decreased, suggesting that the new models will generalize better to datasets of that domain from the same tutor. This translates to having increased accuracy from a knowledge tracing perspective, which is important for intelligent tutoring systems. The improved accuracy will help provide the students the correct amount of problems needed to achieve mastery for a given skill, particularly when this process and improvement is applied to multiple KCs. Having this closer fitting model is key in order to avoid over or under-practice [10]. With a more accurate problem selection, these tutoring systems can help students learn more efficiently and make better use of their limited time. It also models the student learning process better, allowing for the suggestions of next problems that contain only the KCs a student still needs to master and letting them advance through the tutor at the correct pace.

While the Geometry case did not show as much improvement as the Psychology, only a single KC was broken down in that instance. The Geometry dataset came from an ITS used in production, one developed by professionals at a company rather than by a single professor, so the quality of the original model may have been stronger, needing less refinement. This means there may not be as many improvements possible to the model. It is not to say that breaking KCs into multiple ones always leads to an improved model, sometimes the remapping of a KC is required rather than creating a new one, as was the case for the second learning curve in the Psychology example. However, in both studies, only working with a few learning curves led to changes that created new improved models. While we did not get to feed these models back into the host educational technology, OLI and an ITS, several studies support that the improved AIC, BIC, and RMSE scores will result in improved student learning [4, 8, 17]. This is particularly useful for the Geometry case, where the model is used for a tutoring system with knowledge tracing.

Aside from improving the student model, iterative improvements to the educational technology systems are other potential outcomes of such analysis. In addition to adjusting the KCMs, there may be a need to adjust the systems and content to support these model refinements, in order to fully realize the improvements. The data from the Psychology example utilizing OLI had its activities mapped with the original model's KCs by a domain expert, the course instructor. Analysis of the "describe_psychoactive_drugs" learning curve that led to it being broken down two component ones might suggest that new assessments be added for the added component skill. In that case, the original KC still had 18 opportunities, but the added one only had 5. Providing more opportunities for the later could provide a more accurate measure of student learning for that KC and ample opportunities to develop mastery. Similarly, the analysis of the "analyze_intelligence_concepts" learning curve suggested it was assessing a different KC and needed to be remapped. It may be the case that other assessments in the course are actually targeting "research methods" like this KC, but are also mislabeled.

6 Conclusion and Future Work

We presented two cases where novice users of DataShop were able to utilize its features and the categorization of learning curves to assist in identifying potential problem areas within a course. The results of utilizing the learning curve categorization, drilling into the learning curves, and breaking them down into multiple KCs or remapping them, led to improved student models for both cases. Data that created the curves came from two different educational technology systems, yet both benefited from similar methods that utilized the affordances of DataShop. Not only were the technology systems different, but they also represented two completely different domains. However, we were able to apply similar techniques to both in order to improve their corresponding student models. Our study is another step toward showing how novice users can analyze the large amount of data their educational technology systems collect in a way that feeds into the iterative improvement of courses. It supports that by using the learning curve categorizations as a starting point, users can make informed judgements when it comes analyzing KCs. The improved student models these KCs feed into not only better model learning, but can be used to accurately inform course instructors of their students' learning and areas they might target for course improvement.

Continued work should look at applying similar techniques and utilizing the categorizations to find areas of course improvement in even more diverse domains. One such domain we are moving towards is computing principles, which currently has several years worth of data available in DataShop. Such courses often have a mix of questions types, from programming activities to free response. We believe that analysis of the curves for that datasets will reveal the need for similar interventions as the two presented case studies.

Additionally, this process supported by DataShop offers the potential to create an improved model from a semester's worth of data and see how it translates to many other datasets from different semesters of the same course. Finding how generalizable an improved model is suggested by the RMSE, is important in creating a solution that is effective across all student populations. Additionally, future work should look to feed the improved student models back into the educational technology and measure the learning gains students have from the better fitting model. This is key for intelligent tutoring systems or other educational technology systems that utilize knowledge tracing, as the student model is core to the system. Building upon this, there may be a benefit for looking at ways to help users identify when to remap problems steps to KCs, or to breakdown a KC into multiple ones. While the categorization and viewing of learning curves helps to indicate there is a potential problem, it may not be clear to the user how to optimally resolve the problem.

Acknowledgment. The research reported here was supported, in whole or in part, by the Institute of Education Sciences, U.S. Department of Education, through grant R305B150008 to Carnegie Mellon University. The opinions expressed are those of the authors and do not represent the views of the Institute or the U.S. Department of Education.

References

1. Baker, R.S., Inventado, P.S.: Educational data mining and learning analytics. In: Learning Analytics, pp. 61–75. Springer, New York (2014)
2. Murray, T.: An overview of intelligent tutoring system authoring tools: updated analysis of the state of the art. In: Authoring Tools for Advanced Technology Learning Environments, pp. 491–544. Springer, Dordrecht (2003)
3. Fancsali, S.E., Ritter, S., Stamper, J., Nixon, T.: Toward "hyperpersonalized" cognitive tutors. In: AIED 2013 Workshops Proceedings Volume, vol. 7, pp. 71–79 (2013)
4. Stamper, J., Koedinger, K.R.: Human-machine student model discovery and improvement using DataShop. In: Kay, J., Bull, S., Biswas, G. (eds.) Proceeding of the 15th International Conference on Artificial Intelligence in Education (AIED 2011), pp. 353–360. Springer, Berlin (2011)

5. Stamper, J., Koedinger, K., Baker, R.S., Skogsholm, A., Leber, B., Rankin, J., Demi, S.: PSLC DataShop: a data analysis service for the learning science community. In: International Conference on Intelligent Tutoring Systems, p. 455. Springer, Heidelberg (2010)
6. VanLehn, K.: The behavior of tutoring systems. Int. J. AIED **16**, 227–265 (2006)
7. Shepard, L.A.: What policy makers who mandate tests should know about the new psychology of intellectual ability and learning. In: Gifford, B.R., O'Connor, M.C. (eds.) Changing Assessment: Alternative Views of Aptitude, Achievement and Instruction, pp. 301–328. Kluwer, Boston (1992). 10, 978-94
8. Baker, R.S., Corbett, A.T., Aleven, V.: More accurate student modeling through contextual estimation of slip and guess probabilities in Bayesian knowledge tracing. In: Intelligent Tutoring Systems, pp. 406–415. Springer, Heidelberg, June 2008
9. Anderson, J.R., Corbett, A.T., Koedinger, K.R., Pelletier, R.: Cognitive tutors: lessons learned. J. Learn. Sci. **4**(2), 167–207 (1995)
10. Cen, H., Koedinger, K.R., Junker, B.: Is over practice necessary? Improving learning efficiency with the cognitive tutor. In: Proceedings of the 13th International Conference on Artificial Intelligence and Education (2007)
11. Koedinger, K.R., Mathan, S.: Distinguishing qualitatively different kinds of learning using log files and learning curves. In: ITS 2004 Log Analysis Workshop, pp. 39–46 (2004)
12. Koedinger, K.R., McLaughlin, E.A., Jia, J.Z., Bier, N.L.: Is the doer effect a causal relationship?: how can we tell and why it's important. In: Proceedings of the Sixth International Conference on Learning Analytics & Knowledge, pp. 388–397. ACM, April 2016
13. Anderson, J.R., Conrad, F.G., Corbett, A.T.: Skill acquisition and the LISP tutor. Cogn. Sci. **13**(4), 467–505 (1989)
14. Cen, H., et al.: Learning factors analysis – a general method for cognitive model evaluation and improvement. In: ITS 2006, pp. 164–175 (2006)
15. Moore, S., Stamper, J., Soniya, G.: Human-centered data science for educational technology improvement using crowd workers. In: Companion Proceedings 9th International Conference on Learning Analytics & Knowledge, pp. 341–347, March 2019
16. Draney, K.L., Pirolli, P., Wilson, M.: A measurement model for a complex cognitive skill. Cogn. Diagn. Assess., 103–125 (1995)
17. Koedinger, K., Stamper, J., McLaughlin, E.: Using data-driven discovery of better student models to improve student learning. In: Proceedings of the 16th International Conference on Artificial Intelligence in Education (AIED 2013) (2013)
18. Koedinger, K.R., McLaughlin, E.A., Stamper, J.C.: Automated Student Model Improvement. Int. Educ. Data Mining Soc. (2012)
19. Nguyen, H., Wang, Y., Stamper, J., McLaren, B.M.: Using knowledge component modeling to increase domain understanding in a digital learning game. In: Proceedings of the 12th International Conference on Educational Data Mining (EDM 2019), pp. 139–148 (2019)
20. Stamper, J., Koedinger, K., Mclaughlin, E.: A comparison of model selection metrics in datashop. In: Educational Data Mining, July 2013
21. Koedinger, K.R., Baker, R.S.J.D., Cunningham, K., Skogsholm, A., Leber, B., Stamper, J.: A data repository for the EDM community: the PSLC DataShop. In: Romero, C., Ventura, S., Pechenizkiy, M., Baker, R.S.J.D. (eds.) Handbook of Educational Data Mining. CRC Press, Boca Raton (2011)
22. Koedinger, K.R., Baker, R.S., Cunningham, K., Skogsholm, A., Leber, B., Stamper, J.: A data repository for the EDM community: the PSLC DataShop. In: Handbook of Educational Data Mining, vol. 43 (2010)

23. Murray, R.C., Ritter, S., Nixon, T., Schwiebert, R., Hausmann, R.G., Towle, B., Vuong, A.: Revealing the learning in learning curves. In: International Conference on Artificial Intelligence in Education, pp. 473–482. Springer, Heidelberg, July 2013
24. Rohrer, D.: Interleaving helps students distinguish among similar concepts. Educat. Psychol. Rev. **24**(3), 355–367 (2012)

Digital Content Editing System for Smartphones (Athrim 1.0)

Mirtha Idania Gil Rondón, Dionis López Ramos(✉), and Silena Herold García

Universidad de Oriente, Santiago de Cuba, Cuba
{mirtha.gil,dionis,silena.gil}@uo.edu.cu

Abstract. At present there is a large number of users with mobile devices, however, this potential is not used to the maximum. In this regard, it was decided to develop a system with mobile technology that allows the visualization of content about to any topic. With this, students to have a way of studying about a subject or several at all times, from anywhere, without having to be connected to the internet and not install new apps. This work describes the design and implementation of a system with two tools. The first, Athrim 1.0, an application oriented to mobile devices, which allows students to visualize and consult information using different multimedia resources. The second tool is a desktop application MobileDataCreator, which is going to be manipulated by the teachers. This tool allows create different contents and saving them in a file to be visualized in Athrim 1.0, with aim to contribute the learning of diverse topics.

Keywords: Android · Desktop application · Smartphone application

1 Introduction

1.1 Justification

The use of the computer limits us to specific moments where certain circumstances occur (electricity, Internet, etc.). However other elements such as tablets or smartphones have the necessary autonomy to not cause this type of problem. We can say that this technology has great possibilities in relation to its relatives, such as the camera, the touch screen, the motion sensor, or simply the fact that it can be carried more easily than a computer as stated in [1].

The mobiles, as well as the tablets, have become a key piece in the day to day of the people. The constant tendency to socialization makes people need a continuous connection to these media that promote the relationship both personal and certain content. This is also the case with applications for mobile devices, they are so diverse and have multiple functionalities that contribute to society an improvement in the quality of life as stated in [2].

The use of information and communication technologies (ICT) can be seen as a component of innovation that can contribute to the improvement of quality and coverage of education, in this sense, digital educational content or mobile applications for

M. E. Auer and D. May (Eds.): REV 2020, AISC 1231, pp. 762–772, 2021.
https://doi.org/10.1007/978-3-030-52575-0_62

education play a very important role as support for teaching, not only in the classroom, but anywhere through the production of high-quality applications that are found on the network so that they can be used and reused by all actors in the education sector.

Mobile applications focused on education are composed of digital content, and can be categorized as stated in [3]:

- Multimedia: photography, illustration, video, animation, music, sound effect, locution, composite audio, narrative text, hypertext, graphics, integrated media.
- Information system: database, table, graph, conceptual map, navigation map, multimedia presentation, tutorial, digital dictionary, digital encyclopedia, periodical digital publication, web/thematic or corporate portal, wiki, weblog.
- Computer application: multimedia creation, editing tool, web creation, editing tool, office automation tool, programming tool, analysis, information, knowledge organization tool, process, procedure support tool, learning, work management tool individual, cooperative, collaborative.
- Service: multimedia creation, editing service, web creation, editing service, office automation service, programming service, analysis, information, knowledge organization service, process, procedure support service, individual learning, work management service, cooperative, collaborative.
- Didactic content: guided lectures, master lesson, text comment picture, discussion activity, exercise or closed problem, contextualized case, open problem, real or virtual learning scenario, didactic game, webquest, experiment, simulation, questionnaire, exam, self-evaluation.

Something important when it comes to developing mobile applications for education is the process of identification, recovery and description of these applications and their content as stated in [2]. Given this, the application developed in this project is classified in Information System.

The system proposed in this work consists of an application for mobile devices and a desktop application, the latter complements the first one, being able to create and modify the contents shown in the mobile application. Its objective is the creation of a system that consists of two tools: a desktop application that allows creating different contents for an application of a smartphone, allowing users of the latter to view and consult the creative information with the aforementioned application. Priority, which contributes to learning with various topics such as medicine, sports, information of places, etc., through the integration of multimedia resources such as images, texts, videos and questionnaires.

Due to these facilities, the tool allows the interaction between student and teacher, because the teachers create the information that the students through the mobile application will consult, which facilitates to carry at all times with teaching contents that the teacher needs that the Students consult, therefore, the system supports the teaching-learning process.

At present there is a large number of users with mobile devices, however teachers do not take advantage of these potentialities to guide activities through the use of mobile applications, on the other hand it has been detected from the study that there is no application that allows the processing of information to be viewed in an affordable way for its use in the preparation of users. In this regard, it was decided to develop a system that allows the use of mobile technology the visualization of content applied to any context, taking into account the possibilities it has.

2 Material and Method

2.1 Programming Tools

Android operating system was used to implement the application for mobile devices (Athrim 1.0). Android is a complete open source software solution for phones and mobile devices. It is a package that includes an operating system based on the Linux kernel, a Java-based execution routine, a set of low and medium level libraries and an initial set of applications for the end user (all developed in Java) as stated in [3].

Platforms such as Microsoft's Windows Mobile and Apple's iPhone also provide a richer and more simplified mobile development environment. However, these operating systems, unlike Android, are not open source. In some cases, they prioritize native applications over those created by third parties. Android offers new possibilities for mobile applications providing an open development environment. It was based on an open source Linux kernel, providing access to all the functions of the equipment to all applications completely through the use of API libraries as stated in [4].

Android allows developers to write codes that run under the management of a virtual machine in the Java language, controlling the device through Java libraries developed by Google, releasing most of the codes under the Apache license, free license and open source.

For the development of the (Athrim 1.0) application, the IDE Android Studio v 2.1 was used, a computer program composed of a set of multiplatform open source programming tools. It is currently the standard development tool for Android applications as stated in [5].

It was also used, SQLite as Database Management System. The SQLite library is linked to the program and becomes an integral part of it. The program uses SQLite functionality through simple calls to subroutines and functions. This reduces the latency in the access to the database, because the calls to functions are more efficient than the communication between processes as stated in [6].

For the implementation of the (MobileDataCreator) desktop application, the NetBeans Integrated Development Environment (IDE) was used, which is a free integrated development environment, made primarily for the Java programming language. There is also a significant number of modules to extend it. NetBeans IDE is a free and free product with no restrictions on use.

The NetBeans platform allows applications to be developed from a set of software components called modules. Applications based on the NetBeans platform can be easily extended by other software developers. The NetBeans IDE is an open source IDE written entirely in Java using the NetBeans platform as stated in [6].

To design the graphical interface, the XML language was used, the tags being the visual components used in the graphic interface. This language was the propeller of modern HTML, in addition, allows all the visual components present in each activity to be organized efficiently, managing to adjust to each screen different from mobile devices, as well as adapting to the landscape position (the device is in vertical position and the content of the app also) or portrait (the content of the application is located horizontally as the position of the device) as the case may be as stated in [9]. It allows declaring the properties of each component within the same label, which greatly helps the programmer since reading the code becomes much easier. It was also used to save the content in text format, in the desktop application. To declare the classes, as well as the implementation of the actions performed by the application, the Java language was used (typical of Android applications), which has many libraries and classes already implemented that accelerate the work of the programmer and ensure the effectiveness of the code as stated in [8].

2.2 Development Methodology

The agile methodologies are specially oriented for projects that need a customized solution, with a high simplification without leaving aside the assurance in the quality of the product. They focus on the human factor and the software product; that is, they give greater value to the individual, to the collaboration of the client and to the incremental development of the software with very short iterations as stated in [8]. The methodology used for this project was the XP methodology (Extreme Programming). This methodology is a software engineering approach. It is the most widely used of agile software development processes. Like these, extreme programming differs from traditional methodologies mainly in that it places more emphasis on adaptability than on predictability as stated in [9].

This methodology considers that the changes of requirements on the march are a natural, inevitable and even desirable aspect of the development of projects; being able to adapt to changes in requirements at any point in the life of the project is a better and more realistic approach than trying to define all the requirements at the beginning of the project and investing efforts later in controlling changes in the requirements as stated in [10], unlike the robust methodologies, which required too much documentation, meetings and planning, and time was wasted considerably compared to XP.

The XP methodology fits perfectly with the type of project, the conditions of development, as well as the idea of the system as stated in [10]. Below, the fundamental reasons that were taken into account when choosing this methodology.

1. Start small and add functionality with continuous feedback: The development of the system starts from the basic requirements and from there, are added functionalities that both the developer and the client understand necessary.
2. Few roles: This methodology is aimed at small development groups with few roles like this case.
3. The client or the user becomes a member of the team.

3 Planning and System Design

There is a desktop application (MobileDataCreator), which will allow the insertion of the information to be displayed in the application for mobile devices (Athrim 1.0); this information (text, images and questions and answers) will be saved in a compressed file, which will contain the images and an XML file with the texts, this compressed file will be taken to the mobile device and loaded by the application through of the configuration provided by it. The diagram of system deployment of the is the one shown in Fig. 1.

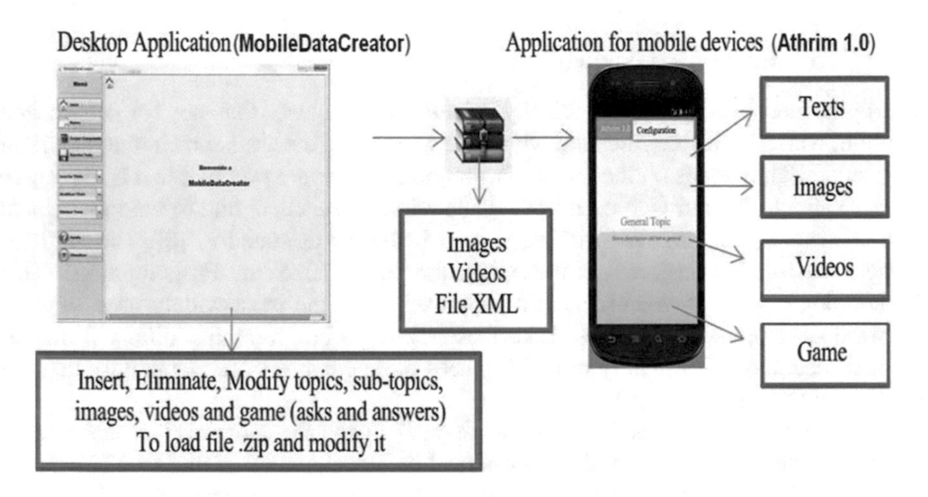

Fig. 1. System deployment diagram

3.1 Application Oriented to Mobile Devices With Android Operating System (Athrim 1.0)

The architecture shows the way in which the application is designed, where it is necessary to separate the responsibilities, which allows the specification of the work forces. The architectural style defines the general rules of organization in terms of a pattern and the restrictions in the form and structure of a large and varied group of software systems, more specifically. For the development of the system an architectural style was used in layers, specifically in two layers: presentation layer and business logic layer. They are described below and are shown in Fig. 2.

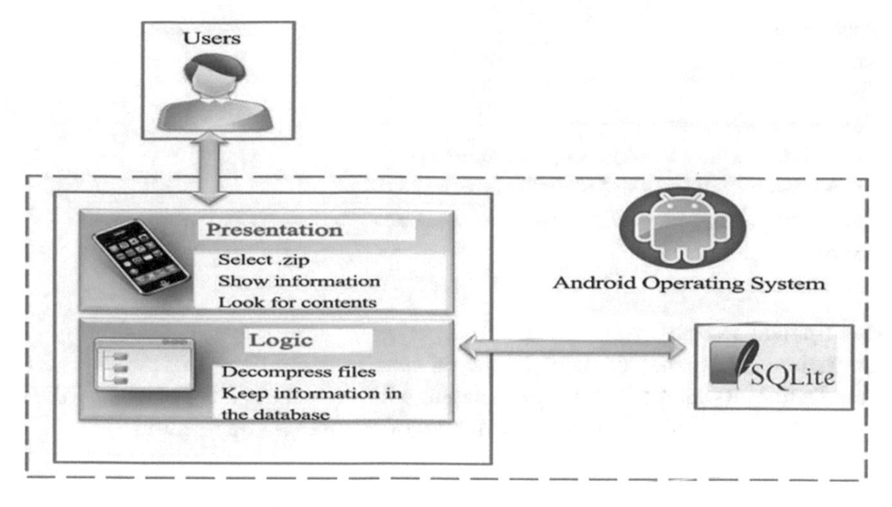

Fig. 2. Architecture of the android application

Presentation layer: It communicates only with the business layer, transporting the necessary data or records. In it are the graphic interfaces of the program, which will allow the user to interact with the application. Some of these interfaces are: load compressed file, search contents, among others.

Logic layer: This layer is where the functionalities that are going to be executed are, the user's requests are received and the answers are sent after the process. Some of the activities are: decompress files, read XML file, save information in the database, among others. This layer uses the SQLite database manager that is present in the Android operating system to store all the information in an orderly manner.

For the zip file to be considered correct, it must contain an XML file, and a set of images and videos (images and videos are not required). The XML file must have the following structure:

```
<Presentation>
<Name > </Name>
<Text > </Text>
<AuthorName > </AuthorName>
<CorreoAutor > </CorreoAutor>
<ContrasenaAutor > </ContrasenaAutor>
<Theme>
<Name > </Name>
<Body > </Body>
<Image > </Image>
<Video > </Video>
<SubTema>
<Name > </Name>
<Body > </Body>
<Image > </Image>
```

```
</SubTema>
</Theme>
<Game>
<Question > </Question>
<Correct Response > </Correct Response>
<AnswerIncorrecta1 > </AnswerIncorrecta1>
<AnswerIncorrecta2 > </AnswerIncorrecta2>
</Game>
</Presentation>
```

Label Specifications:
Presentation: Main label, contains all the others.
Name: Defines the name that the presentation will have, that is, the one that will show the application, it is shown in the first screen of the mobile application.
Text: It is a brief description of the title.
AuthorName: Name and surname of the user, defined when the content is saved and compressed.
CorreoAutor: Email of the author, defined when the content is saved and compressed.
ContrasenaAutor: Password defined by the author when the content is saved and compressed.
Theme: This label contains the topics that the application will present, that is, topics derived from the title, these will contain name (name of the theme), body description of the theme), images and a video.
Subtopic: It is a lower level than the theme, they will be content derived from the theme that will have the same characteristics.
Game: This tag is going to present a didactic question and answer game, where the question is going to define the question that will be asked to the user of the mobile application and the incorrect and correct answers, previously defined.

If the XML file is defined with the structure described above and the database has been updated with the information contained therein, then the mobile application can already show in its different views all the content stored in the tablet. These tags are created through the tool (MobileDataCreator).

Functionality of (Athrim 1.0)
The main functionalities of the application (Athrim 1.0) are the following:

- Unzip the .zip file.
- Store the information found in the compressed file in the database.
- Load the information stored in the database in the different views of the application.
- Browse between the different topics, and if there are subtopics, navigate among them.
- Correctly visualize the images and videos.
- Search subtopic by name, through a search filter by letter word or phrase.
- Evaluate the user's responses in the game and give the results.

For the correct functioning of both applications it is necessary to take into account e following requirements:

Requirements for (Athrim 1.0)

- Software Requirements: To install the application you need the Android V 4.0 or higher operating system.
- Hardware Requirements: The device must allow the installation of applications from sources other than the Play Store (found in the security options of the device).

3.2 Desktop Application (MobileDataCrcator)

For the development of the system a layered architectural style was used, as described below in Fig. 3:

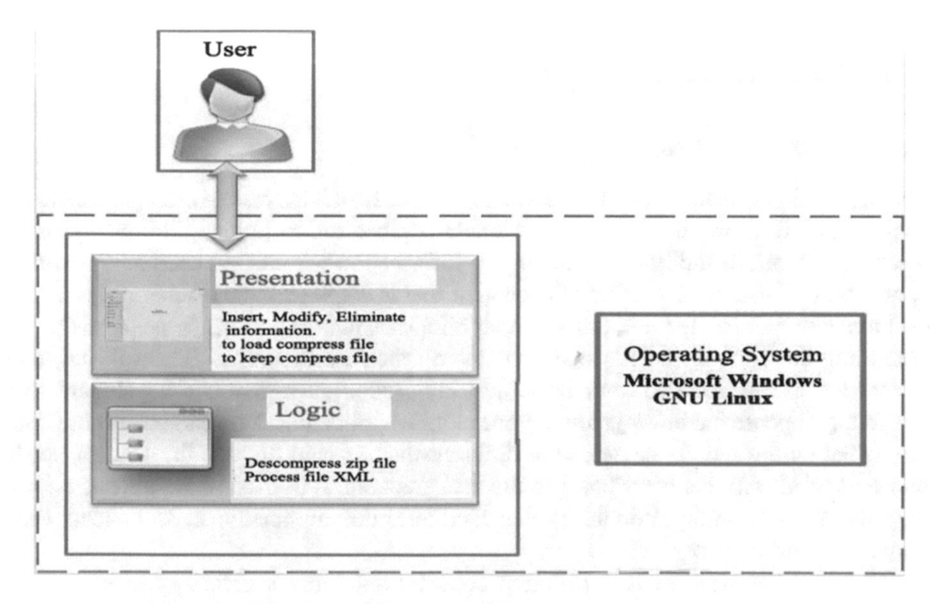

Fig. 3. Architecture of (MobileDataCreator)

Presentation layer: It communicates only with the business layer, transporting the necessary data or records. In it are the graphic interfaces of the program, which will allow the user to interact with the application. Some of these interfaces are:

Insert, Modify and Delete information, display content, load compressed, save compressed, among others.

Logic layer: This layer is where the functionalities that are going to be executed are, the user's requests are received and the answers are sent after the process. Some of the activities are: decompress zip file, process XML file, among others.

Functions of (MobileDataCreator)

- Insert, Modify and Delete Theme, Subthemes, Images, Videos and Didactic Game of questions and answers.
- Save the information inserted in an XML file.
- Compress the XML file, the images and videos inserted in a .zip file.
- Show preview of the information stored in the XML file.
- Load .zip file to modify the stored information, validating the password of the user who created the tablet.

Requirements for (MobileDataCreator)

- Software Requirements: You must have Java SE Development Kit 8 Update 66 in an Operating System versions of Windows from XP or GNU/Linux.
- Documentation and Help Requirements: The software has a help, which explains all the services offered by the application and the different ways to access them.
- Requirements of Hardware: Minimum requirement Pentium IV Processor to 700 MHz with 512 Mb of RAM.

3.3 Results and Discussion

The application oriented to mobile devices (Athrim 1.0), has great advantages due to its adaptability. It allows the student, in a single application, to upload different types of teaching content, in addition to updating the information without the need of an Internet connection. If you want to update the content that is displayed, the teacher accesses the desktop tool created for this purpose (MobileDataCreator), which is responsible for compiling the compressed file that contains all the information to be displayed, and creates the update that You want the subject in question, then give it to the student, and this in turn, update the information in your mobile application. You can have more than one tablet on the mobile device, with different themes, and the one the student needs will be loaded. All this contributes to the self-preparation in dissimilar subjects with a high degree of usability, portability that facilitates the application in any social, economic, industrial context, etc.

This tool has been used in different contexts obtaining satisfactory results. Such was the case, in the Allergy clinic of the Provincial Teaching Hospital “Saturnino Lora Torres”, where the application was used (Athrim 1.0), to improve the learning of the disease in question by patients and relatives. The system allows self-preparation of patients’ relatives efficiently, provides a tool that provides clarification at all times to frequent questions and guidance in some cases, allows the user to verify their level of knowledge through the game of questions and answers and possesses. All this has a positive effect on the attention the allergic patient receives from his family, and in a certain way it reduces the costs of specialized care given the probability of decreasing the assistance of patients to consultation. In parallel increases the quality of life of the patient, and also the family to be able to cope with more preparation a certain action with these patients. As for the specialists, it provides a tool that can serve as a bibliography to support their preparation in terms of facing the process of orientation of the relatives, which increases their experience in the care of the patient and their family.

Fig. 4. Adaptation of the application for mobile devices (Athrim 1.0) to different topics (Allergies and Computing).

The efficiency of the tool was checked through a survey of 50 patients and relatives who had a mobile device with the necessary features to install the mobile application. The results of the survey were satisfactory.

In addition, (Athrim 1.0) has been applied in the university educational process, being employed in the subject Programming Oriented to Mobile Devices, contributing to the preparation of the students in the matter in question. Adaptation of the mobile application for the Android operating system (Athrim 1.0) to different themes. See Fig. 4.

References

1. Expert, A.C.: Android Cookbook [Internet]. USA: Android Comunity Expert, pp. 1–688 (2011)
2. Beck, K.: Extreme Programming Explained: Embrace Change. Addison-Wesley Professional, Boston (2000)
3. Cantillo, C., Roura, M., Sánchez, A.Y.: Tendencias actuales en el uso de dispositivos móviles en educación. La Educación digital. **147**, 1–21 (2002)
4. Fowler, M.: The new methodology. Wuhan Univ. J. Nat. Sci. **6**, 12–24 (2001)

5. Gijones, T.: El gran libro de Android (Segunda Edición). Grupo Editor Alfaomega. Barcelona, España (2012)
6. Jackson, W.: Android Apps for Absolute Beginners. Apress, New York (2011)
7. Núñez, C.F.M.: Dispositivos móviles en la educación médica. Teoría de la Educación. Educación y Cultura en la Sociedad de la Información. **11**(2), 28–45 (2010)
8. Niemeyer, P.: Learning Java, A Bestselling Hands-On Java Tutorial, 4th edn. pp. 1–1010. O'Reilly Media, Sebastopol (2013)
9. Orjuela, A.: Las metodologías de Desarrollo Ágil como una oportunidad para la ingeniería de software educativo. Avances en Sistemas e Informática **5**(2), 159–171 (2008). Medellín. Colombia
10. Baumeister, H., Lichter, H., Riebisch, M.: Agile processes in software engineering and extreme programming. In: 18th International Conference, XP, pp. 22–26 (2017)

Virtual Hotel – Gamification in the Management of Tourism Education

Petra Poulova[1(✉)], Miloslava Cerna[1], Jana Hamtilova[1], Filip Malý[1], Tomáš Kozel[1], Pavel Kriz[1], Jan Han[2], and Zdenek Ulrych[3]

[1] Faculty of Informatics and Management, University of Hradec Kralove, Rokitanskeho 62, 500 03 Hradec Kralove, Czech Republic
{petra.poulova, miloslava.cerna, jana.hamtilova, filip.maly, tomas.kozel, pavel.kriz}@uhk.cz

[2] Institute of Hospitality Management in Prague, Svídnická, 181 00 Prague, Czech Republic
han@vsh.cz

[3] University of West Bohemia, Univerzitni 8, 306 14 Pilsen, Czech Republic
ulrychz@kpv.zcu.cz

Abstract. We are experiencing increasing interactions between individuals and smart environments, whether it is the interaction of smart devices during ordinary activities or teaching/learning. There is growing emphasis on teachers in passing their knowledge and experience to students to get students more engaged because students expect more interaction and communication in the online world. The virtual world is the type of online community, which is primarily a form of computer-simulated environment that users can use, create, and interconnect certain objects. Currently, there are some schools and educational institutions around the world providing the opportunity to attend part of the training courses using simulation games. Simulators and their possibilities will be described in this work together with examples of utilization of virtual reality for educational purposes.

Keywords: Application · Blended learning · E-learning · Education · Game · Process · Simulation · Smart device

1 Introduction

Virtual Reality is a technology that enables a user to interact with a simulated environment. Virtual reality technologies create the illusion of the real world (e.g., combat training, piloting, learning or the fictional world of computer games). It is a visual, auditory, tactile or other experience creating a subjective impression of reality using computer imaging equipment. Special audio-visual helmets, glasses, motion sensing, and stimulating touch or other techniques evoking perception and sensation are utilized.

There is a rising emphasis on teachers in passing their knowledge and experience to students to get students more engaged into the process of education. Current students expect more interaction and communication in the online world [1]. The virtual world is the type of online community, which is primarily a form of computer-simulated environment that users can use, create, and interconnect certain objects [2].

M. E. Auer and D. May (Eds.): REV 2020, AISC 1231, pp. 773–781, 2021.
https://doi.org/10.1007/978-3-030-52575-0_63

The computer simulation of the world is perceived as predominantly perceptual stimuli that allow the user to manipulate the elements of the model world and create a sense of realism. These simulations are now used in many industries, whether it's the business sector or the activities of different organizations [3].

Simulation is used in education, training and analysis in the hotel industry, as well. It is primarily designed to help participants in the learning process to develop skills in identifying problems, detecting potential problems and finding solutions. It also helps learners develop social skills in team building and teamwork.

In summary it can be said that the hotel simulations enable students to develop the ability to plan, analyse and apply knowledge in the market for strategic business planning process [4].

2 Computer Games

Computer games and simulations are used at all levels of education. Most of the educational games are trying to introduce students to new knowledge and helps them acquire and improve their skills and abilities [5].

Computer hotel simulations were generally designed for college students in the hotel industry. The simulation should provide managerial experience in accounting, problem solving and decision making.

Significant development of computer systems that help manage production and logistics in restaurants and hotels could be seen during the last decade. It is essential that potential hotel managers were informed about the principles and rules involved in the development and use of computer applications, both in accommodation and food sectors of the hotel industry.

It is of key importance that potential hotel managers are familiar with the policies that are involved in the development and the use of computer applications, both in the accommodation and food sectors of the hotel industry.

Scope and possibilities of these specific applications will be demonstrated on practical examples.

2.1 Cesin Hospitality - Simulation of Hotel and Restaurant Management

The interactive game Cesim Hospitality ranks among widely utilized educational simulations. Its aim is to improve the business skills of students in the hotel business. It is used primarily by universities in tourism and hospitality programs.

The goal is to achieve success within teams. The game is focused on the management of operating profit, net profit, return on assets and cash flows. Customer satisfaction is a key factor for success in both game and reality. The simulation includes all major hotel industry specific situations. The simulation game develops participants' ability to identify, analyse, and influence key operational processes that affect hotel and restaurant operations in a competitive environment. In addition, the simulation enhances analytical decision-making and highlights the financial implications of decisions by linking decisions with cash flows and performance [6] (Fig. 1).

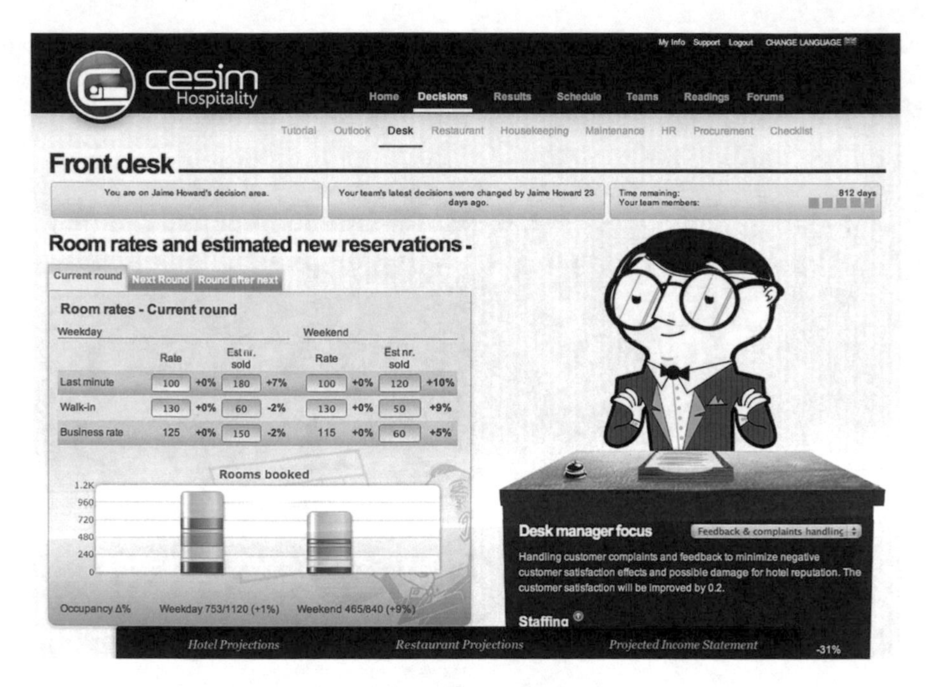

Fig. 1. Cesim Hospitality

2.2 HOTS – Hotel Simulation

Another representative of the interactive game is the HOTS hotel simulation based on the management of a large hotel. Players control a virtual environment reflecting the real world. It is a competitive environment in which participants work individually or in teams. Performances in the hotel's Hots simulation can be targeted at different educational goals, which include, for example: strategic management, finance, risk management, social media, revenue management, and many others [7].

This hotel simulation is often used as part of training courses, management training and team training activities. Companies use it as part of learning and development activities.

2.3 Virtual Business Hotel

Virtual Business Hotel is a simulation that allows students to take control of a complete hotel. This simulation of hotel processes is focused mainly on modelling the internal functioning of the hotel.

Students get familiar with everyday business decisions that lead to the successful operation of a top hotel. The simulation includes: price and revenue management, marketing, customer service (reception), social media feedback, restaurant management, gastronomy, cleaning and financial reports [8].

This "game" consists of ten lessons. Students gain experience in all key roles in the hotel business. The system records individual information about its guests and their stays. In addition, the system monitors information on how guests have rated the accommodation on a simulated website similar to TripAdvisor. The main goal of the game is to keep the guests feel good and satisfied (Fig. 2).

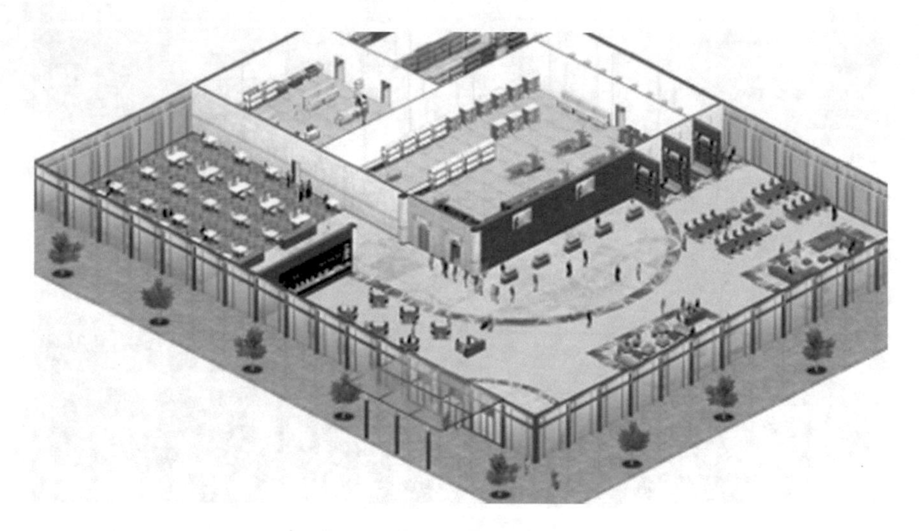

Fig. 2. Virtual Business Hotel

2.4 Hotel Giant

The aim of the Hotel Giant game is to develop and manage a pre-made hotel, attract a large number of guests and thrive as a hotel professional.

In this tycoon game there are several popular tourist destinations and 26 kinds of hotels to choose from. For example, a hotel on the beach will have more demanding customers than a hotel located in the city centre, where predominantly business people go. An important factor in the construction of the hotel is the number of floors, because as the number of floors increases, the costs of furnishing it increase.

The architecture is fully up to the player. He/she can really arrange everything from common rooms, Internet cafes, bars, swimming pools to the detailed design of the room equipment. Player designed room is automatically saved as a template for the future use. This step eliminates the need for designing terraced rooms. After opening the hotel the player just follows the wishes and complaints of guests and accommodates them [9].

This game specializes more in designing and simulating hotel environment, less in complex hotel operation (Fig. 3).

Fig. 3. Hotel Giant

2.5 The Change Game

Organizational leadership is seen as increasingly important in strategic management as it is an essential element of the company's ability to survive and remain competitive on the market. But very little attention has been focused on how to teach managers to accept change within their organizations. Simulation called "The Change Game" is useful for managers as support to implement changes [10].

The benefit of this game is to practice three key points aimed at the arts of adopting strategic changes in the organization [10].

The first point is to describe the various phases of innovation (awareness, assessment, adoption). The second point concerns the definition of individual differences (as it was before, as it is now). The third point focuses on the practice of application of the appropriate tactics and strategies depending on the stage of adoption of innovations.

3 PROTUR Hotel Simulator

The application PROTUR is developed by the authors′ team for training students of the hotel school and hotel staff.

The following diagram schematically captures the frame algorithm simulator (Fig. 4).

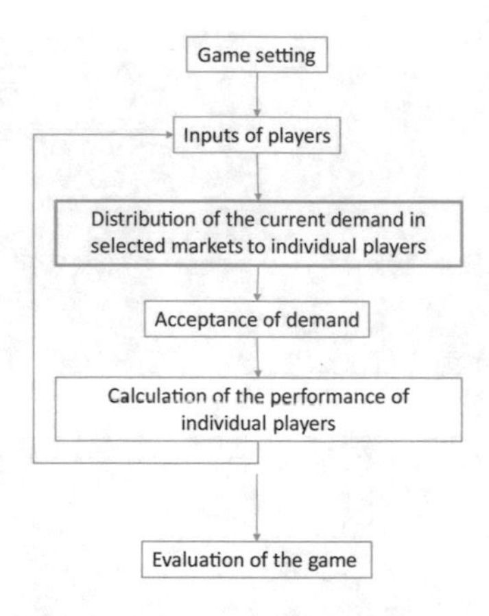

Fig. 4. Frame algorithm simulator

The simulator is prepared in two basic modes. In the case of on-line mode, players compete with each other on the market for current demand and players also control the operating side of the hotel.

In the off-line mode, players do not compete with each other in the markets, everyone has the full amount of demand regardless of the games of the others, and as for the hotel's operational side, it is firmly assigned by the game manager. In this case, players decide only in revenue management processes.

4 Further Use of Information Technology in the Hotel Industry

Information technologies offer operators, employees and clients unprecedented possibilities. Students studying in the tourism and hospitality programmes who are preparing for their business should be familiar with the tools which the latest technologies offer in their business field, e.g., Virtual reality, Blockchain technology or the use of robots.

4.1 Virtual Reality in the Hotel Industry

The virtual hospitality reality aims to attract guests to the hotel. Virtual reality allows potential visitors to get acquainted with the environment of the hotel or with the given destination before their arrival. The customer who has the opportunity to know the destination in advance will book the hotel with a greater amount of trust [11] (Fig. 5).

Fig. 5. Virtual reality

For example, the Best Western Hotels & Re-sort, which has a 360-degree high-resolution view of hotel processors, can serve as an example of virtual reality. Individual videos offer an interactive tour of the hotel swimming pool, lobby, fitness center and guest rooms. Thus, before making a reservation, the guest can first explore everything from the lobby to the type of chair they will be sitting on.

There are a number of ways in which hotels can incorporate virtual reality into a guest experience. Virtual reality can ensure that the guest arrives at the hotel with expectations that can be fulfilled and leaves satisfied.

4.2 Blockchain

Blockchain technology is one of the most interesting innovations that may in the future fundamentally change the way business transactions are conducted.

Blockchain is basically a public list of digital records where transactions are recorded anonymously and permanently. The records or "blocks" in the database are secured by cryptography, and once the information has been recorded in the block series, it is resistant to modification, which means that the data cannot be changed. Blockchain technology is decentralized (information is not stored in one place). This prevents data manipulation. Event records can be shared among many people in the network [12] (Fig. 6).

In the future, blockchain could ensure safer payments, loyalty programs and even luggage tracking.

Fig. 6. Blockchain

4.3 Robots in Hotels

Modern robots can be either autonomous or semi-autonomous and can use artificial intelligence and speech recognition technology. Most robots are programmed to perform specific tasks with great precision. The reason why robots have become a relatively quite popular technology trend in the hotel business is that automation and self-service ideas play an increasingly important role in customer experience. The use of robots may eventually lead to improvements in speed and cost-effectiveness [12].

As an example can be mentioned chatbots allowing hotels to provide 24/7 support to clients through online chat (Fig. 7).

Fig. 7. Robots in hotels

5 Conclusion

Hotel simulators offer excellent tools that can be used in educational activities. The aim is to provide students with a gentle and playful way technique enabling them understand the basic principles and improve their decision making skills.

Currently there are on the market a large number of all kinds of games that relate to the hotel industry. Development of the game market constantly moves forward, it is important to be able to accept and work with new information. Automated world forces us to invent new innovations that ultimately make life easier for us.

Augmented reality has emerged as an important concept in the hotel industry in recent years by enabling hotels and other related businesses to improve the physical environment they sell or improve their cognition experience of the environment.

Among the current trends in the hotel industry rank: interactive hotel rooms, gamification and enlarged the hotel environment.

Acknowledgments. This study is supported by the TACR project TL01000191 and SPEV project 2019, run at the Faculty of Informatics and Management, University of Hradec Kralove, Czech Republic.

References

1. Oblinger, D.: Boomers, Gen-Xers a Millennials: understanding the new students. EDUCAUSE Rev. **38**(4), 37–47 (2003)
2. Bishop, J.: Enhancing the understanding of genres of web-based communities: the role of the ecological cognition framework. Int. J. Web-Based Communities **5**(1), 4–17 (2009)
3. Hán, J.: Modelování a simulace procesů v hotelnictví a gastronomii [online]. COT (2017). https://www.icot.cz/modelovani-a-simulace-procesu-v-hotelnictvi-a-gastronomii. Accessed 09 Nov 2018
4. Gursoy, D.: Welcome to hotel business management training simulation (2018). http://www.hotelsimulation.com/. Accessed 09 Nov 2018
5. Bouki, V., Mentzelopoulos, M., Protopsaltis, A.: Simulation game for training new teachers in class management. In: SIGDOC '11 Proceedings of the 29th ACM International Conference on Design of Communication. ACM, New York (2011)
6. Cesim Oy: Hotel and Restaurant Management Simulation (2018). https://www.cesim.com/. Accessed 09 Nov 2018
7. Rusell Partnership Technology: HOTS - The Hotel Simulation (2017). http://www.thetotalsimulator.com/hots—hotel-simulation. Accessed 09 Nov 2018
8. Knowledge Matters: Virtual Business Hotel (2018). https://knowledgematters.com/highschool/hotel/. Accessed 09 Nov 2018
9. Dobrovsky, P.: Hotel Giant. Tiscali Media (2002). https://games.tiscali.cz/. Accessed 09 Nov 2018
10. Lyles, M., Near, J., Enz, C.: A simulation for teaching skills relating to organizational self-renewal. J. Manage. Dev. **11**(7), 39–47 (1992)
11. Hotel-Online: Virtual Reality – The Next Game Changer for Hotels (2018). https://www.hotel-online.com/press_releases/release/virtual-reality-the-next-game-changer-for-hotels. Accessed 09 Nov 2018
12. Revfine: Augmented reality hospitality industry (2018). https://www.revfine.com/augmented-reality-hospitality-industry/. Accessed 09 Nov 2018

How to Motivate Students? The Four Dimensional Instructional Design Approach in a Non-core Blended Learning Course

Małgorzata Gawlik-Kobylińska(✉), Dorota Domalewska, and Paweł Maciejewski

War Studies University, Warsaw, Poland
ggawlik2000@yahoo.com,
{d.domalewska,p.maciejewski}@akademia.mil.pl

Abstract. The four dimensional design approach focuses on cognitive, emotional, social, and psychomotor dimensions of learning. It relies on balanced differentiation of educational activities. Such a differentiation aims at increasing learners' motivation. In this paper, we examine the adoption of the four dimensional design to a non-core blended course designed and led at the War Studies University in Poland. Data collected from 59 respondents through structured interviews and the analysis of learning outcomes revealed that the four dimensional instructional design approach applied to a blended course keeps students motivated and engaged in the learning content. Our results suggest that university instructors should incorporate the four dimensional approach into instructional strategies to maximise learners' motivation and learning outcomes.

Keywords: Blended-learning · Motivation · Instructional design · Security

1 Introduction

In recent years, the interest in information and communication technologies (ICTs) in didactics has been gaining in popularity. Researchers and educators consider them as tools for improvement and enhancement of didactic endeavours. However, their application has to be purposeful and boosts the attractiveness of learning. ICTs as versatile didactic tools can help to motivate learners, especially in cases when subjects or titles of courses are seen as irrelevant or unattractive. The ongoing question is about motivation; how can students be motivated to learn specific issues, especially in a situation when they have to start studying side subjects? At some universities, the number of non-core subjects that can be freely chosen is limited. Sometimes, students have to choose subjects which are not in line with their primary interests. Therefore, the role of a teacher is to design a course which not only triggers students' curiosity and pushes them in new directions, but is also outstanding and led in an innovative way. It can be realized through the implementation of innovative didactic tools (ICTs) in a blended form of teaching. A blended course can rely on a specific approach to instructional design which assumes maintaining a balance between the four dimensions

M. E. Auer and D. May (Eds.): REV 2020, AISC 1231, pp. 782–794, 2021.
https://doi.org/10.1007/978-3-030-52575-0_64

of learning: cognitive, emotional, social, and psychomotor. In our research, we concentrated on boosting student motivation through the application of the blended learning form and using the four-dimensional instructional design approach. The motivation assessment will be described with three categories: e-learning component assessment (the opportunity to deepen knowledge on a specific issue, effective time management, and unlimited access to teaching materials); enjoyment of learning (friendly methods of tests, motivating content, attractive style); and attitude to blended learning forms (more classes can involve blended strategies, adequacy of using the blended form in non-core courses). We assume that a well-designed course may trigger students' interests from a particular area and build an interdisciplinary view on a specific discipline which is being studied [1]. We also think that the use of the four dimensional instructional design in a blended form of teaching will improve learning outcomes. The article can stand as a proposal on how to encourage students to learn new, unknown issues. It can be useful for teachers who would like to convince their learners to discover new topics, areas for studying.

2 Motivation in a Blended-Learning Environment

Blended learning addresses some of the difficulties of classroom instruction, such as large and diverse student cohort [2]; it offers efficiency, convenience and learning outcomes [3, 4]. Blending synchronous and asynchronous learning is also superior to online learning as it provides motivation through human interaction, diversity of learning activities and prompt feedback. In fact, motivation is a key contributor to the effectiveness of blended learning as it is a strong predictor of academic success. It determines students' engagement in the learning process and the effort they are willing to put in to attain the goal. Several studies have investigated how motivation contributes to blended learning outcomes. Technology provides more opportunities to construct meaning through interaction [5] enhancing active learning and creating a community of inquiry, i.e. the "context to conceptually and operationally define and operationalise metacognition in a socially shared environment" [6] in which members offer each other social and emotional support [7, 8]. Engagement in a peer community supports active learning; students become more engaged in learning and more creative [9, 10]. Furthermore, technology-supported classrooms enhance learning and develop higher order thinking skills [11, 12] and sharing ideas online helps to develop critical thinking and facilitates integration of ideas and construction of meaning [13]. López-Pérez, Pérez-López, & Rodríguez-Ariza [4] found the correlation between blended learning and objective outcomes: blended learning reduced dropout rates and improved exam marks. Similarly, Wentao, Jinyu and Zhonggen's [14] longitudinal study showed that the dropout rate during the study's 4-year period reduced and the pass rate significantly increased, for both men and women. On the other hand, Sugahara and Boland [15] reported a high rate of dropouts in blended learning, which exerted a negative effect on learning outcomes. A crucial factor that reduces the dropout rate is, on the one hand, creating a positive learning environment and increasing student satisfaction from the learning experience. On the other hand, research shows that the proportion of online and classroom instruction also correlates with students' academic success.

Horn and Staker [10] argue that technology-supported learning can adopt one of the four models of integrating online and face-to-face instruction: (1) students switch to different modes of learning; at least one of them is online or digital (rotational model); (2) students are mostly engaged in online learning, which is done at school under teacher supervision (flex model); (3) the entire course is online with an online instructor (a la carte model); (4) students firstly take one-on-one lessons with their teacher, next they proceed to the online component of the course that they complete independently. Thus, teachers can adopt one of the models to meet the course requirements; however, the models produce different academic outcomes. Owston and York [16] examined the correlation between the ratio of online to face-to-face learning on students' perceptions and performance. The study proves that students enrolled in the Medium (36% to 40% online) and High (50%) blends have the most positive perceptions of the course compared to learners in the Low (27% to 30% online) and Supplemental blends (100% classroom instruction enriched with online tutorial sessions). Furthermore, students attending the High and Medium blends significantly outperformed learners in the Low and Supplemental blends. Therefore, in order to make blended learning most beneficial, the online component should take up at least one-third of classroom instruction.

Stein and Graham [3] and Van Der Merwe [17] maintain that the online component boosts students' intrinsic motivation. A large body of studies point at motivation fluctuation in blended learning. Schober and Keller [18] found that motivation rises during certain periods but the overall level is lower than expected on a long term scale. Students tend to engage in interaction with their peers shorty before and shortly after the workshops and fall to a minimum between them. Smirnova and Katashev [19] note that students delay doing their independent assignments until right before the deadline. Therefore, they either do the assignment improperly or fail it, which prolongs the term of learning.

Schober and Keller [18] distinguished several determinants of motivation in blended learning: features and usability of the software, local parameters at school (such as Internet bandwidth, hardware and software facilities) and the general workload that the students had to cope with. Technical problems experienced during study time annoys students and result in faltering motivation because students expect high performance from a Learning Management System on a comparable level to professional platforms they use in their spare time. Therefore, improving the performance of software and eliminating technical obstacles might directly affect the greater motivation of students in blended courses, which will result in a lower dropout rate. Apart from the software, the teacher plays a crucial role in motivating students in blended learning environments. MacDonald [20] investigated the tutors' perceptions of their role. The researcher has compiled a list of tutor qualities, i.e. high level educational interventions that ensure quality in the learning process:

1. catering for the needs of students, i.e. boosting their confidence and developing a working relationship with individual learners;
2. being dialogic, i.e. tailoring to individual student needs and refining support offered to them;
3. helping students to focus on studying and harnessing distractions;

4. encouraging students to reflect on their contribution and offering them flexibility to accommodate learning into their work schedule;
5. focusing on timely and relevant issues;
6. reversible in offering feedback both to individual students and the whole group;
7. being accessible and providing support to all students regardless of their level of competence and technological limitations.

Tutor-mediated support plays a vital role as it develops learners' social and emotional skills that underpin effective learning. However, interactions with classmates also play a role as social learning in a blended course consists of the following interconnected components: instructor involvement, community cohesion, interaction intensity, affecting association, as well as knowledge and experience [21]. Technology-supported instruction can, in fact, stir various emotions that will enhance learning. The emotions experienced during study sessions, resulting mainly from random interactions with classmates or the teacher, help to engage students in meaningful learning [22].

The discussion so far focused on cognitive, social and emotional components of motivation as these components can activate, direct and sustain behaviour. Blended learning caters for all of them as technology-supported instruction can support building students' cognitive skills, can enrich interactions between a teacher and learners, and arouse emotions as well as intrinsic motivation. Finally, online instruction develops psychomotor skills to some extent, as learners may virtually manipulate content and explore objects or places. Even though many psychomotor skills which rely on the use of specific equipment cannot be practiced online (e.g. carpentry), online simulations may be introduced to develop skills at a distance [23].

3 The Review of Instructional Strategies and Approaches

Blended learning offers a diversified and integrated approach that accommodates students with different learning needs, styles and interests [24]. Teachers can apply a series of instructional strategies to pave the way for personalised and creative learning [25] and to overcome cross-cultural challenges [26] because even though learning strategies are culture-dependent [27], students, regardless of their culture, select strategies that help them achieve academic excellence [28]. Moreover, if implemented correctly, blended methodology can boost learning outcomes. It is then worth considering what factors contribute to the effectiveness of blended learning. First, learners engaged in technology-based instruction actively participate in their learning. Even though blended learning can be teacher- or content-centered, if technology-based courses focus on the learner and provide content that is relevant and linked with what is already known, students' performance can be boosted [12, 29, 30]. Active learning, which follows constructivist learning pedagogy [31], encourages learners to build their knowledge in a process of exploration, collaboration [32], negotiation of meaning and scaffolding [24]. Constructivist web-based instruction invites students to participate in critically reflective practices, encourages them to question prior knowledge and beliefs and practice problem-solving skills [33]. In addition, blended learning engages every student and fosters social interaction [6, 21]. Furthermore, learners are more autonomous

[34] and responsible for their learning [35] as blended methodology requires the use of self-regulated learning strategies and time management strategies [32].

Web-based instruction paves the way for flipped classroom strategies, which restructure the classroom [36] placing learners at the centre of their own learning [37]. When instructional content is assigned as homework, the class time can be devoted to problem-based instruction and collaborative learning [38]. The flipped classroom model provides accessible technology to design a course in economics [39], ICT Management course [31], General Health courses [35], Advanced Medical courses [40], English for Specific Purposes [41], and Military Training [42], among others. However, research shows than not all courses can be transformed according to the flipped classroom methodology. Strayer [43] found that students enrolled in an introductory statistic course were dissatisfied with their new pedagogical approach.

Another challenge of blended learning relates to students' being accustomed to traditional lecture-based instruction as most teaching and learning practice, especially in higher education, is carried out with transmissive rather than interactive strategies [29, 44]. Furthermore, since students are required to switch to different learning environments, they may feel confused about the learning objectives and have difficulty selecting and using an effective learning approach or strategy [29].

4 The Four Dimensional Instructional Design Approach

The starting point for considerations on instructional design is a theory of Danish scientist, Kund Illeris, who developed an approach to learning which integrates key educational theories [45]. In other words, *this theory has been constructed as a sort of umbrella, offering an overview and a structure of the landscape of learning which can be applied in both analysing and planning learning processes, both inside and outside of the educational system* [45].

According to Illeris, the process of learning should be understood in terms of three dimensions: cognitive, emotional, and social.

The cognitive dimension refers to knowledge and skills acquired by a learner and includes both knowledge and motor learning. These are also opinions, insight, meaning, attitudes, values, types of behaviour, methods, strategies, etc. [46]. Originally, Illeris uses two German words for the English term experience: Erlebnis (life experience) and Erfahrung (effected consciousness). The cognitive dimension of learning concerns Erfahrung, an event of understanding. It includes reflection, meta-learning and transformative learning [47]. All the content which is learnt builds the understanding and capacity of the learner.

The emotional dimension stresses the importance of psychic energy transmitted by feelings, emotions, attitudes, motivations, and volition [45, 48]. This is also known as the incentive dimension. It is vital to stress that incentives are always influenced by the content, e.g. new information can change the incentive condition [46].

According to Illeris, these two dimensions are always initiated by impulses from the interaction processes and integrated in the internal process of elaboration and acquisition [46].

The social dimension of learning emphasises that it is a social process taking place in the interaction between the individual and its surroundings [46, 48]. Social interactions, collaborative work, and participation in group work contribute to the description of the dimension.

The theory can be used in any instructional design activities as it provides a balance between three dimensions: the content (cognitive aspect), emotions (visual aspect), and social aspect (design of interactions). While scenario and learning activities development, different techniques and methods of learner's engagement (such as gamification, storytelling, etc.), an instructional designer can revise or confront the ideas with the learning theory. Such an approach to instructional design helps in achieving a balance between the scenario components, e.g. the content is varied (text, tabs containing engaging illustrations, videos), visual aids are relevant to the content and their number is not exceeded (fireworks effect is avoided), the number of tasks for students (collaborative work) is relevant, purposeful and does not overwhelm other types of tasks, which requires an individualised approach and self reflexivity. It can be noticed that instructional design also requires planning motor activities, which should be of different types to avoid boredom. In the theory of Illeris, a motor aspect of learning belongs to a cognitive domain. However, taking into account the specificity of activities performed in cyberspace, which requires operating a mouse, joystick, and intelligent accessories, a new dimension – a psychomotor one, can be distinguished [49, 50]. In this dimension, the activities such as "drag-and-drop", "click on the spot", "mark the object", "grab the object" should be considered with special attention. It is vital to stress that a psychomotor domain of learning was described by Bloom [51] as cognitive learning and affective learning [52]. As a result, the four dimensional approach to instructional design: cognitive, emotional, social, and psychomotor aspects, was proposed [50].

5 Research Methodology

The research aimed to prove that the four dimensional instructional design approach applied to a blended course keeps students motivated and engaged in the blended learning process. It encompassed the realisation of a blended course on Pedagogy (with the four dimensional instructional design) for students of the first year of the Faculty of Command and Management. The course was conducted at the War Studies University in Poland in the winter semester 2018/2019. The quantitative research methods were applied (analysis of learning outcomes and the results of structured interviews).

5.1 Research Procedure

The four dimensional design for a non-core blended course was proposed and agreed with the departments' council. 59 students enrolled the course (meanwhile other 65 students participated in a traditional, stationary course).

Upon the course completion, eight-item structured interviews and learning outcomes analysis were conducted. The interview questions fell into three categories: e-learning component assessment, enjoyment of learning, and attitude to blended learning form. The first category concerned only the e-learning part, while the second and third – the whole course.

E-learning component assessment included three statements.

1. E-learning module provided me with the opportunity to deepen specific knowledge.
2. Thanks to the e-learning component, I could effectively manage my time.
3. I had permanent access to materials and, therefore, I could learn as much as I needed.

Similarly to the previous category, the second, enjoyment of learning, also relied on three statements:

1. Assessment (online tests) was student-friendly.
2. The content of the b-course was motivating.
3. I found the b-course attractive in its form.

Attitude to blended learning form, the last category, involved two statements:

1. Since the subjects are not obligatory, the b-learning form is highly adequate.
2. In the future I would like to see more b-learning classes.

The data were collected from structured interviews (students participating in a blended course) and analysis of learning outcomes (comparison of students' results from blended-learning and traditional courses).

5.2 Description of the Pedagogy Course and Its Four Dimensional Instructional Design

The objectives of the course were to convey knowledge on social processes from a pedagogical perspective; to foster understanding of the interactive dimension of education and acting in social situations, and the challenges of modern education as well as the pedagogical aspects of the work situation. Students had to gain the ability to work in small social groups (project-based tasks), to work for their own educational and professional development, and also to explain the problems of the functioning of an adult person in the organisation. The structure of the course involved topics such as social processes, challenges of modern education, adult development and education, pedagogical aspects of the work situation (lectures); types, rules and stages of creating an educational project, interactive dimensions of education and activities in social situations, work in small social groups, adults in organisations, and project-based activities.

Regarding the four dimensional instructional design, the content, visual aspects, interactions, and psychomotor tasks were planned. The aim was to maintain a balance between all four dimensions of instructional design.

The cognitive aspect concerned the preparation of pills of knowledge – short, informative snippets of text. The content design was in line with general didactic principles, specifically, the principle of affordability in teaching (grading difficulties), activity, visibility, and accessibility [53]. It meant using hypermedia, hypertext solutions, bookmarks, and interactive graphics showing the content.

The emotional aspect relied on engaging activities for real-time meetings as well as organising online contests and games. In this dimension, a particular attention was paid to visual aspects of the course, appealing illustrations, and using discriminable elements.

Regarding the social dimension, project-based tasks were applied in both real and virtual parts of the course.

The last, psychomotor dimension, the applied tasks were related to "drag-and-drop" exercises, "marking the area", "typing phrases", and "marking an option". During the classroom meetings, the psychomotor aspect was restricted to typical behaviour presented in the classroom.

5.3 Participants

The participants who took part in the study were 59 freshmen students of aviation, logistics, and management (three majors offered by the Faculty of Command and Management). This was a heterogeneous group of full-time and part-time students. They mostly perceived Pedagogy as a minor and generic subject associated with coaching, mentoring, cooperation, problem solving, and negotiations skills. When asked about their past experience of participation in blended courses or e-learning, only a few students confirmed that they knew and took part in these forms of learning.

Their participation in the research involved students' full consent and was anonymous.

5.4 Limitations

The study was limited to a small group and, therefore, the study should be treated as preliminary. Moreover, the analysis of learning outcomes (blended learning course vs. traditional course) can be too general and may lead to bias in the assessment of the learning outcomes.

5.5 Research Results

Analysis of Learning Outcomes

The analysis of learning outcomes included the project-based tasks as well as test results achieved at the end of the semester. Both groups (receiving the blended learning course and the traditional one) had the same final tests. It was noticed that the results obtained by 59 students enrolled in the blended course (4.25) exceeded the average grade obtained by 65 students who participated in a full-time Pedagogy course (3.78) in the 2018/2019 academic year.

Structured Interviews

The perception of the blended learning course was investigated. On the five point Likert scale, learners had to mark the number which applied to the level of agreement with eight statements.

The statements fell into three categories: e-learning component assessment, enjoyment of learning, and attitude to blended learning form (Tables 1, 2, 3).

Table 1. E-learning component assessment

Statements	Agreement scale				
	1	2	3	4	5
E-learning component provided me with the opportunity to deepen specific knowledge	–	.02	.15	.34	.49
Thanks to the e-learning component, I could effectively manage my time	–	.05	.07	.05	.83
I had permanent access to materials; therefore, I could learn as much as I needed	.02	–	.13	.08	.76

As can be seen, the dominant value was 5 in all statements concerning the e-learning component assessment. Students appreciated the e-learning component mostly because of the possibility of time management and permanent access to didactic materials. During the interviews students admitted that they relied on deadlines to control their progress.

Table 2. Enjoyment of learning

Statements	Agreement scale				
	1	2	3	4	5
The assessment (online tests) was student-friendly	–	.02	.10	.08	.79
The content of the b-course was motivating	–	–	–	.20	.80
I found the b-course attractive in its form	–	–	–	.22	.78

Regarding the category "Enjoyment of learning", all statements (method of knowledge checking, motivating content course and attractiveness) were highly assessed.

Table 3. Attitude to a blended learning form

Statements	Agreement scale				
	1	2	3	4	5
Since the subjects are not obligatory, the b-learning form is highly adequate	–	.07	.22	.20	.51
In the future I would like to see more b-learning classes	–	.05	.12	.08	.74

In the last category, the students indicated that they would like to have more b-learning courses at the university. However, it seems that the fact that the course is non-core, is not a prerequisite for having it in a blended form.

6 Conclusions

Consistent with prior research [3, 9, 10, 17], the analysis of learning outcomes and structured interviews revealed that the four dimensional instructional design approach applied to a blended course keeps students motivated and engaged in the learning content. This conclusion was based on the analysis of three categories: the e-learning component assessment (students can deepen their knowledge on a specific issue, can effectively manage their time, have unlimited access to teaching materials); enjoyment of learning (friendly methods of tests, motivating content, attractive form); attitude to blended learning form (more classes can involve blended strategies). Furthermore, the e-learning component was assessed as very good. In the interviews, the students emphasised that the blended environment not only provided opportunities for more effective and enjoyable learning but also encouraged them to employ a variety of instructional and metacognitive strategies. The present findings confirm that as other researchers have reported [19], a structured blended course with its focus on deadlines and regular quizzes helps to keep students on track with their learning. It must be noted that the respondents indicated that the status of the course (core or non-core) did not determine their preference for its form (traditional or blended).

In line with previous studies [3, 4], these findings show that blended learning has the proven potential to improve students' outcomes. Therefore, our results suggest that university instructors should incorporate the four dimensional approach to instructional strategies to maximise learners' motivation and learning outcomes.

7 Discussion and Further Implications

Nowadays, in the digital era, maintaining university students' motivation to learn non-core courses has become a challenge. However, both blended approaches to instructional design and digital teaching aids can be effectively implemented to boost learners' motivation to gain new knowledge. Especially the knowledge on topics which aim to improve personal and social skills necessary to deal with contemporary challenges [54, 55] This study has proven the potential of using the e-learning component to design motivating non-core courses offered to students majoring in aviation, logistics and management. Furthermore, the findings show the online component with the inherent structure of regular assessment eliminates procrastination and helps students manage their learning. Thus, the findings fill in the gap in research on self-directed motivation among students, instructional strategies as well as innovations in education.

Further research may involve exploration of strategies employed by students in the blended courses, specifically time management and self-discipline. Further exploration of the four dimensional instructional design (4D ID) approach could be vital to prove its effectiveness.

Acknowledgement. Funding for the present work was supported by the Ministry of National Defense (Republic of Poland), Research Grant No. GB/4/2018/208/2018/DA (2018–2020).

References

1. Ivanitskaya, L., Clark, D., Montgomery, G., Primeau, R.: Interdisciplinary learning: process and outcomes. Innov. High. Educ. **27**, 95–111 (2002)
2. Garrison, D.R., Vaughan, N.D.: Blended Learning in Higher Education: Framework, Principles, and Guidelines. Jossey-Bass, San Francisco (2008)
3. Stein, J., Graham, C.R.: Essentials for Blended Learning: A Standards-Based Guide. Routledge, New York, London (2014)
4. López-Pérez, M.V., Pérez-López, M.C., Rodríguez-Ariza, L.: Blended learning in higher education: Students' perceptions and their relation to outcomes. Comput. Educ. **56**, 818–826 (2011). https://doi.org/10.1016/J.COMPEDU.2010.10.023
5. Domalewska, D.: Technology-supported classroom for collaborative learning: blogging in the foreign language classroom. Int. J. Educ. Develop. Inf. Commun. Technol. **10**(4), 21–30 (2014)
6. Garrison, D.R.: E-Learning in the 21st Century: A Community of Inquiry Framework for Research and Practice, 3rd edn. Routledge/Taylor and Francis, London (2017)
7. Lee, J., Bonk, C.J.: Social network analysis of peer relationships and online interactions in a blended class using blogs. Internet High. Educ. **28**, 35–44 (2016). https://doi.org/10.1016/J.IHEDUC.2015.09.001
8. Rourke, L., Anderson, T., Garrison, R.D., Archer, W.: Assessing social presence in asynchronous text-based computer conferencing. J. Distance Educ. **14**, 50–71 (2007)
9. Al-Zahrani, A.M.: From passive to active: the impact of the flipped classroom through social learning platforms on higher education students' creative thinking. Br. J. Educ. Technol. **46**, 1133–1148 (2015). https://doi.org/10.1111/bjet.12353
10. Horn, M.B., Staker, H.: Blended: Using Disruptive Innovation to Improve Schools. Jossey-Bass, San Francisco (2017)
11. Domalewska, D.: Blogs as means for promoting active learning: a case study of a Thai University. In: Smyczek, S., Matysiewicz, J. (eds.) New Media in Higher Education Market. University of Economics in Katowice, pp. 278–288 (2015)
12. Roehl, A., Reddy, S.L., Shannon, G.J.: The flipped classroom: an opportunity to engage millennial students through active learning strategies. J. Fam. Consum. Sci. **105**, 44–49 (2013)
13. Angelaina, S., Jimoyiannis, A.: Analysing students' engagement and learning presence in an educational blog community. EMI Educ. Media Int. **49**, 183–200 (2012). https://doi.org/10.1080/09523987.2012.738012
14. Wentao, C., Jinyu, Z., Zhonggen, Y.: Learning outcomes and affective factors of blended learning of english for library science. Int. J. Inf. Commun. Technol. Educ. **12**, 13–25 (2016). https://doi.org/10.4018/IJICTE.2016070102
15. Sugahara, S., Boland, G.: The effectiveness of powerpoint presentations in the accounting classroom. Account. Educ. **15**, 391–403 (2006). https://doi.org/10.1080/09639280601011099
16. Owston, R., York, D.N.: The nagging question when designing blended courses: does the proportion of time devoted to online activities matter? Internet High. Educ. **36**, 22–32 (2018). https://doi.org/10.1016/J.IHEDUC.2017.09.001
17. Van Der Merwe, A.: Using blended learning to boost motivation and performance in Introductory Economics modules. South African J. Econ. **75**, 125–135 (2007). https://doi.org/10.1111/j.1813-6982.2007.00109.x
18. Schober, A., Keller, L.: Impact factors for learner motivation in Blended Learning environments. Int. J. Emerging Technol. Learn. (iJET), **7**(Sp. Iss. 2: FNMA), 37–41 (2012)

19. Smirnova, G.I., Katashev, V.G.: A study module in the logical structure of cognitive process in the context of variable-based blended learning. Eur. J. Contemp. Educ. **4**, 102 (2017). https://doi.org/10.13187/ejced.2017.1.4
20. MacDonald, J.: Blended Learning and Online Tutoring: Planning Learner Support and Activity Design. Gower, Aldershot (2008)
21. Whiteside, A.L.: Introducing the social presence model to explore online and blended learning experiences. Online Learn. **19**, 2 (2015)
22. Parlangeli, O., Marchigiani, E., Guidi, S., Mesh, L.: Disentangled emotions in blended learning. Int. J. Hum. Factors Ergon. **1**, 41–57 (2012)
23. Zirkle, C., Fletcher, E.C.J.: Utilization of distance education in Career and Technical Education (CTE) Teacher Education. In: Wang, V.C.X. (ed.) Handbook of Research on e-Learning Applications for Career and Technical Education: Technologies for Vocational Training, pp. 1–13. Information Science Reference, Hershey, New York (2009)
24. Okaz, A.A.: Integrating blended learning in higher education. Procedia Soc. Behav. Sci. **186**, 600–603 (2015)
25. Nauman, S., Yun, Y., Sinnappan, S.: Emerging web technologies in higher education: a case of incorporating blogs, podcasts and social bookmarks in a web programming course based on students' learning styles and technology preferences. J. Educ. Technol. Soc. **12**, 98–109 (2009)
26. Parrish, P., Linder-VanBerschot, J.: Cultural dimensions of learning: addressing the challenges of multicultural instruction. Int. Rev. Res. Open Distrib. Learn. **11**, 1 (2010). https://doi.org/10.19173/irrodl.v11i2.809
27. Oxford, R.L.: Language Learning Strategies Around the World: Cross-cultural Perspectives. University of Hawaii, Honolulu (1996)
28. Domalewska, D.: Approaches to studying across culturally contrasting groups: implications for security education. Secur. Def. Q. **16**, 3–19 (2017)
29. Bonk, C.J., Graham, C.R.: The Handbook of Blended Learning: Global Perspectives, Local Designs. Pfeiffer, San Francisco (2005)
30. Koller, V., Harvey, S., Magnotta, M.: Technology-Based Learning Strategies (2006)
31. Al-Huneidi, A.M., Schreurs, J.: Constructivism based blended learning in higher education. Int. J. Emerg. Technol. Learn. **7**, 4–9 (2012). https://doi.org/10.3991/ijet.v7i1.1792
32. Broadbent, J.: Comparing online and blended learner's self-regulated learning strategies and academic performance. Internet High. Educ. **33**, 24–32 (2017). https://doi.org/10.1016/J.IHEDUC.2017.01.004
33. Cooner, T.S.: Dialectical constructivism: reflections on creating a web-mediated enquiry-based learning environment. Soc. Work Educ. **24**, 375–390 (2005). https://doi.org/10.1080/02615470500096902
34. Snodin, N.S.: The effects of blended learning with a CMS on the development of autonomous learning: a case study of different degrees of autonomy achieved by individual learners. Comput. Educ. **61**, 209–216 (2013). https://doi.org/10.1016/J.COMPEDU.2012.10.004
35. Melton, B.F., Graf, H., Chopak-Foss, J.: Achievement and satisfaction in blended learning versus traditional general health course designs. Int. J. Scholarsh. Teach. Learn. **3**, 1–13 (2009). https://doi.org/10.20429/ijsotl.2009.030126
36. Strayer, J.: The effects of the classroom flip on the learning environment: a comparison of learning activity in a traditional classroom and a flip classroom that used an intelligent tutoring system. Dissertation. The Ohio State University (2007)
37. Bergmann, J., Sams, A.: Flip Your Classroom: Reach Every Student in Every Class Every Day. International Society for Technology in Education, Alexandria (2012)
38. Tucker, B.: The flipped classroom. Educ. Next **12**, 82–83 (2012)

39. Lage, M.J., Platt, G.J., Treglia, M.: Inverting the classroom: a gateway to creating an inclusive learning environment. J. Econ. Educ. **31**, 30 (2000). https://doi.org/10.2307/1183338
40. Sonesson, L., Boffard, K., Lundberg, L., et al.: The potential of blended learning in education and training for advanced civilian and military trauma care. Injury **49**, 93–96 (2018). https://doi.org/10.1016/J.INJURY.2017.11.003
41. Whittaker, C.: A military blend. In: Tomlinson, B., Whittaker, C. (eds.) Blended Learning in English Language Teaching: Course Design and Implementation, pp. 175–183. British Council, London (2013)
42. Tyler, K.M, Dolasky, K.C.: Educating warrior diplomats. Blended and unconventional learning for special operations forces. In: Blended Learning: Research Perspectives, pp. 235–248. Routledge, New York, London (2014)
43. Strayer, J.F.: How learning in an inverted classroom influences cooperation, innovation and task orientation. Learn. Environ. Res. **15**, 171–193 (2012). https://doi.org/10.1007/s10984-012-9108-4
44. Marmah, A.A.: Students' perception about the lecture as a method of teaching in tertiary institutions. Views of students from College of Technology Education, Kumasi (COLTEK). Int. J. Educ. Res. **2**, 601–612 (2014)
45. Illeris, K.: Towards a contemporary and comprehensive theory of learning. Int. J. Lifelong Educ. **22**, 396–406 (2003)
46. Illeris, K.: A comprehensive understanding of human learning. Contemporary Theories of Learning, pp. 1–14. Routledge, New York, London (2018)
47. Poscente, K.: The Three Dimensions of Learning: Contemporary Learning Theory in the Tension Field between the Cognitive, the Emotional and the Social (Author: Illeris, K.). Int. Rev. Res. Open. Distrib. Learn. **7** (2006). https://doi.org/10.19173/irrodl.v7i1.305
48. Illeris, K.: What do we actually mean by experiential learning? Hum. Resour. Dev. Rev. **6**, 84–95 (2007)
49. Brown, D., Bell, B., Goldberg, B.: Authoring adaptive tutors for simulations in psychomotor skills domains. In: Proceedings of MODSIM World 2017. NTSA, Virginia Beach, VA (2017)
50. Gawlik-Kobylińska, M.: The four dimensional instructional design in the perspective of human-computer interactions. In: Petkov, N., Strisciuglio, N., Travieso-González, C.M. (eds.) Applications of Intelligent Systems: Proceedings of the 1st International APPIS Conference 2018, vol. 310. IOS Press, Amsterdam, Berlin, Washington DC (2018)
51. Bloom, B.S. (ed.): Taxonomy of Educational Objectives. Cognitive Domain, vol. 1. David McKay, New York (1956)
52. Rovai, A.P., Wighting, M.J., Baker, J.D., Grooms, L.D.: Development of an instrument to measure perceived cognitive, affective, and psychomotor learning in traditional and virtual classroom higher education settings. Internet High. Educ. **12**, 7–13 (2009). https://doi.org/10.1016/J.IHEDUC.2008.10.002
53. Kupisiewicz, C.: Dydaktyka. Oficyna Wydawnicza Impuls, Kraków (2012)
54. Świerszcz, K., Bożejewicz, W., Jędrzejko, M.: Człowiek w ponowoczesności - postęp czy zagrożenie? In: Jędrzejko, M., Malinowski, J.A. (eds.) Młode pokolenie w zderzeniu cywilizacyjnym. Studia – badania – praktyka. Toruń: AKAPIT (2014)
55. Świerszcz, K., Bożejewicz, W., Jędrzejko, M.: "Inżynieria społeczna" człowieka w epoce postmodernizmu i jej implikacje. In: Jędrzejko, M. (ed.) "Zwariowany" świat ponowoczesności. ASPRA-JR, Warszawa-Milanówek (2015)

Model for Educational Free Software Integration into Artificial Intelligence Teaching and Learning

Yuniesky Coca Bergolla(✉) and María Teresa Pérez Pino

University of Informatics Science, Havana, Cuba
{ycoca, mariatpp}@uci.cu

Abstract. Artificial Intelligence is a branch of Computer Science with complex subjects. A trend in the teaching and learning of Artificial Intelligence is to use software where students modify the source code. However, teachers do not use the freedoms of free software in their classrooms. Based on a previous review, it is difficult to see methodological indications to use or create new programs with these characteristics for other teaching-learning processes. The goal of this work is to develop a model of integration of educational free software to the teaching-learning process of Artificial Intelligence. The authors use several research methods: historical-logical, analysis and synthesis, functional-structural-systemic and modeling. The model has three fundamental components. The conceptual component represents the principles and their theoretical foundations. The structural component represents the main elements that intervene in the integration. The instrumental component constitutes the materialization of the model and contains three stages: preparation, execution and evaluation. The authors used the focus group technique, a pre-experiment and Iadov's satisfaction technique to validate the model. These methods made it possible to verify the relevance of the model, its good acceptance by the teachers and the acceptable satisfaction of the students after the first application of the model.

Keywords: Artificial Intelligence · Educational free software · Teaching and learning

1 Introduction

Artificial Intelligence (AI) is the branch of Computer Science responsible for developing systems with rational behavior [1]. The goal of AI is to solve problems for which there is no algorithm, or this is computationally intractable [2]. For this reason, it addresses advanced elements of computer programming.

Since the 90s, several articles addressed the use of computer applications as teaching-learning means where students program the algorithms and techniques discussed in class [3–6]. This type of software is visually attractive and has methodological elements that guide teachers and students for program the source code. Those software characteristics allow following trends in education technologies [7] such as blended learning, mobile learning and makerspace.

M. E. Auer and D. May (Eds.): REV 2020, AISC 1231, pp. 795–810, 2021.
https://doi.org/10.1007/978-3-030-52575-0_65

These applications allow the source code modification. Scientifics know the benefits of the free software movement, in terms of technological sovereignty and free distribution. The Free Software Foundation [8] defends educational free software as the software based on the freedoms of free software that uses this featuring for some educational purpose.

The AI diversity of topics and rapid advance [9] implies a constant updating of each didactic components. The teaching-learning means need update systematically, for which the teachers must be able to create, modify and update them. The traditional educational software uses the interface to interact with the students; however, educational free software adds the source code to interact with the student. This characteristic allows the student to analyze, modify or reuse the software source code in the solution of teaching tasks. The professors need theoretical knowledge and methodological elements to be able to select, update, build or use these means according to the characteristics of their classes.

Several related to educational technologies address the creation of models or conceptions [10–12] and strategies or methodologies [13–15]. These researches discourse about the creation and management of educational resources, the teacher improvement and the design of educational software. The analysis of these investigations allowed identifying some elements of interest for the use of educational free software in Artificial Intelligence subjects:

The use of educational software, where students modify the source code from the teaching-learning process, provides advantages to the teaching-learning process.

Lack of a specific pedagogical treatment of free software.

Few methodological indications to use or create new software with these characteristics for specific teaching-learning processes.

Gaps in the treatment of the particularities of educational software to teaching-learning process of Artificial Intelligence.

Insufficient treatment of the relationships between educational software and the context where the teaching-learning process takes place.

Lack of a specific treatment to educational software based on free software and the use of their freedoms in a teaching-learning process, specifically Artificial Intelligence.

The researches approach the development of educational software independently of its use in a specific teaching-learning process.

The study shows the need for the development and use of educational software that takes advantage of the freedoms of free software in the teaching-learning process of Artificial Intelligence; however, there are theoretical and methodological shortcomings for the integration of educational free software to the teaching-learning process of Artificial Intelligence. This work aims to present a model of integration of educational free software to the teaching-learning process of Artificial Intelligence.

To carry out the research, the authors proceeded to analyze bibliographic references on the teaching of Artificial Intelligence, on educational software, free software and the integration of technologies to teaching-learning processes. Then they synthesized the essential elements to obtain an operational definition of the integration of educational free software to the teaching-learning process of Artificial Intelligence.

The model design took into account the indicators for the integration of educational free software to the teaching-learning process of Artificial Intelligence. The model includes a conceptual component that supports theoretically each of defined indicators, includes a structural component that addresses each essential element of integration and its fundamental relationships, and it include an instrumental component that guides methodologically to contribute to the improvement of defined indicators.

Finally, theoretical and empirical methods and techniques were applied to obtain valuations on the model. We applied the technique of focus groups and applied a pre-experiment in the Faculty 4 of the University of Computer Science. The pre-experiment evaluated the initial state of integration, following the specific stages and actions of the instrumental component. In addition, the students showed satisfaction with the software when applying Iadov's technique.

2 Related Works and Definitions

The study of Artificial Intelligence (AI) prepares the student to determine when an AI approach is appropriate for a given problem, identify the appropriate representation and reasoning mechanism, and implement and evaluate it [9].

Artificial Intelligence appears since the early 90s of the last century, in computer related careers curricula. The development of these curricula was vertiginous in the first years [16]. The Computing Curricula 2001 [17] contributed to consolidate a basic number of contents in AI, which has been refined over the years with the development of computer science.

An AI teachers group surveyed professors and AI researchers to contrast they propose to include in classes [18]. The study highlights the desire of researchers to achieve in students skills in systems engineering, this element did not appear in the responses of educators. The educators consider it important to treat the issues from a vision of games, while researchers would like to expose students to systems with a strong engineering component. There are already examples of courses that try to bring real world problems of high complexity.

The consulted bibliography reflects several trends in the use of software for the teaching of AI (Table 1).

The analysis of these references allowed ratifying the following problematic:

The use of software not designed to teach Artificial Intelligence, limits the possibilities of working as an integrated system.

The tasks to the students are generally limited to adding code, but assigning tasks to study, modify or reuse the source code that has already impregnated the software.

The assignment of tasks to solve by the students with the software, not taken into account the pedagogical context.

Table 1. Kinds of software used in the teaching-learning of Artificial Intelligence.

Kinds of software	Distinctive features
For algorithm visualization [19, 20]	The software visualize the algorithms to facilitate students understanding. Some software allows interaction with the visualization to modify it. This software has closed the sourcecode
For the algorithms implementation in simulators or graphic visualization systems	Teachers guide students in the development of algorithms in an open source software. This software show the results in a pleasant way
For develop virtual players	The game is already developed and students implement the intelligence of virtual players. It allows interaction through competition among students. There can be open source or not
For the implementation of games or parts of them	Students implement the algorithms and techniques to give intelligence of video games or games partially implemented on open source platforms

2.1 Educational Free Software Integration to a Specific Teaching-Learning Process

Educational software appears in the bibliography from two different perspectives. One of them covers computer applications designed to achieve various purposes in the educational field; second is limited to applications used in the context of the teaching-learning process. This perspective also has divergent positions. Some authors consider educational software as any computer program used in teaching learning process. Other authors ponder educational software as any computer application designed with a specific educational purpose and assumed as mean in a pedagogical process [11].

The movement of "open knowledge" has great reach and impact in various spheres of society. This movement is present in education as part of educational technology. International organizations such as UNESCO have contributed significantly to its expansion. This organization introduced and defended the term Open Educational Resource (OER).

Several investigations promote the creation of OER in institutions [10, 21]. However, the creation of OER has been limited to the construction of materials that make up online courses or learning objects. The creation of educational software as open educational resource has been poor theoretical treatment.

The term "open source" derive from "open knowledge" concept. It refers to the computer programs that allow access to the source code.

The term 'open source' reached notoriety since 1983 under the name of Free Software Movement [8], establishing four freedoms:

- The freedom to run the program as you wish, for any purpose (freedom 0).
- The freedom to study how the program works, and change it so it does your computing as you wish (freedom 1).
- The freedom to redistribute copies so you can help others (freedom 2).
- The freedom to distribute copies of your modified versions to others (freedom 3).

The Free Software Foundation establishes the term educational free software for some software used by educational institutions. This type of software puts into practice and effectively apply the freedoms that free software grants. We assume educational free software as any computer application designed with the intention of making the most of the freedoms of the free software movement in a teaching-learning process.

One of the most important theoretical positions in educative technology is the Technology Integration in Education (TIE) [22]. The distinctive element is the transformation of the teaching learning process to include TIE.

The TIE is a contextualized, systemic, continuous, planned and reflective process, oriented to the pedagogical practice transformation, with the purpose of harmoniously incorporating technologies of information and communications to satisfy educational goals [23].

We consider the educational free software integration (EFSI) to a specific teaching-learning process as a contextualized, systemic, continuous, reflective and planned process, with the purpose of harmoniously incorporate computer applications, designed with the intention of making the most of the freedoms of the free software movement, for a teaching-learning process transformation.

3 Educational Free Software Integration Model

In order to make effective the EFSI, we modeled it. We assume the following general methodology:

1. Define the educational free software integration.
2. Determine dimensions, subdimensions and indicator to evaluate the model (Table 2).
3. Model the process: First, we analyzed the theoretical foundations to define a conceptual component. This component contains the model's principles. Second, we determined the elements that conform the integration and we analyzed its relationship to represent the structural component. Finally, we proposed actions to contribute at each indicators defined in the variable operationalization. Stages and phases delimited these actions to conform the model instrumental component (Fig. 1).
4. Obtain criteria about the model to improve it.
5. Apply the model.

Table 2. Dimensions, subdimensions and count of indicators

Dimension	Subdimension	Count of indicators
Educational free software didactic design	Didactic	6
	Technologic	5
	Spatial	3
	Management	2
Didactic-technologic relation	Contextualized	2
	Systemic	2
	Continued	3
	Reflexive	3
	Planned	6

Three general components conform the model:

Conceptual Component: It represents the principles that govern the model and its theoretical foundations.

Structural Component: The conceptual component determines this component. It represents the main elements that intervene in the integration of educational free software to the Artificial Intelligence teaching-learning process and its fundamental relationships.

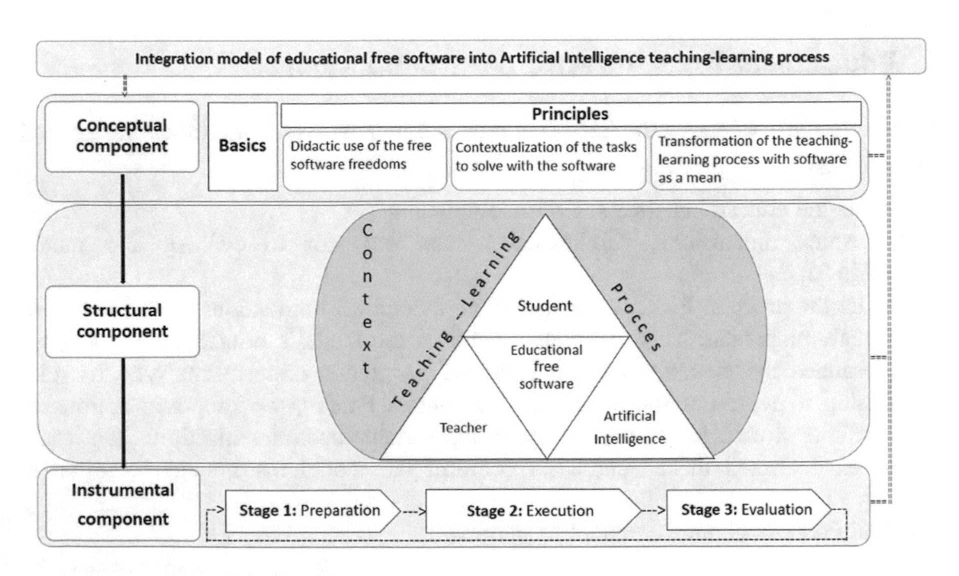

Fig. 1. Integration model of educational free software

Instrumental Component: It constitutes the materialization of the structural component. A didactic strategy establishes this component and it is expressed in the following stages:

- Stage I: Preparation (Phase I: Diagnosis, Phase II: Overcoming and Phase III: Planning)
- Stage II: Execution (Phase I: Didactic design and Phase II: Use of the SLE)
- Stage III: Evaluation

3.1 Conceptual Component

Dialectical materialist approach support the model components. We assume the investigation and therefore the model starts from an objective reality, this reality arrives to the mental plane and it is materialized transforming the reality. We consider real and objective the relationship between theory and practice. We assume the internal interactions and the causal interdependence between the different components of the research object. In addition, we analyze the internal contradictions of each component and between them, contributing to its dialectical transformation.

Basics

The model assumes the Leninist theory of knowledge. The dialectical path of knowledge of truth transits from living contemplation to abstract thought and from thought to practice. The proposal starts from the observation of the teaching-learning process to conform a theoretical model allowing the study of objective reality and contributes to its transformation in practice.

From the sociological point of view, we assume the education must respond to the needs and interests of the environment and students. The active role of the teacher and students in the process must be contribute to the socialization of information in order to satisfice the society demands. The school constitutes an organized social institution capable of perpetuating, reproducing and transmitting ideology and culture. This institution has the purposes of achieving autonomous individuals, capable of thinking and acting for themselves. We defend the problems resolution in the pedagogical context, through activities with the software. We paid special attention to the technological or instrumental mediation of software.

The model assume the historical-cultural approach from the Vygotsky and followers conceptions. Learning is a social process expressed in the context in which individuals perform [10]. The concepts “zone of proximal development” and “development of concepts” [24] support a conception of the integration of AI favoring the development of students, from their active participation in the educational free software transformation to solve contextualized professional problems.

The pedagogy is a science having as essential features the social projection, the humanistic orientation and a transformer character. The relationship between education and instruction, forming a dialectical unit. All pedagogical processes should materialize this relation, with an organized manner and with an appropriate direction.

The didactic is a practical theory where its main theoretical realization is the development of the principles about the contextualization and intercommunication between different theories. Contextualization is a logical development process, it allow locating in concrete relevant situations in the subject, in a branch of curriculum and in society. It is use as motivational framework and thematic conductor for the presentation, development and evaluation of the contents, for learning purposes [25].

We assume technology as a social practice that involves knowledge, skills, organizational problems, values and ideologies. Science and technology are social processes that should contribute to the development of society. Educational software is a mean within a teaching-learning process increasingly influenced by new technologies. We recognize and support the possibility of introducing current trends in technology such as blended learning, mobile learning, makerspace. The EFS is a mean useful in online courses to assign students task solving in personal computers, laptops mobiles or other student devices. The EFS flexibility and freedom help to develop makerspace, we assume the EFS as a virtual makerspace.

The free software movement is a liberating and social technology conception. It defends the knowledge socialization with shared benefits, playing a role not only instructive but also educational and to the value formation.

From the legal point of view, the licenses created for the free software development support the software created as part as the model. The copy left and creative commons licenses offer the possibility of studying, modifying and reusing the software.

Principle

A principle is a guiding idea, a fundamental conduct rule. The didactic principles have a transformative function; they determine the content, methods, procedures, organizations forms and evaluation, having as a guiding category the objectives [26]. We assume these didactic principles based on the historical-cultural approach. We propose three principles to the educational free software integration model to the Artificial Intelligence teaching-learning process.

- Didactic use of the free software freedoms

We start at the needed to develop a computer application under the terms of free software. The principle proposes the exploitation within the teaching-learning process of the freedoms that free software movement defends. We suggest the assignment of activities to students, related to the study, modification and reuse of the source code of the educational free software.

Within the teaching-learning process, the freedom to study the code allows to use parts of it as example in conferences and perform written tasks in practical classes. The freedom to modify allows assigning familiarization tasks in laboratory classes. The reuse can be support extra class tasks of greater complexity. The teachers should be use the freedom of distribution of the improvements and modifications in the exchange of results among the students. This contributes to the development of values of solidarity, as the student provides the results of their work for the benefit of their peers.

- Contextualization of the tasks

The historical-cultural approach support the contextualization. The students would be work in activities with educational free software related with the subject or a branch of curriculum or with the research line where they receive professional practice. Also, we suggest take into account the students interests.

The subject limit the contents and specifics skills, however, the professional practice mounts the scope of the students tasks. The professional practice gets the professional problems to work with the educational free software. The diagnosis facilitates the necessary characterization of the students for a personalized software tasks assignment taking into account the context.

- Transformation of the teaching-learning process

The purpose of integration is to transform the teaching-learning process. The software for the present investigation is a teaching-learning mean. The mean is a didactics category and is a facilitator of the process. It is not going to replace the teacher, but it is for help in the appropriation of the content, complementing the method, for the achievement of the objectives. In this research, the means increases its importance, by enhancing the ideas of developer learning, enhancing the student's work to acquire knowledge and develop skills.

The integration process nourishes the teaching-learning process; achieving a dialectic transformation. The teaching-learning process maintain its essential characteristics but incorporating software as a mean to transform each of the didactic components, including the entire means system.

3.2 Structural Component

We direct the attention to the student and teacher personal components within the teaching-learning process. We assume the educational free software as a mean and as center of the model. However, we analyze all the non-personal components, goals, contents, methods, means, organization forms and evaluation. Additionally, we proposed the context as an essential element inside the model.

Essential Elements

Student: The students are the main element; the final goal of the teaching-learning process is to contribute to their formation. The teacher need to know the fundamental characteristics of students to face the process, both their personal interests with respect to the subject and their connection to the professional practice. We suggest applying a diagnosis, based on the cultural historical approach, to provide the interest elements.

Teacher: The teachers play a decisive role; they design the software to the teaching-learning process. They must have a good preparation to promote conditions and create learning situations in which the student takes ownership of knowledge and skills that enable a responsible and creative action. They require knowledge of the programming languages and paradigms, as well as software engineering and educational technology. The teachers drive the process, assuming a flexible position to exploit each student develop area.

Subject Artificial Intelligence: Artificial intelligence is the branch of computer science responsible for applying representation, processing and extraction of knowledge methods, through multi-paradigm programming, in the development of computer systems with rational behavior. In the curricula, AI is a subject, a branch of curriculum or group of assignments within a branch. It includes elements of applied and discrete mathematics. In all the analyzed cases, one subject includes the essential elements of Artificial Intelligence.

The subject defines the specific objectives to the student. This objective derive from the objectives of the professional model of the career and the objectives of the academic year. The definition of artificial intelligence frame the contents and objectives determine it. We suggest active methods that allow working with educational free software. However, we consider all possible methods to work with the software, including the expository in the presentation of contents.

Software is the main mean to use, however it needs other means such as the computer. We propose a system of traditional means and new technologies, such as mobile telephones or learning platforms to enhance work with educational free software. The use of this should be not only in laboratory classes or in independent work. The study of the code is included, even in lecture classes or other types of classes.

Finally, we consider the evaluation during all the process must be integrative and take into account the development of the student.

Context: The context refers to the student characteristics, includes the characteristics of the center or research group where he performs his professional practice. In addition, other subjects that propose tasks and jobs including features that propitiate solutions by using Artificial Intelligence methods.

Educational Free Software: Educational free software is a computer application, designed with the intention of taking advantage of the freedoms of the free software movement, in a specific teaching-learning process transformation. From this definition, we consider necessary a didactic design. In this design, one dimension is the didactic, responsible for specifying the relationship between software and each component of the didactics. The spatial dimension determines the spaces where educational free software will work. The management dimension ensures the conditions in the use of software, whether technological or distribution conditions, including software licenses. Finally, the technological dimension specifies the elements of software engineering for the application development.

We assume elements for the development of educational software. As well as some specific elements for free software:

- Developed under license according to the movement of free software.
- Modular and easily extensible architecture.
- Developed from a programming multiparadigm vision, with readable code.

3.3 Instrumental Component

The instrumental component allows materialize the model. Three general stages conform this component.

Stage I: Preparation

This stage prepares the conditions to design and use the educational free software in the Artificial Intelligence teaching-learning process. Newly three phases compose the stage.

- **Phase I:** Diagnosis

This stage allows diagnosing the integration of educational free software to the specific teaching-learning process. We determine two dimensions and several indicators for the research. We define two specifics task, in relation with each dimensions:

- Diagnosis of the didactic design level of the software used in the subject. We propose 16 indicators into 4 sub-dimensions, related to didactic, technological elements, pedagogical spaces and the software management.
- Diagnosis of the relationship level between the software and the Artificial Intelligence teaching-learning process. We propose 16 indicators into 5 sub-dimensions, contextualized, systemic, continuous, reflective and planned.

- **Phase II:** Upgrade

We propose a phase to upgrade the teachers' skills in educational free software integration. This phase include:

- Organize the necessary improvement activities.
- Execute the upgrading activities.

- **Phase III:** Planning

Planning plays an important role in to materialize the integration. The results depend on the complexity of the design and development of the educational free software. This phase may involve more than one academic year. The specific actions are:

- Student diagnosis preparation.
- Calendar plan adequacy.
- Planning of the didactic design of the software.
- Planning feedback of the process.

Stage II: Execution
The execution phase consists of two phases: the didactic design of the software and its use on the Artificial Intelligence teaching-learning process.

- **Phase I.** Didactic software design

This phase defines concrete actions from the sub-dimensions defined for the study of the integration of educational free software to a teaching-learning process. The specific actions are:

- Describe the relationships between educational free software and each didactics components.
- Determine kinds of activities in the different available pedagogical spaces.
- Evaluate the conditions to distribute and use educational free software in the teaching-learning process.
- Analysis, design and computational development of educational free software.

- **Phase II.** Use of educational free software in the Artificial Intelligence teaching-learning process

- Initial diagnosis to students.
- Presentation of educational free software.
- Assignment the tasks to be performed with the software.
- Monitoring the tasks assigned to students
- Final evaluation.

Stage III: Evaluation

When teaching learning process finish, the teacher must executes actions to renovate the process and the model.

- Analysis of student results.
- Evaluate if necessary software redesign.
- Analysis of the process results.
- Feedback of the model.

4 Results and Evaluation

We execute three general actions to evaluate the model. Each of them providing specific elements that complement the evaluations of the model:

- Valuation by focus groups.
- Model's application.
  - Preexperiment.
  - Iadov satisfaction technique.
- Methodological triangulation.

We conform two focus groups to the following general composition:

G1 Methodological: This group was conforming by 11 teachers from Programming Techniques Department of Faculty 4. The aims was evaluate the applicability of the models in a specific faculty.

G2 Specific Methodological: This group was conforming by 8 teachers from Artificial Intelligence collective in the university. The aims was obtain the criteria about the possible benefits of the model to Artificial Intelligence teaching-learning process.

We define a guide of subjects and a set of operational criteria for a better development of the workshops.

The results of the workshops was positive. The groups said the model benefit the artificial intelligence teaching-learning process and they consider the model is applicable in the current institution conditions. Both groups expressed ideas to improve the model and put it into practice.

To apply the model we used a pre-test/post-test preexperiment design. We apply the model from August 2017 to February 2019 in the Faculty 4 of the University of Informatics Sciences. The preexperiment involves as units of analysis, software, teachers, students and subjects. We followed the stages of the model's instrumental component to carry up the application.

We determined as objectives to work with the software:

- Represent knowledge about a specific domain to develop intelligent systems.
- Apply techniques to deal with the uncertainty present in knowledge datasets.
- Apply supervised and unsupervised learning algorithms to solve data mining tasks.

The specific contents to work with the software will be:

- For representation of knowledge
  - Logic (Rules in Prolog)
  - Cases (Facts in Prolog)
- For uncertainty treatment
  - Probabilistic analysis (Naive Bayes)
- For data mining:
  - Classification algorithms available in weka (Software platform developed in Java for data mining).
  - Clustering algorithms available in weka.

The analysis of the technological conditions and the preliminary requirements of the software led to proposal two software. The software called Logical Agent [27] incorporate the knowledge representation and uncertainty treatment with the use of the swi-prolog. The other software called SmekDB include machine learning with the use of weka as algorithms library.

The teachers used both software on 4505 group and 4503 group, in the first semester of the 2018–2019 academic year.

The teachers applied the diagnosis to students, they present the software in the classroom and they assign initial tasks for software modification.

Some examples of assigned tasks related to studying the code:

- Analyze how the software constructs the rules from the visual interface.
- Analyze how the software performs the treatment of the uncertainty in case-based reasoning.
- Analyze how educational free software makes the call to the classification and grouping methods implemented in weka library.

Some examples of assigned tasks related to modifying the source code:

- Add a new rule in the Prolog code, recompile and analyze the results from the interface.
- Apply a new method of classification of those incorporated by weka, with their respective configurations, from the source code.

For the code reuse, the teachers assigned a specific task to each student based on initial diagnosis.

Some examples of assigned code reuse tasks:

- Develop an application that keeps a dataset about music and allows consult that knowledge from a java interface.

- Develop an application that reads certain information from a dataset and applies some of the data mining algorithms that the weka library includes.

In order to follow up on the final task assigned to the students, the teachers used laboratory classes.

To corroborate the results, we applied the same instruments used at the beginning of the investigation about defined dimensions. The results reflects in all the indicators more favorable results than at the beginning of the pre-experiment. This method offers two important signs, specifically to improve the actions in the execution stage and to carry out a redesign to obtain a single integrated software.

At the end of the course, we applied an Iadov Satisfaction Technic to the 23 students of the 4505 group. The questioner contained ten questions. Fifteen students obtain the best index of satisfaction (1), six obtain value two (2) and three obtain value three (3). We calculate the group satisfaction index (GSI) by the following formula:

$$\mathrm{GSI} = \frac{A(1) + B(0,5) + C(0) + D(-0,5) + E(-1)}{N}$$

The value of GSI was 0.78, showing good satisfaction. The open questions allowed enrich the criteria about the process. The most important sign was the few time of students to solve the reuse source code task. That is why we propose integral design of the branch in the curriculum including the educational free software integration as part of the program.

Finally, we proceeded to a methodological triangulation to establish the divergences and coincidences between the theoretical and empiric techniques.

The focus group technique provided theoretical and methodological elements. This technique revealed the theoretical feasibility of the model, its components and relationships, as well as the effective possibility to implement it. Also exposed the relevance and coherence of the model.

The application of the model allows get criteria to improvement the educational free software integration to the Artificial Intelligence teaching-learning process. In addition, it allows demonstrate the applicability of the model.

The Iadov technique verified the criteria of students, the most important element in the model. Theirs criteria are significant to improvement the model.

The results of each technique get us elements to estimate as positive the model.

5 Conclusions

The current development of educational technology requires flexible means adapting to new learning conditions. Free software and the use of its freedoms can contribute to the development of trends such as blended learning, mobile learning and makerspace.

The use of the freedoms of free software is necessary in the teaching-learning process of Artificial Intelligence. That is why it is necessary a correct didactic design of the software used and planning a set of actions that allow organizing the relationships between educational free software and the teaching-learning process.

The model presents three fundamental components. The conceptual component represents the principles and its theoretical foundations. The structural component represents the main elements that intervene in the integration and its fundamental relationships. Finally, the instrumental component constitutes the materialization of the model and contain three stages: preparation, execution and evaluation.

Pedagogical theory support the model; the model application in a faculty of the University of Informatics Science showed favorable preliminary results, demonstrating the effectiveness of the model.

Rcferences

1. Russell, S., Norvig, P.: Artificial Intelligence. A Modern Approach, 3rd edn. Prentice Hall, Upper Saddle River (2010)
2. Bello, R., García, Z.Z., García, M.M., Reynoso, A.: Aplicaciones de la Inteligencia Artificial. Universidad de Guadalajara, Jalisco (2002)
3. Sosnowski, S., Ernsberger, T., Cao, F., Ray, S.: SEPIA: a scalable game environment for artificial intelligence teaching and research. In: AAAI Symposium Educational Advances in Artificial Intelligence (2013)
4. Mark, O.R.: A Python engine for teaching artificial intelligence in games. arXiv e-prints arXiv:1511.07714 (2015)
5. Renz, J.: AIBIRDS: the angry birds artificial intelligence competition. In: AAAI Conference on Artificial Intelligence (2015)
6. Veliz, O., Gutierrez, M., Kiekintveld, C.: Teaching automated strategic reasoning using capstone tournaments. In: Proceedings of the Sixth Symposium on Educational Advances in Artificial Intelligence, Phoenix (2016)
7. EDUCAUSE: NMC Horizon Report Preview. 2018 Higher Education Edition, Horizon Project (2018)
8. FSF Homepage. https://www.fsf.org/. Accessed 11 May 2019
9. ACM-IEEE: Computer Science Curricula 2013. Curriculum Guidelines for Undergraduate Degree Programs in Computer Science (2013)
10. Cabrera, J.F.: Modelo de Centro Virtual de Recursos para contribuir a la integración de las TIC en el Proceso de Enseñanza Aprendizaje en el Instituto Superior Politécnico José Antonio Echeverría, La Habana (2008)
11. Rodríguez, L.: Concepción didáctica del software educativo como instrumento mediador para un aprendizaje desarrollador, La Habana (2010)
12. Area-Moreira, M., Hernández-Rivero, V., Sosa-Alonso, J.J.: Models of educational integration of ICTs in the classroom. Comunicar **24**(47), 79–87 (2016)
13. Lombillo, I., Valera, O., Rodríguez, I.: Estrategia metodológica para la integración de las TIC como medio de enseñanza en la didáctica universitaria. Apertura **3**(2) (2011)
14. Álvarez, A.: Estrategia pedagógico-tecnológica para la integración de las tecnologías de la información y las comunicaciones en el proceso de enseñanza-aprendizaje desde la producción de materiales educativos digitales en el ISPJAE, La Habana (2014)
15. Ramos, N.: Una metodología para el proceso pedagógico de desarrollo de software educativo de Química en la Educación General cubana, La Habana (2016)
16. Urretavizcaya, M., Onaindía, E.: Docencia universitaria de Inteligencia Artificial. Inteligencia artificial: Revista Iberoamericana de Inteligencia Artificial **6**(17), 23–32 (2002)
17. ACM-IEEE: Computer Science Curricula 2001. Curriculum Guidelines for Undergraduate Degree Programs in Computer Science (2001)

18. Wollowski, M., Selkowitz, R., Brown, L., Goel, A., Luger, G., Marshall, J., Neel, A., Neller, T., Norvig, P.: A survey of current practice and teaching of AI. In: Proceedings of the Sixth Symposium on Educational Advances in Artificial Intelligence (2016)
19. Osella, G.L., De Vito, C., Russo, C.C., Ramón, H.D.: AutoPython: una herramienta para la automatización de sesiones interactivas de Python. In: XI Congreso de Tecnología en Educación y Educación en Tecnología (2016)
20. Asensio, J.J., García, P., Mora, A.M., Fernández, A.J., Merelo, J.J., Castillo, P.Á.: Progamer: aprendiendo a programar usando videojuegos como metáfora para visualización de código. ReVisión, 93–104 (2014)
21. Allendes, P.A., Chiarani, M.C., Noriega, J.E.: Desarrollo de recursos educativos abiertos en la universidad pública. In: XI Congreso de Tecnología en Educación y Educación en Tecnología (2016)
22. Bernard, R.M., Borokhovski, E.F., Schmid, R., Tamim, R.M.: Gauging the effectiveness of educational technology integration in education: what the best-quality meta-analyses tell us. In: Spector, M., Lockee, B., Childress, M. (eds.) Learning, Design, and Technology, pp. 1–22. Springer, Cham (2018)
23. Cabrera, J.F., Álvarez, A., Herrero, E.: Contribución del Centro Virtual de Recursos a la integración de las TIC en la CUJAE. Referencia Pedagógica, 39–50 (2013)
24. Vygotski, L.S.: Historia del desarrollo de las funciones psíquicas superiores. Científico-Técnica, La Habana (1987)
25. Addine, F.: !Didáctica! ¿Qué Didáctica? In: Didáctica. Teoría y práctica, 8–26. ISPEJV, La Habana (2002)
26. Zilberstein, J., Silvestre, M.: Didáctica desarrolladora desde el enfoque histórico cultural. Ediciones CEIDE, México (2004)
27. Coca, Y., Rosell, L.B., Velazquez, A.: Modelo de agente lógico con inferencia basada en hechos. Revista Cubana de Ciencias Informáticas **11**(2), 29–45 (2017)

Integration of Work and Study in Computer Science

Robert Pucher(✉)

University of Applied Sciences – Technikum Wien, Vienna, Austria
robert.pucher@echnikum-wien.at

Abstract. Universities of Applied Sciences in Austria offer study courses especially designed and optimized for today needs of companies and students. During the last 20 years, several approaches have been worked out. Feedback of companies employing graduates and feedback of students are among the main driving factors in the process of optimizing these study programs. An important need of adaption of study programs arises from the preferences of students [1]. Today almost all students of computer science start to work in their field while still studying. In many cases, this comes from the fact that companies are intensively trying to attract students, as the labor market in the IT field increasingly demands personnel. This in turn leads to many dropouts of students who are not going to finish their studies. One approach to overcome this problem is to integrate the education experienced while working in companies with the education of students in computer science. At the University of Applied Sciences Technikum Wien, such a program started on the bachelor level in computer science [2]. In the present paper, the authors describe a program on the master's level in software engineering, which is scheduled to be introduced in fall 2021.

Keywords: Work · Study · Computer science · Education · Curriculum

1 Introduction

1.1 Job Vacancies in Computer Science and the Consequences

"4000+ Software Engineer jobs in Munich, Bavaria, Germany. New Software Engineer jobs added." Statements like these you find on any platform, which offers jobs in computer science. In the fall term, 2018 students of the master's program Computer Science at the University of Applied Sciences Technikum Wien asked the administrative staff to stop forwarding job offers per email. "We all do have a job and are not interested in getting new offers twice a week", they said.

Today almost all students of computer science start to work in their field while still studying. In many cases, this is because companies are intensively trying to attract students. In the third year of the bachelor's program computer science at the University of Applied Sciences Technikum Wien, around 90% of all students already work besides their studies.

In the master's program, Software Engineering the percentage of working students is almost 100%. This can lead to drop outs of students who are not going to finish their

M. E. Auer and D. May (Eds.): REV 2020, AISC 1231, pp. 811–816, 2021.
https://doi.org/10.1007/978-3-030-52575-0_66

studies, as in many cases the workload in the job simply does not leave enough time to study. One approach to overcome this problem is to integrate education at the workplace with the education of students in the regular of computer science to form a joined curriculum. This requires an approach that allows combining learning experiences at the workplace and teaching at the university. At the University of Applied Sciences Technikum Wien, such a program was introduced in 2018 at the bachelor level in computer science [2]. To continue this program, a curriculum is being developed for the master's program computer science.

1.2 Computer Science at the University of Applied Sciences Technikum Wien

At the University of Applied Sciences Technikum Wien several study programs exist in the field of computer science. The most important study programs are briefly listed in Tables 1 and 2.

Table 1. University of applied sciences Technikum Wien, faculty of computer science, bachelor's degree programs.

Study program	Details
Business informatics	Six semester bachelor's degree program
Computer science	Six semester bachelor's degree program

The bachelor's degree programs are organized in various ways and are offered as full-time and part-time study degree programs. Business Informatics was offered as Distant Learning Program, however due to many dropouts in the program this program is no longer offered as distant study program.

Table 2. University of applied sciences Technikum Wien, faculty of computer science, machelor's degree programs.

Study program	Details
Software engineering	Four semester masters's degree program
Game engineering and simulation technology	Four semester masters's degree program
Information management and IT security	Four semester masters's degree program
Information systems management	Four semester masters's degree program

Master's degree programs at the University of Applied Sciences Technikum Wien consist of four semesters, 30 ECTS each semester are earned. As the programs are designed to fulfill the needs of companies seeking skilled personnel in the field of computer science, graduates of theses master's degree programs predominantly start working. Only a very small percentage (<1%) continues to study in a doctoral program.

2 The Master's Program in Computer Science

The University of Applied Sciences Technikum Wien was one of the first Universities which introduced the Bachelor Master system in Austria.

The master's program in computer science was introduced in 2004 at the University of Applied Sciences Technikum Wien. This program was continuously developed based on a feedback process from companies and students. Teaching is done in a blended format. Face to face lessons are combined with distant study elements. Today the program is a typical standard two years' master's program with a workload of 120 ECTS in total. In average one ECTS equivalents approximately 25 h of work, including all overhead activities. However, workload is a relative term, any individual student will have to do a different amount of work to finally be awarded with the degree of "Master of Science in Engineering". For a brief explanation of the ECTS system see [3, 4].

The study program computer science is designed to enable students to work beside their study. Around 50% of all lessons are done in face to face mode, around 50 percent are thought in distant learning modules. These distant modules can be studied at any location and are offered in the learning management system (LMS) Moodle. Students appreciate this, but still feel an enormous workload, as 120 ECTS equivalent to around 1500 h per year.

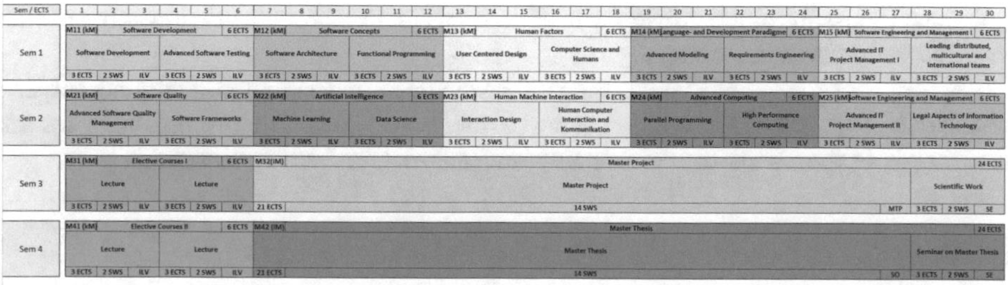

Fig. 1. Four semester master's program in computer science.

Figure 1 shows the structure of the master's program computer science. The first year predominately consists of lessons related to computer science. The second year, semester three and four just consist of a large block of work, the master's project and the master's thesis, plus two lessons which can be chosen of a pool of lessons, according to the individual interest of students.

Most students who work report, they really like the first year. The major reason for this is the content of the first year. In many lessons theoretical knowledge fitting to the practical work students do is being taught.

The second year was perceived by students in a very different way. Some students reported they could use working time to work on the master's project and on the master's theses, while other students reported they had to work in the company and on top of that work they had to do the large modules. Therefore, this second year of study was evaluated and the author tried to find a better way to integrate study and work life.

3 Learning at the Workplace Combined with Academic Teaching

In the last years a special type of study programs became increasingly popular in Austria. This type of study programs is called "integrated work/study degree program" [5]. These programs are designed to combine work and study. Teaching takes place at the University and in specific subjects teaching takes place in partner companies. Students need to get employed at these partner companies.

Especially for computer sciences this model does have numerous advantages. The most important of them are briefly summed up here:

- Theoretical knowledge from the university is combined with practical applications in a managed process. This has positive influence on students' performance.
- Companies work together with university and communicate their expectations of students' skills. This helps universities to continuously improve the curriculum and to adapt to the changing needs of the software industry.
- Students do learn while working in companies. In the case of such joined programs the learning in the company is optimized to add to the learning at the university.

In the case of the master's program software engineering a balance between theoretical knowledge and practical skills must be carefully obtained. If students work in the IT field, most of them practice during work how to apply theoretical knowledge. This means, lessons at the university need to focus on the theoretical knowledge needed to fulfil the tasks at the company.

A group of teachers, experts from IT companies and students identified subjects which can be taught while working at the partner companies. The responsibility for this teaching is in the hands of the company and only supervised by academia.

4 Results

In the planned curriculum "Computer Science - Integrated Work/Study Degree Program" the following lessons are carried out at the partner companies involved (Table 3).

Table 3. Joined teaching with companies.

Semester	Subject	ECTS
1	Advanced IT Project Management I	3
2	Advanced IT Project Management II	3
3	Master Project	21
4	Master Thesis	21
	Total	**48**

In total 48 ECTS out of 120 will be thought in a company, these equivalents to 40% of the total workload of the master's program.

Please note, the two lessons master's project and master's thesis, 21 ECTS each, in the third and fourth semester will be accompanied by 3 ECTS taught at the university to assure the academic level.

Figure 2 shows the structure of the curriculum "Computer Science - Integrated Work/Study Degree Program". Lessons marked in red, are carried out at partner companies.

It can be seen, that in the first year only 10% of the workload of students is transferred in partner companies (6 ECTS out of 60 ECTS), whereas in the second year around 60% of the workload is being taught in companies. This accounts mainly on the large amount of applied practical work in the second year. The result of this work can be used by companies which employ the students. The companies benefit from the academic supervision of this work. For students this means a significant improvement, as the work they do counts for the companies needs and the study program.

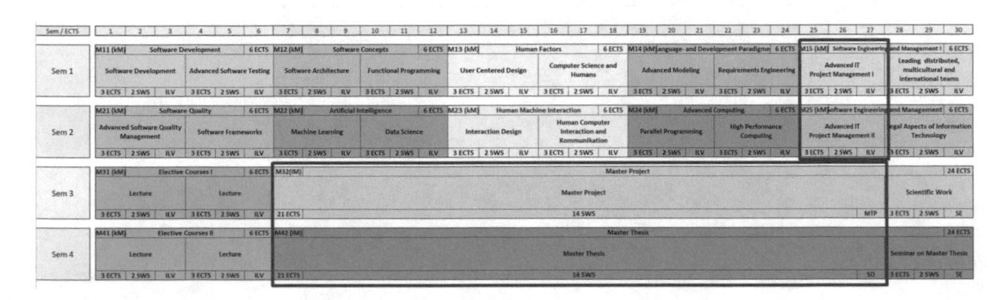

Fig. 2. Structure of the curriculum "Computer Science - Integrated Work/Study Degree Program". Lessons that will be taught at partner companies are marked in red.

5 Conclusions

Students in the area of computer science very often start to work besides their studies. In many cases this leads to a phenomenon referred to as "job outs". The students increasingly start to shift their engagement away from study towards work. This finally leads to quitting the study. Therefore, in Austria in universities of applied sciences much effort was done to develop study programs that allow students to work and study at the same time. One of the most successful models emerged in the last years, the "Integrated Work/Study Degree Program" [5]. This model integrates work at a company in a managed way into the study program.

References

1. Pucher, R., Holweg, G., Mandl, T., Salzbrunn, B.: Optimizing Higher Education for the Professional Student–The Example of Computer Science Education at the University of Applied Sciences Technikum Wien Conference: X International GUIDE Conference, Vienna, Austria, September 16-18, 2015, At Vienna, Austria, In: Proceedings of the X International GUIDE Conference, Vienna, Austria, 16–18 September 2015 (2015). (http://www.guideassociation.org/proceedings/Guide_2015_Vienna/Proceedings_Vienna.pdf)
2. https://www.technikum-wien.at/duales-studium-bachelor-studiengang-informatik/, 1 June 2019
3. https://en.wikipedia.org/wiki/European_Credit_Transfer_and_Accumulation_System, 1 June 2019
4. https://europass.cedefop.europa.eu/sites/default/files/ects-users-guide_en.pdf, 1 June 2019
5. https://www.dualstudieren.at/dual-studieren/warum-dual-studieren/ , 1 June 2019
6. https://www.researchgate.net/post/What_are_the_reasons_for_drop-out_of_CS_IT_studies_courses, 1 June 2019
7. Bahl, A., Dietzen, A. (eds.): Work-Based Learning as a Pathway to Competence-based Education – A UNEVOC Network Contribution. Verlag Barbara Budrich, Leverkusen (2019)

Digital Citizenship and Life Long Learning

Olga Bombardelli(✉)

University of Trento, Trento, Italy
olga.bombardelli@unitn.it

Abstract. Our world becomes increasingly digital. Technology is changing the way we live, work, learn, participate in public life. This paper deals with strategies for enabling young people and adults for competent Digital Citizenship. Citizenship competence is the ability to act as responsible citizens and to fully participate in civic and social life, based on understanding of social, economic, legal and political concepts and structures, as well as global developments and sustainability (EU Key competences 2018). This also involves the ability to access, and interact with both traditional and new forms of media. Digital literacy includes digital etiquette for digital communication, critical thinking, digital security. Education for digital citizenship is to be embedded into the curriculum at school for youngsters and adults, in a continuing process of life long learning, using active methods and participative evaluation, developing new forms of teacher training.

Keywords: Digital citizenship · Competence · Education · Digital literacy

1 Introduction

Our world has been deeply transformed by the rise of new information and communication technologies (ICT) and has become increasingly digital. The Internet shapes almost every sphere of our life: Finance, Public life, Tourism, Entertainment, Commerce, Work and Social relationships.

Digital tools and the World Wide Web, one of the main services of Internet, make it possible to connect to the rest of the world. Thanks to digital technologies, today we can bank, read the news, write e-mails and chat with people across the globe, study for a degree, all without leaving the comfort of our homes. The Internet has provided basic information of countless political subjects accessible to all citizens skilled and motivated enough to seek it out; it allows individuals to post, at a minimal cost, messages and images that can be potentially viewed instantly by global audiences.

The current use of technology is changing our rights, and responsibilities, the way we live, work and participate in political life. A new relationship can be developed between municipalities and local residents, and between parliamentarians and citizens. Internet facilitates new forms of political organization, improves communication by and to governing bodies, bottom up approach initiatives such as flash mobs etc.; political strategies are in a new era, changing the role of traditional actors [3] such as political parties, governing bodies, trade unions, NGOs, and digital citizenship is becoming more and more important in the digitally mediated world.

M. E. Auer and D. May (Eds.): REV 2020, AISC 1231, pp. 817–826, 2021.
https://doi.org/10.1007/978-3-030-52575-0_67

All politicians, economic groups, national and international organisations and citizens use social networks, among others, for political purposes. It has provided basic information on countless political subjects accessible to all citizens skilled and motivated enough to seek it out.

This paper aims to analyse the potentials of digital citizenship and deals with Life Long Learning (LLL) to increase the advantages and reduce the dangers of modern digital communication that citizenship can have.

I will deal with the situation of digital citizenship, including its potential and risks for democracy, considering citizenship competences and digital competence, focussing on the LLL for inclusive education, in order to support people in achieving high competences and fighting discrimination, removing at least the educational gap.

The methodology includes the use of updated studies and official documentation by International bodies; I adopt a value-based and vision-driven approach to innovation in education.

Digital online tools seem to open new paths for active citizenship. Given the ubiquitous and pervasive character of digital online technologies in mediating communication, and forging cultural identities, digital citizenship can be seen as a continuation and broadening of the core concept of citizenship.

It is not clear if a significant number of previously inactive citizens have been recruited into political activism by online communication; the question requires new in-depth research projects to answer this.

Online communication can potentially support great opportunities of information and participation, at the same time it can increase distance, indifference, reduce the sense of belonging, decrease one's feeling of community attachment, depersonalizes communication, all factors that are detrimental to civic participation.

Individuals who consume political information online are more likely to participate in political discussions, have higher levels of political knowledge, and have more acute political awareness. Although the empirical studies do evidence a correlation between Internet use, political interest, and civic engagement, findings are overstated by asserting a causal relationship. Internet merely reflects politics as usual, failing to increase civic engagement among the disengaged and merely provides instant, cost-free information to those already politically motivated, because political enthusiasts are more likely to seek political information online as compared to the politically apathetic. Highly debated is the question of the possible power of citizens' participation online in affecting political and economic decisions.

I focus on active participation, and on education for inclusive democratic digital citizenship through LLL.

2 Digital Citizens

Being a digital citizen means using regularly digital technologies to participate in society, with the skills, and knowledge to effectively communicate with others, create and consume digital content for citizenship, being aware both that digitalization is an extraordinary opportunity, and that it is not even a phenomenon that we can take lightly.

According to the Council of Europe, digital citizenship is the competent and positive engagement with digital technologies (creating, socializing, investigating, playing, communicating and learning), participating actively and responsibly in communities (local, national, global) at all levels (political, economic, social, cultural and intercultural), being involved in a double process of Life Long Learning, and defending human dignity [7]. A digital citizen has to maintain an attitude relying on critical thinking as a basis for meaningful and effective participation in his/her community [7].

The Internet penetration rate is high by global regions; as of January 2019, North America and Northern Europe were both ranked first with an online penetration rate of 95%, followed by Western Europe with 94% [15]. The global average penetration rate was 57%, an increase from 35% since 2013 [15]. As of January 2019, East Asia accounted for 1 billion internet users, followed by Southern Asia with 803 million internet users. Young adults (aged 18 to 29) use the Internet more frequently than the general population. 1.56 billion people on average log onto Facebook daily and are considered daily active users [15] for March 2019. A huge and vastly growing number of Facebook users are active and consistent in their visits to the site, making them a promising audience for marketing efforts.

Political civil society activities include a broad range of engagement forms, mostly digitally mediated, from coordinating activities in formal and informal structured NGOs, to turning up at a protest or being member of societal organizations, like political parties, religious and cultural groups, trade unions, volunteering, work for electoral campaigns, representativeness in the working place, consumer protection, attendance in demonstrations, citizens' initiatives, communication among citizens and with the institutions, use of the media, interactive communication (off-line and online, chats, e-mail, online news).

New forms of political participation, outside of the classical ones, are established and play a pivotal role in almost all forms of active citizenship even crowdfunding, pirate parties, proposals, from formal politics to community activities.

There are different fields of relationship among who takes decisions and the citizens: the open access to politically relevant information; communication, which implies the possibility to express their own voice, being heard and debating, contributes in building public opinion; contribution to the general decision making, control, and consultation, especially competent political and electoral choices [2].

Platforms and open data have the potential to promote transparency and collaboration in cities for public benefit and innovation. Open government is expected to create dialogue among citizens and public administration; it requires understandable information, skills to use in a proper way, however, significant segments of the population are still excluded from digital citizenship, undermining the democratic principle of openness to all.

Digital online networks not only bring great opportunities; feeds and social networks are instruments, which create widely unseen barriers and exclusive mechanisms. The link structure of the web is critical in filtering what content citizens see, using forms of Artificial Intelligence within its newsfeed. The software code, the search engines which guide most users' online search behaviour. Giants of the web are likely to be owners of public opinion. As the news we are fed is curated for us by remote algorithms, customizing information according to our interests, there is a danger that it undermines the objectivity of our judgments and predetermines our decisions. The erosion of democratic systems is not far behind as long as predetermination by algorithms leads to a brainwashing of the people downgrading individuals to mere receivers of commands, who automatically respond to stimuli.

There are big problems in the democratic governance of the digital space: security questions online, espionage, hackers, hate speech, cyberbullying, messages of hate, obscenity, extremism, criminality, terrorism, difficulties in the protection of safety, intrusion in identity (e.g. scanning photos and providing facial recognition), questions of intellectual property, privacy protection, rights as freedom of expression, risks of manipulation, of homogenisation of society and fake news, and the digital divide.

Therefore, requires interventions by Governments, International Organisations, and the NGOs; very difficult tasks, such as monopolist global properties are not easily submitted to the sovereignty of States, in a situation of planetarium alliances and fights, and new strong markets.

The responsibility of users is to be encouraged. Ethics is at the core of digital citizenship for users too; just like people who should be good citizens in society, they should also be good citizens online, being careful of contacts through the Internet with unknown people, sharing of compromising pictures, downloads and use of music, movies, and pictures, copy and paste texts, behaviour to malicious or rude online messages.

The 2018 World Development Report 'Learning to Realize Education's Promise' suggests that digital platforms and virtual convening could be transformative in terms of citizen engagement, as long as it allows provision for digital inclusion. There are barriers and exclusions, both because of the functioning mechanisms of the Internet, and because of technological and social inequalities [16, 17], with heavy consequences in information, digital participation, and democracy. Large differences are visible across countries, and between people of different age groups and educational backgrounds in access to informatic technology/internet, and in skills. Despite the current progress, a "digital divide" remains between those with and without Internet access, between those with and without digital competences.

"Digital divide" refers to systematic disparities in access to computers and the Internet, affecting people who are poorer (low household income), under-educated (low school degree), elderly, belonging to some minority groups (gender, and ethnicity). Throughout the globe, countries and people are unequally prepared to seize the benefits of digital transformation, and that makes digital citizenship a privilege for the elite, not a democratic process, when citizens either are having no access, or are not competent enough and are credulous and easily manipulated. Participatory forms of citizenship, face to face and online, require political literacy, which incorporates civic and general knowledge, skills, attitudes and values for the capability to exercise rights. Bennett stresses the importance of the student's experiences of engagement and active participation within the educational system to shape the outcomes of citizenship for future generations, practicing and experiencing digital citizenship [1].

3 Citizenship Competence and Digital Competence

According to the Recommendation of the European Union on the key competences for Life Long Learning 2018 'Digital competence involves the confident, critical and responsible use of, and engagement with, digital technologies for learning, at work, and for participation in society. It includes information and data literacy, communication and collaboration, media literacy, digital content creation (including programming), safety (including digital well-being and competences related to cybersecurity), intellectual property related questions, problem solving and critical thinking' [8].

Essential knowledge, skills and attitudes related to this competence. Individuals should understand how digital technologies can support communication, creativity and innovation, and be aware of their opportunities, limitations, effects and risks.

They should understand the general principles, mechanisms and logic underlying evolving digital technologies and know the basic function and use of different devices, software, and networks. Individuals should take a critical approach to the validity, reliability and impact of information and data made available by digital means and be aware of the legal and ethical principles involved in engaging with digital technologies [8].

Individuals should be able to use digital technologies to support their active citizenship and social inclusion, collaboration with others, and creativity towards personal, social or commercial goals. Skills include, the ability to use, access, filter, evaluate, create, programme and share digital content. Individuals should be able to manage and protect information, content, data, and digital identities, as well as recognise and effectively engage with software, devices, artificial intelligence or robots.

Citizenship competence is the ability to act as responsible citizens and to fully participate in civic and social life, based on understanding of social, economic, legal and political concepts and structures, as well as global developments and sustainability. Skills for citizenship competence relate to the ability to engage effectively with others in common or public interest, including the sustainable development of society [8].

High competences help to overcome obstacles and risks. The European key competences for Life Long Learning 2018, state that citizenship competence involves '… critical thinking and integrated problem-solving skills, as well as skills to develop arguments and constructive participation in community activities, as well as in decision-making at all levels, from local and national to the European and international level. This also involves the ability to access, have a critical understanding of, and interact with both traditional and new forms of media and understand the role and functions of media in democratic societies. Respect for human rights as a basis for democracy lays the foundations for a responsible and constructive attitude' [8].

4 Life Long Learning for Inclusive Education to Competent Digital Citizenship

Participatory forms of Citizenship require political literacy, which incorporates civic and general knowledge, skills, attitudes and values and reducing risks. Civic engagement, political interest, political influence in discussion and in decision making have to build upon general and political knowledge. Education is central to the process of developing competences both for general culture, for citizenship and for digital competences.

According to UNESCO, LLL includes all learning activity undertaken throughout life, with the aim of improving qualifications for personal, social and/or professional reasons. The European ET strategy 2020 promotes LLL in formal and non-formal, informal education, and includes digital literacy [10] too.

Besides the traditional "hard" skills, such as reading, writing, and arithmetic, disciplinary contents, transversal competences, "soft" skills are needed [12, 13].

The Survey OECD 2019 considers only 58,3% of the OECD citizens able to use the Internet in a complex way [12]. Education for digital citizenship is not simply a matter of information, knowledge and know-how. It is also a matter of interpersonal and inherently ethical relations [12]. Citizens are expected to develop documented opinions, and the awareness of the need of improving individual and common decision making.

Measures for fostering inclusion in digital citizenship are to be implemented both in short and long term, supporting people in all stages of their life to achieve high competences of citizenship and fighting discrimination, removing at least the educational gap in the use of digital tools for citizenship. Educational efforts for political literacy by the school and by social bodies should improve knowledge and cognitive skills, media literacy, habits of selecting information, fostering awareness and responsibility. Human rights have to be respected on the *Net* too.

The capabilities of digital citizenship depend not entirely on educational institutions (school, universities etc.); they are developed in family, formal and informal groups, peers, media and the political world as well, through a constant exchange with the environment around us.

A promising opportunity lies on the horizon, such as Local Municipalities. City mayors and local leaders are expected to invest in social infrastructure to help individuals, families and communities to better participate in decision making processes about society and the economy, also promoting transparency and collaboration. Government bodies should work to ensure that all who want to access the Internet are granted it potentially, expanding broadband access points, lowering the cost of broadband access, as well as addressing educational and technological disparities, giving the necessary support in place, adopting user friendly platform models and choosing the right tools. However, those who are genuinely not interested in being online should not be discriminated against by governments.

Local platforms can enable citizens to find information, to submit ideas, rank priorities for allocating public resources according to more meaningful political participation, improving the quality and legitimacy of decision-making. Democratic countries and cities (examples: the 'smart city', and 'the networked city') invite a bottom-up approach participation via online platforms on different data feeds: from transport and mobility to urban planning, to energy and waste, allowing sometimes participated policy making to plan and participatory budgeting; that can unlock the bottom-up approach knowledge of communities, experiences, skills, and resourcefulness of citizens.

The education and training system must continue to be focused on the development of hard and soft skills that allow personal development and the innovative integration of youngsters in a society that is constantly changing.

Traditional learning must be combined with innovative work. Education for Digital citizenship, and regular use of ICT are to be embedded into the curriculum at school for youngsters and adults, in an ongoing way, not only as separate areas. Some stand-alone lessons can certainly be useful, but digital citizenship can be practiced in the educational institutions and in the social environment, in order to empower learners to participate effectively in a culture of democracy.

This is a rather complex process, which entails a concerted action involving the adoption of educational methodologies that stimulate the development of these competences in the classroom, choosing topics corresponding to the expectations, and needs of the students, caring for good human relations and solidarity.

I consider five propositions: active teaching/learning methods, practicing the use of ICT, peer tutoring, participatory assessment and evaluation, participation, teacher recruitment and training, to contribute in developing the following aspects: take a critical approach to the validity, reliability and impact of information and data (media literacy, fact checking, debunking against fake news regularly quoting the sources), using criteria and models for evaluating online information and to select from the myriad of available sources, the promotion of activities geared towards the real world and the needs of learners; diversification of pedagogical methods and strategies according to the aims; the fostering of the students' abilities with problem-solving, adaptive, resilience and social competences; and the promotion of peoples' motivation to engage in Life Long Learning.

For this teaching process to be successful, the student's active role in their learning is critical. Besides Citizenship education, Philosophy for children can be introduced to foster coherent reasoning; the habit of dealing with controversial issues [14] is a strategy to face difficult tasks in a balanced form, to develop arguments and practice respect.

Useful examples of active teaching/learning methods are work-based projects, project work, case-and-problem-based projects, which foster and boost the development of not only technical-scientific, but also transversal competences in students. The work at reality-based projects helps to achieve the objectives of 'real life' areas of practice; project-based learning allows the development of several transversal competences, such as the ability to work as a team, problem-solving, acceptance of different perspectives and critical analysis.

One of the teaching/learning strategies is virtually recreating (to a certain extent) simulated situations, that are close to the real context, using ICT when appropriate, for finding answers and solving problems.

Important is the adoption of clearly defined assessment criteria, namely through the use of rubrics, adopting participatory evaluation criteria [11], interactive and well-known to the students, who, in this way, will have a clear perception of the evolution of their learning regarding the competences that they are expected to attain.

Digital Citizenship is something that all teachers need to be aware of. With the growing amount of online activity that students are involved in, educators need to provide students with clear guidance on 'best practice' when working online, and when practising digital citizenship. There is a need for special initial and in-service training (communities of practice and international exchange) of teachers; their competence and motivation shape a successful education for citizenship, fostering interest and advancement without indoctrination. They need to learn new working methods, besides frontal teaching; through the use of active methodologies, they have the possibility of enhancing interaction with their students, thus fostering close interaction and a more individualized assessment of the progress and competences, but also the students' difficulties and weaknesses.

5 Conclusion

Digital citizenship is based on e-participation through new media. Online participation has the potential to benefit society as a whole, even though it includes serious risks as well.

Citizens should become competent enough to give constructive contributions, to avoid manipulation and misbehaviour online; the digital divide is a danger for democratic digital citizenship.

In this paper I highlighted competences of digital citizenship and focussed on LLL enabling people to contribute to the common development. Empowering citizens for digital citizenship in view of democratic representativeness and inclusion.

Education is responsible for the future, putting the 'why-ness' of design, values and visions before the 'how-ness', and the 'what-ness', as it carefully needs pondering over the foundations and society, which is not reduced to training.

Digital democracy is possible only with the inclusion of those with less access and fewer digital and interpretational skills, because disadvantaged groups, digitally excluded for socioeconomic, geographic, cultural reasons lag' behind in their access and use of the Net, or/and have poor critical skills.

Educational measures for developing competent citizenship cover competences, like properly managing personal and other people's information shared online and skills for political participation. Engagement with digital technologies and content requires a reflective and critical, yet curious, open-minded and forward-looking attitude to their evolution. It also requires an ethical, safe and responsible approach to the use of these tools' [8].

In terms of limitations, this manuscript critically reflected on the topic of the relevance of attaining transversal competences for the education and training of individuals, seeking to focus on the role of LLL; it does not have an empirical component. It is a field of wider, further research, as comparable longitudinal studies are still at the beginning. Although it is difficult to calculate intangible changes, like the relation between increased Internet use, and civic engagement, studies should compare in a longitudinal way the impact of the Internet on the phenomenon of low civic engagement and voter turnout, in similarly situated communities (per size, economic wealth, average income, ethnic and educational backgrounds of its citizens).

I hope that this work may be a basis for future studies and whose purpose is to assess the relevance of the attainment of transversal competences in the individuals' digital citizenship. Digital citizenship is potentially a good answer to a sustainable future, as long as a high competence is developed in a Life Long Learning process.

References

1. Bennett, W.L.: Changing citizenship in the digital age. In: Lance Bennett, W., The John, D., Catherine, T. (eds.) Civic Life Online: Learning How Digital Media Can Engage Youth. MacArthur Foundation Series on Digital Media and Learning, pp. 1–24. The MIT Press, Cambridge, MA (2008)
2. Bombardelli, O.: Cittadinanza digitale e educazione. In: Mascia, M. (ed.) Democrazia rappresentativa e partecipazione politica. Cacucci, Bari (2014); Bombardelli, O., Ferreira, P.: Digital tools and social science education. J. Soc. Sci. Educ. **1**(2016), 2–5 (2016)
3. Calisee, M., Musella, F.: Il principe digitale. Laterza, Bari (2019)
4. ComScore. https://www.comscore.com/Insights/Data-Mine/Finnish-Internet-Users-are-Most-Avid-Consumers-of-Online-News
5. ComScore. https://www.comscore.com/Insights/Data-Mine/Nearly-7-million-Internet-Users-in-Italy-Access-Government-Sites
6. Commission of the European Communities: Making a European area of Lifelong Learning a reality. Communication from the Commission. COM (2001) 678 final, Brussels (2001). http://viaa.gov.lv/files/free/48/748/pol_10_com_en.pdf
7. Council of Europe Digital Citizenship Education (DCE) (2019). https://rm.coe.int/10-domains-dce/168077668e
8. Council Recommendation of 22 May 2018 on key competences for Lifelong Learning (Text with EEA relevance) (2018/C 189/01). https://eur-lex.europa.eu/legal-content/EN/TXT/PDF/?uri=CELEX:32018H0604(01)&from=EN
9. Darnis, J.P., Polito, C. (eds.) La geopolitica del digitale, Roma: Edizione Nuova Cultura (2019). https://www.iai.it/sites/default/files/iaiq_20.pdf
10. Digital Agenda for Europe. http://europa.eu/legislation_summaries/information_society/strategies/si0016_en.htm
11. Gómez-Gasquet, P., Verdecho, M.J., Rodriguez-Rodriguez, R., Alfaro-Saiz, J.J.: Formative assessment framework proposal for transversal competencies: application to analysis and problem-solving competence. J. Ind. Eng. Manag. **2018**(11), 334 (2018)
12. OECD: OECD Skills Outlook 2019 Thriving in a Digital World. OECD Publishing, Paris (2019). https://doi.org/10.1787/df80bc12-en
13. Sà, M.J., Serpa, S.: Transversal competences: their importance and learning processes by higher education students. Educ. Sci. **8**(3), 126 (2018)
14. Schiele, S., Schneider, H. (eds.): Das Konsensproblem in der Politischen Bildung, Stuttgart (1977). https://www.lpb-bw.de/beutelsbacher-konsens.html

15. Statista (2019). https://www.statista.com/statistics/269329/penetration-rate-of-the-internet-by-region/
16. Warschauer, M.: Technology and Social Inclusion: Rethinking the Digital Divide. The MIT Press, Cambridge, MA (2004)
17. World Development Report: Learning to Realize Education's Promise (2018). http://www.worldbank.org/en/publication/wdr2018

Blended Learning

Designing a Mentoring System for Pre-training Preparation in a Blended Digital Badge Program

Kei Amano(✉), Shigeki Tsuzuku, Katsuaki Suzuki, and Naoshi Hiraoka

Research Center for Instructional Systems, Kumamoto University, Kumamoto, Japan
{keiamano,tsuzuku,ksuzuki,naoshi}@kumamoto-u.ac.jp

Abstract. The purpose of this study was to redesign a mentoring system and verify its effects. The research field involved a blended instructional design (ID) course conducted as a university extension course composed of prior online assignments, face-to-face workshops, and post-learning assignments. The results demonstrated that the redesign of the ID course may improve the completion rates of online pre-training and post-learning assignments altogether. Additionally, the findings showed that motivational e-mails were useful for facilitating participants' access e-learning. Furthermore, some participants' reports suggested that the syllabus provided on the website gave participants a perspective of the course and motivated them to learn during the course. In addition, motivational e-mails were helpful in making participants feel a sense of belonging in the ID course. Based on these results, we proposed three design principles for supporting participants' completion of lifelong learning activities that do not provide incentives such as university credits and job promotions.

Keywords: Mentoring support · Pre-training preparation · Blended digital badge program

1 Introduction

Since 2011, Kumamoto University has offered an instructional design (ID) course as part of its university extension courses in Japan. This course was an open lecture for working adults. Thus far, a broad variety of professionals from various fields, such as university faculties, medical doctors, nurses, human resource professionals in companies, and Japanese language teachers, have attended the course. In 2015, these courses were redesigned from being just one-day face-to-face workshops to a blended style comprising online prior assignments, one-day face-to-face workshops, and online post-learning assignments [1]. The purpose of this improvement was to ensure that the time used in face-to-face workshops is invested for establishing interactions between participants and instructors, as well as among participants, and that participants spend less time listening to the instructors and more time applying the learning contents in some

M. E. Auer and D. May (Eds.): REV 2020, AISC 1231, pp. 829–841, 2021.
https://doi.org/10.1007/978-3-030-52575-0_68

cases, by acquiring the basic knowledge during prior online learning phases. The aim was also to confirm each participant's learning outcomes during the online post-learning phases; this could not be realized in face-to-face workshops before redesign due to time constraints. In addition, participants who met the passing criteria for post-learning assignments received a digital badge as a certificate of completion.

Although this redesign was expected to provide a better learning experience, not many participants worked on the online prior-learning assignment due to the following problems: 1) Participants were unable to identify the requirements of the assignment because the pre-assignments were not set very often during the seminar or training; 2) Some participants experienced difficulty in accessing the online learning environment due to a lack of IT literacy; 3) Participants were not motivated to work on assignments. Therefore, the purpose of this study is to design a mentoring system to improve participants' work on pre-training preparation. The study also discusses their reactions to the improved design of the mentoring system. Further, the lessons learned are demonstrated to guide the introduction of pre-training preparation in the blended learning program through reflection on our experience.

2 Literature Review

In this research, mentoring is defined as a support method for the learning process of participants by those who are familiar with the learning contents and a variety of learning methods. The method of mentoring has been made more sophisticated to prevent e-learning dropouts [2]. The person responsible for mentoring is called a mentor. Mentors answer questions from learners, and sometimes help learners by advising about study methods and time management [3]. In other words, the role of a mentor is to provide support to the participants, except for the learning contents and instructional design, for which subject matter experts and lecturers are responsible. Chang (2004) proposed that a mentor has the following three roles [4]:

1) Teaching assistants: supporting the faculty's role as subject matter experts; mentors provide students with extra help toward clarifying and comprehending course contents.
2) Social connectedness: helping students develop an e-learning connectedness.
3) Technical supporters: solving participants' technology-related problems.

The timing of need for help from mentor is found in previous findings. For example, a UK open university found that participants are most likely to drop out of e-learning before the first assignment [5]. Specifically, the participants drop out at the timing of registration for e-learning or before working on the first assignment. From these findings, it is suggested that support at the beginning of the learning is important. In this study, we refer to these findings as the UK open university's retention model (Fig. 1).

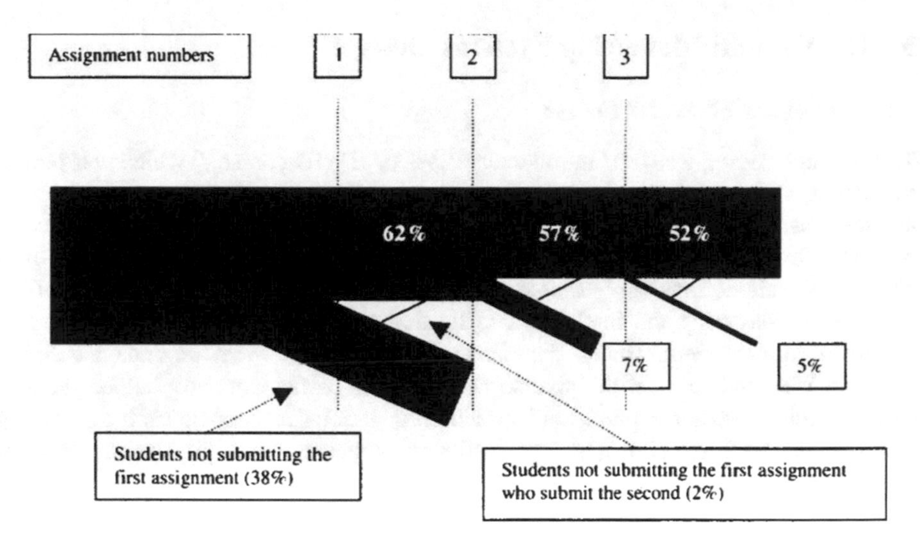

Fig. 1. UK open university's retention model (Retrieved from Figure 4.1. of Simpson 2003 [5])

It is recommended that motivational messaging is effective in facilitating the learning of participants. Visser (1998) proposed the design principles of message design to motivate participants to complete the educational courses based on the ARCS motivational model [6], which involves four steps for promoting and sustaining motivation during the learning process: Attention, Relevance, Confidence, and Satisfaction (ARCS) [7]. The design principles are collectively called MMSS (Motivational Message Support System). Visser indicated that motivational messages are needed to increase the sense of belonging to the course and to make the participants feel empathized with while facing difficulties in working on the assignments. Furthermore, Visser's research demonstrated that personalized messages are not effective compared to collective messages that are delivered to a group of participants. If the message is designed and systemized collectively, collective message might be collective. This enables the efficiency of education. The effects of Visser's MMSS were only verified in a university formal education setting, and not in lifelong learning settings.

There is little research regarding how mentoring can facilitate participants' completion of lifelong learning activities that are not prerequisite for university credits or job promotions. Based on the UK open university's retention model and Visser's MMSS, we redesigned the mentoring system to support participants' completion of the ID workshop.

3 Course and Mentoring Systems Design

3.1 Overview of the ID Course

The research setting was the "Introductory class of ID (ID course)," which was part of the extension courses provided at Kumamoto University. In this course, learners acquire basic ID skills and consider improvement proposals for educational cases presented by lecturers. The learning objectives and evaluation method are shown in Table 1. We aimed to not only enable participants in memorizing the knowledge of ID, but also in mastering the intellectual skills that allow them to apply ID theory to education improvement. The ID theories taught in the ID course are demonstrated in Table 2. We tried to meet the diverse needs of the participants and helped them in solving their educational problems by including various ID models such as learning evaluations, teaching strategies, motivational strategies, and process models for improving education.

Table 1. Instructional design course objectives and evaluation methods

Introductory course
■ Learning objectives
• Demonstrate how to use basic instructional design (ID) skills to improve education programs
• Identify the problems in specific cases of educational programs and select appropriate solutions based on the Attention, Relevance, Confidence, Satisfaction model
■ Evaluation methods and criteria
• A score of 80% or more on the comprehension test
• Submitting the final report and scoring 80% or more based on the grading criteria. Based on the course content, summarizing the following items in the final report:
1) Analysis and improvement proposal from ID course participants 2) Action plans to use learning outcomes 3) Three things that participants learned most in the course
• Comments on other participants' final reports in the discussion forum on Moodle

Table 2. 10 ID models learned in the workshop

- ADDIE Model
- Carroll's Model of School Learning
- ARCS Model
- Gagne's Nine Events of Instruction
- Mager's Three Questions
- Gagne's Five Categories of Learning Outcomes
- Kirkpatrick's Four Levels of Evaluation
- Andragogy
- First Principles of Instruction
- TOTE Model

3.2 Blended Design of the ID Course

This course consists of prior learning activities (online), a face-to-face program (one day), and post-learning activities (online). The entire composition of the ID course is demonstrated in Fig. 1. The learning management system (LMS) used for the e-learning courses was Moodle. The online phases allowed participants to effectively utilize the day of the face-to-face program. These online activities were required for submitting coursework items; therefore, they were a prerequisite for evaluation, enabling participants to select the best learning path. Finally, participants would acquire a digital badge if they fulfilled the evaluation criteria along with the prerequisites for evaluation in the course (Amano, Suzuki, Tsuzuku, & Hiraoka, 2017). The badge was designed to display the accomplishment of learning objectives, with evidence such as online report assignments and an asynchronous record of discussion forum posts created during a blended educational program (Fig. 2). By introducing the digital badge, the researchers improved the certification of program completion from seat-time-based to mastery-based, allowing them to verify whether the learners reached their learning objectives at the end of the program, and to ensure the quality of the program.

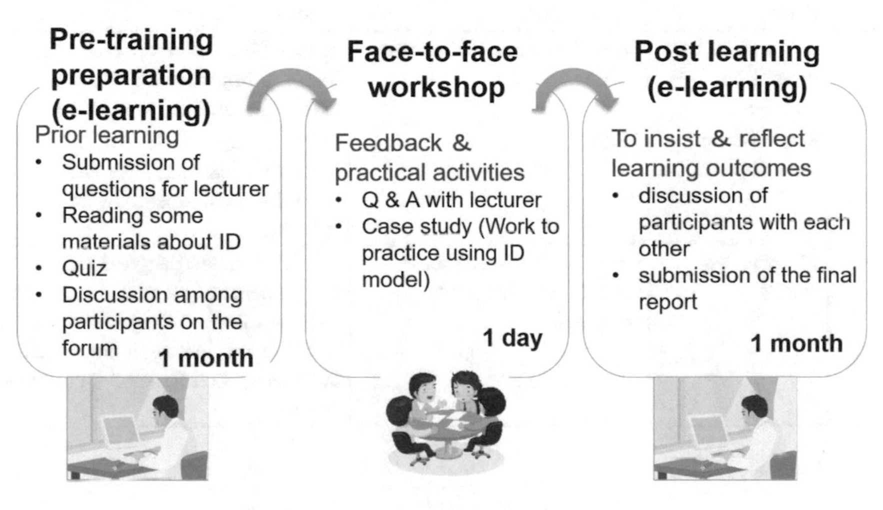

Fig. 2. The composition of the ID course

3.3 Improvement of Mentoring Systems to Support Participants' Online Pre-training Preparations

We first introduced the blended digital badge program in 2015, but because problems still remained, and we gradually improved them. The strategy introduced for supporting participants' work on online prior assignments are demonstrated in Table 3.

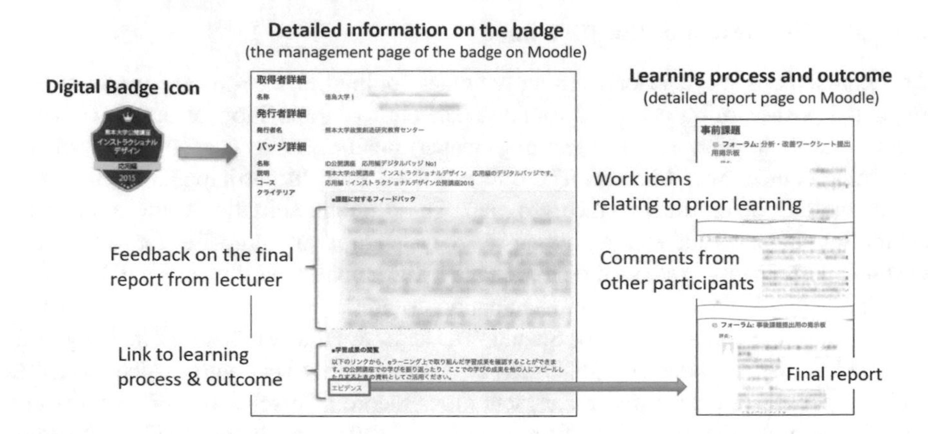

Fig. 3. The digital badge that was linked with the learning portfolio

Table 3. The improvements in the mentoring system

No.	Items	Before (2015)	After (2016)
1	Description of promotion websites	Only the course outline was written.	A syllabus, which specified the learning objectives, evaluation, and learning methods, was added
2	Step-by-step manuals	None	Step-by-step manuals on how to login to e-learning sites and work on online assignments were introduced
3	Motivational emails	Only the announcements regarding assignments were delivered	In addition, the progress on assignments of all participants, such as how many participants were working on their assignments and how many had finished them, was shared

In 2015, the participants were given two weeks for the pre-training assignment. However, not many participants worked on the pre-training assignment. One of the causes of this problems is considered to be the short period given for pre-training assignment. Therefore, the time for working on the assignment was extended from two weeks to one month. This constituted the logistical improvement of the ID course.

Additionally, we improved the way of providing information related to the ID course design. In 2015, we only announced information about the online pre-training assignment. As this seemed insufficient, we tried to inform the participants about the course design intent. For example, the detailed syllabus was delivered to participants through promotion websites (Fig. 3). Step-by-step manuals on how to login to e-learning sites and work on online assignments were also prepared for participants. These manuals were created in the PDF format. In addition, e-learning screen capturing and operation instructions were provided visually (Fig. 4). Furthermore, motivational e-mails were delivered to participants. Motivational information included the progress of all participants in the assignments, including the numbers of participants still working on their assignments and those who had finished them. In order to make this possible, information on who was logging in to the e-learning site and who was working on which assignments was recorded on the LMS. The requirements of the assignments were confirmed and encouragement messages were sent through e-mail (Table 3). These emails were delivered to students one week before and one day before the deadline for the assignment. The purpose of the e-mails was to improve the pacesetting of the learning process and inform the participants that they were not alone, and many other participants were going through the same experience as them. In this way, an effort was made to inform participants about the requirements of the assignments and the complex learning flow of the blended ID course (Table 4).

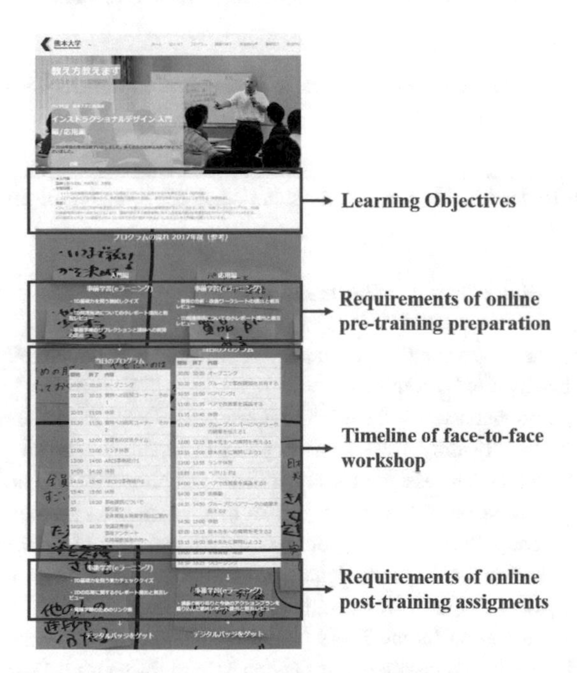

Fig. 4. The promotion websites of the ID course that explained the course syllabus

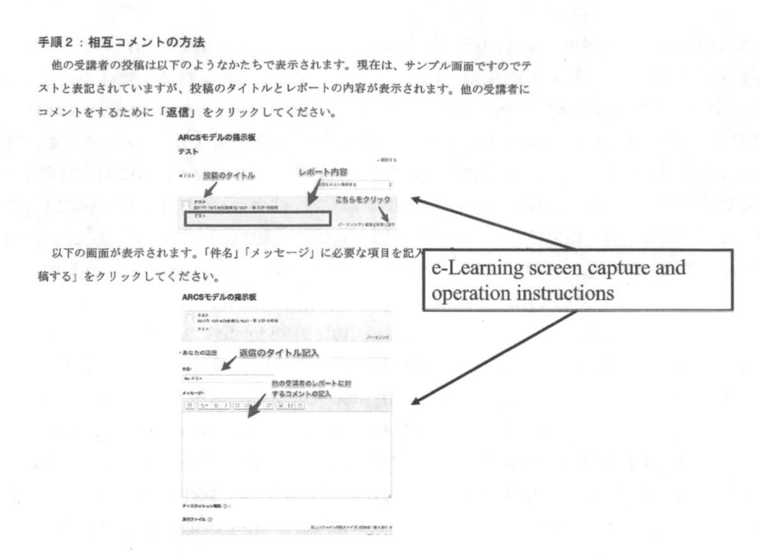

Fig. 5. Sample pages of the step-by-step manuals

Table 4. A sample of the Motivational E-mails. ※(1) showed empathy with the participants' difficulties while working on assignments. ※(2) showed facilitating a sense of belonging to the community.

Subject：【10/16 deadline!】 Request for submission confirmation of the prior assignment of the ID course

Dear participants who will be attending the ID course,

Thank you for your cooperation. My name is Kei Amano from Kumamoto University. I am responsible for managing the ID course.

(1) The deadline for submission of the prior assignments is October 16 (Tuesday), next week. As you might be busy with work, I think some of you will be able to take the time this weekend to work on the assignment. The assignment is not difficult because it is related to your job, but requires some time to work on. E-learning activities in advance are a prerequisite for attending the face-to-face workshop. Please confirm your work on the pre-training assignment.

(2) Regarding e-learning for the pre-assignment, 148 out of 180 participants from all venues are currently participating. A total of 154 people have taken the quiz, and 57 have submitted their short reports on the discussion forum. There is a lively discussion occurring on the discussion board. Please participate in this discussion. It may be helpful for your professional development as you can learn about other participants' educational cases.

If you have already finished the task, please provide us your review comments on other participants' posts on the discussion forum. This may be helpful for knowledge sharing in the community of the ID course.

If you face any troubles while working on the assignments, don't hesitate to contact me. I look forward to seeing you at the face-to-face workshop. Thank you.

---

4 Method

The effectiveness of improvement of the mentoring system was determined by comparing the data from before (2015) and after (2016) the interventions. The participants and the data used in this research was described.

4.1 Participants

Blended ID workshops, which combine face-to-face seminars with online and at-home preparations, were held in several districts both in 2015 (Tokyo, Osaka, Nagoya, Fukuoka) and 2016 (Tokyo, Osaka, Nagoya, Fukuoka, Kumamoto). Although their locations varied, each workshop followed the same program, and all participants participated in and discussed the same e-learning course. A total of 180 participants in 2015 and 202 participants in 2016 attended the ID course. The ratio of participants' attribution is shown in Fig. 5. Participants from various backgrounds attended the ID course, with some engaging in the course or training as part of their jobs. Although the total number of participants varied, the trends in the ratio of their job types were almost similar between 2015 and 2016.

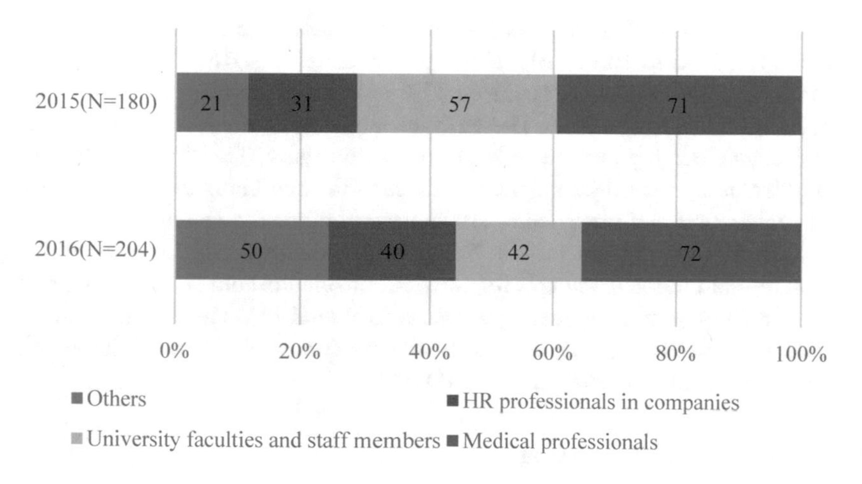

Fig. 6. Ratio of participants' attribution

4.2 Data

Three research datasets were gathered. The first data was regarding the completion rates of online prior assignments and post-learning assignments in order to verify the overall improvement of the mentoring system. The second included logs recorded in the LMS. These datasets were collected to determine the trigger that leads participants to participating in the learning process. The third comprised free descriptions of the final reports. The participants were required to write their impressions of the ID course based on the ARCS motivational model [6]. The purpose of this data was to examine the perceptions of the improvement strategy used for the ID course. The ARCS model may be suitable to determine participants' perceptions from the viewpoint of motivation.

5 Results and Discussions

Detailed results from 2015 to 2017 will be reported in proceedings of this presentation. The completion rates of online prior assignments increased from 66% (118 out of 180 participants) in 2015, 100% (202 out of 202 participants) in 2016 to 100% (181 out of 181 participants) in 2017. The completion rates of the entire program also increased from 35% (63 out of 180 participants) in 2015, 71% (144 out of 202 participants) in 2016 to 77% (140 out of 181 participants) in 2017. These results suggested the effectiveness of the strategy for learning accessibility introduced in this study. A log stored on the LMS would be shown as a result of the reaction to each strategy in detail. Some participant comments in the questionnaire and final reports that were part of the post-learning assignments also demonstrated positive reactions to some interventions.

5.1 Completion Rates of Online Pre-training Assignments and Post-learning Assignments

The completion rates of online prior assignments increased from 66% (118 of 180 participants) in 2015 to 100% (202 of 202 participants) in 2016. The completion rates of the entire program also increased from 35% (63 of 180 participants) in 2015 to 71% (144 of 202 participants) in 2016. These results suggest the effectiveness of the strategy for learning accessibility that was introduced in this study. The strategy introduced in this study included support during the processes at the beginning of the course, such as account registration and pre-training assignments before the face-to-face workshops. These findings are in agreement with the previous findings, which demonstrated that support at the start of the learning is important to enable the completion of assignments, as shown in the UK open university's retention model [5]. The results in this study confirmed that the strategy for supporting the process at the start of the learning is effective in the mentoring system of the ID course.

5.2 Access Logs of E-Learning

The access logs for e-learning were verified as a result of the interventions. Figure 6c shows the transition of access to e-learning websites before the face-to-face workshops. While stable e-learning access was demonstrated in this duration, we found that access was increasing when we sent motivational e-mails to participants. Some of this timing was included when we sent the participants motivational and reminder e-mails, as shown by the red dot in Fig. 6. These results suggest that motivational e-mails facilitate participants' sense of community and feeling empathy for other participants facing difficulties in working on the assignment, as shown in Table 3. In addition, the e-mails were effective for encouraging pace-setting of the participants' learning. These results imply that the motivational e-mail design based on Visser's MMSS [6] may also be effective in supporting completion of lifelong learning opportunities in addition to formal education (Fig. 7).

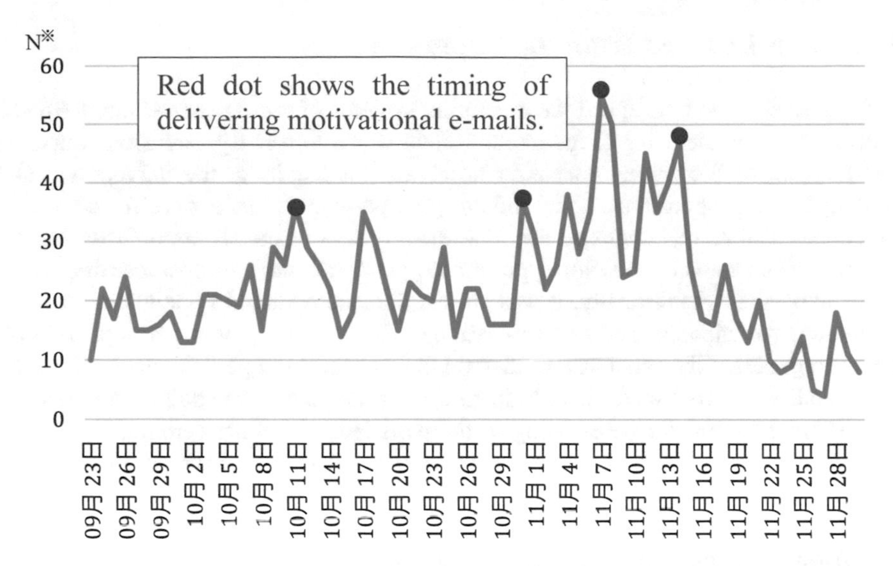

Fig. 7. Access logs to e-learning in the LMS in 2016 ※N showed the numbers of those who accessed e-learning website.

5.3 Participants Comments in the Final Reports

Participants were required to post short reports on the discussion forum, which included an analysis and improvement proposal from the viewpoint of motivation, based on ARCS model. Here, we introduce some participants' comments regarding the mentoring system in the ID course.

Some participants reported that the promotion websites encouraged them to learn in the ID course: "The learning objectives were clearly shown in the syllabus of the promotion websites. It was also possible for me to recognize the learning requirements of what I was required to do through the ID course"; "The phrase 'teach how to teach' written on the websites was attractive for me because I want to learn teaching methods for training. In addition, the evaluation method and course program are displayed on the websites, enabling me to imagine what I should do in the course. This made me feel familiarity with the course."

In this manner, the syllabus provided on the websites gave participants a perspective of the course and motivated them to learn during the course.

Furthermore, other participants commented the following regarding motivational emails: "I did not forget the pre-training assignment because the motivational e-mail reminded me to work on the assignment. The progress of other participants helped me feel that I am not alone as there are other participants."; "The motivational e-mails reminded me of the purpose of attending the course. Thanks to the e-mail, I knew there were many participants who encountered the same problems with their jobs as me."

These comments suggest that motivational e-mails were helpful in enabling participants to feel a sense of belonging to the course in the online learning platform.

6 Lesson Learned from the Experience

In this research, we redesigned the mentoring systems of the ID course and verified its effects. The characteristics of the research field was a university extension course that was not part of the formal education necessary for credits at the university, but of lifelong learning opportunities that did not give participants incentives for completing the course. The results demonstrate that the redesign of the ID course may improve the completion rates of the online pre-training assignments and post-learning assignments altogether. Additionally, it was shown that motivational e-mails are useful for facilitating participants' access to e-learning. Furthermore, some participants' comments suggest that the syllabus provided on the website gave participants a perspective of the course and motivated them to learn during the course; in addition, motivational e-mails were helpful for pace-setting of the participants' online learning.

Based on these results and reflections on our practices, we present our conclusions. The features of this research setting were that the workshops were held as open seminars that did not offer incentives such as university credits or promotion at work. The learning contents also followed an instructional design, targeting those who had engaged in some education. The following are the lessons learned and implications for those who attempt a similar study in similar settings, to improve the mentoring support of pre-training preparation in a blended digital badge program.

1. Create step-by-step manuals to work on online assignments, clarifying the learning path. It might be effective for supporting participants' work and increase the completion rates of online pre-training and post-learning assignments (UK open university's retention model [5]).
2. Provide explicit learning requirements through syllabus on the promotion website before application. This gives participants a perspective of the learning process and clearly informs them about the learning requirements (UK open university's retention model [5]).
3. Deliver motivational e-mails for sharing other participants' learning progress to facilitate pacing. This increased the access to e-learning and enabled participants to feel a sense of belonging (Visser's MMSS 1 [6]).

These findings show that effort to inform the participants of the learning processes facilitated their working on the assignments. However, these design principles were not examined in educational cases other than the blended digital badge in this study. Further investigation of this point is therefore recommended for refining the design principles through trials in various training programs.

Acknowledgment. This work was supported by JSPS KAKENHI Grant Number JP17K12948.

References

1. Amano, K., Suzuki, K., Tsuzuku, S., Hiraoka, N.: Designing a digital badge as a reflection tool in blended workshops. J. Inf. Syst. Educ. **16**(1), 12–17 (2017)
2. Simpson, Ormond: Student Retention in Online, Open and Distance Learning. Routledge, London (2003)
3. Kasami, N., Takeuchi, T., Matsuda, T., Sudo, K., Saito, Y.: Redesign and development of e-mentor short training program for e-learning courses based on '3C + C' model. In: McFerrin, K., Weber, R., Carlsen, R., Willis, D. (eds.) Proceedings of SITE 2008–Society for Information Technology & Teacher Education International Conference, pp. 5113–5118. Association for the Advancement of Computing in Education (AACE), Las Vegas (2008). https://www.learntechlib.org/primary/p/28081/
4. Chang, S.: The roles of mentors in electronic learning environments. AACE J. **12**(3), 331–342 (2004). Norfolk, VA: Association for the Advancement of Computing in Education (AACE). Retrieved May 1, 2019 from https://www.learntechlib.org/primary/p/4881/
5. Simpson, O.: Supporting Students in Online, Open and Distance Learning, 2nd edn. Kogan Page, London (2002)
6. Visser, L.: The Development of Motivational Communication in Distance Education Support. University of Twente, Enchede (1998)
7. Keller, J.M.: Motivational Design for Learning and Performance: The ARCS Model approach. Springer Science + Business Media LLC, New York (2010)

Natural Language Processing and Deep Learning for Blended Learning as an Aspect of Computational Linguistics

Marcel Pikhart(✉)

Faculty of Informatics and Management, University of Hradec Kralove, Hradec Kralove, Czech Republic
Marcel.pikhart@uhk.cz

Abstract. Machine learning has been used for several years already in information science to create algorithms which enable computers to solve problems without giving them particular tasks and instructions to perform these tasks. This paper attempts to concentrate on possibilities of this AI (artificial intelligence) subfield inasmuch it could prove helpful for blended learning. It brings possible questions which are connected to e-learning, blended learning and machine learning and its utilization in university courses. It also brings several questions of computational linguistics which could prove extremely helpful in blended learning processes in which it could bring big data analysis. These phenomena are relatively new, therefore, neglected by blended learning scholars, thus this paper brings these new ideas together and wants to present new ideas and concepts which will be utilized in blended learning area. It also suggests new approach to blended learning, coined by the term blended learning 2.0, which implements modern approaches such as computational linguistics and corpus linguistics into the utilization of e-platforms in the educational process.

Keywords: Blended learning · E-learning · Computational linguistics · Corpus linguistics · Natural language processing · Deep learning

1 Computational Linguistics in Blended Learning

Computational linguistics is now a classic linguistics discipline which utilizes statistical data collected from a computational perspective and then analysed giving thus answers to traditional linguistics questions [1–3]. The paper brings these traditional questions and attempts to prove that they may be used efficiently in blended and hybrid learning.

Computer aided learning has been ubiquitous for a few years and so is computational linguistics, however, we don't see much utilization of the latter in blended learning yet and also the literature on the issue is basically non-existent despite the vast literature on the use of mobile apps in the learning process [4–9]. The paper, therefore, stresses the importance of this use of modern approaches which are already present around us in business, marketing and sales [17–28], but still somehow neglected by educators and curricula creators for unknown reasons. In business and marketing all these modern tools are used vastly to analyse customer behavior, to predict their future behavior and motivate them so that their shopping performance is improved and

M. E. Auer and D. May (Eds.): REV 2020, AISC 1231, pp. 842–848, 2021.
https://doi.org/10.1007/978-3-030-52575-0_69

therefore also the performance of the company. The same algorithm can be used in the learning process, i.e. both the institution and the user will experience enhancement of their performance and involvement in the learning process [11, 14, 16].

What are the aspects of computational linguistics and big data analysis and which can be utilized both in blended learning processes and applied linguistics? The paper tries to stress how much computational linguistics, corpora analysis, big data mining, etc., could help when designing and implementing blended learning courses into curricula [29, 30]. Thus, it brings basic ideas which are still missing in the area of blended learning and could be as an inspiration for further research and implementation of these modern approaches to the learning process [12, 31, 36].

2 Research

The research was conducted to prove that even a very basic tool which uses elementary artificial intelligence algorithm or some kind of computational linguistics or corpora linguistics will be very efficient in blended learning process and can bring very quick but long-term gains.

2.1 Research Methodology

The research was conducted in students of Faculty of Informatics and Management of the university of Hradec Kralove, the Czech Republic, in the winter semester of the academic year 2018/2019 into the issues and questions which arise in vocabulary retention process after the use of an online course of the English language which uses basic artificial intelligence algorithm. The number of respondents was 42, divided randomly into two groups. Computational linguistics methodology was used to obtain data on their efficacy of learning new vocabulary and vocabulary retention in the users of the tool.

There were two groups of students - one which used only traditional methods of learning vocabulary with the use of Blackboard platform (scanned materials and lists of new words presented by the Blackboard), and the other group which used on-line tool to improve their vocabulary. The on-line tool used basic artificial intelligence process, i.e. the tool which is able to analyse the performance based on the previous failures of the student and derives further new processes based on this performance. If the student answers incorrectly, then the future test incorporates all these failures into the future tests so that the student is permanently tested on the vocabulary they are not able to retain until the retention of the new vocabulary is reached, then, later on, the student is systematically tested on the retention again and again.

The data were collected based on the tests of vocabulary comprehension and retention in the two groups. Also, the use of new vocabulary was tested in the use of them in essay writing. The students had to write an essay at the beginning of the semester and another one at the end of the semester and were tested on their vocabulary improvement – i.e. corpora analysis was used, which means that their vocabulary improvements were tested statistically by a computer, so that it was objectively possible to measure and quantify the vocabulary retention.

In all cases computational linguistics, i.e. collecting and analyzing data using big data analysis was used.

2.2 Research Question and Hypothesis

The research question was whether there will be any significant improvement in vocabulary retention in the group which uses on-line tools to improve their vocabulary compared to the group which uses only traditional approaches.

The hypothesis was that there will be a significant improvement in the group which will use the new approach.

3 Findings

It is clearly visible that big data analysis is an extremely powerful tool which can be used in many theoretical and practical areas of human expertise and blended learning is one of them. The results of the research clearly showed better progress in the group of the students who use online tools to vocabulary learning. The first group, i.e. using only Blackboard platform, compared to the second group proved much worse performance in the students' new vocabulary retention. Therefore, the hypothesis was confirmed.

The aim of the semester was to improve the vocabulary of the students by 300 new words from the area of finance and management. The second group proved to have better results at the end of the semester by 39% compared to the group one in the vocabulary test at the end of the course. In the essay analysis, the second group was able to incorporate and used new vocabulary by 23% compared to the parallel group. These simple facts bring us to a clear finding that this kind of blended learning enhanced by artificial intelligence proves to be much more efficient even if we take into account possible statistical deviation and research limitations.

Big data analysis proved to be very helpful mostly in the essay part, where it is almost impossible for the teacher to analyse manually the use of newly acquired vocabulary - only big data analysis can manage such a large corpus which is used. Statistically, it is visible that corpus linguistics and computational linguistics can prove helpful in blended learning as it provides us with invaluable data which would not be obtainable through traditional research methods.

4 Limitations of the Research

It must be noted that it is possible that the research sample was not large enough and the research was minimalistic, however, it can be postulated that the research provides enough valid data which can lead to our conclusions. Of course, it is also necessary to say that various students were tested and their initial knowledge and learning performance as a group is not perfectly the same so that in the second group we may have had better students or vice versa.

We also don't know anything about the learning styles and learning preferences of the respondents so it was impossible to create statistically equalized and balanced groups of respondents, which could give us much more statistically relevant results.

Therefore, further research must be conducted with much larger group of respondents, and the author of the paper is already working on it so that we obtain more relevant and objective results which could be further processed with higher statistical relevance. However, the results were so clear that the limitations of the research might be neglected when we want to receive at least some basic picture of the benefits of corpora analysis and computational linguistics in blended learning.

5 Discussion

The conducted research clearly showed, despite its limitations, that there is a huge potential of the use of artificial intelligence and all these afore-mentioned approaches in blended learning. It must be added that there is basically no research into this topic and blended learning professionals are ignoring this approach in their research, therefore, this paper is a daring attempt to change the situation by showing and highlighting the importance and benefits of these modern methodologies.

The modern era of blended learning is only feasible if we take into account these changes in technology and will implement them into on-line courses which are used in universities and other higher education institutions [10, 13, 33, 34]. It is a question of future success of educational institutions because more and more young users call for the use of modern technologies in education. A lot of research proves that this use of modern mobile devices is inevitable, however, it will not be enough if we only stick to the traditional, i.e. old-fashioned approaches of presenting print materials on-line. This approach cannot, by any means, be called on-line or blended learning.

Corpora linguistics and computational linguistics can also be a very useful tool for the teachers because of the objectivity of the test results as it gives very systematic and quantifiable results which are almost impossible to be collected by a human brain.

6 Future Outlook of Blended Learning: Blended Learning 1.0 vs Blended Learning 2.0

Traditional blended learning utilizes electronic platforms as a repository of various texts and materials used by the students, such as pdf scans, videos, short films, audio files, podcasts, etc. It also uses various platforms for testing, submitting essays of the students and, last but not least, communication tools for the tutor and the students and also the students with each other. Various group tools are used to create presentations and other group assignments. All these approaches and tools are very useful and have been used widely in all kinds of e-learning, blended learning and hybrid learning approaches, however, they are not sufficient and should be considered obsolete if they are not enriched by afore-mentioned approaches. We can coin the term blended learning 1.0 for all these traditional approaches.

Blended learning 1.0 utilizes e-learning platforms in a way that is rather old-fashioned inasmuch it does not count on the findings of computational linguistics, corpora linguistics, deep learning, etc. All these mentioned are modern tools which can move blended learning to a much higher level of efficiency.

Blended learning 2.0 will use all these methodologies used by blended learning 1.0, however, will add a few very useful and effective tools already known in ITC departments. Thus, blended learning should use data analysis, big data mining and other modern useful tools already used by large companies to analyse customer behaviour and on their previous purchases build new more customized offers and increase thus customer traffic. In blended learning 2.0, it should work in the same way. The mistakes of the students should be analysed using large databases of their previous mistakes so that the tests should be created automatically based on their previous performance, so that the test itself does not only work as a test but as a very efficient educational tool which is able to compare previous performances of the student, see their progress, and suggest improvements. Students' essays can be analyses by corpus linguistics methodology, i.e. computational comparison of vast databases of the texts produced by the students, and again, the progress of, for instance, vocabulary development and syntax improvement, etc., can be observed and compared to previous performance.

The possibilities of blended learning 2.0 are vast and it is almost incredible how it is possible all these information technologies are not used to their fullest potential in blended and hybrid learning [15, 32, 35, 37]. This paper is not only a summary of modern approaches which *could* be implemented, but rather an inspiration that something *should* be implemented into our attempts to create successful and efficient tools for educational purposes. Blended learning 2.0 is a challenge we all must face and try to make it real, otherwise we can never succeed as global educators.

Acknowledgements. The paper was created with the support of SPEV 2019 at the Faculty of Informatics and Management of the University of Hradec Kralove, Czech Republic. The author would like to thank the student Jan Sprinar for his help when collecting the data of the research.

References

1. Alpaydin, E.: Machine Learning. The New AI. MIT Press, Cambridge (2016)
2. Buckland, M.: Information and Society. MIT Press, Buckland, M.: Information and Society. MIT Press (2017)
3. Clark, A., et al.: The Handbook of Computational Linguistics and Natural Language Processing. Blackwell, Chichester (2010)
4. Klimova, B.: Teacher's role in a smart learning environment—a review study. In: Uskov, V., Howlett, R.J., Jai, L.C. (eds.) Smart Innovation, Systems and Technologies, vol. 59, pp. 51–59. Springer (2016)
5. Klimova, B.: Assessment in the eLearning course on academic writing – a case study. In: Wu, T.T., Gennari, R., Huang, Y.M., Xie, H., Cao, Y. (eds.) Emerging Technologies for Education, SETE 2016. LNCS, vol. 10108, pp. 733–738 (2017)

6. Klímová, B., Berger, A.: Evaluation of the use of mobile application in learning English vocabulary and phrases – a case study. In: Hao, T., Chen, W., Xie, H., Nadee, W., Lau, R. (eds.) Emerging Technologies for Education, SETE 2018. LNCS, vol. 11284, pp. 3–11 (2018)
7. Klimova, B., Poulova, P.: Mobile learning and its potential for engineering education. In: Proceedings of 2015 I.E. Global Engineering Education Conference (EDUCON 2015), pp. 47–51. Tallinn University of Technology, Estonia, Tallinn (2015)
8. Klimova, B., Poulova, P.: Mobile learning in higher education. Adv. Sci. Lett. **22**(5/6), 1111–1114 (2016)
9. Klimova, B., Simonova, I., Poulova, P.: Blended learning in the university English courses: case study. In: Cheung, S., Kwok, L., Ma, W., Lee, L.K., Yang, H. (eds.) Blended Learning. New Challenges and Innovative Practices, ICBL 2017. Lecture Notes in Computer Science, vol. 10309, pp. 53–64. Springer (2017)
10. Lopuch, M.: The effects of educational apps on student achievement and engagement (2013). http://www.doe.virginia.gov/support/technology/technology_initiatives/e-learning_backpack/institute/2013/Educational_Apps_White_Paper_eSpark_v2.pdf. Accessed 2 Mar 2018
11. Luo, B.R., Lin, Y.L., Chen, N.S., Fang, W.C.: Using smartphone to facilitate English communication and willingness to communicate in a communicative language teaching classroom. In: Proceedings of the 15th International conference on Advanced Learning Technologies, pp. 320–322. IEEE (2015)
12. Males, S., Bate, F., Macnish, J.: The impact of mobile learning on student performance as gauged by standardised test (NAPLAN) scores. In: Issues in Educational Research, vol. 27, no. 1, pp. 99–114 (2017)
13. Mehdipour, Y., Zerehkafi, H.: Mobile learning for education: Benefits and challenges. Int. J. Comput. Eng. Res. **3**(6), 93–101 (2013)
14. Miller, H.B., Cuevas, J.A.: Mobile learning and its effects on academic achievement and student motivation in middle grades students. Int. J. Sch. Technol. Enhanced Learn. **1**(2), 91–110 (2017)
15. Muhammed, A.A.: The impact of mobiles on language learning on the part of English foreign language (EFL) university students. Procedia Soc. Behav. Sci. **136**, 104–108 (2014)
16. Oz, H.: Prospective English teachers' ownership and usage of mobile device as m-learning tools. Procedia Soc. Behav. Sci. **141**, 1031–1041 (2013)
17. Pikhart, M.: Sustainable communication strategies for business communication. In: Soliman, K.S. (ed.) Proceedings of the 32nd International Business Information Management Association Conference (IBIMA), 15–16 November 2018, Seville, Spain, pp. 528–553. International Business Information Management Association (2018). ISBN: 978-0-9998551-1-9
18. Pikhart, M.: Intercultural business communication courses in european universities as a way to enhance competitiveness. In: Soliman, K.S. (ed.) Proceedings of the 32nd International Business Information Management Association Conference (IBIMA), 15–16 November 2018, Seville, Spain, pp. 524–527. International Business Information Management Association (2018). ISBN: 978-0-9998551-1-9
19. Pikhart, M.: Multilingual and intercultural competence for ICT: accessing and assessing electronic information in the global world (MISSI 2018). In: Advances in Intelligent Systems and Computing, 11th International Conference on Multimedia and Network Information Systems, MISSI 2018, Wroclaw, Poland, 12 September 2018 through 14 September 2018, vol. 833, pp. 273–278 (2018). https://doi.org/10.1007/978-3-319-98678-4_28. ISSN 2194-5357

20. Pikhart, M.: Technology enhanced learning experience in intercultural business communication course: a case study. In: Hao, T., Chen, W., Xie, H., Nadee, W., Lau, R. (eds.) Emerging Technologies for Education. Third International Symposium, SETE 2018, Held in Conjunction with ICWL 2018, Chiang Mai, Thailand, 22–24 August 2018, Revised Selected Papers. Book Series: Lecture Notes in Computer Science. Springer (2018). Print ISBN: 978-3-030-03579-2, Electronic ISBN: 978-3-030-03580-8
21. Pikhart, M.: Communication based models of information transfer in modern management - the use of mobile technologies in company communication. In: Soliman, K.S. (ed.) Proceedings of the 31st International Business Information Management Association Conference (IBIMA) 25–26 April 2018, Milan, pp. 447–450. International Business Information Management Association (IBIMA) (2018). ISBN: 978-0-9998551-0-2
22. Pikhart, M.: Current intercultural management strategies. The role of communication in company efficiency development. In: Proceedings of the 8th European Conference on Management, Leadership and Governance (ECMLG), pp. 327–331 (2012)
23. Pikhart, M.: Communication based models of information transfer in modern management – the use of mobile technologies in company communication. In: Innovation Management and Education Excellence through Vision 2020, IBIMA 2018, pp. 447–450 (2018)
24. Pikhart, M.: Electronic managerial communication: new trends of intercultural business communication. In: Innovation Management and Education Excellence through Vision 2020, IBIMA 2018, pp. 714–717 (2018)
25. Pikhart, M.: Managerial communication and its changes in the global intercultural business world. In: Web of Conferences (ERPA 2015), vol. 26 (2016)
26. Pikhart, M.: Intercultural linguistics as a new academic approach to communication. In: Web of Conferences (ERPA 2015), vol. 26 (2016)
27. Pikhart, M.: Implementing new global business trends to intercultural business communication. In: Procedia Social and Behavioral Sciences, ERPA 2014, vol. 152, pp. 950–953 (2014)
28. Pikhart, M.: New horizons of intercultural communication: applied linguistics approach. In: Procedia Social and Behavioral Sciences, ERPA 2014, vol. 152, pp. 954–957 (2014)
29. Simonova, I., Poulova, P.: Innovations in data engineering subjects. Adv. Sci. Lett. **23**(6), 5090–5093 (2017)
30. Simonova, I., Poulova, P.: Innovations in enterprise informatics subjects. Adv. Intell. Syst. Comput. **544** (2017)
31. Sung, Y.T., Chang, K.E., Liu, T.C.: The effects of integrating mobile devices with teaching and learning on students' learning performance: a meta-analysis and research synthesis. Comput. Educ. **94**, 252–275 (2016)
32. Tayan, B.M.: Students and teachers' perceptions into the viability of mobile technology implementation to support language learning for first year business students in a Middle Eastern University. Int. J. Educ. Literacy Stud. **5**(2), 74–83 (2017)
33. Teodorescu, A.: Mobile learning and its impact on business English learning. Procedia Soc. Behav. Sci. **180**, 1535–1540 (2015)
34. Tingir, S., Cavlazoglu, B., Caliskan, O., Koklu, O., Intepe-Tingir, S.: Effects of mobile devices on K–12 students' achievement: a meta-analysis. J. Comput. Assist. Learn. **33**(4), 355–369 (2017)
35. Wu, Q.: Learning ESL vocabulary with smartphones. Procedia Soc. Behav. Sci. **143**, 302–307 (2014)
36. Wu, Q.: Designing a smartphone app to teach English (L2) vocabulary. Comput. Educ. (2015). https://doi.org/10.1016/j.compedu.2015.02.013
37. Wu, Q.: Pulling mobile assisted language learning (MALL) into the mainstream: MALL in broad practice. PLoS ONE **10**(5), e0128762 (2015)

Infret: Preliminary Findings of a Tool for Explorative Learning of Information Retrieval Concepts

Aleksandar Bobić[1(✉)], Christian Gütl[1], and Christopher Cheong[2]

[1] Graz University of Technology, Graz, Austria
bobic.aleksandar92@gmail.com, c.guetl@tugraz.at
[2] RMIT University, Melbourne, Australia
christopher.cheong@rmit.edu.au

Abstract. To help students better understand concepts in an information search and retrieval (ISR) class using the Motivational Active Learning (MAL) pedagogical approach, a tool called Infret was developed. It serves as the interactive experimentation and visualization component of MAL. The design of Infret is based on feedback collected from a testing session with a first proof of concept tool built in Java a year earlier. Infret was developed using Web technologies in order to have a simpler update process and be accessible from any Internet connected device with reasonable screen sizes (PC or tablet) without the need for manual installation. It was used and evaluated as part of a text statistics exercise in which students explored different properties of a text-based document collection. When students completed the exercise, they filled out a multi-part survey about Infret. Findings revealed that Infret helped participants gain a better understanding of text statistics and that the activity was well received by them. The usability score of Infret was 76.9, which indicates a usability level that is slightly above average and the emotions scale results were mostly positive. Additionally, the most commonly experienced emotion by students was happiness followed by negative emotions such as sadness, anxiety and anger which were experienced almost none of the time. There were multiple improvement suggestions such as improving the responsiveness of the UI on small screens, offering the option to see the content of text collections and the addition of formula explanations. These suggestions will be considered for improving Infret in future versions. Additionally, students expressed they would like to use Infret in other areas of information retrieval and also in other subjects.

Keywords: Information retrieval · Active learning · Interactive visualisation tool · Web-based learning · Explorative learning

1 Introduction

Existing studies [4, 12] have identified some of the common issues students face when learning in class. These issues include losing focus during face-to-face activities which inhibits students' ability to understand important concepts during class, low knowledge

M. E. Auer and D. May (Eds.): REV 2020, AISC 1231, pp. 849–865, 2021.
https://doi.org/10.1007/978-3-030-52575-0_70

retention after class or after the course has finished and limited understanding of abstract concepts.

An example of an area in which the understanding of abstract concepts is critical is information retrieval (IR) in which many concepts rely on strong understanding of text statistics and mathematical concepts. For details on these concepts refer to [20]. Therefore, students should be actively engaged during classes and have the opportunity to explore concepts taught in class.

An idea that describes learning as a continuous process built on gaining experience and reflecting on that experience is experiential learning [14]. However, experiential learning does not specifically take into account virtual environments. That is why exploratory learning [8] expands its definition and introduces the concept of exploration which helps learners expand their knowledge. Active learning [9] can be seen as one of the popular pedagogic approaches that attempts to address the issue of loss of focus during class. Multiple evaluation studies concluded that engaging or active delivery of knowledge during class increases students focus during class. To attempt to address other issues such as low knowledge retention and having problems understanding abstract concepts new approaches built on active learning had to be created.

Two approaches that try to tackle the issues mentioned above are Technology-Enabled Active Learning (TEAL) and Motivational Active Learning (MAL). TEAL attempts to improve active learning and address the issue of low knowledge retention and understanding of abstract concepts with the use of computer-based interactive visualisations [3]. MAL expands the concepts of TEAL by introducing gamification elements in an attempt to motivate students and lower their negative feelings towards the new approach [16].

MAL was applied and evaluated in an information search and retrieval (ISR) course at Graz University of Technology for several years. The experiences in the ISR course indicate there is a need for an interactive tool for experimentation and visualisation that would help students understand the abstract concepts taught at the course. Thus, this research paper focuses on the conceptualisation, development and evaluation phases of developing an interactive exploration tool for understanding IR concepts.

The remainder of the paper is organised as follows; Sect. 2 introduces background and related work on some common approaches that attempt to address the previously mentioned issues students experience while studying. The conceptual design and architecture of our exploration tool are described in Sect. 3. The evaluation process and its results are discussed in Sect. 4. Finally, the paper is concluded and possibilities for future work are presented in Sect. 5.

2 Background and Related Work

2.1 Selected Learning and Teaching Approaches

Students in traditional face-to-face lessons tend to have common learning issues. These include low attention span and low knowledge recall. Students interest in face-to-face classes needs to be raised by actively engaging them and offering them possibilities for exploration of concepts taught in the class [4, 12].

One well known approach to overcome the aforementioned issues is experiential learning [14], which is based around the idea that knowledge is created through the transformation of experience and that learning is a continuous process. The approach models learning as a cycle of concrete experience, followed by observation and reflection on the experience. These observations lead to the formation of abstract concepts which have certain implications. These implications are then tested and lead to new experiences. Exploratory learning [8] expands the definition of experiential learning by taking into account interaction in virtual learning environments. Furthermore, it introduces the concept of exploration which can occur in either specifically designed activities or in open-ended activities and in physical or virtual environments and helps learners expand their knowledge.

A popular learning approach that attempts to address the issues of low attention span and low knowledge recall is active learning [9], which engages students in activities during a lecture. These are typically instructional activities completed by students in which they actively think about what they are doing [9]. A simple example of active learning is introducing a short 2-min pause every 15 min to let students clarify their notes [17]. This approach was evaluated and shown to be effective in increasing student recall [18]. A different study of 6,000 students found that interactive engagement increases students' performance [10]. It was also found that active learning effectively increases the examination performance of students in STEM classes and lowers the possibility of failing a class [7].

One of the approaches that attempts to build on and further improve the concepts of active learning is TEAL [3]. It combines aspects of active learning with technology aid in order to decrease the failure rate of undergraduate students in a physics course and to change the way electromagnetism is taught at MIT. It leverages visualisations and animations to help students understand abstract concepts. TEAL classes are structured from smaller lectures or recitations, each followed by hands-on experiments and discussions. This type of activity also requires students to work in groups in a specially designed classroom. The evaluation of students attending the TEAL class indicates they had a substantial increase in conceptual understanding and lower failure rates compared to non-TEAL students in a similar course. In addition, an evaluation of students who attended the TEAL class 12 to 18 months after the course has ended found that these students had higher knowledge retention of the course content than the non-TEAL students [4]. Although TEAL was successful in its goal to help students improve knowledge retention the students showed a lot of resistance as the new approach required them to adjust from the traditional lectures.

An approach that attempts to help students transition from traditional lectures into a more engaging environment is Motivational Active Learning (MAL) [16]. It combines the active and interactive aspects of TEAL with gamification approaches to motivate students. It includes various methods such as group work and assignment of points based on performance in order to create a competitive but also teamwork enabling environment. The basic structure of MAL consists of short face-to-face lecture activities, each followed by engaging tasks and quizzes. At the end of each task students immediately receive points for the task and are encouraged to further improve. This enables students to immediately react on feedback on their performance and improve it in case they are not happy with the result. The overall progress of each of the class and course performance can be transparently followed by each student. In order to master the course, it is expected from students to master each topic with more the 50%. However, repeated attempts to improve the final score and students' knowledge are highly encouraged.

MAL was first used in 2013 in an Information Search and Retrieval course (ISR) in a class of 28 students. The general attitudes towards the new methodology were assessed in this first use and the results indicate that students enjoyed receiving points instead of grades. They were also motivated to complete further assignments to achieve more points. Students liked the interactivity of the classes, however, there are possibilities for improvement, such as the introduction of a tool for experimentation comparable with hands-on labs in physics education. MAL proved to be effective in motivating students. In order to increase the course interactivity and make the abstract content of the ISR course more understandable, an interactive tool for experimenting and visualising related concepts is in the focus of the next improvement cycle.

2.2 Existing Interactive Tools

Among existing tools there is the Information Retrieval Game [11], which is a web-based tool for improving learners' query forming capabilities by enabling them to write queries, administering the queries on textual and image test collections and evaluating the results of these queries using precision/recall curves and relevance bars. Another solution is the IR-Toolbox, an experiential teaching tool which helps non-technical students understand the processes and algorithms of IR systems. It presents the document analysis, searching and evaluation steps of IR. However, there are no detailed information about the tool to our knowledge [5].

Another student-oriented tool [15] attempts to improve students' understanding of fuzzy information retrieval systems (FIRS). It enables students to create custom test collections or use existing ones, define weighted Boolean queries, use the custom queries on the test collections and evaluate and visualise their performance. Findings revealed it increased students' understanding of FIRSs models, their motivation and their marks in examinations.

The tool VIRLab enables users to implement and evaluate custom retrieval functions, configuring a custom search engine by selecting a test collection and a retrieval function and compare their retrieval functions with other users retrieval functions through leaderboards [6].

Finally, Xtrieval Web Lab is a web-based tool structured from multiple configurable components with a focus on helping users with no programming knowledge understand the IR process [21]. It provides gamification through a level (new levels unlock new locked components) and achievement system (new achievements are unlocked by exploring components which are not part of the level system).

The aforementioned approaches described attempts to support students learning of query relevance concepts, models of search engines and enable them to write retrieval functions. However, the main focus is not on basic concepts of IR and assume the students have already the appropriate pre-requisite knowledge.

Despite the existence of tools which support the understanding and experimentation of IR concepts, the course layout and activities based on the MAL approach (see also 2.1) is not fully supported. Thus, to foster students in their understanding of basics and also more advanced concepts of IR following the MAL approach, a first proof of concept of an interactive visualisation tool ISRapp has been developed that implements basic text statistics functions and a simple search engine [19]. Although the tool is useful for the purpose of providing an interactive visual experience, possibilities for further improvements have been identified, such as web-based accessibility to be used by various end user devices, higher flexibility for adding new features and for gathering user interaction information. To solve these issues a new web-based tool called Infret was researched, designed and developed. These are described and discussed in the remainder of this paper.

3 Application Design and Architecture

3.1 Aims and Foundational Concepts

Based on the findings of Sect. 2 and the pedagogical concepts of MAL, we were motivated to research an overall concept and start developing a tool for explorative learning of information retrieval concepts, called Infret.

The Infret concept aims to be used in a MAL lecture environment supporting students in learning or better understanding IR concepts by providing an engaging and exploratory interactive visualisation environment with a focus on text documents. The concept also includes the accessibility from any computer or tablet independent of the operating system or the browser. Infret should support a connection to learning management systems such as Moodle and offer short tasks, quizzes and point assignment based on how well they complete a task through them. In addition, students would benefit from a personalised learning path inside of Infret which would guide them based on their progress. Moreover, it would be useful for the lecturer to have an overview of which topics are most commonly viewed in Infret and what the students' progress are. Furthermore, Infret will provide functionalities related to topics of IR such as text statistics and term weighting, enable students to use NLP approaches and to write their own code for example for custom weighting functions. Finally, it should offer the option to crawl web pages and collect data from them, support custom document collections and comparison of different document collections.

3.2 Requirements and Architecture Overview

To create the Infret prototype, we focus only on a small subset of the aforementioned overall concept. Thus, the high-level requirements are:

- Provide an engaging and exploratory environment
- Provide interactive visualisations
- Support multiple text statistics
- Focus on text documents
- Be accessible from any computer or tablet

- Enable multiple developers to work on the project at the same time
- Have low entry barrier for new developers
- Be easily extensible and modular

To satisfy these requirements, a simplified architecture, which can be seen in Fig. 1, has to be defined. The main components in the figure are presented as blocks and the sub-components or details of implementation are presented as sub-blocks. Additionally, most main components include a block with "…" which signifies possible further extensions.

In order to support text statistics, Infret first has to pre-process the text document collections and convert them into an internal representation. The data is first read using the file reader. Once the files are read the parser selector selects the appropriate parser for the files based on the file extension. Currently Infret only supports files with the "cran" file extension. The data is pre-processed using the file parser component which extracts metadata, such as authors, title, and the full text from each file. Finally, the extracted metadata is stored as internal representation in a data storage for later access.

The cached internal representations can then be used by the processing component to extract text statistics. Once the text statistics are calculated, they are cached in the data store in order to be quickly accessible by other components and with that also to provide a faster experience for users.

To provide an engaging and exploratory environment, Infret offers a variety of ways on how to explore and interact with the data. It provides these options by including multiple data visualisations and controls for those visualisations on the client. The implemented visualisations include a bar chart component, a line chart component and a table component for a better overview of the data. To make it easier for users to explore the data it offers a side panel with filters (such as selecting the number of results to be shown and in which format to show the results) which can be used to manipulate the visualisations. Furthermore, the side panel enables users to select a text statistic function and the subset of the text collection they would like to explore. All of these components are hosted in a cloud and accessible through a web-browser in order to satisfy the requirement of being accessible from multiple computers and tables at any time.

3.3 Technology Comparison and Final Decisions

To be accessible from any computer or tablet, Infret runs on a remote server and can be accessed using a web browser. Therefore, the client-side framework choices were limited to JavaScript frameworks popular at the time of the Infret planning phase in 2017. The frameworks considered were Angular[1] version 2 and React[2] version 15 as they proved to be efficient for building large single page web applications on the client-side and were well documented. To enable multiple developers to work on Infret at the same time and have a low entry barrier for new developers on the client side, the framework, we choose had to enforce strict coding styles and enable easy component-based code structuring. Both React and Angular offered modular and extensible

[1] https://angular.io/.

[2] https://reactjs.org/.

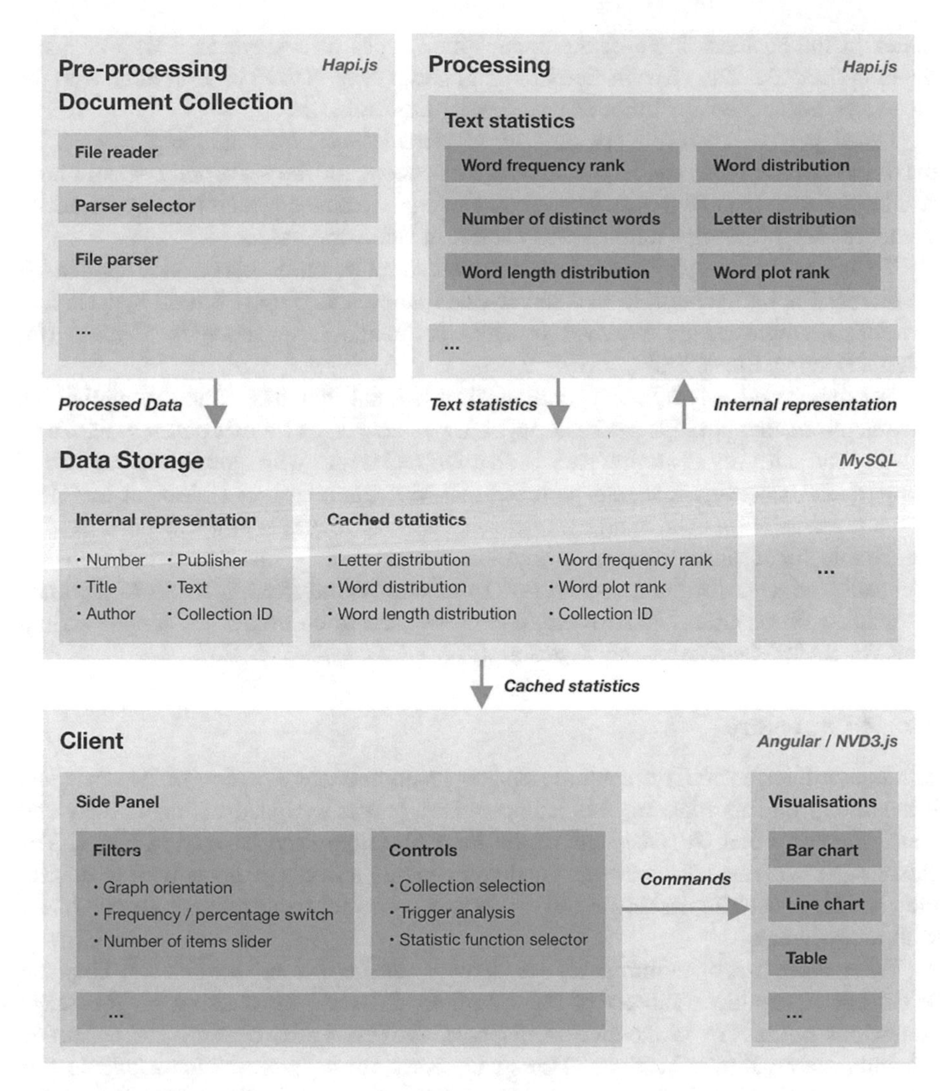

Fig. 1. Simplified architecture of Infret. Connections between components are displayed with arrows. The black italic text next to a connection represents the data passed via that connection. The blue italic text represents the main technologies used in the component.

structuring, however, only Angular enforced a strict coding style and had a code style guide. Thus, Angular was chosen as the client-side framework for Infret.

TypeScript[3] was chosen as the development language of the Infret client-side application as it is the preferred language for Angular. It adds types and advanced language features which are natively not supported by JavaScript. This helps with the development, especially in bigger teams as it is much easier to avoid bugs and find

[3] https://www.typescriptlang.org/.

issues in the business logic of the application. Types also serve as a way of documenting the code. The visualisation library of choice was NVD3.js[4] as it offered all the visualisations needed for Infret and was easily customisable.

Developers will likely work on both server-side and client-side applications. To make the development easier, the language of choice for the server-side should have similarities with the client-side. Therefore, Node.js[5] version 6 which is an event-driven asynchronous JavaScript runtime was chosen for the server-side.

To achieve an easily extensible and modular structure on the server-side application we choose Hapi.js[6] version 15 as a server-side framework. Hapi.js is built by a team of developers employed by Walmart in order to handle large amounts of users (for example during Black Friday[7]).

We compared MySQL[8], a traditional relational database and MongoDB[9] a document-oriented NoSQL database. MySQL was chosen as Infret's database due to its stability and safe way of storing data. MongoDB had issues with data loss at the time of comparison. However, it should be noted that MySQL is also not ideal for text document representation even though it offers an easy way to retrieve data and is proven to be reliable for a much longer. In addition to the already mentioned technologies, Webpack[10] is used for the transpiling of TypeScript to JavaScript, SASS to CSS and automatization of tasks. Furthermore, linters are used to enforce a set of code styling rules for SASS, JavaScript and TypeScript.

3.4 UI Structure

Infrets user interface (UI) structure resembles its component structure in the sense that it is split up into multiple regions (components). It was designed with simplicity and consistency in mind. A screenshot of the Infret prototype can be seen in Fig. 2. The sidebar (A) contains all the navigational components and components used to control the visualisations. The dashboard (B) serves as a container for visualisations and tables with analysis results.

When users first open Infret they see an empty dashboard and the sidebar. They can select one or multiple subsets of the Cranfield dataset[11] by clicking on the select collections button (F) in the sidebar displayed in Fig. 4 and opening the collection selection dialog shown in Fig. 3. Users can select one or more of the available text collections (H). During the selection process the number of files from all selected collections is displayed in the bottom left corner (J) of the dialog. Users can close the

[4] http://nvd3.org/.

[5] https://nodejs.org/en/.

[6] https://hapijs.com/.

[7] The Friday after Thanksgiving in the United States (US) which marks the start of the Christmas shopping season in the US.

[8] https://www.mysql.com/.

[9] https://www.mongodb.com/.

[10] https://webpack.js.org/.

[11] A dataset consisting of 1400 text documents used for testing and evaluation purposes.

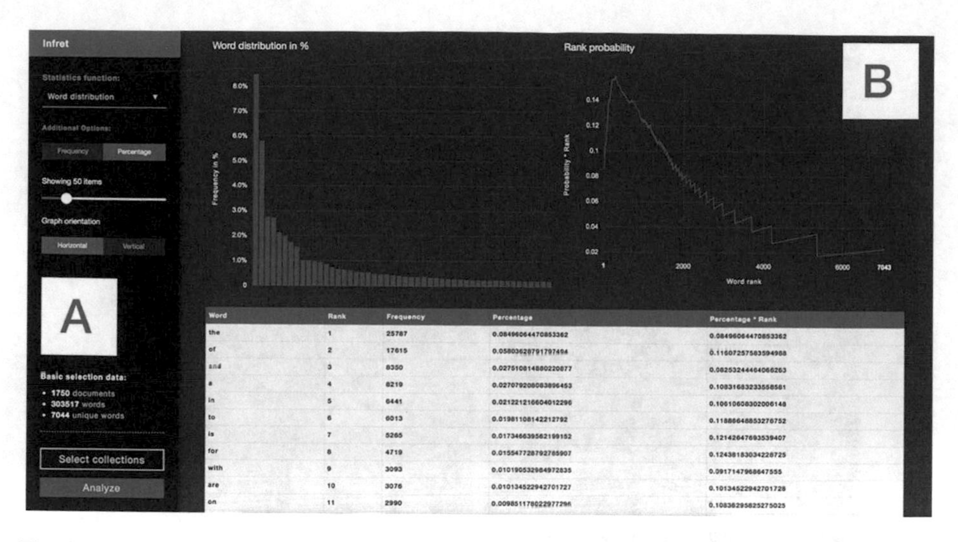

Fig. 2. The Infret prototype screenshot. On the left side (A) is the sidebar containing controls and filters. On the right side (B) is the visualisation dashboard containing the visualization of results.

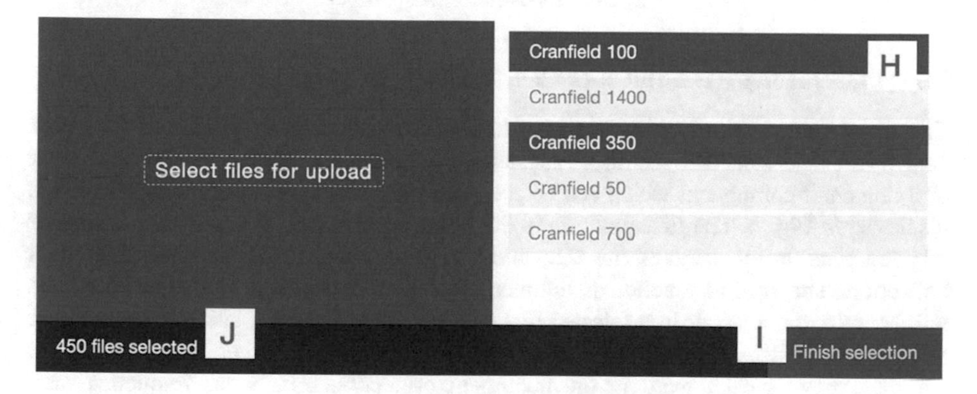

Fig. 3. Collection selection dialog.

dialog by clicking the "Finish selection" button (I). Once a collection is selected, they chose a text statistics function which they would like to use on the selected collection from the dropdown (C) at the top of the sidebar. Once they select a statistic function, they trigger the analysis by clicking on the "Analyze" button (G) at the bottom of the sidebar. When the analysis is completed, its results are displayed in the dashboard. In addition, users can see useful information about the selected collections (E) at the bottom of the sidebar. Users can then either chose to manipulate the visualisations and explore the results by using the controls in the middle of the sidebar (D) or select a different text statistics function.

Fig. 4. The sidebar component.

3.5 Implemented Text Statistics Concepts

The Infret prototype offers students six text statistics functions to choose from and analyse the data with. The results of the statistic function processing are shown in a bar or a line chart component which can be seen in Fig. 5 and a table component which is displayed in Fig. 6. The first function is the letter distribution. When users administer this function, Infret displays the frequency of each letter in the selected document collection. The second function is number of distinct words which displays how the number of distinct words in a selected text collection changes in relation to the number of words in the selected collection. The word distribution function enables users to see the frequency of each word in the document collection. The word frequency rank displays the number of words with a certain frequency (for example 503 words have a frequency value of 3). Users can also see the distribution of word lengths by using the word length distribution function and get an insight into the relationship between the word frequency and word rank values in a logarithmic scale by using the word plot rank.

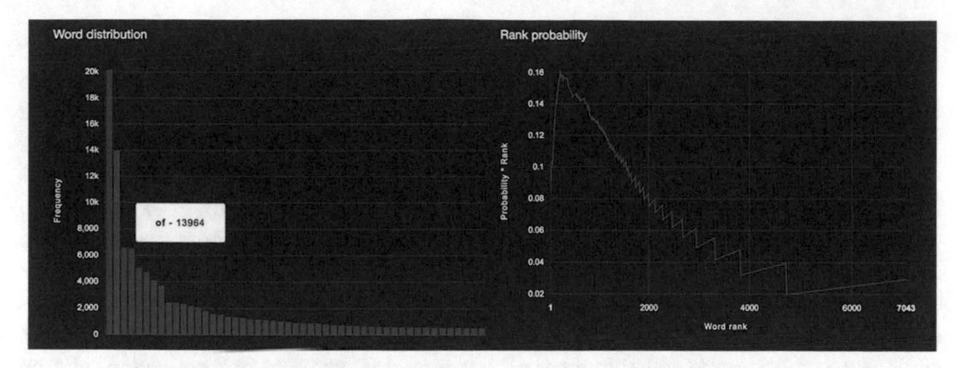

Fig. 5. Example of a bar chart and a line chart component.

Word	Rank	Frequency	Percentage	Percentage * Rank
the	1	5611	0.08708405761112491	0.08708405761112491
of	2	3651	0.05666439036503601	0.11332878073007202
and	3	1768	0.027439781475043456	0.08231934442513036
a	4	1713	0.026586168363546065	0.10634467345418426
in	5	1407	0.02183697541594239	0.10918487707971195
to	6	1299	0.020160789669729326	0.12096473801837596
is	7	1151	0.0178637943878818	0.1250465607151726
for	8	1015	0.01575304196672461	0.12602433573379687
flow	9	692	0.010740004966476285	0.09666004469828657
on	10	661	0.010258877576359572	0.10258877576359572
with	11	653	0.010134715669232679	0.11148187236155946
are	12	648	0.01005711447727837	0.12068537372734045
at	13	565	0.008768934690836852	0.11399615098087908
boundary	14	498	0.007729078718649119	0.10820710206108766
layer	15	470	0.007294512043704991	0.10941768065557468
by	16	463	0.007185870374968959	0.11497392599950335

Fig. 6. Example of a table component.

4 Evaluation

4.1 Study Design

The focus of the study is the evaluation of the Infret prototype. Additionally, we want to see if Infret can help students understand or deepen their knowledge of text statistics concepts. Moreover, we want to get an insight into what is the overall students' experience of Infret's first version like. We do this by comparing the usability scores and emotions experienced during the usage of Infret to the results of the ISRapp evaluation.

To answer the aforementioned questions, we first provide students with instructions for predefined text-statistic-related learning activities which require the usage of Infret, Excel, or Python, and knowledge gained during the ISR course. Students are given one week to complete activities outside of class and submit their answers. Once the activities are completed students are provided with a link to an online questionnaire. Finally, the results of the questionnaire are evaluated in order to reach final conclusions.

4.2 Setting and Instruments

The five activities given to students are divided based on the concepts they were designed to teach students. The first activity is exploratory and asks students to explore the overall properties of the selected document collections. The second activity focuses on letter distribution and asks students to explore values in Infret and provide use cases for letter distribution in IR. The third activity instructs the students to explore word distribution of the 40 words with the highest term frequency. Additionally, it asks them to use their knowledge from the ISR course and apply selected formulae on the acquired data. The fourth activity focuses on the exploration of single word occurrence distribution for the 15 most common words. Finally, the last activity instructs the students to compare the actual number of words in a vocabulary of a collection with an estimated number calculated by using a formula. All activities except the first one, require students to apply formulae and approaches they learned in the ISR lectures to the results acquired from Infret. They may do this by using Excel or Python.

Once students complete the activities, they are provided with a link to an online questionnaire divided into multiple sections. The first section are demographics questions. The second section is the System Usability Scale (SUS) [2] which enables the comparison of usability over multiple versions of Infret and identification of major issues in usability. It is particularly useful due to its small size (10 Likert scale questions) and proven reliability. The third section contains questions from the Computer Emotion Scale (CES) [13] which displays what emotions students felt while using or learning to use Infret. The Scale uses 12 questions to evaluate the intensity of 4 distinct emotions: happiness, sadness, anxiety and anger. The questionnaire is concluded with general feedback questions used to further improve Infret. To analyse the results of the questionnaire we used basic statistical approaches in Python and Excel.

4.3 Study Participants

The participants of the study were students studying in the ISR course in the winter semester of the school year 2017/2018 at the Graz University of Technology. The 27 students enrolled in the course were studying basic concepts of information retrieval during the semester and were given the activities at the end of the semester. This means the students already had experience with the concepts presented in Infret and were improving their existing knowledge and expand their understanding of various concepts.

From the 27 enrolled students, 25 filled out the questionnaire. 22 (88%) students were male and 3 (12%) were female. 15 (60%) students were between 20 and 25 years old, 9 (36%) students were between 26 and 30 years old and 1 (4%) student had over 30 years. The majority studied computer science or software development and business management.

4.4 Findings and Discussion

When asked about experiments and hands-on activities 88% of the students agreed that they like experiments and hands-on activities and that experiments and hands-on activities help them better understand the theory and methods. Furthermore, most students agreed that Infret helped them better understand text statistics with 28% staying neutral and 16% disagreeing. Similar feedback was received for the ISRapp which was used in the same course and with the same set of student activities [19]. This further supports the idea of having an interactive visualisation element in the class as support for teaching abstract concepts. Detailed answer statistics for the aforementioned statements can be seen in Table 1.

Table 1 Answers to the following statements: S1 – I like experiments and hands-on activities; S2 – Generally experiments and hands-on activities help me to better understand theory and methods; S3 – This tool has helped me to better understand and reflect aspects of text statistics.

	Strongly agree		Agree		Neutral		Disagree		Strongly disagree	
S1	7	28%	15	60%	1	4%	2	8%	0	0%
S2	12	48%	10	40%	2	8%	1	4%	0	0%
S3	2	8%	12	48%	7	28%	1	4%	3	12%

When it came to the SUS score the mean value is 76.9 and the standard deviation 11.8. The average SUS score is 68, which indicates that Infret has a slightly above average usability [1]. In addition, it has somewhat better usability than the ISRapp which had a mean SUS score of 73.8 with a standard deviation of 11.1. This increase in the mean SUS score could be attributed to Infret's less cluttered UI. As it does not have a search engine it requires less UI elements which could potentially cause confusion in users' minds.

The mean intensity of the four CES emotions can be seen in Fig. 7. A value of 1 indicates the students never experienced that emotion during the usage of Infret and a value of 4 that they experienced an emotion all of the time. The prevailing emotion students experienced while using Infret was happiness with a mean value of 2.32 (SD: 0.54). Although, students did occasionally experience negative emotions. The prevailing negative emotion was sadness (Mean: 1.34, SD: 0.4), followed by anger (Mean: 1.27, SD: 0.47) and anxiety (Mean: 1.18, SD: 0.41). Similar results were observed during the evaluation of the ISRapp. However, students who used the ISRapp reported a higher happiness occurrence with a mean value of 2.76. These results might further support students' statement that they like interactive tools and it could also be an indication that they enjoyed using Infret.

When asked about what they liked students reported that they liked the visualisations and animations, interactivity of the UI, the interactive learning process, the representation of the data, ease of use and the overall design of the UI: *"The animated graphs and the easy navigation were quite good", "Easy to understand, clean design, responsiveness", "I liked the design a lot. The usage of the software was also very intuitive. Everything was structured well and easy to understand", "interactive and dynamic adaptions depending on which option was selected"*. However, there were also some aspects that students did not like and that could be improved in the future.

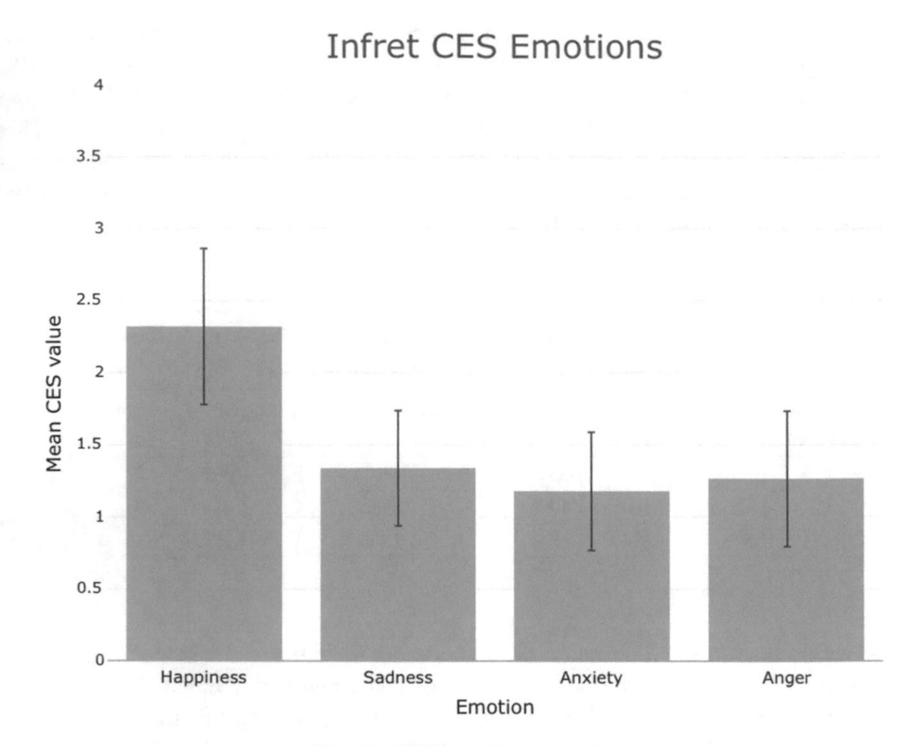

Fig. 7. CES emotion intensity

Some of these aspects include not being able to select single columns in the result table, the slider element for setting the number of elements shown in the chart not working well, lack of more options for manipulating the chart visualisations and some also did not like the animations: *"That you could not select single columns for copying the values. Aside from that there was not anything else that I did not like at all", "The slider for setting the amount of elements shown in the graph was very finnicky and generally a nightmare to control", "I didn't like that you couldn't alter the graph's representation more. The only parameter you could change was the number of items showing"*. Students suggested the following recommendations for improvements: the option to view the collection documents, addition of soring functionality to the table columns, adaptation of the UI responsiveness, customise additional chart properties, create custom collections, improving the slider functionality and more: *"Adding sorting functionality to the table columns would be very useful", "With a smaller device (such as my laptop) it was sometimes a bit frustrating to scroll in the tool, because there were 2 scrollbars. and when I open the tool I can't directly see the analyze button", "seeing a list of documents from the collection, possibility to add/remove columns from analyzed results, customizable graphs (label, color, aspect ratio), add possibility to change the data of the collection to experiment a little with it"*.

Although there is a lot of room for improvement it seems that Infret was well received by students and it helped them understand the concepts of text statistics. Most students agreed that they would like to use such a tool for other subjects in ISR. Examples of IR areas and subjects where students would like to use such a tool include digital libraries, term weighting concepts, image queries and performance measures.

A limitation of the study is that the survey results represent a subjective view. This could be addressed in the future by measuring grades or achieved points and comparing them between user groups. Additionally, students filled out the survey at various times after the exercise completion. Some students might have forgotten the details of the experience by the time they filled out the survey. Providing students with the survey in class and letting them fill it out during the class might address the aforementioned issue in the future. Finally, the students could have copied the question results from their peers and may not have explored concepts of IR by themselves. This could be detected in future versions of Infret by providing an anonymous unique identifier in the tool with which it would be possible to track usage statistics and detect how many students used the tool.

In conclusion, the evaluation results indicate that Infret was well received by the students and that it helped them learn the aspect of text statistics. In addition, it achieved its goal of supplying the students with an interactive visualisation experience which was mentioned as one of the aspects of the tool that students liked. Based on the SUS scores, the emotions of students and the general student feedback it can be concluded that the tool does an equally good job of supporting students in learning text statistics concepts while offering the advantages of web-based tools.

5 Summary and Future Work

In this paper, we describe the conceptualisation and development of Infret, a prototype of an interactive exploration tool for IR. The main goal of the tool is to aid students in a MAL-based ISR class in understanding text statistics, the basic concepts of many IR topics. Infret is composed of a server-side and a client-side application which was built using modern web technologies. Infret was evaluated in a MAL ISR course in a class of 27 students. The students were given a take-home assignment composed of five learning activities which required the use of Infret in combination with the knowledge previously learned in the course. Most students agreed that Infret helped them better understand text statistics and mostly experienced happiness during usage. The usability of the tool was measured using the SUS score and was also slightly above average which can be attributed to the simplistic UI of Infret. Furthermore the students' experienced emotions were measured using CES. Students experienced mostly happiness, and rarely experienced any negative emotions. Most students liked the UI design, visualisations, animations and interactivity. However, there were several possible improvements identified which will be considered in the future development of Infret. Some of the improvements are improving the responsiveness of the UI on small screens, offering the option to see the content of text collections and the addition of formula explanations for more complex formulae used in the tool.

Infret will be expanded with various weighting concepts and potentially to other areas of information retrieval in the future. Future versions of Infret will also support advanced features such as web crawling, in-application coding of IR concepts and more. Moreover, Infret will be evaluated in larger user studies where the variety of user groups will be extended.

Acknowledgment. We would like to acknowledge and thank Graz University of Technology for funding and RMIT University's School of Business IT and Logistics for hosting Aleksandar Bobić. We would also like to thank for participants for volunteering in our research project.

References

1. Brooke, J.: SUS: a retrospective. J. Usability Stud. **8**(2), 29–40 (2013)
2. Brooke, J.: SUS-A quick and dirty usability scale. Usability Eval. Ind. **189**(194), 4–7 (1996)
3. Dori, Y.J., Belcher, J.: How does technology-enabled active learning affect undergraduate students' understanding of electromagnetism concepts? J. Learn. Sci. **14**(2), 243–279 (2005)
4. Dori, Y.J., Hult, E., Breslow, L., Belcher, J.W.: How much have they retained? Making unseen concepts seen in a freshman electromagnetism course at MIT. J. Sci. Educ. Technol. **16**(4), 299–323 (2007)
5. Efthimiadis, E.N., Freier, N.G.: IR-Toolbox: an experiential learning tool for teaching IR. In: Proceedings of the 30th Annual International ACM SIGIR Conference on Research and Development in Information Retrieval - SIGIR 2007, p. 914 (2007). https://doi.org/10.1145/1277741.1277982
6. Fang, H., Wu, H., Yang, P., Zhai, C.: Virlab: a web-based virtual lab for learning and studying information retrieval models. In: Proceedings of the 37th International ACM SIGIR Conference on Research & Development in Information Retrieval, July 2014, pp. 1249–1250. ACM (2014)
7. Freeman, S., Eddy, S.L., McDonough, M., Smith, M.K., Okoroafor, N., Jordt, H., Wenderoth, M.P.: Active learning increases student performance in science, engineering, and mathematics. Proc. Natl. Acad. Sci. **111**(23), 8410–8415 (2014)
8. De Freitas, S., Neumann, T.: The use of 'exploratory learning' for supporting immersive learning in virtual environments. Comput. Educ. **52**(2), 343–352 (2009)
9. Frost, S.H.: Academic advising for student success: a system of shared responsibility. In: ASHE-ERIC Higher Education Report No. 3, 1991. ASHE-ERIC Higher Education Reports, The George Washington University, One Dupont Circle, Suite 630, Washington, DC 20036 (1991)
10. Hake, R.R.: Interactive-engagement versus traditional methods: a six-thousand-student survey of mechanics test data for introductory physics courses. Am. J. Phys. **66**(1), 64–74 (1998)
11. Halttunen, K., Sormunen, E.: Learning information retrieval through an educational game. Is gaming sufficient for learning? Educ. Inf. **18**(4), 289–311 (2000)
12. Johnson, R.T., Johnson, D.W.: Active learning: cooperation in the classroom. Ann. Rep. Educ. Psychol. Jpn. **47**, 29–30 (2008)
13. Kay, R.H., Loverock, S.: Assessing emotions related to learning new software: the computer emotion scale. Comput. Hum. Behav. **24**(4), 1605–1623 (2008)
14. Kolb, D.A.: Experiential Learning: Experience as the Source of Learning and Development. Prentice Hall, Englewood Cliffs (1984)

15. Lopez-Herrera, A.G., Alonso, S., Cabrerizo, F.J., Porcel, C., Cobo, M.J., Herrera-Viedma, E.: Using a visual tool to guide students in the complex process of learning fuzzy weighted information retrieval systems. In: INTED2010 Proceedings, pp. 361–370. IATED (2010)
16. Pirker, J., Riffnaller-Schiefer, M., Gütl, C.: Motivational active learning: engaging university students in computer science education. In: Proceedings of the 2014 Conference on Innovation & Technology in Computer Science Education, June 2014, pp. 297–302. ACM (2014)
17. Rowe, M.B.: Getting chemistry off the killer course list. J. Chem. Educ. **60**, 954 (1983)
18. Ruhl, K.L., Hughes, C.A., Schloss, P.J.: Using the pause procedure to enhance lecture recall. Teach. Educ. Special Educ. **10**(1), 14–18 (1987)
19. Ziessler, F.: A Framework for E-Learning in the field of information retrieval, p. 145 (2018)
20. Baeza-Yates, R., Ribeiro-Neto, B.: Modern Information Retrieval, vol. 463. ACM press, New York (1999)
21. Wilhelm-Stein, T., Kahl, S., Eibl, M.: Teaching the information retrieval process using a web-based environment and game mechanics. In: Proceedings of the 40th International ACM SIGIR Conference on Research and Development in Information Retrieval, August 2017, pp. 1293–1296. ACM (2017)

"Schools of Education as Agents of Change: Coping with Diversity in India and Germany Through a Collaborative, Interactive and Blended-Learning Environment – a Pre-test Study"

Christoph Knoblauch(✉), Jörg-U. Keßler, and Minke Jakobi

Ludwigsburg University of Education, Ludwigsburg, Germany
{christoph.knoblauch,kessler,
jakobi}@ph-ludwigsburg.de

Abstract. This paper discusses a multi-method Pre-test focusing on students' attitudes and preferences towards collaborative, interactive and blended-learning courses in an intercultural setting. These courses were developed and taught in the context of a project funded by the *Baden-Württemberg-STIPENDIUM for University Students – BWS plus* programme, a programme of the *Baden-Württemberg Stiftung*. The project promotes collaboration between Dr. Ambedkar University (AUD) in Delhi, India, and Ludwigsburg University of Education (LUE), Germany. In this paper, we present a critical reflection on the initial stages of the collaborative, interactive and blended courses taught as part of the project, using a triangulation pre-test that combines the methods of quantitative evaluation research and qualitative interviews. By doing so, our study assesses current practice and analyzes the above-mentioned courses closely and in detail, thus seeking to investigate how students deal with a collaborative and blended-learning environment, especially in an intercultural, international and digital setting.

Keywords: Collaborative and interactive blended learning · Intercultural and borderless education · Multi-method research

1 Background and Context of the Study

The process of learning and teaching in intercultural and digital settings has become increasingly important and has changed greatly in recent years. Diversification and new information technologies have become integral features of societies and their educational systems all over the world. The world of learners and teachers has become highly

This publication "Schools of Education as Agents of Change: Coping with Diversity in India and Germany through a collaborative, interactive and blended learning environment – a pre-test Study" was produced as part of the "Schools of Education as Agents of Change. Coping with Diversity in the Digital Age" project, which is part of the *Baden-Württemberg-STIPENDIUM for University Students – BWS plus* programme, a programme of the *Baden-Württemberg Stiftung*.

M. E. Auer and D. May (Eds.): REV 2020, AISC 1231, pp. 866–876, 2021.
https://doi.org/10.1007/978-3-030-52575-0_71

diverse and interconnected as a result of migration, globalization and digitalization. This is reflected in the research by many academics whose interests lie within the scope of intercultural, collaborative and blended learning: In this regard, the development of intercultural competence is often discussed as one of the most important skills in teacher training,[1] which can be acquired effectively by means of collaboration between students studying in different cultural settings.[2] Blended-learning approaches seem to have proven particularly relevant in such intercultural and collaborative environments, as they can offer tools for communication, interaction and shared learning between students and teachers in different countries.[3] Consequently, there is a strong demand for the development of collaborative and interactive blended courses in a diverse and digital-learning environment. Schools of Education play a major role in these developments and must be ready to deal with the known challenges in order to prepare students for learning and teaching in new contexts. However, well-researched resources offering guidance on how to design and implement such courses, with students studying in different countries and specializing in different disciplines, remain rare.

This study reports on a long-term cooperative project between Dr. Ambedkar University Delhi (AUD) in India and Ludwigsburg University of Education (LUE) in Germany. The project and cooperation between the two universities develop, implement and evaluate multiple types of blended and interactive collaboration between Indian and German students in joint courses. Participants in all courses were students enrolled in education programs, i.e. teacher-training programs or programs for early childhood development. Students were enrolled at either of the two universities. Most students were female, their ages ranging between 21 and 26 years of age. However, the data presented in this paper was mostly collected from the German participants only. In a follow-up study, more data concerning the Indian students will be analyzed. The results presented in this study can be considered to be a pilot study that will be extended as the project itself develops. In follow-up studies, more data from India and Germany will be collected and analyzed: During a summer school program for students from both universities held at LUE in October 2019, we will collect new data, implementing questionnaires and carrying out interviews with participating students and teaching staff both from Delhi and Ludwigsburg.

The courses presented, evaluated and discussed in this paper are developed and implemented in the context of a *BWS plus-project* funded by the *Baden-Württemberg Stiftung*[4]. The project also promotes collaboration between AUD and LUE. The key question of the project is how Schools of Education in India and Germany can perform as agents of change in diverse and digitally connected societies. The courses discussed have therefore been developed as collaborative, interactive and blended-learning courses focusing on the theme of diversity in educational and social contexts in both countries. The focus of this paper is the introduction of collaborative blended-learning scenarios that support the participating students in a two-fold way: On the one hand,

[1] cf. Byram (1997); Yeh, Jasiwal-Oliver and Posejpal (2017): 32–51; Ilse and Keßler (2017).

[2] cf. Thanh (2014): 11–23.

[3] cf. So and Bonk (2010): 189–200.

[4] https://www.bw-stipendium.de/en/students/bws-plus/ (last accessed 2019/05/03).

participants should develop their own learning skills and expertise in blended learning as such; here, our aims are not only to introduce them to blended-learning activities furthering their own content learning but also to make future teachers aware of and capable of blended-learning skills they might transfer to their own future teaching in classroom settings. On the other hand, all activities covered in our blended-learning environments were intended to contribute to the intercultural learning and ICC (in Byram's sense) of the participating students, who come from very different cultural backgrounds.

In order to facilitate an insight into the learning strategies of students in the described settings, our paper discusses a multi-method pre-test focusing on the attitudes and preferences of students to collaborative, interactive and blended-learning courses in an intercultural setting. In this paper, we present a critical reflection on the initial stages in these collaborative, interactive and blended courses, using a triangulation pre-test that combines the method of quantitative evaluation research and qualitative interviews. By doing so, our study assesses current practice, analyzes the above-mentioned courses closely and in detail, thus seeking to investigate how students deal with a collaborative and blended-learning environment, especially in an intercultural, international and digital setting.

2 Description of Course Design and Digital Content

The courses serving as the basis for this study were conducted at LUE, Germany, in winter semester 2018/2019. The online content was designed at both AUD and LUE. The main foci of the courses are Diversity Education, Resilience and Multilingual Education. Approximately 50% of the course sessions were conducted in classroom settings, whereas the other 50% were carried out online, individually or in collaborative teams consisting of students from India and Germany. In this context, the term 'blended learning' is understood as the integration of traditional teaching methods and eLearning, both using hybrid learning arrangements and encouraging collaboration.[5] Additionally, colleagues from the two universities worked together in classroom settings either via Skype-sessions or through in-person visits to each other's classrooms. Some of these joint sessions were also video-recorded.

The classroom sessions were designed to deal with matters of organization, to discuss key terminology and key concepts and, finally, to present and discuss collaborative tasks and projects.

The online content was designed to (1) present course topics from an Indian and German perspective; (2) to offer input from proven experts; (3) to establish the foundations for collaboration; and (4) to encourage interaction between Indian and German students. To these ends, (1) videos of Indian and German professors teaching in teams were used; (2) a podcast with experts was made available; (3) common issues were presented through a selection of pertinent literature; and (4) teams of Indian and German students were set up in order to develop joint projects. In a collaborative

[5] cf. Buran and Evseeva (2015): 178.

setting, students were asked to interact online, using communication tools of their own choice. The participating universities and professors established a common ground for the discussions, providing contacts and communication technologies.

The courses are now part of 'Educational Studies' in the English Department at LUE. Students can gain up to three credit points (as defined by the European Credit Transfer and Accumulation System – ECTS) for each course.

3 Design of the Pre-test Study

The study was designed as a pre-test study, focusing on the initial stages of the cooperative, interactive and blended courses. It uses a creative multi-method design in order to create a flexible research process and to produce various findings. However, the design of the pre-test is also meant to serve as the basis for a follow up study, which uses the first set of findings and reflects the research methods.

In the context of the complex research focus – students' attitudes and preferences to collaborative, interactive and blended-learning courses in an intercultural setting – a multi-method test, using quantitative and qualitative methods, was designed: The study looks for both reliable quantitative data as well as individual, subjective feedback which offers a more comprehensive view.[6]

First of all, two evaluations of the pilot-courses, carried out by means of quantitative questionnaires, illustrate trends about attitudes towards and usage of certain online learning methods, helping to develop advanced structures and methods for any following courses and further research. All participating students took part in this quantitative evaluation, completing two questionnaires about (1) blended-learning courses in general; and (2) the blended-learning approach of these courses in particular. Whereas the first questionnaire focuses on the balance of online and classroom learning, resulting learning strategies and attitudes towards the blended learning setting in general, the second questionnaire evaluates online learning preferences, specific forms of digital collaboration and the concrete use of online tools (i.e. podcasts, videos and interactive options).

The subsequent series of semi-structured qualitative interviews, conducted after having analyzed trends present in the quantitative study, identify individual attitudes and preferences towards collaborative and blended learning in this unique setting. The questions focus on the use of the online content provided, options for interaction, attitudes towards collaboration, reflections on learning strategies, and perceptions of change within learning strategies and attitudes.

The experience of multi-method pretesting notably helped to identify strengths and weaknesses in the conducted courses by evaluating students' responses in different empirical settings.

[6] cf. Creswell (2015): 15.

4 Findings of the Multi-method Pre-test[7]

4.1 The Quantitative Perspective

The quantitative perspective consists of two surveys, which are completed by participants of the courses (n(1) = 16/n(2) = 13). The surveys aim to (1) show trends in attitudes and preferences of students concerning collaborative, interactive and blended learning courses in an intercultural setting; and (2) to help develop items for the following qualitative evaluation.

The first set of questions focuses on attitudes towards blended learning and the online content provided. It shows that the blended structure and the online content of the courses were generally well accepted by the participating students: ***61% strongly agree*** with the statement "I would take a blended learning class with this structure again." Both the online content and the classroom sessions are genuinely appreciated by the sample and show similar results: ***31% strongly agree and 54% rather agree*** with the statement "I enjoyed working with the online content." In comparison, ***38% strongly agree and 54% rather agree*** with the statement "I enjoyed working in class." The blending of online and classroom learning also resulted in very positive attitudes: ***42% strongly agree and 29% agree*** with the statement "The balance between online and classroom learning was good." The findings indicate that students exhibit a general satisfaction with the blended-learning design. Against this backdrop, the question arises as to what the students specifically appreciate and which improvements can be made – questions the following qualitative evaluation incorporates.

The second set of questions focuses on experiences with and attitudes towards collaboration. The results show that the study group truly appreciates the various forms of collaboration offered in the courses: ***23% strongly agree and 54% rather agree*** with the statement "The collaboration with others was very helpful for my learning." The results are even more pronounced when the sample is asked to rate classroom interaction and collaboration with international peers: ***59% strongly agree and 35% rather agree*** that "The classroom discussions were very helpful for their learning", while in point of fact ***76% strongly agree and 24% rather agree*** that "Collaboration with internationals was very helpful for their learning." The findings indicate that the students value highly collaboration on a national and, particularly, international level. In this regard, the question arises as to which specific traits of collaboration are valued and why – again, questions the following qualitative evaluation incorporates.

The third set of questions focuses on changes in the learning behaviour of the participating students. The results show that a high number of students ***(54% strongly agree/38% rather agree)*** share the opinion that they "learned new techniques of learning in the courses." ***23% strongly agree and 61% rather agree*** with the statement that "they developed their learning skills during the courses." The participating students especially appreciate "the independent ways of learning" ***(61% strongly agree/31% rather agree)*** and the "flexibility of the blended design" in the courses ***(69% strongly***

[7] All results, quotations, charts, transcriptions and analysis phases of the pre-test study are available through the authors.

agree/23% rather agree). The findings indicate that students develop their learning skills and acquire new learning techniques in a blended and collaborative setting. Independent and self-regulated methods of learning and the flexibility of the blended design seem to be the main reasons for this. The qualitative study can be used to gain a deeper insight into these findings.

4.2 The Qualitative Perspective

The qualitative study focuses on (1) attitudes towards blended learning; (2) use of and attitude towards the online content provided; (3) experiences with and attitudes towards collaboration; and (4) changes in learning. The items were developed through the analysis of the preceding quantitative tests and a thematic analysis of the qualitative data.[8] The qualitative testing thus produces highly interesting insights towards individual attitudes, preferences and use of the blended, collaborative and interactive approach of the courses.

(1) Attitudes towards blended learning: The interviewed students (n = 5) in this pretest seem to value blended learning in general, mostly discussing the benefits of self-regulated learning, interaction and the diversified working options. Additionally, the students cite specific arguments as to why they appreciate the blended design of the courses. One important factor seems to be the potential for reflection generated by the mix of online and classroom sessions: *"...I liked the mix of both (...) so when I watched the videos it was more thinking by myself and making my own ideas, but it was also very important to be back in class (...) I got new ideas how they (other students) understood the tasks, how they answered the questions and how they think about this and this was really interesting."* In addition, the mix of team-teaching videos as a foundation of and the podcasts as a source for deepening knowledge is mentioned several times. Students report that they appreciated the clear structure and the tasks of the online content provided *"... because they were very helpful as a kind of guidance through the content and stirred my attention to the central themes of the videos."* Furthermore, some students put an emphasis on the fact that the learners could influence both digital and classroom learning: *"...and (the) best part about it was that students were also taking the lead. It was not completely driven by the teacher."* Finally, the group discussions held after individual (online) learning sessions seem to be an important factor for the interviewed students: *"... (the course was) open for group discussions (...) and these encouraged independent thinking and the reflection of one's own ideas and thoughts."* In general, the sample shows a positive attitude towards blended learning in general and the designed courses approach in particular: *"... the blended learning approach arouses students interests and combines ways of (...) modern teaching with interesting in-class sessions. So, it was a good mix of watching the videos and coming together in class and talking about them."* The answers show that the online content and associated tasks have to be adjusted carefully in order to prepare further interaction. Furthermore, the mix of

[8] See Braun and Clarke (2006): 77–101.

online content and classroom discussion, the opportunity for self-regulated learning and the different means of interaction seem to play a major role for the interviewed students.

(2) Use of and attitude towards the online content provided: The interviewed students seem to value the mix of videos, podcasts and tasks, and the fact that they can use the content repeatedly for their own reflection. Additionally, some students were adamant that educational videos are a valuable option and should be used more often, especially when looking at the workload some texts require. A problem however seems to be that the courses use the online content only as a preparation for classroom discussion and collaboration but not as an option to develop new (digital) content – this could be changed in future courses.

(3) Experiences with and attitudes towards collaboration: All interviewed students highly appreciate the collaboration both online and in class. The reassessment and reconfiguration of one's own opinions and ideas seems to be the main reason for this feedback: "*… for me it was getting to know new numerous different points of views and insights on opinions of other students (...) this was particularly interesting for me (...) because it required me to reassess my own ideas and enabled me to make additions and changes accordingly.*" As a special focus in this collaboration, some students mention the online-interaction between Indian and German students. This interaction takes place via student-controlled tools like Skype and WhatsApp – the students did not ask for any support from the participating universities. This online interaction seems to be especially productive when it is prepared for and guided by online content and tasks: *"... to first think so what do I know, what do I think and then also to listen to the others' answers. Because so I could compare and I also got more ideas (…) sometimes it was also a little "wow-effect". So "wow" what did the others think about this questions and how different it was. Probably it was because (...) different countries, so between India and Germany for me."* Student perspectives like these underline the importance of international collaboration in interactive blended-learning settings. The close interlinking of students in a borderless intercultural setting can use this instance of alterity as a new opportunity for learning. *"The best was that through the media it was possible to work with people from different countries and to do that instantly."* All interviewed students show a positive attitude towards collaboration, both in class and online. Again, the mix of self-regulated learning and interaction seems to be a favourable means of learning for the sample. An emphasis is put on the international collaboration as it often confronts students with alterity and challenges consolidated positions in a special way.

(4) Changes in learning: The changes in learning reported by the students interviewed mostly concern learning strategies and collaboration. The changes can be traced back to the blended-learning design and interactive tasks: "*…we had the chance to work on our own, completely independent of time and place, which was good. And afterwards we could share in class our solutions and our thoughts (…).*" All the interviewed students welcomed the fact that the online content served as a preparation for collaboration and could be used independently. The value of the combination of self-regulated learning and interactive collaboration was mentioned emphatically several times. One of the most impressive perspectives

referenced by the students linked the blended and collaborative approach to one of the main themes of the courses: *"I think the blended and collaborative learning approach did change my way of learning to a certain degree since it illustrated how important diversity in education actually is, as it doesn't only give you multiple perspectives but it also shows how much oneself can profit from the ideas and perspectives of others. And it further underlines the fact that there isn't just one fixed answer to a question (...) which I thought became very clear in our seminar and I really liked that."* Diverse methods of learning in a blended-learning approach are correlated with the theme of diversity in education in this answer. The chance to deal with and learn about multiple perspectives is positively valued by all students interviewed and is often associated with the international collaboration dimension of the project: *"... (the blended approach) further enabled me to take notes of the given ideas and thoughts of the Indian students, while I could reflect on (...) them and my own concurrently."* Several students report that their methods of learning changed because of the use of online content, the connection between self-regulated learning and classroom discussion, as well as the opportunity to interact in an intercultural digital setting (which enables students to reflect on multiple perspectives). *"I feel that looking at the problem from a wider perspective helps in respecting all cultures in the world and hence finding the solution."*

4.3 Analysis and Interpretation Through Triangulation

The presented pre-test uses a multi-method approach and is looking for both reliable quantitative feedback as well as individual, subjective perspectives. The quantitative perspectives (see Sect. 4.1) were mainly used to gain general feedback about the courses, to show trends in attitudes and preferences, and to help develop items for the following qualitative evaluation. The qualitative perspective was then included to gain deeper knowledge about the individual attitudes and preferences of the participating students, and to learn about the demands of a collaborative blended-learning environment in intercultural settings in general (see Sect. 4.2). The methods employed largely focus on the same phenomena and are discussed in this final analysis, using the method of triangulation.[9]

Three main foci could be established during the analysis process: (1) attitudes towards blended-learning designs and traits; (2) attitudes towards collaboration and interaction in intercultural settings; and (3) development of learning skills.

(1) In general, the participating students show positive attitudes towards the blended-learning design of the courses. This seems largely due to the mix of digital learning and classroom discussion, as well as the structure of the courses, which usually used the online content to prepare subsequent classroom interaction. Within the blended setting of the courses, the participating students especially appreciated the opportunity for self-regulated learning. This is explicitly mentioned several times in the conducted interviews, and it leads to a development of

[9] cf. Denzin (1978): 291.

learning skills. In relation to this, the students seem to value the independence the blended design offers: The online content can be retrieved and reused whenever and wherever the students choose to; online interaction can take place at individually fixed dates; and the collaboration enables students to share perspectives outside the classroom with individually chosen partners. These findings correlate with the high appreciation which students show towards the flexible structure of the courses, which enable them to work in an independent, self-regulated manner, marked by responsibility for oneself.

(2) In general, the findings indicate that students value highly the collaborative and interactive approach of the courses. Students report that they enjoy the blending of individual reflection via online contents, the interaction with other international students and, finally, the collaboration – online and in class – in small groups. Many participating students put a special emphasis on the fact that the structure of the courses enabled them to reflect upon different perspectives on both a national and international level. The interactive and collaborative design of the courses led to the discussion of multiple experiences and opinions – a trait most students especially appreciate. In this context, students greatly value the international and intercultural approach of the courses, which enables them to learn about other educational systems and concepts, and at the same time reflect on familiar ideas. In this context, students report that consolidated positions have been challenged by alterity.

(3) In general, the findings indicate that students develop their learning skills and acquire new learning techniques in the blended and collaborative setting of the courses. Students report that they use the provided online content and tasks in a flexible and largely independent manner. This marks a method of learning which was new to many of them. In addition to this, many students place emphasis on the fact that the digital interaction with international students is a means of collaboration which changed their learning habits. Finally, self-regulated learning with a focus on responsibility for oneself seems to embody a challenge which offers special learning potential via a blended and collaborative design.

5 Outlook

Blended collaborative learning environments in intercultural settings are in need of a structure which enables students (1) to find themes and prepare topics individually through online content and tasks; (2) to share their perspectives in class and via online interaction; (3) to develop shared knowledge through collaboration in class and online; and (4) to develop new content and tasks for themselves and others. The course designers should present a rough outline which describes the balance between digital and classroom learning, and between individual learning and interaction. This outline, however, should allow a certain degree of flexibility for and self-regulation by the students. Within the described structure, the collaboration should especially focus on international, intercultural perspectives concerning common themes. The discussion of topics which are relevant to all partners in an intercultural setting seems to offer a large learning potential. In this context, blended learning and intercultural collaboration can function as links between students from different countries and cultural backgrounds. At the same time, such learning scenarios can connect educational systems of different countries, sharing different experiences, perspectives and concepts. It is clear that this learning potential, on a student and an institutional level, is particularly encouraged by the concept of alterity. Against this backdrop, Humboldt's idea of alienation from ones' own lifeworld as a constructive means of learning plays an important role: In modern educational theory, the understanding of education often refers to Humboldt's fragment "Theorie der Bildung des Menschen" [Theory of the Education of Man].[10] Learning potential evolves when the human being discovers a world which is different to his own lifeworld. *"Our encounters with the world are transformative in that they are mediated by "self-alienation," that is, alienation from our taken-for-granted and habitual self-understandings. (...) Bildung refers to the active and receptive self-other relation implicit in educational process."*[11] This alienation is additionally supported by the use of a language that is neither the mother tongue of the majority of the Indian nor of the German students. Using English as a lingua franca in our project has proven to be another constructive opportunity, rather than a challenge, to raise the intercultural awareness of the participants and to sensitize them to the necessary openness for both the topics dealt with in the courses as well as the ideas and concepts of their international fellow students.

In both the quantitative and qualitative study, students comment on the process of collaboration as being a crucial factor to their learning experiences. By exchanging ideas, especially in an intercultural setting, students compare and discuss different perspectives in active and receptive self-other relations. These encounters, which are made possible by the collaborative and blended design of the courses, might be regarded as transformative (in the spirit of Humboldt) and should therefore play a major role in collaborative and interactive blended-learning environments in general.

References

Braun, V., Clarke, V.: Using thematic analysis in psychology. Qual. Res. Psychol. **3**, 77–101 (2006)

Buran, A., Evseeva, A.: Prospects of blended learning implementation at Technical University. In: Social and Behavioral Sciences, vol. 206, pp. 177–182 (2015)

Byram, M.: Teaching and Assessing Intercultural Communicative Competence. Multilingual Matters, Clevedon (1997)

Creswell, J.W.: A Concise Introduction to Mixed Methods Research. SAGE, Los Angeles (2015)

Denzin, N.K.: The Research Act. A Theoretical Introduction to Sociological Methods, 2nd edn. McGraw-Hill, New York (1978)

English, A.R.: Discontinuity in Learning. Dewey, Herbart and Education as Transformation. Cambridge University Press, Cambridge (2013)

von Humbodt, W.: Theorie der Bildung des Menschen (1793)

[10] cf. von Humboldt (1793).

[11] English (2013): 13.

Ilse, V., Keßler, J.-U.: Living interculturalism instead of teaching interculturalism. In: Aicher-Jakob, M., Marti, L. (eds.): Education - Dialogue - Culture. Migration and Interculturalism as Educational Responsibilities, pp. 139–152. Schneider Verlag Hohengehren, Baltmannsweiler (2017)

Keengwe, J.: Handbook of Research on Promoting Cross-cultural Competence and Social justice in Teacher Education. IGI Global (2017)

So, H.-J., Bonk, C.: Examining the roles of blended learning approaches in computer-supported collaborative learning (CSCL) environments: A delphi study. Educ. Technol. Soc. **13**(3), 189–200 (2010)

Thanh, P.T.H.: Implementing cross-culture pedagogies. Cooperative learning at Confucian heritage cultures. In: Education in the Asia-Pacific Region, vol. 25, pp. 11–23. Springer (2014)

Yeh, E., Jasiwal-Oliver, M., Posejpal, G.: Global education professional development: A model for cross-cultural competence. In: Handbook of Research on Promoting Cross-cultural Competence and Social Justice in Teacher Education, pp. 32–51. IGI Global (2017)

Enhancement of Engineering Design Competencies Through International Engineering Collaboration: A Pedagogical Approach Used in SEPT. Part 1

Dan Centea(✉) and Lucian Balan

McMaster University, Hamilton, ON, Canada
{centeadn, balanl}@mcmaster.ca

Abstract. A recent trend in engineering education related to the expected design competencies of graduating engineers is a shift from prescribing what the design requirements of a program should include toward examining the corresponding learning outcomes. One of the approaches used in implementing this trend is the adoption of outcome-based assessment criteria in the engineering design curriculum. Graduates should be able to apply knowledge of mathematics, science and engineering by designing systems, components and processes in learning environments that include both traditional and modern approaches, such as project-based learning, and should be able to apply their engineering design knowledge in open-ended real-life projects. The paper describes an approach to develop engineering design competencies using open-ended projects by providing detailed expected outcomes of the designed product and providing limited design specifications. Students are expected to go through all the steps of the engineering design process from defining the specifications of the designed product that addresses the expected outcomes to testing the final product. The purpose of the study is to find the engineering design competencies that are enhanced through international design collaboration.

Keywords: Engineering collaboration · International engineering design collaboration · Engineering competencies · Engineering design competency

1 Introduction

To meet the challenges of globalization, the education system needs to prepare graduates for a workplace where responsibilities are constantly changing, where information passes through multiple channels, where networking is needed and initiative-taking is valued, and where strategies are complex because of the expansion of markets beyond national borders [1, 2]. Education must help individuals to perform tasks for which they were not originally trained. These tasks combine personal skills such as creativity, improvisation and critical thinking with team skills in multinational environments. The role of education is therefore not only to provide students with knowledge and skills but also to prepare them for the new global society.

M. E. Auer and D. May (Eds.): REV 2020, AISC 1231, pp. 877–885, 2021.
https://doi.org/10.1007/978-3-030-52575-0_72

Internationalization is one of the ways in which higher education is responding to the challenges of globalization. It recognizes national boundaries and the uniqueness of individual societies and cultures in the face of the forces of globalization and urges international understanding and cooperation [3]. Internationalization includes academic linkages, open online courses with no border restrictions, foreign students, student and faculty mobility, and various forms of cooperation.

Universities form academic linkages to be able to compete in the global higher educational market. Linkages help newer universities to gain visibility and market share and help older or stronger universities to maximize their advantageous positions. Linkages can take the form of bi-lateral agreements, academic associations, academic consortiums and institutional networks [3].

The open online courses with participants from multiple countries are modern ways to educate students located in multiple countries. Their popularity increased significantly in the last 7 years; currently they represent the highest form of academic internationalization. For example, by the end of 2018, over 900 universities around the world have announced or launched more than 11,400 Massive Open Online Courses (MOOCs). In 2018 alone, 101 million students enrolled in MOOCs [4].

The foreign students attending a university outside their country's borders represent the second largest form of internationalization. For example, if we consider only the courses taught in English, statistics show that in 2018 there were more that 2.8 million foreign students in the English-speaking countries (1.094 million in the US, 690,000 in Australia, 572,000 in Canada, and 458,000 in the UK [5, 6]. This number does not include the large number of foreign students enrolled in courses taught in English and offered in countries where English is not the main language.

Other forms of internationalization are constantly developed by the academic community; student and faculty mobility based on international inter-institutional exchange agreements has become an important goal of most universities. Various mobility agreements are currently in place for students, faculty and staff, and they include teaching, learning, research, and training. These mobilities, based on academic alliances, enable students to gain international exposure and credentials that will help in their careers.

Cooperation between universities takes different forms. The most frequently used ones include double and/or joint degrees, PhD cotutelle, common curriculum development, common research projects, summer internships, common course offering, international design cooperation, and so on. Most of these are started and led by faculty members who agree to invest time and effort in preparing students for today's complex, diverse and challenging world through the use of innovative methodologies that promote multiculturalism.

This paper focuses on one form of international cooperation, namely international engineering design collaboration. This cooperation includes significant dedication and effort from both students and mentors as it generally involves frequent conference calls between students from different countries, conference calls between mentors, and annual competitions.

2 Engineering Graduate Attributes and Competencies

Looking at the outcomes of higher education, many universities are preparing their graduates for employment by focusing the graduate attributes – generally defined as the understandings, academic abilities, personal qualities, and transferrable skills that students should develop during their time with the university. Engineering accreditation boards from different countries have defined the graduate attributes that an engineering graduate should possess. Taking Canada as an example, the graduate attributes for engineering include a knowledge base for engineering; problem analysis, investigation; design; use of engineering tools; individual and team work; communication skills; professionalism; impact of engineering on society and the environment; ethics and equity; economics and project management; and life-long learning [7].

In addition to the theoretical and technical knowledge, the engineering profession requires reflective and practical knowledge and competencies to deal with complex, open ended problems. Considering engineering design as an example, the graduate attribute can be defined as "*an ability to design solutions for complex, open ended engineering problems and to design systems, components or processes that meet specified needs with appropriate attention to health and safety risks, applicable standards, and economic, environmental, cultural and societal considerations*" [7]. This graduate attribute is accomplished through the processes carried out between identifying a problem and creating a product that solves the problem. These processes include: generate design specifications and carry out iterative activities to establish objectives and criteria; brainstorm, evaluate and chose a solution; generate alternatives; analyze, develop and prototype a solution; construct, test, and evaluate the solution; and, when the solution meets the requirements, communicate the results.

The International Engineering Alliance defines in [8] the graduate attributes as a set of individually assessable outcomes that are components indicative of the graduate's potential to acquire competence to practice at appropriate level. The document describes the 12 graduate attribute profiles defined by the Washington Accord. The professional competence is described by the same document as a set of attributes corresponding largely to the graduate attributes, but with different emphasis. The document describes 12 professional competency profiles that a person must demonstrate in order to be able to practice competently in his/her practice area.

Unlike graduate attributes, professional competence is more than a set of attributes than can be assessed individually; competence must be assessed holistically [8]. Engineering competency represents a set of capabilities, skills, aptitude, proficiency, and expertize required to perform professional duties effectively. The published literature offers many descriptions of engineering competencies [8–24].

Although the engineering competencies needed to solve an engineering problem may vary from one application to another, these competencies can be rated in terms of importance. A statistical analysis [9] shows that the 11 competencies defined by the U.S. Accreditation Board for Engineering and Technology (ABET) can be rated in terms of mean importance into three clusters. The top cluster includes three competencies: problem solving, communication, and teamwork; the intermediate cluster includes ethics; life-long learning; math, science and engineering knowledge;

engineering tools; experiments and data analysis; and design. The bottom cluster includes contemporary issues; and understanding impacts. The placement of engineering design in the intermediate cluster is debatable considering the importance given by the current engineering accreditation policies and trends.

3 Engineering Design Competency

Professional competency profiles reflect the essential elements needed to assess evidence of competence. There are no universally accepted engineering competency profiles, each institution having its own approach of grouping the competencies into categories. NC State University, for instance, defines in [10] the competency profile of an engineer through a set of six competencies that include professional knowledge; program/project management; engineering review, decision making and analysis; communication; engineering design and analysis; and leadership.

The assessment of each competency considered separately can be accomplished by defining performance indicators. These indicators represent the actions taken by a professional to demonstrate competence. In the case of the engineering design competency, which is the focus of this paper, competency is defined by different authors with different levels of granularity. The authors of reference [8] define five major performances to assess design competency: identify and analyze design requirements and draw up detailed requirements specifications; synthesize a range of potential solutions to problem or approaches to projects execution; evaluate potential approaches against requirements and impact outside requirements; fully develop design of selected option; and produce design documentation for implementation. The authors of paper [11] define 28 design competencies, and group them in eight categories [12]. Other authors [13] identified multiple competencies and grouped them into eight affinity groups.

The published literature and examples presented above show that there is no generally accepted consensus on defining and assessing the engineering design competencies. Demonstrating that an engineering design competency is enhanced through any means is not trivial. This paper analyzes a set of competencies that are needed to perform design processes in an international environment and suggest an assessment approach that can indicate if some competencies are enhanced or not through international engineering collaboration.

4 Analysis of Engineering Design Competencies in an International Engineering Collaboration

Engineering education is evolving in response to the technological advances, political and economic realities, engineering job markets, and public interests. Consequently, the engineering accreditation is also evolving. The following trends have been observed in engineering accreditation [13]: examining the outcomes of engineering education; emphasizing engineering design as a key component for engineering accreditation; developing international agreements on engineering education;

accreditation of innovative engineering programs; and creating partnerships in education between engineering educational institutes and outside stakeholders such as engineering associations, industry and businesses.

The trends in engineering accreditation listed above place engineering design as a key component of the engineering accreditation. Ensuring that the graduates of engineering programs master the competencies needed to carry out the processes included in engineering design is an important goal of engineering education. Although many competencies can be acquired through the courses included in engineering curricula, they can be reinforced and probably enhanced through extracurricular activities such as engineering clubs, engineering collaboration, and engineering competitions.

The study carried out in this paper focuses on an international engineering design collaboration that involved 65 universities from 10 countries grouped in eight teams, five major industrial partners, and more than 20 industrial contributors. The purpose of the student collaboration was the design of personal assisted mobility devices by teams of students from different countries located in several times zone. The process involved faculty advisors from each university, input and mentoring provided by major industrial partners, and the use of several engineering design software tools provided at no- or low-cost by industrial contributors. The two-year long design process started from a design problem with specific constraints provided by the industry partners and finished with two annual competitions, one for the design and building of a prototype, and one for the manufacturing and testing of the final product.

This paper uses the definition of the design engineering competency proposed in Canada by the Natural Sciences and Engineering Research Council (NSERC) Chairs in Design Engineering [13]. The proposed definition includes the following competency categories: general knowledge, specific knowledge in a professional environment, knowledge of procedures, operational skills, experiential skills, social/personal skills, and cognitive skills. The paper analyses the first three groups of competencies. A paper to following will analyze the remaining four groups.

The **general knowledge** competency group is defined through knowledge of mathematics, basic science, and engineering sciences. The analysis of the work done by students show that, through their weekly collaboration with their international partners, they have put limited effort in applying these skills to proper technical design analysis. They relied very much on the software tools, and a similar approach was also seen in their engineering partners from abroad. No evidence has been found to state that the international collaboration enhanced the general knowledge competency.

A second competency group is related to the **specific knowledge in a professional environment**. This competency is defined through the students' awareness and use of technologies, regulations, safety, liability, ethics, and role in society. The group analysed in this papers included students from different disciplines working together to reach a common goal – the design of a mobility device. Students belonged to several departments such as mechanical, electrical, computing and mechatronic. They initially felt that they were in silos and were able to communicate only with their peer from the same discipline. However, due to the multidisciplinary aspect of the design many of them started to be aware and have some technical understanding of a broad range of technologies beyond the technologies of their engineering discipline. Due to this awareness they could understand most engineering systems and components of the design product and were effective in working in cross-disciplinary teams.

The propulsion system of the designed mobility device was required to be electrical. Most teams decided to use Li-ion batteries. Because of the possible risks associated with these batteries, all group members became aware of the regulation of manipulating, storing and transporting these batteries. Safety being an important competency element, the group analysed it in detail. The group members from different countries had different opinions on the minimum level of safety that could be accepted for their design, and they have different possible users of the device based on their market research. Design criteria such as maximum acceleration and deceleration and the maximum speed on curves were seen differently based on the age group of their potential customers. They reached a compromise only after analysing, in detail, safety and liability implications, and discussed the ethics related to each of the alternative designs. Safety, liability and ethics are important competencies related to the specific knowledge in a professional environment group of competencies, and the collaborative work in an international team enhanced their awareness and skills to apply these competencies.

Another important competency that belongs to the specific knowledge in a professional environment group is the role of the design engineer in the society. The international collaboration placed all students in a multicultural environment, which is a clear fact in today's global society. Furthermore, by working with their industrial mentors from multinational companies, students were made aware of a corporate culture and the global nature of business. By following the milestones and deadlines imposed by the industrial mentors, students learned to accept responsibility, to develop their interpersonal skills, to coordinate their activities with their international partners, to assess the capabilities of others, and to develop strong team building and collaboration skills. Although some of these skills could be developed in any industrial-led project, working in an international cooperation enhanced their competency to work and compete in multinational teams.

A third group of competencies, termed **knowledge of procedures** by [13], is defined as knowledge of the product development process, engineering design process, and engineering design tools. In the two years of international collaboration, the students understood the product development process by taking their designed product from conception to market. With different components of the product designed in different countries, students experienced the challenges of meeting intermediate milestones in the conditions of different academic and cultural work gaps such as academic years, term tests, exams, work terms, statutory and religious holidays.

The collaborative work taught the students how the transition from a common problem to a proposed design approach, which initially was different for different countries, and through a common approach to define the technical specifications, to propose and develop a solution, to implement the solution, to be involved in prototyping, manufacturing, and testing. Although most of these steps of an engineering design process competency are not necessary related to an international team, the process of agreeing on a single approach included the use of personal skills that involved international negotiation addressed to different cultural backgrounds can be considered as an enhancement of the engineering design process competency.

In order to follow the steps required by the engineering design process, the students needed to learn and be proficient in the use of several engineering design tools. They collected relevant design information from their countries to define the problem; used in their weekly conference calls engineering drawing, from sketches to CAD; and used engineering calculations using statics (to determine the position of the centre of mass), CAD (for the shape and mass and interference of parts) stress and thermal analysis (using FEA), and multi-body dynamics (ADAMS to calculate acceleration and interference between moving parts). An interesting aspect was that some teams were more experienced than others in using some of the tools based on the focus and importance given by each specific university in using some of these tools. The students helped each other in improving their skills in using most of these engineering tools. This could not be done in a group of students from the same university. As a consequence, it is possible to assume that the related competency was enhanced through student-centred international engineering cooperation.

The development of the design includes the development of a business-related competency through the use of project management skills, market research, and cost estimation at various stages for prototyping, manufacturing and transportation between universities and at competitions. In the development of the business-related competency the students were made aware of needs of different markets, the assumed quantities to be produced based on the individual market researches, the suggested selling cost that was acceptable by the customers from different countries, and the associated costs related to the manufacture and transportation in different countries.

The analysis presented above indicates that all the components of the knowledge and procedures competency group are enhanced through student-focused international engineering cooperation.

5 Summary

This paper describes an engineering design collaboration that took place between 50 universities grouped in several teams with the purpose of designing a mobility device, manufacturing and building it, and participating in annual competitions. The current trends of education towards internationalization are presented, and the current internationalization approaches in education are described. International engineering design collaboration being one of the trends, the paper aims to study if student-centered international engineering collaboration provides benefits by enhancing the engineering competency of the participants. The graduate attributes and engineering competencies published in the literature are analyzed, and several proposed definitions of engineering competency are provided. The paper focuses on the engineering design competency, uses a definition of this competency proposed in Canada by the NSERC Chairs in Engineering Design, and analyses three of the seven groups of engineering design competencies included in this definition. The analysis concludes that most of the engineering design competencies are enhanced through a two-year design process carried out by groups of students from several countries located in different time zones.

Acknowledgements. The authors would like to acknowledge the mentoring and financial help provided to the team that included McMaster University by the Partners for the Advancement of Collaborative Engineering Education (PACE) comprised of General Motors, Autodesk, HP, Oracle and Siemens as PACE Partners, and several PACE contributors that provided software tools: ANSYS, MathWorks, and MSC Software.

References

1. Zhao, Q., Zheng, X., Zhou, S.: Exploration on education model of international engineering competencies for undergraduate students through Project-Based Learning: a case study from China. In: 2018 IEEE 10th International Conference on Engineering Education (ICEED 2018), Kuala Lumpur, Malaysia, 8–9 November 2018, pp. 10–14 (2018). https://doi.org/10.1109/ICEED.2018.8626957
2. Sharafi, S., Bassak Harouni, G., Torfi, S., Makenalizadeh, H., Sayahi, A.: Studying implication of globalization on engineering education. Int. J. Educ. Pedagogical Sci. **5**(10), 1277–1280 (2011)
3. Chan, W.W.Y.: International cooperation in higher education: theory and practice. J. Stud. Intern. Educ. **8**(1), 32–55 (2004). https://doi.org/10.1177/1028315303254429
4. https://www.classcentral.com/report/mooc-stats-2018/
5. [D3]. https://www.iie.org/Why-IIE/Announcements/2018/11/2018-11-13-Number-of-International-Students-Reaches-New-High
6. http://monitor.icef.com/2019/02/canadas-foreign-student-enrolment-took-another-big-jump-2018/
7. Engineers Canada Consultation Group on Engineering Instruction and Accreditation, Webinar, 7 January 2016 (2016). https://engineerscanada.ca/sites/default/files/Graduate-Attributes.pdf. Accessed 10 May 2019
8. International Engineering Alliance. Graduate Attributes and Professional Competencies. Version 3: 21 June 2013 (2013). http://www.ieagreements.org/assets/Uploads/Documents/Policy/Graduate-Attributes-and-Professional-Competencies.pdf. Accessed 10 May 2019
9. Passow, H.J., Passow, C.H.: What competencies should undergraduate engineering programs emphasize? a systematic review. Res. J. Eng. Educ. **106**(3), 475–526 (2017). https://doi.org/10.1002/jee.20171
10. NC State University. ENGINEER Competency Profile. https://ts.hr.ncsu.edu/wp-content/uploads/sites/14/2016/06/Engineer.pdf. Accessed 10 June 2019
11. Trevisan, M.S., Davis, D.C., Crain, R.W., Calkins, D.E., Gentili, K.L.: Developing and assessing statewide competencies for engineering design. J. Eng. Educ. **87**(2), 185–193 (1998). https://doi.org/10.1002/j.2168-9830.1998.tb00340.x
12. Davis, D.C., Crain, R.W., Jr., Calkins, D.E., Gentili, D.E., Trevisan, M.S.: Competency-based engineering design projects. In: Proceedings, 1996 ASEE Annual Conference, ASEE, Session 1608, pp. 1.108.1–1.108.17 (1996)
13. Angeles, J., Britton, R., Chang, L., Charon, F., Gregson, P., Gu P., Lawrence, P., Stiver, W., Strong, D., Stuart, P., Thompson, B.: The Engineering Design Competency (2004). https://ojs.library.queensu.ca/index.php/PCEEA/article/download/3991/4061
14. Goff, R.M., Terpenny, J.P.: Engineering design education - core competencies. In: Industrial and Manufacturing Systems Engineering Conference Proceedings and Posters, vol. 11 (2012). http://lib.dr.iastate.edu/imse_conf/11, https://doi.org/10.2514/6.2012-1222

15. Robinson, M.A., Sparrow, P.R., Clegg, C., Birdi, K.: Design engineering competencies: future requirements and predicted changes in the forthcoming decade. Design Stud. **26**(2), 123–153 (2004). https://doi.org/10.1016/j.destud.2004.09.004. ISSN 0142-694X
16. Woollacott, L.C.: Taxonomies of engineering competencies and quality assurance in engineering education. In: Patil, A., Gray, P. (eds.) Engineering Education Quality Assurance, pp. 257–295. Springer, Boston (2009)
17. Passow, H.J.: Which ABET competencies do engineering graduates find most important in their work? J. Eng. Educ. **101**(1), 95–118 (2012). https://doi.org/10.1002/j.2168-9830.2012.tb00043.x
18. Downey, G.L., Luicena, J.C., Moskal, B.M., Parkhurst, R., Bigley, T., Hays, C., Jesiek, B. K., Kelly, L., Miller, J., Ruff, S., Lehr, J.L., Nichols-Belo, A.: The globally competent engineer: working effectivcly with people who define problems differently. J. Eng. Educ. **95**(2), 107–122 (2006). https://doi.org/10.1002/j.2168-9830.2006.tb00883.x
19. Walther, J., Kellam, N., Sochacka, N., Radcliffe, D.: Engineering Competence? An Interpretive Investigation of Engineering Students' Professional Formation. J. Eng. Educ. **100**(4), 703–740 (2011). https://doi.org/10.1002/j.2168-9830.2011.tb00033.x
20. Charyton, C., Jagacinski, R.J., Merril, J.A., Clifton, W., DeDios, S.: Assessing creativity specific to engineering with the revised creative engineering design assessment. J. Eng. Educ. **100**(4), 778–799 (2011). https://doi.org/10.1002/j.2168-9830.2011.tb00036.x
21. Kishline, C.R., Wang, F.C., Aggourne, E.M.: Competency-based engineering design course development. In: Northcon/98 Conference Proceedings (Cat. No.98CH36264), Seattle, WA, USA, 1998, pp. 202–207 (1998). http://doi.org/10.1109/NORTHC.1998.731537
22. Strong, D.S., Stiver, W.: Engineering design competency: perceived barriers to effective engineering design education. In: Brennan, R., Yellowley, I., (eds.) Proceedings of the 2nd CDEN Design Conference. Design education. University of Calgary (2005). http://citeseerx.ist.psu.edu/viewdoc/download?doi=10.1.1.452.5883&rep=rep1&type=pdf\
23. May, E., Strong, D.: Is engineering education delivering what industry requires. In: Proceedings of the Canadian Design Engineering Network (CDEN) Conference, Toronto, Canada, July 24–26, 2006, pp. 204–212 (2006). https://doi.org/10.24908/pceea.v0i0.3849
24. Dym, C.L., Agogino, A.M., Eris, O., Frey, D.D., Leifer, L.J.: Engineering design thinking, teaching, and learning. J. Eng. Educ. **94**, 103–120 (2005)

Blended-Learning Experiences in the Elective Course "Scientific Writing"

María Teresa Pérez Pino(✉), Liliana Argelia Casar Espino, Leonid Rodríguez Basabe, Ailec Granda Dihigo, Tito Díaz Bravo, Rosa Adela González Noguera, and Pedro Castro Álvarez

University of Informatics Sciences, La Habana, Cuba
{mariatpp, lily, lioni, agranda, tdiaz, rosygonzan, pcastro}@uci.cu

Abstract. Information and communication technologies are a reality in university classrooms. Young people live connected. Blended learning is becoming a most relevant approach in university contexts that use these environments as alternatives for developing lessons. In face-to-face teaching, technology is viewed as a complement not as a real toll for a methodological change. The purpose of this study is to increase students' motivation and academic efficiency in the elective course "Scientific Writing" at the University of Informatics Sciences. The research methods used were the historical-logic; the documentary analysis and the pre experiment. Wicolxon or ranked sign test. B-learning is assumed from a perspective oriented towards planning its use as a changing strategy in terms of organizational, methodological and didactic aspects. Teaching processes must be conceived taking into account the need for flexibility and continuous change. It was also confirmed the advantage of using virtual communication resources, new ways of participating and constructing information.

Keywords: Blended-learning · Elective course · Scientific writing

1 Introduction

1.1 Context

Teaching has been transformed, at the beginning of the new century, with communication technologies. They have become accelerators of change that influence the main actors of the process, both teachers and students, who use technologies to interact, communicate, carry out activities and tasks, teach and learn [1]. Computer-mediated teaching-learning models have led to new ways of teaching and learning beyond spatialtemporal coordinates. E-Learning, blended learning and mobile learning offer a new universe of interactivity for the didactic relationship [2].

Virtual teaching has been the maximum representation of the use of technology for teaching, but in classroom teaching technology was seen as a complement rather than as a real tool for methodological change [3]. Blended Learning is acquiring a fundamental relevance in university contexts that use these and other environments as

M. E. Auer and D. May (Eds.): REV 2020, AISC 1231, pp. 886–897, 2021.
https://doi.org/10.1007/978-3-030-52575-0_73

alternatives to develop the classes [4]. One of the most effective patterns of modern higher education is the learning mix that combines traditional classroom activities with elements of distance learning, and makes extensive use of information technology [5].

The pedagogical experience was carried out in the elective course Scientific Writing taught to third year students at the University of Informatics Sciences. This course is designed for a face to face modality and students must fulfill the learning objectives and acquire the content knowledge stated in the curriculum. However, this is a traditional conception and does not meet the students' characteristics and needs. Students are not motivated and their academic performance as shown in the first topic is low.

1.2 Purpose or Goal

The aim of this work is to contribute to raising the motivation and academic performance of students in the elective course Scientific Writing with the application of the b-learning modality from a perspective oriented to the planning of its use as a strategy of change of organizational, methodological and didactic character.

Research problem: How to contribute to raise the academic performance of students in the elective course scientific writing?

Hypothesis: If b-learning is applied from a perspective oriented to the planning of its use as a strategy of change of organizational, methodological and didactic character in the elective course Scientific Writing then it will help to raise the academic performance of the students.

Specific objectives:

1. Systematize the main theoretical and methodological foundations about b-learning and its incidence in the university environment.
2. Diagnose the initial state of the academic performance of students in the first topic of the course Scientific Writing.
3. Apply b-learning from a perspective oriented to the planning of its use as a strategy of change of organizational, methodological and didactic character in the elective course Scientific Writing.
4. Check through a pre experiment the effectiveness of the design and application of blearning in the elective course Scientific Writing on the academic performance of students.

2 Development

2.1 Blended Learning

B-learning is a learning approach that combines in a harmonious way face-to-face as well as virtual components. A review of the literature of recent years was carried out to define the concept of b-learning and its incidence in the university context.

The blended concept, since its appearance, has had different names: hybrid, blended, mixed and mixed learning [6–10]. And also different visions or meanings a) blended-learning (b-learning from now on) as a combination of face-to-face learning;

b) b-learning as a combination of distribution systems or training distribution technologies; c) b-learning as a combination of strategies or learning models [11]. The first of the trends has been consolidated and is widely accepted, while the terminology is still diffuse and the terms hybrid, mixed and blended are used interchangeably [12].

Blended Learning is understood as a system in which face-to-face and non-face-toface situations are mixed, resorting to the most appropriate technologies for each need [13]. Regarding its diversity, it is stated that there are not two identical Blended Learning designs, because they have their specificities [14].

Collaborative teamwork is very important in Blended Learning designs, the group is the space where it is naturally possible to develop the competences (interpersonal, intercultural and social, and civic competence) recommended by the Council of Europe on key competences for lifelong learning [13] and [15].

Mixed learning refers to the use of both face-to-face and non-face-to-face technological resources to optimize the result of the training. While the educators talk about "blended learning", a learning in which face-to-face and virtual resources are mixed, the students begin to come from another environment, live "blended lives", lives in which face-to-face and virtual reality are mixed [13].

While teachers consider the possibility of introducing elements of e-learning, and through blended learning to optimize learning, students live combining both realities.

This is a point that teachers should be clear about: a Blended learning design is not going to be used because it is more effective, but because it is the usual way to communicate, to access information, to manage social networks, to build knowledge. "… that is precisely BL: choosing the most appropriate resource according to their potential and the needs of the subject" [13].

This author defends the idea of a model "… perhaps closer to the learning communities, model in which the presence and virtuality are mixed in a continuous way". [13], a criterion that is shared.

The student requires new competences, such as:

- Self-regulated learning ("Self-regulated learning").
- Digital competence as described by the European Commission (2005) [15] including the critical capacity to manage information.
- Knowledge of other languages to access other sources of resources [13].

It is recommended to keep in mind that much of the information is already on the Internet or in books and magazines. The need for the teacher to focus his work on the activities that allow students to develop the required competencies is highlighted. Only when it is known that students need some information not available should be provided.

You should not limit the development of your ability to search, assess, select, structure information.

Bartolomé, (2008) highlights some features of the blended learning model that he proposes:

- The environment as a space in which the subject develops basic skills such as the ability to self-regulate learning or digital skills.

- Offer the student a resource-rich environment so that he can determine his training needs, find the resources that can help him solve them and apply them effectively.
- Do not neglect the potential of audiovisual language.
- Provide the environment with flexibility, so that teachers and students are comfortable in it, that they can use it adapting it to their needs and characteristics.
- Do not try to offer all the resources: there are many options on the Internet. Using them will not only save costs and effort, but also prepare students to continue using them when they finish their training.
- Starting from a flexible and focused on the subject curriculum [13].

It is suggested that b-learning fosters a favorable attitude, both on the part of professors and students, towards online methodologies, but, more especially, towards mixed or blended models, of which the usefulness was highlighted, the flexibility allowed, and the increase of the involvement and participation of the students and that contributes to the increase of the academic performance of the students through the implementation of blended learning activities [16]. The b-learning provides teaching flexibility in educational times and spaces, access to multiplicity of resources in addition to those offered by the teacher, new ways of interaction between student-teacher and students, increased autonomy and student responsibility in their own process [17] as elements of educational improvement, in addition to facilitating the development of digital competence [16]. The b-learning is assumed from a perspective oriented to the planning of its use as a change strategy of an organizational, methodological and didactic character.

2.2 The Classroom Modality in Cuban Higher Education

In the face-to-face mode, the group is a necessary natural environment, both because of the teacher's action and because of the pace imposed by the face-to-face meeting in class sessions. The emotional aspects are present naturally through the use of non-verbal languages: a phrase of the teacher always includes an emotional aspect since it is always accompanied by a facial expression, a tone, and a body position [13].

The face-to-face session with the group is a very valuable resource if it is achieved that it is not only used as a vehicle for transmitting content. Unfortunately, many teachers still identify this function as fundamental in the classroom. This also explains the massive use and type of use of "PowerPoint" presentations, a real threat to university teaching [13] and [18].

The design of the curriculum based on objectives encourages many teachers to focus their teaching on transmitting content. It would be advisable to reflect on the convenience of a design based on competencies and the use of much more adequate resources to transmit information. A curriculum focused on the activities for the development of the same and not in the transmission of content, even when knowing is part of the process of its acquisition.

The recommendations of tasks carried out by Bartolomé, (2008) for the face-to-face sessions are interesting and valuable:

- Present a topic globally.
- Give guidelines for a job.

- Encourage students, help them to be motivated.
- Show the relationship of a topic with others.
- Present the fundamental elements of a topic in a succinct way.
- Suggest important aspects to study.
- Generate group dynamics that help learning.
- Carry out group tutoring.
- Supervise simultaneous individual or group activities.
- Present small and precise information packages (short duration).
- Show the practical application of a theoretical aspect. Present devices, examples, experiences [13].

2.3 Material and Method

The 27 students of the third year group were taken as units of analysis, the program of the course, the result of the systematic evaluation of the first topic of the course, the study plan. The logical historical for the systematization of the theoretical - methodological foundations about the b - learning and its incidence in the university environment were used as methods in its realization. The documentary analysis for the analysis of the program of the subject, the normative documents and the evaluative results of the first subject of the course Scientific writing. The pre experiment to check the hypothesis. The percentage analysis for the data processing and the test of the ranked or Wicolxon signs to check if significant changes are obtained in the academic achievement of the students after applying the b-learning.

2.4 Results of the Analysis of the Program of the Course and the Diagnosis of the Academic Achievement of Students in Topic 1

The course Scientific writing is located in the curriculum as an elective course for third and fourth year students.

Table 1 shows the distribution of the topics according to the types of traditional classroom classes.

The result of the application of the method of documentary analysis to the program of this course shows that it is a traditional program, conceived only for the classroom modality.

The class predominates as a form of organization and its types include the lecture, the seminar and the practical lesson. The normative documents offer the possibility of making changes, provided that the number of class hours and the contents to be addressed are respected. In the Teaching Regulations for Higher Education, Resolution 2/2018 [19], in Article 99, it is stated that in the different academic periods an adequate balance of the time that students dedicate to face-to-face and non-face-to-face activities must be guaranteed, as a way to promote their autonomous learning under the guidance and control of teachers. In article 125 it is emphasized that the first priority in the teaching work is the correct application of the integral approach for the educational work in the universities, which is specified in all the teaching activities that are carried out.

Table 1. Distribution of the topics of the subject Scientific writing

General data				
Subject:	Scientific writing			
Discipline:	Professional Practice			
Specialty:	Informatics Science Engineering			
3rd and 4th Academic Years	2nd Semester	Academic Course: 2017– 2018		
	L	PL	S	Total
Topic 1	4	4	2	10
Topic 2	4	4	2	10
Topic 3	2	2	2	6
Topic 4	2	2	2	6
Total	12	12	8	32

Note: L: Lecture PL: Practical Lesson S: Seminar
Topic 1: Introduction to Scientific writing
Topic 2: Information management. Citations and references
Topic 3: The scientific article
Topic 4: The thesis or report

In article 126 it is stated that the organizational form of the teaching work is the structuring of the activity of the professor and the students, in order to achieve in the most efficient and effective way the fulfillment of the objectives foreseen in the study plans. In the development of the different organizational forms it is essential that the teacher guarantees the activity and communication of the students in an affective climate and manages to awaken interest in the content object of learning, so that they feel committed to the achievement of the objectives to reach.

Article 128 states that the class is one of the organizational forms of teaching work, whose objectives are the acquisition of knowledge, the development of skills and the formation of values and cognitive and professional interests in students, through the realization of activities of an essentially academic nature. Classes are classified on the basis of the objectives to be achieved and their main types are: the lecture, the practical lesson, the seminar, the meeting class, the laboratory practice and the workshop.

The results of the academic achievement of students in topic 1. The students show little interest in the subject. There are low results in academic achievement in the first subject, which are reflected in Table 2.

Table 2. Results in the academic achievement of students in Topic1.

			Evaluative categories					
Group	Enrollment	Students that Passed	2	3	4	5	% Passed/Enrollment	% Quality
RTC	27	14	13	4	6	4	52	38

As you can see, the results were low in terms of the percentage of students that passed attending enrollment: 52%, as well as the quality: 38%.

The research problem is confirmed and the need to intervene in the educational reality is demonstrated. It is based on the hypothesis: if b-learning is applied from a perspective oriented to the planning of its use as a strategy of change of organizational, methodological and didactic character in the elective course Scientific Writing then it will help to raise the academic performance of the students.

The proposed redesign of the b-learning program is applied as a possible solution from a perspective oriented to the planning of its use as a change strategy of organizational, methodological and didactic character and its application to the other topics of the course with the use of virtual communication resources and new modes of participation and construction of information.

2.5 Teaching Experience in the Redesign and Application of the Program of the Elective Course Scientific Writing

In the redesign of the program of the elective subject Scientific writing is part of a strategy of organizational change. Other types of classes are introduced in the planning and control model of the teaching process as it appears in Table 3.

Table 3. Redesign of the program of the elective course Scientific writing

General data							
Subject:			Scientific Writing				
Discipline:			Professional Practice				
Specialty:			Informatics Science Engineering				
3rd and 4th Academic Years			2nd Semester	Academic Course: 2018–2019			
	L	GL	PL	S	LP	EW	Total
Topic 1	2	2	0	2	2	0	10
Topic 2	2	2		0	4	2	10
Topic 3		2	2	2	0	0	6
Topic 4	2		2	2	0	2	6
Total	6	6	4	6	6	4	32

Note: L: Lecture GL: Guiding Lesson PL: Practical Lesson S: Seminar LP: Laboratory Practice EW: Evaluative Workshop

In the redesign of the program of the subject other types of lessons were included: guiding lessons, laboratory practices and evaluation workshops. These contribute to favor the active role of the students in the learning process.

2.6 Methodological Recommendations

The change must have a methodological and didactic character with the use of virtual communication resources and new modes of participation and construction of information. The methods, means, organizational forms should favor the integration of ICT and an active position of the students towards learning. The learning tasks should promote self-preparation of students for the development of their individual potential and their autonomous learning.

Keep in mind the work with curricular strategies: the widespread use of information and communication technologies, the use of the mother tongue, communication in English, the humanistic education of the student, among others.

In lectures, students should be given the most up-to-date scientific-technical foundations, through the appropriate use of scientific and pedagogical methods, in order to help them to integrate the knowledge acquired and the development of skills and values for a better professional performance.

In the seminars, the students should be guided by learning tasks that encourage them to consolidate, expand, deepen, discuss, integrate and generalize the oriented contents, by using the methods of the branch of knowledge and scientific research. They should encourage the development of their oral expression, the logical ordering of contents and skills in the use of different sources of knowledge.

In practical classes the teacher must plan teaching tasks that require intra, inter and transdisciplinary approaches. This type of tasks allow students to execute, expand, deepen, integrate and generalize work methods characteristic of subjects and disciplines, which allow them to develop skills to use and apply, independently, knowledge.

In the guiding lessons the teacher will clarify the doubts corresponding to the contents and activities previously studied by the students. These contents will be discussed and exercised and their compliance will be evaluated. The essential aspects of the new content will be explained and the independent work that the students must carry out in order to achieve an adequate command of these will be clearly and precisely oriented. The most important mission that the teacher has in the guiding lesson is to contribute to the development of the cognitive independence of the students and to favor the development of values that enhance it.

In the workshops students will apply the knowledge acquired in the different disciplines to solve problems of the profession, from the link between the academic, investigative and labor components. The workshops should contribute to the development of skills for the integral solution of professional problems in groups, for the group and with the help of the group, where interdisciplinary relations prevail.

It must be guaranteed that in the laboratory practices the students acquire the proper skills of the methods and techniques of work and scientific research; expand, deepen, consolidate, generalize and check the theoretical foundations of the subject or discipline through experimentation, using the necessary means. As a rule, in this type of class the individual work of the students in the execution of the planned tasks must be guaranteed.

In the workshops students will apply the knowledge acquired in the different disciplines to solve problems related to the profession, from the link between the academic, investigative and labor components. The workshops should contribute to the development of skills for the integral solution of professional problems in groups, for the group and with the help of the group, where interdisciplinary relations prevail.

The research work of the students should be encouraged, so that it contributes to develop the skills of the technical and scientific research work, through tasks that require the use of elements of the methodology of scientific research. It contributes to the development of the initiative, the cognitive independence and the creativity of the students. It fosters the development of skills for the efficient and up-to-date use of information sources, of foreign languages, of computer methods and techniques and of the national standardization, metrology and quality control system.

Virtual communication resources: the use of the Zera platform is recommended, communication tools such as: chat, email, forum …; digital educational resources, cell phones and the digital portfolio.

Make available to students on the platform materials for their training in different types of resources: textual, audiovisual and multimedia. The materials can be consulted by students at the time they consider appropriate, which allows a flexibility of the training action.

Communicate with students, both synchronously and asynchronously and develop more successfully the tutoring and discussions through communication systems. It is important to ensure that students design their own learning strategies, work as a team, possess interpersonal communication skills and know how to search for reliable information on the web.

In the application of the pre experiment, the results of the evaluative report of topic 1 were taken into account as an initial test. The program was given taking into account the redesign carried out and the final evaluation of the subject was made.

Once the experience of redesigning the program and its application in the classroom was completed, it was possible to confirm the results of the final evaluation of the subject that appear in Table 4.

Table 4. Results of the final evaluation of the course

			Evaluative categories					
Group	Enrollment	Students that Passed	2	3	4	5	% Passed/Enrollment	% Quality
RTC	27	26	1	6	11	9	96	74

Results of the Mann-Whitney W Test (Wilcoxon) to compare medians
Null hypothesis: median1 = median2
Alt. Hypothesis: median1 <> median2
Average range of sample 1: 16,375
Average range of sample 2: 32,625
W = 483.0 P-value = 0.0000200273
The null hypothesis is rejected for alpha = 0.05.
It was used as a statistician: STATGRAPHICS version 15.2

This option executes the Mann-Whitney W test to compare the medians of two samples. This test is constructed by combining the two samples, ordering the data from least to greatest, and comparing the average ranks of the two samples in the combined data. Because the P-value is less than 0.05, there is a statistically significant difference between the medians with a confidence level of 95.0%.

It is statistically verified that there is a significant difference between the initial and final tests with respect to the academic achievement of the students in the Scientific writing course.

The students were asked to indicate the positive and negative aspects of the subject that they studied with the application of b-learning elements. As positive aspects pointed out: the amount of teaching materials put at their disposal in different formats, the possibility of working independently of the class schedule and access to information from anywhere. More contact with the teacher and with her classmates. As negative aspects point out: the problem of the availability of internet connection and the malfunctioning of the platform.

The advantages of the application of b-learning in the academic achievement of students in the subject Scientific writing are demonstrated [5, 16]. The incorporation of virtual resources was positive because it is the usual way for young people to communicate, to access information, to build knowledge [13]. Interactivity is favored for the didactic relationship [2]. The use of technology is a real tool for methodological change [3]. The b-learning is verified from a perspective oriented to its use in the subject as a change strategy of organizational, didactic and methodological character. The convenience of the use of virtual communication resources and new modes of participation and construction of information is confirmed. It shows that it is necessary to guide the training processes in a flexible way [12]. The students are involved, participate more and are more active in the training action.

It is recommended to continue investigating the possibilities of b-learning as a strategy to overcome the vision of the classroom and the face-to-face class as exclusive possession of teachers and face-to-face evidence as the main source of information for the students' learning process and its evaluation.

3 Conclusions

The b-learning or mixed learning is approached from a perspective oriented to its use in the elective subject. Scientific writing at the University of Informatics Sciences as a change strategy of organizational, didactic and methodological nature. Training processes must be conceived based on the need for flexibility and continuous change. The convenience of the use of virtual communication resources and new modes of participation and construction of information is confirmed.

We offer a series of recommendations oriented to pedagogical planning and the integration of virtual and face-to-face modalities, taking into account the orientation of the training process based on the need for flexibility and continuous change. The pre experiment is applied. The advantages of the application of b-learning in the academic achievement of students in the subject Scientific writing are demonstrated.

References

1. Islas Torres, C.: El B-learning: un acercamiento al estado del conocimiento en Iberoamérica, 2003–2013. Apertura **6**(1), 1–12 (2014). http://www.udgvirtual.udg.mx/apertura/index.php/apertura3/article/view/500/357
2. Bartolomé Pina, A.R., García Ruiz, R., Aguaded Gómez, J.I.: Blended learning: panorama y perspectivas. RIED: revista iberoamericana de educación a distancia, vol. 21, n. 1, pp. 33–56 (2018). https://doi.org/10.5944/ried.21.1.18842. http://rabida.uhu.es/dspace/handle/10272/14430
3. García-Peñalvo, F., Ramírez Montoya, M.: Aprendizaje, Innovación y Competitividad: La Sociedad del Aprendizaje. Revista De Educación a Distancia, (52) (2017). https://revistas.um.es/red/article/view/282141
4. Castaño, R., Jenaro, C., Flores, N.: Percepciones de estudiantes del Grado de Maestro sobre el proceso y resultados de la enseñanza semipresencial -Blended Learning-. Revista De Educación a Distancia, (52) (2017). https://revistas.um.es/red/article/view/282161 http://dx.doi.org/10.6018/red/52/2
5. Shurygin, V.Y., Sabirova, F.M.: Particularities of blended learning implementation in teaching physics by means of LMS Moodle. Revista espacios **38**(40), 39 (2017). http://www.revistaespacios.com/a17v38n40/17384039.html
6. Bartolomé Pina, A.: Blended learning. Conceptos básicos. Pixel-bit **23**, 7–20 (2004). https://idus.us.es/xmlui/handle/11441/55455
7. Llorente, M.C.: Formación semipresencial apoyada en Red (Blended Learning). Diseño de acciones para el aprendizaje, Eduforma, Alcalá de Guadaira (2009)
8. Moran, L.: Blended-learning. Desafío y oportunidad para la educación actual. Edutec, Revista Electrónica de Tecnología Educativa **39** (2012). http://dx.doi.org/10.21556/edutec.2012.39.371
9. Picciano, A.: Introduction to blended learning: research perspectives. In: Picciano, A., Dziuban, C.R., Graham, C.R. (eds.) Blended Learning: Research Perspectives. Routledge, New York (2014)
10. Valverde-Berrocoso, J., Balladares Burgos, J.: Enfoque sociológico del uso del b-learning en la educación digital del docente universitario. Sophia, Colección de Filosofía de la Educación, (23), 123–140 (2017). https://dx.doi.org/10.17163/soph.n23.2017.04
11. Tayebinik, M., Puteh, M.: Blended learning or e-learning? Int. Magazine Adv. Comput. Sci. Telecommun. **3**(1), 103–110 (2012). http://doi.org/10.1016/j.iheduc.2012.12.001
12. Salinas Ibáñez., J., de Benito Crosetti, B., Pérez Garcies, A., Gisbert Cervera, M.: Blended learning, más allá de la clase presencial. RIED. Revista Iberoamericana de Educación a Distancia **21**(1), 195–213 (2018). https://doi.org/10.5944/ried.21.1.18859
13. Bartolomé Pina, A.: Entornos de aprendizaje mixto en educación superior. RIED. Revista Iberoamericana de Educación a Distancia **11**(1), 15–51 (2008). https://doi.org/10.5944/ried.1.11.955
14. Garrison, D., Kanuka, H.: Blended learning: Uncovering its transformative potential in higher education. Internet Higher Educ. **7**, 95–105 (2004). http://www.sciencedirect.com/
15. Comisión de las comunidades Europeas (2005). Recomendación del Parlamento Europeo y del Consejo sobre las competencias clave para el aprendizaje permanente. COM (2005) 548 final. http://ec.europa.eu/education/policies/2010/doc/keyrec_es.pdf

16. Cabero, J., Llorente, M.C., Morales, J.A.: Aportaciones al e-learning desde un estudio de buenas prácticas en las universidades andaluzas. Revista de Universidad y Sociedad del Conocimiento (RUSC) **10**(1), 45–60 (2013). http://rusc.uoc.edu/ojs/index.php/rusc/article/view/v10n1-cabero-llorente-morales/v10n1-cabero-llorente-morales-es http://dx.doi.org/10.7238/rusc.v10i1.1159
17. Adell, J., Area, M.: eLearning: Enseñar y aprender en espacios virtuales. En J. De Pablos (Coord.), Tecnología Educativa. La formación del profesorado en la era de Internet. Aljibe, Málaga, pp. 391–424 (2009)
18. Aliaga, F., Bartolomé, A.: El impacto de las nuevas tecnologías en Educación, en: Escudero, T.; Correa, A. Investigación en Innovación Educativa, pp. 55–88. Madrid: La Muralla (2006)
19. Ministerio de Educación Superior. (2018) Reglamento docente metodológico. Resolución 2/2018

Intelligent System Tutorial for Distance Learning the Computer Science Engineering Career

Maidelis Milanés Luque(✉), Bienvenido Hanley Roque Orfe, and Natalia Martínez Sánchez

Informatics Sciences University, Havana, Cuba
mmilanes@uci.cu

Abstract. With the advance of technology, there are new ways of teaching that break the barrier of time and place; distance education can be considered as the most advanced model at this time, however you may bring difficulties if you need support an advisory and/or guardian by the student at a given time and this is not available. To solve this problem a technological tool that behaves like a teacher in both the teaching strategy and mastery of subject teaching is required, guiding the student on how to apply this knowledge in their individual characteristics. Intelligent tutoring systems simulate the behavior of a human tutor, using Artificial Intelligence techniques in order to provide the student the required cognitive support for learning.

In this work, the creation of an Intelligent Tutorial System for the teaching of Engineering in Computer Science and distance many students at once using Artificial Intelligence techniques is proposed.

Keywords: Distance education · Intelligent tutoring system · Teaching and learning

1 Introduction

The current era and the foreseeable future are influenced by a need where information and knowledge are the main pillars to achieve development, Cuba as an underdeveloped country must seek the necessary means to achieve the advancement of society in pursuit of a better future. In this context Higher Education must play an active role, achieving transformation for the good of society, a mission that must be fulfilled by adapting to the frequent changes of today.

The University of Informatics Sciences (UCI), plays a fundamental role in the accelerated development of the country and the positive transformation of the Cuban community and society in general, therefore it must take full advantage of its technological component to fulfill this mission. One of the main actions that can be carried out would be to be able to teach the degree of Engineering in Computer Science remotely, but this could cause several disadvantages among: time available to possible distance learners to attend a meeting with teachers and clarify necessary doubts, time differences by geographical conditions between the applicant to pursue the career and

M. E. Auer and D. May (Eds.): REV 2020, AISC 1231, pp. 898–908, 2021.
https://doi.org/10.1007/978-3-030-52575-0_74

trained teachers, among others. The first barrier could perhaps be overcome with the development of today's technologies, even with the transformations that are being carried out in the country for the computerisation of the same, but the second may be more difficult due to the fact that trained personnel would be needed 24 h a day, depending on the students number.

Therefore, it would be interesting to have a system that, without neglecting the importance of the teacher, could, when the teacher is not available, guide the student in his or her learning process in a way that simulates the teacher, and contact the teacher when his or her direct intervention is already necessary. This is made possible with the development of Artificial Intelligence techniques and is concreted in the creation of Intelligent Tutorial Systems (ITS). This paper proposes the creation of a STI for the teaching of Engineering in Computer Science.

2 Development

Intelligent Tutor Systems (ITS) began to develop in the 1980s with the idea of being able to impart knowledge using some form of intelligence in order to assist and guide the student in the learning process. The aim was to emulate the behaviour of a human tutor, i.e. through a system that could adapt to the student's behaviour, identifying the way in which the tutor solves a problem in order to be able to provide cognitive aids when required (Humanante Ramos 2016).

"It is a software system that uses artificial intelligence (AI) techniques to represent knowledge and interacts with students to teach it to them (Burns and Capps 1988).

"Systems that model teaching, learning, communication, and mastery of specialist knowledge and student understanding of that mastery (Wolf 1984).

"A system that incorporates AI (Artificial Intelligence) techniques in order to create an environment that takes into account the diverse cognitive styles of the students who use the program" (Giraffa et al. 1997).

According to (Paquette et al. 2010). "An ITS uses AI techniques, mainly to represent knowledge, and to direct a teaching strategy; and it is capable of behaving like an expert, both in the domain of knowledge it teaches (showing the student how to apply that knowledge), and in the pedagogical domain, where it is capable of diagnosing the situation in which the student finds himself, and according to it, offering an action or solution that allows him to progress in learning.

As it can be appreciated in the previous definitions, Artificial Intelligence techniques are used for the simulation of the human tutor in the interaction with the student.

2.1 Artificial Intelligence in Intelligent Tutorial Systems

- Rule-Based Systems

According to (Enrique Castillo), two important elements are involved in rule-based systems: The knowledge base and data. Data consists of evidence or facts known in a particular situation.

Rule 1: If you score > 9, then grade = outstanding.

Rule 2: If < 20 or note > 7, then Admit = yes and Notify = yes.

Each of the above rules relates two or more objects and is made up of the following parts:

- The premise of the rule, which is the logical expression between the keywords if and then. The premise may contain one or more object-value statements connected to logical operators and/or not. For example, the premise of Rule 1 consists of a single object-value statement, while the premises of Rule 2 consist of two object-value statements connected by a logical operator.
- The conclusion of the rule, which is the logical expression after the keyword then. A rule is usually written as "If premise, then conclusion".

- Neural Networks

Neural networks are "massive parallel interconnections of simple elements that respond to a certain hierarchy by trying to interact with real objects as a psychological neural system would" (Boussabaine and Kaka 1998). Neural networks have the characteristic of assimilating knowledge based on experiences through the generalization of cases (Haykin 1999).

They're models that try to reproduce the brain's behavior. In the same way as the former, it makes a simplification, finding out which are the relevant elements of the system. An adequate choice of its characteristics, plus a convenient structure, is the conventional procedure used to build networks capable of carrying out a certain task. This model has elemental process devices, the neurons, as in the biological model. Together these can generate specific representations, for example a number, a letter, or any other object. (Chavez Huapaya and Contreras Ochoa 2018).

According to (Ruiz-Shulcloper 2012). For example, to predict academic performance, a back propagation type neural network can be used taking as input data the results of partial evaluations disaggregated in two ways.

- Taking the case of resolution by exercises.
- Taking exercises based on cognitive achievement.

Using the data from the students' partial evaluations, future performance can be predicted.

- Data Mining

It seeks to generate information similar to that which could be generated by a human expert, which also satisfies the principle of comprehensibility. The objective of this is to discover interesting knowledge, such as patterns, associations, changes, anomalies and significant structures from large amounts of data stored in databases, data warehouses, or any other means of storing information.

Data Mining is a special case of Automatic Learning, it uses its methods to find patterns, with the difference that the observed scenario is a database. In an Automated Learning scheme, the real world is the environment in which learning takes place, these are translated into a finite set of observations or objects that are encoded in some legible format. The set of examples constitutes the necessary information for the training of the system. (Chavez Huapaya and Contreras Ochoa 2018).

2.2 Examples of Intelligent Tutorial Systems

- Intelligent Tutorial System for Auto Training in Tuning Control Systems (STISIN).

The STISIN was conceived under the premise of developing a system that was as close as possible to a form of teaching according to the theory of the cognitive field. This meant that the student could interact with the system without it always being imposed on the teaching-learning process. The STISIN is oriented to the instruction of techniques of tuning of control systems. It is also a prototype that meets important specifications for an intelligent tutorial system. The system functions as a query system and as a tutorial system. (Veitía Rodríguez 2010).

- Intelligent Tutor System for the Diagnosis and Treatment of Sexually Transmitted Infections (STIITS)

STIITS as an application is the first Cuban Intelligent Tutorial System for the diagnosis and treatment of ITS developed in the province of Cienfuegos. STIITS responds to a social need but also has employment possibilities in teaching. There is no other software in the country with similar features for the same purpose. The goal of STIITS as an STI is to provide greater flexibility to the computer-driven tutorial and to make it allow for better interaction with the student. STIITS aims to capture the knowledge of experts in the specialties of gynecology and obstetrics and create interactions in a dynamic way, for a better understanding and development of these topics by students. (Veitía Rodríguez 2010).

The main problem of intelligent tutorial systems is that they are defined on a particular field of science, however it would be interesting to have an integral system capable of allowing the simulation of a teacher in the teaching-learning process on several sciences at the same time, in addition to allowing experts in the field to create the conditions to increase the knowledge of ITS. The main problem to be solved is the following: Will it be possible to create an intelligent tutorial system capable of teaching a complete career?

3 Initial Proposal for an Intelligent Tutorial System for the Distance Teaching of the Engineering Career in It Sciences

According to (Humanante Ramos 2016) ITS allow the emulation of a human tutor to determine what to teach, how to teach and who to teach through:

- Domain Module: defines the domain of knowledge
- Student Module: is able to define the student's knowledge at each point during the work session
- Tutor Module: generates learning interactions based on discrepancies between specialist and student
- User Interface Module: allows the interaction of the student with an ITS in an efficient way

The proposed ITS will contain the modules defined according to Zulma Cataldi:

In the domain module will be stored the necessary knowledge represented in an appropriate way according to the possible types of students identified so that it can efficiently meet the required objectives, in addition this module must be able to generate the best way to evaluate the student and give answers to problems presented through the intervention of the tutor module.

The student module will be in charge of characterizing the student's needs, as well as keeping the student's status updated as he or she progresses in the subject taught. In a first version will have identified the types of students that can be obtained, so it will use the supervised learning as a technique of Artificial Intelligence, it is desired in the future to ensure that the STI is able to characterize the student without predefined criteria.

In the tutor module will be defined the objectives that each student needs to meet according to the subject matter being studied, supervise and establish the pedagogical strategy to follow based on the characteristics identified by the modeling of the student, be able to identify the needs of the same as it progresses in the matter and redefine the strategy to follow if necessary, order and update the student's trajectory, as well as communicate with real teaching advisors if necessary.

In the user interface module, the interaction of the student with the STI will be defined, having as premises that this is as interactive as possible, allowing the student to consult what is referred by other members of the group, as well as the interaction in real time with them, simulating a virtual class.

Initially the first topic that was taken into account to include the STI was the English subject for specific purposes established according to the above reflected that:

The student's module was in charge of:

- Initially make a diagnosis to classify students according to selected learning styles and learning needs, which was prepared by(Guzmán Páez and German Campos 2009) in his master's thesis. See Annexes 1 and 2.
- Do a diagnostic test to establish the level of communicative-informational competence of the student.
- Inform you of the level of informational communicative competence according to the result of the diagnosis.
- Inform the student about the teachers who will be their advisors in the CASIE, as well as the schedule that they work in person in the same depending on the subject chosen and the available schedules of the English teachers and the specialty that are in the database.
- Maintain the student's file updated according to his or her trajectory.

This module through the use of Case Based Reasoning allows to obtain the level of knowledge associated to the student, to know future performances and to determine learning patterns, elements that will be taken into account by the virtual tutor to establish the learning strategy to follow.

The tutor module is able to guide the student through pedagogical strategies in the successful development of informational communicative competence using Pattern Recognition techniques to identify in the activities developed by the student if he made definitions, comparisons, generalizations and/or descriptions depending on the established dimension of informational communicative competence.

Three dimensions of informational communicative competence were instituted:

1. Access to information
2. Understanding information
3. Information production

Each dimension is defined by a set of descriptors and indicators:
INDICATORS AND DESCRIPTORS OF THE ACCESS TO INFORMATION DIMENSION

- Information Search: Uses reliable and relevant sources to find the information according to the determined need to solve the task in question.
- Selection of information: Choose digital documents to meet your information needs.
- Critical evaluation of the information: (date of publication, recognized source on the subject matter and topicality of the content of the article.
- Explanatory note: The student must determine an item from those assigned in the library of the subject, here ends the first dimension of informational communicative competence (Access to information)

INDICATORS AND DESCRIPTORS OF THE INFORMATION COMPREHENSION DIMENSION

- Determines the general idea of the article
- Identifies acts of speech: definition, classification, general comparison, description appearing in it.
- Underlines the essential paragraphs of the article

INDICATORS AND DESCRIPTORS OF THE INFORMATION PRODUCTION DIMENSION

- Summary in Spanish/English or one of the two (the application must give a notification for this task to be performed)
- Write the task solution in English language
- Establish the level of informational communicative competence for the activities developed.

4 Conclusions

The intelligent tutorial systems have allowed since their emergence to have one more opportunity to take teaching to different levels, simulating the presence of a human tutor thanks to the advances achieved in the area of artificial intelligence, although it is not possible from the point of view of the authors of this paper to replace the results achieved in terms of quality and efficiency of a real teacher, without doubt these are an extremely important help in the area of pedagogy.

Although doubts still persist in the development of intelligent tutorial systems referring to: the magnitude of the subjects in which they can be specialists, the affective component that the teacher takes into account in his teaching-learning process, the formation of values and the educational work that can be carried out in man-to-man training, it would undoubtedly be interesting to investigate how far we can go to improve the quality of teaching and differentiated work by breaking down the barriers of space and time. This is an issue that we must exploit in light of the accelerated advance in technology development and the need for good education and instruction in today's troubled world.

Annexes

Annex 1: Learning Styles Questionnaire

Have you ever thought about how to optimize your learning? Knowing your preferences for one way or another to perceive and process information, you can better guide your efforts to achieve school effectiveness

Answer the following questionnaire honestly:

A lot, if the situation fully fits your preferred learning style.

Little, if the situation fits only in a few cases to your preferred learning style.

None, if the situation does not fit in any way with your preferred learning style.

No	My preferred learning channels	M	P	N
1	I remember something better if I see it written down			
2	In class I like to follow the teacher's explanations attentively and to write little			
3	I learn better and more from what I see than from what I hear			
4	For the tests I like to study with someone who reads me their class or consulting notes			
5	I understand my teachers better when they write and graph information on the blackboard or on a transparency			
6	I don't like to study in absolute silence, I prefer to study with someone with whom I can exchange or at least with background music			
7	To engrave a new word or term well, I need to represent it in my mind first			
8	I can easily follow the thread of the teacher's explanation even if the teacher doesn't write anything on the board			
9	In order to learn something well, I must first repeat it several times for myself and, if possible, aloud			
10	I like the information presented through graphs, tables, diagrams, etc			

(*continued*)

(*continued*)

No	My preferred learning channels	M	P	N
My preferred ways of processing information				
11	When I consult the bibliography I concentrate more on concrete facts and global aspects than on detailed descriptions			
12	In class I don't like to miss a single detail with which the teacher deals with facts and phenomena			
13	When I read I usually do it by jumping around looking for the relevant and trying to avoid the insignificant and the details			
14	I have a fondness for Mathematics, statistical processing, logic, etc			
15	In the solutions to problems, I like to theorize as much as possible, without neglecting any of the specific characteristics of the issue I address			
16	When processing information from a text, or listening to my teacher, I like to activate my intuition, my experiences and previous experiences			
17	In class I prefer to take brief notes synthesizing the most relevant aspects that the teacher presents, rather than writing down everything he says			
18	I like to be concrete and realistic in the analysis of a phenomenon or in the search for a solution to a problem			
19	I am meticulous in my analysis, and I like not to overlook even the slightest detail			
20	Many times I am slow to read, because I like to capture with depth every idea and detail			
My preferred ways of orienting myself towards the fulfillment of my goals as an apprentice				
21	I am very careful to be late for classes or to deliver out of time the assigned work			
22	I'm quick to make decisions			
23	I'm used to being very organized for my class notes			
24	Instead of following a pre-established plan of activities, I like to be spontaneous and open in my personal and academic life			
25	In class I like to be actively involved in the search for solutions to problems raised by the teacher			
26	I am usually slow to make decisions and require a lot of time to analyze each variant and arrive at a final decision			
27	I'm not used to thinking too much about my answers			
28	When concentrating on a problem or task, I like to do it cautiously, thinking through every possible solution and every step			
29	To solve a task I like to have the methodology to follow			
30	I am usually always willing to take on an academic task or problem in class			

(*continued*)

(*continued*)

No	My preferred learning channels	M	P	N
My preferred ways of socially orienting myself in learning				
31	I like to study as a team, even for the tests			
32	In class I like the teacher to give me responsibilities to my classmates			
33	Of all, I prefer individual mental concentration activities such as reading, computing, etc			
34	I usually enjoy debates, group discussions and activities in which I can express my views to other colleagues			
35	I like to be open and sociable			
36	For practical classes, seminars and tests, I like to prepare myself and study alone			
37	I prefer written rather than oral evaluations in which I must express my views to the group			
38	I'm bored studying and preparing myself			
39	In class, I'm usually quiet			
40	When I work in a team, I am easily distracted and I am always disturbed by cooperation			

To know the level of preferences for the auditory visual or verbal learning channel, add the items: 1, 3, 5, 7, 10 and 2, 4, 6, 8, 9 respectively.

To know the level of preferences for the global or analytical way of processing information, add the items: 11, 13, 16, 17, 18 and 12, 14, 15, 19, 20 respectively.

To know the level of preferences for the planned or spontaneous way of orienting towards the fulfillment of your goals, add the items: 21, 23, 26, 28, 29 and 22, 24, 25, 27, 30 respectively.

To find out the level of preference for the cooperative or individual way of socially orienting oneself towards learning, add items: 31, 32, 34, 35, 38 and 33, 36, 37, 39, 40 respectively.

Annex 2: Learning Needs Questionnaire

User Name: ——————— E-mail:———————————

The purpose of this questionnaire is to help you as a user and your tutors to give you better advice. Your valuations will be very useful, because they will allow the creation of the center's database and record your needs and demands that you as a user pursue when you visit us. For which we thank you in advance.

Answer the following questions:

1-How many years have you studied English?
2-UD. use the Internet as a way to improve your knowledge of English?
3-On a scale from 1 to 5 (1 is the lowest level and 5 is the highest).
What is your level in each of these skills?

Grammar
Vocabulary
Writing
Reading
Oral expression
Hearing
Pronunciation
4-On a scale from 1 to 5 (1 is the lowest level and 5 is the highest).
Which of the following would you be interested in practicing at our center?
Reading of scientific articles.
Creation of ovcrviews.
Exhibition in workshops and seminars.
Presentation of projects.

5-List on a scale from 1 to 5 the following aspects to determine the degree of importance they represent for you.

Reading scientific articles
Making Overviews
Exhibition in workshops and seminars
Presentation of projects

6-How would you like to use our university's Center for Self-study and Foreign Language Services (CASIE)?

Individually
Co-Learning (with other students) Tutor them.
Combined
7-How would you like to be evaluated?
Self-evaluation
Co-evaluation
Assisted by a virtual tutor.
Assisted by a teacher-adviser
8-In what form would you like to be evaluated?
Written
Oral
Auditory
Combined

9-On a scale of 1 to 5 assess the expectations pursued in the interdisciplinary course offered at CASIE with the assistance of a computer tool:

_______ The contents and topics dealt with in the English language will be relevant for my future professional training.

_____ A Motivation to learn the English language to solve problems in the subjects of the specialty.

_____ At the level of analysis and critical appraisal of the texts dealt with in the subjects of the speciality.

Access to EVA's existing English-language specialty _______literature.

_______ To the level of updating given by access to Internet sites in English language recommended by the professors of the specialty.

References

Boussabaine, A.H., Kaka, A.P.: A neural networks approach for cost flow forecasting. Constr. Manage. Econ. **16**(4), 471–479 (1998)

Burns, H.L., Capps, C.G.: Foundations of Intelligent Tutoring Systems: An Introduction. Lawrence Erlbaum Associates, Hillsdale (1988)

Chavez Huapaya, S.M., Contreras Ochoa, C.Y.: Implementación de Business Intelligence, utilizando la Metodología de Ralph Kimball, para el proceso de toma de decisiones del área de ventas. Empresa Yukids (2018)

Giraffa, L.M.M., Nunes, M.A., Viccari, R.M.: Multi-ecological: an learning environment using multi-agent architecture. In: Proceedings of the Masta 1997 (1997)

Guzmán Páez, M.G., German Campos, M.A.: Incidence of didactic material for the English language learning of tenth year students of basic education at TELMO HIDALGO DIAZ high school, in SAN PEDRO SANGOLQUÍ, third trimester, school year 2008-2009. In: QUITO/ESPE-IDIOMAS/2009 (2009)

Haykin, S.: Neural Networks, A Comprehensive Foundation, pp. 161–175. Prentice-Hall Inc., Upper Saddle River (1999)

Humanante Ramos, P.R.: Resumen de la Tesis Doctoral: "Entornos Personales de Aprendizaje Móvil (mPLE) en la Educación Superior" (2016)

Paquette, L., Lebeau, J.-F., Mayers, A.: Authoring problem-solving tutors: a comparison between ASTUS and CTAT. In: Nkambou, R., Bourdeau, J., Mizoguchi, R. (eds.) Advances in Intelligent Tutoring Systems. SCI, vol. 308, pp. 377–405. Springer, Heidelberg (2010)

Ruiz-Shulcloper, J.: User Profile Module of the Virtual Tutor for the Evaluation of Self-Language Learning. Universidad de las Ciencias Informáticas, La Habana (2012)

Veitía Rodríguez, M.: Analysis and Design of the Virtual Evaluation Tutor for Autonomous Language Learning. Universidad de las Ciencias Informáticas, La Habana (2010)

Wolf, B.: Context Dependent Planning in a Machine Tutor. University of Massachusetts. Tesis Doctoral (1984)

Computational Tool to Support the Process of Teaching and Learning in the Discipline of Artificial Intelligence

Bienvenido Hanley Roque Orfe(✉), Maidelis Milanés Luque, Yunia Reyes González, Natalia Martínez Sánchez, Wilber Luna Jiménez, and Madelin Haro Perez

University of Computer Science, Havana, Cuba
bhroque@uci.cu

Abstract. The University of Computer Science is an educational-production centre that develops computer applications and services in different areas of knowledge. Because of the added value that the developed applications get when Artificial Intelligence techniques are used for decision making, the academic formation of the students in the different disciplines of the profession with emphasis on Artificial Intelligence is of high interest. This work describes a computational tool to support teaching that uses the conceptual algorithms of logical combinatorial recognition to determine the architecture and the set of initial weights of an artificial neural network, which contributes to the temporal efficiency of the learning process of the network and the efficiency of the classification. Experiments using Friedman's test and cross validation method demonstrate the applicability of this hybrid model in a Multilayer Perceptron.

Keywords: Artificial intelligence · Teaching-learning · Artificial neural networks · Conceptual algorithms · Combinatorial logical pattern recognition

1 Introduction

Information and Communication Technologies (ICT) have brought to the field of education innovative aspects that represent a qualitative improvement in the ways of teaching and learning with a multidisciplinary development involving Pedagogy, Psychology, computing, among other sciences.

Consequently, learning environments have been designed and implemented, which are articulated with the characteristics of the teaching-learning processes in an effective way, which favors the diffusion of different types of Computer-Assisted Teaching (EAC) media (Camejo 1992), according to its structure: Tutorials, Virtual Laboratories, Simulators, Trainers, Information Seekers, Intelligent Learning Teaching System (Martínez et al. 2007), among others.

This work describes a computer tool to support the process of teaching to learn the subjects of Artificial Neural Networks and Pattern Recognition of of the Artificial Intelligence Discipline. The objective of the tool is to support the student in the entire engineering process with the knowledge implicit in the design of the model of an

M. E. Auer and D. May (Eds.): REV 2020, AISC 1231, pp. 909–921, 2021.
https://doi.org/10.1007/978-3-030-52575-0_75

artificial neural network: network topology, set of initial weights and training set, using conceptual algorithms of Combinatorial Logical Pattern Recognition.

On the other hand, it is used in the software development process of the development centers of the university, in those products that are applicable techniques to give an added value.

The centers have the mission to develop and produce applications and computer services to part of the link of study work serving as support to the Cuban industry of the computer science.

2 Artificial Neural Networks

Artificial Neural Networks (ANNs) is defined as a mathematical-computational model that tries to imitate the structure and functionality of biological neural networks (Rosenblatt 1962).

The basic elements that make up ANNs are:

- A set of processing units, called neurons.
- The architecture or topology; the way neurons are organized.
- A learning rule.

According to the author (Montaño Moreno 2017) ANNs are information processing systems whose structure and functioning are inspired by biological neural networks. They consist of a set of simple processing elements called nodes or neurons connected to each other by connections that have a modifiable numerical value named weight.

ANNs are capable of detecting complex and non-linear relationships between variables. Variables are divided into input and output variables, related by some type of correlation or dependency. Neurons can be arranged in different layers, defining the type of layer according to the neuronal model you are facing.

Criterion 1: In relation to their layered structure, single-layer and multi-layer networks exist. Single-layer networks are those composed of a layer of neurons. Layered networks are those whose neurons are organized in several layers.

Criterion 2: Based on the flow of data in the neural network, there are unidirectional networks and recurrent networks. In unidirectional networks (feedforward), information circulates in a single direction, from the input neurons to the output neurons. In recurring or feedback networks, information can flow between layers in any direction, including outbound-inbound.

ANNs have two modes of operation, operation mode is the way in which the network can be used:

Learning or convergence: it is closely related to the defined learning rule and is charged from a set of entry patterns, it teaches the network a specific domain of application, by readjusting their weights.

Remembrance or execution: it begins when the system has already been trained and the learning ends (the weights and the architecture are fixed), being ready to solve the problem corresponding to the application domain on which it was trained.

A neural model is a mathematical model that attempts to reproduce some of the problem-solving capabilities of biological neural networks. Among the different types of artificial neural network models are Back-Propagation, radial base, probabilistic, McCulloch-Pitts and Multilayer Prceptron. The architecture of these models has more than two layers, which will be used in this research.

3 Combinatorial Logic Pattern Recognition

The RLCP can be applied to solve problems in most areas of knowledge: character recognition, medical diagnosis, remote sensing of the earth, identification of human faces and fingerprints, prognosis of breakages in equipment and machinery, analysis of biomedical signals and images, automatic inspection, blood cell count, analysis of well logs, archaeology, prognosis of mineral deposit, analysis of seismological activity, classification of documents, among others (Reyes González 2014).

According to (Ruiz-Shulcloper 2009) the problems of pattern recognition focus on feature selection, supervised classification, unsupervised classification and semi-supervised classification. The RLCP highlights the representation of objects in terms of a set of features, which allows the resolution of problems presented in the so-called nonformalised sciences, where the description of these objects can be mixed and incomplete.

Within the RLCP are the conceptual grouping algorithms that are composed of two fundamental tasks, which do not necessarily have to be independent or be performed in a particular order (González Kings 2014).

1. Extension structuring or determination: the process of grouping entities, in which groups are determined from a collection of objects.

Intentional characterization or determination: the concept of each structuring group is determined, the properties that characterize the grouping.

Conceptual grouping algorithms can be divided into two large groups, incremental and non-incremental algorithms (Pascual Belda 2018). Incremental algorithms base their operation on the adaptation of the groupings (or concepts) with the new objects presented to them, that is, each time a new object arrives by means of a certain strategy, it is classified in the already existing groupings or new groupings are created. On the other hand, non-incremental algorithms structure a sample of objects without assuming that they arrive one at a time (Ruiz-Shulcloper 2009).

In the works of (Ruiz-Shulcloper 2009), (Reyes González 2014), (Pons Porrata 2004) a critical analysis of different conceptual grouping algorithms is done, taking into account their characteristics and functioning, highlighting the following among their main results:

The main disadvantage of incremental type conceptual algorithms is the dependence of the result (the structuring) according to the order of presentation of the objects to the algorithm while in non-incremental type the number of groupings is determined randomly, which can be difficult in real life, due to ignorance of possible groupings in certain types of problems.

Algorithms that do not belong to the combinatorial logical approach to pattern recognition construct their concepts on the basis of probabilistic or statistical criteria, so their interpretation can be cumbersome for people not specialized in these areas of knowledge.

The LC-Conceptual and RGC algorithms (both belonging to the RLCP) construct their concepts based on logical properties, based on the features of the objects being studied. In addition, they do not require the number of clusters to be specified a priori. The difficulty that these algorithms have is the computational complexity in the calculation of typical testers.

4 Computer Tool

The computer tool is used in topic II of the subject of artificial neural networks that is taught in the University of Computer Science with the title Multilayer Networks, fulfilling the objective of training a multi-layer neuronal model applying conceptual algorithms. It also serves as a support to fulfill the final task of the subject where they are integrated with the different neuronal models that impact the subject with the combinatorial logical recognition of patterns applying algorithms of conceptual groupings.

The tool contains the Back-Propagation, Radial Base, Probabilistic, McCulloch-Pitts and Multilayer Perceptron neuronal models, their architecture being multi-layered and its neurons being connected with heavy connections to each other. In the case of the design of the Multilayer Perceptron, which is composed of three layers: an input layer, a hidden layer and an output layer with their respective heavy connections to each other, because of these characteristics it is first necessary to determine the number of neurons corresponding to each of the layers and then the two sets of initial synaptic weights; those between the input layer and the hidden layer and those between the hidden layer and the output layer.

The finding of the elements to design a network with the characteristics described above is carried out using the conceptual algorithms LC and RGC, taking into account their two stages. The extensional determination stage and the second intentional determination. The computer tool has four modules shown in Fig. 1.

The first module "Calculation of concepts" begins with the extensional determination of the Conceptual Algorithms in case of working with an unsupervised problem, grouping the objects in classes, then based on the intentional stage the concepts are calculated, which represent the distinctive characteristics of the objects that conform the groups. If the problem is supervised, it is directly based on the intentional determination since the classes are already grouped together.

The second module selects the artificial neuronal model to be used for the configuration of the network.

On the third module according to the type of neuronal model the neurons are selected by layer in correspondence with the results obtained after applying the conceptual algorithms.

Finally, the fourth module "Determination of initial synaptic weights", as its name suggests, defines the weights of the connections between the layers if there are more than one.

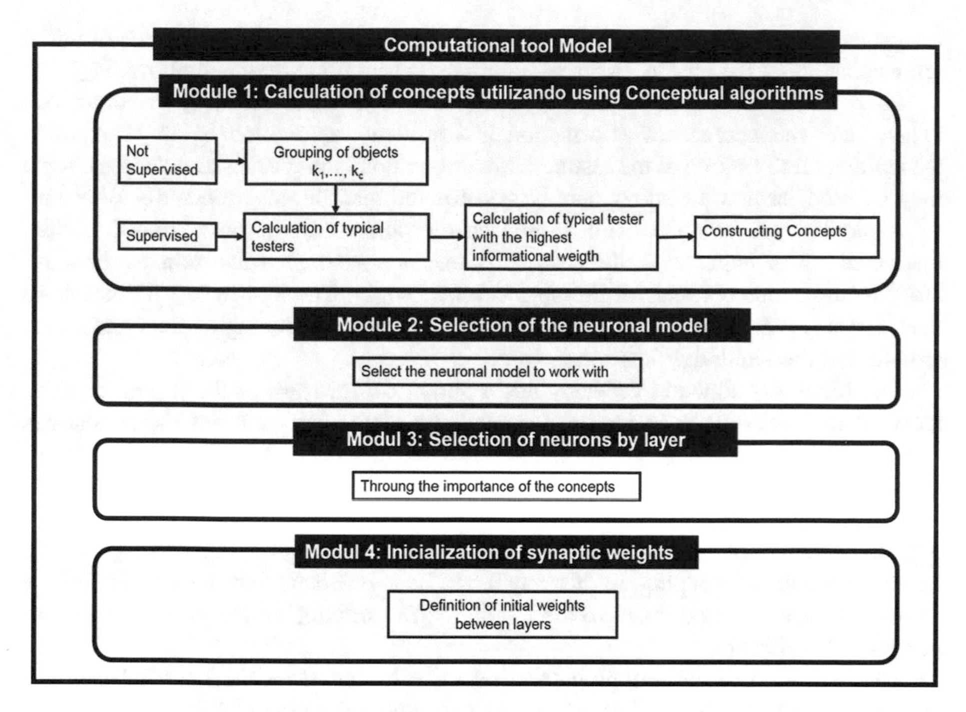

Fig. 1. Computational tool module

4.1 Module 1: Calculation of Concepts Using Conceptual Algorithms

Figure 2 shows the concept construction module where the possibility of doing it step by step as a explanatory guide of the concepts construction process is offered, or manually, which is made with the expert's knowledge, and lastly, 'complete', which allows to do the concepts calculations directly.

In this phase the algorithm LC-Conceptual or RGC is applied to take the typical testor of greater informational weight and the number of concepts formed.

Objective: To learn how conceptual grouping algorithms work through the calculation of typical testers and the selection of the testor with the greatest informational weight and the concepts associated with each class.

Regardless of the type of problem to be solved, it is the formal approach of the same, which aims to specify the objective and identify the sources of information with which you work. The presence of the specialist in the area of application is of vital importance, since his criteria largely decide several of the aspects to be taken into account, such as: what are the features and their importance, how are the variables and objects compared, determine the way in which the objects of study are represented, whether they are in classes, whether they are dissected or not, hard or diffuse. For this process it is proposed to adopt the methodology described in (Ruiz-Shulcloper 2009).

From the logical-combinatory approach to problems of classification without learning, it is known that objects can be represented in an initial M_i matrix of the set of descriptions of objects in a universe. For each x_i that are the features, there is associated a set of admissible values M_i i = 1,…, n, consequently the space of initial representation of the objects. A value comparison criterion is defined above M_i.

Each trait is a random variable with discrete (nominal or ordinal) or continuous values, and the absence of information is admitted, represented by the symbol *. Depending on the nature of the feature, different comparison criteria studied in the topic may be used, or new functions may be constructed that are not necessarily Boolean.

A similarity function is defined between the object descriptions: From the initial matrix and the comparison criteria a matrix can be constructed that reflects the similarity relationships between all the objects under study. The tool makes it possible to visualize this whole process in order to understand the functioning of the comparison criteria and the similarity function.

The Similarity function β determines a numerical measure of the degree of similarity of one object with respect to the other by taking into account the similarities between the features. This function linked to the IM is used to find the matrix of likeness.

The grouping criterion linked to the similarity function and the existence of other objects is the reason why an object will belong to a grouping or why two objects will belong to the same grouping. In this way it can be appreciated that the selection of the criterion to use, is determinant in the quality of the solution of the problem of unsupervised classification.

The definition of the grouping criterion must be based on the knowledge of the particular problem being addressed, in order to define the type of behaviour between objects based on their similarities, which is significant, according to the particular problem. Therefore, in selecting some grouping criteria, given a set of objects and the function of similarity, the family of groupings has been defined indirectly. The formal approach to the structuring of universes, of unsupervised classification, consists in finding a grouping criterion that responds to the interests of the problem in question.

For the construction of concepts, it is necessary to first select the determining features in the problem, this is done by calculating the testers, which allows the reduction of the number of features describing the objects. From the Training Matrix (ME) is obtained the Difference Matrix (MD) and then the Basic Matrix (MB), through which the testers are found. The typical testers are obtained, then the l-complexes of each grouping are obtained with the operator Conditioned Refunion or Extended Refunion which are the ones that apply the conceptual algorithms LC and RGC respectively (Martínez Trinidad and Ruiz Shulcloper 1999). The operator Generalization (GEN) and finally the concepts are formed for each class belonging to the problem.

Figure 2 visualizes the module of construction of the concepts where it offers the possibility of doing it step by step as explanatory guide of the process of construction of the concepts or manually that is with the knowledge of the expert to do the entire procedure and finally complete that allows to do the calculation of the concepts directly.

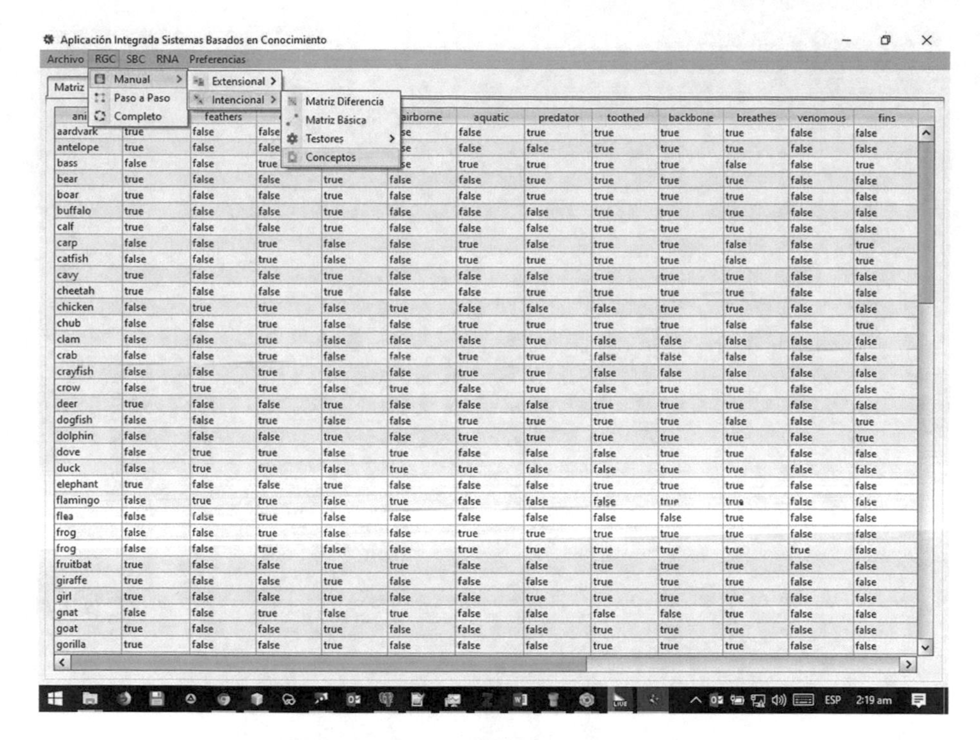

Fig. 2. Concepts construction module

4.2 Module 2: Selects the Artificial Neuronal Model

The phase is in charge of selecting the different neuronal models that are implemented in the computational tool. The models implemented are: Back-Propagation, Radial Base, Probabilistic, McCulloch-Pitts and Multilayer Perceptron their architecture being composed of three and more layers, allowing to establish synaptic weights between the entry layer and the hidden and between the hidden and the exit layer. Allowing to solve problems with a higher level of complexity in the different areas of knowledge (See Fig. 3).

4.3 Module 3: Selection of Neurons by Layer

This phase corresponds to the physical structure of the model to be used as a function of the number of processing units (neurons) corresponding to the layers (See Fig. 4).

Objective: To learn how to design the neuronal models chosen by selecting the number of neurons per layer.

- Input layer: the neurons coincide with the features that make up the concepts built on the basis of the typical testors used.
- Hidden layer: neurons are associated with constructed concepts.
- Output layer: the neurons coincide with the number of classes initially defined in the problem.

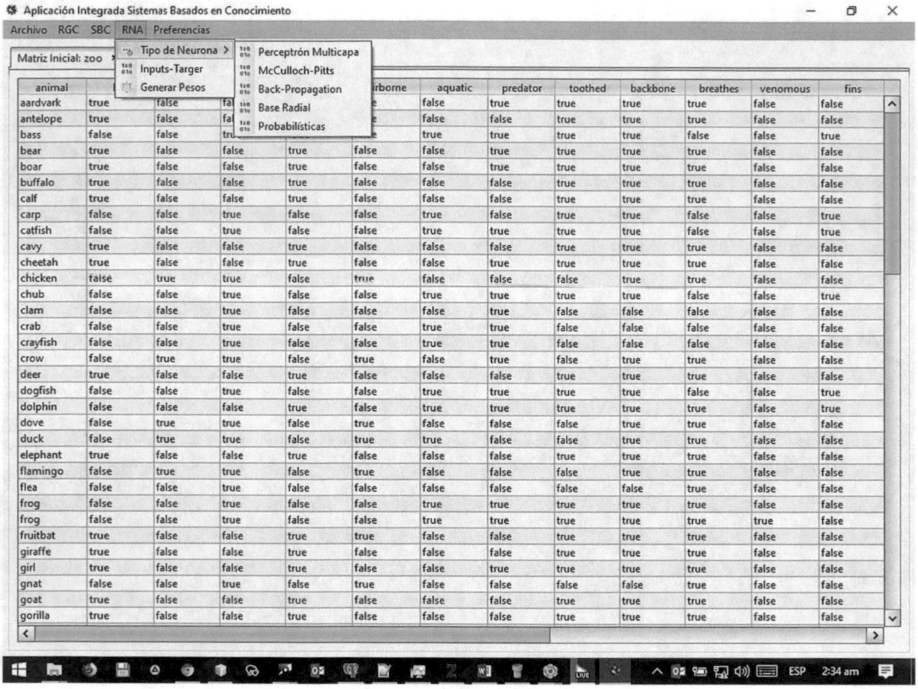

Aplicación Integrada Sistemas Basados en Conocimiento

Archivo RGC SBC RNA Preferencias

Matriz Inicial: zoo

Tipo de Neurona >
Inputs-Targer
Generar Pesos

Perceptrón Multicapa
McCulloch-Pitts
Back-Propagation
Base Radial
Probabilísticas

animal					irborne	aquatic	predator	toothed	backbone	breathes	venomous	fins
aardvark	true	false	fal		e	false	true	true	true	true	false	false
antelope	true	false	fal		e	false	false	true	true	true	false	false
bass	false	false	tru		e	true	true	true	true	false	false	true
bear	true	false	false	true	false	false	true	true	true	true	false	false
boar	true	false	false	true	false	false	true	true	true	true	false	false
buffalo	true	false	false	true	false	false	false	true	true	true	false	false
calf	true	false	false	true	false	false	false	true	true	true	false	false
carp	false	false	true	false	false	true	false	true	true	false	false	true
catfish	false	false	true	false	false	true	true	true	true	false	false	true
cavy	true	false	false	true	false	false	false	true	true	true	false	false
cheetah	true	false	false	true	false	false	true	true	true	true	false	false
chicken	false	true	true	false	true	false	false	false	true	true	false	false
chub	false	false	true	false	false	true	true	true	true	false	false	true
clam	false	false	true	false	false	false	true	false	false	false	false	false
crab	false	false	true	false	false	true	true	false	false	false	false	false
crayfish	false	false	true	false	false	true	true	false	false	false	false	false
crow	false	true	true	false	true	false	true	false	true	true	false	false
deer	true	false	false	true	false	false	false	true	true	true	false	false
dogfish	false	false	true	false	false	true	true	true	true	false	false	true
dolphin	false	false	false	true	false	true	true	true	true	true	false	true
dove	false	true	true	false	true	false	false	false	true	true	false	false
duck	false	true	true	false	true	true	false	false	true	true	false	false
elephant	true	false	false	true	false	false	false	true	true	true	false	false
flamingo	false	true	true	false	true	false	false	false	true	true	false	false
flea	false	false	true	false	false	false	false	false	false	true	false	false
frog	false	false	true	false	false	true	true	true	true	true	false	false
frog	false	false	true	false	false	true	true	true	true	true	true	false
fruitbat	true	false	false	true	true	false	false	true	true	true	false	false
giraffe	true	false	false	true	false	false	false	true	true	true	false	false
girl	true	false	false	true	false	false	true	true	true	true	false	false
gnat	false	false	true	false	true	false	false	false	false	true	false	false
goat	true	false	false	true	false	false	false	true	true	true	false	false
gorilla	true	false	false	true	false	false	false	true	true	true	false	false

ESP 2:34 am

Fig. 3. Neuronal model selection module

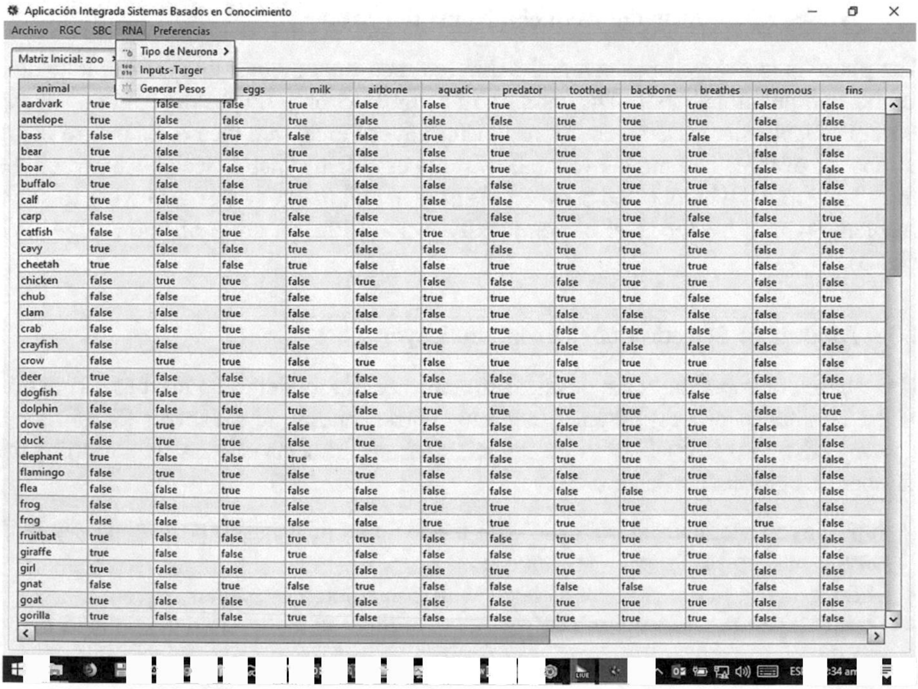

Aplicación Integrada Sistemas Basados en Conocimiento

Archivo RGC SBC RNA Preferencias

Matriz Inicial: zoo

Tipo de Neurona >
Inputs-Targer
Generar Pesos

animal			eggs	milk	airborne	aquatic	predator	toothed	backbone	breathes	venomous	fins
aardvark	true	false	false	true	false	false	true	true	true	true	false	false
antelope	true	false	false	true	false	false	false	true	true	true	false	false
bass	false	false	true	false	false	true	true	true	true	false	false	true
bear	true	false	false	true	false	false	true	true	true	true	false	false
boar	true	false	false	true	false	false	true	true	true	true	false	false
buffalo	true	false	false	true	false	false	false	true	true	true	false	false
calf	true	false	false	true	false	false	false	true	true	true	false	false
carp	false	false	true	false	false	true	false	true	true	false	false	true
catfish	false	false	true	false	false	true	true	true	true	false	false	true
cavy	true	false	false	true	false	false	false	true	true	true	false	false
cheetah	true	false	false	true	false	false	true	true	true	true	false	false
chicken	false	true	true	false	true	false	false	false	true	true	false	false
chub	false	false	true	false	false	true	true	true	true	false	false	true
clam	false	false	true	false	false	false	true	false	false	false	false	false
crab	false	false	true	false	false	true	true	false	false	false	false	false
crayfish	false	false	true	false	false	true	true	false	false	false	false	false
crow	false	true	true	false	true	false	true	false	true	true	false	false
deer	true	false	false	true	false	false	false	true	true	true	false	false
dogfish	false	false	true	false	false	true	true	true	true	false	false	true
dolphin	false	false	false	true	false	true	true	true	true	true	false	true
dove	false	true	true	false	true	false	false	false	true	true	false	false
duck	false	true	true	false	true	true	false	false	true	true	false	false
elephant	true	false	false	true	false	false	false	true	true	true	false	false
flamingo	false	true	true	false	true	false	false	false	true	true	false	false
flea	false	false	true	false	false	false	false	false	false	true	false	false
frog	false	false	true	false	false	true	true	true	true	true	false	false
frog	false	false	true	false	false	true	true	true	true	true	true	false
fruitbat	true	false	false	true	true	false	false	true	true	true	false	false
giraffe	true	false	false	true	false	false	false	true	true	true	false	false
girl	true	false	false	true	false	false	true	true	true	true	false	false
gnat	false	false	true	false	true	false	false	false	false	true	false	false
goat	true	false	false	true	false	false	false	true	true	true	false	false
gorilla	true	false	false	true	false	false	false	true	true	true	false	false

Fig. 4. Module of selection of neurons by layers

4.4 Module 4: Determination of Initial Synaptic Weights

At this stage, the initial synaptic weights of the model being worked on are established.

Objective: To select the set of initial weights using metrics defined as a heuristic based on the informational importance of traits and concepts.

For the initialization of the weights it is proposed to calculate the weights referring to the neurons by layers, between the entry layer and the hidden layer and between the hidden layer and the exit layer. This division is due to the fact that between the first two layers the one that governs the process is the trait, and therefore the informational weight of the same is used in relation to the interpretative association that it has with the concepts that define the neurons of the hidden layer. While between the neurons of the hidden layer and those of exit the relation is between the concepts and the classes (See Fig. 5), being directed by the informational importance given by the concepts.

Archivo RGC SBC RNA Preferencias

Matriz Inicial: zoo

Tipo de Neurona
Inputs-Targer
Generar Pesos

animal			eggs	milk	airborne	aquatic	predator	toothed	backbone	breathes	venomous	fins
aardvark	true	false	false	true	false	false	true	true	true	true	false	false
antelope	true	false	false	true	false	false	false	true	true	true	false	false
bass	false	false	true	false	false	true	true	true	true	false	false	true
bear	true	false	false	true	false	false	true	true	true	true	false	false
boar	true	false	false	true	false	false	true	true	true	true	false	false
buffalo	true	false	false	true	false	false	false	true	true	true	false	false
calf	true	false	false	true	false	false	false	true	true	true	false	false
carp	false	false	true	false	false	true	false	true	true	false	false	true
catfish	false	false	true	false	false	true	true	true	true	false	false	true
cavy	true	false	false	true	false	false	false	true	true	true	false	false
cheetah	true	false	false	true	false	false	true	true	true	true	false	false
chicken	false	true	true	false	true	false	false	false	true	true	false	false
chub	false	false	true	false	false	true	true	true	true	false	false	true
clam	false	false	true	false	false	false	true	false	false	false	false	false
crab	false	false	true	false	false	true	true	false	false	false	false	false
crayfish	false	false	true	false	false	true	true	false	false	false	false	false
crow	false	true	true	false	true	false	true	false	true	true	false	false
deer	true	false	false	true	false	false	false	true	true	true	false	false
dogfish	false	false	true	false	false	true	true	true	true	false	false	true
dolphin	false	false	false	true	false	true	true	true	true	true	false	true
dove	false	true	true	false	true	false	false	false	true	true	false	false
duck	false	true	true	false	true	true	false	false	true	true	false	false
elephant	true	false	false	true	false	false	false	true	true	true	false	false
flamingo	false	true	true	false	true	false	false	false	true	true	false	false
flea	false	false	true	false	false	false	false	false	false	true	false	false
frog	false	false	true	false	false	true	true	true	true	true	false	false
frog	false	false	true	false	false	true	true	true	true	true	true	false
fruitbat	true	false	false	true	true	false	false	true	true	true	false	false
giraffe	true	false	false	true	false	false	false	true	true	true	false	false
girl	true	false	false	true	false	false	true	true	true	true	false	false
gnat	false	false	true	false	true	false	false	false	false	true	false	false
goat	true	false	false	true	false	false	false	true	true	true	false	false
gorilla	true	false	false	true	false	false	false	true	true	true	false	false

ESP 2:35 am

Fig. 5. Module for determining the initial synaptic weights

After the second phase, the neural network is ready to be trained. Figure 6 shows the architectural design after applying the conceptual algorithm.

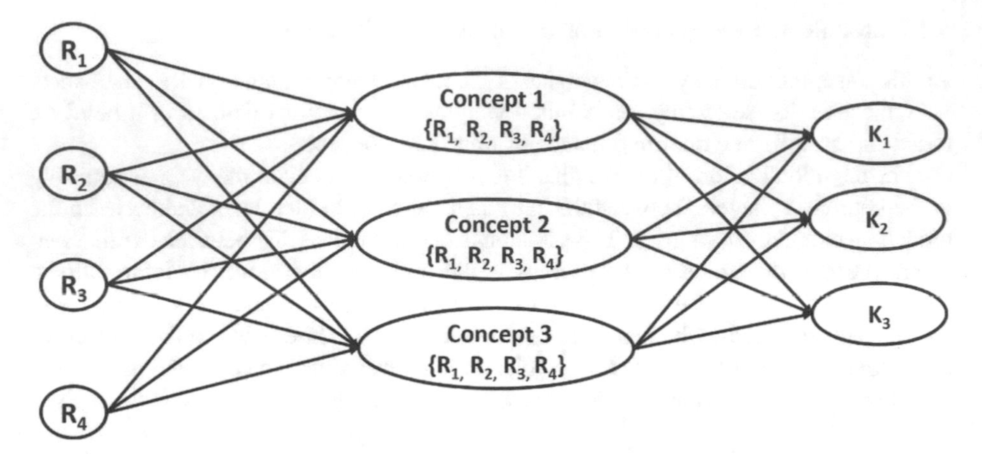

Fig. 6. Design of the architecture of an MLP after applying the conceptual algorithm

5 Results and Discussion

For the analysis of the proposed solution, the computational tool is used to support the subject by applying the grouping algorithms to the design of the architecture of a Multilayer Perceptron using 10 internationally recognized databases available in the repository for automated learning at the University of Irvine, California (Merz 1998). The selection considers data sets with varied characteristics such as: the presence of numerical and non-numerical traits, the absence of information, and diverse in terms of the number of traits and objects (Table 1).

Table 1. Databases

Database	Number of objects	Number of features	Number of classes	Type of BD
Wine	178	13	3	Supervised
Annealing	798	16	38	Supervised
Zoo	101	18	7	Supervised
Iris	150	5	3	Supervised
Glass	214	10	7	Supervised
Tae	151	6	–	Unsupervised
Sonar	207	60	–	Unsupervised
Solarflare	324	13	–	Unsupervised
Lymph	148	19	–	Unsupervised
Liver-disorders	345	8	–	Unsupervised

The international databases are run in the tool for the Multilayer Perceptron neuronal model, one design is the MLP without applying any conceptual grouping (MLP1), the other is the design of the MLP applying the LC-Conceptual algorithm (MLP-LC) and the last design is the MLP applying the RGC algorithm (MLP-RGC).

For the training and testing of the designs, the k-fold cross validation method was applied. This method consists of dividing the master data into two parts; one part is used as a training set to determine the parameters of the neural classifier and the other part, called the validation set, is used to estimate the generalization error, i.e., the incorrect classification rate of the classifier with data different from those used in the training process. 85% of the data is selected for network training and the remaining 15% for simulation. The Friedman test as suggested in (Berlanga and Rubio Hurtado 2012) is also used to check whether there are significant differences between the effectiveness of each model.

Databases are tested by applying the cross validation method to check the effectiveness of the classification of the computational tool created (See Table 2 and Fig. 4).

Table 2. Experiment data for supervised and unsupervised databases

Database	Number of variables in MLP	Number of variables in MLPLC	Number of variables MLP-RGC	MLP Effectiveness %	MLP-LC efficiency %	MLPRGC efficiency %
Wine	13	3	3	95.02	95.60	100
Annealing	38	11	11	98.51	84.43	98.99
Zoo	18	7	7	90.23	92.60	93.33
Iris	5	3	3	96.34	96.34	100
Glass	10	4	4	98.32	99.32	100
Tae	6	4	4	92.50	96.24	97.33
Sonar	60	3	3	78.33	90.23	92.14
Solarflare	13	12	12	89.50	90.33	88.60
Lymph	19	11	11	85.20	88.21	99.11
Liverdisorders	8	3	3	98.24	90.34	99.45

After validating MLP1, MLP-LC and MLP-RGC, Friedman's test is applied for the percent variable of correct classifications, with a 95% confidence interval. The null hypothesis is to assume that all algorithms are statistically equivalent. The test is performed with a p-value = 0.0008781 as this value is below 0.05 the null hypothesis must be rejected and therefore it can be stated that there are significant differences between neural network designs. To detect them, the post hoc test is applied with Finner's correction (Berlanga and Rubio Hurtado 2012), which results in the MLP-RGC model being statistically significant in comparison with the MLP1 neuronal model. In Fig. 7 it can be observed that the critical distance that marks the limit of significance does not include the MLP1 model, on the other hand, the differences are not significant with respect to the model created applying the LC-Conceptual algorithm, although the results are higher.

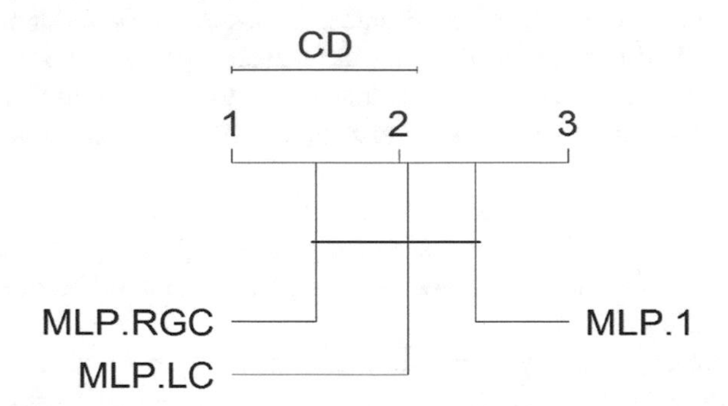

Fig. 7. Graph showing significant differences between neural models

6 Conclusions

The development of the computer tool allowed the teaching of artificial neural networks in the University of Computer Science to achieve a better understanding in the learning of the subject where the different neural models are integrated with the incremental algorithms of conceptual groupings of pattern recognition based on the adaptation of objects. It allowed to establish comparisons between the designs of the neuronal models Back-Propagation, Radial Base, Probabilistic, McCulloch-Pitts and Multilayer Perceptron to select the most suitable model for the different types of problem.

The scientific methods used for the validation of the computational tool allowed to verify that the effectiveness in the classification in the neuronal models raised in the research.

References

Berlanga, V., Rubio Hurtado, M.J.: Clasificación de pruebas no paramétricas. Cómo aplicarlas en SPSS. REIRE. Revista d'Innovació i Recerca En Educació **2**, 101–113 (2012)

Camejo, S.: Proyecto de investigación efecto del uso del computador en la elaboración del proyecto de trabajo de grado de los estudiantes de la UNA: revisión bibliográfica. Informe de Investigaciones Educativas, **6**(1–2), 101–136 (1992)

Martínez, N., Ferreira, G., García, Z.Z.: Sistema de enseñanza/aprendizaje inteligente para grafos (2007)

Martínez Trinidad, J., Ruiz Shulcloper, J.: Algoritmo LC conceptual duro. Presented at the IV Simposio Iberoamericano de Reconocimiento dePatrones, La Habana, Cuba (1999)

Merz, C.J.: UCI repository of machine learning databases. http://www.Ics.Uci.Edu/~ Mlearn/MLRepository.Html (1998)

Montaño Moreno, J.J.: Redes neuronales artificiales aplicadas al análisis de datos (2017)

Pascual Belda, A.: Caracterización del trastorno del espectro autista basado en técnicas de aprendizaje automático a partir de características extraídas de la conectividad funcional del cerebro en estado de reposo (2018)

Pons Porrata, A.: Desarrollo De Algoritmos Para La Estructuración Dinámica De Información Y Su Aplicación A La Detección De Sucesos (Tesis Doctoral) (2004)

Reyes González, Y.: Modelo para la adaptación de las soluciones en un Sistema Basado en Casos utilizando el agrupamiento conceptual (Tesis de Maestría). Universidad de las Ciencias Informáticas (2014)

Rosenblatt, F.: Perceptions and the theory of brain mechanisms. Spartan books (1962)

Ruiz-Shulcloper, J.: Reconocimiento lógico combinatorio de patrones: teoría y aplicaciones (Tesis en opción al grado científico de Doctor en Ciencias). Universidad Central de Las Villas, Santa Clara (2009)

Smart Study

Blended Learning in a Merger of Vocational and Higher Education

Harald Jacques(✉) and Reinhard Langmann

Hochschule Duesseldorf University of Applied Sciences, Duesseldorf, Germany
{jacques,langmann}@ccad.eu

Abstract. The 4th industrial revolution with cyber physical systems (CPS) entering automation and production processes, so called Industry 4.0 (I40), forces universities to establish new educational models. To fulfill the upcoming demands of the industry, there has to be a permanent transfer of knowledge between university and industry and use of digital media in the learning/training process, called *Smart Study*. The paper gives an overview on the merging of vocational and higher education in Germany to a Smart Study and shows some examples for blended learning in this study.

Keywords: Smart Study · Dual Study · Vocational training · Blended learning

1 Introduction

What is *Smart Study*? It is a model of integrated learning at different learning areas, such as industry and university, using blended learning methods. Usually, blended learning is defined as "combination of face-to-face instruction with computer mediated instructions" [1]. This definition only show, how information from the teacher is better driven into knowledge of the learners, but missing the competency to solve problems in an unknown environment. Combining an engineering study at a university with working and learning in industry simultaneously, the theoretical knowledge from university can be directly used in industrial applications. Thus gives a high impact on communication and collaboration skills, needed for the society of the 21st century [2].

In the first part is shown, how to manage the cooperation between university study program and vocational training in industry. The second part shows some examples for blended learning in a Smart Study.

2 Part I: Merger of Vocational and Higher Education

2.1 Requirements of the Future

The education of highly qualified engineers is more and more related to the development of technologies for industrial purposes. The start of the 3rd industrial revolution (~1970) moving in microelectronics into industrial processes, forces Germany to

M. E. Auer and D. May (Eds.): REV 2020, AISC 1231, pp. 922–934, 2021.
https://doi.org/10.1007/978-3-030-52575-0_76

reform the higher education system. There was high need for practical oriented engineers for the industry besides theory based academics. Therefore Germany established a new type of universities: the "Universities of Applied Sciences" (short: UAS). The main improvement: after vocational training and some years of professional experience, skilled workers get the matriculation standard for UAS. Hence, the education at UAS formed professional experienced engineers.

Today Germany is well known for his skilled workers and engineers. But are we prepared to educate engineers for the demands of the 4th industrial revolution? Do we need a new education system, e.g. Education 4.0 (called E40)?

Following a study of Karlsruhe Institute of Technology (KIT) in 2013 [3], we have to expect a total change in our system of industrial work. Usually we had in the past a pyramid with lots of unskilled workers as basis, less skilled workers and only a few engineers and managers at the top (see grey pyramid in Fig. 1). In the future, this will transform into a rhombus (see red rhombus in Fig. 1) with only a few unskilled workers left but a maximum at the interface between skilled workers and academics.

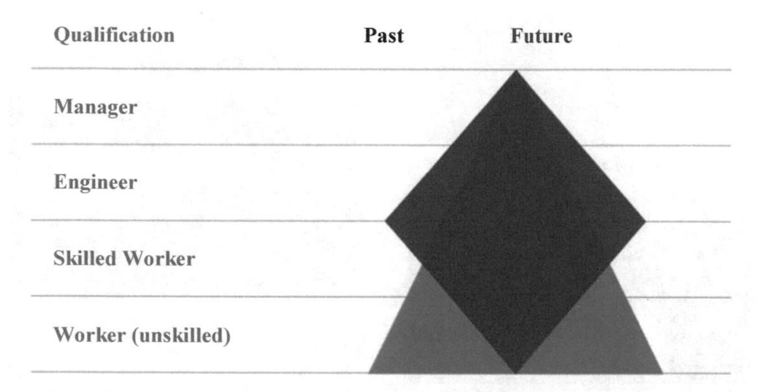

Fig. 1. Transformation of occupational qualifications from a pyramid (grey) in the past to a rhombus (red) in the future [3]

A study of the Association of German Engineers (VDI) in 2018 [4] postulates, that the digital transformation needs interdisciplinary and transdisciplinary collaboration skills. Thus the engineers of the future have to have cooperation competences, the ability to collaborate in social, technical and business matters, to achieve consensus and mutual acceptance. As a consequence for engineering education there should be a dialogue between institutes of higher education and companies, associations and unions offering the opportunity of mutually coordinating the skills required by digital transformation. Experiences from integrated degree programs can be helpful [4].

But where will the engineers get their skills and experience from? Are small projects and simulations in the higher education institutes enough? Competence development is a constant challenge, which not only conveys knowledge but also has to be processed emotionally and also leads to immediately recognizable consequences in a real decision-making process [5]. In the real environment of an industrial enterprise, the required competences can best be developed. And not just after graduation in the first working phase but best parallel to the study program.

Some years ago, universities of applied sciences in Germany started to collaborate with industrial companies in a new study model called "***Dual Study***". This study programs support the process of transformation (see Fig. 1) by combining the vocational training of skilled workers with a bachelor study course, thus forming special educated engineers with practical skills at the interface of vocational and higher education.

2.2 Dual Study

The main idea of the Dual Study program was the combination of the university curriculum with the program of vocational training. Institutes of higher education can teach the basic principles, deeper points of focus and learning strategies while companies, associations and unions can meet specific requirements like "Industry 4.0" through training and further education opportunities [5].

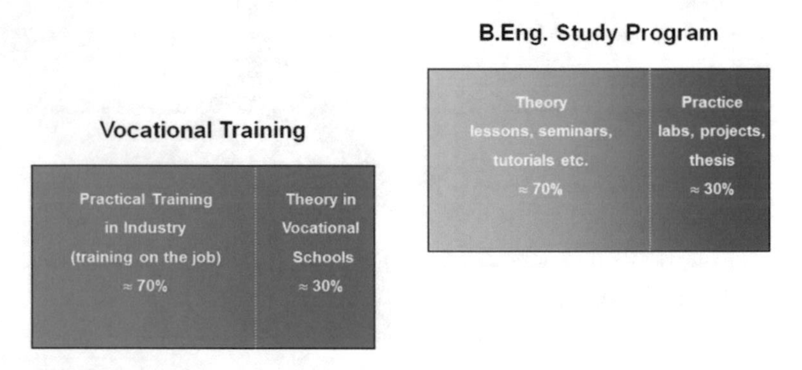

Fig. 2. Schematic drawing of the practical and theoretical part of vocational training (left) and engineering study program (right).

A brief overview of the program of vocational training and engineering study curricula is shown in Fig. 2. The vocational training in Germany includes about 30% of theoretical learning content, which is taught in the vocational schools. In addition, there are approximately 70% practical in-plant-training and learning by doing in an apprenticeship training company. In opposite, the study programs for engineers at universities of applied sciences (UAS) include about 70% theoretical lessons and 30% practical training such as laboratories, projects and industrial internship.

A closer look to the curricula of the two training methods shows, in some respects, overlapping parts (see Fig. 3). The theoretical part of the vocational training is widely congruent with the basics of the engineering study program while part of the practical training in the engineering study program is done more extensively in the industrial training. Therefore the theoretical training in vocational schools can be replaced by the theory at the UAS while, on the other hand, the practical applications in the apprenticeship training companies can compensate most of the laboratory exercises at the UAS and the industrial internship. The synergetic combination of occupational and academic training thus can form highly skilled, professional engineers saving a lot of time by simultaneous training.

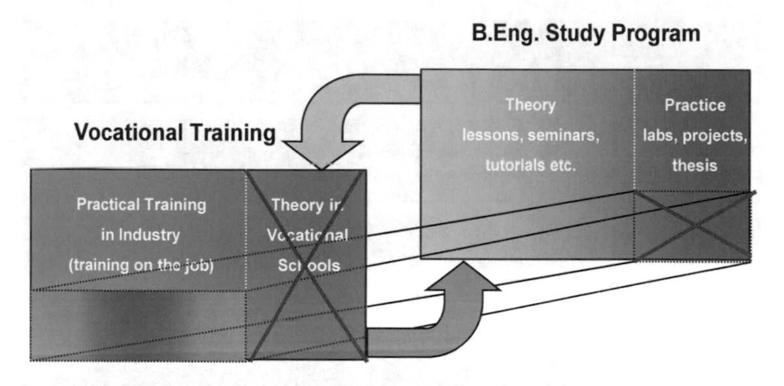

Fig. 3. Schematic drawing of the practical and theoretical part of vocational training and engineering study program with overlapping parts.

There are various models for the collaboration of higher education institutes and industry:

- The standard version is the integrated model, combining a study course with a vocational education, thus ending into two professional qualifications, the bachelor degree and the certificate of apprenticeship. This model accounts for 45% of *Dual Study Programs* [6].
- The second model only integrates work experience in the industry without vocational training. This model is forced by the *Universities of Cooperative Education* (like *Duale Hochschule Baden Württemberg DHBW*), now counting for 37% [6].
- Moreover, mixed models have increased to 18%. The latest is a *Triple Degree Course*, combining an apprenticeship in the crafts with advanced master craftsman training and a bachelor degree course [6].

The first two models will be exemplified in more detail:

Figure 4 shows for example a bachelor course with normal duration of 6 semesters with an integrated vocational training with a normal duration of 3 years too. In the first two years, the trainees are half the time of the week in the vocational training in their apprenticeship training company (without vocational school) and the other half in the UAS as part-time-student. The in-plant-training normally takes about 70% of the time, but the dual students/trainees have to fulfill the vocational training in 50% of the time, thus making it ambitious. In the first two years as part-time-students they fulfill the normal study program of the first two semesters. In the 3rd year the trainees are full-time-students and in their apprenticeship training companies only in the lecture free times of the university. The examination at the chamber of commerce or chamber of crafts for the certificate of apprenticeship ends the vocational training. The 4th year is a normal study year with lessons and the bachelor thesis, ending with the degree Bachelor of Engineering (B.Eng.). By that schedule, the length of both training programs together (normally 3 + 3 = 6 years) can be decreased by 2 years to 4 years in total.

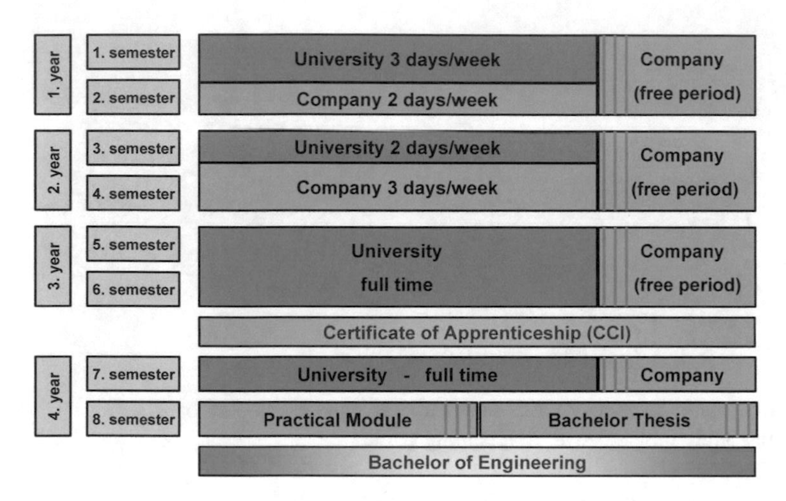

Fig. 4. Time-table for a "dual study program" with an integrated vocational training (example).

The second model without vocational training but regular practical work in industry in between is shorter (see Fig. 5), but has only one approved certificate, the B.Eng. The UAS don't know, if students are trained in industry or still working and learning by doing. Thus making the study requirements much stronger, because the students have to perform the engineering courses in 12 weeks per semester instead of 15–16 weeks usually.

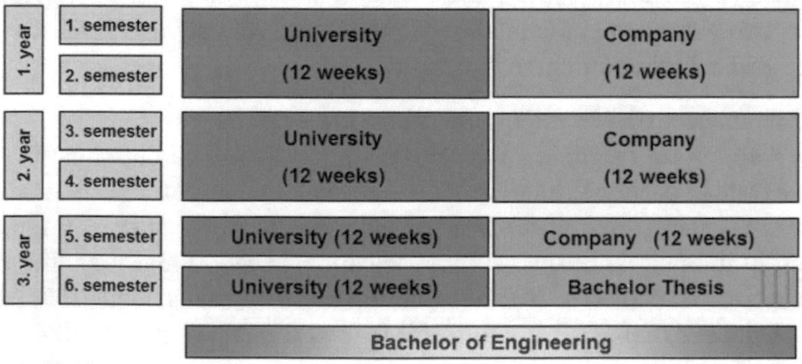

Fig. 5. Time-table for a "dual study program" without vocational training, but practical work experience (example).

The implementation of a *Dual Study Program* is for both models the same: First, if a company shows interest in professional graduates, the conditions such as the curriculum of both training center, the time schedule and the cross approval of training courses are fixed in a cooperation agreement between the company and the university. Second, interested applicants for the "cooperative education" must first apply to the industrial company and sign a contract of apprenticeship or of work. Thereafter, the trainee is applying at the UAS for a university place (assuming the university entrance qualification) and is enrolled as a student. This is the so called "Triangle of Dual Study Programs" in Germany (see Fig. 6).

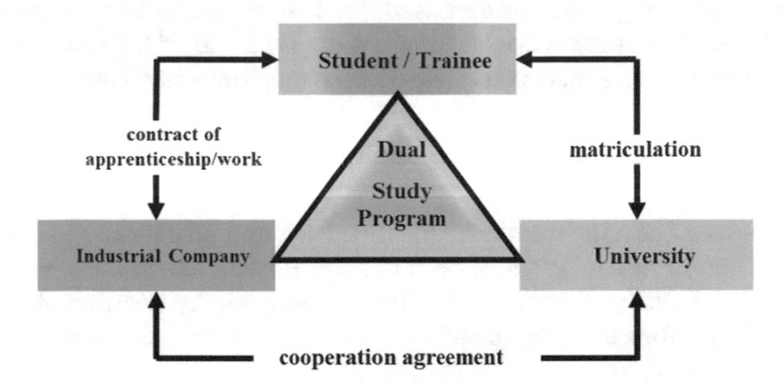

Fig. 6. The "Triangle of Dual Study Programs" in Germany

Because of the parallelism of the two training paths, close coordination between the partners is necessary. Partly, course units are offered by a training partner, which are carried out spatially and temporally from home and also within the course times in the respective other training center. This is only possible through the massive use of digital media in the context of blended learning (see part II).

2.3 Benefits

The *Dual Study Program* brings benefits to all three parties:

For training companies:

- attractive for graduates from grammar schools, so they can select out of a great number of applicants with high potential - "searching for the best minds"
- Qualification of students/trainees can be observed and promoted for a long time
- Getting high qualified engineers with professional experience – short period of job-adjustment after graduation.

For trainees/students:

- Two qualified and internationally recognized (certified) qualifications (in the model with integrated vocational training)
- Vocational training and a university study will need less time in total
- Financial security during training/studying and secure job after graduation.

For universities:

- Very good and highly motivated students, preselected by the training company
- Intensification of cooperation and know-how transfer between universities and companies by industrial internship, R&D projects, bachelor thesis etc.

3 Part II: Blended Learning for Automation Technology

Following three application examples of blended learning in a *Smart Study* for Automation Technology will be described. The Smart Study is implemented in a Bachelor course "Automation Engineering" of the Duesseldorf University of Applied Sciences. 50% of the students in this course are coming from the above described Dual Study program.

All examples use as a key element of the blended learning scenario remote labs or remote experiments. Thereby you can distinguish two types [7]:

- ***Fixed Remote Lab:*** The remote lab presents itself as a predefined didactic learning environment, in which the student has to solve predetermined tasks remotely on a real system. Usually, the setup of a fixed remote lab environment requires high development efforts and corresponding know-how, which means that a teacher will not create and modify such a remote lab by himself.
- ***Open Remote Lab:*** You need for an open remote lab an open tool environment for the configuration of the concrete lab system. As a learning environment, this is comparable to a freely accessible classical laboratory workplace. The student has all possibilities to create a required functional system. The learning task is described in corresponding didactic learning documents. A tutor chaperones the student during the remote work in the system, either through a remote collaboration tool or via face-to-face consultations.

As a tool environment for all described examples is used the Internet-of-Things (IoT) platform FlexIOT [8].

3.1 Example I: Fixed Remote Experiment with a Rotary Table

The example uses a processing and test station with a rotary table (Fig. 7).

In the experiment, a student is asked to operate the position-controlled rotary table, and record and evaluate the angular speed and acceleration as well as determine the positions of processing modules in the station.

The process data access to the system occurs via a web connector for Modbus TCP. The example is implemented as a project in the IoT portal FlexIOT. A student needs only a user account to work with the experiment. He can operate the experiment but he can't change the structure of the experiment.

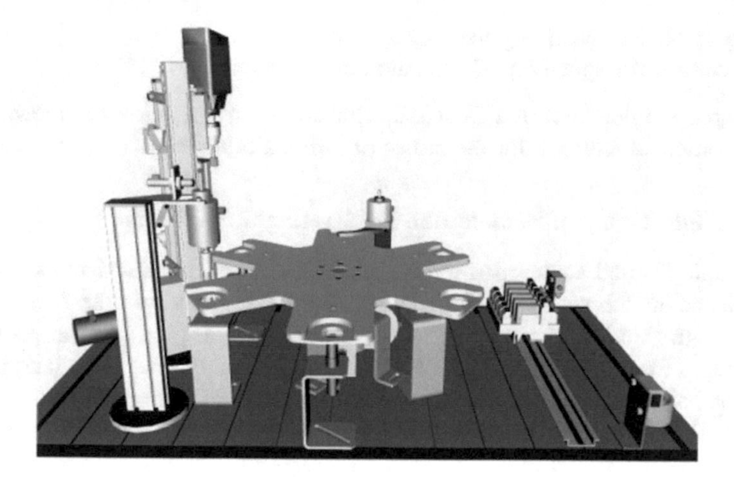

Fig. 7. Processing and test station with position-controlled rotary table

Figure 8 shows the remote experiment in the web browser.

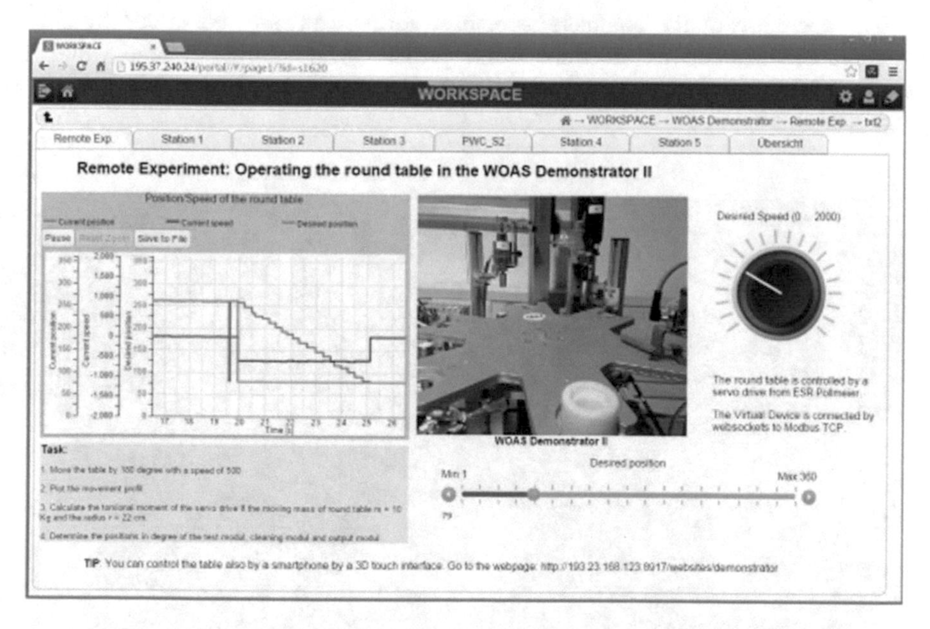

Fig. 8. Fixed remote experiment for a position-controlled rotary table

The following lab functionalities are available to the student on the website:

- Video image of the rotary table.
- 2-channel real-time plotter. The plotter enables the saving of displayed values on the client PC.

- Rotary knob for specifying the angular speed.
- Slide control for specifying the position target value.

The example is integrated in a Dual Bachelor course Automation Engineering and is used in a practical exercise for the basics of Drive Technology.

3.2 Example II: Open Remote Lab for an Assembly Station

This example is used as an open remote lab in a course Human-Machine Communication (Bachelor, 5th semester, Automation Engineering). Within the framework of a project task, the student must configure and test an operator panel for the stations of an assembly line for model cars. Figure 9 shows the assembly line. Each learning group implements an operator panel for testing and operation of stations 2–5.

The work on the real system is exclusively carried out remotely via corresponding accesses to the IoT portal FlexIOT. Each project group consists of four students, who collectively have access to one "Administrator" login. In the IoT portal itself, it is stipulated that a maximum of four administrators may work simultaneously (but at different stations) using this access login. In order to work on the real stations the students have to make a time reservation before by a special web-based reservation system.

The operation of the complete assembly process can only be solved by closed collaboration between all students in the project group during remote work. To accomplish this, a chat service integrated as a part of the IoT portal, can be used (Fig. 10).

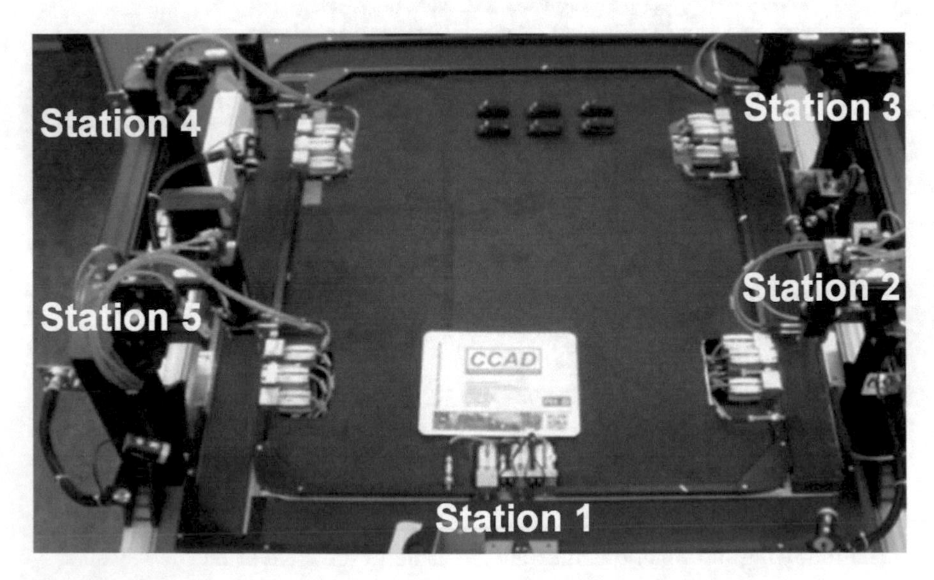

Fig. 9. Assembly line for model cars with five stations (Station 1: Test station for the assembly parts, Station 2: Assembly of the axis modules, Station 3: Assembly of the body, Station 4: Disassembling the body, Station 5: Disassembling of the axis modules)

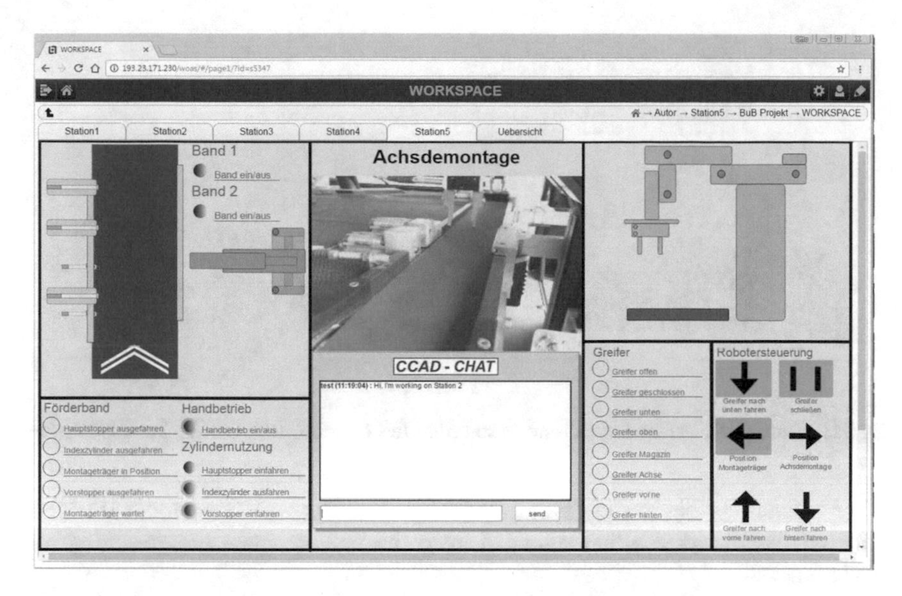

Fig. 10. Operator Panel for Station 5 in the manual mode from a student group in the WS 2015/16

Every winter semester, the learning task is carried out successfully by approximately 35 dual students. As an example, Fig. 5 shows the result of a project group from the winter semester 2015/16. More information about the example you can find in [9].

3.3 Example III: Cloud-Based Blended Learning Lab for PLC Education

The Cloud-based blended learning lab (CBLL) combines training on a real on-site PLC with technology models from the cloud. The technology models are loaded into the learner's web browser from the cloud where they are run. The models are linked directly with the inputs and outputs of the PLC via a special CloudIO adapter based on Raspberry Pi to enable a PLC program to control the technology models. Management of the technology models is performed via the IoT platform FlexIOT where model use is offered as a service.

Figure 11 shows the principle diagram of a CBLL for PLC training. The detailed design and implementation of the example is described in [10] and the video in [11] illustrates the work within this lab.

The CloudIO is built onto a DIN A4 board within the didactic learning system Eduline created by Phoenix Contact, which ensures that the cloud adapter can easily replace the simulation board in this PLC training system. The connection to the inputs and outputs of the Eduline-PLC ILC 191 ME/AN or ILC 131 ETH is performed via two 24-pole D-SUB cables. The CloudIO can also be linked to any other digital/analogue process signals (e.g. Siemens PLC) for the purposes for connecting these to the Internet/Intranet via a web protocol.

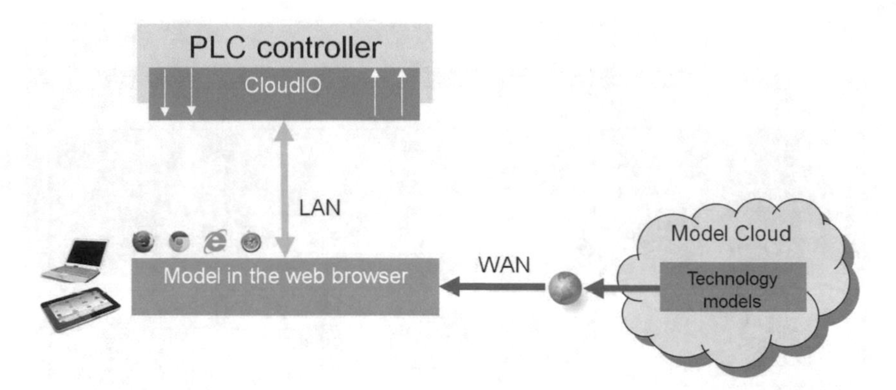

Fig. 11. Principle diagram for a cloud-based blended learning lab (CBLL) for PLC training

The technology models use web technologies, such as SVG and/or X3D for their generation. JavaScript can be embedded in to the models when programming model behaviour.

Since the Internet can be accessed via the PLC's input/output signals via the CloudIO, all appropriate systems available online can generally be controlled using the on-site PLC. This results in a variety of options for potential technology models. For instance, any real stations which is accessible via Internet can be also used a technological model for PLC training.

An evaluation of the CBLL system showed that technology models with process times of ca. 50 ms can be used via this method. This makes the models applicable for most training tasks in PLC programming.

On-site, learners no longer require stand-alone installations of software or licenses. Execution of the technology models can be performed by the learner on every PC or post-PC (smartphone, tablet, etc.).

Thus example is applicated in a basic course for PLC programming for different Smart Study courses (Bachelor) and uses the Eduline education system from Phoenix Contact with different technology models (2D/3D models) from the Cloud.

4 Conclusions

Smart Study as a model of integrated education of engineers at different places (UAS/industry), with different challenges (academic/occupational) and different learning techniques (face-to-face/blended learning) is a possible answer to the demands of the future. Experiences so far show that the use of blended learning scenarios for the learning/training of automation engineers in dual study has many advantages.

- The *Dual Study* is the fastest growing part of the tertiary education system in Germany with a growth rate of approximately 10% pa [6].
- The drop-out-rate in the Dual Study program is less than 8%, which is much lower than in normal engineering study programs with about 40% [12].

- Graduates have a very high employability (nearly everyone changes directly from university to work place).
- New developments and future trends in technology are passed on directly from the cooperating companies to the university, so that a timely response with changes in training and education are possible.
- The training using practice-oriented blended learning labs with consideration of remote labs or remote experiments significantly increases the efficiency of the training. In particular, human resources for the practical laboratory support can be saved to a considerable extent. The remote labs are available to students around the clock and so the dual students can learn also on the workplace in their companies (if time is available).
- Dual students are very purposeful, motivated and result-oriented compared to classical students (without vocational training, only high school graduation). These students value the independency and self-management in performing the tasks in the described learning scenarios.
- In specially designed learning scenarios, such as in example II, cooperation between students within the group is possible regardless of the location of students. Students can work across local boundaries on practice-oriented tasks, and at the same time learn to work in distributed and global teams. This kind of collaborative learning work prepares the students for the future networking of machines and people in the context of Industry 4.0.

Overall, it is estimated that combining dual study with modern media and learning methods into a *Smart Study* can be an essential pillar of a future education for engineers 4.0.

References

1. Bonk, C.J., Graham, C.R.: The Handbook of Blended Learning Environments: Global Perspective, Local Designs. Jossey-Bass/Pfeiffer, San Francisco (2006)
2. Beers, S.Z.: 21st Century Skills: Prepare Students for the Future. http://cosee.umaine.edu/files/coseeos/21st_century_skills.pdf
3. Landmesser, M.: Herausforderungen für die akademische Weiterbildung aus Unternehmersicht, KIT, März 2013
4. VDI Discussion Paper: Smart Germany – Engineering Education for the Digital Transformation. VDI Publishing, March 2018
5. Kuhlmann, A., Sauter, W.: Innovative Lernsysteme. Springer Publishing, Heidelberg (2008)
6. BIBB (Federal Institute for Vocational Education and Training): Dual higher education study – upwards trend carries on. AusbildungPlus BIBB Database, Bonn, 26 September 2017. https://www.bibb.de/en/pressemitteilung_67829.php
7. Langmann, R.: A CPS integration platform as a framework for open remote labs in automation engineering. In: Azad, A.K.M., et al. (eds.) Cyber-Physical Laboratories in Engineering and Science Education. Springer International, pp. 305–330 (2018)
8. FlexIOT: http://www.flexiot.de
9. YouTube video: https://www.youtube.com/watch?v=7CIp4n72lAU

10. Langmann, R., Coppenrath, M.: A cloud-based, blended learning lab for PLC education. In: Auer, M.E., Langmann, R. (eds.) Smart Industry and Smart Education. Lecture Notes in Networks and Systems, vol. 47, pp. 3–13. Springer International (2018)
11. YouTube video: https://www.youtube.com/watch?v=Nu5PmZVsUyg
12. acatech (National Academy of Science and Engineering): Student drop-out in the engineering sciences; position paper (2017). www.acatech.de/Publikation/student-drop-out-in-the-engineering-sciences-multi-university-analysis-and-recommendations/

20 Years' of eLearning and Blended Learning at Czech Universities

Petra Poulova(✉), Blanka Klimova, and Lenka Tvrdikova

Faculty of Informatics and Management, University of Hradec Kralove,
Rokitanskeho 62, 500 03 Hradec Kralove, Czech Republic
{petra.poulova,blanka.klimova,lenka.tvrdikova}@uhk.cz

Abstract. 20 years ago the European Commission announced eLearning to be a strategic plan for developing new trends in education. This was also important for the Czech Republic. In the last decades, blended learning and eLearning have become common part of the tertiary education. The article describes the research of the development of eLearning and blended learning at Czech universities, as it is reflected in annual reports of these institutions of higher learning. The findings from the monitored annual reports of public universities indicate the importance of the use of eLearning in university education all over the Czech Republic. Almost 20 years of the existence of eLearning education, as well as numerous conferences on this topic, prove its effectiveness for higher learning.

Keywords: Blended learning · eLearning · eLearning polices · Infrastructure · Learning management system · Research · University education

1 Introduction

Current society is often seen as digital. Like the invention of book print or the industrial revolution of the 19th century, digital technology is changing many areas of human life. And, of course, they also concern education.

This need of the use of digital technology in education has been also reflected in the strategic documents of the European Union for almost 20 years. The most important was then to develop key competences especially in the field of the information and communication technologies (ICT) and foreign languages [1–3] with special focus on media literacy, critical thinking and foreign language communication.

Therefore, also the Czech Republic has tried to implement these key competences in university educational system. In addition, since 1990 eLearning courses have been used; firstly, for distant education and then as support for face-to-face classes [4].

Digital technology, especially the computer support in education has been widely and frequently applied, researched and evaluated for last three decades. eLearning is approached from two main directions. First, it is an educational process, which involves information and communication technologies. Second, it is perceived as a set of technological tools applied in education. In the wide context, eLearning is the educational process supported by information and communication technologies [6].

M. E. Auer and D. May (Eds.): REV 2020, AISC 1231, pp. 935–943, 2021.
https://doi.org/10.1007/978-3-030-52575-0_77

2 eLearning at Czech Universities

In 2001 in the Czech Republic, a strong emphasis was put on education and improving competences in the field of ICT and foreign languages [5].

The process of eLearning implementation could be structured into three basic steps: [6]

- First, getting new hardware, software and other class equipment was essential.
- Second, forming digital competence followed. It means that all possible participants of the process had to learn how to work with computers and become computer literate.
- Third, applying the general computer literacy in education which is the most difficult and demanding part, crucial for the process of education. It covers didactic training aimed at both teachers and students so that they know how to teach and study being supported or managed by ICT.

At the very beginning of this process, Step Zero was usually introduced at several universities, which meant, for example, presenting study materials in shared directories, using e-mail for communication between teachers and students, creating university web pages, editing electronic journals etc. [7].

3 The Process of eLearning and Blended Learning Implementation at Czech Universities

Altogether, there exist 27 public universities, two state colleges and more than 40 private (non-state) institutions of higher learning in the Czech Republic. Nevertheless, these private institutions of higher learning have only a small number of students. Thus, the authors of this article involved only public universities in the research. Annual reports of the Czech public universities from 1999 to 2019, which are available at their websites, were the main source of information for this research. In total, more than 600 annual reports were included in the research.

Twenty-seven public universities were included in the research:

- Academy of Art, Architecture and Design, Prague, (VSUP)
- Academy of Performing Arts, Prague, (AMU),
- Academy of Fine Arts, Prague, (AVU)
- Charles University, Prague, (UK)
- College of Technology and Economics, České Budějovice (VSTE)
- College of Polytechnics, Jihlava, (VSP)
- Czech Technical University, Prague, (CVUT)
- Czech University of Life Sciences, Prague, (CZU)
- Institute of Chemical Technology, Prague, (VSCHT)
- Jan Evangelista Purkyně University, Ústí nad Labem, (UJEP)
- Janáček Academy of Performing Arts, Brno, (JAMU)
- Masaryk University, Brno, (MU)
- Mendel University, Brno, (MENDELU)

- Palacký University, Olomouc, (UP)
- Silesian University, Opava, (SU)
- Technical University, Brno, (VUT)
- Technical University of Liberec, (TUL)
- Tomas Bata University, Zlín, (UTB)
- University of Economics, Prague, (VSE)
- University of Hradec Králové, (UHK)
- University of Ostrava, (OU)
- University of Pardubice, (UPa)
- University of South Bohemia, České Budějovice, (JU)
- University of Veterinary and Pharmaceutical Sciences, Brno, (VFU)
- University of West Bohemia, Plzeň, (ZCU)
- VŠB - Technical University of Ostrava. (VSB)

Information on eLearning and blended learning was searched, both in the form of single chapters and data mentioned in the text. The collected data underwent the process of critical examination and evaluation (Fig. 1).

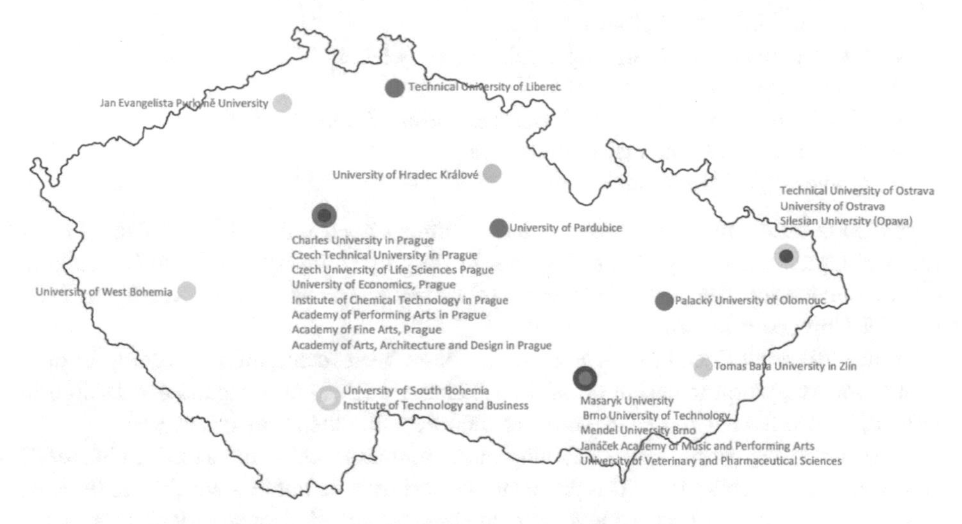

Fig. 1. Locations of public universities in Czech Republic (https://managingtheuniversitycampus.nl)

3.1 Chronological Development

The first university, which started with electronic education, was the University of Ostrava in 1999. In 2000, it was followed by the University of Hradec Králové, which began to implement the virtual learning environment in courses for distant (part-time) students. Furthermore, in the same year, the University of Economics in Prague developed an eLearning course for its academic staff.

One year later, in 2001, eLearning was mentioned in ten annual reports:

- piloting eLearning projects
  - Czech Technical University, Prague,
  - University of Hradec Králové
- established the eLearning department or built the infrastructure
  - University of Ostrava
  - Jan Evangelista Purkyně University, Ustí nad Labem
  - Institute of Chemical Technology, Prague
- training for academic staff to prepare them for online instruction and distance education methodology
  - University of Ostrava
  - University of Hradec Králové
  - University of Economics, Prague,
- made decision about which LMS would be used at the university
  - University of Ostrava,
  - Silesian University, Opava,
- started the process of designing e-subjects for students
  - University of Hradec Králové
  - University of Veterinary and Pharmaceutical Sciences, Brno
  - VŠB-Technical University of Ostrava
  - University of Economics, Prague

In 2002, eLearning was implemented at other Czech universities, which were as follows: Czech University of Life Sciences in Prague, University of South Bohemia in České Budějovice, Technical University of Liberec, Tomas Bata University in Zlín, Mendel University in Brno.

Since 2010, all Czech public universities have been mentioning eLearning in their annual reports, with the exception of the Academy of Art, Architecture and Design in Prague, whose annual reports have never included this theme in recent years.

The occurrence of the term eLearning in the annual report is not a clear indicator of how well the university uses its tools to support education, but in a way, it can indicate how important it is for each university. In this respect, the imaginary winner is the Czech University of Life Sciences in Prague, which mentions the term eLearning in its annual reports from 2010–18 in total 80 times. This university has a large number of study programs for part-time studies and therefore also a high number of part-time students for whom eLearning support for learning is very useful. Otherwise, the occurrence of the term eLearning in the annual reports of individual universities is relatively balanced. Please consult Fig. 2 below.

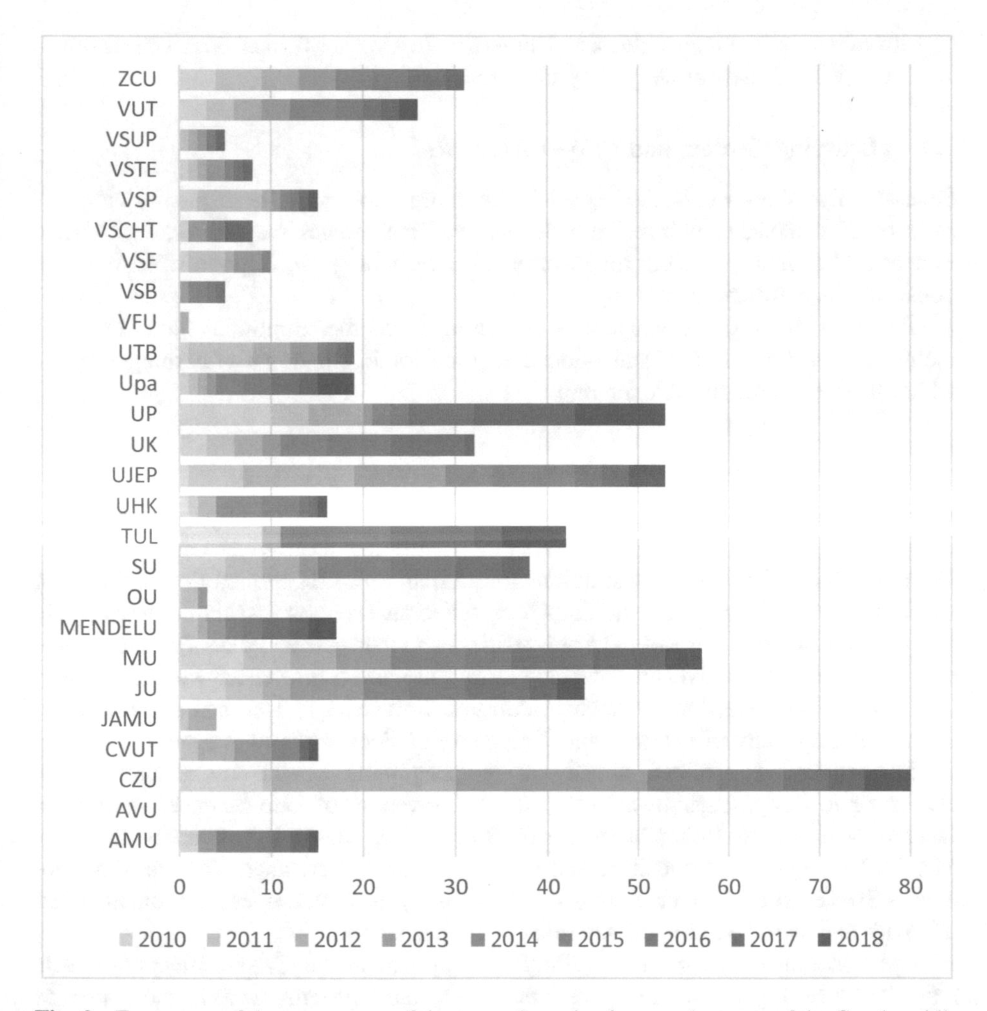

Fig. 2. Frequency of the occurrence of the term eLearning in annual reports of the Czech public universities in the years between 2010 and 2018

3.2 Academic Staff Education

The effective running of university eLearning courses requires relevant skills and competences of academic staff. Therefore, many trainings were carried out as far as technical, pedagogical and communication skills were concerned. The results of these trainings were then again described in annual reports.

The first university, which started with such trainings, was the University of Economics in Prague in 2000. Followed by the University of Hradec Králové and the University of Ostrava in 2001.

In 2002, another five universities began methodological trainings for their academic staff. These included the Czech Technical University in Prague, the Czech University of Life Sciences in Prague, Silesian University in Opava, Tomas Bata University in Zlín, and VŠB-Technical University of Ostrava.

3.3 eLearning Centers and Project Activities

Several universities established special eLearning centers or departments, either as parts of IT centers, or working independently. Their names may differ, e.g. Virtual Education Laboratory, Centre for Electronic (or technological) Support of Instruction, eLearning Department, etc.

Financial funding and support of eLearning is another important field. 17 of 27 monitored institutions (63%) mentioned national or international eLearning projects which were a source of their funding.

3.4 Learning Management System

The effectiveness of eLearning in educational process is significantly connected with the relevant Learning Management System (LMS) that is used at each university. Twenty-two of twenty-seven institutions (more than 80%) mentioned the used type of LMS in the annual report. The findings from the annual reports reveal that they usually chose one or two types of LMS (16 universities) and then selected the more appropriate one. However, some universities decided to try out more types. These were as follows: University of Ostrava, Jan Evangelista Purkyně University in Ústí nad Labem, VŠB-Technical University of Ostrava, and University of Economics in Prague.

Six universities decided already at the beginning of the implementation of eLearning to design their own LMS (Czech University of Life Sciences in Prague, University of South Bohemia in České Budějovice, Masaryk University in Brno, Mendel University of Agriculture and Forestry in Brno, Technical University of Liberec, VŠB-Technical University of Ostrava). Usually, however, over the years have left this path and switched to LMS Moodle or another system.

LMS Moodle is the most frequently used system in the Czech universities. It is reported by twenty universities. Other frequently used systems were in the course of past 20 years Tutor 2000 or iTutor (3 universities), LearningSpace (3 universities), EDEN (2 universities), WebCT or BlackBoard (2 universities), eDoceo (2 universities), MS Class Server (2 universities), Unifor (1 university), IBM Workplace Collaborative Learning (1 university) and Oracle eLearning (1 university).

4 The Process of eLearning and Blended Learning at FIM UHK

The development history of blended learning at the Faculty of Informatics and Management of the University of Hradec Kralove (FIM UHK) began just after the foundation of the faculty in 1993 by using shared directories where study materials were presented. Gradually, the importance of electronic mail increased for communication between students, and students and teachers, then other services followed: electronic administration of credits and examinations, displaying syllabi, timetables, entrance exams results, university websites were designed and e-journal Telegraph published. Teachers′ websites supporting instruction appeared, and in 1997 nearly 25% of teachers used them. After that, the professional virtual learning environment Learning Space was bought. In 2001, it was replaced by WebCT [8]. In 2019, more than 300 e-courses are accessible in LMS Blackboard.

Development of computer assisted education leading to sophisticated eLearning and blended learning strategy at FIM UHK has gone through several phases likewise development at other universities involved in eLearning from shared study materials on the shared disc place to virtual mobility of students enrolled in e-courses, run in the virtual learning space of several learning management systems [9–16].

- 1991 –> Utilization of shared information sources on a shared disc space,
- 1994 –> Utilization of the Internet to support the process of education via e-mail communication and web pages,
- 1996 –> Development of multimedia applications (CDs) supporting the process of education,
- 1996 –> Support of the process of education with electronic tools: Information system and Timetable on the Internet,
- 1996 –> Project cooperation with prominent European centres focused on eLearning (European project support TEMPUS, SOCRATES/MINERVA, GRUNDVIG, eLearning),
- 1997 –> First attempts with creation and implementation of on-line courses: Internet in the Process of Education, Modern Presentation and Education.
- 1999 –> eLearning courses for further education, Choice of professional LMS (LearningSpace)
- 2001 –> Purchase of WebCT license, creation and implementation of supportive e-subjects for students of present and combined forms of study
- 2002 –> eLearning as a strategic priority at the FIM,
- 2005 –> Interuniversity study, virtual mobility on national and international basis,
- 2007 –> Move from LMS WebCT to LMS Blackboard,
- 2008 –> REKAP Project – education of academic staff in creation of eLearning courses,
- 2009 –> Research project Evaluation of the modern technologies contributing towards forming and development university students′ competences,
- 2010 –> Number of e-courses supporting the process of education exceeded 200 subjects,
- 2011 –> Research project A flexible model of the ICT supported educational process reflecting individual learning styles,
- 2015 –> Testing and implementation of mobile learning.
- 2018 –> Upgrading LMS Blackboard to a new version and a comprehensive transformation of all e-subjects in accordance with a unified structure.

5 Conclusion

The findings from the monitored annual reports of public universities indicate the importance of the use of eLearning in university education all over the Czech Republic. Almost 20 years of the existence of eLearning education, as well as numerous conferences on this topic, prove its effectiveness for higher learning.

Acknowledgments. This study is supported by the SPEV project 2019, run at the Faculty of Informatics and Management, University of Hradec Kralove, Czech Republic.

References

1. European Commission. Growth, competitiveness, and employment. The challenges and ways forward into the 21st century. Brussels: European Commission (1993)
2. European Commission. White Paper on Education and Training - Teaching and Learning - Towards the Learning Society. Brussel: European Commission (1995)
3. European_Commission. The eLearning Action Plan: Designing tomorrow's education. Brussel: European Commission (2001)
4. Šimonová, I.: Learning styles in foreign language instruction. Interactive collaborative learning (ICL2011) – 11th International Conference Virtual University, VU′11, pp. 595–601 (2011)
5. Šimonová, I., Poulová, P., Šabatová, M., Bílek, M., Maněnová, M.: On Contribution of Modern Technologies Towards Developing Key Competences. M. Vognar, Hradec Králové (2009)
6. Poulová, P., Šimonová, I.: eLearning at Czech universities in 1999 – 2010. eLearning: proceedings of the 9th European Conference, pp. 512–520. Academic Publishing, Reading (2010)
7. Poulová, P., Šimonová, I.: The traditional versus ICT-supported instruction within the tertiary education: comparative study. In: i-Society 2011: International Conference: Proceedings, London, pp. 321–326 (2011)
8. Klimova, B., Simonova, I., Poulova, P.: Blended learning in the university english courses: case study. blended learning. new challenges and innovative practices. In: ICBL: Lecture Notes in Computer Science, vol. 10309. Springer, Cham (2017)
9. Černá, M., Poulová, P., Šrámková, H.: eLearning strategy - case study. In: Information and Communication Technology in Education 2005. Ostravská univerzita: Ostrava, pp. 255–259 (2005)
10. Manenova, M., Simonova, I., Poulova, P.: Which one, or another? comparative analysis of selected LMS. Procedia Soc. Behav. Sci. **186**, 1302–1308 (2015)
11. Sokolova, M.: Analysis of the effectiveness of teaching with the support of eLearning in the course of Principles of Management I - performance analysis. Procedia Soc. Behav. Sci. **28** (2011)
12. Simonova, I., Poulova, P., Kriz, P.: Reflection of intelligent ELearning/Tutoring - the flexible learning model in LMS Blackboard. In: Transactions on Computational Collective Intelligence XVIII. Lecture Notes in Computer Science, vol. 9240, pp. 20–43 (2015)
13. Poulova, P., Cerna, M., Svobodova, L.: University network - efficiency of virtual mobility. In: Proceedings of the 5th International Conference on Educational Technologies (EDUTE09), pp. 87–92. WSEAS Press, La Laguna (2009)

14. Zimola, B., et al.: Building a virtual learning and teaching community in the Czech Republic and in Europe. In: 4th International Conference on Education and Information Syst/2nd International Conference on Social and Organizational Informatics and Cybernetics, pp. 77–81. Int Inst Informat & System, Orlando (2006)
15. Greener, S.: How are web 2.0 technologies affecting the academic roles in higher education? a view from the literature. In: 11th European Conference on eLearning, ECEL 2012, Groningen, pp. 124–132 (2012)
16. Poulová, P., Hynek, J. Application of information and communication technologies. In: Education and Information Systems: Technologies and Applications, pp. 80–85. International Institute of Informatics and Systemics, Orlando (2003)

Learning Professional Skills

Students' Attitudes Towards the Mobile App Used in English Classes

Blanka Klímová(✉)

University of Hradec Králové, Hradec Králové, Czech Republic
blanka.klimova@uhk.cz

Abstract. Mobile devices have penetrated all spheres of human activities, including education. Especially the mobile phones/smartphones are becoming widely used in learning thanks to their proved benefits such as improved knowledge retention and increased student engagement. The purpose of this article is to discuss students' attitudes towards the mobile app used for learning vocabulary and phrases in English classes at a university level. The methods included a literature review and a questionnaire survey. The results showed that the positive aspects of the mobile app prevailed over the negative ones. The mobile app helped students prepare for the final achievement test, learning was less stressful, interacting with the app helped students retain the English vocabulary better, as well as become more confident in their learning. In addition, using the app gave students confidence knowing that they had all necessary resources at hand and could access them at any time when moving around. They also enjoyed using the app and would opt for its implementation in other courses taught at the faculty. Thus, the findings of this study revealed that students have positive attitudes towards the use of mobile apps in classes and they consider mobile apps suitable additional learning tools, which can thanks to their interactivity, ubiquity, or portability, motivate them to learn.

Keywords: Mobile app · Students · Attitude · English

1 Introduction

Today, young people cannot imagine their life without mobile phones, respectively smartphones. They consider using mobile phones as natural as breathing. Moreover, these mobile devices have penetrated all spheres of human activities, including education. As statistics [1] points out, 64% of students consider accessing their learning material from a mobile device essential. Smartphone learners complete course material 45% faster than those using a computer and 89% of smartphone users download apps, 50% of which are used for learning. Especially the mobile phones/smartphones are becoming widely used in learning thanks to their proved benefits such as improved knowledge retention and increased student engagement [2, 3]. This is also true for English language learning. Such language learning enhanced with the help of handheld mobile devices, especially smartphones, is called Mobile Assisted Language Learning (MALL) [4]. It is considered to be part of the so-called Computer Assisted Language Learning (CALL) [5]. However, as Kukulska-Hulme and Shield [6] state, MALL is different from CALL because it

M. E. Auer and D. May (Eds.): REV 2020, AISC 1231, pp. 947–953, 2021.
https://doi.org/10.1007/978-3-030-52575-0_78

emphasizes spontaneity of access and interaction across different contexts of use. In addition, MALL attempts to promote learning, which is personalized, situated, authentic, spontaneous, collaborative and with the learner at the center of the learning process [7]. Furthermore, the classroom environment is no longer the only learning environment as students become part of the context and interact with their peers [8]. Thus, interaction in both formal and informal settings is available.

The purpose of this article is to discuss students' attitudes towards the mobile app used for learning vocabulary and phrases in English classes at a university level.

2 Methods

2.1 Methodology

The methodology was based on the description and evaluation of the findings of a literature review of the available sources found on the research topic in the world's acknowledged databases such as Scopus and Web of Science.

Furthermore, a method of a questionnaire survey was used to discover students' attitudes towards the used technology, in this case a mobile app, used for learning and retention of new words and phrases. The app was tailored to their needs (learning and retaining new words and phrases from their study, i.e., from the field of management of tourism) in order to enhance their learning outcomes.

The questionnaire consisted of 15 questions, out of which two were open-ended questions. The questionnaire and its findings are presented in the section Results and discussion below. Five scale ranging from "strongly disagree" to "strongly agree" was used. The questionnaire was filled in by 14 students of Management of Tourism in their third year of study, by those who used the app in the winter semester of 2018.

2.2 Study Design

The mobile app was used by students of Management of Tourism in their third year of their study at the Faculty of Informatics and Management in Hradec Králové as an additional support for learning English in the winter semester of 2018. The semester lasted for 13 weeks and mobile app was used outside the contact English classes. All the contact classes lasted 90 min per week and were held from the end of September to the mid of December. Altogether, there were 20 students out of which 14 were using the mobile app and 6 who decided not to use it.

The content of the app was aimed at practicing and retaining new words and phrases, which proved to be students' main weakness when learning English. The mobile app is called Angličtina (English) TODAY. It is a software architecture that uses the server part, the web application and the mobile application [9]. The mobile app was developed both for the Android and iOS operating systems. There are ten lessons of vocabulary and 10 lessons of phrases. The students must translate the word or the phrase from their native language into English. Each vocabulary lesson is done as a test and comprises 15–18 new words on average. The same is true for the phrases lessons, which include 10 new phrases on average. In addition, students received notifications sent to them on their smartphones at least twice a week by their teacher in order to remind them to study on a regular basis. Figure 1 below provides a picture of the teacher's interface and Fig. 2 demonstrates the mobile app screen.

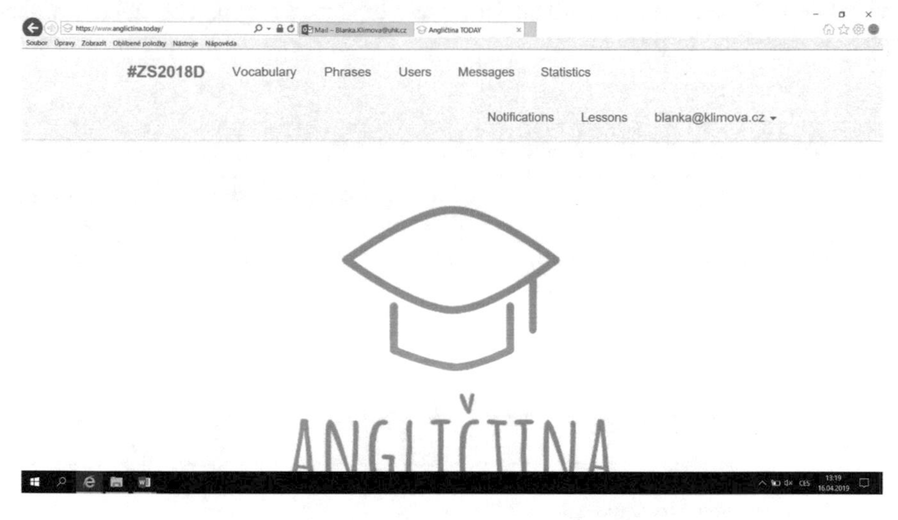

Fig. 1. Teacher's interface

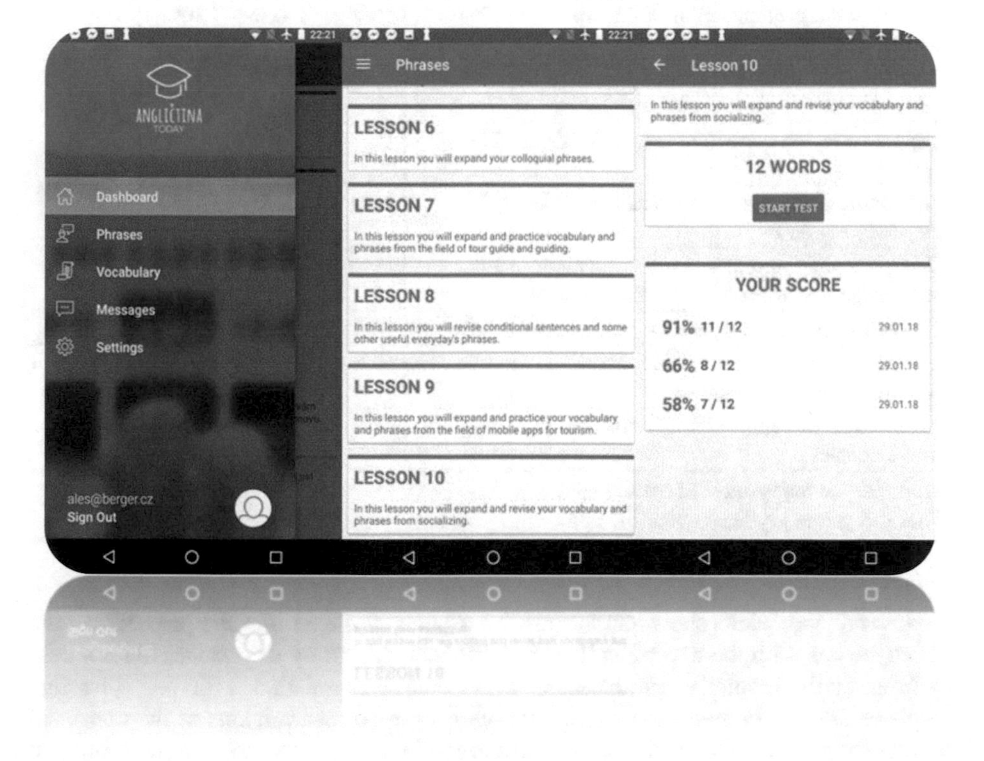

Fig. 2. Mobile app screen

3 Results and Discussion

Table 1 below presents the results of the questionnaires, which were filled in by 14 students (13 female students and 1 male student) using the mobile app. Their age range was between 21 and 25 years. All students were of Czech origin. The amount of their replies is described in percentages.

Table 1. Questionnaire results on students' attitudes towards the mobile app

	Strongly disagree	Disagree	Neutral	Agree	Strongly agree
I enjoyed using a mobile app to learn	0%	14%	43%	29%	14%
Using the app helped me become more confident in my learning	0%	14%	29%	57%	7%
The app was more accessible than books when I was moving around	0%	0%	21%	29%	50%
The app had a positive effect on my study behavior	0%	14%	50%	36%	0%
The app gave me confidence knowing I had my resources at hand and could access it at any time	0%	7%	29%	43%	21%
I checked the pronunciation of the words I was learning on the app	7%	36%	21%	36%	0%
Interacting with the app helped me remember my English vocabulary better	0%	7%	29%	57%	7%
I appreciated the corrective feedback of the app	21%	14%	29%	29%	7%
The notifications sent by the teacher helped me study regularly	7%	36%	36%	21%	0%
Using a mobile app to test my vocabulary knowledge was more fun and less stressful	14%	14%	14%	58%	0%
The app helped me prepare for the final test	0%	0%	0%	71%	29%
Using the app helped me enhance my communication performance	0%	43%	43%	14%	0%
I would like the app to be implemented in future courses	14%	7%	14%	51%	14%

What did you enjoy most about the app?
What did you enjoy least about the app?

Overall, the results show that the positive aspects of the mobile app prevail over the negative ones with the exception that the notifications sent by the teacher did not help students study regularly. Furthermore, they were not satisfied with the corrective feedback, which, in their opinion, did not offer more correct versions of the translated word(s) or phrases. This fact was also mentioned in their answer to what they enjoyed least about the app. However, the correctness of the words and phrases was carefully analyzed and evaluated by the teacher, which means that students might not have been right in this statement (please consult Fig. 3 below). For instance, instead of writing infinitive of *look forward to*, they wrongly wrote *looking forward* to.

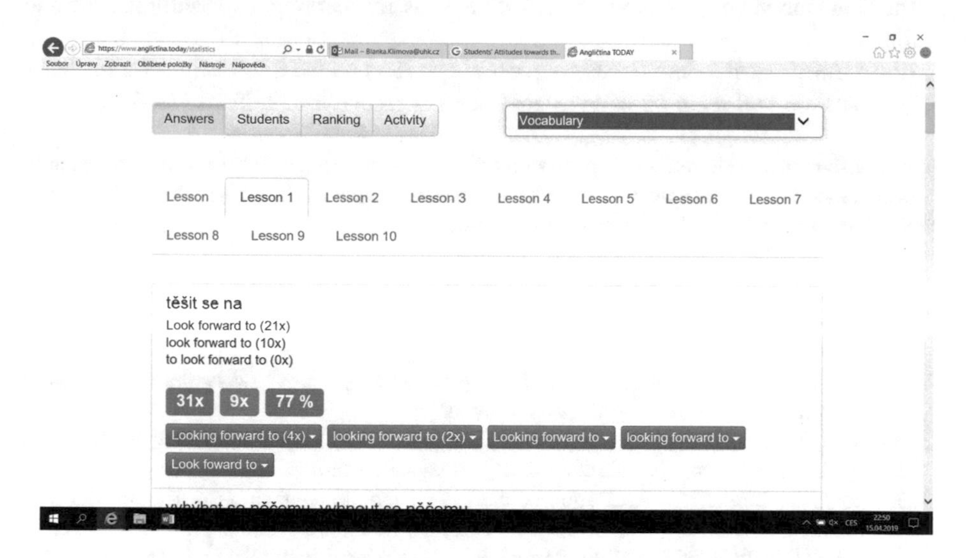

Fig. 3. Students' performance and their mistakes

On the contrary, the results of the questionnaire reveal that the mobile app helped them prepare for the final achievement test, learning was less stressful, interacting with the app helped them retain the English vocabulary better, as well as to become more confident in their learning. In addition, using the app gave them confidence knowing that they had all necessary resources at hand and could access them at any time when moving around.

Last but not least, they enjoyed using the mobile app and would opt for its implementation in other courses taught at the faculty. In the open-ended question they also pointed out that they were learning faster and more effectively since they could use it and access it at any time and anywhere, on the way home, for example, on the bus or train. The aspect of ubiquity is also reflected in the research by Yurdagul and Oz [10], who in their descriptive study revealed that their respondents had appreciated the instant and easy access to information in language learning.

These findings have been also confirmed by other research studies, e.g. [11–17], which illustrate that students using mobile apps seem to be more stimulated to learn both in formal and informal settings. Furthermore, students seem to be more confident in their learning and they exhibit less anxiety [18].

The limitations of this article consist in a small sample of respondents. Therefore, further survey is needed to provide more reliable results.

4 Conclusion

The findings of this study reveal that students have a positive attitude towards the use of mobile apps in classes and they consider mobile apps suitable additional learning tools, which can thanks to their interactivity, ubiquity, or portability, motivate them to learn. Thus, mobile apps seem to be effective tools for learning, especially in informal settings.

For further research into the use of mobile apps and all the implications for business communication and learning of languages see the research of Pikhart [19–21].

Acknowledgment. This study is supported by the SPEV project 2104/2019, run at the Faculty of Informatics and Management, University of Hradec Kralove, Czech Republic. The author thanks Aleš Berger for his help with data collection.

References

1. 10 Stats that Prove Mobile Learning Lives up to the Hype. https://www.docebo.com/blog/10-stats-prove-mobile-learning-lives-up-to-hype/
2. Klimova, B.: Mobile phones and/or smartphones and their apps for teaching English as a foreign language. Educ. Inf. Technol. **23**(3), 1091–1099 (2017)
3. Klímová, B., Berger, A.: Evaluation of the use of mobile application in learning English vocabulary and phrases – a case study. In: Hao, T., Chen, W., Xie, H., Nadee, W., Lau, R. (eds.) SETE 2018. LNCS, vol. 11284, pp. 3–11. Springer, Cham (2018). https://doi.org/10.1007/978-3-030-03580-8_1
4. Chinnery, G.: Going to the MALL: mobile assisted language learning. Lang. Learn. Technol. **10**(1), 9–16 (2006)
5. Yang, J.: Mobile assisted language learning: review of the recent applications of emerging mobile technologies. Engl. Lang. Teach. **6**(7), 19–25 (2013)
6. Kukulska-Hulme, A., Shield, L.: An overview of mobile assisted language learning: from content delivery to supported collaboration and interaction. ReCALL **20**(3), 271–289 (2008)
7. Kukulska-Hulme, A.: Will mobile learning change language learning? ReCALL **21**(2), 157–165 (2009)
8. Wang, B.T.: Designing mobile apps for English vocabulary learning. Int. J. Inf. Educ. Technol. **7**(4), 279–283 (2017)
9. Berger, A., Klímová, B.: Mobile application for the teaching of English. In: Park, James J., Loia, V., Choo, K.-K.R., Yi, G. (eds.) MUE/FutureTech -2018. LNEE, vol. 518, pp. 1–6. Springer, Singapore (2019). https://doi.org/10.1007/978-981-13-1328-8_1
10. Yurdagul, C., Oz, S.: Attitude towards mobile learning in English language education. Educ. Sci. **8**, 142 (2018)
11. Teodorescu, A.: Mobile learning and its impact on business English learning. Procedia Soc. Behav. Sci. **180**, 1535–1540 (2015)
12. Balula, A., Marques, F., Martins, C.: Bet on top hat – challenges to improve language proficiency. In: Proceedings of EDULEARN15 Conference, 6–8 July 2015, Barcelona, Spain, pp. 2627–2633 (2015)
13. Tayan, B.M.: Students and teachers' perceptions into the viability of mobile technology implementation to support language learning for first year business students in a Middle Eastern University. Int. J. Edu. Literacy Stud. **5**(2), 74–83 (2017)

14. Davie, N., Hilber, T.: Mobile-assisted Language Learning: Student attitudes to using smartphones to learn English vocabulary (2015). https://www.researchgate.net/publication/270550041_Mobile-assisted_Language_Learning_Student_Attitudes_to_using_Smartphones_to_learn_English_Vocabulary
15. Heil, C.R., Wu, J.S., Lee, J.J., Schmidt, T.: A review of mobile language learning applications: trends, challenges, and opportunities. EuroCALL Rev. **24**(2), 32–50 (2016)
16. Glahn, C., Gruber, Marion R., Tartakovski, O.: Beyond delivery modes and apps: a case study on mobile blended learning in higher education. In: Conole, G., Klobučar, T., Rensing, C., Konert, J., Lavoué, É. (eds.) EC-TEL 2015. LNCS, vol. 9307, pp. 127–140. Springer, Cham (2015). https://doi.org/10.1007/978-3-319-24258-3_10
17. Goz, F., Ozcan, M.: An entertaining mobile vocabulary learning application. EPESS **7**, 63–66 (2017)
18. Luo, B.R., Lin, Y.L., Chen, N.S., Fang, W.C.: Using smartphone to facilitate English communication and willingness to communicate in a communicate language teaching classroom. In: Proceedings of the 15th International Conference on Advanced Learning Technologies, pp. 320–322. IEEE (2015)
19. Pikhart, M.: Communication based models of information transfer in modern management – the use of mobile technologies in company communication. In: Innovation Management and Education Excellence through Vision 2020, pp. 447–450. International Business Information Management Association (2019)
20. Pikhart, M.: Electronic managerial communication: new trends of intercultural business communication. In: Innovation Management and Education Excellence through Vision 2020, pp. 714–717. International Business Information Management Association (2019)
21. Pikhart, M.: New horizons of intercultural communication: applied linguistics approach. Procedia Soc. Behav. Sci. **152**, 954–957 (2014). ERPA

Critical Success Factors in Computer Aided Intercultural Business Communication Course

Marcel Pikhart(✉)

Faculty of Informatics and Management, University of Hradec Kralove, Hradec Kralove, Czech Republic
Marcel.pikhart@uhk.cz

Abstract. The paper describes the introduction of the course of intercultural business communication into the curriculum of the Faculty of Informatics and Management of the University of Hradec Kralove, Czech Republic, five years ago as an obligatory subject for foreign students visiting the university. Initially, the subject was taught using traditional methods like lectures of the tutor and the seminars. However, after a few years blended learning methodology was introduces into the course. The students' satisfaction with the course was tested after each year and the paper compares the results before and after the introduction of blended learning into the subject. Computer aided course clearly proves to be assessed by the students more positively and the paper thus stresses the important of necessary introduction of hybrid and blended learning into traditional curricula as it shows significant improvements in students' satisfaction, motivation and performance levels.

Keywords: Blended learning · Computer aided learning · Intercultural business communication · Intercultural communication

1 Introduction

The paper deals with the advantages and disadvantages of the utilization of computer aided learning design after the introduction of the course of intercultural business communication into university curricula supported by e-learning platforms.

The course of intercultural business communication was introduced for foreign and Czech students of the Faculty of Informatics and Management of the University of Hradec Kralove five years ago. Initially, the subject was taught using merely traditional methods of lectures and seminars. The student's satisfaction with the classes was tested each year using traditional questionnaires testing their overall satisfaction.

Two years ago, after careful consideration, literature review and best practice research, blended learning methodology was introduced and implemented into the teaching process and the student's satisfaction was tested again after each year of the course.

2 Literature Review and Main Topics

There has been enormous literature on blended learning, computer aided learning, m-learning, etc., in the past few years [4, 6, 9, 13]. Current research into the use of these new methodologies generally shows significant improvement of performance levels [7,

M. E. Auer and D. May (Eds.): REV 2020, AISC 1231, pp. 954–960, 2021.
https://doi.org/10.1007/978-3-030-52575-0_79

10, 12] and users' satisfaction [5, 14, 15, 32, 34, 35] with these new means of information transfer and learning process in modern education institutions. Further research into machine learning, corpus linguistics and computational linguistics is also necessary to understand the issue [1–3].

Therefore, after the review of current literature and research, the tutor of the intercultural business communication course felt the urgent need to improve the course by implementing new and modern methods into the course (following the trend [30, 31]) which is usually in European universities taught by using traditional methods (the tutor has the experience with the course or similar courses from many European universities which he had visited before the research as a visiting lecturer, the universities included countries like Finland, Italy, Germany, Spain, Portugal, Turkey, Romania, the Netherlands, and the UK [18–29]).

The current situation regarding the use of mobile technologies can be described as follows. Younger adults (i.e. university students) use mobile devices on a daily basis and the natural increase in the quantity of time they spend with mobile devices has increased unprecedentedly in the past few years [10, 16, 17, 32, 33]. Therefore, naturally, the creators of university curricula have to take these aspects into their account when creating new curricula and syllabi if they want to be attractive and create sufficient impact on the new generation of the students. It is not only gaming and social media which are being used massively, but also various utilities and apps which are somehow connected to natural human desire to obtain information and learn new things.

A lot of research clearly shows that by using blended learning the performance of the students is enhanced, in language education the retention of newly acquired vocabulary is improved, the motivation rises and the users of these new platforms are willing to spend more time using the devices while learning [36–38]. The benefits of computer aided learning process are so clear that it goes without saying that modern learning process is basically unimaginable without it [7, 8, 11]. The optimization of the learning process is so obvious that further research into it is basically unnecessary and we should more focus on the possible pitfalls and drawbacks which could potentially arise when using these e-learning platforms.

Moreover, further research proves that smart devices and mobile learning (m-learning) brings many new benefits which are useful for optimized usability and utilization of our time while we study, such as portability, possibility to study anytime and almost anywhere, and the possibility to implement various communication tools both used by the students when they communicate with each other or with the tutor of the course [8, 16].

The last but not least, it must be noted that e-learning also enables the students who are suffering from various physical disabilities and with various special needs such as dyslexia to participate in the learning process in an easier way and university studies are thus more viable option for them because of an easier access, both physical and mental as well [4, 6].

3 Blended Learning Platforms and Critical Factors in Computer Aided Learning

The described course of intercultural business communication started to use Blackboard after three years of its introduction into the university curriculum as the template platform for blended learning, i.e. first three years of the course were taught without any computer aided learning platforms and the last two years with the help of the computer aided learning, namely the Blackboard.

After the implementation of the Blackboard environment, the students had an opportunity to use the platform as a repository of various texts to be studied, a basic tool for communication in the group and with the tutor, they could also share their projects and presentations there and could use various collaborative tools in their project work as well. The Backboard environment was fully used by all the participants of the course and there were no technical issues with it during the course whatsoever.

The presented paper attempts to focus on critical success factors in computer aided learning comparing previous situation, i.e. without any computer aided learning, with the new one, i.e. with the utilization of blended learning, because this is the opportunity to highlight not only the benefits of the bended learning but also flag up potential bottlenecks which may prove crucial and possibly very risky when creating university courses supported by computer aided learning.

4 The Research Format

Quantitative research was conducted each year after the course was introduced in five subsequent years, data analyzed and compared and final conclusions were created. The students of the course of intercultural business communication at the Faculty of Informatics and Management of the University of Hradec Kralove, Czech Republic, were given online questionnaires after the course was finished.

The questionnaires were anonymous, asking a few simple questions about the overall satisfaction with the course and also contained a few open questions for further recommendations of the participants, what to improve, what to add to the course and what are the parts of the course which are unnecessary. The data were collected each year and compared in two groups, i.e. before the implementation of computer aided learning (first three years) and after the implementation (last two years of the course).

The data of the research were collected from various nationalities, as the group of respondents was created by the students from all over the world, such as China, Taiwan, many European countries and a few countries from Latin America. The number of respondents was around 50 each year, i.e. the total number of the course participants. Male-female ratio was nearly 50-50 and the age group was 20–25 years.

5 The Importance of the Research

The findings of the research clearly show the benefits of blended learning, however, also stresses potential pitfalls. The results of the research are very important because they can help us when we try to flag up potential bottlenecks when designing any e-learning courses.

E-learning is generally considered very beneficial in language learning and other kinds of educational process, however, we still lack ideas about limitations of e-learning so that we would be able to fight with the drawbacks which are necessarily connected to the use of new methods in education, such as blended learning in the modern learning process.

The paper brings a five-year research comparison and will prove helpful both for theory and practice of blended learning and computer aided learning because it brings findings which could be used in our everyday teaching practice, when creating curricula and when trying to implement computer aided design into the learning process.

6 The Findings of the Research

The most important findings are that the respondents showed their interest in new methods of teaching/learning compared to traditional approaches. The vast majority of the students were very satisfied with the implementation of the new means of transferring information and learning new things by using modern communication tools, such as PC's and smart phones.

The research clearly shows that the new generation of students is willing to use new technologies to enhance their learning experience, however, they also showed lack of motivation when the information transferred through electronic media could be transferred otherwise - the time they spend with mobile devices in their free time is so long that any time spend with mobile devices which is not necessary is considered as lost time. This finding is very important for us educators and creators of e-learning courses and must be taken into consideration because it is crucial when developing and implementing new courses.

There are many crucial aspects for computer aided educational process and the findings clearly show that its implementation into the teaching process is not straightforward as it seems in the first place. The drawbacks and possible pitfalls are present and somehow neglected by the current research which is available now, therefore, this paper tries to highlight the cons which this kind of teaching/learning brings, rather than its benefits which have already been described and depicted in detail by many scholars. Thus, it can help us to design, create and implement this extremely efficient tool with eliminating its drawbacks.

7 Limitations of the Research

The limitations of the research are the relatively small sample of the respondents, even if all the participants answered the questionnaire. However, the author is convinced the data from this small sample are statistically relevant and could be generalized and used on a larger scale. The research sample was different each year, which may have influenced the results dramatically as well.

Moreover, further research is planned on a larger scale at another Czech university into the same topic, i.e., the potential of e-learning (blended or hybrid learning) in university curricula and the potential benefits and improvements of the learning process while utilizing these platforms.

8 Conclusions

The paper highlights the importance of the implementation of computer aided learning in any available way, such as blended learning, hybrid learning, m-learning, etc., so that the modern educational process will be more attractive for the users, will become more efficient and will thus bring better performance and enhance our competitiveness in the global educational market.

On the other hand, it attempts to show certain limits to the utilization of these modern platforms inasmuch as they can bring several serious issues to the learning process which must not be ignored. Otherwise, the creators of these courses will not succeed in the long-run, and therefore, will not be able to create efficient and motivating platforms for further educational process. The conducted research is just an attempt to highlight the importance of realizing these potential pitfalls because the current literature on computer aided education somehow lacks this view.

The designers of curricula should take these facts into account and then they will be able to create successful and motivating courses which will have an opportunity to bring synergy effect of bended learning into our university education.

Acknowledgements. The paper was created with the support of SPEV 2019 at the Faculty of Informatics and Management of the University of Hradec Kralove, Czech Republic. The author would like to thank the student Jan Sprinar for his help when collecting the data of the research.

References

1. Alpaydin, E.: Machine Learning. The New AI. MIT Press, Cambridge (2016)
2. Buckland, M.: Information and Society. MIT Press, Cambridge (2017)
3. Clark, A., et al.: The Handbook of Computational Linguistics and Natural Language Processing. Blackwell, Chichester (2010)
4. Klimova, B.: Teacher's role in a smart learning environment—A review study. In: Uskov, Vladimir L., Howlett, Robert J., Jain, Lakhmi C. (eds.) Smart Education and e-Learning 2016. SIST, vol. 59, pp. 51–59. Springer, Cham (2016). https://doi.org/10.1007/978-3-319-39690-3_5

5. Klimova, B.: Assessment in the eLearning course on academic writing – A case study. In: Wu, T.-T., Gennari, R., Huang, Y.-M., Xie, H., Cao, Y. (eds.) SETE 2016. LNCS, vol. 10108, pp. 733–738. Springer, Cham (2017). https://doi.org/10.1007/978-3-319-52836-6_79
6. Klímová, B., Berger, A.: Evaluation of the use of mobile application in learning english vocabulary and phrases – A case study. In: Hao, T., Chen, W., Xie, H., Nadee, W., Lau, R. (eds.) SETE 2018. LNCS, vol. 11284, pp. 3–11. Springer, Cham (2018). https://doi.org/10.1007/978-3-030-03580-8_1
7. Klimova, B., Poulova, P.: Mobile learning and its potential for engineering education. In: Proceedings of 2015 I.E. Global Engineering Education Conference (EDUCON 2015), pp. 47–51. Tallinn University of Technology, Estonia, Tallinn (2015)
8. Klimova, B., Poulova, P.: Mobile learning in higher education. Adv. Sci. Lett. **22**(5/6), 1111–1114 (2016)
9. Klimova, B., Simonova, I., Poulova, P.: Blended learning in the University English Courses: Case study. In: Cheung, S.K.S., Kwok, L., Ma, W.W.K., Lee, L.-K., Yang, H. (eds.) ICBL 2017. LNCS, vol. 10309, pp. 53–64. Springer, Cham (2017). https://doi.org/10.1007/978-3-319-59360-9_5
10. Lopuch, M.: The effects of educational apps on student achievement and engagement. http://www.doe.virginia.gov/support/technology/technology_initiatives/e-learning_backpack/institute/2013/Educational_Apps_White_Paper_eSpark_v2.pdf (2013). Accessed 2 March 2018
11. Luo, B.R., Lin, Y.L., Chen, N.S., Fang, W.C.: Using smartphone to facilitate English communication and willingness to communicate in a communicative language teaching classroom. In: Proceedings of the 15th International Conference on Advanced Learning Technologies, pp. 320–322. IEEE (2015)
12. Males, S., Bate, F., Macnish, J.: The impact of mobile learning on student performance as gauged by standardised test (NAPLAN) scores. Educ. Res. **27**(1), 99–114 (2017)
13. Mehdipour, Y., Zerehkafi, H.: Mobile learning for education: Benefits and challenges. Int. J. Comput. Eng. Res. **3**(6), 93–101 (2013)
14. Miller, H.B., Cuevas, J.A.: Mobile learning and its effects on academic achievement and student motivation in middle grades students. Int. J. Sch. Technol. Enhanced Learn. **1**(2), 91–110 (2017)
15. Moher, D., Liberati, A., Tetzlaff, J., Altman, D. G.: The PRISMA group. Preferred reporting items for systematic review and meta-analysis: The PRISMA statement. PLoS Med. **6**, e1000097 (2009)
16. Muhammed, A.A.: The impact of mobiles on language learning on the part of English foreign language (EFL) university students. Procedia – Soc. Behav. Sci. **136**, 104–108 (2014)
17. Oz, H.: Prospective English teachers' ownership and usage of mobile device as m-learning tools. Procedia-Soc. Behav. Sci. **141**, 1031–1041 (2013)
18. Pikhart, M.: Sustainable communication strategies for business communication. In: Soliman, K. S. (ed.) Proceedings of the 32nd International Business Information Management Association Conference (IBIMA), 15–16 November 2018, Seville, Spain, pp. 528–553 (2018). International Business Information Management Association. ISBN: 978-0-9998551-1-9
19. Pikhart, M.: Intercultural business communication courses in European Universities as a way to enhance competitiveness. In: Soliman, K.S. (ed.) Proceedings of the 32nd International Business Information Management Association Conference (IBIMA), 15–16 November 2018, Seville, Spain, pp. 524–527 (2018). International Business Information Management Association. ISBN: 978-0-9998551-1-9
20. Pikhart, M.: Multilingual and intercultural competence for ICT: Accessing and assessing electronic information in the global world. In: Choroś, K., Kopel, M., Kukla, E., Siemiński, A. (eds.) MISSI 2018. AISC, vol. 833, pp. 273–278. Springer, Cham (2019). https://doi.org/10.1007/978-3-319-98678-4_28. ISSN 2194-5357

21. Pikhart, M.: Technology enhanced learning experience in intercultural business communication course: A case study. In: Hao, T., Chen, W., Xie, H., Nadee, W., Lau, R. (eds.) SETE 2018. LNCS, vol. 11284, pp. 41–45. Springer, Cham (2018). https://doi.org/10.1007/978-3-030-03580-8_5. Print ISBN: 978-3-030-03579-2, Electronic ISBN: 978-3-030-03580-8
22. Pikhart, M.: communication based models of information transfer in modern management - The use of mobile technologies in company communication. In: Soliman, K.S. (ed.) Proceedings of the 31st International Business Information Management Association Conference (IBIMA), 25–26 April 2018, Milan, pp. 447–450 (2018). International Business Information Management Association (IBIMA). ISBN: 978-0-9998551-0-2
23. Pikhart, M.: Current intercultural management strategies. The role of communication in company efficiency development. In: Proceedings of the 8th European Conference on Management, Leadership and Governance (ECMLG), pp. 327–331 (2012)
24. Pikhart, M.: Communication based models of information transfer in modern management – The use of mobile technologies in company communication. In: Innovation Management and Education Excellence through Vision 2020, IBIMA 2018, pp. 447–450 (2018)
25. Pikhart, M.: Electronic Managerial communication: New trends of intercultural business communication. In: Innovation Management and Education Excellence through Vision 2020, IBIMA 2018, pp. 714–717 (2018)
26. Pikhart, M.: Managerial communication and its changes in the global intercultural business world. In: Web of Conferences (ERPA 2015), vol. 26, p. 01013 (2016)
27. Pikhart, M.: Intercultural linguistics as a new academic approach to communication. In: Web of Conferences (ERPA 2015), vol. 26, p. 01005 (2016)
28. Pikhart, M.: Implementing new global business trends to intercultural business communication. In: Procedia Social and Behavioral Sciences, ERPA 2014, vol. 152, pp. 950–953 (2014)
29. Pikhart, M.: New horizons of intercultural communication: Applied linguistics approach. In: Procedia Social and Behavioral Sciences, ERPA 2014, vol. 152, pp. 954–957 (2014)
30. Simonova, I., Poulova, P.: Innovations in data engineering subjects. Adv. Sci. Lett. **23**(6), 5090–5093 (2017)
31. Simonova, I., Poulova, P.: Innovations in Enterprise Informatics Subjects. Advances in Intelligents Systems and Computing, 544, (2017) Springer, Cham
32. Sung, Y.T., Chang, K.E., Liu, T.C.: The effects of integrating mobile devices with teaching and learning on students' learning performance: A meta-analysis and research synthesis. Comput. Educ. **94**, 252–275 (2016)
33. Tayan, B.M.: Students and teachers' perceptions into the viability of mobile technology implementation to support language learning for first year business students in a Middle Eastern university. Int. J. Educ. Literacy Stud. **5**(2), 74–83 (2017)
34. Teodorescu, A.: Mobile learning and its impact on business English learning. Procedia Soc. Behav. Sci. **180**, 1535–1540 (2015)
35. Tingir, S., Cavlazoglu, B., Caliskan, O., Koklu, O., Intepe-Tingir, S.: Effects of mobile devices on K–12 students' achievement: A meta-analysis. J. Comput. Assist. Learn. **33**(4), 355–369 (2017)
36. Wu, Q.: Learning ESL vocabulary with smartphones. Procedia Soc. Behav. Sci. **143**, 302–307 (2014)
37. Wu, Q.: Designing a smartphone app to teach English (L2) vocabulary. Comput. Educ. (2015). https://doi.org/10.1016/j.compedu.2015.02.013
38. Wu, Q.: Pulling mobile assisted language learning (MALL) into the mainstream: MALL in broad practice. PLoS ONE **10**(5), e0128762 (2015)

Formative Assessment in a Blended Learning English Course

Blanka Klímová(✉)

University of Hradec Králové, Hradec Králové, Czech Republic
blanka.klimova@uhk.cz

Abstract. At present, formative assessment is widely used since it seems to significantly improve student achievement results. Most recently, it has become an inseparable part of the so-called blended learning (BL), which currently combines the use of mobile learning and face-to-face instruction. The implementation of formative assessment in a BL environment is increasing because it can enhance the learning quality. Moreover, research reveals that 87% of the formative assessment in BL is conducted automatically because it provides more flexible, timely and prompt feedback to students. The purpose of this article is to demonstrate how the formative assessment in a blended learning English course taught at the Faculty of Informatics and Management of the University of Hradec Kralove, Czech Republic, can help students to improve their English proficiency. The methods include a literature review of available sources found on the research topic in the world's acknowledged databases such as Scopus and Web of Science, a method of analysis of students' performance in a mobile app, and a method of analysis of students' final test results in order to illustrate the effectiveness of the formative assessment and discuss its implementation in the English course. The findings of the statistical analysis and students' performance in the mobile app reveal that the formative assessment has a positive impact both on students' learning and teacher's instruction.

Keywords: Formative assessment · Blended learning · English · Mobile phones/smartphones · Effect

1 Introduction

Teachers usually use in their classes both formative and summative assessment. The summative assessment is most often conducted at the end of the course in order to provide learners with a grade on their overall course performance. On the contrary, the formative assessment is carried out throughout the whole course to give learners feedback on their progress and guide their instruction effectively [1]. In addition, formative assessment significantly improves student achievement results [2, 3]. The key characteristics of formative assessment is that agents in the classroom (teacher, peer and learner) collect evidence of student learning and, based on this information, adjust teaching and/or learning [4]. In this way, new emerging technologies can help enhance the learning process and they are additional support to traditional, face-to-face teaching.

M. E. Auer and D. May (Eds.): REV 2020, AISC 1231, pp. 961–967, 2021.
https://doi.org/10.1007/978-3-030-52575-0_80

In fact, the use of technologies was combined with contact classes and blended learning came into being.

Blended learning (BL) is defined as a combination of eLearning courses and face-to-face instruction [5–10]. However, nowadays, with the emergence of mobile devices, especially smartphones, BL is becoming more a combination of mobile learning and face-to-face instruction. In comparison with eLearning, mobile learning enables students to learn ubiquitously from anywhere and at any time, on their own pace. The main advantages of the use of mobile devices are their easy portability, interactivity and informal learning settings [11].

Research shows that assessment is improved if BL is used [12, 13]. As Febriani and Abdullah [14] point out, the implementation of formative assessment in a BL environment is increasing since it can enhance the learning quality. Moreover, their research reveals that 87% of the formative assessment in BL is conducted automatically because it provides more flexible, timely and prompt feedback to students. Furthermore, the BL approach enhances the effectiveness and consistency of the whole assessment process [13].

The purpose of this article is to demonstrate how the formative assessment in a blended learning English course taught at the Faculty of Informatics and Management of the University of Hradec Kralove, Czech Republic, can help students to improve their English proficiency.

2 Methods

2.1 Methodology

The methods include a literature review of the available sources found on the research topic in the world's acknowledged databases such as Scopus and/or Web of Science in order to describe the recent findings on this topic, as well as a method of analysis of students' performance in a mobile app, used as an additional informal support tool for their learning. Furthermore, a method of analysis of students' final test results was applied in order to illustrate the effectiveness of the formative assessment and discuss its implementation in the English course.

2.2 Study Design

The mobile app was used outside the regular classes by students of Management of Tourism in their third year of their study at the Faculty of Informatics and Management in Hradec Králové as an additional support for learning English in the winter semester of 2018. The semester lasted for 13 weeks and mobile app was used outside the contact English classes. All the contact classes lasted 90 min per week and were held from the end of September to the mid of December. Altogether there were 28 students out of which 22 were in an experimental group, using the mobile app, and 6 students in a control group, not using a mobile app. Their level of English according to Common European Reference Framework for languages (CERF) [15] was B2 (upper-intermediate level of English).

The content of the app focused on the assessment and retention of new words and phrases, which proved to be students' main weakness when learning English. This finding was based on students' needs analysis carried out during the first lesson. The mobile app is called Angličtina (English) TODAY [16]. It is a software architecture that uses the server part, the web application and the mobile application. The mobile app was developed both for the Android (Fig. 1) and iOS operating systems (Fig. 2). There were ten lessons of vocabulary and 10 lessons of phrases. The students had to translate the word or the phrase from their native language into English. Each vocabulary lesson was done as a test and consisted of 15–18 new words on average. The same was true for the phrases lessons, which include 10 new phrases on average. After completing each lesson, students could immediately see the results of their performance, i.e. continuous assessment conducted through a mobile device. In addition, students received notifications sent to them on their smartphones at least twice a week by their teacher in order to remind them to study on a regular basis. In this way, the formative assessment was stimulated.

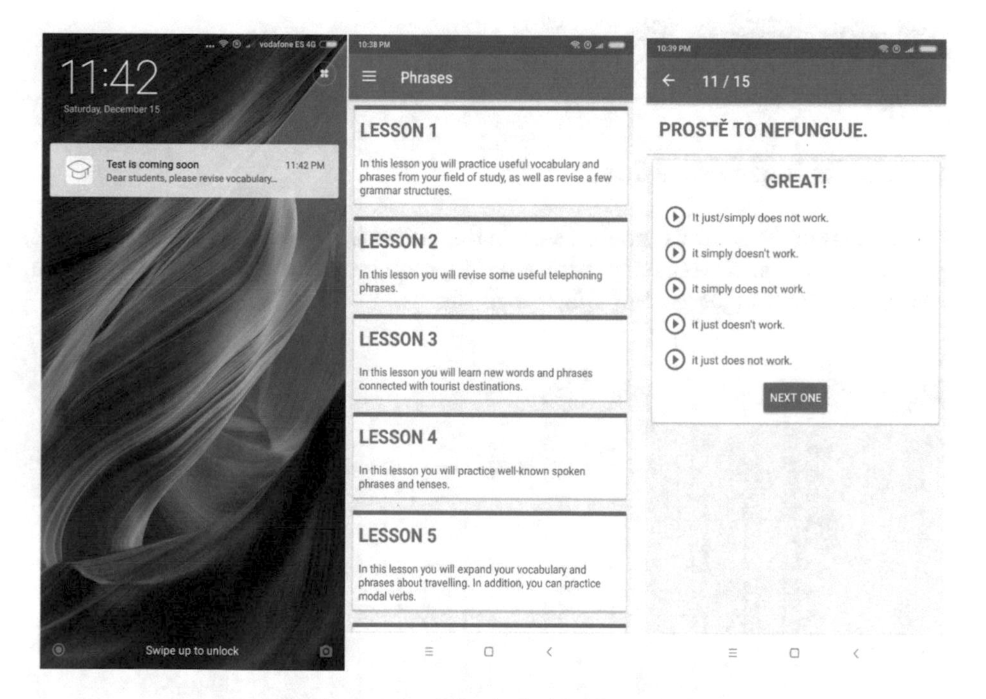

Fig. 1. Screen of the Android OS application

The methods also involved a method of analysis and evaluation of the results of students' achievement tests, including a statistical analysis. The pass mark for doing the final achievement test was 50%, i.e., 30 points. All the results were recorded and statistically analyzed.

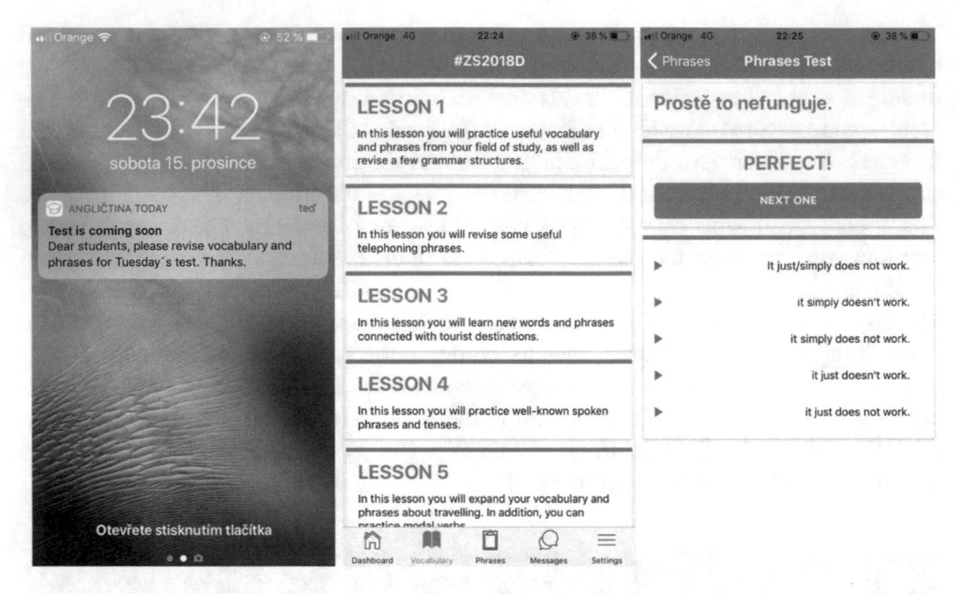

Fig. 2. Screen of the iOS application

3 Results and Discussion

The findings of the students' performance in the mobile app reveal that students studied on a regular basis. Nevertheless, before the final achievement test their performance was several times higher as Fig. 3 below illustrates.

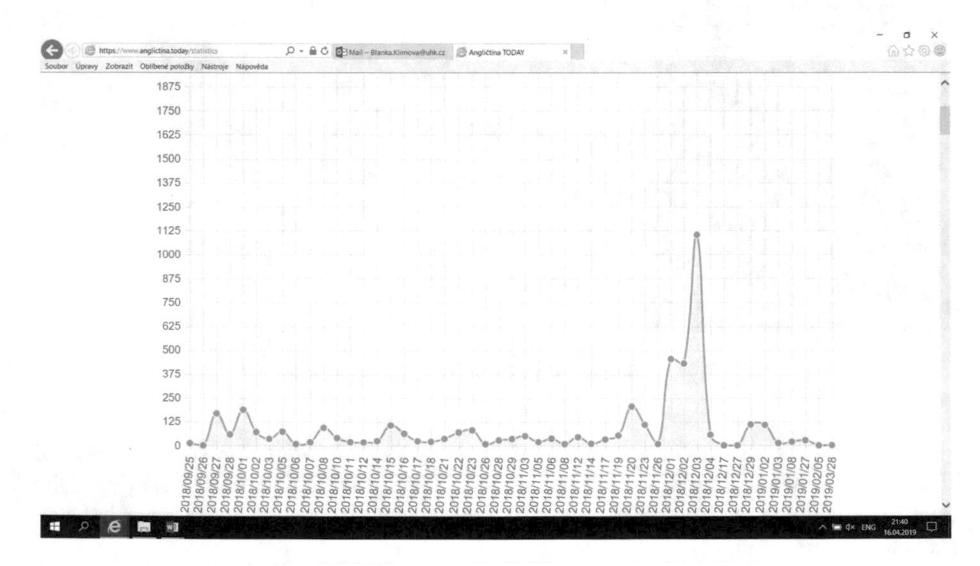

Fig. 3. Students' performance in the mobile app

The statistical analysis of students' results in the final achievement test show that the students in the experimental group, using the mobile app for the formative assessment, achieved the higher scores than the students in the control group, not using the mobile app. Table 1 below provides the results of descriptive statistics. The results show that the students in the experimental group achieved 39 points on average, while the students in the control group only 31.9 points.

Table 1. Descriptive statistics for variable results of the final test in points

Year	Mean	95% Confidence Interval for Mean		Median	Std. deviation	Minimum	Maximum
		Lower bound	Upper bound				
2018/19	35.518	31.906	39.130	37.000	9.316	17.0	54.0

This is in line with study by Mehmood et al. [17], whose research findings conducted among secondary school students also show that the formative assessment had a positive effect on students' achievement results in the experimental group. In addition, other research studies confirm that formative assessment have a considerable effect on higher academic achievement levels [18–20]. For instance, the study by Das et al. [20] revealed that formative assessment inspired students for deep learning and regular study. Furthermore, 78% of students reported that the feedback collected from formative assessment remained important for them as it helped to fill their learning gaps. They also agreed that formative assessment helped the faculty to identify student's weak point.

Overall, the results indicate that formative assessment helps students to enhance their learning performance. In addition, it gives students more chances to practice learning material, which in the case of the described BL English course contributes especially to a more permanent retention of new words and phrases [21–26]. Furthermore, such an assessment makes students more motivated to study since they can see a gradual progress of their learning.

The main limitations of this study include a small sample of students, as well as the fact that the students who did not use the app were deprived of the opportunity of continuous assessment. For further research into implications of blended learning in communication and learning process see research of Pikhart [27–29].

4 Conclusion

The findings of this study reveal that the formative assessment has a positive impact both on students' learning and teacher's instruction since the teacher can immediately detect, specifically through a feedback provided by the mobile app what students need to improve in their English learning and modify the learning material, both in formal and informal settings, to their needs.

Acknowledgment. This study is supported by the SPEV project 2104/2019, run at the Faculty of Informatics and Management, University of Hradec Kralove, Czech Republic. The author thanks Aleš Berger for his help with data collection.

References

1. Klimova, B.: Assessment methods in the course of academic writing. Procedia Soc. Behav. Sci. **15**(1), 2604–2608 (2011)
2. Black, P., Wiliam, D.: Assessment and classroom learning. Assess Educ. Principles Pol. Pract. **5**(1), 7–74 (1998)
3. Hattie, J., Timperley, H.: The power of feedback. Rev. Educ. Res. **77**(1), 81–112 (2007)
4. Andersson, K., Palm, T.: Characteristics of improved formative assessment practice. Educ. Inq. **8**, 104–122 (2017)
5. Frydrychova Klimova, B.: Blended learning. In: Mendez Vilas, A., et al. (eds.) Research, Reflections and Innovations in Integrating ICT in Education, pp. 705–708. FORMATEX, Spain (2009)
6. Klimova, B.: Evaluation of the blended learning approach in the course of business English – a case study. In: Huang, T.-C., Lau, R., Huang, Y.-M., Spaniol, M., Yuen, C.-H. (eds.) SETE 2017. LNCS, vol. 10676, pp. 326–335. Springer, Cham (2017). https://doi.org/10.1007/978-3-319-71084-6_37
7. Klimova, B., Simonova, I., Poulova, P.: Blended learning in the university English courses: case study. In: Cheung, S.K.S., Kwok, L.-F., Ma, W.W.K., Lee, L.-K., Yang, H. (eds.) ICBL 2017. LNCS, vol. 10309, pp. 53–64. Springer, Cham (2017). https://doi.org/10.1007/978-3-319-59360-9_5
8. Hubackova, S., Semradova, I.: Comparison of on-line teaching and face-to-face teaching. Procedia Soc. Behav. Sci. **89,** 445–449 (2013)
9. Simonova, I., Kostolanyova, K.: The blended learning concept: comparative study of two universities. In: Cheung, S.K.S., Kwok, L.-F., Shang, J., Wang, A., Kwan, R. (eds.) ICBL 2016. LNCS, vol. 9757, pp. 302–311. Springer, Cham (2016). https://doi.org/10.1007/978-3-319-41165-1_27
10. Frydrychova Klimova, B., Poulova, P.: Forms of instruction and students' preferences - a comparative study. In: Cheung, S.K.S., Fong, J., Zhang, J., Kwan, R., Kwok, L.-F. (eds.) ICHL 2014. LNCS, vol. 8595, pp. 220–231. Springer, Cham (2014). https://doi.org/10.1007/978-3-319-08961-4_21
11. Yurdagul, C., Oz, S.: Attitude towards mobile learning in English language education. Educ. Sci. **8**, 142 (2018)
12. Gimeno-Sanz, A.: Intermediate online English: an attempt to increase learner autonomy. Teach. Engl. Technol. **10**, 35–49 (2010)
13. Sejdiu, S.: English language teaching and assessment in blended learning. J. Teach. Learn. Technol. **3**(2), 67–82 (2014)
14. Febriani, I., Abdullah, M.I.: A Systematic review of formative assessment tools in the blended learning environment. Int. J. Eng. Technol. **7**, 33–39 (2018)
15. CERF. https://www.coe.int/en/web/common-european-framework-reference-languages
16. Anglictina TODAY. https://www.anglictina.today/
17. Mehmood, T., Hussain, T., Khalid, M., Azam, R.: Impact of formative assessment on academic achievement of secondary school students. Int. J. Bus. Soc. Sci. **3**(17), 101–104 (2012)

18. Ozan, C., Remzi, Y.: The effects of formative assessment on academic achievement, attitudes toward the lesson, and self-regulation skills. Educ. Sci. Theory Pract. **18**(1), 85–118 (2018)
19. Carrillo-de-la-Pena, M.T., Bailles, E., Caseras, X., Martinez, A., Ortet-Fabregat, G., Perez, J.: Formative assessment and academic achievement in pre-graduate students of health sciences. Adv. Health Sci. Educ. **14**(1), 61–67 (2009)
20. Das, S., Alsalhanie, K.M., Nauhria, S., Joshi, V.R., Khan, S., Surender, V.: Impact of formative assessment on the outcome of summative assessment – a feedback based cross sectional study conducted among basic science medical students enrolled in MD program (2017). https://www.researchgate.net/publication/318077434_Impact_of_formative_assessment_on_the_outcome_of_summative_assessment_-_a_feedback_based_cross_sectional_study_conducted_among_basic_science_medical_students_enrolled_in_MD_program
21. Klimova, B.: Mobile phones and/or smartphones and their apps for teaching English as a foreign language. Educ. Inf. Technol. **23**(3), 1091–1099 (2017)
22. Klímová, B., Berger, A.: Evaluation of the use of mobile application in learning english vocabulary and phrases – a case study. In: Hao, T., Chen, W., Xie, H., Nadee, W., Lau, R. (eds.) SETE 2018. LNCS, vol. 11284, pp. 3–11. Springer, Cham (2018). https://doi.org/10.1007/978-3-030-03580-8_1
23. Chinnery, G.: Going to the MALL: mobile assisted language learning. Lang. Learn. Technol. **10**(1), 9–16 (2006)
24. Yang, J.: Mobile assisted language learning: review of the recent applications of emerging mobile technologies. Engl. Lang. Teach. **6**(7), 19–25 (2013)
25. Wang, B.T.: Designing mobile apps for English vocabulary learning. Int. J. Inf. Educ. Technol. **7**(4), 279–283 (2017)
26. Berger, A., Klímová, B.: Mobile application for the teaching of English. In: Park, J.J., Loia, V., Choo, K.-K.R., Yi, G. (eds.) MUE/FutureTech -2018. LNEE, vol. 518, pp. 1–6. Springer, Singapore (2019). https://doi.org/10.1007/978-981-13-1328-8_1
27. Pikhart, M.: Multilingual and intercultural competence for ICT: accessing and assessing electronic information in the global world. In: Choroś, K., Kopel, M., Kukla, E., Siemiński, A. (eds.) MISSI 2018. AISC, vol. 833, pp. 273–278. Springer, Cham (2019). https://doi.org/10.1007/978-3-319-98678-4_28
28. Pikhart, M.: Technology enhanced learning experience in intercultural business communication course: a case study. In: Hao, T., Chen, W., Xie, H., Nadee, W., Lau, R. (eds.) SETE 2018. LNCS, vol. 11284, pp. 41–45. Springer, Cham (2018). https://doi.org/10.1007/978-3-030-03580-8_5
29. Pikhart, M.: Communication based models of information transfer in modern management - the use of mobile technologies in company communication. In: Soliman, K.S. (ed.) Proceedings of the 31st International Business Information Management Association Conference (IBIMA), April 25–26, pp. 447–450. International Business Information Management Association, Milan (2018)

Didactic Strategy with the Application of Technologies to Contribute to the Development of Modelling Skills from Operations Research Subject

Lester González López[1](✉), Valentina Badía Albanés[2], and Yudeisy Valdés Fernández[1]

[1] University of Informatics Sciences, Havana, Cuba
lester@uci.cu
[2] University of Havana, Havana, Cuba

Abstract. The knowledge society demands new ways to develop the learning - teaching process, where Information and Communication Technologies (ICT) plays a major role with the use of educational technology as a didactic tool for the development of skills. The University of Informatics Sciences offers a degree in Informatics Sciences Engineering, which has Operations Research subject within the Applied Mathematics discipline. This program is in charge of developing in the students the skill of modelling problems of linear programming, necessary for a professional of this area. Through the review of semester reports, pedagogical observations, interviews and surveys applied to students and teachers; several insufficiencies were evidenced in the learning teaching process of this subject; that indicates a low level of skill development of modeling linear programming problems. For this reason, this research objective is to elaborate a didactic strategy that makes use of the ICT for contributing to the development skills to model problems from Operations Research course in students of Informatics Sciences Engineering. As part of the strategy, a set of actions were defined to be executed by students and teachers in each of the four stages developed: diagnosis, planning, execution and evaluation. In this stage, a group of indicators was defined to measure the level of the development of the skill, according to a three levels scale. A collection of exercises was also made to model linear programming problems to work in classes, according to the current socioeconomic context. Finally, the results of the validation of the strategy are shown, through a pre-experiment, and the application of Iadov's satisfaction test.

Keywords: Modeling · Skill · Didactic strategy · Educational technology

1 Introduction

Scientific and technological development is one of the most influential factors in the knowledge society in which we live, where new forms of teaching and learning processes are increasingly in demand, since the role played by information, technologies

M. E. Auer and D. May (Eds.): REV 2020, AISC 1231, pp. 968–981, 2021.
https://doi.org/10.1007/978-3-030-52575-0_81

and the use of applications in the development of all fields of human activity cannot be ignored: in health, in education, in commerce, in public services, and so on.

Education demands specific forms of learning and the use of ICTs as didactic tools that contribute to the achievement of meaningful learning by students. In this context, the challenge for Cuban universities is the formation of a competent professional, committed to the Revolution and capable of describing and solving, with a scientific character, the real problems that arise in their area of action. It is necessary to transform the conception of the teaching learning process (TLP), in order to achieve a more active role of the student as the center of the process, where he will play a leading role in the development of his own knowledge, through the assimilation and management of the skills necessary for this.

The career in Informatics Sciences Engineering (ICI for the Spanish initials), which is studied at the University of Informatics Sciences (UCI for the Spanish initials), aims to form "integral professionals, committed to the homeland and to the development of the Cuban socialist model, whose function is associated with the development of computerization of Cuban society from three important edges: the development of the national software industry, the transformations of processes in the entities to assume its computerization and the necessary support for its maintenance. These needs are in accordance with the level reached in the computerization of society, the objectives proposed by the country, current and future international trends (University of Informatics Sciences 2014). In order for graduates to perform well in their profession, students must incorporate during the career a system of knowledge and skills necessary to achieve this goal. One of the disciplines that contributes significantly to this is Applied Mathematics, which is responsible for studying at a basic level the probabilistic and statistical modeling of processes to characterize them through their variability, as well as mathematical models and methods of operations research, which help build skills in the analysis and processing of data for decision making. In the model of the professional it is defined that one of the objectives of the discipline is: "to model mathematically the structure of a problem and the relations of the data to discover useful information, to arrive at the knowledge of the problem and to represent problems of decision making". Also within the skills to be developed is "modeling and applying methods of solution to problems of Service Systems".

Within the Applied Mathematics discipline is the subject Operations Research (OR), which has among its general objectives "to formulate mathematical models associated with problems of Linear Programming and Discrete Programming, and network problems", being mathematical modeling one of the main skills to overcome by students to solve the problems of the subject and real life, understood as the way to describe or represent the interrelation between the real world and mathematics.

Some meanings of mathematical modeling are found in the literature and it is concluded that there are multiple definitions that differ in theoretical and epistemological aspects. Among the definitions found, some are identified with the intention of constructing mathematical models through a structural process (Bassanezi and Biembengut 1997; Borromeo Ferri 2010; Hein and Biembengut 2006; Nieto 2004; Villa-Ochoa et al. 2009), and other contributions are related to the proposal and solution of problematic situations of context in school classrooms (Araújo 2009; Suárez and Cordero 2010; Bassanezi and Biembengut 1997; Berrío 2011; Biembengut and Hein 2006; Bossio 2014; Muñoz Mesa et al. 2014; Huapaya Gómez 2012).

The starting point for modeling must be a real problem situation, which must be simplified, idealized, subject to conditions and assumptions, and must be specified according to the interests of the problem solver. This leads to a formulation of the mathematical model that responds to the original situation, which must be solved through the solution methods in order to obtain certain mathematical results. These must be validated, that is, they must be transferred back to the real world, to be interpreted in relation to the original situation. When the model is validated, discrepancies may occur that lead to having to change or modify the model, but a model may not be found, perhaps because the original problem is not accessible to the mathematical treatment. Quite the opposite happens when you get a satisfactory model, which would serve for decision making. This means that it is important to interpret the model and the solutions to analyse whether it fits the original situation.

However, in all these aspects, shortcomings have been observed in previous courses. The analysis of the six-monthly reports of the subject Operations Research of the last 4 years (2013–2014, 2014–2015, 2015–2016, 2016–2017) made it possible to identify that certain problems have arisen in the TLP of the OR in the university, in addition to the author's own experience and the results of the surveys carried out with teachers and students who have worked on the subject, the following difficulties were noted:

- Limitations on being able to translate from the natural language in which the problems are described, to mathematical language.
- Difficulties in identifying the different elements of the mathematical model.
- Low understanding of problem statements, as well as for the selection of information that may be relevant and necessary to model the problem.
- Limitations on verifying and interpreting the results obtained in the context of the problem.
- More than 40% of students in the last 2 academic years did not develop modeling skills in the subject.

The deficiencies found corroborate that there is a contradiction between the insufficient development of problem modeling skills and the need for students to be prepared to solve real problems that present themselves as future professionals, where it will be essential to master this skill.

This contradiction leads to the following scientific problem: How to contribute to the development of the ability to model problems from the subject Operations Research in the students of the career of Engineering in Informatics Sciences?

The object of study of this research is the teaching-learning process of the subject Operations Research in the university, framed in the field of action, development of the ability to model from the TLP of the subject OR with the use of technologies.

In order to respond to this problem, the general objective is to develop a didactic strategy that makes use of Information and Communication Technologies in order to contribute to the development of the ability to model problems from the subject Operations Research in the students of the degree in Informatics Sciences Engineering.

2 Materials and Methods

The following theoretical methods were used for the development of the research:

- Analysis-synthesis: to analyse the object of the research and to carry out a synthesis of that which is central to the definition and implementation of the strategy. In addition, it allowed, through logical thought processes, the foundation of the thesis, as well as the reaching of partial and final conclusions.
- Historical-logical: to analyze the evolution of theoretical conceptions about the ability to model problems, as well as the essence of the laws of the TLP of the OR in the university.
- System approach: to establish the links between the diagnosed problem, its theoretical-methodological references and the proposed didactic strategy. In addition, it made it possible to establish the structuring and interrelationship of the components of the strategy.
- Induction-deduction: to arrive at conclusions on the characteristics of the didactic strategy to be elaborated, based on the observed facts.
- Modeling: for the conception of the didactic strategy in its different stages, and to interpret it and adjust it to the reality of the educational practice.

The following empirical methods were used:

- Documentary analysis: for the study of the different normative documents on the TLP, the model of the professional, analytical program of the subject, reports of partial, six-monthly and end of course tests.
- Interview: teachers were exchanged to ascertain the characteristics of the students in terms of their mathematical modeling skills.
- Survey: students and teachers were surveyed in order to identify possible inadequacies in the TLP of OR in the university, mainly those related to the development of problem modeling skills.
- Pedagogical test: to determine the level of development of the ability to model problems in the OR subject.
- Pedagogical observation: focused on clarifying students' difficulties in developing problem modeling exercises independently within the framework of practical OR classes.
- Pre-experiment: to verify the validity of the didactic strategy presented, through its implementation.

Also, as statistical methods:

- Descriptive statistics: to show the results obtained from applying the different instruments that were designed. The 87 students of the third year of the Informatics Sciences career of the Faculty 2 of the university of the 2017–2018 academic year were selected as population, and as it shows the 19 students of the 2304 group. This sample is intentional non-probability, which was selected because it is the group to which the researcher gives classes.

The research is based on the postulates of the historical-cultural approach of Vigotsky and his followers, as it focuses on learning in the educational field, putting the student as the protagonist of the process, and defends the influence of the environment in the assimilation of knowledge by the individual. Its design is not based on a pure research paradigm, but on a mixed approach, combining quantitative and qualitative. Research is also educational, since different ways are used to obtain information about the object and the current state of the ability to model problems in the OR subject is characterized. It is applied, according to the objective that is tried to reach, directed to the solution of existing difficulties that slow down the development of the ability modeling in the students.

3 Results

The main result of the work is the didactic strategy that is designed, for which the author assumes in the research, for its comprehensive character the definition proposed by Silva (2016), who in his doctoral research defines didactic strategy as: "the system of actions that makes it possible to transform the real state of the teaching-learning process into the desired state, based on the objectives and actions that are carried out in the stages of diagnosis, planning, execution and evaluation during the treatment of the content of the subject directed at the actions of the teacher, the student and the group in order to promote a developing learning with an interdisciplinary approach…".

In every strategy a set of actions is developed to transform the current state into a desired state, and that these are always intentional and aimed at solving problems in practice. The actions to be defined must integrate the preparation of both the students and the teachers who will participate in their development, since there are insufficiencies in both senses in the process to be perfected.

A system of actions was planned to transform the current state of TLP from content modelling of linear programming problems of OR at higher levels and according to concise and concrete objectives, where a dialectical relationship between TLP components is fostered. In addition, the following were defined as general characteristics of a didactic strategy for the development of skills: systemic character, objective-subjective duality, subordination to content, search for meaningful learning, planned and flexible character, integration of individual and group work, and the coexistence of systematic control and control by results.

The aforementioned elements served as a basis for the design of the didactic strategy whose mission is to contribute to the general training of the Informatics Sciences Engineering from the TLP of the OR subject with the use of ICT. The general objective of the teaching strategy is to contribute to the development of the ability to model linear programming problems from the OR subject with the use of technologies. The elements that compose it are the set of actions of the different stages (Fig. 1), as well as the teaching tasks with the use of ICT where the students deploy the necessary operations and procedures in the modeling of problems of the subject.

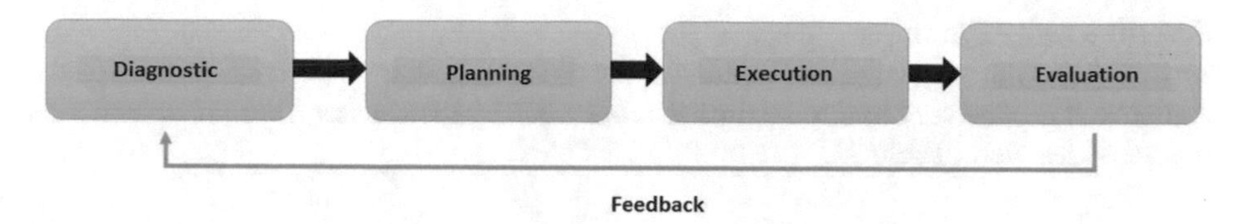

Fig. 1. Stages of the strategy

3.1 Strategy Actors

There are two fundamental actors: the teacher who implements the strategy and the students. The first is in charge of diagnosing the needs of the TLP in order to plan and orient the activities according to the needs detected. It must also prepare students for the use of ICT, specifically for work in the virtual learning environment (EVEA for Spanish initials), as well as motivate and encourage the active participation of the student. Among other functions it must ensure the correct evaluation and control of the activities, as well as make any adjustments in the implementation of the strategy. The students execute the actions and tasks guided by the teacher, maintaining an active action during all the development of the activities. In addition, they must be able to evaluate themselves from the development in the realization of the tasks, as well as to verify solutions from the practical guide of the teacher mediated by the technology.

3.2 Step-by-Step Strategy Actions

This section describes the actions that are conceived for each stage of the strategy, and suggests some ideas for their practical execution. Actions can be enriched, based on the principle of flexibility, from the creativity and experience of the teachers who put it into practice.

First Stage - Diagnosis

This stage is the starting point of the strategy and its main objective is to determine the real state of the TLP. The potentialities and difficulties of the students must be identified in the contents preceding the modeling of problems. The good execution of this stage makes it possible to guide in an adequate and precise way the actions of the teacher and the students when conceiving, planning and directing the TLP (Table 1).

Table 1. Actions carried out by the teacher and the student in the first stage

Teacher actions	Student actions
– Analyzes the governing documents – Elaborates and analyzes the pedagogical test and the initial survey of students – Assesses the results obtained – Diagnose the ways to use educational technologies – Assesses the types of activities to be prioritized in the EVEA	– Actively participates in the conduct of the diagnostic test and the initial survey – Reflect on the results obtained

Second Stage - Planning

In this stage the objective is to prepare the actions that will be carried out by the students. It is necessary to devote time to table work by the teacher to organize and plan the activities well (Table 2).

Table 2. Actions carried out by the teacher in the second stage.

Teacher actions– Definition of indicators to measure the development of modeling skills in linear programming problems – Elaboration and design of teaching tasks – Classify teaching assignments in terms of level of complexity – Preparation of help materials on the work in the EVEA and on the computer tools for the modeling of problems – Selection of the educational resources to be used for each teaching task – Preparation of the necessary resources for the self-learning of the students in the problems of modeling – Organized and coherent planning of activities in the EVEA

Third stage - Execution

The objective of this stage is to apply in practice the actions conceived in the planning stage, which allow contributing to the development of the ability to model problems in the TLP of OR (Table 3).

Table 3. Actions carried out by the teacher and the student in the third stage.

Teacher actions	Student actions
– It guides the students in the development of the activities – Provides different levels of assistance to students – Place in the EVEA short audiovisuals with the solution of modeling problems – Coordinate and have time in the forum to exchange solutions with students – It attends to students in a personalized way through videoconference – Review and monitor the activity of each student in the EVEA – Assesses the performance of students in the execution of activities to issue an evaluation	– Actively participates in the realization of the activities guided by the teacher – Request levels of help through the forum – Review the modeling exercises through short audiovisuals – Ask the teacher for personal attention – Evaluates the performance of other students in solving activities – Proposes to the teacher a self-evaluation of his performance in the activity

Fourth Stage - Evaluation

The objective of this stage is to evaluate the level of effectiveness of the actions implemented as part of the didactic strategy to develop the ability to model problems.

This evaluation must be systematic and continuous, so that the difficulties presented in the process can be perfected and corrected, depending on the fulfillment of the stated objective (Table 4).

Table 4. Actions carried out by the teacher and the student in the fourth stage.

Teacher actions	Student actions
– Reflect and exchange with students about the actions developed – It detects the main deficiencies and corrects them – Performs frequent evaluations – Design and apply partial and final evaluations – Design and apply survey to students	– It issues criteria on the actions of the strategy – Performs frequent, partial and final evaluations – Responds to a survey, offering ratings on the strategy

4 Discussion

The assessment of the proposed strategy was carried out through the implementation of a pre-experiment, and the application of Iadov's test to verify the degree of student satisfaction.

4.1 Evaluation of the Results Based on the Pedagogical Pre-experiment

The main purpose of the pre-experiment was to assess the effectiveness of the didactic strategy to develop the ability to model problems in the OR subject with the use of technologies. The same was done with the teaching group that was receiving the OR course, taught by the author of the research in the academic year 2017–2018, so it can be considered as a natural pre-experiment.

Diagnostic Stage
From the review and analysis of the guiding documents of the OR course, it was identified that the development of the ability to model problems contributes to the modes of action proposed in the Professional Model. In order to obtain a vision of the didactic treatment given to the modeling of problems of linear programming in the subject and to be aware of the most frequent difficulties presented by students when modeling the proposed problems, a survey was applied to 9 professors who teach the subject, and to 32 students in the third year of Faculty 2. In addition, a documentary review of examinations in the last academic years was carried out in order to identify the main difficulties presented by the students.

The results of the processing and analysis of the applied instruments, allowed a correct planning of the actions in the following stages of the strategy, in order to fulfill its objective.

Planning Stage
At this stage, a set of indicators was defined to measure the development of the ability to model linear programming problems achieved by each student. These are grouped

into three fundamental dimensions that respond to the ability to model linear programming problems: model preparation, model formulation, and model testing.

This was followed by the development of a set of teaching assignments for use in ICT-assisted TLP. The tasks were classified by level of complexity, from the simplest to the most difficult for the students to understand.

In order to prepare students for work in an environment of educational technologies, the use of different collaborative tools was oriented to carry out initial familiarization tasks, not only with the contents that were assimilated, but also with the computer tools themselves. The students appropriated the previous knowledge and skills necessary to develop the experimentation.

Based on identifying the resources that are most accepted by the students, in addition to being better dominated by them, each task is associated with the different resources. Other resources were then prepared so that the students could evaluate themselves in solving the modeling problems. In this case the most used technology was the audiovisual. Each problem was divided into parts according to the different situations to be modeled, and for each one a video of how the teacher executes the modeling was made. Finally, the virtual learning environment was used to place all the teaching tasks previously associated with educational resources.

In addition, a collection of modeling problems was elaborated to be solved in the EVEA, taking into account different degrees of complexity, and that the context of the exercises was related to social demands.

Execution Stage

From the orientations offered by the teacher the students were carrying out the tasks in the EVEA in a systematic way, during the teaching of the subject. At the beginning, the students had difficulties in carrying out the tasks that were oriented to them, mainly due to the insufficiencies they had in aspects related to the previous mathematical contents and sometimes due to lack of knowledge of how to interact with the platform.

A good discussion environment was provided at all times in certain situations regarding the different criteria of the students for the modeling of the same problem, as well as co-evaluation and self-evaluation. For each task, a time was spent in the analysis of the answer, in addition, with the audiovisuals the students checked the solution of the problems of modeling, and to offer different levels of help the forum was used. Even when more personalized attention was needed, a videoconference or consultation was arranged.

In general, a change of attitude was observed in the students in front of the modeling of problems of linear programming, feeling more motivated by the activities of the subject.

Evaluation Stage

At all times, reflections prevailed, starting from qualitative criteria to quantitative ones, but in such a way that everyone accepted their difficulties and assumed them as a challenge to overcome them. Respect for the criteria emitted by other students was encouraged, which determined the presence of criticism and self-criticism.

Evaluations of the EVEA itself, those issued by other students, and the self-evaluation of each were taken into account. Frequent classroom assessments were also conducted to measure students' level of development. A partial test was done on the topic of modeling, and it was also evaluated in the final exam. Evaluations were made with the students on the strategy, applying a survey, which served to determine the level of satisfaction.

Figure 2 shows a comparative graph with the results achieved in the partial test in which the ability to model was evaluated, where it can be seen that the results in the pre-experiment group were much higher with respect to the rest of the groups:

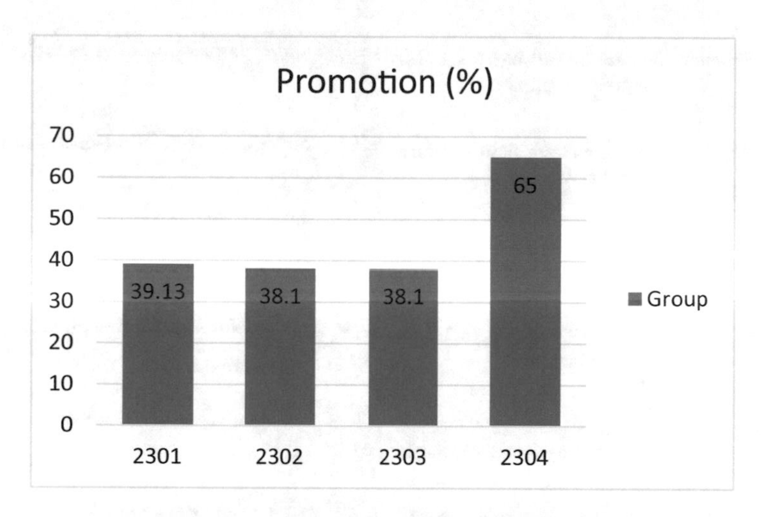

Fig. 2. Comparison of promotion results among teaching groups

In the final exam the skill could be measured through a question. On this occasion the results are compared with those of group 2303, the results of the group in which the didactic strategy was applied being superior. The results were accounted for through the indicators defined at the planning stage (Fig. 3). The information provided indicates the proportion of students in percentage grades who are placed on the different scales defined for each indicator. The percentage difference between the average of those evaluated in the high scale for each indicator of group 2304 (84.8%), with respect to this average in group 2303 (52.3%), amounts to 32.5%, which is considered very significant.

Considering that the results in the pre-experiment group were better with a large percentage margin in most of the indicators, it can be concluded that the proposed strategy contributes to the development of problem modeling skills in students of the university.

4.2 Iadov Satisfaction Test

In order to know the degree of satisfaction of the students after having applied the didactic strategy, Iadov's satisfaction test was carried out, which in its original version was created for the study of satisfaction in pedagogical careers. After applying Iadov's logical framework to each of the respondents, the following results were obtained (Table 5):

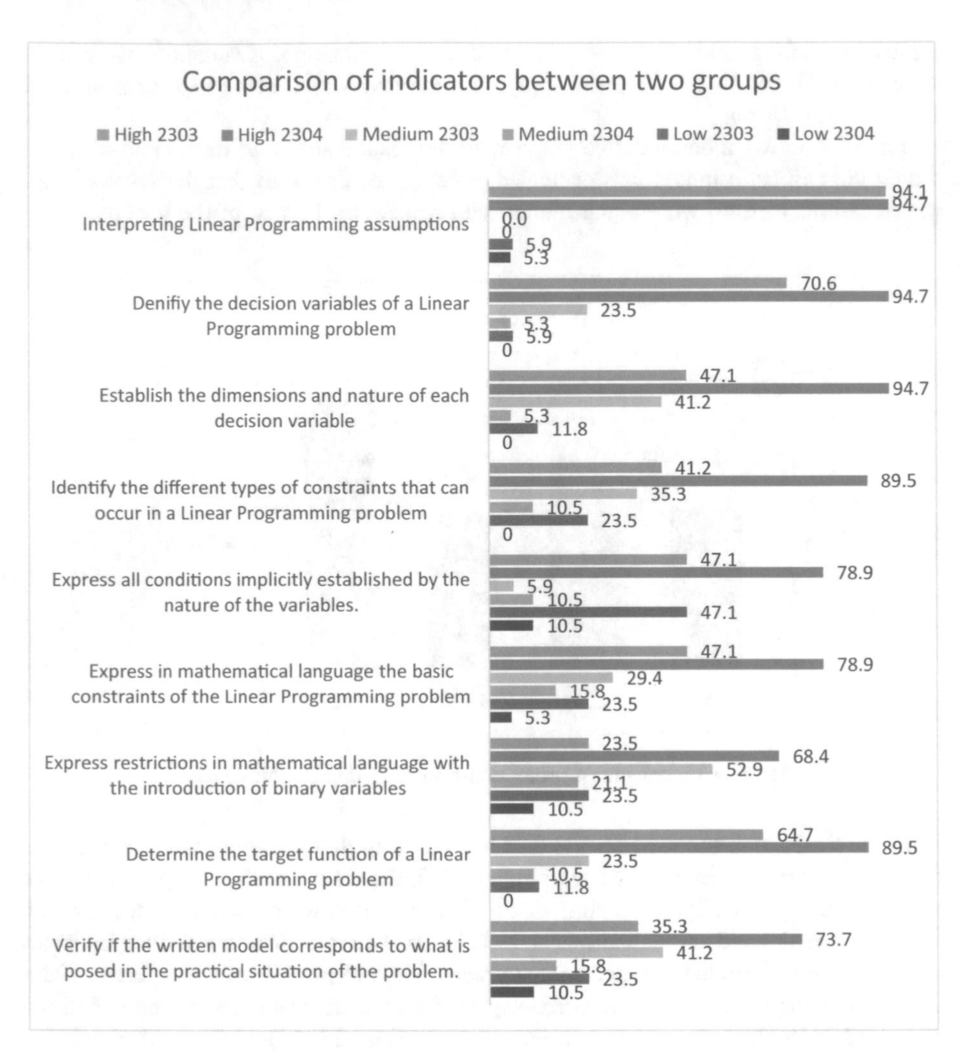

Fig. 3. Comparison of the level of development of the modeling skill for each indicator between two groups (2304 and 2303).

Table 5. Number of students per satisfaction scale

Satisfaction scale	Respondents: 19	
	Quantity	%
Maximum satisfaction	13	68.4
More satisfied than dissatisfied	4	21.1
Not defined	2	10.5
More dissatisfied than satisfied	0	0
Maximum dissatisfaction	0	0
Contradictory	0	0

In view of this result, the present investigation has to,

$$ISG = \frac{13(1) + 4(0,5) + 2(0) + D(-0,5) + E(-1)}{19} = 0,789$$

It is evident that there is a high level of satisfaction in the group where the pre-experiment took place. In the satisfaction survey there were also other questions that helped to know some considerations about the TLP of the OR subject. The most important and majority criteria are set out below:

- The students consider the audiovisual as the most used resource that motivated them and helped in a significant way in the development of skills for problem modeling.
- The students reflect that the modeling exercises were related to practice, and that their precise description, as well as their disaggregation from small to high complexity, helped them to execute a solution plan.
- Among the main elements for which students say they like to use technologies in learning are: they feel motivated and attracted, they can study anytime and anywhere, they have an immediate response to their concerns and doubts, they achieve greater concentration, and they can listen to the teacher's explanation as many times as they need it.
- Among the aspects to be improved in the subject, according to the students, are: that videos are also used for other subjects besides modeling, that more practical classes are given, and that the exams can be taken with the software.

The theoretical contribution is in the characterization that is done on the ability modeling of problems, the use of the technologies for the development of cognitive abilities, as well as of the current state of the subject OR as far as these edges are concerned.

The practical contribution is given by the set of actions that were defined in the implementation of the strategy in order to contribute to the development of the ability to model problems in the OR subject, in addition to a collection of exercises of modeling problems of linear programming in correspondence with the demands of the master strategy of ideological political work of the university as for the approach of the content of the texts, to contribute in the formation of values and of an economic culture in the students. It also makes available to the faculty various educational resources such as tasks and audiovisuals for the subject, which are the result of applying the didactic strategy, as well as a software for solving problems of linear programming to test the models.

The current state of research is given by the growing demand to train professionals who are capable of taking the problems from reality to a model, in order to apply methods for the search for solutions. It is based on the need to transform the learning style of students in the current university context, to develop skills progressively with an intensive use of technologies.

5 Conclusions

As a result of the study carried out, the general objective defined is fulfilled, for which reason the author of this research paper considers the following:

- The theoretical-methodological references of the investigation, support the importance of the development of the ability modeling of problems with the use of the technologies.
- The analysis of the results from the characterization of the modeling ability in the TLP of OR, allowed to verify that the students present a low level of development of the ability, and that it is necessary to elaborate a strategy to transform the current state of the process.
- The analysis of the object of study allowed the foundation and design of a didactic strategy with the use of technologies to contribute to the development of the ability to model linear programming problems from the OR subject in students.
- The validation of the results of the didactic strategy, through a pre-experiment and the application of Iadov's satisfaction test, made it possible to verify that the use of technologies in the TLP of the OR subject, contributes to develop in the students the ability to model problems, making it possible for the student to perform better when he or she enters working life as a professional.

References

Araújo, J.D.: A socio-critical approach to mathematical modeling: the perspective of critical mathematical education. Revista de Educaçao em Ciência e Tecnologia (2009)

Bassanezi, R., Biembengut, M.S.: Mathematical modeling: an old form of research - a new teaching method. J. Didactics Math. (1997)

Berrío, M.d.: Elements that intervene in the construction that students do as opposed to mathematical models. The case of coffee cultivation. Medellín (2011)

Biembengut, M., Hein, N.: Mathematical modeling as a research method in mathematics class. V International Mathematics Festival, Costa Rica (2006)

Borromeo Ferri, R.: On the influence of mathematical thinking styles on learners´s modeling behavior. J. Math. Didakt. 31, 99–118 (2010)

Bossio, J.L.: A mathematical modeling process from a situation in the context of banana cultivation with tenth grade students by generating linear models. Medellín (2014)

Hein, N., Biembengut, M.S.: Mathematical modeling as a research method in mathematics classes. In: Proceedings of the V International Mathematics Festival, Murillo, M (Chairman) (2006)

Huapaya Gómez, E.: Modeling using quadratic function: teaching experiments with 5th grade high school students. Pontificia Universidad Católica del Perú (2012)

Muñoz Mesa, L.M., Londoño Orrego, S.M., Jaramillo López, C.M., Villa-Ochoa, J.A.: Authentic Contexts and the production of school mathematical models. Universidad Católica del Norte Virtual Magazine (2014)

Nieto, M.: The role of basic sciences in engineering education. Congress on Engineering Education, Mexico (2004)

Suárez, L., Cordero, F.: Theoretical elements to study the use of graphs in the modeling of change and variation in a technological environment. Electron. J. Res. Sci. Educ (2010)

University of Informatics Sciences.: Model of the Professional Engineer in Informatics Sciences. Havana (2014)

Valiente Mesa, R.: Didactic strategy with the use of participative methods to contribute to the development of the ability to solve problems from the subject operations research in the students of the career of engineering in informatics sciences. Havana (2015)

Villa-Ochoa, J., Bustamante Quitero, C.A., Berrio Arboleda, M.D., Osorio Castaño, J.A., Ocampo Bedolla, D.A.: Sense of reality and mathematical modeling: the case of alberto. revista de educaçâo em ciência e tecnologia. (2009)

Assessing Digital Skills and Competencies for Different Groups and Devising a Conceptual Model to Support Teaching and Training

Xhelal Jashari[1(✉)], Bekim Fetaji[2], Alexander Nussbaumer[1], and Christian Gütl[1]

[1] Graz University of Technology, Graz, Austria
xhelal.jashari@student.tugraz.at,
{alexander.nussbaumer,c.guetl}@tugraz.at
[2] Mother Teresa University, Skopje, North Macedonia
bekim.fetaji@unt.edu.mk

Abstract. The assessment of digital skills and competencies required to make use of digital resources poses several challenges. In order to evaluate the current situation and trends, this research study initially presents a review of history, existing trends and digital competence frameworks, learning and teaching methodologies for digital literacy and provides insights and recommendations. The review of existing framework and methods used to assess digital skills of different groups has revealed that self-assessment is an assessment category that most often results in an overestimation of own digital skills by respondents. To address such deficiencies, a digital competency model (DigComp) was created in 2013, as part of the EU Education and Training Agenda 2020. This model describes which skills and competencies are needed to use digital technologies in a confident, critical, collaborative and creative way and to accomplish the goals related to work, learning, and leisure in a digital society. Starting from the existing framework, this research study devises a conceptual model to support teaching and training. Each user group requires a different approach in order to accurately evaluate their digital competences; therefore, the use of a flexible and integrated approach is necessary. The model proposed in this study describes which skills should be learned, how they should be taught, and how they can be assessed.

Keywords: Digital skills · Digital competencies · Assessment · DigComp framework · Conceptual model

1 Introduction

Nowadays data, information and knowledge have been widely digitalized which makes it easier to be accessed by almost everyone. According to [2] in 2019 there are more than 4.3 billion internet users, including 3.4 billion active social media users and approximately 3 billion mobile social media users. With an increase of 3.1% from the

M. E. Auer and D. May (Eds.): REV 2020, AISC 1231, pp. 982–995, 2021.
https://doi.org/10.1007/978-3-030-52575-0_82

previous year, there are more than 2.8 billion people purchasing goods via e-commerce, where by penetration of consumer goods is 37% of total population.

To acquire and make use of digital resources, one must possess the appropriate skills and digital competencies needed. However, it has also been observed that there is a deficiency of those competencies. In order to assess the current situation, a literature review of current trends and situations have been performed, which revealed that some people cannot adequately assess their digital skills. Respondents incorrectly evaluate their competences and usually over-estimate their skills. For example, in Austria, 94% of survey participants assessed their general computer skills as 'average' to 'very good'. However, in the practical test, only 39% of them scored on the presumed level [1]. Obviously, the method of self-assessment of the digital skills and competencies realized currently is not appropriate. As part of EU Education and Training Agenda 2020, in 2013, a digital competence model DigComp was created. This model is a framework for developing and understanding digital competency in Europe, whose aim is beside other things to create a common understanding of digital competence at European level, that would promote social inclusion. DigComp describes which skills and competencies are needed to use digital technologies in a confident, critical, collaborative and creative way thus accomplish the goals related to work, learning and leisure in digital society. In order to do a thorough analysis of these processes, this research study has been focused on a literature review of current trends and situations related to the assessment of digital skills and competencies. Based on the review performed, issues were identified and insights and recommendations have been stated, discussed, and argued. The assessment of the digital skills cannot be performed through a single component, and be tested with just one type of assessment, and this is crucial in this process. Therefore, a flexible and integrated approach is quite necessary.

The aim of this study is to explore the required digital skills and competencies for specific groups as youth, vocational training students, and elderly people, and devise a conceptual model to support teaching and the training. It is based on the literature review of current trends and situations used for the assessment of digital skills and competences.

This paper is structured as follows. It starts with the introductory section, followed by Sect. 2 that reports on the history, current trends of the digital competence and related aspects. Section 3 summarizes three digital competence frameworks that have been considered throughout the paper. Section 4 consists of learning and teaching methodologies for digital literacy. Section 5 considers the instruments used to assess digital literacy and competence. Section 6 provides a conceptual model. Finally, Sect. 7 concludes the paper.

2 Background and Related Work

In 2006, recommendations on the crucial competences for lifelong learning that included digital competence were published by the European Parliament and Council. The recommendations implied that competence was "a combination of knowledge, skills and attitudes", stressing out the fact that key competences are those "which all individuals need for personal fulfilment and development, active citizenship, social inclusion and employment" [3].

In the Communication provided by European Parliament and Council on the Key Competences for Lifelong Learning, the following definition of digital competence was proposed: *"Digital competence involves the confident and critical use of Information Society Technology (IST) for work, leisure and communication. It is underpinned by basic skills in ICT: the use of computers to retrieve, assess, store, produce, present and exchange information, and to communicate and participate in collaborative networks via the Internet"* [3]. On the other hand, according to [4] digital literacy is defined as the ability to access, manage, understand, integrate, communicate, evaluate and create information safely and appropriately through digital technologies for employment, decent jobs and entrepreneurship. It includes competences that are variously referred to as computer literacy, ICT literacy, information literacy and media literacy.

According to [3] The concept of 'digital competence' (first acknowledged as a Key Competence in December 2006, Official journal of the European Union) has been described as a human right: "*A multifaceted moving target, covering many areas and literacies and rapidly evolving as new technologies appear. Digital competence is at the convergence of multiple fields. Being digitally competent today implies the ability to understand media (as most media have been/are being digitalized), to search for information and be critical about what is retrieved (given the wide uptake of the Internet) and to be able to communicate with others using a variety of digital tools and applications (mobile, internet). All these abilities belong to different disciplines: media studies, information sciences, and communication theories*".

3 Digital Competence Frameworks

Nowadays an increasingly higher number of governmental and non-governmental organizations and companies are creating different frameworks that are used to describe, categorize, and enhance digital skills, literacy and competencies. Below, we have summarized three Digital Competencies' Frameworks: The European Commission's Digital Competence Framework; UNESCO's Digital Literacy Global Framework (DLGF); and the Digital Intelligence (DQ) Framework.

DigComp Framework

The European Digital Competence Framework for Citizens, also known as DigComp, is used as a tool to describe, categorize, and improve citizens' digital competences. According to [6] DigComp 2.0 Framework competences are divided into the following 5 areas as shown in Table 1.

Table 1. DigComp competence areas [6]

Digital competence area	Competencies
1. Information and data literacy	1.1 Browsing, searching and filtering data, information and digital content 1.2 Evaluating data, information and digital content 1.3 Managing data, information and digital content
2. Communication and collaboration	2.1 Interacting through digital technologies 2.2 Sharing through digital technologies 2.3 Engaging in citizenship through digital technologies 2.4 Collaborating through digital technologies 2.5 Netiquette 2.6 Managing digital identity
3. Digital content creation	3.1 Developing digital content 3.2 Integrating and re-elaborating digital content 3.3 Copyright and licenses 3.4 Programming
4. Safety	4.1 Protecting devices 4.2 Protecting personal data and privacy 4.3 Protecting health and well-being 4.4 Protecting the environment
5. Problem solving	5.1 Solving technical problems 5.2 Identifying needs and technological responses 5.3 Creatively using digital technologies 5.4 Identifying digital competence gaps

UNESCO Global Framework of Reference on Digital Literacy Skills

UNESCO proposed a framework that included seven competency areas. Through the mapping process that analyzed 20 existing frameworks in fourteen countries from four different regions, it also identified two areas of digital literacy competences in addition to the existing DigComp 2.0 framework. It also proposed the addition of one competence under area 5, Problem solving [4].

According to UNESCO proposed competence areas and competences for the Digital Literacy Global Framework that should be included in addition to the competencies in DigComp 2.0 are (see Table 2):

Table 2. UNESCO proposed competence areas and competences [4]

Digital competence area	Competencies
0. Devices and software operations	0.1 Physical operations of digital devices 0.2 Software operations in digital devices
5. Problem solving	5.5 Computational thinking
6. Career-related competences	6.1 Operating specialized digital technologies for a particular field 6.2 Interpreting and manipulating data, information and digital content for a particular field

Competence Area 0 - Devices and software operations result from the consultation showed general agreement that a digital literacy framework should include competences needed for basic operations of devices and software, particularly in the context of low-income and developing countries. The proposed addition of CA0 (Devices and software operations) comprises two additional competences, 0.1 Physical operations of digital devices and 0.2 Software operations in digital devices.

Competence Area 5 - DigComp 2.0 (solving technical problems) also involves the operation of devices and software, while competence 5.5 refers to higher levels of understanding and skills involved in the problem-solving process.

Competence Area 6 - Career-related competences frameworks which target adults include career-specific competences, such as the use of digital technology in engineering (e.g. computer-aided design/computer-aided manufacturing equipment) and in education (e.g. use of Learning Management Systems). Career-specific knowledge, skills and attitudes should be included depended on the specific socioeconomic context, which also change over time. CA6, Career-related competences, is an addition to DigComp 2.0, comprising competences 6.1 (Operating specialized digital technologies for a particular field) and 6.2 (Interpreting and manipulating data, information and digital content for a particular field). One advantage of the openness and flexibility for this competence area is that countries can identify the competences required for economic growth and development in targeted fields and in specific contexts.

Digital Intelligence (DQ) Framework (Global Standards for Digital Literacy, Skills, and Readiness)
The Digital Intelligence (DQ) was formed by a coalition of a cross-sector cooperative network of organizations comprised of Organization for Economic Cooperation and Development (OECD), the IEEE Standards Association, and the DQ Institute in association with World Economic Forum and launched on September 26, 2018 [7].

Digital Intelligence (DQ) is a comprehensive set of technical, cognitive, metacognitive, and socio-emotional competencies grounded in universal moral values that enable individuals to face the challenges of digital life and adapt to its demands. The DQ Framework offers a holistic set of digital competencies with a systematic structure as a reference framework. The aim is to enable any organization to adopt the DQ Framework, and to be able to practically tailor the framework to meet their needs.

The DQ Framework is structured around two categories: “areas” and “levels” of digital intelligence. Eight broad areas of one’s digital life have been identified: Digital Identity, Digital Use, Digital Safety, Digital Security, Digital Emotional Intelligence, Digital Communication, Digital Literacy, and Digital Rights. The competencies within these eight areas can be further differentiated by three different “levels” of maturity - Digital Citizenship, Digital Creativity, and Digital Competitiveness.

4 Learning and Teaching

Learning and teaching for digital literacy consist of the ability to access digital media and ICT, to understand and critically evaluate different aspects of skills: to create and access contents, and to communicate effectively in a variety of contexts. Main elements of learning and teaching that the learner will acquire at the end of the process are categorized in three important categories: knowledge, digital skills and competencies (see Table 3).

The Knowledge that the learner will obtain can be theoretical which involves different concepts such as; scientific concepts, mathematical concepts, theorems and axioms. The knowledge obtained can also be factual which is fact-based.

Digital skills that the learner will gain can be cognitive, which is accessed by acquiring knowledge and understanding through thought and reflection. The other form of digital skills can be practical skills which are accessed through applied methods and experience.

The competences are divided into two forms: individual autonomy type and group type. The first form being self-experience and self-directed activities and the latter, group, guided and team-managed learning processes.

Table 3. Main elements of digital competences: knowledge, digital skills and competence type

Main elements	Aspects/Forms	
Knowledge	Theoretical	Factual
Digital skills	Cognitive	Practical
Competence type	Individual autonomy type and socially responsible level	Group (team) type and socially responsibility level

Lifelong learning strategies need to answer to the growing need for advanced digital competence for all jobs and for all learners including the digital natives and older generations and address the gap in between them. Learning digital skills not only needs to be addressed as a separate subject but also embedded within teaching in all disciplines and fields of study, scientific and art. According to [8] Building digital competence by embedding and learning ICT should start as early as possible, i.e. in primary education, by learning to use digital tools critically, confidently and creatively, with attention paid to security, safety, and privacy. Teachers, educators in general need to be equipped with the new digital skills and competence themselves, in order to support this process. There is a need for Digital Skills and Competencies passport for educators as well as for learners.

Educational policies should make sure that digital literacy in its widest sense is included in educational curricula in primary and secondary education. Education should start building digital competences as early as possible in primary education, through learning to use digital tools confidently, critically and creatively.

According to [9] currently, the concept of digital competence is re-shaped by the emergence and use of new social computing tools, which give rise to new skills related to collaboration, sharing, openness, reflection, identity formation and also to challenges such as quality of information, trust, liability, privacy and security.

However, as technologies and their usages evolve, and new knowledge, digital skills and competences arise with them. According to [10] the development of information literacy and critical thinking skills it important that students successfully gain domain knowledge in the designed digital classrooms. The students' information literacy and critical thinking skills should be fostered. Teacher professional development related to digital classrooms is needed in future.

It is often assumed that young people are 'digitally native' whose skills with digital technology far surpass those of their 'digital immigrant' parents and teachers. According to [11] It is true that many young people are confident in using a wide range of technologies and often turn to the internet for information. They seem able to learn to operate unfamiliar hardware or software very quickly and may take on the role of teaching adults how to use computers and the internet.

However, several important qualifications are needed to the 'digital natives' concept. For one thing, digital skills and knowledge are not evenly spread amongst all young people. Their distribution is affected by class, race, gender and nationality, creating a 'participation gap' [12].

According to [13] young people's confidence with technology can also be misleading. Students frequently struggle when applying ICT to research tasks, and teachers sometimes complain of 'copy and paste syndrome'. Students can find it difficult to work out whether information on an unfamiliar website is trustworthy, with many of them relying on their chosen search engine's rankings for their selection of material. According to [14] many have little understanding of how search terms work or of the powerful commercial forces that can result in a particular company being top of the search engine's list.

According to [14] for general education (levels 1 to 4), the same three dimensions are grouped as follows:

(a) knowledge:
    (i) language and communication;
    (ii) mathematics and natural sciences;
    (iii) social functioning;
(b) skills:
    (i) language and communication;
    (ii) mathematics and natural sciences;
    (iii) social functioning;
    (iv) learning;
(c) social competence:
    (i) language and communication;
    (ii) health and the environment;
    (iii) social functioning.

Educators therefore have a crucial role to play in ensuring that students are digitally literate across a set number of dimensions of learning.

5 Assessment and Feedback

In order to assess digital literacy or digital competence, nowadays there are numerous instruments developed by national, regional, international and commercial agencies.

Based on [4] the mapping of cross-national and national ICT and digital literacy frameworks, has found that the competences described in these frameworks can all be mapped to the DigComp 2.0 framework. Therefore, these instruments can be mapped as part of the proposed DLGF.

According to [4] the digital literacy assessment instruments that already exist were developed to serve different purposes, starting from certification until the evaluation of individuals or population groups, to research, and other purposes, and are structured into four major assessment categories (see Table 4).

Table 4. Assessment categories and descriptions [4]

Assessment categories	
1. Performance assessment	Performance assessment requires the individual to demonstrate how he/she performs certain tasks
2. Knowledge-based assessment	Knowledge-based assessment requires the individual to explain how he/she would perform certain tasks
3. Self-assessment	Self-assessments are subjective evaluations of one's own competence and may not really reflect a person's competence in real-life situations
4. Secondary data-gathering and analysis	Secondary data-gathering and analysis may provide some information about competence at the group or population level but not at the individual level

According to [4] performance assessment is used by the International Computer and Information Literacy Study 2013 to investigate, in a range of countries, the ways in which young people are developing computer and information literacy (CIL) to support their capacity to participate in the digital age [15]. It is used for certification or comparison of digital literacy achievements (such as in the ICIL Study 2013 and 2018) [4].

According to [16] the DigComp proposal comprises two different interrelated outputs:

- a framework identifying, for each area, all the related competences, and providing for each competence a general description, descriptors on three levels, examples of the knowledge, attitudes and skills, and examples of applicability for different purposes.
- a self-assessment grid that proposes the areas of Digital Competence and descriptors for three proficiency levels;

The self-assessment grid could also be used as a tool for each of the citizens to tell about their own level of digital competence to the interested third parties and to understand the ways and methods to improve their own digital competence. The self-assessment grid can be used as a communication tool too, because it represents the model in a brief and easy-to-grasp way [16].

According to [16] curricula and initiative developers can use this framework to develop the digital competence of a specific target group. The level of abstraction of the competences that are foreseen in the framework allows stakeholders to refine and specify sub-competences in the terms they consider most appropriate for the target groups or context. The framework could also be used as a reference tool to compare existing frameworks and initiatives, in order to map which areas and which levels are taken into account by a currently existing framework (or certification scheme, or syllabi).

The shell of the DigComp framework is structured in five dimensions. These dimensions reflect different aspects of the descriptors and different stages of granularity.

These five dimensions are:

Dimension 1: Competence areas identified to be part of digital competence;
Dimension 2: Competence descriptors and titles that are pertinent to each area;
Dimension 3: Proficiency levels for each competence;
Dimension 4: Knowledge, skills and attitudes applicable to each competence;
Dimension 5: Examples of use, on the applicability of the competence to different purposes.

Table 5. The overview and differences between DigComp 2.0 and DigComp 2.1. [17]

DigComp 2.0 (year 2016)		DigComp 2.1 (year 2017)	
Competence areas (dimension 1)	Competences (dimension 2)	Proficiency levels (dimension 3)	Examples of use (dimension 5)
Five competence areas	21 competences	Eight proficiency levels for each of the 21 competences	Examples of use of the eight proficiency levels applied to learning and employment scenario in the 21 competences

As shown in Table 5, DigComp 2.1 has eight proficiency levels for each of the 21 competences and provides examples of using them.

DigComp 1.0 Framework had three proficiency levels in Dimension 3 (1. Foundation, 2. Intermediate and 3. Advanced). According to [17] these have now been increased to eight levels in DigComp 2.1. A wider and more detailed range of proficiency levels supports the development of learning and training materials. It also helps in the design of instruments for assessing the development of citizens' competence, career guidance and promotion at work.

Eight proficiency levels for each competence have been defined through learning outcomes using action verbs, following Bloom's taxonomy and inspired by the structure and vocabulary of the European Qualification Framework (EQF). Moreover, each level description contains knowledge, skills and attitudes, described in one single descriptor for each level of each competence (see Table 6); this equals to 168 descriptors (8 × 21 learning outcomes) [17].

Table 6. Eight proficiency levels in DigComp 2.1 [17]

Level	Complexity of tasks	Autonomy	Cognitive domain
L1	Simple tasks	With guidance	Remembering
L2	Simple tasks	Autonomy and with guidance where needed	Remembering
L3	Well-defined and routine tasks, and straightforward problems	On my own	Understanding
L4	Tasks, and well-defined and non-routine problems	Independent and according to my needs	Understanding
L5	Different tasks and problems	Guiding others	Applying
L6	Most appropriate tasks	Able to adapt to others in a complex context	Evaluating
L7	Resolve complex problems with limited solutions	Integrate to contribute to the professional practice and to guide others	Creating
L8	Resolve complex problems with many interacting factors	Propose new ideas and processes to the field	Creating

Programming, as one of the 21 competencies, is thoroughly explained in (see Table 7 below) through a given scenario, as stated on [17]. The level of autonomy of the example is as seen in the Table 6. L4: independent and according to my needs.

Table 7. Learning scenario example [17]

Area	3: Digital content creation
Competence	3.4: Programming
Example:	*Learning scenario:* *Prepare a presentation on a certain topic that I will make to my classmates*
Scenario developed:	*Using a simple graphical programming interface (e.g. Scratch Jr), I can develop a smartphone app that presents my work to my classmates* *If a problem appears, I know how to debug the programme and I can fix easy problems in my code*
Proficiency level	L4

Areas and Competences

According to [16] the areas of digital skills' competences can be summarized as follows:

1. **Information:** identify, locate, retrieve, store, organize and analyses digital information, judging its relevance and purpose.
2. **Communication:** communicate in digital environments, share resources through online tools, link with others and collaborate through digital tools, interact with and participate in communities and networks, cross-cultural awareness.
3. **Content-creation:** Create and edit new content (from word processing to images and video); integrate and re-elaborate previous knowledge and content; produce creative expressions, media outputs and programming; deal with and apply intellectual property rights and licenses.
4. **Safety:** personal protection, data protection, digital identity protection, security measures, safe and sustainable use.
5. **Problem-solving:** identify digital needs and resources, make informed decisions on most appropriate digital tools according to the purpose or need, solve conceptual problems through digital means, creatively use technologies, solve technical problems, update own and other's competence.

Areas 1, 2 and 3 are rather for linear assessment while areas 4 and 5 are more transversal. This means that while areas 1 to 3 deal with competences that can be re-traced in terms of specific activities and uses, areas 4 and 5 apply to any type of activity that is been carried out through digital means.

6 Conceptual Model

This section proposes a conceptual model for teaching and learning digital skills and competences. This model strives for integrating two strands of research in technology-enhanced learning. According to [5] Technology Enhanced Learning (TEL) refers to the support of teaching and learning through the use of technology and can be used synonymously with e-learning.

First, work and frameworks on digital skills and competences as described in the previous sections are taken into account. This work describes which skills should be learned, how they should be taught, and how they can be assessed. The research second strand addresses research and practice of adaptive e-learning and personalization. This strand provides models for learning general domain-related skills, such as mathematics or science.

The conceptual model (see Fig. 1) integrates the digital skills framework with an adaptive learning strategy. Traditional concepts for adaptive learning consist of user models, content and domain models, and tutoring model [18, 20]. The domain model is consisted of digital skills and domain knowledge. The content model includes information about the available content that can be presented to the learner. The user (or student) model includes characteristics and information about the student by whom the content should be adapted. The tutoring model provides an adaptive strategy that calculates learning paths out of the content and user model. An assessment (e.g. knowledge tests) as a strategy that includes self-reporting and formal and informal assessment, before and during the learning process gathers information for the user model. More recent learning analytics initiatives include non-invasive assessment and behavior tracking methods to gather user information that are used for direct feedback to the learner [19].

In order to design a general framework for learning, teaching, and assessing digital skills, we propose to enrich the traditional adaptation concepts with the digital skills framework. The domain content would include either only content for learning digital skills or a combination for digital skills and a different domain to be learned. This allows both thc tcaching of digital skills in combination with or without a different domain. The digital skills framework includes skills to be learned and also has references to content objects. The user model includes information about currently available digital skills, as well as other user characteristics needed for the teaching strategy. The learning and teaching strategy includes a model that selects appropriate content based on current available skills. The learning environment presents the selected content and provides a user interface for the student. The assessment strategy module includes different strategies to assess the current available skills. Based on Sect. 5, it provides knowledge tests, self-assessment, and non-invasive strategies. The results are stored in the user model.

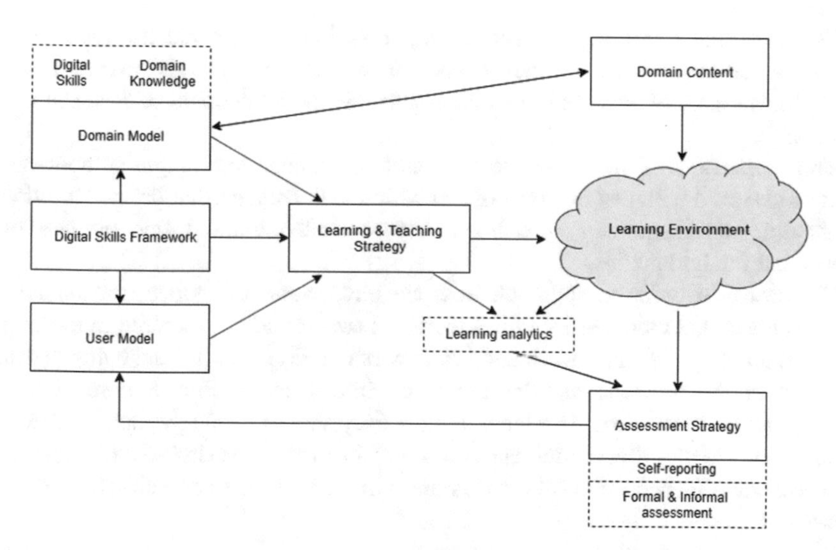

Fig. 1. The conceptual model for learning, teaching, and assessing digital skills and competences.

For exemplifying the conceptual approach, we present a use case that deals with teaching digital skills in primary schools. The goal is to teach basic skills to understand data. To this end, several content modules are created that allow the pupils to interact with weather data. In particular, modules are created that can be used to get an overview of all data, to filter data, to represent data with different graphical diagrams, and to perform simple statistical analyses. Digital skills are defined as skills that are trained with each module. These skills are also related to the DigComp framework. A pedagogical model that provides a story line for the usage of the content modules. For example, a student could start with filtering temperature data within a certain time frame, viewing the filtered data as bar chart and line chart, and calculate the mean value of the filtered data. After such a learning sequence, the pupil is asked to answer some questions about the performed actions. In addition, the log data of the interactions with the modules are analyzed by assessment component. Both assessment types lead to an update of the user model regarding the available digital skills. Based on this update, the teaching strategy can focus on presenting certain content modules to learn missing skills or improve existing ones (e.g. by selecting certain diagram types).

7 Conclusion

Digital competence stands as an important challenge for the educational systems in Europe. From the literature review realized and analyzed frameworks we have come to conclude that there are many issues concerning the assessment of digital skills in EU. Respondents incorrectly evaluate their digital competences, most often by overestimating. This varies and also depends based on the targeted respondents. Each group requires a different approach.

Self-assessment is always a challenging task because people always attempt to conduct a self-assessment but they do not always succeed in this regard. The main reason for the lack of success in self-assessment is the inadequate self-esteem or self-awareness.

The assessment of the digital skills cannot be reduced to a single component, nor can it be assessed with just one type of test which is the real crucial point. The adoption of a flexible and integrated approach is therefore needed, without renouncing to define criteria and methodologies.

Therefore, in order to approach this, we have created a conceptual model. The devised conceptual model strives for integrating two strands of research in technology-enhanced learning. Initially, work and frameworks on digital skills and competencies as described in the previous sections are taken into account. Furthermore, the model describes which skills should be learned, how they should be taught, and how they can be assessed. Finally, this model can contribute in further research to this field which will investigate in detail the differences and similarities between self-assessment and assessment.

References

1. ECDL Foundation: Perception and Reality - Measuring Digital Skills in Europe. ECDL Foundation (2016)
2. Kemp, S.: Digital 2019: Global Digital Overview. Datareportal.com, 31 January 2019. https://datareportal.com/reports/digital2019globaldigitaloverview

3. European Parliament and Council: Recommendation of the European Parliament and of the Council of 18 December 2006 on key competences for lifelong learning. Official Journal of the European Union, L 394/10 (2006)
4. Law, N.W.Y., Woo, D.J., de la Torre, J., Wong, K.W.G.: A global framework of reference on digital literacy skills for indicator 4.4.2 (2018)
5. IGI Global (n.d.): Technology enhanced learning (TEL). https://www.igi-global.com/dictionary/technology-enhanced-learning-tel/29510
6. Vuorikari, R., Punie, Y., Gomez, S., Van Den Brande, G., et al.: DigComp 2.0: The digital competence framework for citizens. Update phase 1: The conceptual reference model. Publications Office of the European Union, Luxembourg (2016). https://doi.org/10.2791/11517
7. DQ Institute: DQ Global Standards Report 2019 Digital Intelligence (2019)
8. CareerBuilder. Press release 10 September 2008. http://www.careerbuilder.com/share/aboutus/pressreleasesdetail.aspx?id=pr459&sd=9/10/2008&ed=12/31/2008&cbRecursionCnt = 2&cbsid = 2ece1ba5ca224298adebb8cdc4994e70-279121665-J2-5&ns_siteid = ns_xx_g_CareerBuilder.com_res. Accessed 5 May 2019
9. Ala-Mutka, K., Malanowski, N., Punie, Y., Cabrera, M.: Active Ageing and the Potential of ICT for Learning. Office for Official Publications of the European Communities, Luxembourg (2008). ftp://ftp.jrc.es/pub/EURdoc/JRC45209.pdf. Accessed 12 June 2019
10. Ala-Mutka, K.: Social Computing: Use and Impacts of Collaborative Content. IPTS Exploratory Research on Social Computing. Office for Official Publications of the European Communities, Luxembourg (2008). http://ftp.jrc.es/EURdoc/JRC47511.pdf. Accessed 7 May 2019
11. ComScore. Press release 10 October 2007. https://www.comscore.com/Insights/Press-Releases/2007/10/UK-Social-Networking. Accessed 5 Mar 2019
12. Cachia, R.: Social Computing: Study on the Use and Impact of Online Social Networking IPTS Exploratory Research on the Socio-economic Impact of Social Computing. Office for Official Publications of the European Communities, Luxembourg (2008). https://publications.jrc.ec.europa.eu/repository/bitstream/JRC48650/jrc48650.pdf
13. Cachia, R., Kluzer S., Cabrera, M., Centeno, C., Punie, Y.: ICT, Social Capital and Cultural Diversity. Report on a Joint IPTS-DG INFSO Workshop. EUR 23047 EN (2007). http://ipts.jrc.ec.europa.eu/publications/pub.cfm?id=1534
14. Chou, C., Chan, P.-S., Wu, H.-C.: Using a two-tier test to assess students' understanding and alternative, conceptions of cyber copyright laws. Br. J. Educ. Technol. **38**(6), 1072–1084 (2007)
15. Fraillon, J., Schulz, W., Ainley, J.: International computer and information literacy study: assessment framework. IEA, Amsterdam (2013). https://www.iea.nl/sites/default/files/2019-04/ICILS_2013_Framework.pdf
16. Ferrari, A.: DIGCOMP: A framework for developing and understanding digital competence in Europe. Publications Office of the European Union, Luxembourg (2013). http://doi.org/10.2788/52966
17. Carretero, S., Vuorikari, R., Punie, Y., et al.: DigComp 2.1: The Digital Competence Framework for Citizens with eight proficiency levels and examples of use. Joint Research Centre. Publications Office of the European Union, Luxembourg (2017). https://publications.jrc.ec.europa.eu/repository/bitstream/JRC106281/web-digcomp2.1pdf_(online).pdf
18. Brusilovsky, P.: Methods and techniques of adaptive hypermedia. J. User Model. User-Adap. Interact. **6**, 87–129 (1996)
19. Siemens, G.: Learning analytics: the emergence of a discipline. Am. Behav. Sci. **57**(10), 1380–1400 (2013). https://doi.org/10.1177/0002764213498851
20. Shute, V., Towle, B.: Adaptive e-learning. Educ. Psychol. **38**(2), 105–114 (2003)

C^2ELT^2S- A Competitive, Cooperative and Experiential Learning-Based Teamwork Training Strategy Game: Design and Proof of Concept

Matthias Maurer[1], Christopher Cheong[2], France Cheong[2], and Christian Gütl[1,3](✉)

[1] Graz University of Technology, Graz, Austria
maurer.matthias@gmx.at, c.guetl@tugraz.at
[2] RMIT University, Melbourne, Australia
{christopher.cheong, france.cheong}@rmit.edu.au
[3] Curtin University, Perth, WA, Australia

Abstract. Playing provides many opportunities for learning as learners can take charge and make choices about their learning. However, since all play is not learning, it makes sense to combine learning and playing activities. In the area of teamwork training, despite the existence of a number of learning and leisure games, there is still room for a holistic approach to design such games. Thus, a game design is introduced based on well-proven and documented aspects of teamwork training. Based on this design, an initial proof of concept prototype is created and evaluated with 18 participants. The results are promising as the relationship between the prototype and teamwork was recognized, the prototypes training and playing potential was confirmed, and even its usability was rated as good. However, a more advanced prototype and artwork and a longer term study are required to further establish the training capabilities of the game design.

Keywords: Teamwork training · Game design · Serious game · Experiential learning cycle

1 Introduction

There is increasing interest in serious games [10, 33] due to their desirable features, especially their engaging and motivating characteristics [62]. However, there are a number of issues with using serious games which are not all fully addressed [62]. These include the over-simplification of concepts and generalization of approaches without consideration of nuances and subtleties in the initial situation which lead to less realistic and accurate results.

A very interesting and, hence, often considered content area for learning games is teamwork. It is an important soft-skill and an important influencing factor in various fields of application and research [32]. This is an especially popular topic in environments where consequences of rarely occurring errors are high, such as health care

M. E. Auer and D. May (Eds.): REV 2020, AISC 1231, pp. 996–1015, 2021.
https://doi.org/10.1007/978-3-030-52575-0_83

[39]. In business areas, the success of companies depends on the teamwork capabilities of the team members, as is the case in many STEM focused (Science, Technology, Engineering, and Mathematics) companies [57]. A lot of attention is directed towards the crucial field of team research to examine the complex structures involved.

Due to the interest in teamwork focused serious games, a considerable number of serious games concerning teamwork emerged recently. The majority of these games are designed for teamwork research, not for learning and training teamwork skills [44]. Other game designs, focusing on teaching teamwork, are often described vaguely and lack a strong scientific justification [7], are not accessible [8], or focus only on specific stages during team construction [24]. Serious games focusing on aspects relevant for teamwork training are discussed at conferences such as FGD [25], CSCW [17], CHI [13] and CHI Play.

At this point, it should be stressed that teamwork training is distinct from team building [36]. While team training is skill-focused, team building focuses on social and interpersonal aspects. As a result, teamwork training games and team building games both aim at improving teamwork but they used different approaches and, hence, should not be compared directly.

Although a good start on serious games for teamwork has been made, what is lacking are more well-designed, scientifically-based, and well-documented teamwork training game for various context areas. This motivated us to initiate the C²ELT²S game research and development project. The idea of the game design is grounded on competitive and cooperative game elements, as well as experiential learning as the base of its learning design. Teamwork research is analyzed in order to incorporate the key teamwork aspect into this strategic game design, targeting teamwork beginners. A discussion of related work, concept and design details can also be found in [44].

In this paper, we first analyze the related work in this field. Then, the proposed serious game design for teamwork training is introduced, followed by a proof of concept and the related evaluation. We conclude our paper by discussing future work.

2 Background and Related Work

In this section, we discuss three categories of related work, namely teamwork research, serious games for skill training, and serious games for communication skills.

2.1 Teamwork Research

There are a number of theories and frameworks introduced to model teamwork, each of them targeting and explaining the topic from a different angle. Well known ones are concerned with team processes, team knowledge, adaptive team performance, team roles, and teamwork components.

Team processes describe team behavior over time, namely the three main processes of (1) transition, (2) the action process, and (3) the process associated with interpersonal matters [42]. Alternative frameworks can also be found, of which one of the most accepted was introduced by Tuckman [59]. He introduced four sequential processes through which groups progressed: forming, storming, norming, and performing.

Team knowledge is a very broad topic [15, 16], defined as the assortment of the task- and team-related knowledge held by teammates, and their collective understanding of the situation confronting the team.

Adaptive performance in teams, describes the need for teams to adapt to changing circumstances [52]. This adaptation is especially important for teams operating in dynamic environments, to keep the plan for reaching the team's objective up to date. It is modeled using different elements, such as individual team members' characteristics, an adaptive assessment-planning-learning-feedback cycle, and a change of team information.

The theory of team roles was first rigorously investigated by Belbin [6] who describes the different behavioural tendencies for different persons on a team. This idea was picked up by Mathieu et al. [43], introducing a classification similar to the one from Belbin, but related to the measures of the well researched 'Big 5' personality construct [4].

Components of teamwork are defined as the basic elements of which teamwork is composed, which also separates individual tasks from teamwork [15]. When comparing the different frameworks describing components of teamwork [22, 46, 54, 55], five components can be found repeatedly, namely leadership, monitoring, adaptability, orientation, and communication.

2.2 Serious Games for Skill Training

Skill training based on serious games can be found in a variety of areas and discipline, such as health care, and the security sector.

Training for surgery, or other health care interventions where an error might cost a life, are particularly suitable for training methods that include serious games. A review lists seventeen serious games specifically designed to train professionals in medicine, the majority of which are highly relevant to surgical trainees [27], showing the relevance of such a training. Two advantages of this approach are improved surgical operation performance and realistic performance not biased by anatomical variation, compared with practicing on animals [28].

Another health care application of serious games is given with so-called exergames [30]. These kinds of games bring physical, social, and cognitive benefits to patients [58], but also mental issues can be addressed with exergames. For example, it has been shown that an exergame intervention was able to achieve significant improvements in depressive symptoms, mental health-related quality of life, and cognitive performance [53].

Another high-risk application area, where lives are at stake, is fire safety. It is difficult to provide the required skills training using traditional methods; however, a serious game approach brings along several advantages [14]. The biggest of these advantages include the opportunity to receive training in a variety of emergency situations, which is more challenging with traditional methods. A simulation of a fire emergency, for example, can be used due to the unavailability of a comparable and harmless situation in real life, but there are also other reasons for simulating certain training situations, such as high costs.

2.3 Serious Games for Teamwork Skills

There are also serious games with a focus on teamwork introduced in the literature. However, most of these games are designed for teamwork research, not for training teamwork skills [44]. Table 1 lists a selection of teamwork games including their purpose and teamwork focus. The purpose, in this context, is either training which describes the game's aim to convey knowledge to the players or analysis. Games with an analysis purpose are used as tool for team research as they prove useful for this purpose [16]. A clearly smaller share of these games is intended to train teamwork skills; two representative examples are selected and discussed in the following.

Table 1. Selection of teamwork games.

Game	Purpose	Teamwork focus
Gazebo [50]	Analysis	Team performance, contributions, collaboration
Tinsel Town [21]	Analysis	Relationship: team effectiveness and diversity
eScape [9]	Analysis	Influence of individuals to group interaction
Infiniteams [35]	Analysis	Leadership in online environments
DREAD-ED [31]	Analysis	Decision making, communication
GaMeTT [20]	Analysis	Level of presence: virtual team-building
Let's team! [29]	Analysis	Confidence (communication, commitment)
4-C Model [49]	Analysis	Coordination, cohesion, communication, cognition
ColPMan [48]	Analysis	Decision making
Zoom [11]	Analysis	Perceived team cohesiveness
Artemis [40]	Analysis	Motivation, communication skills, efficacy
Crossing Ravine [24]	Training	Team building
SimVenture [8]	Training	Lasting impact of decision making, communication
TeamUp [7]	Training	Not described

One representative, a game for teamwork skills training is TeamUp!, the product of a master's thesis which focuses on the implementation aspect [7]. As such it targets a commercialization. This adventure game requires four players to solve puzzles in order to reach a shared goal. This goal is either to reach a certain area in the game or the next level. The in-game barriers placed in abandoned temples can be overcome by using switches and buttons, whose functionalities are not explained to the players but can be grasped very quickly. The theoretical framework used is not strongly described and it lacks scientific justification as the work is focused on the implementation of the game. The game does not focus on teamwork aspects, it rather focuses on quantitative feedback to allow the players and facilitators to draw their own conclusions.

The last teamwork training game discussed in this section, SimVenture, is an enterprise business simulation intended to promote self-efficacy and employability through teamwork, presentation skills, and greater self-confidence [8]. The players set up, manage and grow a business by making strategic decisions based on the market, competitor research, customer research, and response to performance feedback. SimVenture has been used in a university context in order to investigate the lasting impact of decision making and communication with main focus on engage friendships among students, in order to foster resilience and prevent isolation. Its game design, however, is not accessible, which limit the transferability and makes an evaluation of its features impossible.

As discussed in this section, there are existing serious games that focus on teamwork, including those that focus on teamwork training, however, these games are not described in detail or lack scientific justification regarding their learning approach and teamwork background based on the current literature. To address this, we present our work on a game designed for teamwork training. This game design is based on teamwork research, as well as learning theories proving its potential in a small scale preliminary study.

3 Serious Game Design

In this section, the serious game design of the C^2ELT^2S game is outlined. It is designed to be a game to train teamwork skills of teamwork beginners in a content independent manner. Due to the nature of serious games, this design phase includes multiple disciplines, including the learning content, the learning design, and the game design [19]. Since these three elements of serious game design are interdependent a cyclic design approach seems natural and has been applied, which is outlined in detail in the following subsections. Subsequently, the in-game story is discussed.

3.1 Learning Content Design - Skill Dimensions

Based on discussion in Sect. 2.1, we decided to build upon the components of teamwork: leadership, orientation, monitoring, adaptability, and communication [15, 22, 46, 54, 55]. This is motivated by the fact, that team processes, team roles, and team knowledge rather describe theoretical knowledge about teamwork which can hardly be transformed for practical skill training. Also, adaptive performance in teams is embedded in the teamwork components.

Team leadership is defined by the ability to direct and coordinate the activities of other team members, assess team performance, assign tasks, develop team knowledge, skills, and abilities, motivate team members, plan and organize, and establish a positive atmosphere [55]. The team orientation of a team member is defined by her or his willingness to note and consider other team members' contributions, as well as prioritization of team goals versus individual goals [55]. Mutual performance monitoring is concerned with developing a common understanding of the environment and applying strategies to monitor team members' performance [55]. Feedback is also part of this teamwork component. Adaptability is characterized by the ability to adjust strategies based on new or changing information gathered from the environment [55]. Closed-loop communication is concerned with exchanging information [55]. This especially includes precautions that the information sent is the same as the information received to prevent miscommunication.

As one result of the iterative design approach, four of these five components, all but leadership are used in the presented serious game design. Leadership is excluded, due to the inherent nature of leadership, requiring one team member to behave differently to the rest of the team. This is unsuitable for the symmetric game design of our game focus in terms of responsibilities and operating principles and, hence, we made the game design decision to exclude this aspect of teamwork. This leaves us with the four components of teamwork, namely orientation, monitoring, adaptability, and communication, which represents the final focus to construct the game and serve as a basis for the further serious game design.

3.2 Design for Skill Training

As it can not be assumed that playing such serious game, including a chosen training content, will automatically trigger a learning process, an intentional effort is made to foster learning [45]. Thus, for training skills in a playful environment, we focus on the well-accepted learning approach for this purpose, namely experiential learning [18]. This approach is strongly based on exploratory learning patterns [47] and the constructivist learning theory [26].

Basic requirements for using the experiential learning circle are (1) the presence of a problem, (2) the possibility to apply a solving-strategy to the problem, and (3) time to think about the problem to refine a solving-strategy or find it in the first place [5, 23, 37].

Another aspect appearing in the game design, due to its close relationship to teamwork, is collaborative learning. It is *'an educational approach to teaching and learning that involves groups of learners working together to solve a problem, complete a task, or create a product'* [38]. The close relationship is obvious, when it is compared to teamwork - defined as working in a team, which *'consists of two or more individuals, who have specific roles, perform interdependent tasks, are adaptable, and share a common goal'* [2].

3.3 Game Design

Based on the considerations in Sects. 3.1 and 3.2, our aim is to define the game design for training the team skills orientation, monitoring, adaptability, and communication by considering an experiential learning design. For the game design, we are following the mechanics, dynamics, and aesthetics (MDA) framework [34], which is an often used tool in this context [60].

The MDA framework is based on mechanics, dynamics, and aesthetics. Aesthetics are the anticipated emotional responses of the player when playing the game. These emotional responses are a result of the game's dynamics, as the run time behaviour of the game mechanics acting on player input over time. Game mechanics are particular components of the game, at the level of data representation and algorithms. When designing a game one normally starts with the aesthetics (user experience viewpoint) or game mechanics (game designer viewpoint).

The aesthetics categories we are focusing on in the game design are: challenge (game as obstacle course) and fellowship (game as social framework). These aesthetics translate into a multiplayer game with teams of four. The challenge between the four teams is an important intrinsic source of motivation [41, 56, 61]. Furthermore, playing in a team makes the incorporation of teamwork aspects into the game easier. These considerations are also in line with the strategic game genre [51] and follows the ideas of the of the experiential learning cycle (see Sect. 3.2). This is especially supported by strategic game implementations used for browser games, like Ikariam [1]. The choice for this genre can not be argued scientifically, it is a pure game design decision and grounded in the easy incorporability of the experiential learning cycle and the four teamwork aspects which will be discussed below.

A typical browser-strategy game features basic game mechanics and dynamics, which includes the assignment of (a fixed total number of) work units to collect different resources and spending resources to erect or expand buildings. These buildings grant bonuses and advantages over other players. Strategic decisions on which resources to collect and which buildings to expand decide the game's outcome. This genre allows an integration of the experiential learning cycle and its three requirements. The first requirement, the existence of a problem, is given with the competition among players. For this purpose, a score is introduced, comparing the performances with each other. A strategy needs to be found in order to reach a higher score than the competitors. This strategy can be followed for a certain time and its effects on the score can be evaluated. This means that the problem-solving strategy can be applied to the problem at any time during the game which satisfied the second requirement. Also, the third requirement, having enough time to think about the problem, is met by this approach. This type of game is played over a very long time span, possibly multiple months, for a short amount of time per game session.

Not only the experiential learning circle is supported by the strategic game genre, as argued above, but also the four teamwork aspects which will be shown in the following. The orientation aspect is reproduced by introducing two conflicting scores. The team score is the same for all team members and can be increased by spending resources for team objectives. The individual score is different for the team members and increases if resources are spent for a purpose not contributing to the team objectives. This aspect is shown in Fig. 1.

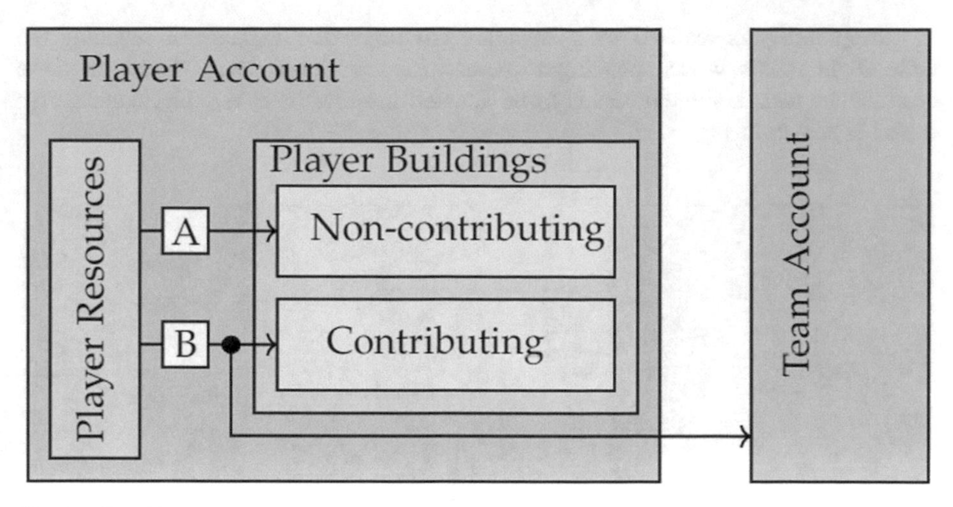

Fig. 1. Graphical representation of the game mechanic 'team versus individual goal'. It is used to evaluate the orientation dimension of a player's teamwork skills. Resources spent on non-contributing buildings (A) and resources spent on either contributing buildings or put towards the team account (B). A simple orientation metric can be calculated as $B/(A + B)$.

Monitoring is considered by introducing player controlled working conditions for workers collecting resources. If a player matches the workers' unobservable desired working conditions with the controllable working conditions, the workers will collect more resources. A team member is able to observe the desired working conditions and give feedback to the concerned player. This aspect is shown in Fig. 2.

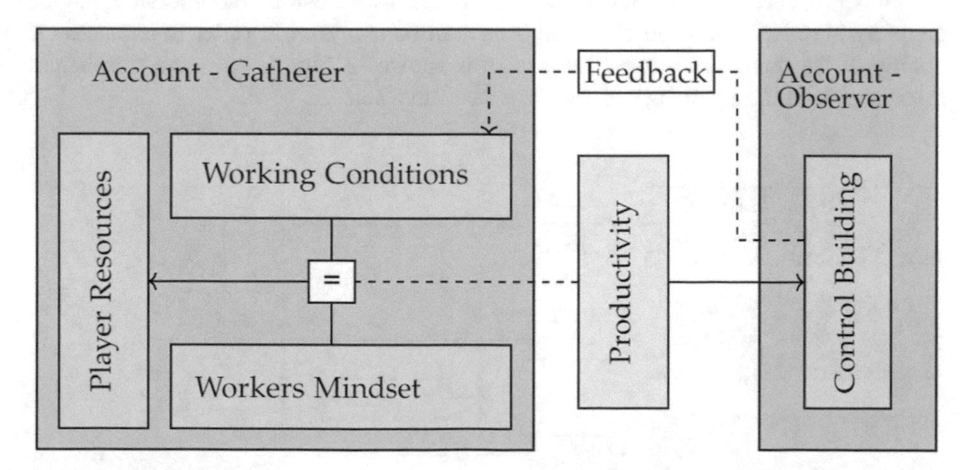

Fig. 2. Graphical representation of the game mechanic 'working conditions'. It is used to evaluate the monitoring dimension of a player's teamwork skills. The productivity depends upon the working conditions and the workers mindset, which need to coincide for a high resource gathering performance. The productivity can be monitored by a team member (observer), which can report the productivity back to the player (gatherer). It can then be adjusted to gain higher productivity. Each player is a gatherer and observer to exactly one team member.

Adaptability is realized by a changing environmental variable, influencing the increase in points when spending resources for team objectives. Based on these changes the team's strategy needs to be adapted in order to gain a high score. This aspect is shown in Fig. 3.

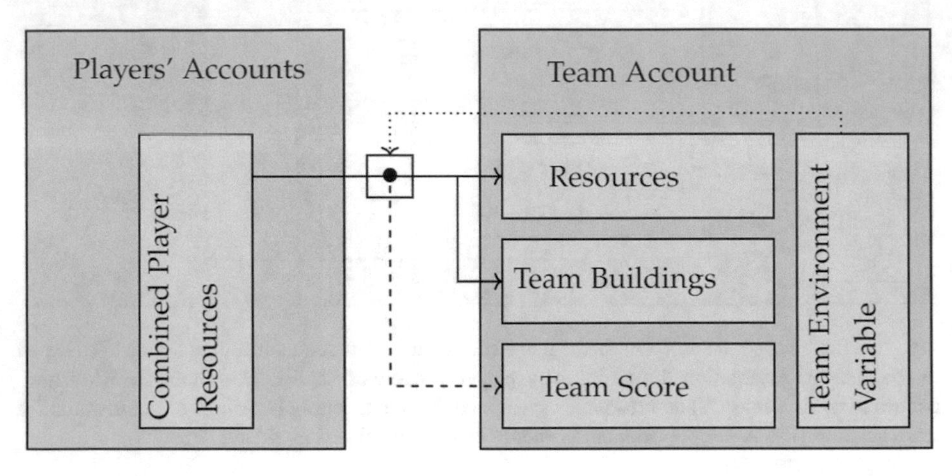

Fig. 3. Graphical representation of the game mechanic 'bonus scores'. It is used to evaluate the adaptability dimension of the team's teamwork skills. The players on a team can transfer resources to the team account, which increases the team score. Depending on the team environmental variable (visible to all players), bonuses are granted in addition to the points received by the team.

The omnipresent **communication** aspect of teamwork can be incorporated into the game by introducing a team chat. This chat can be analyzed in order to evaluate the quality of the communication. This aspect is shown in Fig. 4. All game mechanics associated with the teamwork aspects are listed in Table 2.

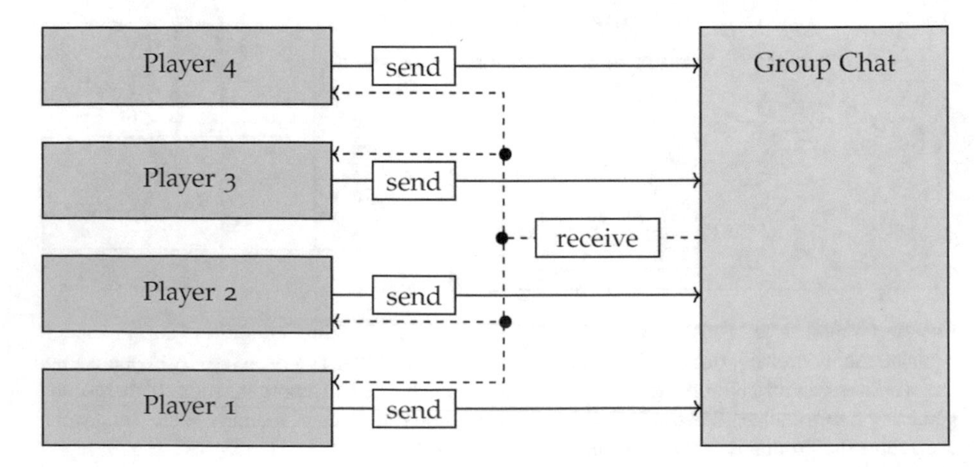

Fig. 4. Graphical representation of the game mechanic 'group channel'. It is used to evaluate the communication dimension of the player's teamwork skills.

Table 2. Game mechanics used.

Component	Game mechanic	Metric
Orientation	Team versus individual goal	Tracking the number of resources spent for the team account and resources spent without contributing to reaching the team goal
Monitoring	Working conditions	Tracking of the times at which the control building is visited. Tracking of sent or missing feedback
Adaptability	Bonus scores	Tracking of the bonuses gained by the team
Communication	Group channel	Tracking the number of words and the number of messages for each player

3.4 In-Game Story

The in-game story is used to wrap the game mechanics, dynamics, and aesthetics into an appealing and reasonable appearance, and is a very central and important game element. Although it is not of primary importance in a strategy game, it can be used to motivate the available in-game actions and provides reasons for them [51]. In this section we will discuss different options for the story, as well as their advantages and disadvantages to show that there are numerous ways to shape a game based on the introduced game design.

The story needs to explain multiple game mechanics, such as resource gathering, building creation, player grouping, the existence of the team account, the individual goal, and the team goal. This list can be supplemented by considering more and more fine details of the game design and depends strongly on the chosen level of detail.

The decision for a specific story can be based on its ability to justify the game mechanics, but also on other considerations, such as its motivating capabilities and its level of related controversy. Adventure stories with pirates might be more appealing in comparison to accounting activities. Controversial stories, including sex, violence, religion, and/or politics need to be handled carefully. Of course, motivation and the level of controversy also depend on the target user group.

A suitable story can be depicted by a religious war among different gods. In this setting, each player is the leader of a town worshiping a certain god. The scenario is set in the past, in a certain culture, such as the Viking, Roman, Egyptian, Greek, Maya or even an imaginary culture. In this context, gathering resources and creating buildings is natural. The teams are grouped based on the god they worship.

Based on the common in-game beliefs of a team, the team goal is to strengthen their god as much as possible. This can be done by building temples for their god and sacrificing resources to them, which explains the team account and the team goal. The individual goal can be defined as becoming the ruler of the most important town.

This in-game story includes religion and war, which can both be considered as controversial areas. The level of controversy is reduced by placing the game in the past and using an outdated religion or even an imaginary one. A certain level of controversy is present in many successful leisure games, such as Assassins Creed, Final Fantasy, and Zelda to name but a few, which may indicate their motivating capabilities.

4 Proof of Concept

A visual proof of concept prototype realizing the elaborated serious game design was created and is described in the following. It is important to emphasise that this prototype as a first work-in-process step towards a fully featured game, containing only the basic game mechanics (leading to dynamics and, in the best case, aesthetics following the MDA framework). As such, it does not implement an in-game story.

4.1 Architecture and Implementation Overview

The prototype is based on a server-client architecture, communicating with network sockets. The server is covering the game logic and dynamic. The client, functioning as the representation layer, can be seen as the acting entity in the game world, following the game world's rules and acting according to them. A short overview is given next, a more detailed description can be found in [44].

The client is used by the player to get a visual representation of the player's current game state. It also allows the player to perform game world manipulations and interactions. These manipulations trigger a message, which is sent to the server to manipulate the game world. The client is also responsible to change its local game state and the visualization based on messages received from the server.

The server stores the game state and provides a player specific version of the game state to each player. It is also responsible for changing the game state and updating the player specific game state when necessary. This is done based on regular events happening in the game and based on game world manipulations done by the player.

The prototype was implemented using Unity3D and Node.js. The game engine Unity3D was used for the game client. For the server component, Node.js (including the packages express, socket.io, and http) was used. The communication between the components takes place by using a JSON notation.

4.2 Player's Viewpoint

In this section, different viewpoints of the implemented proof of concept are shown in order to get a better understanding of the game (see Fig. 5).

The default viewpoint of the player is shown in the top left corner of Fig. 5. It includes an overview of the player's game statistics on top and a representation of the game world including all possessed buildings. By clicking on a building, a new viewport appears as shown in the bottom left corner of Fig. 5. This focus also gives access to the building's information and manipulation methods, allowing the reassignment of the player's workers. It also displays the set working condition of the player's workers.

In the bottom and top right corner of Fig. 5, different states of a building are shown. The big cone represents a certain building, identified by color. Small cones represent the expansion level of the building.

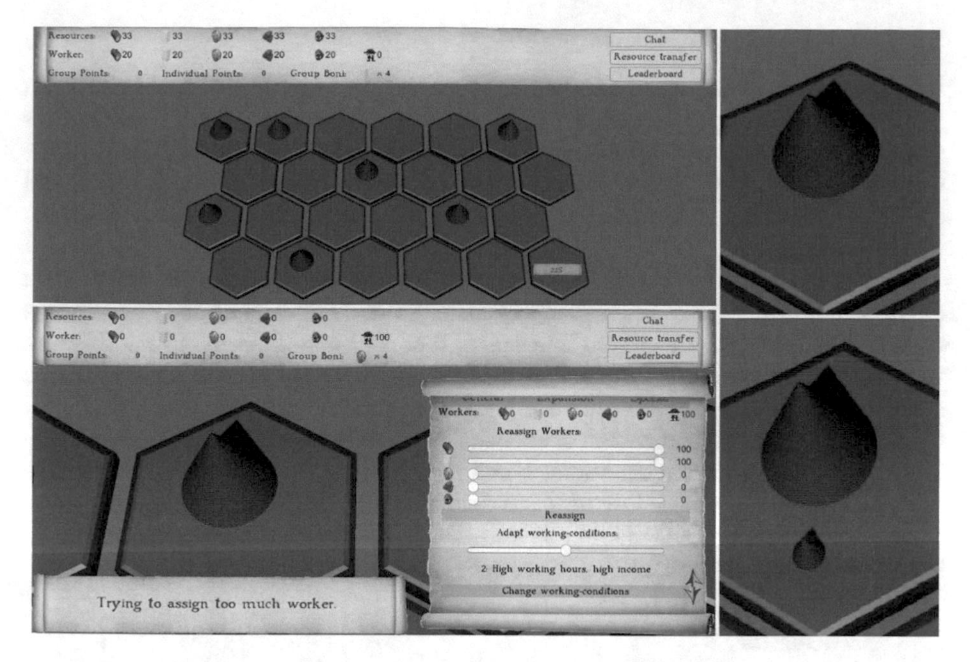

Fig. 5. Different player viewpoints. The default viewpoint of the player (top left corner), the building focused viewport (bottom left corner), and different building states (right-hand side) are shown.

5 Evaluation

This section discusses the evaluation, consisting of research design, settings and instruments, participants of the study, as well, as the study's results.

5.1 Research Design

The study is intended to evaluate the prototype and its game design. We, therefore, focus on three aspects of the prototype, namely its relationship to teamwork, the quality of its user interface, and its potential to be expanded to a fully-fledged serious game for teamwork training.

In order to cover all three aspects, the evaluation is divided into two parts, the expert evaluation and the user experience evaluation. The expert evaluation examines the pedagogical aspects of the prototype. The user experience part of the study evaluates aspects related to the potential for fun and handling of the prototype. The three types of experts considered in this study are experts in the area of learning, computer games, and cognitive research.

5.2 Setting and Instruments

Both parts of the study were conducted in a supervised and controlled environment. At first, the participants engage with the game (watch a video and play it) and then answer questionnaires. One standardized questionnaire, the system usability scale (SUS) questionnaire [12], and three self-created questionnaires focusing on the learning and entertaining qualities of the prototype were used [44]. The procedure differs between the two parts of the study.

The user experience evaluation consists of five parts. First, a demographic data questionnaire is answered. Second, an explanatory video of the game and its features is shown. Third, the game is played in teams of four for approximately 15 min. Fourth, the SUS questionnaire for evaluating the usability is answered. Fifth and last, a concept-specific player questionnaire is answered.

The expert evaluation consists of six parts. First, a demographic data questionnaire is answered. Second, a foreknowledge questionnaire is answered to determine the type of expert. Third, an explanatory video of the game and its features is shown. Fourth, eight tasks are performed in the game. Fifth, a first content-specific expert questionnaire is answered. Sixth and last, a second concept-specific expert questionnaire is answered.

5.3 Participants

The participants of the study were recruited differently, depending in which part of the study they were participating. For the user experience part, students were recruited by advertising the study on different channels. Experts were recruited by specifically requesting their participation.

For the user experience evaluation, three test runs were carried out with four participants for each test run. Four of the participants were in the 18–25 age bracket, and eight were in the 26–30 age bracket. The highest degree or level of schooling completed by the participants included five master's degrees, one bachelor's degree, five higher education certificates, and one secondary school certificate. There were two female and ten male participants.

For the expert evaluation, six test runs were carried out. Each test run included one expert. At least two of them were experts in teaching/learning, at least two of them were computer game experts, and at least two of them were cognitive science experts. Four of them were in the 26–30 age bracket, one in the 36–40 age bracket, and one in the 41–50 age bracket. The highest degree or level of schooling completed by the participants were four master's degrees and two doctorate degrees. There were two female and four male participants.

5.4 Results and Discussion

The SUS score is a linear score yielding values between 0 (poor usability) and 100 (good usability) evaluating the usability of a system. The scores of the twelve participants (user and experts) can be found in the interval between 37.5 and 90 with a sample mean of 73.5 and a sample standard deviation of 14.4. This means that the software has good usability [3].

We now group items, investigating one aspect of the study, into one standardized score, reaching from 0 (not in favour of the prototype/design) to 100 (in favour of the prototype/design). The aspects under investigation are the prototype's relationship to (A1) teamwork in general and (A2) the four teamwork dimensions communication, orientation, monitoring, and adaptability. Furthermore, we investigated the prototype's potential to become a fully-fledged game in terms of (A3) teaching teamwork skills and (A4) entertaining the player. The grouping of items and the mean score can be found in Table 4.

The results of the software-specific expert questionnaire, part 1, and the user experience questionnaire are given in Fig. 6 and Fig. 7. In order to reach a standardized score, each answer to these item gets an integer value, from zero for strongly disagree to four for strongly agree. Only item three of the user experience questionnaire has an inverse scale, meaning that zero stands for strongly agree and four represents strongly disagree. As a result, high numbers can be interpreted as in favour of the prototype or design. Summing up the items of a category and multiplying them with $\frac{100}{ms}$ with ms the maximum possible points as the number of items multiplied with 4.

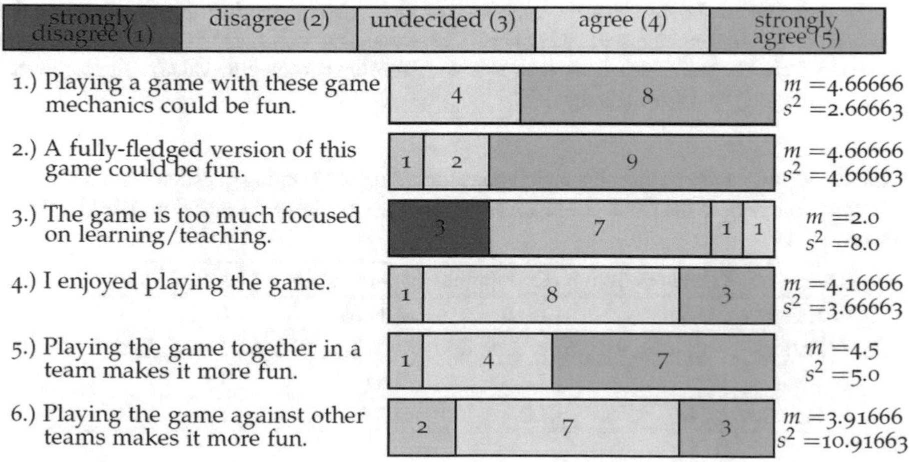

strongly disagree (1) | disagree (2) | undecided (3) | agree (4) | strongly agree (5)

1.) Playing a game with these game mechanics could be fun. 4 8 $m = 4.66666$ $s^2 = 2.66663$

2.) A fully-fledged version of this game could be fun. 1 2 9 $m = 4.66666$ $s^2 = 4.66663$

3.) The game is too much focused on learning/teaching. 3 7 1 1 $m = 2.0$ $s^2 = 8.0$

4.) I enjoyed playing the game. 1 8 3 $m = 4.16666$ $s^2 = 3.66663$

5.) Playing the game together in a team makes it more fun. 1 4 7 $m = 4.5$ $s^2 = 5.0$

6.) Playing the game against other teams makes it more fun. 2 7 3 $m = 3.91666$ $s^2 = 10.91663$

Fig. 6. Answers to the Likert-scale questions of the software-specific user experience questionnaire. Numbers in rectangles represent quantities. The mean (m) and standard deviation (s^2) are calculated for each item. Each item answer is therefore transformed to values ranging from 1 (strongly disagree) to 5 (strongly agree).

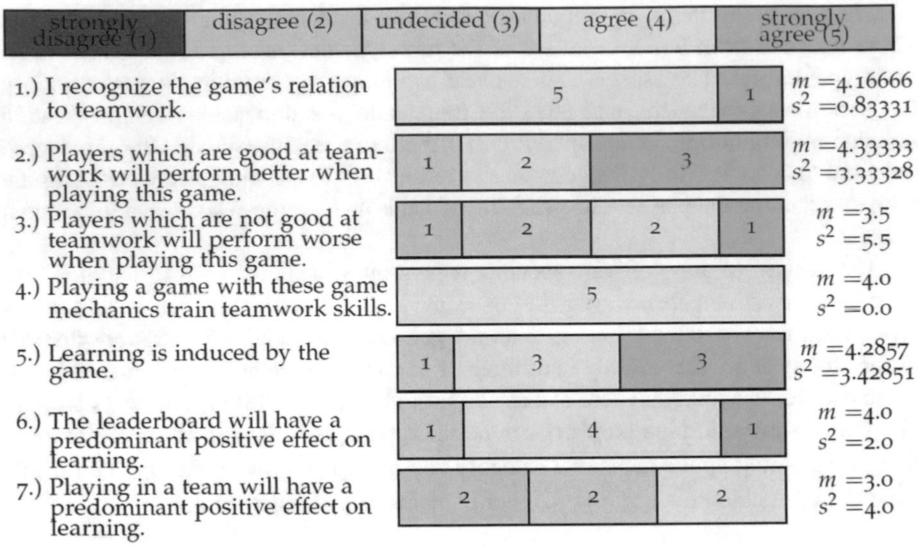

strongly disagree (1)	disagree (2)	undecided (3)	agree (4)	strongly agree (5)

Item	Answers (quantities)	Statistics
1.) I recognize the game's relation to teamwork.	5, 1	$m = 4.16666$, $s^2 = 0.83331$
2.) Players which are good at teamwork will perform better when playing this game.	1, 2, 3	$m = 4.33333$, $s^2 = 3.33328$
3.) Players which are not good at teamwork will perform worse when playing this game.	1, 2, 2, 1	$m = 3.5$, $s^2 = 5.5$
4.) Playing a game with these game mechanics train teamwork skills.	5	$m = 4.0$, $s^2 = 0.0$
5.) Learning is induced by the game.	1, 3, 3	$m = 4.2857$, $s^2 = 3.42851$
6.) The leaderboard will have a predominant positive effect on learning.	1, 4, 1	$m = 4.0$, $s^2 = 2.0$
7.) Playing in a team will have a predominant positive effect on learning.	2, 2, 2	$m = 3.0$, $s^2 = 4.0$

Fig. 7. Answers to the Likert-scale questions of the software-specific expert questionnaire. Numbers in rectangles represent quantities. The mean (m) and standard deviation (s^2) are calculated for each item. Each item answer is therefore transformed to values ranging from 1 (strongly disagree) to 5 (strongly agree).

Table 3. Result - software-specific expert questionnaire, part 2, rating questions. Representation of teamwork aspects in the game (1 - low representation, 10 - high representation) rated by the six experts.

	Communication	Orientation	Monitoring	Adaptability
Range	1–10	1–10	1–10	1–10
Ratings	10, 4, 8, 9, 5, 6	8, 7, 5, 8, 7, 8	9, 2, 8, 8, 6, 8	7, 4, 6, 7, 6, 5
Mean	7	7.2	6.8	5.8
Std. Dev.	2.4	1.2	2.6	1.2

The results of the rating questions of the software-specific expert questionnaire, part 2, are given in Table 3. In order to reach the standardized score, we sum up the answers of all four categories, subtract four and multiply it with $\frac{100}{36}$. The grouping of items and the mean score can be found in Table 4.

Table 4. Evaluation scores overview.

Aspect	Questionnaire	Items	Mean-score	Standard deviation
SUS	SUS	1–10	73.5	14.4
A1	Expert 1	1–4	75.0	11.2
A2	Expert 2	1–4	63.4	16.8
A3	Expert 1	4–7	67.7	4.7
A4	User	1–6	83.0	9.2

Qualitative feedback collected from users during the study mainly addresses issues concerning usability and the depth of gameplay. They suggest improvements in the user interface, as well, as in the illustration of the in-game units. Also, the number of buildings and possibilities to manipulate the game world is assessed as too low. This information is, although very helpful on were to continue work for the next stage prototype, expected due to the nature of the prototype.

The expert evaluation's remarks focus on how to improve or enhance the teamwork training functionality of the prototype. There are three extremely valuable suggestions among them. The first talks about different scenarios for the game, the second recommends the introduction of teamwork activities to unlock buildings or resources. The last suggests explicitly talking about the different teamwork aspects. These three possible improvements will now be examined.

The first suggestion (different scenarios), is especially helpful in situations where learners play the game very often. The need to find a strategy and discuss possible problem solving strategies vanishes after playing this game a few times. By introducing different scenarios that brings along different building costs, different goals, and different in-game possibilities, players are again forced to discuss a common strategy.

The second suggestion (including teamwork activities to unlock buildings or resources) can be used to directly enhance the teamwork training capabilities of the game. These activities can be incorporated as mini-games or mini-challenges to unlock in-game advantages. They can be realized as single player activities or even as multiplayer activities.

The last suggestion (to talk explicitly about teamwork) is interesting but needs to be discussed critically. Although this approach could raise the teamwork training capabilities of the game, it might also transform it to a great extent from a fun game into learning software. As this depends very strongly on the concrete implementation and the subjective perception of the players, tests need to be carried out to evaluate this approach.

When looking on the three aspects evaluated, the prototype's relationship to teamwork (A1 and A2), the quality of its user interface (SUS), and its potential to be expanded to a fully-fledged serious game for teamwork training (A3 and A4), all three aspects speak in favour of the prototype.

The relationship between the prototype and teamwork was rated in average 63.4 with 0 meaning there is a low representation of the four teamwork aspects and 100 that there is a high representation of teamwork the teamwork aspects represented in the prototype. The perceived relationship to teamwork was rated on average with 75 out of 100.

The game's potential to be expanded into a fully-fledged game was rated in average 67.7 for the teaching component and 83 for the entertaining component out of 100 of the prototype.

Also, the usability was rated as good, with a mean score of 73.5.

These results approve the prototype's potential to be expanded into a fully fledged teamwork training game. Its relationship to teamwork, its usability, and its potential for both, learning and playing, was shown.

6 Summary and Future Work

This paper introduces a game design for training teamwork skills. It is based on well-researched teamwork- and learning theories. Included theories are the experiential learning circle and components of teamwork, namely adaptability, orientation, monitoring, and communication.

A proof of concept is implemented as a visualisation of the game design using a server-client architecture. It is also used to conduct a study to evaluate the prototype's potential for expansion into a fully-fledged teamwork training game, its usability, and its relationship to teamwork. The findings support the introduced game design and its prototype in all three evaluated areas.

These evaluation results suggest that the game should be expanded into a fully-fledged teamwork training game. This requires the extension of game mechanics and improvements of the graphics. Also, different scenarios, implicit teamwork activities, and explicit communication about teamwork as suggested by the experts should be incorporated into the updated game design. A next step would then be to conduct a long term study over multiple month to evaluate the training abilities of the game design.

Acknowledgment. This work was partly conducted at RMIT University, including the design and implementation of the prototype system. Funding support was provided by University of Graz.

References

1. Ikariam (2017). https://en.wikipedia.org/wiki/Ikariam
2. Baker, D.P., Day, R., Salas, E.: Teamwork as an essential component of high reliability organizations. Health Serv. Res. **41**(4p2), 1576–1598 (2006)
3. Bangor, A., Kortum, P., Miller, J.: Determining what individual SUS scores mean: adding an adjective rating scale. J. Usability Stud. **4**(3), 114–123 (2009)
4. Barrick, M.R., Mount, M.K.: The big five personality dimensions and job performance: a meta-analysis. Pers. Psychol. **44**(1), 1–26 (1991)
5. Barrows, H.S.: A taxonomy of problem-based learning methods. Med. Educ. **20**(6), 481–486 (1986)
6. Belbin, R.M.: Management teams-why they succeed or fail. In: Proceedings of the 2005 ASEE/AaeE 4th Global Colloquium on Engineering Education. Butterworth Heinemann (1981)
7. Bezuijen, A.: Teamplay: the further development of teamup, a teamwork focused serious game. Master thesis, TU Delft (2012)
8. Bhardwaj, J.: Evaluation of the lasting impacts on employability of co-operative serious game-playing by first year computing students: an exploratory analysis. In: 2014 IEEE Frontiers in Education Conference (FIE), pp. 1–9. IEEE (2014)
9. Bluemink, J., Hmlinen, R., Manninen, T., Jrvel, S.: Group-level analysis on multiplayer game collaboration: how do the individuals shape the group interaction? Interact. Learn. Environ. **18**(4), 365–383 (2010)

10. Boyle, E.A., Hainey, T., Connolly, T.M., Gray, G., Earp, J., Ott, M., Lim, T., Ninaus, M., Ribeiro, C., Pereira, J.: An update to the systematic literature review of empirical evidence of the impacts and outcomes of computer games and serious games. Comput. Educ. **94**, 178–192 (2016)
11. Bozanta, A., Kutlu, B., Nowlan, N., Shirmohammadi, S.: Effects of serious games on perceived team cohesiveness in a multi-user virtual environment. Comput. Hum. Behav. **59**, 380–388 (2016)
12. Brooke, J., et al.: SUS-a quick and dirty usability scale. Usability Eval. Ind. **189**(194), 4–7 (1996)
13. CHI 2019: Conference on Human Factors in Computing Systems, 01 May 2019. https://chi2019.acm.org/
14. Chittaro, L., Ranon, R.: Serious games for training occupants of a building in personal fire safety skills. In: Conference in Games and Virtual Worlds for Serious Applications, VS-GAMES 2009, pp. 76–83. IEEE (2009)
15. Cooke, N.J., Kiekel, P.A., Salas, E., Stout, R., Bowers, C., Cannon-Bowers, J.: Measuring team knowledge: a window to the cognitive underpinnings of team performance. Group Dyn. Theory Res. Pract. **7**(3), 179 (2003)
16. Coovert, M.D., Winner, J., Bennett Jr., W., Howard, D.J.: Serious games are a serious tool for team research. Int. J. Serious Games **4**(1), 41–55 (2017)
17. CSCW 2019: Conference on Computer-Supported Cooperative Work and Social Computing, 01 May 2019. https://cscw.acm.org/2019/
18. De Freitas, S.: Learning in Immersive Worlds: A Review of Game-Based Learning. JISC (2006)
19. De Gloria, A., Bellotti, F., Berta, R.: Serious games for education and training. Int. J. Serious Games **1**(1) (2014)
20. De Leo, G., Goodman, K.S., Radici, E., Secrhist, S.R., Mastaglio, T.W.: Level of presence in team-building activities: gaming component in virtual environments. arXiv preprint arXiv:1105.6020 (2011)
21. Devine, D.J., Habig, J.K., Martin, K.E., Bott, J.P., Grayson, A.L.: Tinsel town: a top management simulation involving distributed expertise. Simul. Gaming **35**(1), 94–134 (2004)
22. Dickinson, T.L., McIntyre, R.M.: A conceptual framework for teamwork measurement. In: Team Performance Assessment and Measurement, pp. 19–43 (1997)
23. Ellenbogen, J.M., Hu, P.T., Payne, J.D., Titone, D., Walker, M.P.: Human relational memory requires time and sleep. Proc. Natl. Acad. Sci. **104**(18), 7723–7728 (2007)
24. Ellis, J.B., Luther, K., Bessiere, K., Kellogg, W.A.: Games for virtual team building. In: Proceedings of the 7th ACM Conference on Designing Interactive Systems, pp. 295–304. ACM (2008)
25. FDG 2019: Foundation of digital games, 01 May 2019. http://fdg2019.org/
26. Fosnot, C.T., Perry, R.S.: Constructivism: a psychological theory of learning. Constructivism Theory Perspect. Pract. **2**, 8–33 (1996)
27. Graafland, M., Schraagen, J.M., Schijven, M.P.: Systematic review of serious games for medical education and surgical skills training. Br. J. Surg. **99**(10), 1322–1330 (2012)
28. Grantcharov, T.P., Kristiansen, V., Bendix, J., Bardram, L., Rosenberg, J., FunchJensen, P.: Randomized clinical trial of virtual reality simulation for laparoscopic skills training. Br. J. Surg. **91**(2), 146–150 (2004)
29. Guenaga, M., Eguluz, A., Rayn, A., Nez, A., Quevedo, E.: A serious game to develop and assess teamwork competency. In: 2014 International Symposium on Computers in Education (SIIE), pp. 183–188. IEEE (2014)

30. Gbel, S., Hardy, S., Wendel, V., Mehm, F., Steinmetz, R.: Serious games for health: personalized exergames. In: Proceedings of the 18th ACM International Conference on Multimedia, pp. 1663–1666. ACM (2010)
31. Haferkamp, N., Kraemer, N.C., Linehan, C., Schembri, M.: Training disaster communication by means of serious games in virtual environments. Entertain. Comput. **2**(2), 81–88 (2011)
32. Hoegl, M., Gemuenden, H.G.: Teamwork quality and the success of innovative projects: a theoretical concept and empirical evidence. Organ. Sci. **12**(4), 435–449 (2001)
33. Hollins, P., Westera, W., Iglesias, B.M.: Amplifying applied game development and uptake. In: European Conference on Games Based Learning, p. 234. Academic Conferences International Limited (2015)
34. Hunicke, R., Leblanc, M., Zubek, R.: A formal approach to game design and gameresearch. In: Proceedings of AAAI Workshop on Challenges in Game AI, vol. 4, p. 1722 (2004)
35. Kaplancali, U.T., Bostan, B.: Gaming technologies for learning: virtual teams and leadership research in online environments. In: 3rd International Future-Learning Conference (2010)
36. Klein, C., DiazGranados, D., Salas, E., Le, H., Burke, C.S., Lyons, R., Goodwin, G.F.: Does team building work? Small Group Res. **40**(2), 181–222 (2009)
37. Kolb, D.A.: Experiential Learning: Experience as the Source of Learning and Development. FT Press, Upper Saddle River (2014)
38. Laal, M., Ghodsi, S.M.: Benefits of collaborative learning. Procedia-Soc. Behav. Sci. **31**, 486–490 (2012)
39. Leonard, M., Graham, S., Bonacum, D.: The human factor: the critical importance of effective teamwork and communication in providing safe care. Qual. Saf. Health Care **13** (suppl 1), i85–i90 (2004)
40. Luu, S., Narayan, A.: Games at work: examining a model of team effectiveness in an interdependent gaming task. Comput. Hum. Behav. **77**, 110–120 (2017)
41. Malone, T.W.: Toward a theory of intrinsically motivating instruction. Cogn. Sci. **5**(4), 333–369 (1981)
42. Marks, M.A., Mathieu, J.E., Zaccaro, S.J.: A temporally based framework and taxonomy of team processes. Acad. Manag. Rev. **26**(3), 356–376 (2001)
43. Mathieu, J.E., Tannenbaum, S.I., Kukenberger, M.R., Donsbach, J.S., Alliger, G.M.: Team role experience and orientation: a measure and tests of construct validity. Group Organ. Manag. **40**(1), 6–34 (2015)
44. Maurer, M.: Using the experiential learning cycle to design a pure virtual serious game for developing teamwork skills. Master thesis, TU Graz (2018)
45. Maurer, M., Nussbaumer, A., Steiner, C., van der Vegt, W., Nadolski, R., Nyamsuren, E., Albert, D.: Efficient software assets for fostering learning in applied games. In: International Conference on Immersive Learning. pp. 170–182. Springer (2017)
46. McEwan, D., Ruissen, G.R., Eys, M.A., Zumbo, B.D., Beauchamp, M.R.: The effectiveness of teamwork training on teamwork behaviors and team performance: a systematic review and meta-analysis of controlled interventions. PLoS ONE **12**(1), e0169604 (2017)
47. McGrath, R.G.: Exploratory learning, innovative capacity, and managerial oversight. Acad. Manag. J. **44**(1), 118–131 (2001)
48. Nonaka, T., Miki, K., Odajima, R., Mizuyama, H.: Analysis of dynamic decision making underpinning supply chain resilience: a serious game approach. IFAC PapersOnLine **49**(19), 474–479 (2016)
49. Ramachandran, S., Presnell, B., Richards, R.: Serious games for team training and knowledge retention for long-duration space missions. In: 2016 IEEE Aerospace Conference, pp. 1–11. IEEE (2016)
50. Roberts, D., Wolff, R., Otto, O., Steed, A.: Constructing a Gazebo: supporting teamwork in a tightly coupled, distributed task in virtual reality. Presence **12**(6), 644–657 (2003)

51. Rollings, A., Adams, E.: Andrew Rollings and Ernest Adams on Game Design. New Riders, Indianapolis (2003)
52. Rosen, M.A., Bedwell, W.L., Wildman, J.L., Fritzsche, B.A., Salas, E., Burke, C.S.: Managing adaptive performance in teams: guiding principles and behavioral markers for measurement. Hum. Resour. Manag. Rev. **21**(2), 107–122 (2011)
53. Rosenberg, D., Depp, C.A., Vahia, I.V., Reichstadt, J., Palmer, B.W., Kerr, J., Norman, G., Jeste, D.V.: Exergames for subsyndromal depression in older adults: a pilot study of a novel intervention. Am. J. Geriatr. Psychiatry **18**(3), 221–226 (2010)
54. Rutherford, J.: Monitoring teamwork: a narrative review. Anaesthesia **72**(S1), 84–94 (2017)
55. Salas, E., Sims, D.E., Burke, C.S.: Is there a big five in teamwork? Small Group Res. **36**(5), 555–599 (2005)
56. Schlütz, D.: Bildschirmspiele und ihre Faszination: Zuwendungsmotive, Gratifikationen und Erleben interaktiver Medienangebote. Fischer, München (2002)
57. Smith, K.A., Douglas, T.C., Cox, M.F.: Supportive teaching and learning strategies in stem education. New Dir. Teach. Learn. **2009**(117), 19–32 (2009)
58. Staiano, A.E., Calvert, S.L.: Exergames for physical education courses: physical, social, and cognitive benefits. Child Dev. Perspect. **5**(2), 93–98 (2011)
59. Tuckman, B.W.: Developmental sequence in small groups. Psychol. Bull. **63**(6), 384 (1965)
60. Vegt, N., Visch, V., de Ridder, H., Vermeeren, A.: Designing gamification to guide competitive and cooperative behavior in teamwork. In: Gamification in Education and Business, pp. 513–533. Springer (2015)
61. Vorderer, P., Hartmann, T., Klimmt, C.: Explaining the enjoyment of playing videogames: the role of competition. In: Proceedings of the Second International Conference on Entertainment Computing, pp. 1–9. Carnegie Mellon University (2003)
62. Westera, W.: Games are motivating, aren't they? Disputing the arguments for digital game-based learning. Int. J. Serious Games **2**(2), 3–17 (2015)

Virtual and Face-to-Face Teaching Practices for Dissertation Writing: Current Challenges and Future Perspectives

María Isabel Pozzo(✉)

National Scientific and Technical Research Council, Rosario Institute for Research in Educational Sciences, Rosario, Argentina
pozzo@irice-conicet.gov.ar

Abstract. Dissertation writing constitutes a personal and professional challenge that has a strong impact at an institutional and even national level. Indeed, the completion of the thesis is the crowning moment of postgraduate studies. Simultaneously, graduation rate is one of the main indicators of effectiveness of postgraduate programs, with regional and international repercussions.

Due to the importance of dissertation writing, there are several useful strategies to mitigate difficulties. In this regard, virtual and face-to-face teaching practices related to thesis writing are presented: gains and limitations based on a contextualized cases study in Argentina. After the presentation of the different aspects, current challenges and future perspectives about virtual and face-to-face teaching practices for thesis writing are posed for discussion.

Keywords: Virtual teaching practices · Face-to-face teaching practices · Dissertation writing · Postgraduate studies

1 Introduction

The integration of technologies and education has opened a broad range of possibilities for innovating teaching practices at different levels and disciplines. However, this integration is not a magical resource, but it demands reflections to make the most appropriate decisions in what is the goal of every teacher: to achieve more genuine learning. This challenge requires constant 'epistemological vigilance' [1], whatever our object and approach are. In this context, this paper refers to the integration of technologies and education in a very specific field: the training of high-level professionals as postgraduate students (Specialization, Master's and Doctorate) in their facet of authors of a thesis. Although it may seem to be a very particular issue, it involves a numerous population, considering the expansion of postgraduate studies during the last decades worldwide, in Latin America [2], and also in Argentina (Fig. 1).

Likewise, the problem concerns an even greater population that involves thesis supervisors, project, scholarships, and thesis evaluators, and professors in charge of courses related to thesis writing. After this enumeration, it is easy to notice the wide scope of this issue. Taking care of the tasks related to the research process of postgraduate careers also requires training. And a growing number of postgraduate students demands in turn more human resources trained to forge researchers.

M. E. Auer and D. May (Eds.): REV 2020, AISC 1231, pp. 1016–1032, 2021.
https://doi.org/10.1007/978-3-030-52575-0_84

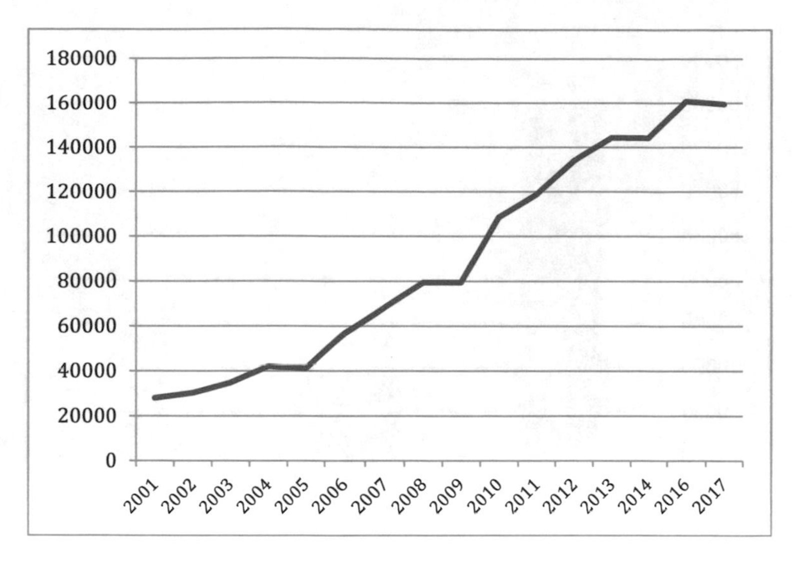

Fig. 1. Number of postgraduate students in Argentina (2001–2017) Source: own elaboration adapted from [3]

There is still another reason that highlights the importance of this topic: Whether we are postgraduate students, thesis supervisors, or professors, dissertation writing represents a great professional challenge that has a strong impact at personal, institutional and even national level. Indeed, the completion of the thesis constitutes the greatest challenge for those professionals who decide to undertake a postgraduate study, it is the crowning moment of their careers. Simultaneously, graduation rate is one of the main indicators of effectiveness of postgraduate programs, with regional and international repercussions.

To sum up, the importance of the topic lies on:

- the expansion of postgraduate studies in recent decades at a global scale,
- the need of a larger body of trained professors-researchers to cover tasks associated to the research process, and
- the importance of concluding the thesis at a personal, institutional and national level.

The negative aspect of the increasing number of postgraduate studies is a very low rate of students' completion at the global level. Even having passed all the mandatory courses, many leave their postgraduate careers unfinished for not completing their thesis. A very resonant case in this regard was the guitarist of the English band Queen, Brian May, who being an astrophysicist, took 30 years to complete his doctorate [4]. Beyond this extreme case of public knowledge, there are numerous cases of postgraduate studies postponement or dropout. Figure 2 shows the situation in Argentina in 2017 in terms of the difference regarding the number of students and graduation rate at postgraduate level.

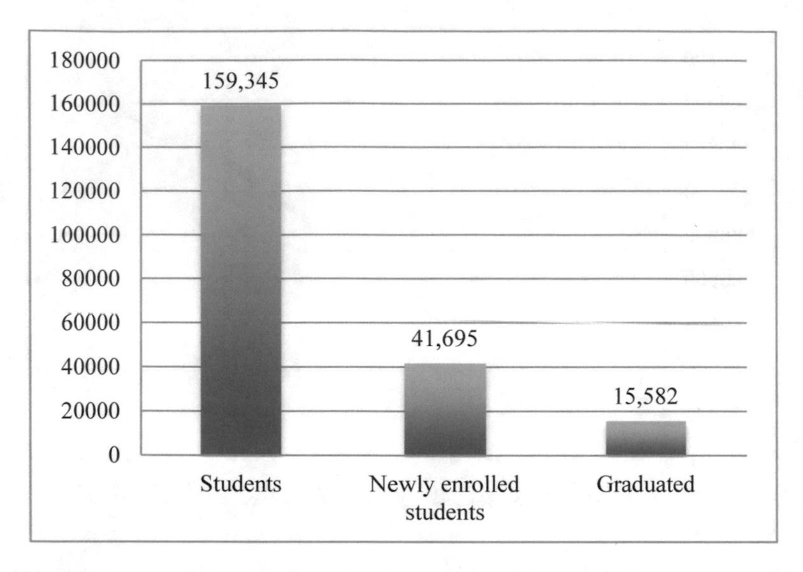

Fig. 2. Graduation rate in postgraduate students (Argentina, 2017) Source: own elaboration adapted from [5]

In this context, training researchers became an issue of the university agenda [6]. Thesis workshops, courses or seminars, with their different denominations, aim to deploy strategies to promote their completion. The implementation of this kind of courses mediated by platforms -although still less frequent- brings in new perspectives and opportunities. However, virtual media implies challenges and demands decisions to make the most of technology. These challenges and decisions concern multiple aspects: the production of teaching materials, the follow-up of student learning, the evaluation of every process, and teacher training. Here we could recall the question: "If technology is the answer, what was the question?" [7]. Within the scope of this paper, this question is inspired by the name of ICBL Conference: how do we achieve interactive collaborative learning in computer mediated teaching on dissertation writing?, what can computer mediated instruction add to face-to-face dissertation writing instruction on an interactive collaborative basis?

In this regard, it should be noted that research on thesis writing teaching through virtual courses is scarce compared to face-to-face, but there are already some systematizing efforts. On the one hand, online tutoring of postgraduate writing has been described in general terms [8]. On the other hand, a more specific approach studies teacher feedback for learning in virtual thesis writing workshops [9], while other works focus on particular areas of research training, such as the teaching of statistics through platforms [10].

Due to this vacancy, the purpose of this paper is to systematize some of the challenges relating to pedagogical mediation, defined as the way to deal with contents in order to foster students' knowledge construction, through virtual teaching practices designed for postgraduate students. Afterwards, some comparisons with face-to-face courses in traditional environments are presented as well: gains and limitations of teaching practices related to thesis writing, their peculiarities and points in common of both environments. After this presentation, current challenges and future perspectives about virtual and face-to-face teaching practices of thesis writing will be posed for discussion.

The analysis is done through a case study focused on a virtual Seminar on post-graduate thesis writing carried out as a participant observer in the role of the professor (Fig. 3). The Seminar is offered by the National University of Rosario, Argentina, which is in fifth place in the ranking of fifty five National Universities. The Seminar is delivered through the university virtual campus in a Moodle platform. It offers numerous and varied resources and activities, out of which web page, folder, home-work and forum are used.

Fig. 3. Screenshot of the course portal.

This paper aims to contribute to two lines of inquiry: research training in general, and research training in blended and online environments in particular. Due to the underlying importance of writing in cognition, a Didactic based on Psychology perspective is adopted, following the Swiss psychologist, Aebli [11], as it will be described in the following sections.

2 Development

2.1 The Didactic Triad in Virtual Educational Environments

Virtual teaching environments preserves the didactic triad (Fig. 4): teacher-student-contents.

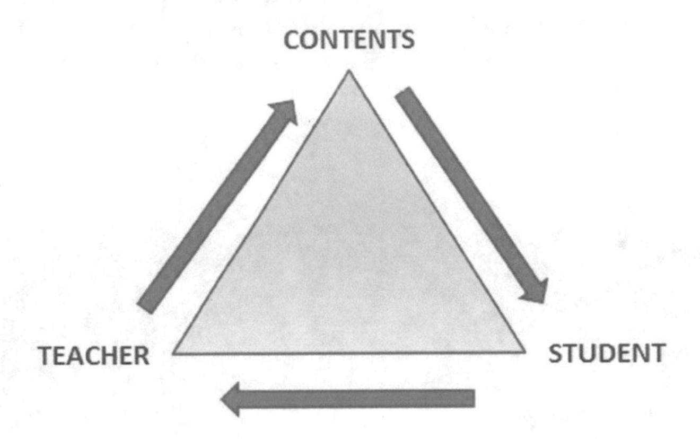

Fig. 4. The didactic triad

The history of didactics has recorded the transition from a teacher-centered tradition to a student-centered position. That is, non-directive teaching has surpassed directive. However, a retrospective analysis indicates that neither extreme is good. The teacher retains the responsibility to choose the most suitable way for contents' teaching, and selecting the most appropriate bibliography. Therefore, access to the information available in the web is not equivalent to knowledge achievement. In thesis writing courses, the availability of digitized and open access bibliography, such as research methodology handbooks or thesis uploaded in institutional repositories, does not assure the concretion of learning. This premise could be extended to the mindless use of Internet in education, as if the digital media could guarantee learning achievement. "The most modern technology does not assure the value of the proposal; the quality of the materials is not linked to the medium, but to the contents and the activities that are developed as long as they generate good learning", says Litwin [12], an Argentinean pedagogue, and there, teachers have a fundamental role that will show in the teaching practices.

2.2 Teaching Practices in Virtual Educational Environments

Teaching practices comprises the stages of design, implementation and evaluation [13]. Within this framework, the visible activities of the class are just the tip of an iceberg.

The planning stage is reflected in a course syllabus, which not only comprises (as sometimes happens) a list of contents, but also the rationale, objectives, activities and evaluation criteria.

In the thesis writing seminar of the case study, the syllabus is shown in abbreviated form on the website, and it is fully displayed within the platform, only accessible for the course students. The contents are structured around the sections of a thesis, which are precisely the steps of the scientific method [14]. This selection and sequencing criteria may be the same in virtual and face-to-face learning environments, but they differ in the distribution and grouping of contents because they differ in the conception of time:

In face to face learning environments, the limit of the class is given by the time table: once finished, a class may have been more or less productive, depending on the quantity and quality of content that the teacher managed to develop, but there will be no doubts regarding time limit. In virtual environments, time has no limit, so it is necessary to create discrete units of teaching. These units should be balanced according to two variables: the density of the contents and the extent of their development.

According to these criteria, the contents of the virtual course are distributed in ten units (Fig. 5) coinciding with the ten-week course duration, assigning each unit a similar complexity. Each week a unit is uploaded, so as to create some surprise effect, together with the security of a plan conceived beforehand portrayed in the Syllabus.

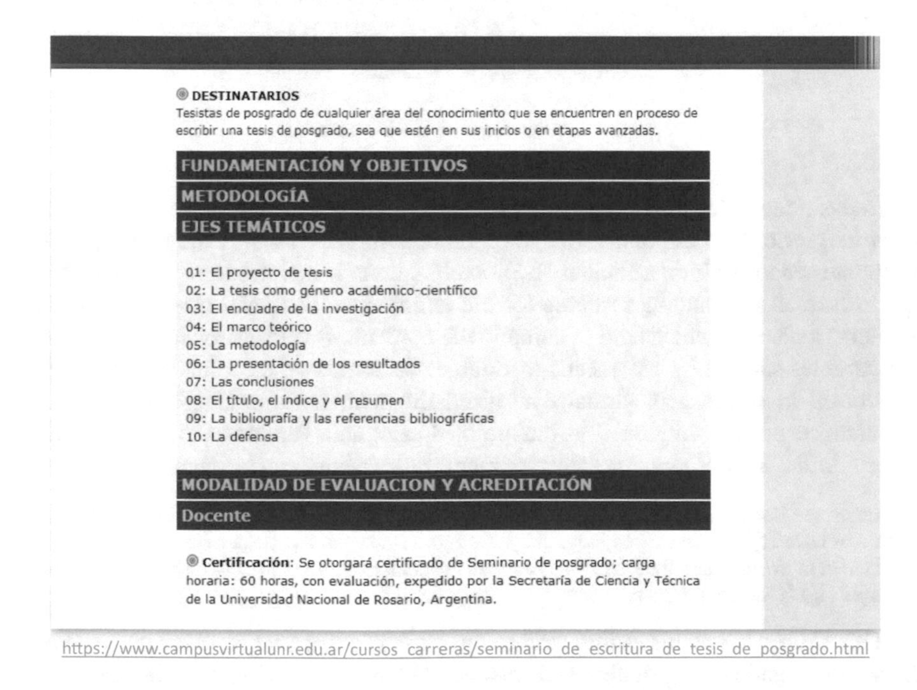

Fig. 5. The design stage of the course: the Syllabus

The presentation of the contents is carried out through self-made systematizing texts, of moderate length and medium difficulty (Fig. 6). This decision is especially appropriate for heterogeneous groups, so that such input would be -as the case may be- the first access to the contents, a review of something already known, or an arrangement of scattered knowledge.

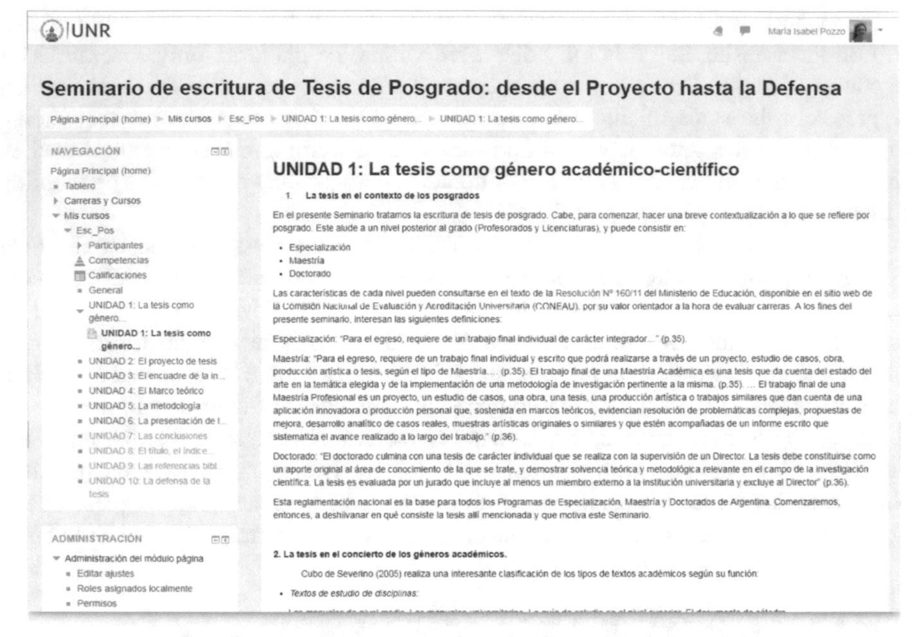

Fig. 6. An excerpt of Unit 1 core text

Indeed, one of the differences of virtual environment teaching is the need to portray in written form the usual oral verbiage of the classroom. Although the platform allows to upload videos (within a limited size), explanations in written form are unavoidable and in turn an outstanding resource for the learning process. Not even the traditional teachers' materials, which tend to compile the contents developed during the classes, is comparable, since they are meant to double, to reinforce the oral language of the classroom. In contrast, in virtual education, the written communication is the most prevalent, if not the only one. At this point, Olson and Torrance [15], two Canadian experts in the area of cognitive development and writing, can be recalled:

> The natural human being is not a writer or a reader, but a speaker and a listener. This must be as true of us today as it was seven thousand years ago. Literacy at any stage of its development is in terms of evolutionary time, a mere upstart, an artificial exercise, a work of culture, not nature, impost upon the natural man.

This expression confirms that writing the class is a real challenge for the teacher, in the sense of anticipating doubts and misunderstandings, in order to avoid them. This challenge can be achieved with the teacher's expertise in the subject, since some doubts and errors are recurring. However, the activities in which the students apply knowledge and exchange perspectives are the instances par excellence to detect them. In virtual environments, the appropriate means are the forums (Fig. 7), which have the advantage of preserving the exchanges to be read more than once, meditated and answered carefully. Screen shots are especially useful for research purposes, allowing to keep record of the exchanges for later analysis. Data collection becomes more complex in face-to-face classroom observation, in which there is the problem of unclear expressions, voices overlapping, the interference of the observer, etc.

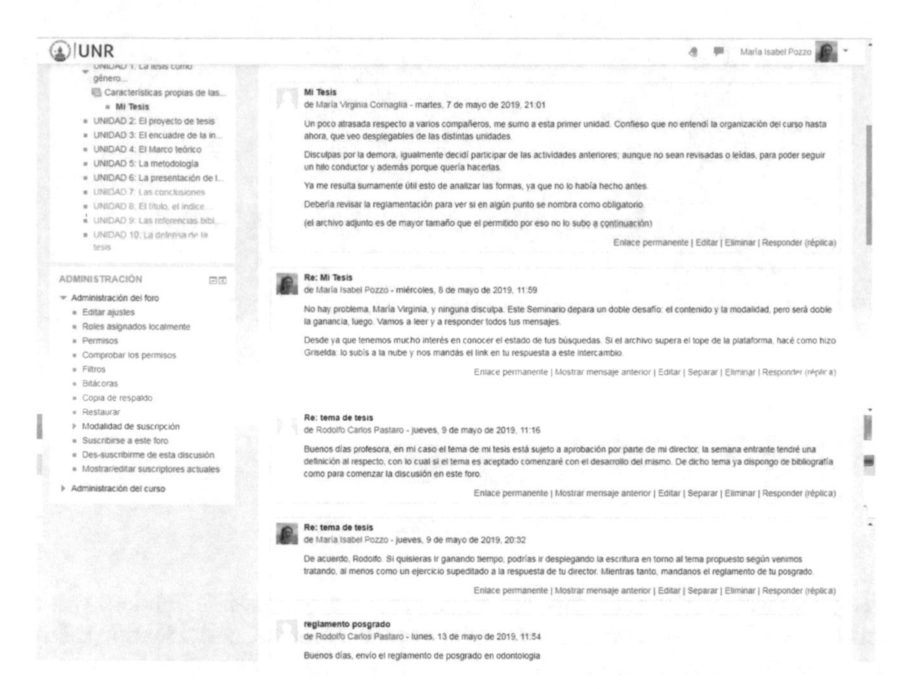

Fig. 7. An excerpt of the Forum

2.3 Verbal Written Interaction as a Vehicle for Pedagogical Mediation

Since verbal written language is the vehicle for pedagogical mediation in virtual environments, it is necessary to decide the tone of verbal interaction. Academic language is characterized by a high degree of formality and distance. Paradoxically, while the content of the course aims to achieve this style, it is necessary to gain the confidence of the students so that they are not afraid or embarrassed to share their productions, raise their doubts or show their uncertainties publicly. It should be born in mind that postgraduate students are professionals. Thus, to achieve this mutual trusting atmosphere, it is necessary to use a fluid, friendly, colloquial language, without losing the rigor of the academic environment and the precision of scientific research. "A clear, direct and expressive language gives the student the idea that he/she is the permanent interlocutor of the teacher and that both participate jointly in the construction of knowledge." [16].

Even the illustration of the platform can also contribute to an atmosphere of mutual confidence (Fig. 8).

Fig. 8. An illustrated menu

Together with an agreeable tone and even a cute illustration, interventions will have to follow one of the main characteristics of academic discourse, and scientific behavior: order. Even the personal presentation of the group, which aims to be a moment of relaxed sociability, can be channeled into a rubric following a pattern:

> In this forum we will introduce ourselves indicating: 1) degree; 2) postgraduate career and institution in which you are enrolled; 3) why you enrolled in this course; 4) previous experience in distance education.
> We listen to each other….

The teacher's presentation is also structured following the same items, even the one that aims to elicit the reasons to register for the Seminar (adapted to the professor status, of course) by providing them with a similar and coherent input:

> Welcome to the Seminar. It is a pleasure for me to accompany you in the complex process of writing a thesis. I begin by introducing myself:
> My studies: I am a Professor, Licenciate and Doctor in Educational Sciences graduated at the National University of Rosario and Magister graduated at the University of Barcelona, Spain. This Seminar relates in general to my teaching and researcher jobs; specifically, as co-director of the Research Project "Teaching practices of writing postgraduate thesis: comparative study of experiences in face-to-face and virtual environments".
> I have already given distance courses, but on other subjects. I find the modality extremely flexible and favorable, so I am very excited to implement it to promote the writing of postgraduate thesis.
> Now I invite you to introduce yourselves at the Forum below …

In this way, the teacher operates as a promoter of cooperative learning networks that turns the physical loneliness of virtual education and the individual enterprise of thesis writing into collaborative work through virtual interaction.

2.4 Advantages of Virtual Environment for Thesis Writing

One of the most obvious practical advantages of virtual education is the possibility of "attending" class overcoming the difficulties of coinciding in space and time with the teacher and the other students. This is a huge advantage for graduate students, since they are full-time professionals, with very little time availability. However, there are some more substantive advantages that concern the learning process:

- It provides general learning strategies.
- It contributes to introspective work.
- It provides tools for writing as a recursive process.

2.4.1 The Promotion of Learning Strategies and Metacognitive Reflection

In agreement with [17], good proposals in virtual education assume the commitment -not very assumed by face-to-face education- of providing students general learning strategies.

In the case of postgraduate thesis writing, virtual environments contribute to introspective work, leading to metacognitive and metalinguistic skills. The possibility of individual reflection guarantees more elaborate, thoughtful productions, which in the traditional classroom cannot be done.

2.4.2 Writing as a Recursive Process

Writing a thesis is usually dealt with linearly, throughout the different stages of the scientific method. However, it is really carried out in a spiral process with the constant expansion and reformulation of its chapters. Indeed, academic writing is a slow, recursive process, with fluctuations. For this reason, the chance to go back to previous units from the platform helps the students in this non linear process.

The availability of digitalized versions from the platform to be read carefully is another advantage provided by virtual environments. It allows the teacher to track the students' improvements along the course. This can be done in two ways: 1) by requesting periodic deliveries in files identified by the date or section to which they correspond (Fig. 9); or 2) by uploading updatable shared files. The first way is more rigid, but allows the teacher to track students' progress along the course. The second (one updated file) does not allow the teacher to follow the learning process longitudinally but it guarantees access to the latest version.

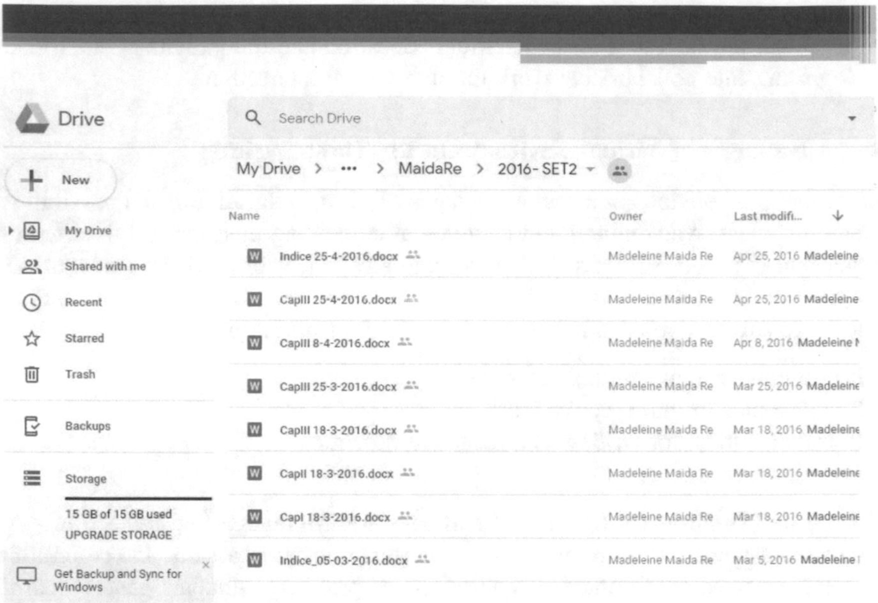

Fig. 9. A screenshot of the manuscript files uploaded periodically in a blended learning course.

2.5 The Socialization of Progress

Introspective work becomes even more valuable if it is socialized. In this way, the students notice recurrences, common mistakes, and alternative strategies. Thus, after the students have introduced themselves, the course begins sharing what everyone has written at that point.

Accordingly, the forum of the first unit asks to share the thesis manuscript in an attached file, to have a "photograph" of the starting point of each student.

> Forum of unit 1: Our productions
> Regarding the identification of the parts within the textual structure of the thesis: which one are you working on? Together with your comments please attach what you have written of your thesis so far. It doesn't matter how scarce it is, or if it is messy. We are interested in keeping it to review it:
>
> – throughout the Seminar in the light of the topics that have being dealt with, and
> – when we have finished the seminar, to see how much progress you have made, in what aspects, etc.

However, whereas the forum has the advantage of illustrating the comments with the attached documents, the forum thread may become very long and cumbersome to look for attachments. In order to easily retrieve them in future instances, they are compiled in a folder, while the forum threads illustrated by the files are also preserved. This 'resource box' works as an internal repository of the class. It resembles the portfolio of architects, where they keep their work. It is in turn a learning strategy that allows us to keep track of the theses advancements throughout the course or at the end of it and recall the progress.

Thus, in unit 1 devoted to the research framework, the rubric is shown not only in the forum but also in the folder:

> Folder of unit 1: Our productions
> In this folder of Unit 1, we compile what you have written of your thesis at the beginning of the Seminar. It doesn't matter how scarce it is, or if it is messy. We are interested in keeping it to review it:
>
> – throughout the Seminar in the light of the topics being dealt with, and
> – when we have finished the seminar, to see how much progress we have made, in what aspects, etc.

In turn, an epigraph announces this availability, since a virtual course requires to clearly explicit the internal organization.

The importance of being acquainted with the production of the students justifies this files' duplication (Fig. 10). Thus, benefits and disadvantages of technological means and resources need to be examined.

Another more dynamic option for students is to write or to paste their writing in a wiki page (Fig. 10). This is particularly appropriate for those students who have just begun writing, and the text to be pasted is still short, simple and prone to numerous changes.

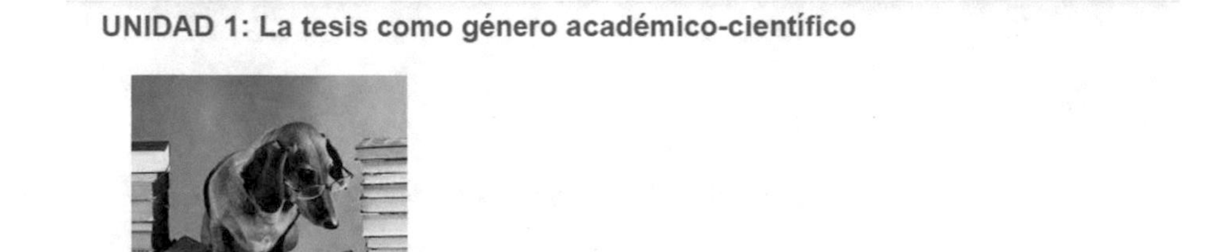

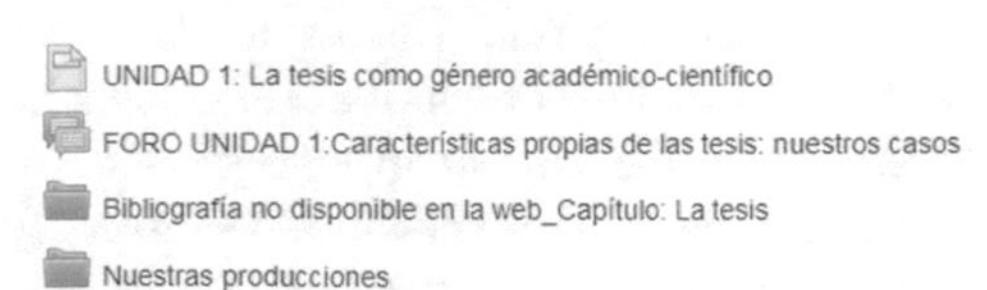

Fig. 10. Menu with files duplicated in Forum and Folder, and Wiki pages.

On the other side, the only possibility of the traditional classroom is to narrate what is being written, but the oral reported speech is far from the thesis as a product. Therefore, virtual education provides resources that allow postgraduate students to deal with the theses themselves. Sharing the thesis files in the cloud (Fig. 11), also possible in a blended learning approach, allows every student to have access to the productions of each classmates. However, this teaching strategy depends on the technological availability in the classroom. In other words, if students cannot take their notebooks to class or if it cannot take place in computer rooms (as it is usually the case in Latin American universities), access to the files can only be done at home, after the class.

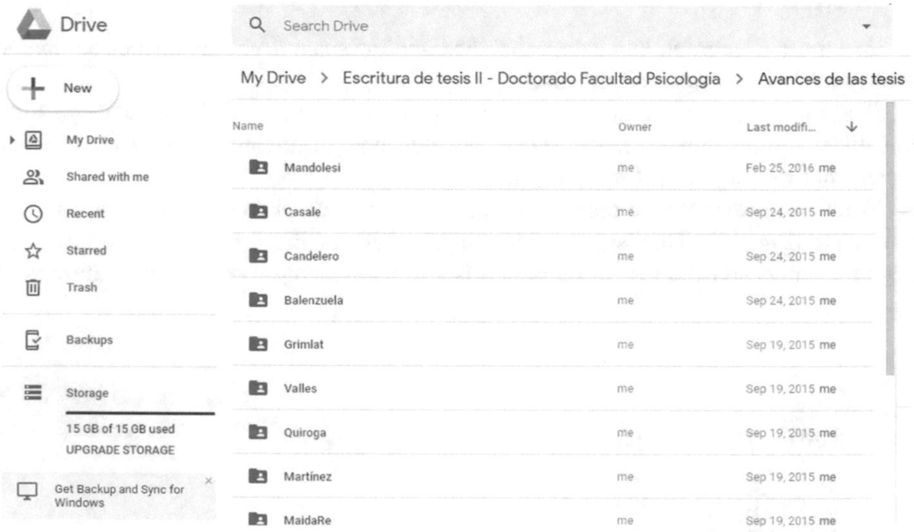

Fig. 11. Sharing the thesis files in a blended learning course

2.6 The Teacher as Manager of the Virtual Classroom

Virtual education started with a kind of social division of labor between the teacher as an expert in the disciplines, the content writer as a specialist in the development of teaching materials for virtual education and the tutor as the teaching assistant in charge of accompanying and enriching the pedagogical proposal. Thanks to the growing teachers' familiarity with ICTs and the technical simplification of the educational LMS, this division of roles disappeared, converging on the teacher in charge of a virtual course. From this centralizing role, the teacher has the responsibility of diversifying teaching practices, in order to avoid reproducing the monotony of university face-to-face classes.

Besides technological skills, the teacher who undertakes a virtual course is aware of the need to create activities for each unit, whereas a face-to-face course does not necessarily require it. Traditional classes on thesis writing are more bound to lecture whereas virtual teaching tends more to laboratory in order to provide students the feedback they need.

Again, following a pattern for every unit helps the students to predict what comes next. They will no longer have to see what the activity of each unit is about, but they may anticipate it in some way, as it only differs the section of the thesis that is being dealt with.

Unit 3 > Forum: The framework of our thesis
In this Forum, please make comments about the initial section of the theses uploaded in the respective folder. You can refer to your progress and/or those of your partners.
Unit 5 > Forum: The methodology of our thesis
In this Forum, please make comments about the methodology of the theses uploaded in the respective folder. You can refer to your advances or those of your classmates.

At the same time, it is a relief for the teacher who may feel overloaded crcating many activities and correcting the outcomes.

2.7 Other Complementary Textual Resources: The Exemplification

Undoubtedly, exemplification constitutes a resource of vital importance in knowledge construction. Together with a clear and coherent explanation of each topic, a thesis excerpt -or its reduced version, an article-, illustrating a certain topic helps to understand in a concrete way what each section of the thesis implies. However, several questions arise:

- Where should the example be placed: after the core text or on a Resource folder?
- Which would be the better way to exploit the example: as an illustration or as a recognition activity?
- How should the activity be presented: as an individual task or in the forum?

An apparently irrelevant decision, which in fact is very important, is where to place the examples in the teaching sequence. Locating them after the expository text provide comfortable immediacy, but at the same time leaves them too exposed in the central area of the unit. While the core text may require several readings, the example will not deserve so many. The need to separate the example from the core text is even greater when the example is in fact an "anti-example", that is to say, the example of what should *not* be done. For example: in the unit devoted to the thesis abstract, its characteristics are illustrated with erroneous decisions throughout the writing of successive versions.

In order to separate the example from the expository text, the former can be placed in another section, such as a folder devoted to upload files with examples. Due to its role in the learning process, it can be called "Resource Box", a name that is broader than simply "Examples." In order to obtain the complementary effect it is necessary to indicate very clearly for which segment of the main text each example corresponds to. In turn, in the examples folder, each file has to be clearly linked with its corresponding segment of the core text. Thus, in the folder of the unit devoted to the Abstract, there are:

- a file with the different versions in which an Abstract was modified by its author based on the corrections received, starting with the antiexample previously mentioned, and
- a file with an example of an Abstract correctly written.

Another decision to be taken regarding the examples concerns the best way to exploit them for learning. Thus, a fragment of a thesis can be used as an illustration of a certain topic, or, if presented without any additional comments, it can be exploited by students as a recognition activity. For example, a fragment of the methodology chapter can be used to illustrate or to recognize some methodological elements. Given this double possibility, it is convenient to start presenting an example to illustrates the theoretical explanation and then in a second step, engage students in a more active task of recognizing methodological elements (such as the description of the case in a case study, the data collection techniques used and the detail of the field work carried out). Each topic should be exemplified with fragments of different theses, since once the students have given the correct answer, the exercise will no longer make sense. And this decision in turn challenges the teacher to find examples of similar content and difficulty. In its general concept, this proposal follows the principles of general teaching [11]. Now, as for the specificity of thesis writing virtual teaching, it raises the question on how to distribute the example spatially: When presenting an example, the activity should be dealt with as an individual task and not as a topic of discussion in the forum, because as soon as the first participant answers, the rest would only have to confirm. At a later instance, and once everyone has answered, the forum is a good medium where to justify the individual answers, even to closed questions.

3 Final Comments

This paper has reviewed some didactic decisions that arise in the process of designing and implementing virtual teaching practices in the university postgraduate level aiming at collaborative learning. These decisions are translated into very specific actions that involve challenges about the best way to mediate pedagogically in virtual environments. In this sense, the role of oral and written verbalization, the temporal distribution of contents and activities, the progression of contents, the creation of activities, the cross-references, and the exemplification, are debated within two crucial dimensions: the online instruction of the Seminar, and the specificities of academic discourse as a vehicle for scientific research, within the framework of the postgraduate thesis genre.

At this point it is worth asking: do teachers meditate about all these questions in a traditional classroom course? As a participant observer of this case study, the answer would be negative. In higher education didactics, it is not frequent to reflect so much about the activities, not at least with so much detail. Virtual education poses educators in a position to ask themselves about these aspects, and surely the students perceive this dynamism.

Then it is worth to reflect in which situations each resource can promote the writing of a thesis, and which solutions they offer for each specific need. These conclusions imply deciding: do we adopt a technological approach (what is called in English 'technology driven approach'), which implies that teachers integrate ICTs into class simply because they are available or because they are considered as the solution to all problems of teaching? Or do we adopt a didactic approach (or 'pedagogy driven approach'), which advises us to start from the objectives we want to achieve and then see if ICTs can help us to achieve these objectives more effectively? Of course this second position was adopted hereby. Virtual education is not exempt from risks, however, given the economic bias of a knowledge market that tries to make attractive products to be sold, in order to provide new sources of financing the scientific communities. Likewise, virtual education faces the risk of becoming a means of cultural homogenization through the use of technologies disconnected from the reality of the users. In this circumstance, it is necessary to generate pluralistic proposals that address and value diversity, recognizing the multiple ways in which each culture attributes meanings and solves its problems" [18].

Even within the fixed procedures of the academy, virtual teaching practices for thesis writing should tend to promote creative, reflective and self-critical postgraduate students. The systematization carried out in this paper can contribute to encourage teachers to explore virtual environments for teaching thesis writing and provide tools for those who are considering starting a similar enterprise. Future studies should investigate the perception of the students about these topics and, subsequently, the learning outcomes obtained in different circumstances.

Acknowledgment. This paper reflects part of an ongoing research project funded by the National Scientific and Technical Research Council and the National Agency for Scientific and Technological Promotion of Argentina.

References

1. Bachelard, G.: The Formation of the Scientific Mind. Clinamen, Manchester (2002)
2. Mollis, M.: Imágenes de posgrados: entre la academia, el mercado y la integración regional. En: Mollis, M., Núñez Jover, J., García Guadilla, C. Políticas de posgrado y conocimiento público en América Latina y el Caribe: desafíos y perspectivas. CLACSO, Buenos Aires (2010)
3. Secretariat of University Policies of Argentina (SPU): University statistics consultation system (2018). http://estadisticasuniversitarias.me.gov.ar/#/seccion/2
4. May, B.: Official Biography (2018). https://brianmay.com/brian/biog.html
5. Department of University Information. Ministry of Education, Culture, Science and Technology of Argentina: University Statistics 2017–2018. Synthesis of Information, pp. 31–32 (2018). https://www.argentina.gob.ar/educacion/universidades/sintesis-de-informacion-universitaria-2016-2017
6. Wainerman, C.: Consejos y advertencias para la formación de investigadores en ciencias sociales. En: Wainerman, C., Sautu, R. (comp.) La trastienda de la investigación. Manantial, Buenos Aires, pp. 27–51 (2011)
7. Moravec, et al.: Manifiesto15. Aprendizaje en evolución (2015). https://manifesto15.org/es/
8. Difabio De Anglat, H., Alvarez, G.: Alfabetización académica en entornos virtuales: Estrategias para la promoción de la escritura de la tesis de posgrado. Traslaciones. Revista Latinoamericana de Lectura y Escritura **4**(8), 97–120 (2017). http://revistas.uncu.edu.ar/ojs/index.php/traslaciones/article/view/1066
9. Alvarez, G., Difabio De Anglat, H.: Retroalimentación docente y aprendizaje en talleres virtuales de escritura de tesis. Apertura. Revista de innovación educativa, Méjico **10**(1), 8–23 (2018). http://www.udgvirtual.udg.mx/apertura/index.php/apertura/article/view/996

10. Camarero, L.A., García De Cortázar, M., Del Val, C.: La enseñanza de la estadística y de las técnicas de investigación social a distancia. Empiria. Revista de Metodología de Ciencias Sociales (1), 203–212 (1998)
11. Aebli, H.: Doce formas básicas de enseñar. Una didáctica basada en la Psicología. Narcea, Madrid (2000)
12. Litwin, E. (comp.): La educación a distancia. Temas para el debate de una nueva agenda educativa, p. 26. Amorrortu Editores, Buenos Aires (2000)
13. Zabala, A.: La práctica educativa. Cómo enseñar. Graó, Barcelona (2010)
14. Samaja, J.: Epistemología y Metodología. Elementos para una teoría de la investigación científica. Eudeba, Buenos Aires (1994)
15. Olson, D., Torrance, N. (Comps.): Literacy and Orality, p. 37. Cambridge University Press, New York (1991)
16. Soletic, A.: La producción de materiales escritos en los programas de educación a distancia: problemas y desafíos, p. 113. En: Litwin, E. (comp.) La educación a distancia. Temas para el debate de una nueva agenda educativa, pp. 105–134. Amorrortu Editores, Buenos Aires (2000)
17. Maggio, M.: El tutor en la educación a distancia. En: Litwin, E. (comp.) La educación a distancia. Temas para el debate de una nueva agenda educativa, pp. 135–160. Amorrortu Editores, Buenos Aires (2000)
18. Coiçaud, S.: La colaboración institucional en la educación a distancia, p. 84. En: Litwin, E. (comp.) La educación a distancia. Temas para el debate de una nueva agenda educativa, pp. 73–104. Amorrortu Editores, Buenos Aires (2000)

Author Index

M. E. Auer and D. May (Eds.): REV 2020, AISC 1231, pp. 1033–1036, 2021.
https://doi.org/10.1007/978-3-030-52575-0

Printed in the United States
By Bookmasters